INTERNATIONAL SOCIETY FOR ROCK MECHANICS

SOCIETE INTERNATIONALE DE MECANIQUE DES ROCHES

INTERNATIONALE GESELLSCHAFT FÜR FELSMECHANIK

International Congress on Rock Mechanics

Congrès International de Mécanique des Roches

Internationaler Kongress der Felsmechanik

PROCEEDINGS / COMPTES-RENDUS / BERICHTE

VOLUME / TOME / BAND 2

Editors / Editeurs / Herausgeber
G.HERGET & S.VONGPAISAL

Montréal / Canada / 1987

Proceedings
Comptes-rendus
Berichte

Sixth International Congress on Rock Mechanics

Sixième Congrès International de Mécanique des Roches

Sechster Internationaler Kongress der Felsmechanik

Montréal / Canada / 1987

VOLUME 2 Theme 4 Underground openings in overstressed rock
TOME 2 Thème 4 Souterrains en massifs rocheux sous grandes contraintes
BAND 2 Thema 4 Untertägige Hohlräume im überbeanspruchten Gebirge

A.A.BALKEMA / ROTTERDAM / BOSTON / 1987

and   Canadian Rock Mechanics Association / CIM / CGS
et   L'Association canadienne de mécanique des roches
und   Kanadische Vereinigung für Felsmechanik

For the complete set of three volumes, ISBN 90 6191 711 5
For volume 1, ISBN 90 6191 712 3
For volume 2, ISBN 90 6191 713 1
For volume 3, ISBN 90 6191 714 X
© 1987 by the authors concerned
Published by A.A.Balkema, P.O.Box 1675, Rotterdam, Netherlands
Distributed in USA and Canada by: A.A.Balkema Publishers, P.O.Box 230, Accord, MA 02018

Contents / Contenu / Inhalt
Volume 1 / Tome 1 / Band 1

**1 Fluid flow and waste isolation in rock masses
Ecoulement de fluides et enfouissement de déchets dans
les massifs rocheux
Flüssigkeitsbewegung und Abfallisolierung im Fels**

2 Rock foundations and slopes
Fondations et talus rocheux
Felsgründungen und Böschungen

3 Rock blasting and excavation
Sautage et excavation
Sprengen und Ausbruch

Contents / Contenu / Inhalt
Volume 2 / Tome 2 / Band 2

4 Underground openings in overstressed rock
Souterrains en massifs rocheux sous grandes contraintes
Untertägige Hohlräume im überbeanspruchten Gebirge

Author index
Index des auteurs
Authoren Verzeichnis

4

Underground openings in overstressed rock
Souterrains en massifs rocheux sous grandes contraintes
Untertägige Hohlräume im überbeanspruchten Gebirge

State of stress in rock and rock-burst proneness in seismicactive folded areas
Etat de contrainte et appréciation de la susceptibilité aux coups de toit du massif rocheux dans des régions plissées de l'activité séismique
Spannungszustand und Bewertung der Gebirgsschlaggefahr für Gebirgsmassive in tektonisch aktiven Gebieten

I.T.AITMATOV, Institute of Physics & Mechanics of Rocks, Academy of Sciences of Kirghiz SSR, Frunze, USSR
K.D.VDOVIN, Institute of Physics & Mechanics of Rocks, Academy of Sciences of Kirghiz SSR, Frunze, USSR
K.CH.KOJOGULOV, Institute of Physics & Mechanics of Rocks, Academy of Sciences of Kirghiz SSR, Frunze, USSR
U.A.RYSKELDIEV, Institute of Physics & Mechanics of Rocks, Academy of Sciences of Kirghiz SSR, Frunze, USSR

ABSTRACT: Evaluation of rock stress state in seismic active folded areas is performed. Application of the direct boundary element method for calculating rock-burst proneness of rock mass is considered.

RESUME: On cite les dependances, permettantes apprecier l'état de contrainte et la susceptibilité pour coup de terrain dans les regions plissés de l'activité séismique. On discute la possibilité de l'application du methode numerique pour la définition de la susceptibilité potentielle de coup de terrain.

ZUSAMMENFASSUNG: Die empirische Abhängigkeiten sind forlegen, die der Spannungszustand des Gebirgsmassives in der seismotätig Gebirgfaltgebiete zubewerten gestatten. Man besprecht die Möglichkeit der Anwendung der numerischen Methode zur Bestimmung von potentiellen Gebirgsschlaggefahr.

1 IN SITU STATE OF STRESS

Results of in situ rock stress measurements indicate heterogeneity of the global stress field in the upper part of the earth crust. It would be logical to assume that the heterogeneity of the global stress field is caused by different history of rock development in different regions of the Earth which resulted in various quantitative and qualitative changes in rock under the influence of tectonic , magmatic and other geological processes. However, one may also assume that bodies of rock situated in different places should have identical or similar stress state under the influence of similar geotectonic conditions, mechanical characteristics of the rock being alike.

In accrdance with the assumption above we have analyzed published data on seismic active folded areas in different regions of the Earth: the Alps (Nothern Italy, Switzerland, Austria, France), Japanese island arcs, Malaysia, Paleozoic folded structures of South-East Australia, the Cordilleras, the Rocky Mountains and the Appalachians in North America.

Basing on this analysis and results of numerous in situ stress measuments in highlands and foothills of Middle Asia and South-East Kasakhstan (Aitmatov et al. 1976, 1978,1980, Aitmatov 1981) the following relations between the average horisontal stresses and the depth are established:

1. In hard rock with Young's modulus ranging from 5×10^4 MPa to 1.1×10^5 MPa (skarns, diorites, granodiorites, listwanites, quartz porphyries, etc.)

$$\sigma_x + \sigma_y = 9.5 + 0.075 \gamma h \quad MPa$$
$$\sigma_x = 4.5 + 0.045 \gamma h \quad MPa \qquad (1)$$
$$\sigma_y = 5.0 + 0.030 \gamma h \quad MPa$$

2. In moderate strength rock with Young's modulus ranging from 2×10^4 MPa to 5×10^4 MPa (marmorized limestones, breccias, silt-stones, etc.)

$$\sigma_x + \sigma_y = 5.0 + 0.058 \gamma h \quad MPa$$
$$\sigma_x = 3.0 + 0.030 \gamma h \quad MPa \qquad (2)$$
$$\sigma_y = 2.0 + 0.028 \gamma h \quad MPa$$

3. In weak rock with Young's modulus ranging from 10^4 MPa to 2×10^4 MPa (coal-clay shales, serpentinites, sandstones, etc.)

$$\sigma_x \approx \sigma_y \approx \gamma h \quad or \quad \sigma_x \approx \sigma_y \approx \lambda \gamma h \qquad (3)$$

where h is the depth below surface, γ is the unit weight of the rock, λ is the lateral pressure ratio.

In weak and hard rock vertical stress correlates in most cases with the gravity load, i.e.

$$\sigma_v = \gamma h \quad MPa$$

At the same time in seismic active folded areas high residual stress has been revealed in certain kinds of massive rock, especially in metamorphic and igneous rock bodies free from faults, shear zones, etc. Rock stress distribution in Sarydshas area (the Central Tien Shan) is the most demonstrative example. Experimental stress investigations using stress relief, core discing and static indentor tests were carried out in the massive hard rock. The stresses in the rock appeared to be much greater than those expected in accordance with relation (1). The high stresses mainly result from the intensive process of granitization and consolidation of the rock in the bounded tectonic zone subjected to general compression. In this zone rock deformation and rock fracturing are not severe.

Thus for design purpose, in seismic active folded areas initial prediction of stress state of any rock should be realized according to relations (1-3). In some cases specific regional geological and tectonic patterns and peculiarities of ore bodies are to be taken into consideration, which permits to make more exact evaluation of actual rock stress state.

2 EVALUATION OF ROCK-BURST PRONENESS

The results obtained can be usefull to evaluate rock-burst proneness in openings. Solution of this problem depends upon correct determination of the initial stress field and a rock failure criterion. The energy criterion proposed by Petukhov and Linkov (1974) is one of modern rock-burst proneness criteria. According to this energy criterion a rock burst can be expected at that part of the opening where

$$-\frac{dW}{dA} > g \qquad (4)$$

Here $\frac{dW}{dA}$ is the energy flow into the rock as the opening surface increases per unit of area, g is the maximum energy absorbed by fractured material during the increase.

The quantity g is commonly adopted in fracture mechanics as a characteristic of material. It is called "effective surface energy" and for many rock types it has the order of $1 \ J/m^2$ (Petukhov and Linkov 1983). Relatively small value of g is the specific feature of rock.

One can calculate the value $\frac{dW}{dA}$ using numerical methods of rock mechanics when the stress field in rock mass is determined. The direct boundary element method (DBEM) is the most effective for determining the stresses at any point around the underground opening (Banerjee and Butterfield 1981, Ryskeldiev 1986). The method permits to reduce the dimension of the problem in comparison with other numerical methods, e.g. the finite element method.

The main equation of the DBEM for a three-dimensional body with the surface S can be written as

$$\tfrac{1}{2} u_j(x) = \int_{S} t_i(\xi) \, U_{ji}(x,\xi) \, d\xi -$$
$$\int_{S} u_i(\xi) \, T_{ji}(x,\xi) \, d\xi \qquad (5)$$

where x, ξ are the points of the surface S, $u_i(x)$ is the displacement vector and $t_i(x)$ is the force vector in the point x, $T_{ji}(x,\xi)$, $U_{ji}(x,\xi)$ are Kelvin's tensors, $i,j = 1,2,3$.

Defining boundary conditions for equation (5) in accordance with (1-3) and performing the usual procedure of dividing the surface S into boundary elements one can obtain a system of algebraic equations. From this system of equations one can determine the vectors $t_i(x)$, $u_i(x)$. Then one can calculate the stresses in rock mass around the opening and finally the stress intensity factor k_1 in given points. In terms of k_1 the expression for the energy flow can be written as

$$-\frac{dW}{dA} = \frac{1-\nu^2}{E} \, k_1^2$$

where ν is Poisson's ratio, E is Young's modulus. Thus considering g as a given value one can use the energy criterion for the prediction the rock-burst proneness. The whole algorithm of the problem solution is computerized.

The combination of the energy criterion and the DBEM features the proposed method for the prediction the rock-burst proneness. Using this combination gives a reduction of labour consumption and computer time for the problem solution.

As an example consider a thick ore body at the depth h from the surface, α being a dip angle. Chamber mining with backfill is used. In the Figure a cross-section of the opening is shown when three chambers have been worked out.

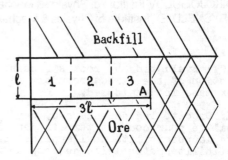

Figure 1. Sequence of minig.

Broken lines and figures show the sequence of mining, l is the height of the chamber. Initial stresses at the depth of the ore body are determined from relation (1).

The numerical analysis of rock-burst proneness for different sectors of the excavation is performed using the DBEM, elastic constants of the ore and the fill are assumed equal. The analysis indicates that Sector A is the most rock-burst prone as mining proceeds.

3 CONCLUSIONS

1. Evaluation and prediction of rock stress state should be performed according to relations (1-3).
2. Direct boundary element method with the energy criterion is efficient for determining rock-burst proneness in underground openings.

REFERENCES

Aitmatov, I.T., K.D.Vdovin & K.Ch.Kojogulov 1976. Some results of stress measurements in Kurusay-Turanglin ore deposit.(In Russian). In E.I.Shemiakin (ed.), Stress measurements in rock mass, p.2, 32-34. Novosibirsk: Inst.of Mining (IGD SO AN SSSR).

Aitmatov, I.T., K.D.Vdovin, K.Ch.Kojogulov, K.T.Tilegenov, F.S.Hudaybergenov & I.S. Jasykov. State of stress in rocks of Middle Asia ore deposits and evaluation of rock-burst proneness. (In Russian). In M.V.Kurlenia (ed.), State of stress in rock mass, 112-115. Novosibirsk: Inst. of Mining (IGD SO AN SSSR),1978.

Aitmatov, I.T., K.D.Vdovin, K.Ch.Kojogulov & V.A.Melikhov 1980. Specific features of stress state of steep dipping ore deposits in Middle Asia. (In Russian). In M.V.Kurlenia (ed.), Diagnostics of stress state and properties of rock, 65-69. Novosibirsk: Inst. of Mining (IGD SO AN SSSR).

Aitmatov, I.T. 1981. Stress state of rock
  in the upper part of the earth crust in
  seismic active regions of Middle Asia
  and South-East Kasakhstan. (In Russian).
  In V.J.Stepanov (ed.), Geomechanical conditions and rock pressure dynamic manifestations in   Middle Asia ore mines,
  3-20. Frunze: Ilim.
Petukhov, I.M. & A.M.Linkov 1974. On stability energy criterion in rockburst
  theory.(In Russian). Transaction of the
  VNIMI Institute (Trudy VNIMI) 91: 182-
  194.
Petukhov, I.M. & A.M.Linkov 1983. Rockburst
  and outburst mechanics. (In Russian).
  Moscow: Nedra.
Banerjee, P.K. & R.Butterfield 1981. Boundary element methods in engineering
  science. New-York: Mc-Graw Hill.
Ryskeldiev, U.A. 1986. Application of
  boundary integral equation method to the
  problem of rock-burst proneness evaluation. (In Russian). In I.T.Aitmatov (ed.)
  Stress state and fracture of rock,
  49-53. Frunze: Ilim.

Seismic stability of rock masses
Stabilité séismique des massifs rocheux
Erdbebensicherheit von Felsmassiven

E.ARKHIPPOVA, 'Hydroproject' Institute, Moscow, USSR

ABSTRACT: The dynamic spectral method is proposed to calculate the seismic loading on the rock mass. "An elastic body on elastic foundation" model is realized. The finite element method is used. The influence of the different factors on the seismic loading is analysed. The application of the method proposed for evaluation of the seismic stability of the slope of the Dnestrovskaya pump storage station is described.

RESUME: Pour les calculs de la charge sismique agissant sur les massifs rocheux on propose d'utiliser la méthode dynamique spectrale. On utilise le modèle "corps élastique sur la fondation élastique" dont l'exécution s'effectue à l'aide de la méthode des éléments finis. L'analyse de l'influence des divers facteurs sur la valeur de la charge sismique, est présentée. La mise en pratique de la méthode proposée pour l'estimation de la résistance aux séismes du talus de la centrale hydroélectrique de pompage Dniestrovskaya, est décrite.

ZUSAMMENFASSUNG: Zur Berechnung der seismischen Belastung des Felsmassivs wird die Ausnutzung der dynamischen spektralen Methodik vorgeschlagen. Es wird das Modell "elastischer Körper auf dem nachgiebigen Untergrund" benutzt, das mit der FEM realisiert wird. Es wird der Einfluß verschiedener Faktoren auf die Größe der seismischen Belastung analysiert. Die Anwendung der empfohlenen Methodik zur Einschätzung der Erdbebensicherheit der Böschung des PSW Dnestrowskaja wird beschrieben.

1 INTRODUCTION

While analyzing the stability, the rock masses as a rule are assumed to be rigid disks interacting with the stiff space through friction and cohesion. Therefore the static theory of seismic resistance is applied to determine seismic loading on them. In accordance with the theory: 1) horizontal seismic force is assumed to be equal to the product of the rock mass weight and the seismic coefficient for this region; 2) vertical component of the seismic force is considered to be small and is neglected in the stability analysis.

The paper views the estimation of seismic loading using the dynamic spectral method that is widely applied for calculations of seismic stability of industrial, civil and hydrotechnical structures. This method takes into account dynamic properties of the rock mass, i.e. frequencies and natural modes. The method of finite elements is used as well. The result obtained in the course of studies are in a good agreement with the results of H.Ito and T.Watanabe (1985).

2 FORMULATION

The horizontal component of the seismic loading at K point of the slope corresponding to mode i can be found as:

$$S_{ik} = A\ K_1\ Q_k\ K_2\ K_\gamma\ B_i\ \eta_{ik} \qquad (1)$$

where Q_k - weight of the slope element; A - seismic coefficient; K_1 - coefficient dependent on admissibility of local fissuring in the slope; $K_2 = 1$, $K_\gamma = 1$; B_i - dynamic response factor of the rock mass according to mode i, η_{ik} - coefficient dependent on the rock mass deformation form with its natural vibrations according to mode i.

For the static theory of seismic resistance the product $B_i\ \eta_{ik}$ is assumed to be equal 1 and the equation (1) is as follows:

$$S_k = A\ K_1\ Q_k \qquad (2)$$

where S_k - horizontal component of the loading at K point of the rock mass.

The basic provisions for the design model are as follows: 1) the rock mass is thought to be separated by more or less distinctly defined sliding plane from the surrounding medium considered as the rock mass foundation; 2) the rock mass is thought to be elastic located on the elastic foundation; 3) the rock mass is believed to have the weight, but the foundation is thought to be inertialess (Fig. 1).

3 APPLICATION OF METHOD OF FINITE ELEMENTS FOR CALCULATION OF FREQUENCIES AND NATURAL MODES OF ROCK MASSES

Two-dimensional problem is being solved. The rock mass together with the part of foundation are approximated by the finite element mesh (Fig. 2). The differential equation for calculations of the natural modes of the system rock mass - foundation without taking the damping into account takes the following form:

$$[M]\{\ddot{U}\} + [K]\{U\} = 0 \qquad (3)$$

where $[M]$ - weight matrix of the system rock mass - foundation, $[K]$ - stiffness matrix, $\{U\}$, $\{\dot{U}\}$, $\{\ddot{U}\}$ - vectors of displacement, speed, acceleration of the rock mass points. Introducing the notion of circular frequency ω and proper values $\lambda = 1/\omega^2$, some manipu-

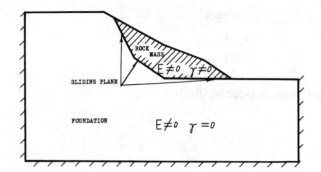

Figure 1. Calculation scheme.

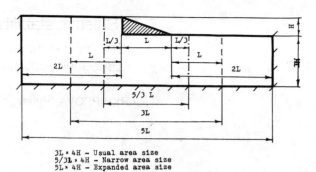

3L × 4H - Usual area size
5/3L × 4H - Narrow area size
5L × 4H - Expanded area size

Figure 3. Selection of foundation dimensions.

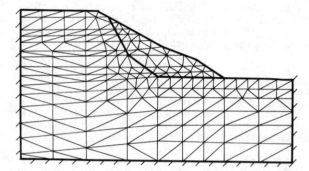

Figure 2. Finite element mesh.

lation yield, basing on equation (3), the particular problem for proper values:

$$\{U\} = \omega^2 [D][M]\{U\} \qquad (4)$$

or

$$[D][M]\{U\} = \lambda\{U\} , \qquad (5)$$

where $[D] = [K]^{-1}$ - matrix of compliance.

Basing on the proper values λ_i calculated are the periods of natural modes of analysed slope

$$T = 2\pi \sqrt{\lambda_i} \qquad (6)$$

and frequencies

$$\omega_i = \sqrt{1/\lambda_i} \qquad (7)$$

Dynamic response factors B_i are determined according to the soil category by seismic properties but not less than 0.8.

The coefficient η_{ik} is calculated by the formula

$$\eta_{ik} = \frac{X_{ik} \cdot \sum_{j=1}^{n} M_j \cdot X_{ij}}{\sum_{j=1}^{n} M_j X_{ij}^2} , \qquad (8)$$

where X_{ik} - normalized displacements of the rock mass along the axis X at natural modes i at the points (at the K point under the consideration, in particular), where the concentrated weights M_j are located, n - number of elements in the analysed area.

The value of seismic acceleration at K point along the rock mass is equal to:

$$W_k = A K_1 g \sqrt{\sum_{i=1}^{n} (B_i \eta_{ik})^2} \qquad (9)$$

where g - gravitational acceleration.

4 METHODICAL CALCULATIONS

The rock slope is an infinite half-plane. Therefore the application of the method of finite elements requires some restriction

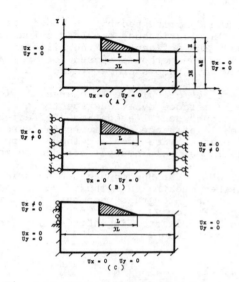

Figure 4. Selection of boundary conditions.

on the dimension of the area to be analysed. Three areas are considered to analyse the influence of the foundation width of the study area on the seismic loading (Fig. 3).

"The usual area" is characterized by the foundation dimension, generally accepted in the statistic analyses using FEM. Table 1 shows average values of $B_i \eta_{ik}$ for three study areas in the rock mass.

Table 1.

Area size	$(B_i\eta_{ik})$ aver. horizontal	$(B_i\eta_{ik})$ aver. vertical
Usual	0.91	0.41
Expanded	0.92	0.44
Narrow	0.82	0.39

The area of "usual" dimensions was considered under three boundary conditions to analyse the influence of boundary conditions along the outlines of the foundation on the seismic loading value (Fig. 4) (Table 2).

The analysis of the influence of the relationship between the modulus of deformation of the rock mass proper and the foundation was made. The dimensions of the analysed area are "usual", the boundary condition - overall embedment around the whole outline. The rock slopes with the ratio of modulus of deformation $E_{found.}/E_{rock\ mass}$ = 1.0; 1.25; 2.5; 5.0; 10.0; 20.0 were considered (Fig.5,6).

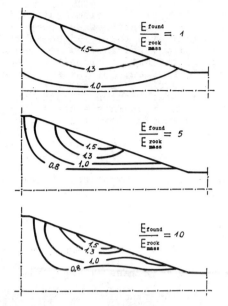

Figure 5. Isolines Biηik in the horizontal direction.

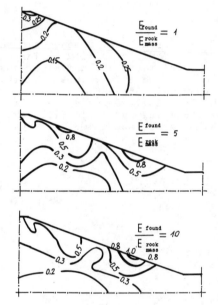

Figure 6. Isolines Biηik in the vertical direction.

Table 2.

Type of boundary condition	(Biηik) aver. horiz.	(Biηik) aver. vert.
Complete embedment around the whole outline	0.91	0.41
Embedment along the lower part of the outline. The right side and the left side is fixing along axis X	0.91	0.41
Embedment along the lower part of the outline and the right side. The left side is fixing partially along axis Y	0.92	0.41

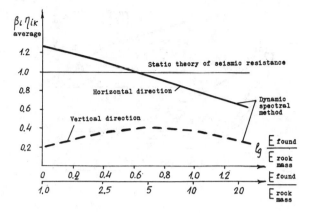

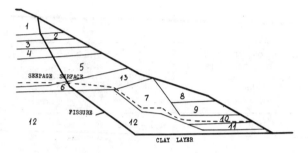

Figure 7. (Biηik) aver. versus foundation stiffness. Diagram.

Figure 8. Slope of the Dnestrovskaya pump storage station. Physical-mechanical properties of rocks: 1-13 at Table 3.

Basing on the results of calculations the diagram of (Biηik) average in the horizontal and vertical directions against the foundation stiffness is plotted (Fig. 7). Figures 5, 6, 7 show:

1. Seismic loads along the rock mass height are extremely nonuniform. Maximum accelerations occur near the surface approximately in the middle part of the slope.

2. In the slope more or less uniform height-wise horizontal seismic forces calculated by dynamic spectral methods exceed that determined by the static theory of seismic stability almost by 20%.

3. Vertical seismic forces are rather high and are comparable with the horizontal ones.

5 DETERMINATION OF SEISMIC STABILITY OF
   SLOPE AT DNESTROVSKAYA PUMP STORAGE STATION

The slope at the Dnestrovskaya pump storage station is composed of disintegrated argillite, marls, limestones alternating with clay interlayers. Dynamic moduli of elasticity of these rocks differ considerably from each other (Fig. 8, Table 3).

The part of the slope separated from the foundation by the steeply dipping fissure and horizontal clay layer is considering (Fig. 8). The analysis was made at the dead load and seismic load.

This region is characterized by the seismic acceleration 7 points (0.025 g). The diagrams of analytical horizontal seismic acceleration for three cross-section lengthwise the slope are given in Fig. 9a. This suggests that:
1) maximum seismic accelerations are equal to 0.17 g in the lower part of the slope in the zone of rocks with minimum dynamic modulus of

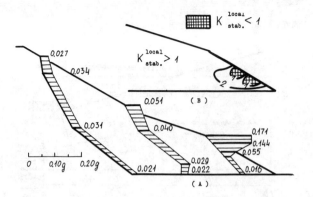

Fig. 9. Slope of the Dnestrovskaya pump storage station: (A) diagrams of seismic accelerations; (B) isolines of $K^{local}_{stab.}$.

Table 3. Mechanical and physical properties of rock mass.

Number of zone	Dynamic modulus of elasticity MPa	Poisson's ratio	Unit weight KN/m³	
			Dry rock	Saturated rock
1	4400	0.30	21.3	-
2	500	0.25	17.8	-
3	5250	0.38	18.8	-
4	12100	0.38	17.7	20.3
5	5250	0.38	21.7	22.9
6	34000	0.34	25.7	25.9
7	12000	0.38	24.0	24.1
8	280	0.38	19.4	-
9	280	0.38	19.4	20.3
10	380	0.48	21.5	22.
11	11500	0.25	24.	24.1
12	14000	0.38	-	25.9
13	30000	0.31	25.7	-

elasticity (E = 280 MPa); 2) in the upper part of the slope the distribution of the acceleration is rather uniform heightwise.

Basing on the values of the compressive and tangential stresses in each element of the approximating mesh the local stability of the slope was estimated using the formula (Lee & Roth 1977):

$$K^{loc.}_{stab.j} = \frac{(\sigma^{stat.}_{y_j} + \sigma^{dyn.}_{y_j}) \, tg \, \vartheta^{loc.}_j + c^{loc.}_j}{\tau^{stat.}_{xy_j} + \tau^{dyn.}_{xy_j}} \quad (10)$$

where σ_{y_j} - compressive strain in element j, τ_{xy_j} - tangential strain in element j, $tg \, \vartheta^{loc.}_j$ $c^{loc.}_j$ - characteristics of local rock stability.

The calculations of $K^{loc.}_{stab.}$ indicated that the whole slope is stable exception of the lower part (Fig. 9b).

The total shear strength of the slope as a rigid disk (using the limit equilibrium theory) was determined. The Maslov - Berer horizontal force technique was used. The seismic acceleration was presented by diagrams obtained by FEM basing on the dynamic spectral method. In this case the sliding safety factor of the slope equals 1.03.

The calculations of the seismic stability of the slope of the Dnestrovskaya pump storage station indicated that: 1) the earthquake of intensity 7 points does not cause displacement of the entire slope but leads only to local disturbance of the slope stability near the surface at the lower part; 2) the sliding safety factor computed on the basis of representation this slope as a rigid disk is the lowest estimation of the slope stability. The results of the analysis qualitatively agree with conclusions of Japanese specialists Hiroshi Ito and Toshiyuki Watanabe (1985) from the results of experimental and analytical investigations of dynamic and static failure of slopes.

6 CONCLUSIONS

1) The static theory of the seismic resistance may be applied to evaluate the seismic stability of the rock mass at the early stage of designing.
2) The use of the dynamic spectral method is required to determine the seismic stability of the slope for important structures.

REFERENCES

Ito, H. & T. Watanabe 1985. Some considerations on the seismic stability of large slopes, by model tests and numerical analysis. Proc. of the 5 Int.Conf.Num.Meth. Geomech. (Nagoya), 989-996.
Lee, K. & W. Roth 1977. Seismic stability analysis of Hawkins Hydraulic Fill Dam. J.Geotech.Engng.Div.ASCE, 103, 627-644.

Anchorage performance and reinforcement effect of fully grouted rockbolts on rock excavations
Performance de l'ancrage et effet de renfort de boulons cimentés durant l'excavation
Verankerungsverhalten und Verstärkungseffekt von vollständig einbetonierten Felsankern in Felsabtragungen

Ö.AYDAN, Nagoya University, Japan
T.KYOYA, Nagoya University, Japan
Y.ICHIKAWA, Nagoya University, Japan
T.KAWAMOTO, Nagoya University, Japan

ABSTRACT: Anchorage mechanism of bolts is clarified through an experimental program and a numerical representation for grouted rock bolts is proposed. This model is then applied to simulate rockbolts in the analyses of pull-out tests and their reinforcement effects on discontinuous rock pillars and a tunnel excavated in a weak rock formation and the results of the analyses are compared with the actual measurements. The analyses have shown that the proposed model is capable of simulating the bolts and enable one to evaluate the reinforcement of bolts quantitatively

RESUMÉ: Le mechanism de l'ancrage des boulons est élucidé par une etudie experimentale et une représentation numérique est proposé dans les cas les boulons scellés. Ce modéle est appliqué en vue de simuler les boulons lors des analyses de tests de traction et de l'effet de renfort sur piliers de roche discontinue et un tunnel creusé dans une formation faible. Les resultats des analyses sont comparés avec les données experimentales. Les analyses ont démontré que le modéle proposé est capable de simuler les boulons et permettent l'évaluation quantitative leur effet de renfort.

ZUSAMMENFASSUNG:Ankernmechanismen von Felsankern werden anhand eines experimentellen Programmes geklärt und eine numerische Repräsentation von eingeklenten Felsankern in vorgeschlagen. Das Modell wird dann angewendet auf die Simulation von Felsankern in Zugversuchen, auf die Simulation der Verstärkungseffekte von Felsankern in diskontinuierlichen Felssäulen und in einem Tunnel in einer weichen Felsformation. Die Resultate der Analyse werden mit tatsächlichen Messungen verglichen.Die Analyse hat gezeigt, dass das vorgeschlagene Modell in der Lage ist, die Felsankern zu simulieren und einen in die Lage versetzt, die Verstärkung durch Felsankern quantitativ auszuwerten.

1. INTRODUCTION

Rockbolts have become one of the widely used supporting members in geotechnical engineering practices in recent years. This probably results from the easiness of their transportation, storage and installation of and quickly developing reinforcement effects as compared with other conventional supporting members such as steel ribs with wooden lagging. This has initiated numerous theoretical, numerical and experimental researches all over the world in order to enlight and evaluate the qualitatively well known superior reinforcement of bolts in engineering practices for the optimum design of support systems for geotechnical engineering structures in quantitative terms. These research works on rockbolts can be broadly classified into two main groups:
1. works on their reinforcement effects, and
2. works on their pull-out and shear resistances.
According to these works, the reinforcement effects of bolts are grouped into the followings:
1. Suspension effect,
2 Beam building effect,
3. Arch action,
4. Radial confinement,and
5. Utilising the inherent structural properties of the surrounding rock
depending upon the location and ground conditions in which they are installed.

When the reinforcement effects of bolts are considered, there is a tendency to associate that with the tensile and/or shear strength of steel bar with the condition, that is, bolts have a sufficiently long anchorage.

On the other hand, the works on pull-out resistance of bolts mainly involve the effect of factors such as the kind of host rock, bolt-borehole diameter ratio, surface roughness of steel bar and/or hole walls, and curing time of resin or cement based mortar, and others.

These works have clarified some of several important aspects of bolts qualitatively and quantitatively up to some extent. Nevertheless, the uncertainty still exists on the role of rockbolts as a permanent supporting member for rock excavations as most of these works involved with the resultant

effects of the bolts rather than the fundamental ones. It is the opinion of the authors of this paper that if any mathematical model incorporates the axial and shear behaviour of steel bars with due considerations of behaviour of grout annulus and interfaces between grout and steel bar, and grout and rock, it will be possible to evaluate the reinforcement effect of bolts in quantitative terms and remove the uncertainty on the permanency of bolts as a supporting member in securing the stability of rock excavations. Though it may be possible to obtain some closed form solutions for anchorage capacity and reinforcement effects of bolts (Aydan et al 1985, 1986), a general modelling of rockbolts for rock engineering structures with complex geometry and material behaviour will require a numerical model such as a finite element representation. Among all the finite element representations, there is only one representation proposed by John and Van Dillen (1983) which deserves mentioning herein. This representation has attempted to incorporate the behaviour of grout annulus as well as the axial and shear behaviour of steel bar. However, it fell short of considering the most fundamental behaviour of interfaces which is felt to be the most important element on overall anchorage behaviour of the bolts.

In this paper, experimental and theoretical studies on the anchorage performance of grouted rockbolts are described and a numerical representation has been proposed. This model is then applied to simulate the bolts in pull-out tests and their reinforcement effects on discontinuous rock pillars and a tunnel excavated in weak rock. The calculations are then compared with those of actual measurements and a good agreement has been found.

2. BEHAVIOUR AND ANCHORAGE MECHANISM OF GROUTED ROCKBOLT SYSTEM

The grouted rockbolt system can be visualised as consisting of mainly two materials; steel bar and grout annulus, and two interfaces between grout and steel bar, and grout and rock. The steel bar is the principal element within the system to resist to axial and shear forces and the grout annulus functions to transfer these forces to/from the surrounding medium.

2.1 Behaviour of steel bar

The behaviour of steel bar in simple axial and shear is very well known and is described as elastic-strain hardening plastic. A typical stress-strain curve for a steel bar subjected to tension is shown in Figure 1.

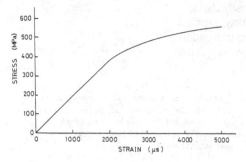

Figure 1. Stress-strain curve for a steel bar in tension

2.2 Behaviour and anchorage mechanism

As the stress transfer between steel bar and surrounding medium is made through the grout annulus and interfaces, their behaviour is of paramount importance. A push-out and pull-out testing program was initiated at our institute with the objective of investigating the anchorage mechanism of bolts and the parameters influencing that. Thus, a special triaxial cell capable of applying a uniform pressure on rock samples of 120 mm in diameter and 200 mm long was developed for this purpose. A steel was instrumented with axially directional e.r.s. gauges spaced on a machined surface.

A rapid hardening cement with fine sand was used as a grouting material and steel bars were embedded into centrally drilled holes in a weak rock samples, called Oya tuff. Bars were either pushed out or pulled out by applying loads in increments. Test results of pull-out and push-out tests are summarised in Figure 2.

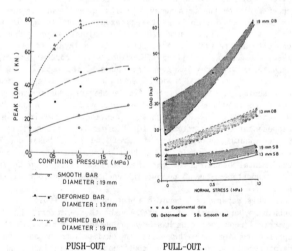

PUSH-OUT   PULL-OUT.

Figure 2. Load bearing capacity of rockbolts in laoratory tests

Although it is difficult to describe the findings of the testing program in detail within this limited space, some of those will be briefly outlined here as follows:

1. Bolt-borehole diameter ratio has a great influence on the location of failure and consequently pull-out resistance in the case of deformed bars, but has no influence in the case of smooth surface bars.

2. Application of confining pressure results in higher pull-out or push-out resistance.

3. Rough surface configurations of bars and borehole walls also results in higher pull-out resistance.

4. Steel bars pulled or pushed-out at one of interfaces by shearing except some splitting failures of rock in the case of deformed bars at low confining pressures. Shearing firstly starts at the loaded end and propagates towards the unloaded end. Interfaces were weakness surfaces in tests.

5. Stress distributions along the bolts are non-uniform and depend upon the load level and shearing along one of interfaces and its propagation.

6. It is extremly difficult to, if not impossible, determine the parameters for a mathematical modelling of mechanical behaviour of grout annulus and/or interfaces. As a result, pull-out or push-out tests are not suitable for such a purpose.

In view of these experimental facts, a direct shear testing program on the interfaces and grouting material was undertaken. Typical shear stress-displacement and normal displacement-shear displacement curves for each interface and grouting material are shown in Figure 3. As noted from Figure 3, the shear resistance of the interfaces less than that of the grouting material, and indicates that interfaces are the weakness surfaces within the system.

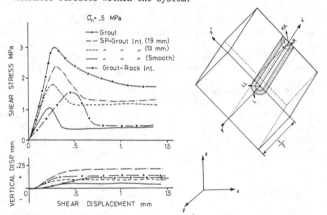

Figure 3. Shear behaviour of various interfaces and grout under normal stress of 0.5 MPa

Figure 4. Perspective view of bolt element

3. MATHEMATICAL MODELLING

A bolt element which can count the actual mechanism involved with bolts must be a three dimensional one and be also capable of allowing the slippage of the interfaces which will require the use of joint or interface elements may cause many complexities and be uneconomic in the analysis of large geotechnical engineering structures. Therefore, to keep this element as simple as possible while taking into account the facts pointed out in previous sections results in a coupled form of a bar element and axisymmetric interface element with a finite thickness similar to the one proposed by Ghaboussi et al.(1973). In regard with the mathematical modelling of the bolt element, the following assumptions are made; the steel bar has an axial stiffness within the continuum and additionally shear stiffness at only discontinuities. The bolt is further assumed to have no resistance against bending and the displacement of bolts perpendicular to its axis is the same as that of rock mass. The grout annulus is considered to be an axisymmetric body around the bar and its shear strain is evaluated in terms of relative displacements of the interfaces and dimensions of the bar and borehole depending upon the possible failure location.

A three dimensional perspective view of the element is shown in Figure 4.

Since the full description of the derivation of the stiffness matrix of the bolt element is given elsewhere (Aydan et al. 1986) the interested reader is advised to refer the above article for details. The finite element form of the stiffness matrix is given as:

$$\underset{\sim}{K}^e = \int_\Omega \underset{\sim}{B}^T \underset{\sim}{D} \underset{\sim}{B} \, dV \qquad (1)$$

with the use of the following expressions for matrices B and D

$$\underset{\sim}{B} = \begin{bmatrix} 0 & 0 & -\frac{dN_{KK}}{dx} & \frac{dN_{LL}}{dx} & 0 & 0 & 0 & 0 & 0 & 0 & 0 & 0 \\ -A_r N_{KK} & -A_r N_{LL} & A_r N_{KK} & A_r N_{LL} & 0 & 0 & 0 & 0 & 0 & 0 & 0 & 0 \\ 0 & 0 & 0 & 0 & -\frac{dN_K}{dx} & \frac{dN_L}{dx} & 0 & 0 & 0 & 0 & 0 & 0 \\ 0 & 0 & 0 & 0 & 0 & 0 & 0 & -\frac{dN_K}{dx} & \frac{dN_L}{dx} & 0 & 0 \end{bmatrix} \quad (2)$$

$$N_K = N_{KK}, \quad N_L = N_{LL}, \quad N_{KK} = \frac{x'_{LL} - x'}{x'_{LL} - x'_{KK}}, \quad N_{LL} = \frac{x' - x'_{KK}}{x'_{LL} - x'_{KK}}$$

$$\underset{\sim}{D} = \begin{bmatrix} E_b & 0 & 0 & 0 \\ 0 & G_{ga} & 0 & 0 \\ 0 & 0 & G_b & 0 \\ 0 & , 0 & , 0 & , G_b \end{bmatrix} \quad (3)$$

one can easily obtain the following expressions for the stiffness matrix in local coordinates:

$$\underset{\sim}{K}^e = \begin{bmatrix} 2K_{ga} & K_{ga} & -2K_{ga} & -K_{ga} & 0 & 0 & 0 & 0 & 0 & 0 & 0 & 0 \\ K_{ga} & 2K_{ga} & -K_{ga} & -2K_{ga} & 0 & 0 & 0 & 0 & 0 & 0 & 0 & 0 \\ -2K_{ga} & -K_{ga} & 2K_{ga}+K_a & K_{ga}-K_a & 0 & 0 & 0 & 0 & 0 & 0 & 0 & 0 \\ -K_{ga} & -2K_{ga} & K_{ga}-K_a & 2K_{ga}+K_a & 0 & 0 & 0 & 0 & 0 & 0 & 0 & 0 \\ 0 & 0 & 0 & 0 & K_s & -K_s & 0 & 0 & 0 & 0 & 0 & 0 \\ 0 & 0 & 0 & 0 & -K_s & K_s & 0 & 0 & 0 & 0 & 0 & 0 \\ 0 & 0 & 0 & 0 & 0 & 0 & 0 & 0 & 0 & 0 & 0 & 0 \\ 0 & 0 & 0 & 0 & 0 & 0 & 0 & K_s & -K_s & 0 & 0 \\ 0 & 0 & 0 & 0 & 0 & 0 & 0 & -K_s & K_s & 0 & 0 \\ 0 & 0 & 0 & 0 & 0 & 0 & 0 & 0 & 0 & 0 & 0 \\ 0 & 0 & 0 & 0 & 0 & 0 & 0 & 0 & 0 & 0 & 0 \end{bmatrix} \quad (4)$$

where $\quad K_a = \frac{E_b A}{L} \quad K_s = \frac{G_b A}{L} \quad L = x'_{LL} - x'_{KK}$

The specific forms of Kga in the above expressions differs depending upon the chosen form of Ar. These are:

$$K_{ga} = \pi G_{ga} L / (3 \ln(r_h/r_b)) \quad (5a)$$

$$K_{ga} = \pi G_{ga} L (r_h^2 - r_b^2) / (6(r_b \ln(r_h/r_b))^2) \quad (5b)$$

$$K_{ga} = \pi G_{ga} L (r_h^2 - r_b^2) / (6(r_h \ln(r_h/r_b))^2) \quad (5c)$$

The stiffness matrix in local coordinates is then assembled by using the coordinate transformation.

4. APPLICATIONS AND DISCUSSIONS

In this section, the applications of the bolt element described in the previous section are presented with the objective of illustrating the fundamental reinforcement effects of bolts and enlighting some certain aspects of the interaction between grouted rockbolts and rock mass with particular reference to actual situations.

The first applications of the bolt element are made to pull-out tests carried out in our laboratory as it has some practical implications to the suspension effect of the bolts as well as to support the statement, that is, the pull-out tests can not be the right method of testing the bonding behaviour of rockbolts.

The material properties of interfaces, grouting material, rock and steel bar used in analyses were obtained from our laboratory tests. Two of pull-out test simulations are presented herein and calculated axial strain distributions and shear stress distributions at bolt-grout interface are shown in Figure 5 together with the measured distributions at various levels of applied pull-out load. As noted from the figure, a shear stress concentration occurs at the loaded end. As the debonding starts to take place along bolt-grout interface, the shear stress concentration region moves away from the proximal end. The good agreement between calculated and measured axial strain distributions at various pull-out load levels is quite apparent.

We, next, simulate the shear reinforcement effect of rockbolts on a rock pillar with a discontinuity inclined at an angle of 45° to the direction of applied load. This analysed example has a direct relevance to such a pillar tested and reported by Egger and Fernandes (1983). The material properties used in the analysis are those reported in the referred paper with an exception that two bolts used in actual tests was replaced by a rockbolt with an equivalent cross

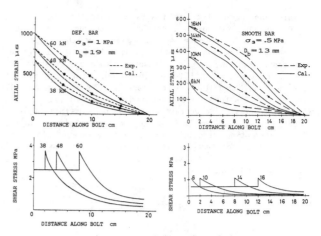

Figure 5. Distributions of axial strain and shear stresses in/along rockbolts during pull-out tests

section. Figure 6 shows the layout of the model and load and displacement curve. In this analysed case, the bolt makes an angle of 45° with the normal of discontinuity. Discontinuity was represented by an interface element of Ghaboussi type. The load-displacement curve calculated follows the same path as that measured up to 60 kN. However, a discrepancy between two curves starts to take place thereafter. This discrepancy is attributable to the phenomena of rupturing of rock at the vicinity where the bolt crosses the discontinuity. This was not considered in the numerical model. Nevertheless, the magnitude of ultimate applied loads for both cases almost the same. The figure also shows the development of axial and shear stresses in the bolt element crossing the discontinuity during the application of loads.

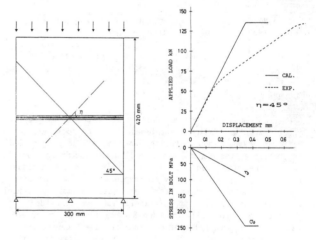

Figure 6. Layout of discontinuous rock pillar model and load-displacement curve

The third example of applications is on the effects of radial confinement and improvement in rock mass properties due to the rockbolts in the case of a section of a highway tunnel passed through a weak rock formation with a uniaxial strength of 1.7-4.0 MPa, and 400 m below the ground surface. As the strength and elastic modulus (250 MPa) of the rock mass were very low, plastification of ground and large displacements were expected. The initially designed support system for this section consisted of 6 m long rockbolts of 24 mm in diameter and two layers of shotcrete; the first layer was 50 mm thick, and the second layer 200 mm thick with 4-6 slits in order to allow to large deformations. The final unreinforced concrete lining was about 400 mm thick. A number of measurement stations were selected and instrumented with the objective of observing the deformations of ground by extensometers and convergence, axial forces in rockbolts, and radial and tangential stresses in shotcrete.

This support system was unable to maintain the allowed amount of deformations of rock mass and the deformations did not cease even the face was far away from the respective sections. At various locations, the head of bolts and shotcrete were ruptured. As a result, additional rockbolts of 9 and 13.5 m long had to be installed. Then, the support system was altered. These alterations are the increase of bolt length from 6 m to 9 m and decrease the number of slits and to fill them up after a certain amount of deformations. At some locations, no slit was left in the secondary shotcrete layer. These alterations resulted in success. Figure 7 shows the observed range of tunnel wall displacements for two types of support system. As for the sections where no slits were left, the observed tunnel wall displacement was less than 50mm.

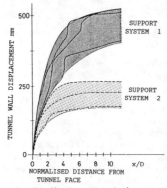

Figure 7. Tunnel wall displacement during tunnel face advance for two different support system.

Five cases given in Table 1 were analysed. The behaviour of rock mass is assumed to be elastic-perfectly plastic. The supports are also assumed to be installed as soon as the excavation was made.

Table 1. Details of cases

Case No.	Bolt		Shotcrete	Comments
	Length m	No	Thickness mm	
C-1	–	–	–	Unsupported
C-2	6	48	–	Rockbolt only
C-3	6	48	50	
C-4	9	60	50	
C-5	9	60	250	

Figure 8 shows the ground reaction curve and development of plastic zone with allowed displacement of the rock mass for 5 different cases. When rockbolts 9 m long and no slit is left in shotcrete, the maximum tunnel wall displacement is about 50 mm. On the other hand, if slits are left in the secondary shotcrete layer which simply corresponds to a case, that is, shotcrete is 50 mm thick. For this case, the maximum tunnel wall displacement is approximately 150 mm, provided that shotcrete does not fail. The Case 3 is similar to the previous case. However, the bolts of 6 m long in Case 2 can not resist to excessive straining beyond that at the ultimate failure load, and all bolts fail. As a result, the displacement of tunnel wall starts to approach that in the unsupported case. The calculated axial force distributions for various cases for 6 and 9 m long bolts are shown in Figure 9 together with the

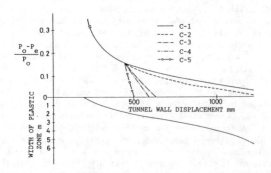

Figure 8. Ground reaction curves for various supporting cases

Table 2. Material properties and in-situ stress

Rock				Bolt			Shotcrete			In-situ
E GPa	ν	σ_C MPa	Φ °	E GPa	ν	σ_C MPa	E GPa	ν	σ_C MPa	σ_0 MPa
250	0.4	2	35	210	.3	450	10	.2	35	10

measure ments of some rockbolts around the tunnel. As it is apparent from figures, the axial forces in bolts are approximately equal or greater than yield load of the steel bars. The straining in bolts was far greater when no or very thin shotcrete layer is applied. The analyses clearly shows why the inital support system was not successfull. The second interesting point to note from calculated ground reaction curves is the decrease of deformability of rock due to bolting. This effect of bolts will probably be more apparent in weak rocks.

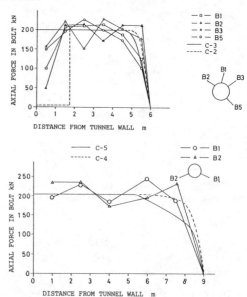

Figure 9. Axial force distributions along rockbolts

5. CONCLUSIONS

The anchorage mechanism of a grouted rockbolt has not yet been clarified. Firstly, we here presented results of triaxial pull-out and push-out tests in laboratory, then it was understood that the surface geometry of the bolt and the confining pressure have large influences on the behaviour of rockbolts.

Based on these results, we have proposed a new rockbolt element with interface for the finite element analysis. Applications of the proposed model were discussed on the above pull-out and push-out tests, a rock pillar test with a discontinuity, and an actual tunnelling problem in a weak rock formation. It was observed that the proposed rockbolt model is very effective in simulating their reinforcement effects.

REFERENCES

Aydan,Ö., Y.Ichikawa & T.Kawamoto 1985.Load bearing capacity and stress distributions in/along rockbolts with inelastic behaviour of interfaces.Procs.Int. Conf. Num.Meths. in Geomechanics, Nagoya.
Aydan,Ö., Y.Ichikawa & T.Kawamoto 1986.Reinforcement of geotechnical engineering structures by grouted rockbolts. Int.Symp. Engng. Complex Rock formations,ISRM,Beijing.
Egger,P. & H. Fernandes 1983.Nouvelle presse triaxiale - etude de modeles discontinus boulones. Procs. Int. Conf. Rock Mechs.,ISRM, Melbourne.
Ghaboussi,J., E.L.Wilson & J Isenberg 1983. Finite element for rock joint and interfaces. Procs. ASCE, Vol. 99, SM10.
John.C.M. & D.E.Van Dillen 1983. Rockbolts: a new representation and its application to tunnel design. 24th U.S. Symp. Rock Mechs.

Méthode intégrale de dimensionnement d'ancrages cimentés dans le rocher
Integral method for the design of grouted rock anchors
Ein integrales Verfahren zur Bemessung von Injektionsankern im Fels

G.BALLIVY, Département de génie civil, Université de Sherbrooke, Québec, Canada
B.BENMOKRANE, Département de génie civil, Université de Sherbrooke, Québec, Canada
A.LAHOUD, Département de génie civil, Université de Sherbrooke, Québec, Canada

ABSTRACT: This paper presents the results of theoretical and experimental studies of the uplift behaviour of grouted anchors in a homogeneous rock. The originality of the theoretical model by the finite element method lies in the fact that the geotechnical characterization of the bonding at the grout/tendon interface, the type of failure observed here, are obtained from an uplift test conducted on short anchors injected under the same conditions. In doing so, the roughness of the borehole, the strength of the grout cured in the same environment and the geometry of the anchor are taken into account in this model. The results obtained from this study show that the loads and the types of failure predicted by the model agree with those observed in the field. The field tests on anchors instrumented with strain gauges show a good correlation between stress distributions and the progress of debonding.

RESUME: Les résultats d'une analyse théorique et expérimentale du comportement à l'arrachement d'ancrages injectés dans une roche homogène par utilisation, en particulier de la méthode des éléments finis sont présentés ici. L'originalité de ce modèle théorique réside dans le fait que les paramètres géotechniques caractérisant l'adhésion à l'interface tige-coulis, mode de rupture observé ici pour les roches très compétentes, sont déduits d'un essai d'arrachement d'ancrage court scellé dans les mêmes conditions que les ancrages projetés. Ainsi il est possible de prendre en compte la rugosité de la paroi du trou de forage, la résistance du coulis pour un mûrissement dans l'environnement concerné et la géométrie de l'ancrage. Les résultats ainsi obtenus montrent que les charges et les modes de rupture prédits par le modèle concordent avec ceux obtenus sur des ancrages installés en chantier. Aussi, il a été observé une bonne corrélation entre les distributions des contraintes et l'évolution de la décohésion; celles-ci furent déduites expérimentalement en chantier sur des ancrages instrumentés à l'aide de jauges de déformation.

ZUSAMMENFASSUNG: Eine theoretische und experimentale Untersuchung mit Resultaten über die Eigenschaften der Injektionsanker in homogenen Fels wird in der vorliegenden Schrift unter Benutzung der Methode der endlichen Elemente behandelt. Dieses theoretische Model ist original, wobei die geotechnischen Parameter welche die Adhesion an der Grenzschicht Mörtel-Stahl, und die Bruchform in harten Fels characterisieren, wurden mittels einem Abhebungstest für injektierten Kurzanker unter denselben Bedingungen als für einen projektierten Anker ausgeführt. So ist est möglich mit dr Rauhigkeit der Wand der Bohrung sowie mit dem Widerstand des Mörtels und der Geometrie des Ankers zu rechnen. Die Resultate zeigen, dass die Afuladungen und die vorgesagte Bruchweise ausdem Model mit denen an der Baustelle gut miteinander entsprechen. Dies bezieht sich auch auf eine gute Korellation zwischen der Verteilung der Spannungen welche von der Baustelle mit Hilfe eines Deformationstandes entnommen wurden.

1. INTRODUCTION

La technique des ancrages injectés dans les massifs rocheux, connaît un développement très important, que ce soit pour retenir des structures sollicitées en traction ou stabiliser des ouvrages tels que murs de soutènement, excavations souterraines, culées de ponts, pylônes hydro-électriques, appuis de barrage, talus rocheux...

Si, cependant, l'emploi des ancrages injectés dans le rocher s'intensifie et se généralise, les techniques de prévision de leur résistance ont par contre peu évolué. Pour pallier à cette déficience, on utilise des longueurs ancrées élevées, en plus des essais sur le chantier sont effectués d'une façon systématique pour vérifier la compétence de l'ancrage une fois qu'il est installé.

Un examen de la littérature révèle que les méthodes de dimensionnement actuelles sont très approximatives et fortement empiriques et qu'il existe d'importantes incertitudes particulièrement en regard du mécanisme de transfert des charges du tirant d'acier au massif rocheux, à la distribution des contraintes le long de l'ancrage et au mécanisme de décohésion du coulis de scellement. Ces lacunes sont d'autant plus importantes quand on sait que plusieurs millions d'ancrages sont installés chaque année dans le monde et qu'il y a une demande sans cesse grandissante d'ancrages à très haute capacité: charge supérieure à 10 000 kN.

On rapporte ici les résultats d'une étude théorique qui comprend une simulation par éléments finis du comportement à l'arrachement d'ancrages injectés dans une roche homogène. Ces résultats sont comparés à des résultats obtenus lors d'essais en chantier sur plus de 40 ancrages scellés avec différents produits de scellement (BALLIVY et al. (1986)). La comparaison est faite particulièrement aux niveaux des charges et des modes de rupture, des distributions des contraintes et de l'évolution de la décohésion du scellement.

2. MODES DE RUPTURE D'UN ANCRAGE INJECTÉ DANS LA ROCHE

La rupture d'un ancrage injecté dans une roche peut se faire de quatre façons (Figure 1):
- Rupture du tirant d'acier
- Rupture du massif rocheux
- Rupture du scellement au niveau du contact tige-coulis
- Rupture du scellement au niveau du contact roche-coulis

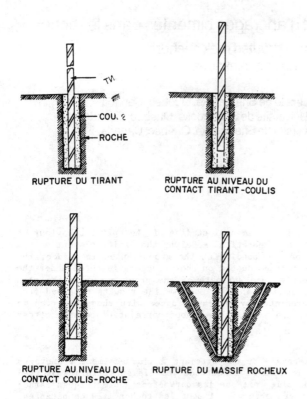

RUPTURE DU TIRANT

RUPTURE AU NIVEAU DU CONTACT TIRANT-COULIS

RUPTURE AU NIVEAU DU CONTACT COULIS-ROCHE

RUPTURE DU MASSIF ROCHEUX

Figure 1. Illustration des quatre modes de rupture d'un ancrage injecté

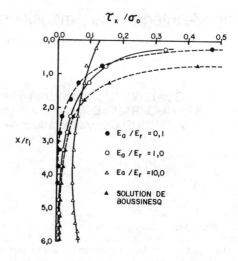

Figure 2. Variations de la contrainte de cisaillement en fonction de la profondeur de l'ancrage pour diverses valeurs de E_a/E_r (COATES et YU (1970))

Le dimensionnement est basé sur ces quatre modes de rupture. La rupture du tirant d'acier est un mode de rupture très facile à vérifier; les autres modes de rupture sont caractérisés ainsi:

- La rupture du massif rocheux ne se produit généralement que dans le cas des ancrages courts injectés dans un massif rocheux très fracturé. La surface de rupture dépend des discontinuités géologiques et elle est souvent assimilée à un cône inversé;

- La rupture du scellement se fait soit au niveau des contacts tige-coulis soit au niveau des contacts roche-coulis et ce dépendamment des caractéristiques mécaniques du coulis et de la roche, de la géométrie de l'ancrage (diamètres de la tige et du trou de forage) et des caractéristiques des surfaces de la tige et du trou de forage (HANNA, 1982)). En général, pour un ancrage injecté dans une roche massive, la rupture se fait dans le coulis au niveau du contact tige-coulis, tandis que dans le cas d'une roche tendre la rupture se fait dans la roche au niveau du contact roche-coulis; dans les deux cas, la rupture est caractérisée par un cisaillement.

3. ÉTUDES THÉORIQUES PAR ÉLÉMENTS FINIS

Les résultats de plusieurs études théoriques par éléments finis sur le comportement à l'arrachement des ancrages injectés dans la roche ont été analysés. Il s'agit en particulier des travaux des auteurs suivants: COATES et YU (1970), HOLLINGSHEAD (1971) et YAP et RODGER (1984).

COATES et YU (1970) ont étudié de façon théorique, par la méthode des éléments finis, la répartition des contraintes le long d'un ancrage. Ils n'ont cependant modélisé l'ancrage que par un ensemble de deux matériaux élastiques, le tirant et la roche. Cette étude montre que la distribution des contraintes de cisaillement le long de l'ancrage dépend du rapport entre le module d'élasticité du tirant, E_a, et le module d'élasticité de la roche, E_r (Figure 2). Ainsi, plus la valeur de E_a/E_r est faible, cas d'une roche dure, plus il y a concentration de contrainte à l'extrémité chargée de l'ancrage; les valeurs élevées de E_a/E_r, cas d'une roche tendre, sont par contre associées avec une distribution de contraintes de cisaillement plus ou moins uniforme.

HOLLINGSHEAD (1971) analysa le problème des ancrages injectés en considérant les trois matériaux impliqués soit la tige d'acier, le coulis de scellement et la roche. De plus un comportement élastoplastique a été introduit. Les trois matériaux ont un comportement élastique parfaitement plastique et l'écoulement plastique est conforme au critère de Tresca. Cette étude montre (Figure 3) que l'écoulement plastique du coulis débute au niveau de l'extrémité chargée de l'ancrage et progresse vers l'autre extrémité entraînant un transfert de contrainte à la tige d'acier de plus en plus important à mesure que la contrainte de chargement dans l'ancrage augmente.

La différence principale entre l'étude présentée par YAP et RODGER (1984) et celle de HOLLINGSHEAD se situe au niveau du choix des critères de l'écoulement plastique pour les matériaux impliqués dans le système d'ancrage. YAP et RODGER ont fait appel au critère d'écoulement plastique de Von Mises pour l'acier de la tige et à un critère développé spécialement pour le béton, appelé critère d'écoulement octaédrique, pour le coulis et la roche.

Ces études théoriques permettent les observations suivantes:

- Le comportement à l'arrachement d'un ancrage injecté et le mécanisme de transfert des charges demeurent un problème très complexe;

- L'hypothèse de l'adhésion parfaite considérée pour les contacts tige-coulis et roche-coulis est satisfaisante et l'approche par éléments finis pour la prédiction du comportement à l'arrachement d'un ancrage injecté constitue un outil très satisfaisant que l'on doit vérifier et compléter par des essais en chantier (YAP et RODGER). En fait, aucun des paramètres mécaniques utilisés dans les deux modèles élastoplastiques ne tient compte des conditions en chantier du coulis de scellement et tout particulièrement des caractéristiques de surface et de la qualité des contacts tige-coulis et roche-coulis. La détermination des paramètres mécaniques impliqués dans ces deux modèles se fait à l'aide d'essais de traction ou de compression en laboratoire sur des échantillons de coulis et de roche.

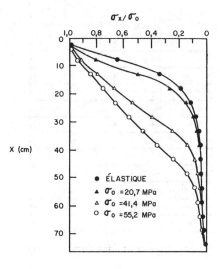

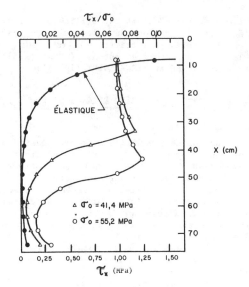

Figure 3. Distribution des contraintes axiales dans la tige et des contraintes de cisaillement au contact tige-coulis (HOLLINGSHEAD (1971))

Or, il a été montré (BALLIVY et al (1986)) qu'un produit de scellement présentant les meilleures caractéristiques mécaniques en laboratoire ne signifie pas forcément une meilleure résistance à l'arrachement.

4. ÉTUDE PAR ÉLÉMENTS FINIS DU COMPORTEMENT À L'ARRACHEMENT

Pour se pourvoir d'une approche plus fondamentale au dimensionnement d'un ancrage injecté dans une roche, une analyse par éléments finis du comportement à l'arrachement fut entreprise; celle-ci a abouti au développement d'un programme informatique (BENMOKRANE (1986)).

4.1 Formulation par éléments finis

Le type d'élément fini utilisé pour représenter le système d'ancrage (tige d'acier, coulis de scellement et roche) est un élément isoparamétrique quadrilatéral à huit noeuds tel qu'illustré dans la figure 4a. L'élément est également axisymétrique car la géométrie et le chargement d'un ancrage vertical considéré ici sont axisymétriques. Les contacts au niveau tige-coulis et roche-coulis furent modélisés par des éléments de faible épaisseur respectivement du côté coulis et du côté roche (figure 4b), car en général, dans un ancrage injecté, la rupture "tige-coulis" est caractérisée par une rupture dans le coulis, tandis que la rupture "roche-coulis" est caractérisée par une rupture dans la roche. Dans cette modélisation, on a considéré une adhésion parfaite entre les deux matériaux d'un interface. Cette hypothèse de l'adhésion parfaite au niveau des contacts tige-coulis et roche-coulis, utilisée par plusieurs auteurs, évite le recours à des éléments d'interface spéciaux et s'avère très satisfaisante pour ce type d'étude.

4.2 L'élément fini élastique parfaitement plastique

Dans la formulation de ce modèle, on a considéré que les trois matériaux impliqués dans l'ancrage (tige d'acier, coulis de scellement et roche) se comportent tel un milieu élastique parfaitement plastique avec perte de rigidité pour le coulis de

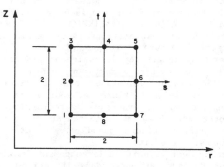

(r,z) : AXES DE COORDONNÉES GLOBALES

(s,t) : AXES DE COORDONNÉES LOCALES

a) ÉLÉMENT ISOPARAMÉTRIQUE À HUIT NOEUDS

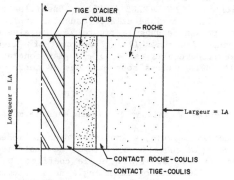

b) REPRÉSENTATION DE L'ANCRAGE

Figure 4. Élément isoparamétrique à huit noeuds et la représentation de l'ancrage utilisée dans la modélisation par éléments finis

scellement et la roche. Cette perte de rigidité survient une fois que l'élément en écoulement plastique atteint une déformation de cisaillement prise égale à 0.4%. La matrice de rigidité de l'élément est alors réduite à 1% de sa valeur initiale; cette matrice de rigidité réduite est utilisée dans l'assemblage de la matrice de rigidité globale à chacun des incréments de charge subséquents.

Le critère de Mohr-Coulomb est utilisé pour indiquer l'écoulement plastique dans le coulis de scellement et la roche. Ce critère utilisé pour les sols, les roches et les bétons (CHEN (1983)) s'écrit ainsi (NAYAC et ZIENKIEWICK (1972)):

$$F = \sigma_m \sin\phi + \sqrt{J_2}(\cos\theta - \frac{\sin\theta \sin\phi}{\sqrt{3}}) - S_o\cos\phi \quad (1)$$

avec F = fonction décrivant l'écoulement plastique; σ_m = contrainte moyenne; J_2 = deuxième invariant de la contrainte déviatrice; ϕ = angle de frottement interne; S_o = cohésion; θ = angle de Lode.

Pour l'acier, on a utilisé le critère de Tresca (cohésion seulement). L'utilisation d'un critère d'écoulement pour l'acier de la tige n'est en réalité pas nécessaire pour ce type d'étude, car on considère que l'on a atteint la rupture dans l'acier une fois que la charge totale appliquée a atteint la charge de limite d'écoulement plastique de la tige d'ancrage.

La méthode de la contrainte initiale avec une loi d'écoulement non associée fut utilisée pour résoudre ce problème non linéaire. Une fonction de la même forme que celle décrivant l'écoulement plastique a été retenue pour le potentiel plastique, l'angle de frottement étant remplacé par un angle de dilatation (NAYAC et ZIENKIEWICZ (1972)); une valeur égale à zéro est attribuée à cet angle pour un changement de volume plastique nul.

Un élément ayant atteint l'écoulement plastique, les contraintes excédentaires, en dehors de la surface d'écoulement plastique, sont considérées comme des contraintes initiales et sont alors converties en forces. Le processus de redistribution de la force s'arrête, et l'incrément de charge extérieure suivant débute quand un des critères ci-dessous est satisfait:

1. Le changement relatif de la composante maximum du vecteur déplacement entre deux itérations successives est inférieur à 1%;

2. Le nombre d'itération maximum fixé est atteint.

L'organisation et les séquences détaillées du programme informatique mis au point ont été publiées ailleurs (BENMOKRANE (1986)). Une illustration du maillage typique utilisé dans les applications est présentée dans la figure 5. L'épaisseur des éléments finis modélisant les contacts tige-coulis et roche-coulis est de 5 mm; les contraintes et les déformations sont évaluées au centre des éléments finis.

4.3 Paramètres mécaniques impliqués dans le modèle

Les paramètres mécaniques impliqués dans le modèle proposé sont la cohésion c, l'angle de frottement ϕ, le module d'élasticité E et le coefficient de Poisson ν. La cohésion et l'angle de frottement interviennent dans le calcul élastoplastique et permettent de savoir à l'aide du critère de Mohr-Coulomb si l'élément atteint l'écoulement plastique.

Les ruptures "tige-coulis" et "roche-coulis" s'effectuent par cisaillement. Les valeurs de la cohésion du coulis et de la roche à appliquer dans le modèle proposé peuvent donc être déterminées à partir d'essais de cisaillement direct. La conduite de tels essais en laboratoire est cependant compliquée (MARTIN (1981)) et en plus ces essais ne permettent pas de tenir compte rigoureusement des conditions en chantier, telles la rugosité (LADANYI et DOMINGUE (1980)). Il est proposé ici de réaliser plutôt des essais en chantier; et d'attribuer à la cohésion du coulis la valeur de la résistance au cisaillement ultime tige-coulis, τ_{t-c} ultime, déterminée à partir d'essais d'arrachement d'ancrages courts scellés dans les mêmes conditions que celles

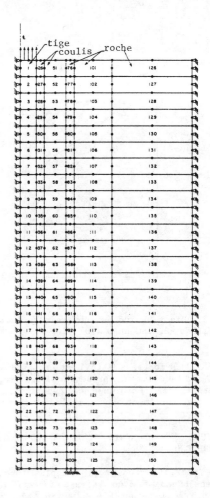

Figure 5. Configuration du maillage typique utilisé pour les applications

envisagées pour le projet considéré. Cette méthode permet ainsi de tenir compte de la rugosité de la paroi du trou de forage, de la résistance du coulis dans l'environnement concerné, et des géométries respectives de la tige et du forage. Le scellement étant très court, longueur scellée inférieure à 10 fois le diamètre de la tige, l'hypothèse d'une répartition uniforme des contraintes est alors justifiée (BALLIVY et MARTIN (1984)).

5. RÉSULTATS DU MODÈLE COMPARÉS AUX RÉSULTATS EXPÉRIMENTAUX

5.1 Essais d'arrachement sur des ancrages en chantier

Les résultats expérimentaux ont été obtenus à partir d'une étude en chantier (BALLIVY et al. (1984) et (1986)) qui a porté sur plus de 40 ancrages scellés avec différents produits de scellement à base de coulis de ciment. Les ancrags furent scellés en carrière à Sherbrooke, dans une roche volcano-détritique massive et compétente appartenant à la formation des Appalaches. Les barres d'acier utilisées, de type Dywidag, ont un diamètre de 36 mm et une charge ultime de 1 030 kN; les trous de forage, de diamètre égal à 76 mm, furent forés par rotation et percussion. Plusieurs ancrages furent instrumentés avec des jauges de déformation, collées sur les barres d'acier, afin d'évaluer les contraintes de cisaillement et la progression de la décohésion le long de la zone injectée durant un

essai de chargement. Une représentation schématique des ancrages est illustrée sur la figure 6.

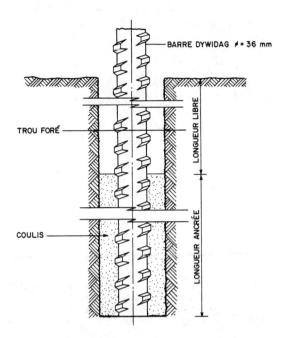

Figure 6. Représentation schématique des ancrages

5.2 Prédiction de la charge et du mode de rupture

L'évolution de la rupture de l'ancrage, en fonction
de la charge d'arrachement, est décrite analytique-
ment par la rupture successive des éléments finis
constituant le maillage de l'ancrage. L'incrément
de charge utilisé, pour suivre cette évolution jusqu'à atteindre la charge de rupture de l'ancrage, fut fixé à 20 kN; cette valeur représente
environ le 1/50 de la charge de rupture maximum des
ancrages considérés soit 1 030 kN.

Deux modes de rupture furent observés:

- Rupture au niveau du contact tige-coulis, lorsque les 25 éléments finis modélisant le coulis le
long de tout le contact tige-coulis (figure 5) ont
subi la rupture;

- Rupture dans la tige, lorsque la charge totale
appliquée, qui est égale à la charge incrémentale
cumulée, a atteint la limite d'écoulement plastique
de la tige soit 1 030 kN.

Un sommaire des valeurs des paramètres mécaniques
introduits dans le programme informatique est montré dans le tableau 1.

Le tableau 2 illustre les charges de rupture et
les modes de rupture, tels qu'obtenus d'une part
avec le modèle par éléments finis, d'autre part
avec les résultats obtenus sur les ancrages installés en chantier. On observe que le modèle prédit
des charges de rupture et des modes de rupture
semblables à ceux obtenus lors d'essais d'arrachement en chantier. Cette concordance entre les
résultats théoriques et expérimentaux se retrouve
pour l'ensemble des ancrages injectés avec les
différents produits de scellement utilisés.

La variation des charges de rupture théoriques et
expérimentales dans les ancrages injectés avec le
coulis de ciment usuel (Type C2) est reportée dans
la figure 7 pour différentes longueurs ancrées.
·Cette figure montre ainsi que la charge de rupture
théorique augmente de façon linéaire avec la longueur ancrée jusqu'à une longueur ancrée égale à
630 mm, comparativement à 650 mm pour les résultats
expérimentaux. Au-delà de cette longueur (630 mm),
la charge de rupture demeure constante et égale à
la charge limite de rupture de la barre.

Tableau 1. Sommaire des valeurs des paramètres
mécaniques introduites dans le modèle par éléments
finis

	Module d'élasticité (GPa)	Coefficient de Poisson	Angle de frottement (degré)	Cohésion (MPa)
Tige d'acier	200	0,3	0	505
Roche	59,8	0,25	50	38,1
Coulis C1	17,6	0,19	31	12,6*
Coulis C2	11,5	0,18	30	14,5*
Coulis C3	16,4	0,20	27	15,8*
Coulis C4	17,9	0,17	28	19,0*

C1 = Coulis de ciment usuel (E/C = 0,4)
C2 = C1 + poudre d'aluminium
C3 = C2 + fumée de silice
C4 = C3 + sable
* Résistance au cisaillement ultime tige-coulis, τ_{t-c}
ultime, telle qu'obtenue en chantier à l'aide
d'essais d'arrachement sur des ancrages.

Tableau 2. Résultats théoriques, obtenus par le
modèle, comparés aux résultats expérimentaux, obtenus d'essais d'arrachement en chantier sur des
ancrages

Type de coulis	Longueur ancrée (mm)	RESULTATS THEORIQUES		RESULTATS EXPERIMENTAUX	
		Charge de rupture (KN)	Mode de rupture	Charge de rupture (KN)	Mode de rupture
C1	360	500	t-c	492	t-c
C1	385	520	t-c	541	t-c
C2	115	180	t-c	185	t-c
C2	260	420	t-c	410	t-c
C2	350	560	t-c	550	t-c
C2	415	660	t-c	685	t-c
C2	500	800	t-c	760	t-c
C2	630	1 030	t-c	---	---
C2	660	1 030(1)	t	> 860(2)	p.a.
C2	710	1 030	t	1 030	t
C2	780	1 030	t	> 860	p.a.
C2	1 060	1 030	t	> 860	p.a.
C2	1 510	1 030	t	> 860	p.a.
C3	440	780	t-c	780	t-c
C3	470	820	t-c	800	t-c
C4	380	780	t-c	795	t-c
C4	400	820	t-c	870	t-c
C4	430	900	t-c	891	t-c
C4	440	920	t-c	910	t-c
C4	460	980	t-c	940	t-c
C4	470	1 000	t-c	1 000	t-c

(1) Charge de limite de rupture de la tige
(2) Charge de limite élastique de la tige
t-c Rupture tige-coulis
t Rupture de la tige
p.a. Ancrage pas arraché

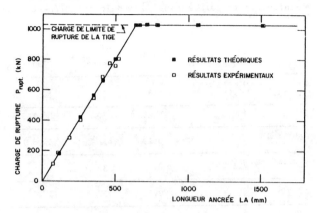

Fig. 7 Charges de rupture théoriques comparées aux charges de rupture expérimentales

5.3 Évolution de la rupture et des contraintes en fonction de l'augmentation de la charge d'arrachement

L'évolution de la rupture en fonction de l'augmentation de la charge d'arrachement pour deux ancrages, l'un court (LA = 260 mm) et l'autre long LA = 1510 mm), est illustrée sur la figure 8. La rupture des ancrages est due alors à une rupture du coulis situé le long des contacts tige-coulis. La rupture du coulis commence en haut de la zone ancrée et progresse vers le bas de l'ancrage à mesure que la charge augmente; ce qui rejoint donc les observations antérieures (BALLIVY et al. (1984) et BENMOKRANE (1986)) faites à partir des profils des contraintes de cisaillement obtenus sur les ancrages instrumentés, comme celui indiqué ici sur la figure 9.

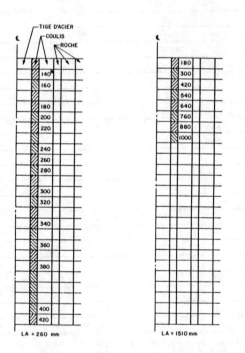

Fig. 8 Exemple de l'évolution de la rupture d'ancrage prédite par le modèle

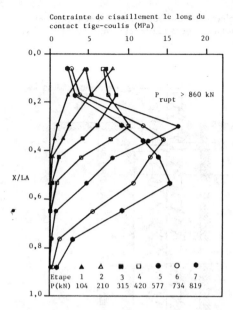

Fig. 9 Exemple de distributions de la contrainte de cisaillement le long du contact tige-coulis en fonction de la charge d'arrachement pour un ancrage instrumenté de longueur ancrée égale à 1060 mm

Les distributions des contraintes et des déplacements de la tige à l'extrémité tendue de l'ancrage, telles qu'obtenues par le modèle, sont montrées à la figure 10. Les courbes de contraintes sont présentées pour trois niveaux de chargement; le premier niveau correspond au comportement élastique, absence de rupture du coulis, tandis que les deuxième et troisième niveaux correspondent au comportement avec rupture du coulis. La forme des courbes demeure la même quelque soit la charge à laquelle les contraintes élastiques sont déterminées. La figure 10 conduit aux observations suivantes:

- L'allure générale des diverses courbes concorde avec les données expérimentales obtenues à partir des essais en chantier sur les ancrages instrumentés. Avant la rupture du coulis, une concentration de contraintes de cisaillement apparaît à l'extrémité tendue de l'ancrage, avec une diminution progressive vers l'autre extrémité de la zone ancrée;

- La rupture du coulis engendre un transfert de contraintes vers le bas de l'ancrage à mesure qu'elle progresse; le niveau de contrainte dans la tige se retrouve ainsi accru de façon très significative, comparativement au comportement élastique avant la rupture, comme le montrent les courbes des contraintes verticales et des déplacements;

- La rupture du coulis modifie non seulement les contraintes de cisaillement le long du contact tige-coulis mais aussi celles agissant le long du contact roche-coulis. Dans les deux cas, les courbes de distribution des contraintes de cisaillement perdent leur forme exponentielle dans la zone rupturée. Ceci peut s'expliquer par le fait que dans cette zone, le transfert de contraintes tige-coulis-roche devient plus faible (BALLIVY et MARTIN (1984));

- Avant la rupture du coulis, le déplacement vertical de la tige à l'extrémité tendue augmente linéairement avec la charge. Après l'initiation de la rupture, la courbe des déplacements devient non linéaire, le déplacement augmentant fortement avec la charge à mesure que la rupture du coulis progresse vers le bas de l'ancrage. Ceci est le résultat de l'accroissement relatif du niveau de contrainte verticale dans la tige, comparativement au comportement élastique.

766

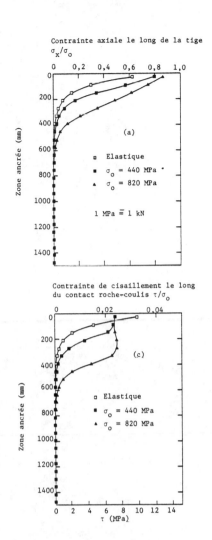

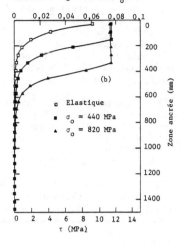

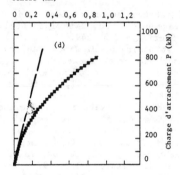

Figure 10 Un sommaire des résultats obtenus à l'aide du modèle pour un ancrage de longueur ancrée égale à 1 510 mm

6. CONCLUSIONS

Le modèle de calcul proposé, basé sur la méthode des éléments finis, peut être utilisé pour étudier le comportement à l'arrachement d'ancrages injectés dans une roche homogène. Il permet ainsi de prédire pour des ancrages de géométrie donnée, l'évolution de la rupture, le déplacement en tête de l'ancrage et les contraintes à mesure que la charge d'arrachement augmente. Les charges et les modes de rupture prédits concordent avec les charges et les modes de rupture observés sur les ancrages installés en chantier.

Les paramètres mécaniques impliqués dans le modèle sont le module d'élasticité, le coefficient de Poisson, l'angle de frottement et la cohésion. Les valeurs de la cohésion du coulis et de la roche à appliquer dans le modèle sont déterminées à partir d'essais d'arrachement sur des ancrages courts installés en chantier (rapport entre la longueur ancrée et le diamètre de la tige inférieur à 10). On attribue à la cohésion du coulis la valeur de la résistance au cisaillement ultime tige-coulis, τ_{t-c} ultime, et à la cohésion de la roche la valeur de la résistance au cisaillement ultime roche-coulis, τ_{r-c} ultime; l'utilisation d'un ancrage court rend négligeable l'effet de la contrainte normale, σ_n, et ainsi dans la relation $\tau = S_o + \sigma_n \tan \phi$ on peut dire que la résistance au cisaillement ultime n'est représentée que par la cohésion.

En l'absence d'essais triaxiaux l'angle de frottement, ϕ, peut être évalué par l'angle de frottement résiduel; celui-ci se détermine assez facilement par des essais de cisaillement direct.

Ce modèle pourrait être appliqué aussi aux massifs rocheux fracturés avec différents espacements et orientations pour les discontinuités. Une démarche analogue pourrait être entreprise par des travaux en chantier et sur modèles mathématiques; le massif rocheux fracturé étant simulé à l'aide d'un modèle du type proposé par LADANYI et ARCHAMBAULT (1970, 1972).

REMERCIEMENTS

Cette étude a été rendue possible grâce à l'aide financière du Conseil National de la Recherche, Ottawa (octroi no 1-42-48), du Ministère de l'Éducation du Québec (FCAR, 84 EQ 2051) et la participation active du personnel du Laboratoire de mécanique des roches de l'Université de Sherbrooke.

RÉFÉRENCES

Ballivy, G., Benmokrane, B. et Aïtcin, P.C. 1984. Les ancrages et la consolidation des parois rocheuses", Colloque international sur le renforcement en place des sols et des roches, Ecole Nationale des Ponts et Chaussées, Paris, 213-217.

Ballivy, G., Benmokrane, B. et Aïtcin, P.C. 1986. Le rôle du scellement dans les ancrages actifs injectés dans le rocher, à paraître dans Revue Canadienne de Géotechnique, Vol. 26, No 4.

Ballivy, G. et Martin, A. 1984. The dimensioning of grouted anchors, Proc. Int. Symp. on Rock bolting, ABISKO, Suède, Septembre 1983, édité par O. Stephanson, A.A. Balkema, pp. 353-365.

Benmokrane, B. 1986. Contribution à l'étude du comportement à l'arrachement et au fluage d'ancrages actifs injectés dans la roche, Thèse de Doctorat, Université de Sherbrooke, Département de génie civil, Sherbrooke, Québec, Canada, 311 p.

Chen, W.F. 1983. Plasticity in reinforced concrete, Mc Graw Hill Book Company, 475 p.

Coates, D.F. et Yu, Y.S. 1970. Three dimensional stress distributions around a cylindrical hole and anchor, Proceedings of the 2nd International Conference on Rock Mechanics, Belgrade, pp. 175-182.

Hanna, T.H. 1982. Foundations in tension ground anchors, Trans. Tech. Publications, Serie on Rock and soil Mechanics, First Edition, 574 p.

Hollingshead, G.W. 1971. Stress distribution in rock anchors, Canadian Geotechnical Journal, V. 8, No 4, 588-592.

Ladanyi, B. and Archambault, G. 1970. simulation of shear behaviour of a jointed rock mass, Proc. 11th Sympo. on Rock Mechanics, (AIME) p. 105.

Ladanyi, B. and Archambault, G. 1972. Evaluation de la résistance au cisaillement d'un massif rocheux fragmenté, Proc. 24th Int. Geol. Cong., Montreal, Section 13, p. 249.

Ladanyi, B. et Domingue, D. 1980. An analysis of bond strength for rock socketed piers, Proceedings of the Int. Conf. on Structural Foundations on Rock, Sidney, Mai 1980, édité par P.J.N. Pells, A.A. Balkema, pp. 363-373.

Martin, A. 1981. Contribution à l'étude de la répartition des contraintes le long d'un ancrage scellé dans le rocher. Mémoire de maîtrise, Université de Sherbrooke, Département de génie civil, Sherbrooke, Québec, Canada, 181 p.

Nayac, G.C. et Zienkiewicz, O.C. 1972. convenient form of stres invariants for plasticity, Proc. ASCE, V. 98, No ST4, pp. 949-954.

Yap, L.P. and Rodger, A.A. 1984. A study of the behaviour of vertical rock anchors using the finite element method, Int. J. of Rock Mech. and Min. Sci., V. 22, No 2, pp. 47-61.

Zienkiewicz, O.C. 1977. The finite element method in engineering sciences, Mc Graw-Hill Company, London, 3rd edition.

Three-dimensional stress state and fracturing around cavities in overstressed weak rock

Etat des contraintes tri-dimensionnelles et rupture autour de cavités en roche faible surpressionnée
Dreidimensionales Druckspannungsfeld und Bruch um Hohlräume im überbelasteten Weichgestein

STAVROS C.BANDIS, Norwegian Geotechnical Institute, Taasen, Oslo
JON LINDMAN, Norwegian Geotechnical Institute, Taasen, Oslo
NICK BARTON, Norwegian Geotechnical Institute, Taasen, Oslo

ABSTRACT: Investigations of the stress-state and the failure modes of unlined cylindrical cavities in poor quality ground were conducted by utilizing instrumented 3-D physical models. Stress redistribution around an opening in overstressed porous material seems to be influenced by an associated compaction/strain-hardening factor. The intermediate (axial) stress possesses a significant stabilizing potential. For the present type of material, shearing is the dominant failure mode. The failure planes develop in a pattern resembling logarithmic spiral slip lines. There are indications that a tensile stress field could develop locally under anisotropic stresses. Closed-form solutions, based on conventional failure criteria, show a moderate approximation to the physical reality.

RESUME: Des études de l'état de contrainte et du mécanisme de rupture de cavités cylindriques non tubées dans un rocher faible ont été menées en utilisant des modèles tri-dimensinnels instrumentés. La redistribution des contraintes autour d'une ouverture dans un matérial poreux en surpression semble être influencée par un facteur associé compactage/durcissement. La contrainte intermédiaire (Axiale) pesséde un important potentiel de stabilization. Pour le matériel étudié, le cisaillement est le mécanisme de rupture dominant. Les plans de rupture se forment selom un réseaus ressemblant à des lignes de glissement en spirale logarithmique. On trouve des indications qu'un champ de contrainte anisotropes. Les solutions analytiques, basées sur des critéres de rupture conventionnels, représentant des approximations modérées du phénomène physique.

SUZAMMENFASSUNG: In instrumentierten 3-D Modellversuchen wurden Untersuchungen des Spannungs entstandes und Bruch verhaltens nicht gesicherer zylinderischer Hohlräume in Gesteinen schlechter Qualität untersucht. Die Spannungsumlagerung in der Ungebung eines Hohlraumes in überbelasteten porösem Gestein Scheint von einem assozierten Verdichtungs/Verfestigungs Fakta abzuhängen. Die mittlere (axiale) Hauptspannung bestizt einen wesentlichen stabilisierenden Einfluz. Für das im Augenblick verwendete materiel ist der Scherbruch das vorherrschende Bruch verhalten. Das sich entwickelnde Bruch flächenmuster ähnelt logarithmisch-spiralförmigen scher linien. Es gibt Anzeichen, dass sich unter einem anisotropen Spannungszustand ein Zugspannungs feld entwickeln kann. Exakte lösungen, welche auf den bekannten Bruch Kriterien beruhen, sind mittelmässige Annäherungen an die physikalische Wirklichkeit.

1 INTRODUCTION

The problem of designing complex underground structures is increasingly focused on the concern for "...the inability of present analytical models (elasto-plastic with all kinds of sophistication) to predict apparition and discretization of lines of failure..." (1984 Meeting of the ISRM Commision on Rock Failure Mechanisms on Underground Openings). Diverse opinions are often heard about the basic mechanism of tunnel wall failures - shearing, extension or combinations of the two. More complications arise when excavating in an overstressed environment. Current engineering applications falling in the latter category are sub-sea tunnels, oil wells, deep-sited caverns for toxic waste storage and traditional deep-level mining.

The literature reveals that usually uniaxially and biaxially loaded models have been employed by experimentalists, to investigate the stability conditions, deformations and failure mechanisms around simulated tunnels. These models have employed various types of material (natural or model) with or without tunnel linings; e.g. Rummel (1971), Abel and Lee (1973), Worotnicki et.el. (1976), Kennedy and Lindberg (1978), Casarin and Mair (1981), Dhar et.al. (1981), Kaiser and Morgenstern (1981, 1982), Kaiser et.al. (1985). A few 3-dimensional test cases are also found, e.g. Stillborg et.al. (1979), Bakthar et.al., (1985). Finally, two studies by Sauer and Fornano (1979) and Konda et.al. (1981) are the only experimental studies known to the authors, where stress changes were monitored during excavation, under a 3-D primary stress field.

A systematic study, currently in progress at NGI, has been designed to investigate the mechanical performance of cylindrical openings in poor quality ground, by utilizing 3-dimensional physical models.

The technique has been applied successfully to investigate the stability of deep boreholes (Bandis and Nadim, 1985, Bandis and Barton, 1986) and unsupported tunnels in overstressed ground (Bandis, 1985, Bandis et.al., 1986). This paper presents recent findings on the 3-D stress redistribution and the failure modes around circular tunnels excavated in porous, weak rock under various conditions of overstressing.

2. DESIGN OF THE MODEL EXPERIMENTS

2.1 General

Artificial rock simulants were used in model blocks measuring 30 x 30 x 40 cm. The latter were pressurized in a polyaxial ($\sigma 1 \neq \sigma 2 \neq \sigma 3$) test facility. A special feature of the tests was the use of embedded miniature pressure cells for monitoring the primary and secondary stresses around the model tunnel. Furthermore, a line of small diameter holes drilled in a plane perpendicular to the excavation axis and grouted with dyed model material, helped in identifying the failure modes within the distributed zone.

2.2 Test facility - Instrumentation

The polyaxial test chamber and peripherals are shown in Fig. 1. The 3-D stress state was accomplished by using three pairs of steel flat jacks. Use of specially designed friction-reducing plattens enabled the elimination of undesirable end-effects. Excavation was simulated by manual operation of a circular cuttings tool (ϕ35 mm). Excavation debris was removed by application of a vacuum tube. The pressure sensors were of a commercially available type, circular in shape (12 mm in diameter and ~ 3 mm in thickness) and with a max. capacity of 5 bars.

Test chamber with model block in place.

FIGURE 1. Photographs of the three-dimensional pressure chamber for testing model tunnels.

Test rig assembled for three dimensional stressing.

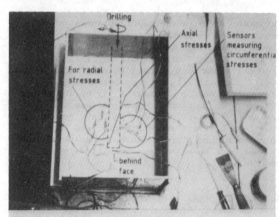

FIGURE 2. Emplacement of pressure sensors at suitable location for controlling the initial stress state and monitoring the stress perturbation upon drilling in 3-D.

The response of the sensors due to a deflection of a silicon beam, relies upon the deformation path of the matrix. This creates an inherant limitation in obtaining very precise measurements of the stress during load cycling. A technique involving "dedicated" calibration of each sensor has been devised, to allow for these effects. The sensors were installed along a section of the block to record the radial, tangential and longitudinal stresses (Fig. 2). The sensors were approached and finally passed in the drilling process, by at least three excavation diameters.

2.3 Modelling principles - Equivalent materials

The fundamental scaling relationship states that

$$\zeta = \lambda \cdot (\rho_p/\rho_m) \qquad (1)$$

where ζ = stress scale factor, λ = geometric scale factor and ρ_p and ρ_m are the the densities of the prototype rock and the model material respectively. Once λ is chosen, all of the prototype properties with specific force dimensions must be reduced by equation (1).

Multi-component rock-like materials were developed by varying the relative proportions of the wet and dry constituents of the gypsum cemented sand-based frictional/dense filler mixtures and selecting appropriate curing temperatures. Thus, a range of prototype rocks could be simulated - weak sandstone, chalk, marlstone etc. (Bandis, 1985).

The prototype tunnel was assumed to have a diameter of 5,0 m and, hence a scale of 1 $\simeq$ 145. Assuming a density ratio of 1,2, the stress scale factor ζ became $\simeq$ 170. The material used for the experiments reported herein, had a uniaxial compressive strength $(\sigma_c)_m \simeq$ 100 kPa and a cohesion $c_m \simeq$ 25 kPa. Conversion to the prototype scale gives $(\sigma_c)_p \simeq$ 15,0 MPa and $c_p \simeq$ 4,0 MPa. The modulus ratio was ~ 200-300, and $\phi \simeq$ 18o - 20o, hence, the material could be characterized as a marl or a shale.

2.4 Test procedure

Initially, the blocks were subjected to a consolidation phase under the desired primary stress field. A period of approximately 12 hours was usually adequate for stabilization of the block/flat jack system. Tunnel excavation (usual length $\simeq$ 30 cm i.e. 43,5 m at prototype scale) was completed within $\simeq$ 3 hours. After relaxation, the tunnel was loaded to collapse. Data relevant to the initial stress state of each block and the pressure monitor location are presented in Table 1.

Table 1. Initial stress levels and configuration of pressure cells in model tunnel expriments.

MODEL NO.	INITIAL STRESS STATE	INITIAL STRESS LEVEL	LOCATION OF SENSOR INSTALLATION
1	HYDRO-STATIC	$\sigma_v \simeq \sigma_H \simeq \sigma_h \simeq$ 200kPa$\simeq$2,0 + σ_c	axial ... max σ_ϑ ... behind face
2		$\sigma_v \simeq \sigma_H \simeq \sigma_h \simeq$ 400kPa 4,0 + σ_c	σ_r ... 1,2 ... 2,0 ... 4,0*$_r$
3	ANISO-TROPIC	$\sigma_H \simeq$ 300kPa, $\sigma_h \simeq$ 300kPa $\sigma_v \simeq$ 400kPa$\simeq$4,0 * σ_c σ_h : $\sigma_v \simeq$ 0,75	min σ_ϑ ... 3,6 1,7 ... 4,6 5,5*$_r$ ... axial max σ_ϑ

3. EXPERIMENTAL RESULTS

In all the experiments, the excavation was performed in small steps, approximately 1,0 - 1,5 times the diameter of the tunnel. The first changes in the initial stresses (assumed as the stress state after consolidation) were detected at different distances ahead of the advancing face, depending on the degree of stress anisotropy. The maximum effect on the tangential (σ_θ) and radial (σ_r) stresses was seen, when the excavation was adjacent to, and beyond the depth of the instrumented section. A summary of the experimental results is presented in Table 2. For each tunnel, the secondary stresses correspond to those measured at the end of excavation (stable condition) and at the instant of collapse respectively. The stress magnitudes in Table 2 have been corrected for the hysteretic response of the sensors, wherever applicable, following post-test "dedicated" calibrations.

Table 2. Summary of approximate magnitudes of secondary stresses (model scale) at various experimental stages (r = tunnel radius, x = distance from tunnel centre).

Model	Initial average stresses (kPa)	x/r	SECONDARY STRESSES (kPa) End of excavation			Instant of collapse		
			σ_ϑ	σ_r	σ_a	σ_ϑ	σ_r	σ_a
1	$\sigma_v \simeq \sigma_h \simeq 200$ $\sigma_H \simeq 180$	1,2	230	65		190	180	
		2,0	200	100	150	410	350	–
		4,5	160	140		300	360	
2	$\sigma_v \simeq \sigma_h \simeq$ $\sigma_H \approx 400$	1,2	480	80		260	80	
		2,0	470	–	180	600	–	80
		4,0	380	220		400	400	
			Max σ_ϑ	Min σ_ϑ		Max σ_ϑ	Min σ_ϑ	
3	$\sigma_v \approx 400$ $\sigma_H = 300$ $\sigma_h = 200$	1,7	380	–		280	–	
		2,1	–	290		–	320	
		9,7	–	280	370	–	345	100
		3,6	500	–		–	–	
		4,6	420	–		445	–	
		5,5	420	–		400	–	

σ_θ = tangential, σ_r = radial, σ_a = longitudinal (axial) stresses.
σ_v = overburden pressure, σ_H and σ_h = maximum and minimum horizontal stresses.

4. STRESS REDISTRIBUTION

The excavation of a tunnel relieves the radial stresses (σ_r) and the load is transferred around the circumference as a hoop stress (σ_θ). Overstressing of the wall material, will create a fractured zone around the tunnel. Both experiments and field experience suggests that, under a wide range of stress combinations, the fractured zone retains part of its load carrying capacity, while its deformation modulus decreases greatly and causes stress transfer away from the opening.

The experimental stress values σ_θ, σ_r, and σ_a were normalized w.r.t. the primary field stresses σ_v, σ_h, and σ_H, respectively. The measured stress redistribution curves around the modelled tunnels No. 1, 2 and 3 are presented in Figures 3, 4 and 5. Also included are the stresses in fractured zones, as predicted by a 2-D elasto-plastic closed form solution for each experimental case. The numerical model utilizes the linear Coulomb criterion of material behaviour and has been described elsewhere (Bandis et al. 1986).

4.1 Radial stresses (σ_r)

Readjustment of σ_r agrees reasonably well with the numerical prediction from the assumed 2-D plane strain solution (Figs. 3a and 4a). However, considerable difference in the magnitudes of σ_r can be seen at the instant of failure of tunnels No. 1 and 2 (Figs. 3b and 4b), which can be attributed to the different extent and nature of the predicted "plastic" and the real fractured zones.

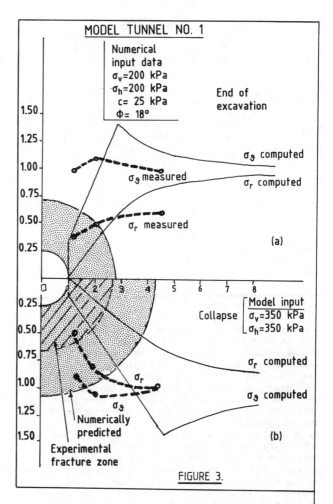

FIGURE 3.

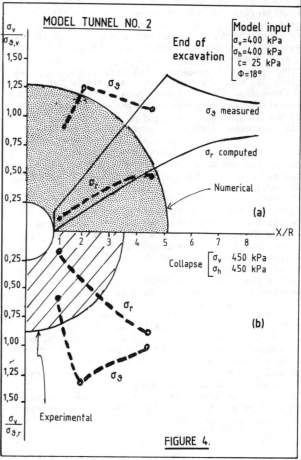

FIGURE 4.

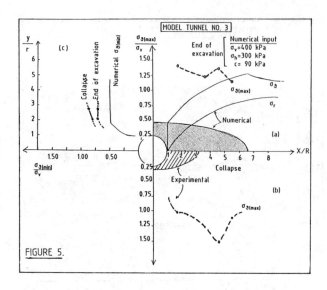

FIGURE 5.

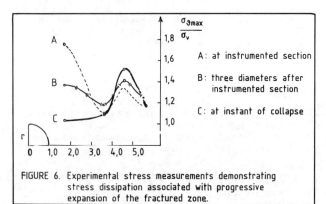

FIGURE 6. Experimental stress measurements demonstrating stress dissipation associated with progressive expansion of the fractured zone.

4.2 Tangential stresses (σ_θ)

A marked peaking in the value of maximum σ_θ is dominant in both the highly stressed tunnels (Figs. 4a and b, 5a and b). By contrast, only a minor peaking was recorded in the case of tunnel No. 1 (Fig. 3). A plausible explanation is that the excavation-induced σ_θ caused material compaction, thus, volume reduction, which enabled absorption of the near wall redistributed stresses. Material contraction and the "hardening" factor associated with porous materials, was clearly detected in our experiments.

The later may be responsible for limiting the extent of the fractured zones. For the present material type and stress levels, the threshold point of the σ_θ curves, (conceptually denoting transition from the "plastic" to the "elastic" region) appeared at 2-4 radii from the tunnel centre (at the instant of collapse). The corresponding numerical predictions were 3-7 radii. The progressive expansion of the fractured zone, (indicated by shifting of the σ_θ peak) up to kinematic initiation of failure, is clearly demonstrated in Fig. 6.

The $\sigma_\theta$max peak concentrations were between approximately 1.1 and 1.9. Variations were due to the different stress levels, stress anistropy and stage of excavation (see Fig. 7). With the exception of tunnel No. 1, the $\sigma_\theta$max magnitudes were not unrealistically different from those predicted numerically.

The observations of Sauer and Fornano (1979), were also qualitatively similar. These authors also pointed out the development of altering stressed and destressed zones. The latter is seen in Fig. 5 for our model No.3.

Intuitively circumferential compaction of a zone under tangential compression could create a ring of increased stiffness acting as a stress absorbing bearing shell, thus creating an understressed surrounding zone within a certain radius of influence. Upon failure of the stiff near-wall zone, the adjacent understressed ring would act as the new bearing shell, with the mechanical process being repeated interactively in diminishing fashion with increasing distance from the tunnel wall.

Important stress changes were detected on the plane of the minimum tangential stresses. The maximum stress concentration ($\sigma_\theta$min/σ_h) at 2-3 radii from the centre was between 1.3-1.4 and occurred upon the end of excavation, when the tunnel walls suffered the maximum anisotropic yielding. The result was a radial decompression of the roof and a build-up of compressive stresses. Tensile cracks above the tunnel roof parallel to the minimum horizontal stress (σ_h) have been identified in post-failure tunnel sections (see Fig. 8b).

4.3 Axial stresses (σ_a)

The intermediate principal stress (σ_a) plays an important stabilizing role (Bandis et.al. 1986). Under the anistropic stresses of tunnel test No.3, σ_a had increased by 20-30% at the end of excavation. The effect could be related to shear fracturing, and associated dilation phenomena. No such change was monitored in the case of the hydrostatically stressed tunnel No. 2.

4.4 Stresses ahead of the tunnel face

Under anisotropic field stresses (tunnel No. 3) a strong anisotropic compression field was detected at a distance of 2-3 diameters ahead of the excavation face (Fig.7). Abel and Lee (1973) observed stress changes at 2-4 diameters ahead of simultated tunnels under a uniaxial compressive field. Under hydrostatic primary stresses, the disturbed zone appears restricted to less than one diameter ahead (tunnel No. 2). Konda et.al. (1981) also detected a region of maximum compression at 1.0-2.0 diameters in front of a simulated tunnel excavation in a cohesive earth simultant under isotropic stress field. Sauer and Fornano (1979) also observed strain peaks at 1.0 diameter distance.

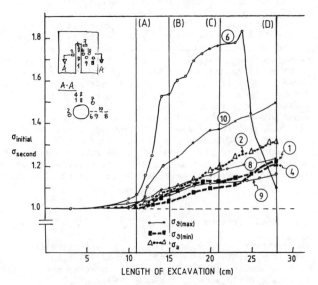

FIGURE 7. Relative changes in the 3-D stress field ahead and behind the excavation face of tunnel no. 3 (anisotropic stresses).

(A) - 3 diam. before instrumented section
(B) - 1,5
(C) - at instrumented section
(D) - 2,3 diam. after instrumented section

5. FRACTURED ZONES - FAILURE MECHANISM

The fractured zones formed after tunnel collapse were traced on sectioned planes and are presented for comparison in Fig. 8. Geometrically, the curved fracture planes resemble logarithmic spiral slip lines. Conceptually, a slip line is a curve whose tangent at every point denotes the shear direction at that point. (Janikis, 1983). Assuming that at limit equilibrium, two possible rupture surface curves exists through every point in the rock, it is possible to derive an axisymmetric fracture zone, as shown in Fig. 9a and b. A comparison of the fractured zone around tunnel No.2 with a computed slip line pattern shows a remarkable agreement (Fig.9c)

The above concepts appear compatible with physical reality. The line-drilled grouted holes demonstrated that shearing is the main failure mechanism within the circular (Fig. 8b and 10a) and elliptical (Fig.8c and d) fracture zones, depending on the stress anisotropy (Fig. 10a).

Numerical investigations by Deaman and Fairhurst (1972) and Kobayashi et.al. (1981) indicate inclination of the fracture zone to the tunnel axis near the tunnel face. The effect is clearly seen in the photograph of Fig. 10b. Mechanically it can be attributed to the relaxation of the axial stress near the face and reorientation of the maximum shear stress between the tangential and axial stresses.

The observations of model tunnel performance have clearly indicated that the degree of compaction should be treated as an integral element of models during the analysis of tunnel stability in poor quality ground. As a preliminary step to allow for compaction effects, an approximate zoning of the material surrounding tunnel No. 3 has been attempted as shown in Fig. 11.

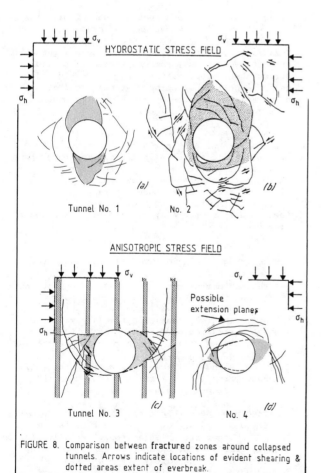

FIGURE 8. Comparison between fractured zones around collapsed tunnels. Arrows indicate locations of evident shearing & dotted areas extent of everbreak.

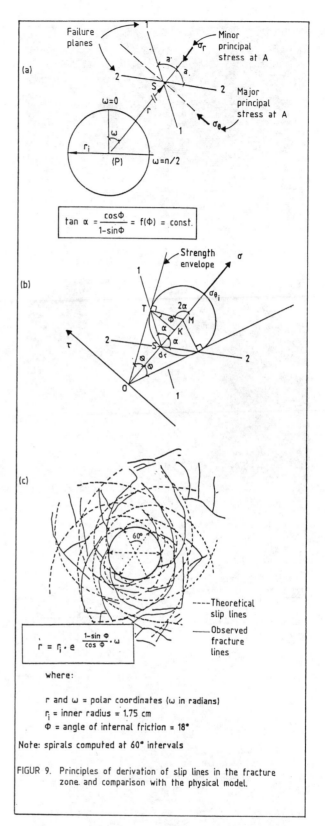

$$\tan \alpha = \frac{\cos \Phi}{1 - \sin \Phi} = f(\Phi) = \text{const.}$$

$$r = r_i \cdot e^{\frac{1 - \sin \Phi}{\cos \Phi} \cdot \omega}$$

where:

r and ω = polar coordinates (ω in radians)
r_i = inner radius = 1.75 cm
Φ = angle of internal friction = 18°

Note: spirals computed at 60° intervals

FIGUR 9. Principles of derivation of slip lines in the fracture zone and comparison with the physical model.

6. CONCLUSIONS

1. Interpretation of physical measurements of stress changes around tunnels excavated through weak, porous rocks, suggests that the associated compaction/strain-hardening factor may play a significant role in the mechanics of stress redistribution and failure.

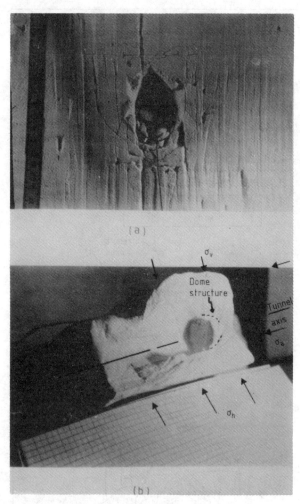

FIGURE 10. View of fractured zone around the failed model tunnel no. 2.

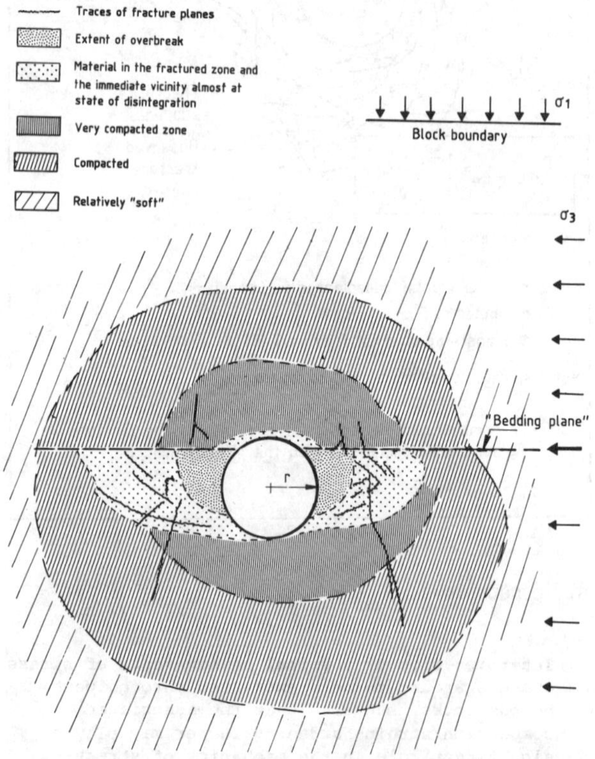

2. If the compressive stresses around the tunnel are high enough compared to the strength of the rock material, compaction of the latter will occur causing volume reduction. Thus, the initial build-up of tangential stresses will be accomodated without significant lateral straining. In addition, the compacted zone will acquire higher stiffness and will form an effective load-bearing shell, surrounded by an understressed ring of material extending within the zone of mechanical influence of the former.

3. The latter stress state will probably apply until overstressing reaches the level required for crack development with the gradual creation of a fractured zone. As the deformation modulus of the zone decreases, quick redistribution of the stresses will occur outwards. The fractured zone will retain part of its load-bearing capacity and may expand in area to a maximum critical size. The latter can be expected to be influenced by material properties (viz. friction), the degree of stress anisotropy, the degree of compaction and the axial stress; since the effect of the collapse is clearly three dimensional and depends also on the degree of freedom in the longitidiual direction.

4. The mechanics of fracturing appear qualitatively similar for both isotropic and anisotropic stress states. Differences concern the magnitudes of the secondary stresses. Under both states the main failure mode in the fractured zone is shear. Anisotropic stress may cause a "tertiary" stress state at the roof of the ellipse, strongly tensile in nature.

5. It is possible to distinguish zones of material with extreme differences in consistency, ranging from hard and compacted to nearly disintegrated. This is of course, the end effect of a sequence of events involving compaction → volume reduction → stress absorption → stiffening → formation of bearing shells → cracking → stress dissipation → propagation of cracks → extension of compacted zone → yielding → final collapse.

6. The fracture planes in 2-D can be approximated by logarithmic spiral slip lines. The outer boundary of the fractured zone may be circular or elliptical depending on stress anisotropy. The extent of the fracture zone may range from 2-4 radii from the tunnel centre. Classical closed-form solutions generally give an overestimated fracture zone.

7. For the range of field stresses applied in the present tests ($\sigma_v \approx 4 \cdot \sigma_c$, $\sigma_v/\sigma_h = 0.75$ or $\sigma_v = \sigma_h$) the maximum tangential stress concentrations (σ_θ/σ_v) at the instant of tunnel collapse were 0.55 - 1.00 near the wall and 1.25-1.50 at the "plastic-elastic" transition boundary with the highest stress ratio corresponding to the anisotropic stress field. The highest concentration was $1.8 \cdot \sigma_v$ and occured approximately 1.0 diameter behind the face. The minimum tangential stresses were 1.4 - 1.7 times higher than the minimum horizontal field stresses σ_h (≈ 0.8 to $0.9 \cdot \sigma_v$). The axial stress (σ_a) remained unchanged when excavating under hydrostatic field stresses, but increased by a factor of 1.2 - 1.3 under stress anisotropy. Finally, stress changes ahead of the excavation were located at distances of approximately 1.0 - 4.0 radii depending on stress anisotropy.

Traces of fracture planes

Extent of overbreak

Material in the fractured zone and the immediate vicinity almost at state of disintegration

Very compacted zone

Compacted

Relatively "soft"

FIGURE 11. Approximate zoning of the material around a tunnel under anisotropic stresses, according to the degree of material compaction.

8. Comparisions between the physical model experiments and predictions from a 2-D elasto-plastic closed-form numerical model using the linear Coulomb criterion offer as expected, only a moderate agreement.

9. It is important that future efforts are directed to a detailed investigation of the compaction/strain-hardening mechanism, and its effects on the stress state, in parallel with the search for appropriate constitutive models of weak rock behaviour.

ACKNOWLEDGEMENTS

The work described in this paper was funded by Statoil, Norsk Hydro, Norsk Agip, Conoco Norway, Saga, Fina Exploration: Permission to publish is gratefully acknowledged.

REFERENCES

Abel, J.F. and F.T. Lee (1973)
Stress changes ahead of an advancing tunnel. Int. J. Rock Mech. Min. Sci. and Geom. Abstr., Vol. 10, pp. 673-697.

Bakhtar, K., Black, A. and R. Cameron (1985)
Load response of modelled underground structures. Proc. 26th U.S. Symp. of Rock Mechs., Rapid City, pp. 1255-1260.

Bandis, S.C. (1985)
Physical models of tunnels in overstressed rock. Fjellsprengningsteknikk - Bergmekanikk - Geoteknikk - Proceedings, Trondheim, Tapir, pp. 33.1 - 33.17.

Bandis, S.C. and F. Nadim (1985)
Stability of a deviated wellbore. NGI Contract Report No. 85619, p. 48.

Bandis, S.C. and N. Barton (1986)
Failure modes of deep boreholes. Proc. 27th U.S. Symp. on Rock Mechs., Alabama, pp. 599-605.

Bandis, S.C., Nadim, F., Lindman, J. and N. Barton (1986)
Stability investigations of modelled rock caverns at great depth. Proc.Int.Symp. on Large Rock Caverns, Helsinki, Vol.2, pp. 1161-1170.

Casarin, C. and R.J. Mair (1981)
The assessment of tunnel stability in clay by model tests. In soft Ground Tunnelling - Failures and Displacements, Rotterdam, Balkema, pp. 33-44.

Daemen, J.J.K. and C. Fairhurst (1972)
Rock failure and tunnel suppport loading. Proc. Int. Symp. for Underground Construction, Luzern, pp. 356-369.

Dhar, B.B., Ratan, S, Sharma, D.K. and P.M. Rao (1981)
Model study of fracture around undergoing excavations in weak rocks. Proc. Int. Symp. on Weak Rocks, Tokyo, pp. 267-271.

Jumikis, A.R. (1983)
Rock Mechanics, Trans. Tech. Publ., 3rd ed., pp. 613.

Kaiser, P.K. and N.R. Morgenstern (1981) (I, II), 1982 (III)
Time-dependent deformation of small tunnels - I. Experimental facilities, II Typical test data, III Pre-failure behaviour. Int. J. Rock. Mech. Min. Sci. and Geom. Abstr., Vol. 18, pp 129-140, 141-152, Vol. 19, pp. 307-324.

Kaiser, P.K., Guenot, A, and N.R. Morgenstern (1985)
Deformation of small tunnels - IV. Behaviour during failure. Int. J. Rock. Mechs. Min. Sci. and Geom. Abstr., Vol. 22, pp. 141-152.

Kennedy, T.C. and H.E. Lindberg (1978)
Model tests for plastic response of lined tunnels. Jl.of the Eng. Mechs. Div., Proc. ASCE, Vol.104, No. EM2, April, pp. 399-420.

Kobayashi, S., Tamura, T., Nishimura, N. and Y. Mochida (1981)
Stresses and deformations around tunnel face in soft rock. Proc. Int. Symp. on Weak Rock, Tokyo, pp. 813-818.

Kanda, T. Inokuma. A, and K. Kanto (1981)
Three-dimensional model test on soft ground tunnels. Proc. Int. Symp. on Weak Rock, Tokyo, pp. 807-812.

Rummel, F. (1971)
Uniaxial compression tests on right angular rock specimens with central holes. Proc. Rock fracture, Int, Symp. Rock. Mech., Nancy 1921, Vol. 1, pp. 90-101.

Sauer, G. and M. Fornano (1979)
Techniques and materials for modelling a tunnel driving operation at moderate depth. Proc. Int. Colloquium on Physical Geomechanical Models, ISMES, Bergamo, Italy, pp. 83-95.

Stillborg, B., Stephansson, O. and G. Swan (1979)
Three dimensional physical model technology applicable to the scaling of underground structures. Proc. 4th Int. Cong. of ISRM, Montreux, Vol. II, pp 655-662.

Worotnicki, G., Wold, M., Enever, J. and R. Walton (1976)
The application of physical modelling techniques to problems in tunnelling. Proc. 2nd Australian Tunnelling Conf., Melbourne, pp. 25-40.

Characterization of rock mass by geostatistical analysis at the Masua Mine
Caractérisation du massif rocheux par analyse géostatistique à la mine Masua
Charakterisierung von Gesteinsmassen mit Hilfe der geostatistischen Analyse in der Masua Grube

G.BARLA, Structural Engineering Department, Technical University of Turin, Italy
C.SCAVIA, Structural Engineering Department, Technical University of Turin, Italy
M.ANTONELLIS, Mining Italiana S.p.A., Roma, Italy
M.GUARASCIO, Mining Institute, University of Trieste, Italy

ABSTRACT: The purpose of the present paper is to show the application of Geostatistical Analysis to rock mass characterization, with the Masua Mine (Sardinia, Italy) being the case study. Following a brief description of geostatistical methods and use of "Variograms" and "Kriging", the criteria applied for quantification of rock mass properties are discussed. A comprehensive engineering geological study of the rock mass, including joint mapping, borehole and core logging, furnishes the data needed for the characterization. The parameters describing rock discontinuities are treated as "regional variables" in statistics terms. At the same time, the spatial correlation and variability within the rock mass are represented. The analysis of rock mass data by geostatistical procedures allows one to infer the variability of rock quality such as defined by the RMR index. Additionally, the deformability and strength parameters are obtained, with the purpose to use them in numerical modeling of mine structures. A comparison of predictions of rock mass parameters, based on Geostatistical Analysis and Monte-Carlo simulation is given. It is stated that the use of geostatistical methods such as "Kriging" is more likely to reflect the expected rock mass mechanical behavior.

RESUME: La communication illustre l'application de techniques du type géostatistique à la caractérisation du massif rocheux, en se référant en particulier à la mine de Masua (Sardaigne - Italie). Après une brève description des méthodes géostatistiques et des modalités d'emploi du "variogramme" et du "Kriging", on présente les critères utilisés pour la quantification des propriétés caractéristiques du massif rocheux. Une étude géostructurale de détail du massif rocheux fournit les données nécessaires pour la caractérisation. Les paramètres caractéristiques des discontinuités sont traités, en termes statistiques, comme des "variables régionalisées" en étudiant leur corrélation spatiale et leur variabilité à l'intérieur du massif rocheux. L'analyse des données du massif rocheux par les méthodes du type géostatistique permet d'obtenir la variabilité spatiale de l'indice de qualité de la roche RMR et donc des paramètres de deformabilité et de résistance à utiliser dans le modèle numérique des structures de la mine. On mentionne enfin les valeurs des paramètres du massif rocheux évaluées à travers l'analyse géostatistique et la technique de simulation de Monte-Carlo; il ressort, de la comparaison, que l'utilisation de méthodes géostatistiques telles que le "Kriging" permet de mieux représenter le comportement mécanique du massif rocheux.

BERICHT: Der Zweck dieser Arbeit ist, die Anwendung von geostatistischen Techniken für die Charackterisierung der Gesteinsmassen zu erläutern, wobei sich hierbei insbesondere auf das Bergwerk in Masua (Sardinien - Italien) bezogen wird. Nach einer kurzen Beschreibung der geostatistischen Methoden und der Anwendungsmodalitäten des "Variogramms" und des "Kriging" werden die für die Quantifizierung der charakteristischen Eigenschaften der Gesteinsmassen verwendeten Kriterien näher betrachtet. Eine detaillierte geo-strukturelle Untersuchung der Gesteinsmassen liefert die für die Charakterisierung notwendigen Daten. Die charakteristischen Diskontinuitätsparameter werden im statistischen Begriff "variabel regionalisiert" behandelt, und ihre räumliche Wechselbeziehung, sowie die Variabilität im Innern der Gesteinsmassen werden untersucht. Die Analyse der Daten der Gesteinsmassen, ausgeführt durch die geostatistische Methode, gewährt die Gewinnung der Raumvariabilität des Qualitätsindexes des Gesteins RMR und somit der Umformbarkeits - und Widerstandsparameter, die für das Verfahren der numerischen Modellierung der Bergwerksstruktur zu verwenden sind. Zum Schluß werden noch die Werte der Parameter der Gesteinsmassen über die geostatistische Analyse ermittelt, sowie die Monte-Carlo simulationstechnike. Aus dem Vergleich geht hervor, daß durch die Anwendung von geostatistischen Methoden, wie "Kriging", eine bessere Darstellung des mechanischen Verhaltens der Gesteinsmassen erreicht wird.

1 INTRODUCTION

Quantification of significant properties of rock mass (i.e. strength and deformability characteristics) is one of the most important aspects of site characterization for input to design of underground excavations. In general, the values of these properties are assumed to fit the normal distribution and the arithmetic mean is considered as representative value.

This procedure may or may not be adequate for design, depending on the actual distribution that fits the data, dispersion of the data about the mean, the size of the sample (number of tests), and the specific sensitivity of the rock mass behavior to a variation in the properties. Methods for projecting (forecasting) properties from one location to other locations need be developed, so as to be able to incorporate the rock mass variability into a design analysis.

One possible scheme for quantification of significant properties is the use of a simulation method, such as Monte-Carlo. With this method, the rock mass is subdivided in a number of regions. On the basis of the histogram distribution of a property, needed for design, a characteristic value is given to each region. Monte-Carlo simulation is applied by assuming that each variable is independent of the location in the rock mass. This implies that the spatial variation in rock mass properties is being neglected (La Pointe, 1980).

The application of geostatistical analysis to rock mass characterization is considered in the present paper. This technique, that holds promise in actual design, allows one to account for the spatial variation of variables. Following a few remarks on geostatistical analysis, the techniques used for the characterization of rock mass at the Masua Mine (Sardinia, Italy) are discussed. The variability of rock mass ratings around mine openings is estimated in conjunction with the strength and deformability characteristics.

2 GEOSTATISTICAL ANALYSIS

In conventional probabilistic analyses, all the stochastic variables are assumed to be independent and are defined to fit a known distribution. In rock mass characterization, this leads to properties that are not dependent on a specific location in the rock mass.

The variability of rock mass properties, such as strength and deformability characteristics, is only "partially random". These properties are such that, for two values of a given quantity, located in near proximity, there is a likelihood to be equal. Similarly, two samples of the same property taken at two different locations in the rock mass, may result in a greater possibility of them being given different values.

The uncertainty in observing the three dimensional distribution of each variable does not permit one to evaluate its representativeness for purpose of engineering design. Also, the most reliable scheme for sampling a given property within the rock mass is not easily defined.

"Geostatistics", based upon the theory of "Regionalized Variables" (Matheron, 1965), allows one to overcome some of the difficulties stated above, which characterize conventional probabilistic analyses. The following is achieved:

1) study of the spatial variation of rock mass characteristics, by means of a limited site exploration program: the tool of "Variogram" is used to define the appropriate "spatial correlation" between sets of data;

2) define a complete sampling scheme, given that the "Variogram" indicates the existence of an allowable spatial variation of the most significant variables in geological-engineering terms; the purpose is to obtain the information needed to characterize the site for design, at a minimum cost;

3) determine the mean value of design variables within "regions" which subdivide the rock mass; the minimum error in the estimate is obtained by the "Kriging" method;

The "Variogram" concept and the "Kriging" method are briefly reviewed in the following. For a complete study of the subject, reference is made to the textbooks of Matheron (1965), Journel (1973), and Guarascio et al. (1975).

2.1 "Variogram"

Given a set of data for the "Regionalized Variable" $z (\vec{x})$, sampled at a number of locations in the rock mass, the variance 2γ is defined as:

$$2 \dot{\gamma} = \frac{1}{n} \sum_{i}^{n} \left(z (\vec{x}) - z (\vec{x} + \vec{h}) \right)^2$$

where the n values are taken at a distance $\vec{h}$, one from another one.

The set of variance 2γ, obtained by changing the interval h between two data points, is defined as a "Variogram". Plotting of 2γ in a x - y plot and fitting to the n data points a mathematical model allows one to illustrate the variation in $z (\vec{h})$, in the direction of interest $\vec{h}$. With reference to a spherical model and the variogram shown in Figure 1, the following most important characteristics are noted:

1) $2 \gamma (h)$ increases with h increasing, up to the "Sill" $2 \gamma (h_s)$, i.e. the variance of all data that have been sampled;

2) the limit of h, defined as "Range" (h_s), stands for the interval above which the spatial variation is no longer applicable;

3) by extrapolating the variogram back to the y axis, an intercept is obtained representing the "Nugget"; this is a measure of that portion of variance which is completely random and is not defined according to a spatial variation; this variation which leads to a spatial correlation is as good as small is the nugget and as large is the range.

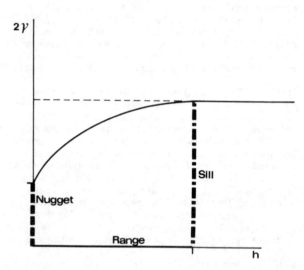

Figure 1. Example of a typical "Variogram".

2.2 "Kriging"

"Kriging" is to permit the estimate of the mean of the "Regionalized Variable" $z (\vec{h})$, in each region used to subdivide the rock mass. This estimate is effected on the basis of the group of available data and the variogram so as to make the error minimum.

The procedure is to account for:

1) the spatial variation of $z (\vec{h})$, according to the variogram;

2) the relative location of the samples taken with respect to the region where the estimate is to be effected;

3) the geometry of each region;

4) the sampling scheme.

On the basis of the range determined by the variogram, the limit for the validity of the correlation is established in each direction. Also, the most desirable sampling scheme is chosen. As an example, Figures 2 and 3 show a scheme proposed by La Pointe (1980) to sample the discontinuities and estimate the spacing between them.

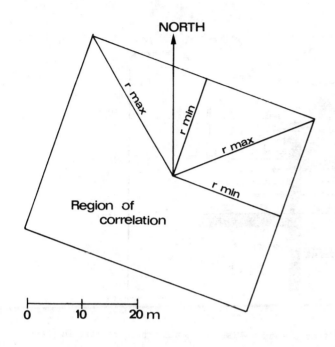

Figure 2. Average region of influence of the fracture frequency at Lannon (La Pointe, 1980).

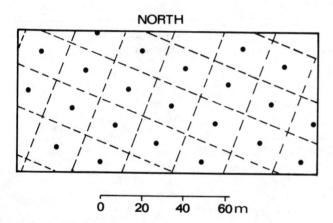

Figure 3. Sampling pattern suggested by the geostatistical analysis of Lannon. Dashed parallelograms define the region of influence for a single sample (La Pointe, 1980).

3 MASUA MINE - AVAILABLE DATA

The methods of geostatistical analysis have been applied to rock mass characterization at the Masua mine, located on the southwestern coast of Sardinia, 70 km West of the city of Cagliari (Italy). The Masua mine, in the mining district of Iglesiente, is considered to be the largest Italian lead and zinc center.

The production of sulphides has attained a yearly level of $7 \cdot 10^5$ t of raw material with a Pb - Zn grade of about 5 per cent. The Pb - Zn sulphide mass strikes North - South, with the footwall and hangingwall consisting of a hard dolomitic limestone. A schist formation is also present at the footwall, to the West of the main orebody (Figure 4). The mass, mined by the sublevel - stopping method, is approximately 370 m high, between levels 320 and -50 (Figure 5). The length ranging from 100 to 350 meters, the width lies between 30 and 60 meters.

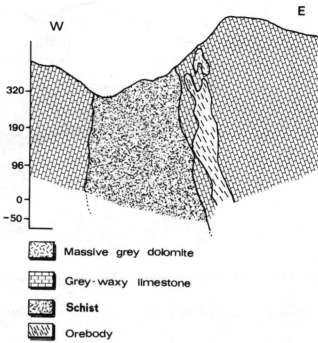

Massive grey dolomite

Grey-waxy limestone

Schist

Orebody

Figure 4. East-West geological section of the Marx Deposit, Masua Mine.

A typical cross section of the orebody, taken in the East - West direction and where the same orebody extends to its maximum height, is shown in Figure 6. At present, two mine pannels, between levels 320 and 191 and between levels 191 and 96, have been excavated. A vertical pillars is being left in place to support the walls (Figure 5), as crown pillars are being created, in order to separate between them the open-stopes which are being formed.

3.1 Rock material properties

Laboratory tests were performed on rock samples obtained by diamond drilling of both limestone and mineralized limestone, so as to obtain their most important physical and mechanical properties as shown in Table 1 (Barla, 1984).

3.2 Rock mass data

In order to characterize the rock mass for the purpose of numerical modeling (°) it was decided to gather the necessary information by: (1) diamond drilling and careful borehole and core logging; (2) detailed drift mapping at specific sites, in the orebody and in the adjacent rock.

A total of 400 m rock cores were drilled at various levels. Also, a systematic discontinuity mapping was carried out at different locations in the mine. The most attention was devoted to the rock mass above level 96 (Table 2).

(°) Numerical modeling of the mine structure was carried out first by a Two-Dimensional Finite Element Model (Barla, 1984). Also a Three-Dimensional Model was applied more recently, based upon the Displacement Discontinuity Method (Barla, 1986). In all cases, the mean value of each mechanical property of rock mass was considered as representative for use in design.

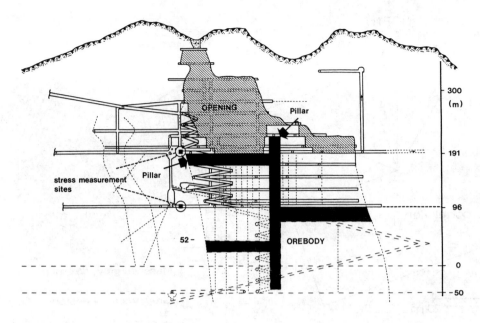

Figure 5. North-South cross section of the Marx Deposit at the Masua Mine. The pillars to be left in place are also shown.

Table 1. Results of laboratory tests on limestone and mineralized limestone.

	limestone			mineralized limestone		
	n	$\bar{x}$	s	n	$\bar{x}$	s
Bulk density	11	2.65	0.70	16	2.76	0.84
Compressive Strength (MPa)	6	87.8	25.0	8	93.5	16.0
Tensile Strength (MPa)	19	5.6	1.6	25	5.0	1.2
Tangent Modulus (10^3 MPa)	6	78.0	16.0	8	74.0	11.0
Tangent Poisson's Ratio (-)	6	0.26	0.07	8	0.26	0.06
Ultrasonic Modulus (10^3 MPa)	11	90.0	12.6	16	95.5	17.3
Basic Friction Angle (°)	9	39.8	1.4	9	38.3	1.5
Cohesion (MPa)	12	31.0	-	12	29.0	-
Peak Friction Angle (°)	12	45.0	-	12	44.0	-

Note: n = number of tests; $\bar{x}$ = mean value; s = standard deviation

The overall area covered by the geomechanical mapping of the wall drifts at each site was approximately 30 x 5 square meter. Three scan lines, each approximately 30 meter in length, were generally laid parallel to the floor, the first one being just 1.5 m above it. The data were collected by analyzing the discontinuities intersecting the scan line.

For every discontinuity, the following data were recorded:
(a) Distance along the scan line, from some datum, to the point of intersection with each discontinuity.
(b) Strike and dip of each discontinuity.
(c) Whether or not the discontinuity is open and by how much.
(d) Whether or not infilling material is present in the open discontinuities.

(e) Whether or not water is present.
(f) Estimate of the nature of the discontinuity surfaces. Waveness and surface roughness of the same surfaces.

4. GEOSTATISTICAL ANALYSIS OF ROCK MASS DATA

The design parameters for numerical modeling of the mine structure were determined by the geostatistical analysis of the available rock mass data. It was decided to infer the deformability and strength characteristics on the basis of the methods which allow for scaling laboratory - measured quantities to in situ values (Hoek and Brown, 1983; Bieniawski, 1984).

Field data were obtained by performing plate tests in an exploratory adit (in the limestone rock mass) and by deformation monitoring of mine drifts during excavation. These data made one accept the scaling procedures for the choice of parameters for the final design stage.

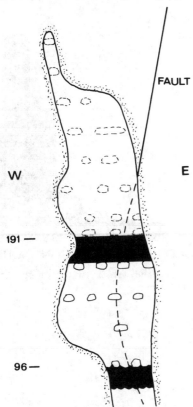

Figure 6. Typical East-West cross section (to be used for two-dimensional modeling) of the Marx Deposit, Masua Mine.

Table 2. Locations for drift mapping and core logging

Level	Drift Mapping	Trend of scan lines	Core logging
191	●	North-South	●
170	●	East-West North-South	●
130	■	–	
110	●	North-South	●
96	●	North-East South-West	■

| ● | Yes | ■ | No |

Given the similarity of behavior of limestone (footwall and hangingwall) and mineralized limestone (orebody) in the laboratory, together with the very limited variability of the most significant rock mass parameters in situ, the available data were analysed as pertaining to the same rock mass.

The following characteristics entering RMR classi-

fication:
- Rock Quality Designation (RQD)
- Spacing of discontinuities
- Orientation of discontinuities (Dip direction and Dip)
- Condition of discontinuities (in accordance with the RMR scheme, a numerical coefficient was evaluated for each discontinuity based upon: persistence, aperture, rock alteration, weathering, and type of infilling)

were treated to obtain their frequency distribution as shown in Figure 7 to 11.

The rock material uniaxial compressive strength was considered to be a deterministic parameter (the mean value obtained in the laboratory as given in Table 1 was used), in conjunction with the groundwater conditions (water was found to be absent in most cases). Then, the importance ratings of various classification parameters were established and summed to yield the frequency distribution of RMR as illustrated in Figure 12.

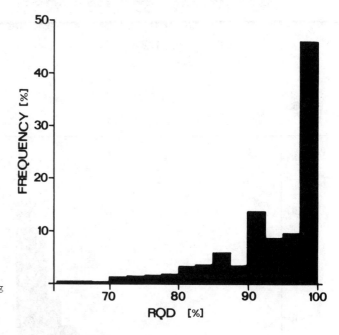

Figure 7. Frequency distribution of RQD.

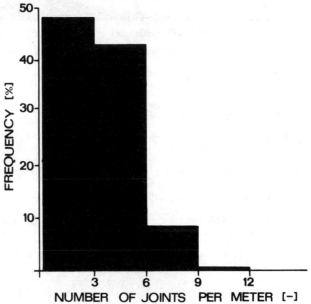

Figure 8. Frequency distribution of average number of discontinuities per m.

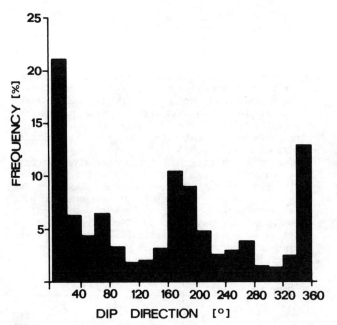

Figure 9. Frequency distribution of dip direction.

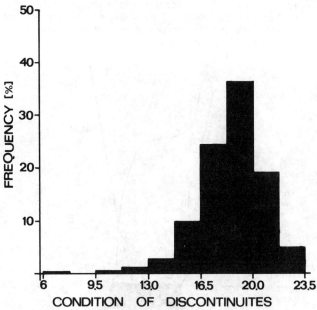

Figure 11. Frequency distribution of conditions of of discontinuities.

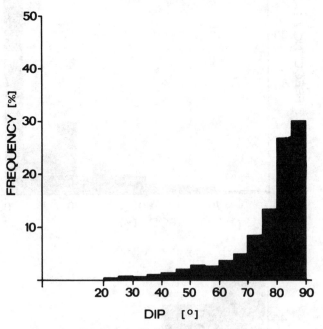

Figure 10. Frequency distribution of dip angle.

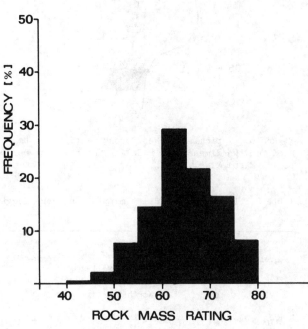

Figure 12. Frequency distribution of RMR index.

4.1 Variograms

Given the values assigned to the five classification parameters above and the location of
measurements in the rock mass, where these same
parameters were determined, the variograms of
each variable were constructed as depicted in
Figures 13 to 18. The following points seem clear:

1. A significant random variability of all the
    parameters is observed, mostly due to the
    presence of microstructures, as is often the
    case for rock characteristics (Matheron, 1963).
2. The range of all the variables being analysed,
    except for the number of discontinuities per
    meter (i.e. the discontinuity spacing) and
    the condition of discontinuities, is of very
    limited size.

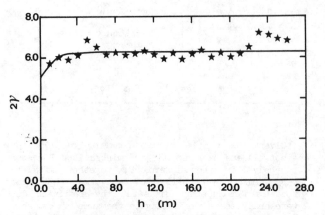

Figure 13. Variogram of RQD index.

3. The variograms illustrated in Figures 13 to 18 are representative of the distribution of the most significant parameters in the rock mass, independently of the direction considered within it (the variogram function does not appear to change significantly as the sampling direction is changed during mapping of discontinuities).

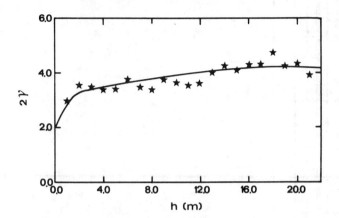

Figure 14. Variogram of average number of discontinuities per m .

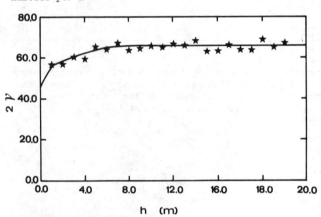

Figure 15. Variogram of dip direction.

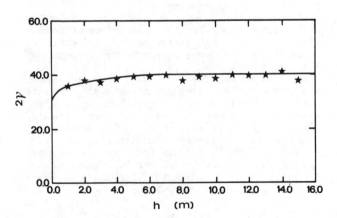

Figure 16. Variogram of dip angle.

4.2 RMR distribution by the "Kriging" method

The rock mass data, obtained according to the scheme described above, are affected by the following limitations:

1. The scan lines were not orderly distributed along the axis of each drift, where the discontinuities were mapped.
2. The rock mass (orebody and host rock) was mapped mostly along scan lines, located in mine drifts, directed either North-South or East-West.
3. The choice of structural regions where to collect the discontinuity data and/or to describe the most significant rock mass characteristics was affected by the need to gain easy access.

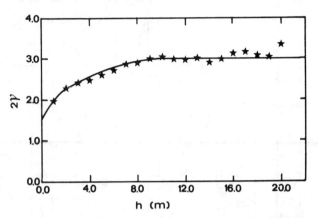

Figure 17. Variogram of condition of discontinuities.

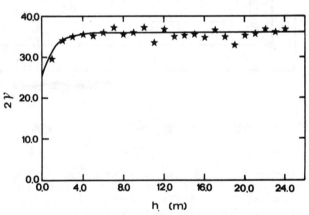

Figure 18. Variogram of RMR index.

With these constraints in mind, and by considering the limited "range" of the rock mass parameters, as shown by each variogram (Figures 13 to 18), a number of assumptions were introduced in order to obtain the distribution needed for geostatistical analysis. In fact, the location of each scan line in the rock mass was varied as shown in Figure 19, so as to concentrate the available data in the near proximity of the cross section of interest, later to be used for design.

On the basis of the variograms of Figure 13 to 18, and according to the "Sampling Scheme" assumed as in Figure 19, the "Kriging" method was applied to a three-dimensional model, by obtaining the mean value of each variable (rock mass parameters) of interest, including the RMR coefficient. The rock mass around the open stopes was subdivided in 1710 regions (elements). Then, the classification parameters pertaining to them were determined. As an example, Figure 20 illustrates the distribution of the RMR indices in one of the cross section of interest. The

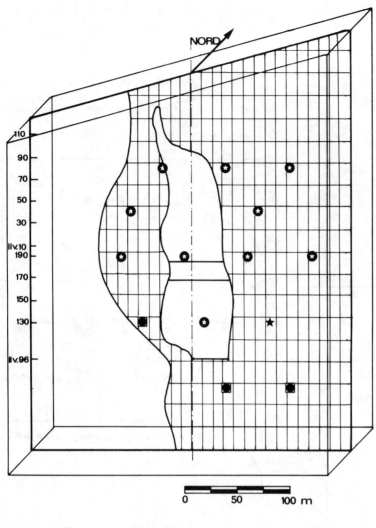

Figure 19. Rock mass of interest for geo-
statistical analysis and sampling scheme.
Trend of sampling: North-South (NS), North
-East (NE), East-West (EW).

○ :NS   ■ : NE   ★ :EW

distribution of this index, as evaluated for the same cross section according to Monte-Carlo simulation, is reported in Figure 21.

The following remarks can be made:

1. The distribution of the RMR index, as illustrated in Figure 20, gives the expected variability of rock mass quality around the open stopes. The predictions based on geostatistical analysis is shown to exhibit a very limited variability of this index, when moving from one rock volume to another one, located in its near proximity.

2. As a comparison, the predictions based on Monte-Carlo simulation give marked differences in the RMR parameters evaluated at neighboring locations, within the rock mass. This result is considered to be unrealistic and inconsistent with the most likely distribution of the desired rock mass characteristics.

4.3 Distribution of deformability and strength characteristics

The distribution of the RMR index obtained by the Kriging method (Figure 20) and Monte-Carlo simulation (Figure 21) allowed to evaluate, for the various elements in the cross section of interest, the following deformability and strength characteristics:

1) the in-situ modulus of deformability E_d by applying the correlation between the RMR index and E_d, as proposed by Bieniawski (1978) for use for better quality rock masses;

2) the m and s empirical constants (depending on the properties of rock and the degree of fracturing of rock mass), entering the criterion of failure of rock mass strength, as proposed by Hoek and Brown (1980).

The three-dimensional distribution of these parameters in the rock mass obviously reflects the same trend in the distribution of the RMR index, as obtained for the rock mass quality, assessed by using the results of Geostatistical Analysis and Monte-Carlo simulation. In order to give a "feel" for the rock mass as assessed by the two methods above, typical values of the deformability and strength characteristics are summarized in the following Table 3.

Table 3. Deformability and strength characteristics of rock mass obtained by geostatistical analysis and Monte-Carlo simulation.

	$\bar{x}$	$\bar{\sigma}$	max	min	$\bar{x}$	$\bar{\sigma}$	max	min
E_d (GPa)	27.23	7.68	40.00	5.00	30.15	14.70	59.00	5.00
m	1.38	0.18	1.71	0.90	1.49	0.39	2.45	0.71
s	0.019	0.007	0.036	0.005	0.028	0.023	0.049	0.002
	Kriging Method				Monte-Carlo Simulation			

E_d = modulus of deformability $\bar{\sigma}$ = standard deviation
m,s = empirical constants   max = maximum value
$\bar{x}$ = mean value   min = minimum value

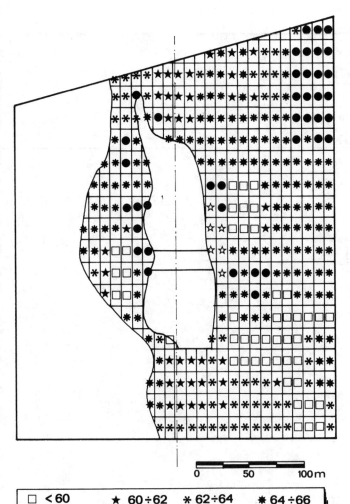

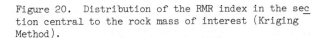

□ < 60	★ 60÷62	✳ 62÷64	✱ 64÷66
✻ 66÷68	● 68÷70	☆ > 70	

Figure 20. Distribution of the RMR index in the section central to the rock mass of interest (Kriging Method).

5. CONCLUDING REMARKS

A theoretical method based on the principles of geostatistical analysis has been presented, with the interest to assess the spatial variation of rock mass characteristics, for design and analysis of mine structures. The rock mass at the Masua Mine, explored on the basis of engineering geological mapping along mine drifts, and borehole and core logging, provided the data for analysis. The following concluding remarks are made.

• The spatial variation of the rock mass quality (assessed by the RMR index) and of the deformability (modulus of deformability) and strength (empirical constants entering the Hoek and Brown rock mass failure criterion) characteristics is found to be quite small. This fact is also well illustrated by the variograms of the most significant rock mass parameters. It reflects a prevailing homogeneity of the rock mass being investigated, where the heterogeneity in mechanical properties is found to be mostly random.

• The small spatial variability in rock quality allows however the clustering of regions in the rock mass, where the deformability and strength characteristics attain representative values. The distribution of parameters given by geostatistical analysis (Kriging method) has been compared with the distribution, in the same cross section of the mine structure, as estimated by Monte-Carlo simulation. The numerical results show the estimate by the Kriging method to be more likely to reflect the rock mass mechanical behavior. Monte-Carlo simulation gives marked differences and inconsistencies in the estimated parameters, for regions which are nearly located or even contiguos.

• The spatial variability of rock mass mechanical parameters has been referred to a two dimensional cross section of the mine, so that the stability analysis of open stopes and mine structures can be effected by numerical methods, when accounting for the expected distribution of deformability and strength properties. This will allow to associate with the mobilized strength estimate in the rock mass around mine openings a "confidence limit". It is expected to obtain by this approach a better estimate of the performance of underground excavations, than it is possible with the notion of "Factor of Safety", in defining in engineering terms the amount of over design.

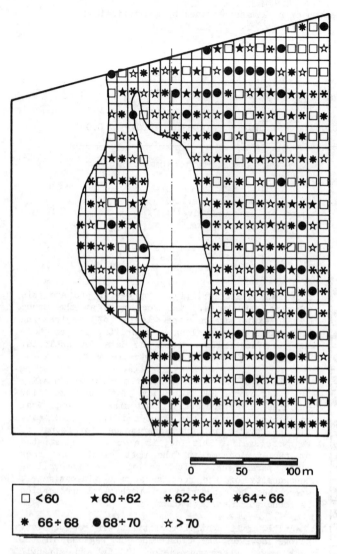

0   50   100 m

| □ <60 | ★ 60÷62 | ✳ 62÷64 | ✳ 64÷66 |
| ✳ 66÷68 | ● 68÷70 | ☆ >70 | |

Figure 21. Distribution of the RMR index in the sec
tion central to the rock mass of interest (Monte-Car
lo simulation).

6. REFERENCES

BARLA G., 1984. Methods Applied for Stability
    Studies of Underground Openings at the Masua
    Mine. I.S.R.M. Symp., Cambridge.
BARLA G., 1986. General Report - Session C2. Proc.
    Symp. on Large Rock Caverns. LRC '86. Helsinki.
BENIAWSKI Z.T., 1976. Rock Mass Classifications
    in Rock Engineering. Proc. Symp. on Exploration
    for Rock Engineering. Balkema Press., Rotterdam
    97-106.
GUARASCIO M., DAVID M., HUIJBREGTS C., 1975.
    Advanced Geostatistics in the Mining Industry.
HOEK E. and BROWN E.T., 1980. Underground
    Excavations in Rock. The Institution of Mining
    and Metallurgy, London.
JOURNEL A., HUIJBREGTS C., 1978. Mining
    Geostatistics.
LA POINTE P.R., 1980. Analysis of the Spatial
    Variation in Rock Mass Properties through
    Geostatistics. 21st u.s. Symposium on Rock
    Mechanics. Rolla 570-580.
MATHERON G., 1963. Principles of Geostatistics.
    Econ. Geol. 1246-1266.
MATHERON G., 1965. Les Variables Régionalisées
    et leurs Estimations. Masson, Paris.

Interpretation of tunnel convergence measurements

Interprétation des mesures de convergence en tunnel
Auswertung von Konvergenzmessungen im Tunnelbau

J.P.BARLOW, HARDY-BBT Ltd, Edmonton, Canada
P.K.KAISER, Professor of Civil Engineering, University of Alberta, Edmonton, Canada

ABSTRACT: Monitoring of wall movements in the immediate vicinity of the tunnel face has become an integral part of the design and performance evaluation process. For this reason, a convergence data interpretation procedure was developed to consider all important aspects of tunnel advance, including near face conditions, sequential excavation, and support/ground interaction. The tunnel convergence interpretation method is introduced together with a discussion of practical aspects of field data interpretation. The technique is assessed by comparison with field measurements from a large highway tunnel (Enasan Tunnel, Japan).

Résumé: Le contrôle des déplacements du mur à proximité immédiate de la face d'un tunnel est devenu une partie intégrante des processus de design et d'évaluation du comportement. Pour cette raison, une procédure d'interprétation des données de convergence a été développée pour tenir compte de tous le aspects importants de la progression d'un tunnel, par exemple, les conditions à proximité de la face, l'excavation séquentielle et l'interaction entre le sol et les supports. La méthode d'interprétation de la convergence d'un tunnel est présentée conjointement avec une discussion de considérations pratiques relatives à l'interprétation des données du terrain. La technique est évaluée en comparaison avec des mesures en provenance du chantier d'un tunnel routier de grandes dimensions (Tunnel Ensan, Japon).

Zusammenfassung: Messungen von Tunnelwanddeformationen in der Nähe der Ortsbrust sind heute fuer den erfolgreichen Tunnelbau von grosser Bedeutung. Darum wurde eine Methode entwickelt, die es erlaubt, Konvergenzmessungen unter Berücksichtigung von wichtigen Einflussfaktoren, wie die Verhältinisse Nähe der Ortsbrust, der stufenweise Aushub und das Verhalten des Einbaues, zu interpretieren. Diese Methode wird erlaütert und mit praktischen Anwendungen zusammen diskutiert. Ihre Gültigkeit wird durch einen Vergleich mit Messungen eines grossen Stassentunnels (Enasan Tunnel, Japan) beurteilt.

1. INTRODUCTION

Monitoring of wall movements in the immediate vicinity of the tunnel face has become an integral part of the design and performance evaluation process. The response of a rock mass to excavation should be a good indicator of the rock mass properties and its ability to adapt to the stress change induced by the removal of the rock as well as the interaction of the support with the rock (Kaiser and Korpach, 1986). Tunnel wall convergence measurements are often preferred over extensometer measurements because of space and time restrictions during tunnel advance, eventhough extensometer measurements are more desirable for the purpose of rock property determination, design evaluation and assessment of localized yield processes (Kaiser, 1981).

Many authors have published procedures to determine time-dependent and time-independent properties from convergence measurements. Most of them focus primarily on the component of time-dependent tunnel convergence due to rock deformation and often neglect the influence of stress changes resulting from the tunnel face advance. The interpretation procedure presented by Guenot et al. (1985) is exceptional in this sense because it respects the dependence between time-dependent stress changes due to tunnel face advance and stress-induced creep as well as yield zone propagation. However, several important aspects such as the effects of support/ground interaction were not properly evaluated by them. In a recent study, we have expanded their work and have developed and evaluated a rational, more generally applicable, convergence interpretation procedure by comparison with field measurements.

The approach by Guenot et al. (1985) is briefly summarized below to provide an introduction for the extensions that are presented afterward. The application of this extended method will then be demonstrated on data from the Enasan Tunnel (Japan).

2. APPROACH INTRODUCED BY GUENOT, PANET AND SULEM (1985)

The equation proposed by Guenot et al. (1985) gives the convergence (C) as a function of the position of the face (x) and time (t) as follows:

$$C(x,t) = [C_1(x)] \{C_{x\infty} + A\, C_2(t)\} \qquad (1)$$

where:
$C_1(x)$ = time-independent or loading function, which depends on the position of the face, x;

$C_2(t)$ = time-dependent function, representing the creep behavior of rock;

$C_{x\varphi}$ = ultimate time-independent convergence (elastic + plastic);

A = ultimate time-dependent convergence;

$(C_{x\infty} + A)$ = total ultimate convergence.

The loading function $C_1(x)$ describes the stress increase in the rock surrounding the tunnel excavation as the face advances:

$$C_1(x) = C_{x\infty} \left\{ 1-[\frac{1}{1 + \frac{x}{\overline{X}}}]^2 \right\} . \qquad (2)$$

Panet and Guenot (1982) established the following relationship between the parameter X and the radius of the yield zone, R, as X = 0.84 R.
If valid, this could be of great practical importance, because the radius of the plastic zone can be determined by fitting the above function to observed convergence data.

The time-dependent, or creep component of convergence is described by:

$$C_2(t) = \left\{ 1-[\frac{1}{1 + \frac{t}{\overline{T}}}]^{0.3} \right\} \qquad (3)$$

where:
   T = parameter reflecting the shape of the creep curve for a particular rock.

3. EXTENSION OF CONVERGENCE SOLUTION

3.1 Introduction

The convergence equation by Guenot et al. (1985) was modified to broaden its applicability to more realistic tunnelling conditions (Barlow, 1986) and to include:

1) the interaction between a tunnel support and the ground; and

2) the effects of sequential tunnel excavation procedures.

The approach adopted for modelling tunnels excavated in stages is to treat each excavation stage as a separate opening and to superimpose the individual responses. It is necessary, for this application, to model the convergence that occurs both ahead, and behind the excavation face as each stage passes the section of measurement. A series of axisymmetric, linear elastic, finite element analyses (Barlow, 1986) have been used to aid in the development of this extension to the convergence equation.
The convergence curve that is represented by this extended approach is shown in Fig. 1, plotted with respect to the distance from the face. In this figure, the lined tunnel case is compared with the unlined case (given by Guenot et al., op cit).
Convergence occurs in response to tunnel excavation because of the stress redistribution associated with the removal of rock. The convergence curve has been divided into three zones with distinctly different stress redistribution modes. The convergence ahead of the tunnel face, indicated by Zone A in Fig. 1, is produced by the compression of the core of rock ahead of the face as the radial stress in this region increases. In Zone B, the region between the tunnel face and the support installation point, the mechanism of stress redistribution changes. With the core of rock now removed, much more stress is transferred to the tunnel walls resulting in a sharp increase in convergence rate. The additional stress change must be carried by the tunnel walls, the rock at the

tunnel face and some arching to the delayed lining. Thus, as the face advances, the proportion of the stress change carried by the rock wall increases in Zone B. When the tunnel support is activated, in Zone C, it alters this process of stress transfer, by sharing some load with the lining as the tunnel face advances. This reduces the amount of additional load carried by the rock mass, and decreases the rate of convergence of the tunnel walls.

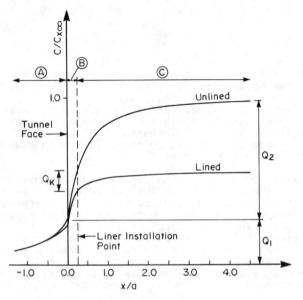

Fig.1 Schematic Convergence Curve

3.2 Development of Equations

Zone A

The convergence that occurs in Zone A can be approximated by the convergence equation if the loading function, $C_1(x)$, is altered to reflect the stress redistribution that occurs ahead of the face, in the following manner:

$$C(x,t)=[Q_1 C_{pf}(x)-Q_k P_{k+}(x)]\{C_{x\infty}+AC_2(t)\} \qquad (4)$$

where:
   Q_1 = portion of normalized convergence ahead of the face due to the total stress change associated with excavation (see Fig. 1);

$$C_{pf}(x) = [\frac{1}{1 + (\frac{x_f-x}{X})}]^{1.2};$$

x_f = value of x at tunnel face;

Q_k = amount of normalized convergence that is inhibited in Zones A and B due to the tunnel support (see Fig. 1);

$$P_{k+}(x) = [\frac{1}{1 + (\frac{x_s-x}{X})}]^{a};$$

$$a = 1 + \frac{a}{L_d};$$

x_s = value of x at support activation point;

a = tunnel radius; and

L_d = distance between tunnel face and liner activation point.

The term $Q_1C_{pf}(x)$ represents the stress increase resulting from the approaching excavation, and $Q_kP_{k+}(x)$ represents the effect that the tunnel support has on decreasing the convergence in this region.

Zone B

The convergence that occurs in Zone B is described by:

$$C(x,t)=[Q_1+Q_2C_1(x)-Q_kP_{k+}(x)]\{C_{x\infty}+AC_2(t)\} \quad (5)$$

where:

Q_2 = portion of normalized convergence due to the total stress change associated with excavation after the face (see Fig. 1).

Note that $C_{x\infty}$ and A now represent the total (pre-face plus post-face) time-independent and time-dependent components of the ultimate convergence. This is the ultimate convergence that would occur if the tunnel remained unsupported.

The value of Q_1 determined by finite element analyses (for linear elastic rock) was 0.27, which corresponds to the value given by Panet and Guenot (1982). However, from their elasto-plastic analyses, they discovered that Q_1 depends on the radius of the plastic zone, R. For example, for R=2.15a, Q_1 was 0.58.

Zone C

After the support is installed, the convergence can be described by:

$$C(x,t)=[Q_1+Q_2C_1(x)-P_s(C(x,t))-Q_kP_{k-}(x)]$$
$$\{C_{x\infty}+AC_2(t)\} \quad (6)$$

where:

C_s = Convergence at the point of liner installation;

$$P_{k-}(x) = [\frac{1}{1 + (\frac{x-x_s}{X})}]^a; \quad (6.1)$$

$$P_s(C(x,t)) = K_s \frac{\Delta C_1}{2a} \quad (6.2)$$

ΔC_1 = Convergence of the tunnel support;

$$K_s = \frac{E_s[a^2-(a-t_s)^2]}{(1+\nu_s)[(1-2\nu_s)a^2+(a-t_s)^2]} \quad (6.3)$$

K_s = ring stiffness for a circular thickwalled support ring;

E_s = Young's modulus of the liner;

t_s = liner thickness; and

ν_s = Poissons ratio of the liner.

Eqn 6 can be rewritten as:

$$C(x,t)=\frac{[Q_1+Q_2C_1(x)+KC_s-Q_kP_{k-}(x)]}{[1+K(C_{x\infty}+AC_2(t))]}\{C_{x\infty}+AC_2(t)\} \quad (7)$$

where:

$K = \frac{K_s}{2ap_o}$.

Thus, the convergence ahead and behind the face of a supported tunnel can now be described by Eqns 4, 5 and 7.

The application of this series of equations requires a knowledge of one additional parameter Q_k. Values of Q_k, derived from the finite element analyses, are shown in Fig. 2. This figure demonstrates that Q_k decreases with a decrease in liner stiffness and with an increase in support delay. The stiffness of the liner is expressed in terms of the relative stiffness of the liner by the compressibility ratio, C^* (Einstein and Schwartz, 1979):

$$C^* = \frac{E\,a\,(1-\nu_s^2)}{E_s\,A_s(1-\nu^2)}$$

where:

E = Young's Modulus of the ground;
ν = Poisson's Ratio of the ground;
A_s = cross sectional area of the support per unit length.

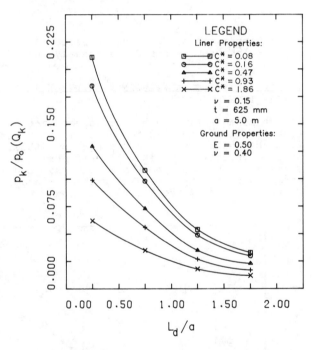

Fig.2 Normalized Pre-liner Equivalent Lining Pressure vs Normalized Support Delay (from Finite Element Analyses)

This chart is valid for elastic ground conditions only.

While the convergence model has grown from a single equation to a set of three equations, it has only gained one additional parameter, K. This parameter reflects the stiffness of the support system, and can be determined directly from the liner properties. If the support system is other than a circular liner, K may be estimated by fitting convergence data. Thus, for a circular liner, the application of the Convergence Solution is no more complex than the original equation. Both involve the determination of four parameters by curve fitting. The expanded set of convergence equations makes it applicable to a wide range of realistic tunnelling conditions.

3.3 Validation of New Equations

The convergence equations (Eqns 4, 5 and 7) were fitted to a convergence curve produced by finite element analyses modelling a liner with a relative stiffness of C^*=0.50, and a support delay of 1.25 radii. The curves presented in Fig. 3, show excellent agreement between the finite element results and the proposed extension of the convergence equation.

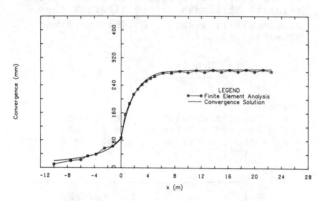

Fig.3 Comparison of Convergence Solution with Results from Finite Element Analysis

4. APPLICATION OF CONVERGENCE SOLUTION TO ENASAN TUNNEL (JAPAN)

The proposed extension of the convergence equation is now applied to measurements taken from the Enasan Tunnel in Japan (Ito, 1983). Some additional data was collected during a site visit in 1984. The Enasan highway tunnel was constructed by the NATM, with an average depth of cover of about 1000 m at the location considered in the following. Because the rock was weak and heavily fractured (Takino et al., 1983), excavation proceeded in three sequential stages, with a top heading, a bench and an invert (see Fig. 4 inset).

Five sets of convergence measurements are plotted against time in Fig. 4, together with the excavation sequence for each heading. The excavation history of the three stages is extremely variable due to many changes in rate of advance and work stoppages. The irregular nature of the convergence curves reflects this variability in excavation history.

A detailed analysis of the entire data set was performed by Barlow (1986) and several simplifying assumptions had to be made to apply the convergence equations to the entire range of data presented in Fig. 4 for all stages of excavation. In the following, only the convergence associated with the heading excavation will be examined and the convergence equations will be fitted to the measurements from Position H2 only (Fig. 4). A comparison of the best fit of the convergence equations to these measurements is shown in Fig. 5.

It is most convenient to plot the convergence rate normalized to the maximum convergence rate measured near the tunnel face. In this manner, the curves become independent of the magnitude of the convergence rate. Barlow (1986) demonstrated that this method of presentation is most useful for the practical purpose of field data interpretation. In this fashion the number of parameters is reduced from four to three, as now only the ratio $A^* = A/C_{x\infty}$ is required. The parameters that correspond to the fit shown in Fig. 5 are: X=9.45 m, T=1.0 days and A^*=1.0. These parameters

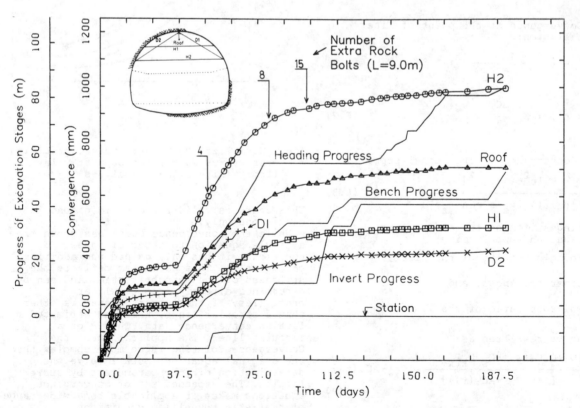

Fig.4 Tunnel Convergence and Face Progress at Station A, Enasan Tunnel (after Ito,1983)

characterize the response of the rock mass to the tunnel excavation and can now be used to predict the ultimate size of the yield zone, the influence of tunnel supports, and the future response of the rock mass.

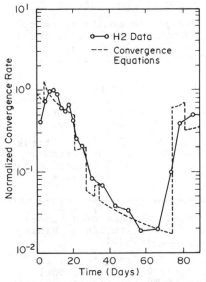

Fig.5 Normalized Convergence Rate of Position H2

4.1 Yield Zone

The convergence equations reveal valuable information about the extent of the yield zone surrounding a tunnel. The value of X determined in the above fitting process is 9.45 m, which corresponds to R=2.5a or 11.25 m (6.75 m from tunnel wall). With the excavation of the second and third stages the yield zone would likely propagate. If the yield zone propagates to maintain a constant R/a ratio, then the final extent of yielding would be R=15 m, or 9 m from the tunnel wall. The rock bolt length and pattern could be selected on the basis of this prediction.

4.2 Influence of Tunnel Support

The two major decisions, with regard to final liner placement, that must be made during the observational design approach concern:

1) the type and dimensions of liner; and
2) the liner installation point and time.

Once the response of the rock mass has been characterized by a set of parameters, the convergence equations may be used to model the response of the rock mass to a variety of potential support scenarios. The optimum support can then be selected on a rational basis, to ensure that the allowable convergence and lining pressures are not exceeded.

An example of this process is given for the parameters determined in the previous section. Fig.6 shows the effect that concrete liners of various thicknesses have on both the convergence and convergence rate curves.

The curves in Fig.6 depict the response of the rock mass at the Enasan Tunnel to an "idealized" excavation, with all work stoppages eliminated. Several simplifying assumptions, such as a constant rate of advance of 1m/day, were required to model the three stage excavation. These assumptions are explained in more detail by Barlow (1986). As

expected, Fig.6 shows that an increase in liner thickness causes a downward translation of the convergence rate curve, and hence less ultimate convergence. Also, an increase in liner thickness causes a substantial increase in the ultimate lining pressure. This pressure may be calculated using Eqn 6.2 and the deflections given in Fig.6.

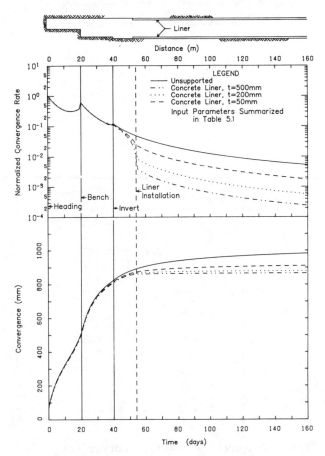

Fig.6 Convergence Solution Curves for Idealized Excavation with Various Liner Thicknesses

The effect that a variation in the liner installation location has on the convergence curves from the same idealized excavation is presented in Fig.7. This figure shows the interesting result that all convergence rate curves eventually merge into one curve independent of their respective installation locations (the ultimate convergence, however, differs). This has implications for the pressure build-up in the liner, as it is proportional to the deformation of the liner. The ultimate rate of pressure increase (eg. after 100 days) in the liner is independent of installation location. Only the initial load increment depends on the location of the liner installation point. This demonstrates that ultimate lining pressure determined by extrapolation based on the initial load build up, must lead to erroneous results.

4.3 Framework for Interpretation of Data

Once the three required parameters (X, A^* and T) of the convergence equations have been determined, by curve fitting to the field measurements, the equations may be employed to model the future behavior of the underground opening. The convergence curve that should occur in response to a given excavation/construction schedule can be predicted. This prediction then provides a framework to interpret future measurements, so that a change in the characteristic rock mass response, such as the propagation of the yield zone, can be readily identified.

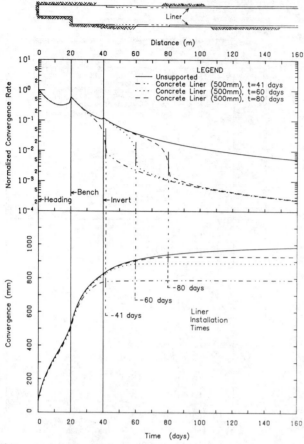

Fig.7 Convergence Solution Curves for
Idealized Excavation with Various Liner
Installation Locations

5. SUMMARY AND CONCLUSIONS

A convergence interpretation procedure has been developed on the basis of the approach given by Guenot et al. (1985). It is now applicable to a wide range of realistic tunnelling conditions. When this procedure is used to interpret convergence measurements, valuable information can be gained about the response of the rock mass to the excavation and support process, the time-dependent behavior and the extent of the yield zone. Once the response of the rock mass has been characterized, both the long term response of the rock mass, and the effect that a variety of support systems would produce can be predicted. This provides valuable guidance for the support design process.

6. REFERENCES

Barlow, J.P., 1986. Interpretation of Tunnel Convergence Measurements. M.Sc.Thesis, Department of Civil Engineering, University of Alberta, 235 p.

Einstein, H.H. and Schwartz, C.W., 1979. Simplified Analysis for Tunnel Supports. Journal of the Geotechnical Engineering Division, American Society of Civil Engineers, 105, pp. 499-518.

Guenot, A., Panet, M., and Sulem, J., 1985. A New Aspect in Tunnel Closure Interpretation. 26th U.S. Symposium on Rock Mechanics, Rapid City, pp. 455-460.

Ito, Y., 1983. Design and Construction by NATM through Chogiezawa Fault Zone for Enasan Tunnel on Central Motorway. (in Japanese) Tunnels and Underground, 14, pp. 7-14.

Kaiser, P.K. , 1981. Monitoring for the Evaluation of the Stability of Underground Openings. First Conference on Ground Control, West Virginia, pp. 90-.97.

Kaiser, P.K. and Korpach, D., 1986. Stress Changes Near the Face of Underground Excavations. Interrnational Symposium on Rock Stress and Stress Measurements, Stockholm, pp. 635-645.

Panet, M. and Guenot, A., Analysis of Convergence Behind the Face of a Tunnel. Tunnelling'82, The Institution of Mining and Metallurgy, pp. 197-204.

Takino, K., Kimura, H., Kamemura, K. and Kawamoto, T., 1983. Three Dimensional Ground Behavior at Tunnel Intersections. International Symposium on Field Measurements in Geomechanics, 2, Zurich, pp. 1237-1246.

Contribution to the study of the behaviour of support pillars in mines

Contribution à l'étude du comportement des piliers de soutènement dans les mines
Beitrag zur Untersuchung des Verhaltens von Stützpfeilern in Bergwerken

M.BARROSO, Laboratório Nacional de Engenharia Civil, Lisboa, Portugal
L.N.LAMAS, Laboratório Nacional de Engenharia Civil, Lisboa, Portugal

ABSTRACT: This paper presents a study, on laboratory scale, of the deformation and strength characteristics of mine support pillars.
The experimental approach used small-scale models of the pillars, shaped in such a way that the uniaxial compression tests, with a very stiff servo-controlled testing machine, were not affected by the contacts between the testing machine and the specimens. An elastic numerical analysis was carried out, in order to investigate the influence of some parameters on the pillar stability

RÉSUMÉ: Cette communication présente une étude, à l'échelle de laboratoire, des caractéristiques de déformation et résistance des piliers de soutènement des mines. L'approche expérimentale usa des modèles des piliers à échelle réduite, ayant une forme telle que les essais de compression uniaxiale, avec une machine d'essais servo-contrôlé très rigide, n'ont pas été affectés par les contacts entre la machine d'essais et les spécimens. Une analyse numérique élastique a été faite, afin de chercher à investiguer l'influence de quelques paramètres sur la stabilité de piliers.

ZUSAMMENFASSUNG: Dieses Referat stellt eine Untersuchung, im Labormaßstab, der Verformungs- und Festigkeits charakteristika von Bergwerksstützpfeilern vor. Der experimentelle Ansatz benutzte Modelle in verkleinertem Maßstab von den Pfeilern, die solch eine Form hatten, daß die einachsigen Druckversuche, mit einer sehr steif en servo-gesteverten Versuchsmaschine, nicht durch die Kontakte zwischen der Versuchsmaschine und den Spezimina beeinflußt wurden. Eine elastische numerische Analyse wurde durchgeführt, um den Einfluß einiger Parameter auf die Pfeilerstabilität zu untersuchen.

1 INTRODUCTION

In practical terms the use of multipurpose pillars has widespread in underground mines. Particularly when pillars provide the main support for the roof, in spite of progress of knowledge in rock mechanics already permitting to approach the design of such pillars rationally, it still seems justified to carry on research to investigate the mechanisms governing the behaviour of pillars from loading to failure.

From the rock mechanics viewpoint, the design of support pillars aims at ensuring that their sizes and those of spans may be enough to prevent failure as long as required. Their design demands a detailed quantitative analysis of failure conditions result - ing from the interactions of the complex matrix of parameters that condition the overall mine design. This matrix has dynamic characteristics which are imparted to the factor of safety of the pillars, accor ding to the development of mining which often introduces changes in load distribution on the supports.

This work presents the overall question of the design of support pillars in mines, the study of the behaviour of models of pillars following experimen - tal and mathematical approaches by using a schistous metamorphic rock from a tungsten mine exploited by a method that sometimes employs overstressed support pillars.

The experimental approach made use of pillars shaped in such a way that deformation and failure conditions were not affected by the contacts between the testing machine and the specimens tested. Besides the tests were carried out with a very stiff servo--controlled testing machine, able to produce fairly reduced strain rates.

Particularly the effects of the dimensional rela - tions of the pillar (width/height ratio) were analysed for the complete load — deformation curves, including the region beyond the pick strength where they have a negative slope.

The numerical analysis was carried out by using a linear elastic three-dimensional boundary element model, in order to investigate which zones are most likely to fail first, and the influence of the Poisson's ratio and the width-heigth ratio on the stability of the pillar.

2 EXPERIMENTAL STUDY

The aim of the study on the basis of this paper was an experimental simulation in laboratory of the usual work conditions of support mine pillars as well as a theoretical approach, in order to contribute to the knowledge of the behaviour of mine pillars, in particular in the period beyond rock elastic behaviour until complete breakdown.

The laboratory testing was carried out in a 4000 KN quick-response servo-controlled testing machine with a stiff frame of 10^{10}N/m. The electronics package has three controllers monitoring three different control variables: position of the piston of the machine, axial displacement and axial load. The outputs are fed to a X-Y plotter which gives the load-deformation curve of the specimen being tested according to the conditions fixed with a ramp generator in which the sign varies monotonically with time.

The different properties of the material of the platens and the rock provoke the wellknown end effects which can influence the results obtained from the tests. In order to overcome this difficulty all the pillar models tested were of the shape represented in Fig. 1. The relative dimensions of the cylindrical pillar models are such as to reduce the possibility of failure by bending of the zone above the top of the pillar. According to the work presented by Mendes (1972), these dimensions should be such that:

793

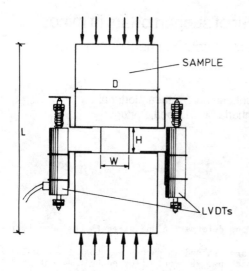

Figure 1. Schema of model pillar

$$\frac{L}{D} = \frac{7}{3} \quad \text{and} \quad \frac{W}{D} = \frac{1}{3}$$

where W is the diameter and H the height of the pillar, D is the diameter of the upper and lower parts of the specimen and L is its total height.

The rock used for the model pillars was a schistous metamorphic rock with high strength, whose schistosity varies slightly around the axes of the model pillars.

All the model pillars were tested in uniaxial compression. The vertical deformation was the only one measured, and in two different ways:

i) by the position of the piston of the testing machine;

ii) by the use of Linearly Variable Differential Transformers (LVDTs) as shown in Fig 1. While with the first system the deformation of the whole specimen was measured, the LVDTs only measured the deformation of the pillar. The tests were carried out with a uniform rate of displacement of the piston of the testing machine. This rate was 2 μm/min at the beginning of the test and, after the first signs of yield were visibles, it was changed to 0.3 μm/min. The strain rate in the pillars was also measured and its value, before yielding, was approximately 0.4×10^{-6}/s (Barroso, 1986).

The total number of tests carried out until now is small but, nevertheless, it has been possible to obtain information about the tendency of the changes of the pillar strength and overall behaviour of the pillar with the width-height ratio. Fig. 2 represents

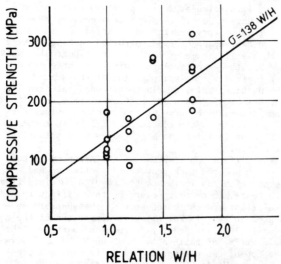

Figure 2. Evolution of the compressive strength with the relation W/H

a graph of peak strength against the width-height ratio. It is possible to establish a linear relation between them:

$$\sigma = 138 \ W/H$$

where σ is given in MPa. This relation implies a faster variation of the peak strength with the width-height ratio than that obtained by Wagner (1984) with models of cylindrical sandstone pillars

The yield occurred for very high loads, corresponding approximatety to 85% of the maximum load. Despite the fact that the width-height ratio was only 1.75 and that a rate of axial deformation was imposed, there was a tendency for the stabilization of the load, at its maximum value, for a certain period of time after the peak. It is, thus, a kind of rock which leads to controlled collapse of the pillars, with high residual strength for width-height ratios greater than one.

Furthermore, for the five model pillars with a width-height ratio of 1.75, two of them failed by punching (Fig. 3 and 4), This justifies the useful

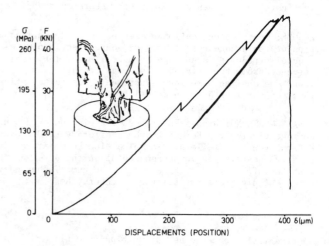

Figure 3. Stress-strain behaviour of a model pillar and sketch of failure mode

ness of some principles of foundation engineering for values of the width to height ratio which, for this kind of rock, do not have to be very high.

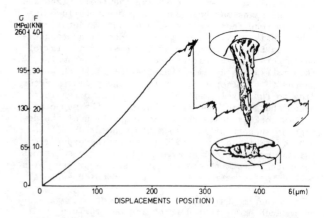

Figure 4. Stress-strain behaviour of a model pillar and sketch of rupture mode .

In the case shown in Fig. 3, failure occurred by punching of one of the pillar bases. This failure was early announced by some local failures at 171 MPa (55% of the maximum load) and at 273 MPa (88% of the maximum load). It is worth noting that the diagrams for the two unloading-loading cycles car

ried out immediately before peak strength, have slo_
pes similar to those recorded before yielding. In the
sketch presented in Fig. 3 some lines of shear pro -
voked by punching are visible.

In the case shown in Fig. 4 the punching effect
was even clearer, which justifies the high value for
the residual strength (150 MPa, 57% of the peak
strength) and can explain why it lasted for a very
long period of time.

The fact that these cases of failure by punching
were obtained, reveals the necessity of simulating
the effect of confinement of the "heads" of the specimens, in order to better approach the actual behaviour of mine pillars.

3 NUMERICAL MODELLING OF PILLAR BEHAVIOUR

A numerical analysis was carried out in order to ana_
lyse the distribution of stresses and the displace -
ments in the test specimens. The rock material was
considered to have a linear elastic, homogeneous and
isotropic behaviour. With a linear elastic model it
is not possible to study the behaviour of the specimens up to failure. However, with the values of the
stresses at different points of the specimen and a
failure criterion, it is possible to identify the zo_
nes which start yielding first. For this numerical
analysis a three-dimensional boundary element model
developed at LNEC (Lamas, 1984) was used. This model
uses a direct formulation of the boundary element me
thod and rectangular parabolic isoparametric boundary elements with 8 nodal points. To model the pillar
specimens a mesh of 104 boundary elements with 378
nodes was used. The stresses and displacements were
obtained at 70 internal points.

Three different situations were studied. In all of
them the specimens had a pillar diameter W = 14 mm
and obeyed the relations between their dimensions
presented earlier. The modulus of elasticity had a
unitary value. The height of the pillar, H, and the
Poisson's ratio, $\vee$, varied in the following way:

Situation	H (mm)	$\vee$
I	14	0.2
II	14	0.4
III	8	0.2

The situation I corresponds to the most frequent
geometry of the specimens tested and the Poisson's
ratio is a usual value for this kind of rocks. In si_
tuation II only $\vee$ was changed. This high value is con_
sidered to be a reasonable upper limit for the values that $\vee$ can take in this kind of rock. Also, it
can give a better idea of the distribution of the
stresses for high axial loads, because, usually $\vee$ in
creases with the axial load. In the situation III an_
other value of H was used, with the same $\vee$ as in I,
in order to investigate the influence of the width -
-height ratio of the pillar on the distribution of
stresses.

For each situation, three cases of loading were con_
sidered in the numerical analysis. All applied loads
were unitary and uniformly distributed as follows:
a) vertical load acting on the tops of the specimens; b) horizontal load acting perpendicular to the
vertical faces of the specimen, excepting the walls
of the pillar; c) horizontal load acting on the walls
of the pillar. Case a) corresponds to the test condi_
tions and simulates unitary vertical stress in the
rock mass. Case b) models a unitary stress field, hy_
drostatic in the horizontal plane, in the rock mass.
With case c) it was intended to study the influence
of a filling material on the stability of the pillar.

The analysis of the results obtained for case of
loading c) directly shows that, in an elastic anal -
ysis, the influence of a small confining pressure on
the walls of the pillar on the distribution of stres
ses is negligible. This result was expected because
the confining pressure is very small when compared
with the stresses that develop in the pillar due to
the field stresses. Only when failure is starting to

occur the filling starts playing its role by limit -
ing the large deformations of blocks of the pillar
and hence increasing considerably its strength.

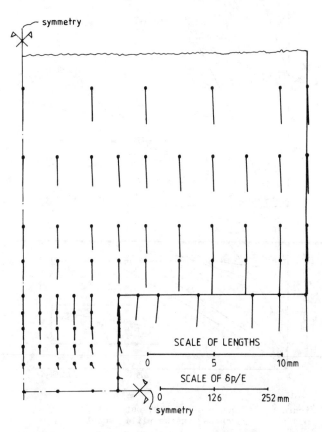

Figure 5. Displacements obtained by the numerical mo_
del

Fig. 5 shows an image of the displacements in a
diametral plane obtained for situation I, i.e., the
test conditions. It clearly shows the bending of
the horizontal surface at the top of the pillar and
the increase in the values of the displacements, in
the zone above the pillar, from the centre to the
periphery. This evidences the advantage of this kind
of specimens for these studies, since they can bet -
ter simulate the actual conditions at the mine.

For situation I, the influence of a horizontal
stress on the stability of the pillar was investigat_
ed. With this purpose, 1/3, 1 and 3 times the stresses obtained for loading case b) were added to the
stresses calculated for loading casa a). Fig. 6 represents the principal stresses at the most signifi-
cant internal points in a σ_I/σ_c Vs σ_{III}/σ_c graph,
where σ_c is the uniaxial compressive strength of the
rock, for a value of the vertical load near to the
failure load of the specimens. The location of these
internal points is shown in Fig. 7. Hoek-Brown's
failure criterion envelopes were drawn relative to
the intact rock and to several qualities of rock
masses (Brady and Brown, 1985), for the type of rock
of the specimens (schistous rock). For all the points,
the addition of a horizontal stress led to improve -
ment on the stability of the pillar. Most tensions
disappear and the stress path with the application
of horizontal stresses has the direction of the fail_
ure envelopes for lower quality rock masses. This
stabilizing effect was already expected and confirms
results obtained in other ways.

Fig. 7 represents the stresses at the internal
points in a σ_I/σ_c Vs σ_{III}/σ_c graph for the three
situations analysed, and only for loading case a).
The points corresponding to internal points in the
same horizontal line were united by lines, in order
to give a better visualization of the evolution of
the stresses inside the specimen. This figure shows

795

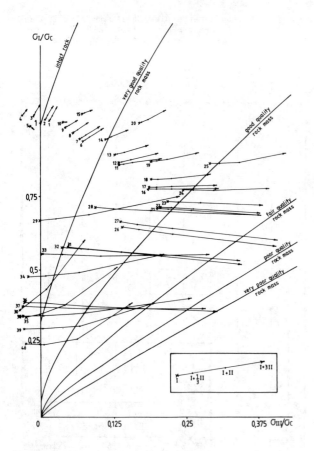

Figure 6. Influence of an horizontal distributed
load on the stresses in the pillar model

that, from situations I to II the change in the values of the stresses in the zone of the pillar is
not very important, though it increases for zones
farther from the pillar. Apart from points 1 to 4,
an increase in Poisson's ratio led to a more favor
able situation, the points moving in the direction
of lower quality rock mass failure envelopes. However for points 1 to 4 σ_{III} is close to zero and
σ_I has a small increase, which means that in this
central zone of the pillar, where failure is likely
to occur first, an increase of Poisson's ratio may
lead to an unfavorable situation.

Comparing the curves for situations I and II it be
comes clear that there is an important improvement on
the stability of the pillar when their height decrease and, consequently, the width to height ratio increases.

4 INFLUENCE OF THE WIDTH-HEIGHT RATIO ON PILLAR
BEHAVIOUR

In order to assess the global behaviour of a mine ex
ploited by using support pillars it is very important to study the strength of the pillars and their ca
pacity of deformation. As a matter of fact, the deformational aspects of pillar behaviour are those
that most influence the redistribution of stresses
associated to any exploitation method. If the flexibility of a pillar is defined as the relation between
the maximum deformation and the corresponding load,
it is clear that the flexibility of the pillars in -
creases markedly with the width to height ratio.
This is a consequence of the change from brittle to
to ductile load-deformation behaviour of the model
pillars, with the increase of the width to height
ratio. Besides, it is noteworthly that the maximum
load that each pillar will withstand depends main-

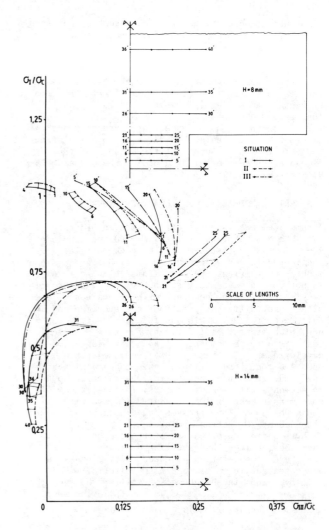

Figure 7. Influence of the Poisson's ratio and W/H
on the stresses in the pillar model

ly on the overburden pressure and on the spacing be
tween the pillars.

In order to obtain the most suitable conditions of
deformation of the pillars, their design should be
preceded by the estimation of the curves representative of the non linear variation of the peak strengths and the maximum deformations with the width-
-height ratio. Then, knowing the load to be supported by the pillar, the most suitable value of the
width-height ratio can be calculated. This is to say
that, knowing the load that the pillar will have to
support, the width-height ratio will control the glo
bal behaviour of the pillar, which can vary from
elasto-brittle to elasto-plastic. A main difficulty,
however, consists of the quantification of the parameters that define this type of behaviour.

The evaluation of the loading conditions has to ta
ke into account the modifications provoked by the
progress of mining. This can be done using mathemati
cal models that make use of the parameters that condition the global behaviour of the mine. The models
have to be calibrated using the information obtained
by monotoring the behaviour of the mine during the
mining activity. Due to the difficulties associated
with stress measurements in rock masses, it is usual
to monitor the deformational behaviour, mainly by
measuring stope convergences. If it is possible to
associate stope convergences with deformability of
pillars, the design of support pillars will be done
in a safer way. That is why the study of pillar be
haviour up to failure is regarded as a field of research of major interest, in order to make the in -

put of quantitative data possible in the calculation methods.

Due to their nature the filling materials used in the mines have very high deformability and very low strength, when compared with the properties of the rock masses in mineral exploitations. Therefore, in an elastic analysis, the contribution of the filling material for the global stability of the mine is negligible. Actually, the effect of the filling material on pillar behaviour only starts to be important when the first local failures occur, hindering from detaching from the pillars the rock blocks that may form. Hence, even if those local failures are of an elasto-brittle kind, the effect of the filling material is that the behaviour of the zones where these local failures occur becomes elasto-plastic. Furthermore, as with the progress of failure the confinement pressures increase and, consequently, the strength of the pillars, it is possible to conclude that the effect of the use of a filling material is very similar to increasing the width-height ratio. If the filling material completely fill the cavities, it has also an important effect of improving the stability of the roof by delaying the onset of failure and reducing its magnitude. This is a field of knowledge worth of deeper studies both by experimental and theoretical ways.

5 CONCLUSIONS

The study developed with model pillars of a schistous metamorphic rock showed that the compressive strength increases rapidly with the width-height ratio. Simultaneously, the load-deformation behaviour changes from brittle to plastic with a relatively small increase of the width-height ratio. The numerical analysis carried out showed that the filling material had no influence on the pillar stability during the elastic stage, and that an increase of the width-height ratio leads to the improvement of the pillar stability. Confining pressure applied on the upper and lower part of the specimens also has a favourable effect.

As the experimental and mathematical approaches clearly showed it is important to test the influence of the confinement of the roof and floor strata. It is also important to study the influence of a fill, cemented or not, on the strength and load-deformation characteristics of the pillars.

REFERENCES

Barroso, M. & F.M. Mendes 1986. Study of the behaviour of the pillars of Panasqueira Mine (in Portuguese). LNEC, Internal Report, Lisbon.

Brady, B.H.G. & E.T. Brown 1985. Rock mechanics for underground mining. London, George Allen and Unwin.

Lamas, L. 1984. Three-dimensional boundary element model for underground structures (in Portuguese). LNEC, Internal Report, Lisbon.

Mendes, F.M. & C.D. da Gama 1972. Laboratory simulation of mine pillar mechanical behaviour. 14th American Symposium on rock mechanics, Pennsylvania State University.

Wagner, H. 1984. Fifteen years experience with the design of coal pillars in shallow South African collierie: an evaluation of the performance of the design procedures and recent improvements. ISRM Symposium, Cambridge, U.K.

Geomechanical mine design approach at Noranda Minerals Inc.

Approche géomécanique pour la conception des mines à Noranda Minerals Inc.
Eine geomechanische Entwurfsmethode für den unterirdischen Bergbau der Noranda Minerals Inc.

W.BAWDEN, Centre de Recherche Noranda, Pointe Claire, Canada
D.MILNE, Centre de Recherche Noranda, Pointe Claire, Canada

ABSTRACT: A design approach is used at Noranda underground mines which incorporates both numerical and empirical modelling techniques. It requires the collection of a comprehensive rock mechanics data base and is backed up by a rational monitoring program. Two case histories from mines with widely differing stability concerns are outlined in this paper.

RESUME: L'approche pour l'ingénièrie de conception des mines souterraines de Noranda comporte des techniques de modélisation numérique et empirique. Elle nécessite la création d'une base de données en mecanique des roches et est supportée par un programme rationnel d'instrumentation des massifs rocheux. Deux histoires de cas de mines ayant des problèmes de stabilité passablement différents sont mis en relief dans cet article.

ZUSAMMENFASUNG: Eine Entwurfsmethode für Noranda's unterirdische Bergwerke wurde entwickelt, welche numerische als auch empirische Modelliertechniken anwendet. Diese Methode erfordert die Zusammenfassung einer umfangreichen felsmechanischen Datenbasis. Das Verhalten der so entworfenen Struktur wird dann durch ein rationelles Überwachungsprogramm bewertet. Zwei Beispiele werden gegben wo die Methode in zwei Bergwerken angewendet wurde, welche grundlegend verschiedene Stabilitätsprobleme hatten.

1 INTRODUCTION

In recent years rock mechanics has finally begun to take its place as a practical tool for underground hard rock mines. The increased acceptance of rock mechanics techniques in mining is largely due to rapid advancement in numerical modelling techniques. Using two case histories, (Norita and Chadbourne Mines, Quebec, Fig. 1), we hope to show how these techniques can be used in design to augment the experience and empirically based methods which will always remain important mine design tools. The Norita Mine will be discussed in detail. Chadbourne will then be covered briefly to show that widely different rock mechanics questions can be addressed using similar techniques.

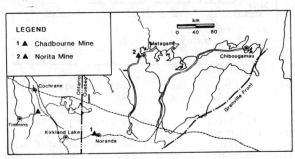

Fig. 1. Location Plan

2 NORITA MINE

The Norita deposit is located on the north limb of a regional anticline. The deposit, a stratiform massive sulphide, is a vertical, tabular ore body located in a thick volcanic sequence (Fig. 2). In the area presently being worked, the ore body is about 25 m wide, has a strike length of about 200 m and extends to a depth of 670 m.

The mine has experienced stability problems in the past using sub-level caving. Present mining is by transverse stoping as shown on the longitudinal section, Fig. 3.

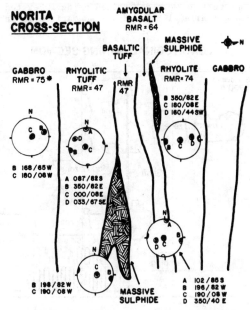

Fig. 2. Norita cross section showing rock characterization.

2.1 Rock mechanics data

In order to evaluate stability and optimize the mining sequence, a rock mechanics study was initiated. A thorough database was developed. To estimate virgin stresses, seven overcoring tests in two HX diamond drill holes were done using CSIR strain cells. The tests were done at depths of 443 m and 714 m below surface in areas removed from major mine excavations. Results were difficult to interpret but suggest vertical stresses about equal to overburden weight and horizontal stress about $2\frac{1}{2}$ times the vertical.

Table 1. Norita mine, mechanical properties of seven rock types (determined with the Bemek rock tester)

ROCK TYPE	YOUNG'S MODULUS Mean (GPa)	S.D.[1]	POISSON'S RATIO Mean	S.D.	RUPTURE MODULUS Mean (MPa)	S.D.	COMPRESSIVE STRENGTH Mean (MPa)	S.D.	TENSILE STRENGTH Mean (MPa)	S.D.
Gabbro (fine-grained)	123.0	13.3	.20	.04	41.7	6.9	219.2	40.8	21.2	3.9
Gabbro (coarse-grained)	109.9	9.1	.18	.02	42.5	7.0	238.7	42.9	24.7	6.4
Rhyolite	91.9	4.6	.18	.02	29.2	5.9	185.3	29.2	32.1	5.4
Amygdular Andesite	82.8	8.8	.22	.04	20.1	4.9	94.5	23.8	21.3	2.7
Basaltic Tuff	95.0	18.9	.26	.08	37.0	10.2	118.0	33.9	11.8	3.2
Rhyolitic Tuff	67.9	24.0	.15	.08	18.1	5.8	98.3	14.0	15.2	1.9
Sulphide	232.2	31.4	.16	.04	32.9	7.7	316.0	34.8	30.4	4.7

[1.] S.D. = Standard Deviation

Key joint fabric and rockmass classification data was collected and is summarized in Fig. 2. Intact rock properties were determined using index texts (Kanduth & Milne 1986) and are summarized in Table 1.

LONGITUDINAL MINING SECTION

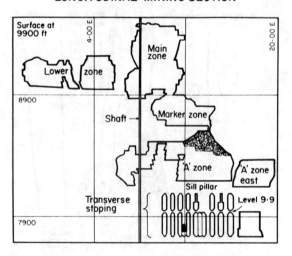

Fig. 3.   Norita Mine longitudinal section

2.2 Recent Instability

Problems were reported at the end of 1985 in the form of ground movement and slabby, unstable back conditions in the 8-8 drift. This drift is located in the 25 m thick sill pillar between the "A" zone and the transverse mining block below (Figs. 3 and 4). Initially problems appeared local and related to weaker stringers included in the massive sulphide. Monitoring in the form of drift closure stations was recommended.

Continued deterioration of ground conditions in the new year included:

(i) Continuing vertical and horizontal closure of up to 2 cm in the 8-8 sill drift after only one month of readings.

(ii) Poor ground conditions in the level 9-9 draw point pillars (Fig. 4). These pillars, located between the 9/10 and 10/11 vertical stopes, required frequent reconditioning.

TRANSVERSE STOPE SEQUENCING & INSTRUMENTATION

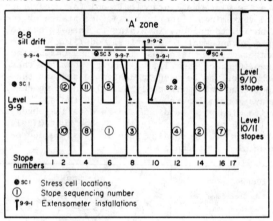

Fig.4. Norita longitudinal section showing stope sequencing and instrumentation

At this time questions concerning longer term mine planning were also raised, resulting in a more comprehensive analysis. A complete modelling program was then started. Also, a comprehensive monitoring package was recommended to assist in interpretation of model results and assessment of overall mine behaviour.

2.3 Modelling

Two numerical programs, a two-dimensional boundary element program (BEA) and a displacement discontinuity program (Mintab) were used.

BEA was used to model transverse and plan sections. Principal stress and stress directions are calculated while failure is estimated based on the Hoek and Brown failure criteria (Hoek and Brown 1980). The Mintab program was used to simulate longitudinal sections and to evaluate the effect of stope sequences on minewide stress distribution.

2.4 Initial model calibration

Before predictive modelling can be done it is essential to back analyse known failure conditions. At Norita, failure in relatively unconfined rock occurs at stress levels predicted by Mintab at about 1/3 to 1/2 of the unconfined compressive strength (UCS) of the intact material.

2.5 Calibration to present conditions

Following the initial calibration, the lower "A" zone and transverse stoping block (Fig. 3) was modelled to simulate present mining conditions (May 1986). Fig. 5 shows the resulting normal stresses acting through the pillars. Stress levels of about 80 to 100 MPa are predicted in the 10/11 stoping block pillars (∿1/3 the UCS of the massive sulphide). It is interesting to note that, in these stopes, over-break in the blast rounds (i.e. 2 m rounds with 3.3 m pull) appears to correlate with the predicted stress levels of about 1/3 the UCS of the rock. Plan views through the 9/10 and 10/11 stoping blocks, modelled using BEA, compare reasonably well with Mintab results. In all cases, individual pillars are predicted to fail at stress levels of about 1/3 UCS.

NORITA-MINTAB RESULTS

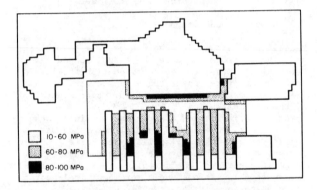

Fig. 5. Norita Mintab results showing normal stresses in pillars

In the area of the level 9/10 draw points, gradual failure of the drawpoint pillars agrees with predicted failure at stresses between 1/3 and 1/2 of the UCS in areas of low confining stress.

It is now possible to relate convergence in the 8-8 sill drift to blasting in a central stope in the 9/10 stoping block. Fig. 6 shows typical results from one closure station with mining activity superimposed. Neither the BEA nor Mintab models predicted the apparent instability in this drift. Predicted normal stresses in this area are between 10 and 60 MPa. Apparent model discrepancies are presumed to be due to linear elastic modelling which does not account for unloading from the transverse pillars and subsequent stress shedding onto the sill.

2.6 Modelling results

Extensive modelling was done with Mintab to investigate the effects of varying the transverse stope sequence. BEA models were also run to act as a check on the induced stresses and to allow application of a failure criteria in the pillar areas. An optimum excavation sequence was recommended taking into account backfilling and other mining constraints and is shown in Fig. 4.

Even with the stress distribution optimized by sequencing, failure conditions are predicted for any isolated secondary transverse pillars. A proposed pillar in the 9/10 stoping block comprising 3 transverse pillars shows induced stresses at or near

CLOSURE STATION

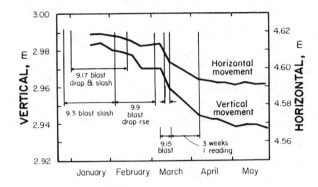

Fig. 6. Typical closure station results from the 8-8 drift

failure by the end of mining. Stresses shown in the sill pillar increased with mining and approach assumed failure loads near mine completion. Induced stresses at the shaft indicate a change of only 20% to 30% between present (May 1986) and final mining conditions. No stress related problems are predicted at the shaft.

2.7 Monitoring

The monitoring of ground response to increased stress is of paramount importance for determining the validity of the modelling and its interpretation. Areas of concern were highlighted by the modelling and appropriate monitoring is being installed. Extensometers in the secondary stope pillars are showing ground movement in response to increasing stress (Fig. 7). Vibrating wire stress cells are being used in conjunction with the extensometers to measure stress change.

EXTENSOMETER 9-9-1 (58°)

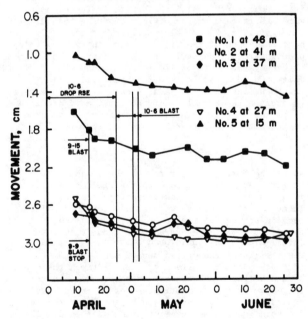

Fig. 7. Movement recorded from extensometer 9-9-1

Fig. 8 shows the preliminary results from 2 of the stress cells. The first cell, installed near the shaft, shows steadily increasing stress indicating the rock has not reached failure. The second cell, installed near the edge of 3 transverse pillars in the 9/10 block, shows decreasing stress with mining. This indicates that, at least near the edge

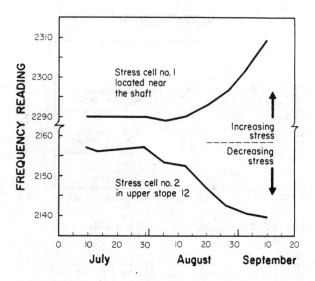

Fig. 8.   Initial results from stress cells

of this pillar, failure has already initiated.
   Additional extensometers and stress cells, as well
as closure stations are being installed in the sill
and shaft pillar areas to monitor ground behaviour
as mining progresses. These results will be
continually correlated with the modelling results.

2.8 Summary

The precise behaviour of the rock following initiation of failure could not be predicted by present
modelling. Monitoring is necessary to show if
stresses are continuing to increase or relax and to
verify the occurrence of large scale movement.

3 CHADBOURNE MINE

The Chadbourne deposit is located in a breccia pipe
intruded through andesite. The ore is funnel shaped
with a surface diameter of 76 m tapering to a diameter of 16 m, 240 m below surface. A cross
sectional view of the mine is shown in Fig. 9. Mine
planning required leaving a sill pillar in the main
ore zone at a depth of 170 m and a study was
initiated to determine the required sill pillar
thickness and support. Additionally it was necessary
to analyse the stability of the lower main ore zone
stope assuming it is pulled completely empty. The
mining method is vertical crater retreat (VCR).

3.1 Rock mechanics data

As at the Norita Mine, a rock mechanics database
program was conducted to obtain virgin stresses,
joint fabric and rockmass classification data.
Intact rock properties were also determined. The
results showed the ore to be very strong and
competent (UCS = 240 MPa and RMR = 80 (Bieniawski
1974)) with 3 well developed orthogonal joint sets.
The results of the stress measurement program
indicated vertical stresses about $2\frac{1}{2}$ times the overburden weight and horizontal stresses twice vertical.
The above very high vertical stresses are not well
understood but are assumed to be local and related to
the point source measurement techniques used. The
need for modelling to predict induced stresses in the
sill pillar is, however, indicated.

3.2 Recent instability

Back failures in overcuts had occured in the past in
the form of large block failures (1800 m³ and 3600 m³).
These failures had occured along the known joint sets
under low stress conditions. The model being used
could not accurately predict these failures.

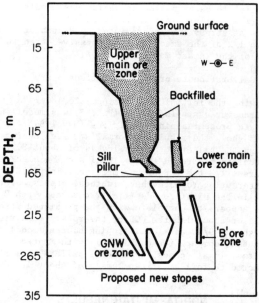

Fig. 9.   Chadbourne mine cross section

3.3 Computer modelling

Modelling was done with the BEA program on vertical
and plan sections. Due to the funnel-shaped geometry of the mine, the vertical section models did
not accurately represent the stress distribution.
Plan views were run to attempt to correct the results
of the vertical views.
   Numerous simulations of progressive mining stages
and sequences were done. Special attention was
concentrated on the sill pillar area where probable
failure was indicated by the end of mining. A high
level of confidence was not placed on the BEA results
from vertical sections and a more empirical approach
was used to augment the computer analysis.

3.4 Empirical modelling

An empirical design approach called the Mathews
method (CANMET 1980) was used to model the stability
of the sill pillar. This is a graphical approach
which plots stability number (N) versus hydraulic
radius (S). The graph is broken into zones labelled
stable, potentially unstable and potentially caving
(Fig. 10). The term N is based on rock classification, induced stress and intact strength, joint
orientation and the orientation of the surface being
studied. The shape factor S is simply the area of
the surface under study divided by the perimeter.
   The method was used to analyse two block failures
located on upper levels of the mine. These failures
each represented well over 1000 m³ of rock. Both
results plotted in the stippled zone between the
potentially unstable and potentially caving zones
(Fig. 10). This indicates that this zone on the
Mathews graph accurately represents the conditions
at which failure could be expected at this mine.
   Present and projected sill pillar stability was
next analysed. With progressive mining below the
sill pillar, only the increase in induced stresses
changes the stability number N used in the Mathews
analysis. The change in induced stresses was
calculated using the BEA model. At stress levels
indicated at the start of mining of the lower zone,
the sill pillar plots well within the stable area of
the Mathews graph (Fig. 10). At the end of mining
of the lower zone much higher induced stresses are
calculated with the BEA model. These higher stresses
shift the plot of the sill pillar stability within

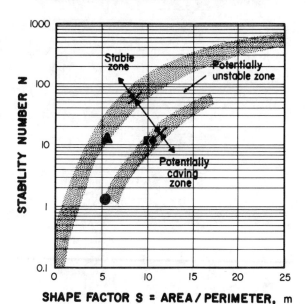

MATHEWS EMPIRICAL DESIGN METHOD

SHAPE FACTOR S = AREA / PERIMETER, m

▲ Actual sill roof

● Sill roof, after the lower main ore zone is extracted

■ Level 6, back failure

◆ Level 5, back failure

Fig. 10. Mathews analysis results for Chadbourne case histories

the expected zone of failure on the Mathews graph (Fig. 10). This agrees with the stability indicated with the BEA model. These methods, however, show that stability is highly dependent on the accuracy of the predicted induced stress in the sill. It was realized that the 2-dimensional BEA model was over estimating the induced stresses in the sill pillar, however, it was not possible to determine by how much. For this reason the following monitoring was proposed.

3.5 Monitoring

Extensometers and stress cells were installed in key areas of the mine. In the sill pillar, stress cells were used to determine the accuracy of the predicted stress increase with mining. Modelling predicted stress to more than double to a maximum of about 75 MPa. Due to the geometry of the deposit, a lower stress change than that predicted was anticipated. With the majority of mining completed, the measured stress change in the sill pillar was less than 10%. Had substantial stress increase been measured in the sill pillar, cable bolts would have been installed. With the monitoring results obtained, no additional support measures were necessary.

3.6 Summary

The 3-dimensionality of this mining problem made it difficult to accurately model using the 2-dimensional models at our disposal. The use of more than 1 analysis technique made it possible to determine the key parameters which effect stability. This enabled instrumentation to be chosen to augment analysis results where they are weakest.

4 CONCLUSIONS

A design approach is used which consists of the following:
1. definition of the problem,
2. collection of a suitable data base, (virgin stress, rock fabric and classification and intact rock properties),
3. modelling to known failure conditions using numerical and empirical design techniques, (back analysis),
4. modelling future mining conditions and interpretation of results and
5. monitoring in key areas to check the validity of the modelling (closing the loop).
Application of this general approach has been described for two mines with very different problems. At Norita problems were stress dominated, whereas Chadbourne's past instability was structurally controlled. At both mines the modelling approach, when coupled with appropriate monitoring, produced good results. Intelligent coupling of modelling and monitoring is essential to optimize interpretation of both.

ACKNOWLEDGEMENTS

The authors wish to thank the engineering staff from both Norita and Chadbourne Mines for their valuable assistance in the rock mechanics studies. Also, appreciation is given to the mines and Noranda Inc. for permission to present this information.

REFERENCES

Hoek, E. and Brown, E.T., 1980. Underground Excavations in Rock, London, Institute of Mining and Metallurgy.
Kanduth, H. and Milne, D., 1986. Assessment of the Bemek Advanced Portable Rock Tester, Proceedings of the 27th U.S. Symposium on Rock Mechanics, Tuscaloosa, Alabama.
CANMET, 1980. Prediction of Stable Excavation Spans for Mining at Depths Below 1,000 Meters in Hard Rock, DSS Serial No. OSQ80-00081, DSS File No. 17SQ.23440-0-9020.

Etude expérimentale du comportement mécanique du sel en extension
Experimental study of mechanical behaviour of rock salt in extension
Experimentalstudie des Dehnungsverhaltens von Steinsalz

J.BERGUES, Laboratoire de Mécanique des Solides, Ecole Polytechnique, Palaiseau, France
J.P.CHARPENTIER, Laboratoire de Mécanique des Solides, Ecole Polytechnique, Palaiseau, France
M.DAO, Laboratoire de Mécanique des Solides, Ecole Polytechnique, Palaiseau, France

ABSTRACT : This study is undertaken in the frame of the research performed on the stability of underground caverns. In this paper, the authors give a description of a triaxial cell with temperature, as well as some experimental results on the mechanical behaviour of salt in extension.

RESUME : Cette étude s'inscrit dans le cadre des recherches entreprises sur la stabilité des cavités souterraines. Les auteurs présentent, d'une part, la description d'une cellule triaxiale en température et d'autre part les résultats expérimentaux relatifs au comportement mécanique du sel gemme en extension.

ZUSAMMENFASSUNG : Diese Studie im Rahmen der Forshungen über die Standsicherheit von unterirdischen Hohlraüme erfolgt. Hier wird zuerst die Beschreibung einer dreiachsigen Anlage mit Temperaturän derungs möglichkeit und, denn, die auf das mechanischen Verhalten von Steinsalz in Streckung bezüglichen experimentellen Ergebnisse vorstellt.

1 INTRODUCTION

Les recherches entreprises ces dernières années sur les cavités souterraines ont pu montrer l'importance du mode de sollicitation sur la stabilité des structures.Ainsi lors de certaines phases de la réalisation d'un tunnel (M. Panet, 1979) ou dans le cas d'un stockage de gaz dans les cavités souterraines, certaines zones du massif rocheux se trouvent être dans un état de contrainte en extension (J. Mandel, 1959), type de sollicitation relativement dangereux pour les structures réalisées au rocher et qui est encore très peu étudié sur échantillon au laboratoire. En effet, il est apparu aussi bien à partir de tests en laboratoire sur structures, (par exemple, des essais sur tubes de craie (J. Bergues et al,1979), ou sur maquettes simulant le chenal non réactif dans le cas d'une gazéification souterraine du charbon (Nguyen Minh D. et al, 1986), que sur des observations in situ d'ouvrages réels, que ce type de sollicitation peut faire apparaitre des phénomènes de rupture comme de l'écaillage ou de la fracturation (M. Panet, 1969). Dans le cas d'un enfouissement de déchets nucléaires dans le sel gemme, les conceptions actuelles de stockage conduisent à une élévation de température de l'ordre de 200° C au voisinage du dépôt dont il faut examiner les effets (dilatation, contraintes thermiques, propriétés rhéologiques). Aussi le Laboratoire de Mécanique des Solides a développé depuis plusieurs années des équipements permettant d'étudier le comportement mécanique de géomatériaux à température élevée et sous sollicitations complexes (J.P. Charpentier, 1985). Un accent particulier a été mis sur l'étude du comportement triaxial du sel gemme sous température en extension dont on présente ici les principaux résultats :
- la cellule triaxiale en température miseau point au laboratoire,
- les résultats d'une première série d'essais sur la rupture du sel en extension,
- une analyse des mécanismes de rupture en compression monoaxiale faite à partir des résultats d'essais sur éprouvettes ayant subi un chargement particulier.

2 DESCRIPTION DE LA CELLULE D'ESSAI TRIAXIAL

La figure 1 représente une coupe détaillée de la cellule qui permet de réaliser des essais triaxiaux cylindriques jusqu'à $\sigma_2 = \sigma_3 = P = 60$ MPa de pression latérale. Ces essais peuvent être :
- de compression $\sigma_1 > P$ ou
- d'extention $\sigma_1 < P$
avec $\sigma_1 = $ contrainte axiale.

La nomenclature placée sous la figure l repère les éléments principaux de la cellule.
- Un four (repère 10) disposé à l'intérieur du conteneur permet la mise en température jusqu'à 150° C. Cette température est régulée par une électronique à partir d'un thermocouple (repère 13) disposé au coeur de l'éprouvette.
- L'éprouvette (repère 12), un cylindre de 70 mm de diamètre et de 180 mm de hauteur est protégée du fluide de confinement par une gaine étanche (repère 11).
- Les déformations au cours du chargement sont déterminées par deux types de capteurs :
a - Les déformations longitudinales, dans la direction de σ_1, sont déduites d'un capteur de déplacement L.V.D.T. placé en tête du piston (repère 2) mesurant les variations de longueur $\Delta \ell$ de l'éprouvette.
b - Les déformations transversales, dans la direction de P, sont déterminées à partir d'un capteur, mis au point au LMS , ancré sur deux points diamétralement opposés de l'éprouvette mesurant les variations de diamètre Δt de l'échantillon. Ce capteur est constitué d'un arc déformable dont la section médiane est affaiblie pour augmenter sa sensibilité; les déformations de cet arc sont mesurées par une jauge extensométrique collée sur l'extrados de l'arc. Le capteur est positionné sur l'éprouvette par l'intermédiaire de deux tétons scellés (repère 14). Ce capteur est préalablement étalonné à l'aide d'un micromètre.

Les essais triaxiaux instantanés sont réalisés à l'aide d'une installation entièrement asservie comprenant :
- une presse d'une capacité de 1 000 KN pour la charge axiale σ_1,
- la cellule triaxiale dont la mise en pression est assurée par un multiplicateur de pression.

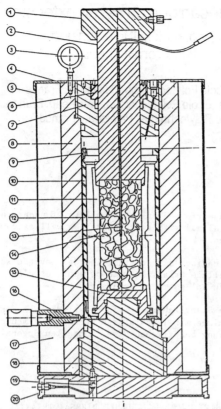

1) Appui du piston avec système de refroidissement
2) Piston avec passage du thermocouple
3) Anneau de manutention
4) et 5) Carter pour isolation thermique
6) Appui pour joint d'étanchéité
7) Bouchon supérieur
8) Corps principal de la cellule
9) Connexions pour sorties électriques
10) Corps de chauffage
11) Jaquette de protection de l'éprouvette
12) Eprouvette
13) Thermocouple
14) Tétons pour mesure des déformations transversales
15) Talon inférieur
16) Arrivée d'huile
17) Laine de roche pour isolation thermique
18) Bouchon inférieur
19) Passages des sorties électriques
20) Support de prises

Figure 1 : Description de la cellule triaxiale.

3 BUT DE L'ETUDE ET PROGRAMME EXPERIMENTAL

Le but de cette étude expérimentale sur la rupture instantanée du sel est double ; elle vise d'une part à proposer pour le matériau un critère de rupture en extension et d'autre part à étudier l'influence du trajet de chargement sur la rupture en extension.

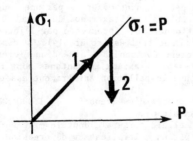

Figure 2 : Trajet de chargement du type A.

Pour cela deux types de chargement ont été étudiés.

. *Chargement du type* A : Représenté dans le plan $[\sigma_1, P]$ sur la figure 2.

Il s'effectue en deux temps :

ler temps : Mise en contrainte hydrostatique $\sigma_1 = \sigma_2 = \sigma_3 = P$ jusqu'à une valeur déterminée.

2ème temps : Mise en contrainte déviatorique d'extension $\sigma_1 - P < 0$ en relachant la contrainte axiale σ_1 et en maintenant constante la contrainte latérale P. Le chargement se poursuit jusqu'à la rupture.

. *Chargement du type* B : Représenté dans le plan $[\sigma_1, P]$ sur la figure 3.

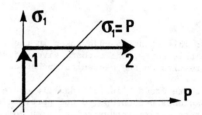

Figure 3 : Trajet de chargement du type B.

Il s'effectue également en deux temps :

ler temps : Mise en contrainte monoaxiale σ_1 jusqu'à une valeur déterminée.

2ème temps : Mise en contrainte latérale $\sigma_2 = \sigma_3 = P > 0$ en conservant σ_1 constant. Le chargement se poursuit jusqu'à la rupture.

Pour les deux types de trajet de chargement et pour les deux temps de mise en charge, l'essai est asservi en contrainte à vitesse constante $\dot{\sigma} = 0,1$ MPa s-1. Les essais ont été réalisés sur du sel de Bresse, à une température de 50° C ; cette mise en température s'effectue progressivement en trois heures.

4 RESULTATS EXPERIMENTAUX ET INTERPRETATION

4.1 Chargement du type A :

Les résultats relatifs à la rupture sont donnés par le tableau 1 et sur la figure 4 (x et trait plein). Le tableau mentionne l'état de contrainte et les déformations correspondant à la rupture.

$\sigma_1, \frac{\Delta\ell}{\ell}$ contrainte et déformation axiales

$P, \frac{\Delta t}{t}$ contrainte et déformation latérales

Les déformations sont comptées à partir de l'état de contrainte nulle (allongement noté moins). La figure 4 donne la rupture dans le plan $[\sigma_1, P]$

Tableau 1

P(MPa)	σ_1 (MPa)	$\Delta\ell/\ell$ 10^{-2}	$\Delta t/t$ 10^{-2}
30	7,5	− 1,2	1,8
40	19	− 0,9	1,8
45	17	− 1,2	1,7
50	26	− 1	1,6

Dans tous les cas l'éprouvette se rompt d'une façon brutale en deux parties suivant un plan parfaitement horizontal qui ne contourne jamais les cristaux de sel qui sont rompus d'une façon nette. On observe également le long de l'éprouvette d'autres amorces de rupture horizontales d'autant plus nombreuses que la pression latérale est élevée.

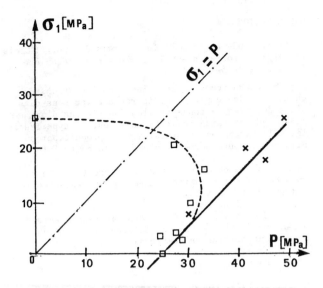

Figure 4 : Critères de rupture dans le plan $[\sigma_1, P]$ pour les deux types de trajet de chargement.

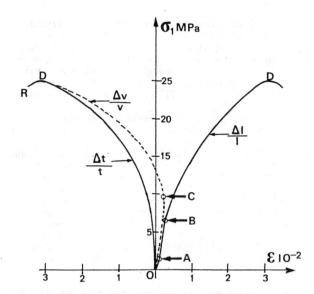

Figure 5 : Courbes "effort-déformation".

Les déformations à la rupture sont sensiblement constantes quelle que soit la contrainte latérale.

Dans ces conditions de chargement, à partir de la figure 4 on constate que le critère de rupture est du type TRESCA :

C = 12,5 MPa
φ = o

avec C = cohésion, φ = frottement interne.

4.2 Chargement du type B

Les résultats relatifs à la rupture sont donnés par le tableau 2 et sur la figure 4 (□ et pointillé).

Tableau 2

σ_1 (MPa)	P(MPa)	$\Delta\ell/\ell$ 10^{-2}	$\Delta t/t$ 10^{-2}
25	0	+ 2,8	- 3
20	26	+ 1,4	- 1
15,6	33	+ 0,6	+ 0,2
9,9	30	- 0,7	+ 0,8
3,9	28	- 1,8	+ 1
3,4	24	- 2,2	+ 2
2,6	28	- 2,5	+ 2,2
0	25	- 3,2	+ 2,1

Le premier essai du tableau correspond à l'essai de compression uniaxiale dont la figure 5 donne les différentes courbes "effort-déformation" :

$$\sigma_1 = f\ (\Delta\ell/\ell)$$
$$\sigma_1 = g\ (\Delta t/t)$$
$$\sigma_1 = h\ (\Delta v/v)$$

avec $\dfrac{\Delta v}{v} = \dfrac{\Delta\ell}{\ell} + 2\ \dfrac{\Delta t}{t}$ = variation volumique.

L'analyse de ces différentes courbes permet de mettre en évidence un comportement mécanique du sel à cinq phases :

- *Première phase* : O A jusqu'à 1 MPa, une phase de serrage qui peut provenir soit de la fermeture des fissures dans le matériau soit également d'un phénomène de compaction du sédiment argileux.
- *Deuxième phase* : AB jusqu'à 7 MPa, une phase élastique.
- *Troisième phase* : BC jusqu'à 10 MPa, une phase de dilatance qui est caractérisée par l'apparition de la fissuration qui se développe dans la direction de la contrainte σ_1.
- *Quatrième phase* : CD jusqu'à 25 MPa, une phase de foisonnement qui correspond a un développement intense de la fissuration. Durant cette phase, l'observation visuelle de l'éprouvette au cours du chargement montre qu'à partir d'un certain niveau du chargement il se développe un réseau de fissuration perpendiculaire à σ_1, comme le montre le dessin de la figure 6.

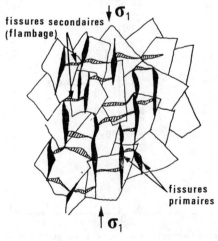

Figure 6: Réseaux de fissuration développés en compression uniaxiale.

Ces observations font penser à un mécanisme d'endommagement du sel en deux périodes :

1ère période : développement de la fissuration dans la direction de σ_1 avec formation de petites colonnettes.

2ème période : instabilité et rupture par flambage
de ces petites colonnettes provoquant un réseau de
fissures perpendiculaire à σ_1. Il est important de
noter que dans le cas d'essais triaxiaux classiques de
compression $\sigma_1 > P$ cette deuxième période est consi-
dérablement retardée voir inexistante ce qui confère
au matériau une déformabilité importante sans atteindre la rupture.
- *Cinquième phase* : DR = Perte de cohésion entrai-
nant la ruine du matériau.

La comparaison des essais de la figure 4 montre que
pour des contraintes axiales comprises entre 0 et
5 MPa environ le trajet de chargement n'a aucune influence sur la rupture. Au delà de cette valeur on
constate que pour le chargement du type B la rupture
survient prématurément par rapport au chargement du
type A. Ce résultat s'explique par l'apparition du
second réseau de fissures par flambage survenant au
cours de la première phase du chargement , qui en-
dommage le matériau dans la direction perpendiculaire
à σ_1.

Ce résultat est également mis en évidence sur la
figure 7 où l'on a reporté les déformations longitudi-
nales $\frac{\Delta \ell}{\ell}$ à la rupture (données dans les tableaux 1
et 2) en fonction de σ_1 pour les essais des deux types de chargement. On peut remarquer sur cette courbe une anomalie (point F sur la figure) liée probablement à l'apparition du second réseau de fissuration.
En effet, si pour les essais du type B l'endommagement se limitait au cours du chargement monoaxial à
un développement des fissures suivant la direction
de σ_1, ces fissures lors de la mise en contrainte latérale P (2ème temps du chargement) auraient tendance
à se refermer et à rétablir la continuité du matériau. De ce fait la rupture devrait se produire pour
des états de déformations similaires à ceux obtenus
pour les essais du type A, ce qui n'est pas le cas
comme cela apparaît clairement sur la figure 7.

5 CONCLUSION

Cette étude expérimentale a permis d'apporter trois
résultats importants sur le comportement mécanique
instantané du sel.

1 - Contrairement à certaines configurations de
chargement en compression, la rupture (au sens classique du terme) a toujours lieu dans le cas d'un chargement en extension ($\sigma_1 < P$). Elle est du type fragile, elle est en effet brutale et se produit par un
plan de fracture perpendiculaire à la contrainte σ_1.

2 - Influence du trajet de chargement sur la rupture du sel en extension causée par une anisotropie
induite de fissuration.

3 - Ces essais ont permis également de préciser le
mécanisme de fissuration du sel dans le cas d'un
chargement initial en compression.

BIBLIOGRAPHIE

Panet, M. 1979. Les déformations différées dans les
   ouvrages souterrains. 4ème congrès international
   de Mécanique des Roches. Montreux. Tome 3
   pp. 291-301.
Mandel, J. Janvier 1959. Revue de l'industrie minérale. pp. 78 à 92.
Berest, P., Bergues, J., Nguyen, M.D. 1979. Comportement
   des roches au cours de la rupture ; application
   à l'interprétation d'essais sur des tubes épais.
   Revue Française de Géotechnique n° 9, pp. 5-12.
Nguyen M.D., Schmitt, N. 1986. Physical modelling for
   stability analysis of the linkage in deep underground gasification. 12th Conf. Saarbrühen.
Panet, M. Décembre 1969. Quelques problèmes de
   mécanique des roches posés par le tunnel du Mont-
   Blanc. Annales ITBTP.
Charpentier, J.P., Berest, P., Août-Septembre 1985
   Fluage du sel gemme en température. Moyens d'essai
   et résultats. Revue générale de thermique,
   tome 24, n° 284-285, pp. 685-687.

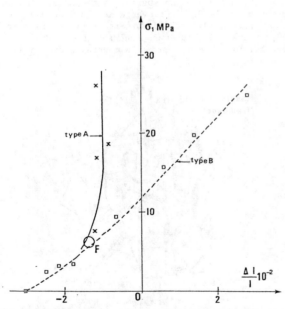

Figure 7 : Déformations à la rupture pour les deux
types de trajet de chargement.

Non-linear and discontinuous stress variation with depth in the upper crust of the Baltic Shield

Variation non-linéaire et discontinue des contraintes en fonction de la profondeur dans la partie supérieure du bouclier baltique
Nichtlineare und unstetige Spannungsveränderung mit der Tiefe in der oberen Kruste des baltischen Schildes

BJARNI BJARNASON, Sweden
OVE STEPHANSSON, Sweden

ABSTRACT: Hydraulic fracturing stress measurements have been conducted at five sites in Sweden and one site in Finland using the multihose system developed at the Luleå University of Technology, Sweden. All test sites are located within the crystalline bedrock of the Baltic Shield. The maximum measuring depth is 500 m. In general, the stresses vary non-linearly with depth. Calculated stresses from hydrofracturing measurements are smaller in magnitude than stresses determined by overcoring methods. A stress discontinuity of more than 10 MPa is probably associated with a major fracture zone at one of the test sites. Thus, rock stresses should still be measured before any large underground construction is undertaken in the shield.

RÉSUMÉ: Des mesures de contrainte par fracturation hydraulique ont été effectuées sur cinq sites en Suède et sur un site en Finlande, en utilisant le systéme de câble multiconducteur hydro-électrique développé par l'Université Technologique de Luleå, Suède. Tous les sites de mesure appartiennent au socle cristallin du Bouclier Baltique. La profondeur maximale des mesures est de cinq cents mètres. Génèralment, les contraintes varient d'une façon non-linéaire avec la profondeur. Les contraintes calculées à partir des mesures par fracturation hydraulique ont des magnitudes inférieures à celles des contraintes déterminées par des méthodes de surcarott-age. Sur un site, une discontinuité de plus de dix Mégapascals dans la régime des contraintes est probable-ment associée à une zone de fracture majeure. En zone cratonique, les contraintes doivent toujours mesurées avant d'entreprendre toute construction souterraine.

ZUSAMMENFASSUNG: Hydraulische Bruchspannungsmessungen sind an fünf Orten in Schweden und einem Ort in Finn-land mit dem bei der technischen Universität von Luleå, Schweden entwickelten Mehrschlauchsystem durchgeführt worden. Alle Versuchsanlagen lagen im kristallinen Muttergestein des baltischen Schildes. Die grösste Mess-tiefe betrug 500 m. Im allgemeinen schwankten die Spannungen nichtlinear mit der Tiefe. Die sich aus Hydro-frakturiermessungen ergebenden berechneten Spannungen sind geringer als die mit Kernmethoden bestimmten Spann-ungen. Eine Spannungsdiskontinuität von mehr als 10 MPa wird an einer der Anlagen mit einer möglichen gröss-eren Bruchzone in Verbindung gesetzt. Darum sollten Felsspannungsmessungen auch zukünftig durchgeführt werden, bevor grössere unterirdische Bauarbeiten im Bereich des Schildes erfolgen.

INTRODUCTION AND THEORY

Hydrofracturing stress profiling in the Baltic Shield has only been conducted during the last few years. However, the results have increased our knowledge of the shallow stress field in the region.

The results presented in this paper are obtained by application of the classical hydrofracturing method, where the borehole wall is fractured by pressurizing water. Principal stresses are assumed to be vertically and horizontally oriented, and the vertical stress, S_v, is assumed to be equal to the overburden pressure. Horizontal stresses are calculated as follows:

$$S_h = P_s$$
$$S_H = 3S_h - P_{c2}$$

where
S_h = minimum horizontal stress
S_H = maximum horizontal stress
P_s = stable shut-in pressure
P_{c2} = fracture reopening pressure

The hydrofracturing tensile strength of the rock is taken to be equal to the difference between the break-down pressure and the reopening pressure, $P_{c1} - P_{c2}$. The pore pressure in the rock is neglected. (For de-tailed information on the application of the method, see Bjarnason. 1986a).

COMPILATION OF STRESSES FOR SWEDISH BEDROCK

The following compilation is based on hydrofracturing results from five sites in Sweden, (Bjarnason 1986a, Bjarnason 1986b, Stephansson and Ångman 1986). Site locations are shown in Figure 1. The data has been treated selectively and only results from vertical, induced fractures are presented. All measurements re-present virgin rock stresses far from man-made open-ings.

Figure 2 illustrates the mean horizontal stress di-vided by the theoretical vertical stress, versus depth for five sites in Sweden. The method of pre-sentation is similar to the well-known compilation by Brown and Hoek (1978). The figure shows that a high ratio of horizontal to vertical stresses for hydro-fracturing is a very shallow phenomenon in Swedish bedrock. Down to 100 m depth, k-values up to 7 are encountered. At 100 m depth the k-value is generally aroun 2, and at 500 m depth the k-ratio is close to unity.

The data in Figure 2, at a given depth z below sur-face, is fitted to the following equation:

$$k = \frac{160}{z} + 0,8$$

Figure 1. Hydrofracturing stress measurements in the Baltic Shield, conducted by the Division of Rock Mechanics, Luleå.

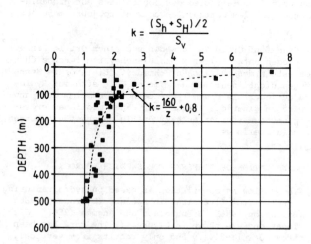

$$k = \frac{(S_h + S_H)/2}{S_v}$$

$$k = \frac{160}{z} + 0,8$$

Figure 2. Ratio of the average horizontal stress and the theoretical vertical stress, $(S_h + S_H)/2S_v$, versus depth for hydrofracturing measurements at five sites in Sweden.

Figure 3 shows the average horizontal stress, $(S_H + S_h)/2$ as a function of depth in Fennoscandian bedrock, determined by different measuring methods. Curves 1, 2 and 3 represent results from overcoring methods, curve 4 results from hydrofracturing measurements and curve 5 is the assumed vertical stress calculated from an average rock unit weight of 27 KN/m^3. Curves 1, 2 and 3 are obtained by a non-selective compilation of data from the recently established Fennoscandian Rock Stress Data Base, (FRSDB), Stephansson et al. 1986.

The overcoring measurements, (with the exception of the SSPB method), are conducted in underground openings, mostly mines. Induced stresses surrounding the excavations and the inhomogenety of the rock mass in mining areas may be partly responsible for the discrepancy between the overcoring and hydrofracturing results. Referring to curve 1, it has been concluded by Bergman (1977) that the one-dimensional character

of Hast's overcoring method may lead to a severe overestimate of the in-situ stresses when the stress tensor is determined by combining results from different measuring points and different boreholes.

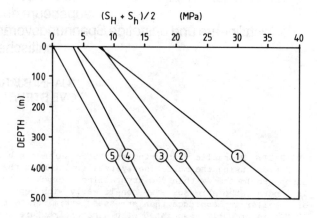

Figure 3. Average horizontal stress versus depht in Fennoscandian bedrock.

(1) Overcoring, Hast: S_{Hav} = 7,2 + 0,063 z MPa
(2) Overcoring, SSPB:* S_{Hav} = 7,7 + 0,037 z MPa
(3) Overcoring, others: S_{Hav} = 3,8 + 0,039 z MPa
(4) Hydrofracturing: S_{Hav} = 3,1 + 0,025 z MPa
(5) Theoretical vertical stress S_v = 0,027 z MPa

* SSPB = Swedish State Power Board

To summarize, it can be stated that the hydrofracturing results presented above do not confirm the relatively large horizontal stresses close to ground surface, reported from overcoring measurements in Sweden and other Fennoscandian bedrock.

BI-LINEAR STRESS GRADIENTS

The multihose hydrofracturing system used, enables measurements at short intervals troughout the length of the borehole without excessive field expenses. Typically, 25-30 valid measurements are conducted in a 500 m deep borehole. This large number of measuring points reveals the details of the stress profiles.

Figure 4 shows shut-in pressures versus depth in borehole Gi-1 at Gideå in Northern Sweden. This is one of the study sites in the Swedish programme for the final disposal of high level radioactive waste. The joint frequency at depths less than 300 m in the rock mass at Gideå is about 4,7 joints/m, falling abruptly to ca 2,7 joints/m at depths greater than 300 m. The gradient of shut-in pressures versus depth is 0,03 MPa m^{-1} from the surface down to approximately 300 m depth. At depths greater than 300 m the gradient is 0,008 MPa m^{-1}. Further details on the results from Gideå are given by Bjarnason (1986a).

Figure 5 illustrates results from borehole BSP-1 at Staverhult test site along the 80 km long fresh water tunnel Bolmen, in Southern Sweden. The results are similar to those obtained from Gideå, with a steep gradient of shut-in pressures from surface down to 70 m, but with a small gradient at depths greater than 70 m. Again, the gradients are reflected by the joint frequency of the rock mass. Above 70 m depth the rock mass is heavily jointed and locally crushed. At depths greater than 70 m the rock is predominantly fresh and of good quality, with intact core lengths up to 8 m. At 230 m depth, the shut-in pressure increases again.

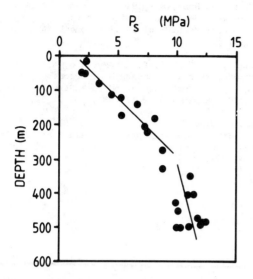

Figure 4. Shut-in pressures versus depht for borehole Gi-1 at Gideå in Northern Sweden.

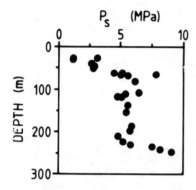

Figure 5. Shut-in pressures versus depth for borehole BSP-1, Staverhult, Bolmen fresh water tunnel in Southern Sweden.

Figures 4 and 5 illustrate that the idealized picture, of a linear variation of stresses with depth, often presented for large regions, (see Figure 3 as an example), is not valid for individual test sites.

DISCONTINUOUS STRESS FIELD AT LAVIA, FINLAND

The test site at Lavia is situated in Western Finland. It is included in the Finnish programme for final disposal of high level radioactive waste, Figure 1. Measurements were conducted down to a depth of 500 m in a 1000 m deep borehole, inclined at 10-15° to the vertical. The main rock type is granodiorite. The results are presented in Figure 6.

Horizontal fractures

The hydrofracturing tests were conducted at five levels in the borehole, at depths of approximately 100, 200, 300, 400 and 500 m. Five tests were conducted at each test level, except at 500 m depth, where the rock quality permitted only 3 test points. At 100, 300, and 400 m levels, only true horizontal hydrofractures were formed. This occurred in spite of exceptionally good rock conditions with intact core lengths up to 20 m. The formation of exclusively horizontal fractures prevented determination of the horizontal stresses at these three test levels. However, shut-in pressures for true horizontal fractures give the vertical stress component, and a

possibility to compare a measured vertical stress, equal to P_s, with the theoretical vertical stress, S_v, calculated from the weight of the overburden. The shut-in pressures are plotted in Figure 6. Table 1 shows the ratio of the measured shut-in pressures P_s to the theoretical vertical stress, S_v.

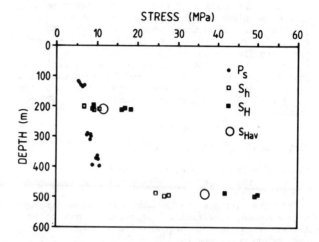

Figure 6. Results from hydrofracturing measurements at Lavia, Finland.

Table 1. Ratio of the measured shut-in pressure P_s to the theoretical vertical stress, S_v from Lavia, Finland.

Depth (m)	P_s/S_v	Depth (m)	P_s/S_v	Depth (m)	P_s/S_v
116,5	1,8	291,6	1,0	362,3	1,1
119,8	1,8	293,0	1,1	370,0	1,0
127,8	1,7	294,6	1,0	373,2	1,1
130,7	2,0	297,5	1,1	391,7	0,9
133,8	1,9	309,1	1,0	396,8	1,0

The measured vertical stress at 100 m depth is almost double the expected value from the weight of the overburden. However, at 300 and 400 m levels the ratio of P_s/S_v is close to unity. The shut-in pressures for the horizontal fractures at Lavia remained stable after repeated pressurization cycles and an increased injection volume. This implies stable horizontal propagation of the fracture away from the borehole wall and, hence, a near vertical orientation of the minimum principal stress.

Large horizontal stresses

At 200 m and 500 m test levels, vertical hydrofractures were formed at all test points in the borehole at Lavia. The shut-in pressures, ($P_s = S_h$ in the case of vertical fractures), the maximum horizontal stress and the average horizontal stress are plotted for all test points at the two levels, Figure 6. The results are summarized in table 2.

Table 2. Hydrofracturing results from 200 m and 500 m test levels at Lavia in Finland. Mean values from all test points at each depth level.

Level (m)	S_v (MPa)	S_h (MPa)	S_H (MPa)	S_H/S_h	S_{Hav} (MPa)	S_{Hav}/S_v
200	5,4	8,6	14,1	1,6	11,3	2,1
500	12,8	26,4	47,2	1,8	36,8	2,9

The borehole intersects a major fracture zone at 440 - 480 m depth. The orientation of the zone is unknown. The test points at 500 m level are located immediately below this fracture zone. The magnitude of horizontal stresses below the fracture zone is more than three times higher than the stresses at 200 m depth. The ratio of the average horizontal stress to the vertical stress at 500 m depth is almost three times higher than the ratio at the same depth obtained from measurements in Sweden, cf. Figure 2. Two valid points, (200 m and 500 m levels), give insufficient information on the details of the stress profile at Lavia. However, the difference in shut-in pressures between 200 and 500 m levels gives a strong indication of a major discontinuity in stress magnitudes across the fracture zone at 440 - 480 m depth.

CONCLUSIONS

Experience from hydrofracturing at the six sites presented in this paper can be summarized as follows:

Hydrofracturing stress measurements in Swedish bedrock have not confirmed the relatively large horizontal stresses close to ground surface reported from overcoring measurements.

The registration of vertical stress profiles has shown that a linear variation of stresses with depth does not apply to individual test sites within the upper crust in Fennoscandia.

The results from Lavia in Finland indicate a pronounced discontinuity in stress magnitudes across a major fracture zone at 440-480 m depth in the borehole. Similar results have been reported from Forsmark in Sweden by Stephansson and Ångman 1986. The vertical stress measured at the 100 m level at Lavia is almost double the calculated overburden pressure.

The deepest hydrofracturing measurements in Sweden and Finland reach only down to 500 m. Hence, the results presented are only valid for the very shallow crust in the region. A new multihose for measurements down to 1000 m depth is now under construction at the Technical University of Luleå.

ACKNOWLEDGEMENTS

The authors wish to thank the following companies and organisations for financial support that made the field measurements possible and for the kind permission to present the data from the measurements: TVO (Inustrial Power Company, Finland), SKB (Swedish Nuclear Fuel and Waste Management Company), BeFo (Swedish Rock Engineering Research Foundation), and DDP (Swedish Deep Gas Project).

REFERENCES

Bergman, S.G.A., 1977. Rock Stress Measurements in Scandinavian Bedrock - Presumptions, Results and Interpretation. Technical Report No. 64, Swedish Nuclear Fuel and Waste Management Company, Stockholm, (in Swedish).

Bjarnason, B., 1986a. Hydrofracturing Rock Stress Measurements in the Baltic Shield. Licentiate Thesis 1986:12 L, Division of Rock Mechanics, Luleå University of Technology, Sweden.

Bjarnason, B., 1986b. Rock Stress Measurements in Borehole 4, Stajsås, Siljan Area, Sweden. Preliminary Report, CENTEK/Luleå University of Technology, Sweden.

Brown, E.T. and Hoek, E., 1978. Trends in Relationship Between Measured In-Situ Stresses and Depth. Int. J. Rock Mech. Min. Sci. & Geomech. Abst., vol. 15, pp. 211-215.

Stephansson, O., Särkkä, P. and Myrvang, A., 1986. State of Stress in Fennoscandia. Proc. of the Int. Symp. on Rock Stress and Rock Stress Measurements, Stockholm 1-3 Sept. 1986, pp. 21-32.

Stephansson, O. and Ångman, P., 1986. Hydraulic Fracturing Stress Measurements at Forsmark and Stidsvig, Sweden. Bull. Geol. Soc. Finland, vol. 58. Part 1, pp. 307-333.

A rock mechanical investigation of a longitudinal opening in the Kiirunavaara Mine
Investigation de mécanique des roches d'une ouverture longitudinale à la mine Kiirunavaara
Eine felsmechanische Untersuchung des längsgehenden Durchbruchs in der Grube Kiirunavaara

TORGNY BORG, Ph.D., (Rock Mechanics), Consultant, Falun, Sweden
ANDERS HOLMSTEDT, Min.Eng.,M.Sc., Rock Mechanics, LKAB, Sweden

ABSTRACT: In the Kiirunavaara mine, longitudinal opening of transverse sub-level caving is tested. The ore along the hanging wall contact is excavated through a longitudinal drift, leaving caved material as contact for following opening for the transverse sub-level caving on the same level. Serious stability problems have been observed around the longitudinal drifts at the hanging wall mining of the above level. An investigation has been performed including FEM-analysis, measurements, observations of fracturing and failures. The study has shown that the stability problems in the longitudinal opening related to areas with high stress magnitudes can be significantly reduced by moderate changes in the current mining practice. Crosscuts should not be driven through to the longitudinal drift at the hanging wall contact but stopped at a distance of one or two rounds. Meeting places of fronts should be placed in competent rock and prereinforced. In areas with very poor rock massblasting is proposed.

RESUME: L'essai d'ouvertures par tracages longitudinales a ete faite dans la mine de Kiirunavaara. Le minerai le long du toit est abattue et evacue par un tracage longitudinal, laissant un contact de sterile pour l'ouverture de l'abattage suivant des tracages transversalles du même niveau. Des graves problèmes de stabilitê du tracage longitudinal a etee observe pendant l'ouverture et l'exploitation de ce niveau. Des mesures, des observations de la formation des fractures et des éboulements, et une analyse-FEM ont été exécutés. Ce travail a demontrê que les problèmes des stabilitê qui existent dans les ouvertures longitudinales où on trouve une grande tension, peuvent être appréciablement rêduits par des changements modérês de la mêthode d'exploitation actuelle. Les tracages transversalles doivent être arrete a une distance du tracage longitudinal La rencontre des fronts d'exploitation doivent se faire dans un rocher solide. Ces emplacements sont renforcees a l'avance. Les sections a mauvaise stabilitee sont fraturees en masse.

ZUSAMMENWASSUNG: Im Teilsohlenbruchbau im Kiirunavaara wurde am Hangenden ein Versuch mit längsgehendem Durchbruch ausgeführt. Für diesen Zweck wurde am Hangeden entlang eine Strecke vorgetrieben und später wurde von dieser Strecke der Durchbruch zu überliegenden Teilsohle gemacht. Dieser längsgehende Durchbruch diente als Öffnung für den queer zum Erz gehenden Bruchbau und sollte auch die Selektivität der verschiedenen Erztypen verbessern. Während des Durchbruchs entstanden jedoch am Hangenden ernsthafte Stabilitätsprobleme. Eine Untersuchung in Bezug auf u.a. FEM-Analyse, Gebirgsmessungen, Beobachtung von Bruchzonen und Erzausbeute wurde durchgeführt. Diese Studie zeigte, dass die Stabilitätsprobleme vor allem in Gebieten mit grossen Spannungen auftreten. Die negativen Einwirkungen können durch geringe Änderungen in der Abbaupraxis erheblichverbessert werden. Die Querstrecken werden ein zu zwei Bohrlochkranzrundevor der Hangendstrecke beendet. Beim Rückabbau der Hangenstrecke wird der Schlusabbau in festes Gebirge verliegt, beigeringer Standfestigkeit werden mehrere Bohrreihengleichzeitig gesprengt.

1. INTRODUCTION

In the middle of the 1970's rock mechanical calculations, showing that the sublevel caving layout had to be altered for the Kiirunavaara mine on larger depths were performed (Knobloch 1976). In 1977 it was decided to increase the drift distance from 11 m to 16,5 m maintaining the slice height of 12 m. Increased drift distances resulted in a more difficult opening blasting which led to the decision that an additional hanging wall drift would be driven.

Since the middle of the 1970's, the demand for high-phosphorous ore has gradually decreased, which has resulted in the study of caving methods giving a larger percentage of low-phosphorous ore than conventional sublevel caving. Within the ore body, the high-phosphorous ore is concentrated to the hanging wall side, and by mining this high-phosphorous ore separately along the hanging wall, the selectivity would be remarkably improved. The method of mining the high-phosphorous ore separately along the hanging wall has been named stair-step mining, and it will at the same time form an opening for the conventional transverse sub-level caving (Schmiedhofer 1981).

When the development work in a mining area has been performed, the stair-step mining is commenced by opening the caving, from the hanging wall drift at one level, towards the above level (Figure 1). Subsequent-ly, the ore at the hanging wall is mined from the longitudinal drift until this is done in the entire area. The mining of the transverse sub-level caving is then commenced, using the longitudinal caving as an opening.

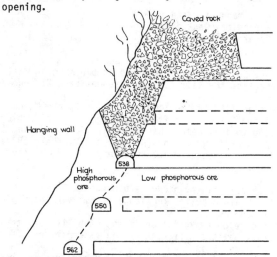

Figure 1. Vertical profile of the stair-step mining method at Kiirunavaara iron ore mine.

The best selectivity is achieved if the longitudinal sublevel caving is located two slices below the transverse caving in a stair-step pattern (Schmiedhofer 1981). Caving of the longitudinal drifts is performed along several fronts at the same time in one level (Figure 2). This process gives several locations for the opening of the hanging wall caving along the level, as well as several places where the fronts will meet.

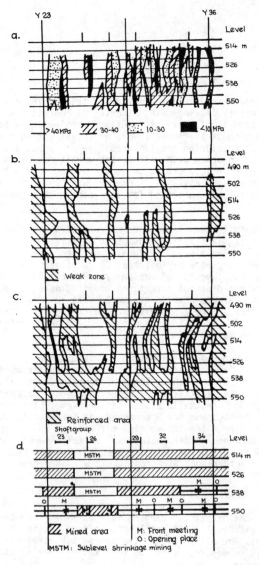

a. Rock mass strength in MPa
b. Geological weak zones
c. Reinforcements
d. Proposed production plan

Figure 2. Longitudinal section at the hanging wall contact of the Kiirunavaara mine between Y23 and Y36.

The stability of the longitudinal drift at the hanging wall contact when shifting from conventional sublevel caving to this method has generaly been poor, and in certain areas very poor (Holmstedt & Liedberg 1982). This has initiated a rock mechanical study of the method.

The objects of the study have been:
- to obtain a better knowledge of the working failure mechanisms
- to develop a prediction of the stability and to study how the different mining alternatives affect the prediction.
- to translate the knowledge to suggestions of changes in the mine layout.

This article states rock mechanical observations, measurements, and production data. We describe the use of extension strain prediction for the interpretation of these data, and we examine how different mining alternatives affect the predictions. The article is concluded with a discussion on experience gained and a description of the suggested changes in the production plan that the work has resulted in.

2. GEOLOGY AND ROCK STRESSES

2.1 Geology

The hanging wall drifts within the examination area Y 28 - 33 (Y2800 - Y3300= 500m in strike direction) have been driven in magnetite ore, and the dominate joint set is repeated domains of structures dipping 45 - 50 S almost perpendicular to the strike of the ore body. The joint spacing is mostly in the range of 10 - 40cm. In most joints there is a coating or filling of weathered calcite and talc. The apatite content in the ore is high and it is often concentrated to laminations or layers. The rock mass is classified as small block size or middle block size (Holmstedt & Liedberg 1982).

2.2 Rock stresses

Rock stress measurements with the LUT-cell developed at the University of Luleå according to the Leeman principle (Leijon 1986) have been performed above the overhead slice, to enable the determination of the virgin stress. The maximum principal stress ( ) is 30.6MPa before the opening of the overhead slice, and the direction is almost perpendicular to the strike of the ore body. After the opening of the overhead slice, the maximum principal stress is 35MPa and its direction has shifted 10 to the north.

3. OBSERVATIONS AND MEASUREMENTS

3.1 Observations in the area Y 31.7 - 32.8

In connection to the mining of the hanging wall drift at level 538m, continuous visual observations at the levels 538m and 550m were performed. Every rock failure and fracturing was noted, and these notes in combination with data of the ore recovery has then been related to the position of the mining front at 538m (Figure 3).

The rock failures and the fracturing are generally dependent on the position of the mining front at level 538m. The last blasting at 538m, starting approximately 35m from the encounter with the north front, has a low ore recovery with bridge formation, that is, remaining parties of intact ore in the blasted rounds. The bridge formation is primarily caused by poorly charged drill-holes (Figure 4).

The rock failures and the fracturing are not as notable in connection to the mining front at the 538m level. Within these areas, strong reinforcements have been made after earlier rock failures at the 538m level (Figure 3).

3.2 Displacement measurements

The area that has been followed up can be divided into two sections, one in the north (Y 29.5 - 30.0) and one in the south (Y 31.7 - 32.8). Distometer measurements in the northern section show that the roof center point has displaced 50mm down (Figure 5) in connection to the encounter of the two fronts at the overhead slice. In the southern section, minor displacements of the drift profiles ranging from 1 to 2 mm were registered. Measurements with borehole extensometer within the same area show no displacements.

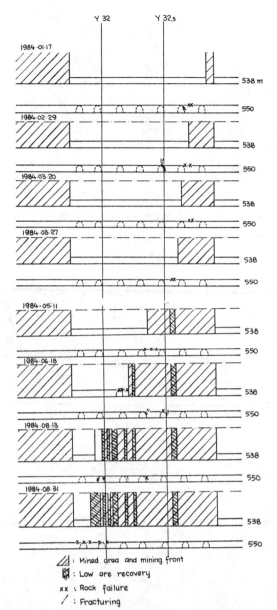

Figure 3. Longitudinalsection between Y31.5 and Y33. Visual observations at different times when two mining fronts are meeting.

Legend:
- 🗆 : Mined area and mining front
- 🗆 : Low ore recovery
- xx : Rock failure
- ⁄⁄ : Fracturing

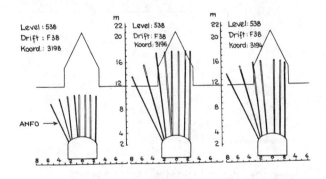

Figure 4. Results of loading fans

3.3 Measurements of rock stress changes

The stress changes caused by the mining has been measured with CSIRO's Hollow Inclusion stress cell. The positions for the Hollow Inclusion stress cells were in the slice between 538m and 550m, above the

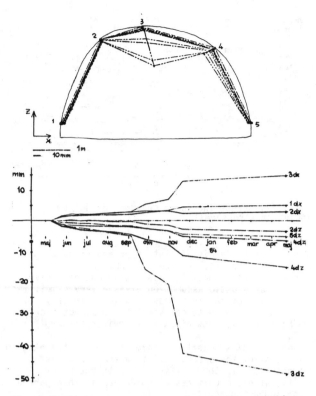

Figure 5. Distometer measurements, northern section.

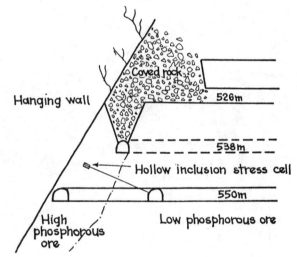

Figure 6. Positioning of the Hollow Inclusion stress cell.

hanging wall drift and in the vicinity of the ore contact (Figure 6).

The monitoring of the stresses started when level 538m was opened 80m south of the position for the stress cells and continued until the front passed above the stress cells. The stress increase registered during of the mining of the overhead slice was 15MPa (Figure 7.)

4. PREDICTION OF ROCK STRESSES

A two-dimensional Finite Element analysis has been performed to establish a rock stress prediction and to analyse some conceivable mining alternatives.

815

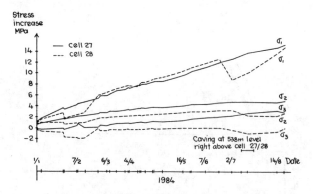

Figure 7. Stress increases measured by Hollow Inclusion stress cells.

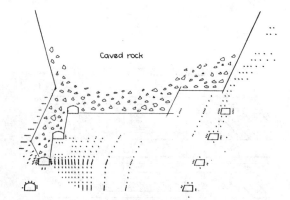

Figure 9a. Major extension strain vectors.

The rock mass is divided into four areas; hanging wall, ore, foot wall and caved rock. An isotropic linearly elastic material assumption has been used. The linear elastic assumption gives a very simplified description of the rock mass behaviour. In spite of this, it has shown to be useful to construct rock mechanic predictions of stress-magnitudes (Borg 1983).

The highest stress levels are located to the hanging wall side of the opening drift, where the stress magnitudes rate 100MPa (Figure 8). In the underlying drift, stress magnitudes of 60MPa appear in a narrow zone above the roof.

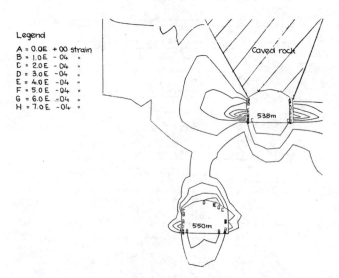

Figure 9b. Iso-curves of major extension strains.

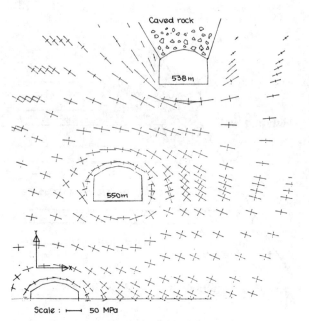

Figure 8. Principal stress vectors.

Predictions of failures can be analysed using an extension strain criterion (Stacey et al 1981).

The highest magnitudes of extension strains appear within the earlier described areas with high magnitudes of rock stresses, but also within areas where the mining is performed by conventional sublevel caving and in the vicinity of the foot wall drifts (Figure 9a and 9b).

The result of a comparison between an alternative where the crosscuts has been driven towards the hanging wall drift and an alternative where a section has been left to the hanging wall drift is shown in Figures 10a and 10b. The results from the two alternatives indicate that the remaining section has considerably decreased the magnitudes of the extensional strains in the vicinity of the drifts.

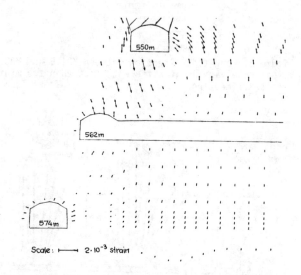

Figure 10a. Major extension strain vectors (The crosscuts connected with the hanging wall drift)

In figure 11 suggestions of measures are stated. If the suggestions are followed, the magnitudes of extension strains can be reduced in critical sections.

816

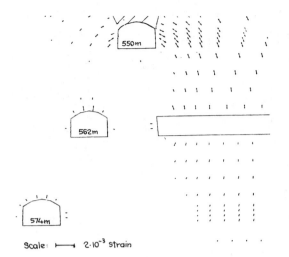

Scale: ⊢———⊣ 2·10⁻³ strain

Figure 10b. Major extension strain vectors (The crosscut not connected with the hanging wall drift).

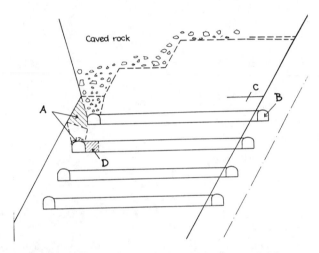

A: Zone with high stress magnitude. The analysis shows that the stress magnitudes increasees when two adjacent mining fronts approaches each other. In those situations an alternative method for openings is recommended, e.g. mass blasting or rock destress blasting.
B: The analysis indicates that a failure in the foot wall drift can be expected when mining advances to the drift level.
C: Increase the distance between ore contact and the foot wall drift.
D: Leave an intact rock bridge between the crosscut and the hanging wall drifts in critical areas.

Figure 11. Suggested measures.

5. DISCUSSION

The measurement results from the longitudinal drift shows that displacements   15mm in drift profile points will not result in fracturing of the rock and also that beginning fracturing has been registered at a movement of 50mm. The Hollow Inclusion cell shows an increase of the stress magnitude of 15MPa during the mining of the overhead slice and when simulating this in the model, a stress magnitude increase of 4MPa over the longitudinal drift is received, a value that coincides with the LUT-cell measurements. An explanation to the higher stress magnitude values indicated by the Hollow Inclusion cell can be shrinking phenomena in the material of the measurement cell.

At the production follow-up several sections with low ore recovery and probable bridge formation have been observed (Figure 3). The bridge formation can be attributed to poor blasting results, which in its turn primarily can be attributed to poor charging of drill holes (Figure 4). The poor drill-hole charging is caused by the fact that many of the drill holes are crushed or sheared off. In the model, it can be observed that the stress magnitudes are high in the upper parts of the drill fans of the hanging wall contact, which may affect the stability of the drill holes.

The rock failures in the hanging wall drift at level 550m within the observed area are relatively few, in comparison to earlier observations of the stability at hanging wall drifts. A probable explanation to this can be the observed bridge formations at level 538m, which reduce the stress magnitudes at the underlying level. This explanation was verified by model simulations.

The results are used to establish a plan for suitable positions of openings of hanging wall drifts and front encounters. The openings will be placed in areas with weak rock. Meeting places of fronts will be placed in competent rock and where the crosscuts have been stopped one or two rounds before the longitudinal drift. A schema like that in figure 2 can be used to determine opening and meeting places for the longitudinal sub-level caving. In areas with very poor ground condition mass blasting is proposed for meeting of fronts. Meeting places are recommended to be reinforced.

Then consideration also has to be taken to the position of the ore passes in the actual area to achieve an acceptable haulage route. All together positioning of openings and meeting places of fronts is an optimization of rock mechanics and efficient production conditions.

As the set-forward of the development work has been relatively large, it will take a long time before the results of taken measures can be observed. The suggestion of measures presented above could be fully used at the construction of a mining layout for level 562m, Y 28 - 38, which will be mined during 1987.

6. CONCLUSIONS

1. The calculated zones with high extension strain values around the hanging wall drifts correspond satisfactory to zones where failures have been observed.
2. It has been possible to use the model for the study of certain mining variations reducing the extension strain levels around the hanging wall drifts.
3. The accumulated knowledge of the rock mass as well as refined model analyses can be used when planning reinforcement works.
4. To reach through with a measure suggestion is very dependent upon if the measures are conceived as realistic by the management, line management and mine operators. The rock mechanical engineer has to spend a considerable amount of time discussing with planners and production people.

7. REFERENCES

Borg, T.1983,Prediction of failures in mines with application to the Näsliden Mine in northern Sweden Doctoral Thesis,Luleå University,Sweden.Report no. 1983:260.
Holmstedt, A and Liedberg, S.1982.Internal report.
Knobloch, H. 1976.Internal report.
Leijon, B.1986,Application on the LUT triaxial overcoring technique in Swedish mines.Proc.of the Int. Symp.on Rock Stress and Rock Stress Measurements, Stockholm,Sweden,569-579.
Schmiedhofer, A. 1981,Internal report.
Stacey, T.R.1981,A simple extension strain criterion for fracture of brittle rock.Internal Journal Rock Mechanics,Mining Science and Geomech.Abstract,vol. 18,no 6,469-474.

Bohrlochkonvergenz und Spannungsrelaxation im Steinsalzgebirge

Borehole convergence and stress relaxation in rock salt
Fluage et relaxation autour d'un trou de sonde percé dans un massif de sel gemme

G.BORM, Institut für Bodenmechanik und Felsmechanik, Universität Karlsruhe, Bundesrepublik Deutschland

ABSTRACT: The rheological behaviour of rock salt around boreholes or shafts is studied by analytical and numerical calculations. As material model a generalized nonlinear Maxwell-body is assumed, where the rate of creep deformation is a power function of deviatoric stress.
    Special emphasis is directed to the interaction of creep and relaxation in the rock. As a result of stress relaxation, the rates of borehole convergency decrease monotonously with time, indicating apparently strain hardening material properties.

RESUME: L'objet de cette étude est l'analyse du comportement rhéologique du sel gemme aux abords d'un trou de sonde libre. Le comportement du matériau sera approché par un modèle de Maxwell généralisé et non-linéaire. Pour ce dernier, le taux de déformation visqueux s'exprime comme puissance de la différence des contraintes.
    On considère en particulier l'influence réciproque du fluage et de la relaxation dans un massif de sel. Le processus de déchargement en contraintes du massif se laisse approcher de façon réaliste par une loi non-linéaire visco-élastique. La différence des contraintes, qui peut être absorbée au maximum, est d'autant plus grande, que la vitesse de fluage est importante. Comme conséquence de la relaxation en contraintes, il résulte que la convergence du trou de sonde décroît avec le temps, de telle sorte qu'on simule un écrouissage à froid du sel.

ZUSAMMENFASSUNG: Die rheologische Bewegungsgleichung für das zeitabhängige Konvergenzverhalten eines Bohrloches oder Schachtes im Steinsalzgebirge wird hergeleitet und semi-analytisch gelöst. Das Stoffgesetz entspricht dem eines verallgemeinerten nicht-linear viskoelastischen Maxwell Körpers, dessen viskose Deformationsrate einem Potenzgesetz folgt.
    Die maximal aufnehmbare stationäre deviatorische Spannungsintensität ist umso höher, je höher die Kriechdehnungsgeschwindigkeit ist. Sie ist nach oben hin durch die Bruchfestigkeit des Gesteins begrenzt. Im Lauf der Zeit nimmt die Konvergenzrate monoton mit der fortschreitenden Spannungsrelaxation des Gebirges ab. Das Gesetz für die zeitabhängige, instationäre Bohrlochkonvergenz wird in geschlossener Form angegeben.

1 EINLEITUNG

Bei der untertägigen Einlagerung von Sondermüll im Salzgebirge hat die Frage nach dem zeitabhängigen Konvergenz- und Festigkeitsverhalten der Befahrungs- und Einlagerungshohlräume eine zentrale Bedeutung. Modellrechnungen und in situ Messungen erlauben erste langfristige Prognosen.

Im folgenden wird die rheologische Bewegungsgleichung für das zeitabhängige Konvergenzverhalten eines Bohrloches oder Schachtes im Salinar hergeleitet und semi-analytisch gelöst. Das Stoffgesetz entspricht bei bruchlosem Kriechen dem eines verallgemeinerten, nichtlinear viskoelastischen Maxwell Körpers, bei dem die viskose Deformationsgeschwindigkeit einem Potenzgesetz folgt.

Die Relaxationskennlinien werden dimensionslos dargestellt, damit sie für beliebige Werte von äußerer Gebirgsspannung, Bohrlochradius, Schubmodul und Kriechparameter gültig sind. Mit ihnen können die Wirkungen der einzelnen Einflußgrößen quantitativ abgeschätzt werden.

Umgekehrt läßt sich damit die Gültigkeit des Kriechgesetzes vom Potenztyp für felsmechanische Berechnungen im Steinsalzgebirge durch Vergleich der theoretischen Resultate mit Ergebnissen von in-situ-Messungen überprüfen.

2 STATIONÄRE SPANNUNGEN UND KONVERGENZRATEN BEI EINEM BOHRLOCH IM SALINAR

2.1 Geometrie, Randbedingungen, Stoffgesetz

In erster Näherung wird das Bohrloch durch eine zylindrische Öffnung im Salzgebirge unter ebenem Formänderungszustand und konstantem Außendruck $p_o = -\delta_o$ (Gebirgsdruck) betrachtet (Fig.1).

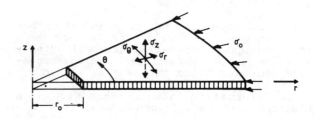

Fig.1. Koordinaten und Hauptspannungungsbezeichnungen am Bohrlochmodell.
Coordinates and notations of principal stresses of borehole model.
Coordonnées es notations des contraintes principales a un modèle du trou de sonde.

Das Materialverhalten des salinaren Gebirgs-
körpers wird nichtlinear viscoelastisch angenommen und durch das einachsige Kriechgesetz (Gl.1) beschrieben:

$$\dot{e} = k \, 6^n \; ; \quad n \geq 1 \qquad \qquad \dots (1)$$

Darin bedeuten:
$\dot{e}$ axiale Kriechdehnungsgeschwindigkeit
6 axiale Spannung
k,n temperaturabhängige Meßgrößen

Die Verallgemeinerung des eindimensiona-
len Kriechgesetzes (Gl.1) auf die mehrachsi-
gen Spannungsverhältnisse am Bohrlochrand
erfolgt nach Odqvist&Hult(1962) für volumen-
treues Kriechen durch

$$\dot{e}_r = 3/4 \; k \; 6_e^{n-1} \; (6_r - 6_\theta) \qquad \dots (2a)$$

$$\dot{e}_\theta = - e_r \qquad \qquad \dots (2b)$$

$$\dot{e}_z = 0 \quad (\text{ebene Formänderung}) \qquad \dots (2c)$$

mit
$$6_e = \sqrt{3/4} \; \text{abs}(6_r - 6_\theta) \qquad \dots (3)$$

Darin bedeuten:
$\dot{e}_r$ radiale Dehnungsgeschwindigkeit

$\dot{e}_\theta$ tangentiale Dehnungsgeschwindigkeit

$\dot{e}_z$ axiale Dehnungsgeschwindigkeit

6_r radiale Spannung

6_θ tangentiale Spannung

6_e effektive Spannung

2.2 Stationäre Spannungen und Konvergenzra-
ten am Bohrlochrand

Die stationären radialen, tangentialen und
axialen Spannungen 6_r, 6_θ und 6_z sowie die
Deformationsraten $\dot{e}_r$ und $\dot{e}_\theta$ können für das
Bohrloch im Salinar unter Annahme gleichför-
miger äußerer Gebirgsspannungen, inkompres-
siblen Materialverhaltens und der Gültigkeit
eines Kriechgesetzes vom Potenztyp (Gl.1)
analytisch entwickelt werden (Nadai 1963):

$$6_r = p_o \left((r/r_o)^{-2/n} - 1 \right) \qquad \dots (4a)$$

$$6_\theta = p_o \left((1-2/n)(r/r_o)^{-2/n} - 1 \right) \qquad \dots (4b)$$

$$6_z = 0.5 \; (6_r + 6_\theta) \qquad \dots (4c)$$

Darin bedeuten
r_o Innenradius
p_o äußerer Gebirgsdruck

In Fig.2 sind die tangentialen und radialen Spannungen am Bohrloch im Salinar (n=5)
in Abhängigkeit von der Stoßtiefe in dimensionsfreier Form aufgetragen. Die Spannungsdifferenz am Bohrlochrand beträgt $(2/n)6_o$.

Die Deformationsraten ergeben sich nach
Einsetzen der Gln.4 in die Gln.(2) zu:

$$\dot{e}_r = a \; p_o^n \; (r/r_o)^{-2} \qquad \dots (5a)$$

$$\dot{e}_\theta = - \dot{e}_r \qquad \qquad \dots (5b)$$

$$a = k \; (3/4)^{(n+1)/2} \; (2/n)^n \qquad \dots (5c)$$

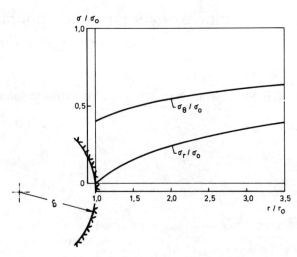

Fig.2. Tangentiale und radiale Spannungen
am Bohrloch im Salinar (n=5).
Tangential and radial stresses of borehole in rock salt (n=5).
Contraintes tangentielles et radiales aux
abords de la paroi interne du trou de
sonde dans un massif de sel gemme (n=5).

Die stationäre radiale Konvergenzgeschwindigkeit $\dot{u}$ des Bohrlochs folgt aus der Deformationsrate $\dot{e}_\theta$ über die Beziehung

$$\dot{u} = r_o \; \dot{e}_\theta = a \; p_o^n / r_o \qquad \dots (6)$$

Die radiale Konvergenz u des Bohrlochs ergibt sich bei stationärem Kriechen des Salinars und kleinen Verformungen durch lineare
Integration der Gl.(6) nach der Zeit t:

$$u(t) = (a/r_o) \; p_o^n \; t \qquad \dots (7)$$

mit der Anfangsbedingung u(t=0) = 0.
Die Konvergenzgeschwindigkeit $\dot{u}$ des Bohrloches ist nach Gl.(6) stationär, wenn sich
der Gebirgsdruck p_o mit der Zeit nicht ändert. Bei großen Verformungen ($e_r \geq 3\%$) muß
die zeitliche Veränderung des Radius r_o in
der Zeitintegration berücksichtigt werden.
Dabei ergibt sich eine monotone Abnahme der
Konvergenzrate mit der Zeit, die als kinema-
tische Dehnungsverfestigung bezeichnet wird.

3 INSTATIONÄRE SPANNUNGEN UND KONVERGENZ-
RATEN BEI EINEM BOHRLOCH IM SALINAR

3.1 Relaxation der Haupt-Differenzspannungen

Die Spannungsdifferenz am Bohrlochrand

$$6: = (6_r - 6_\theta)|_{r=r_o}$$

ist nach den Gleichungen 4 gegeben durch

$$6 = - (2/n) \; p_o \qquad \dots (8)$$

Für den Fall n=1 stellt sich am Bohrloch
eine linear viskose Spannungsverteilung ein,
die nach dem Korrespondenzprinzip gleich der
linear elastischen Spannungsverteilung ist.
Dieser Spannungszustand (Fig.3) wird im folgenden als Anfangsspannungszustand angenommen.
Nach vollständiger Relaxation der elastischen Spannungen (Fig.3) stellt sich asymp-
totisch der stationäre viskose Spannungszustand (Fig.2) ein. Während des Relaxations-
vorgangs nimmt die Kriechgeschwindigkeit mo-

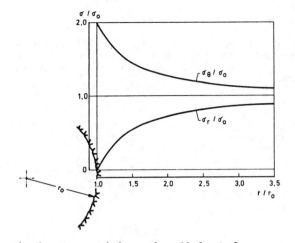

Fig.3. Tangentiale und radiale Anfangsspannungen am Bohrloch (n=1).
Tangential and radial initial stresses of borehole (n=1).
Contraintes initiales tangentielles et radiales aux abords de la paroi interne du trou de sonde (n=1).

noton entsprechend der Abnahme der Hauptdifferenzspannung am Bohrlochrand ab, bis sie schließlich einen stationären Wert nach Gl.5 annimmt.
Eine analytische Lösung der Bewegungsgleichung für die Kriechkonvergenzen eines freistehenden Bohrloches im Steinsalzgebirge bei relaxierenden Tangentialspannungen war bisher nicht bekannt (Nadai 1963). Sie wird im folgenden vorgestellt.

Wir betrachten die Spannungen am Bohrlochrand ($r=r_1$):

$$\sigma_r = 0 \ , \quad \sigma_\theta = \sigma_\theta(t) \ , \quad \sigma_z = \sigma_\theta/2$$

und definieren: $\sigma = \sigma_\theta(r=r_o)$.

Aus numerischen Analysen des Verfassers konnte die Relaxationsfunktion V(t) für die tangentiale Spannung σ_θ am Bohrlochrand bestimmt werden:

$$\sigma = \sigma_\infty \, V(t) \qquad\qquad \dots (9)$$

$$\sigma_\infty = 2 \, \sigma_o/n \qquad\qquad \dots (10)$$

$$V(t) = 1 + (n-1) \, Y(t) \qquad \dots (11)$$

$$Y(t) = (1+\hat{t})^{1/(1-n)} \qquad \dots (12)$$

$$\hat{t} = t/t_o \qquad\qquad \dots (13)$$

$$t_o = (e_o/\dot{e}_o)/(n-1) \qquad \dots (14)$$

$$e_o = \sigma_o/(2G) \qquad\qquad \dots (15)$$

$$\dot{e}_o = A \, \exp(-T^*/T) \, \sigma_{eo}^n \qquad \dots (16)$$

Darin bedeuten:
σ_o primäre Gebirgsspannung
n   Exponent im Kriechgesetz
V(t), Y(t) Relaxationsfunktionen
t   Zeit
t_o Relaxationszeit
G   Schubmodul
T   Kelvin Temperatur
A, T^* Stoffparameter des Kriechgesetzes
$\dot{e}_o, \sigma_{eo}$ Referenzwerte im Kriechgesetz

Fig.4 zeigt in dimensionsfreier Darstellung die Relaxation der tangentialen Randspannungen des Bohrloches für verschiedene Exponenten des Kriechgesetzes von Steinsalz. Bei linearer Visko-Elastizität (n=1: Maxwell Körper) würde keine Spannungsrelaxation am freien Rand ($r=r_o$) stattfinden.

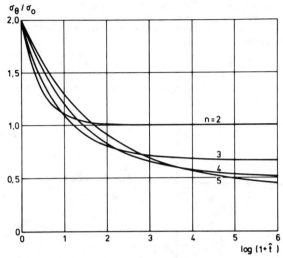

Fig.4. Tangentiale Spannungsrelaxation am Bohrlochrand in Abhängigkeit von der Zeit $\hat{t}$ und verschiedenen Exponenten n des Kriechgesetzes in dimensionsfreier Darstellung.
Tangential stress relaxation at the borehole wall as a function of time $\hat{t}$ and exponent n of the creep law; (dimensionless representation).
Relation entre la constrainte tangentielle adimensionnée subie par la paroi du trou de sonde et le temps adimensionné.

3.2 Zeitliche Abnahme der Konvergenzrate

Die Verformungsraten des Bohrlochrandes sind beim ebenen, volumentreuen Formänderungszustand des Salinars gegeben durch $\dot{e}_r = -\dot{e}_\theta$. Wir definieren: $\dot{e} = \dot{e}_r$.
Die Gesamtverformungsrate setze sich aus einer elastischen und einer viskosen Deformationsrate zusammen (Prandtl-Material):

$$\dot{e} = \dot{e}^{ela} + \dot{e}^{vis} \qquad\qquad \dots (17)$$

Die zeitabhängige elastische Verformungsrate läßt sich aus den Gln.(9-12) berechnen zu

$$\dot{e}^{ela} = -(e_\infty/t_o) \, Y^n(t) \qquad \dots (18)$$

$$\text{mit} \quad e_\infty = e_o/n \qquad\qquad \dots (19)$$

Die Formeln für e_o, t_o, V(t) und Y(t) sind in den Gln.(11-15) gegeben.

Die viskose Verformungsrate folgt dem Potenzgesetz

$$\dot{e}^{vis} = \dot{e}_\infty \, V^n(t) \qquad\qquad \dots (20)$$
mit
$$\dot{e}_\infty = A \, \exp(-T^*/T)(3/4)^{(n+1)/2} \, (2/n)^n \, \sigma_o^n$$

Als Anfangswerte für die Zeit t = 0 findet man
$$\dot{e}_o = \dot{e}_o^{ela} + \dot{e}_o^{vis}$$

$$\dot{e}_o^{ela} = - e_\infty/t_o = - n^{(n-1)} \, \dot{e}_\infty \quad \dots (21)$$

$$\dot{e}_0^{vis} = n^n \, \dot{e}_\infty \qquad \dots (22)$$

$$\dot{e}_0 = \dot{e}_\infty \, n^{(n-1)} \qquad \dots (23)$$

Fig.5 zeigt bezogene Anfangswerte $\dot{e}_0/\dot{e}_\infty$ der Verformungsraten am Bohrlochrand für unterschiedliche Exponenten n des Kriechgesetzes. Beispielsweise ist die anfängliche Konvergenzrate für ein Steinsalzgebirge mit dem Kriechparameter n=6 etwa 10,000-fach höher als die stationäre Kriechrate nach erfolgter Spannungsrelaxation.

Die zeitliche Abnahme der bezogenen Kriechdehnungsgeschwindigkeiten ($\dot{e}/\dot{e}_\infty$) ist in Abhängigkeit von der bezogenen Zeit $(1+\hat{t})$ in Fig.6 in doppelt-logarithmischer Darstellung aufgetragen.

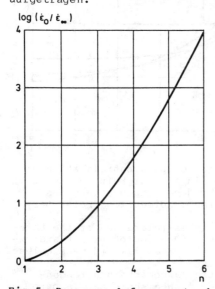

Fig.5. Bezogene Anfangswerte der Konvergenzraten des Bohrlochs für verschiedene Exponenten n des Kriechgesetzes.
Relative initial values of rates of convergency of borehole as a function of exponent n of the creep law.
Relation entre le taux des convergences initiales et la puissance n dans la loi de fluage.

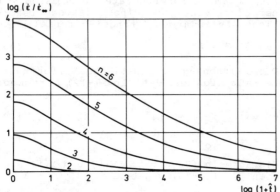

Fig.6. Monotone Abnahme der Bohrlochrand-Verformungsraten als Funktion der Zeit t für verschiedene Werte des Exponenten n des Kriechgesetzes in dimensionsfreier Darstellung.
Monotonous decrease of rates of borehole convergency as a function of time and exponent n of the creep law; (dimensionless representation).
Evolution du taux de déformation visqueux adimensionnée subie par la paroi interne du trou de sonde.

3.3 Bohrlochkonvergenzen

Wie bei den Verformungsraten wird angenommen daß sich die Gesamtverformung aus einem elastischen Anteil und einem viskosen Anteil zusammensetzt

$$e = e^{ela} + e^{vis}$$

Die elastischen und die viskosen Deformationen ergeben sich durch Integrationen der entsprechenden Verformungsraten (Gl.18) bzw. Gl.(20) nach der Zeit t. Die Integrationen lassen sich bei geradzahligen Exponenten des Kriechgesetzes analytisch durchführen; anderenfalls wird man numerische Integration anwenden.

$$e^{ela} = \int_0^t \dot{e}^{ela} \, dt = e_\infty \, V(t)$$

$$e^{vis} = \int_0^t \dot{e}^{vis} \, dt = \dot{e}_\infty \int_0^t V^n(t) \, dt$$

Die Bohrlochkonvergenzen, die nach innen positiv gezählt werden, folgen aus den Verformungen durch Integration über die radiale Koordinate. Für den Fall großer Verschiebungen erhält man $u = r_0(1-\exp(e))$.

Da die sofortige Konvergenz zum Zeitpunkt $t=t_0$ durch $u_0=r_0(1-\exp(e_0))$ gegeben ist, gilt für das Verhältnis von zeitabhängigen Konvergenzen und Anfangsradius die Beziehung $(u-u_0)/(r_0-u_0) = 1 - \exp(-e+e_0)$ ... (24)

Diese relativen zeitabhängigen Konvergenzen nach Gl.(24) sind in Fig.7 für verschiedene Exponenten n des Kriechgesetzes für die Zeit bis zum vollständigen Schließen der Bohrung aufgetragen. Die Kennlinien können durch geeignete Zeittransformation (Multiplikation von t_0 in Gl.14 mit dem Faktor n^n) auf ein enges Band zusammengezogen werden.
In einem doppelt-logarithmischen Diagramm erscheinen die Konvergenzkurven bei kleinen Verformungen als Geraden und bei beidseitig linearer Auftragung als Parabeln.

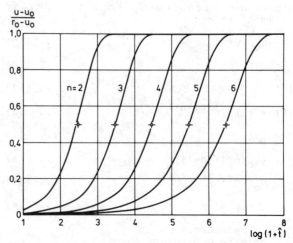

Fig.7. Bohrlochkonvergenzen als Funktion der Zeit t für verschiedene Exponenten n des Kriechgesetzes in dimensionsfreier Darstellung.
Borehole convergencies as a function of time t and exponent n of the creep law; dimensionless representation.
Evolution de la convergence visqueux adimensionnée de la paroi interne du trou de sonde.

In Fig.8 sind als Beispiel der üblichen
linearen Auftragung theoretische Konvergenz-
Zeit-Verläufe für eine 31-cm Bohrung im Salinar für zwei unterschiedliche Teufen dargestellt. Ohne Kenntnis des Relaxationsver-
haltens könnte man aus dem konkaven Verlauf
der Kennlinien ein transientes Kriechgesetz
mit Dehnungs- oder Zeitverfestigung ableiten
und zu einer Fehleinschätzung des langzeiti-
gen Konvergenzverhaltens des Gebirges kommen.

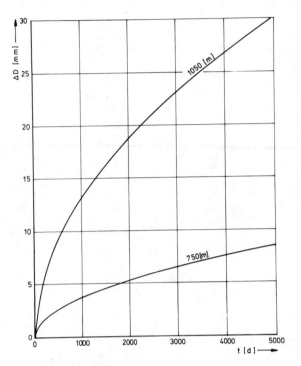

Fig.8. Zeitabhängige radiale Konvergenzen
einer 31cm Bohrung im Salinar.
Time dependant radial convergencies of a
31cm borehole in rock salt.
Evolution de la convergence radial visqueux de la paroi interne d'un trou de
sonde (Ø 31cm).

4 SCHLUSSBEMERKUNGEN

Die Wechselwirkungen zwischen der Konvergenzgeschwindigkeit und der Relaxation der
deviatorischen Spannungen am Bohrlochrand im
Salinar sind nachgewiesen.
Die vorliegenden Relaxationskennlinien in
dimensionsloser Darstellung sind theoretisch
für beliebige Werte von äußerer Gebirgsspannung, Bohrlochradius, Schubmodul und Kriechparameter des Potenzgesetzes gültig. Mit ihrer Hilfe können die Wirkungen der einzelnen
Einflußgrößen quantitativ abgeschätzt werden
Umgekehrt ist es möglich, die Gültigkeit
des Kriechgesetzes vom Potenztyp für felsmechanische Berechnungen im Steinsalzgebirge
zu überprüfen, indem man die theoretischen
Resultate mit den Ergebnissen von in-situ-
Messungen vergleicht.
Schon Davis (1960) stellt gravierende Unterschiede zwischen den rheologischen Stoff-
parametern fest, die man einerseits im stationären Kriechversuch, andererseits im rei-
nen Relaxationsversuch an Metallen gemessen
hat. Für Steinsalz deutet nach experimentellen Labor- und Felduntersuchungen des Lehr-
stuhls für Felsmechanik an der Universität
Karlsruhe vieles auf ebensolche Unterschiede
hin.

Beim bergmännischen Hohlraumbau im Salinar
beeinflussen sich Kriechen und Spannungsrelaxation gegenseitig, so daß die vorliegende
Theorie eine erhebliche praktische Bedeutung
besitzt. Sie läßt sich qualitativ auch auf
andere als zylindrische Hohlräume im Steinsalzgebirge übertragen.
Bei komplizierteren Strukturen muß man auf
numerische Berechnungsverfahren nach der Methode der finiten Elemente zurückgreifen. Es
ist dabei wichtig, daß neben den Ungewißhei-
ten über die richtigen Annahmen des Stoffge-
setzes, des Einflußbereichs und der Randbedingungen nicht noch zusätzliche Ungenauig-
keiten wegen mangelnder Konvergenz des Zeit-
integrationsverfahrens auftreten. Erst eine
analytische Lösung ermöglicht die objektive
Prüfung der Genauigkeit der verwendeten numerischen Rechenverfahren.

5 LITERATUR

Borm, G. 1980. Analyse chronischer Gebirgsverformungen beim Felshohlraumbau. Veröf-
fentlichungen des Institutes für Bodenmechanik und Felsmechanik der Universität
Karlsruhe, Nr.88.

Borm, G. 1985. Wechselwirkung von Gebirgs-
kriechen und Gebirgsdruckzunahme am
Schachtausbau. Felsbau 3, Nr.3: 153-158.

Davis, E.A. 1960. Relaxation of a cylinder
on a rigid shaft. J.Appl.Mech.: 41-44.

Heard,H.C. 1972. Steady State Flow of Poly-
crystalline Halite at Pressure of 2 Kilo-
bars. In: Flow and Fracture of Rocks.
AGU-Monograph No.16, Washington, D.C.

Lurje,A.I.1963. Räumliche Probleme der Elastizitätstheorie. Berlin: Akademie Verlag.

Nadai, A. 1963. Theory of flow and fracture
of solids. New York: Mc Graw Hill.

Odqvist, F.K.G & J.Hult 1962. Kriechfestig-
keit metallischer Werkstoffe.
Berlin, Göttingen, Heidelberg: Springer.

DANK

Meinen Kollegen am Institut für Bodenmechanik und Felsmechanik der Universität Karlsruhe, besonders den Herren Prof.Dr.-Ing. O.
Natau, Dr.rer.nat.Ch.Lempp und Dipl.-Ing. M.
Haupt danke ich für viele klärende Diskussi-
onen zur Rheologie des Salzgebirges.
Der Deutschen Forschungsgemeinschaft(DFG)
danke ich für die Förderung des Grundlagen-
projekts Bo-764 "Standsicherheit von Schächten im Salinar", in dessen Rahmen die vorge-
stellten numerischen und analytischen Untersuchungen durchgeführt werden konnten.

Effects of the extraction of tabular deposits around vertical shafts in deep-level mines

Effets de l'extraction des dépôts tabulaires autour des puits verticaux dans les mines à grande profondeur

Auswirkungen vom Abbau tafelförmiger Ablagerungen um vertikale Schächte in tiefen Gruben

S.BUDAVARI, Professor of Rock Mechanics, University of the Witwatersrand, Johannesburg, Republic of South Africa
R.W.CROESER, Research Officer, University of the Witwatersrand, Johannesburg, Republic of South Africa

ABSTRACT: Leaving shaft pillars, in the deep-level gold mines of South Africa, is becoming prohibitive from both economic and mining points of view. The alternative option is to extract the orebody early in the life of the shaft and limit the ensuing damage by controlling the strata displacement.

On the basis of some simple models, the paper describes the mechanisms of strata displacement along a shaft and the damage that may result from the early extraction of an orebody. The same method of modelling is utilised in generating numerical data to enable a comparison of measured and computed strata displacements brought about by an actual extraction.

RÉSUMÉ: Laissant les pilliers des puits, dans les mines d'or à grande profondeur en Afrique du Sud, devient inabordable au point de vue des aspects financiers et des operations minieres. L'option alternative est d'extraire la masse minerale tôt dans la vie d'un puits et de limiter les dégâts resultants en reglant le déplacement des couches.

Sur le fondement de quelques modèles simples, la communication décrit les mecanismes de déplacement des couches le long d'un puits et les dégâts qui peuvent être provoqués par l'extraction hâtive d'une masse minerale. La même méthode de modelage est utilisée pour produire les données numériques pour faciliter une comparaison des déplacements mésurés et computés des couches provoqués par une véritable extraction.

ZUSAMMENFASSUNG: In den tiefen Goldminen in Südafrika wird es sowohl vom wirtschaftlichen, als auch vom bergbaulichen Standpunkt aus zu aufwendig, die Schachtpfeiler stehenzulassen. Die Alternative wäre, das Erz schon früh in der Lebensdauer des Schachtes abzubauen, und Folgeschäden durch Kontrollierung der Gesteinsschichtverschiebungen zu begrenzen.

Dieser Bericht beschreibt anhand einfacher Modelle die Mechanismen von Verschiebungen in Gesteinsschichten längs des Schachtes und den Schaden, den früher Abbau des Erzes zur Folge haben kann. Die gleiche Modellmethode wird benutzt, um numerische Daten zu gewinnen für einen Vergleich zwischen gemessenen und errechneten Schichtverschiebungen, die in der Praxis bei einem Abbau entstanden.

INTRODUCTION

The most common method of protecting shaft systems from the damaging effects of mining is to leave the ore unmined in an area surrounding the shaft; that is to leave a shaft pillar. The design method currently preferred for the design of shaft pillars in the deep-level mines of South Africa, which exploit narrow, flat-dipping gold-bearing reefs, was suggested by Salamon (1974). According to this method, at depths greater than 2,5 to 3,0 km, protection of vertical shaft systems by shaft pillars becomes prohibitive because of the large sizes of pillars required.

An alternative option to leaving shaft pillars is the removal of the mineral deposit early in the life of the shaft from an area that would normally constitute the shaft pillar area. However, the induced vertical displacements in the hanging- and footwalls are substantial and the associated strata disturbance can cause considerable damage to the shaft.

The scarcity of field observations and the lack of retrospective analysis of these data are the main reasons for the uncertainty in planning the early extraction of an orebody around the shaft. In order to contribute to the solution of this problem, the present study is concerned with the mechanisms of damage involved and the comparison of observed and calculated effects of such an extraction on the hangingwall strata. It seeks information on the applicability of the elastic model in providing quantitative data for design purposes.

EFFECTS OF REEF EXTRACTION ON THE STRATA

Stoping of a horizontal, tabular deposit close to any shaft will have a number of effects on the rock mass containing the shaft. The most obvious consequence is that the immediate hangingwall in the stope moves downwards and the footwall moves upwards. These movements result in relaxation of the rock mass for a distance along the shaft. As the extent of extraction increases, the movements of the foot- and hangingwalls continue and the affected distances up and down the shaft increase until closure occurs. The consequences of these rock mass deformations are that, in some regions of both the foot- and hangingwalls, components of the resultant stresses and induced strains become tensile in nature. Once closure takes place, virgin stress conditions tend to be re-established over the area of contact between foot- and hangingwalls.

The distribution of resultant stresses and induced vertical strain along the centre line of two parallel, diverging longwall stopes can be studied by accepting linearly isotropic elastic rock mass behaviour and using any appropriate numerical method. Such a distribution, in a two-dimensional section across the stopes just before closure, is depicted in Fig. 1. In this diagram the x and z axes are orientated horizontally and vertically respectively. Some additional and relevant information are presented in the diagram. Assuming the shaft to be located along this centre line, the effects of extraction on the vertical shaft can be indicated.

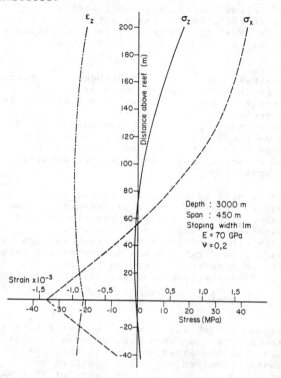

Figure 1. Distribution of resultant stresses and induced vertical strain along part of the shaft just before closure.

Axial extension and subsequent compression of the shaft can affect the integrity of the shaft lining and the shaft steelwork. Fracturing of the shaft lining can cause problems in water-bearing rock masses and dislocation of parts of the fractured lining or loose rock may represent a hazard within the shaft. Relative movements, beyond a certain magnitude, in the shaft steelwork may make high-speed winding dangerous. The induced differential movement could also affect pipe columns located in the shaft.

In order to assess the effects of the computed differential movements and to predict the extent of the affected zones, it is necessary to compare these to some empirically established critical values. Only a few references can be found in the relevant literature to critical tensile strains derived from measurements in, or experience with, concrete lined vertical

shafts. According to Kratzsch (1983), tensile fracture in concrete can occur between strain values of $0,15 \times 10^{-3}$ and $0,30 \times 10^{-3}$. The level of tensile strain that the shaft steelwork can tolerate without damage is recorded, by More O'Ferrall (1983), to be $0,40 \times 10^{-3}$.

FIELD CONDITIONS AND OBSERVATIONS

The in situ observations were carried out in and around the Ventilation Shaft of Harmony Gold Mine in the Orange Free State by Barcza and von Willich (1960). Mining in the area around the shaft was confined to the Basal Reef which intersected the shaft at a depth of 1329 m below surface. The Basal Reef is underlain by massive, competent quartzites and it is overlain by a succession of quartzite and conglomerate beds belonging to the Main-Bird and Kimberley Series. Several shale beds are included in these series. The strata sequence above these successions comprises massive, competent quartzites. The total thickness of the strongly bedded strata between the Basal Reef and the massive quartzite is about 138 m. The dip of the strata is 7 degrees towards the West.

Mining was carried out at a stoping width of 1,2 m using the longwall method of extraction. Three sets of parallel longwalls were mined symmetrically about the shaft at centre-to-centre distances of about 250 m. Initial measurements started with the commencement of mining in 1954 and continued until total extraction was achieved over a sufficiently large area around the shaft. The hangingwall in the worked-out area was supported by timber packs.

Measurements were made of changes in vertical distances between horizontally installed roof-bolts along the full length of the shaft. These bolts were arranged in a vertical line and placed 3 m apart close to the reef. From 15 m upwards, the distance between measuring points gradually increased to a maximum of 60 m. The relative vertical movements were measured by suspended steel tapes. Observations were also made of fractures occurring in the shaft lining and inflows of water. Additional measurements were carried out in haulages, in stopes and on the surface, however, these are not relevant from the points of view of this study.

DETERMINING THE ROCK MASS RESPONSE

The process of mining was modelled by the MINSIM D computer modelling system using three mining layouts spaced equally in time over a period of three years. For each layout the resultant stresses and induced vertical strains were calculated at bench-marks located along the vertical axis of the shaft. This resulted in three different sets of calculated values.

In order to obtain an estimate of the magnitude of in situ modulus, the computed rise of the footwall, in close vicinity of the shaft, was compared to the corresponding measured value. The reason for choosing the footwall rise as a measure of the rock mass response was that the footwall was thought to be free from strata separation and other inelastic deformations. After the first series of computer runs it was obvious that the correlation between measured and calculated displacements was not close when using the widely accepted 70 GPa as the modulus value. As can be seen in Fig. 2, a reasonable correlation was obtained when a modulus value of 50 GPa was accepted. Therefore, in all subsequent computations this value of the modulus was used.

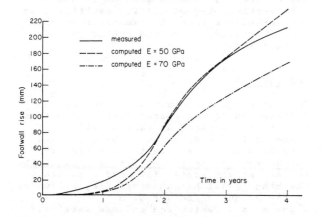

Figure 2. Comparison of measured and computed values of footwall rise.

For comparison purposes, some measured and computed data are presented together in Figs. 3,4 and 5 corresponding to the second, third and fourth year of mining respectively. It will be seen from these illustrations that details of only those quantities which are expected to affect the rock mass response and the shaft are given.

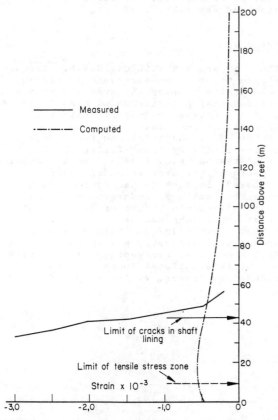

Figure 3. Comparison of measured and computed induced, vertical strains in hangingwall at the end of second year of mining.

COMPARISON OF COMPUTED AND MEASURED DATA

The most striking observation one can make from Figs. 3,4 and 5 is that the extent of strata disturbance in the immediate hangingwall is substantial. The height of this zone increases with the progress of mining, however, after reaching a maximum value, it remains stationary at about 138 m. Beyond this disturbed zone the strata response approximates that of the idealised elastic substance.

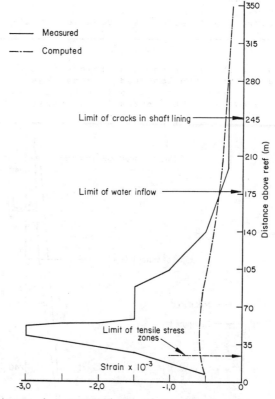

Figure 4. Comparison of measured and computed induced, vertical strains in hangingwall at the end of third year of mining.

One can expect that bed separation and loosening up of the strata would take place in the tensile stress zones. A comparison of the extent of tensile stress zones with the loosened up zones shows, however, that the latter extend well into the compressive stress zones. Admittedly, the magnitude of compressive stresses are low in this transitional region. The fact that the loosening of the strata extends to the base of the massive quartzite bed indicates that the bedded nature of strata below this massive, competent bed has an overriding influence on the conditions brought about.

In the massive quartzite strata, the calculated induced, vertical strain correlates reasonably well with the measured values. Considering the limitations inherent in the instrumentation and measuring systems used, the correlation is excellent. Therefore, provided an estimate of the in situ modulus can be obtained, the use of the elastic model gives acceptable prediction of the vertical induced strain in this region.

To compare the measured and calculated induced strains at which the development of cracks or inflow of water occurred, the relevant data are tabulated below.

Table. Comparison of measured and calculated strains.

		3rd Year	4th Year
Strain at which cracks occurred in shaft lining	measured	$0,20 \times 10^{-3}$	$0,20 \times 10^{-3}$
	calculated	$0,24 \times 10^{-3}$	$0,13 \times 10^{-3}$
Strain at which water inflow occurred	measured	$0,31 \times 10^{-3}$	$0,31 \times 10^{-3}$
	calculated	$0,34 \times 10^{-3}$	$0,20 \times 10^{-3}$

Since the observed data presented in Fig. 5 are within the loosened up zone, the critical strain from this illustration is excluded from the figures in the table.

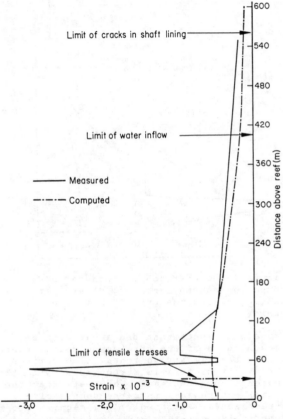

Figure 5. Comparison of measured and computed induced, vertical strains in hangingwall at the end of fourth year of mining.

The figures presented in the table serve to illustrate the following points. Correlation between corresponding measured and calculated data are very reasonable. The critical values for the computed quantities are more affected by their locations above the reef than those of the corresponding measured quantities. The strain effect, to enable the flow of water, is greater than that for cracks to occur. It appears, therefore, that as far as damage to shaft lining is concerned, the intensity of cracking and the width of cracks are more important than the initial occurrence of cracks. Both the measured and calculated

values of strain at which tensile fracturing of the concrete took place falls within the range suggested by Kratzsch.

The amount of strata displacement and the upward migration of the disturbed zone, observed in the illustrations, could have been limited by reducing the effective stoping width. In recent times, a number of methods have been employed to achieve this. One of the most common methods is the introduction of good quality backfill into the worked-out area. A discussion on this topic, however, is beyond the scope of the present study.

CONCLUSIONS

The elastic model gives an acceptable description of the response of rock masses beyond the loosened up hangingwall.

The region in which bed separation and loosening of the strata takes place, may extend beyond the theoretical limits of the tensile stress zones. The bedded nature of the hangingwall has an overriding influence on the height of this region.

The zone in which the induced vertical strain causes undesirable effects in the shaft extends well beyond the tensile stress and loosened up zones.

The critical strains for tensile fracturing of the shaft lining and for inflow of water were found to be $0,20 \times 10^{-3}$ and $0,30 \times 10^{-3}$ respectively. Both these values are below the tolerable figure of $0,40 \times 10^{-3}$ from the point of view of the unimpaired use of the shaft steelwork.

The magnitude of induced strain and the locations of its critical values can be determined and controlled.

REFERENCES

BARCZA, M. and VON WILLICH, G.P.R. (1960). "Strata movement measurements at Harmony Gold Mining Company, Limited". Papers and Discussions, Association of Mine Managers of South Africa 1958-1959. Chamber of Mines, Johannesburg.
KRATZSCH, H. (1983). "Mining subsidence engineering". Springer-Verlag, Berlin.
MORE O'FERRALL, R.C. (1983). "Rock mechanics aspects of the extraction of shaft pillars". In Rock Mechanics in Mining Practice, edit. S. Budavari, chap. 10. The South African Institute of Mining and Metallurgy, Johannesburg.
SALAMON, M.D.G. (1974). "Rock mechanics of underground excavations". In Advances in Rock Mechanics, Proceedings 3rd Congress of the International Society of Rock Mechanics, Denver, v. 1 (B), pp. 951-1099.

Scale effect on the deformability and strength characteristics of rock masses
Effet d'échelle sur les caractéristiques de déformabilité et de résistance des massifs rocheux
Der Massstabseffekt bei den Verformungs- und Festigkeitseigenschaften der Gebirge

J.G.CHARRUA-GRAÇA, Laboratório Nacional de Engenharia Civil, Lisboa, Portugal

ABSTRACT: Scale effect is defined and its cause is discussed. A model for scale effect on deformability is presented and results obtained with this model for some rock masses are shown and discussed. A model is also presented for scale effect on shear strength of joints as well as some results confirming it.

RÉSUMÉ: On définit effet d'échelle et on discute sa cause. On présente un modèle concernant l'effet d'échelle sur la déformabilité, faisant la discussion de quelques résultats obtenus avec ce modèle.
On présente encore un modèle pour l'effet d'échelle sur la résistance au cisaillement de diaclases, aussi bien que quelques résultats qui confirment sa validité.

ZUSAMMENFASSUNG: Der Maßstabseffekt wird definiert und seine Ursache erörtert.
Es wird ein Modell für den Maßstabseffekt bei der Verformbarkeit vorgestellt und einige mit diesem Modell erhaltene Ergebnisse besprochen.
Ein Modell für den Maßstabseffekt bei der Scherfestigkeit von Klüften wird auch vorgestellt, sowie einige Ergebnisse, die die Gültigkeit des Modells bestätigen.

1 - INTRODUCTION

It has been experimentally observed that geometrically homothetic samples of a definite material (rock or rock mass), when submitted to states of stress wich respect the similarity conditions, present variations in their mechanical characteristics, and that these variations depend on the dimensions of the sample.

Normally, sets of samples of the same universe but with different dimensions present distributions with different parameters for the characteristics being studied.

The variation of these characteristics with the dimension of the sample is what we call the "scale effect".

Though different phenomena may be pointed out as causes for the occurrence of the scale effect, it can be assumed that for most of the characteristics, This effect depends mainly on the heterogeneity of the material. This is the reason why in rock masses, which are so heterogeneous in general, the scale effect is so important. The mastering of the scale effect for each property could be based on the knowledge of:
- the law of variation of the mean value of a property with the dimension of the sample.
- the law of the scattering of the values obtained for a certain dimension of the sample.

2 - SCALE EFFECT ON THE DEFORMABILITY

2.1 - An explanatory model

It has been shown (Peres-Rodrigues, 1974) that the deformability modulus (E) of a body made of two materials with deformability moduli (E_1 and E_2) and occurring in volumetric proportions ($\emptyset_1$ and $\emptyset_2$) can be obtained through

$$E = \frac{1}{\frac{\emptyset_1}{E_1} + \frac{\emptyset_2}{E_2}} \qquad (1)$$

If we consider a uniform random distribution for the occurrences and assume that the component with modulus (E_1) occurred in elements with dimension (a), the probability a sample of dimension (na) to contains (k) of these elements is given by:

$$P_{na}^{(K)} = \frac{n!}{K!(n-K)!} \emptyset_1^K (1-\emptyset_1)^{(n-k)} \qquad (2)$$

since:

$$n\emptyset_1 > 4 \qquad (3)$$

$$n(1-\emptyset_1) > 4, \qquad (4)$$

the binomial expression (2) may be approximated by the normal distribution

$$P_{na}^{(K)} = \frac{e^{-\frac{(K-n\emptyset_1)^2}{2n\emptyset_1(1-\emptyset_1)}}}{\sqrt{2\pi\,n\emptyset_1(1-\emptyset_1)}} \qquad (5)$$

Equation (1) may be written as:

$$\frac{1}{E} = \sum_i \frac{\emptyset_i}{E_i} \qquad (6)$$

or in terms of deformability D = 1/E,

$$D = \sum_i \emptyset_i D_i \qquad (7)$$

As, in the case under study

$$\emptyset_1 = K/n$$

$$\emptyset_2 = 1 - K/n$$

the preceding equation becomes

$$D = (D_1 - D_2) K/n + D_2 \qquad (8)$$

The probability of a sample with dimension (na) to have deformability (D) will thus be

$$P_{na}^{(D)} = \frac{\sqrt{ne}^{-\frac{n}{2}\frac{\left[D-D_1\phi_1-(1-\phi_1)D_2\right]^2}{\phi_1(1-\phi_1)(D_1-D_2)^2}}}{\sqrt{2\pi}\,\phi_1(1-\phi_1)(D_1-D_2)^2} \qquad (9)$$

Hence the expression of a normal distribution with the mean value

$$\overline{D} = D_1\phi_1 + D_2(1-\phi_1), \qquad (10)$$

and the standard deviation

$$\sigma = (D_1-D_2)\sqrt{\frac{\phi_1(1-\phi_1)}{n}} \qquad (11)$$

In the case of b dies formed of (i) materials the binomial expression (2) will be replaced by the multinomial expression

$$P_{na}^{(k_1)} = \frac{n!}{K_1!\,K_2!\,\ldots\,K_i!}\,\phi_1^{K_1}\,\phi_2^{K_2}\,\ldots\,\phi_i^{K_i} \qquad (12)$$

This expression may be approximated by the normal expression

$$P_{na}^{(D)} = \frac{\sqrt{ne}^{-\frac{1}{2}\frac{n(D-\sum_i D_i\phi_i)^2}{\sum_i D_i^2\,2\phi_i(1-\phi_i)-2\sum_i\sum_j D_iD_j\phi_i\phi_j}}}{\sqrt{2\pi}\left[D_i^2\,\phi_i(1-\phi_i)-2\sum_i\sum_i D_iD_j\phi_i\phi_j\right]} \qquad (13)$$

to which corresponds the mean value

$$\overline{D} = \sum_i D_i\,\phi_i \qquad (14)$$

and the standard deviation

$$\sigma = \sqrt{\frac{1}{n}\left[\sum_i D^2\,\phi_i(1-\phi_i)-\sum_i\sum_j D_iD_j\phi_i\phi_j\right]} \qquad (15)$$

Expressions (13), (14) and (15) generalize expressions (9), (10) and (11), deduced for the case of bodies formed by only two materials.

The expressions above show that: i) for the same body (in which, therefore, the proportions of the occurrence of the constituting materials and the deformability characterietics remain the same), the dispersion of the deformability of the samples decreases proportionally to the square root of their dimension (n); and ii) for the same body, the mean deformability value of the samples with a given dimension is constant and equal to the mean deformability of the body.

2.2 - Results of the model

The expressions (14 and 15), which synthetize the model allow to foresee the behaviour of a multicomposed body in elastic deformation, and so to analyse the expected results of the deformability tests of a rock mass, for instance.

Consider a rock mass in which 90% is constituted by a rock with a modulus of 40 000 MPa, and 10% by joints filled with a material with a modulus of 200 MPa.

The curves that define the extreme probabilities of 5% of observance as functions of (n) are plotted in Fig. 1, and it can be observed that they are very si-

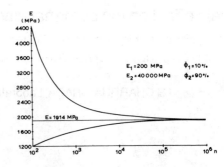

Fig.1- Curves limiting 90 % of the expectable results

milar to the curves obtained in different experimental works.

Similar plots have been drawn for the specific situation of a granitic rock mass. The results can be seen in Fig. 2.

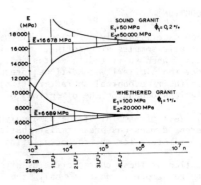

Fig.2 - Curves (5%) for a granitic rock mass

2.3 - Experimental results

LNEC has a large experience of conducting deformability tests with the Large Flat Jack Method, interesting different volumes of rock mass due to different numbers of LFJs used in the test. From recent tests we have results for one site tested with four LFJs, eight sites with three LFJs and eighteen sites with two LFJs. As in each test results from each LFJ and from each combination of LFJ are obtained, it is possible to plot the results versus the volume interested.

In Fig. 3 is shown the global result obtained. In this figure shows in abissae the volumes interested in the test as function of the maximum volume tested on site and in ordinates the values of the deformability modulus as function of the mean deformability modulus obtained for the site.

In what concerns the distribution of values for one certain dimention, in Fig. 4 is represented the cumulative distribution curve of relative deformabilities for V = 1/2 as well as the most likely normal distribution.

As can be seen for both figures a good agreement

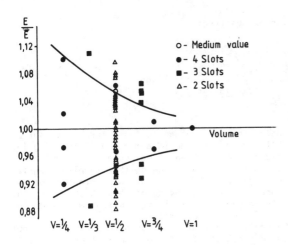

Fig. 3 – Modulus of deformability as function of volume

with the model was obtained in general.

We think much better results could be reached if a parameter related with the heterogeneity of each rock mass (for instance, unit cubic block) was used instead of the volume.

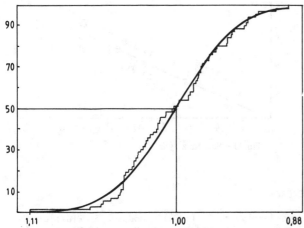

Fig. 4 – Comulative curve for relative deformabilities

3 - SCALE EFFECT ON THE STRENGTH

3.1 - An explanatory model

Let us assume that the irregularity of a surface can be well described as an ergotic aleatory variable. It can be shown that the square of the median of the maximal amplitudes ($\bar{h}$) is a linear function of the logarithm of the surface area (Ω)

$$\bar{h}^2 = K_1 \ln \Omega + K_2 \qquad (16)$$

If the irregularity considered is that displayed after displacement in shearing, the value h corresponds to the overlip occurred, and can be expressed in terms of the overlip angle ($\Delta \phi$), in fact an additional to the friction angle for a plane surface (Fig. 5)

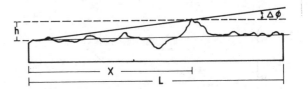

Fig. 5 - Irregularity and $\Delta \phi$ angle

$$\frac{h}{x} = \text{Tan} \, \Delta \phi \qquad (17)$$

As x can equally be any value between 0 and L, for a certain dimension the variable

$$\frac{h}{\text{Tan} \, \Delta \phi} = x \qquad (18)$$

has a uniform distribution with median value L/2, then:

$$\overline{\text{Tan} \, \Delta \phi} = \frac{\bar{h}}{L/2} \qquad (19)$$

and

$$(\overline{\text{Tan} \, \Delta \phi})^2 = K_1 \ln \Omega + K_2 \qquad (20)$$

3.2 - Results of the model

In an earlier work (Peres-Rodrigues, Charrua Graça, 1985) a good correlation was already found between results presented by other authors and the proposed model, namely in what concerns the variation of shear strenght with the dimension of the sheared surface (Fig. 6) and the scattering of results for a certain dimensions of the surface (Fig. 7).

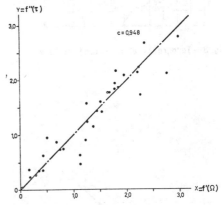

Fig. 6 – Correlation between strength and sheared area

Other tests were conducted on joints of a mudstone-sandstone rock mass both in laboratory (samples ≈ 15x x15 cm) and in situ (blocks ≈ 70x70 cm).

The irregularities of the shear surfaces were measured and analysed on basis of charts like the one in Fig. 8.

The behaviour of the sanstone joint is shown in Fig. 9 where Coulomb's straight-lines for both dimensions of surfaces are presented. In the figure is also indicated the maximum amplitude of irregularities found.

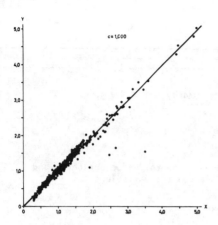

Fig.7 - Interpretation of the scattering of the results as uniform distribution of x

As can be observed:

$$\frac{\dfrac{\overline{h}_{70\times70}}{70}}{\dfrac{\overline{h}_{15\times15}}{15}} \simeq \frac{Tan(32^{o}-25^{o})}{Tan(39^{o}-25^{o})} = 0,492$$

25° being the friction angle found for plane surfaces. Similar tests made in a mudstone joint (Fig.10) ilustrate that such an effect depends on the irregularity of the surface, showing that in plane surfaces such as this the volume of K_1 decreases to near zero and without increasing of $\overline{h}$ with the area, the friction angle remains almost constant with the increase of the sheared area.

The small number of tests unfortunately did not allow to check the fitting of the scattering law.

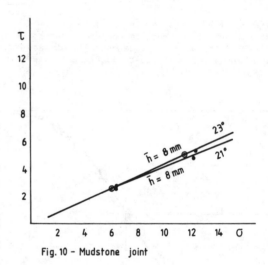

Fig. 10 - Mudstone joint

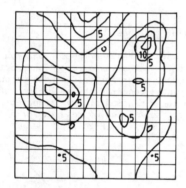

Fig. 8 - Topography of a sheared surface

REFERENCES:

Peres-Rodrigues, F. (1974) - Modulus of Elasticity of a rock obtained from the Moduli of Elasticity of its constituints - 2nd Int. Conf. IAEG.S.Paulo, Brasil.

Peres-Rodrigues, F.; Charrua-Graça,J.G. (1985) - Scale effect on the strength characteristics of rock masses - Int Symp. the role of rock mech. in exc. for mining and civil works. Zacatecas. Mexico.

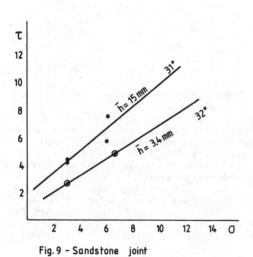

Fig. 9 - Sandstone joint

Back analysis of a major ground caving in a sub level retreat mine in northwestern Quebec

Rétro-analyse d'un effondrement majeur dans une mine en sous-niveaux rabattus dans le nord-ouest du Québec

Rück-Analyse eines hohen Firstbruchs in einem Bergwerk mit rückschreitendem Teilsohlenabbau im nordwestlichen Quebec

P.CHOQUET, Université Laval, Québec, Canada
P.GERMAIN, Centre de Recherche Noranda, Pointe Claire, Canada

ABSTRACT: This paper describes the events which have led to a major ground caving in a 25 m wide vein type orebody mined by a sub level retreat mining method. Various measurements and tests were firstly carried out; they included in situ stress measurements, joint mapping, application of geomechanical classification schemes and laboratory testing for the mechanical properties of rock and of discontinuities. Numerical modelling work is then reported that confirms that unusually high tensile stresses have been the main cause for the initiation and propagation of caving.

RESUME: L'article décrit les événements qui ont mené à un effondrement majeur dans un gisement filonien de 25 m de largeur exploité par une méthode des sous-niveaux rabattus. Divers essais et mesures ont d'abord été réalisés; ils comprennent des mesures de contraintes in situ, des relevés de discontinuité, l'évaluation des classifications géomécaniques des terrains et des essais de laboratoire pour l'obtention des propriétés mécaniques de la roche et des discontinuités. Des simulations numériques ayant servi à confirmer que des contraintes de traction relativement élevées ont été la cause de l'amorce puis de la propagation de l'effondrement, sont ensuite décrites.

ZUSAMMENFASSUNG: Der Artikel beschreibt die Ereignisse die zu einem höheren Grundfall führten, in einem 25 m breit aderlichen Vorkommen dass mit dem rückgefahrenen Teilsohlenabbau gewonnen war. Verschiedene Messungen und Proben wurden zuerst durchgeführt; sie enthalten in situ Spannungsmessungen, Felsrisseverzeichnis, Anwendung der geomechanischen Klassifizierungen und Laborprüfung der mechanischen Eigenschaften der Fels und Felsrisse. Numerische Modellierungen sind dann vorgestellt, die es befestigen, dass ungewohnliche starke Zugspannungen waren der Hauptgrund für die Einführung und die Verbreitung des Grundfalles.

1 INTRODUCTION

The Norita deposit, operated by Noranda inc., Matagami division, is located 2 kilometers north west of the town of Matagami, Quebec. It consists of five massive sulphide lenses containing about 4 175 000 tonnes grading 4,1% zinc and 1,8% copper.

The orebody was discovered in 1957 but was not brought into production before 1976. Exhaustion of reserves is now forecasted in the course of 1988.

The five lenses are interbedded in a precambrian volcanic succession composed essentially of rhyolitic and basaltic tuffs. These formations strike approximately E-SE with a sub vertical dip.

The sulphide lenses have variable dimensions; the smallest (Upper zone) has a height of 100 m, a length of 150 m and a 6 m average thickness; the largest (A zone) which is also the deepest, is 400 m in length, 210 m in height and has a width of 1.2 m in its top portion, rapidly increasing to a width of 25 m. A longitudinal section of the various lenses is presented in figure 1.

2 SUMMARY OF EVENTS

A detailed description of the events has been published elsewhere (Goodier and Dubé 1984).

The first lenses e.g. the Main, Lower and Marker zones were mined successfully by a sub level retreat method (figure 2), with sub level drifts established at 12 m intervals. In 1979, when A zone was due to enter production, it was decided to continue to use the same method of ore extraction. The decision was backed by existing experience acquired during mining of the lenses above, and also by the shape of the top portion of the A zone which was narrow and nearly vertical in dip.

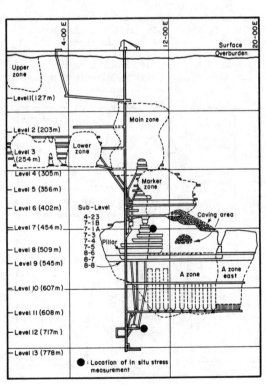

Figure 1. Longitudinal section of the 5 massive sulphide lenses.

Development and production of the first three sub levels were quite normal. It was known from the

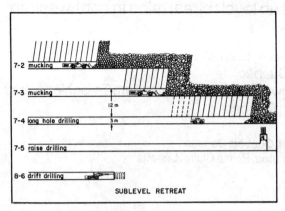

Figure 2. Schematic of the sub level retreat mining method.

original diamond drilling that the increasing width of the A zone below the 7-3 sub-level would begin to affect the mining sequence. It was decided at this time to leave small 3 x 4.5 m pillars to support the back of sub levels in the wider sections.

As the production sequence commenced in these sub-levels ground conditions in the walls and backs of the sub-levels began to deteriorate. Small rock-falls, displacement within the production long holes and sloughing of the walls began to take place, and a program of rehabilitation of the drifts had to be initiated.

To complete the development of 8-6 and 8-7 sub-levels, it was decided to maintain a 6 m wide continuous pillar in the center of each sub-drift and a thin section of ore was left on the north and south contacts to prevent sloughing of the walls into the sub drifts.

Development of the 8-6 sub-level and production faces in 7-5, 7-4 and 7-3 subs were brought to a complete halt in July 1980, when major movements and caving of the backs in each of these subs occured. The extent of the zone affected by caving can be seen in figures 1 and 3. This event which will be referred to as event no. 1, forced the mine to

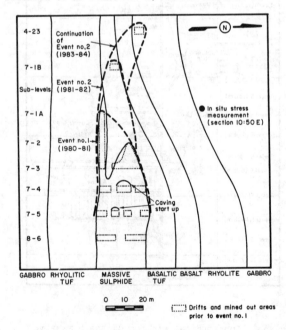

Figure 3. Transversal section 13:00 E with simpli-fied geology and caving events.

modify the mining method to a sub-level caving method of ore extraction with waste fill, mined from an adjacent waste rock stope, being introduced at the top of the A zone to give some support to the walls.

A haulage system was established on the 8-7 sub-level and preparations were made to have a mass blast of the 7-3 to 8-6 sub-levels which took place in July 1981. Furthermore, a modified sub-level retreat method was used to mine the sill pillar between the 8-7 and 8-6 subs.

When the complete sill was blasted, mucking of the draw points commenced. The draw points were mucked in a systematic way in order to pull down the mass of broken ore as one unit, while waste rock was being introduced at the top.

Rapidly however, due to a large mass of ore and waste material being in continuous downward movement, sloughing of north and south walls above the 7-3 sub elevation took place. The result of this sloughing created a large void above the A zone which resulted in deterioration and loss of the waste dumping points at 7-1B sub-level, and the remaining two production sub-levels of the Marker zone. The extent of the zone affected by this event no. 2 can be seen in figures 1 and 3.

Obviously, the need for a more efficient mining method was becoming a necessity as ground control, dilution and costs were becoming excessive. The mine decided then to change to a transverse open stope mining method with delayed backfilling to recover the ore between levels 9 and 11. A description of ground control occurences in these transverse stopes is reported elsewhere (Bawden and Milne 1986).

3 PURPOSE OF THE BACK ANALYSIS

In 1985, a study was initiated jointly with the department of mining and metallurgy at Laval Univer-sity, the mining technology division of the Noranda Research Center and the Norita mine in order to assess the prevailing ground conditions and rock mechanics parameters at the mine and to back analyze event no. 1.

It was decided to concentrate the efforts on event no. 1 since it had entirely taken place in the mas-sive sulphides whereas event no. 2 involved a more complex situation with the caving reaching the con-tact between the sulphide lens and the basaltic tuff and then extending into it. Moreover, as mentioned, caving during event no. 2 was initiated by the flow of broken waste into the stope which caused wall sloughing.

The steps involved in the back analysis work, which are described in the next sections were the following.
- In situ stress measurement.
- Joint mapping and geomechanical classifications.
- Laboratory testing of intact core specimens and shear testing of joints.
- Numerical modelling.

4 IN SITU STRESS MEASUREMENT

Seven CSIR triaxial stress measuring cells were in-stalled in two different boreholes at the 7 and 12 levels and were overcored in 7.6 cm diameter.

The first borehole was in the gabbro formation at a depth of 443 m and had an azimuth of N 100°E and a plunge of +5°. It was located 55 m north of A zone at a position shown in figures 1 and 3. For this reason, the four cells installed into it were almost certainly under the influence of stress migration caused by the adjacent mined A zone stope.

The second borehole was also located in the compe-tent gabbro formation at a depth of 714 m, with an azimuth of N 346° W and a plunge of +5°. It was located 60 m north of A zone at a position shown in figure 1 and was most probably under no influence of stress migration due to transverse stoping above 11 level, since only five stopes were mined out at that time.

A number of verifications were conducted on the overcored cylinders retrieved from the boreholes. Among others, strain gauge readings were checked through the validity test of comparing the sum of readings taken two by two and also through laboratory loading and unloading to assess the quality of gauge bonding to the cylinder walls.

After overcoring in the field, it was noticed that the readings of some of the stress cells could drift during a few hours before stabilizing. This was attributed, even though a compensation gauge was used, to a difference between the rock mass and the test drift which was 4°C warmer, and also to creeping of the strain gauge glue. For this reason, the strain gauge readings which were considered valid were those recorded after complete stabilization, and they were further corrected by a correction factor determined in the laboratory for each cylinder. This latter correction factor proved to be fairly low, varying between -1 and -3 micro-strain per °C increase.

Based on the verifications, two cells were rejected, one for each borehole. The final in situ stresses were then computed and are given below, as a function of the volumic weight $\gamma = 0.027$ MN/m^3 of the overlying rock mass, and the depth z (m).

Table 1. Value of measured in situ stresses.

	7 Level			12 Level		
	Magnitude	Strike	Plunge	Magnitude	Strike	Plunge
Vertical stress	0.42 γ.z	N 13°W	63°	1.02 γ.z	N 30°E	59°
Horizontal stress (N-S)	0.78 γ.z	N 162°E	27°	3.66 γ.z	N 157°E	19°
Horizontal stress (E-W)	4.34 γ.z	N 107°W	2°	3.50 γ.z	N 76°E	23°

It can be noted that stresses for the 12 level are much in agreement with past knowledge on the magnitude of natural stresses in the Canadian Shield (Herget, 1980) with a vertical stress equal to the weight of the overlying material, and horizontal stresses greater than the vertical. Consequently, the initial stresses for the numerical modelling will be those measured at the 12 level. The strong anisotropic stresses measured at the 7 level could be used for calibration purposes of a non linear stress analysis program accounting for the yield of the rock mass, although this aspect will not be discussed in the present paper.

5 JOINT MAPPING AND GEOMECHANICS CLASSIFICATIONS

As already mentioned, the A zone is contained in volcanic formations. From north to south, they are the following: rhyolite, basalt, basaltic tuff, massive sulphide (A zone) and rhyolitic tuff, as described by MacGeehan and al (1981) (figure 3).

Fourteen traverses stretching over 330 m were established in these formations. A total of 294 discontinuities larger than 0.6 m were mapped. Any information needed for rock mass classification was also noted.

For the various formations, the strike and dip of discontinuity sets encountered are given in the table below.

It is worthwhile noting that a flat lying major joint set was observed in every formation. It accounted for 40% to 80% of the mapped joints. Other joint sets were more or less minor, except of the basaltic and rhyolitic tuffs on both sides of the massive sulphides.

Table 3. Geomechanical classification of the basaltic and rhyolitic tuffs and the massive sulphides.

CSIR classification								
Parameter:	Strength	RQD	Spacing	Condition	Water	Orientation	Rating RMR	Description
Tuffs	12	13	10	6	10	-10	41	Poor to fair rock
Sulphides	15	17	25	12	10	-10	69	Good rock

NGI classification							
Parameter:	RQD	J_n	J_r	J_a	J_w	SRF	Rating Q
Tuffs	65	11	1.0	2.33	1.0	2.0	1.26
Sulphides	88	8	1.1	0.85	1.0	2.5	5.7

The geomecanical classifications CSIR and NGI were also determined at nine locations in the basaltic and rhyolitic tuffs and in the massive sulphides. The average parameters of the classification schemes are given above, in accordance with the nomenclature suggested by their authors (Bieniawski, 1974 and Barton & al, 1974).

Finally, during field work, some additional measurements were carried out in order to complement the information needed for evaluating the true shear strength of the discontinuities and foliation of the tuffs. More specifically, Schmidt hammer rebound values were obtained on joint walls for use in Barton and Choubey (1977)'s shear strength equation. Discontinuity roughness was also evaluated by means of repeated measurements of the attitude of several joints with a compass placed on plates of different dimensions (5, 10, 20 and 40 cm), after a method proposed by Fecker and Ranger (1971).

6 LABORATORY TESTING

Table 2. Strike and dip of joint sets in each geological formation.

Joint Set	Rhyolite			Basalt			Basaltic tuff (and rhyolitic tuff)			Massive sulphides		
	Strike	Dip	Importance*	Strike**	Dip	Importance	Strike	Dip	Importance	Strike	Dip	Importance
1	0°	8°E	58 %	0°	8°E	59 %	0°	8°E	50 %	0°	8°E	80 %
2	160°	44°W	24 %	196°	82°W	20 %	87°	82°S	24 %	198°	82°W	20 %
3	350°	82°E	18 %	350°	40°E	11 %	33°	67°E	12 %	—		
4	—			102°	82°S	10 %	269°	35°N	12 %	—		
5	—			—			350°	82°E	12 %	—		

Standard laboratory tests, as recommended by ISRM (1981) were performed on 50 mm diameter intact rock core specimen. The tests included the uniaxial compressive strength Ω, the modulus of elasticity E and Poisson's ratio ν. Results of the tests are given in table below, where anisotropy of the tuffs can also be evaluated.

Testing was also conducted on joints and foliation planes in a

* Percentages provide a basis for comparison of importance of the various joint sets.

** Strike as an azimuth, dip on right hand side

835

Table 4. Results of standard laboratory tests.

	n	E (MPa) $\bar{x}$	E (MPa) s	ν $\bar{x}$	ν s	Q (MPa) $\bar{x}$	Q (MPa) s
Gabbro	7	88 893	3 693	0.277	0.013	221	53
Sulphide	15	144 957	28 582	0.252	0.063	286	44
Rhyolite	6	77 782	17 239	0.251	0.071	103	44
Basaltic tuff (parallel to foliation)	9	85 262	2 982	0.386	0.018	126	22
Basaltic tuff (perpendicular to foliation)	8	63 573	3 957	0.316	0.039	154	33
Rhyolitic tuff (parallel to foliation)	1	94 062	–	0.268	–	85	–
Rhyolitic tuff (perpendicular to foliation)	6	73 829	7 362	0.223	0.028	142	96

n: number of tests; $\bar{x}$: average, s: standard deviation

a scale effect, the polar stereonets representing the repeated measurements of field joint attitudes by the method of Fecker and Jones mentioned previously, were evaluated. An example of the stereonets for a joint in massive sulphides and a foliation plane in the basaltic tuff is presented in figure 4.

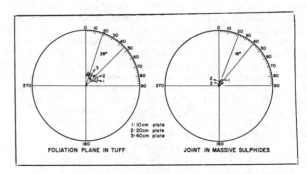

FOLIATION PLANE IN TUFF   JOINT IN MASSIVE SULPHIDES

Figure 4. Stereonets of repeated measurement of discontinuity attitude on different base lengths.

shear apparatus accomodating samples with a cross section up to 15 x 20 cm. However most of the samples tested had a cross section between 4 x 5 cm and 10 x 15 cm. For this reason, it was felt that laboratory values should be corrected for a scale effect to obtain their true field values, by means of various approaches presented below.

Standard shear tests as recommended by ISRM (1981), were performed on a number of samples. A modified procedure was selected for application of the normal load since the test conditions were of the constant normal stiffness type as discussed by Brady and Brown (1985), instead of maintaining a constant normal load. Normal stiffness was made constant by means of shutting the valve of the compensation system which is normally used in the shear apparatus to maintain a constant normal load. Thus, the normal load would increase due to dilation during the shear movement, a condition which is considered more representative for a joint confined in the rock mass at depth, as opposed to shearing in open pit situations. Normal displacements and increase in normal load were recorded during testing for computation of the normal stiffness and to ensure that the magnitude of this latter parameter was similar to the value that could be computed for the stiffness of the rock mass on both sides of the joint walls.

Results of the tests which are described in detail by Germain (1986), are presented in the table below.

From the figure, it can be concluded that the joints present a roughness of $\pm$ 9° for a wave length of 20 cm, while the foliation planes appear to have a much higher roughness of $\pm$ 14° over a wave length of 40 cm.

One method of using the above roughness values is that of Mc Hanon (1985) who suggests, based on back analyses of failed slopes, that the effective friction angle is given by the sum of the residual friction angle measured in a shear apparatus and the mean of relatively large scale roughness angles (e.g. corresponding to Patton's "first order" irregularities), which are the same as the above reported roughnesses. However, this procedure applies to tests performed under the constant normal force condition, and there is not yet any reason to believe that it could directly be applied to results of tests carried out with a constant normal stiffness. Moreover, it is generally recognized that scale effect is reduced for tests in which dilation is inhibited, as it is in the constant normal stiffness test condition.

Alternatively, an attempt was made to estimate the shear strength of the discontinuities by the method suggested by Barton and Choubey. This method which is well documented in ISRM (1981) involves evaluation of three parameters, namely JCS (joint wall compression strength), JRC (joint roughness coefficient) and ϕ_r (residual angle of friction). The first parameter is estimated through a Schmidt hammer test, while the second parameter is estimated by comparison with sketches of typical roughness profiles, or by a

Table 5: Results of shear tests for joints and foliation planes.

	Area of joint wall (cm²)	Number of tests	Normal stress σ (MPa)	Peak shear strength τ_p (MPa)	Residual shear strength τ_R (MPa)	Normal stiffness during shearing (GPa/m)	Increase in normal stress during shearing (MPa)
Joints in massive sulphides	18 to 33	4	0 to 15	$2.1 + \sigma_n \tan 41.9°$	$0.7 + \sigma_n \tan 23.2°$	7.3 to 9.4	1.1 to 5.5
Foliation in basaltic tuff	55 to 128	8	0 to 15	$2.0 + \sigma_n \tan 43.4°$	$0.6 + \sigma_n \tan 29.1°$	3.2 to 3.8	2.2 to 5.4
Foliation in rhyolitic tuff	82 to 147	3	0 to 10	$3.8 + \sigma_n \tan 40°$	$\sigma_n \tan 29°$	1.9 to 8.3	2.4 to 3.9

tilt test. The third paramater comes directly from the laboratory shear tests.

It was thought of interest to conduct separately an evaluation of the parameters in the field and in the laboratory on the samples prior to the shear tests.

Table 6 reports the values of JCS, JRC and ϕ_r to be included in Barton and Choubey's equation for peak shear strength τ_p, namely $\tau_p = \sigma_n \tan [JRC \log_{10} (\frac{JCS}{\sigma_n}) + \phi_r]$.

However, for comparison purposes with the laboratory peak shear strengths presented in table 5, a linearized equation which best fits Barton and Choubey's equation in a range of 0 to 15 MPa for σ_n, is presented in the table.

It should be noted that the cohesion reported in the table for both peak and residual strengths is an apparent cohesion due to the increase in normal stress during testing. For example, tests with a zero normal stress at the beginning showed an increase of about 3 MPa in normal stress after a few millimeters of shearing.

In order to correct the peak shear strengths for

Table 6. Parameters of Barton and Choubey's peak shear strength equation.

LABORATORY				
	JCS* (MPa)	JRC**	ϕ_r	Linearized τ_p (MPa)
Joints in massive sulphide	412	5.5	23.2	$0.2 + \sigma_n \tan 30.7^{\circ}$
Foliation plane in rhyolitic tuff	113	10.6	29.1	$0.6 + \sigma_n \tan 37.3^{\circ}$
Foliation plane in basaltic tuff	102	6	29.0	$0.3 + \sigma_n \tan 33.4^{\circ}$

* Schmidt hammer test on laboratory samples
** Tilt test and typical roughness profiles

FIELD				
	JCS*	JRC**	ϕ_r	Linearized τ_p (MPa)
Joints in massive sulphide	676	8	23.2°	$0.4 + \sigma_n \tan 35.6^{\circ}$
Foliation plane in rhyolitic tuff	188	5	29.1°	$0.3 + \sigma_n \tan 34^{\circ}$
Foliation plane in basaltic tuff	112	6	29	$0.3 + \sigma_n \tan 34^{\circ}$

* Schmidt hammer test
** Field visual evaluation of typical roughness profiles

Considering table 6, there appears to be a scale effect of 5° to be applied to the shear strength of joints in massive sulphides, whereas no strong scale effect is apparent for the foliation planes in the two types of tuff. On the other hand, it is obvious that the peak shear strengths from Barton and Choubey's equation are much lower than those reported in table 5.

The final opinion of the authors is that the laboratory peak shear strengths obtained under normal stiffness test conditions are the most representative of the actual field properties of the discontinuities. This assertion is partly validated by a certain amount of back analysis work which is not reported in the paper (Germain 1986). However it is also the authors' opinion that the information provided could be of some help for future novel research in assessing exact field properties of discontinuities.

7 NUMERICAL MODELLING

A two-dimensionnal boundary element program developed by Hoek and Brown (1980) was used to model transverse sections of the mine levels affected by caving.

The program assumes a homogeneous, isotropic and linear elastic material, whereas the medium to be modelled is composed of formations of different mechanical properties, especially in the case of the massive sulphides and the adjacent tuffs.

However, early simulations showed rapidly that stress distributions computed under the homogeneous and elastic medium assumption were already very significant to explain the onset of caving, as presented below. The authors agree however that computed stress distributions could be refined by using a non linear and yielding stress analysis program allowing for the following situations: different elastic properties for the sulphides and the tuffs, anisotropy of mechanical properties of the tuffs, slip criterion for interfaces between various formations and yielding of rock formations with reduction of rock strength to residual values. Such an analysis would allow a comparison with the in situ stress value measured at level 7.

On the other hand, during the early modelling, various attempts were made to use the peak failure criterions of both the rock and the discontinuities. These attempts proved to be valid in the sense that failure zones detected corresponded quite accurately to the extent of the caving zone observed. However, since these failure criteria are in shear, whereas most of the caving occurence will be explained below by the existence of tension zones, it was not thought of

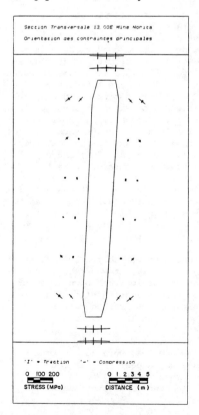

Figure 5. Plot of principal stresses for transversal section 13:00 E (mining stage 1).

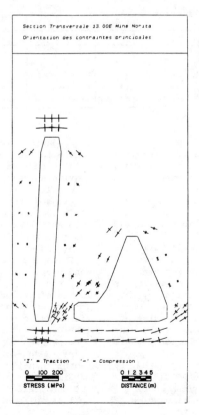

Figure 6. Plot of principal stresses for transversal section 13:00 E (mining stage 2).

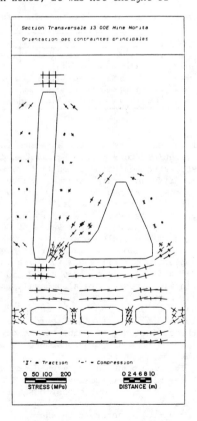

Figure 7. Plot of principal stresses for transversal section 13:00 E (mining stage 3).

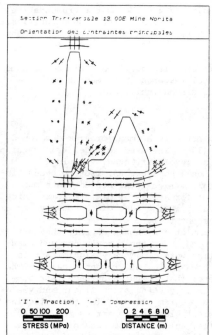

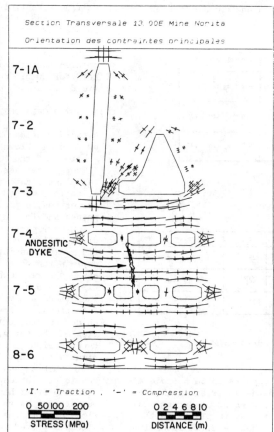

Figure 8. Plot of principal stresses for transversal section 13:00 E (mining stage 4).

Figure 9. Plot of principal stresses for transversal section 13:00 E (mining stage 5)

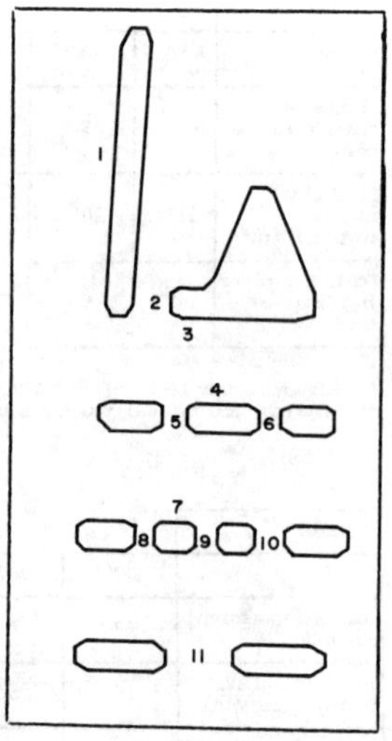

Figure 10. Position of stress values of table 7.

Table 7. Principal stress values for different mining stages, in MPa.

Position:	1	2	3	4	5	6	7	8	9	10	11
Stage 1	0.5, – 4.5	29.5, 1.4	–	–	–	–	–	–	–	–	–
Stage 2	0.2, – 2.7	70.0, – 41.0	66.5, – 7.3	–	–	–	–	–	–	–	–
Stage 3	0.2, – 2.8	72.7, – 41.4	74.2, – 6.4	82.7, 0.4	23.0, 0.03	37.9, – 2.3	–	–	–	–	–
Stage 4	0.2, – 2.9	73.9, – 41.5	76.4, – 6.8	84.9, 0.5	22.0, – 0.15	35.0, – 2.4	70.9, 3.9	28.0, – 0.50	30.0, – 0.65	35.2, 4.0	–
Stage 5	0.2, – 3.0	74.4, – 41.6	77.4, – 7.0	85.7, 0.52	21.3, – 0.26	34.0, – 2.4	72.3, 3.8	25.0, – 0.65	29.6, – 0.82	32.8, 3.9	33.4, 24.5

interest to present the attempts in this paper.

Figures 5 to 9 present graphical plots of principal stress values computed on transversal section 13:00 E at different mining stages. On these figures, stresses are represented by scaled symbols, with tensile stresses represented by a "I" symbol and compressive stresses by a "_" symbol.

Some principal stress values at positions shown in figure 10 are also presented in table 7.

In this table, it is apparent that tensile stresses appear in all pillars which were left in sub-levels 7-5 and 7-4 (positions 5,6,8,9 and 10). Moreover, tensile stresses are also developing in the sill pillar between sub-levels 7-4 and 7-3 (position 3), which has eventually led to the massive caving after sub-level 8-6 was driven. Figure 9 is representative of the situation just prior to the caving while the other figures show the progression of stress build-up until that final stage.

The following caving mechanism can be suggested. After mining of sub-levels 7-1, 7-2 and 7-3 (figures 5 and 6), the two empty stopes were surrounded by tensile stresses causing some wall sloughing which was actually observed in them.

After driving sub-level 7-4 (figure 7), some of the

conditions prior to the caving were already present, with tensile stresses appearing at positions 3 and 6. These conditions were however accentuated after driving sub-level 7-5 since it becomes apparent that all pillars at positions 5,6, 8 and 9 are subjected to tensile stresses oriented to cause pillar split. Such a pillar split condition is a well known cause of roof degradation, a condition which was observed at this stage in the two sub-levels 7-4 and 7-5, and especially at locations shown in figure 3, where the caving actually started.

Finally, sub-level 8-6 was driven. Stress distributions do not seem to be considerably altered after this stage as can be seen in table 7, however it is probable that enough movement and loosening in the rock mass had taken place so that caving eventually took place.

Three more conditions have probably helped to cause and propagate caving. Firstly, it was recognized after the event that an andesitic dyke shown in figure 9 had contributed to the propagation of caving. Secondly, horizontal joints which are the main structural features in every rock formations and especially in the massive sulphides, have also certainly added to the propagation of caving. Additionally, small pillars in sub-levels 7-5 and 7-4, were not so continuous longitudinally as they may appear in the figures and their actual size was most probably smaller than planned, all of this leading to an increase of the pillar split effect.

8 CONCLUSION

A detailed description of a major ground caving associated with the sub-level retreat mining method is provided in the paper.

The method had been preferably chosen to mine the vein-type orebody because of its several advantages, for example the reduction of development work required as compared to some other mining methods, which eventually leads to lower mining costs.

However, presently, the sub level retreat method has probably been brought to its limits because of the wide width required for the sub levels. The understanding of the mechanisms which have led to caving could however be of some help in extending the areas in which the mining method could be applied.

On the other hand, it can be stressed that extensive rock mass recognition and characterization work has been carried out in the mine as it is reported in the paper, although the majority of it has not been fully used yet in explaining the caving occurrence. More refined analyses calibrated to the events which have occurred at the mine could still be conducted.

ACKNOWLEDGEMENTS

The authors wish to thank the Norita mine for financially supporting the project and the engineering staff, namely MM. J. Gagné, A.J. Bonenfant, R. Dubé and A. Goodier for their assistance and contribution at the mine site.

Thanks are also extended to the Centre de Recherche Noranda for a great involvement in the project.

REFERENCES

Barton, N. & Choubey, V. 1977. The shear strength of rock joints in theory and practice. Rock Mechanics (Springer Verlag) 10, 1-54.

Barton, N., Lien, R. & Lunde, J. 1974. Engineering classification of rock masses for the design of tunnel support. Rock Mechanics, Volume 6, no. 4.

Bawden, W. & Milne, D. 1987. Geomechanical mine design approach at Noranda Minerals Inc., 6th Congress of the International Society for Rock Mechanics, Montreal, August 30th - Sept. 3rd, 1987.

Bieniawski, Z.T. 1974. Geomechanics classification of rock masses and its application in tunnelling. Proc. Third International Congress on Rock Mechanics, ISRM, Denver, Volume 11A.

Brady, B.H.G. & Brown, E.T. 1985. Rock mechanics for underground mining, George Allen and Unwin (Publishers) ltd, London, pp. 121-124.

Fecker, E. & Rengers, N. 1971. Measurement of large scale roughnesses of rock planes by means of profilograph and geological compass. Rock Fracture, Proc. of Int. Symp. Rock Mech. Nancy, paper I.18.

Germain, P. 1986. Rétro-analyse d'un effondrement minier majeur dans une mine exploitant par la méthode des sous-niveaux rabattus, M.Sc. thesis, Université Laval, Dept. of mining and metallurgy.

Goodier, A. & Dubé, R. 1984. Changes in mining methods to overcome ground conditions at the Norita mine, Preprints 86th Annual Meeting of the Canadian Institute of Mining and Metallurgy, Ottawa, April 15-19, 1984.

Herget, G. 1980. Regional stresses in the Canadian Shield, 13th Canadian Rock Mechanics Symposium, Toronto, May 28-29, 1980, Underground Rock Engineering (CIM Special Volume 22), Montreal.

Hoek, E. & Brown, E.T. 1980. Underground Excavations in Rock, Institution of Mining and Metallurgy, London, pp. 493-511.

International Society for Rock Mechanics 1981. Rock Characterization, Testing and Monitoring, ISRM suggested methods, Pergamon Press ltd, pp. 3-52 (Suggested methods for the quantitative description of discontinuities in rock masses), pp. 113-117 (Suggested method for determining the uniaxial strength and deformability of rock materials), pp. 135-137 (Suggested method for laboratory determination of direct shear strength).

MacGeehan, P.J., Mac Lean, W.H. & Bonenfant, A.J. 1981. Exploration significance of the emplacement and genesis of massive sulphides in the Main Zone at the Norita mine, Matagami, Quebec, The Canadian Mining and Metallurgical Bulletin, April 1981.

Mc Mahon, B.K. 1985. Some practical considerations for the estimation of shear strength of joints and other discontinuities. Proceedings of the international Symposium on Fundamentals of Rock Joints, Björkliden, September 15-20, 1985, pp. 475-485.

Rock mass classifications for tunnel purposes – Correlations between the systems proposed by Wickham et al., Bieniawski and Rocha

Classifications de massifs rocheux pour tunnels – Corrélations entre les systèmes proposés par Wickham et al., Bieniawski et Rocha

Felsklassifizierung für Tunnel – Beziehungen zwischen dem Wickham et al., und dem Bieniawski und Rocha System

A.S.COSTA-PEREIRA, M.Sc., Universidade do Porto – Portugal, CÊGÊ, Lda, Lisboa
J.A.RODRIGUES-CARVALHO, Ph.D., Universidade Nova de Lisboa, Portugal

ABSTRACT: Most of the supports for tunnels in rock masses are designed based on empirical classifications. Firstly, the authors present a brief description of the classification proposed by Rocha. After that and based upon the results obtained from 15 tunnel sections, the Wickham et al. (RSR), Bieniawski (RMR) and Rocha (MR) classification systems are correlated.

RESUME: Les soutènements pour les tunnels en massifs rocheux sont choisis trés frequemment a partir de classifications empiriques. Les auteurs (commence pour presenter la classification proposé par Rocha. En suite ils) presentent des correlation entre les classifications de Wickham et al. (RSR), Bieniawski (RMR) et Rocha (MR), en se baseant sur les resultats de 15 sections de tunnels etudiés.

1 INTRODUCTION

Most of the supports for tunnels in rock masses are designed on the basis of empirical geotechnical classifications which have been developed for this purpose by several authors. The first of these type of classifications was introduced by Terzaghi (1946) and a number of others appeared more recently, particulary during the 70's, namely those proposed by Wickham et al.(1974), Bieniawski (1974, 1976, 1979), Barton (1974) and Association Française de Travaux en Souterrain - AFTES (1976).

In this kind of classifications, the quality of a rock mass is "quantified" on the basis of a number of geological and geotechnical features and parameters in order to antecipate the behaviour of the rock mass to be tunnelled and to produce some recommendations on the support to be used.

Selecting the most appropriate and relevant features and parameters that control the behaviour of the rock masses as well as their relative importance in order to build up a classification lies,to a great extent, upon the experience of each author. Consequently,it is important to investigate if there is a correlation between the different classification systems. If this correlation exists, one may apply the easiest system according with the parameters available in the situation under study and derive the result that would be obtained by using other. At the end, this would enable a comparison between the recommendations about the support established by the different classification systems.

Also during the 70's, Rocha (1976) proposed his own empirical classification to "quantify" the quality of the rock masses in order to assist the design of tunnel supports. Althoug frequently used in Portugal together with others, namely the ones from Wickham et al., Bieniawski and Barton, Rocha's classification is almost unknown abroad. It was introduced in 1976 during the lessons delivered by that author to the M.Sc. courses in Engineering Geology and Soil Mechanics at the Universidade Nova de Lisboa. By that time Rocha prepared, for their students, a set of written notes under the litle "Underground Structures" where the basis of this classification are explained. Rocha, certainly wanted to carefully check the results of his own classification and this is certainly the main reason why it was never published before his death in 1981.

Within other comparative studies carried out by some authors, as refered by Bieniawski (1979), Rutledge (1978) correlated the classification systems proposed by Wickham et al. (RSR), Bieniawski (RMR) and Barton (Q) by using data from tunnels in New Zealand and established the following correlations:

$$RMR = 13.5 \log Q + 43$$
$$RSR = 0.77 RMR + 12.4$$
$$RSR = 13.3 \log Q + 46.5$$

Firstly, in this paper, the authors summarize the classification developed by Rocha. After that, this classification together with the ones proposed by Wickham et al. and Bieniawski are correlated by using the results obtained from 15 tunnel sections.

2 THE ROCHA'S CLASSIFICATION (MR)

Rocha assumes that the pressure on the support is imposed by a volume of rock that may detach (come off) from the rock mass after the excavation, this being ruled mainly by the characteristics of the existing joints. That volume has the shape presented in Fig. 1 where h_c and h_n are the widths of rock that may come off from the roof and from the walls of the excavation. The magnitude of h_c and h_n for a given rock mass depends upon both L (tunnel span) and k, this being a coefficient derived from the quality of the rock mass (MR).

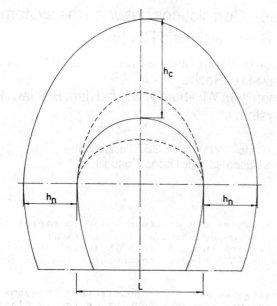

Fig. 1 - Volume of rock that may load the support

Different ratings are allocated to each of the above parameters in accordance with the existing conditions as specified in Table I. The quality of rock mass is then represented by the MR index which is expressed by the total of those ratings and may vary between 0 and 100.

$$MR = Pe + Ps + Pr + Pp$$

The values of k in conexion with the quality of the rock mass may be obtained from the courve represented in Fig. 2.

The load in the roof of the support is given by the equation

$$h_c = kL$$

Concerning the loads on the support walls, Rocha considers that they are dependent upon both, h_c and MR and suggests the following values:

$$h_n = 0 \qquad \text{for } MR > 60$$

$$0 < h_n < \frac{1}{2} h_c \text{ , for } 50 < MR < 60$$

$$\text{and } h_n = \frac{1}{2} h_c \text{ , for } MR < 50$$

In order to quantify the quality of a rock mass Rocha adopts four parameters:

- Joint spacing (Pe)
- Joint systems (Ps)
- Joint shear strength (Pr)
- Water conditions (Pp)

JOINT SPACING

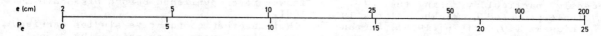

JOINT SETS

SHEAR STRENGTH

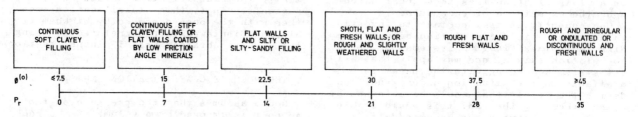

WATER PRESSURE

Table I - Rocha's classification: parameters and their ratings

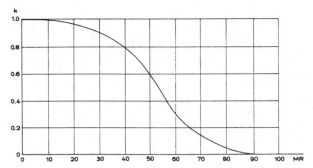

Fig. 2 - Obtaining k from the rock mass
quality (MR)

Table II presents the five classes of
rock masses considered by Rocha together
with the corresponding range of values for
MR and k and contains also some considerations of that author about support needs.

CLASS	MR	k	SUPPORT
I	80-100	0 -0.05	Sporadic support;ex: rock-bolting,in accordance with the observed roof conditions.
II	60-80	0.05-0.3	Systematic support in the roof.
III	50-60	0.3-0.6	Systematic support in the roof. Sporadic support in the walls may also be necessary
IV	30-50	0.6-0.9	Systematic support in both, roof and walls is necessary.
V	0-30	0.9- 1	

Table II - Classes of rock masses; their
ratings and considerations on
supports needs (Rocha, 1976)

4 CORRELATION BETWEEN THE ROCHA, WICKHAM et al. AND BIENIAWSKI CLASSIFICATION SYSTEMS

During the geological and geotechnical
studies carried out for the design of five
tunnels (four in Portugal and one in Macau)
the characteristics of the different rock
masses led to define three geotechnical zones in each case. In sequence of this the
classification systems suggested by Rocha,
Wickham et al. and Bieniawski were applied
to each of those fifteen zones.
The total ratings obtained for each zone
by using the three above mentioned classifications are presented in Table III. The
figures in that table allows to derive the
following correlation between Wickham's
(RSR) and Bieniawski's (RMR) classifications:

$$RSR = 0.7 \; RMR + 23.5$$

TUNNEL AND ROCK TYPE	GEOTECHNICAL ZONNE	RATINGS		
		RMR	RSR	MR
GUIA (MACAU) Granite	I	21	48	21
	II	37	50	53
	III	57	65	63
BELICHE-GAFA Shale+greywack (flysch)	I	17	34	18
	II	36	44	44
	III	46	56	52
ODELEITE Shale+greywack (flysch)	I	28	34	24
	II	45	44	48
	III	62	66	59
CASTELO DO BODE Gneiss	I	17	36	25
	II	36	50	45
	III	60	69	60
ÁLAMOS-ALQUEVA Phillite	I	20	34	25
	II	45	50	43
	III	65	76	67

Table III - Total ratings obtained for the
15 tunnel sections studied

On the other hand, correlating the results between Rocha's (MR) and Bieniawski's
(RMR) classifications the authors arrived
at the following equation:

$$MR = 0.95 \; RMR + 5.4$$

The above correlations together with the
one between RSR and RMR, presented by Rutle
dge (1976) are represented in Fig. 3.

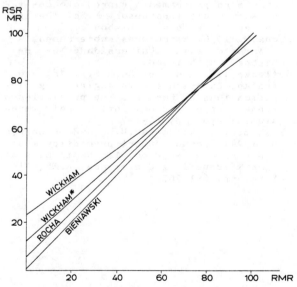

Fig. 3 - Correlations between the classifi-
cations of Bieniawski and those of
Wickham et al. and Rocha
(*) Rutledge (1978)

843

5 CONCLUSIONS

The correlation between Bieniawski's (RMR) and Wickham's (RSR) total ratings for the sections of tunnels studied, as obtained by the authors, and that presented by Rutledge are close to each other. It may be said that, in practical terms, a difference occours only for the rock masses of poorer quality. In situations like this the correlation obtained by the authors led to less conservative marks to the Wickham's ratings than the ones derived by using the correlation of Rutledge.

The ratings derived by using Rocha's clas sification are closest to the Bieniawski's than the ones obtained from Wickham's, clas sification as illustrated in Fig. 3.

By the final phase of engineering geologi cal studies for tunnels in rock masses it is usual to apply different empirical clas sifications of the type dealt with in this paper and to compare the results before making suggestions on the support to be used.

In the authors oppinion the correlations presented in this, paper are good reasons to include Rocha's classification in the number of classifications to work with.

From Fig. 3 it may be pointed out that the differences in the total ratings obtained by means of the different classifications discussed here, decreases as the rock mass quality is better.

LITERATURE

Association Française de Travaux en Souter-
    rain (AFTES) 1976. Texte provisoire des
    recommendations relatives au choix d'un
    type de soutènement en galerie. Tunnels
    et Ouvrages Souterraines, cpecial issue.
Barton,N; Lien,R. and Lunde,J. 1974. Engi-
    neering classification of rock masses for
    the design of tunnel support. Rock Mecha-
    nics, Vol 6, nr.4, pp. 189-236.
Bieniawski,Z.T. 1979. The geomechanics clas
    sification in rock engineering applicati-
    ons. Proc. 4th Int. Cong. Int. Soc. Rock
    Mech., Montreux, Vol II, pp. 41-48.
Costa-Pereira,A.S. 1985. A geologia de enge
    nharia no planeamento e projecto dos tu-
    neis em maciços rochosos. M.Sc. thesis,
    Universidade Nova de Lisboa.
Rocha,M. 1976. Estruturas subterrâneas -
    - Notas de aula - Universidade Nova de
    Lisboa, Unpublished.
Rutledge,T.C. and Preston,R.L. 1978. New
    Zealand experience with engineering clas-
    sifications of rock for the prediction of
    tunnel support. Proc. Int. Tunnel Sympo-
    sium, Tokyo, pp. A3-1-7.
Wickham,G.E.; Tiedemann,H.R. and Skinner,E.
    H. 1974. Ground support prediction model-
    - RSR concept -. Proc. 1st Rapid Excava-
    tion & Tunnelling Conference, AIME, New
    York, pp. 691-707.

Analysis of tunnel behaviour in discontinuous rock masses
Analyse du comportement des tunnels dans des massifs rocheux discontinus
Analyse des Tunnelverhaltens im Fels mit Diskontinuitäten

A.P.CUNHA, Laboratório Nacional de Engenharia Civil (LNEC), Lisbon, Portugal

ABSTRACT: Some parametric studies about the influence of discontinuity properties on tunnel behaviour are reported, considering overstressed joints. Stress and displacement fields are presented, and a comparison is made with isotropic and anisotropic, analytical or numerical continuous solutions.

RESUMÉ: On présente des études par élements finits sur l'influence des propriétés des discontinuités dans le comportement des tunnels en roche où la resistance au cisaillement des discontinuités est dépassée. Une comparaison est faite avec des solutions numériques et analytiques pour massif continu, soit isotropique soit anisotropique.

ZUZAMMENFASSUNG:Es wird von einigen parametrischen FE - Untersuchungen über den Einfluß der Diskontinuitätseigenschaften auf das Verhalten von Tunnels im Felsen berichtet, wobei die Wirkung der Spannungen berucksichtigt wird. Spannungs - und Verschiebungsfelder werden vorgestellt, und ein Vergleich mit isotropen und anisotropen, analytischen oder numerischen Kontinuumlösungen gezogen.

1 INTRODUCTION

The random pattern of the discontinuities, the scattering of their geometric and mechanical characteristics and the heterogeneity and anisotropy of the rock assign to the rock masses a complex structural behaviour, the analysis and forecasting of which has to be done,due to computational and geotechnical data limitations, by means of continuous or discontinuous simplified models.

The verisimilitude and meaning of such approaches are directly dependent on the macroscopic structure of the rock masses, the discontinuous models requiring the definition of the mechanical properties both of the geomechanically important discontinuities and of the equivalent continuous medium between those joints.

For shallow underground openings, with low stress levels, the deformation of the rock blocks can often be negligeable, if compared with the rotation and translation movements which take place along the joints, block models being then commendable for the stability analysis (Lamas, 1986). For deep tunnels in high stress fields, however, both block and joint deformation must be taken into account. Herein, the finite element method is a suitable tool, both in the investigation of the tunnel phenomenology and in design.

In the paper some parametric studies on the influence of discontinuity properties and primary state of stress on tunnel behaviour are reported. A comparison is made with isotropic and anisotropic, analytical or numerical continuous solutions. As overstresses induce joint failure, their effects are evaluated in terms of stress and displacement fields and the safety factors against sliding, along the discontinuities, are investigated.

2 CIRCULAR TUNNEL IN A ROCK MASS WITH A MAIN JOINT SET

A plane strain analysis of a circular tunnel (R=3 m) in a sedimentary rock mass with a main joint set was carried out. The geometrical and mechanical conditions of nine parametrical studies are described in figs. 1 and 2, considering the rock mass between discontinuities as a continuous elastic medium and assigning

both linear and non-linear behaviour to joints. The initial vertical stress was σ_v=2 MPa. Displacements, principal stresses and equal stress lines for all calculations were plotted and can be seen elsewhere (Cunha, 1981).

As the existence of the main joint set gives the rock mass a clear anisotropic behaviour, the definition of equivalent continuous elastic anisotropic media, based on the mechanical properties of the joints and rock matrix,was made for the several calculations. The displacement and stress fields relative to these

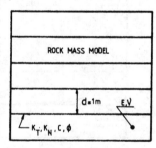

Figure 1. Sedimentary rock mass with a main joint set

CASE	STRESS $\lambda = \sigma_H/\sigma_V$	ROCK E (MPa)	JOINTS K_T(MPa/m)	K_N(MPa/m)	c (MPa)	φ(°)
J1	1	10^4	10^3	$2,5×10^3$	0	15
J2	1	"	"	"	"	30
J3	1	"	"	"	0,15	15
J4	1	"	$2×10^4$	$5×10^4$	0	15
J5	3	"	10^3	$2,5×10^3$	"	15
J6	1/3	"	"	"	"	15
J7	1	"	$2×10^6$	$5×10^6$	"	15
J8	3	10^3	10^3	$2,5×10^3$	"	15
J9	1	10^3	"	"	"	15

Figure 2. Mechanical parameters for the analysis.

equivalent media, when compared with the elastic discontinuous ones, show good similarities as regards the trend of stresses and displacements (Cunha, 1981). Fig. 3 describes the joint yielding lengths for the various calculations, yielding depending on joint position related to tunnel wall and only attaining joints II to VI. It can be observed that the extension of joint yielding is related to the decrease of joint stiffness or rock properties and to the direction of

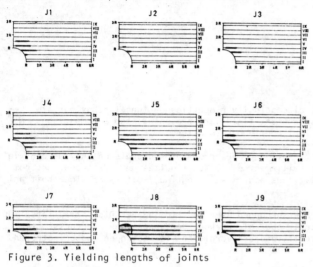

Figure 3. Yielding lengths of joints

the maximum initial principal stress. Joint displacements and yielding extension are greater when this stress is parallel to the joint set. Yielding extension also increases when joint strength decreases or joint stiffness increases. When joint stiffness is equivalent to the deformability of the rock matrix,

joint slidings were nil (J7), although joint strength, computed by the Mohr-Coulomb criterion, was exceeded for some joints. The yielding extension increases from invert to crown, where the maximum stress σ_θ is parallel to the joint set, thus developing a tangential effort on the joints. So the maximum yielding length occurs in joints III and IV, respectively located below and above the tunnel crown, and the same applies for joint displacement.

In fig. 4 displacements and stresses of joints III and IV for all calculations are shown. The displacements do increase as the stiffness characteristics and the strength of the discontinuities decrease, that is, as the greater is the weakness represented by joints as regards the rock matrix. Differences are also strongly influenced by the orientation of the primary stresses. The most unfavorable situation occurs when the maximum principal component, parallel to the joints (calculations J5, J8) brings about a slower vanishing of tangential stresses along the joints and an overall decrease in their safety coefficient against sliding, leading to an increase in the extent of the zones where the joint strength may be exceeded and, consequently, to an increase of joint slidings. The normal stresses on joints are similar, for all calculations, and the normal displacements show a trend to open, near the tunnel crown, related to a bending effect of the strata, the result of which was the release of stabilizing normal stresses in the fractures and consequent deterioration of their safety conditions. A similar or opposite role, depending on whether the joint is being opened or closed during sliding, is played by the dilatancy of the joints, whose effect was not considered in the calculations.

The evolution of the punctual safety factor against sliding for the several joints, and calculations J5, J6, is presented in fig. 5, shearing strength of joints being evaluated through the Mohr-Coulomb envelope. When the greatest initial stress is parallel to the joints

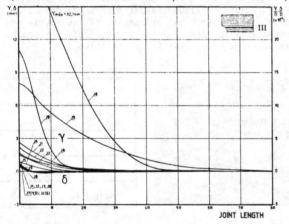

γ, δ – TANGENTIAL AND NORMAL DISPLACEMENTS

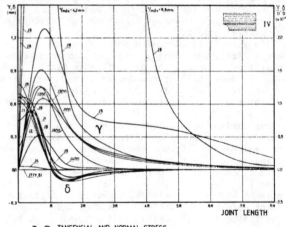

τ, σ – TANGENCIAL AND NORMAL STRESS

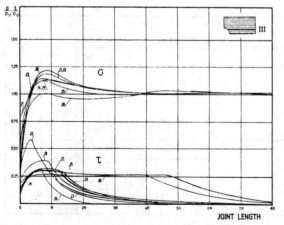

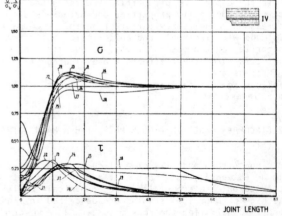

Figure 4. Normal and tangential displacements and stresses of the joints III and IV.

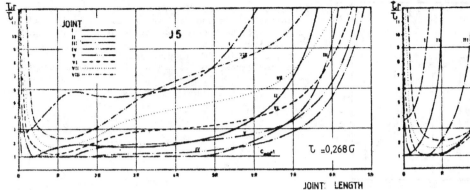
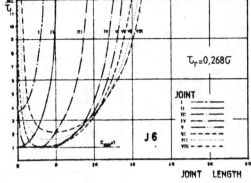

$\tau_r = 0,268\sigma$

Figure 5. Joint safety factors against sliding

the tangential stresses decrease slowly, and the safety factor in each joint increases also slowly with the distance to the wall, depending on joint strength and joint position too. Note that the punctual exceeding of the joint strength does not necessarily determine the collapse of the rock mass — unless these joints are crossed by other fractures, defining blocks which, owing to the existence of an opening and due to their dead weight, accquire an effective possibility of movement — but it may correspond to a significant increase of displacements, as regards the elastic discontinuous or continuous solutions (relationships of ten to one were determined).

Finally, fig. 6 shows the tunnel wall displacements (u-horizontal, v-vertical, δ_r-radial) relative to a quarter of the circunference ($\theta=0°-90°$), for linear and non-linear behaviour of discontinuities (J5, J6). These displacement curves are discontinuous at the wall-fracture intersections ($\theta=30°$, $60°$) and the comparison between the discontinuous elastic displacements and the equivalent continuous anisotropic ones in the same figures shows, besides the concentrated effects of the fractures, quite a similar trend and amplitude. However, the ratio between discontinuous elastic and elastoplastic maximum displacements on the wall and the isotropic continuous ones for the same initial stresses and mechanical properties of the rock has reached the maximum value of eight.

3 CIRCULAR OPENING IN A ROCK MASS WITH TWO JOINT SETS

A plane strain analysis of a circular opening (R=3m), in a rock mass with two perpendicular joint sets with a spacing of 1 m was also carried out. Since a 3rd joint set would exist, normal to the previous ones, and since the study of the kinematic behaviour of each block under its self weight and applied forces was not considered, the analysed situation corresponds better to a cross section of a vertical shaft than to a tunnel.

Two calculations were made, assuming the rock as elastic and a non linear behaviour for the joints. The initial stresses ($\sigma_I=\sigma_{III}$=2 MPa) and mechanical properties are shown in fig. 7.

CASE	σ_I / σ_{III}	\multicolumn{5}{c}{MECHANICAL PROPERTIES}					
		\multicolumn{1}{c}{ROCK}	\multicolumn{4}{c}{HORIZONTAL JOINT SET σ_{III}}	\multicolumn{4}{c}{VERTICAL JOINT SET σ_I}
		E(MPa)	K_T(MPa/m)	K_N(MPa/m)	C (MPa)	φ(°)	K_T(MPa/m)	K_N(MPa/m)	C (MPa)	φ(°)
F1	1	10^4	10^3	25×10^3	0	15	10^3	$2,5 \times 10^3$	0	15
F2	1	10^4	2×10^4	5×10^4	0	15	10^3	$2,5 \times 10^3$	0	15

Figure 7. Mechanical parameters for the analysis

The joint yielding lengths, evaluated by means of the Coulomb criterion, are presented in fig. 8. On the vertical and horizontal simetry axis there is no yielding for both sets in all the calculations, on account of the stress condition τ=0. Since the primary stresses are the same (in case F1 the two joint sets have identical mechanical properties and different ones in case F2), the two diagrams are not very different.

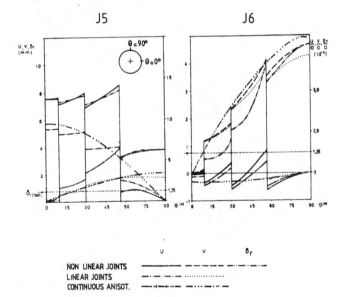

J5   J6

Figure 6. Tunnel wall displacements.

Displacements for linear and non linear joint behaviour have quite a similar trend, the differences between corresponding values increasing with the amount of yielding for each joint and each case.

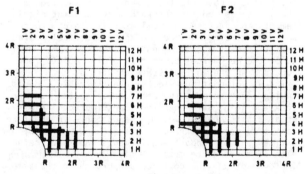

Figure 8. Yielding lengths of joints

Finally, the tunnel wall displacements (u-horizontal, v-vertical, δ_r-radial) relative to a quarter of the circunference — $\theta=0°$ to $\theta=90°$ (crown) — are shown in fig. 9, for elastic and non linear behaviour of discontinuities and for the equivalent elastic anisotropic media wich took into account joint and rock deformability.

For the discontinuous approaches, the diagrams are

847

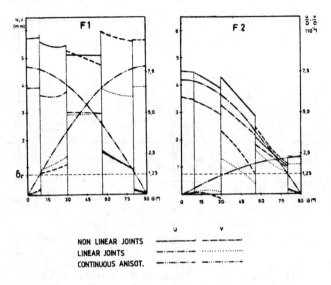

NON LINEAR JOINTS

LINEAR JOINTS

CONTINUOUS ANISOT.

Figure 9. Displacements of tunnel wall

discontinuous at the wall - fracture intersections. The difference bethween the linear and non-linear joint behaviour, due to the amount of yielding occurred (whose extension was presented in fig. 8) is clearly shown, as well as the quite reasonable approach to linear discontinuous behaviour represented, in terms of deformability, by the continuous anisotropic model. It could also be seen, by comparison with the correspondent calculations relative to a circular opening with a main joint set, that as the number of joints increases, their individual influence on the tunnel wall displacements decreases, that is, the discontinuous pattern of wall displacements is smoothed, but the overall deformability of the rock mass is increased by the additional contribution brought by joint displacements.

4 CONCLUSIONS

The presence of discontinuities brings about an increase of the overall deformation of the rock mass and, as it eliminates the continuity, is expressed by a stress pattern in the blocks that can be quite different from that existing in the continuous medium. The differences do increase as the stiffness and strength properties of the joints decrease, that is, as the greater is the weakness represented by the joints as regards the rock matrix. Joint displacements and stresses as well as their safety factor against sliding depend on the geometrical characteristics of the opening and the rock mass, as well as on the mechanical properties of the joints and rock and on the initial state of stress.

The use of continuous elastic anisotropic models, taking into account the deformation modes of the rock and joint sets, when rock behaviour can be assumed as elastic, has proved to be acceptable in the forecast of displacements next to the tunnel wall, for the estimation of support requirements and interpretation of convergence measurements. There is a growing improvement of the continuous approach in the simulation of the deformational behaviour of rock masses as their mechanical characteristics and joint spacing decrease.

5 REFERENCES

Cunha, A.P. 1980. Mathematical modelling of rock tunnels (in Portuguese). Research Officer thesis.
Lamas, L. 1986. Statistical analysis of the stability of rock faces. M.Sc. Dissertation. London, Imperial College.
Pedro, J.O. 1973. Finite element stress analysis of plates, shells and massive structures. Lisbon, CEB.
Rocha, M. 1976. Underground Structures. (in Portuguese). Lisbon, LNEC.
Sousa, L., Teles, M. 1980. A 3-D finite element model for tunnel analysis (in Portuguese).Lisbon, LNEC.

Linear arch phenomenon analysis in the roof of layered rock excavations
Analyse linéaire du phénomène de voûte dans le toit d'excavations en milieu rocheux stratifié
Analyse von Gewölbebildung in der Firste geschichteter Gesteine

O.DEL GRECO, Dipartimento di Georisorse e Territorio del Politecnico di Torino, Italy
G.P.GIANI, Dipartimento di Georisorse e Territorio del Politecnico di Torino, Italy
R.MANCINI, Dipartimento di Georisorse e Territorio del Politecnico di Torino, Italy
L.STRAGIOTTI, Dipartimento di Georisorse e Territorio del Politecnico di Torino, Italy

ABSTRACT: The paper reports a rock roof excavation stability criterium available in stratified rock and when the preexisting or induced by excavation fractures cause a pressure arch between the rock segments of the roof. The stability criterium is set up in the case of stratified rock which has variable spacing and deformability and strength caractheristics which vary layer by layer.

RESUME: Le mémoire décrit une méthode de contrôle de la stabilité du toit des vides creusés en roches strati-fiées, dans l'hypothése de la présence d'un phénomène d'arc linéaire en pression entre les morceaux de roche situés en couronne, déterminés par une fissuration originaire ou par la concentration des pressions des ter-rains. Cette méthode a été mise au point par rapport au comportement des formations stratifiées, ayant des épaisseurs différentes et même pour des roches dont les couches élementaires aient des caractéristiques méca-niques (déformabilité, résistance à la rupture) variables.

ZUSAMMENFASSUNG: Die Anmerkung erläutert ein Kriterium für die Stabilitätsprüfung der Grabenskrone in Schrstgesteine, im Falle das, Drucksbogen wegen des Vorhandenseins von vorher bestehenden oder verursachten Brüchen unter den Felselementen derselber Decke Zeigt. Das Kriterium wird für Schichtungen eingestellt, die verschiedene und Elastizitäts und widerstands eigenschaften haben, die sich von Schicht zu Schicht ändern.

1 INTRODUCTION

The case of mining and civil excavation works car-ried out in rock masses of a sedimentary type in which regular layering is present, is not rare.

The stability conditions in such excavations are a function of various parameters: the depth of the excavations themselves, their geometry, the thickness of the rock layers, and the strength and deformabi-lity characteristics of the rocks and of the discon-tinuities naturally present in them or induced by the excavation procedure.

The methods for checking the stability of this kind of structure relates to the theory of the ela-stic beams as long as the layers maintain a natural continuity. Instead, when breaking arises, the stability of the roof of the excavation can still exist thanks to the rise of the well known "linear arch" or "arching" effect. In this case the checking has an analytical solution different from that adopted for the elastic field.

2 CONDITIONS FOR THE EXISTENCE OF THE LINEAR ARCH OR ARCHING EFFECT

An analytical solution which would allow a stability analysis of the linear arch is here examined and developed as a first approximation criterium. It can be considered valid within certain limited con-ditions relative to the geometry of the structure and to the geomechanical characteristics of the rocks.

Such conditions, refering to a single layer, are here given.

A) The span of the excavation (L) must be such that the elastic equilibrium conditions are exceeded (when tensile fracturation occurs):

$$L > \sqrt{\frac{2 \cdot T_o \cdot t^2}{\gamma}} \tag{1}$$

where:

T_o = rock tensile strength
t = layer tickness
γ = unit weight of the rock

B) The span of the excavation must not be wider than the linear arch bearing capacity, imposed by the rock compressive strength (C_o):

$$L < \sqrt{\frac{1.5 \cdot C_o \cdot t^2}{\gamma}} \tag{2}$$

C) The sag of the linear arch at midpoint (f) must be small in comparison . to the width of the beam (t) because as the f/t ratio grows, the equilibrium tends to become unstable.

The relationship between f and the load remains approximately linear until f does not exceed the value of $0.2 \cdot t$; in this case the first approxima-tion expression is valid:

$$f \cong \frac{\gamma \cdot l^4}{8 \cdot E \cdot t^2} \tag{3}$$

from which one can obtain the condition

$$L < \sqrt[4]{\frac{1.6 \cdot E \cdot t^3}{\gamma}} \tag{4}$$

where: E = rock Young modulus.

D) The slices of rocks which make up the linear arch must not slide vertically because of insufficient lateral contrast, this is checked by:

$$L > \frac{2 \cdot t^2}{tg\,\varphi} \qquad (5)$$

where: φ = friction angle of the rock fracture

3 EQUILIBRIUM CONDITIONS ON THE ROOF OF AN EXCAVATION

When the roof of an excavation is made up of a series of layers for which the free span exceeds the limit conditions 1) one must expect successive breaks to happen in the layers, and the switch to the equilibrium condition of the linear arch as defined by 2). In each layer the forseen sequence of events runs as follows:

- break at the extremes, where the bending moment is maximum ($\gamma \cdot t \cdot L^2/12$);

- switching from the condition where the beam is mortised at the extrems to the condition where the beam is hinged at the extrems, with the maximum bending moment in the middle ($\gamma \cdot t \cdot L^2/8$);

- break in the midpoint, and switch to the "linear arch" equilibrium.

The same succession of events would then be repeated layer by layer in the layers above which are deprived of the support.

If all the median breaks were to occur exactly in the midpoint of the span, all the broken layers would assume the same maximum sag, and the phenomenus of the successive breaking would spread to the whole packet of layers; in this hypothetical case the stability conditions would net be different from those seen for the single layer.

However, this hypothesis is not realistic, it is in fact not very likely that the breaks at the extremes would occur simultaneously and completely. Actually, the break in one of the joints could be c completed before the breaking phenomenus begins in the other and in such a case the position of the maximum central bending momentum moves from the midpoint by 1/8 of the span (right or left wise). If the break in the second joint begins before the break in the other is completed, the intermediate break would occur around the middle within a range of ± 1/8 of the span, from the midpoint.

The progressive breaks sequence of the layers and the variability of the breaking point always reduces the maximum sag allowed to the single layers, layer by layer, as the break moves upward from the roof layer of the excavation.

The graphic scheme of fig.1 represents what has already been stated.

If we indicate the sag of the lower layer with f_a and those of the layers above with f_1, f_2, f_3,..., then f_1/f_a, f_2/f_a ... f_n/f_a are given by the general formula:

$$\frac{f_n}{f_a} = \frac{0.5 + a_2}{0.5} \cdot \frac{0.5 + a_3}{0.5 + a_2} \cdot \ldots \cdot \frac{0.5 + a_n}{0.5 + a_{n-1}} \qquad (6)$$

where the factors a represent the distance of the median fracture from the midpoint (expressed as a

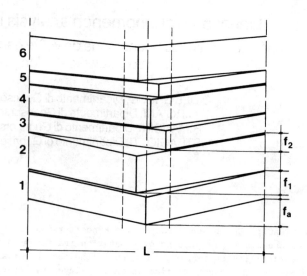

Fig. 1

fraction of the span of the beam, considered to be unitary) and are taken with a positive sign when $a_{n-1} > a_n$ and viceversa with a negative sign.

For an approximate check of the more likely values of the f_n/f_a ratio as a function of n, the breaks are considered to be equiprobable in an interval of ± 1/10 of the span with respects to the midpoint.

The statistical distribution of the values of a sample of 1000 groups of random a values was analyzed using a computer program. The procedure was first applied to a model made up of five layers of equal thickness and then repeated increasing the number of layers until reaching a maximum of 50, so obtaining the values continued in tab. 1.

Tab. 1

Number of layers	Mean value of f_n/f_a	Standard deviation
5	0,589	0,119
10	0,310	0,093
15	0,166	0,065
20	0,085	0,038
25	0,045	0,024
30	0,024	0,015
35	0,0123	0,008
40	0,0065	0,0044
45	0,0034	0,0025
50	0,0018	0,0015

From the tabulated data the following empiric expression, which correlates the most probable value of f_n/f_a to the number of broken layers, was been obtained:

$$\frac{f_n}{f_a} = 1.1319 \cdot e^{-0,129 \cdot n} \qquad (7)$$

4 ROOF STABILITY ANALYSIS

Assuming, therefore that the sag (f_a) which correspond to the maximum bearing capacity developed by the arch effect is reached at the lower layer, the layers above can develop only a progressively smaller fraction of their bearing capacity which would develop if they were singly considered.

Now, disregarding the sliding stability analysis, the stability checking procedure of the packet of layers entails the following steps:

Calculation of the maximum sag allowed for the lower layer. One obtains from 2) and 3)

$$f_a = 0.195 \frac{C_o \cdot L^2}{t \cdot E} \qquad (8)$$

on the condition that $f_a < 0.2 \cdot t$ in order to avoid instability caused by excessive deformation (the value of f_a to be assumed is the smaller of the two).

Calculation of the number of broken layers. The progressive breaking of the layers stops when the sag elastic allowes to the nth layer is less or equal to the maximum elastic deflection relative to the upper layer of the packet, considered as mortised at both extremes. This is expressed as:

$$f_e = 0.0624 \cdot \frac{T_o \cdot L^2}{t \cdot E} \qquad (9)$$

The number of broken layers $(n-1)$ is therefore obtained from (7), written in the form:

$$\frac{f_e}{f_a} = \frac{0.0624 \dfrac{T_{on} \cdot L'^2}{t_n \cdot E_n}}{0.195 \dfrac{C_{o1} \cdot L^2}{t_1 \cdot E_1}} = 1.1319 \cdot e^{-0.129 \cdot n} \qquad (10)$$

Calculation of the cumulated load bearing capacity (C_t) of the packet of broken layers. Is given from the summation of the contributions of every layer of the packet (obtainable from 3):

$$C_t = \sum_{i=1}^{n-1} \frac{8 \cdot E_i \cdot t_i^3}{L^3} f_i = \sum_{i=1}^{n-1} \frac{8 \cdot E_i \cdot t_i^3}{L^3} f_a \cdot$$
$$\cdot (1.1319 \cdot e^{-0.129 \cdot i}) =$$
$$= \frac{1.77 \cdot C_{o1}}{L^3 \cdot t_1 \cdot E_1} \cdot \sum_{i=1}^{n-1} E_i \cdot t_i^3 \cdot e^{-0.129 \cdot i} \qquad (11)$$

If for one of the layers $f_i > 0.2\, t_i$ or $f_i > 0.195.$ $\frac{C_{oi} \cdot L^2}{t_i \cdot E_i}$, the corresponding contribution to C_t is assumed to be equal to zero.

Calculation of the complessive weight (P_t) of the packet of broken layers. Obtained by the summation of the weights of the layers:

$$P_t = \sum_{i=1}^{n-1} \gamma_i \cdot t_i \cdot L \qquad (12)$$

Calculation of the upthrust (N) for the support of the elastically bent layers. Let us recall that, due to the imposed condition (1), the last layer (that is the first unbroken layer) can support only part of its own weight; if not supported, it would disconnect from those above and it would break according to the previously mentioned mechanism. The layer immediately above would find itself in the same conditions and the breaking phenomenus would continue to propagate. The support offered by the packet of broken layers should at least hold up the breaking of the layer under the action of its own weight.

This condition is not however sufficient: the excess of the bearing capacity should in fact be large enough to also hold up the sliding between layers, creating a "monolith" thick enough to be self bearing. The thickness (t') of this is calculated as:

$$t' = \frac{\gamma \cdot L^2}{2\, T_o}$$

and the maximum shear force that would occur for action of its own weight is equal to

$$\tau = \frac{\gamma \cdot L}{2}$$

Assuming that the sliding between the layers is held up only by the friction (φ = angle of the friction on the joints of the layering) the excess of the load bearing capacity necessary to ensure the stability should therefore be at least:

$$N = \frac{\gamma \cdot b \cdot L^2}{2\, tg\varphi} \qquad (13)$$

Still disregarding the risk of downward sliding of the broken strata, the stability of the structure is assured by the condition:

$$C_t > P_t + N$$

and a safety coefficient can be expressed by the ratio

$$\frac{C_t}{P_t + N}$$

5 SLIDING STABILITY ANALYSIS

The phenomenon of the vertical sliding of the packet broken layers can occurs when the value of $P_t + N$ is greater than the friction resistence produced by the lateral thrust of the broken layers.

The lateral thrust (F_i) of each broken layer is given by

$$F_i = \frac{f_i \cdot t_i^2 \cdot E_i}{0.26 \cdot L^2} \qquad (14)$$

The total force which the broken layers packet opposes to the sliding is therefore:

$$T = tg\varphi \sum_{i=1}^{n-1} \frac{t_1^2 \cdot E_i}{0.26 \cdot L^2} \cdot f_a (1.1319 \cdot e^{-0.129 \cdot i}) =$$
$$= \frac{0.85 \cdot tg\varphi \cdot C_{o1}}{t_i \cdot E_i} \cdot \sum_{i=1}^{n-1} E_i \cdot t_i^2 \cdot e^{-0.129 \cdot i} \qquad (15)$$

Therefore the stability condition against sliding is given by:

$$T > P_t + N \qquad (16)$$

6. APPLICATION EXAMPLE

An application example of the proposed procedure to the excavation roof stability problem is here described.

The purpose of this application example is the roof stability analysis of different widths of the excavation and the enduration of the improvement of the statical conditions provided by rock bolts of different length.

The rock bolt action is considered in the stability analysis only for the consolidation effects of several layers and not for the pretension which can be applied to them.

The case examined is referred to a stratified rock formation consisting of alternations of sandstones and mudstones.

The layers have a sub-horizontal or slightly inclined dip; both the sandstone and the mudstone layers have a variable thickness.

The formation statigraphy was obtained by means of several drillings which alloved to know the layer depositional charactheristics.

In the investigated area the sandstone forms about 70% of the formation; the formation is composed of series of sandstone layers, with a cumulated thickness in the 5 to 10 m range, alternating with thinner mudstone layers.

In the examined problem, geometrical correlation between the different stratigraphy, obtained from different drillings, are not found and consequently the variability of the results of the laboratory determinations of deformability and strength refer only to the single litotype and not to the spatial position of the specimen.

The aleatoriety of the geometrical and physico-mechanical parameters of the problem is taken into account by means of a probabilistic analysis. A Monte Carlo simulation is carried out for this purpose utilizing cumulative frequence distributions for the input data of the problem.

Fig. 2 shows the frequence distribution of the thickness of the sandstone and mudstone layers.

Figures 3 and 4 show the uniaxial compressive strength (C_Q), the tensil strength (T_Q) and the secant elastic modulus (E_s) for the two litotypes.

The stability analysis is carried out assuming a sandstone layer as the first layer in the excavated roof, and assuming the sandstone litotype to form the 70% of the formation in each geometrical excavation configuration.

Following the hypothesis above described, stratigraphyes, formed by alternation of sandstones and mudstones having deformability and strength parameters varying in the range of values experimentally determined, are generated.

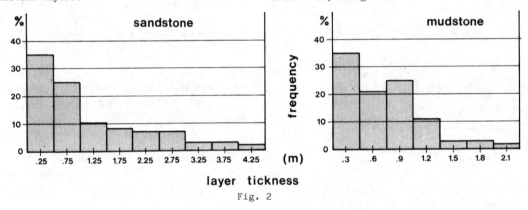

layer tickness

Fig. 2

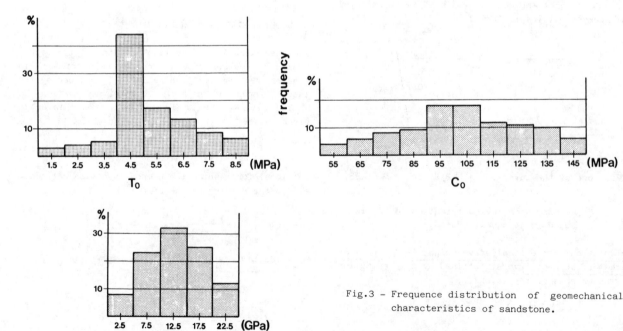

Fig.3 – Frequence distribution of geomechanical characteristics of sandstone.

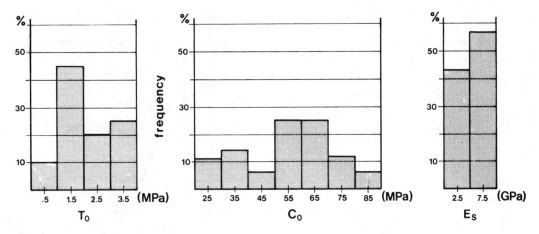

Fig.4 – Frequence distribution of geomechanical characteristics of mudstone.

The stability analyses involve the execution of the following main steps for each generated stratigraphy.

1) Computation of the equilibrium condition of the first layer in the roof of the excavation determining the corrispondent situation (collapse, linear arch field, elastic field);
2) Computation of the number of the layers subjected to the linear arch effect;
3) Computation, for the layers subjected to the linear arch condition, of the layer weight, the bearing capacity and the safety factor.

Repetition steps 1-3 is carried out for a large number of different stratigraphyes until obtain a stable frequence distribution of the safety factor.

The stability analyses are carried out for excavation spans variable between 10 to 26 m.

The same stratigraphies are examined simulating the presence of a rock bolt system. The stability analyses are carried out for lengths of the rock bolts variing between 1 to 4 m.

The computation procedure assumes the rock bolt application as a consolidation phenomenon of a layer packet having a complessive tickness equal to the length of the rock bolts. For the sake of stability the consolidated layer packet must be included between two sandstone layers.

Utilizing this criterium, a mudstone layer placed in correspondence to the rock bolt anchorage is not considered.

The assumed strength and deformability parameters obey the following criteria adopted for the sake of stability.

a) the tensile strenth and the elastic modulus values are the values of the lower bolted layer, where the greater tensions stresses are mobilized and when the rotations assume the maximum values;
b) the uniaxial compressive strength value is the average value of the bolted packet, taking into account the different thickness of the layers.

The results of the stability analyses carried out for the different excavation spans and for the situations corresponding to the absence and the presence of the rock bolt systems are reported in figures from 5 to 9.

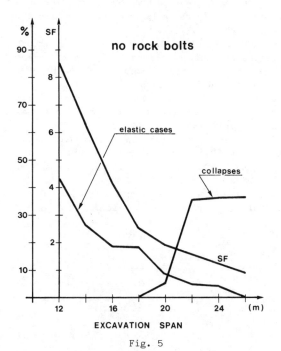

Fig. 5

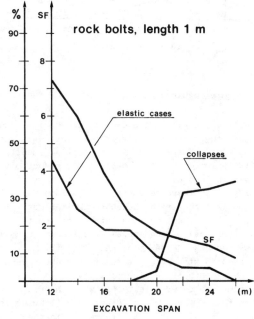

Fig. 6

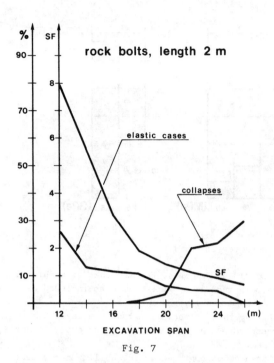

Fig. 7

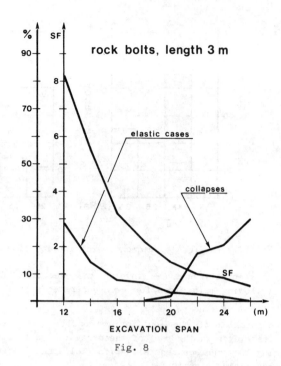

Fig. 8

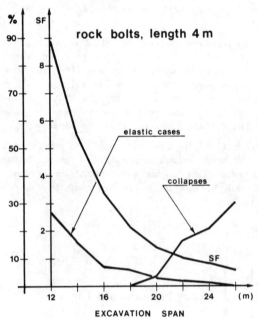

Fig. 9

Fig. 5-9 – Collapses and elastic cases frequence distribution diagrams variation with excavation span.
Average safety factor (SF) diagrams variation with excavation span in the linear arch cases.

LITERATURE

Barker, R.M.: Joint effects in bedded formation roof control. Amer. Soc. of Civil Eng. Pennsylvania State Univer., 11-14 June 1972.

Berry, P., Manfredini, G., Ribacchi, R.: The mechanical properties of the rocks of the Laga formation in the Camposto area. Intern. Symposium "The Geotechnics of structurally complex formations". Capri, 19-21 Sept. 1977.

Del Greco, O., Mancini, R., Stragiotti, L.: Studio analitico del bullonaggio nella corona di camere in formazioni stratificate. XV Convegno Naz. di Geotecnica, A.G.I., Spoleto, 4-7 May 1983.

Innaurato, N.: Un'applicazione della teoria delle travi al caso di meccanica delle rocce. Boll. Ass. Min. Subalpina, a.V, n.1-2, 1968.

Mancini, R., Del Greco, O.: Un'applicazione della teoria dell'effetto arco al calcolo delle armature sospese. Boll. Ass. Min. Subalpina, a. XVI, n. 2, 1980.

Woodruff, S.: Methods of working coal and metal mines. Vol. 1, Pergamon Press, Oxford, 1966.

7. CONCLUDING REMARKS

Rock bolting usage, as the example shows gives just a scarce long term stability improvement, even when long bolts (4 m) are used. In fact stability benefits only consist in a neglettable reduction of collapse incidence.

Bolting can increase the risk of sliding of the entire packet of strata: because, the thicker the broken beam, the smaller the lateral contrast is.

Moreover, as bolting increases the beam thickness, it also reduces the sag too; therefore the roof layer becomes stronger, but the layers above cannot fully develop their bearing capacity.

Bolting can be efficient only when the roof layer remains in the field of the elastic equilibrium (elastic beam), without cracks therefore in the case of comparatively small spans.

Tunnels autoroutiers à Liège: Caractéristiques du massif rocheux et évaluation des pressions de soutènement par diverses méthodes

Highway tunnels in Liège: Geomechanical characteristics of the rock and evaluation of support pressures by different methods
Die Autobahntunnel in Liège: Felskennwerte und Gebirgsdruckbestimmung mit Hilfe verschiedener Methoden

J.DELAPIERRE, Groupement E5/E9, Liège, Belgique
E.POHL, Groupement E5/E9, Liège, Belgique
R.FUNCKEN, S.A.Tractebel, Bruxelles, Belgique
F.LATOUR, Ministère des Travaux Publics, Liège, Belgique
J.NOMÉRANGE, Institut Géotechnique de l'Etat, Liège, Belgique
J.P.MARCHAL, Institut Géotechnique de l'Etat, Liège, Belgique

ABSTRACT: For the construction of two highway tunnels in Liège, the site exploration comprised three shafts, one tunnel and 100 boreholes. Dilatometer readings, shear characteristics and the Bieniawski classification were used for the tunnel calculations

RESUME : Cette étude concerne la construction de 2 tunnels autoroutiers destinés à relier les autoroutes E40 et E25 en passant sous la colline de Cointe, à proximité du centre la ville de Liège. Les principales caractéristiques géomécaniques ont été obtenues grâce à une campagne géotechnique comprenant 3 puits, une galerie et une centaine de forages de reconnaissance. Les principales caractéristiques utilisées pour le calcul des tunnels sont les caractéristiques dilatométriques, les caractéristiques de cisaillement et la classification de Bieniawski. Dans cet article, le calcul des tunnels a été fait par différentes méthodes utilisant chacune les caractéristiques qui leur sont propres. Les résultats obtenus sont comparés et analysés.

ZUSAMMENFASSUNG: Das Explorationsprogramm für zwei Verkehrstunnel in Liège umfasste drei Schächte, einen Tunnel und ungefähr 100 Bohrlöcher. Dilatometermessungen, Schercharakteristiken und die Bieniawski Klassifikation wurden für die Tunnelberechnungen benutzt

1. INTRODUCTION

En Belgique, et plus précisément à Liège, un tronçon d'autoroute urbaine doit être réalisé pour relier les Pays-Bas et l'Allemagne Fédérale au Sud de la Belgique, au Luxembourg et à la France. La construction de ce tronçon d'autoroute nécessite le percement de deux tunnels (1300 mètres de long chacun et 14 mètres de diamètre) à travers la colline de Cointe. Cette colline

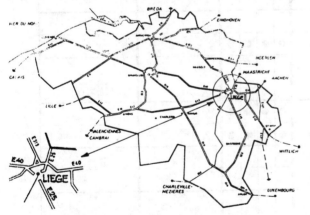

Figure 1. Situation géographique du projet.

est située à proximité du centre-ville et de la station de chemin de fer principale. Les reconnaissances géologiques ont montré que le sous-sol de la colline de Cointe appartient à l'étage westphalien datant de la fin du Primaire. Il s'agit de séries houillères comprenant des schistes, des couches de charbon et des grès. Au cours des temps géologiques, ces formations ont été plissées en une structure complexe; on trouve des plis, souvent cisaillés par des failles, dont certaines très précoces, ont été reprises dans des plissements ultérieurs. A la fracturation d'origine tectonique, se sont surimposées des fissures dues aux exploitations minières profondes. En outre, d'anciens travaux miniers peu profonds et non reconnus, exécutés dans la colline en bordure des versants, ont été mal ou non remblayés et constituent des secteurs particulièrement délicats à traverser par les tunnels. Les sollicitations qui ont servi aux calculs de stabilité du revêtement sont déterminées ci-après par différentes méthodes. Une étude statistique de la stabilité du versant de la colline au droit de l'entrée Nord des tunnels est présentée dans l'article "A probabilistic approach of slope stability in fractured rock".

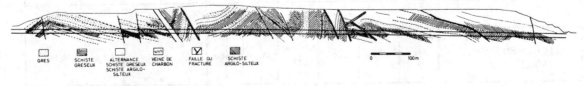

GRES · SCHISTE GRESEUX · ALTERNANCE SCHISTE GRESEUX SCHISTE ARGILO-SILTEUX · VEINE DE CHARBON · FAILLE OU FRACTURE · SCHISTE ARGILO-SILTEUX · 0 100m

Figure 2. Coupe géologique succinte de la colline de Cointe.

2. CARACTERISTIQUES GEOMETRIQUES, GEOMECANIQUES ET RHEOLOGIQUES

2.1. Caractéristiques géométriques

Couverture à la clef : 55 mètres.
Entre-axe des 2 tunnels : variant de 30 à 60 mètres.

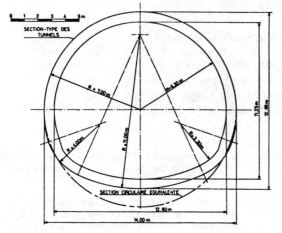

Figure 3. Coupe-type d'un tunnel et section circulaire équivalente.

2.2. Caractéristiques géomécaniques

Poids spécifique du massif rocheux :
$\gamma = 25$ KN/m³
Coefficient de Poisson : $\nu = 0,2$
Dilatance : $\Delta V/V = 0$

Des essais triaxiaux effectués sur des échantillons de roche de la région de Liège (Polo-Chiapolini 1974) ont permis de mesurer l'influence de la lithologie et de la déconsolidation du massif sur la valeur du coefficient m (1)
Pour des grès homogènes non influencés par les exploitations minières : m = 27,4
Pour des grès sur des exploitations minières : m = 19,1
Pour des schistes : m = 10,8
Nous avons retenu :
m = 10,8 pour le schiste et 19,1 pour le grès (1)
s = 1 pour le schiste et pour le grès (1)

Zone élastique	Zone plastique
S_o = 0,1 MPa ϕ = 20° E = 5250 MPa m' = 0,22 pour le schiste (1) m' = 0,36 pour le grès (1) S' = 0,0001 pour le schiste et le grès (1)	S_o = 0,06 MPa ϕ = 13° E = 2600 MPa m' = 0,057 pour le schiste (1) m' = 0,1 pour le grès (1) S' = 0 pour le schiste et le grès (1)

(1) Méthode de Hoek et Brown

2.3. Caractéristiques rhéologiques

Par des essais de fluage au dilatomètre d'une durée de 6 à 8 heures, on a approché les caractéristiques rhéologiques du massif. Le massif étant considéré comme un corps de Burgers, les valeurs obtenues sont les suivantes :
- module de cisaillement à court terme :
G_2 = 1300 MPa
- module de cisaillement à long terme :
G_1 = 2000 MPa

Tableau I

METHODE		CONVERGENCE/CONFINEMENT				STATIQUE				RHEOLOGIQUE	EXPERIMENTALE
DEVELOPPEE PAR		AMBERG-LOMBARDI	EGGER-DESCOEUDRES	GOODMAN	HOEK & BROWN	CAQUOT	TERZAGHI	EGGER	TALOBRE	GOODMAN	BIENIAWSKI
ESSAIS	E_d	●	●	●							
	E_e	●		●	●	●	●	●	●		
DILATOMETRIQUES	G_1, G_2 η_1, η_2									●	
ESSAIS DE CISAILLEMENT	ϕ	●	●	●	●	●	●	●	●		
	c	●	●	●	●	●	●	●	●		
ESSAIS TRIAXIAUX	m				●						
	s				●						
CLASSIFICATION DU MASSIF (Forages, piézo-mètres, etc...)	RMR				●						●

Tableau II

PRESSION DE SOUTENEMENT (2) (MPa)		0,21	0,22	0,20	0,34 (schiste) 0,19 (grès)	0,43	0,19	0,22	0,40		0,14 à 0,20

(2) Calculée pour δ = 5 cm dans les méthodes convergence/confinement.

856

- viscoélasticité à court terme :
  η_2 = 13.10⁴ MPa.min.
- viscoélasticité à long terme :
  η_1 = 6.10⁴ MPa.min.

3. METHODES DE CALCUL

Plusieurs méthodes sont utilisées pour
calculer la pression sollicitant le revêtement des tunnels. Le tableau I donné
ci-avant indique les paramètres pris en
compte suivant les différentes méthodes
utilisées.

3.1. Méthodes convergence/confinement

La figure 4 ci-après fournit les courbes de
convergence obtenues par les différentes
méthodes. Ces méthodes étant devenues classiques, nous ne reprenons pas ici les
formules utilisées. Notons que les courbes
obtenues par les méthodes de
Egger-Descoeudres, Amberg-Lombardi et
Goodman sont des courbes obtenues à partir
de caractéristiques moyennes tandis que les
2 courbes obtenues par la méthode de Hoek et
Brown se rapportent l'une au schiste et
l'autre au grès.

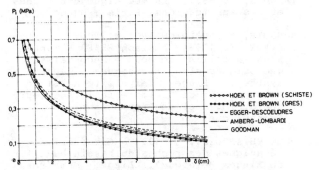

Figure 4. Courbes intrinsèques du massif rocheux
obtenues par les différentes méthodes.

La détermination des coefficients caractéristiques m' et s' utilisés dans la méthode
de Hoek et Brown est faite à l'aide de
l'abaque donné à la figure 5. Cet abaque
permet, via la classification de Bieniawski,
de passer des coefficients m et s déterminés
à partir d'essais triaxiaux exécutés sur des
échantillons de roche intacte (seuls résultats disponibles) aux coefficients m' et s'
correspondant au massif pris dans son

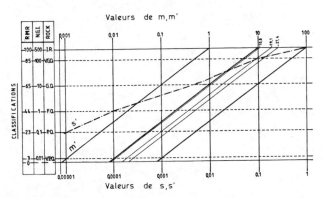

Figure 5. Détermination des coefficients m,m' et s,s'
en fonction de la classification des massifs rocheux.

ensemble et permet d'évaluer l'évolution de
ces coefficients en fonction de
l'altération du massif.

3.2. Méthodes statiques

Les charges de dislocation ont été estimées
par différentes méthodes et ont donné les
résultats suivants :
Méthode de Caquot : 0,43 MPa
Méthode de Terzaghi : 0,19 MPa
Méthode d'Egger : 0,22 MPa
Méthode de Talobre : 0,40 MPa

Les méthodes de Talobre et de Caquot sont
très similaires, elles font intervenir la
pesanteur en intégrant cet élément depuis
l'intrados jusqu'à la surface du sol ou
jusqu'à la limite du rayon plastifié; par
contre, les méthodes de Terzaghi et d'Egger
sont semi-empiriques et font intervenir un
effet-voûte favorable, ce qui explique les
valeurs plus faibles obtenues. On peut
noter l'influence très importante des
caractéristiques c et ϕ sur la charge de
dislocation (voir tableau III) et donc
l'influence d'une dégradation des caractéristiques géomécaniques lors de l'exécution
des travaux par suite de la détente.

Tableau III

Influence de c et ϕ sur la charge de dislocation en utilisant la méthode de Talobre

c MPa	ϕ (°)	P dislocation (MPa)
0,06	13	0,40
0,06	20	0,20
0,10	20	0,10

3.3. Méthode "rhéologique"

La figure 6 fournit les courbes de
convergence du massif en fonction du temps
dans l'hypothèse où un revêtement de 5 cm
ou de 10 cm de béton est placé immédiatement après réalisation de l'excavation.
Le module élastique du béton est pris égal
à 15000 MPa et le coefficient de Poisson du
béton égal à 0,15.

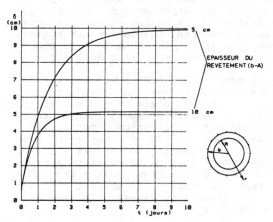

Figure 6. Convergence à la paroi du tunnel en
fonction du temps et de l'épaisseur du revêtement.

3.4. Classification suivant Bieniawski

Sur base des résultats des reconnaissances géologiques et géotechniques, la méthode de Bieniawski a été utilisée pour classifier le massif rencontré au droit des tunnels. Les résultats obtenus sont fournis au tableau IV.

TABLEAU IV

Classe	Répartition (%)	RMR calculé Moyenne	Ecart-type
I	0	-	-
II	3	66	4
III	57	51	6
IV	25	33	5
V	15	(*)	(*)

(*) non représentatif

En appliquant l'abaque de Bieniawski-Unal, on arrive à des valeurs de pression sur le soutènement variant de 0,14 à 0,20 MPa.

4. ANALYSE DES RESULTATS ET CONCLUSIONS

Le tableau II donné ci-avant fournit les pressions sollicitant le revêtement obtenues par les différentes méthodes. Pour les méthodes convergence/confinement, si on se fixe une convergence de 5 cm ou 10 cm, les valeurs des pressions sollicitant le revêtement sont les suivantes (tableau V) :

TABLEAU V

	δ = 5 cm	δ = 10 cm
Amberg-Lombardi	0,21 MPa	0,13 MPa
Egger-Descoeudres	0,22 MPa	0,14 MPa
Goodman	0,20 MPa	0,12 MPa
Hoek et Brown (grès)	0,19 MPa	0,11 MPa
Hoek et Brown (schiste)	0,34 MPa	0,25 MPa

Grâce aux différentes méthodes de calcul des convergences et aux observations faites lors du creusement de la galerie de reconnaissance , une estimation des tassements en surface causés par le creusement et la déformation correspondante du massif a pu être faite. En se basant sur des caractéristiques moyennes, les tassements maxima prévus dans le cas des tunnels autoroutiers de Liège seraient de l'ordre de 3 à 4 cm.

Dans le cadre de l'étude des tunnels sous la colline de Cointe, une campagne de reconnaissance assez complète a permis d'utiliser plusieurs méthodes de calcul faisant appel à des paramètres variés et de comparer les résultats obtenus. On ne constate heureusement pas de discordance majeure. L'instrumentation mise en place pendant la construction des ouvrages doit permettre de confirmer les estimations données ci-avant et de prendre, si nécessaire, les mesures qui s'imposent.

REFERENCES

AMBERG, A. & LOMBARDI, G. 1980. Une méthode de calcul élasto-plastique de l'état de tension et de déformation autour d'une cavité souterraine. Compte-rendus du Troisième Congrès de la Société Internationale de Mécanique des Roches à Denver. Thèmes 3-5. Volume II, tome B : 1055-1060.

BIENIAWSKI, Z.T. 1984. Rock Mechanics Design in Mining and Tunneling. Rotterdam : Balkema.

HOEK, E. & BROWN, E.T. 1980. Underground Excavations in Rock. London : The Institution of Mining and Metallurgy.

DESCOEUDRES, F. 1984. Mécanique des Roches. Lausanne : Ecole Polytechnique Fédérale de Lausanne. Département de Génie Civil. Institut des Sols, Roches et Fondations. Laboratoire de Mécanique des Roches.

GOODMAN, R.E. 1980. Introduction to Rock Mechanics. New York : John Wiley & Sons.

TALOBRE, J. 1967. La mécanique des roches et ses applications. Paris : Dunod.

MONJOIE, A., POLO-CHIAPOLINI, C., DELVAUX, T. 1986. Liaison E5/E9 - Tunnel sous la colline de Cointe-Reconnaissances géologiques et géotechniques. Rapport de synthèse. Liège : Université de Liège. Laboratoires de Géologie de l'Ingénieur, d'Hydrologie et de Prospection Géophysique.

POLO-CHIAPOLINI, C. 1974. Caractéristiques mécaniques des roches du bassin houiller de Liège (Belgique). Liège : Mémoires CERES (Nouvelle Série) n° 47 : Comité Belge de Géologie de l'Ingénieur.

Numerical analysis of coal mine chain pillar stability

Analyse numérique de la stabilité des piliers inter-chambres des mines de charbon
Eine numerische Analyse zur Standsicherheit von Zwischenpfeilern im Kohlenbergwerk

M.E.DUNCAN FAMA, CSIRO Division of Geomechanics, Mount Waverley, Victoria, Australia
L.J.WARDLE, CSIRO Division of Geomechanics, Mount Waverley, Victoria, Australia

ABSTRACT: A numerical model for assessment of stability of inter-panel pillars for longwall mining is described. The model uses realistic coal properties including post-yield strain-softening. Results are compared with Wilson's pillar design method developed for British conditions. The agreement is excellent provided the load deficiency in the caved waste is correctly estimated. But the results suggest that Wilson's method needs adjusting for environments in which pre-mining stresses are not re-established in the caved waste.

RESUME: On décrit un modèle numérique pour l'estimation de la stabilité des piliers se trouvant entre les chambres isolées dans le cas de l'exploitation par longwall. Le modèle emploie des propriétés réalistes du charbon, y compris le ramollissement dû à des contraintes se produisant après l'affaissement. On fait une comparaison des résultats avec les valeurs de la méthode de Wilson pour la construction des piliers, développée pour répondre aux exigences en Grande-Bretagne. L'accord est excellent pourvu que la manque de charge des déblais soit correctement estimée. Les résultats font penser cependant qu'il se peut que la méthode de Wilson ne soit pas valide pour les milieux où les efforts se produisant préalablement à l'exploit-ation minière ne seraient pas rétablis dans les déblais.

ZUSAMMENFASSUNG: Es wird ein numerisches Modell zur Auswertung der Standsicherheit der beim Strebbau zwischen Bauabteilungen verwendeten Pfeilen beschrieben. Das Modell verwendet wirklichkeitsnahe Kohlen eigenschaften, einschliesslich der folglich des Zusammenschubs vorkommenden Spannungsweichung. Die Ergebnisse werden mit Werten der Wilsonschen, für britische Verhältnisse entwickelten Pfeilerplanungsmethode verglichen. Die Über-einstimmung ist ausgezeichnet, vorausgesetzt, dass der Ladungsausfall in den Bruchmassen richtig geschätzt ist. Die Ergebnisse deuten darauf hin, dass die Wilsonsche Methode vielleicht für Umgebungen nicht gilt, wo vor der Abbautätigkeit sich ergebende Spannungen in den Bruchmassen nicht wiederhergestellt worden sind.

1 INTRODUCTION

The design of inter-panel or 'chain' pillars for a coal mine longwall operation requires a balance between two competing criteria. Firstly the pillars should be as narrow as possible to minimize the volume of coal lost in the pillar. Secondly the pillars must be wide enough to protect the roadway from the influence of high abutment stresses. Excessive rib spall and roof stability problems may be encountered in the roadway if the pillars are too narrow.

The aim of this paper is to lay the groundwork for a rational method for assessment of chain pillar stability for longwall operations in a wide variety of conditions. As it is generally recognized that the post-yield behavior of coal has a major influence on pillar stability, our numerical model incorporates strain-softening coal properties. The key features of the model are illustrated using stress distributions predicted from two-dimensional (plane strain) finite element analyses of a number of sample problems.

A comprehensive evaluation of existing pillar design methods is beyond the scope of the present paper. The underlying assumptions and limitations of two commonly used approaches are briefly examined below.

Salamon and Munro (1967) analyzed statistically pillar failures in room-and-pillar mines in South Africa and derived the following formula:

$$\sigma_p = \sigma_1 \, w^{0.46} \, h^{-0.66}$$

where σ_p is the pillar strength;
σ_1 is a constant;
and w and h are the width and height of the pillar.

The formula is generally acknowledged to have two major shortcomings: (i) coal strength variations for different coal seams are not allowed for, and (ii) the strength of squat pillars is underestimated because they were not represented in the statistical analysis (the data for collapsed pillars is for aspect ratio (w/h) in the range 0.9-3.6).

Wagner and Madden (1984) described an extension of the formula to take account of squat pillars and suggested parameter values for the extended formula on the basis of laboratory test data. However, the formula cannot be used confidently until field trials provide realistic parameter values.

Wilson (1983) developed a method for design of longwall inter-panel pillars that primarily is based on considerations of gateroad stability. He used an assumed stress distribution alongside a single panel to determine the distance that a roadway should be placed from the panel ribside to avoid undue damage during panel extraction. The resulting pillar would have a high factor of safety against ultimate failure.

The approximate distribution of stress adjacent to the excavated panel is deduced by considering the 'stress balance' over a section through the panel. The method relies on certain assumptions about the stress distributions in three zones: (i) a yield zone in the coal, (ii) an elastic zone in the coal and (iii) the caved waste or 'goaf'.

The coal is assumed to obey a linear Mohr-Coulomb failure criterion:

$$\sigma_1 = \sigma_0 + k\sigma_3$$

where σ_1 is the maximum principal stress at failure;
σ_3 is the minimum principal stress (confining pressure);
σ_0 is the unconfined compressive strength;
and k is the triaxial stress coefficient (given by $(1+\sin\phi)/(1-\sin\phi)$, where ϕ is the angle of internal friction.)

The unconfined compressive strength in the yield zone is assumed to drop to a substantially smaller value σ_0', typically about 0.2 MPa.

Wilson considers two cases: (i) yield in seam only
and (ii) yield in roof, seam and floor. Case (i) is
more relevant to Australian conditions, and will be
discussed here. Case (ii) will be the subject of a
future publication.

Analytical solutions for simple geometries are used
to approximate the yield zone stresses and an
exponential decay is assumed for the stresses away
from the yield zone.

Using empirical data for United Kingdom conditions,
Wilson assumes that the vertical stress in the goaf
returns to cover load at a distance of 0.3 H from the
rib-side if the face length is greater than 0.6 H
(H is the cover depth).

The stress distribution in the yielded and elastic
coal is calculated using the above assumptions (see
Fig. 6 for a typical example).

A pillar design method ideally should use all
available geomechanical data: elastic moduli, triaxial
strength data, pre-mining stress measurements,
geological data and so on. No general formula can
specifically take into account all of these important
parameters. Although simple design methods may
provide a reasonable first approximation, unusual,
different or new regimes should be studied using more
rigorous methods such as numerical stress analysis.
This issue is usually faced when longwalling is first
introduced to a given country or region, where
external experience is not directly translatable.

2 NUMERICAL MODEL

2.1 Yield model

A key feature of our numerical model is the use of
post-yield strain-softening properties.

The yielded material can be described as a linear
elastic inhomogeneous material, as its hypothetical
modulus of elasticity varies from point to point in
the yield zone. This modulus variation is determined
iteratively so that the final stresses in the yield
zone satisfy a Mohr-Coulomb post failure criterion
(Figure 1). The iterative technique is based on that
developed by Kripakov(1981). The validity of the
material description and the convergence of the
iterative technique has been demonstrated for a
circular opening in a hydrostatic stress-field by
comparison with an analytical solution (Duncan Fama
1984).

The post-failure criterion is taken to be

$$\sigma_1 = k\sigma_3 + \sigma_0^*$$

where $\qquad \sigma_0^* = \chi\sigma_3 + \sigma_0' \quad$ provided $\sigma_0^* \leq \sigma_0$

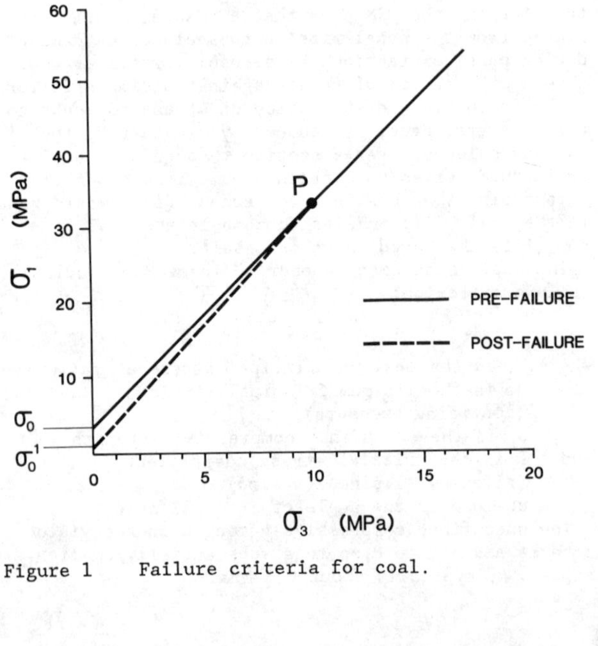

Here χ is a constant. This defines a smooth
variation in σ_0^* across the yield zone, with $\sigma_0^* = \sigma_0'$
at the boundary of the (unsupported) opening. This
variation is similar to that of Brown et al. (1983).
Their post-failure (Hoek-Brown) parameters were taken
proportional to strains in the strain-softening
region.

Wilson's post-failure criterion is given by $\chi = 0$,
and leads to a discontinuity in the major principal
stress across the yield-zone/elastic interface. This
causes numerical difficulties in the iterative
procedure. These difficulties are overcome by a
judicious choice of χ.

Thus, for $\chi \neq 0$ the post-failure criterion has the
form

$$\sigma_1 = k'\sigma_3 + \sigma_0'$$

with $\qquad\qquad k' = k + \chi$

and this line intersects the pre-failure criterion
at a point P, determined by χ (see Fig. 1).

If χ is large, $\sigma_0^* = \sigma_0$ everywhere except in elements
at the coal ribs and the post failure has very little
effect.

If χ is small, $\sigma_0^* = \sigma_0'$ well into the ribs and then
rises to its final value σ_0 (see Fig. 11). The post-
failure criterion can then have a major effect both on
the stresses in the yield-zone and the depth of yield
(see Fig. 4 and 5).

In general χ should be chosen so that the point P is
representative of the stresses at the yield-elastic
interface.

2.2 Model parameters

The following parameters are used for the model
unless otherwise indicated:
Young's moduli: $E_{coal} = 1$ GPa; $E_{rock} = 5$ GPa
Poisson's ratios: $\nu_{coal} = \nu_{rock} = 0.3$
Coal strength parameters: $\sigma_0 = 3.6$ MPa; $k = 3.0$
(friction angle, $\phi = 30°$, cohesion, $c = 1.04$ MPa); σ_0'
$= 0.2$ MPa; $k' = 3.3$; tensile strength, $T_0 = 0.05$ MPa.
Coal seam thickness $= 2.6$ m.
The surrounding rock is assumed to be linear
elastic.
Cover depth $= 150$ m, hydrostatic pre-mining
stresses $= 3.6$ MPa.

Figure 2 is a schematic of the model geometry. The
pillar loading is simulated by following the standard
excavation loading procedure (unload opening and
superimpose pre-mining stresses on the results).

The goaf material is not explicitly included in the
model. Instead, a range of representative goaf
loading conditions is simulated by unloaded openings
of various widths. Results are presented for the
following opening widths: 12 m, 48 m and 100 m. The
loading of the pillar and panel flank is a function of
the 'goaf load deficiency', i.e. the area between the
goaf stress and the pre-mining stress on the stress-
distance diagram (Figure 3) and should be essentially
independent of the detailed stress distribution in the
goaf. The unloaded 100 m opening has the same goaf
load deficiency as a 200 m wide panel with an average
goaf load that is half the cover load. The 48 m
opening approximates Wilson's goaf load deficiency,
which corresponds to a 45 m unloaded opening.

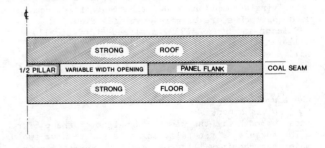

Figure 2   Model geometry.

Figure 1   Failure criteria for coal.

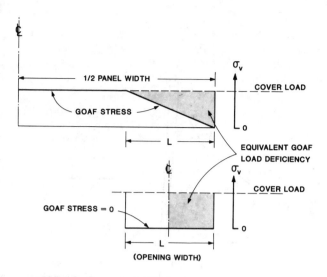

Figure 3   Goaf loadings.
            (a) Wilson's goaf load assumption
                 (L=0.3 cover depth for British
                 conditions)
            (b) Unloaded opening with equivalent
                 goaf load deficiency

Wilson's design method gives a pillar width of
22.7 m. Finite element results are presented for two
pillar widths: 15 m and 25 m. The 25 m pillar
approximates Wilson's pillar design.

2.3   Results

Results from our numerical model are presented in
the form of vertical (σ_v) and horizontal (σ_h) stress
distributions on the coal seam mid-plane. Results are
given for the panel flank and the pillar. The
stresses in the panel flank are essentially
independent of whether the pillar width is 15 m or
25 m.

Figures 4 and 5 highlight the essential differences
between the linear elastic and non-linear finite
element solutions. Figure 4 compares the linear
elastic and non-linear solutions in a 25 m pillar
(100 m wide opening). Figure 5 gives the
corresponding solutions in the panel flank. Allowing
the coal to yield makes the peak abutment stress move
deeper into the coal. Realistic post-failure
properties move it as far again.

The remaining results are for the yield model.

Figure 6 shows the growth in the yield zone for a
25 m pillar as the pillar stress increases due to
widening the opening. Figure 7 compares the flank
stresses for the different opening widths. Wilson's
solutions are also included for comparison. Wilson's
solution agrees well with the finite element solution
for the opening width (48 m) that corresponds to U.K.
goaf load conditions. However, for the 100 m opening
the finite element solution is rather different from
Wilson's solution.

Figures 8 and 9 compare solutions for the 15 m and
25 m pillars for the 48 m opening and 100 m opening
respectively.

Figure 10 illustrates the sensitivity of results to
changes in two of the key parameters, the pre-mining
horizontal stresses and the coal friction angle.
Keeping the cohesion constant, dropping the friction
angle ϕ from 30° to 25.4° (i.e triaxial strength
coefficient k reduced from 3.0 to 2.5) and dropping k'
from 3.3 to 2.8 has a substantial impact on the growth
of the yield zone. Reducing the pre-mining horizontal
to vertical stress ratio from 1.0 to 0.7 has a similar
effect.

Figure 11 shows the variation in unconfined
compressive strength across the yield zone for a
typical problem.

3   DISCUSSION AND CONCLUSIONS

The numerical model allows most of the critical
geomechanical parameters to be taken into account.

Although Wilson's method is not based on rigorous
application of continuum mechanics theory, it is clear
from comparisons with the finite element calculations
that his hypotheses about stress distributions are
basically sound. When the finite element loading is
matched to his goaf load deficiency the resulting
stress distributions are virtually identical - a
remarkable result. However, there are two key factors
in his stress balances. The first is the assumed load
deficiency of the goaf, described above. The second
is the assumption that the horizontal stress in the
yield zone does not exceed the pre-mining value. This
determines the yield zone depth independently of the
load deficiency in the goaf.

The finite element solutions show that the yield
zone horizontal stress can rise above its pre-mining
value. A larger load deficiency than that assumed by
Wilson can lead to a substantially larger yield zone
with much higher abutment stresses than Wilson's
design method would allow. Although Wilson's
hypotheses give a good representation of U.K.
conditions, they need to be adjusted for environments
where the goaf loads are significantly less than cover

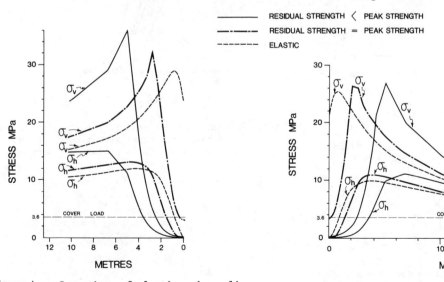

Figure 4   Comparison of elastic and non-linear
            solutions in 25 m pillar (100 m opening).

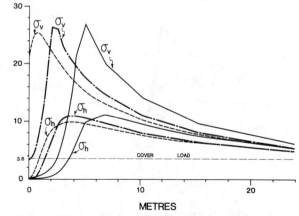

Figure 5   Comparison of elastic and non-linear
            solutions in panel flank (100 m opening).

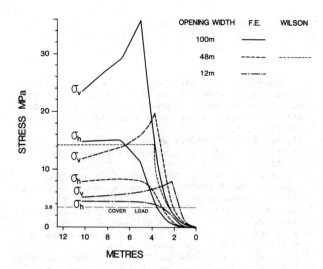

Figure 6   Stress distributions in 25 m pillar for various opening widths.

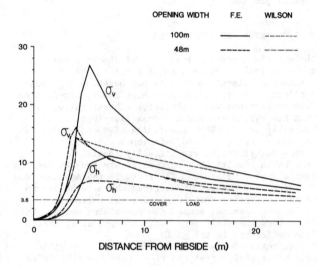

Figure 7   Stress distribution in panel flank for different opening widths.

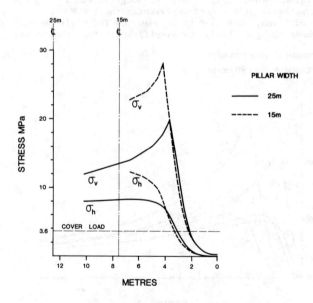

Figure 8   Comparison of solutions for 15 m and 25 m pillars for 48 m opening.

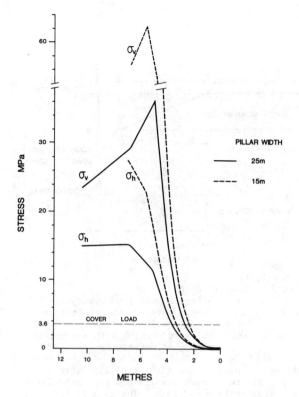

Figure 9   Comparison of solutions for 15 m and 25 m pillars for 100 m opening.

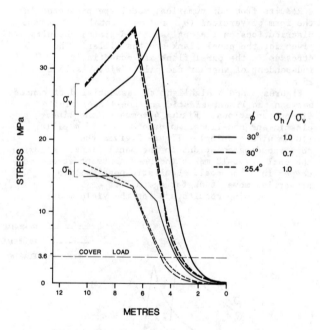

Figure 10   Influence of coal friction angle and pre-mining horizontal stress on yield zone development (25 m pillar, 100 m opening).

862

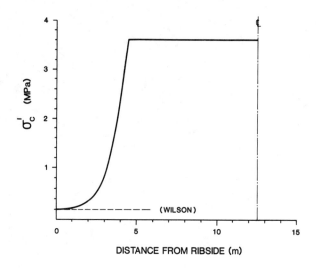

Figure 11 Variation of unconfined compressive strength across yield zone (25 m pillar, 100 m opening).

shallow South African collieries: an evaluation of the performance of the design procedures and recent improvments. Design and Performance of Underground Excavations (ISRM Symposium, Cambridge, U.K.), pp. 391-399.

Wardle, L.J. and McNabb, K.E. (1985). Comparison between predicted and measured stresses in an underground coal mine. Proceedings 26th U.S. Rock Mechanics Symp., Rapid City, pp. 531-538.

Wilson, A.H. (1983). The stability of underground workings in the soft rocks of the Coal Measures, Int. J. Mining Engineering, Vol. 1, pp. 91-187.

load. Back-analysis of stress monitoring results (Wardle and McNabb, 1985) from the Laleham No.1 Colliery in Australia's Bowen Basin suggests that the goaf load deficiency could be about twice that assumed by Wilson.

The high abutment stresses can only be mobilized if the pillar-rock interface can sustain high shear stresses. Finite element modelling has shown that structural discontinuities such as claybands could prevent such stress levels (McNabb and Wardle, 1986).

A change in the ratio of horizontal to vertical pre-mining stress ratio has also been shown to have as marked an effect on yield zone characteristics as a change in friction angle (see Fig. 10). Wilson's design method caters for the latter effect, but not the former. This is an important consideration in design for Australian mining conditions.

Wilson's design philosophy can be extended to cater for a much wider variety of conditions by replacing his assumed yield zone stress distributions by those derived from the strain softening finite element analysis.

ACKNOWLEDGEMENTS

Ken McNabb was responsible for preliminary finite element analyses of chain pillar stability.

REFERENCES

Brown, E.T., Bray, J.W., Ladanyi, B. and Hoek, E. (1983). Ground response curves for rock tunnels. J. Geotech. Engng. ASCE, Vol. 109, pp. 15-39.

Duncan Fama, M. E. (1984). A new constitutive equation for a Coulomb material. Proc. I.S.R.M. Symp. Design and Performance of Underground Excavations, Cambridge, U.K., pp. 139-147.

Kripakov, N.P. (1981). Finite element analysis of yield-pillar stability. Computers and Structures, Vol. 13, pp. 575-593.

McNabb, K.E. and Wardle, L.J. (1986). Numerical modelling of development roadways, German Creek Central Colliery, Central Queensland, CSIRO Aust., Div. Geomech., Geomechanics of Coal Mining Report No. 63.

Salamon, M.D.G. and Munro, A.H. (1967). A study of the strength of coal pillars. J.S. Afr. Inst. Min. Metall., Vol. 68, pp. 55-67.

Wagner, H. and Madden, B.J. (1984). Fifteen years' experience with the design of coal pillars in

Räumliche Modellversuche für geankerte seichtliegende Tunnel
3-D model tests for shallow rock-bolted tunnels
Essais sur modèle à trois dimensions de tunnels ancrés à faible profondeur

P.EGGER, Laboratoire de Mécanique des Roches, Ecole Polytechnique Fédérale de Lausanne, Schweiz

ABSTRACT : A series of 3D geomechanical tests were performed at a scale of 1:20 in order to simulate the execution of shallow tunnels in poor ground. Both passive and prestressed bolts were used for ground support, and the superiority of the latter was observed. Surface settlements and tunnel stability were compared with analytical and numerical forecasts as well with monitoring results of two real tunnels.

RESUME : Dans une série d'essais géomécaniques tridimensionnels, le creusement de tunnels à faible couverture, dans des terrains médiocres, a été simulé à l'échelle 1:20e. Le support a consisté en des ancrages passifs ou précontraints, ces derniers s'avérant supérieurs. Tassements en surface et stabilité du tunnel ont été comparés aux prévisions fournies par des méthodes analytiques et numériques, de même qu'aux résultats d'auscultations de deux tunnels réels.

ZUSAMMENFASSUNG : Es wurde eine Reihe räumlicher geomechanischer Modellversuche im Massstab 1:20 durchgeführt, die das Auffahren seichtliegender Tunnel in Gebirge geringer Festigkeit simulierten. Die Tunnelsicherung erfolgte mittels schlaffer oder vorgespannter Anker, wobei sich letztere als überlegen erwiesen. Standsicherheit des Tunnels und Oberflächensetzungen wurden mit den Prognosen aus analytischen und numerischen Berechnungen sowie mit den Ergebnissen von Messungen an zwei ausgeführten Tunneln verglichen.

1. VORBEMERKUNG

Im städtischen Verkehrswegebau stellen seichtliegende Tunnel in vielen Fällen die technisch beste, und zudem eine umweltfreundliche Lösung dar. Infolge der geringen Tiefenlage ist das mittlere Spannungsniveau des Gebirges gewöhnlich bescheiden, andererseits zeichnen sich die oberflächennahen Schichten meist durch geringe Festigkeit und hohe Verformbarkeit aus. Deshalb verursacht im Normalfall die Dimensionierung des endgültigen Ausbaus weniger Schwierigkeiten als die zweckmässigste Sicherung des Vortriebs und die Begrenzung der Oberflächensetzungen. Der Art der Sicherung sowie dem Abstand zwischen Vortriebsbrust und Sicherungseinbau kommen in diesem Zusammenhang erfahrungsgemäss besondere Bedeutung zu.

2. ZIEL UND ART DER UNTERSUCHUNGEN

Im Rahmen eines an der Ecole Polytechnique Fédérale de Lausanne durchgeführten Forschungsprogramms (Egger, 1985) sollten in einer ersten Phase - worüber hier berichtet wird - vor allem die Einflüsse einer Hohlraumsicherung mit schlaffen bzw. vorgespannten Ankern, sowie der Ueberlagerungshöhe des Tunnels untersucht werden. Gebirge und Tunneldurchmesser wurden nicht variiert. Um die Stützwirkung der Tunnelbrust und die Verhältnisse im Nahfeld der Anker möglichst naturgetreu zu simulieren, wurde das Hauptgewicht des Forschungsvorhabens auf räumliche geomechanische Modellversuche gelegt. Diese wurden durch begleitende ebene Base-Friction-Untersuchungen sowie durch ebene und räumliche numerische Berechnungen ergänzt.

3. VERSUCHSBEHAELTER

Der Versuchsbehälter wurde derart ausgelegt, dass kreiszylindrische Tunnel mit einem Durchmesser von 10 Metern in einem Massstab von 1:20 aufgefahren werden konnten. Der gewählte Modellmassstab stellt einen Kompromiss zwischen technologischen Zwängen und der gewünschten Beschränkung der Modellgrösse dar und führte zu Behälterabmessungen von 3.0 m Breite × 2.0 m Länge × 3.0 m Höhe. Der Behälter besteht aus einer Sohlplatte und drei festen Wänden aus Stahlbeton, sowie einer demontierbaren Stirnwand aus Stahlprofilträgern, welche auch das 50 cm - - Durchmesser - Ansatzstück für die Tunnelauffahrung trägt (Abb. 1). Eine bewegliche Arbeitsbühne ermöglicht den Ein- und Ausbau des Modellmaterials ohne Betreten der Oberfläche. In die Wände des Behälters sind an verschiedenen Stellen Glötzl-Druckmessgeber eingelassen.

Abbildung 1. Modellbehälter mit 18 m^3 Fassungsvermögen. Im Vordergrund Behälter für Vorversuche

4. MODELLMATERIAL

In den Versuchen sollte bindiges Gebirge geringer Festigkeit, z.B. verwitterte tertiäre Mergel oder tonige Moränen - wie häufig in Mitteleuropa angetroffen -, durch ein äquivalentes Modellmaterial dargestellt werden. Bei einem annähernd gleichen Raumgewicht wie das natürliche Gebirge musste das Modellmaterial eine um den geometrischen Massstab 20-fach verringerte Festigkeit aufweisen. Eine Reihe von Vorversuchen führte zur Wahl einer inerten, d.h. wiederverwendbaren Mischung aus Baryt, Zinkoxyd und Paraffinöl (John, 1969), deren Festigkeit je nach Intensität der Verdichtung in weiten Grenzen variiert werden kann.

Das Material wurde in etwa 6 cm dicken Schichten in den Behälter eingebaut und lagenweise mit einer Rüttelplatte verdichtet. Am Ende jedes Versuches wurden in verschiedenen Höhen ungestörte Proben zur Bestimmung der geomechanischen Kennwerte entnommen; deren Mittelwerte sind in der Tabelle 1 zusammengestellt.

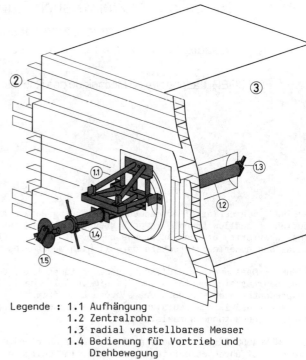

Kennwerte	Versuch Nr.					
	1	2	3	4	5	6
z [m]	0.85	0.85	0.85	0.85	0.65	1.05
Sicherung	ohne	schl.A.	Tüb.	vorg.A.	vorg.A.	vorg.A.
γ [kN/m^3]	17.0	17.4	18.1	19.5	19.5	19.7
q_c [kN/m^2]	5.0	6.0	7.0	12.0	12.0	15.0
E [kN/m^2]	2200	2400	2200	1500	4000	6600
E/q_c	440	400	314	125	333	440
c [kN/m^2]	1.5	1.6	1.6	3.0	3.0	4.0
ϕ [°]	35	35	35	35	35	35

Legende : 1.1 Aufhängung
1.2 Zentralrohr
1.3 radial verstellbares Messer
1.4 Bedienung für Vortrieb und Drehbewegung
1.5 Bedienung für Ausbruchradius
2 Stirnwand des Behälters
3 Modellkörper

Abbildung 2. Ausbruchsvorrichtung

Legende : z ... Tiefe der Tunnelachse unter GOK
schl.A. .. schlaffe Anker
vorg.A. .. vorgespannte Anker
Tüb. ... Tübbinge

Tabelle 1. Geomechanische Kennwerte des Modellmaterials

Die Festigkeiten der letzten drei Versuche liegen deutlich über jenen der ersten, hervorgerufen durch eine unbeabsichtigte Frequenzänderung der Rüttelplatte.

5. AUSBRUCH UND SICHERUNG

5.1 Ausbruchsvorrichtung (Abb.2)

Der Tunnelausbruch geschieht mit Hilfe eines an einem steifen Zentralrohr radial angeordneten Messers, wobei der Ausbruchsradius zwischen 15 und 40 cm verändert werden kann. Drehbewegung und Vorschub werden händisch von aussen gesteuert. Für die Schutterung hat sich ein Industriestaubsauger gut bewährt.

5.2 Schlaffe Anker

Die zur Hohlraumsicherung verwendeten schlaffen Modellanker bestehen aus einem nichtrostenden Stahlrohr, versehen mit einer Kopfplatte und einer Ankerspitze, die V-förmige Drahtbügel zur Erzielung eines mechanischen Verbunds enthält.

Die Länge der Anker beträgt drei Viertel des Tunnelradius. Zum Einbringen der Anker wurde eine ferngesteuerte pneumatische Setzvorrichtung entwickelt, die auf die Ausbruchsvorrichtung geschraubt wird.

5.3 Vorgespannte Anker (Abb. 3)

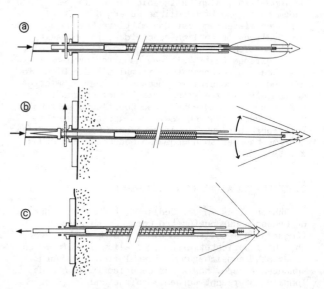

Legende : a .. Anker einbaubereit
b .. Ausklinken der V-Bügel
c .. Vorspannen des Ankers infolge Federwirkung

Abbildung 3. Modell-Vorspannanker : Schema und Einbauphasen

Ankerspitze und Kopfplatte sind durch Zwischenschaltung einer federgespannten Teleskopstange gegeneinander verschieblich miteinander verbunden. Der Modellanker wird im gespannten Zustand mit Hilfe der pneumatischen Setzvorrichtung eingebracht, hierauf werden mittels eines Bolzens die V-förmigen Drahtbügel aus dem Rohr ausgeklinkt, sie spreizen sich und üben über die gespannte Feder eine Vorspannkraft auf die Kopfplatte aus.

5.4 Tübbinge

In Ergänzung zu den beiden Arten von Ankersicherung wurden spreizbare Gelenktübbinge entwickelt. Darauf wird hier jedoch nicht näher eingegangen, da bisher nur ein einziger Versuch damit durchgeführt wurde.

6. MESSVORRICHTUNG

Neben den bereits erwähnten Erddruckmessungen in den Seitenwänden wurden vor allem die Verschiebungen (Setzungen) an der Oberfläche und im Inneren des Modellkörpers gemessen. Für erstere wurden an einem unabhängigen Rahmen befestigte mechanische Messuhren verwendet, während im Innern des Gebirgsmodells elektromagnetische "Bison"-Spulen entlang von drei vertikalen Messstrecken eingebaut wurden. Diese Spulen zeigen - ähnlich wie Extensometer - Abstandsänderungen während des Versuchs an, haben aber den grossen Vorteil, dass sie bereits während des lagenweisen Einbringens des Modellmaterials miteingebaut werden können.

7. VERSUCHSDURCHFUEHRUNG

Der typische Versuchsablauf gliedert sich in folgende Phasen :
- Einbau des Modellmaterials und der Messeinrichtungen;
- Konsolidierung unter dem Eigengewicht;
- Ggf. Vorausbruch mit Durchmesser 30 cm bis zur Tiefe von 1.70 m (Vortriebsgeschwindigkeit 30 cm/Tag);
- Beobachtung des Abklingens der Verformungen;
- Aufweitung auf 48 cm Durchmesser mit Einbringen der Sicherung bis zu einem Vortrieb von 85 cm;
- Weiterführen der Aufweitung, jedoch ohne Sicherung, bis zur grösstmöglichen Tiefe;
- Ausbau des Modells mit Entnahme von ungestörten Proben in verschiedenen Tiefen für Laboruntersuchungen.

Insgesamt wurden sechs Versuche durchgeführt (s. Tabelle 1), wobei zunächst bei gleichbleibender Tiefe des Tunnels das Sicherungssystem variiert wurde (Versuche 1 bis 4), während die beiden letzten Versuche (Nr. 5 und 6) gleich wie Versuch Nr. 4 mit vorgespannten Ankern, doch mit verschiedenen Ueberlagerungshöhen gefahren wurden.

8. VERSUCHSERGEBNISSE

Die bisher durchgeführten Versuche lieferten folgende wichtigsten Ergebnisse :
- Alle Vorausbrüche (betrifft die Versuche Nr. 1, 4, 5, 6 : Stollen mit 30 cm Durchmesser) waren ohne Stützmassnahmen standsicher. Das gilt auch für den im Versuch Nr. 4 im hinteren Teil des Tunnels zwischengeschalteten zweiten Vorausbruch mit 40 cm Durchmesser.
- Der ungesicherte Vollausbruch mit 48 cm Durchmesser (d.h. Gesamtlänge des Versuchs Nr. 1, hinterer Teil des Tunnels in den Versuchen Nr. 3 bis 6) war nur in den Versuchen Nr. 5 und 6 standsicher; alle anderen Tunnel verbrachen (Abb. 4).
- Desgleichen verbrach der mit schlaffen Ankern gesicherte Versuchstunnel Nr. 2 während des Ausbruchs (bei einem Vortrieb von 90 cm).
- Der mit Tübbingen gesicherte Versuch Nr. 3, sowie die mit Vorspannankern gesicherten Tunnelröhren (Versuche Nr. 4 bis 6) waren standsicher (Abb. 5). Der Verbruch des ungesicherten hinteren Teils des Tunnels im Versuch Nr. 4 verursachte zwar ein Absakken der Kalotte im vorderen, ankergesicherten Bereich, doch keinen völligen Verbruch (Abb. 6); dies deutet auf eine wirksame Baugrundverbesserung durch die Anker.

- Die Oberflächensetzungen in Tunnelachse, 90 cm vom Tunnelmund entfernt, sind in den Abb. 7 und 8 eingezeichnet.

Die Abb. 9 zeigt eine Zusammenstellung der in den Versuchen und bei zwei ausgeführten Bauwerken (vgl. § 11) gemessenen Setzungen in dimensionsloser Darstellung. Daraus geht hervor, dass die Setzungen bei einer Tunnelachstiefe von etwa dem doppelten Durchmesser einen Grösstwert erreichten.

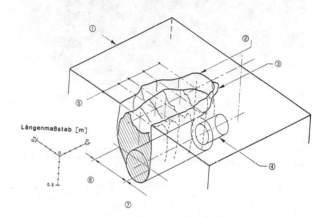

Legende : 1 .. Modellkörper
2 .. Rand des Einbruchsgrabens
3 .. Risse an der Oberfläche
4 .. Tunnel vor dem Verbruch
5 .. Aufgemessene Querschnitte
6 .. Verstürzter Bereich
7 .. Stehengebliebener Bereich

Abbildung 4. Versuch Nr. 1 - Blick auf den Verbruch

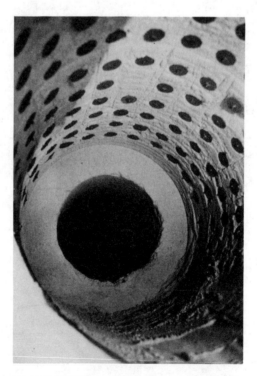

Abbildung 5. Versuch Nr. 6 - Mit Vorspannankern gesicherter Tunnel (ø = 48 cm); im Hintergrund Vorausbruch (ø = 30 cm)

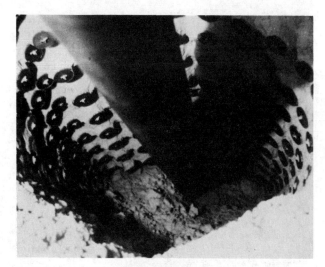

Abbildung 6. Versuch Nr. 4 - Nach Verbruch im ungesicherten Teil (Hintergrund) abgesackte ankergesicherte Kalotte

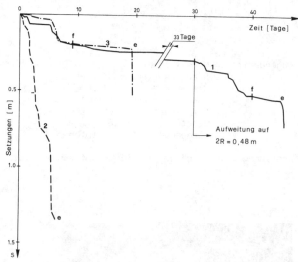

Legende : f .. Ende des Vortriebs; e .. Verbruch

Abbildung 7. Entwicklung der Oberflächensetzungen in Tunnelachse - Versuche Nr. 1, 2, 3

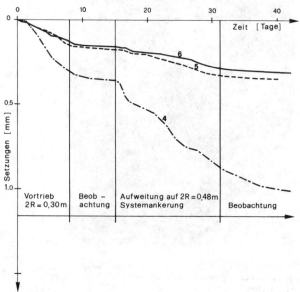

Abbildung 8. Entwicklung der Oberflächensetzungen in Tunnelachse - Versuche Nr. 4, 5, 6

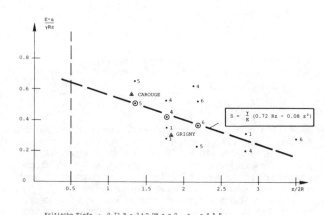

Kritische Tiefe : $0.72\ R - 2 \cdot 0.08\ z = 0$ $z_{cr} = 4.5\ R$

Abbildung 9. Oberflächensetzungen in Abhängigkeit von der Tiefe der Tunnelachse

9. ANALYTISCHE STANDSICHERHEITSUNTERSUCHUNGEN

Zur Abschätzung der rechnerischen Standsicherheit wurden die erforderlichen Gebirgsfestigkeiten nach drei verschiedenen Methoden für einen Reibungswinkel von $\phi = 35^{\circ}$ ermittelt. Für den Vollausbruch (48 cm Durchmesser ergaben sich folgende Werte für $\sigma_{c,erford.}$ $[kN/m^2]$:

Methode Versuch Nr.	A	B	C Achse	C Kämpfer	$\sigma_{c,vorh.}$	
1	5.9	5.4	4.2	10.0	5.0	Verbruch
2	6.1	5.5	4.3	10.3	6.0	Verbruch
3	6.3	5.7	4.5	10.7	7.0	Verbruch
4	6.8	6.1	4.9	11.5	12.0	Verbruch
5	6.5	6.1	4.9	11.5	12.0	stand. sicher
6	7.1	6.3	4.9	11.6	15.0	stand. sicher

Legende : A .. nach Caquot (1966)
B .. nach d'Escatha-Mandel (1974)
C .. nach Egger (1983)

Tabelle 2. Rechnerisch erforderliche und vorhandene Gebirgsfestigkeiten

Die beste Uebereinstimmung zwischen Standsicherheitsprognose und Beobachtung ergibt sich für die nach Egger (1983) im Kämpferbereich erforderliche Gebirgsfestigkeit, während die anderen Methoden zu optimistisch erscheinen.

10. NUMERISCHE BERECHNUNGEN

Alle Versuche wurden mit Hilfe eines ebenen Finite-Elemente-Programms MALINA, das am Laboratoire de Mécanique des Roches der EPFL weiterentwickelt wurde, nachgerechnet. Dabei zeigte sich zweierlei :
- Ein bevorstehender Verbruch kündigt sich deutlich durch eine fortschreitende Ausbreitung der plastifizierten Zonen in der Umgebung des Tunnels an.
- Die rechnerischen Setzungen betragen etwa das Doppelte der im Versuch beobachteten. Dies dürfte darauf zurückzuführen sein, dass die Sicherung im Versuch direkt an der Brust eingebracht wurde, während sie in der Rechnung erst nach erfolgter elastischer Entspannung in Ansatz gestellt wurde.

Zusätzlich wurden zwei Versuche mit einem räumlichen Finite-Elemente-Programm ADINA nachgerechnet (Giani, 1985). Das Netz (Abb. 10) besteht aus 448

Elementen und 850 Knoten; die vollständige Simulation eines Versuchs benötigte 2.5 Stunden CPU an der VAXII-780 der Abteilung für Bauwesen der EPFL. Die Ergebnisse zeigten eine gute Uebereinstimmung der Oberflächensetzungen (Abb. 11) im mittleren Bereich des Versuchsbehälters, während die Randbereiche im geomechanischen Modell offensichtlich durch Wandreibung deutlich verringerte Setzungen erlitten.

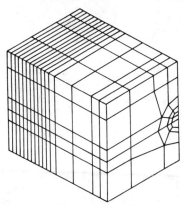

Abb. 10. Netz für räumliches Finite-Elemente-Modell

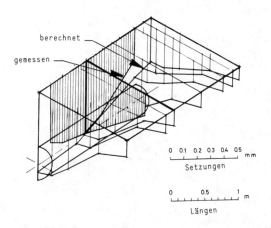

Abb. 11. Versuch Nr. 6 - Gemessene und berechnete Oberflächensetzungen

11. VERGLEICH MIT AUSGEFUEHRTEN BAUWERKEN

Um konkrete Hinweise auf die Aussagekraft der geomechanischen Modellversuche zu gewinnen, wurden Vergleiche mit zwei ausgeführten Bauwerken angestellt, deren Einbettungstiefe und Baugrundverhältnisse in etwa den durchgeführten Versuchen entsprachen.

Der Strassentunnel in Carouge (Odier u.a., 1981) wurde in Moränen aufgefahren, deren wichtigste Bodenkennwerte aus der Tabelle 3 ersichtlich sind; es gelangte die Schalenbauweise zur Anwendung. Dagegen durchörterte der Eisenbahntunnel in Grigny (Egger, 1975) tertiäre Mergel; der Bau erfolgte nach den Prinzipien der N.Oe.T.

In der Tabelle 3 sind die geometrischen und geomechanischen Kennwerte, u.a. das Verhältnis von Ueberlagerungsgewicht zur Gebirgsdruckfestigkeit $(\frac{\gamma \cdot z}{\sigma_c})$ sowie die in dimensionsloser Form dargestellten Oberflächensetzungen $(\frac{E \cdot s}{\gamma \cdot R \cdot z})$ sowohl für diese beiden Bauwerke als auch für die auf Naturgrösse hochgerechneten Modellversuche Nr. 4 bis 6 zusammengestellt. Dabei zeigt sich eine erstaunlich gute Uebereinstimmung der Oberflächensetzungen für vergleichbare Einbettungstiefen.

12. SCHLUSSFOLGERUNGEN

Die bisher durchgeführten räumlichen Modellversuche zur Simulation des Vortriebs seichtliegender Tunnel erlauben, folgende Schlüsse zu ziehen :
- Der gewählte Versuchsmassstab von 1:20 gestattete eine weitgehend naturgetreue Modellierung des Ausbruchs sowie verschiedener Sicherungssysteme (schlaffe und vorgespannte Anker, Tübbinge).
- Dagegen erwies sich die Verdichtungskontrolle des äquivalenten Modellmaterials wegen der angestrebten geringen Festigkeiten (σ_c = 5 .. 15 kN/m^2) als schwierig.
- Die Beobachtung der Standsicherheit der ungesicherten Tunnelröhren und der Oberflächensetzungen erlaubte, die Gültigkeit verschiedener analytischer Rechenverfahren sowie die Zweckmässigkeit üblicher Vorgangsweisen bei ebenen numerischen Verfahren (z.B. Berücksichtigung der vor dem Einbau der Tunnelsicherung erfolgten Verformungen) wirkungsvoll zu überprüfen.

Modell	γ [kN/m^3]	ϕ [°]	c [kN/m^2]	σ_c [kN/m^2]	E [MN/m^2]	E/σ_c	z [m]	$\frac{z}{2R}$	$\frac{\gamma \cdot z}{\sigma_c}$	2R [m]	Sicherung	$\frac{E \cdot s}{\gamma \cdot R \cdot z}$
Versuch 4	19.5	35	60	240	30	125	17	2.83	1.4	6.0	ohne	0.20
								2.13		8.0	ohne	0.62
								1.77		9.6	ohne	0.53
								1.77		9.6	Vorspannanker	0.42
Versuch 5	19.5	35	60	240	80	333	13	2.17	1.1	6.0	ohne	0.23
								1.35		9.6	ohne	0.65
								1.35		9.6	Vorspannanker	0.51
Versuch 6	19.8	35	80	300	132	440	21	3.50	1.4	6.0	ohne	0.28
								2.19		9.6	ohne	0.52
								2.19		9.6	Vorspannanker	0.37

| Carouge | 19.0 | 26 | 10 | 87 | 15 | 172 | 13 | 1.30 | 2.8 | 10.0 | Schalenbauw. | 0.57 |
| Grigny | 19.0 | 15 | 50 | 250 | 10 | 40 | 20 | 1.82 | 1.5 | 11.0 | N.Oe.T. | 0.30 |

Tabelle 3. Vergleich zwischen Modellversuchen und zwei ausgeführten Bauwerken

- Während der mit schlaffen Ankern gesicherte Tunnel
im Versuch Nr. 2 verstürzte, waren die mit
Vorspannankern gesicherten Abschnitte der Röhren in
den Versuchen Nr. 4 bis 6 standsicher. Der Vergleich
zwischen dem Sicherungseffekt schlaffer und vorgespannter Anker wurde jedoch durch die in der zweiten
Hälfte der Versuchsserie beobachteten höheren Gebirgsfestigkeiten erschwert. Eine deutliche Gebirgs-
verfestigung durch die systematische Vorspannanke-
rung zeigte jedenfalls der Versuch Nr. 4 : im
Gefolge des Versturzes des gesamten hinteren, ungesicherten Bereichs der Tunnelröhre sackte die
davorgelegene ankergesicherte Kalotte wohl ab, doch
ging der Gebirgszusammenhalt nicht verloren und die
Setzungsbeträge verringerten sich deutlich gegen den
Tunnelmund hin (vgl. Abb. 6).
- Die beobachteten Oberflächensetzungen hingen deutlich von der Tiefenlage des Tunnels ab (s. Abb. 9);
sie erreichten einen Grösstwert, sobald die Ueberla-
gerung etwa das Doppelte des Tunneldurchmessers
erreichte, und klangen dann wieder ab. Inwieweit
diese Ergebnisse quantitativ auf andere Gebirgsverhältnisse und Bauverfahren übertragen werden können,
muss noch durch zusätzliche Untersuchungen geklärt
werden.

13. SCHRIFTTUM

CAQUOT, A. 1966. Traité de mécanique des sols.
4e éd., Paris Gauthier-Villars.
D'ESCATHA, Y. et J. MANDEL 1974. Stabilité d'une
galerie peu profonde en terrain meuble. Revue
Ind. Minér. : 45-53.
EGGER, P. 1975. Erfahrungen beim Bau eines seicht-
liegenden Tunnels in tertiären Mergeln. Rock
Mech. Suppl. 4, Springer, Wien : 41-54.
EGGER, P. 1983. Roof stability of shallow tunnels in
isotropic and jointed rock. Proc. 5th Int. Congr.
Rock Mech. Melbourne, vol. 1 : C295-301.
EGGER, P. 1985. Stabilité des tunnels à faible
profondeur et tassements en surface. Schlussb.
Forsch.-Vorh. 38/80, EVED-BAS.
GIANI, G.P. 1985. Comportement d'un tunnel à faible
profondeur au voisinage du front. Ing. et Arch.
suisses No 8 : 124-128.
JOHN, K.W. 1969. Festigkeit und Verformbarkeit von
druckfesten, regelmässig gefügten Diskontinuen.
Veröff. Inst. Boden-u.Felsm. Univ. Karlsruhe,
H. 37.
ODIER, M., P. EGGER et F. DESCOEUDRES 1981.
Auscultation d'un tunnel sous faible couverture.
C.R. X ICSMFE Stockholm, Vol. 1 : 329-334.

Evaluation of the stability of mine tunnels in changing field stresses
Evaluation de la stabilité des tunnels de mine dans les tensions de champs changeants
Einschätzung der Beständigkeit bergmännischer Tunnel unter veränderlichen Gesteinsspannungen

G.S.ESTERHUIZEN, Department of Mining Engineering, University of Pretoria, Republic of South Africa

ABSTRACT: The field stresses through which mine tunnels are developed seldom remain constant over the life of the tunnels. Mining activities may cause large changes in the stresses. This paper describes practical methods of evaluating the effects of changing field stresses on mining tunnels in low stress and high stress environments. The approach is based on considering rock loosening in low stress fields and on numerical modeling of the initially failed rock around a tunnel to asses further failure in high stress environments.

RESUME: Les tensions de champ par lesquelles les tunnels de mine sont développés restent rarement constamment pendant la vie des tunnels. Les activités de mine peuvent causer des larges changements des tensions. Cet étude décrit des méthodes pratiques pour évaluer les effets des tensions de champ changeants sur les tunnels de mine dans les environnements de tension haute et basse. Cet approche est fondé en considérant le desserrement de roche dans les champs de tension basse et le modelage numérique de la roche défaillie au début autour d'un tunnel pour estimer la défaillance secondaire dans les environnements de tension haute.

ZUSAMMENFASSUNG: Die spannungen in Gebirgen, die von Tunneln durchörtert werden, bleiben selten während der Lebensdauer der Tunnel unverändert. Abbautätigkeit kann grosse Veränderungen des Spannungszustandes verursachen. Das vorliegende Referat beschreibt praktische Methoden zur Analyse der Wirkungen sich ändernder Spannungen in der Umgebung bergmännischer Tunnel in Spannungsfeldern niedriger und hoher Intensität. Die Methode für den Fall niedriger spannungsintensität beruht auf Betrachtugen der Lockerung des Gebirges, während die Methode für den Fall hoher Spannungsintensität auf numerischer Modellierung bereits zerrütteten Gebirges in der Tunnelumgebung basiert, um weiterschreitende Zerrüttung zu erfassen.

1 INTRODUCTION

In mining situations the field stresses through which tunnels are developed are seldom constant over the productive life of the tunnels. Large increases in one component of the stress may be accompanied by a decrease in another component. The changes in the field stresses are normally induced by stoping operations. When designing support for mining tunnels it is necessary to predict the effect of changes in the stress field on the stability of the tunnels. Tunnels often suffer considerable damage as a result of stoping operations, although the total stresses are well below the rock mass strength. Conversely, in high stress situations, a zone of failed rock forms around tunnels when they are developed, this changes the effective shape of the tunnels and hence changes the way in which they react to a new stress field.

This paper discusses methods of assessing the effect of stress and stress changes in mine tunnels. The methods are based on the behaviour of mining tunnels in brittle hard rock environments.

2 MODES OF STRESS-RELATED FAILURE IN TUNNELS

Stress-related failure of the rock mass around a tunnel may take place in three ways:
1. Movement along pre-existing joint planes
2. Joint movement and fracturing of intact rock
3. Fracturing of intact rock.

The first two modes of failure are common in low to medium stress environments. The latter mode of failure is typical in the deep level gold mines in South Africa, where stress levels are high and joints are widely spaced.

The mode of failure will depend to a large extent on the level of stresses, the strength of the intact rock and on the characteristics of jointing in the rock mass. For the purpose of this paper it is assumed that support pressures are negligible in relation to the stress levels and the strength of the rock.

When assessing the effect of stresses on the stability of a tunnel it is necessary to determine the mode of failure. Kirsten (1983) has simplified and rationalised the determination of the stress effect on tunnels. The method is based on the Q-System of classifying tunneling conditions by Barton et al (1974) where the stress reduction factor (SRF) of the Q-system is calculated as follows:

$$SRF = 0,244 \ K^{0,346}(H/\sigma_c)^{1,322} + 0,176(\sigma_c/H)^{1,413} \qquad \ldots\ldots\ldots\ldots\ldots(1)$$

A stress reduction limit is defined, below which the SRF cannot fall, this is calculated as follows:

$$SRF = 1,809 \ Q^{-0,329} \qquad \ldots\ldots\ldots\ldots(2)$$

where K is the ratio of the principal field stresses, H is the depth in metres which corresponds to the maximum principal stress, σ_c is the uniaxial compressive strength of the intact rock and Q is the tunnelling quality based on the SRF calculated in equation 1. If the value of SRF calculated by equation 1 is lower than the SRF calculated by equation 2 then the rock mass will be controlled by the stresses and rock mass strength. Otherwise the rock mass behaviour will be dominated by the jointing.

3. EVALUATION OF THE EFFECT OF STRESS CHANGES IN LOW STRESS ENVIRONMENTS

When stresses are low relative to the rock mass strength, and jointing is indicated to be the controlling factor in the rock mass behviour, it has been found that loosening of the rock mass due to reductions in the field stresses is the main cause of damage. A practical method of assessing the potential damage to such tunnels is to calculate the amount of loosening of the rock mass in the plane perpendicular to the axis of the tunnel. A rock loosening factor (RLF) may be calculated as follows:

$$RLF = (\Delta\sigma_1 - \nu.\Delta\sigma_2) + (\Delta\sigma_2 - \nu\Delta\sigma_1) \dots\dots\dots(3)$$

where $\Delta\sigma_1$ and $\Delta\sigma_2$ are the changes in the maximum and minimum principal stresses in the plane perpendicular to the axis of the tunnel, ν is the poissons ratio of the rock.

The RLF will return a positive value for an increase in stresses and a negative value for a decrease. When the RLF is positive it is important to check whether stress-related failure could occur.

It is necessary to find the correlation between the calculated values of the RLF and the observed damage in tunnels. The system could be refined to include the rock mass rating as a normalising factor.

An example of the application of the above concept is the stability of sublevel development in a diamond mine where a pipelike kimberlite orebody was mined. The rock mass was classified as a 'good quality' rock mass and had continuous joint planes at random orientations. The uniaxial compressive strength of intact rock samples was in the order of 120 MPa. The tunnels were initially supported with 1,8m long grouted steel rods spaced approximately 3m apart. Large open stopes were made, and it was necessary to predict the damage to tunnels which will be situated in the pillars between future open stopes. The stress levels in these tunnels would not be high enough to cause fracture of the rock mass. Several damaged tunnels were available for inspection and analysis. Figure 1 illustrates observed damage and RLF values for tunnels adjacent to one of the open stopes.

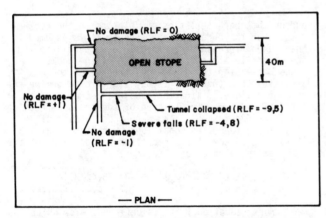

Figure 1. Observed damage and calculated values of rock loosening factor for tunnels adjacent to a large open stope.

The correlation between the RLF and observed damage to tunnels in Kimberlite adjacent to large open stopes is shown in table 1. Further stress analyses were carried out to estimate the stress changes to which the drives in the pillars would be subject. It was found that the horizontal field stresses will decrease, and the vertical stresses will increase. The values of the RLF varied between -5 and -15, indicating that the tunnels will experience severe damage unless they are adequately supported.

4. EVALUATION OF THE EFFECT OF STRESS CHANGES IN HIGH STRESS ENVIRONMENTS

When stresses are high relative to the rock mass strength and failure through intact rock is possible,

Table 1. Relationship between rock loosening factor (RLF) and observed damage in tunnels in kimberlite.

RLF	Additional damage due to stress change
÷ 0	no additional damage
0 to -2	opening of joints, no additional support required
-2 to -4	minor falls of roof and side-walls
-4 to -9	moderate falls requiring additional or resupport of tunnel
-9 to -12	major falls - possible collapse of tunnel

it becomes more difficult to predict the effect of stress changes. Empirical methods are commonly used such as rock classification methods. The Q-system (Barton et al. 1974) may be used as well as the adapted geomechanics classification by Laubscher (1976). These methods account for stress effects in a qualitative way. When stresses change, the stress effects cannot readily be assessed.

Stress analysis using finite element or boundary element techniques may be used to establish the extent of rock mass failure around a tunnel through the use of appropriate failure criteria. However, when the field stresses change it is necessary to account for the already failed rock mass in the model. There are several ways of achieving this. A simple method, which has provided satisfactory results, is to conduct a stress analysis to determine the rock mass failure due to the initial sate of stress, using a boundary element model. The initially failed rock is then modelled as a zone of 'soft rock' around the excavation and the analysis is repeated for the changed state of stress. The additional failure may then be assessed. The depth of failure predicted by the stress analysis may be used as an indication of the amount of deformation that a tunnel will experience.

Analyses were carried out to assess the effect of changing stresses on highly stressed tunnels in hard rock. The base case was a tunnel developed into highly stressed ground with a vertical stresses of 110 MPa and horizontal stresses of 55 MPa. The rock mass was assumed to be 'good quality' quartzites and the empirical failure criterion of Hoek and Brown (1980) was used to evaluate failure of the rock mass. The amount of initial failure predicted by the model is shown in figure 2A. This agrees well with observed failure around highly stressed tunnels, More O'Ferrall (1983).

A second model was prepared in which the failed zone was modelled as a softer rock type. The properties of the failed rock were varied until no additional failure of the unfailed rock mass was indicated for the initial state of stress. Both the horizontal and vertical field stresses were then increased by 50 percent. The amount of additional failure is shown in figure 2B. The results again agree well with observations of increased dilation of sidewalls with increasing stresses, Hepworth (1984).

The effect of a change in the orientation of the major principal stress is shown in figure 2C. Due to the major stress becoming more favourable to the orientation of the failed ellipsoid, little additional damage is predicted.

A study of the effect of a change of the ratio of principal stresses showed that tunnel stability is

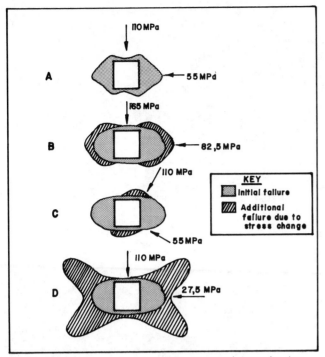

Figure 2. Failure around highly stressed tunnels due to changes in the field stresses.

very sensitive to this parameter. Figure 2D illustrates additional failure when the vertical stress remains constant, but the horizontal stress is decreased. The ratio of vertical to horizontal stresses was 0,33.

5. CONCLUSIONS

When stress levels are low relative to the rock mass strength, the change in field stress is likely to result in stability problems if loosening of the rock mass will occur. The amount of loosening may be assessed from the results of numerical models and may be correlated to actual tunnel behaviour. Appropriate support measures may be installed at an early stage to prevent deterioration of the rock mass.

In highly stressed rock masses the amount of initial failure which occurs around a tunnel during development will determine the effect of further stress changes. Simple numerical models may be used to assess the amount of additional failure which will take place due to changes in the stress field. Results show that a tunnel with a zone of failed rock around it will be most sensitive to a change in the ratio of the principal stresses, whilst changes in the magnitude or the orientation of the principal stresses are less serious to tunnel stability.

REFERENCES

Barton N, Lien R & Lunde J, Engineering classification of rock masses for the design of tunnel support, Rock Mechanics, Vol 6, No 4, 1974: 189-236.
Hepworth N, Correlation between deformation observations and support performance in a tunnel in a high stress, hard rock, mining environment. MSc Dissertation, University of the Witwatersrand, 1985.
Hoek E & Brown ET, Underground excavations in rock. Inst of Min & Metall, London 1980.
Kirsten HAD, The combined Q-NATM system for the design and specification of primary tunnel support. South African Tunnelling, Vol 6 No 1, 1983.
Laubscher DH & Taylor HW, The importance of geomechanics classification of jointed rock masses in mining operations, Proc Symp Exploration for rock engineering, Johannesburg, 1976: 119-178
More O'Ferrall RC & Brinch GH, An approach to the design of tunnels for ultra deep mining in the Klerksdorp district. Symp Rock Mechanics in the Design of Tunnels, South African National Group on Rock Mechanics, Johannesburg, 1983: 61-68.

Generation and analysis of stable excavation shapes under high rock stresses
Génération et analyse de formes d'excavations stables sous fortes contraintes
Entwicklung und Analyse stabiler Hohlraumformen für hohe Druckspannungen im Fels

RUSSELL T.EWY, Department of Materials Science and Mineral Engineering and Earth Sciences Division, Lawrence Berkeley Laboratory, University of California, USA
JOHN M.KEMENY, Department of Materials Science and Mineral Engineering and Earth Sciences Division, Lawrence Berkeley Laboratory, University of California, USA
ZIQIONG ZHENG, Department of Materials Science and Mineral Engineering and Earth Sciences Division, Lawrence Berkeley Laboratory, University of California, USA
NEVILLE G.W.COOK, Department of Materials Science and Mineral Engineering and Earth Sciences Division, Lawrence Berkeley Laboratory, University of California, USA

ABSTRACT: Failure criteria based on micromechanical models of splitting failure and shear failure are implemented in boundary element programs to analyze the stability of various two-dimensional excavation shapes under high stress. A technique for modelling the failure processes of progressive spalling, shear, and tensile failure is developed and is used to simulate the formation of stable shapes through progressive failure. The shapes resulting from initially circular openings are strikingly similar to those recognized widely as well-bore breakouts. All the shapes studied lead to the formation of pointed breakout regions, and these final shapes are stable with respect to splitting, shear, and tensile failure. The size of the failed region is smaller for the case of gradually increasing field stress around a preexisting opening than for an opening made instantaneously in rock with preexisting stress.

RESUME: Des critères de rupture basés sur des modèles micromécaniques de ruptures par éclatement et par cisaillement sont utilisés avec un programme d'équations intégrales pour analyser la stabilité en deux dimension de diverses formes d'excavation sous forte contrainte. Une technique pour modéliser les phénomènes de rupture par lamellation et flambage progressif, rupture par cisaillement, et rupture en tension est exposée et utilisée pour simuler la formation de formes stables par rupture progressive. Les formes provenant d'ouvertures initialement circulaires sont similaires à celles communément reconnues comme "breakouts" ou rupture en paroi de puits. Toutes les géométries étudiées révèlent la formation de zones de ruptures en forme de pointe, il en résulte des ouvertures stables en ce qui concerne l'éclatement, le cisaillement, et la rupture en tension. L'étendue de la zone de rupture est moindre dans le cas d'un champ de contraintes progressivement croissant autour d'une ouverture déjà existante, et est plus large dans le cas d'une ouverture faite instantanément dans la roche sous contraintes déjà établies.

ZUSAMMENFASSUNG: Bruchkriterien, die auf mikromechanischen Modellen von Spalt-und Scherbrüchen beruhen, werden in ein Grenzelementprogramm einbezogen um die Stabilität verschiedener zweidimensionaler Hohlraumformen unter hohen Spannungen zu untersuchen. Eine Methode wird entwickelt um die Bruchvorgänge von fortschreitendem Splitter-, Scher-und Zugbruch in einem Modell darzustellen. Dieses wird benützt, um die Bildung stabiler Formen durch fortschreitenden Bruch zu simulieren. Die Formen, die von ursprünglich kreisrunden Öfnungen ausgehen, ähneln den wohlbekannten Ausbrüchen in Brunnenbohrlöchern. Alle untersuchte Formen führen zur Bildung spitzer Ausbruchzonen; diese Formen sind schliesslich stabil im Hinblick auf Splitter-, Scher-und Zugbruch. Die Bruchzone ist kleiner für den Fall wo die Spannung um eine vorher existierende Öffnung allmählich erhöht wird; als für den Fall einer Öffnung, die schlagartig in einen Fels unter vorher bestehender Spannung eingebracht wird.

1 INTRODUCTION

Fracture of rock adjacent to the surfaces of underground excavations is of considerable interest in connection with safety in underground mining, the performance of mined geologic repositories for the disposal of nuclear wastes, the strength of deep excavations for defense, and the stability of deep boreholes for gas and oil production or scientific observation.

Boreholes are the most ubiquitous form of underground excavation. Under conditions where the magnitudes of the stresses in the rock are of the same order as the strength of the rock, it has been established that the walls of boreholes fracture, elongating the cross section of the borehole in the direction of the minimum stress orthogonal to its axis (Gough and Bell 1982; Hickman et al. 1985; Plumb and Hickman 1985; Zoback et al. 1985).

Both Gough and Bell (1982) and Zoback et al. (1985) have analyzed the process of fracture that produces well-bore breakouts by applying a Mohr-Coulomb failure criterion to the distribution of stresses given by the Kirsch solution for a circular hole in an elastic material. Gough and Bell (1982) propose that breakouts are produced by two pairs of shear fractures whose orientations define the pointed, "dog ear" shape of a typical breakout cross section. Zoback et al. (1985) define regions adjacent to a circular borehole within which the elastic stresses exceed the strength of the rock, and then define these broad, flat-bottomed regions to be the shape of the initial breakout cross section. Using a numerical model for breakout growth, they found that the stresses in the rock adjacent to the initial, and succeeding, predicted cross sections become

increasingly severe, so that breakouts should propagate indefinitely. They suggest that inelastic processes must provide the stability of breakout cross sections observed in the field.

In this paper we show that progressive spalling of the borehole wall can lead to a stable breakout shape.

Boreholes are certainly not the only underground excavations subject to progressive spalling failure. This behavior, commonly referred to as sidewall failure, is the most common and most severe mode of failure occurring around excavations in highly stressed brittle rock (Hoek and Brown 1980). It has been documented not only in the deep mines of South Africa (Ortlepp et al. 1975) but also in tunnels in Scandinavia (Selmer-Olsen and Broch 1982), where highly anisotropic stresses are common in the steep valley sides of the fjords. This progressive slabbing results from the formation of cleavage fractures parallel to the excavation surface and parallel to the maximum principal stress (Ortlepp 1983).

Under severe stress conditions or extreme stope dimensions, as for longwalls, tensile and shear fractures are also observed. In this paper we show that progressive spalling can lead to stable shapes for excavations, and we also illustrate the formation of tensile and shear fractures.

2 FAILURE OF ROCK IN COMPRESSION: AXIAL SPLITTING AND SHEAR FAULTING

Many rocks are extremely brittle in uniaxial and triaxial compression. Wawersik and Fairhurst (1970) and Wawersik and Brace (1971) measured complete stress-strain curves on a variety of rocks in uniaxial and triaxial compression. In uniaxial compression, deformation is essentially elastic up to near the maximum ordinate of the stress-strain curve, or the strength of the rock. Beyond this point many rocks are intrinsically unstable (Class II, Wawersik and Fairhurst 1970), in that the stored energy in the rock at the maximum stress exceeds the energy needed to fracture the rock, whereas in other rocks the residual strength diminishes very rapidly with small increases in strain (Class I). These researchers observed that failure in uniaxial compression is characterized by fracturing predominantly parallel to the direction of loading and parallel to the free sides of the sample.

In contrast to the axial splitting mode of failure in uniaxial compression, shear faulting is the dominant mode of failure in rocks subjected to confining stress. Recent work with high-resolution scanning electron microscopes (Tapponnier and Brace 1976; Evans and Wong 1985; Wong and Biegel 1985) has revealed details of the formation of shear faults under triaxial compression. In the initial portions of the stress-strain curve, microcracks grow in the direction of maximum loading, as in uniaxial compression tests. As loading progresses, however, these cracks appear to stabilize, as "shear localization" processes dominate, which include the growth of new cracks in the zones of high shear stress, microbuckling of slender columns formed by the initial cracks, kinking of mineral grains, and rotation and crushing between mineral grains. Thus the shear localization process leading to shear faults is not a simple crack growth phenomenon as with axial splitting. Moreover, shear localization produces post-failure stress-strain curves that are more gently sloped, and thus more stable, than the steeply sloping stress-strain curves due to axial splitting.

Fracture models for axial splitting and for shear localization have been developed by Horii and Nemat-Nasser (1985), Kemeny and Cook (1987), and others, using linear elastic fracture mechanics. In these models the solid rock is always linearly elastic. Macroscopic nonlinear behavior results only from crack growth and interaction. These models result in macroscopic complete stress-strain curves that agree with those from laboratory uniaxial and triaxial compression tests. They can be used to understand the mechanisms controlling different modes of macroscopic fracture.

The axial splitting and shear failure models of Kemeny and Cook (1987) are shown in Figures 1 and 2, respectively. The axial splitting model consists of a two-dimensional column of "sliding cracks" under axial and confining stresses. The sliding cracks are inclined, sliding, source cracks that propagate in the direction of maximum principal stress as the loading is increased (Brace et al. 1966; Fairhurst and Cook 1966). The sliding cracks are driven by the Mode I stress intensity factor, and thus the important material parameter is the fracture toughness, K_{IC}. Ultimate failure in this model occurs by axial splitting as the individual sliding cracks coalesce to form a macroscopic axial crack. Stress-incremental strain curves for the axial splitting model are shown for different values of confining stress in Figure 1. Stresses and strains have been normalized with respect to the initial length of the sliding cracks, $2l_o$, the pitch between vertically aligned sliding cracks, $2b$, as well as the fracture toughness, K_{IC}. Note also that the strains in Figure 1 are crack strains; i.e., the elastic strain due to the model with no cracks is left out. Figure 1 shows that the initial growth of the sliding cracks results in nonlinear strain hardening and that interaction between the vertically aligned sliding cracks results in highly unstable Class II behavior. Figure 1 also shows a large, linear increase in the ultimate strength with small increases in confining stress, indicating that axial splitting is strongly inhibited by confining stress. This agrees with the experimental results of Wawersik and Brace (1971). The model also predicts a similarly dramatic decrease in ultimate strength when small values of tension are applied normal to the splitting cracks.

The shear failure model shown in Figure 2 consists of a two-dimensional, collinear row of shear cracks at some angle to horizontal that propagate in their own plane. The shear failure model is a simplification of several mechanisms known to occur during the shear localization process, as described above. Even though isolated cracks do not

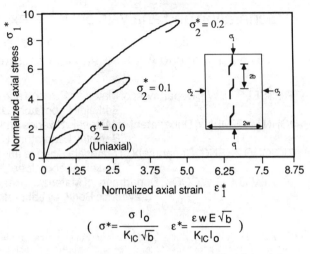

$$\left(\quad \sigma^* = \frac{\sigma\, l_o}{K_{IC}\sqrt{b}} \qquad \varepsilon^* = \frac{\varepsilon\, w\, E\sqrt{b}}{K_{IC}\, l_o} \quad \right)$$

Figure 1. Stress-incremental strain curves for the axial splitting model at different values of confining stress.

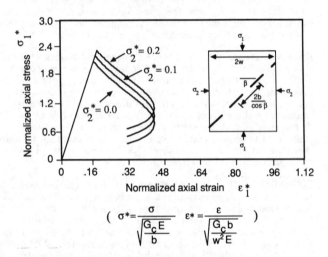

$$\left(\quad \sigma^* = \frac{\sigma}{\sqrt{\dfrac{G_c E}{b}}} \qquad \varepsilon^* = \frac{\varepsilon}{\sqrt{\dfrac{G_c b}{w^2 E}}} \quad \right)$$

Figure 2. Stress-incremental strain curves for the shear failure model at different values of confining stress.

grow in shear, the cracks in the shear model can be thought of as shear zones that propagate in a crack-like manner due to the shear localization processes. The shear cracks are driven by the Mode II stress intensity factor, and the important material parameter is the critical value of the energy release rate, G_C. Normalized results of the shear failure model, for different values of confining stress, are presented in Figure 2. This shows that the initial growth of the shear cracks defines the ultimate strength and immediately produces nonlinear strain-softening slopes, and these slopes are initially shallower and more stable than the strain-softening slopes produced by the axial splitting model. The linear increase in strength with increasing confining stress is much less dramatic than with the axial splitting model. Thus, at higher values of confining stress, these models predict that failure should occur by shear failure rather than by axial splitting.

The maximum ordinates of the stress-strain curves in Figures 1 and 2 define the strengths of rock in the splitting and shear failure modes, respectively. Since these strengths increase linearly with confining stress, the following failure criteria can be prescribed for failure by splitting, shear, and tensile fracture.

splitting: $\quad \sigma_1 \geq Q + m\sigma_3 \qquad$ for $\sigma_3 > T \qquad$ (1)

shear: $\quad \sigma_1 \geq kQ + n\sigma_3 \qquad$ for $\sigma_3 > T \qquad$ (2)

tension: $\quad \sigma_3 \leq T \qquad\qquad\qquad\qquad\qquad$ (3)

where: $\quad \sigma_1$ = major principal stress,
$\qquad\qquad \sigma_3$ = minor principal stress,
$\qquad\qquad Q$ = compressive stress associated with onset of

spalling at a free boundary,
T = tensile strength (negative),
k,m,n = constants, $k \geq 1$.

The splitting and shear failure criteria correspond to Mohr-Coulomb criteria expressed in terms of the principal stresses. The constants m and n are related directly to the angles of internal friction. In accord with the models and with physical observation, the value of m should be greater than that of n. Hoek (1983) also prescribes a high instantaneous friction angle when the effective normal stress is low and a decrease in friction angle as normal stress increases.

Because the slope of the σ_1 vs. σ_3 failure criterion for shear is less than that for splitting, it should intercept the σ_1 axis at an effective uniaxial strength greater than Q, i.e., $k > 1$. Hoek (1983) prescribes an instantaneous cohesion that increases with increasing effective normal stress, which would increase the instantaneous (effective) uniaxial strength. In addition, K_{IC} depends on the velocity of crack propagation (Atkinson 1984). The long-term splitting strength, associated with slow crack growth, could therefore be much less than that derived from short-term laboratory tests. These failure criteria are used in conjunction with a numerical boundary element analysis (Crouch and Starfield 1983) to examine the failure of excavations and the stability of the resulting shapes.

3 FAILURE OF CIRCULAR HOLES

When a borehole is drilled, the tangential stress in the rock around the hole is increased near the ends of a diameter parallel to the minimum stress in the rock mass, and the radial stress is diminished to zero in the case of an empty hole or, otherwise, to the fluid pressure in the hole, as given by the well-known Kirsch solution. In the absence of drilling mud or casing, the rock in this region adjacent to the borehole is subjected essentially to a biaxial, compressive effective stress. If this stress is sufficiently great in relation to the strength of the rock given by the maximum ordinate of the stress strain curve in Figure 1 (Q in eq. (1)), the rock should fail by extensile splitting subparallel to the wall of the borehole. Note that such splitting must occur quite close to the free surface of the borehole, because the radial stress increases rapidly with radial distance into the rock away from the circumference of the borehole, and it has been shown that such confinement inhibits axial splitting very strongly. Also, the extensile splitting becomes intrinsically unstable just beyond the peak of this curve because of its Class II character. Comparison of the stresses around the borehole with the shear failure criterion (eq. (2)), using the conservative assumption that $k = 1$ and $m = n = 3.7$ (angle of friction = 35 degrees), reveals that only a thin skin of rock close to the boundary meets the failure criterion. Because these failure zones correspond exactly to those regions of the boundary subject to splitting failure, it is assumed that the more unstable splitting behavior will be the mode of failure. The tensile failure criterion, using $T = -Q/10$, is not met anywhere around the boundary for the values of field stresses used in this analysis.

As a result of the formation of the first spall by extensile splitting, a new free surface is created, the shape of the cross section of the borehole is changed, and the tangential stress is increased. The new free surface diminishes the radial stress in the rock adjacent to it, so that another extensile fracture should form subparallel to the circumference of the elongated portion of the new cross section. Well-bore breakout is assumed to be the result of a series of episodes of spalling by extensile splitting, such as described above.

To model this sequence of changes in cross section by successive spalling, a method is required for systematically changing the cross section of the borehole as a result of each episode. This has been done by dividing the rock around an initially circular borehole into a large number of small blocks defined by angular increments of 3 degrees and radial increments of 0.067 of a borehole radius. Initially, the boundary element analysis (Crouch and Starfield 1983) is used to determine the tangential stress around a circular borehole for values of the rock mass stresses of interest. An

D = Breakout Depth/Original Radius; θ = Breakout Angle; ① = Cycle Number

Figure 3. Cross sections through a borehole illustrating the steps by which a stable breakout cross section evolves.

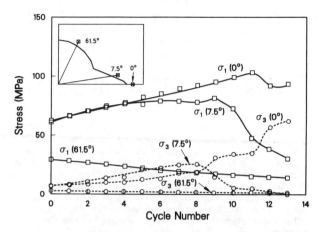

Figure 4. Changes in the principal stresses at selected positions around a borehole as the cross section changes from circular to a stable shape. Cycle number is as illustrated in Figure 3.

annular arc of rock is removed that has a radial thickness of 0.067 of the borehole radius and which includes that portion of the borehole circumference within which the tangential stress exceeds the strength. The new distribution of stress in the rock around the elongated cross section resulting from this change is now calculated using the boundary element analysis, and another layer of rock of the same thickness as in the previous cycle is removed from anywhere on the circumference of the now noncircular cross-section, wherever the tangential stress exceeds the compressive strength. The final breakout shape of the borehole is allowed to evolve by a succession of such episodes until the tangential stress is everywhere less than the strength of the rock. Changes in the angular and radial increments by factors of 0.5 to 2 do not change the final breakout shape.

If at any stage of the breakout process the tangential stress anywhere on the boundary becomes more negative than the tensile strength, a tensile fracture is allowed to form normal to the boundary at that point to relieve the stress. This occurs at the ends of a diameter parallel to the maximum field stress for highly unequal field stress values in conjunction with large breakouts. The formation of these fractures has little or no effect on the stresses near the breakout regions. Tensile fractures of this nature have been noted by others (Gay 1976; Hoek and Brown 1980), who also observe that these fractures simply relieve the excess tensile stresses but have little effect on the stability of an excavation in unjointed rock.

Once the final breakout cross section has been found, the stability of this cross section is verified in two ways. First, the tangential stress around the circumference of the breakout must be everywhere less than the compressive splitting strength of the rock (eq. (1)). Second, the distribution of the stresses everywhere in the rock outside the breakout cross section is compared with equation (2) for the triaxial strength of the rock. If the distribution of elastic stresses is everywhere less than that needed to produce failure, either by extensile splitting or Mohr-Coulomb shear, the breakout cross section should indeed be stable. The rock will then be in an elastic condition everywhere, so that the stress analysis using boundary elements in an elastic material is valid.

The episodes by which an initially circular borehole

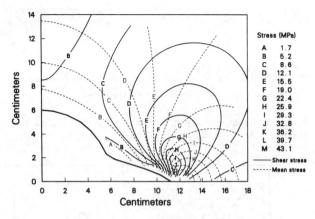

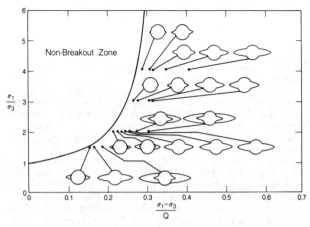

Figure 5. Contours of maximum shear stress and mean stress around the stable breakout shape of Figures 3 and 4.

Figure 6. Simulated stable breakout cross sections for different values of field stress and biaxial compressive strength. Larger breakouts are for a hole created instantly in rock with preexisting stress. Smaller breakouts are for gradually increasing stress around a preexisting hole.

develops a stable breakout cross section, as determined by the numerical simulation, is illustrated by the sequence of cross sections shown in Figure 3. Clearly, the key to this process is the nature of the stress redistribution that results from the changing cross-sectional shape. For a vertical borehole where the horizontal stresses orthogonal to the hole axes are 30.3 MPa and 15.17 MPa, respectively, and the compressive strength is 69 MPa, the changes in the principal stresses at critical positions ahead of the circumference of the borehole and at a distance of 0.133 of a borehole radius into the rock are illustrated in Figure 4. In general, the magnitudes of both local principal stresses near the unfailed part of the boundary diminish monotonically as the stress concentration zone moves away from the original borehole. The principal stresses near the elongated part of the borehole first increase as the result of the increase in stress concentration caused by the changing cross-section, and then decrease as the concentration zone passes by the point and moves farther away. The maximum principal stress directly in front of the ends of the elongating diameter does not start decreasing until the ends of the breakout become pointed. At the same time the minimum principal stress, which provides confinement, increases rapidly. This reduces the stress difference and raises the mean stress, both of which contribute to a stable state of stress.

Detailed contours of the maximum shear stress (half the principal stress difference) and the mean stress (half the stress sum) in the region near the tip of a stable breakout are shown in Figure 5. It can be seen that the mean stresses are high enough so that the rock in this region is far from failure in terms of a Mohr-Coulomb criterion.

Using the methodology described above, stable wellbore breakout shapes have been determined for a range of ratios between the principal stresses in the rock mass orthogonal to the axis of a borehole and a range of the magnitudes of these stresses relative to the strength of the rock, as is illustrated in Figure 6. The region on the top left of this figure represents conditions where the tangential stress around a circular borehole is not sufficient to result in spalling and produce breakouts, which is determined from equation (1). To the right of this are shown stable breakout shapes corresponding to a variety of stress conditions. The larger breakouts result from applying the stresses first to the rock mass and then excavating the borehole. To some extent this represents what occurs when a borehole is drilled, although the full stress concentrations around the hole must actually develop progressively as the hole is drilled beyond the plane where the cross section is being considered.

In laboratory tests of well-bore breakouts it is much easier to increase the stresses in a specimen of rock in which a pre-drilled borehole exists (Haimson and Herrick 1985). We have used a similar methodology to that described above to simulate well-bore breakouts under these conditions of progressively increasing stress. The stresses in the rock are increased in small, regular steps from critical values necessary to initiate fracture at the ends of a diameter parallel to the direction of the minimum principal stress in the rock to the desired final values while

the stress ratio is kept the same. After each stress increase the failed region of the boundary is removed to change the cross section, as before.

Stable well-bore breakout cross sections for these conditions of progressively increasing stress on the rock mass are also shown in Figure 6 for some of the stress and strength values. These appear as smaller breakouts within the larger one. The difference between stable breakout cross sections for the conditions of a borehole instantly made in a rock mass subjected to preexisting stresses and cross sections for a borehole in a rock mass where the stresses are progressively increased to their final values is remarkable. The amount of breakout needed to develop a stable cross section in the former case is much larger than in the latter case. Further, the differences between the principal stresses a short distance into the rock from the breakout for the case of steadily increasing stress, although less than those needed to cause failure, are greater than for the larger breakouts associated with drilling into rock under a preexisting state of stress. The real case of drilling into rock with a preexisting stress probably lies somewhere between these two cases.

Breakout cross sections predicted by the methodology described above have a very pointed shape in the direction of the minimum principal stress orthogonal to the borehole axis. The shapes in Figure 6 resemble those shown by Haimson and Herrick (1985) and Mastin (1984) for laboratory tests.

That such different breakout cross section shapes can both be stable under the same final far-field stress condition (Figure 6) is attributed to the shape of the breakouts. The pointed tip of a breakout limits the stress concentration zone to a very small area and therefore reduces the stresses elsewhere around the borehole except for a very small region adjacent to the tip. In this region, both principal stress values are increased (Figure 4), but the higher mean stress together with the lower stress difference ensures stability.

4 FAILURE OF NONCIRCULAR EXCAVATIONS

An excavation with a more rectangular cross section (Figure 7) causes a stress distribution quite different from that around a circle and provides an excellent comparison case. This opening could represent a tunnel or the early stages of a longwall stope.

The directions and magnitudes of the principal stresses in the surrounding rock are shown in Figure 7 for the case of a vertical field stress equal to 50 MPa and a horizontal field stress equal to 25 MPa. In contrast to a circular excavation it is seen that zones of slight tension are formed above and below the opening, especially near the corners, where these small tensile stresses are associated with orthogonal compressive stresses of significant magnitude. The extent

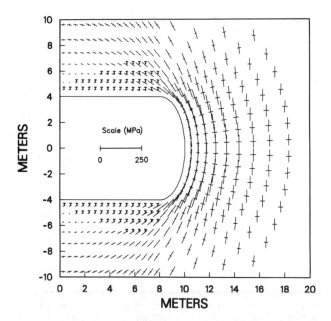

Figure 7. Vectors showing direction and magnitude of principal stresses around an 8 × 20 m opening subject to vertical and horizontal field stresses of 50 and 25 MPa. 'T' indicates tensile stress.

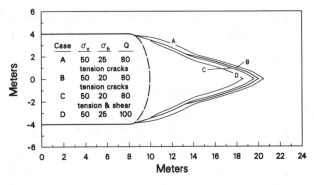

Figure 8. Stable shapes resulting from the initial shape of Figure 7 under different values of field stress and biaxial compressive strength. Fractures included in the analysis are noted.

and severity of these tensile zones increases with increased width to height ratio or increased vertical to horizontal stress ratio. Another significant feature of the stress distribution is the presence of high tangential boundary stress (145–155 Mpa) along almost the entire lengths of the excavation ends.

If equations (1), (2), and (3) are applied to the rock surrounding the excavation in Figure 7, setting $Q = 100$ MPa, $T = -Q/10$, $k = 1.5$, $m' = 10$ ($\phi = 55°$), and $n = 4.6$ ($\phi = 40°$), it is found that it is everywhere safe against both tensile and shear failure. However, along almost the entire lengths of the curved ends, a thin skin of rock is unsafe with regard to splitting failure.

Just as in the analysis of borehole breakouts, a progressive spalling model is implemented within a boundary element code, in which each segment of the boundary subject to splitting failure is moved outward along a radius line extending from the center of the opening. The final stable shape is shown in Figure 8 (Case D). The rock surrounding this final shape is stable everywhere with respect to splitting, shear, and tensile failure.

Although the initial shape of this excavation is vastly different from a circle, the similarity between the final shape and that for a circular hole subjected to the same ratios of stresses and strength (Figure 6) is striking. Both have developed long triangular breakout regions with small-angle tips. Contours of maximum shear stress and mean stress in the rock surrounding the stabilized shape exhibit a remarkable similarity to those for a stable wellbore cross section (Figure 5). The same pattern of stress redistribution that acts to stabilize a well-bore breakout also acts to stabilize this broken out shape, which originates from a very different initial shape.

If the value of Q is set to 80 MPa rather than 100, a similar breakout sequence is observed. In accord with the observations presented earlier for initially circular holes (Figure 6), this lower strength results in a greater breakout depth (Figure 8, Case A). With $Q = 80$ MPa the tensile strength of the rock is exceeded during the breakout process along the top and bottom boundaries near the center of the opening. Vertical tensile fractures grow from these boundaries to a final length of 2 m, and the formation of these fractures ensures that all remaining tensile stresses above and below the opening are insufficient to cause failure.

The final shape for the above stress and strength values is stable everywhere with respect to tensile failure and stable everywhere on the boundary. There is a small zone just above and below the pointed tip in which splitting or

shear could occur. Because this zone is small and movement of the rock is restricted by the geometry, any fracture in this region is not expected to have any significant effect.

If the horizontal field stress is reduced to 20 MPa and the vertical field stress is left at 50 MPa, the pattern of principal stresses is very similar to that in Figure 7, except the zone of tensile minor principal stress is slightly more extensive. With $Q = 80$ MPa the only area subject to shear failure is a very thin skin of rock along each of the curved ends. These areas are so close to the boundary, however, that failure is actually expected to occur by spalling, which is more critical at low confining stresses.

There is also a zone above and below each corner in which the factor of safety against splitting is less than one (Figure 9). This is due to the high values of σ_1 in association with slightly tensile values of σ_3. The high ratio of field stress values (2.5) is not a requirement to form these zones of splitting failure near the corners. A lower ratio (2.0) as before, but with a lower Q value, will also result in these unsafe zones. These splitting zones can also be caused by increasing the width-to-height ratio of the opening, as in a longwall slope. Under many conditions, the corner zones of splitting are seen to connect with the splitting zones along the ends of the excavation.

Different types of fractures are possible in this zone. In one scenario, splitting cracks will form and coalesce near the excavation boundary at the corners. As a crack extends into the tensile zone just above the corner, the stress intensity factor at the crack tip will increase very rapidly as the crack lengthens. A splitting fracture near the excavation boundary could therefore propagate parallel to the major principal stress above the corner as a tensile fracture. Fractures of this nature would have an orientation relative to horizontal of 35 to 55 degrees, depending upon the starting point of the fracture, the horizontal-to-vertical stress ratio, and the width-to-height ratio of the opening.

In another scenario, small splitting cracks will form parallel to the major principal stress throughout the unsafe zone above the corner, in accord with the model presented in this paper. Because there is no free surface, the confining stress acting across these small splitting cracks increases rapidly and the growth of the cracks is arrested. The high compressive stresses are still present, and the small slip interfaces that initially gave rise to the small splitting fractures can now coalesce through a weakened zone of rock to form a macroscopic shear fracture. A fracture of this nature would probably be inclined about 25 to 35 degrees steeper than the direction of major principal stress, or about 60 to 90 degrees relative to horizontal.

Several kinds of fractures have been observed around the deep level longwall slopes in South Africa. Significantly, Pretorius (1966) noted a prevalent set of fractures dipping at 60 to 90 degrees and another set dipping at 20 to 40 degrees. Recently, Adams et al. (1981) described fractures dipping at 60 to 75 degrees and others dipping at 30 to 40 degrees.

In laboratory experiments on model openings, Gay (1976) observed fractures with dip angles ranging from 60 to 80 degrees. Also carefully noted in these experiments was the formation of tensile cracks in the roof and floor and

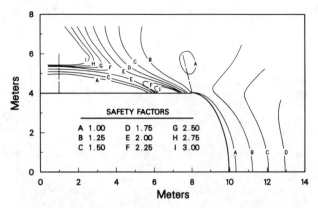

Figure 9. Safety factors for the opening shape of Figure 7 under the Case C conditions of Figure 8, showing unsafe zones with respect to splitting above the corner and along the end and with respect to tension directly above the opening. Dashed lines indicate locations of tensile and shear fractures used in the analysis.

significant spalling at the ends of the openings.

The effect of fracturing in these zones above and below the corners is investigated in this numerical model by including steep-angle shear fractures at each corner, since these fractures will have the greatest redistribution effect on the stresses and are also potentially the more dangerous. For the opening shape previously described, with vertical and horizontal stresses of 50 and 20 MPa and a Q strength of 80 MPa, these fractures are placed at an angle of 76 degrees from horizontal, emanating from the ends of the straight boundary segments (8 m from center) and extending 2 m above and below the excavation (Figure 9). This angle places the shear fractures within the unsafe zone and also provides a logical orientation relative to the principal stress directions. Tensile fractures are also included above and below the center of the opening (Figure 9).

With the above model, the processes of progressive spalling, tensile fracture, and shear fracture are modeled simultaneously. When the initial configuration is subjected to the field stresses, the tensile fractures immediately crack to a length of 2 m to relieve the tensile stresses, and the shear fractures undergo inelastic slip along their entire lengths. The tangential stresses along the curved ends remain high, however, and these surfaces fail by spalling.

The initial slip of the shear fractures relieves the excess shear stress across them. As the excavation geometry changes, however, it alters the stress distribution around the opening. As the concentration of stresses moves farther away, the shear stress acting across the fractures decreases and eventually switches direction. Tensile stresses also develop across the fractures. A process of incremental tension cracking and reverse slip occurs as the excavation geometry widens. At the final breakout depth (Figure 8, Case C), the shear fractures have cracked in tension to a height of 1.75 m. As in the previous case, the horizontal tensile stresses above and below the excavation increase, and the tensile fractures grow to a final length of 5 m when the final breakout depth is reached.

Except for small zones above and below the pointed tip, the rock surrounding this final shape is stable with respect to splitting, shear, and tensile failure, even around the fractures. If the above analysis is repeated without allowing any shear failure, the final shape (Figure 8, Case B) is similarly stable. Thus the shear fractures appear to have little effect on the progressive spalling and, although they relieve stresses around the initial shape, are not important to the stability of the final broken out shape.

The final shape for a similar initial shape subject to the Case D conditions of Figure 8 is shown in Figure 10. Tension cracks near the center of the opening reach a final length of 2 m. This shape is stable everywhere on and off the boundary with respect to splitting, shear, and tensile failure. Its final breakout depth is greater, but not significantly greater, than that for the original shape of Figure 7 under the same stress and strength values (Figure 8).

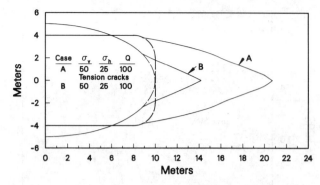

Figure 10. Stable shapes resulting from a rectangle with curved corners and a 2:1 ratio ellipse subject to the Case D conditions of Figure 8. Fractures included in the analysis are noted.

A classic opening shape of interest is the ellipse. An ellipse with a width equal to 20 m and a height of 10 m subject to vertical and horizontal field stresses of 50 Mpa and 25 Mpa has a maximum tangential stress of 225 MPa at the ends of its major axis. If this initial shape is assigned a Q strength of 100 MPa and allowed to spall progressively, it reaches a final breakout depth of 4.22 m (Figure 10). The only unsafe zone is, as in some of the previously described cases, a small area just above and below the pointed tip.

If this ellipse is oriented with its major axis vertical, it becomes a harmonic excavation, and the tangential stress has a constant value of 75 Mpa everywhere on the boundary. As long as the splitting strength Q remains above 75 MPa this orientation is stable. If this strength becomes less than 75 MPa, either through excavation in weaker rock or through time-dependent weakening, the entire surface of the excavation is subject to spalling, which could lead to very large breakouts before a stable shape is achieved.

If the ellipse is oriented horizontally and the Q strength is 75 MPa, it reaches a final breakout depth of 6.14 m. Of greater interest, however, is that the final shape shown in Figure 10, which was arrived at by setting $Q = 100$ Mpa, is also stable for a Q of 75 Mpa, even though the breakout depth is only 4.2 m. The well-bore breakout analysis reveals that the smaller breakouts achieved by gradually increasing the field stresses are stable under the final state of stress (Figure 6). It is quite probable, therefore, that a horizontal ellipse with an intentionally excavated breakout of even less than 4.2 m would be stable under the above stress and strength values. This same logic should apply equally well to almost any initial excavation shape.

5 DISCUSSION

Using linear elastic fracture mechanics, micromechanical models have been developed for the failure of rock both by extensile splitting parallel to the direction of maximum compressive stress and by inclined shear fracture. Utilizing failure criteria based on these models, a methodology has been developed to generate breakout cross sections from excavations of various initial shapes in elastic brittle rocks using numerical simulation. Included in this simulation is the formation of tensile and shear fractures, in addition to progressive spalling.

There are several interesting results. First, it is shown that the breakout cross section redistributes the stresses in an elastic rock, so that these stresses become everywhere less than the biaxial compressive strength (splitting strength), the tensile strength, and the shear strength given by a Mohr-Coulomb criterion, thus indicating that the final cross section must be stable. Second, it is shown that the resulting cross section depends upon the stress path. Breakouts are much smaller if they result from a gradual increase in stress than if an excavation is created instantly in rock already subjected to stress, yet both cross sections result in stress distributions that do not exceed the strength of the rock. Third, the formation of tensile and/or shear fractures does not appear to have a significant effect on the spalling behavior or the final shape, and the changed cross section removes zones which could otherwise be subject to shear failure.

The predicted breakout cross section should be exact for truely brittle rocks. Many softer rocks exhibit some degree of Class I stable strain softening. It is likely that in such rock the residual strength of the rock at strains in excess of that corresponding to peak strength may prevent complete removal of the rock by spalling, as assumed in the above model. Indeed, Plumb and Hickman (1985) have detected regions of enhanced electrical conductivity at the base of borehole breakouts that could correspond to porosity caused by incomplete spalling and dilatation of the rock in this region. In horizontal excavations in which the sides spall as a result of high vertical stress, much of the failed material remains in the breakout region and acts to increase the confinement and prevent further breakout. When the dominant stress is horizontal, the removal of spalled rock in the roof is assisted by gravity, and in these cases a definite pointed breakout shape is observed (Brekke 1970; Selmer-Olsen 1970).

Finally, spalling may be a time dependent phenomenon, as has been observed in the laboratory by Mastin (1984) and in the field by Plumb and Hickman (1985). If the spalling strength decreases with time and approaches a lower bound, this has the same effect as gradually increasing the stresses in the numerical model and will result in small stable breakouts.

ACKNOWLEDGEMENT

The authors wish to acknowledge support from the following sources: The Office of Civilian Radioactive Waste Management through the Office of Geologic Repositories, the U.S. Bureau of Mines for Grant No. G1164106 to the California Mining and Mineral Resources Research Institute, and the Jane Lewis Fellowship Fund at the University of California, Berkeley.

REFERENCES

Adams, G.R., A.J. Jager & C. Roering 1981. Investigations of rock fracture around deep level gold mine stopes. Proceedings 22nd U.S. Symposium on Rock Mechanics, p. 213—218. Cambridge: MIT.

Atkinson, B.K. 1984. Subcritical crack growth in geological materials. J. Geophys. Res. 89(B6):4077—4114.

Brace, W.F., B.W. Paulding & C. Sholtz 1966. Dilitancy in the fracture of crystalline rocks. J. Geophys. Res. 77:3939.

Brekke, T.L. 1970. A survey of large permanent underground openings in Norway. In T. Brekke & F. Jörstad (eds.), Large permanent underground openings, p. 15—28. Oslo: Universitetsforlaget.

Crouch, S.L. & A.M. Starfield 1983. Boundary Element Methods in Solid Mechanics. London: George Allen & Unwin.

Evans, B. & T.-F. Wong 1985. Shear localization in rocks induced by tectonic deformation. In Z. Bazant (ed.), Mechanics of geomaterials. New York: John Wiley and Sons.

Fairhurst, C. & N.G.W. Cook 1966. The phenomenon of rock splitting parallel to a free surface under compressive stress. Proceedings First Congress of International Society of Rock Mechanics, Lisbon 1:687—692.

Gay, N.C. 1976. Fracture growth around openings in large blocks of rock subjected to uniaxial and biaxial compression. Int. J. Rock Mech. Min. Sci. 13:231—243.

Gough, D.I. & J.S. Bell 1982. Stress orientations from borehole wall fractures with examples from Colorado, east Texas, and northern Canada. Canadian J. Earth Sci. 19:1358—1370.

Haimson, B.C. & C.G. Herrick 1985. In situ stress evaluation from borehole breakouts experimental studies. Proceedings 26th U.S. Symposium on Rock Mechanics, p. 1207—1218. Rotterdam: Balkema.

Hickman, S.H., J. H. Healy & M.S. Zoback 1985. In situ stress, natural fracture distribution, and borehole elonga-tion in the Auburn geothermal well, Auburn, New York. J. Geophys. Res. 90(B7):5497—5512.

Hoek, E. 1983. Strength of jointed rock masses. Geotechnique 33(3):187—223.

Hoek, E. & E.T. Brown 1980. Underground excavations in rock. London: Institute of Mining and Metallurgy.

Horii, H. & S. Nemat-Nasser 1985. Compression-induced microcrack growth in brittle solids: Axial splitting and shear failure. J. Geophys. Res. 90(B4):3105—3125.

Kemeny, J. & N.G.W. Cook 1987. Crack models for the failure of rock in compression. To appear, Proceedings of the Second International Conference on Constitutive Laws for Engineering Materials, Tucson, Arizona.

Mastin, L. 1984. The development of borehole breakouts in sandstone. M.S. Thesis, Stanford University.

Ortlepp, W.D. 1983. Considerations in the design of support for deep hard-rock tunnels. Preprints Fifth Congress of International Society for Rock Mechanics, Melbourne D:179—187.

Ortlepp, W.D., R.C. More O'Ferrall & J.W. Wilson 1975. Support methods in tunnels. Symposium on Strata Control & Rockburst Problems of the S.A. Goldfields, Association of Mine Managers of South Africa 1972—73, p. 167—194. Johannesburg: Chamber of Mines of South Africa.

Plumb, R.A. & S.H. Hickman 1985. Stress-induced borehole elongation: A comparison between the four-arm dipmeter and the borehole televiewer in the Auburn geothermal well. J. Geophys. Res. 90(B7):5513—5521.

Pretorius, P.G.D. 1966. Contribution to rock mechanics applied to the study of rockbursts. J. South African Inst. Min. & Metall. 66:705—713.

Selmer-Olsen, R. 1970. General discussion: experience with using bolts and shotcrete in area with rock bursting phenomena. In T. Brekke & F. Jörstad (eds.), Large permanent underground openings, p. 275—278. Oslo: Universitetsforlaget.

Selmer-Olsen, R. & E. Broch 1982. General design procedure for underground openings in Norway. In Norwegian hard rock tunnelling, p. 11—18. Norwegian Soil and Rock Engineering Association, Trondheim: Tapir.

Tapponnier, P. & W.F. Brace 1976. Development of stress induced microcracks in Westerly Granite. Int. J. Rock Mech. Min. Sci. 13:103—112.

Wawersik, W.R. & W.F. Brace 1971. Post-failure behavior of a granite and diabase. Rock Mechanics 3:61—85.

Wawersik, W.R. & C. Fairhurst 1970. A study of brittle rock fracture in laboratory compression experiments. Int. J. Rock Mech. Min. Sci. 7:561—575.

Wong, T.-F. & R. Biegel 1985. Effects of pressure on the micromechanics of faulting in San Marcos gabbro. J. Struct. Geol. 7:737—749.

Zoback, M.D., D. Moos, D. & L. Mastin 1985. Wellbore breakouts and in situ stress. J. Geophys. Res. 900(B7):5523—5530.

Le calcul du soutènement dans les terrains fissurés
The design of roof support in jointed rock masses
Ausbauberechnung für zerklüftete Gesteinsmassen

JACQUES FINE, Centre de Mécaniques des Roches, Ecole Nationale Supérieure des Mines de Paris, France
THOMA KORINI, Université de Tirana, Albanie

ABSTRACT: A method for the design of roof support in the jointed rock masses is described. Using data about the joints distribution provided from a drift we identify the tetrahedron-shaped blocks and with criterions based on dilatancy and sliding, we estimate the probability of blocks fall by simulations of the spatial distribution of joints. The roof bolting is calculated according to the assumption of the hanging support.

RESUME: On propose une méthode pour calculer le soutènement dans les terrains affectés par une fissuration naturelle. A partir du relevé de fissures dans une galerie, on cherche les blocs en forme de tétraèdres délimités par les fissures et au moyen de critères basés sur la dilatance et le glissement on estime la probabilité de chutes de blocs en effectuant des simulations de la distribution des fissures dans l'espace. Le boulonnage est alors calculé selon le principe du soutènement suspendu.

ZUSAMMENFASSUNG: Es wird eine Methode zur Berechnung vom Ausbau von zerklüfteten Gesteinsmassen vorgestellt. Auf der Grundlage der Aufnahme der Zerklüftung in einer Strecke, sucht man tetraedrische Blocken, die durch die klüfte begrenzt werden. Mit hilfe von Kriterium die sich auf die Dilatanz und Gleitfähigkeit gründen, schätzt man die Wahrscheinlichkeit von Blockenstürzen, indem Simulationen über die Verteilung der Klüfte im Raum durchgeführt werden. Der Ankerausbau wird dann nach dem Prinzip des hängenden Ausbaus, errechnet.

INTRODUCTION

Le calcul du soutènement d'une excavation souterraine nécessite la connaissance du comportement des terrains. Dans les terrains affectés par une fissuration naturelle préexistante avant tout creusement de cavité, l'étude de ce comportement est généralement analysée de deux manières différentes:
- dans les terrains très fissurés où il n'est guère possible de faire un relevé précis de chaque fissure, on aborde le problème sous l'aspect de la mécanique des milieux continus: la roche est remplacée par un matériau globalement continu dont on cherche à définir le comportement,
- dans les terrains où il est possible de faire un relevé des fissures, le massif rocheux est considéré comme un assemblage de blocs dont on cherche à déterminer les conditions d'équilibre.

C'est ce second cas qui est examiné dans cette note. On supposera que les contraintes qui s'exercent dans le toit des excavations minières sont insuffisantes pour créer de nouvelles fissures par excès de compression. L'objet de l'étude n'est pas d'étudier la cinématique d'un effondrement de toit, mais de définir un schéma de boulonnage susceptible d'empêcher toute chute de bloc.

MODELISATION DE LA STRUCTURE DU MASSIF

Il est nécessaire que le massif rocheux ait été reconnu par des travaux souterrains, par exemple par un ensemble de galeries.
Dans ces galeries, on procédera à un relevé spatial de chaque fissure, c'est à dire on notera:
- l'azimut et le pendage de la fissure,
- la position du point où ces paramètres ont été relevés. Ce point est défini par ses trois coordonnées X,Y,Z dans un repère liée à la galerie.

Pour modéliser le massif rocheux, on fera les hypothèses suivantes:
- une fissure est assimilée à un plan. La connaissance des paramètres azimut, pendage et point de relevé permet d'écrire l'équation de ce plan dans un repère orthonormé lié au massif rocheux.
- ce plan de fissure est infini. Il serait aisé d'introduire d'autres paramètres caractérisant l'extension des fissures, mais la connaissance réelle de ces paramètres risquerait d'être illusoire. Admettre des fissures très étendues conduit probablement à des résultats pessimistes, en considérant le massif plus fracturé qu'il ne l'est.

Le relevé de n fissures se traduit donc par n équations de plans. Ces n équations caractérisent la structure du massif au voisinage de la galerie.

Il faut noter que le relevé des fissures doit faire l'objet d'une méthodologie précise. On peut, par exemple, effectuer un relevé sur chaque parement, mais il ne faut pas caractériser le massif par la somme des deux relevés car on risquerait de définir deux fois les mêmes fissures, les fissures d'un parement apparaissant sur l'autre parement. Il est donc recommandé de comparer les systèmes de fissures relevées sur chaque parement, ce qui permet de voir si l'hypothèse de continuité des fissures est justifiée. Dans certains cas les relevés dans une seule galerie peuvent être insuffisants pour définir entièrement la fissuration du massif. En effet:
- s'il existe des fissures dont l'azimut est voisin de la direction de l'axe de la galerie, ces fissures échapperont vraisemblablement au relevé,
- si le problème à résoudre consiste à déterminer le soutènement d'une future galerie plus ou moins éloignée de la galerie d'observation, le système de fissuration peut ne pas être strictement le même.

Dans ces cas là, on a jugé intéressant d'établir la répartition spatiale des fissures par simulation.
La méthodologie suivante est proposée:
- on effectue des relevés dans deux galeries, dont les directions devront si possible être orthogonales afin d'être certain de recouper chaque système de fissure. L'idéal serait bien sûr de disposer d'une troisième galerie dont l'axe serait perpendiculaire au plan des deux autres, pratiquement d'une cheminée verticale. Cette opportunité ne se présente que très rarement.
- on définit les familles de fissures en utilisant un diagramme azimut-pendage adapté à un traitement informatique. On trouvera sur la figure 1 le type de diagramme employé. L'espace azimut-pendage

est divisé en cases élémentaires. On compte le nombre de fissures entrant dans chaque case. Ce diagramme constitue un histogramme à deux dimensions. On met ainsi en évidence les différents systèmes de fissuration

- pour chaque système de fissures, on établit un histogramme donnant la répartition de la distance entre deux fissures. Cette distance est calculée le long d'un axe passant dans la galerie d'observation. La figure 2 reproduit un tel histogramme.

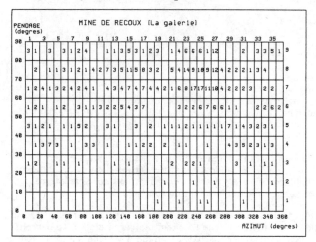

Fig. 1. Diagramme azimut-pendage

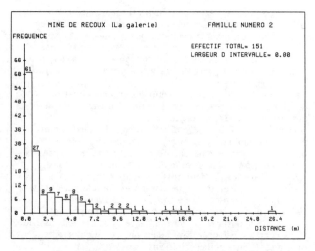

Fig. 2. Diagramme des intervalles entre fissures

Pour engendrer un réseau de fissures simulant le réseau réel, on parcourt successivement dans un repère orthonormé lié au massif, des axes ayant la même direction que les galeries d'observation et par tirage au sort d'abord dans l'histogramme des distances puis dans l'histogramme azimut-pendage, on définit les paramètres nécessaires pour obtenir l'équation d'un plan. La distribution de ces plans dans l'espace suit donc la même loi que la distribution des fissures réelles relevées dans le massif rocheux.

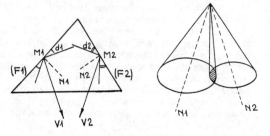

Fig. 3. Critère de dilatance

EQUILIBRE DES BLOCS

Les n plans de fissure caractérisant la fissuration du massif découpent des blocs qui peuvent avoir des formes diverses. On se limitera dans cette étude aux blocs ayant la forme de tétraèdres, c'est à dire aux blocs découpés par quatre plans.
Dans une première phase, on identifiera tous les tétraèdres ayant une base sur le toit de la galerie, donc formés par trois plans de fissure et par le toit.
Cette recherche se fait en combinant trois à trois les n plans de fissure.
Dans une seconde phase, on cherchera à déterminer si ces tétraèdres sont en équilibre.
Deux critères de stabilité ont été retenus: d'abord un bloc peut rester en équilibre parce qu'il est parfaitement emboîté dans le toit; pour le faire tomber, il faudrait le tirer avec une force suffisante pour provoquer des ruptures de cisaillement le long des plans de fissure. Ce critère sera appelé critère de dilatance.
Ensuite, un bloc peut rester en équilibre parce que les forces de frottement à vaincre le long des plans de fissure sont trop importantes. Ce critère sera appelé critère de glissement.

- Critère de dilatance
Les plans de fissure qui affectent les massifs rocheux ne sont pas en général des surfaces parfaitement planes et polies. Il en résulte que si on impose à un bloc de glisser d'une quantité u parallèlement à un plan de fissure, on observera également un déplacement v dans la direction perpendiculaire au plan de fissure. Autrement dit, le bloc se décollera de la fissure. Ce phénomène appelé dilatance est caractérisé par l'angle d définissant la tangente à l'origine de la courbe $v=f(u)$.
En se décollant le long d'une face, le bloc étant supposé indéformable, les autres faces risquent, à l'inverse, d'être comprimées. Le bloc reste alors encastré dans le toit et ne peut pas tomber.
Mathématiquement, la condition d'encastrement se traduit de la façon suivante: si un bloc glisse sur une de ses faces F1, le vecteur déplacement V1 d'un point M1 de cette face devra se situer à l'intérieur d'un cône de révolution ayant pour demi angle au sommet $\pi/2-d1$ (voir figure 3). De même, si le bloc glisse sur la face F2, le vecteur déplacement V2 d'un point M2 situé sur cette face se trouvera à l'intérieur d'un cône de révolution de demi angle au sommet $\pi/2-d2$.
Un mouvement quelconque du bloc peut se décomposer soit en une translation, soit en une rotation. Si on envisage une translation, les vecteurs déplacements V1 et V2 doivent être parallèles et égaux, la distance M1 M2 restant constante puisque le bloc est indéformable. Cela s'exprime par le fait que les deux cônes de révolution doivent avoir une partie commune. Si on considère la troisième face, le même raisonnement permet d'aboutir à la condition nécessaire pour que l'encastrement du bloc ne soit pas possible: les cônes de révolution définissant les directions de mouvement admissible sur les trois faces doivent avoir une partie commune.
Si on envisage une rotation, on peut démontrer que si cette rotation est possible, alors une translation est également possible et on peut donc se contenter de l'examen de la seule translation.
Cette analyse montre la nécessité de caractériser les fissures par un coefficient de dilatance. Bien que rien n'empêche numériquement d'affecter un coefficient de dilatance à chaque fissure, on s'est contenté ici de définir un coefficient unique. Des études et essais sont en cours pour pouvoir mesurer directement in situ ce coefficient au moyen d'un rugosimètre (Mines d'Alès P. WEBER).

- Critère de glissement
D'une manière générale, l'équilibre d'un tétraèdre soumis à un champ de contraintes s'analyse comme

suit: chaque face A_i est soumise à une force normale N_i et une force tangentielle T_i que l'on calcule en résolvant les équations d'équilibre:

$$\Sigma N_i + \Sigma T_i + \Sigma \text{forces extérieures} = 0$$

Il y a équilibre sur la face A_i si:

$$N_i \, tg\, \varphi_i + S_i > T_i \qquad (1)$$

$tg\, \varphi_i$ étant le coefficient de frottement et S_i la force de cohésion de la face A_i.

Dans le cas contraire, on remplace T_i par $N_i \, tg\, \varphi_i + S_i$ sur la face A_i et on recommence le calcul avec pour inconnues la force normale N_i sur la face A_i et les forces tangentielles T_i sur les autres faces.
On itèrera ce processus autant de fois que nécessaire.
Il y a équilibre s'il y a au moins une face sur laquelle l'inéquation (1) est vérifiée.
Ce processus jugé trop long a été simplifié:
- en écrivant que les seules forces extérieures étaient une contrainte horizontale isotrope et le poids du bloc,
- en considérant comme nulle la cohésion des fissures,
- en utilisant un critère pondéré sur les trois faces:

$$\Sigma A_i \frac{N_i\, tg\, \varphi_i}{T_i} < 1.5\, \Sigma A_i$$

pour les tétraèdres dont la projection du sommet sur le toit tombe à l'intérieur de la base, tous les autres types de tétraèdres étant considérés comme stables.

CALCUL DU BOULONNAGE

Le principe adopté pour le calcul du boulonnage est celui du soutènement dit suspendu: un bloc susceptible de tomber doit être accroché par des boulons à une zone saine du toit. Pour cela, les conditions suivantes doivent être remplies:
- un boulon doit être ancré solidement dans le toit supérieur; la longueur d'ancrage doit donc être suffisamment longue surtout s'il s'agit d'un ancrage réparti à la résine ou au ciment,
- le nombre de boulons doit être suffisant pour supporter le poids du bloc.

Les boulons étant dans la pratique implantés suivant une maille régulière, la méthode utilisée ici consiste à passer en revue plusieurs schémas de boulonnage obtenus en faisant varier la densité de boulons (côté de la maille carrée), la longueur des boulons et leur résistance à la rupture. On retiendra un schéma qui garantit l'absence de chute de blocs.
Dans le cas d'une simulation de la structure du massif, les résultats seront fournis sous un aspect probabiliste.

APPLICATIONS

On présentera ici une application de cette méthode de calcul effectuée dans la mine du RECOUX.

La mine du RECOUX (VAR France) exploite un gisement de bauxite en forme d'amas. Les relevés de fissures ont eu lieu dans une galerie principale de 3.5m de large creusée sous un recouvrement de 80m dans la masse de bauxite ainsi que dans une recoupe de 2m de large perpendiculaire à la galerie principale. La galerie est fortement boulonnée, la recoupe n'est pas boulonnée. L'examen des plans de fissures dont la surface n'a ni un faciès conchoïdal ni un faciès lustré a conduit à adopter un coefficient de dilatance de 20 degrés et un coefficient de frottement de 40 à 45 degrés, correspondant aux valeurs que l'on a pu mesurer en laboratoire sur des surfaces de cisaillement.
Quant à la contrainte horizontale σ_h, elle a été prise égale à $0.5\, \sigma_v$ soit 1 Mpa.
Une première analyse a été faite en utilisant directement les relevés faits in situ. On a cherché à vérifier que les résultats du calcul étaient conformes aux observations des exploitants.
On a d'abord traité les données recueillies sur le parement droit de la galerie principale. La longueur du tronçon étudié était de 192m.

Le logiciel fournit les résultats suivants:
- nombre de plans de fissure: 225
- nombre de tétraèdres situés au toit de la galerie:3368
- nombre de blocs ne pouvant tomber parce que les deux critères (dilatance et glissement) ne le permettent pas: 2714
- nombre de blocs ne pouvant pas tomber par suite du seul critère de glissement: 575
- nombre de blocs ne pouvant pas tomber par suite du seul critère de dilatance:0
On a pu vérifier que le critère de dilatance commençait à jouer un rôle très important pour des valeurs de dilatance supérieures à 35 degrés.
- nombre de blocs purgés: 807. On estime que les blocs de faible hauteur tombent immédiatement soit au tir soit au purgeage. Il n'est donc pas besoin de les soutenir. La hauteur de purgeage adoptée dans ce calcul était de 0.40m.
- nombre de blocs tombant sans soutènement: 32 Le poids de ces blocs varie de 9 kN à 130 KN. La hauteur du bloc le plus haut est de 1.15m
- avec des schémas de boulonnage à la maille de 1m ou 1.5m, il subsiste toujours des chutes de blocs. Le schéma à la maille de 0.5m avec des boulons à ancrage ponctuel de 1.5m de long permet d'éviter toute chute. Ce résultat est donc conforme à la réalité, cette galerie étant très boulonnée, parfois avec treillis métallique.

Un calcul analogue effectué à partir des données recueillies sur le parement gauche de la galerie indique 277 plans de fissures, 6342 tétraèdres dont 33 peuvent tomber. Le même schéma de boulonnage que précédemment peut les retenir. Les résultats obtenus avec les deux parements sont donc bien concordants. Les calculs concernant la recoupe n'indiquent aucune chute de blocs, aussi bien avec le relevé du parement droit que le relevé du parement gauche. Là aussi, la confrontation avec la réalité est satisfaisante puisque la recoupe n'est pas boulonnée.

Une seconde analyse a été faite au moyen d'une simulation. Les fissures ont été classées en quatre familles. Trois familles correspondent à des fissures clairement identifiées en azimut et pendage (voir figure 4), la quatrième famille rassemble des fissures très diverses, à azimut et pendage variés qui ne peuvent pas être rattachées à l'une des trois premières familles. Pour chacune de ces familles, on a défini un histogramme azimut-pendage et un histogramme fréquence. La figure 2 reproduit l'histogramme fréquence de la famille 2.
La simulation a consisté à analyser l'équilibre des blocs au toit d'un tronçon de galerie d'une longueur de 100m en effectuant 100 tirages de distribution des fissures dans l'espace.

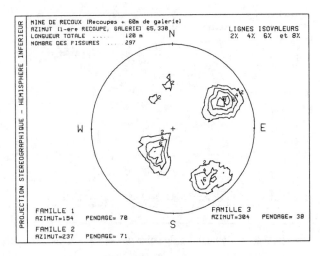

Fig 4. Projection stéréographique (hémisphère inférieur)

Les résultats principaux sont condensés sur les diagrammes de la figure 5. Le schéma 0 représente la probabilité de chute d'un nombre donné de blocs. On voit par exemple que l'on peut obtenir au pire une chute de 51 blocs, mais avec une probabilité de 1%. Au mieux, on n'observera pas de chute de blocs et cela avec une probabilité de 4%.

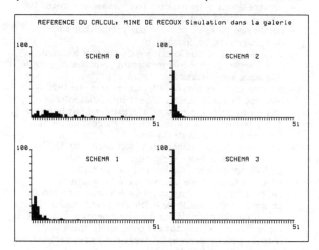

Fig. 5. Probabilité de chutes de blocs

La figure 6 fournit la répartition des hauteurs de blocs qui tombent: on voit qu'il n'existe pratiquement aucune chance d'observer la chute de blocs de plus de 2m de hauteur.

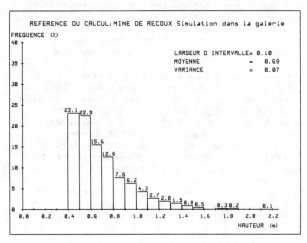

Fig. 6. Répartition de la hauteur des blocs

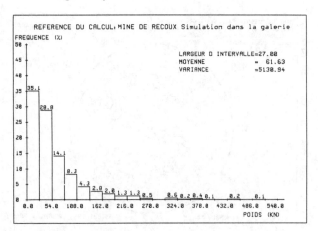

Fig. 7. Répartition du poids des blocs

La figure 7 donne le diagramme analogue en ce qui concerne le poids des blocs.

Les schémas 1,2 et 3 de la figure 5 permettent de juger de l'efficacité du boulonnage. Les caractéristiques de ces schémas sont:
- schéma 1   maille de 1.5m boulons de 1.8m
- schéma 2   maille de 1.0m boulons de 1.8m
- schéma 3   maille de 0.5m boulons de 1.5m.
On voit que seulement le schéma 3 permet de garantir l'absence de chute de blocs selon une probabilité de 100%.

A la suite de ces tests, on en conclut que les résultats de la simulation sont en accord avec la réalité. C'est pourquoi la simulation a été utilisée pour proposer le soutènement des futurs travaux miniers. Ceux-ci consisteront à élargir les recoupes pour en faire des chambres d'exploitation. Leur largeur sera portée à 8.5m.

La simulation montre que dès que la largeur des recoupes dépasse 3.5m, le soutènement devient indispensable. Le schéma 3 de boulonnage est suffisant. Il ne serait pas nécessaire d'augmenter la longueur de boulons malgré l'élargissement de la galerie. En pratique, on pourrait se contenter du schéma 2, à condition d'utiliser un grillage destiné à retenir les blocs échappant à la maille de boulonnage.

CONCLUSIONS

L'objectif à atteindre est de fournir aux exploitants miniers une méthode et un outil informatique aptes à prévoir le soutènement du toit de galeries dans des terrains fissurés. En abordant ce problème par la mécanique des blocs, on risque de tomber dans deux extrêmes:
- ou bien on s'en tient à une analyse sommaire ne dépassant guère l'étude structurale du massif et alors les renseignements que l'on peut en tirer sont surtout d'ordre qualitatif,
- ou bien on analyse en détail les conditions d'équilibre de chaque bloc et la cinématique des effondrements et alors les données nécessaires sont telles qu'il sera illusoire de pouvoir les recueillir: la méthode ne sera pas applicable dans la pratique.

En menant ces recherches, on a voulu surtout mettre en évidence et quantifier les principales raisons pour lesquelles un bloc peut tomber. La démarche adoptée permet de fournir des indications très précises aux mineurs concernant le choix du schéma de boulonnage. Son application doit néanmoins se faire avec beaucoup de prudence et le maximum de vérifications préalables sur les travaux souterrains existants sur le site.

BIBLIOGRAPHIE

BRADY,B.H.G.,BROWN,E.T. 1985. Rock mechanics for underground mining. London:George Allen and Unwin
CRAWFORD,A.M., BRAY,J.W. 1983. Influence of the in situ stress field and joint stiffness on rock wedge stability in underground openings. Can.Geotech J.20,pp276-287.
DOUGLAS,T.H.,RICHARDS,L.R., and ARTHUR,L.J.1979 Dinorwic Power Station:Rock support of underground caverns. Proceedings,4th International Congress on Rock Mechanics, Montreux, Switzerland,Vol.1,pp361-369
GOODMAN,R.E., SHI Gen-hua, BOYLE,W. 1982.Calculation of support for hard jointed rock using the keyblock principle. Proc.23rd Symp.on Rock Mechanics,pp.883-898. Berkeley.

Optical studies of cavity failure in weak sedimentary rocks

Etudes optiques de la rupture d'une cavité dans les roches sédimentaires de faible résistance

Optische Studien der Hohlraum Deformation in weichen sedimentären Gesteinen

ERLING FJÆR, IKU (Continental Shelf and Petroleum Technology Research Institute), Trondheim, Norway
ROLF K.BRATLI, RockMech A/S, Trondheim, Norway
JAN T.MALMO, SINTEF (The Foundation for Scientific and Industrial Research), Trondheim, Norway
OLE J.LØKBERG, Physics Department, NTH (The Norwegian Institute of Technology), Trondheim
RUNE M.HOLT, IKU (Continental Shelf and Petroleum Technology Research Institute, Trondheim, Norway

ABSTRACT: Cavity failure in weak sedimentary rocks have been studied both experi mentally and theoretically. Visual inspection of the cavity during the failure process revealed sand producing zones and cracking prior to failure. Detailed information about the deformation/displacement of the cavity walls was obtained by TV-holographic studies. Cavity deformation was also measured by a cantilever strain gauge system. The stress/strain state around the cavity was studied theoretically by means of a finite element method, revealing the localization of plastified, hence potentially sand producing zones prior to failure.

RESUME: Une étude expérimentale et théorique a été éffectuée sur la rupture d'une cavité dans les roches sédimentaires de faible résistance. L'inspection visuelle de la cavité pendant la rupture a révélé des zones de production de sable et de la fracturation avant la rupture compléte. L'information detaillée apropos la déformation/déplacement de l'interieur de la cavité a été obtenue par les études TV-holographique. La déformation de la cavité a été aussi mésurée avec une systéme de jauges de déformation du type "Cantilever". L'état théorique des contraintes et déformations autour de la cavité a été étudié par les élément finis. Ces résultat théoriques indiquent l'éxistence des zones plastifiées que pourraient don produire de la sable avant la rupture compléte.

ZUSAMMENFASSUNG: Experimentelle und theoretische Undersuchungen der Hohlraumstabilität wurden in weichen sedimentären Gesteinen durchgeführt. Beobachtungen während des Zusammenbrucks zeigten Zonen, aus welchen Sand austritt, und Brüche vor dem eigentlichen Kollaps. Detallierte Informationen über die Deformation und Vershiebung der Hohlraumwände wurden durch TV-Holographie und zusählich mit Deformationsmessgeräten ermittelt. Mit Hilfe der "Finite Element Method" wurde anf theoretischen weg der Stress/Strain-zustand um den Hohlraum beschrieben, um plastisch deformierende Zonen, welche potentiell vor dem Zusammenbruch sand produzieren wurden, zu lokalizieren.

INTRODUCTION

When petroleum is being produced in weak sedimentary rocks, sand production may often be a problem. This happens when the producing cavities no longer are able to withstand the pressure difference and the fluid flow rate in the cavity.

In this paper we describe a series of experiments designed to investigate how a cavity fails due to an increasing pressure difference. The objectives for doing these experiments have been to identify the failure mode(s), to study the cavity deformation prior to and during failure, and to monitor the sand production as pressure increases.

The tests are performed on cylindrically shaped samples having a cylindrical cavity drilled out at one end (Figure 1). During a test, the sample is subject to increasing confining and axial pressure, leaving only the cavity unloaded.

The pressure is increased until the cavity fails.

The loading towards failure situation has been modelled theoretically, using a stress simulation program developed at IKU/SINTEF based on the finite element technique.

The simulation predicts the development and extension of a plastic zone around the cavity (which has also been predicted by analytical methods, see f.i. Risnes et. al., 1981). It has been one of the objectives of these experiments to test the validity of these predictions.

The experimental techniques that have been used to study the cavity failure process are: cantilever strain gauge/LVDT measurements - to monitor the displacement of certain points within the cavity, continuous visual inspection - to reveal information about the sand production at various stress levels as well as the progress of the failure process, and TV-holography - to measure with high accuracy displacements/ deformations of larger parts of the cavity interior.

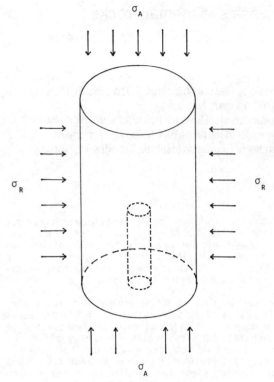

Figure 1 Geometry of sample and stresses.

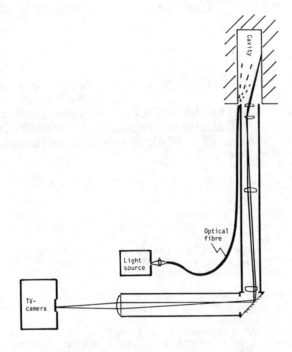

Figure 2 Optical system for visual inspection of the cavity during failure tests.

The investigating techniques are described in the next section of this paper. Some examples of the results that have been obtained are presented in sec   3.

INSTRUMENTATION

The samples to be tested are cut as cylinders, 20 cm in diameter and 25-30 cm long. A cylindrical cavity, 2 or 4 cm in diameter, is drilled out in one end of the sample (see Figure 1). The depth of the cavity is usually taken to be 3 times the cavity diameter.

When prepared, the sample is mounted in a high pressure cell. The sample is then standing on a steel plate with the cavity opening downwards. A hole in the steel plate exactly matches the cavity opening. A plastic sleeve is surrounding the sample, while a steel plate is resting on the top of the sample. Confining pressure σ_R is produced by pressurised oil surrounding the plastic sleeve, while axial stress σ_A is produced by a steel piston resting on the upper steel plate.

Tests are normally run with the relation σ_R/σ_A equal to one (hydrostatic pressure) or equal to a fixed number less than one.

A cantilever strain gauge system was constructed to measure the change in diameter of the cavity during a failure test. The system is capable of monitoring two cavity diameters at orthogonal directions at a given depth of the cavity. This depth may be varied within certain limits. For 4 cm diameter cavities, an LVDT system may simultaneously monitor the displacement of the end surface of the cavity.

An optical system has been especially designed and constructed for the visual inspection and TV-holography tests. The purpose of this system is to transfer a picture from the sample cavity to the target of a TV camera positioned outside the pressure cell. The main requirements to the system are: wide-angle viewing, narrow transmission line, module construction with option for routinely disassembling and reassembling, operational under dirty conditions (sand, oil, dust), extreme stability, and - of course - optimal picturing qualities. The solution was a set of lenses and a mirror mounted in perspex tubes (see Figure 2). The front end lense is protected by a glass plate that is replaced after each experiment.

The front end of the optics is mounted in the hole of the steel plate upon which the sample is resting. Thus the direction of observation is upwards, into the cavity.
To prevent sand and dust from covering the protecting glass plate and thus blocking the view, the plate can be flushed with air from a separate tube. Illumination of the cavity is produced by a white light source positioned outside the cell. A bundle of optical fibres guides the light into the cell. The picture produced by the optical system is focussed on the target of a TV camera. The entire course of all experiments are routinely recorded on videotape.

The optical system is also used in the TV-holography tests (this is why the system has to be extremely stable). The white light source and the TV camera is then replaced with a complete TV-holography system (also called ESPI - Electronic Speckle Pattern Interferometry system), including a He-Ne laser as a light source.

TV-holography as a measuring technique is used to study - with extremely high resolution - displacements and deformations of surfaces. We shall here briefly outline the basic principles of the technique (For a more comprehensive description, see f.i. Løkberg 1980): A picture (in fact a hologram) is detected by a TV camera and stored electronically for later use as a reference. Every subsequent picture (25 per second) is electronically subtracted from the reference, effectively producing a situation where the current picture appears to be optically interferring with the reference picture. Thus, if the object has been deformed since the recording of the reference, the resulting picture will be coated with a fringe pattern. The difference in displacement between two neighbouring fringe lines is 0.31 μm. The resolution is therefore at least as good as 0.3 μm; even better if the picture quality is reasonably good. The technique is sensitive to displacements in one direction only, dependent on the geometry of illumination and viewing. With the geometry used in our experiments, the sensitive direction is along the direction of observation.

The reference picture may be updated at any time (takes a fraction of a second); usually it is done when the interference picture is too overcrowded with fringes. Thus, the technique can be used to study very large displacements, without relaxing on the high resolution.

Since the TV-holography technique reveals information about displacements within a surface, the amount of information revealed is enormous. Clearly, one wishes to use as much of this information as possible, with the least possible effort.
Our solution to this problem has been:

1. Identify a complete set of displacement/deformation "modes" occuring in the experiment

2. Identify the lowest number of point displacements necessary to describe each mode

3. Stepwise analysis of the experiment, assigning each step to a specific mode. It is normally convenient to choose as one step the interval between two subsequent updatings of the reference picture

At present, the technique requires a certain amount of manual interpretation. It is quite clear, however, that this work will be taken over by computers in the future.

Results

A series of experiments has been performed on two types of weak sandstones. These are an artificial silicate cemented sandstone and a fine-grained fluvial deposit from the lower triassic (sampled in Great Britain) which we have called "Red Wildmoor" sandstone. The typical failure mode observed in these experiments is a mirrorsymmetric rupture (see Figure 3) starting at the upper end of the cavity and proceeding downwards. (A similar mode observed in failure tests on coal is described by Kaiser et. al., 1985).

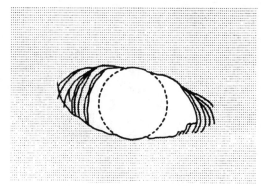

Figure 3 Cross section of a cavity after failure. The section is located half-way between the opening and the end of the cavity. Red Wildmoor sandstone.

Also, when the cavity has a flat end surface, a disc-like piece of the cavity end often falls out, producing a curved end of the cavity, which is apparently more stable.

The theoretical simulation predicts a plastic zone in the cavity end, which is very similar to this disc-like piece. Also, the initiation and extension of plastic zones in the cavity walls, as predicted by the theoretical simulation, confirms to some extent the localization of the observed rupture zones with respect to the depth in the cavity.

The progress of the failure process can be illustrated by the results from a cantilever strain gauge experiment, shown in Figure 4. It is seen that the cavity deformation is small (and probably elastic) up to about 300 bar. Then a large, nonsymmetric deformation (obviously non-elastic) occurs within a very small pressure region. This large deformation is due to the rupturing.

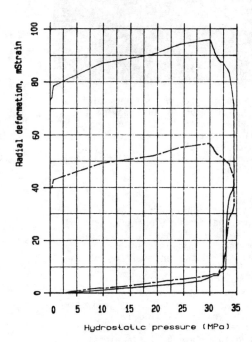

Figure 4 Radial deformation of the cavity versus external hydrostatic pressure. Artificial sandstone.

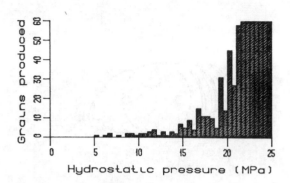

Figure 5 Sand production from cavity versus hydrostatic pressure. Artificial sandstone.

Using the visual inspection technique, it is possible to monitor the sand production as pressure increases.

Figure 5 shows as an example the sand production versus pressure in the lower pressure region, from a test on artificial sandstone. From Figure 5 it is seen that sand starts to drissle at fairly low pressures. However, on the basis of the figures, we suggest that the real sand production starts at 18-19 MPa in this experiment.

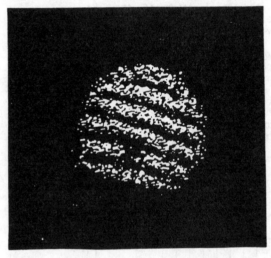

Figure 6 Example of TV screen picture recorded during an experiment on Red Wildmoor sandstone using the TV-holographic technique. The picture shows the circular end-surface of the cavity tilting surface.

The TV-holography technique is used to study displacements and deformations of parts of the cavity surface. We here present as an example how the technique is used to study the deformation of the end surface of the cavity.

During the experiment, a disc constituting the entire end surface is gradually loosening and finally falls down. A typical screen picture obtained during the experiment is shown in Figure 6. The equidistant, straight fringes show that the disc is tilting at this stage. The direction of tilting is revealed by the direction in

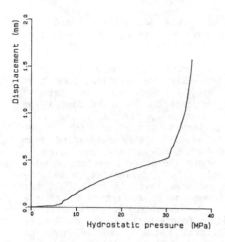

Figure 7 Displacement at the centre of the end surface of the cavity, measured by TV-holographic technique. Red Wildmoor sandstone.

which the fringes are moving. The accompanying displacement rate is found by counting the periods of dark fringes at a reference point (for instance the centre of the disc) and relate it to the pressure increase since the last updating of the reference picture. The displacement at the centre of the end surface, obtained by the TV-holographic technique, is shown in Figure 7. This curve shows only a minor part of the available information, however. In fact, similar curves may have been presented for any point at the end surface. We prefer, however, to present the results as a set of perspective plots, as in Figure 8. Here we see that the surface is bulging and finally tilting as the disc is loosening, in addition to the displacement revealed in Figure 7. Notice that in Figure 8 the displacement has been enhanced by a factor of 50 to make the effect visible.

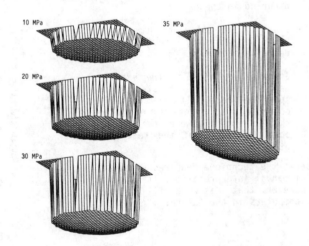

Figure 8 Perspective plots showing the displacement and deformation of the end surface of the cavity, at various hydrostatic pressures. The underlaying data (obtained by TV-holographic technique) are the same as those from which Figure 7 is extracted. The displacement at each surface point is enhanced by a factor of 50, for visibility.

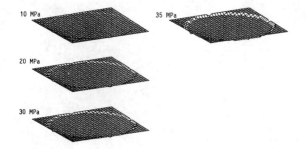

Figure 9 Same as Figure 6, without enhancement of the displacements.

By comparison, the same plots are shown in
Figure 9 without any enhancement of the
displacements. The difference between the
Figures 8 and 9 clearly illustrates the
power of the TV-holographic technique.

Summary

The set of experimental techniques that has
been described, offers possibilities for
extensive studies of cavity failure. In
particular, it is important that the very
sensitive and informative TV-holographic
technique has proved to be operational under
the rather tough conditions of a rock
mechanical experiment. It is our belief that
this technique has a large potential as a
tool for detailed studies of rock mechanical
behaviour.

ACKNOWLEDGEMENT

This work has been funded through IKUs
research program "Formation Mechanics" by
the following companies: Arco Norway Inc.,
Conoco Norway Inc., Elf Aquitaine Norge A/S,
Fina Exploration Norway Utenl. A/S, Norsk
Agip A/S, Norsk Hydro a.s, Statoil, and
Unocal Norge A/S.

References

Kaiser, P.K., Guenot, A. and Morgenstern,
N.R. (1985):
"Deformation of Small Tunnels - IV.
Behaviour During Failure", Int.J.Rock Mech.
Min.Sci. & Geomech.Abstr., 22, 141-152.

Løkberg, O.J. (1980):
"Electronic Speckle Pattern Interferometry"
Phys. in Technol., 11, 16-22.

Risnes, R., Bratli, R.K. and Horsrud, P.
(1982):
"Sand Stresses Around a Wellbore",
SPEJ December.

Inverse problem of design of temporary concrete lining with the aid of measured displacements

Problème inverse du calcul du boisage provisoire en béton selon le mouvement de terrains mesuré

Berechnung des vorläufigen Betonausbaus nach gemessenen Verformungen

NINA N.FOTIEVA, Tula Polytechnical Institute, USSR
N.S.BULYCHEV, Tula Polytechnical Institute, USSR

ABSTRACT: An experimental-analytical technique of tunnel temporary concrete lining design with the application of the data of lining internal surface displacements full-scale measurements is given. The processes of hardening and creepage of concrete taking place simultaneously at the concrete being charged at an early age is taken into account in designing. The design is based upon the solution of the inverse problem of estimating lateral rock pressure coefficient in the intact body by the values of the lining displacements measured.

RESUME: On examine méthode expérimental-analytige du calcul du boisage provisoire en bétone du tunnel selon les mesures du mouvement de sa surface interne. On prend en considération le durcissement du béton frais et son fluage. La méthode du calcul se répose sur la solution du problème inverse déstiné à trouver le coefficient de force latérale du massiv vierge selon les mesures du deplacement du boisage.

ZUSAMMENFASSUNG: Es wird die experimentell-analytische Methode der Berechnung des vorläufigen Betonausbaus des Tunnel auf dem Grunde der Messungen der Verschiebungen der inneren Flache des Ausbaus in natura dargelegt. Bei der Berechnung werden gleichzeitig vollziehenden Prozesse der Erhartung und des Kriechens des Freihalter beladenen Betons in Betracht gezogen. Die Berechnung ist auf der Lösung der Ruckaufgabe über die Bestimmung des Nutzefekts des Seitendrucks des Gesteins auf dem nicht verletzten Massiv nach gemessenen Verschiebungen des Ausbaus gegrundet.

1 FORMULATION OF THE PROBLEM

Processes of hardening and creep of concrete being charged at an early age taking place simultaneously greatly influences the temporary lining stressed state with tunnels being erected at the opening of the face upon a full section with the application of temporary concrete lining, the temporary lining forming later on an element of constant casing.

The technique based upon the solution of N.N. Fotieva (1974) the elasticity theory flat contact problem for the ring of arbitrary form (with a single symmetry axis) supporting the hole in linearly deformed medium from other material, simulating rock massif having an initial stress field called forth by the weight of the rock itself is at present applied for designing temporary lining.

Registration of hardening and creep of concrete is made with the aid of the inverse modulii method by summing up stresses (displacements), arising at moments of overlappings consequently fulfilled. Simultaneously stresses and displacements at the moment of every i-th everlapping is designed at the corresponding $E_1^{(i)}$ value of concrete deformation modulus estimated on the base of S.N. Silvestrov, K.P. Bezrodniy eds. (1982) and at the multiplier α_i^* being introduced into initial stress values, its dependence upon the relative distance of the lining section being designed till the working expressed by formula

$$\alpha_i^* = exp(-1.3\, \ell_{i-1}/R) - exp(-1.3\ell_i/R), \quad (1)$$

where ℓ_0 is the distance from the section where concrete is performed till the working, $\ell_i - \ell_{i-1}$ is the length of overlapping, R is the mean working radius.

However with such a design the results being acquired depend to a great extent upon the value of the coefficient of the lateral rock pressure in an intact massif which is given as a rule approximately at the design stage applying Academician A.N. Dinnick's formula. That is why measurements of displacements of temporary lining internal surface points are made with the face advancing to control the state of the temporary lining.

The technique giving the possibility to estimate the coefficient value of lateral rock pressure in an intact massif applying full-scale measurements data of displacements in internal surfaces of temporary lining (of vertical displacement of the arch upper point and converging of the lateral wall points at a different height from the foot of the working is stated in the paper. Alongside with the further lining design at the estimated coefficient value of lateral pressure the technique presented forms the base of the experiment-analytical design method of the tunnels temporary lining, erected with the base being opened on a full section.

The solution of the inverse problem (with the application of the designed scheme corresponding) of estimating lateral rock pressure coefficient on condition that the lining cross-section internal outline displacements received by method of designing approach the measured one as near as possible in the sense of the least quadratic deviation is put on the base of the technique.

893

2 SOLUTION OF THE INVERSE PROBLEM

The initial elasticity theory contact problem is solved in the book by N.N. Fotieva (1974) with the application of the apparatus of the analytical function theory of the complex variable, conformal reflection, complex series characteristics of the Cauchy type integrals. Formulae for estimating displacements of temporary lining cross-section internal outline points have been received in the paper by N.N. Fotieva, N.S. Bulychev eds. (1986), hardening and creep of concrete being taken into consideration. The formulae acquire the following appearance:

$$V_{x,N} = \frac{\gamma HR(1+\nu_1)}{8(1-\nu_1)} \sum_{i=1}^{N} \frac{\alpha_i^*}{E_i^{(i)}} \left\{ V_x^{(i)}(\bar{I}a) + V_x^{(i)}(\bar{I}b) + \lambda [V_x^{(i)}(\bar{I}a) - V_x^{(i)}(\bar{I}b)] \right\},$$

$$V_{y,N} = \frac{\gamma HR(1+\nu_1)}{8(1-\nu_1)} \sum_{i=1}^{N} \frac{\alpha_i^*}{E_1^{(i)}} \left\{ V_y^{(i)}(\bar{I}a) + V_y^{(i)}(\bar{I}b) + \lambda [V_y^{(i)}(\bar{I}a) - V_x^{(i)}(\bar{I}b)] \right\}, \quad (2)$$

where $V_{x,N}$, $V_{y,N}$ are the vertical and horizontal displacements of the lining section internal outline points at the moment of the $N_{(i)}$-th overlapping correspondingly; $V_x^{(i)}(\bar{I}a)$, $V_y^{(i)}(\bar{I}a)$, $V_x^{(i)}(\bar{I}b)$, $V_y^{(i)}(\bar{I}b)$ are the displacements constituent appearing from loading upon every i-th overlapping estimation algorithm taken from the solution of direct problem at values $\lambda = 1$, and $\lambda = -1$ is given in the paper by N.N. Fotieva, N.S.Bulychev eds. (1986).

The solution of the inverse problem by measured displacements of the upper point of the arch $\tilde{V}_{N_1}$, $\tilde{V}_{N_2}$,..., $\tilde{V}_{N_S}$ (N_1, N_2,..., N_S are the numbers of overlappings after which measurements were made) coming after several overlappings and by wall converging values $\tilde{\Delta} b_{j,N_1}$, $\tilde{\Delta} b_{j,N_2}$,..., $\tilde{\Delta} b_{j,N_S}$ in the j-th points ($j=2,3,...,m$), at the height of h_j from the longwall to estimate the coefficient value of the lateral rock pressure λ is made on the base of formulae (2). Assuming to be basic it is possible to design the stresses and forces in lining.

For that purpose it is necessary to solve a redistributed system of linear algebraic equations in reference to λ. The system has the form:

$$V_{x,1,N} = \frac{\gamma HR(1+\nu_1)}{8(1-\nu_1)} \sum_{i=1}^{N} \frac{\alpha_i^*}{E_1^{(i)}} \left\{ V_{x,1}^{(i)}(\bar{I}a) + V_{x,1}^{(i)}(\bar{I}b) + \lambda [V_{x,1}^{(i)}(\bar{I}a) - V_{x,1}^{(i)}(\bar{I}b)] \right\} = \tilde{V}_N,$$

$$(N=N_1,N_2,...,N_S ; j=2,3,...,m) \quad (3)$$

$$\tilde{\Delta} b_{j,N} = 2 V_{y,j,N} = \frac{\gamma HR(1+\nu_1)}{4(1-\nu_1)} \sum_{i=1}^{N} \frac{\alpha_i^*}{E_1^{(i)}} \left\{ V_{y,j}^{(i)}(\bar{I}a) + V_{y,j}^{(i)}(\bar{I}b) + \lambda [V_{y,j}^{(i)}(\bar{I}a) - V_{y,j}^{(i)}(\bar{I}b)] \right\} = \tilde{\Delta} b_{j,N}.$$

Introducing significations

$$B_{1,N} = \sum_{i=1}^{N} \frac{\alpha_i^*}{E_1^{(i)}} [V_{x,1}^{(i)}(\bar{I}a) - V_{x,1}^{(i)}(\bar{I}b)],$$

$$C_{1,N} = \frac{8(1-\nu_1)\tilde{V}_N}{\gamma HR(1+\nu_1)} - \sum_{i=1}^{N} \frac{\alpha_i^*}{E_1^{(i)}} [V_{x,1}^{(i)}(\bar{I}a) + V_{x,1}^{(i)}(\bar{I}b)],$$

$$B_{j,N} = \sum_{i=1}^{N} \frac{\alpha_i^*}{E_1^{(i)}} [V_{y,j}^{(i)}(\bar{I}a) - V_{y,j}^{(i)}(\bar{I}b)],$$

$$(j=2,3,...m) \quad (4)$$

$$C_{j,N} = \frac{4(1-\nu_1)\tilde{\Delta} b_{j,N}}{\gamma HR(1+\nu_1)} - \sum_{i=1}^{N} \frac{\alpha_i^*}{E_1^{(i)}} [V_{y,j}^{(i)}(\bar{I}a) + V_{y,j}^{(i)}(\bar{I}b)],$$

we approach the system of the form

$$\lambda B_{j,N} = C_{j,N} \quad (N=N_1,N_2,...,N_S; j=1,2,...,m). \quad (5)$$

The solution of system (5) leads to the formula for estimating lateral pressure coefficient in an intact rock body

$$\lambda = \frac{\sum_{j=1}^{m} \sum_{N=1}^{N_S} {}^* B_{j,N} C_{j,N}}{\sum_{j=1}^{m} \sum_{N=1}^{N_S} {}^* B_{j,N}^2}, \quad (6)$$

where the sign * at the sums means that N assumes the values only equal to N_1, N_2, ..., N_S.

As the problem of the elasticity theory being solved with the application of the conformal reflection, the position of the lining cross-section internal outline points is estimated by angle θ in the reflected field, the operations estimating θ_j ($j=2,3,...,m$) corresponding the points of lining walls converging measurements (angle θ_1 =0 corresponds the upper arch point) is added to the computing algorithm given in the paper by N.N. Fotieva, N.S. Bulychev (1986). Those operations consist of estimating coordinates x_p of the lining cross-section internal outline points with the angle θ being changed by 1°, i.e. at $\theta_p = (p-1)\pi/180$ ($p = 1,2,...,181$), estimating values $h_p = x_p - x_{181}$ and finding angles θ_j ($j=2,3,...,m$) where $h_j \approx \tilde{h}_j$.

The algorithm described of processing the data of full-scale measurement has been programmed for the computer.

3 EXAMPLE OF DESIGN

Examples estimating lateral pressure coefficient in an intact rock massif with the aid of temporary lining internal surface measured displacements are given as illustrations below.

The shape and sizes of the lining cross-section are shown in fig. 1.

Initial data for design:
- depth of tunnel embedding H = 180 m;
- weight of rock volume unit γ = 27 kN/m³;
- deformation characteristics of rock E = 2700 MPa, ν_0 = 0.28;
- the Poisson ratio coefficient of concrete ν_1 = 0.2;
- unit strain of concrete linear creep C_N = 0.148·10⁻⁴;
- relative dampness at concrete hardening w = 0.8;
- parametre of concrete composition influence ρ = 0.25;
- characteristics of creep speed a_0 = 135 days;
- length of everlapping ℓ = 2m;
- distance from the face to the concrete lining edge ℓ_0 = 0.5 m;

- time passing till the first loading
  (charge) t_{θ} = 0.5 days;
- time of working cycle T = 1 day.
  In working the tunnel the employees of
the construction organization made measurements of the temporary lining arch
upper point vertical displacements and lining walls converging at a height of 1.5 m
from the working foot. Values of the measured displacements in seven sections along
the tunnel lenght at the definite moments
of time are given in table 1.

Table 1. Values of measured displacements

Numbers of sections	Days	Vertical displace-ments in arch	Converging of walls in lining
1	4 9	- 2.2 - 6.8	4.9 0.9
2	4 9	13.1 11.7	- 3.0 - 4.3
3	1.5 9	0 0	- 1.2 - 2.6
4	3 7	- 3.6 - 3.7	- 4.8 3.0
5	4 12	1.5 - 3.6	- 3.1 - 3.8
6	4 10	5.8 3.1	1.0 4.4
7	3 8	5.6 11.7	- 1.9 3.5

As a result of the design described above
values of lateral rock pressure coefficient
for every section have been received. The
values are given in table 2.

Table 2. Values of lateral rock pressure
coefficient λ

Numbers of sections	λ	Numbers of sections	λ
1	0.28	5	0.83
2	1.46	6	0.70
3	0.81	7	1.02
4	0.54		

Thus, the mean values of lateral rock pressure coefficient is λ = 0.8. Results of
the lining design at the value of λ received are given in fig. 1, where epures
of normal contact stresses σ_ρ , normal
tangential stresses upon an internal outline, bending moments M and longitudinal
forces N at the time moment t = 1.5 days
are shown. The values of those very stresses and forces on the nineth day.

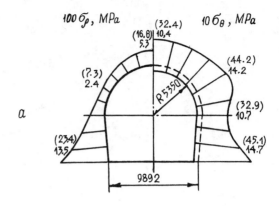

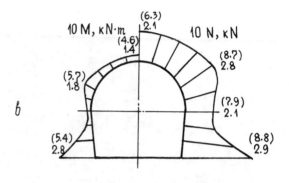

Figure 1. Epures of stresses and forces in
lining

In conclusion we note that the technique
offered may also be developed for processing the results of other measurements,
for example, displacements of several arch
points or lining walls, or wall points converging not placed at the same height from
the foot, etc. In particular a similar method may be applied while measuring tangential deformation of lining.

REFERENCES

Fotieva, N.N. 1974. Design of tunnel li-
    nings of non-circular shape. Moscow:
    Stroyizdat.
Fotieva, N.N. & Bulychev, N.S. (eds.) 1986.
    Ways of increasing reliability of linings
    and reducing material consumption of tun-
    nel linings. Proceedings of "Underground
    structure mechanics", p. 3-17. Tula, Tula
    Polytechnical Institute.
Silvestrov, S.N. & Bezrodniy, K.P. (eds))
    1982. Investigation of interaction of
    massif with concrete lining loaded at an
    early age. Proceedings of "Underground
    structure mechanics", p.67-70. Tula, Tula
    Polytechnical Institute.

Application of stress control methods to underground coal mine design in high lateral stressfields

Application de méthodes de contrôle des contraintes à la conception de mines de charbon souterraines où existent de fortes contraintes latérales
Die Anwendung von Spannungskontrollmethoden im Entwurf von Kohlegruben in einem starken seitlichen Spannungsfeld

WINTON J.GALE, Australian Coal Industry Research Laboratories Ltd, Corrimal, Australia
JAN A.NEMCIK, Australian Coal Industry Research Laboratories Ltd, Corrimal, Australia
ROBERT W.UPFOLD, University of Wollongong, Australia

ABSTRACT: Methods of stress control include roadway orientation to optimise stress redistributions during driveage and utilisation of rock failure about roadways to provide stress relief to adjacent driveage. These strategies allow versatility in mine layout and pillar size whilst obtaining significant driveage optimisation.

RESUME: La Method de controle des constraintes consiste a orienter les tracages afin d'optimiser la distribution des constraintes autour du tracage et prendre l'advantage de la fracturation et la detente des terrains adjacents.

AUSZUG: Methoden der spannungskontrolle beinhalten streckenfuehrung zur optimierung der spannungsverteilung waehrend der streckenvortriebe und machen gebrauch von gebirgsversagen (nachgeben) um ienen spannungsabbau neben den strecken zu erhalten. Dieses vorgehen erlaubt vielseitigkeit in der untertaegigen auslegung der grube und der pfeilerabmessungen bei gleichzeitiger optimierung des streckenvortriebes.

1 INTRODUCTION

Coal measure strata are relatively weak engineering materials which require artificial support and reinforcement to maintain the utility of openings during mining operations (Gale 1986 a). Rock failure in the form of shear fracture through the material, tension fracture and shear along bedding planes and joints readily occurs about roadways created in highly stressed environments. Lateral stress components have been identified as the main contribution to rock failure about roadways driven under these conditions (Gale et al 1984, Gale 1986b).

The existence of high in-situ lateral stresses and relatively weak rock materials requires mine design based on efficient rock reinforcement systems and stress control strategies to optimise mining operations.

The potential application of stress redistributions which occur about rock falls (Agapito et al 1980), extraction panels (Enever et al 1980) and softened ground about roadways (Roest 1982, Karwoski et al 1983, Gale et al 1984, Gale 1986 b) have been noted in other studies.

This paper presents an overview of the stress control strategies currently being applied to reduce lateral stress components together with parameters which have been found to affect their field performance.

2 STRESS CONTROL STRATEGIES

2.1 Roadway orientation based on optimum stress redistribution

Rock failure typically occurs about roadways in highly stressed areas as a result of stress redistributions occurring during development driveage. Minimisation of the stress concentrations about roadways can be achieved by orientation of main roadway developments with respect to the in-situ stressfield geometry. Stressfields measured in the Illawarra Coal Measures have distinct directional characteristics whereby σ_1 is essentially horizontal and the ratio of lateral stress components is (1.3-2):1 (Gale 1986b).

The effect is summarised in Figure 1 which shows factors of safety based on coulomb shear failure criteria normalised as a percentage of the maximum value and plotted relative to the angle at which σ_1 intersects the roadway direction. The data were obtained from three dimensional modelling of stress distributions about the face of a rectangular mine roadway (Gale and Blackwood, in press).

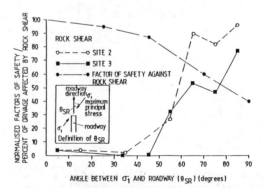

FIGURE 1. THEORETICAL FACTORS OF SAFETY (COULOMB) AND THE OCCURRENCE OF ROCK SHEAR FRACTURING AT THE FACE OF ROADWAYS PLOTTED RELATIVE TO θ_{SR} (AFTER GALE 1986B).

The occurrence of shear fracture about the roadways driven in various orientations to σ_1 is also plotted and indicates the operational benefits of stress minimisation which can result from roadway orientation with respect to the stressfield within the rock mass. These aspects are elaborated elsewhere (Gale 1986b; Gale and Blackwood, in press) and have been noted by others (e.g. Parker 1973).

2.2 Lateral ground destressing utilising rock softening about openings

The occurrence of rock failure in the roof and floor of mine roadways induces a stress redistribution about the failed material. The stress redistribution is affected by the extent of failure about the opening, residual strength and stiffness of the failed material and the nature of the "intact" rock about the excavation. The post failure properties of coal measure rock at low confining pressures (Hassani et al 1984) indicate that failed rock about a mine opening would be ineffective in the transmission of field stresses across the opening. The result of such rock failure would be to enlarge the effective excavation dimensions. This effect has been delineated by stress measurements about mine roadways subject to significant roof/floor failure. Figure 2 shows the results of stress measurements conducted at different roof heights about a roadway. The roadway was driven perpendicular to the in-situ σ_1 and as a result of roof failure, this component (σ_1) was totally relieved in the lower roof. In the upper roof area σ_1 was measured as being in the same direction as the in-situ σ_1 but was oriented to be consistent with redistribution about a significantly larger excavation (Gale et al 1984).

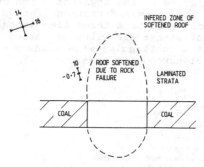

FIGURE 2. STRESS MEASUREMENTS ABOUT A SOFTENED ROADWAY INDICATING THE LATERAL DESTRESSING EFFECT AND STRESS REDIRECTION ABOUT THE SOFTENED MATERIAL (TENSION (-VE)). ROADWAY 5.5 M x 2.5 M.

The effect of such lateral stress redistributions on mining operations is to allow horizontal stress relief for subsequent roadways driven in close proximity. Roadways driven in the laterally stress relieved ground are typified by good driveage conditions (i.e. significantly reduced rock failure), long term stability and reduced reinforcement requirements.

The practical effect of lateral destressing about a softened roadway is summarised in Figure 3 which shows the roof displacements and the incidence of shear fracture above roadways driven in proximity to the softened roadway. In these examples, the stress redistributions about the softened roadway created improved ground conditions up to 60 m adjacent to the initially driven roadway.

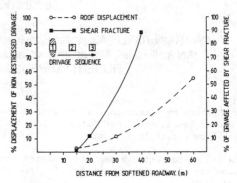

FIGURE 3. PRACTICAL EXPRESSION OF DESTRESSING ADJACENT TO SOFTENED ROADWAYS. (ROADWAY 4.8 M - 5.5 M WIDE.)

The application of this technique is also adopted to 'protect' key excavations of particular application. Lateral stress relief procedures are adopted during driveage of longwall face installation roadways. In high lateral stressfields the driveage of stable and wide roadways (6 m - 7 m) requires the use of stress relief protection roadways driven prior to the wide installation roadway. This technique has proved successful and reduced roof displacements in protected roadways in the order of 50 - 100 fold relative to non-protected equivalent roadways (Figure 3). Other instances where these general principles apply include pillar extraction on the advance (Nicholls 1978) and Wongawilli extraction techniques.

Purposeful roadway sequencing and pillar geometry can allow panel development to occur based on ground destressing strategies and can have significant advantages in optimised driveage conditions, roadway stability and reinforcement requirements.

2.3 Yield pillar methods

The use of narrow failed (yielded) coal pillars as a method to control vertical stress abutments about extraction panels is well established in practice, however, the technique has not been extended to panel design to control lateral stress components.

The existence of practical stress relief in excess of 40 m adjacent to a softened but fully utilised roadway allows considerable versatility in pillar geometry and roadway sequencing. Under these conditions a yielding pillar approach to lateral stress relief is not necessarily appropriate, as ground destressing can be achieved without the requirement of "failed" (yielded) pillars. In this manner the vertical bearing capacity of pillars can be maximised whilst still obtaining practical horizontal stress relief.

3 FACTORS AFFECTING LATERAL STRESS RELIEF ABOUT SOFTENED ROADWAYS

3.1 Dimensions of roof-floor softening

The effect of softening dimensions on the areal extent and magnitude of lateral stress relief as an independent parameter is indicated in Figure 4 for coal seam enclosed by an isotropic massive medium.

Under the conditions modelled, stress relief is limited to a lateral extent of less than 30 m and the magnitude of relief is strongly influenced by the vertical dimensions of roof - floor softening.

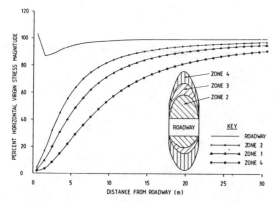

FIGURE 4. EFFECT OF SOFTENED ZONE DIMENSIONS IN A COAL SEAM SURROUNDED BY ISOTROPIC MATERIAL (AFTER GALE 1986B).

3.2 Effect of stiffness contrast within the surrounding strata

The composition of strata about the coal seam is a major factor affecting the magnitude and extent of stress relief. Key parameters affecting lateral stresses adjacent to softened roadways are the stiffness properties and the disposition of stiff and soft rock types. Of particular importance is the existence of bedding planes and bands of material exhibiting stiffness contrasts with the surrounding strata.

The general effects are indicated in Figure 5 which shows the magnitude and extent of lateral stress relief for three strata composition "end members" for constant roof softening geometry. The cases modelled are -
i) A 2.5 m thick coal seam surrounded by massive isotropic material.
ii) A 2.5 m thick coal seam surrounded by stiff strata having thin bands with low shear stiffness along weak bedding planes.
iii) An 8.0 m thick coal seam sequence having thin bands with high shear stiffness along strong bedding planes.

The results show that lateral destressing is most effective in heterogeneous materials particularly where bedding planes or thin bands exhibit low shear stiffness. The existence of bands with high shear stiffness tend to reinforce the soft rock mass. The effectiveness of lateral stress relief is strongly dependent upon rock mass properties and height of softening.

Lateral stress relief is optimised in interbedded stiff strata with discrete bedding planes of low shear stiffness. These conditions are commonly satisfied in coal measure sequences.

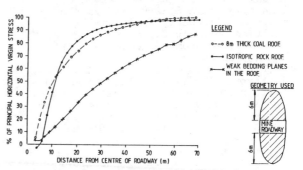

FIGURE 5. EFFECT OF STRATA COMPOSITION ABOUT A ROADWAY ON LATERAL DESTRESSING. ROADWAY WIDTH 5.25 M, HEIGHT 2.5 M.

The rock properties considered in these models are given in Table 1.

Table 1 Rock and Bedding Plane Properties

	Roof Rock	Coal	Weak Bedding Planes	Strong Bedding Planes
E (GPa)	16	4		
v	.25	0.3		
C			0.3	0.5
Ø			15	30
(KT) MPa/m Shear Stiffness			0.1	0.5
(kN) Normal Stiffness			1 GPa	1 GPa

4 CONCLUSIONS

Stress control methods adopted in mining operations within high lateral stressfields are undertaken to optimise roadway driveage and panel sequencing. The methods include roadway orientation to optimise the stress redistributions occurring during development, and the use of softened roadways to provide lateral ground destressing on a panel wide basis.

Lateral destressing can provide practical stress relief in excess of 40 m from a securely reinforced yet softened roadway.

The lateral extent of destressing adjacent to a softened roadway is affected by the composition of strata about the coal seam and the extent of rock failure about the roadway.

The use of lateral destressing strategies allows versatility in mine layout, panel sequencing, pillar sizes and pillar bearing capability, whilst obtaining substantial driveage optimisation.

REFERENCES

Agapito,Jose F.,Scott,Mitchell J.,Hardy,
Michael P.&Hoskins,William N.1980.
Determination of in-situ horizontal
rock stress on both a mine-wide and
district-wide basis. United States
Bureau of Mines O.F.R.143-80.
Enever,J.R.,Shepherd,J.,Cook,C.E.,
Creasy,J.W.,Rixon,L.K.,Crawford,G.,
Dean,A.&White,A.S.1980. Analysis of
factors influencing roof stability at
Wallsend Borehole Colliery. CSIRO Aust.
Div. Applied Geomech.,Geomech. Coal
Mining Rep.No.15.
Gale,W.J.,Rawlings,C.D.,Cook,C.E.,Stone,
I.,Rixon,L.K.,Enever,J.R.,Walton,R.J.&
Litterbach,N.1984. An investigation of
in-situ stressfield and its effect on
mining conditions at Tahmoor Colliery
NSW Australia. CSIRO Aust.,Division of
Geomechanics, Geomechanics of Coal
Mining Report No.49.
Gale,Winton J.1986a. Design consider-
ations for reinforcement of coal mine
roadways in the Illawarra Coal
Measures. Proc.Aus.I.M.M. Illawarra
Branch, Ground movement and control
related to coal mining, Symp.
Wollongong:82-92.
Gale,Winton J.1986b. The application of
stress measurenments to the optimisation of coal mine roadway driveage
in the Illawarra Coal Measures. Proc.
Int.Symp. on Rock Stress and Rock
Stress Measurements, Stockholm:551-
560.
Gale,W.J.&Blackwood,R.L.(in press).
Three dimensional computational
modelling of stress and rock failure
distributions about the face of a
rectangular underground opening and
its correlation with coal mine roadway behaviour. Int.J.Rock.Mech.&
Mining Sci.&Geomech.Abstr.
Hassani,Faramarz,P.,White,Martin,J.&
Branch,David 1984. The behaviour of
yielded rock in tunnel design. Proc.
2nd Int.Conf.on stability in under-
ground mining A.I.M.E.Lexington:126-
142.
Karwoski,William,J.,McLaughlin,William,
C.,Blake,Wilson 1983. Rock pre-
conditioning to prevent rock bursts -
Report on a field demonstration. United
States Bureau of Mines,Report of
Investigations 8381.
Nicholls,B.1979. Pillar extraction on the
advance at Oakdale Colliery. Proc.1st
Int.Symp. on stability in coal mines.
(Edited by C.O.Brawner and
I.P.F.Dorling) Miller Freeman, San
Francisco:182-196.
Parker,J.1973. How to design better mine
openings. Engineering & Mining Journal
Dec.:76-80.
Roest,J.P.A.1982. Conventional deep
mining research. Rock support by a
destressed ring of rock around a
gallery under severe stress. In:J.J.
Dozy(ed) research on the coal beneath
The Netherlands-Geol.Mijnbouw 61:367-
372.

The practice of rock mechanics in South Africa
La pratique de la mécanique des roches en Afrique du Sud
Felsmechanik in Südafrika

N.C.GAY, South African National Group on Rock Mechanics, Johannesburg
B.MADDEN, South African National Group on Rock Mechanics, Johannesburg
W.D.ORTLEPP, South African National Group on Rock Mechanics, Johannesburg
O.K.H.STEFFEN, South African National Group on Rock Mechanics, Johannesburg
M.J.DE WITT, South African National Group on Rock Mechanics, Johannesburg

ABSTRACT: The International Society of Rock Mechanics has been in existence for almost 25 years and the South African National Group has been a member for much of that time.

It is believed that an increased knowledge of the activities and objectives of individual National Groups could further the understanding and community of interest that the parent society seeks to promote. With this purpose in mind, the size and structure of rock mechanics practice in South Africa is briefly reviewed. This gives an indication of the main problems in hard- and soft-rock mining and in surface and sub-surface civil engineering and the emphasis and approach which is adopted in these areas.

RESUME: La Societé Internationale de Mécanique des Roches existe depuis presque 25 ans et le Groupe National sud-africain en est un membre pendant beaucoup de ces années.

On croit qu'une connaissance augmentée des activités et des objectifs de chaque groupe national pourrait augmenter l'entendement et la solidarité d'interêts que la societé paternelle quête d'avancer. Ayant cet objectif en vue, on examine en bref la dimension et la structure de mecanique des roches en Afrique du Sud. Cette communication donne une indication des problèmes principaux en exploitations minières des roches dures et fragiles et en génie civil de surface et souterrain et l'insistance et la méthode que l'on adopte dans ces domaines.

ZUSAMMENFASSUNG: Die 'International Society of Rock Mechanics' besteht seit nahezu 25 Jahren, und die 'South African National Group' ist seit geraumer Zeit Mitglied.

Es besteht die Meinung, dass ein besseres Wissen um die Arbeit und die Ziele der einzelnen 'National Groups' Verständnis und gemeinsame Interessen, wie sie von der Muttergesellschaft angestrebt wird, fördern könnte. Im Hinblick hierauf werden Umfang und Struktur angewandter Felsmechanik in Südafrika kurz betrachtet. Daraus ergibt sich ein Umriss der Hauptprobleme beim Abbau harten und weichen Gesteins, und im Hoch- und Tiefbau, sowie der Zweckbetonung und der Zielrichtung, die auf diesen Gebieten verfolgt werden.

1 INTRODUCTION

The profession of rock mechanics is a relatively small, specialized community which has perhaps a disproportionate responsibility for the safety and the welfare of the society in which it functions. Its working environment is the earth's crust which can be both hostile and unforgiving of error. Tectonic processes from time to time cause the largest natural disasters. However, inadequate understanding of even its more quiet moods can result in design errors which may lead to incidents causing major loss of life.

This great responsibility requires that the profession should draw closer together in the search for improved understanding. International congresses are expressly intended to provide opportunities for the sharing of knowledge and the strengthening of bonds of professional friendship. It is with the hope of encouraging this process that the South African National Group on Rock Mechanics (SANGORM) presents the following description of its structure and a brief outline of the areas of interest of the main groups which make up its membership.

2 MEMBERSHIP

At the time of writing, the total individual membership of SANGORM was 333, in addition to which there are three institutional bodies who are supporting members of the ISRM. South Africa is a country rich in mineral resources with an economy which is, to a large extent dependent on their exploitation. Gold is the largest single contributor with much of the production coming from deep mines with many rock mechanics-related problems. Consequently it is perhaps not surprising that two-thirds of the membership of SANGORM are directly involved with mining. The great majority of these (80 %) are based on the Gold Mines while most of the remainder are from the Coal Mining Industry. Civil Engineers and Engineering Geologists represent the next largest group, together making up 22 % of the total membership; while the remaining 12 % includes people from universities, research organizations and other geotechnical activities. Figure 1 analyses the mining-related membership of SANGORM, together with information on the revenue derived from the various mining activities.

3 GOLD MINING

The total annual SA gold production of 610 metric tons comes from some 30 mines which together process 106 million metric tons of ore, hoisted from mining depths of as much as 3 500 m. At an average working height

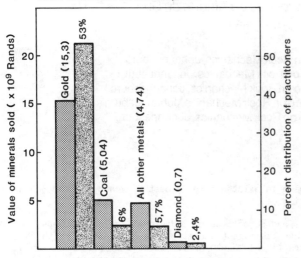

Total mining production value R25,7 x 10⁹
Total SANGORM membership:333

Figure 1. Histogram comparing the value of the various minerals produced with the percentage of rock mechanics practitioners in the South African Mining Industry.

of 1,4 m, this represents an area of about 28 square kilometres of rock extracted each year.

The major rock mechanics problems in the deep South African gold mines arise from high rock pressures. As a result of these pressures, rock around excavations fractures and deforms. This rock behaviour is affected by geological structures, but generally fracturing occurs in a stable manner. On occasions however, rockmass instabilities can lead to rock-bursts; the sudden and violent collapse of part of this fractured rock or closure of the excavations. These rockbursts, together with simple gravity-driven falls of ground, are responsible for a large propor-tion of the accidents occurring in the mining environ-ment.

The occurrence of rockbursts is perhaps the most pressing problem at this time. To alleviate this hazard, improvements are being made to methods of mining, to the design of excavations and their layouts and to methods of supporting the fractured rock. Furthermore, fundamental research is being carried out to obtain a better understanding of the causes of unstable rock behaviour.

Improved, practical procedures for designing mines are under development. These include the design of regional support systems for medium and deep level mining by the proper use of stabilizing pillars or backfill, the siting and development of large service excavations and tunnels, the protection of deep level shafts and the layout of stopes in faulted ground. For this purpose, it is necessary to establish design criteria which allow for the effects of geological discontinuities, such as faults and dykes. Thus, advanced computing methods are continually being developed to model and analyse the behaviour of excavations, their interaction with one another and with geological faults, so that the best layout can be determined.

The effective support of the fractured rock around mining excavations and in varying geological condi-tions, during the rapid deformations which accompany a rockburst, is a major aspect of the problem. The performance of various support systems for stopes and tunnels has been studied in the laboratory and under-ground, and some of the causes of falls of ground have been identified.

Measurements show that the rates of closure in excavations range from less than 1 mm/day to several

m/sec, indicating that support units have to accommo-date large strains over a wide range of strain rates. The fractured nature of the rock being supported also requires that support systems provide adequate areal coverage, so as to maintain the integrity of this rock mass during both normal and rockburst conditions. An example of an inadequately supported tunnel is shown in Figure 2. Current efforts are focussing on enhanc-ing the yieldability (energy-absorbtion capability) and areal coverage of tunnel and stope support systems.

Fundamental research is concerned mainly with under-standing the causes of, and developing methods for the control of rockbursts. From seismic studies, it is known that on average one rockburst occurs for every eight seismic events, although, for events of large magnitude, the frequency of rockbursts is signifi-cantly greater. Thus to reduce rockburst occurrence, the number and size of seismic events must also be reduced. It is also known that most seismic events probably originate close to geological structures, as a result of movement across these zones of potential weakness, when mining-induced stresses perturb the pre-mining state of stress. Unfavourable tectonic stresses would facilitate this movement and thus need to be further investigated. Also needed are guide-lines for the design of mining layouts in geologically disturbed areas.

The identification of potentially hazardous areas of anomalous stress, the early warning of seismic event concentration, and the possiblity of relaxing high or unstable stress conditions are areas of research which are being addressed in the on-going endeavour to control the rockburst/seismic event problem.

Research work on the rock pressure problem is being carried out in various research organizations, includ-ing the Chamber of Mines Rock Mechanics Laboratory, the Research and Development Services of Anglo American Corporation, and Rock Mechanics Departments of various mining groups, consulting companies, and at Universities. In rockburst-prone mining districts, regional or minewide seismic networks provide coverage of seismically hazardous geological structures. On individual mines, close-in networks are being developed as management tools for the evaluation of the potential rockburst hazard in specific areas.

Applied rock mechanics is pursued on almost all the mines at a level of intensity which varies according to the size and depth of the mine and to some extent on the attitudes of management. Mostly this takes the form of a design and advisory service to mine manage-ment in matters such as mine layouts, support methods for excavations, and mining sequences.

Most of the mining activity in South Africa is under the technical control of one or other of 6 major mining groups. Most of these groups were originally founded on gold mining but are now involved in mining activities which range through the complete spectrum of minerals. Consequently, the rock mechanics depart-ments within these groups usually deal with a wide range of problems corresponding to their broad range of activities.

In one group, the rock mechanics effort is structured separately within each of its Gold and Uranium, Diamond, and Coal divisions. Some 80 techni-cally trained people make up the rock mechanics department of the Gold and Uranium division; one-third of these, most of whom have academic qualifications, are involved with research of a more fundamental nature while the remainder provide a routine service of an advisory nature directly to production manage-ment.

Most of the other groups have somewhat smaller complements of rock mechanics staff who generally concentrate on providing an advisory service. The Rock Mechanics Laboratory of the Chamber of Mines Research Organization has the responsibility for carrying out the fundamental and more general applied research which is of relevance to all mining groups. The Laboratory has a staff of about 40 professional

Figure 2. An old arch-supported tunnel at a depth of 2 700 m after a
rockburst of Richter magnitude 2,1. Use of a modern tendon, mesh and
lace support system would probably have protected this tunnel adequately.

Figure 3. Sidewall spalling of a 2,0 m high squat
coal pillar with a width to height ratio of 14:1 at a
depth of 500 m.

Figure 4. Tensile crack damage and block dislocation
on the surface of a bituminous road due to the
approach of 3,0 m high longwall extraction panel 180 m
below surface.

and technical personnel. The direction of research is
controlled by various committees established by the
Mining Industry to ensure that the research programmes
meet the needs of, and are adequately funded by the
Industry.

4 COAL MINING

The total South African coal production is about 180
million metric tons of which some 30 million tons come
from 7 large open-cast operations and the remainder
from underground workings which are mostly of shallow
depth. The aim of the coal mining industry in South
Africa is to increase the percentage extraction of the
reserves in a safe and economic manner. To achieve
this aim rock mechanics research is being undertaken
into the conditions specific to the South African
mining industry.

The two broad areas of rock mechanics investigation
cover, firstly, potential problem areas where rock
mechanics input can achieve a rational and economic
mining sequence and, secondly, areas where rock
mechanics can improve on current methods of extraction
which already produce satisfactory results.

For example, the application of a safety factor

formula has been very successful in designing stable
bord and pillar workings during the past 18 years.
However, this formula is now being reviewed to take
account of individual seam strengths, and to extend it
to the design of pillars at greater depths of mining.
This extended formula takes into account the rapid
increase in strength of squat pillars with large width
to height ratios. Monitoring of the performance of
squat pillars is continuing over a range of conditions
(Figure 3) and is being conducted concurrently with
investigations into the most cost-effective support
methods required to maintain roadway stability at
greater depths.

The presence of dolerite sills, sometimes in excess
of 100 m in thickness, presents problems in panel
layout as well as in the selection of face support in
longwall operations. Monitoring of load variation in
face supports caused by overhanging dolerite sills is
being conducted to assist in the design of future
longwall face support systems.

Total extraction at shallow depths has a profound
effect on the surface. Monitoring of surface
structures such as roads, electric power lines and
buildings (Figure 4) during total extraction has

yielded valuable information on the tolerances that these structures can sustain. As a result of these experiments, restrictions imposed on the undermining of surface structures have now, in some cases, been altered to allow for total extraction. Further investigations are being conducted into this aspect.

Rock mechanics problems associated with mining thick seams on two horizons, or two seams in close proximity to each other, are receiving attention at several collieries. Computer modelling and field trials to monitor the performance of the parting between the seams and the performance of pillars, are being conducted to improve on the currently used multi-seam design guidelines.

The research discussed above is controlled by two committees who determine coal rock mechanics research problem areas, allocate priorities and provide funds to conduct research.

The Strata Control Advisory Committee of the Coal Mining Research Controlling Council comprises members of the mining industry, government departments, research organizations and the Universities. This Council was formed to help solve rock mechanics problems following the Coalbrook disaster of 1960 and continues to oversee problems affecting safety.

The Collieries Research Advisory Committee consists of representatives of the mining houses belonging to the Chamber of Mines of South Africa. This committee allocates funds and directs the research conducted by the Chamber's Coal Mining Laboratory on behalf of mining groups which are members of the Chamber.

Major mining houses also employ their own rock mechanics personnel, the number of which varies widely, depending on group policy and extent of coal-mining interests. The range extends from one person who may occasionally address coal-mining problems, to a department of five rock mechanics personnel actively involved in coal problem areas with their own labora-tory and field testing facilities.

No collieries have rock mechanics personnel actually stationed on the mine, but all do have access to rock mechanics advice either through in-house expertise, the Chamber of Mines of South Africa, or private consultants. The Chamber of Mine's Coal Mining Laboratory and the mining groups together employ a total of 20 people who are devoted exclusively to coal mining rock mechanics problems.

5 SURFACE MINING

Outside of the wide and rather dominant arena of the underground mining world, rock mechanics practice in South Africa extends diversely through the civil engineering fields of slope stability, foundation design, tunnelling, etc..

An area of some overlap is that of surface mining which, in South Africa, takes the form of open-cast mining in the coal industry with fewer but often larger open-pits in iron ore, base metal and diamond mines and overburden stripping in alluvial diamond mines.

A quantitative assessment of the quality of geologi-cal data remains the main concern in ensuring reli-ability in slope design. Very large slope surfaces are continuously being exposed in the mining opera-tion. Only limited information can be gathered with the available resources and time. The most cost-effective utilization of such resources demands that a highly selective programme of data-gathering be followed. A decision criterion is required to ensure that such a process results in a quantitative assess-ment of conditions rather than a subjective judgement.

Application of rock mass classification techniques has been developed to assist in the reduction of the large data bases obtained from exposures, into manage-able design numbers. Use of these numbers with kine-matic models for stability assessments has dominated the design process. Aerial photography has become a valuable method for identifying potentially hazardous areas for spoil piles in large open-cast mining opera-tions. Probability techniques are favoured as a means

for optimizing haul-road layouts.

Successful controlled blasting procedures and more advanced survey surveillance equipment has resulted in the mining of very steep slopes, in excess of 60°.

Almost invariably open-pit mining companies make use of the larger geotechnical consulting firms during both the design and operational phases of a pit, rather than develop in-house expertise.

6 CIVIL ENGINEERING

South Africa is a relatively large country with a generally somewhat arid climate and a population that is mainly concentrated around a few industrial centres and ports. Consequently, the provision and mainte-nance of rail and road networks are of major economic strategic importance. Although there is very little hydro-electric power potential, the impoundment and distribution of water to industrial centres is a vital concern. The demand for rock mechanics understanding in the routine maintenance of these activities is largely provided by the civil engineering firms them-selves, often with assistance from engineering geology consultants when required. From time to time major state-funded projects make extraordinary demands which can only be satisfied by the involvement of the larger rock mechanics consultancy groups from overseas, as well as those locally-based. Under such circumstanc-es, the civil engineering industry has to draw, to some extent, upon the resources and experience of the mining industry for the construction of underground works. Nevertheless, unique problems arise in civil engineering because the work is executed under contract rather than by in-house resources, because the geological environment is often quite different, and because there is a fundamental difference in design philosophy between underground excavations for mining and civil engineering works.

Mining philosophy often conflicts with civil engineering requirements and consequently personnel with a mining background need to be weaned away from emphasis on production to emphasis on quality of the excavated opening. In this respect the civil engineering industry has benefited from working in consortia with foreign contractors whose experience is primarily or wholly in civil engineering construction.

The problem of rock mass characterization for the purposes of defining rock support requirements for tunnel excavations has enjoyed considerable atten-tion. Contractual difficulties have arisen from the use of numerical rock mass classification as a means for defining ground conditions, theoretical support requirements and payment under the contract. This, together with the inherent inability of existing numerical classification systems to accurately predict support requirements, particularly in weak, degrad-able, bedded, sedimentary rock typical of many formations encountered near surface, has given impetus for the development of site specific classification systems in which emphasis is placed on the unambiguous description of ground reference conditions. It is hoped that in this way the potential for conflict which can arise because of different interpretations of the meaning of numerical classes will be avoided and a coherent, identifiable link between geology and support requirements maintained.

Satisfactory definition of ground conditions is dependent upon adequate site investigation. An inte-grated approach progressing through remote sensing techniques (primarily aerial photography), field mapp-ing, core drilling and, less frequently, exploratory excavations, is normally employed. In more homo-geneous strata such as the basalts and sandstones, stratigraphic continuity is vague and therefore opportunities to optimize tunnel alignments in rela-tion to geology are limited. General rock mass quality can be assessed from relatively few borings, and drilling investigations tend to be concentrated at discontinuities (dykes, shear zones). The intensity of drilling typically averages about 100 m of hole per kilometre of tunnel, with a spacing between holes

Drakensberg Basalt

Clarens Sandstone

Interbedded sandstones & mudrocks Elliot formation

Figure 5. Karroo Sequence at the Caledon River Lesotho Highlands Water Project.

ranging from 1 000 to 3 000 m.

Stratigraphic continuity is a feature of the commonly encountered flat lying sedimentary strata (Figure 5). Geological conditions are more predictable and tunnel alignments can with some confidence be optimized in relation to more competent horizons. Hence core drilling to identify the stratigraphic sequence tends to be more intensive, typically averaging about 350 m/km, with the average spacing between holes varying from 250 to 450 m. The intensity of core drilling for tunnels in folded strata is highly variable depending on cover over the tunnel, the opportunities for geological definition and the incidence of discontinuities.

Problems which have arisen because of the presence of weak, moisture sensitive mudrocks (claystones) which degrade rapidly on exposure have been successfully overcome by immediate application of shotcrete in conventionally excavated tunnels. Refined tunnelling techniques will need to be developed for the proposed Lesotho Highlands Water Project where integration of machine boring with shotcrete protection and rapid installation of heavy rock support in a relatively highly stressed environment will be required. Water tunnels through these rocks will have to be concrete lined. In pressure tunnels the potential for erosion occurs if unreinforced concrete linings crack. Solutions adopted in South Africa include the provision of plastic erosion barriers on the rock/concrete interface and prestressing of the lining by grout injection at the interface.

The presence of swelling clay minerals causes certain dolerites and basalts to exhibit rapid weathering properties. This property of the basalts is of particular concern to the Lesotho Highlands Water Project where their presence will influence decisions to concrete line the transfer tunnel and also affect the selection of concrete and rockfill materials. Basic research at the feasibility stage into mechanisms of deterioration will be followed by

further laboratory and in situ investigation at the design stage.

Apart from these directly applied research requirements, fundamental on-going research is largely conducted by State institutes such as the CSIR and Department of Water Affairs and at Universities.

6 CONCLUSION

Rock mechanics as an engineering discipline in South Africa probably had its origins in the formal research effort mounted against the rockburst problems of the gold mining industry early in the nineteen-fifties.

From those early times its growth was slow but generally steady in the mining industry with substantial growth impulses provided from time to time by major civil engineering projects such as the Orange-Fish water transfer scheme, several pump-storage projects, tunnels on the Richard's Bay Coal Line and now the Lesotho Highlands Water Project.

Aided to some extent by the relatively predictable elastic behaviour of the rock mass and the mathematical simplifications afforded by the tabular nature of the ore deposits, substantial contributions to the science of rock mechanics have been made in the way of numerical models. Development of sophisticated seismic monitoring systems has provided considerable understanding of fracturing processes in rock masses.

In other ways and in other fields of application, rock mechanics has become a valuable and even indispensable part of all rock engineering practice. This development has been paralleled by the growth of SANGORM which has, through the years, played an important role in fostering and promoting it. The organization of annual symposia and bi-monthly technical discussion evenings, is the principal way in which this growth is encouraged. With the help and guidance of the parent body, ISRM, this growth and nurturing will continue for many years yet.

7 BIBLIOGRAPHY : SANGORM sponsored publications on
Rock Mechanics in South Africa:

(1) Rock Mechanics in the Design of Tunnels.
    I.S.R.M. N.G. S.AFRICA, Johannesburg, August 1983.

(2) Monitoring for safety in Geotechnical Engineer-
    ing. I.S.R.M. N.G. S.AFRICA, Johannesburg, August
    1984.

(3) Rockbursts and Seismicity in Mines. Eds. N.C. Gay
    and E.H. Wainwright. S.A.I.M.M. Symposium series
    No. 6. (in collaboration with South African
    National Group on Rock Mechanics) Johannesburg
    1984.

(4) Backfill in a scattered mining environment.
    I.S.R.M. N.G. S.AFRICA (Klerksdorp Regional
    Group), Klerksdorp, May 1985.

(5) Rock Mass Characterization. I.S.R.M. N.G.
    S.AFRICA, Johannesburg, November 1985.

(6) The effect of underground mining on surface.
    I.S.R.M. N.G. S.AFRICA, Johannesburg, October
    1986.

Tangential stresses occurring on the boundary of square openings
Contraintes tangentielles en périphérie de cavités carrées
Tangentielle Spannungen am Umfang quadratischer Hohlräume

HASAN GERÇEK, Hacettepe University, Zonguldak, Turkey

ABSTRACT: A formula has been developed for calculating elastic tangential stresses occurring on the boundary of square openings located in a general biaxial in-situ stress field. The formula is relatively simpler and can be applied to five different square shapes with a radius of curvature at the corners varying between 0.06 and 0.007 times the width of the opening.

RESUME: Une formule a été développée pour calculer les contraintes élastiques tangentielles formeés aux surfaces des ouvertures en forme carrée, crées dans le champs d'une contrainte biaxiale in-situ. Etant relativement simple, la formule peut appliquer aux cinq différent carreaux dont les rayons de corbure de ses angles varient entre 0.06 et 0.007 fois de largeur des ouvertures.

ZUSAMMENFASSUNG: In einer allgemeinen zweiachsigen in-situ Spannungsfeld wurde eine Formel entwickelt, die die Berechnung der elastisch tangentialen Spannungen auf der quadratförmigen Oberfläche einer Hohlraum ermöglicht. Diese Formel ist relativ einfach und kann für fünf verschiedene Quadrate angewandt werden, deren Kurvenradius in den ecken sich zwichen der 0.06 und 0.007 fache Länge der seite ändert.

1. INTRODUCTION

The overall aim in engineering desing is to ensure thet the structure or system fulfills it's function without losing it's stability under service conditions and during it's planned lifetime. Stress analysis methods provide valuable information about the stability of structures; therefore, they are indispensable tools of the design engineers. As a matter of fact, the engineers who are involved with the design of underground openings utilize the stress analysis methods along with other design approaches such as empirical methods.

The emergence of computer technology with high-speed and large-storage capacities, and improvements in numerical solution techniques have facilitated the use of computational methods of stress analysis in many areas of modern engineering, including the design of underground openings. Some of these methods (finite element method, boundary element method, etc.) have been successfully used in the design problems of underground excavations involving complex opening geometries, complicated boundary conditions, or nonlinear constitutive behaviour of the medium that may include heterogeneous and anisotropic materials.

However, there may be situations in which the necessary software and/or hardware facilities for using the computational methods are inadequate or the users may lack the sufficient knowledge and experience required for the successful application of these specialized methods. In such cases and especially in the preliminary stages of the design process, some of the avaliable mathematical solutions, most of which include only elastic behaviour, are used for estimating the induced stresses around underground openings. It is an undeniable fact that, when studied with engineering judgement, the distribution of elastic stresses occurring around underground excavations can yield valuable information for the design although rock masses, particularly those which cause instabilities around excavations, rarely behave elastically.

One of the important duties of the design engineer is to predict the potential instabilities due to the induced stresses. In this respect, the first step in the logical methodology for excavation design in massive elastic rock involves determination of elastic stresses occurring at the excavation boundary where they attain their critical values (Brady and Brown, 1985: 185). Then, these critical stress concentrations are compared with the uniaxial tensile and compressive strengths of the rock mass.

2. TWO-DIMENSIONAL ELASTIC SOLUTIONS

In plane elasticity theory involving isotropic and homogeneous materials, when the body forces are absent, the same two-dimensional stress distribution is obtained for both plane stress and plane strain by the solution of boundary value problems, in which only the surface tractions are specified (Malvern, 1969: 512). Therefore, the plane-strain elastic stress distribution around a two-dimensional underground opening (e.g. a tunnel) located at a considerable depth (i.e. in a homogeneous, biaxial in-situ stress field) can be approximated by the stress distribution around a hole of the same shape in an elastic, infinitely lage plate under edge loads parallel to the plane of the plate. Indeed, the classical solutions of the problems dealing with the stresses around holes were applied to the earlier studies related to the stresses around underground openings (Terzaghi and Richart, 1952; Obert et al., 1960). However, the majority of these solutions are limited to the opening shapes, the boundaries of which can be approximated by a simple mathematical expression. In fact, very few cavity shapes, other than circle and ellipse, have been studied, and approximate squares are among those shapes pertinent to mining practices.

The closed-form solutions for the stresses around approximate square holes with rounded corners have been given by several researchers (Greenspan, 1944; Brock, 1958; Heller et al., 1958; Savin, 1961). However, the author believes that these solutions are somewhat impractical for engineers since they involve rather tedious computational steps or lack generality in loading conditions. Therefore, a relatively simpler formula which can be used for calculating the elastic boundary stresses around square openings has been developed.

3. CONFORMAL REPRESENTATIONS

In the theory of two-dimensional linear elasticity, a very useful approach for the solution of boundary value problems involving awkwardly shaped regions is to transform the region into one of simpler shape. Although, in general, the transformation produces more cumbersome boundary conditions, the difficulties of handling these are outweighted by the simpler geometry that results (England, 1971: 129).

Muskhelishvili (1953) developed a very powerful solution method for the boundary value problems of plane elasticity by using the complex variable theory. The method has found important applications to the solutions for stress concentrations around holes in an infinite region which can be mapped conformally onto the exterior or interior of a circle. As a matter of fact, except for the solution by Greenspan (1944), all of the solutions for square openings used this approach. A detailed account of the method can be found in several notable texts (Muskhelishvili, 1953; Sokolnikoff, 1956; Savin, 1961; England, 1971). Therefore, only some important aspects of the method will be treated here.

In the problems involving an infinitely large plate with a hole, the primary concern is mathematical representation of a plane (including also infinitely distant points) from which a finite part having a certain shape (i.e. the shape of the hole) is removed. For this purpose, a conformal mapping function which transforms the region surrounding the hole in the complex z-plane, i.e. $z = x + iy$, onto the interior or exterior of the unit circle in the ζ-plane, i.e. $\zeta = \zeta + i\eta$ or $\zeta = \rho \cdot \exp(i\theta)$, is used (Fig. 1). If the origin of the coordinate system in z-plane is located inside the hole, the function $z = \omega(\zeta)$, which maps the boundary of the hole on the unit circle ($|\zeta| = 1$) in ζ-plane and conformally transforms the region outside the hole onto the interior of the unit circle, will be of the following type (Savin, 1961):

$$z = \omega(\zeta) = R(1/\zeta + c_1\zeta + c_2\zeta^2 + \cdots + c_n\zeta^n) \qquad (1)$$

where R is a constant merely affecting the scale and orientation of the hole and $c_1, c_2, \cdots, c_n$ are the coefficients determining the shape of the hole. As noted earlier, the mapping function $z = \omega(\zeta)$ continuously transforms points in the $\zeta(\rho,\theta)$-plane into points in the z(x,y)-plane. Holding ρ constant and varying θ defines a circle of radius ρ in the ζ-plane and this circle transforms into a closed curve in the z-plane. Holding θ constant and varying ρ defines a radial straight line in the ζ-plane and this line transforms into a curve which is orthogonal to the closed curves (obtained by keeping ρ constant and varying θ) in the z-plane. In other words, the complex conformal mapping function $z = \omega(\zeta)$ transforms the families of concentric circles and radial lines in the ζ-plane into an orthogonal curvilinear net in the z-plane. In the ζ-plane, the unit circle (obtained by setting $\rho = 1$ and varying θ) is used to represent the boundary of the hole in the z-plane.

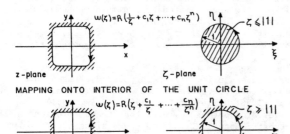

MAPPING ONTO INTERIOR OF THE UNIT CIRCLE

MAPPING ONTO EXTERIOR OF THE UNIT CIRCLE

Figure 1. Basic conformal representations for the problems involving infinite plate with a hole.

In this case, the following parametric equations define the boundary of the hole in the z-plane.

$$x = R(\cos\theta + c_1.\cos\theta + c_2.\cos 2\theta + \cdots) \qquad (2a)$$

$$y = R(-\sin\theta + c_1.\sin\theta + c_2.\sin 2\theta + \cdots) \qquad (2b)$$

Since the mapping function given by Eq.(1) transforms the region surrounding the hole onto the interior of the unit circle, counterclockwise rotation of a point $\zeta = \exp(i\theta)$ on the unit circle corresponds to clockwise rotation of the point $z = x + iy$ along the contour of the hole. If the transformation is made onto the exterior of the unit cincle, the direction of rotation is the same for both ζ- and z-planes.

In the case of conformal transformation, the boundary of the given area, i.e. the contour of the hole, and the contour of the unit circle merge unequivocally, provided that the boundary of the given area is a smooth curve with a continuously changing tangent. However, if the contour of the given hole is a polygon, the change in tangent becomes discontinuous at the corners. An example for this situation is the conformal mapping function which transforms the region surrounding a hole having the shape of a regular polygon with "m" sides, on the inside of a unit circle (Savin, 1961). It is

$$\omega(\zeta) = R\{ 1/\zeta - 2\zeta^{m-1}/[m(m-1)]$$
$$+ 2(m-2)\zeta^{2m-1}/[2m^2(2m-1)]$$
$$- 2(m-2)(2m-2)\zeta^{3m-1}/[6m^3(3m-1)]$$
$$+ 2(m-2)(2m-2)(3m-2)\zeta^{4m-1}/[24m^4(4m-1)]$$
$$- \cdots\} \qquad (3)$$

where m = 3, 4, 5, $\cdots$ (i.e. number of sides). If only a finite number of terms of the series given by Eq.(3) is taken into consideration, approximate polygons with curvilinear sides and rounded-off corners are obtained. Since the series given by Eq.(3) converges rapidly, it is always possible to define a regular polygon with the desired accuracy by using a sufficiently large number of terms in the mapping function (Savin, 1961).

When the mapping function $\omega(\zeta)$ is known, the boundary conditions of the problem given in the z-plane are simultaneously transformed to an appropriate form for the ζ-plane. The problem is solved in the ζ-plane by using two complex potential functions. Finally, the resulting mathematical expressions for stress (or displacement) distributions are inverted to obtain those for the actual problem in the z-plane.

4. TANGENTIAL BOUNDARY STRESSES

In this study, the general mapping function (for regular polygons) given by Eq.(3) is used to obtain a mapping function which transforms the region surrounding a square hole (with rounded corners) onto the interior of the unit circle. The same function was also used by Savin(1961) but he considered only the first four terms and uniaxial loading conditions. It is given below as

$$z = \omega(\zeta) = R(1/\zeta - \zeta^3/6 + \zeta^7/56 - \zeta^{11}/176$$
$$+ \zeta^{15}/384 - 7\zeta^{19}/4864 + \cdots) \qquad (4)$$

The formula developed in this study can be applied to five different cases of approximate squares, starting from the one in which only the first two terms of the function are used to the one in which first six terms of the function are considered. The terms and coefficients appearing in the formula vary as the number of terms used in the above function is changed. Obviously, the relative radius of curvature at the corners and the deviation of the sides from a straight line decrease as the number of terms retained in the mapping function is increased. For example, when first six terms are used, the radius of curvature at the corners is about 0.007 times the side length of the square opening. The parametric equations of the boundary of the square opening with rounded corners are as follows:

$$x = R_n \left[\cos \theta + \sum_{j=2}^{n} c_{j-1} \cdot \cos (4j-5)\theta \right] \quad (5a)$$

$$y = R_n \left[-\sin \theta + \sum_{j=2}^{n} c_{j-1} \cdot \sin (4j-5)\theta \right] \quad (5b)$$

where n is the number of terms used in the mapping function; R_n is the value assigned to coefficient R to obtain a square of unit width (i.e. the distance between the midpoints of opposing sides), and R_n varies with n. The coefficients denoted by "c" are the same ones appearing in the mapping function given by Eq.(4), e.g. for $n = 3$, $c_1 = -1/6$ and $c_2 = 1/56$. Finally, θ is the value of the curvilinear coordinate component orthogonal to the boundary of the hole. As explained earlier, θ increses with clockwise rotation along the boundary. The shapes of the boundary between $\theta = 0$ and $\pi/4$, for different number of terms retained in the mapping function; the radius of curvature at the corners (ρ_C) and at the midpoint of the sides (ρ_A); and the values of coefficient R_n are shown in Figure 2.

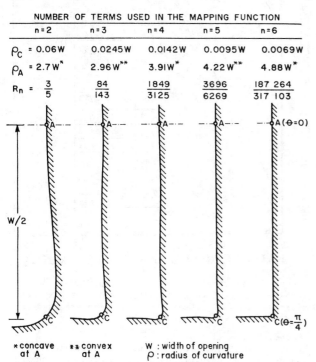

NUMBER OF TERMS USED IN THE MAPPING FUNCTION

	n = 2	n = 3	n = 4	n = 5	n = 6
ρ_C =	0.06W	0.0245W	0.0142W	0.0095W	0.0069W
ρ_A =	2.7W*	2.96W**	3.91W*	4.22W**	4.88W*
R_n =	$\frac{3}{5}$	$\frac{84}{143}$	$\frac{1849}{3125}$	$\frac{3696}{6269}$	$\frac{187264}{317103}$

* concave at A   ** convex at A   W : width of opening ρ : radius of curvature

Figure 2. Types of squares obtained by using different number of terms in the mapping function.

According to the problem for which a solution is presented, the square opening is located in a biaxial in-situ stress field defined by the principal components "P" and "k.P" (Figure 3). The direction in which the component P is acting makes an angle (α) with the horizontal axis ($\theta=0$) of the opening in counter-clockwise direction. Also, the polar angle (γ) of a point on the opening surface (see Figure 3) is given as $\gamma = \arctan (y/x)$. Except for the cases when θ is an integer multipla of $\pi/4$, the absolute values of θ and the polar angle (γ) are not equal.

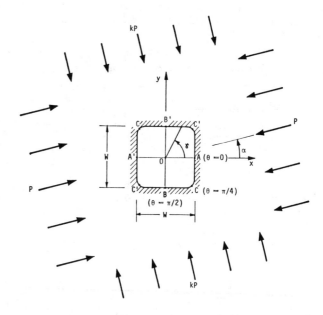

Figure 3. Problem geometry and conditions.

The tangential stress concentration factor at a point on the surface of the opening is given below,

$$\sigma_\theta/P = 4\left[(A.C + B.D) + k(E.C + F.D)\right]/(C^2 + D^2) \quad (6)$$

where the terms A to D can be easily calculated by using the following:

$$A = -0.25 + \sum_{j=2}^{n} \left[a_{2j-3} \cdot \cos 2\alpha . \cos 2(2j-3)\theta \right.$$
$$- b_{2j-3} \cdot \sin 2\alpha . \sin 2(2j-3)\theta$$
$$\left. + a_{2(j-1)} \cdot \cos 4(j-1)\theta \right] \quad (7a)$$

$$B = \sum_{j=2}^{n} \left[a_{2j-3} \cdot \cos 2\alpha . \sin 2(2j-3)\theta \right.$$
$$+ b_{2j-3} \cdot \sin 2\alpha . \cos 2(2j-3)\theta$$
$$\left. + a_{2(j-1)} \cdot \sin 4(j-1)\theta \right] \quad (7b)$$

$$C = -1 + \sum_{j=2}^{n} (4j-5).C_{j-1} \cdot \cos 4(j-1)\theta \quad (7c)$$

$$D = \sum_{j=2}^{n} (4j-5).C_{j-1} \cdot \sin 4(j-1)\theta \quad (7d)$$

where n is the number of terms used in the mapping function. Also, E and F can be found by exchanging "α" with "$\alpha+\pi/2$" in A and B, respectively. The coefficients a,b and c are listed in Table 1 for different square shapes considered in this study.

As expected, for a given state of in-situ stresses, the tangential stresses occurring at the corners are greatly affected by the smallness of radius of curvature of the boundary at these corners. However, this effect is localized around the corners since the stresses occurring at the midpoints of the sides differ slightly with the number of terms retained in the mapping function. At the midpoints of the sides, the slight fluctuation of the tangential stresses becomes insignificant as the deviation of the sides from a straight line gets smaller.

The variation of tangential stress concentration

Table 1. The values of coefficients (a, b and c) to be used with Eq. (7).

Number of Terms Retained in Mapping Function

	n=2	n=3	n=4	n=5	n=6
a_1	3/7	4704/11045	0.4254	0.4253	0.4252
a_2	1/8	59/424	0.1429	0.1444	0.1453
a_3	---*	84/221	0.0428	0.0442	0.0448
a_4	---	-1/32	-0.0419	-0.0458	-0.0478
a_5	---	---	-0.0218	-0.0257	-0.0271
a_6	---	---	1/64	0.0233	0.0268
a_7	---	---	---	0.0144	0.0175
a_8	---	---	---	-5/512	-0.0129
a_9	---	---	---	---	-0.0104
a_{10}	---	---	---	---	7/1024
b_1	3/5	4704/7741	0.6099	0.6109	0.6114
b_2	---	-420/7741	-0.0650	-0.0691	-0.0713
b_3	---	---	0.0312	0.0400	0.0441
b_4	---	---	---	-0.0207	-0.0278
b_5	---	---	---	---	0.0150
c_1	-1/6	-1/6	-1/6	-1/6	-1/6
c_2	---	1/56	1/56	1/56	1/56
c_3	---	---	-1/176	-1/176	-1/176
c_4	---	---	---	1/384	1/384
c_5	---	---	---	---	-7/4864

*not required

factors at the corners, with the ratio and orientation of in-situ stresses, is shown in Figure 4 for the case in which first six terms of the mapping function are used. According to Figure 4, there seems to ba a critical orientation of the in-situ stresses, for which the tangential stresses occurring at opposing corners become constant and equal to the one that occurs in a hydrostatic in-situ stress field, i.e. they are not affected by the ratio of in-situ stresses (k). For example, in the case when only first two terms are used (i.e. n=2), this critical value is given by $\alpha = 0.5 \arcsin(5/8)$. In this case, the tangential stress concentration factor at corner C (i.e. $\theta = \pi/4$) is constant and equal to 6 while it becomes 6k at corner C' (i.e. $\theta = -\pi/4$).

5. CONCLUSION

The formula presented in this study is relatively simpler than those developed previously. This is partly due to the fact that it can be used for calculating only the tangential stresses occurring on the boundary of the opening. Yet, it can be easily utilized by using a simple desk-top calculator or even by hand.
  The formula can be applied to five different, approximately square-shaped openings. The radius of curvature of the opening boundary at the corners varies between 6 and 0.7 percent of the opening width. It can be used in the preliminary stages of the desing of square openings for predicting critically-stressed parts of the boundary, as well as for determining the effect of the orientation and relative magnitudes of the biaxial in-situ stresses on the tangential stress concentrations occurring at the periphery of the opening.

REFERENCES

Brady, B.H.G. and E.T. Brown 1985. Rock mechanics for underground mining. London: George Allen and Unwin.
Brock, J.S. 1958. The stresses around square holes with rounded corners. J. Ship Research 2(2): 37-41,
England, A.H. 1971. Complex variable methods in elasticity. London: Wiley.
Greenspan, M. 1944. Effect of small hole on the stresses in a uniformly loaded plate. Quarterly

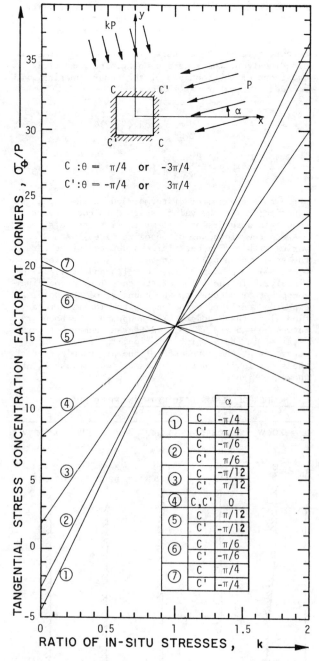

Figure 4. Effect of ratio and orientation of in-situ stresses on the tangential stresses at corners when first six terms of the mapping function are used.

Appl. Math. 2(1): 60-71.
Heller, S.R., J.S. Brock and R. Bart 1958. The stresses around a rectangular opening with rounded corners in a uniformly loaded plate. In Trans. 3rd US congr. appl. mech., p. 357-368.
Malvern, L.E. 1969. Introduction to the mechanics of a continuous medium. Englewood Cliffs: Parentice Hall.
Muskhelishvili, N.I. 1953. Some basic problems of the mathematical theory of elasticity, 3rd edn. Trans. J.R.M. Radox. Gronigen: Noordhoff.
Obert, L., W.I. Duvall and R.H. Merrill 1960. Design of underground openings in competent rock. US Bur. Mines Bull. 587.
Savin, G.N. 1961. Stress concentration around holes. Trans. E. Gros. Oxford: Pergamon.
Sokolnikoff, I.S. 1956. Mathematical theory of elasticity. 2nd edn. New York: McGraw-Hill.
Terzaghi, K. and F.E. Richart 1952. Stresses in rock about cavities. Géotechnique 3(2): 57-90.

Relations between the shear strength of rock joints and rock material
Relations entre le résistance au cisaillement d'un joint rocheux et du matériau rocheux
Verhältnis zwischen der Scherfestigkeit von Felsklüften und des Gesteinsmaterials

C.GERRARD, CSIRO Division of Building Research, Melbourne, Australia
L.MACINDOE, CSIRO Division of Building Research, Melbourne, Australia

ABSTRACT: At high levels of normal stress, an 'equality point' is reached when the shear strength of a rock joint approaches that of the rock material. This point can provide a valuable reference in the development of empirical relations for rock joint strength. Several writers have given methods of estimating this 'equality point'. The different results produced from the application of these methods are compared from the viewpoint of practical consequences.

RESUME: Au niveau élevé de contrainte normal un 'point d'égalité' est atteint quand la résistance au cisaillement d'une diaclase rocheuse se rapproche de celle d'une matrice rocheuse. Ce point peut fournir une référence précieuse dans le développement des relations empiriques de la puissance de la disclase rocheuse. Plusieurs écrivains ont déjà donné le procédé de calculer ce 'point d'égalité'. Les différents résultâts provenant de l'application de cette méthode sont d'un point de vue comparés aux conséquences pratiques.

ZUSAMMENFASSUNG: Wenn die Normalspannung hohe Werte andeutet wird eine 'Gleichheitspunkt' erreicht wobei die Scherfestigkeit einer Felskluft diejenige des ursprünglichen Festgestein annähert. Dieser Punkt kann eine wertvolle Rolle spielen bei der Entwicklung von empirischen Verhältnissen für die Festigkeit der Felskluft. Methoden für die Bewertung dieser 'Gleichheitspunkt' werden schon von mehreren Experten beschrieben. Die praktischen Konsequenzen, die aus einem Vergleich der verschiedenen Ergebnissen bei der Anwendung der obergennanten Methoden folgen, werden gezeigt.

1 INTRODUCTION

In developing empirical relations for the shear strength of rock joints, Gerrard (1986) has emphasised the need to incorporate the physical constraints that apply at very low and very high levels of normal stress. When the normal stress approaches zero, the peak angle of dilation will be maximal and the total peak friction angle will consist of the sum of this dilation angle (Ψ) and the basic friction angle (ϕ_b). At the other extreme, when the normal stress reaches a sufficiently high value, Ladanyi and Archambault (1970, 1980) suggest that the shear strength of the joint will approach that of the rock material at the 'equality point'. This high level of normal stress is referred to as the 'equality stress'. The 'equality stress' falls well outside the stress range of interest in practical rock mechanics applications, in most cases being well in excess of the uniaxial compressive stress(c). However, it does enable the position, and to a large extent the slope, of the $\tau-\sigma$ strength envelope for the rock joint to be fixed for the extreme case of highly elevated stress. In addition, when taken together with the constraints that apply to the location and slope of this envelope at very low levels of normal stress, it provides the opportunity to make an 'educated' interpolation into the practical stress range.

This is shown in Figure 1 where the 'equality point' and the corresponding 'equality stress', f, for the shear strength envelope of a typical rock material (τ_m) are marked. In addition, the constraints for $\sigma \rightarrow$ O are shown together with the interpolated shear strength envelope for the rock joint (τ_j).

The estimation of the position of the 'equality point' can be approached in two stages: (a) definition of the line that represents the shear strength envelope for the rock material, and (b) determination of the location along this line that represents the 'equality point'. These two stages are discussed in subsequent sections of this paper.

By drawing attention to the location of the 'equality point' this paper provides a necessary prerequisite to the further development of empirical relations for the shear strength of rock joints.

2 ROCK MATERIAL STRENGTH

A comprehensive analysis of the data for the shear strength of rock material has been conducted by Hoek and Brown (1982). They found that, to a reasonable degree of approximation, the range of rock types could be grouped into five categories, referred to here as Groups I to V. For each group the shape of the failure envelope is independent of the uniaxial compressive strength, c.

Expressed in terms of shear stress and normal stress, the Hoek and Brown equation is given as,

$$\tau/c = q \ (\sigma/c + t/c)^p,$$
OR
$$\tau^* = q \ (\sigma^* + t^*)^p, \tag{1}$$

where o < p < 1, t is the uniaxial tensile strength, and the starred quantities indicate division by c and hence they become relative to c. For example, t^* is referred to as the relative uniaxial tensile strength.

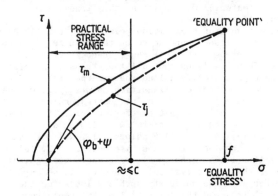

Figure 1. 'Equality point' and strength envelopes for rock material and rock joint.

Hoek and Brown suggest that the parameters q, p, and t* depend only on the Rock Group that a particular rock belongs to.

The physical significance of the quantity q is that it is equal to the value of the relative shear strength (τ^*) corresponding to a relative normal stress (σ^*) of $1-t^*$. The exponent p has the effect of altering the curvature of the shear strength envelope, with increases in p producing an increased tendency toward linearity. The above points are illustrated in Figure 2 where the values of q and t* are fixed and the effect of varying the exponent p is shown.

Group I rocks consist of carbonate rocks with well developed crystal cleavage (e.g. dolomite, limestone and marble) and have the values of q(0.816), p(0.658), and t*(0.140). At the other extreme, Group V rocks are represented by coarse-grained polymineralic igneous and metamorphic crystalline rocks (e.g. amphibolite, gabbro, gneiss, granite, norite, and quartzdiorite). The corresponding values of the parameters are q(1.220), p(0.705), and t*(0.040).

It has been shown by Gerrard and Macindoe (1986) that a relationship exists between the three parameters q, p, and t* so that only two of them can be considered independent. Hence, in subsequent analysis only the exponent, p, and the relative uniaxial tensile strength, t*, are considered as variables. In these analyses the exponent, p, varies over the range of 0.5 to 0.9 while t* varies between 0.04 and 0.20.

3 COMPARISON OF METHODS OF ESTIMATING 'EQUALITY POINTS'

Based on the work of previous authors, there are several methods of estimating the position of the 'equality point' along the shear strength envelope of the rock material. Each of these methods will have different influences on the interpolation of the shear strength envelope for the rock joint since this depends on the 'equality point' position.

Ladanyi and Archambault (1970, 1980) suggest that the position of the 'equality point' for a rock joint will be approximately the same as the position of the 'transition point' for the rock material within which the joint lies. The concept of a 'transition point', and a corresponding level of normal stress known as the 'transition stress', has been developed by Mogi (1965) and Byerlee (1968) to describe the point at which the failure of the rock material changes in character from brittle to ductile. At the 'transition stress' the rock material will undergo shear deformation without significant change in shear resistance. Such deformation can be continued until a continuous fracture is formed. This sensible equality between the shear strengths of the rock material and the rock fracture (joint) is the basis of the assertion that the 'equality point' is equivalent to the 'transition point'. In this context it is of interest to note that peaked behaviour in the shear strength of rock joints disappears as the 'equality stress' is approached and, similarly, peaked behaviour for rock material disappears as the 'transition stress' is approached.

The analyses of 'transition points' by Mogi (1965) and Byerlee (1968) were based on the results of extensive testing over the full range of rock material types. They concluded that all the 'transition points' can be represented by a line that is therefore referred to as the loci of 'transition points', τ_t. On the basis of each 'transition point' corresponding to the 'equality point' for that particular rock, it follows that all the 'equality points' will lie on a line, i.e. the loci of 'equality points', τ_e, will be equivalent to the loci of 'transition points', τ_t. This concept is illustrated in Figure 3 where the equivalence of the respective loci, τ_e and τ_t, is shown together with the shear strength envelopes for the material (τ_m) and the joint (τ_j) of two rock types.

In the analysis of 'transition stress' by Mogi (1965), it was found that for silicate rocks, the loci of 'transition points' was given by the straight line,

$$\tau_t = 0.8 \, \sigma. \tag{2}$$

For carbonate rocks, Mogi suggests lower 'transition stresses' apply, thereby giving the plot of the loci of 'transition points' a concave downwards characteristic. Byerlee (1968), in analysing test results with normal stress levels up to 700 MPa, concludes that this concave downwards characteristic applies to all rocks.

In order to make use of this curved plot of Byerlee (1968), a curve fitting technique has been used. Over the range of normal stress from 0 to 500 MPa, a cubic relationship is used, with a straight line taking over when the normal stress exceeds 500 MPa. Hence,

$$\tau_t = 1.058\sigma - 0.001026\sigma^2 + 0.000000708\sigma^3,$$
$$\text{for } 0 < \sigma < 500 \text{ MPa} \tag{3}$$
$$\tau_t = 79.5 + 0.563\sigma, \text{ for } \sigma > 500 \text{ MPa}.$$

This fit of the curved characteristic of Byerlee is shown as 'Byerlee A' in Figure 4, for the ranges of normal stress 0-500 MPa (Figure 4a) and 0-2000 MPa (Figure 4b).

In a later study, Byerlee (1978) conducted a detailed examination of the frictional properties of rocks. He implies that the characteristic for maximum friction is equivalent to the loci of 'transition points' applicable to the rock material. The wide scatter in the frictional data that occurred at low levels of normal stress, i.e. up to 5 MPa, was attributed to the effects of differing degrees of surface roughness. At moderate to high levels of stress, i.e. up to 100 MPa, and for extremely high levels of stress, i.e. up to 2000 MPa, the scatter was significantly reduced and there seemed to be little dependence of friction on rock type. Byerlee qualified his conclusions by indicating that the friction would

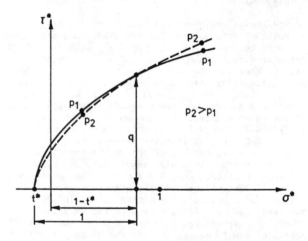

Figure 2. Parameters describing the shear strength of rock material.

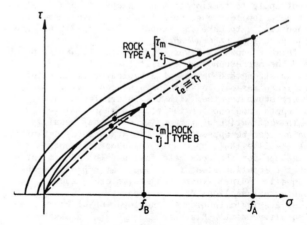

Figure 3. Loci of 'equality points'.

912

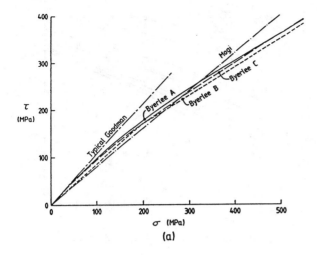

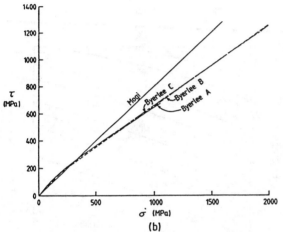

Figure 4. Suggested loci of 'equality points':
(a) 0-500 MPa range; (b) 0-2000 MPa range.

be much lower in cases where the sliding surfaces were
separated by large thicknesses of gouge material.

Byerlee (1978) expresses his maximum friction
characteristic, and, by implication, the loci of
'transition points', in the following bi-linear form,

$$\tau_t = 0.85\sigma, \qquad 0 < \sigma < 200 \text{ MPa}$$
$$\tau_t = 50 + 0.6\sigma, \qquad 200 < \sigma < 2000 \text{ MPa.}$$ (4)

This is shown in Figure 4 as 'Byerlee B'.

For some analyses it is desirable to have a
continuous function to represent the Byerlee results
over the range of normal stresses 0-2000 MPa. Hence, a
hyperbola was fitted to approximate to the Byerlee A
characteristic at low levels of normal stress and to
the Byerlee B characteristic at high levels of normal
stress,

$$\tau_t = 74 \left\{ \frac{0.005608\sigma + 0.00003505\sigma^2}{1 + 0.005608\sigma + 0.00003505\sigma^2} \right\} + 0.585\sigma. \quad (5)$$

This hyperbola is shown as 'Byerlee C' in Figure 4.
One of the features of this curve is that its gradient
is unity when $\sigma = 0$. This compares with a gradient in
excess of unity for Byerlee A and a gradient of 0.85
for Byerlee B.

Mogi's (1965) estimate of the loci of 'transition
points', and hence loci of 'equality points', is given
by Equation (2). When used in conjunction with the
material strength relation, Equation (1), predictions
can be made of the relative transition stress,
$f^*(= f/c)$. Contours of f^*, for ranges of the exponent
p from 0.5 to 0.9 and of the relative uniaxial tensile
strength, t^*, from 0.04 to 0.20 are shown in Figure 5.
In the corner corresponding to low values of exponent
p and high values of t^*, the relative 'equality

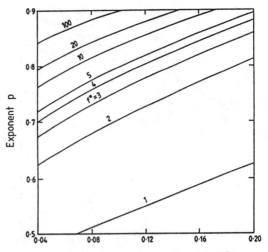

Figure 5. Estimates of relative 'equality stress' (f^*)
based on Mogi (1965).

stress' (f^*) is minimal and less than unity. In the
opposite corner, i.e. high p and low t^*, the relative
'equality stress' (f^*) is maximal with values in
excess of 100. This is unrealistically high and
suggests an impractical combination of p and t^*
values.

Because of the linear nature of the Mogi
relationship, the values of the relative 'equality
stress' (f^*) depend only on the exponent p and the
relative uniaxial tensile strength (t^*). They are
independent of the uniaxial compressive strength (c).
On the other hand, the estimates of the relative
'equality stress' provided by Byerlee (1968, 1978) are
functions of c. This results from the curvilinear
(lines A and C on Figure 4) or bi-linear (line B on
Figure 4) nature of the Byerlee characteristic.

Estimates of the relative 'equality stress' (f^*)
based on Byerlee, have been provided in the form of
contour plots in Figure 6 for c values of 50, 100, 200
and 400 MPa. Values of exponent p ranging from 0.5 to
0.9 and of the relative uniaxial tensile strength (t^*)
ranging from 0.04 to 0.20 have been used in these
plots. The bi-linear form of the Byerlee
characteristic, i.e. Equation (4), was assumed.

In general, in Figure 6, the contours fall into two
regions, separated by a discontinuity. The position of
this discontinuity is along $f^* = 4$ for c = 50 MPa, f^*
= 2 for c = 100 MPa, and $f^* = 1$ for c = 200 MPa. The
discontinuity arises because of the relative 'equality
stress' (f^*) being specified by the intersection of
two downwardly concave curves, i.e. the material
strength curve and the 'equality point' curve. The
lower curve, i.e. the 'equality point' curve,
undergoes a maximal change of gradient at a value of
about $\sigma \approx 200$ MPa, thereby producing an ambiguity
regarding the intersection, and hence causing the
discontinuity. This can be seen from Figure 7 where
values of exponent p and t^* have been chosen so that
the shear strength envelope of the rock material just
touches the 'knee' in the bi-linear 'equality stress'
curve. For a very small increase in exponent p the
rock material envelope will clear the 'knee', with the
intersection occurring at a relatively large distance
above the 'knee'. For a very small decrease in
exponent p the rock material envelope will have one
intersection below the 'knee' and two above it. For
further decreases in exponent p there will be only one
intersection and this will occur below the 'knee'.

It is important to note how very small changes in
exponent p can result in very large changes in the
'equality stress' (f) and hence the 'relative equality
stress' (f^*).

By comparing Figures 5 and 6, it can be seen that,
below the discontinuity in the contours of the
relative 'equality stress' (f^*), the Byerlee estimates

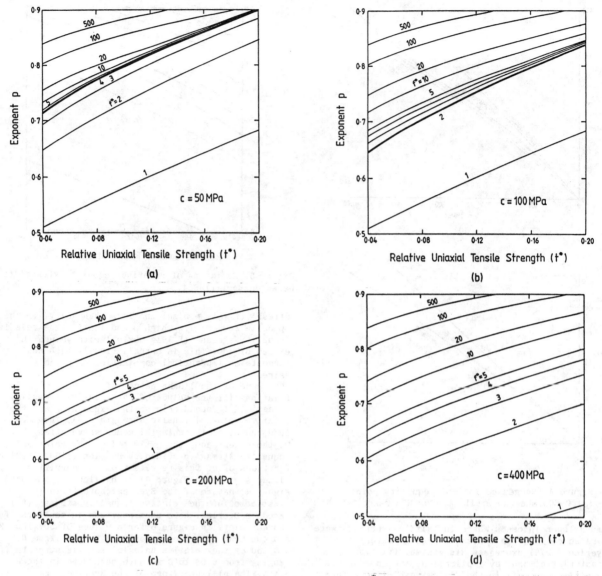

Figure 6. Estimates of relative 'equality stress' (f^*) based on Byerlee (1978).

of f^* are lower than those of Mogi. This tends to occur for relatively low values of c, relatively low values of exponent p, and relatively high values of t^*. This phenomenon is simply due to the gradient of 0.8 in Equation (2) (Mogi) being lower than the gradient of 0.85 in Equation (4a) (Byerlee).

The results shown in Figures 5 and 6, whilst covering general ranges of exponent p and relative uniaxial tensile strength, t^*, can be applied to those values relevant to the Hoek and Brown (1982) Rock Groups. For these, p and t^* both vary monotonically from 0.658 and 0.14 for Group I to 0.705 and 0.04 respectively for Group V. The corresponding values of the relative 'equality stress' (f^*) are shown in Table 1.

Table 1 Values of Relative 'Equality Stress' (f^*)

	Rock Group I	Rock Group V
Mogi (1965) (see Figure 5)	1.3	4.3
Byerlee (1968, 1978) (see Figure 6)		
c = 50 MPa	1.1	3.5
c = 100 MPa	1.2	7.5
c = 200 MPa	1.5	10.0
c = 400 MPa	2.1	11.0

Table 1 shows that the range of f^* values varies from 1 up to about 11, with higher values applying to Rock Group V and, in the case of Byerlee (1968, 1978), to higher values of uniaxial compressive strength, c.

The estimation of the 'equality stress' from Mogi's and Byerlee's results is based on the assumption of equivalence between the 'transition stress' and the 'equality stress'. A different approach has been adopted by Goodman (1976) who suggested that the 'equality stress' will be approximately the same magnitude as the uniaxial compressive strength (c). In contrast with the Mogi and Byerlee estimates of 'transition stress', which are independent of rock type, this assumption directly links the 'equality point' with the strength of the particular rock type.

For the Goodman approach, specification of values for exponent p and the relative uniaxial tensile strength (t^*) will mean that the 'equality points' for different values of unaxial compressive strength will lie on a straight line of gradient μ_G. This gradient changes for different values of exponent p and t^*. For the ranges of exponent p from 0.5 to 0.9 and t^* from 0.04 to 0.20, contours of μ_G are plotted in Figure 8. It can be seen that the values vary from about 0.7 to about 2.0 and, for most cases, tend to predict lower values of the 'equality stress' than either Mogi or Byerlee. For exponent p = 0.7 and t^* = 0.08 the 'typical Goodman' estimate of a line of 'equality

914

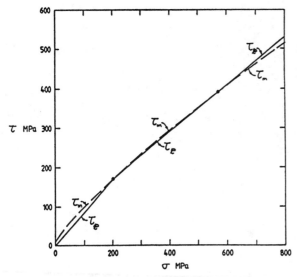

Figure 7. Intersection regime between the shear strength envelope for the rock material (τ_m) and the loci of 'equality stresses' (τ_e).

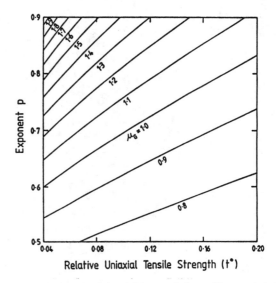

Figure 8. Gradients of lines of 'equality points' based on Goodman (1976).

points' is shown in Figure 4, corresponding to μ_G = 1.04.

In considering Goodman's estimate of 'equality points' it is of interest to note that an 'equality stress' in excess of the uniaxial compressive strength can only be approached when there is significant lateral constraint on the rock joint.

4 PRACTICAL CONSEQUENCES

(a) In general, differences in the estimates of the 'equality stress' (f) will have little effect on the estimated rock joint strength in the civil engineering range, i.e. up to 5 MPa. Here, the strength is mainly controlled by a combination of basic friction and dilation.

(b) However, for the mining engineering range, i.e. up to about 100 MPa, differences in the estimation of the 'equality stress' will have a significant effect on the estimated rock joint strength.

(c) The larger the estimate of the 'equality stress' (f), the lower the estimated rock joint strength will be at these intermediate to high values of normal stress, i.e. higher estimated values of f corresponds to more conservative estimates of rock joint strength.

(d) Goodman's estimate of 'equality stress' is always less than Mogi's or Byerlee's, except in the zone of low values of the exponent p and high values of the relative uniaxial tensile strength (t^*). In the case of Byerlee, this zone disappears when the uniaxial compressive strength (c) is large.

(e) For relatively low values of c, the Byerlee estimate of f is always lower than the Mogi estimate. However, for relatively high values of c, the reverse is true.

(f) For the Hoek and Brown (1982) Rock Groups I to V the estimated values of the relative 'equality stress' (f^*) are 1.3 to 4.3 for Mogi (1965) and 1.1 to 11 for

Byerlee (1968, 1978). Higher values of f^* are associated with Rock Group V and, in the case of Byerlee, with higher values of c.

(g) In general, the consequences of the estimation of the 'equality stress' (f) on the estimation of the rock joint strength (r.j.s.) will be,

r.j.s. based > r.j.s. based > r.j.s. based
on Goodman   on Mogi   on Byerlee

However, for the special case when the exponent p is low, the relative uniaxial tensile strength (t^*) is high, and the uniaxial compressive strength (c) is low, the above order is reversed.

REFERENCES

Byerlee, J.D. 1968. Brittle-ductile transition in rocks. J. Geophys. Res., 73(14):4741-4750.
Byerlee, J.D. 1978. Friction of rocks. Pure Appl. Geophys., 116:615-626.
Gerrard, C. 1986. Shear failure of rock joints: appropriate constraints for empirical relations. Int. J. Rock Mech. and Min. Sci. (in press).
Gerrard, C. & L. Macindoe 1986. Empirical relations linking the shear strength of rock joints and rock material. CSIRO Division of Building Research, Internal Paper No. 86/40.
Goodman, R.E. 1976. Methods of geological engineering in discontinuous rocks. West, New York.
Hoek, E. & E.T. Brown 1982. Underground excavations in rock, 2nd Ed. Inst. of Mining & Metallurgy, London.
Ladanyi, B. & G. Archambault 1970. Simulation of shear behaviour of jointed rock mass. Proc. 11th Symp. on Rock Mech., California, 105-125.
Ladanyi, B. & G. Archambault 1980. Direct and indirect determination of shear strength of rock mass. Preprint 80-25, Soc. of Min. Engrs of AIME, Annual Meeting.
Mogi, K. 1965. Pressure dependence of rock strength and transition from brittle fracture to ductile flow. Bull. of Earthquake Res. Inst. 44:215-232.

Time-dependent ground response curves for tunnel lining design
Courbes caractéristiques différées pour le calcul du revêtement des souterrains
Zeitabhängige Gebirgskennlinien für die Bemessung von Tunnelauskleidungen

D.E.GILL, Department of Mineral Engineering, Ecole Polytechnique, Montréal, Québec, Canada
B.LADANYI, Department of Civil Engineering, Ecole Polytechnique, Montréal, Québec, Canada

ABSTRACT: It has been shown recently (Gill and Ladanyi, 1983;, Ladanyi and Gill, 1984)that the conventional convergence-confinement method, based on isochronous characteristic lines, generally underestimates both the ground pressure on the lining and the tunnel wall convergence. By applying that method to a ground showing a Zener-type creep, and using an exact solution as a basis, the authors show in this paper that this underestimate decreases with time, tending to zero at long term, when the time tends to infinity. A simple numerical method for calculating isochronous characteristic lines, which takes into account the loading history, and requires only a pocket calculator, is finally proposed.

RESUME: On a démontré récemment (Gill et Ladanyi, 1983; Ladanyi et Gill, 1984) que la méthode convergence - confinement conventionnelle, basée sur les lignes caractéristiques isochrones, tend à sousestimer, tant la pression des terrains, que la convergence de la paroi du tunnel.En appliquant cette méthode à un terrain caractérisé par le fluage du type Zener, et en se basant sur une solution exacte du problème, les auteurs montrent que cette sousestimation diminue avec le temps, tendant vers zéro lorsque le temps tend vers l'infini. On propose finalement une simple méthode numérique, n'utilisant qu'une calculatrice de poche, permettant de déterminer approximativement les lignes caractéristiques correctes, qui tiennent compte de l'histoire de chargement.

ZUSAMMENFASSUNG: In zwei vorläufigen Arbeiten (Gill und Ladanyi, 1983; Ladanyi und Gill, 1984) wurde gezeigt, dass die übliche Gebirgskennlinienmethode, die isochrone Kennlinien benutzt, im Algemeinen zu einer Unterschätzung, sowohl des Gebirgsdruckes auf die Tunnelauskleidung, als auch zu einer Unterschätzung der Ausbruchrandverschiebungen führt.. In der vorliegenden Veröffentlichung wurde diese Methode auf ein viskoelastisches Gebirge angewandt, um zu beweisen, dass diese Unterschätzung mit der Zeit abnimmt und gänzlich verschwindet, wenn t unendlich wird. Es wird ebenfalls am Ende der Veröffentlichung ein einfaches numerisches Verfahren vorgeschlagen, das die Belastungsgeschichte berücksichtigt und das auf einem einfachen Taschenrechner ausgeführt werden kann.

INTRODUCTION

The design of tunnel linings has been extensively discussed in the recent literature, and many different design methods have been proposed for that purpose, as can be seen from the surveys made by Ladanyi(1980) and Brown et al.(1983). Among these methods, the characteristic-line ground support design approach, or the convergence-confinement method, has received a considerable attention over the last two decades, even though this concept implies relatively simple conditions. The popularity of this method is mainly due to its clarity and versatility. Used primarily for the design of artificial ground supports, the technique has been recently applied also to the design of mine pillars (Sḁkkḁ, 1984).

The method, introduced originally by Lauffer and Seeber (1961) and Pacher (1964), has since then been used and discussed by numerous authors. A review of that literature reveals that, when the concept is used for designing artificial tunnel supports, the load increase on the lining is usually considered to be due to one or both of the following two processes:

(i) The wall convergence at a given cross-section of an underground opening, due to the advance of the driving face. This convergence can be controlled by the use of air pressure at the face(Panet,1976).

(ii) The long-term convergence of the opening, due either to the time-dependent deformation (creep) of the rock mass, or to its time-dependent deterioration (E.g., Ladanyi, 1974, 1980).

One of the first attempts to introduce the long - term strength concept in the determination of ground pressure on tunnel linings by the characteristic line method was made by Ladanyi (1974). In that approach only two limiting characteristic lines were considered: One for the short-term response ($t \to 0^+$), and another for the long-term response ($t \to \infty$). For intermediate times, it was assumed that the surrounding rock mass undergoes a continuous deterioration, involving a decrease in its modulus of deformation and a gradual loss of strength.

In a subsequent paper (Ladanyi, 1980), the author attempted to fill the time gap between the short - term and the long-term rock mass response, by assuming that the rock around the tunnel creeps according to a non-linear Maxwell (power law) model. This made it possible to calculate any isochronous characteristic line and to find the time-dependent tunnel wall convergence and ground pressure increase.

However, since that solution did not take into account the effect of the loading history on the rock response, but only the rock deterioration with time, independent of the lining pressure, it falls clearly in the class of solutions based on the "aging theory of creep". It is noted that nearly all "conventional" characteristic line methods presently in use around the world, are based on the same "rock aging" assumption.

The necessity of incorporating the loading history into the modelling of the tunnel wall displacement curves was recognized by Kaiser (1980), who applied it to the case in which the support loading process resulted from the advance of the driving face, in a tunnel driven through an imperfectly elastic rock mass.

More recently, Gill and Ladanyi (1983) and Ladanyi and Gill (1984) have shown that the conventional characteristic line concept, when applied to the design

of tunnel lining in a ground showing creep of a Maxwell type (both linear and non-linear), generally
underestimates both the ground pressure on the lining,
and the convergence of the tunnel wall.

In the present paper, an attempt is made to show
that, when the rock mass creeps according to the
Zener-type model, and does not fail, the long-term
strength concept used by Ladanyi (1974), although
based on the aging creep assumption, leads nevertheless to a correct response concerning the ground pressure on the lining. The point is demonstrated in the
paper by comparing an aging creep solution with the
rigorous one, obtained by Gill (1970) and based on
the Zener model.

SOLUTION IGNORING THE LOADING HISTORY

The rock deterioration process to which Ladanyi (1974)
refered to is, in fact, characterized by a slow evolution of the mechanical properties of the rock mass
from their short-term, to their long-term values, as
shown schematically in Fig. 1. In 1980, the same author has shown how to calculate the isochronous characteristic lines, based on the power-law type aging
creep theory, and excluding failure. Alternatively,
such isochronous characteristic lines can also be
calculated for Zener-type linear elastic rock behaviour. In the latter case, if failure does not occur,
any characteristic line can be obtained from the expression:

$$u_i/b = (p_o - p_i)[\frac{1}{k_1} + \frac{1}{k_2} f(t)] \qquad (1)$$

where $f(t)$ denotes the Zener time function, given by:

$$f(t) = 1 - \exp(-tk_2/\eta_2) = 1 - \exp(-t/t_R) \qquad (2)$$

where $t_R = \eta_2/k_2$ is the relaxation time.

In Eqs. (1) and (2), u_i is the time-dependent radial
displacement of the tunnel wall, p_o is the far-field
(pre-excavation) stress, p_i is the internal pressure
acting on the tunnel wall, b is the tunnel radius
(Fig. 2), and t is the time. For the deviatoric
creep considered here, the elastic spring parameters
k_1 and k_2, and the viscosity η_2 in Eqs. (1) and (2)
should be replaced by:

$$k_1 \rightarrow 2G_o = E/(1 + \nu) \qquad (3)$$

$$k_2 \rightarrow 2G_oG_f/(G_o - G_f) \qquad (4)$$

$$\eta_2 \rightarrow 2\eta \qquad (5)$$

where G_o and G_f are the instantaneous (short-term)
and long-term values of the shear modulus of the rock,
respectively, while η denotes its shear viscosity.

In fact, an expression similar to Eq. (1) was found
by Salustowicz (1958), but Eq. (1) is considered to
be its more correct form.

It is noted that, for $p_i = p_o$, Eq. (1) defines the
unrestricted radial displacement of the tunnel wall,
as calculated by Gnirk and Johnson (1964).

The set of dashed lines in Fig. 3, represents the
conventional isochronous characteristic lines calculated from Eq. (1), using the following parameters:

G_o = 2.069 GPa; G_f = 1.0345 GPa;
η_2 = 41.38 GPa.y; p_o = 7.586 MPa.

These lines will be compared in the following with
those obtained by considering the loading history.

SOLUTION CONSIDERING THE LOADING HISTORY

The case in which an elastic lining is installed at

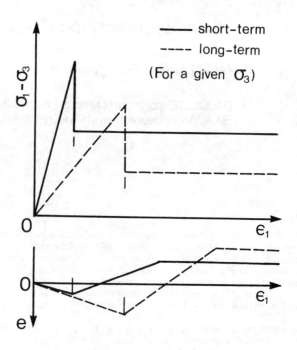

Figure 1. Principal assumptions on rock behaviour
made in Ladanyi (1974).

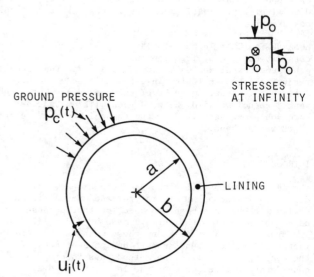

Figure 2. Tunnel cross-section showing the symbols
used in the calculation.

any time $t = t_S$ in a tunnel driven at time $t = 0$,
into a rock mass showing creep properties, has been
solved by Gill (1970), for the Zener-type viscoelastic
case. In the following, that solution will be used
for obtaining the exact isochronous characteristic
lines for a Zener-type rock mass.

The solution is based on the following assumptions
regarding the rock mass, the tunnel, the state of
stress, and the lining installation (Gill, 1970):

(I) The underground opening is a circular cylindrical
tunnel, and the considered section is sufficiently
far from the face, so that plane strain conditions
are valid.
(II) The pre-driving state of stress in the ground
is hydrostatic.

(III) The rock mass is homogeneous and isotropic. Its stress-strain-time relationship is of a linear-visco-elastic type (Zener Substance).
(IV) The strain components remain infinitely small.
(V) The rate of the advance of the tunnel face is large, relative to the creep rate of the ground.
(VI) The ground support system consists of a concrete lining, assumed to be linear-elastic.
(VII) The lining is installed at any time, $t = t_s$, after the driving of the tunnel.
(VIII) There is no failure of rock mass anywhere around the tunnel.

Figure 2 is a section normal to the tunnel axis, showing the symbols used in the solution.

This closed-form solution (Gill, 1970) yields at the wall of the tunnel (Fig. 2):

For $t < t_s$:

$$p_c = 0, \text{ and} \tag{6}$$

$$\frac{u_i}{b} = \frac{p_o}{k_1}[1 + \frac{k_1}{k_2} f(t)] \tag{7}$$

For $t \geq t_s$:

$$p_c = p_o F_1(t) F_2(t) \quad, \text{ and} \tag{8}$$

$$\frac{u_i}{b} = \frac{p_o}{k_1}[1 + \frac{k_1}{k_2} f(t_s)] + p_c D \tag{9}$$

In these equations, $f(t)$, k_1 and k_2 are given by Eqs. (2) to (4), while $F_1(t)$ and $F_2(t)$ are defined by:

$$F_1(t) = \exp(-t_s/t_R) - \exp(-t/t_R) \tag{10}$$

$$F_2(t) = \frac{k_1/k_2}{k_1/k_2 + 1 + Dk_1}\left\{\frac{1 - \exp[-A(t - t_s)/t_R]}{1 - \exp[-(t - t_s)/t_R]}\right\} \tag{11}$$

where

$$A = 1 + \frac{k_1/k_2}{1 + Dk_1} \quad, \text{ and} \tag{12}$$

$$D = \frac{1 + \nu'}{E'}\left[\frac{1 + (1 - 2\nu')b^2/a^2}{1 - b^2/a^2}\right] \tag{13}$$

is the inverse of the rigidity modulus ($K_s = 1/D$ in Ladanyi, 1980) of the annular concrete lining.

In Eqs. (6) to (13), p_c is the time-dependent radial pressure at the ground-lining interface when $t \geq t_s$, i.e., the ground pressure, p_o is the pre-driving ground stress, t_s is the time when the lining becomes effective, u_i is the time-dependent radial displacement at the tunnel wall when $t \geq t_s$, i.e., the tunnel wall convergence, E' is the Young's modulus of the lining material (concrete), ν' is its Poisson's ratio, and a and b are the internal and external radii of the lining.

The preceding Eqs. (6) to (13) define both the ground pressure on the tunnel lining and the tunnel wall convergence. They represent therefore the equations of the coordinates of points of equilibrium on ground-lining interaction diagrams, and, consequently, they define the exact characteristic lines. The full lines in Fig. 3 show a set of such lines obtained for the parameters given previously, and for $t_s = 3 \times 10^7$ sec = = 0.95 years. They are found to lie everywhere above the conventional lines, but the difference decreases with time.

DISCUSSION ON THE EFFECT OF LOADING HISTORY

A detailed examination of Figure 3 leads to the conclusion that any point of equilibrium obtained using the conventional characteristic lines yields generally a lining pressure and a wall convergence lower than those given by the exact characteristic lines.

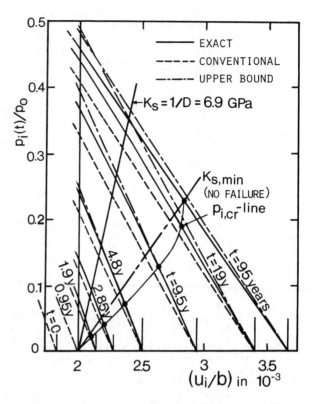

Figure 3. Lower portions of characteristic lines, according to three different calculation methods.

The observed discrepancies are due to the fact that the loading history assumed in the aging theory of creep differs from that actually taking place. This is illustrated in Fig. 4. The curve A, which is a step function, starting at $t = 0$, is implicitly assumed when calculating the conventional characteristic lines. This curve should be compared with the curve B in the same figure, which is deduced from the closed-form solution, Eq. (9). The support loading history defined by this curve corresponds to an exponential function, starting at time $t = t_s$.

However, it can be seen from Fig. 3 that the difference decreases with time, tending to zero for the long-term isochronous characteristic lines. This can be readily demonstrated as follows:

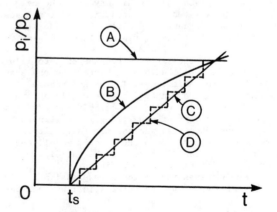

Figure 4. Ground pressure histories. Line A: Assumed for obtaining conventional characteristic lines; Line B: As obtained from the closed-form solution; Line C: Assumed in the proposed approximate method; Line D: Approximation of line C.

Equations (8) and (9) give for the long term conditions ($t \to \infty$) the following results:

$$p_{c\infty} = p_o \frac{k_1/k_2}{k_1/k_2 + 1 + Dk_1} \exp(-t_s/t_R) \qquad (14)$$

and

$$\left(\frac{u_i}{b}\right)_\infty = p_o[\frac{1}{k_1} + \frac{1}{k_2} f(t_s)] + p_{c\infty} D \qquad (15)$$

On the other hand, if one follows the displacement summation method used in Ladanyi (1980) for obtaining the lining pressure p_c, i.e.,

$$[(u_i/b)_o + (u_i/b)_t]_{p_i=p_c} = (u_i/b)_{0,p_i=0} + \frac{\Delta}{b} + (u_{ec}/b)_{(p_i=p_c)} \qquad (16)$$

which, with the Zener's creep model, noting that the gap $\Delta/b = (p_o/k_2)f(t_s)$, gives:

$$(p_o - p_c)[\frac{1}{k_1} + \frac{1}{k_2} f(t)] = \frac{p_o}{k_1} + \frac{p_o}{k_2} f(t_s) + p_c D \qquad (17)$$

one gets finally:

$$p_c = p_o \frac{k_1/k_2}{k_1/k_2 + 1 + Dk_1}[f(t) - f(t_s)] \qquad (18)$$

which, for $t \to \infty$, reverses to Eq.(14), while $(u_i/b)_\infty$, given by the right hand side of Eq.(17), reverses to Eq.(15).

In other words, while at intermediate times the two solutions give different results, they become equal at long term, when $t \to \infty$, at least for a material creeping according to the Zener's model.

For example, for the parameters given previously, and assuming a lining of rigidity $K_s = 1/D = 6.897$ GPa, installed at $t_s = 3\times10^7$ sec $= 0.9513$ years, one gets from both sets of equations:

$p_{c\infty}/p_o = 0.3498$, and $(u_i/b)_\infty = 2.384 \times 10^{-3}$.

However, as seen in Fig.5, for other times, less than about 40 years, the conventional method underpredicts the true lining pressure by up to 35%.

APPROXIMATE UPPER BOUND SOLUTION

In two previous papers (Gill and Ladanyi, 1983; Ladanyi and Gill, 1984), in addition to the complete analytical solutions, some approximate methods for calculating more correct characteristic lines, taking into account the loading history, were also proposed. In the first one (Gill and Ladanyi, 1983), it was shown how any given loading history or sequence could be approximated by a succession of step-loads. The second one (Ladanyi and Gill, 1984) made it possible to approximate any given non-linear lining response.

In this paper, the first method was used to illustrate the effect of the loading history on the shape of the characteristic lines. The method enabled to take into account a linear increase of lining pressure with time (Line C in Fig.4), by transforming it into a finite number of equal step loads (Line D in Fig.4). Each step load was assumed to result in a creep of rock according to the Zener's model, and the final summation was based on the validity of the Boltzmann's principle of superposition.

This method gives the following expression for calculating an isochronous characteristic line, at time $t = \tau$:

$$\frac{u_i}{b} = p_o[\frac{1}{k_1} + \frac{1}{k_2} f(t)] - \frac{p_i}{n} \sum_{i=1}^{n} [\frac{1}{k_1} + \frac{1}{k_2} f(t_i)] \qquad (19)$$

where k_1 and k_2 are given by Eqs.(3) and (4), $f(t)$ is the time function defined by Eq.(2), and t_i denotes:

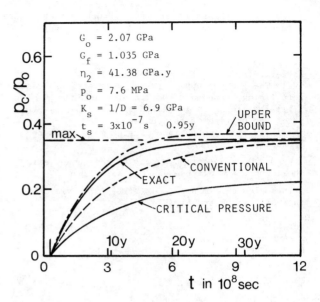

Figure 5. Lining pressure increase with time, according to three different calculation methods.

$$t_i = t - t_s - (2i - 1)\frac{\Delta t}{2} \qquad (20)$$

where

$$\Delta t = (\tau - t_s)/n \qquad (21)$$

is the time interval, obtained by dividing the total time, $(\tau - t_s)$, by the selected number of steps, n.

It is noted that information on $f(t)$ in Eq.(19) can also be obtained from in-situ observations of tunnel wall convergence, because,

$$f(t) = [u_i(t)/bp_o - 1/k_1]k_2 \qquad (22)$$

where $u_i(t)$ denotes the unrestricted radial wall displacement.

The pressure p_i appearing in Eq.(19) represents the selected internal pressure, needed for generation of characteristic lines. Thus, to obtain any particular isochronous characteristic line, it is sufficient to calculate, for a given set of data, the values of $u_i(t)$ for various values of p_i, between 0 and p_o. The dash-dotted lines in Fig.3 represent the isochronous characteristic lines obtained by the described approximate method, and for the previously given parameters. They are seen to lie everywhere above the exact characteristic lines, and represent therefore an upper bound to the latter.

CRITICAL INTERNAL PRESSURE

Since in this paper the rock failure is assumed not to occur, it is necessary to define the limits of validity of that assumption. According to Ladanyi (1974), when the Coulomb's law of failure applies, the rock around the tunnel will start failing only when the lining pressure falls below a critical value given by:

$$p_{i,cr}(t) = \frac{2p_o - C_o(t)}{f_s + 1} \qquad (23)$$

where

$$f_s = (1 + \sin \Phi_s)/(1 - \sin \Phi_s) \qquad (24)$$

is the passive pressure coefficient of the intact rock mass, and

$$C_o = 2c_s\sqrt{f_s} \qquad (25)$$

is the uniaxial compression strength of the mass,

c_s being the corresponding cohesion intercept.

The value of $p_{i,cr}$ can be made time-dependent by assuming that the value of Φ_s remains constant with time, while that of C_0 varies according to a Zener type law, such as

$$C_0(t) = C_0 + (C_f - C_0)f(t) \qquad (26)$$

with $f(t)$ given by Eq.(2).

Assuming, for example, that the initial (short-term) value of $C_0 = 15.86$ MPa, and its final (long-term) value is $C_f = 8.28$ MPa, while $\Phi_s = 30^o$, and $p_0 = 7.60$ MPa, one can determine on each isochronous characteristic line in Fig.3 a point where $p_i = p_{i,cr}$, below which the rock starts failing, and the present theory becomes invalid.

In other words, as shown in Fig.3, in order to prevent the failure of rock to occur after the lining installation, the rigidity of the lining, $K_S = 1/D$, should in the present case be at least equal to about 2 GPa.

Figure 5 shows, for comparison, the curves of the ground pressure increase with time, calculated from the conventional method, the exact solution, and the above approximate upper bound method, respectively. The variation of the critical pressure with time is also shown in the same figure.

CONCLUSIONS

The considerations and analysis shown in this paper lead to the conclusion that, for a rock mass creeping according to the Zener's model, the isochronous characteristic lines are functions of the loading history, which also affects the value of the ground pressure on the lining, as determined by the convergence - confinement method. When compared with the exact solution, shown in the paper, the conventional characteristic line method is found to underestimate the lining pressure up to 35% at short times, but this difference decreases with time to completely disappear, when $t \to \infty$. This gives a justification for the use of the long-term strength concept for tunnel lining design, proposed by Ladanyi (1974).

ACKNOWLEDGEMENTS

The financial support provided for this study by the Natural Sciences and Engineering Council of Canada, is gratefully acknowledged.

REFERENCES

Brown, E.T., Bray, J.W., Ladanyi, B. & Hoek, E. 1983. Characteristic line calculations for rock tunnels. ASCE J.of Geotechnical Engineering, Vol.109, No.1, pp. 15-39.

Gill, D.E. 1970. A mathematical model for lining design in linear viscoelastic ground. Ph.D. Thesis, McGill University, 306 p.

Gill, D.E. & Ladanyi, B. 1983. The characteristic line concept of lining design in creeping ground. Proc. CIM Symp. on Underground Support Systems, Sudbury, Ont., CIM Special Volume, 1986.

Gnirk, P.F. & Johnson, R.E. 1964. The deformational behavior of a circular mine shaft situated in a viscoelastic medium under hydrostatic stress. Proc. 6th U.S. Symp.on Rock Mechanics, Rolla, MO, pp. 231-259.

Kaiser, P.K. 1980. Effect of stress history on the deformation behaviour of underground openings. Proc. 13th Canadian Rock Mechanics Symposium, "Underground Rock Engineering", CIM Spec. Vol. 22, pp. 133-140.

Ladanyi, B. 1974. Use of the long-term strength concept in the determination of ground pressure on tunnel lining. Proc. 3rd Congress of the ISRM,

Denver , Vol. IIB, pp. 1150-1156.

Ladanyi, B.1980. Direct determination of ground pressure on tunnel lining in a non-linear visco-elastic rock. Proc. 13th Canad. Symp. on Rock Mechanics, "Underground Rock Engineering", CIM Spec. Vol. 22, pp. 126-132.

Ladanyi, B. & Gill, D.E. 1984. Tunnel lining design in a creeping rock. Proc. ISRM Symp."Design and Performance of Underground Excavations", Cambridge, U.K., Vol. I, pp. 19-26.

Lauffer, H. & Seeber, G. 1961. Design and control of linings of pressure tunnels and shafts, based on measurements of the deformability of the rock. Proc. 7th Congress on Large Dams, Q.25, Rome, pp. 679-707.

Pacher, R. 1964. Deformationsmessungen in Versuchs-stollen als Mittel zur Erforschung des Gebirgs-verhaltens and zur Bemessung des Ausbaues. Fels-mechanik und Ingenieurgeologie, Suppl. 1, pp. 147-161.

Panet, M. 1976. Analyse de la stabilité d'un tunnel creusé dans un massif rocheux en tenant compte du comportement après la rupture. Rock Mechanics, Vol. 8, No.4, pp. 209-223.

Särkkä, P.S. 1984. The interactive dimensioning of pillars in Finnish mines. Proc. 2nd Int. Conf. on Stability in Underground Mining, pp. 71-84.

Salustowicz, A. 1958. Rock masses as visco-elastic media. Arch. Gorn., Vol.3, No.2, (in Polish), pp. 141-172.

Analysis of underground excavations incorporating the strain softening of rock masses

Analyse d'ouvrages souterrains en tenant compte du radoucissement du massif rocheux
Die Analyse von unterirdischen Hohlräumen, die den Festigkeitsabfall der Felsmasse einschliesst

M.C.GUMUSOGLU, Imperial College of Science and Technology, London, UK
J.W.BRAY, Imperial College of Science and Technology, London, UK
J.O.WATSON, Imperial College of Science and Technology, London, UK

ABSTRACT: A finite element program is developed for the analysis of underground openings excavated at considerable depth or in weak rocks. A rock mass response model based on the Hoek-Brown failure criterion dependent on rock type and rock mass quality is incorporated into the program. Any desired post-failure stress-deformation behaviour, including strain softening, can be considered. The practical implications of this research are demonstrated by analysing the northern section of the St. Gotthard highway tunnel and its safety gallery.

RESUME: Un programme d'éléments finis est développe' pour l'analyse des ouvrages souterrains excave's a' grande pronfondeur ou dans des roches fragiles. Un modèle de comportement pour le massif rocheux qui est base' sur le critère de rupture de Hoek et Brown et qui dépend du type de roche et de la qualite'du massif rocheux est incorpore' au programme. Tous les types de comportement contrainte-déformation peuvent être pris en compte; y compris le radoucissement. Les implications pratiques de cette recherche sont de'montrées par une analyse de la section nord du tunnel routier du St. Gotthard et de sa galerie de se'curite'.

ZUSAMMENFASSUNG: Ein "finite Element" Programm wird, für die Analyse von unterirdischen Ausgrabungen die sehr tief oder in schwachen Felsgesteinen ausgebohrt werden, entwickelt. Ein Felsmassemuster, basierend auf das Hoek-Brown Kriterium, abhängig von der Felskategorie und von der qualität der Felsmasse, ist in dem Programm einbegriffen. Alle gewünschten "post-failure" Druckdeformationen können berücksichtigt werden, einschliesslich des Festigkeitsabfall. Die praktischen anwendungen dieser Forschung werden durch die Analyse des nördlichen Teiles des St. Gotthard Tunnels und seines Sicherheitstollens dar gestellt.

1 INTRODUCTION

In analyses involving the stability, safety and economy of underground excavations, it is essential to account for the probable response of the surrounding rock mass as realistically as possible. In this research, attention is given to the case of openings excavated at considerable depth or in weak rocks, in which the induced stresses reach the available strength in some part of the rock mass surrounding the excavation. Failures controlled by the geological structure of the rock mass are not considered. The finite element method is suited to the treatment of such problems, because any shape and sequence of excavation, initial stress field and rock mass behaviour, as well as the interactive nature of the load-deformation characteristics of both rock mass and support system can be considered. The post-failure behaviour of rock masses accounted for in previous finite element analyses is generally over-simplified and unrealistic. With the improved knowledge of mechanical behaviour of rock masses, it is now evident that a realistic rock mass behaviour model should be based on a yield criterion dependent on rock type and rock mass quality, such as the Hoek-Brown criterion (Hoek and Brown, 1980), and also be able to account for any experimentally determined post-failure stress-strain relationship including strain softening. Thus the principal aim of this research is to develop a finite element program incorporating such a rock mass behaviour model for more realistic and economic design of underground excavations. In order to model correctly most excavation sequences it would be necessary to carry out a three-dimensional analysis, but since this would be very expensive it is supposed here that the rock mass is in a state of plane strain. The techniques discussed here can in principle be applied to the three-dimensional case.

2 ROCK MASS BEHAVIOUR MODEL

For the complete description of the elasto-plastic behaviour of rocks, a yield criterion, post-failure strength-strain relationship and treatment of plastic volumetric strains must be specified:

i) Yield criterion: The failure criterion proposed by Hoek and Brown(1980) is adopted. This is an empirical, nonlinear criterion dependent on rock type and rock mass quality and given by

$$F = \sigma_1 - \sigma_3 - \sqrt{m\sigma_c\sigma_3 + s\sigma_c^2} = 0$$

in which F is the yield function, σ_1 and σ_3 are the major and minor principal stresses respectively, σ_c is the uniaxial compressive strength of the intact rock material and m and s are empirical parameters, which depend on the properties of the rock mass and on the extent to which it had been broken before being subjected to induced stresses. It is assumed that the rock mass is reasonably continuous, isotropic and homogeneous with either very widely or closely spaced joints.

ii) Post-failure behaviour: The progressive failure of rock masses is modelled by linear variation of the empirical parameters m and s with maximum principal plastic strain, ε_1^p - a convenient scalar measure of plastic straining and an index of the amount of structural change (Figure 1). In particular the reduction of s has strong physical significance, since it can simulate the transition of intact rocks (s = 1.0) to completely broken rock masses (s = 0.0), i.e. the complete stress-strain history. In order to provide every possible type of post-failure behaviour one may obtain from experimental results, the model is designed to account for strain hardening, perfectly plastic and strain softening rock behaviours. Only the values of the empirical strength parameters m and s at yield, peak and residual strength levels and ε_{1y}^P, ε_{1p}^P and ε_{1r}^P must be specified.

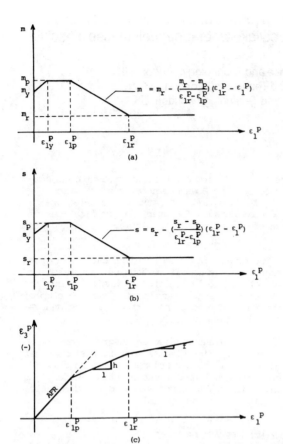

Figure 1. The rock mass behaviour model.

The associated flow rule (AFR) is assumed for strain hardening and perfectly plastic behaviours as shown in Figure 1 ($\varepsilon_1^P \leqslant \varepsilon_{1p}^P$). This assumption is generally supported by experimental data. However, the normality condition seems to disappear when the rock is fractured or poorly interlocked (Brown et al., 1982, 1983). This implies that using the associated flow rule in the strain softening and residual strength regions may give unrealistically large displacement predictions. Therefore non-associated flow rules with constant volume change deformation behaviour (gradient h in Figure 1) in the strain softening regime and a smaller (gradient f) or zero volume change deformation behaviour in the residual strength region are adopted. The ratios h and f of plastic strain increments are to be determined experimentally. In general these are not constant, but they can be approximated as constant fairly realistically.

3 NUMERICAL ANALYSES

The rock mass behaviour model is incorporated into the finite element program, CGFEP, which is a modified and extended version of the program given by Owen and Hinton (1980). Extensive information on the numerical implementation of the rock mass behaviour model and validation of the finite element program is reported by Gumusoglu (1986). CGFEP is systematically verified against various theoretical solutions, including a semi-closed form solution for the analysis of axisymmetric tunnel problems. This solution is developed on the basis of the proposed rock mass behaviour model. In the solution a stepwise procedure is adopted in which stresses and strains are successively calculated on the boundaries of a number of annular rings into which the plastic zone is divided. The solution takes into account variation of the elastic modulus (E) and Poisson's ratio (ν), as well as the strength parameters, in the strain softening region. For the sake of comparison a similar solution based on the Mohr-Coulomb yield criterion, in which cohesion and internal friction angle vary linearly with ε_1^P, is also developed. In finite element analyses a modified initial stress algorithm, in which the global stiffness matrix is updated only in the first iteration of each load increment, is used. In the numerical analyses serious convergence difficulties are encountered, similar to those experienced by Manfredini et al. (1976). The drawbacks of the finite element program are identified and necessary improvements are made. These involve the correction of yield surface drift and overcoming the oscillation of the plastic flow between planes parallel and perpendicular to the excavation face. After these improvements reasonably accurate and convergent solutions are obtained.

3.1 Analysis of the St. Gotthard tunnel

The finite element program and the semi-closed form solution are used to analyse the northern section of the St. Gotthard highway tunnel in Switzerland. Extreme difficulties were encountered in the calcerous sediments of the Mesozoic period and therefore extensive laboratory investigations of the mechanical properties of the rock mass and in situ measurements were carried out (Bourquin, 1975; Pfister, 1976; Lombardi, 1977). Thus there is a reasonable amount of published information, which enables the authors to analyse this section of the tunnel by methods developed in this research. In the Mesozoic zone the depth of the tunnel is 300m below the surface and the in situ stress field is hydrostatic, i.e. $P_o = 7.8$MPa. Based on laboratory investigations and the description of the rock mass at the site, the following material property data are chosen for the rock mass: $E = 10000.0$MPa, $\nu = 0.2$, $\sigma_c = 57.0$MPa, $m_p = 0.2$, $m_r = 0.056$, $s_p = 0.0001$, $s_r = 0.00001$, $h = f = 2.1$, $\varepsilon_{1y}^P = \varepsilon_{1p}^P = 0.0$, $\alpha = 10.0$, $\varepsilon_{1r}^P = \alpha \varepsilon_{1y}^e = 0.0042$ (α is a constant which defines the rate of strength reduction and ε_{1y}^e is the maximum principal elastic strain at the yield strength). It is assumed that body forces are negligible. Since the initial state of stress is hydrostatic and the tunnel shape is approximately circular (radius: a = 6.4m), the finite element mesh illustrated in Figure 2 is used. There are eight 8-noded isoparametric elements and the ratios of their lengths are 1:1:2:4:8:16:28:56. Displacements in the z direction are prescribed to be zero as demanded by the plane strain conditions and the rock mass is free to move in the radial direction. The excavation is simulated in 105 increments by stepwise reduction of stresses equilibrating the initial stress field on the tunnel boundary.

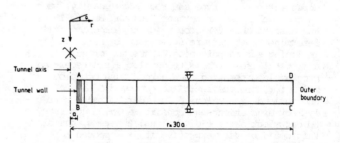

Figure 2. Finite element mesh for the analysis of the St. Gotthard tunnel.

Fair agreement is obtained between the results of the semi-closed form and finite element solutions, as illustrated in Figure 3. In this figure ground response curves are presented in a dimensionless (normalised) form and show the development of radial convergence, u, at the tunnel wall while the internal pressure is decreasing to zero. The maximum difference between the two sets of results is less than 1%. Existing errors in the finite element analysis are mainly due to the simulation of the infinite domain by a finite mesh and discretization of the continuum. In Figure 3 points A and B denote the onset of strain softening and the attainment of residual strength at the tunnel wall respectively. The strain softening model is capable of simulating the long term response of the St. Gotthard tunnel. Consequently good agreement is achieved with the visco-plastic analysis of Lombardi (1977) which agrees well with in situ convergence measurements (Gumusoglu, 1986).

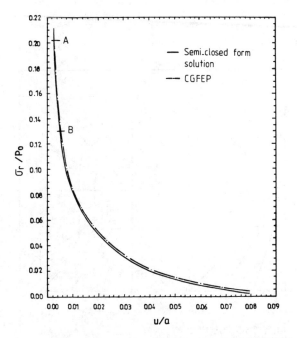

Figure 3. Ground response curves obtained by the semi-closed form and finite element solutions (note that σ_r is the radial stress at the first Gauss point from the tunnel wall and u is the radial convergence).

The semi-closed form solutions are then used to carry out parametric studies in order to check the sensitivity of the solution to several input parameters. Two parallel solutions based on the Hoek-Brown and Mohr-Coulomb yield criteria are performed (H-B and M-C solutions). In the M-C solution, cohesion and internal friction angle at peak and residual strength are chosen as c_p = 0.25MPa, c_r = 0.04MPa, $\emptyset_p$ = 30.0°, $\emptyset_r$ = 28.0° and h = f = 1.85. Other input data are as stated above. In the parametric analyses it is found that the most sensitive input parameters are the internal friction angle and uniaxial strength of the intact rock. Even very small variations of these parameters can influence the support design considerably. It appears that the solution based on the Hoek-Brown criterion is more sensitive to the flow rule and rate of strength reduction, but less sensitive to strength parameters. The variation of normalised radial convergence (u/a) with respect to that of various input parameters is shown in Figure 4. From the results presented one can decide which parameters are most important for determining by suitable laboratory or field tests and which are of secondary importance. Although these results may vary according to the problem and nature of the rock mass, they still do provide useful trends.

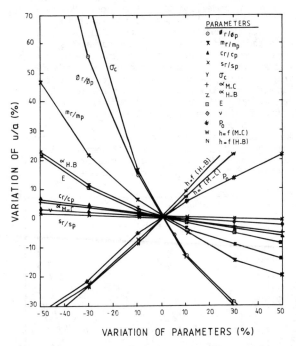

Figure 4. Variation of radial convergence with respect to various input parameters.

As illustrated in Figure 5, the extent of the plastic zone varies considerably with the post-failure behaviour of the rock mass. The influence of modelling strain softening is significant only when the rate of strength reduction, α, is within a specific range of values. Outside this range the rock behaviour can be represented by either perfectly plastic or perfectly brittle models. This result is consistent with the findings of Borsetto and Ribacchi (1979), who have shown that the limits of this range depend upon the strength and deformation behaviour of the rock as well as the initial stress field. It is particularly important to note that, if the residual strength is not reached, the strain softening of the rock mass need not be modelled.

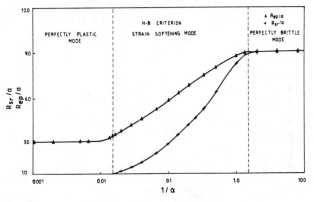

Figure 5. Variations of radii of elastic/strain softening and strain softening/residual strength zone interfaces with respect to $1/\alpha$.

It has been observed in uniaxial cycled loading tests that the strain softening behaviour is accompanied by a reduction of elastic modulus and an increase of Poisson's ratio (Bieniawski, 1969; Elliott, 1982). Although no quantitative justification is made, this concept is generally neglected in numerical modelling of rock masses. However, in this research the influence of the variation of elasticity parameters during the strain softening on the radial convergence is investigated by the semi-closed form solutions. As can be seen from Figures 6 and 7, the solution based on the Hoek-Brown criterion is more sensitive to the variation of elasticity parameters than that based on the Mohr-Coulomb criterion. It appears that, if the elastic modulus decreases to less than half of its original value, its residual value (E_r) should be taken into account in the rock mass behaviour model. On the other hand the influence of increase in Poisson's ratio is of secondary importance and may be neglected for many practical purposes.

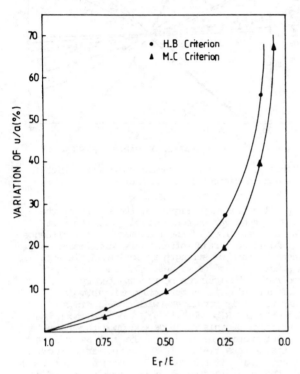

Figure 6. Variation of radial convergence with respect to E_r/E.

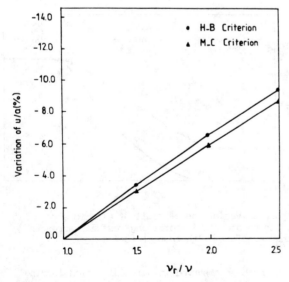

Figure 7. Variation of radial convergence with respect to ν_r/ν.

3.2 Analysis of the safety gallery

A horse-shoe shaped safety gallery, driven parallel with the axis of the St. Gotthard tunnel, is analysed by the finite element program CGFEP. The finite element mesh used in the analysis is shown in Figure 8. The mesh consists of 80 8-noded isoparametric elements, 8 in the radial and 10 in the circumferential direction. The boundaries AF and FG are free to slide in the vertical and horizontal directions respectively and GH and HA are free boundary surfaces. The in situ stress field and the material property data are the same as described in Section 3.1. The safety gallery has been excavated full face with drilling machines in a continuous working cycle. As a temporary support, steel support arches with stiff blocking had to be installed immediately after the excavation to stabilize the rock mass (Bourquin, 1975). In the finite element analysis linear elastic plane strain elements of thickness 0.2m are used to model the support system. The elastic modulus and Poisson's ratio of support elements are E_s = 4500.0 MPa and ν_s = 0.25.

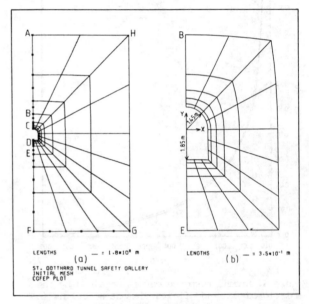

Figure 8. Finite element mesh for the analysis of the safety gallery.

The excavation is simulated in 185 increments by stepwise reduction of stresses equilibrating the initial stress field on the boundary of the gallery. It is determined that the failure of the rock mass initiates at the roof and then at the bottom corner of the gallery, finally enveloping the excavation. The depth of the plastic zone extends approximately to 65% of the width of the gallery. As can be seen in Figure 9 a stress concentration zone develops around the bottom corner, whereas lower stress values are observed in residual strength zones. The development of stresses and displacements in the support system and surrounding rock mass are shown in Figures 10 and 11 respectively. Higher support pressures are predicted at the roof and bottom corner of the gallery. On the contrary larger deformations are obtained at the floor and side-wall of the excavation, where the rock mass is more fractured. Results presented in Figures 9-11 can be used to confirm the adequacy of the support system or to indicate that it is necessary to adjust the support design. Unfortunately, there is not enough field data to verify these numerical results. It was only reported by Bourquin (1975) that the actual support

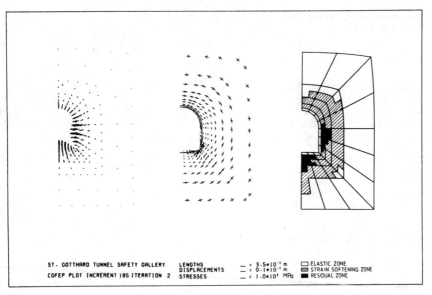

Figure 9.   Final distributions of displacements, stresses and plastic zones around the safety gallery.

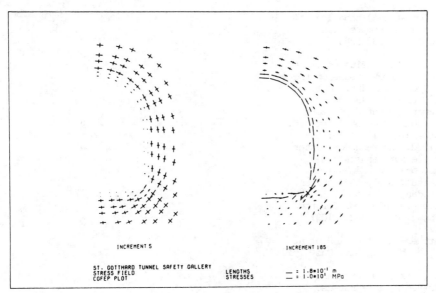

Figure 10.   Development of stresses in the support system and surrounding rock mass.

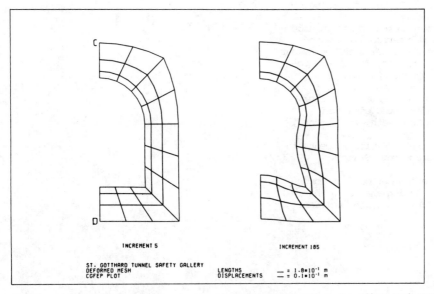

Figure 11.   Development of displacements in the support system and surrounding rock mass.

system was able to support the rock mass, though at some sections plastic hinges had developed.

4 CONCLUSIONS

A finite element program, capable of accounting for the strain softening of rock masses, is developed. This program is verified against a semi-closed form analysis of the St. Gotthard tunnel. In parametric analyses of this tunnel, the significance of the proposed rock mass behaviour model and relative importance of various input parameters are demon-strated. It is shown in the analysis of the safety gallery that the finite element program is capable of assisting in the design of support for non-axi-symmetric underground excavations in strain soften-ing rock masses. Further efforts in this area of research should be directed to improvement of the nonlinear equation solution techniques in order to reduce the computing cost, and refinement of the rock mass behaviour model with the aid of monitoring of underground excavations.

REFERENCES

Bieniawski, Z.T. 1969. Deformational behaviour of fractured rock under multiaxial compression. Proc. Civ. Engng. Mat. Conf., p.589-598. Southampton.

Borsetto, M. & Ribacchi, R. 1979. Influence of the strain-softening behaviour of rock masses on the stability of a tunnel. Proc. 3rd. Int. Conf. on Num. Meth. in Geomech., p.611-620. Aachen.

Bourquin, M. 1975. La franchissement de la zone Mesozoique par le tunnel routier du St. Gotthard. Tunnels et Ouvrages Souterrains. 11: 173-185.

Brown, E.T. & Bray, J.W. 1982. Rock-support inter-action calculations for pressure shafts and tunnels. Proc. Int. Symp. on 'Rock Mechanics: Caverns and Pressure Shafts', p.555-565. Aachen.

Brown, E.T., Bray, J.W., Ladanyi, B. & Hoek, E. 1983. Characteristic line calculations for rock tunnels. J. Geotech. Engng. Div., ASCE. 109: 15-39.

Elliott, G.M. 1982. A further investigation of a yield criterion for rock. Ph.D. Thesis, Univ. of London.

Gumusoglu, M.C. 1986. Analysis of underground excavations in strain softening rock masses. Ph.D. Thesis, Univ. of London.

Hoek, E. & Brown, E.T. 1980. Empirical strength criterion for rock masses. J. Geotech. Engng. Div., ASCE. 106: 1013-1035.

Lombardi, G. 1977. Long term measurements in under-ground openings and their interpretation with special consideration to the rheological behaviour of the rock. Proc. Int. Symp. on Field Measure-ments in Rock Mech., p.839-858. Zurich.

Manfredini, G., Martinetti, S., Ribacchi, R. & Riccioni, R. 1976. Design criteria for anchor cables and bolting in underground openings. Proc. 2nd. Int. Conf. on Num. Meth. in Geomech., p.859-872. Virginia, USA.

Owen, D.R.J. & Hinton, E. 1980. Finite elements in plasticity: theory and practice. Swansea: Pineridge Press.

Pfister, R.E. 1976. Excavation methods for long highway tunnels and ventilation shafts in the Swiss Alps. Proc. 3rd Rapid Exc. and Tun. Conf., p.377-397. Las Vegas.

Die Sicherung der Bruchkante mit dem Alpine Breaker Line Support (ABLS)

Securing the caving edge with the Alpine Breaker Line Support (ABLS)
Le soutènement de l'arête de foudroyage par l'Alpine Breaker Line Support (ABLS)

H.HABENICHT, VOEST-ALPINE AG, Werk Zeltweg, Österreich

ABSTRACT: Four predominant methods for full extraction in flat coal seams are discussed. Longwall and Short-wall have extensive artificial support for the caving edge. In Pillar Extraction (P.E.) and Rib Pillar Extraction (R.P.E.) the caving edge is majorily carried by the coal.

P.E. and R.P.E. need artificial support only in the relatively small working area. Their geometry and mechanical conditions appear to be more complicated. An approach for a clarification is given.

The newly developed, self advancing, support, i.e. ABLS, improves safety, roof control, and economy of P.E. and of R.P.E. Therefore, these become more competitive. Keystones for their success are the rock mechanical and support interrelationships.

As an onset the approximation of the rock load is presented. It involves practically observed parameters. Irregularities of the rock mass, of loading conditions, of configuration, and of the mode of operation, as well as time dependent features and shock loads can cause severe variations in the stability conditions. These require further investigations.

RESUME: Quatre méthods prédominantes d'exploitation complete des couches de charbon sont discutées. Pour grands fronts et pour petites tailles l'arete de foudroyage a le plus extensive soutènement. Pour l'épilage et pour nervures de charbon l'arête de foudroyage est supportée par le charbon.

L'epilage et la récupération des nervures demandent de soutènement seulement dans la petite zone d'extraction. Leurs géometrie et conditions mechaniques sont plus compliquées.Un effort d'éxplanation est présenté.

Le soutènement neuf, i.e. l'ABLS, équipement mobile, peut améliorer la sûreté, la régulation du toit et l'économie. Donc cettes méthodes devient plus competitives. Le clef pour leur succés est la rélation entre la méchanique de roches et le soutènement.

Comme commencement une approximation du charge du roche est présentée. Elle est basée sur paramètres observés dans les mines. Aussi il-y-a des autres effects additionels, p.e., irregularitée de roche, des contraintes, de configuration, de mode d'operation, l'influence de temps, et des contraintes au choc. Ils peuvent causer des variations de stabilitée et donc deservent plus des recherches.

ZUSAMMENFASSUNG: Die vier wichtigsten Abbauverfahren zur Vollausbeute in flachen Kohleflözen werden kurz erörtert.

Beim Strebbau und Kurzfrontabbau wird die Bruchkante umfangreich und weitgehend künstlich gestützt. Beim Pfeilerbau und Langpfeilerbau trägt der Kohlenstoß die größten Längenanteile der Bruchkante. Diese benötigen einen künstlichen Stützausbau nur im eigentlichen verhältnismäßig kleinen Arbeitsbereich. Dessen Geometrie und mechanische Zusammenhänge erscheinen komplizierter, wobei hiemit ein erster Ansatz für eine zweckentsprechende Klärung vorliegt.

Der neu entwickelte selbstfahrende Ausbau ABLS verbessert Sicherheit, Hangendbeherrschung und Wirtschaftlichkeit des Pfeilerbaus und des Langpfeilerbaus. Er gibt daher diesem Verfahren eine bessere Wettbewerbsbasis. Zu deren Erfolg ist jedoch die Kenntnis der gebirgsmechanischen und ausbautechnischen Zusammenhänge von großer Bedeutung.

Ein Ansatz zur angenäherten Berechnung der Gebirgslast wird vorgestellt, der von praktisch beobachteten Parametern ausgeht. Unregelmäßigkeiten im Gebirgsaufbau, im Auslastungszustand, in der Konfiguration und Betriebsweise sowie zeitabhängige Erscheinungen und stoßartige Belastungen können jedoch starke Schwankungen in den praktischen Verhältnissen mit sich bringen. Hiezu bedarf es noch weiterer Untersuchungen.

EINLEITUNG

Zu den wichtigsten und meist angewandten Bruchbauverfahren im untertägigen Steinkohlenbergbau zählen weltweit der Pfeilerbau, der Langpfeilerbau, der Strebbau und der Kurzfrontabbau. Diese unterscheiden sich nicht nur in der Geometrie der Baufelder, in der Ausrüstung und Betriebsweise sowie in der Leistung und Wirtschaftlichkeit, sondern auch in der Beherrschung des Hangenden und im Zubruchwerfen desselben. Für den Strebbau und den Kurzfrontabbau tritt eine sehr regelmäßige Bruchfront bei verhältnismäßig gleichmäßigem Fortschritt ein, für deren örtlichen Bereich auch eine sehr gute Sicherung der Dachschichten durch den beinahe vollkommenen Ausbau gegeben ist. (Arauner 1984, Herwig 1984, Jacobi 1981). Die wissenschaftliche und technische Behandlung der Zusammenhänge zwischen Gewinnungsfortschritt, Dachschichtensicherung, Bruchvorgängen und Ausbaumitteln ist in der Vergangenheit in vielen und umfassenden Arbeiten behandelt worden (z.B. Habenicht 1969, Singhal 1984, Wagner e.a. 1979). Daraus leitet sich ein Wissensstand ab, der weitgehend befriedigende wenn auch nicht völlige Klärung der Zusammenhänge und Möglichkeiten gebracht hat.

Die anderen beiden Verfahren Pfeilerbau und Langpfeilerbau (Crous 1985, Fauconnier e.a. 1982, Habenicht e.a. 1986) waren keiner so weitgehenden Bearbeitung unterzogen, jedoch weisen Bruchverlauf und Bruchkantenausbildung kompliziertere Eigenheiten auf. Die Beherrschung der Vorgänge war eher das Ergebnis praktischer Erfahrung und handwerklicher Geschicklichkeit. Auch die Ausbautechnik war bisher fast ausschließlich durch Einsatz von manuell eingebauten Holzstempeln (Kauffmann e.a. 1980) gegeben.

Erst in den letzten Jahren wurden erfolgreiche Anstrengungen zur Mechanisierung der Ausbauarbeit unternommen (Habenicht e.a. 1986, N.N. 1985), woraus die Entwicklung eines fahrbaren Ausbaus hervorging. Der als Alpine Breaker Line Support (ABLS) bezeichnete Ausbau verändert nicht die prinzipielle Gestaltung

des Bruchvorgangs und der Bruchkante. Er verbessert jedoch die Wirksamkeit des Ausbaus in vieler Hinsicht.
Dadurch gibt er auch den einschlägigen Abbauverfahren mehr Attraktivität und könnte in Richtung einer vermehrten Anwendung derselben wirken. Für eine solche Entwicklung erscheint jedoch die konsequente Behandlung der gebirgsmechanischen Fragen von großer Bedeutung, da die Wirksamkeit und seine Verwendungsweise auf den besten Erfolg abgestimmt werden sollten.

DIE BRUCHKANTE BEI STREBBAU UND KURZFRONTABBAU

Strebbau und Kurzfrontabbau sind jene Verfahren, bei welchen die Bruchkante einerseits vergleichsweise regelmäßig ausgebildet ist und anderseits in höchstem Maße beherrscht wird. Die beim Schneiden der Kohle freigelegte Firste wird in Richtung der Streblänge völlig und in Richtung des Abbaufortschritts zu sehr hohem Anteil unterfangen. Die Bruchkante bildet sich programmgemäß an der rückwärtigen Kante der Schildkappe. Sie verläuft in einer verhältnismäßig geraden Linie und weist nur an den Strebenden eine Abweichung von der stoßparallelen Orientierung auf. Dort schwenkt sie generell ca. 6 - 15 m vor dem Strebende in einen gekrümmten Verlauf, durch welchen sie sich annähernd tangential an die Abbaustreckenrichtung anlegt. Diese Erscheinung ist ein Ergebnis der dort herrschenden zweiseitigen Einspannung.

Diese Bruchkante verlagert sich intervallmäßig in Richtung des Abbaufortschrittes mit dem Vorrücken der generell in einer geraden Linie angeordneten Ausbaueinheiten. Ein Zerbrechen der Dachschichten über oder vor den Kappen kann auftreten, in welchem Fall die flächenartige Unterfangung durch die Ausbaukappen dafür sorgt, daß die Sicherheit der Belegschaft und die Betriebsabläufe möglichst nicht eingeschränkt werden.

Die mechanischen Zusammenhänge für die Stabilitäts- bzw. Bruchbedingungen der Dachschichten und für die Berechnung der Belastung des Strebausbaus (oder Kurzfrontausbaus) sind zwar noch nicht so weit geklärt, daß geschlossene Berechnungen zur Bemessung des Ausbaus führen könnten, doch sind Untersuchungen in umfangreichem Maße durchgeführt worden, welche es erlauben, den Ausbau und damit die Sicherung des Abbauhohlraumes weitgehend effizient zu gestalten.

Ein weiteres wesentliches Merkmal des Strebbaus und Kurzfrontabbaus bildet die Abbaukonvergenz, die zumindest im europäischen Steinkohlenbergbau als Begleiterscheinung geduldet wird. Damit können eine gewisse Mobilisierung der Hangendschichten, ihre Entfestigung und ihr Zerfall eingeleitet werden. In manchen überseeischen Gruben jedoch wird die Konvergenz mit ihren nachteiligen Folgen durch Wahl eines hohen Ausbauwiderstandes verhindert.

Wegen des weitreichenden Bekanntheitsgrades dieser zwei Abbauverfahren wird hier nicht weiter auf die Einzelheiten eingegangen.

DIE BRUCHKANTE BEI PFEILER- UND LANGPFEILERABBAU

Bei den hier genannten zwei Abbauverfahren vermeidet man das Arbeiten an jenem Kohlenstoß, der direkt dem Alten Mann zugewandt ist. Anstatt Personal, Maschinen und Ausbau gegen den Alten Mann zu exponieren, arbeitet man mit Angriffsstellen, die im Zwickel zwischen dem letzten noch nicht in Angriff genommenen Pfeiler der im Abbau befindlichen Pfeilerreihe und der für den Abbau nächstfolgenden Pfeilerreihe liegen (Abb. 1).
Es bleiben auf diese Weise immer Kohlepfeiler oder Pfeilerteile (Kohlenbeine) zwischen dem Arbeitsbereich und dem Alten Mann, welche teilweise das Niederbrechen der Dachschichten im Arbeitsbereich verhindern. Wie in Abb. 1 und 2 jedoch ersichtlich, gestaltet sich damit die Bruchkante in der Form einer abgesetzten Linie.
Beim Pfeilerbau verläuft sie vor und hinter der Knickstelle entlang der bruchseitigen Stöße der jeweils noch nicht abgebauten Pfeilerreihe. Beim Langpfeilerabbau erstreckt sie sich entlang des Stoßes des noch unverritzten Flözteils bis zur Gewinnungsstelle und ab

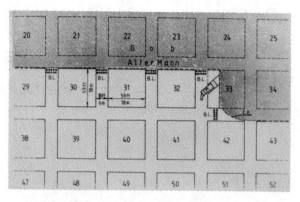

Abb. 1. Pfeilerabbau mit eingetragenen Breaker-Line-Positionen bei Holzstempelausbau (BL) und mit Verlauf der Bruchkante im Firstniveau.
Pillar Extraction with positions of breaker lines shown for timber posts (BL). The caving edge in the roof contour is indicated.
Epilage avec les positions d'arête de foudroyage indiquées pour boisage (BL).

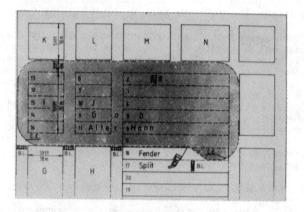

Abb. 2. Langpfeilerabbau mit Position der Holzstempel (BL) und mit Verlauf der Bruchkante.
Rib Pillar Extraction with positions of timber posts (BL) and with location of the caving edge.
Récupération des nervures avec des positions du boisage (BL) et d'arête de foudroyage.

dort entlang des altmannseitigen Stoßes des im Abbau befindlichen Langpfeilers. Die Versetzung der beiden Teile um den Betrag der Breite eines Pfeilers und einer Strecke (Querschlag) erfolgt an der Gewinnungsstelle.
Hier löst sich die Bruchkante vom Kohlenstoß der Pfeiler oder des unverritzten Flözes ab und verläuft ideell gesehen diagonal zum Kohlenstoß hinüber zur anderen Pfeilerreihe. In der Praxis ist an dieser Stelle jedoch häufig auch eine Krümmung im Verlauf festzustellen, die entweder konkav in den Zwickel gerichtet sein kann oder konvex in den Alten Mann. In Abb. 3 ist die Gewinnungsstelle mit konkav in den Zwickel gerichteter Krümmung dargestellt.
Es ist diese Stelle, welche durch technische Maßnahmen mittels natürlicher Bedingungen und künstlicher Hilfsmittel offengehalten werden muß und zwar in jenem Ausmaß, wie es für Personal, Maschinen und Ausbau zum Zwecke der Gewinnung (und anderer Teilaufgaben) erforderlich erscheint.
Ein Ausbau, der die Sicherung dieser Örtlichkeit zu leisten hätte, braucht also nicht entlang der gesamten Bruchkante aufgestellt zu werden, sondern nur an den Schlüsselstellen des Arbeitsbereichs. Hier jedoch

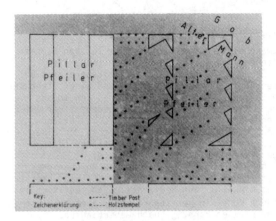

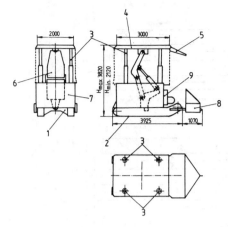

Abb. 3. Ausbauplan für Holzstempel im Fall des Pfeilerbaus bei verminderter Firststabilität oder höherer
Druckhaftigkeit.
  Support plan for timber posts in case of Pillar Extraction at reduced roof stability or at high pressure conditions.
  Plan de boisage en cas d'épilage en conditions du
toif faible ou de pressures elevées.

Abb. 4. Überblick über Gestalt und Komponenten des
ABLS.
  Survey of shape and components of the ABLS.
  Exposé de la forme et des parties constituantes du
ABLS.
1. Grundrahmen, Base Frame, Chassis
2. Fahrraupen, Crawler Tracks, Chenilles
3. Hydraulikstempel, Hydraulic Props, Béquille
Hydraulique
4. Firstkappe, Roof Canopy, Rallonge
5. Vorderkappe, Front Canopy, Rallonge Frontale
6. Lemniskatengetriebe, Lemniscate System, Systéme
de Guidage
7. Bruchschutz, Gob Protection, Protection Contre
Foudroyage
8. Pflug, Plough, Charues
9. Antriebsblock, Power Pack, Groupe Moteur

besteht eine etwas kompliziertere geometrische Konfiguration, für die die Formulierung und insbesondere
die Erkundung der mechanischen Zusammenhänge schwie-
riger erscheint.
  Ein weiteres Beispiel für einen noch mehr aufwen-
digen Holzstempel-Ausbau ist in Abb. 3 gegeben
(Kauffmann e.a. 1980).

DER FAHRBARE AUSBAU UND DIE FRAGE DER BELASTUNGS-
ANNAHME

Beispiele für den Stützausbau des Arbeitsbereiches im
Fall der Verwendung von Holzstempeln sind in Abb. 1
und 2 dargestellt (Fauconnier e.a. 1982, Habenicht
e.a. 1986). Da der grundrißliche Bereich der Pfeiler-
fläche nicht begangen werden darf, bleibt er unaus-
gebaut und nur kleinere Kohlenbeine verbleiben in ihm.
Das Aufstellen der Holzstempel ist vom eigentlichen
Aufwand her umständlich und teuer, aber auch ein Hin-
dernis für die Gewinnung, weil der Continuous Miner
während der Ausbauarbeit keine neue Kohle schneiden
darf. Erst nach Fertigstellung der planmäßig nächst-
liegenden Stempelreihe kann die Gewinnung fortgesetzt
werden.
  Daher besteht ein Verlangen nach Mechanisierung und
Beschleunigung der Ausbauarbeit. Der ABLS erfüllt die
Voraussetzungen für eine Verbesserung der Arbeits-
und Betriebsbedingungen weitgehend (Habenicht e.a.
1986). Er ist ein auf Raupen fahrender hydraulischer
Ausbau (Abb. 4), dessen vier Stempel eine rechteckige,
starre und vollflächige Kappe von großer Abmessung
tragen. Seine Nennlast liegt bei 540 t und seine Setz-
last bei 80 % davon. Es werden in einem Abbau jeweils
vier solche Ausbaueinheiten eingesetzt. Jeweils zwei
davon begleiten die Gewinnungsmaschine beim Rückbau
der Pfeilerteile (Abb. 5 und 6) und jeweils zwei stehen
an der Bruchkante jener Abbaustrecke, auf welche hin
der Pfeiler zurückgebaut wird. Diese vier Ausbaueinheiten mit ihrer Nennlast von zusammen ca. 2080 t ersetzen die herkömmlicherweise aufzustellenden Holzstempel im Arbeitsbereich, deren Anzahl nach dem Beispiel in Abb. 3 je Pfeiler 108 Stück ausmachen kann.
  Da jedoch jeweils nach Fertigstellung einer neuen
Stempelreihe die vorangegangenen programmgemäß geraubt
werden müssen, damit der Hangendbruch unverzögert folgen kann, ist im Regelfall nicht damit zu rechnen, daß
alle im Bild ersichtlichen Stempel zugleich wirken.
Vielmehr wirken immer nur die der Gewinnungsmaschine

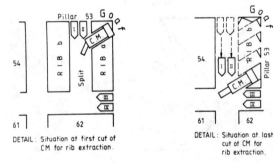

Abb. 5. Gewinnungsschritte beim Abbau eines Pfeilers
mit Ausgangs- und Endstellungen des ABLS.
  Cutting sequence during recovery of a square pillar
with initial and final positions of the ABLS.
  Succession de récupération d'un pilier carré et positions initiales et finales des ABLS.

am nächsten eingebauten Stempel, deren Zahl für die
überwiegenden Fälle mit 16 Stk. anzusetzen ist.
Schreibt man jedem dieser Stempel im Mittel eine Trag-
last von 20 to zu, so erreichen alle 16 Stempel zusammen eine Ausbaustützkraft von 320 t. Die Gegenüber-
stellung mit dem ABLS ergibt also sehr ungleiche Verhältnisse.
  Berücksichtigt man, daß die betreffenden Abbauverfahren mit dem Holzstempelausbau bisher erfolgreich
durchgeführt werden konnten, so können Zweifel an der
Berechtigung der hohen Nennlasten des ABLS aufkommen.
Hiezu sind jedoch zwei Gesichtspunkte zu beachten:
- Der mechanisierte Ausbau soll langlebig und mit
hoher Zuverlässigkeit den Abbau sichern
- Ein Verlust im Alten Mann bei vorzeitigem Niederbruch, wie er bei den Holzstempeln leicht eingegangen
werden kann, soll weitestgehend vermieden werden.

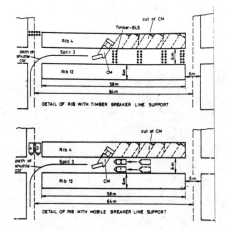

Abb. 6. Gewinnungsschritte beim Abbau eines Langpfeilers mit begleitenden Stellungen des ABLS.
Cutting sequence during recovery of a rib pillar with related positions of the ABLS.
Succession de récupération d'une nervure et positions associées des ABLS.

Für diese Gesichtspunkte wurde der ABLS als äußerst widerstandsfähiges Gerät konzipiert und gebaut, wobei die Verhältnisse von solchen Niederbrüchen als Maß dienen können, welche in ungünstigen Extremfällen auftreten. Es eignet sich hiefür die Untersuchung der Gebirgskörper, welche im Arbeitsbereich niederbrechen. Aus der praktischen Erfahrung mit solchen Fällen kann hier auf Merkmale verwiesen werden, die die Ermittlung

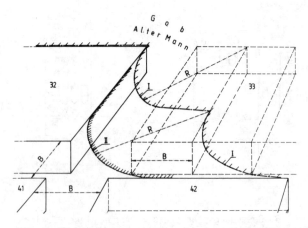

Abb. 7. Räumliche Darstellung des Arbeitsbereiches bei Pfeilerabbau mit eingetragener Bruchkante im Firstniveau.
Threedimensional illustration of the working area in Pillar Extraction with the cave edge shown in the roof level.
Illustration tridiménsionelle de la zone d'extraction en cas d'épilage, avec la position d'arête de foudroyage au niveau du toit.

der Lasten zumindest in einem ersten Ansatz und in Form einer Annäherung ermöglichen.
Zu diesen Merkmalen gehören:
- Der Fall des weitesten Fortschritts der Bruchkante in der Firste tritt ein, wenn der letzte Schnitt in einem Pfeiler erfolgt ist, wodurch also mehr oder weniger keine Kohle mehr zur Stützung der Dachschichten im Arbeitsbereich ansteht.
- Die Bruchkante verläuft dann gekrümmt und ist konkav in den Zwickel hereingezogen (s.Abb. 7). Ihre Gestalt gleicht näherungsweise einem Viertelkreis.

- Die dem vorausgegangene Position der Bruchkante lag im Normalfall näher beim Alten Mann an jener Stelle, an der die gekrümmte Linie gerade den altmannseitigen Rand des vor dem letzten Schnitt bestehenden Kohlenbeins berührte (Abb. 7).
- Auf Grund der Geometrie des Arbeitsbereiches bei einer Pfeilergröße von 18x18 m und einer Streckenbreite von 6 m unter Einsatz eines Continuous Miners mit 3,3 m Schnittbreite bedeutet der zwischen den beiden Positionen liegende Abstand (= Bruchfrontfortschritt) den Betrag von einer Streckenbreite plus einer Schnittlänge (= angenähert gleich einer Streckenbreite).
- Die Abrißfläche in den Dachschichten über dem Arbeitsbereich steigt von der Firste weg unter einem Winkel α zur Horizontalen an, der sich in größere Höhe hin verringern kann, aber auch konstant bleiben kann, je nach Gebirgsaufbau.
Beispiele für zwei von diesen Merkmalen abgeleitete Gebirgskörper sind in Abb. 8 dargestellt. Nimmt man an, daß im Extremfall die gesamte Last aus einem dieser Bruchkörper vom ABLS abzufangen ist, zumindest im Moment des Niederbruchs, so kann man das dem Bruchkörper zuzuschreibende Gewicht für die Ermittlung der Größenordnung der statischen Last ansetzen. Da bei dieser Berechnung nur der Gebirgskörper oberhalb des Arbeitsbereiches berücksichtigt wird, und kein Überhang der Dachschichten in den Alten Mann hinein ange-

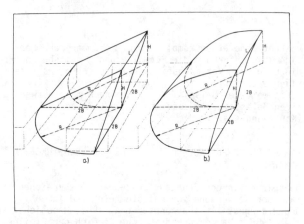

Abb. 8. Prinzipskizze des möglichen Belastungskörpers am Beispiel des Pfeilerabbaus für a) prismatische und b) zylindrische Gestalt.
Basic sketch of the possible loading body in case of Pillar Extraction for a) prismatic, and b) cylindric shape.
Illustration principielle du corps du roche chargeant le soutènement en cas d'épilage pour a) forme prismatique, et b) forme cylindrique.

nommen wird, scheint dieser Ansatz eher die niedrigeren Werte für eine Lastannahme zu ergeben.
Das Gewicht für die in Abb. 8 dargestellten Körper ergibt sich im vereinfachenden Fall, daß die Pfeilerteilbreite gleich der Streckenbreite ist, aus:

Prisma: $G = L \cdot B \cdot H \cdot \gamma$ (Gl. 1)

mit L ... Länge des Prismas (m)
B ... Streckenbreite (m)
H ... Höhe des Querschnitts (= Höhe der Bruchkaverne über Arbeitsbereich)
γ ... spezif. Gewicht der Dachschichten (t/m³) (m)

Zylinder: $G = L \, B^2 \pi \, \gamma$ (Gl. 2)

mit L ... Länge des Zylinders (m)
B ... Streckenbreite (m)
π ... 3,1415
γ ... spezif. Gewicht der Dachschichten (t/m³)

Setzt man in diese Gleichungen die Abmessungen der Abb. 1 ein und nimmt γ = 2,5 an sowie α = 45°, so erhält man beispielsweise Gewichte für das Prisma von

2160 t und für den Zylinder von 3393 t.

Es ist zur Zeit noch unerkundet, welche Beiträge im
Niederbruchvorgang aus der Restreibung, Restfestig-
keit, teilweisen Einspannung und event. aus der zeitlichen Aufeinanderfolge der Mobilisierung in Teilen
u.v.a. noch mitwirken, welche auf die errechneten
Lasten mindernd einwirken können.

Die Minderung besteht jedoch nicht alleine als mög-
licher Modifikator für die Berechnung. Wie schon erwähnt, wäre event. auch ein Überhang zu berücksichtigen. Dazu kommt noch die äußerst bedeutungsvolle
Erfahrung aus Feldeinsätzen mit dem Ausbau, daß Stoß-
belastungen auftreten, deren Intensität die ohnehin
reichlich bemessene Festigkeit des ABLS übersteigen
kann. Der Einsatz von schnell reagierenden Überdruck-
ventilen und von regulären Druckminderungsventilen hat
es erlaubt, daß der Ausbau auch solchen Belastungen
schadenfrei standhalten konnte.

SCHLUSSBETRACHTUNG

Das oben gegebene Beispiel der verhältnismäßig genau
umrissenen und berechneten Bruchkörper und ihre Interpretation als Belastungskörper dürfen nicht darüber
hinweg täuschen, daß im täglichen Betriebsablauf die
Bedingungen und Merkmale starken schwanken können.
Die sich nach einem Niederbruch einstellende neue
Bruchkante und der Winkel der Abrißfläche können unregelmäßig oder stark vom Beispiel abweichend ver-
laufen. Der Überhang, die Mitwirkung zurückgelassener
Kohlenbeine, die aufgebrauchte Feldeslänge und der mit
ihr im Zusammenhang stehende Druckzustand in Firste
und Kohlenstoß bedeuten Faktoren, deren Auswirkung
auf die Stabilität sowie auch auf die Ausbauerfordern-
isse in einem größeren Bereich variieren kann, als
es der Größenordnung der hier berechneten Beträge entspricht. Extreme Entwicklungen können daher zu Verbrüchen des Arbeitsbereichs führen. Die darin bestehen-
de Gefährdung von Personal, Geräten und Betriebserfolg
muß zurückgedrängt werden, wofür insbesondere die
gebirgsmechanischen Zusammhänge ausschlaggebend sind.
Für ihre Klärung wären weiterreichende Untersuchungen
notwendig.

Die mit dem neu entwickelten fahrbaren Ausbau erreichte Verbesserung der Hangendbeherrschung und der Sicherung des Arbeitsbereiches bildet einen Schritt in diese
Richtung. Durch diese neue Technik eröffnen sich neue
Einblicke und Erkenntnisse, deren Umfang hier nur angeschnitten werden konnte.

LITERATURSTELLEN

Arauner, H. Neue Entwicklungen der Ausbautechnik.
Glückauf 120 (1984) Nr. 3, S. 123 - 124.
Crous, P.C. Pfeilerbau auf dem südafrikanischen Stein-
kohlenbergwerk Sigma. Glückauf 121 (1985) Nr. 13,
S. 1026 - 1029.
Fauconnier,C.J. u. Kersten, R.W.O. Increased under-
ground Extraction of Coal. The South African Insti-
tute of Mining and Metallurgy, Monograph Series No.4,
Johannesburg (1982) 345 pp.
Habenicht, H. und Urschitz, E. Rib Pillar Extraction -
An Alternative to Longwalling and Shortwalling.
115th AIME-Annual Meeting, New Orleans (1986) Pre-
print No. 86-65, 14 pp.
Habenicht, H. Systematic Development of a Powered Sup-
port for Faces in Weak Rock. Proc. Fifth Internat.
Strata Control Conference. London (1972).
Herwig, H. Planung des Strebausbaus aus gebirgsmechani-
scher Sicht. Glückauf 120 (1984) Nr. 3, S. 125 - 127.
Jacobi, O. Praxis der Gebirgsbeherrschung. 2. Auflage.
Verlag Glückauf GmbH, Essen (1981) 576 pp.
Kauffmann, P.W., Hawkins, S.A. und Thompson, R.R. Room
and Pillar Retreat Mining - A Manual for the Coal
Industry. United States Department of the Interior,
USBM, Information Circular 8839 (1980), 228 pp.
N.N. Mobile Roof Support For Retreat Mining. USBM-
Technology News No. 232, Nov. 1985, US-Department
of the Interior, Washington, D.C., 2 pp.
Wagner, H. u. Steijn, J.J. Effect of Local Strata Con-
ditions on the Support Requirements on Longwall
Faces. Proc. Internat. Soc. for Rock Mechanics,
4th Congress, Montreux (1979) pp. 557 - 564.
Singhal, R.K. Roof Supports for all Seams. World
Mining Equipment, May 1984, p. 32 - 42.

Design of semi-circular support system in mine roadways
Design de supports d'acier semi-circulaires dans les galeries de mines
Entwurf eines Ausbausystems aus halbrundem Stahl in der Strecke

F.P.HASSANI, Department of Mining and Metallurgical Engineering, McGill University, Montreal, Quebec, Canada
A.AFROUZ, Department of Mining and Metallurgical Engineering, McGill University, Montreal, Quebec, Canada

ABSTRACT: The results of experimental and theoretical investigations into the behaviour of the Semi-Circular steel support under fractured roof loading is presented.

RESUME: Les résultats d'études expérimentales et théoriques relatives au comportement de supports d'acier semi-circulaires avec des conditions du chargement d'un toit fracturé sont présentés.

Zusammenfassung: Die Ergebnisse einer experimentalen und theoretischen Erforschung in das Verhalten von halbrundem Stahlausbau unter gebrochener Deckenbelastung sind vorgestellt.

1 INTRODUCTION

Steel sets are utilized commonly as a passive support in the main access roadways and gateroads in coal mines. They serve to:
 (a) control the strata around the roadways,
 (b) reduce closure of the opening,
 (c) increase safety along the roadways.
Although considerable capital expenditure is involved in the support and associated repairs along roadways in coal mines, there is no standard rule in the coal mining industry for the determination of the support specifications. The design of support systems has tended to be based on experience and availability of support material.

This paper stems from part of an overall research program into rock support interaction in soft and hard rock mining excavations. It is devoted to determining the size and collapse load of semi-circular steel H-section as a roadway support by utilizing Castigiliano's theorem (Timoshenko, 1968) which experimentally has proved to give a good approximation to physical modelling (Jukes et al., 1983).

2 THEORETICAL ANALYSIS

Considering a semi-circular steel support and a point load acting on its crown downwards (Fig. 1), the load carrying capacity of the support can be expressed as follows:

$$\delta_y = M_p/Z_p = R(p - h)/M_s - f \qquad (1)$$
where: δ_y = Yield stress of the support (kPa)
 M_p = Plastic moment $R(p - h)$
 Z_p = Plastic modulus (m^3)
 M_s = Section modulus (m^3)
 R = Radius of the support material (m)
 f = Shape factor = Z_p/M_s (between 1 and 1.6)
 p = Vertical reaction of the support legs, (kN)
 h = Horizontal component of the load acting on the support at its legs (kN)
The bending moment at the crown of the support (M_c) can be expressed as:

$$M_c = p.R - h.R \qquad (2)$$

The bending moment at any point along the support can be expressed as:

$$M = p.x - h.y \qquad (3)$$

According to Jukes et al, (1983):

$$p = P/2$$

$$h = P/\pi$$

$$x = R(1 - \cos\theta) \qquad (4)$$

$$y = R \sin\theta$$

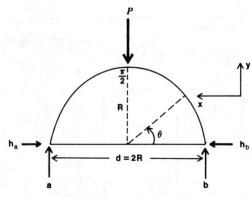

Fig. 1 – Schematic of a semi-circular support, representing forces and notations.

Since it is known that a plastic hinge will form at the crown of the support under the load, then at this point utilizing equations (2) and (4):

$$M_c = M_p = (P_c.R/2) - (P_c.R/\pi)$$

$$= P_c.R(1/2-1/\pi) \qquad (5)$$
where: P_c = Critical load to cause yield of the support, kN
Hence the yield stress at the crown of the support using equations (1) and (5) is:

$$\delta_y \text{ at the crown} = [P_c.R(1/2-1/\pi)]/M_s.f$$

$$= [P_c.R(\pi-2)]/2\pi M_s.f$$

$$= 0.18\ P_c.R/M_s.f$$

Table 1 - Specifications of steel H section supports and fractured rock load (W)

Item No.	Support Radius (R) in m.	Sectional dimensions in mm. (Fig. 4)	Section Modulus (M_s), m³×10⁻⁶ (eqn.9)	Mass per unit length of the support (kg/m)	Load of the fractured roof rock on the support (W), in kN/0.9m			
					At height R (eqn.10)	At height 2R (eqn.11)	At height R/tanϕ (eqn.12)	Terzaghi's formula (eqn.14)
1	1.52	z = 76			51.06	102.02	65.89	91.60
2	1.67	w = 76	45.0	14.67	61.63	123.27	79.54	100.63
3	1.82	t = 8.9			73.20	146.41	94.47	109.67
4	2.42				129.43	258.85	167.02	145.83
5	1.52	z = 89			51.06	102.02	65.89	91.60
6	1.67	w = 89	67.4	19.35	61.63	123.27	79.54	100.63
7	1.82	t = 9.5			73.20	146.41	94.47	109.67
8	2.42				129.43	258.85	167.02	145.83
9	1.52	z = 102			51.06	102.02	65.89	91.60
10	1.67	w = 102	97.0	23.06	61.63	123.27	79.54	100.63
11	1.82	t = 9.5			73.20	146.41	94.47	109.67
12	2.42				129.43	258.85	167.02	145.83
13	1.52	z = 114			51.06	102.02	65.89	91.60
14	1.67	w = 114	132.0	26.78	61.63	123.27	79.54	100.63
15	1.82	t = 9.5			73.20	146.41	94.47	109.67
16	2.42				129.43	258.85	167.02	145.83
17	1.52	z = 127			51.06	102.02	65.89	91.60
18	1.67	w = 114	158.0	29.76	61.63	123.27	79.54	100.63
19	1.82	t = 10.2			73.20	146.41	94.47	109.67
20	2.42				129.43	258.85	167.02	145.83

Table - 2 - Empirical and theoretical collapse load of steel supports (P_C) relevant to the specifications noted in Table 1

Item No. (from Table 1)	Collapse load (P_C) at the crown, in kN			Collapse load (P_C), in kN at $\phi = 45°$, (eqn. 8)	
	Empirically (eqn. 16)	Theoretically (eqn. 6)			
		f = 1.15	f = 1.55	f = 1.15	f = 1.55
1	49.53	48.67	65.74	113.39	152.76
2	45.08	44.39	59.83	103.20	139.04
3	41.36	40.73	54.90	94.70	127.58
4	31.11	26.96	41.29	71.22	95.95
5	74.19	73.05	98.44	169.83	228.80
6	67.52	66.49	89.60	154.57	208.25
7	61.96	61.01	82.21	141.83	191.09
8	46.60	45.88	61.83	106.67	143.71
9	106.77	105.13	141.67	244.41	329.29
10	97.18	95.69	128.94	222.46	299.71
11	89.17	87.81	118.32	204.12	275.01
12	67.06	66.03	88.98	153.51	206.82
13	145.29	143.07	192.79	332.60	448.10
14	132.24	130.22	175.47	302.73	407.85
15	121.35	119.49	161.01	277.78	374.24
16	91.26	89.86	121.09	208.91	281.45
17	173.91	171.25	230.76	398.12	536.37
18	156.37	155.87	210.03	362.36	488.19
19	145.24	143.02	192.72	332.49	447.95
20	109.23	107.56	144.94	250.06	336.89

936

$$P_c = 2\Pi\,\delta_y.M_s.f/R(\Pi-2)$$

$$= 5.56\,\delta_y.M_s.f/R \qquad (6)$$

The yield stress at any other point along the support, from equaitons (1), (3) and (4) is:

δ_y - at any point $= [(P_c.R(1-\cos\Theta)/2) -$

$$(P_c.R\,\sin\Theta/\Pi)]/M_s.f$$

$$= P_c.R(\Pi-\Pi\cos\Theta-2\sin\Theta)/2\,\Pi M_s.f$$

$$P_c = 2\Pi\,\delta_y.M_s.f/R(\Pi-\Pi\cos\Theta-2\sin\Theta) \qquad (7)$$

One of the most likely plastic yield point along the semi-circular steel support is at $\Theta = 45^o$. Therefore, values of δ_y and P_c at this point, using equation (7) is:

$$\delta_y \text{ at } \Theta \text{ of } 45^o = -0.08\,P_c.R/M_s.f$$

$$P_c \text{ at } \Theta \text{ of } 45^o = -12.82\,\delta_y.M_s.f/R \qquad (8)$$

The negative sign in the above expressions indicate compressional failure. All of the above noted expressions are applicable for the supports erected along horizontal or near horizontal level roadways. For roadways dipping with an angle of β^o to the horizontal, the P_c-values should be divided by $\sin\beta$. In this case, the support material may not fail under the load equivalent to the vertical collapse load, but it will be tilted, resulting in the reduction of support efficiency.

3 DESIGN PROCESS

To select a support system, five main steps should be considered. These are as follows:
(a) - The effect of section modulus (M_s) and shape factor (f) on the yield stress of the support (δ_y) and its collapse load (P_c) is seen from the equations (6) and (7). The values of M_s and f are directly proportional to P_c and indirectly proportional to δ_y. The value M_s is determined from the following expression:

$$M_s = A.z/3 \times 10^{-9} \text{ (m}^3\text{)} \qquad (9)$$
where: A = Cross-sectional area of the H-section
support (mm^2)
z = Thickness of the H-section support (m.m) as shown in Fig. 2.

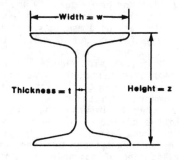

Figure 2. Dimensional notations of H-section supports used in Table 1.

The M_s-values for various dimensions of the H-section supports is given in Table 1. The f-values depend on the cross-sectional shape of the support. Since the steel supports in mines have generally H-shape, f-value is a constant and commonly accepted to be 1.15 (Jukes et al., 1983). This value is conservative in comparison with the value of 1.55

also found experimentaly (Sadler et al., 1965).
(b) - The support should have sufficient load capacity to hold the envelope of fractured strata caused by the strata pressure redistribution. This fractured zone is usually in the shape of a dome, cone or sinosoidal. Fig. 3 shows 4 main loading conditions described as follows:
(i) Fractured envelope extending upto half the diameter of the support (d/2 = R) is shown as curve A. This is most commonly observed where either the roof strata is strong or the roadway is machine-cut, well formed or a good system of packing is employed along the roadway. In this case, the load of fractured roof rock on the support (W_A) and its corresponding pressure (P_A) can be determined from:

$$W_A = \gamma.d.R/2 = 9.81\,\gamma.R^2 = 24.5\,R^2 \text{ (kN/m)}$$

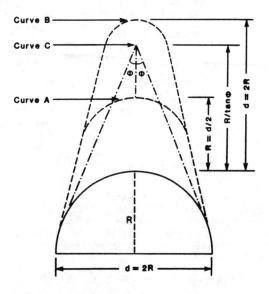

Figure 3. Three main conditions of fractured roof rock above the roadway supports.

However, most roadways driven in coal measure rocks utilize steel supports at 0.8 metre intervals. Therefore, the above equation in terms of kN/0.9 m can be expressed as follows:

$$W_A = 24.5\,R^2 \times 0.9 = 22.1\,R^2 \text{ (kN/0.9m)}$$

$$P_A = 9.81\,\gamma.R^2 = 24.5\,R^2 \text{ (kN/m}^2\text{)} \qquad (10)$$
where: γ = Average density of the fractured roof rock, excluding the coal.
= 2.50 (tonnes/m^3)
Variation of the load (W_A) and pressure (P_A) per support diameter (d) is shown as curve A in Fig. 4.
(ii) Fractured roof envelope extending upto a distance equal to the diameter of the support (d = 2R). This condition prevails when either the roof strata is weak, or intense pressure persists around the roadway. In this case, the load of fractured roof rock on the support (W_B) and its corresponding pressure (P_B) can be determined from:

$$W_B = 9.81\,\gamma.d.R = 2 \times 9.81\,\gamma.R^2 = 49.0\,R^2 \text{ (kN/m)}$$

$$= 0.9 \times 49.0\,R^2 = 44.2\,R^2 \text{ (kN/0.9 m)}$$
$$\qquad (11)$$

$$P_B = 2 \times 9.81\,\gamma.R^2 = 49.0\,R^2 \text{ (kPa)}$$

Variation of the load (W_B) and pressure (P_B) per support diameter (d) is shown as curve B in Fig. 4.
(iii) Fractured roof envelope bond by the average angle of internal friction of the roof

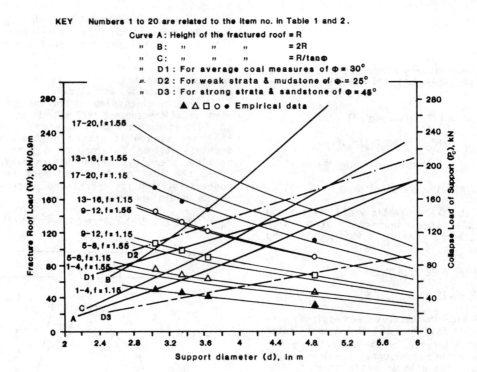

Figure 4.   Load capacity of semi-circular H-section steel supports at the crown.

rocks to the vertical (φ) shown in as curve C. This condition exists when highly stratified strata is over stressed. In this case the load of fractured roof rock on the support (W_c) and its corresponding pressure (P_c) is determined from the following expression:

$$W_c = 9.81\gamma \ (R/\tan\varphi) \ 2R/2 = 9.81 \ \gamma.R^2/\tan \ , \ (kN/m)$$

For characteristic coal measure rocks $\varphi = 30°$, therefore,

$$W_c = 9.81 \ \gamma.R^2/\tan \varphi = 42.5 \ R^2 \ (kN/m)$$

$$= 38.3 \ R^2 \ (kN/0.9m) \qquad (12)$$

$$P_c = 42.5 \ R^2 \ (kPa)$$

Variation of the load (W_c) and pressure (P_c) per support diameter (d) is shown as curve C in Fig. 4.

(iv) Adaption of Terzaghi's formula for the vertical load on the beam shaped supports (Terzaghi, 1943). The formula can be expressed in the following form:

$$W_D = 9.81\gamma(2a+2b \ \tan(45-\varphi/2))/2 \ \tan\varphi$$

$$= 24.5(a+b \ \tan(45-\varphi/2))/\tan\varphi, \ (kN/m)$$

$$= 22.1(a+b \ \tan(45-\varphi/2))/\tan\varphi, \ (kN/0.9m) \qquad (13)$$
where: a = Half width of the support, (m)
b = Height of the support = a, (mm)
For an average angle of internal friction, $\varphi = 30°$, expression (13) can be written as:

$$W_D = 42.46(a+0.577b), \ (kN/m)$$

$$= 38.21(a+0.577b), \ (kN/0.9m)$$

$$= 60.26a, \ (kN/0.9m)$$

$$P_D = 42.46(a+0.577b) = 66.96a, \ (kPa) \qquad (14)$$

Variation of W_D and P_D per support diameter (d) for typical friction angles (φ) is shown as curves

D_1, D_2 and D_3 in Fig. 4.
(c) - Calculation of the collapse load (P_c) - The theoretical value of P_c is given by equations (6-8). Empirically the P_c-value can be derived for the section modulus (M_s) of 132 x 10^3 mm^3 as follows (2).

$$P_c = 45.10/w \qquad (15)$$
where: P_c = Collapse load on the crown, (tonnes)
w = Width of the semi-circular support
= 2R = d, (metre)
Equation (15) can be expressed for f = 1.15 in terms of kN and section modulus (M_s) in the following form:

$$P_c = (45.10 \times 9.806/1.15 \times 260 \times 10^3 \times 132 \times 10^{-6} \times 2) \ (\delta_y.M_s.f/R)$$

$$= 5.6 \ \delta_y.M_s.f/R, \ (kN) \qquad (16)$$

The P_c-values for prime quality grade 43A of H-section structural steel of yield stress (δ_y) = 260 x 10^3 kPa, with various dimensions, section modulus (M_s) and shape factors of f = 1.15 and 1.55 is given in Table 2 and Figure 4.
(d) - Effect of strutting on the collapse load - The H-section steel supports have about 3 times lower load capacity in the direction of its minor axis, in comparison with its major axis (Lewis, 1970). In order to improve the strength of the support in the direction of its minor axis, two steps may be considered:
(i) closer spacing between the supports, and
(ii) utilization of suitable strutting, thereby to strengthen the support system as a whole.

The stability of struts under loads is related to slenderness ratio 1/K (Jukes et al., 1983).
where: 1 = Effective length of the strut, (m)
K = The least radius of gyration of the strut section, (m)
For a solid circular strut (rod): K = r/2
where: r = Radius of strut section (cross-section of the rod).
For a thin-walled tabular strut:

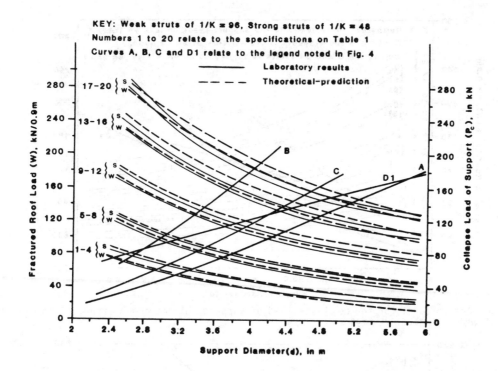

KEY: Weak struts of 1/K = 96, Strong struts of 1/K = 48
Numbers 1 to 20 relate to the specifications on Table 1
Curves A, B, C and D1 relate to the legend noted in Fig. 4
———————— Laboratory results
— — — — Theoretical-prediction

Figure 5 Design chart for the empirically and theoretically determined collapse load of supports at the crown, with f = 1.15, using weak and strong struts.

$$K = \left[r_0^3/(r_0+r_i) \right]^{1/2} \underline{\Omega} \, r_0/\sqrt{2}$$

where: r_0 = Outer radius of the tube section, (mm)
        r_i = Inner radius of the tube section, (mm)

Laboratory tests on scaled model steel supports (Jukes et al., 1983) have shown that the collapse load of supports at the crown is dependent on the slenderness ratio (1/K) and the number of struts (N), when strutting is adopted. The relevant empirical formulae projected for the full size supports in terms of P_{cs} and kN, yields the following expressions:

In general:

$$P_{cs} = 124.10 - 0.17 \; 1/K, \; (kN) \tag{17}$$

For weak struts of 1/K = 96:

$$P_{cs} = 84.95 + 2.94N, \; (kN) \tag{18}$$

For strong struts of 1/K = 48:

$$P_{cs} = 84.07 + 4.71N, \; (kN) \tag{19}$$

The difference between the constants in the equations (18) and (19) is very small. It can be averaged to (84.95+84.07)/2 = 84.51, kN. This is the value of collapse load (P_c) when there is no strutting on the support system (N=0). Hence, the equations (17), (18), and (19) can be expressed in the following forms:

In general:

$$P_{cs} = 1.47 P_c - 0.17 \; 1/K, \; (kN) \tag{20}$$

For weak struts of 1/K = 96:

$$P_{cs} = P_c + 2.94N, \; (kN) \tag{21}$$

For strong struts of 1/K = 48:

$$P_{cs} = P_c + 4.71N, \; (kN) \tag{22}$$

Giving actual values of 1/K, N and P_c (from Table 2) the P_{cs}-values are determined from the equations (20), (21) and (22) and shown graphically in Figs. 5 to 9.

(e) - The support should accommodate the initial roof lowering caused by compression of the gateside pack. This can be done by measuring the roadway convergence onto the packs, and then accounting for an equal amount of convergence to take place in the support by provision of wooden foot blocks, wood chocks, stilts or selecting appropriate support height.

4 CONCLUSIONS

In-situ observations, empirical and thoretical expressions utilized in this investigation suggest the followings:
- The weakest point of the semi-circular steel support lies on its crown.
- Curves A to D in Fig. 4 provide a prediction of the fractured roof load (W) on a semi-circular steel support system for various rock types, strata conditions and method of analysis.
- The collapse load of semi-circular steel supports (P_c), without on the radius (R), sectional modulus (M_s) and shape factor (f) of the support expressed in the following forms:

a) At the crown of the supports:

Theoretical P_c = 11.11 $\delta_y.M_s.f/R$, (kN)

Empirical P_c = 11,65 $\delta_y.M_s.f/R$, (kN)

b) At any other point along the supports:

P_c = 12.56 $\delta_y.M_s.f/R(\Pi - \Pi \cos\theta - 2\sin\theta)$, (kN)

For θ = 45° the above equation can be expressed as:

$$P_c = 25.63 \; \delta_y.M_s.f/R$$

- Figure 4 provides design charts for the semi-circular, H-section, steel supports, without strutting, along mine roadways, by considering crown and shoulders of the supports, respectively.

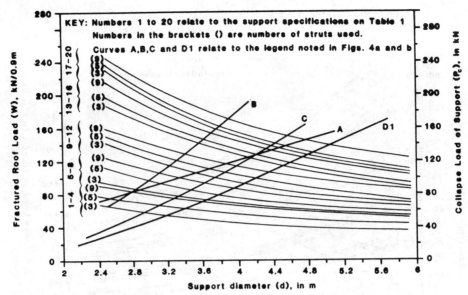

Figure 6. Design chart for the collapse load of supports at the crown with f = 1.15 and varying number of weak struts of l/k = 96.

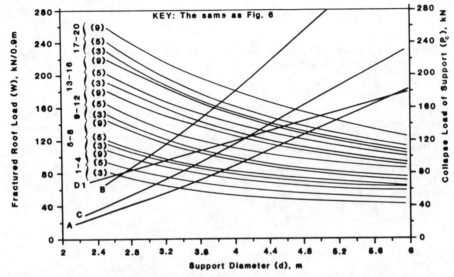

Figure 7. Design chart for the collapse load of supports at the crown with f = 1.15 and varying number of strong struts of l/k = 48.

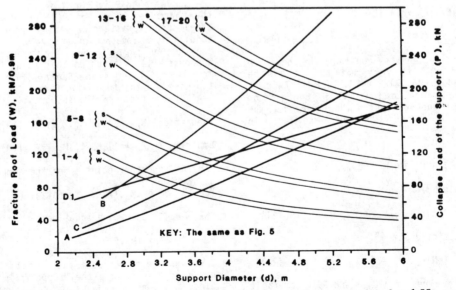

Figure 8. Design chart for the collapse load of supports at the crown with f = 1.55, using strong and weak struts.

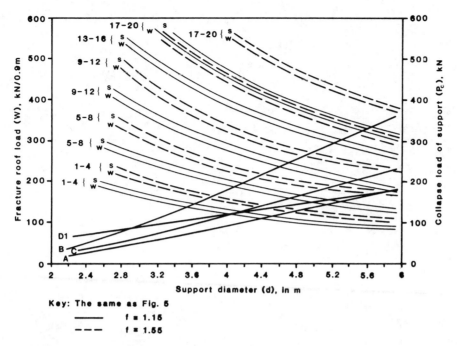

Figure 9. Design chart for the collapse load of supports at the shoulders, with f = 1.15, using weak and strong struts.

- Figures 5 to 9 provide design charts for the above noted supports with strutting.

- High quality and well placed struts can increase the load bearing of steel supports by upto 39%, dependent upon the slenderness ratio (1/K) and number of struts (N) utilized, according to the following expressions:

a) In general:

$$P_{cs} = 1.47 \; P_c - 0.17 \; 1/K, \; (kN)$$

b) For weak struts of 1/K = 96:

$$P_{cs} = P_c + 2.94 \; N, \; (kN)$$

c) For strong struts of 1/K = 48:

$$P_{cs} = P_c + 4.71N, \; (kN)$$

- The above noted conclusions are valid for semi-circular, H-section, steel supports. Work is on hand to extend the investigation towards various type of steel arched supports under different loading conditions.

ACKNOWLEDGEMENT

The authors are grateful to the E.M.R. and N.S.E.R.C. for supporting this project.

REFERENCES

Jukes, S.G., Hassani, F.P. and Whittaker, B.N., 1983/84. Characteristics of Steel Arch Support Systems for Mine Roadways, Parts I and II, Mining Science and Technology, Elsevier, 43-58.
Lewis, S., 1970. Systems of Roadway Supports, Conf. on Strata Control, in Mine Roadways, University of Nottingham, 33-49.
Sadler, G.W., Campbell, S.G., 1965. Approximate Equations for the Yield and Collapse of H-Section Roadway Arches, N.C.B. Laboratory Testing Branch, Report No. 652.
Terzaghi, K., 1943. Theoretical Soil Mechanics, J. Wiley, pp. 194-197.
Timoshenko, S., 1968, Elements of the strength of materials, 5th Ed., Van Nostrand Reinhold.

Simulation of grouting in jointed rock

Simulation d'injection en roche fissurée
Simulierung des Injizierens in rissigem Fels

LARS HÄSSLER, M.Sc., Royal Institute of Technology, Department of Soil and Rock Mechanics, Stockholm, Sweden
HÅKAN STILLE, Ph.D., Royal Institute of Technology, Department of Soil and Rock Mechanics, Stockholm, Sweden
ULF HÅKANSSON, M.Sc., Royal Institute of Technology, Department of Soil and Rock Mechanics, Stockholm, Sweden

ABSTRACT: In order to transform grouting from black magic to science, theoretical understanding and development are needed. A measuring technique for evaluation of the rheology parameters of grouts is presented together with some preliminary results. Results from a numerical model is presented and compared with results from a physical model.

RESUME: Afin de transformer l'art de l'injection de la magie noir en science, le developpement de la théorie est nécessaire. Une technique de mesure pour l'évaluation des parametres de rhéologie des matériaux d'injection est présentée ainsi que quelques résultats préliminaires. Les résultats obtenus utilisant un modele numérique sont présentés et comparés avec les résultats obtenus utilisant un modele physique.

ZUSAMMENFASSUNG: Um das Injizieren aus einer Hexerie in eine Wissenchaft zu vervandeln bedarf es theoritisher Ableitungen. Eine Messtechnik für die rheologishe Parametern von Injektionsgut wird zusammen mit einigen vorläufigen Resultaten beschrieben. Ein numerishes Modell wird dar gestellt und mit resultaten von einen physichen Modell vergleichen.

INTRODUCTION

Reducing water flow through jointed rock is important for different reasons. Water seepage under dams cause energy loss and erosion of the damcore. Water leakage into tunnels, with lowering of the watertable and consolidation settlements of clay layers as a result, can cause damage to buildings on the surface. Water flow through storage of waste material cause an undesired contamination of the ground water.

In order to reduce the water flow, grouting of the jointed rock is a commonly used method. Grouting is today normally based on empirical relationships and on the experience of the grouting personel.

To meet increasing demands of reducing water flow in jointed rock a project was started by the Swedish Rock Engineering Research Foundation (BeFo). The aim of the project is to develop the theory of grouting and the testing procedure of the grout. Some preliminary results from the project will be presented below.

The problems of grouting in jointed rock have been discussed in several books and papers and on different symposiums. Most of the presented material is dealing with case history of grouting or grouting procedure based on experience see for example Evert, Cambefort, Bruce.

Other authors are discussing different grouts, see for example Littlejohn.

In the articles more practical aspects are given on the grout than on theoretically based rheological models. Only a few authors are dealing with mathematical modelling and numerical calculations of grouting see for example Wallner, Lombardi, Chien.

1 PROPERTIES OF GROUTS

The development and use of mathematical models simulating grouting needs knowledge about the properties of the grouts. The grouting process is extended over a timeperiod, therefore the change of these properties with time must be known.

1.1 Measuring technique for grouts

Grouting fluids have a more complicated behavior than Newtonian fluids. In order to describe their behavior, more than one rheological parameter is needed. To find these parameters at least two measurements, at different shearrates, must be made and to evaluate the rheological model many measurements are required. By repeting these series of different shearrates during a time period, the time dependency of the rheological parameters can be found (see fig.1).

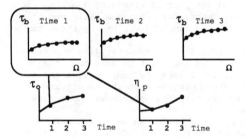

Figure 1: Evaluation of rheological properties and their change with time (Bingham example). τ_b is the shear stress at the surface of the rotating cylinder that is submerged into the grouting fluid and Ω is the angular velocity of the cylinder.

This leads to a rather complicated measuring sequence. To accomplish this a computerized Brookfield Rheoset rotational viscometer is used and connected to a personal computer (fig.2). The program in the viscometer governs the measuring sequence and the personal computer is used for storage, evaluation and report of results. The personal computer can also change the measuring sequence if desired.

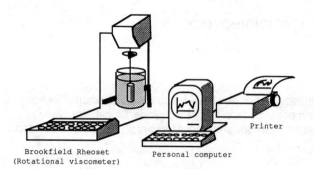

Brookfield Rheoset   Personal computer
(Rotational viscometer)   Printer

Figure 2: Instrumentation for rheological
measurments and evaluation of properties.

1.2 Bingham fluids

According to the state of the art concerning cement
based grouts the Bingham fluid model seems appropriate, see for example Littlejohn. The Bingham
fluid model is similar to the Newtonian with respect
to a linear relationship between shear stress and
shear rate. The difference between the two models is
that, for a Bingham fluid, a yield stress τ_o must be
exceeded in order to initiate flow. In fig.3. the
two models are presented.

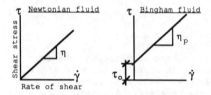

Figure 3: The rheological parameters of Newtonian
and Bingham fluids.

The slope of the curve, in the Bingham model, is a
constant known as the plastic viscosity, η_p. The
yield stress and the plastic viscosity are the
rheological parameters for a Bingham fluid and the
rheological behavior can be described by these two
parameters.

When using a rotational viscometer like the Brookfield rheoset, it is important to note that the
stress distribution across the gap, between the bob
and the cup, is not dependent on the properties of
the fluid. Instead the shear stress will decrease
with the square of the radius.

For a Bingham fluid there can be an area in the
annular space of the cup where the yield stress is
not exceeded and consequently no shearing occurs
(fig.4).

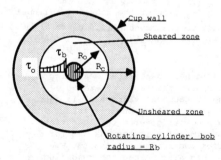

Figure 4: Principles for rotational viscometry of
yield-value fluids

When the stress equals the yield value, the radius
R_o of the sheared zone, is given by:

$$R_o = R_b \sqrt{\tau_b / \tau_o} \qquad (1:1)$$

A criterion for this varying gap size with respect
to τ_b is that $\tau_b > \tau_o > \tau_c$, where τ_c is the stress at the
cup-wall.

In order to evaluate the properties equation (1:2)
is used:

$$\tau_b = \tau_c + \tau_o \ln \left(\frac{R_c}{R_b} \right)^2 + 2\,\eta_p\,\Omega \qquad (1:2)$$

This equation is known as the Reiner-Riwlin equation
and it is valid when all the material in the annulus
is sheared. The expression for any intermediate
radius, i.e an unsheared zone appears, is obtained
by substituting τ_c for τ_o, R_c for R_o and $(R_o/R_b)^2$ for
τ_b/τ_o and this leads to:

$$\tau_b = \tau_o \left(1 + \ln\frac{\tau_b}{\tau_o}\right) + 2\,\eta_p\,\Omega \qquad (1:3)$$

A criterion for equation (1:1) through (1:3) is that
there is no slip between the rotating cylinder and
the fluid or between the cup-wall and the fluid.
For more details concerning rotational viscometry
see for example Van Wazer et al.

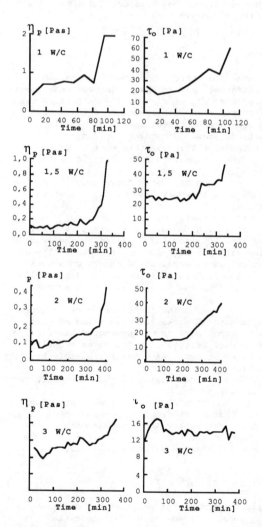

Figure 5: Examples of preliminary results from
measuring of the properties of cement-bentonite
mixtures with 1-3 W/C by wieght and 3% bentonite of
the water weight.

944

1.3 Results from viscometer tests

At present a few cement-bentonite mixtures with W/C (Water-cement ratio by weight) varying from 1-3, and a bentonite content of 3% of the water weight have been tested. The results show very little change of the properties with respect to time for the first hours but after that a dramatical change usually occurs (see fig 5). The result naturally depends on what kind of cement that has been used and on other added material.

2 NUMERICAL MODEL

Very few analytical relations are usable for predicting grouting results. The complexity of the problem can only be satisfactorily solved by using numerical methods. Even when computers are available the problem remains difficult. Therefore at this early stage of the project, a rather simple model has been constructed. The model takes into consideration the effects caused by ground water flow in the jointed rock. At this stage the model is developed for Bingham fluids but it can easily be adjusted to suit other rheological models.

2.1 Philosophy of the model

The model is based on the assumption that a joint plane in a rockmass can be described as a mesh of one-dimensional channels (fig.6). The impervious area between the channels corresponds to, either contact points in the rock, or to filling in the joints.

At present the model can compute the flow in one jointplane at a time but the future aim is to develop the model, in order to simulate grouting of several intersected jointplanes, affected by groundwater flow.

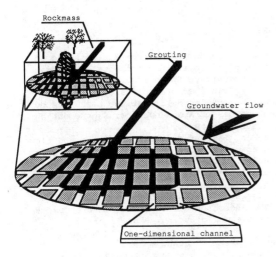

Figure 6: Each jointplane is simulated as a mesh of channels affected by groundwater flow.

2.2 Governing equations for Bingham flow in a channel mesh

Each channel is regarded as one-dimesional with a small opening compared to the width of the channel. The mean velocity of the grouting front in a single channel with no water present can then, according to Wallner, be expressed as:

$$\frac{dX}{dt} = \frac{\gamma b^2}{12 \, \eta_p} \frac{dh}{dx} \left(1 - 3\frac{z_{gr}}{b} + 4 \left(\frac{z_{gr}}{b}\right)^3\right) \qquad (2:1)$$

Where z_{gr} is half the Bingham plug thickness (fig.7), γ is the density of the grout times the acceleration due to gravity, dh/dx is the gradient and dX/dt is the velocity of the front.

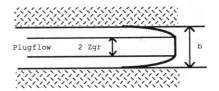

Figure 7: The Bingham plug flow in a one-dimensional channel

Assumptions made in deriving equation (2:1) are presented by Wallner.

Grouting is used to reduce water flow, so consequently water flow should be considered in the equation. The equation for the case when the channel is waterfilled can be expressed as:

$$\frac{dX}{dt} = \frac{\gamma b^2}{12} \frac{(h_o - h_L)}{\zeta} \qquad (2:2)$$

where:

$$\zeta = x(t) \left(\frac{\eta_p}{1 - 3\frac{z_{gr}}{b} + 4 \left(\frac{z_{gr}}{b}\right)^3} - \eta\right) + L\eta$$

$$z_{gr} = \min\left(\frac{\tau_o}{\gamma(h_o - h_x)} \, x \, , \, \frac{b}{2}\right)$$

$$h_x = \frac{dX}{dt} \frac{12 \, \eta}{\gamma b^2} (L - X) + h_L$$

For explanation see fig.8.

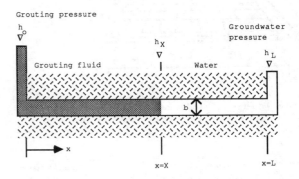

Figure 8: Grouting in a waterfilled channel.

To apply the equation from one channel to a mesh of many channels is simply a matter of mass balance.

$$\sum_{j=1}^{m_i} Q_{ij}(t) = 0 \qquad i = 1, \ldots, n \qquad (2:3)$$

Here m_i is the number of channels that meet in node i and n is the number of nodes in the entire channel mesh. $Q(t)$ is the massflow at the time t. These relations makes it possible to calculate the pressures at the nodes at a given grouting stage if the boundary pressures are known.

2.3 The structure of the numerical program

The numerical program consists of three parts, a preprocessor, a main processor and a postprocessor.

The preprocessor translates the geometry of the jointplane into numerical data that are compatible with the computer.

The main processor calculates and stores all intermediate results for later treatment.

The postprocessor does the printing and plotting of the sampled intermediate results.

The program makes it possible to study how the simulated grouting front proceeds in the jointplane.

The program and the theory of grouting is in detail presented in a report by Hässler et al.

2.4 Special problems

When water and grout are mixed in an intersection it is difficult to calculate the properties of the mixed material. At present the grout is regarded as destroyed and the resulting material is given the same properties as water. This is a rough simplification that needs to be corrected.

3 MODEL TEST RESULTS COMPARED WITH NUMERICAL RESULTS

In order to develop and calibrate the numerical model, physical model tests are made. The joint model consists of two parallel acrylresin panels with a number of square plates inbetween. This simulates a jointplane in a rockmass corresponding to the one used in the numerical model. It is possible to apply different water gradients over the model (see fig.9).

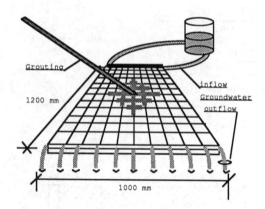

Figure 9: Physical model of a jointplane. Each channel has dimensions 5mm x 1mm and a length of 100mm.

The grouting propagation is continuously registred with a videocamera. Tests have so far only been made with pure bentonite-water mixtures as grout.

3.1 Results

The results from the numerical model shows good agreement with the physical model results when no watergradient is applied over the jointplane (see fig.10). Due to symmetry only half of the propagation is shown.

When watergradients are applied the problem with assuming the properties of watermixed grout so far distorts the numerical calculations.

Time (seconds)

2.5   6   22   65   1426

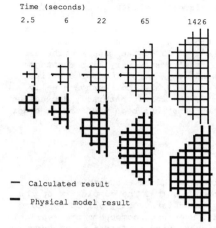

— Calculated result

▬ Physical model result

Figure 10: Physical model results of grouting in a waterfilled jointplane compared with numerical calculations. Bingham grout with τ_o=3 Pa and η_p=0.035 Pas. Grouringpressure equal to 0.48 m of water. No watergradient applied.

4 CONCLUSIONS

Determination of the properties of grouts is important in order to compare different materials and for further development in the field of grouting science. The method described here can be used for evaluation of what fluidmodel is the most accurate and how the rheological properties change with time.

It is possible to calculate, with god accuracy, the groutingprogress in a jointplane simulated by a channelmesh, if there is no watergradient present. Further development needs to be done where the effects of watergradients are better accounted for.

It is also possible to expand the numerical model to three dimensions that simulates an entire rockmass containing intersections of several jointplanes. This tool can be of great use when developing new grouting technics and for choosing the right method and grout for a specific case.

REFERENCES

Bruce N-E. Hellgren A., Bäckström S. Grouting of tunnels in order to avoid detrimental groundwater lowering. Subsurface Space 1980.

Cambefort H. Principes et applications de l'injection. Annales de l'institut technique du batiment et des travaux publics. Supplement au nr:353 septembre 1977.

Chien Sze-Foo. Laminar flow pressure loss and flow pattern transition of Bingham plastics in pipes and annuli. Int.J.Rock Mech.Min.Sci vol7,pp333-356 Perganamon press 1970 Great Britain.

Evert F-K. Rock grouting with emphasis on dam sites. Springer verlag 1985.

Hässler L., Andersson J., Stille H. Theoretical and model studies of grouting, Phase 1 basic relations. Swedish Rock Engineering Research Foundation, BeFo. Stockholm 1986.

Littlejohn G.S. Design of cement based grouts. Grouting in geotechnical engineering. Proceedings. American Society of Civil Engineers 1982 New Orleans.

Lombardi G. The role of cohesion in cement grouting of rock. Quinziéme Congres des Grandes Barrage. Lausanne 1985.

Van Wazer J.R., Lyons J.W., Kim K.Y. and Colwell R.E. Viscosity and flow measurement, a laboratory handbook of rheology. Monsanto chemical company. St.Louis,Missouri. Interscience publishers 1963.

Wallner, M. Propagation of sedimentation stable cement pastes in jointed rock. Rock Mechanics and Waterways Construction. University of Achen, BRD 1976.

Anchorages with high deformation capacity for the support of underground excavations
Tirants d'ancrage à haute capacité de déformation pour le soutènement du rocher dans les excavations souterraines
Nachgiebige Verankerungen für die Sicherung von Hohlraumauffahrungen

THOMAS F.HERBST, Dyckerhoff & Widmann AG, Munich, FRG

ABSTRACT
Deformations of different magnitude follow the excavations in the underground. If anchors are chosen for securing the rock mass surfaces they influence the deformations and herewith the arching arround the opening. Highly yielding anchors gain importance where a semi-rigid support cannot create enough bearing capacity of the rock mass and technical and economic reason limit the quantity, force and length of anchors. Requirement, testing and anchor designs are discussed.

RESUME
Le creusement d'excavations souterraines entraîne des déformations d'ampleur extrêmement variable. Si l'on adopte des tirants d'ancrage afin d'assurer la sécurité de la surface rocheuse, ceux-ci influent sur les déformations et par conséquent sur la formation de la voûte autour de l'ouverture. Les tirants d'ancrage à haute capacité de deformation gagnent en importance lorsqu' un second-oeuvre semi-rigide ne développe pas une capacité portante suffisante et que des considérations tant techniques qu'économiques limitent le nombre, la force et la longueur des tirants d'ancrage. Le présent article décrit les exigences requises des tirants d'ancrage à haute capacité de déformation, leur contrôle et les différents types de tirants.

ZUSAMMENFASSUNG
Verformungen verschiedenster Größe folgen auf den Ausbruch von Hohlräumen im Untergrund. Werden Anker zur Sicherung der Gebirgsoberfläche gewählt, so beeinflussen sie die Verformungen und damit die Gewölbebildung um die Öffnung. Stark nachgiebige Anker gewinnen an Bedeutung, wo ein halbsteifer Ausbau nicht ausreichende Tragfähigkeit des Gebirges entwickeln kann und technische wie wirtschaftliche Gesichtspunkte Anzahl, Kraft und Länge der Anker beschränken. Anforderung, Prüfung und Bauart werden beschrieben.

1. INTRODUCTION

For underground excavations two different and contrary philosophies have governed the requirements for support systems. For civil engineering purposes tunnel lining had to be so rigid that convergencies and subsidence of the roof could be avoided in general. For railway tunnels the geometric shape of the tunnel lining had to be kept within close tolerances.

In mining the support systems had and still have to secure the safety of the mine workers and a required open space for transport vehicles and conveyor belts. As the costs of the support shall be minimized convergencies are accepted under above conditions in particular if the service time of a tunnel is short.

With the construction method of NATM the understanding of the bearing system around a tunnel which is secured with anchors and shotcrete has increased and the existence and necessity of deformations of the rock mass as part of its bearing characteristics are well accepted. Consequently the support systems have to be more flexible (Fig. 1).

If the surfaces of an underground excavation are secured by rock anchors and rock bolts and by a thin lining as e.g. shotcrete, zones of different rock stress are created around the opening. They may reach from a practically unaffected region at big distance from the excavation to elastically deformed and plastified zones and to a post-failure state zone around the opening. If all these states occur the used rock anchors and bolts have to be able to follow considerable deformations without failure and to transfer the loads to the safe rock areas.

Figure 1 Dense pattern of anchor bolts with flexible lining

Figure 2 Anchors in failed rock surface
(Courtesy Bergbauforschung)

For the elastically and plastically deformed zones anchor technology provides designs
which are able to scope with the prevailing
strains and stresses.

2. ANCHORS WITH HIGH DEFORMATION CAPACITY
   AND ROCK MASS PERFORMANCE

Fractured and failed rock zones which form
with weak mechanical properties of the rock
mass and with high stress levels, cause the
greatest deformations in order to release
stresses. Even if the rock mass fractures
it is still able to contribute to the stabilization of a mine roof by supporting adjacent zones, as long as they are held in
place. Their performance may be compared
with that of loose soils. However, anchors
which are applied under such conditions may
not be compared with so-called soil anchors
as the loose soil condition of the post-
failure state rock is mostly created whilst
the anchors have been already installed.
Hence the anchors are submitted to deformations and shear forces in a way which has
not yet been considered as a standard requirement for anchors in underground application.

It may be open for discussion if such high
deformations are good support practice even
for mines. Beside of economic reasons, however, it may be the only way to prevent
heavy rock bursts with high rate of damage
by releasing the stresses in the rock mass
slowly.

Anchors which shall control a rock mass in
its post-failure state or during deforming-
to-failure cannot be considered independently: Rock mass, anchor and lining act as
unit. Anchor head and lining have to converge or yield together at similar rates,
otherwise the anchor heads will punch the
rock mass or the rock mass flows and fails
around the anchor head (Fig. 2).

Lining may consist in its most flexible example of wire mesh, in different cases an
additional net of e.g. used mine rope connect the anchor heads and increases the
bearing capacity of the wire mesh (Fig. 3).
Other linings consist of strips of 2 reinforcement bars connecting two anchor heads.
Thin shotcrete with reinforcement mats are
appropriate for such an application as long
as deformation joints are provided. Advantage of the shotcrete is a complete contact
to the rock surface which prevents even
small rock falls.

Figure 3 Flexible lining with rope reinforcement

The acting part of this composite system
is the rock mass.

Its deformation and failure modes have to
be observed and investigated more closely.
Spalling, shear failure and joint displacement may cause different sollicitations of
the anchor. The failure mode has to be either predictible or the anchor design and
lining has to be able to cover all types
of deformation. This paper limits itself
to discuss anchor designs which provide a
high degree of flexibility and elongation.

3. PROPERTIES OF ANCHORS WITH HIGH DEFOR-
   MATION CAPACITY

The anchor deformation depends on the post-
failure state of the rock mass. The strength
of bedding of the anchor shaft by the rock
material is decisive for its deformation.
If high bedding is combined with a failure
of the rock mass along a joint the anchor
may deform with two sharp bends (Fig. 4).
With a high angle the tendon is pulled in
the direction of the joint, there it may
elongate to its maximum resistance. The angle of the anchor axis to the displacing
joints strongly influences its ultimate
bearing capacity (see Bjurström, Haas).
Low bedding resistance due to weak rock
strength smoothens the deflection of the
anchor to a single or double bend. As this
solicitation is less rigorous than in the
first case all efforts concentrate on anchor designs which are able to withstand
those strong deformations.

Anchors with the property to follow sharp
bends and to provide high elongations under
certain conditions and to allow considerable rock displacements in joints are the
purpose of all developments. They should
follow the displacements of the rock in any
direction axially and perpendicular to the
axis (Fig. 5). Such "intelligent" anchors
have to remain within certain cost limits
if they are to be accepted by the employ-
er, mainly the mines. These limitations restrict the possibilities of design.

Some of the desirable characteristics are
given by the following anchor load - elongation curve (Fig. 6). Curve A shows the

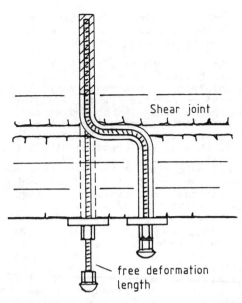

Figure 4   Anchor deformation by shearing

performance of anchors which have special devices to provide a free deformation at a defined load level, the sliding load. It can be predetermined by the anchor design and is taken from an additional length of the anchor shaft. If it is stored at the anchor head, provision has to be made that the required deformation is transferred to the area of required high deformation e.g. joint, without or with low friction losses. As its position cannot be predicted the entire anchor length except bond length has to be able to transfer such a deformation. Curve B shows the characteristics of another design principle. It takes advantage of highly yielding tendon materials as e.g. austenitic steels. The available elongation depends entirely on the steel tendon characteristics, the stress along the tendon and the length of the anchor. The free deformation, however, depends on the ability to transfer the tendon elongation to the required section of the anchor where deformation is needed due to a joint displacement. For case B the characteristic curve of an anchor is plotted where the entire tendon length can be stressed to the indicated load level which occurs, however, only scarcely.

From these considerations it becomes obvious that the friction acting along the anchor shaft especially due to bendings essentially influence the elongation characteristic and the entire performance of the anchor.

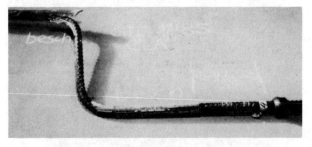

Figure 5   Shear deformation of Dywidag thread-bar for mine anchors after 90 deg. shearing in one shear plane

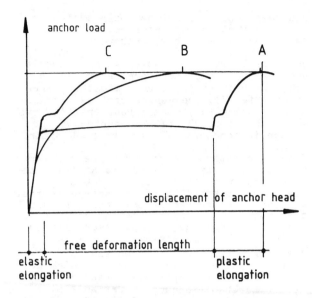

Figure 6   Characteristic curves of anchors with high deformation capability

For reason of comparison curve C shows the characteristic curve of a freely stressable normal anchor bolt.

Anchor bolts with such requirements are basically used for pattern bolting of rock faces liable to post failure condition. Even a very close anchor distance can be assumed. The mean characteristic curve of all anchors in a certain area may be taken into consideration for any kind of model of rock mass failure in the same way as the rock failure mechanism itself is a mean value of many small failures and can only be estimated roughly.

4. TESTING METHODS FOR ANCHORS WITH HIGH DEFORMATION CAPABILITY

For anchors which are mainly axially stressed the properties of the tendon are those as required for reinforcing steel or prestressing steel. These are tensile strength, yield strength, elongation values a_5, a_{10}, a_G and reduction of area z as well as bending angles and straightening after bending around a pin with defined diameter.

Mining authorities in Germany have introduced another property relating to toughness. Notched bar impact toughness in as delivered or aged condition according to one of the standards has to exceed certain limiting values. In certain cases a yield to ultimate strength ratio has to be observed. All these specifications aim at tendon properties with high deformation properties and ductility under load. A brittle failure in mining applications shall be avoided in any case. The above mentioned properties must not be altered by sliding or similar devices during deformation between the two anchorage ends to the extent that basic requirements are not any longer fulfilled.

The testing method influences to a great extend the bearing and deformation characteristics of anchors.
Testing the properties of anchors with great length of deformation and shearing at angles of 90 deg. or other requires special testing

equipment which is not commonly available in material testing laboratories. Only few exist which may test high anchor forces.

One possibility is a set-up where a double ended anchor is placed longitudinally in a concrete beam which is cut into three sections (Fig. 7). The center part is displaced perpendicular to its axis thus creating two shear planes of 90 deg. The center of the anchor is bonded to the center part of the beam.
Free deformation length has to be recovered from the sliding device placed at the outer stressing ends. Load cells, micrometer gages are attached there. Advantage is that 2 anchors are tested at the same time and more commonly available testing components may be used for the set-up. However it is limited to 90 deg. angles.

A testing frame for different angles between anchor axis and shear plane, and which deforms in only a single plane has been built by the Bergbauforschung, a German Mining Research Institute. Angle of 30, 50, 70 and 90 deg. may be chosen. The rock mass is simulated by concrete blocks with 75 MPa strength. The shear force is applied at the joint between 2 blocks.

5. DESIGN OF ANCHORS WITH HIGH DEFORMATION CAPABILITY

Three typical examples of anchors providing high elongation shall be described briefly. For case B as per Fig. 6 a typical example is an anchor bolt the rod of which consists of an austenitic steel with very high ultimate elongations of the freely movable shaft in between bond length and anchor head. Otherwise such highly yielding anchors have a standard rock bolt design with an additional sheathing in the free length.

For case A two designs may be typical. One type are anchors with a telescopic device underneath the bearing plate where the anchor shaft end is equipped with a special friction tool. It slides inside a tubular sleeve which is suspended to the bearing plate. The constant friction is created by metal deformation or by scratching through a hardened material which is placed inside the sleeve. The advantage of such recessed sliding devices is compensated by the need

for increased borehole diameters over the whole anchor length or two staggered borehole diameters.

The other solution is given by an anchor head which yields at a certain sliding load, then maintaining this resistance. It shall be explained for the Dywidag yielding anchor as at present this anchor design meets the requirements for a yielding anchor as described before in a satisfying way (Fig.8). Typical is Dywidag threadbar with a continuous, coarse thread, hot rolled onto the bar during fabrication. A screw-on anchorage device allows the placing of the anchor like a conventional rock bolt. It may even be tightened with an impact wrench. The holding force between bar and anchor head changes at a certain unlocking load into a sliding resistance in order to release the available free deformation. The sliding force is an oscillating value within a maximum and minimum resistance value. It allows the rock mass to arrange its pressure release in a dynamic way. Once the stored free deformation is consumed a stopping end brings the threadbar to its ultimate bearing capacity. The length of free deformation can be chosen in advance. By visual control its reduction by pulling into the anchorage device gives indication about the failure deformation of the rock mass. In the free anchor length the threadbar is covered with a sheath for the purpose of transferring free deformation from the anchor head to the failure plane in which the anchor is highly deformed by shear displacement. For axial deformation the free deformation length may lead to an equal distribution of strains and stresses along the tendon.

6. CONCLUSION

Anchors with high and controlled deformation capacity open a new field in securing rock surfaces under difficult stress conditions. They originate from the needs in mines. Their further development and application will provide useful data on the interaction with the rock mass. They open up a new field of rock securing technique in which stability is connected to high deformation and post-failure states or where the stability is even limited by time due to continuous rock movement. Such a technology

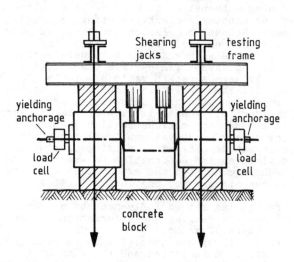

Figure 7   Testing set-up for 90 deg. shearing

Figure 8   Dywidag high deformation anchor bolt

may also be useful to control stress release
which otherwise may occur as rock burst with
high damage. A wide spread application of
these anchors may become reality if theory
is able to explain and calculate stresses
in highly deformed rock mass and to make
above described systems predictable and
controllable.

REFERENCES

Bjurström S., (1974)
  Shear strength of hard rock joints reinforced by grouted untensioned bolts.
  Proc. 3rd ISRM Congress, Denver, Colorado,
  2-B, 1194-1199

DIN 21521
  Gebirgsanker für den Bergbau und
  Tunnelbau, Teil 1 Draft 06.86

Haas C.J., (1976)
  Shear resistance of rock bolts.
  Transactions, Vol. 260, July 1976

Herbst T.F., (1983)
  Application of anchor securing in mines.
  Proc. 5th ISRM Congress, Melbourne, Austr.
  D17-21

Ludwig B., (1983)
  Shear tests on rock bolts.
  Proc. Int. Symp. on Rock Bolting, Abisko,
  Sweden

Götze W.; Stephan P.; Wiegand, (1982)
  Anwendungsgrenzen, Einsatzbereiche und
  künftige Entwicklung der Ankertechnik
  Glückauf 118 S. 1083/91

Schubert P., (1984)
  Das Langzeitverhalten von SN-Ankern im
  Untertageeinsatz.
  Felsbau 2 Nr. 4

Mining induced stresses in Saskatchewan potash
Contraintes induites par minage dans la potasse en Saskatchewan
Druckspannungsverteilung um Hohlräume im Kalibergbau in Saskatchewan

G.HERGET, Mining Research Laboratories, CANMET, Energy, Mines and Resources Canada, Ottawa
A.D.MACKINTOSH, Cominco Fertilizers, Potash Operations, a Division of Cominco Ltd

ABSTRACT: A five room entry system in potash was instrumented in a new section at a depth of 1100 m to determine convergence and stress distribution. Roof, floor, and wall strata were instrumented to a depth of up to 30 m with pressure cells and extensometers. Approximately 150 days were required for creep rates and stress levels to reach a close to steady state condition. After 500 days total horizontal closure was about 45 cm and vertical closure about 25 cm. This appears to be supported by wall stresses exceeding floor and roof stresses by about 100%. Calculation of a viscosity constant from field data gave a value of 0.104 x 10^6 MPa day/ϵ (15 x 10^6 psi day/ϵ).

RÉSUMÉ: Un système de cinq voies parallèles a été instrumenté dans une nouvelle section d'une mine de potasse à une profondeur de 1100 m, afin de mesurer la convergence et la distribution des contraintes. Il a fallu 150 jours avant que les taux de convergence et les niveaux de contrainte atteignent un palier quasi-stationnaire. Après 500 jours, la convergence horizontale était de 45 cm et la convergence verticale de 25 cm environ. Ces données semblent cohérentes avec des contraintes murales excédant celles du plancher et du toît par à peu près 100%. Le calcul d'une constante de viscosité à partir de lectures d'extensomètres et d'un moniteur de pression a donné une valeur de 0,104 x 10^6 MPa .jours/ϵ (15 x 10^{-6} lb/po^2 .jours/ϵ).

ZUSAMMENFASSUNG: In 100 m Tiefe wurden Konvergenz und Druckspannungen in fünf Parallelstrecken im Kalisalz gemessen. Ungefähr 150 Tage waren notwendig bis Kriechgeschwindigkeiten und Druckspannungen annähernd konstant wurden. Nach 500 Tagen erreichte die Horizontalkonvergenz 45 cm und die Vertikalkonvergenz 25 cm. Druckspannungen im Stoß waren doppelt so hoch wie die in der Firste oder Sohle. Die Berechnung der Viskositätskonstante mit Hilfe von Druckmessungen und Extensometerdaten ergab 0,104 x 10^6 MPa day/strain (15 x 10^6 psi day/strain).

INTRODUCTION

Potash mining in Saskatchewan/Canada at depths below 1000 m presents a number of challenges which are not trivial. Stresses acting on mine strata result in continuing deformation which leads to failure of roof, floor, and walls of opennings. Significant amounts of water exist in a number of horizons located between the mining zone and surface. Therefore, it is necessary to avoid excessive roof deformation which might fracture the salt strata acting as a water seal in the hanging wall of the potash ore.

Based on many years of experience in the Prairie evaporite formation, a number of room and pillar layouts have been developed which permit overall extractions of between 28 to 42%. The potential exists to increase extraction ratios by using waste salt as backfill. However, at the present time too little is known in regard to material properties and application of numerical modelling procedures to predict strata behaviour of evaporite formations with a reasonable measure of confidence for various layouts.

To assist in the development of a more quantitative assessment of material behaviour during potash mining, a study is being carried out at the Cominco Ltd. mine near Saskatoon in which pillar deformation, stress levels, and absolute room convergence are being determined in an isolated area of the mine (Fig. 1).

MATERIAL PROPERTIES OF POTASH

Mining of salt and potash at depth has been going on for more than one thousand years but the use of field and laboratory data in analytic studies to predict quantitatively excavation behaviour in mines is still rather limited. Part of the difficulty in undertaking such analytic studies results from the fact that stress distributions in potash evaporite deposits are difficult to obtain. In addition deformation properties of evaporite deposits vary considerably from site to site due to changes in material composition (e.g. type of salts, impurities, size of crystals, inclusions of beds of marl, clay, and anhydrite), and the strong influence by depth of burial (stress) and temperature.

Deformation properties are difficult to quantify, even without the above variations, because deformation behaviour is composed of an elastic element, a time dependent element and if failure occurs, by a post-failure "fracture deformation" or "clastic" element. The time dependent element (creep) is strongly influenced by stress magnitude and thus viscous behaviour is significant. Strain rates can be seen as a composite of (Wallner 1981, 1983): $\dot{\epsilon}_t = \dot{\epsilon}_{el} + \dot{\epsilon}_{th} + \dot{\epsilon}_{cr} + \dot{\epsilon}_f$. ($\dot{\epsilon}_t$ = total strain rate, $\dot{\epsilon}_{el}$ = elastic, $\dot{\epsilon}_{th}$ = thermal, $\dot{\epsilon}_{cr}$ = creep, $\dot{\epsilon}_f$ = fracture). During testing in the laboratory with stresses closely controlled, a cylindrical sample can be made to proceed through stages of primary creep (decelerating or transient creep phase), secondary creep (steady state creep), and tertiary creep (accelerating phase, failure).

These three phases of creep can be detected in field records of excavation convergence. Due to continuing mining activity, many records display significant primary creep, where strains per unit time decrease with time because of an ongoing stress redistribution.

A number of investigators have attributed the decrease of strain rate with time to strain hardening occurring in potash strata but laboratory studies have provided only limited proof.

True secondary creep or steady state creep is difficult to observe in the field but is considered the phase for which stress and creep rates are nearly constant. Most convergence occurs during this time and this is the condition most desirable for mine excavation in potash. Semilog plots of log strain rate/time provide a line parallel to the time axis.

$$\log \dot{\varepsilon}/t = \text{const.} \qquad (1)$$

$\dot{\varepsilon}$ = strain rate, t = time

A simple expression for creep rate is:

$$\dot{\varepsilon} = \frac{\sigma}{\eta} \qquad (2)$$

σ = stress, η = viscosity constant

To predict creep magnitude, the following time power law can be used:

$$\varepsilon = A\sigma^m t^p \qquad (3)$$

where A = factor
m = stress exponent
p = time exponent

Tertiary creep and failure occur when stresses are excessive. An increase in confining stresses will allow higher strain rates without resulting in fracture. This significant point is illustrated in Figure 2 after data by Wallner 1981:744.

SITE DESCRIPTION AND MONITORING PROGRAM

The Cominco potash mine is close to Vanscoy located about 30 km west of Saskatoon/Saskatchewan, Canada. Description of the typical stratigraphy of the Devonian evaporite formation is given in Jones and Prugger (1982) and a cross-section of the test site is given in Fig. 1.

The openings at the test site were completed as follows: 2105 - Jan. 1984, 2101 - Feb. 1984, 2104 - April 1984, 2102 - May 1984, 2103 - 5.5 m - May 1984, 2103 - 7 m - June 1984. The location plan shows the distance to present mining activity. On May 10, 1984, the roof was stabilized in 2105 with a 3 m deep cut in the centre for 50 m and the installation of cribs. This was followed by drilling of holes for extensometers and pressure cells in the east wall, west wall and floor. The temperature at the site is 30.5°C.

Stresses were determined with copper flatjacks (5.4 x 30.4 x 0.3 cm) encased in epoxy resin (colma dur) to achieve a close fit in Bx size [5.4 cm (2 1/8 in.)] drillholes. All pressure cells were prestressed to about 10 MPa (1500 psi) and no grouts or glues were used during installation.

Horizontal convergence was obtained at mid height in openings 2103, 2104, and 2105. Vertical convergence was obtained in the centre and at the rib side. Vertical convergence readings were related to an absolute datum in the mine. Extensometers were installed in horizontal and vertical drillholes.

Monitoring will continue over a number of years; therefore only a preliminary analysis using available data is given below.

MONITORING OF STRESSES

Stress determinations in potash strata have been attempted with a number of procedures: Overcoring, compensation tests by flatjacks, hydraulic fractur-

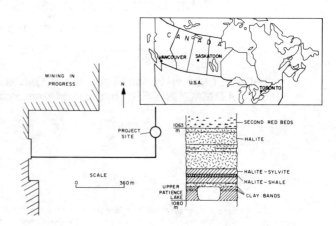

Figure 1. Location of project site at 1100 m (Cominco Ltd.).

ing and fluid filled stress cells. All measurements based on elastic theory provide results with large error margins. The most reliable results have been obtained with sealed fluid filled cells embedded in boreholes. In potash the borehole converges around the cell until fluid pressure and formation pressure are equal. Measurements of stress directions are possible if a flatjack type cell is used with a favourable thickness to width ratio.

In most potash formations it is justified to assume a hydrostatic stress field with overburden load existing in all directions because the amount of shear stress that a potash formation can sustain over a long period of time is quite small.

From the literature it is evident that the creep limit is reached when the "critical shear stress" is exceeded. A typical value of "critical shear stress" has been quoted as 0.6 MPa and corresponds to a normal stress of about 1.2 MPa (Dreyer 1967:37-39). Above this very low value of normal stress, measurable flow occurs. Field observations indicate that creep rate is related to stress gradient.

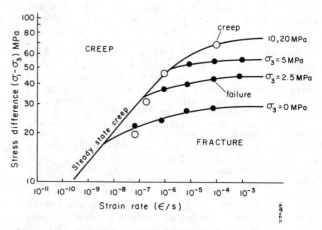

Figure 2. Example of brittle behaviour and creep boundaries based on effective stress and creep rate for salt (Wallner 1981).

At the Cominco project site stresses were monitored in the floor, back and side walls as shown in Figures 3,4,5. Cells in vertical holes were nondirectional, whereas horizontal cells were oriented to measure vertical stress.

Wall Pressures

The highest pressures measured were with cells located at about 3 m (10 ft) depth into the walls and reached for short periods of time 32 MPa (4700 psi). Similar peak values were measured in the yield pillars after 20 to 50 days which reduced to 22 MPa (3200 psi) with time. Stresses in the east wall increased more slowly. The 2105-10 cell reached and leveled off at a pressure level of 22 MPa (3200 psi) after 150 days.

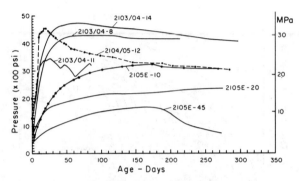

Figure 3. Wall pressure cell readings.

Floor pressures

Floor pressures are generally much lower than wall pressures. The maximum pressure of about 14 MPa (2000 psi) was observed at a depth of 15 m (50 ft) in room 2103. Constant pressures are reached quickly but many cells failed after 60 days, possibly due to bed separation.

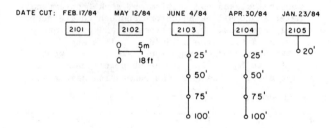

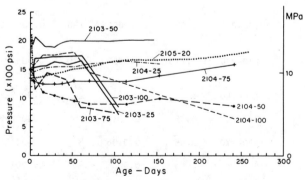

Figure 4. Floor pressure cell readings.

Back pressures

Measurements in the back indicated similar readings as those in the floor. Pressures at 3 m (10 ft) into the back are about 12 MPa (1700 psi) after about 200 days of measurement. More recent data suggest that the pressure cells at 3 m are indicating a reduction in pressure and pressure cells deeper in the back show increases in pressure with time.

The relatively low and constant pressure in the back and floor appears to indicate that a destressed zone develops above the back and below the floor very soon after excavation.

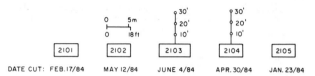

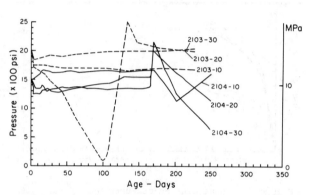

Figure 5. Back pressure cell readings.

MONITORING OF DEFORMATION

Wall and roof convergence was monitored in rooms in vertical (back to floor) and horizontal direction (wall to wall). Boreholes extensometers were installed as shown in Figure 6.

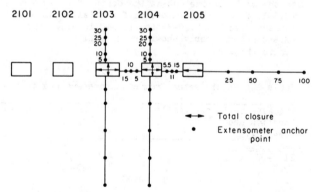

Figure 6. Deformation monitoring.

Room closure

Stress monitoring established that much higher stress levels existed parallel to walls compared to those in roof or floor strata. Therefore, it was to be expected that horizontal closure would be significantly higher than vertical closure. This is indicated in the closure data for rooms 2103 and 2104 (Fig. 7).

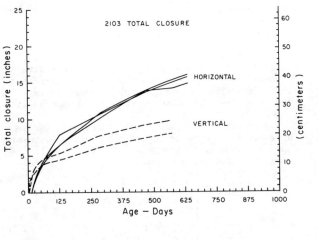

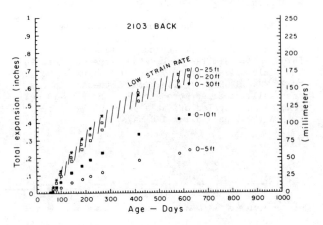

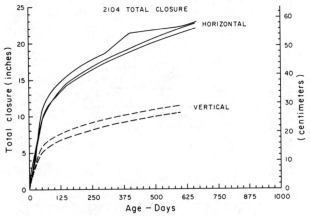

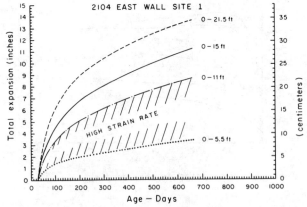

Figure 7. Total closure measurements.

Figure 8. Wall and back extensometer readings.

Wall and roof extensometers

Again the deformation is significantly higher for
readings obtained with horizontal extensometers
than for vertical extensometers as shown in
Fig. 8. For the east wall site it is interesting
to see that the maximum strain does not occur close
to the wall surface but at the depth of 1.7-3.4 m
(5.5-11 ft). In the roof, differential movement
between 6 to 10 m (20-30 ft) is negligible. The
extensometer clearly shows areas of high and low
strain (Fig. 8).

STRAIN RATE AND IN SITU STRESSES

From the graphs it is apparent that high strain
rates relate to high stress levels.
    A number of pressure cell locations were avail-
able with extensometer anchors close by. Strains
were calculated for sections of rock from extensometer anchors located before and behind pressure
cell locations. Deformations of anchor points
covered a time span of 50 days before and 50 days
after pressure readings were taken. These were
fitted to Eq 1. For the seven points calculated
(Table 1), a straight line regression gave a value

Table 1. Strain rates from extensometers readings and stress levels

Pressure cell location and time (days)	σ_V (psi)*	σ_H (psi)	Extensometer spread (ft.)	Anchor separation (in.)	Time (days)	Deformation (in.)	ϵ	days	ϵ/day
Back Locations									
2103-20 (220)		2000	20-30	120	140-270	0.033	0.00025	130	0.000002
2103-30 (220)		2100							
2103-10 (220)		1700	5-20	180	140-270	0.175	0.000972	130	0.000007
Wall Locations									
2105-10 (120)	3000		0-25	300	100-140	0.32	0.00107	40	0.000027
2105-20 (250)	2400		10-45	420	230-300	0.45	0.00107	70	0.000015
2103/4-8 (250)	4250		5-10	60	190-270	0.65	0.0108	80	0.000135
2103/4-11 (250)	3200		10-15	60	80	0.48	0.008	80	0.000100
2104-12.5 (500)	2800		5.5-15	114	300-580	1.8	0.01579	280	0.000056

* 1 MPa = 145 psi; 1 inch (in.) = 2.54 cm; 1 foot (ft.) = 30.48 cm.

of 0.104 x 10^6 MPa day/ϵ (15 x 10^6 psi day/ϵ) with a correlation coefficient of 0.92.

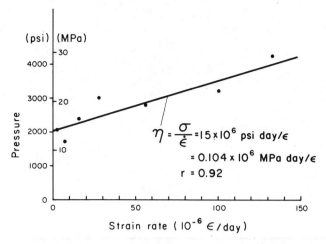

Figure 9. Viscosity constant from field data.

CONCLUSIONS

(1) Stress levels around mine excavations in walls are nearly double those of those measured in roof and floor strata.

(2) The highest stress levels reached 32 MPa (4700 psi) for a short period of time and then stabilized at about 22 MPa (3200 psi).

(3) From extensometer strain rate calculations and in situ pressure readings a field viscocity constant of 0.104 x 10^6 MPa day/ϵ (15 x 10^6 psi day/ϵ) was established in the formation.

REFERENCES

Wallner, M. 1981. Analysis of thermomechanical problems related to the storage of heat producing radioactive waste in rock salt; Proc. 1st Conf Mech Behaviour of Salt, Transtech-Publications 1984; Clausthal-Zellerfeld.

Wallner, M. 1983. Stability calculations concerning a room and pillar design in rock salt; Proc. 5th ISRM Congress Melbourne; D9-D15.

Dreyer, W. 1967. Die Festigkeitseigenchaften natürlicher Gesteine, insbesondere der Salz-und Karbongesteine, Bornträger; quoted from Baar, C.A. (1977:87); Applied salt-rock mechanics 1; The in situ behaviour of salt rocks, Elsevier Scientific Publishing Company, Amsterdam.

Jones, P.R. and Prugger, F.F. 1982. Underground mining in Saskatchewan potash, Mining Engineering; 1677-1683.

Anisotropic mechanical properties of weakly consolidated sandstone
Anisotropie des propriétés mécaniques d'un grès mal consolidé
Anisotrope Materialeigenschaften eines schlecht konsolidierten Sandsteins

RUNE M.HOLT, IKU (Continental Shelf and Petroleum Technology Research Institute), Trondheim, Norway
JENNY BERGEM, NTH (The Norwegian Institute of Technology), Trondheim
TOR H.HANSSEN, SINTEF (The Foundation for Scientific and Industrial Research), Trondheim, Norway

ABSTRACT: An experimental study of a low strength fluvially deposited triassic sandstone has been performed. The investigation correlates the anisotropy in elastic (static and dynamic) moduli with that in strength parameters. Also, permeability anisotropy is measured. Microscopic studies have been performed to relate the observed mechanical anisotropy to the rock's microlayering. The results are compared with theoretical models. Some consequences of rock mechanical anisotropy for borehole or cavity stability are briefly commented.

RESUME: Une étude expérimentale d'un faible grès d'âge Trias a été éffectuée. L'étude fait une correlation entre l'anisotropie des paramètres élastiques (statique et dynamique) et les paramètres de résistance. L'anisotropie de perméabilité est aussi mésurée. Des études microscopiques ont été éffectuées afin de faire une relation entre l'anisotropie méchanique et la nature des microcouches de la roche. Les résultats sont comparés avec les modéles théoriques. Quelques effects de cette anisotropié méchanique sur la stabilité d'un puits ou d'une cavité sont brévement discutés.

ZUSAMMENFASSUNG: Anhand von Experimenten mit fluvialen, triassischem Sandstein von geringe Festigkeit wird versucht, eine Verbindung zwischen der Anisotropie der statischen und dynamischen Moduli und verschiedenen Festigkeitsparametern unabhängig vom Umgebungsdruck herzustellen. Mikroskopische Undersuchungen sollen die Beziehung zwischen der beobachteten mechanischen Anisotropie und der Feinschichtung des Gesteins beleuchten. Die Resultate werden mit theoretischen Modellen vergleichen und einige Konsequenzen der mechanischen Anisotropie für Bohrloch- und Kavernenstabilität werden kurz erwähnt.

INTRODUCTION

Rock mechanical problems associated with weak sedimentary rocks are often encountered by the petroleum industry. Instabilities may occur during drilling and production and may be located to the well itself (breakouts, sand production etc.) or the entire reservoir (reservoir compaction, surface subsidence). In addition, also evaluation of stability of excavations and tunnels in sedimentary strata onshore or offshore call for an improved understanding of the mechanical behaviour of weak rocks. There is also a need for procedures by which mechanical properties can be evaluated from borehole measurements.

Most petroleum reservoirs produce from sandstone formations. When evaluating well stability, current models (Coates and Denoo, 1981; Geertsma, 1985) are based on empirical correlations established from a limited set of experiments with rocks from a limited geographical area. Most of the materials used in these correlations are significantly stronger than the formations causing for instance sand problems. Also, isotropy is generally assumed. At least for sandstones, anisotropy is thought to be of minor importance.

Rock bedding may, however, have consequences for the stability of boreholes. Breakouts are formed in collapse situations, elongating the hole in the direction of the minimum in-situ horizontal stress. Clearly, strength anisotropy may influence on the breakout direction. In particular, if the horizontal stress components are equal, the direction of the breakout will be given by the bedding planes. When dealing with deviated holes (Aadnoy and Chenevert, 1987), the borehole inclination will have to be chosen with care in order to optimize the stability. When drilling through a horizontally bedded formation, the most sensitive range of inclination with respect to borehole collapse is 10-40^{0}, when the rock will fail along the bedding plane.

This paper presents results from a series of rock mechanical experiments performed on a weak sandstone in order to characterize its anisotropic mechanical behaviour with respect to strength parameters and static as well as dynamic elastic moduli.

MATERIAL CHARACTERIZATION

The sandstone used in this study was sampled from an outcrop in Great Britain. We have given it the name "Red Wildmoor sandstone" (RW). The rock is of Triassic origin, and was fluvially deposited. The mineralogy is quite homogeneous throughout the sampled material. The homogeneity is confirmed by He porosity measurements, which give a more or less constant porosity of 26±1%. The main constituents are quartz with some feldspar and fragments of crystalline quartzites and gneisses. Small amounts of mica and clay minerals like illite and kaolinite can be found. The colour is red due to ferrous oxide, possibly hematite mixed with clay minerals, surrounding the grains as a thin film. Neither carbonate nor quartz cement have been observed.

The texture may change significantly from block to block. Figures 1 and 2 illustrate two different structures: In Figure 1, a thin section photograph of material type A, is shown. Areas with a clearly visible lamina-

Fig. 1

Photograph of a sample of type A material showing poorly to moderately sorted silt lamina (diagonal) separating a poorly to moderately sorted layer and a moderately to well sorted layer.

tion can be found. The grains are subrounded to rounded, intermixed with some few angular grains. The sphericity is higher than average, while the orientation of elongated grains seems to be random. The grain to grain contacts are primarily tangential.

Grain sorting is poor to moderate. Both fine (grain size 80-100 μm) and coarser grained regions (bimodal grain size distribution, the larger grains being 220-230 μm and the finer fraction 60-100 μm in diameter) can be found. These regions are separated by lamina of silt having the same mineralogy as the sand fractions, but with a richer content of clay minerals.

Figure 2 shows thin section photographs of material type B. Here, the texture is not as clearly visible as for material A, and some crossbedded areas exist.

It may be seen that elongated grains to a large extent are oriented parallel to each other.

As a part of the petrophysical material characterization, gas (air) permeability was measured. The results for material A are shown in Figure 3. The Klinkenberg correction was quite small. The angle β denotes the orientation of the flow direction with respect to the bedding planes. As expected, the flow parallel to the lamina (β=0) is faciliated with respect to that in the perpendicular (β=90°) direction. The difference corresponds to a factor of $\sim$ 2.

- Material type B had a much lower permeability than A ($\sim$ 40%), but the orientational dependence follows the same pattern as in Figure 3. In case of B, the anisotropic pore/crack size and shape distribution is responsible for the permeability anisotropy.

STRENGTH ANISOTROPY

Uniaxial compression tests

Uniaxial compression tests were run with a large number of dry core plugs of both material types described above. The specimens

Fig. 2   Photographs showing examples of structure and texture in a Red Wildmoor sandstone sample of another type than in Figure 1. Note the higher abundance of elongated grains and that these grains are parallel oriented, introducing more long grain contacts.

The mineralogy of this sample is the same as of that shown in Figure 1

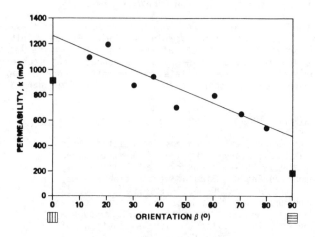

Fig. 3

Air permeability vs. angle between flow direction and bedding orientation in Red Wildmoor Sandstone.
●   Material A, block 1
■   Material A, block 2

were prepared as plane parallel cylinders of $1\frac{1}{2}$" diameter and $\sim$ 3" length. During the tests, the axial stress vs. axial strain curves were recorded. All the curves show a nonlinear behaviour. This will be elucidated further in the next Section. Figure 4 shows a plot of the peak stress (uniaxial compressive strength) vs. orientation for sample types A and B. It should be noticed that no significant difference between the strength of the two textures can be seen. There is a strength minimum at an angle close to 20°.
Notice the large spread in data at β=60°. The strength measured at 90° is approximately the double of that at β=0.
The measured strength data were fitted to a variety of anisotropic failure criteria. Figure 5 shows the best fits to the measurements with material A. The simplest approach is Jaeger's plane of weakness theory (Jeager and Cook, 1979). In this case, the failure of

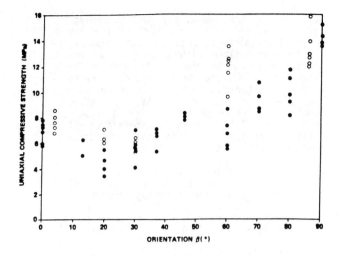

Fig. 4

Uniaxial compressive strength vs. sample orientation for Red Wildmoor Sandstone.
● Material A
o Material B

the matrix material is modelled by a linear Mohr-Coulomb theory:

$$\tau = S_o - \sigma \tan\phi \qquad (1)$$

Here τ is the shear stress and σ is the normal stress acting at an arbitrary oriented plane in the rock. S_o is the cohesion and $\tan\phi$ represents the internal friction (ϕ: friction angle). Failure along the plane of weakness is described by

$$\tau = S_o' - \sigma \tan\phi' \qquad (2)$$

Depending on the angle β between the maximum principal stress and the plane of weakness, shear failure will take place along the plane of weakness when τ as calculated by Eq. (2) becomes smaller than that calculated by Eq.(1).

This is a good description in the case of extremely laminated rocks. Jaeger (1960) has modified this simple approach to account for a continuously varying cohesion:

$$S_o = A - B \cos 2(\beta' - \beta) \qquad (3)$$

Here A and B are constants. β' is the orientation of β for which the cohesion has its minimum. This usually usually occurs at $\beta=\beta' \sim 30^o$ (McLamore and Gray, 1967). They suggested, from experimental data, a generalized anisotropic Mohr-Coulomb criterion, where both cohesion and friction angle were allowed to vary with orientation.

$$S_o(\beta) = A_{1,2} - B_{1,2} (\cos2(\gamma-\beta))^n \qquad (4)$$

$$\tan\phi(\beta) = C_{1,2} - D_{1,2} (\cos2(\gamma' - \beta))^m \qquad (5)$$

Here indices 1 and 2 apply to $0 \le \beta \le \gamma$ and $\gamma < \beta \le 90^o$, respectively, where γ is replaced by γ' when the friction coefficient is concerned. By introducing γ', the friction angle is allowed to have a minimum at an angle different from γ, which was observed in

slates and shales (but not in sandstones). n and m are parameters to be determined from experiments. Values ranging from 1 to 6 are suggested, depending on the type and degree of anisotropy.

Hoek and Brown (1980) suggested another purely empirical failure criterion:

$$\sigma_1 = \sigma_3 + \sqrt{rQ_i\sigma_3 + sQ_i^2} \qquad (6)$$

σ_1 and σ_3 are the major and the minor principal stresses (at failure), Q_i is the uniaxial compressive strength of "intact rock" (in the case of anisotropy determined from tests at $\beta=90^o$). r and s are constants depending on rock type and the degree of fracturing prior to testing. The uniaxial strength Q of a rock sample is related to that of the intact rock:

$$Q = \sqrt{s} \quad Q_i \qquad (7)$$

Thus for fractured rock, s < 1. - In the case of anisotropy, an orientational dependence of r and s may be included in the criterion.

- Hoek and Brown successfully applied this formulation to experimental studies of shale, slate and sandstone materials.

Going back to Figure 5, one is now led to conclude:

i) Jaeger's plane of weakness theory is not valid as a description of the observed behaviour.

ii) Introducing a spatially varying cohesion, this result is improved. There is, however, a discrepancy at $\beta=90^o$, where the model underestimates the strength.

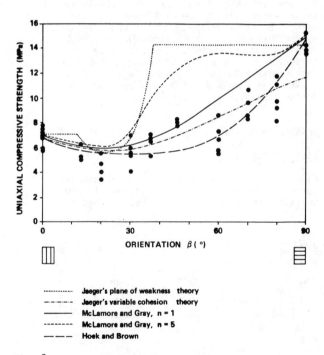

.... Jaeger's plane of weakness theory
–·–·– Jaeger's variable cohesion theory
——— McLamore and Gray, n = 1
- - - - - McLamore and Gray, n = 5
– – – Hoek and Brown

Fig. 5

Best fit to measured uniaxial strength data using some anisotropic failure criteria.

iii) The McLamore-Gray criterion has been tested using a constant $\tan\phi$. The resulting fits are not convincing. $n=1$ fits better than $n=5$, quite contrary to what McLamore and Gray proposed. They assumed $n=1-3$ to be good values for shales, while $n=5$ or 6 was supposed to be valid for bedded sandstones.

iv) The Hoek-Brown criterion gives the best overall fit to the data. This may not be surprising, since the number of adjustable parameters is as large as 9.

Triaxial tests

Triaxial tests were carried out at confining pressures ranging from 2.5 to 40 MPa, using samples of 4 different orientations ($\beta = 0$, 30, 60 and 90°).

The principal stress plot obtained from these triaxial tests shows that the failure envelope is curvilinear. In tests performed with confining pressure above 20 MPa, the failure surface of the broken samples was always covered with a fine grained flour-like powder. This shows that grain crushing takes place and probably is the physical reason for the flattening of the failure envelope at high stress levels.

The anisotropy in strength is preserved under loaded conditions. The largest shear strength is obtained when the maximum principal stress is perpendicular to the bedding plane, and the measurements at $\beta=30^o$ yield the lowest strength. The ratio between maximum and minimum strength decreases with confining pressure, while the strength difference remains more or less constant.

When attempting to fit these data to the failure criteria above, the only nonlinear criterion is that of Hoek and Brown. However, using their formula (Eq.(6)) leads to an overestimation of the strength at high stresses for all orientations. Thus, another criterion has to be sought. This could for instance be a bilinear form of Jaeger's plane of weakness approach. This will be discussed further in later publications.

ELASTIC ANISOTROPY

Static behaviour

Figure 6 shows axial stress vs. axial strain curves obtained for a selection of the uniaxial tests discussed in the preceeding

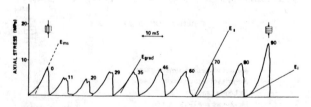

Fig. 6
Uniaxial stress-strain curves for different angles between load direction and rock bedding in Red Wildmoor Sandstone. The horizontal scale is indicated above the curves. The dashed lines drawn indicate how the different elastic moduli are estimated.

Section. As mentioned, all the recordings indicate a nonlinear behaviour, with a low value of the slope during the first stage of loading, associated with closing of microcracks existing in the sample before the experiment is started. Near failure, there is a tendency of reduced slope, associated with the growth of new microcracks. - Elastic moduli are normally defined for a linear elastic material, and it is therefore not strictly correct to attach this term to a nonlinear stress-strain relation.

However, in order to be able to compare results of different tests, we will distinguish between the following "moduli":

i) The initial slope E_i taken at axial stress $\sigma_z=0$ (see the $\beta=0$ curve in Figure 6).

ii) The secant modulus E_s (see the $\beta=70^o$ curve in Figure 6).

iii) The gradient of the stress-strain curve at a fixed stress level, E_{grad} , here chosen at $\sigma_z = 5$ MPa (shown for the $\beta=35^o$ curve in Figure 6).

iv) The maximum slope of the stress-strain curve, E_{ms} (the line drawn for the $\beta=0^o$-curve in Figure 6).

The four moduli defined above are measured for one test at each angle of orientation. There is a large spread in numerical values, and the orientational dependence is quite different from E_i and E_s on one side to E_{ms} and E_{grad} on the other. For the latter, the obtained at $\beta=90^o$, which is not true for E_i and E_s. This shows that the stress influences the anisotropy, in this case through a spatially nonhomogeneous microcrack closure.
- Inserted in Figure 6 is also the uniaxial compressive strength for the same set of samples. The angluar dependence of the strength is seen to be very similar to that of E_{ms} and E_{grad}.

Dynamic behaviour

Sound velocities of P- and S-waves have been measured at atmospheric conditions using ultrasonic (0.5 MHz) transducers attached directly to the sample with a thin layer of acoustic glue. Figure 7 shows the P- and S-wave velocities vs. orientation for dry samples of material types A and B (shear waves with polarization in the bedding plane). The scatter from sample to sample is quite large, which is not at all unexpected for a weak rock with cracks and textural inhomogeneities. Nevertheless, some clear tendency can be read from the measurements: The P-wave velocity changes with orientation, being larger with propagation along the bedding plane than when perpendicular to it. This is expected for a bedded rock, where the velocity parallel to the microlayering is given primarily by the velocity in the faster layers. In the perpendicular direction, it is, however, given (again qualitatively) through an average over the transit times in the different lamina.

- As a total average over all samples, one finds:

$$\langle v_p(\beta = 0^0) \rangle = 1730 \text{ m/s}$$

$$\langle v_p(\beta = 90^0) \rangle = 1560 \text{ m/s}$$

The velocities in material A are slightly larger than in B. - The S-wave velocity shows a smaller angular variance, with an average of ~ 1080 m/s.

Using well established formulations from the theory of elasticity and acoustic wave propagation in anisotropic media (see for instance Amadei, 1983 and Auld, 1973), and assuming that the investigated material is transversely isotropic, one can calculate the elasticity tensor. From this, the effective dynamic E-modulus can be computed for each orientation, i.e.

$$\frac{1}{E(\beta)} = s_{11} \cos^4\beta + s_{33} \sin^4\beta +$$

$$+ (2s_{13} + s_{44}) \sin^2\beta\cos^2\beta \qquad (8)$$

where s_{ij} are components of the inverse elasticity matrix, $\overset{\leftrightarrow}{C}$,

$$\overset{\leftrightarrow}{s} = \overset{\leftrightarrow}{C}^{-1} \qquad (9)$$

and the components C_{ij} are given from the sound velocities by

$$2\rho \, v_p^2(\beta) = C_{11}\cos^2\beta + C_{33}\sin^2\beta + C_{44}$$

$$+ \sqrt{((C_{11} - C_{44})\cos^2\beta - ((C_{33} - C_{44})\sin^2\beta)^2}$$

$$\overline{+ \ 4(C_{13} + C_{44})^2\sin^2\beta\cos^2\beta} \qquad (10)$$

ρ is the mass density.

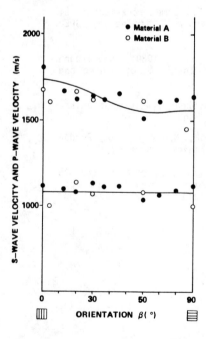

Fig. 7

P- and S-wave velocities vs. orientation of Red Wildmoor Sandstone at non-stressed conditions. Data points represent average over a large number of experiments.

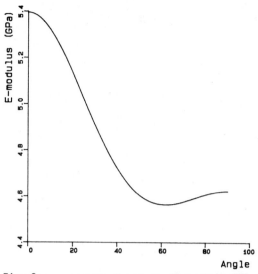

Fig. 8

Calculated dynamic E-modulus vs. orientation based on acoustic measurements on unconfined samples.

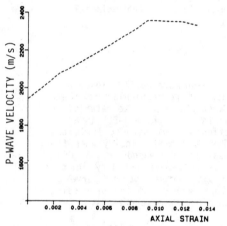

Fig. 9

P-wave velocity parallel to the loading direction during a uniaxial test with Red Wildmoor Sandstone ($\beta=78^0$).

The shear wave velocity with polarization parallel to the bedding plane is:

$$\rho v_s^2(\beta) = C_{44}\sin^2\beta + C_{66}\cos^2\beta \qquad (11)$$

ρ is the material density.

Figure 8 shows the calculated dynamic E-modulus vs. orientation, based on the data above, for unconfined samples. Notice that the dynamic E-modulus is several times larger than any of the statically measured moduli. This is a common observation in rocks with cracks, particularly in weak rocks.

The orientational dependence is, however, dependent on the stress state at which the measurements are performed.

During uniaxial loading there is a strong increase in velocities parallel to the stress up to the failure point (Figure 9). This is related to microcrack closure. This does not affect all orientations in the same way. At $\beta=90^0$, there is obviously more horizontal microcracks present than in any other direction.

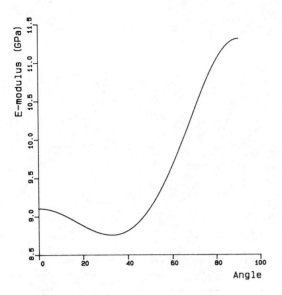

Fig. 10

Orientational dependence of the dynamic
E-modulus based on acoustic measurements
near the maximum slope of the uniaxial
stress-strain curves.

These cracks are also more easily closed when
a stress is applied perpendicular to them.
This leads to a stiffening of the material in
this direction. In fact. while $\beta=90^\circ$ is a low
velocity direction at zero stress, it is the
direction of largest sound velocity when the
uniaxial stress is increased at 3 MPa.
Measuring the axial P-wave velocity in the
same region of the stress- strain curve as
the maximum slope modulus of the proceeding
section is determined from, one finds:

$$v_p(\beta=0^\circ) \mid_{max,slope} = 2210 \text{ m/s}$$

$$v_p(\beta=90^\circ) \mid_{max,slope} = 2440 \text{ m/s}$$

The shear wave velocity also increases with
axial loading. At the same stress level as
above, it has an average of 1450 m/s. Using
these values, the dynamic E-modulus calculated as above according to Eqs. (8)-(10) is
shown in Figure 10. - Thus, although the
dynamic modulus is still much larger than the
static, it follows a similar pattern as E_{ms}
and E_{grad} of Figure 6.

CONCLUDING REMARKS

The experimental work presented in this paper
allows us to draw the following conclusions:

A significant strength anisotropy has been
observed in a weak sandstone. Though the flow
characteristics are related to the details of
rock texture (lamination or oriented elongated grains), this does not seem to be the
case to a similar extent with rock mechanical
parameters. No single failure criterion is
able to describe the behaviour at both
uniaxial and triaxial conditions. While the
Hoek and Brown criterion gives a good fit to
the uniaxial strength data, a nonlinear
extension of Jaeger's varying cohesion theory
seems more appropriate as a generalization.

The elastic anisotropy is strongly influenced
by the stress state, due to the nonlinearity
induced by microcracks which are present at
sampling. Dynamically measured moduli are
much larger than their static counterparts,
but both show the same angular dependence
when referred to identical stress conditions.
This is an important result. Quite often,
correlations are attempted between unconfined
acoustic velocities and static elastic
modulus in the linear part of the stress-
strain curve. This is not likely to give a
positive result, since the measurements refer
to different stress conditions. - Our work
shows a good correlation between the maximum
slope E-modulus (E_{ms}) and uniaxial compressive strength.

ACKNOWLEDGEMENTS

The authors wish to ackowledge Johannes
Stavrum (IKU), for his valuable assistance in
the rock mechanical tests, and Vidar
Fjerdingstad (IKU), who did the thin section
analysis.

LITERATURE

Aadnoy, B.S. & Chenevert, M.E., 1987: Stability of highly included boreholes. SPE/IADC
paper no. 16052.

Amadei, B., 1983: Rock anisotropy and the
theory of stress measurements.
Springer-Verlag.

Auld, B.A., 1973: Acoustic fields and waves
in solids. Vol. I. John Wiley & Sons.

Coates, G.R. & Denoo, S.A., 1981: Mechanical
properties program using borehole analysis
and Mohr's circle. SPWLA 22nd Annual Logging
Symposium Transaction.

Geertsma, J., 1985: Some rock-mechanical
aspecte of oil and gas well completions. SPEJ
25: 848-856.

Hoek, E. & Brown, E.T., 1980: Underground
excavation in rock. Institute of Mining and
Metallurgy.

Jaeger, J.C., 1960: Shear fracture of anisotropic rocks. Geol. Mag. 97: 65-72.

Jaeger, J.C. & Cook, N.G.W., 1979: Fundamentals of rock mechanics. Chapman and Hall.

McLamore, R.T. & Gray, K.E., 1967: A strength
criterion for anisotropic rocks based upon
experimental observations. SPE paper no.
1721.

Comportement mécanique des roches en compression à differentes températures

Mechanical behaviour of rocks in uniaxial compression tests at different temperatures
Das mechanische Verhalten der Gesteine im Druckversuch unter dem Einfluss von verschiedenen Temperaturen

F.HOMAND-ETIENNE, Centre de Recherches en Mécanique et Hydraulique des Sols et des Roches, Ecole Nationale Supérieure de Géologie, I.N.P.L., Nancy, France
R.HOUPERT, Centre de Recherches en Mécanique et Hydraulique des Sols et des Roches, Ecole Nationale Supérieure de Géologie, I.N.P.L., Nancy, France

ABSTRACT : Results about temperature mechanical behaviour of rocks without confining pressure are presented. The rocks used were representative of the principal structures (granites, sandstones and carbonate rocks). We studied the direct influence of temperature by performing experiments at temperature ranging from 20°C to 600°C, on mechanical properties.

RESUME : Les résultats présentés concernent le comportement mécanique des roches en fonction de la température en l'absence de pression de confinement. Les roches testées sont représentatives des principaux types de structure (granites, grès et roches carbonatées). On étudie l'influence de la température entre 20°C et 600°C sur les caractéristiques mécaniques.

ZUSAMMENFASSUNG : Die vorliegende Arbeit betrifft das mechanische Verhalten der Gesteine unter Einfluss der Temperatur in einachsigen Druckversuchen. Die Versuchstemperaturen liegen zwischen 20°C und 600°C. Die untersuchten Gesteine (Granit, Sandsteine, Kalksteine) stellen die wichtigsten Felsstrukturtypen dar.

1 INTRODUCTION

Les recherches sur l'évolution du comportement des matériaux rocheux en fonction de la température, à des pressions de confinement faibles sont peu nombreuses. Par contre, beaucoup d'études se situent dans le domaine des hautes pressions de confinement (> 20 MPa) et elles ont été réalisées en vue de la compréhension de certains processus géologiques et de plus sur de très petits échantillons. Les projets de stockage de déchets radioactifs ou de gazéification ne concernent que de faibles pressions de confinement. Nous présentons des essais réalisés en compression simple ; dans ce cas, nous maximisons les effets de la température puisque, d'une façon générale, la pression de confinement tend à atténuer les effets des hétérogénéités.

Le rôle de la température dans le comportement mécanique des roches est fonction de leur nature minéralogique et de leur structure (texture, surfaces de discontinuités et vides). Les variations de température peuvent affecter la composition minéralogique (départ de certains constituants) et le réseau cristallin (dilatations). Les changements de structure sont liés à la modification des discontinuités (fissures et pores).

Tableau 1. Principales caractéristiques des roches testées.
(D : granularité ; n : porosité ; C : composition du minéral principal).

	D (mm)	n (%)	C (%)
Granite de Senones	1-15	0,2	Quartz = 14-20
Granite de Remiremont	1-2	0,6	" 25-27
Granite de La Clarté	1-20	0,6	" 28-30
Quartzite de Gourin	0,1-0,3	4	" 98
Marbre de Carrare	0,1-0,5	0	Calcite 94
Calcaire d'Euville	2-3	16	" 96
Calcaire de Bazoilles	<0,01	3	" 93
Grès à Voltzia	0,05-0,5	20	Quartz 70-75
Grès de Champenay	0,1-1	14	" 70-80
Grès de Merlebach	<0,2	2,7	" 60

Nous abordons successivement l'influence de températures (de 20°C à 600°C) sur le comportement mécanique de roches granitiques, de grès et de roches carbonatées. Leurs caractéristiques essentielles sont rappelées dans le tableau 1, la description complète de ces matériaux se situe dans une étude de Homand-Etienne (1986).

2 ROCHES GRANITIQUES

D'une façon générale, les caractéristiques du comportement mécanique d'une roche en compression simple peuvent être exprimées par les paramètres suivants : module de déformation, coefficient de Poisson, résistance à la compression simple et limite élastique, comportement post-maximum. Nous exposons ci-après l'évolution de ces différents indicateurs de comportement en fonction de l'augmentation de température, sauf en ce qui concerne le coefficient de Poisson pour lequel les difficultés de mesure des déformations à des températures supérieures à 200°C font obstacle à l'acquisition de données fiables.

2.1 Module de déformation et résistance à la compression

Les variations de module de déformation E/E_0, en fonction de la température pour les granites de Senones, de Remiremont, de La Clarté et le quartzite de Gourin, sont représentées sur la figure 1. Le module calculé correspond à la pente de la partie linéaire de la courbe contrainte-déformation ; les déformations sont mesurées entre les plateaux de la presse et la courbe est corrigée de la déformation des cales d'acier interposées entre le plateau et l'éprouvette. Les courbes de la figure 1 sont assez semblables à celles présentées par Heuzé (1983) dans une étude bibliographique. La chute la plus spectaculaire de module concerne le granite de La Clarté ; à 200°C, celui-ci est inférieur à la moitié du module initial. Ce granite est riche en quartz mais a surtout la plus forte granularité, ce qui semble être le facteur déterminant pour la variation de E/E_0 dans ce type de matériau.

D'une façon générale, la résistance à la compression diminue avec l'augmentation de température, comme le montrent les résultats expérimentaux de la figure 2 ;

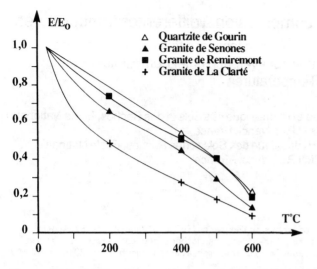

Figure 1. Variation du rapport des modules de déformation E/E_0 en fonction de la température pour les roches granitiques (E : module relatif à la température T ; E_0 : module de la roche témoin à 20°C).

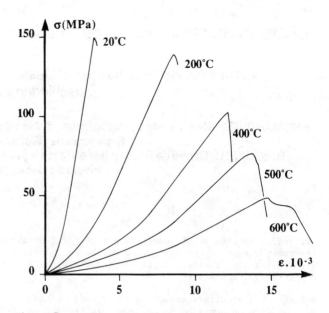

Figure 3. Courbes contrainte-déformation du granite de Senones à différentes températures ; $\dot{\varepsilon} = 10^{-5}s^{-1}$.

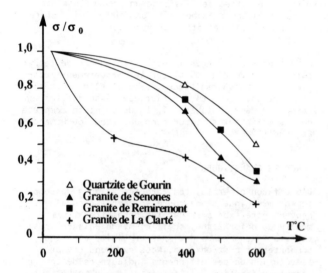

Figure 2. Variation du rapport des résistances à la compression σ/σ_0 en fonction de la température pour les roches granitiques (σ : résistance à la température T ; σ_0 : résistance de la roche témoin à 20°C).

La résistance σ_0, prise comme référence, est celle à la température ambiante du laboratoire. Il faut rappeler que cette résistance à 20°C n'a rien d'absolu dans la mesure où la résistance augmente encore lorsque la température diminue (Houpert, 1973) ; c'est cependant une référence commode. La chute de la résistance, en fonction de l'augmentation de température, est moins forte que celle du module de déformation, jusqu'à 400°C. Au-delà, les pentes des courbes de variation de résistance et variation de module en fonction de la température sont assez semblables.

2.2 Comportement post-maximum

Les essais de compression présentés ont été réalisés avec une machine d'essai asservie ; la vitesse de déformation axiale est de $10^{-5}s^{-1}$. Les courbes contrainte-déformation des granites de Senones et de Remiremont présentent à peu près la même allure (fig. 3 et 4). Les caractères précédemment mis en évidence peuvent être notés sur ces courbes : diminution de la résistance à la compression, du module de déformation avec

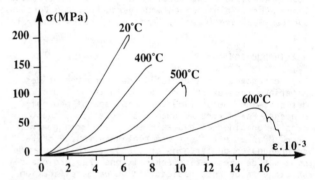

Figure 4. Courbes contrainte-déformation du granite de Remiremont à différentes températures ; $\dot{\varepsilon} = 10^{-5}s^{-1}$.

l'augmentation de la température. La limite de linéarité des déformations axiales, rapportée à la résistance à la compression, diminue légèrement avec la température ; ceci indique une augmentation de la plasticité du matériau. Le comportement post-maximum est, bien sûr, radoucissant, en l'absence de confinement. Les granites de Senones et de La Clarté (roches de classe I), montrent une évolution sensible des déformations post-maximales. Celles-ci peuvent être caractérisées par le rapport de la déformation axiale totale sur la déformation axiale relative à la charge maximale (phase de serrage éliminée) ; ce rapport augmente de 1 à 1,4 lorsque la température varie de 20 à 600°C. Le granite de Remiremont est un matériau généralement de classe II. L'augmentation de la température modifie son comportement à la rupture ; celui-ci est nettement de classe I pour des températures de 500 à 600°C. Cependant, les déformations post-maximales restent faibles.

Le faciès de rupture macroscopique évolue en fonction de la température dans la mesure où le comportement post-maximum est plus développé. Dans des granites à texture hétérogène et à gros grains comme ceux de La Clarté et de Senones, le développement des fissures et des microfractures axiales augmente avec la température. L'éprouvette testée à 600°C présente de très nombreuses fractures ; les cristaux peuvent même être déchaussés ; la forme globale est celle d'un tonnelet à l'intérieur duquel se trouvent individualisés deux cônes. Le développement des microfractures et la forme du tonnelet sont plus amplifiés dans le granite de La Clarté que dans celui de Senones. Dans le granite de Remiremont témoin, on ne voit pas de microfractures axiales dans la mesure où celles-ci ne dépassent pas la taille du grain. Par contre, à des températures plus élevées, on observe le développement de microfractures centimétriques. L'éprouvette est légèrement en forme de tonnelet à l'intérieur duquel les deux cônes sont bien visibles.

3 LES GRES

Des essais sous température ont été réalisés sur des grès : le grès à Voltzia et le grès de Champenay, qui sont des grès vosgiens et le grès schisteux de Merlebach (toit et mur du charbon des Houillères du Bassin de Lorraine).

3.1 Module de déformation et résistance à la compression

Les résultats des essais de compression sont reportés sur la figure 5 pour la variation de module de déformation E/E_O et sur la figure 6 en ce qui concerne les résistances à la compression. Les courbes représentant la variation de E/E_O indiquent une diminution de ce rapport en fonction de la température. La diminution de module semble s'amortir et est plus faible que dans le cas des granites.

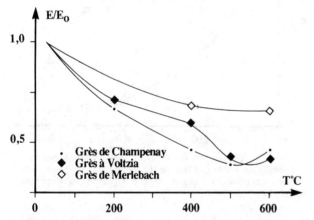

Figure 5. Evolution du rapport des modules E/E_O en fonction de la température pour des grès (E_O : module relatif à la roche témoin).

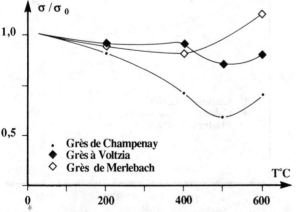

Figure 6. Evolution du rapport des résistances à la compression σ/σ_O en fonction de la température pour des grès (σ_O : résistance relative à la roche témoin).

La résistance à la compression ne chute pas brutalement et il y a même une tendance à l'augmentation entre 500°C et 600°C.

3.2 Courbes contrainte-déformation

Avant de discuter ces résultats, nous pouvons examiner les courbes contrainte-déformation axiale enregistrées sous température du grès de Champenay (fig. 7). Ces courbes sont assez semblables et varient peu en fonction de l'augmentation de température, seule la déformation correspondant à la résistance maximale augmente.

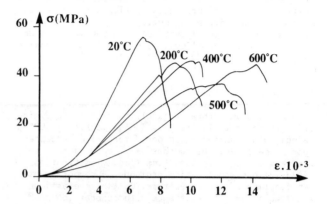

Figure 7. Courbes contrainte-déformation du grès de Champenay à différentes températures ; $\dot{\varepsilon} = 10^{-5}s^{-1}$.

Le contrôle de la rupture est moins facile que sur le matériau témoin. Macroscopiquement, le faciès de rupture est le même quelle que soit la température ; on observe des cônes tronqués par une grande fissure oblique. Les courbes relatives au grès à Voltzia présentent encore moins de différences en fonction de la température. Les courbes $\sigma-\varepsilon_a$ concernant le grès de Merlebach ne présentent pas de différences notables les unes par rapport aux autres, si ce n'est une légère diminution du module avec la température. Il n'y a jamais de contrôle de la rupture ; macroscopiquement, celle-ci se traduit par des fissures verticales découpant l'éprouvette en colonnettes. Le matériau témoin est intermédiaire classe I - classe II.

L'évolution du comportement des grès sous température est difficile à appréhender dans la mesure où plusieurs phénomènes peuvent intervenir. De plus, il faut noter que la dispersion des résistances ne permet pas d'apprécier avec certitude l'évolution des paramètres ; celle-ci paraît plus importante pour les grès que pour les granites. Les grès à Voltzia et de Champenay ont approximativement la même structure ; leur porosité est importante (tableau 1) ; les inégalités de dilatation thermique des différents grains ne conduisent pas à la formation de fissures. La courbe de dilatation thermique (Homand-Etienne, 1986) du grès à Voltzia, montre que ce matériau se dilate très peu sous température ; le coefficient de dilatation thermique calculé est toujours supérieur à celui mesuré. Il existe également des modifications d'ordre minéralogique allant dans le sens d'une consolidation de la structure. De plus, les grès sont les roches dont le point de fusion est le plus élevé ; les expérimentations réalisées le sont donc relativement à des températures plus basses que dans le cas des granites. Le grès schisteux de Merlebach présente une porosité très faible et se dilate cependant très peu. Les grains sont de petite taille, ce qui ne provoque que de faibles inégalités de dilatation.

4 LES ROCHES CARBONATEES

L'étude de ces roches est du plus haut intérêt pour les géologues, du point de vue de la compréhension de la tectonique dans la croûte terrestre. Les travaux de Griggs et al. (1960) et de Heard (1960) concernent des températures entre 25°C et 800°C pour des pressions de confinement atteignant 500 MPa. L'analyse de ces études sur le marbre de Yule et sur le calcaire de Solenhofen conduit à remarquer en premier lieu que le comportement de ces roches se rapproche du comportement élasto-plastique parfait lorsque la température s'élève.

Nous avons choisi d'étudier le comportement sous température de trois roches carbonatées de structure différente : le marbre, roche entièrement cristallisée et de pososité quasi nulle, le calcaire de Bazoilles, variété de micrite dont la porosité est de 3 % et le calcaire d'Euville, biosparite de porosité moyenne égale à 16 %.

4.1 Courbes contrainte-déformation

Dans le cas du marbre (fig. 8), on relève une diminution nette de la limite élastique à partir de 400°C, ainsi qu'une augmentation des déformations acceptées par le matériau. Les courbes σ-ε_a concernant le calcaire micritique de Bazoilles (fig. 9) mettent en évidence une modification profonde du comportement en fonction de la température. La limite élastique confondue avec la résistance à la compression pour la roche témoin, diminue très fortement. Nous observons un passage entre un matériau de classe II à un matériau dont les déformations plastiques augmentent. A partir de 500°C, une phase ductile légèrement radoucissante se développe.

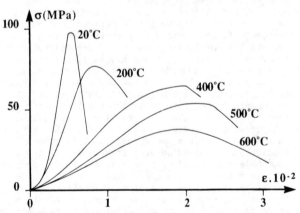

Figure 8. Courbes contrainte-déformation du marbre à différentes températures ; $\dot{\varepsilon} = 10^{-5}s^{-1}$.

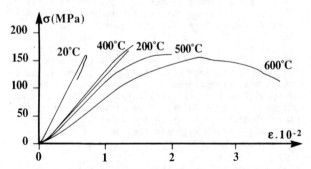

Figure 9. Courbes contrainte-déformation du calcaire de Bazoilles à différentes températures ; $\dot{\varepsilon} = 10^{-5}s^{-1}$.

4.2 Module de déformation et résistance à la compression

Pour ces roches carbonatées, le module de déformation chute très fortement dès 200°C (fig. 10). Les diminutions de module du marbre et du calcaire d'Euville sont ensuite très voisines, tandis que le module de la micrite augmente légèrement à 400°C pour diminuer ensuite. La résistance à la compression s'abaisse en fonction de la température (fig. 11) pour le marbre et le calcaire d'Euville. La variation de résistance de la micrite est moins régulière. Nous retrouvons, en fait, les résultats de Henry (1978), c'est-à-dire une résistance à la compression à 400°C supérieure à celle de la roche témoin et un passage d'une rupture brutale (classe II) à un comportement ductile. Manifestement, à 400°C, la fracture devient plus stable, ce qui entraîne une rupture plus progressive et conjointement des déformations plus importantes. Il est probable que cette stabilisation de la fissuration est liée à un fort développement des zones plastiques en tête de fissure, induites par des transformations

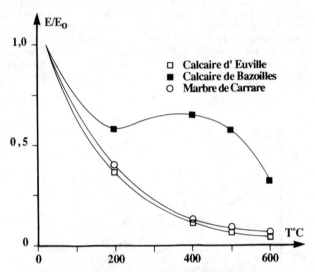

Figure 10. Evolution du rapport des modules de déformation E/E_O en fonction de la température pour des roches carbonatées (E_O : module relatif au matériau témoin).

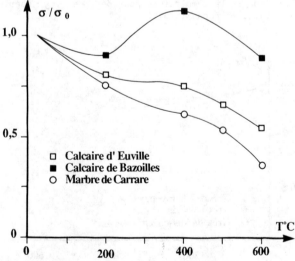

Figure 11. Variation du rapport des résistances à la compression σ/σ_O en fonction de la température pour des roches carbonatées (σ_O : résistance relative au matériau témoin).

minéralogiques (début de fusion). Les observations
que nous avons réalisées au M.E.B. sur des fragments
correspondant à des surfaces de rupture sont assez
révélatrices. Le calcaire témoin ne présente pas de
fissures plurimillimétriques dans la mesure où celles-
ci ne se sont limitées qu'aux grains dans la micrite;
leur taille est donc très réduite. Dans le matériau
chauffé et testé à 600°C, des fissures plus longues
existent. Celles-ci peuvent se développer sur plusieurs millimètres à travers des grains de micrite.

5 CONCLUSION

Le comportement mécanique des roches en compression
sous température dépend étroitement de leur composition minéralogique et de leur structure initiale.
Pour les roches de la famille des granites, la diminution des caractéristiques mécaniques dépend de la
granularité de la roche. Les roches de la famille
des grès montrent un comportement difficile à globa-
liser ; le module d'élasticité diminue mais la résistance à la compression varie peu. Il ne semblerait
pas impossible de passer pour des températures plus
élevées à un comportement ductile. Les roches carbonatées montrent une grande diversité de comportement:
le marbre et le calcaire d'Euville ont globalement
les mêmes réactions, mais celle du calcaire micritique est fondamentalement différente ; on observe
le passage d'un comportement fragile à un comportement ductile.

REFERENCES

Griggs, D.T., Turner, F.J. and Heard, H.C. 1960. Deformation of rocks at 500° to 800°C. In Griggs, D.T.
and Handin, J., Rock deformation, Geol. Soc. Amer.,
mém. 79, pp. 39-104.
Heard, H.C. 1960. Transition from brittle fracture to
ductile flow in Solenhofen limestone as a function
of temperature, confining pressure, and interstitial
fluid pressure. In Griggs, D. and Handin, J., Rock
deformation, Geol. Soc. Amer., mém. 79, pp. 193-226.
Henry, J.P. 1978. Mécanique linéaire de la rupture
appliquée à l'étude de la fissuration et de la fracture des roches calcaires. Thèse Doct. ès Sci.,
U.S.T. Lille, 182 p.
Heuzé, F.E. 1983. High temperature mechanical, physical and thermal properties of granitic rocks - a
review. Internat. J. Rock Mech. Min. Sci., vol. 20,
pp. 3-10.
Homand-Etienne, F. 1986. Comportement mécanique des
roches en fonction de la température. Mém. Sci. de
la Terre, n° 46, 261 p.
Houpert, R. 1973. Comportement mécanique des roches
cristallines à structure quasi isotrope. Thèse Doct.
ès Sci., Univ. Nancy 1.

Rock burst and energy release rate
Coups de toit et taux de relâchement d'énergie
Gebirgsschläge und die Geschwindigkeit der freigesetzten Verformungsenergie

HUA ANZENG, China Institute of Mining & Technology, Xuzhou, Jiangsu

ABSTRACT: The strain energy release procedure in rock mass induced by the change of stress state is studied in this paper. The stable condition of rock is put forward, that is, the actual strain energy is smaller than or equal to the maximum retainable energy. Based on this condition and the affecting factors such as the energy release rate,etc., the reason of sudden spalling off of rock walls of mine openings, abrupt burst of mine headings and unexpected collapse of rock pillars are analysed in connection with cases. Consequently, measures for preventing and controlling rock burst are presented.

RÉSUMÉ: Cet article a recheché le processus de relâchment d'energie de déformation apparaissant par la variation d'éfat de contraintes et a analysé des causes qui ont provoqué l'écroulement soudain de la murette, des
roches de front et des piliers dáprès la vitesse de relâchment d'énergie, l'exemple et les conditions de stabilité de la masse de roche --- l'énergie réelle de déformation étant moins ou égale que l'énergie de déformation quélle permet de laisser. On a raconté encore des prinzipes de mesures de lutte contre ces phénomènes dynamiques.

ZUSSAMENFASSUNG: Dieser Beitrag behandelt den Vorgang der Freisetzung von Verformungsenergie infolge einer Änderung der Gebirgsspannung. Gemäß der stabilitätsbedingung---die gespeicherte Verformungsenergie ist gleich
oder kleiner als die mögliche bleibende Verformungsenergie---und der Geschwindigkeit der Energiefreisetzung
werden das Gebirgszerbersten des Stoßes, das Herausschleudern von Gestein aus dem Streb und der Zusammenbruch
der Pfeiler untersucht. Das Prinzip von möglichen Verhütungsmaßnahmen wird auch erläutert.

1. INTRODUCTION

Since the first record of rock burst at Shengli colliery in China in 1933, many rock bursts have happened at other collieries, metal mines and railway tunnels. Some of them known in metal mines as "rock
bump" are of the slightest harmfulness and occur as
the sudden spalling of the hard rock surrounding mine
workings in a quantity of ten tons or less. The severe
ones induced by the sudden collapse of rock pillar
result in sudden convergence of openings, abrupt roof
fall in a large area and mine seism, etc. For example,
during the recovery of a coal pillar at Chengzi
colliery the remanent parts collapsed suddenly on Oct.
25, 1974. The broken coal rushed in with a floor heave
over 64.5 m and accompanied by a seism of magnitude
3.4 and epicentral intensity 7. What is particularly
severe is the disastrous harzard induced by the collapse of the rock pillar with a large area of roof
rock hanging over it. An example is that occured at
Pangushan colliery where a roof rock, 100m long, 200-
250m wide and 400-500m thick moved down visibly in
several hours and resulted in a surface subsidence of
1.8m. Such a harzard as it occurs in soft rock tends
to induce a rock burst of up to decades of tons. However, when the rock thrown out contains pressurized
gas, it forms a rock and gas burst of over 10,000 t of
rock. For example, a sandstone and carbon dioxide burst
at Yingcheng colliery occurred on June 13, 1975 with
1005 t of sandstone thrown out from a driven heading.
Similarly, a coal and gas burst occurred at Huayingshan
colliery on August 8, 1975 when disclosing a coal seam
and 11,400 t of coal and 140,000m^3 of gas were thrown
out. In order to prevent such a harzard, we have to
investigate the mechanism of rock burst. By doing so,
we will be able to locate the potential burst are;
find effective prevention measures against it.

2. ENERGY RELEASE OF ROCK

2.1 Concept

Although rock burst is commonly considered as a result
of the sudden release of energy, there are different
theories on it. Some theories are based on the energy
release of rock specimens which does not represent
the energy change of rock in situ. For instance, they
often explain the rock burst by the sudden rupture of
specimen blocks in the test machine. But such sudden
rupture rarely occurs in a machine of higher rigidity.
On the contrary, a hard sandstone roof is just the
factor causing the sudden collapse of coal pillars.
Apparently, these two phenomena are substantially
different. In the case of test, the ram of test machine does work on the specimen and deforms it when
the applied load is less than the failure strength of
rock. Strain energy is stored in specimen when it deforms (area A in figure 1). It is worthy to note that
the test machine frame deforms at the same time when
the specimen deforms (area B in fig. 1). As the pressure exceeds the failure strength, the resistance of
the specimen drops its strain energy. The released
energy subsequently does work on the specimen and
thus increases its strain energy (area C in fig. 1)
  Whenever the energy released by the frame (as the
B in fig. 1) is larger than the energy absorbing
capability of the specimen (C in fig. 1) or the
releasing rate is faster than the increasing rate
of the strain energy of the specimen, the spicemen
will break suddenly even though the test machine does
no more work. The excessive released energy is shown
as the shaded part in figure 1. However, being subjected to multiaxial stress, the rock mass has stored
strain energy already. As a driven heading approaches
it, its minor principal stress and elastic limit decrease gradually and the strain strengthening character converts into softening. Consequently, the
retainable strain energy reduces with the reduction
of minor principal stress, and part of the strain
energy in the rock mass will be released. The area
oBb in figure 2 is the strain energy under multiaxial stress. It is going to be oCc, oDd down to oEe

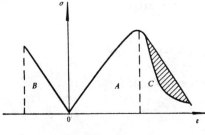

Figure 1

along with the reduction of minor principal stress. The difference between them is the energy released. It is evident that the rock itself releases energy as a result of the variation of stress state with the approaching of the heading. It is quite different from the rebound of the test machine frame.

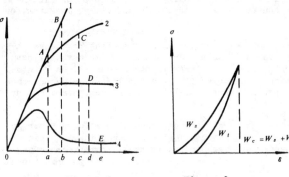

Figure 2   Figure 3

Some scholars consider that the less the difference W_s between the strain energy stored by loading W_c and the energy released by unloading W_t, that is (fig. 3)

$$W_s = W_c - W_t ,$$

the easier the rock break will be or rock burst is liable to occur when $W_t/W_s > 5$. If this were true, metal which is good in elastic recovery would break suddenly whenever it recovers from deformation. Surely not. Elastic recovery does not result in any break at all. Therefore, it is not the real cause of rock burst.

Some scholars refer to the strain energy in the prefailure stage as release energy and that in the postfailure stage as consumed energy. They attribute the sudden break of rock mass to that the pre-failure energy exceeds the post-failure energy. Actually, both of them are only parts of internal energy at different stages. The former produces deformation and the latter results in failure. None of them releases and they do not convert to each other either. Therefore, their relative magnitude has no influence on the break of rock. Nevertheless, the absolute value of the post-failure energy does show the manner of failure, i.e., a small value implies a brittle failure.

2.2 Energy release of the rock surrounding a heading

The intact rock mass subjected to primary stress stores some elastic energy U. By the elastic formula

$$U = [\sigma_1^2 + \sigma_2^2 + \sigma_3^2 - 2\nu(\sigma_1\sigma_2 + \sigma_2\sigma_3 + \sigma_3\sigma_1)]/2E \quad (1)$$

Where $\sigma_1, \sigma_2, \sigma_3$ — principal stresses of intact rock,
    E — elastic modulus,
    ν — Poisson's ratio.
Assuming that the principal stresses $\sigma_1 = \sigma_2 = \sigma_3 = P_0$ and substituting them into (1), we get the stored energy

$$U_y = 3(1 - 2\nu) P_0^2 / 2E \quad (2)$$

It is the sum of areas oAa, oBb, oCc in fig. 4. So long as the heading is driven, the stress in the intact rock surrounding it concentrates. Then the stress in the rock surrounding a round heading becomes:

$$\left.\begin{array}{l} \sigma_1 = P_0(1 + R_0^2 / r^2) \\ \sigma_2 = P_0 \\ \sigma_3 = P_0 (1 - R_0^2 / r^2) \end{array}\right\} \quad (3)$$

where R_0 — radius of the heading,
    r — distance between any point in the rock and the axis of heading.
Substituting (3) into (1), we get the stored energy of a round heading:

$$U_w = P_0^2 \{ 3(1 - 2\nu) + 2(1 + \nu)R_0^4/r^4\} / 2E \quad (4)$$

The stress concentration leads to an accumulation of strain energy. So $U_w > U_y$. The increment is

$$\triangle U = U_w - U_y = P_0^2(1 + \nu) R_0^4 /r^4 E \quad (5)$$

Let the stored energy of any point on the periphery, where $r = R_0$, be U_{w1}, then:

$$U_{w1} = P_0^2 \{ 5 - 4\nu \} /2E \quad (6)$$

From figure 4 we get that

$$U_{w1} = \triangle oDd + \triangle oBb.$$

Besides, the stress-strain curve changes with the stress state. As the rock at the periphery is softened, its stress-strain curve changes into the form shown as oEF in figure 4. Its retainable strain energy is oEFf. When the stress-strain curves in three principal stress directions are nonlinear, the retainable energy of the rock in situ will be

$$U_1 = \int \sigma_1 d\varepsilon_1 + \int \sigma_2 d\varepsilon_2 + \int \sigma_3 d\varepsilon_3 \quad (7)$$

The difference of the actual stored energy U_w and the retainable energy U_1 is the released energy U_s,

$$U_s = U_w - U_1 . \quad (8)$$

The released energy on the periphery U_{s1} can be got from figure 4:

$$U_{s1} = U_{w1} - U_{11} = \triangle oDd - oEFf$$

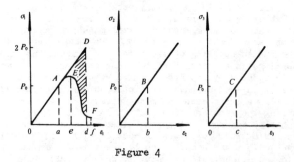

Figure 4

On the basis of the above calculations the relationship between the strain energy and r is got, as shown in figure 5. Based on figure 2 we may deduce that since the stress changes with r, the retainable strain energy changes with r as well, as shown by the dotted line in figure 5. The area where $U_w > U_1$ is in an unstable state.

After the excessive energy is released the actual strain energy is equal to or less than the retainable energy, that is $U_w < U_1$ and then the rock becomes stable. U_w depends on the stored energy and U_1 on the rock mechanical property. Here we take the case of axial symmetry as an example. Other cases are similar. And the energy release ahead of a heading is the same as well.

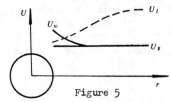

Figure 5

972

2.3 Energy release of rock pillar

Pillars are unconfined on its two or four sides. Hence, according to the preceding statement, the retainable energy U_1 and the actual stored energy U_S can be shown as figure 6. With the reduction of the pillar size the stress state changes, that is, U_1 gradually decreases and U_S increases. Energy only releases at the periphery of the rock pillar, when curves U_1 and U_S nearly coincide with each other, the actual stored energy of the whole pillar approaches a critical state. Then any more disturbance will lead to the release of energy and thorough failure of whole pillar.

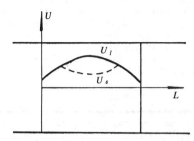

Figure 6

2.4 Energy release rate

A slow release of energy does not result in any harzard. However, a rapid release of a large amount of energy will lead to rock burst and mine seism. The energy release rate V_u may be expressed by the following equation:

$$V_u = d/dt. \int (U_0 - U_t) \, dV$$

where U_0——the initial strain energy of a unit volume;

U_t——the strain energy of the unit volume at the end of a time interval t;

t ——the time interval during which the energy changes;

dV——an elementary volume of the deformed rock.

The line joining points B,C,D,E in figure 2 reflects the law of strain energy release and may be called an energy release curve. According to the mechanical properties of rock, it is evident that the steeper the slope of the energy release curve (absolute value), the higher the energy release rate; the steeper the slope of the strain softening curve (absolute value; EF in figure 4) and the smaller the residual strength, the higher the energy releasing rate. The volume of the rock releasing energy depends on the stress distribution.

3. APPEARENCES OF ENERGY RELEASE

3.1 Sudden spalling off or rock walls

The sudden spalling off accompanied with a loud pop at a working face or periphery of openings is of the slightest type of rock burst, and is referred as rock bump. It often occurs in the hard rock of metal mine. For instance, it has occurred in Jinchuan mine when mining ores at a depth of 400-500m and in Pangushan mine when mining quartzite vein, metamorphic sandstone, etc. at a depth of 400-600m. Though it is relatively seldom in collieries, a number of rock bumps occurred recently in Taiji colliery when mining tuffaceous sandstone at a depth of 700-800m. So far as we know, the stress in the strata of those districts is generally lower than the uniaxial compressive peak strength of the rock, namely, that of the quartzite vein of Pangushan is 124 MPa while the maximum principal stress measured in situ (σ_1) = 32MPa: and that of the ore of Jinchuan mine is 75 MPa while σ_1= 34MPa. In general, the stored energy in surrounding rocks does not exceeds the pre-failure strain energy and the openings are stable there. However, in some stress concentrated zones resulted from geological structure (about 66% of the rock bumps in Pangushan mine occur-

red near faults) or from excavation disturbance (34% of the rock bumps in Pangushan and 100% of Jinchuan mine occurred in stress concentrated zones) and/or in weakened zones where the strength of rock has been reduced, the stored energy in surrounding rocks will exceed the retainable energy, as shown in figure 5. In addition, owing to the fact that the rock is very hard there and thus can store more elastic energy, it can quickly release energy under certain conditions to form rock bumps. It is clear from equation (4) that the stored energy in surrounding rocks is inversely proportional to r^4, namely, stored energy decreases with increasing r. But on the other hand, we can see from equation (3) that the stress state transfers from biaxial to multiaxial, that is, the retainable energy is going to increase with r (as shown by the dotted line in figure 5). In the range of intermediate depth, the stored energy of surrounding rock no longer exceeds the ratainable energy, therefore the extent of rock bump is relatively small.

3.2 Rock burst

In weak strata, rocks surrounding mining excavations are in a plastic state, hence elastic energy can not be accumulated to a large amount in rock walls. Consequently, no rock burst occurs there. Plastically deformed area is essentially an energy release area, but the energy releases progressively with the advance of the plastic zone. Generally, hence, rock burst does not occur there. But once when the working face penetrates the plastic zone instantaneously, the energy will release suddenly and burst rock out. This is the reason why shock blasting can induce rock burst, though seldom. According to statistics, over 80% of rock bursts occur near weak strata. Based on a dimensionless numerical modelling method the writer of this paper proves that discontinuous surfaces, such as faults and beddings, obstruct the continuous change of strain energy and block the progress of the strain energized zone. Along with the advance of the heading end and with the thinning of the rock ahead of it, the rock in that zone will be subjected to a stress up to its critical strength and will release energy at once to form a rock burst. The more the actually stored energy and the less the retainable energy, the more the quantity of rock burst out. Especially serious is the case that the other side of the interface is a coal seam just in a critical state as shown in figure 7. In such a case, once the rock adjacent to the coal bursts out, the energy stored in that coal seam releases abruptly and bursts the coal out violently. Furthermore, when that coal seam contains highly pressurized gas, a vast quantity of gas energy releases at the same time and consequently leads to a rock and gas burst on a large scale.

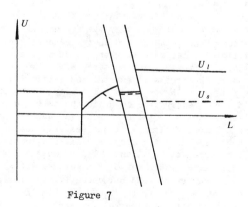

Figure 7

When a crosscut at Nantong colliery disclosed a coal seam, 86 t of coal and 20 m³ of rock burst out. As the crosscut passed through the seam and reached the bottom of it, a second burst with 1473 t of coal and 80 m³ of rock occurred subsequently. Similar events are not seldom. It is difficult to explain such an event by the release of gas, because the coal seam has been disclosed and gas can release gradually instead of suddenly. However, it can be explained reasonably and easily as a sudden strain energy release in front of the interface by the preceding theory. When a heading is driven approaching the roof or floor of an inclined coal seam, the effect of interface leads to a sudden release of strain energy. For the roof, it occurs in the form of a sudden broken of the roof rock which offers a chance for the energy in the coal seam to release, while for the floor, it manifests as a sudden broken of the coal seam itself.

3.3 Sudden collapse of pillars

Many rock burst phenomena or appearences such as sudden convergence of walls, heave up of floor, sudden cave of roof in a large area as well as mine bumps are often resulted from the sudden collapse of pillars. An example is the rock burst occurred at Chengzi colliery on Oct. 25, 1974 (figure 8), where walls

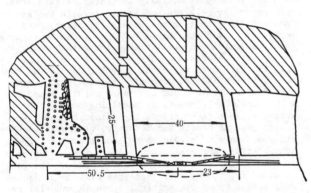

Figure 8

converged, floor heaved up, and 64.5m roadway destroyed. Tracing the reason, we can find that the roof rock of that coal seam is relatively hard, so it kept overhanging instead of caving after the coal seam was extracted and thus a roof of 520,000 m² was supported by pillars with a total area of about 16% of the roof. It is evident that the actual stored energy in coal pillars had already come to the limit of retainable energy or to a critical state. As the stub pillar was being extracted, all pillars suddenly released energy at once with a high speed. Therefore, a rock burst was formed.

Sometimes, instead of the reduction of the pillar size, the retainable energy may reduce with the lapse of time. Hence, the stored energy reaches its critical state easier than ever and leads to the sudden collapse of pillar. Furthermore, after the rupture of pillars, the roof will subside and weight the remaining pillars with its potential energy which converts into strain energy and leads to the collapse of other pillars, and so forth. Once a chain reaction occurs more and more pillars will successively collapse in a short time to bring about a disastrous destroy like that occurred at Majiliang colliery where a large area of roof caved with a seism of magnitude 3.2 and an under-ground storm, and subsequently, the surface subsided over an area up to 70,000 m². Similar events happend at metal mine as well. Their common character is that their roof rocks are too hard to cave along with extraction. Rock burst resulted from collapse of pillars will not occur provided that we induce the

roof to cave with extraction and have the goaf space packed.

4. PRINCIPLES OF PREVENTING ROCK BURST

Till now, the shock blasting method has been extensively used to induce precautionary rock burst. Nonetheless, it badly affects the normal production, so it should be abolished. Instead, we have introduced the releasing hole method to release the strain energy and gas energy at Luling shaft in 1986. By means of that, we have safely passed through three burst prone coal seams one of which is 11m in thickness and 500m in depth. Using the method of drilling rows of releasing holes in weak rock, we have been able to release the strain and gas energy effectively and uniformly beforehand, because the space produced by uniformly distributed boreholes provides a condition of uniform deformation for the rock mass. When the deformation is larger than the convergence of the opening, strain and gas energy will release slowly. According to our experience safety may be ensured by drilling out 1-3 thausandth in volume of intact coal in the protective zone. Whereas for ensuring the safety of pillars, more coal must be removed from them in accordance with the potential roof subsiding quantity. Similarly, to shift the peak stress towards the depth of the rock mass by drilling boreholes or cutting slots may prevent hard rock from burst. After all, extracting a releasing seam is an effective method for regional protection. Unfortunately, it is not unconditionally usable everywhere.

5. CONCLUSION

.1 The change of stress in underground rock mass induced by excavation results in an increase of strain energy and a decrease of retainable energy.

.2 The stable condition of rock mass is that the stored strain energy is smaller than the retainable energy.

.3 Rock burst is formed by rapid release of energy which breaks the rock and throws out the broken rock.

.4 A rock bump is a sudden release of locally accumulated elastic energy in rock walls.

.5 When the actual stored energy in the rock mass between an excavation and a fault or a weak stratum reaches the retainable energy, any additional disturbance will induce a rock burst.

.6 When the strain energy in a pillar reaches its critical state, that pillar will collapse suddenly.

.7 The sudden collapse of pillars will result in an abrupt convergence of walls, heave up of floor, large area roof fall and mine seism.

.8 Shock blasting should be abolished.

.9 Coal burst can be prevented by means of drilling and hollowing a certain space in coal which permits the coal to deform by a quantity larger than the convergence of surrounding rocks.

Tracing of sliding lines in media with planes of weakness and stability analysis of underground rooms

Lignes de glissement dans un milieu contenant des plans de faiblesse et analyse de stabilité de chambres souterraines

Verschiebungslinien im Material mit Schwächeflächen und die Stabilitätsanalyse von Untertagehohlräumen

MLADEN HUDEC, Professor, Civil Engineering, Faculty of Mining, Geology and Petroleum Engineering, Zagreb, Yugoslavia

LIDIJA FRGIĆ, Assistant, Civil Engineering, Faculty of Mining, Geology and Petroleum Engineering, Zagreb, Yugoslavia

ABSTRACT: The sliding lines in rock material with planes of weakness can be drawn by a computer procedure. The stress values and principal directions are computed according to the Theory of elasticity. The Mohr-Coulomb failure criteria applied to the rock itself and to the planes of weakness give the criterion for a prognosis of the sliding propagation. The sliding can be expected in the direction of plane with minimum ratio of the critical and actual shear. The illustrative drawing of the expected sliding lines can be used for the preliminary design of anchors or supporting system.

RÉSUMÉ: A l'aide de la théorie de l'élasticité, il est possible de calculer les niveaux de contraintes et leurs directions préférentielles selon une procédure informatisée. Le critère de rupture de Mohr-Coulomb permet de pronostiquer la propagation des glissements, dans la direction du plan ayant le rapport minimal de cisaillement critique à actuel.

ZUSAMMENFASSUNG: Mit Hilfe eines elektronischen Rechners werden Verschiebunglinien gezeichnet. Druckspannungsgrössen und Hauptspannungsrichtungen werden mit der Elastizitätstheorie berechnet und das Mohr-Coulomb-Kriterium bestimmt den Ansatz für die Verschiebung. Die Analyse kann für die vorläufige Berechnung von Ausbausystemen benutzt werden.

Several approaches exist to the problem of underground rooms stability, first in designing stage and then during the stage of excavation, supporting and lining realisation.

The oldest assumption considers the creation of the supporting arches above the roof. Starting from a simple formulation of the geotechnical rock properties, the volume of the instable rock over the roof can be estimated. (Kommerell, Terzaghy a.o.). Proctor and White claim the same based more on the condition of discontinuities than on the rock strength.

The strain and stress analysis around the underground rooms performed according to the rules of theory of elasticity, gives the stress concentration factors, but only for ideal elastic continuum.

The theory of plasticity, connected with Mohr-Coulombe failure theory, can give the strain and stress distribution with the assumption of plastic flow. The numerical data about the plastificated zone dimensions and some data about the influence of the supporting system can be calculated. Rabchewich illustrated the failure mechanism by the spiral failure lines, based on central symetry of loading and geometry.

Finite element analysis give the deformations, strains and stresses, whereby the arbitrary mechanical properties can be assumed. The output of the computing gives great number of numerical data, but the results are essentially depending on the accuracy of the input data on the rock properties. On the other hand the considerable efforts and money are necessary for the input and execution.

In the block theory the apsolute stiff blocks, limited by the discontinuities are supposed. The form and dimensions of critical polyhedral blocks will be estimated by their geometrical properties. Thereby the state of strains and stresses is completely neglected, what is not acceptable for the underground conditions.

The rock clasifications are based on the mechanical rock properties, density and orientation of the planes of weakness, water content, etc. The estimation of necessary support and lining is based on the experience from previously performed underground works.

The contemporary empiric method diminishes the importance of the previous assumption, but the in situ measurements of the convergence and strains are included. Respecting the security and economy

as the primar postulate, the essential engineering work moves to the working site, where considerable measurements and their interpretation have to be performed. Using the rock itself as the best comuter for
the calculation of the rock properties and response,
the stabilisation of deformations or of the deformations velocities can be estimated and the stability of the
room proved.

The problem of the first approximative design of
the supporting system is not uniformly solved, so it
will be attractive to publish a new attempt of quantification of the events around the underground openings.
In our previous papers we tried to find the shape of
potential sliding lines around the underground rooms.
At the very first (1967), the tracing was based on the
results of a photoelastic procedure, but later we used
a computer procedure based on the solutions of the theory of elasticity, limited to the circular and elliptic
room cross-section. The commonly used linear assumption of the Mohr -circles envelope (fig. 6) is accept-
ed. So the sliding on the plane $\left(\frac{\pi}{4} - \frac{\rho}{2}\right)$ or a crack,
perpendicular to the tension direction, are to be expe-
cted.

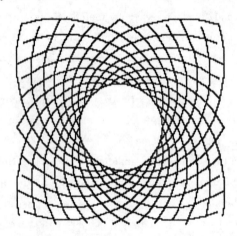

Fig. 1 Sliding lines for a central symetric load

In the simplest case of circular simetry of stress
state and geometry the sliding lines have the form of
spirals (fig. 1). They make visible the wedges demonstrated by Rabchewich, but also the sliding lines described by Terzaghy. Some of the lines are like the contours of the self supporting arches, supposed by Kommerell, Protodiaknoff a.o.

In the fig. 2 the stress trajectories are traced by
a personal computer for the case of a circular hole,
near the surface, for the ratio of horizontal and verti-

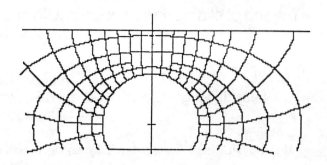

Fig. 2 Stress trajectories for gravity load

cal pressures 1/2.   In the fig. 3 the sliding lines
for this case are visible.

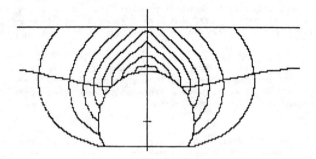

Fig. 3 Sliding lines for example on fig. 2

Principally the procedure is a very primitive one.
In every point of a trajectorie or of a sliding line the
stress components and the direction of principal stresses are calculated. In the direction of propagation of
the trajectories or of the sliding line a segment will
be drawn and the end point of segment will be used for
the next step.

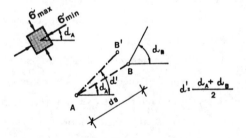

Fig. 4 Tracing a segment of the stress trajectories

In the areas with greater stress direction gradient, or
near isotropic points, the procedure is a little bit more complex (fig. 4). In starting point the direction of
the line propagation will be calculated, and the new
point B will be found by extrapolation.

In this point the new direction will be estimated,
and the definitive segment will be drawn with the arit-

metic mean value of both directions.

The elliptic form of a opening is also included, but without the gravity loads. (fig. 5).

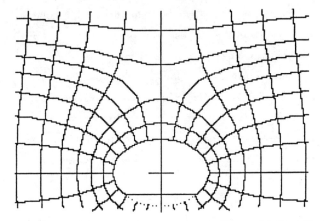

Fig. 5 Stress trajectories around an elliptic opening

The shape of trajectories and sliding lines are very instructive, giving only the qualitative illustration. In this paper we add some quantitative extension, by including the planes of weakness or planes of discontinuities in rock, and the relations between the critical shear in rock and the shear in the planes of weakness.

The calculation of the stress components has to be done to find the direction of principal stresses in any case, and it is only necessary to include the criteria of failure for the plane of weakness, and find the criteria for prognosis of the sliding propagation. A relative factor for both failure possibilities is included in the procedure, as well as the possibility to include more than one system of joint surfaces.

It is questionable if the elastic state of deformations and stresses is appropriate and opportune for a common nonlinear behavior of the rock. An analytical solution for plastic state exist only for the simplest case of central symetry. Thereby, the radial stress component has in both elastic and plastic areas the same range, but the stresses in circular direction significantly change their values, particularly near the free opening boundary. On the other hand, only in the case of a dominant anisotropy the principal stress directions change remarkably. So we can conclude that the trajectories for elastic state can be also accepted, with much probability, for the plastic range.

Failure Criteria and Safety Factors

Rock behavior can be expressed by the Mohr-Coulombe critical shear law:

$$\tau_{cr} = c + \sigma_n \cdot \tan \rho \qquad (1)$$

Here is:

τ_{cr} – critical shear

c – cohesion

σ_n – normal stress

ρ – envelope gradient

We include supstitutions:

$$S = \frac{\sigma_{max} + \sigma_{min}}{2}$$

$$D = \frac{\sigma_{max} - \sigma_{min}}{2} \qquad (2)$$

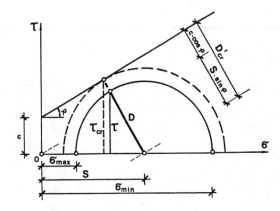

Fig. 6 Failure criteria and safety factor
a – keeping the center of the circle fix

a – Keeping constant the Mohr's circles center, it means the sum of principal stresses, we can define the critical shear and its relation to the existent shear (fig. 6a)

$$s = \frac{\tau_{cr}}{\tau} = \frac{D_{cr}}{D} \qquad (3)$$

From the circle geometry we come to:

$$D'_{cr} = c \cdot \cos \rho + S \cdot \sin \rho \qquad (4)$$

The basic assumption is not acceptable because thereby both of principal stresses have to change their values to reach the condition of critical shear.

b – When the plastic flow around a circular hole occur, only the circular (tangential) stress component change its value, but the radial (maximal) remain quite at the same value. As visible in the fig. 6b the principal stress difference can be increased keeping the point of maximal stress fix. With auxiliary value

$$c' = c + \sigma_{max} \cdot \tan \rho \qquad (5)$$

we obtain:

$$D_{cr} = \frac{c'}{\tan\left(\frac{\pi}{4} - \frac{\rho}{2}\right)} \qquad (6)$$

We can accept the second criteria as more adequate to the stress conditions around the underground opening.

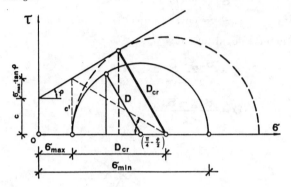

Fig. 6b - keeping σ max fix

For the jointed rock, we can suppose a triaxial test (see fig. 7) where the normal on the plane of weakness and the direction of maximal stress close the angle β.

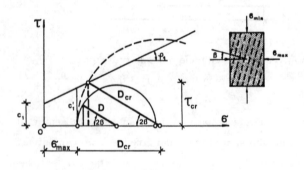

Fig. 7 Failure criteria and safety factor for the rock with plane of weakness

The failure criteria has the previous form, but with numerically other values:

$$\tau_{cr} = c_1 + \sigma_n \cdot \tan\rho_1 \qquad (7)$$

The sliding occurs if the shear on the joint plane reaches the critical value. The normal stress shear may be expressed:

$$\sigma_n = \sigma_{max} + D\left(1 - \cos 2\beta\right)$$

$$\tau_n = D \cdot \sin 2\beta \qquad (8)$$

The critical Mohr's circles radius D can be found as:

$$D_{cr} = \frac{c_1 + \sigma_{max} \cdot \tan\rho_1}{\sin 2\beta - (1 - \cos 2\beta) \cdot \tan\rho_1} \qquad (9)$$

In each point of the plane we can find both of safety fa-

ctors and suppose that the sliding line will propagate in the direction of lower safety factor. There is no problem to include more than one joint system.

In reality the calculated safety factors have a relative value, with the purpose to make the decision for the direction of sliding.

These relative safety factors are calculated for every point of the sliding line, and so a region can be enclosed where this factor is less than a given value. In the procedure the tracing of a sliding line will be broken when the factor reaches the before determinated value.

To avoid the listing of numerical data, they are not easy to survey, a graphic output with drawn lines will be more illustrative.

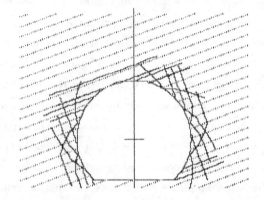

Fig. 8 The sliding lines in joint rock   s= 2.0

In the fig. 8 the sliding lines obtained by developed comupter procedure are represented. The joint inclination 20°, the depth of 200 m, and a cohesion of 3000 kN/m^2 were given. The cohesion in plane of weakness is assumed only a 1/3 of the rock cohesion, and the cohesion perpendicular to joints is 2/3 of this value. The envelope slope is 35°. To get the propagation of sliding lines far from free boundary, a safety factor of 2.0 was given.

In the fig. 9 the safety factor is taken 1.4 with the same other data.

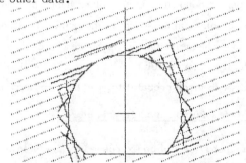

Fig. 9 The sliding lines in the sample   s = 1.40

The authors have not yet the experiences of application. It is evident that the results are some place between the qualitative and quantitative solution, depending on the accuracy of the geotechnical input data. In fig. 10 a similar example is treated, but with very low cohesion on the plane of weakness.

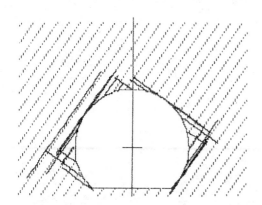

Fig. 10 The sliding lines in rock with low cohesion

Without in situ measurements, the input data, especially the data about the friction and cohesion on the joint planes cen be only rough. But the approximation, even with so given data, represents a useful illustration.

In next step the procedure will be extended to calculate the forces necessary to stabilize and stop the sliding.

Literature:

Papkovič : Teorija uprugosti, Moskva, 1939,

Mindlin: Stress distribution around a tunnel,
    Proc. ASCE, 1939,

Rodin: Influence of circular holes on the stress
    distribution, Vladivostok, 1940,

Savin: Stress concentration around holes,
    Pergamon Press, 1961,

Hudec : Sliding trajectories and their application
    to rock mechanics /in Croatian/, Proceeding
    at The II Yugoslav Congress on Rock mech. soc.
    Skopje 1967,

Kastner : Statik des Tunnel und Stollenbau,
    Berlin, 1971,

Rabchewich : Principles of dimensioning the support-
    ing system for the New Austrian Tunneling,
    Water Power, 1973,

Goodman, Gen-hua Shi : Block theory and its application to rock engineering, Prentice-Hall,
    New Jersey, 1985,

Frgić, Hudec : Stress trajectories and sliding lines
    around circular and elliptic opening /in Croat./
    Proceeding at The IV Yugoslav Congress on
    Rock mech. soc., T, Velenje 1986.

Kaiser effect gauging: The influence of confining stress on its response
Mesure de l'effet Kaiser: L'influence de la contrainte de confinement sur les résultats
Kaisereffektmessungen: Der Einfluss des allseitigen Druckes auf die Ergebnisse

D.R.HUGHSON, Research Associate, Department of Civil Engineering, University of Toronto, Ontario, Canada
A.M.CRAWFORD, Associate Professor, Department of Civil Engineering, University of Toronto, Ontario, Canada

ABSTRACT: The acoustic emission from a rock specimen subjected to increasing uniaxial restress, reveals a Kaiser Effect recollection of the stress that existed in its native environment. If confining stress was a component of that stress environment, the Kaiser Effect reveals the maximum amount by which the major principal stress had exceeded the minor principal stress, even for a brief period.

RESUME: L'émission acoustique provenant d'un échantillon de roche, soumise à un champ de contraintes uniaxial de plus en plus élévé, révèle une mémoire de Kaiser Effect de la contrainte qui existait dans son environment originel. Si la contrainte latérale était un composant de ce champ de contraintes, Kaiser Effect révélerait la déviation la plus grande entre la contrainte principale maximum et la contrainte principale minimum, même pour une brève période.

ZUSAMMENFASSUNG: Die akkustische Emission einer Felsprobe, die einer ansteigenden einachsigen Wiederbeanspruchung unterworfen ist, gibt eine Kaiser Effect Abbildung der Spannung wieder, die in der urgprunglichen Umgebung geherrscht hatte. Falls eine Komponente jener Spannungsumgebung eine Zwangsspannung war, so verrät der Kaiser Effect den Maximalwert, um den die grössere Hauptspannung die kleinere überschritten hatte, selbst für einen kurzen Zeitraum.

1 INTRODUCTION

Kaiser Effect Gauging is a technique which is being developed towards providing a practical method for retrieving the Kaiser Effect recollection of the maximum previous stress to which a sample had been exposed in its native environment (Hughson and Crawford 1986). This technique is anticipated to enable the rapid and economical determination of the pre-existing in-situ stress in rock, concrete and other materials. It is based on the natural phenomenon of Acoustic Emissions (AE).

Acoustic Emissions are the spontaneously generated ultra-sonic pulses that can be detected from certain materials (examples listed in Table 1) when subjected to increasing stress. The wave form is similar to that of a seismic event with duration in the order of 2 milliseconds. In this research project, the AE detection has been filtered to limit the frequency range to between 50 and 500 kHz and is quantified in terms of the cumulative "ring-down count above a threshold" (Nishijima et al 1985).

Dr. Joseph Kaiser (Kaiser 1953) noted that if the stress on a material is relaxed and then gradually increased, there is a significant increase in the AE output as the stress passes from the region of past-experience into the level of new-experience. This increase in AE rate at the level of previous maximum stress has been referred to as the Kaiser Effect.

There is evidence to support the possibility that AE is an acoustic manifestation of micro-sized inelastic strain and fracture occurrences within the material.

These authors have also previously confirmed or established (Hughson and Crawford 1986) that:

1.1 a sample carries in it a Kaiser Effect recollection of its stress environment prior to its extraction from that environment;

Table 1. Some types of materials shown to exhibit Kaiser Effect:

Material	Reported by
rock	Hughson and Crawford 1986
concrete	Cheng 1986, Mlakar et al 1980
metal	Kasier 1953, Baram and Rosen 1979 Rong et al 1983
ceramics	Pollock 1979
plastics	Pollock 1979
glass	Pollock 1979
building materials	Pollock 1979
bone	Fischer Arms et al 1984
snow	Bradley and St. Lawrence 1975
soil	Lord and Koerner 1984
glassy polymers	Reuyuan and Tiangui 1983
superconducting magnets	Wang and Wang et al 1983
fibre-reinforced composites	Gray and Summerscales 1985

1.2 their developed technique of Kaiser Effect Gauging can be an effective and economical procedure for recalling the Kaiser Effect memory held in a sample;

1.3 the Kaiser Effect does not necessarily occur at the exact previous maximum stress. There can be a variation between the two, relative to such factors as duration of the stress and the proportion of applied stress to ultimate stress;

1.4 different materials exhibit their own characteristic pattern of AE response to stress and this has been referred to as their AE Signature Profile.

2 OBJECTIVE

This phase of the Kaiser Effect Gauging research programme reported herein, is to determine by uniaxial testing, the effect of confining stresses on the fidelity of Kaiser Effect recollection of the sample's stress environment. If there were to be found a consistent relationship between the recalled stress and the triaxial stress state of the environment then such a testing procedure could be employed in gauging the in-situ stress field of a rock mass.

This research has found that the Kaiser Effect occurs approximately at the amount of deviatoric stress on the sample.

3 EXPERIMENTAL METHOD

To achieve the experimental objective, it was necessary to determine the responses obtained from many samples, each of which had been exposed to a different combination of axial and confining stresses. Initial uniformity of material selected, including any previous stress history, should be optimised. For this reason, a single block of Berea sandstone was cored into many samples each of which was 54 mm diameter with all axes parallel and being normal to the bedding plane. Sample lengths were trimmed to about 80 mm by the removal of approximately equal amount of material sawed and ground from each end.

Each sample was individually subjected to a different combination of axial and confining stresses. The increasing of stresses up to, and the reduction of stresses from the triaxial stress state, was controlled to follow a self-similar path. Incremental steps of axial and confining stress increase and decrease were kept to within 2 MPa of this path. The stress combination was maintained for a period of about 2 hours in order to achieve stability indicated by substantial decay of residual AE during dwell. When prestress was completed, the sample was sub-cored to produce 7 specimens parallel to the major axis and 4 from each of two directions in the plane of the confining stress as shown in Figure 1. Each specimen was then tested by the Kaiser Effect Gauging procedure (Hughson and Crawford 1986). The output data of cumulative ring-down AE count versus stress was electronically assembled and filed for later retrieval and processing.

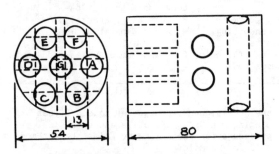

Figure 1. Sub-coring layout of pre-stressed sample. Dimensions in mm of finished core.

4 DATA

The acquired data from each specimen was then processed into suitably scaled but uniform format plots of stress (x axis) versus AE count (y axis) depicted in both linear and logarithmic scales. The Kaiser Effect point is interpreted, by definition, to be the stress level at which there is a rapid (sometimes abrupt) increase in the slope of both plots (linear and semi-log) which does not subsequently recede. This occurs at the transition from past-experience stress into the new-experience range. Figures 2 and 3 show composite examples of these two plots.

The raw data is plotted in both linear and semi-log formats in order to obtain corroboration of the Kaiser Effect point which may occasionally appear indistinct. It is noteworthy that the semi-log plots conform quite closely to two linear regression lines intersecting at the Kaiser Effect point. This suggests that the linear plots are substantially exponential with the past-experience region being at a lower rate than the new-experience region.

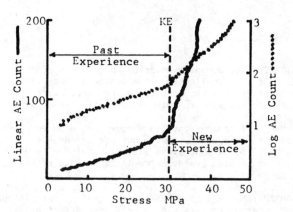

Figure 2. Composite of linear and log values of AE count plotted against stress showing typical slope change at Kaiser Effect point between past-experience and new-experience ranges. Some AE activity is shown in past-experience range.

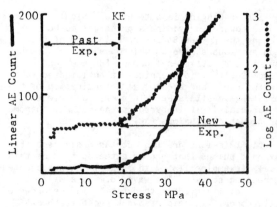

Figure 3. Composite of linear and log values of AE count plotted against stress showing typical slope change at Kaiser Effect point between past-experience and new-experience ranges. Low AE activity is indicated in past-experience range.

On occasion there may be an initial surge of AE output from a specimen, which later recedes to a lower slope of plot. This is likely to be caused by "seating noise" from misalignment in the testing machine or a faulty endcap (Hughson and Crawford 1986). If the slope following this initial surge is lower than the first new-experience slope, a Kaiser Effect point can be identified and that point remains valid. Figure 4 shows a typical example of initial "noise" which might have been caused by seating inaccuracy.

The Kaiser Effect point does not always appear. A sample which has had no previous stress, would produce a substantially straight line semi-log plot, and a linear plot which appears to have a uniformly continuous increase of slope. Figure 5 shows a composite example of such plots.

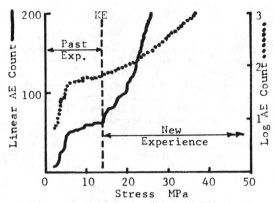

Figure 4. Composite of linear and log plots of AE count versus stress showing effect of initial AE surge, possibly caused by seating noise, which subsided prior to Kaiser Effect point.

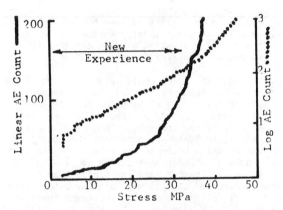

Figure 5. Composite of linear and log plots of AE count versus stress of a specimen which was not exposed to stress.

This could be expected to occur if the rock mass had not been exposed to significant previous stress such as in a relatively shallow sandstone bed or virgin concrete. It could also occur in a specimen extracted from a stressed environment.

It is generally recognized that the strength and elastic properties of all regions within a sample are not necessarily identical. Some regions may be stronger or weaker, stiffer or softer than the norm, even in the same direction. If a region of a sample is substantially surrounded by relatively stiffer material, the prestressing force on the sample will be carried predominantly by the stiffer portions leaving the softer region less highly stressed, or even unstressed. When that softer region becomes an independent specimen subjected to the Kaiser Effect Gauging procedure, it is likely to produce a plot indicative of a very low previous stress, or a virgin state such as is shown in Figure 5.

5 INTERPRETATION OF RESULTS

For the reasons discussed in the previous section, it is unlikely that the presetress load is shared uniformly throughout the sample. Thus the subsequent specimens taken from that sample would reveal a variety of recollections of their stress environment, but each would be valid for its own region. The estimation of overall stress imposed on the sample should therefore be based on the arithmetic average of the results obtained from all specimens relative to a particular direction in the sample. Thus the 7 specimen results for the axial direction, should be averaged to interpret the Kaiser Effect recollection

of the axial stress. The results from all 8 specimens in the plane of the confining stress should be averaged to assess the sample's reaction to it. If no Kaiser Effect could be identified, then that specimen should be interpreted as showing zero stress. It would be illogical to invalidate a specimen's results based on a subjective expectation of what a consensus should produce.

6 FINDINGS FROM THE TEST PROGRAMME

The plots of each of the many specimens produced and tested for this programme, were individually interpreted for of their indicated Kaiser Effect point. The recalled stress level thus derived from each specimen is tabulated and shown on Table 2 along with the average amount for the axial direction (Dir 1) and confining directions (Dir 2 and 3) for each sample. The prestress applied to the sample is also shown.

Table 2 Interpretation of Kaiser Effect Point of Specimens

Stress σ_1 σ_3	Sample No.	Test Dir.	A	B	C	D	E	F	G	Avg
0 0	14	1	0	0	0	0	0	0	18	3
		2		0	0					
		3		0		19	0	15		6
10 0	42	1	0	0	13	18	7	20	16	11
		2	11	18		23	19			
		3	19	13		25	15			18
20 0	16	1	20	20	24	21	20	21	24	21
		2	7	0		14	23			
		3		23	27		14	25		17
30 0	17	1	29	28	28	25	20	26	31	27
		2	22	17		25	22			
		3	28		15	15		18		20
40 0	18	1	33	34	36	32	29	28	30	32
		2	24	0		18	26			
		3	26	13		20	26			19
10 10	43	1	0	0	0	10	0	0	14	3
		2	23	22		14	18			
		3	6		22	26	0			16
20 10	21	1	0	0	0	0	0	0	27	4
		2		20	18		0	12		
		3		12	23		27	18		16
30 10	45	1	28	36	0	20	24	27	20	22
		2	24	24	24	23				
		3	20		19	18		25		22
40 10	46	1	22	27	30	26	25	26	28	26
		2	32	26	21	25				
		3			35	30	34	26		29
50 10	41	1	27	31	27	32	32	24	32	29
		2	19	10			11			
		3			26	28	23	25		20
30 20	27	1	10	8	27	20	20	6	15	15
		2	29	26	24		30			
		3	10	20	24			23		23
40 20	28	1	19		22	10	12	25	18	18
		2		11	25		19	22		
		3			20	23	25			21
50 20	31	1	26	26	29	23	24		17	24
		2	25	31			28	23		
		3	23		31	19		29		26
30 30	32	1	8	0	0	5	0	28	29	10
		2					25	22		
		3			25	26	27	19		24
40 30	33	1	21	14	0	0	20	0	18	10
		2		0	0		12	0		
		3			0	24	20	31		11
40 10 $-\sigma_3$	47	1	36	36	30	30	31	28	33	32
		2	32	25		25	32			
		3	30		31	33		20		29
40 10 $-\sigma_1$	48	1	26	21	31	34	27	28	27	28
		2	30	31		28	29			
		3		30	30		28			29

Consistent with the observations described in sections 4 and 5 above, there can be broad variations in the recollection of specimens in the same direction from the same sample. However the recalled average axial stress for each sample, appears to form a consistent pattern. Figure 6 compares the applied deviatoric stress to the recalled stress in the axial direction. Also shown is a broken line representing the equality between recalled and deviatoric stresses scales. There appears to be substantial agreement of the plotted points with the broken line.

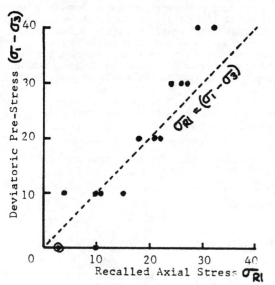

Figure 6. Comparison of the recalled axial stress (σ_{R_1}) to the difference between axial and confining prestresses $(\sigma_1 - \sigma_3)$, the deviatoric stress.

7 FELICITY FACTOR

As has been previously noted by these authors, and others (Gray and Summerscales 1985), the Kaiser Effect point is not necessarily exactly coincident with the level of previous stress. It has been observed that in the case of Berea Sandstone, the recollection can be substantially higher than the actual previous stress if the previous maximum is relatively small. The Kaiser Effect point and the actual deviatoric stress are in approximate agreement at about half range. In the upper region of the materials strength, the recalled stress tends to be significantly lower than the actual previous stress. This variation between recalled and actual previous maximum stresses has been termed the Felicity Factor (Gray and Summerscales 1985). The ratio of the stress at the Kaiser Effect point, to the previous maximum stress has been called the Felicity Ratio (Lord and Koener 1984). Although the magnitudes of this ratio have not been previously established, in the course of our research we have observed this ratio to be as large as 1.5:1 for Sandstone at very low stress levels and in the order of .85:1 near failure. We have further observed that different types of rocks show different amounts of this variation. This suggests that the Felicity Factor relationship to stress level could be a function of each material's individuality, similar to its AE Signature Profile.

Although no quantifying of the Felicity Factor has been established, to our knowledge, if an allowance for this factor were included in considering the results depicted in Figure 6, it is likely that the interpolated locus of recalled stress would become quite similar to the deviatoric stress line.

8 SELF-SIMILAR LIMITATIONS

As described above in Section 3. care was taken to raise and lower both axial and confining prestresses along a self-similar path. This however might be an unrealistic constraint. In a natural rock environment, the development and changes in the triaxial stress field within the recallable time span, might not have been confined to this degree of uniformity. Indeterminable stress paths may have been followed in the in-situ environment. Thus a recollection of the stress field might be relevant to the maximum deviation that had existed between the major principal and the minor principal stresses over the recollection time span.

Furthermore, there is the possibility that the mere process of extracting the sample from its environment could briefly upset the relationship between major principal and minor principal stresses in that sample. This could distort its recollection of the actual stress state. If the material's "memory mechanism" responds quickly to a change in the stress state, and if the coring process might release the stress in one direction prior to its releasing another, then there could be a further variation between the in-situ actual stress and the recalled amount of stress.

In order to test for response speed when one stress direction is released prior to another, a sample was triaxially stressed to a combination of axial and confining stresses already applied to another sample. After the full stress combination had been maintained until a stability state was apparent, the confining stress was released prior to the release of the axial stress. Subsequent specimens showed an axial stress memory 19 percent higher than the similar sample which had been destressed along a self-similar path. With another sample using the similar procedure there was an 8 percent increase noted in the recalled axial stress following the prerelease of the axial before the confining stress. The details of this experiment are included in Table 2 for samples number 47 and 48, and compared with the results of sample number 46.

This experiment suggests that even a brief isolation from the confining stress while the major principal stress remains effective will cause the recollection to approach the major principal stress amount.

9 POSTULATED RATIONALE FOR THE RESULTS

In a given mass of rock there is, figuratively, a large but finite number of micro-sized structural components each possessing a strength/direction capability. The magnitude of this capability might be represented by a histogram showing a normal distribution of strengths for the whole population of micro-sized structural components (Figure 7). As stress on the mass is raised, the stress on the weaker components might be expected to exceed their individual strength limits and, in some form, undergo inelastic strain. This micro-sized event might have two manifestations: an acoustic emission pulse; and the effective loss to the whole population of the strength contributed, however minor, by that component. The progressive raising of stress might be expected to produce progressively increased number of such micro-damage events, the weaker components failing first followed by the progressively stronger ones. The histogram of strengths for the remaining population of structural components, would no longer be a normal distribution. It would show a distortion by virtue of the truncated lower range of strengths.

If the stress were relaxed and then progressively reimposed, the remaining stronger components of the population would have no difficulty in carrying the total load without damage, until it reaches the previous maximum. At that point, the next weakest

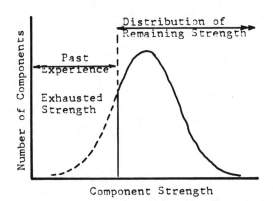

Figure 7. Normal distribution of strengths of micro-sized structural components. The loss of the weaker components results in a distortion of the strength distribution pattern for the remaining population.

components of the population undergo the inelastic strain occurrence. This would result in a significant increase in AE output commencing at the previous maximum level. This is the Kaiser Effect point.

Within practical ranges of stress for the particular rock, a stress field which is equal in all directions, which is an hydrostatic stress field, would do little damage to the micro-sized structural components (Skoblo and Zhigun 1982). If however there was a stress in one direction exceeding that in the other directions, the resultant deviatoric stress could be expected to cause damage to the rock fabric in proportion to magnitude of this deviatoric stress and more or less independent of the magnitude of the hydrostatic component of the stress field.

10 SUMMARY

10.1 The Kaiser Effect response to a triaxial stress field is to indicate the maximum deviatoric stress that had been imposed on a sample's environment, including any maximums which might have existed within the Kaiser Effect retention time span, and also any which occurred in the process of extracting the sample from its stressed environment.

10.2 The Kaiser Effect point is not necessarily precisely at the previous maximum but has a variation pattern relative to stress level.

10.3 The confidence level for derived results can be improved with the number of samplings included in the average.

11 ACKNOWLEDGEMENT

These authors are grateful to Mr. Robert Boothby and Mr. Donald Lau for their assistance in developing the data acquisition and retrieval system, and participation in the specimen preparation and testing, all of which was invaluable to this research.   The work of A.M. Crawford was supported by the Natural Sciences and Engineering Research Council of Canada and EMR. CANMET, Canada.

12 REFERENCES

Baram J. and Rosen, M. 1979. Acoustic emission generated during the tensile testing of aluminum alloys. Materials Science and Engineering 40 p. 21-29.

Bradley, C.C. and St. Lawrence, W., 1975. Kaiser effect in snow. Snow Mechanics - Symp. (Proc. of the Grindelwald Symp., Apr. 1974), IAHS-AISH Publ. No. 114, 1975.

Cheng, E. 1986. Determining characteristic profiles of geological materials and concrete. Thesis for the B.A.Sc. degree, University of Toronto.

Fischer, R.A., Arms, S.W., Pope, M.H. and Seligson. D. 1986. Analysis of the effect of using two different strain rates on the acoustic emission in bone. J. Biomechanics, Vol. 19(2), p. 119-127.

Gray, A.J. and Summerscales, J. 1985. Acoustic emission from cord-reinforced rubber. British J. Non-Destructive Testing, Vol. 27(5), Sept. 1985, p. 300-1, 304-5.

Hughson, D.R. and Crawford, A.M. 1986. Kaiser Effect Guaging: A new method for determining the pre-existing in-situ stress from an extracted core by acoustic emissions. Proceedings, Int. Symp. on Rock Stress and Rock Stress Measurements. Stockholm, Centek

Kaiser, H, 1953. Erkenntnisse und folgerungen aus der messung von gerauschen bie zugbeanspruchung von metallischen werkstoffen. Archiv fur das Eisenhut-tenwesen. Vol. 24, p. 43-45.

Lord, A.E., Jr. and Koerner, R.M. 1984. In-situ stress determination in soil and rock using the acoustic emission method. U.S. Army Research Office, ARO-18687. 6-GS.

Mlakar, P.F., Walker, R.E., and Sullivan, B.R. 1980. Acoustic emission from concrete specimens. WES/MP/SL-81-26 Proj. 4A161101A91D.

Nishijima, S., Iwasaki, H. and Okada, T. 1985. Study of disturbances in superconducting magnests by acoustic emission method. IEEE Trans. Mag. 21(2), p. 388-91.

Pollock, A.A. 1979. An introduction to acoustic emission and a practical example. The Journal of Environmental Science. March/April 1979, p. 39-41.

Renyuan, Q. and Tiangui, W. 1983. Some observations on the acoustic emission of polymers. Eur. Polym. J. Vol. 19, No. 10/11, p. 947-948.

Rong, S.G., Byran, B. and Stephens, R.W.B. 1983. Acoustic emission measurements of a shape-memory alloy. J. Acoust. Soc. Am. 73(4), Apr. 1983, p. 1217-1222.

Skoblo, A.V., Zhigun, A.P., Kolesov, S.A. and Duina, L.P. 1982. Influence of the type of stress state on the nature of acoustic emission signals. Zavodskaya Laboratoriya, Vol. 48(6), p. 91-92.

Wang, E., Wang, B., Xue, Q., Zhu, X., Jiao, Z. and Chen, Z. 1983. Acoustic emission from structural materials of superconducting magnets. Chinese Physics, Vol. 4, No. 2, April-June.

Microseismicity induced by deep coal mining activity

Microséismicité causée par une activité d'extraction de charbon en profondeur
Durch Tiefkohlenbergbau verursachte mikroseismische Bewegungen

YOJI ISHIJIMA, Hokkaido University, Sapporo, Japan
YOSHIAKI FUJII, Muroran Institute of Technology, Japan
KAZUHIKO SATO, Muroran Institute of Technology, Japan

ABSTRACT: Three case studies on the microseismic activities occured in deep coal mining panels are presented with their interpretations. It has been cleared from observation that the sequence of the seismic activities which can be represented by the change of the seismic energy release rate is substantially influenced by the geometry of excavations changing daily with mining activity.

A simulation model has been developed to evaluate the stress distribution and the strain energy release rate which is defined as the total strain energy released from the newly fractured areas during the unit face advance of the mining front.

Since this energy release rate has revealed to be in good correlation with the observed seismic energy release rate for each case study, the newly developed model could be applicable to interprete and to anticipate the fracturing phenomena occuring around the excavation.

RESUME: Trois études de cas concernant les les activités microsismiques produites par l'extraction de charbon en profondeur sont présentées ici, ainsi que leurs interprétatione. Il ressort clairement des observations que la séquence de leure activités, qui peut être représentée par le changement du taux de dégagement d'énergie sismiqué, est influencée de façon substantielle par la géometrie des excavations, qui varient quotidiennement en fonction de l'activité miniére.

Un modéle de simulation a été mis au point afin d'evaluer la distribution des forces et la taux du travail de deformation degage, qui est defini comme le total du travail de deformation dégagé des zones venant d'étre fracturées pondant la progression d'une unité d'exploitation.

Etant donné qu'il a été révélé que ce taux de dégagement d'énergie posséde une bonne corrélation avec le taux de dégagement d'énergie sismique observé dans chacune des études de cas, le modéle qui vient d'étre mis au point pourrait étre utilisé pour interpréter et prévoir les phénoménes de fracturation se produisant aux alentours de l'excavation.

ZUSAMMENFASSUNG: Drei Fallstudien über mikroseismiche Bewegungen in tiefen Kohlengrubenfeldern sowie ihre Interpretationen werden vorgetragen. Aus Beobachtungen hat sich ergeben, daβ der seismichen Aktivitatenablauf, der durch eine Änderung der Freisetzungsrate der seismichen Energie dargestellt werden kann, wesentlich von der Geometrie der sich täglich mit dem Abbau andernden Ausschachtungen abhängt.

Ein Simulationsmodell wurde entwickelt, um die Druckverteilung und die Freisatzungerate der Formenderungsenergie zu bewerten, die als während das Einheitsvortriebs der Abbaufront von den neu durchbrochenen Bereichen freigesetzte Gesamt-Formenderungsenergie definiert wird.

Da diese Energiefreisatzungsrate mit der beobachteten seismichen Energiefreisatzungsrate für jede Fallstudie in guter Korrelation steht, könnte sich das neu entwickelte Modell auf die Interpretation und Prognose von Brucherscheinungen im Umkreis der Ausschechtung anwenden lassen.

1 INTRODUCTION

Working area of Horonai coal mine, Japan, is now located about 1,100 meters below surface and strata control is one of the urgent tasks to be solved. For this purpose the measuring system to monitor the microseismic activities occurring around the excavating panels has been developed and applied in this mine in these ten years. A model study based on the stress analysis to simulate the rock mass fracturings induced around the excavation has also been conducted along this line to interpret the monitoring data.

In this paper, case studies are presented for the three panels on the observed seismic activities and their interpretations using the simulation model. In each panel, two adjacent slightly inclined coal seams, the upper seam and the lower seam separated by 10 meters distance, were extracted by the fully-mechanized longwall mining method. There were no old workings below and above these observed panels, whose condition should often bring high stress state in the vicinity of the excavation and consequently high level of microseismicity.

2 OBSERVATION OF SEISMIC ACTIVITIES

The seismic activities are monitored by using the system developed by Sato et. al (1985) composed of ten velocity-type sensors positioned on the surface and in the ground around the monitored panels. The source location and the maximum amplitude of the ground velocity are determined for each event.

From the latter the Richter's scale magnitude and the amount of seismic energy are evaluated using Gutenberg - Richter's formula and Muramatsu's empirical formula, respectively. The seismic energy data is further supplied to evaluate the seismic energy release rate which is defined as the total sum of its individual energy accompanied with the unit face advance.

These data are processed to deliver graphs and contour maps which visualize the seismic activities in terms of mining progress and/or excavation geometry as shown in chapter 4.

3 EVALUATION OF ENERGY CHANGES REFLECTING THE MINING SEQUENCE

There have been several proposals on the evaluation of energy changes accompanying mining. The most used of these associated with the interpretation of rock burst is the released energy which is defined, in case of tabular mining, as a half of the product of the load which applied to the element just before the excavation and the convergence of that portion removed by the excavation. This quantity was firstly remarked by Cook et. al (1966) and was successfully applied in their research on the rockbursts in South African gold mines.

However poor correlation between this and observed energy in case of Horonai coal mine has inevitably led to introduction of a new energy quantity. Considering the fact that the rock noises are generated accompanying rock mass fracturing, a newly introduced one called strain energy release rate is defined as an amount of strain energy which is released from the newly fractured area when the mining is advanced in unit length. To evaluate this value, however, a three dimensional elasto-plastic stress calculation is required. Since to conduct this calculation is very difficult, approximate method has been devised which can evaluated the quantity only from the elastic analysis. The assumptions used in conducting calculation by this method are the following;
1) The region where the stress condition exceeds a failure criterion is consistent with the fractured region. As for the failure criteria of the rock mass, Paul's criteria which consists of the Coulomb's criteria and the tension cut-off concept is adopted.
2) Stress state in this region immediately drops to a residual strength condition, following the stress-path shown in Fig. 1.

It has been confirmed for the pressurized thick-wall cylinder problem that the analytical solution and the numerical solution obtained by this approximate method is roughly coincident each other, unless the stress drop accompanying the failure is large. Therefore, this method has been applied as a simulation model to interpret the observed data. Three dimensional displacement discontinuity method is used to conduct stress calculation.

4 CASE STUDY FOR THREE PANELS

4.1 Panel 1

As shown in Fig. 2, illustrating the schematic feature of the mining panel, an old working exists sideways of the lower panel, but no old working exists on both side of the upper panel. Only the microseismic activity of the upper seam is paid attention, because it was rather aseismic during mining of the lower seam. This tendency is commonly observed for the three cases, which can be interpreted that the stresses in the lower seam have been substantially subtracted by the over-mining of the upper seam.

In Fig. 3 the three different energy sequences, that are, the measured seismic energy release rate, the simulated energy release rate and the strain energy release rate are shown. The seismic energy release rate is low at the beginning of the mining, then it increases sharply and takes a distinct peak when the face is advanced to about 140 meters from the setup entry. The strain energy release rate shows similar behavior with the observed rate. But the energy release rate behaves in different manners.

The occurrence of peak of the seismic energy release rate at 140 meters from the setup entry might be due to the high stress concentrations at the two adjacent corners formed between the old

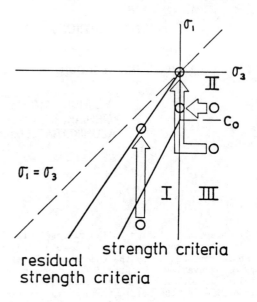

Figure 1. Assumed stress path accompanying rock mass failure.

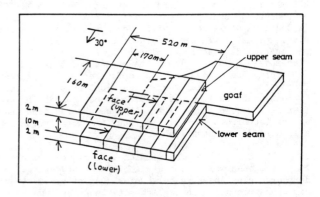

Figure 2. Schematic configuration of the mining panel (panel 1).

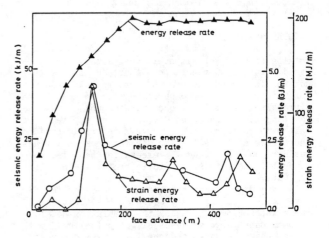

Figure 3. Energy release rates of three types vs. face advance (panel 1).

working and the mined area when the face just passed the former. Correctness of this estimation should be demonstrated by the contour map in Fig. 4 which shows that the seismic energy densities are concentrated at these places.

4.2 Panel 2

A small coal pillar was formed between the tail entry and the old working when the face was advanced as shown in Fig. 5. From Fig. 6 which indicates the sequence of the seismic energy release rate, it is obvious that the mining of the upper seam induced more intense seismicity than that of the lower seam which was worked later as in the case of Panel 1.

The sequence of the seismic energy release rate of the upper seam is as follows; It is rather high at the initial stage of mining. Then it is decreasing during the subsequent period. When the face was just leaving the coal pillar newly formed by the mining, it turns to increase and after taking a peak when the face was advanced to 190 meters, it decreases again.

The reason of high seismicity at the initial stage may be attributed to the existence of two adjacent corners formed between the setup entry and the neighboring old working. On the other hand, the decrease of seismicity when the face advanced more than 200 meters from the setup entry may be due to the decreasing of the corner area at the tail entry.

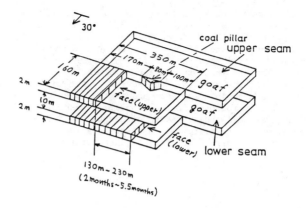

Figure 5. Schematic configuration of the mining panel (panel 2).

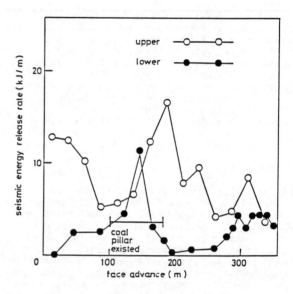

Figure 6. Seismic energy release rate vs. face advance (panel 2).

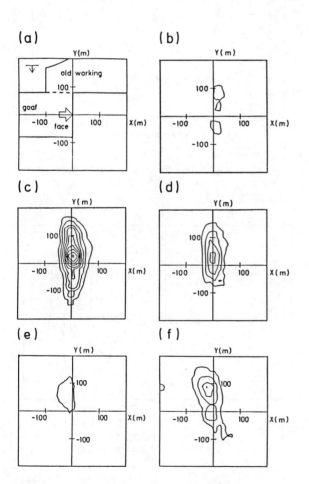

Figure 4. Contour map showing the spatial distribution of the seismic energy for face advance of 0-100 meters (b), 100-200 meters (c), 200-300 meters (d), 300-400 meters (e), 400-500 meters (f), xy-plane is lying on the coal seam and its origin locates at the midpoint of the face line as shown in (a) (panel 1).

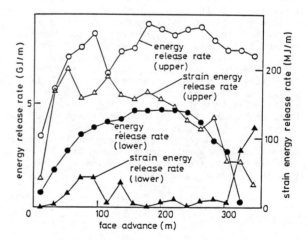

Figure 7. Energy release rates of three types vs. face advance (panel 2).

989

The energy release rate of the lower seam shows a distinct peak when the face reaches the coal pillar formed during the mining of the upper panel. Fig. 7 shows the sequence of the numerically evaluated energy release rate and the strain energy release rate. The existence of distinct peak in the seismic energy release rate vs. face advance curve which are observed both in upper seam and in the lower seam is not clearly followed in the simulation result, although a small-scale peak is distinguished for each seam. On the contrary, there is no such trace in the energy release rate vs. face advance curve.

4.3 Panel 3

The relative position of the panel in relation to the old workings is shown in Fig. 8. Only the behavior of the mining of the upper seam was paid attention due to the same reason as stated in the case of panel 1. The sequence of three rates with mining advance is shown in Fig. 9. The seismic energy release rate has a maximum at the beginning of the mining, then it decreases with the face advance. The strain energy release rate progresses roughly in a similar manner with the observed one, but the energy release rate behaves in a completely different way.

The peak of the seismic energy release rate attained at the initial stage of mining may be due to a high stress concentration occurred in the T-junction region shown by the circle A in Fig. 10, which resembles the initial condition in case 2.

5 DISCUSSION

Studies of the above three case histories have made it clear that, at least for the case of Horonai coal mine, the seismic energy release rate behaves more similarly with the strain energy release rate than with the energy release rate proposed by Cook et. al. It is also cleared that, for three cases, the latter is 10 to 10^2 times larger than the former in magnitude. It should be noticed that only small portion of the total energy released by mining, whose estimation is given by the energy release rate, is used for generation of fracturing.

There is another large discrepancy of magnitude between the measured energy rate and the simulated strain energy release rate; the latter is about 10^3 times greater than that of former; This could be attributed to the following reason; In the seismology, it is known that only a few percent of the strain energy released at the hypocenter of a spontaneous earthquake are used as elastic waves. Measured energy is evaluated based on this elastic wave energy. On the other hand simulated energy is evaluating the released strain energy itself. To assess the strain energy released at a hypocenter of a microseismic event explicitly, the concept of the focal mechanisms of the microseismic event should be introduced, which is left for future study.

The above mentioned considerations should strongly suggest that the strain energy release rate corresponds well to the fracturing induced by mining than the energy release rate.

REFERENCES

Cook, et. al 1966. Rock mechanics applied to the study of rockburst. J. South Afr. Inst. Min. Metal., p.435-528.
Sato, et. al 1985. Microseismic activity induced by mining. Proc. 4th. Conf. AE/MS Activity in Geologic Structures and Materials (In Press).

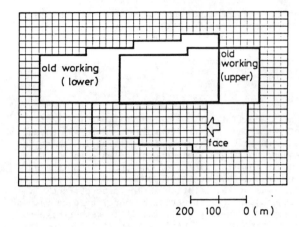

Figure 8. Modeling of the mining panel (panel 3).

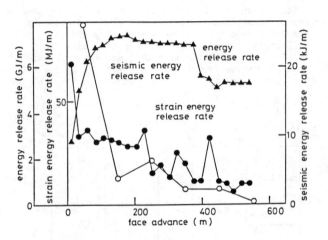

Figure 9. Energy release rates of three types vs. face advance (panel 3).

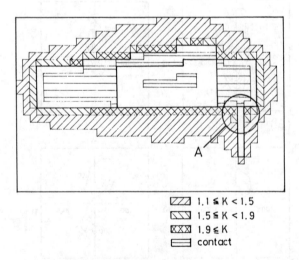

Figure 10. Distribution of vertical stress component in the coal seam, K is ratio of the vertical stress to the initial stress (panel 3).

Rock mechanics aspects of backfill in deep South African gold mines
Aspect de mécanique des roches de remblayage dans les mines profondes sud-africaines
Felsmechanische Aspekte von Hinterfüllung in tiefen südafrikanischen Goldminen

A.J.JAGER, Rock Mechanics Laboratory, Chamber of Mines of South Africa, Research Organization, Johannesburg
P.S.PIPER, Rock Mechanics Laboratory, Chamber of Mines of South Africa, Research Organization, Johannesburg
N.C.GAY, Rock Mechanics Laboratory, Chamber of Mines of South Africa, Research Organization, Johannesburg

ABSTRACT: Rock mechanics related factors have a strong influence on the safety and economics of mining Witwatersrand reefs at great depths where high stresses around stopes cause intense fracturing and deformation of the surrounding rockmass and lead to rockbursts. These problems are countered by introducing regional support to reduce the stope convergence and local stope support to maintain the integrity of the fractured rock. Backfill systems have been developed which can simultaneously fulfil the requirements of both regional and local support. Theoretical studies show that backfill alone can act as the regional support at depths of less than 3 km. Below 3 km the major benefits of backfill are to reduce the stresses acting on stabilizing pillars or to enable an increase in the span between the pillars. Local strata control benefits include improved gully and hangingwall conditions, better stope-width control and reduction in rockburst damage, with the consequence that there is a reduction in dilution of reef and an improvement in safety.

RESUME: Les facteurs liés à la mécanique des roches ont une grande influence sur la sécurité et les aspects financiers de l'exploitation minière des filons à grande profondeur du Witwatersrand ou les extrêmes contraintes autour des gradins provoquent la formation de fissures intenses et la déformation de la masse rocheuse à l'entour et mènent aux coups de terrain. Les problèmes sont opposés par l'introduction d'un soutènement regional pour réduire la convergence du gradin et d'un soutènement local au front de taille pour maintenir l'intégrité des massifs rocheux fissurés. Les systèmes de matériau de remblayage développés peuvent satisfaire simultanément les demandes de soutènement regional et local. Les études théoriques ont montré que le materiau de remblayage seule peut fournir le soutien 'regional jusqu' aux profondeurs de 3 km. Au dessous de 3 km les avantages principaux sont de réduire les contraintes sur les piliers de stabilisation et d'augmenter l'écartement entre les piliers les avantages du contrôle des couches locales comportent les conditions améliorées du couloir et du compartiment supérieur, un meilleur contrôle de la largeur du gradin et une diminution des dégâts par des coups de terrain, par conséquence il y a une diminution en dilution et la sécurité est améliorée.

ZUSAMMENFASSUNG: Felsmechanische Faktoren üben starken Einfluss auf die Sicherheit und Wirtschaftlichkeit des Abbaus dertiefen Witwatersrand Erzgänge aus, denn dort verursachen hohe Spannungen rings um den Abbauort starke Bruchbildung und verformen die umliegende Gesteinsmasse; dies kann zu Gebirgsschlägen führen. Diesen Problemen kann man entgegenwirken, indem man streckenweise Abstützungen vornimmt, um Konvergenz des Abbauorts zu vermindern, und indem man ferner örtliche Abstützungen vor Ort anbringt, um die Vollständigkeit des zerklüfteten Gesteins aufrecht zu erhalten. Es wurden Hinterfüllungssysteme entwickelt, die die Forderungen von sowohl streckenweisen, als auch örtlichen Abstützungen erfüllen. Theoretische Studien haben gezeigt, dass Hinterfüllung allein ausreicht, um als streckenweise Abstützung bis zu 3 km Tiefe zu dienen. Unter 3 km sind die grössten Vorteile, die Spannungen auf stabilisierenden Pfeilern zu reduzieren, und die Spannweite zwischen Pfeilern zu vergrössern. Für die örtliche Beherrschung der Gesteinsschichten ergeben sich als Vorteile: verbesserte Zustände in Bezug auf Schächte und das Hangende, bessere Handhabung der Breite des Abbauortes und eine Verringerung des Schadens durch Gebirgsschlag; das hat zur Folge, dass die Dilatation verringert wird und Sicherheit erhöht wird.

1 INTRODUCTION

Mining extensive thin conglomerate reefs at depths of more than 1 500 m leads to severe rock pressure problems, principal among which are mining-induced seismicity and related rockbursts, intense stress fracturing of the rock surrounding all excavations and large elastic and non-elastic deformations of this rockmass. The combination of these factors within various geological environments results in a wide range of conditions and, thus, requirements for the support of excavations. Until recently these have been fulfilled by an integration of timber props and packs, rapid yield hydraulic props, grout packs, rockbolts, wire mesh and lacing. In addition, mine layouts including stabilizing pillars are designed to improve regional stability.

With increasing depths of mining and greater percentage extraction of the reefs, mining conditions have become more severe. This has had a significant impact on both the safety requirements and economics of mining. Accident statistics show that over the past 10 years the accident rate due to falls of ground and rockbursts has fallen by 38 % while accidents due to other causes have decreased by 52 %. Although these figures show a marked improvement in safety performance, the smaller decrease in rock mechanics related accidents highlights the need for improved support methods. From an economics point of view, labour productivity can vary by as much as 30 % depending on the harshness of geological and mining conditions; in deep mines with potentially high stress conditions more than 15 % of ore reserves may have to be left as stabilizing pillars or as remnants too difficult to mine; the inability to mine to a desired stoping width because of adverse conditions can lead to dilution of ore in excess of 15 % with resultant loss of revenue (Wagner 1986).

Thus from both the safety and economics points of view, the consequences of unfavourable rock conditions are serious and the gold mining industry requires solutions to the problems. There are two main reasons for the unfavourable conditions. Firstly, if high stress levels are allowed to develop at stope faces the rock becomes intensely fractured and is subjected to large elastic and inelastic deformations, and the probability of rockbursts is increased. Secondly, the combination of mining induced fractures with geological structures and poorly cohesive strata leads to numerous key blocks which have to be supported. Effective regional support to reduce convergence and stress levels, and efficient local support to hold key blocks in place is thus required.

Improved engineering technology and a better understanding of support requirements led to the belief in 1977 that backfilling could contribute significantly to solving these problems. However, it is only during the past 5 years that backfill has been widely accepted as a viable support system for deep mines and at present 7 mines have backfill systems of various sizes and types in operation. The amount of backfill being placed constitutes support for less than one per cent of the 270 km of stope face being worked. However, by 1991 it is estimated that backfill will be used in over 25 per cent of stopes.

2 APPLICATIONS OF BACKFILL

Potential rock mechanics applications for backfill in deep mines are as follows:
(i) to provide mine-wide regional support by limiting the amount of elastic deformation of the rockmass affected by mining;
(ii) to reduce the occurrence of seismic events, and hence rockbursts, by reducing energy release rates and by reducing the magnitude of shear stresses acting on planar geological structures which otherwise may fail by shear displacement·
(iii) to reduce the stresses acting on rock stabilizing pillars to below that indicated by pillar failure criteria;
(iv) to increase the percentage ore extraction by allowing greater spacing between stabilizing pillars and by improving conditions to the extent that less reef is left unmined as a result of hazardous conditions;
(v) to provide support during shaft pillar extraction;
(vi) to reduce damage caused by rockbursts by providing total area support close to the working face and access ways in stopes and by providing a much greater resistance to dynamic closure of the stope;
(vii) in conjunction with good temporary support to improve hangingwall control and safety in the working area and access ways of stopes. In particular to support the rock adjacent to geological structures·
(viii) in multi-reef mining situations to support the layers of rock between overlapping stopes - a support function similar to that encountered in shallow mines·
(ix) to improve the support in wide reef stopes, (> 2 m) where the stiffness of conventional support and hence its effectiveness is reduced.
Applications (i) to (iv) relate to the regional support capability of backfill and (v) to (ix) are concerned with the effectiveness of backfill as a local support. In this paper the requirements of backfill to act as both a regional and a local support medium will be defined and the potential of backfilling to fulfil these requirements will be discussed.

3 REQUIREMENTS OF BACKFILL

There are three fundamental requirements of backfill in its application in gold mines. These are, from a regional support point of view, that the ultimate compression that the fill material can experience be minimized. Less than 30 % compression is desirable. For local support, the initial stiffness of the backfill should be as high as possible. Precise requirements have not been defined but a modulus of better than 5 MPa for strains up to 0,02 may prove to be adequate for normal conditions. This implies no shrinkage of the fill which may be difficult to attain in certain fills without the addition of a binder.

The third requirement is that the backfill system must be cost effective with respect to conventional support methods.

The details of these and other requirements will be discussed in the sections on regional and local support.

4 THE TYPES OF BACKFILL USED AND THEIR PHYSICAL PROPERTIES

4.1 Types of backfill

Four main types of backfill material are currently being used on the deep mines. These, in decreasing order of area currently filled, are:
(i) cycloned tailings which are prepared on surface to a relative density of 1,65. The slurry so formed is piped in gravity flow systems to the stopes;
(ii) dewatered tailings, which are prepared by belt filtering or centrifuging the slurry underground close to the stope. The dewatered material is pumped to the stopes as a paste with a relative density of about 1,95;
(iii) cemented dewatered tailings;
(iv) milled waste rock, prepared by crushing underground and pumped to the stopes at a relative density of either 1,6 or 2,1.

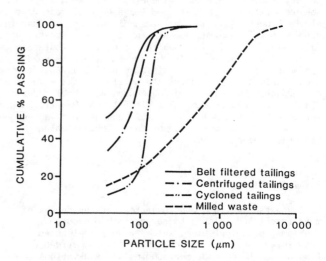

Figure 1. Cumulative particle size distribution of various backfill materials.

The fill is placed in 3 to 6 m wide paddocks or into long bags made of a geotextile material which permits free drainage. The size distributions of the types of fill are shown in Figure 1. The cycloned tailings have about 40 % of the minus 38 μm material removed from the tailings and are therefore coarser but more poorly graded than the dewatered tailings. The milled waste material is coarser still and well graded. As a result of these different grain size distributions the potential minimum placed porosity of the materials is significantly different (Table 1, Stewart et al

1986). However, because of the different treatment and placement techniques resulting in different water contents of the placed material and lack of active compaction during placement the fills initially do not attain their minimum porosity but are 12 % to 20 % above this figure at placement as shown in Table 1. These higher porosities at placement have a deleterious effect on both the initial stiffness and also the ultimate strain in the fill. In applications where backfill is used to provide local support, and little convergence is expected, a binder is added to improve the initial stiffness. Cement addition provides limited improvement to the ultimate strain value and, therefore, uncemented backfills are used for regional support applications.

Table 1. Optimum and placed properties of backfills.

Type	Optimum		At Placement		Pumping Density
	η %	H_2O %	η %	H_2O %	R.D.
Cycloned tailings	40	20	52	23	1,63
Dewatered tailings	35	16	47	24	1,92
Milled waste	22	9	42	6	1,65

η – porosity
H_2O – water content
R.D. – relative density

4.2 Stress-strain properties of backfill

The porosity of the placed fill is of paramount importance to its stress-strain performance. Laboratory test stress-strain curves for the various types of fill are shown in Figure 2. These curves show that there is a significant difference between the strains to which each fill asymptotes. This value of strain correlates closely with the initial porosity of each fill and is of major importance in determining the regional support capability of the fill. The curves also show that all the fills have a low initial stiffness.

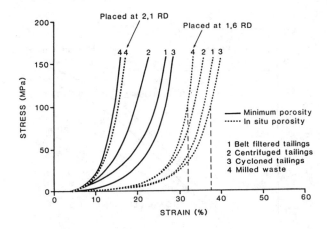

Figure 2. Stress-strain curves for various backfill materials tested in confined compression at their placed and potential minimum porosities.

4.3 Constitutive equations for backfill

A widely-used approximation to the stress reaction behaviour of backfill under normal compression is the so-called 'hyperbolic' model (Ryder and Wagner 1978):

$$\sigma = \underline{a}\varepsilon/(\underline{b} - \varepsilon)$$

where σ and ε are the fill stress and strain, $\underline{a}$ is a stress parameter and $\underline{b}$ is a critical strain asymptote.

This model appears to fit the behaviour of high-quality, cohesive, well-graded fills of low porosity very well. High porosity 'cohesionless' materials, typical of those currently being placed, show very slow early rates of load build-up and are much better fitted by the following 3-parameter 'quadratic' model:

$$\sigma = \underline{a}(\varepsilon - \varepsilon_t)^2/(\underline{b} - \varepsilon) \quad (\underline{a} = o \text{ for } \varepsilon < \varepsilon_t)$$

where $\underline{a}$ is a modulus, ε_t is a transition strain at which significant fill reaction first manifests, and $\underline{b}$ is the critical strain asymptote ($\underline{b} \approx$ initial porosity η_o). (Piper and Ryder, 1986).

These constitutive equations are currently being used in the computer modelling of backfill.

4.4 Monitoring the performance of backfill underground

Great difficulty has been experienced in the development of instrumentation to monitor the performance of backfill placed in stopes. Not only are large differential stresses built up in the fill as the stope closes but also large lateral shear displacements occur. Together the high forces and displacements are capable of breaking welds and high pressure tubing or flattening and stretching armoured electric cables to such an extent that they are inoperable. Strong, telescoping instruments to measure stope closure have been bent and jammed by the fill. However, some instrumentation, viable at low stresses at least, has been developed and installed in backfill underground.

Initial results indicate that the placed performance is similar to that obtained in the laboratory but that in some instances the early stress build-up in the fill underground is less than that obtained in the laboratory at the same strain, suggesting that a certain amount of shrinkage takes place underground, Figure 3. Nevertheless, the limited stress-strain results obtained from underground measurements suggest that laboratory determined stress-strain characteristics of the various types of backfill can be used in computer simulations to assess the regional support benefits of backfill.

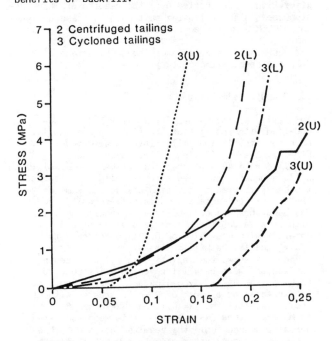

Figure 3. A comparison of stress-strain results obtained underground (U) and in the Laboratory (L).

5 BACKFILL AS A REGIONAL SUPPORT

Rockbursts experienced in the South African gold mines are associated with the energy changes brought about by mining tabular deposits at depth (Cook et al 1966), although by no means on a strict one-to-one basis (Salamon, 1983). These energy changes, expressed in terms of the Energy Release Rate (ERR), are related directly to the volumetric convergence of the mined-out area. Other workers have found direct correlations between seismicity levels (Salamon and Wagner, 1979) and general deterioration of mining conditions (Hodgson and Joughin 1967) with increasing levels of ERR. Consequently, it has been proposed that mining conditions can be improved by designing mining layouts for minimum volumetric convergence and hence low ERR levels. Practical experience has shown that in conventionally supported production areas where the geology is not complex, acceptable conditions are achieved if ERR values are kept below about 40 MJ/m^2. Conjecturally, the use of backfill as a support may permit mining in areas with higher ERR values, because of its large areal coverage.

A number of gold mines have introduced regional support systems consisting of regularly spaced strip pillars, known locally as stabilizing pillars, to control the volumetric convergence and hence the rockburst hazard. The effectiveness of stabilizing pillars for reducing the incidence of seismic events has been demonstrated (Salamon and Wagner, 1979) and they are now accepted as a successful means of providing regional support. However research has shown that the foundation of a highly stressed pillar can fail if the average pillar stress (APS) exceeds three times the uniaxial compressive strength (UCS) of the rock (Wagner and Schumann, 1971). The UCS of typical footwall and hangingwall rocks is usually less than 250 MPa. Therefore, 750 MPa is considered to be the maximum acceptable value of average pillar stress.

The obvious disadvantage of stabilizing pillars is that between 10 % and 15 % of the mine's ore reserves are lost. In addition, their use in deep mines has revealed other problems, such as poor environmental and hangingwall conditions and material handling difficulties in the vicinity of the pillars.

Recent advances in backfill technology make it feasible to use backfill for regional support instead of stabilizing pillars. To assess the potential of backfill as a regional support, a series of computer simulations were carried out; the results of which are discussed in the following section. In these analyses the MINSIM-D stress analysis program (Napier, 1985) was used to calculate the values of ERR and APS for various mining layouts involving combinations of backfill and stabilizing pillars.

5.1 Computer simulations

Initially, a mined area of 1 280 m x 1 280 m was modelled with and without backfill. Subsequently, numerous stabilizing pillar layouts were modelled. Finally, the stabilizing pillar layouts were remodelled with 40 % of the area between the stabilizing pillars backfilled. The stress-strain behaviour of the backfill used in the simulations was similar to that of milled waste with a relative density of 1,8. The behaviour was defined by a hyperbolic function with an a value of 5 MPa and a b value of 0,3. A material with a b value of 0,3 is considered to be 'fair' quality backfill.

Figure 4 shows the variation in ERR with depth below surface and with backfill covering different proportions of the area. It can be seen that if this area is extracted at a depth of 2 km or greater without the use of backfill the ERR will be in excess of 40 MJ/m^2. Since ERR values of this magnitude are normally unacceptable the results indicate that a regional support system of either backfill, stabilizing pillars or combinations of these is required at depths of greater than 2 km. For example filling 20 %

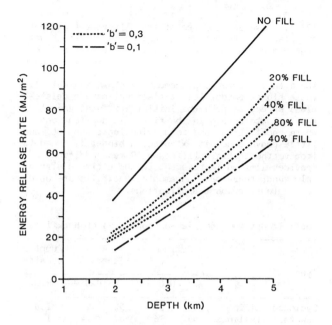

Figure 4. The influence of quantity and quality of backfill on ERR for depths of 2 to 5 km.

of the mined out area at 2 km reduces the ERR to 24 MJ/m^2. Alternatively, an ERR of 40 MJ/m^2 can be achieved at a depth of 2,8 km if 20 % of the area is backfilled, and at a depth of 3,3 km, if 80 % of the area is backfilled. Therefore, in the depth range of 2 to 3,3 km 'fair' quality backfill has the potential to replace stabilizing pillars as a means of providing regional support. Also shown in Figure 4 is the effect of improving the quality of the fill. Reducing the b value to 0,1 enables an ERR of 40 MJ/m^2 to be achieved at a depth of 3,6 km if 40 % of the area is backfilled.

At depths greater than 3,3 km (or shallower if a lower value of ERR is required) stabilizing pillars must be introduced. An analysis was carried out to determine the benefits to be derived from placing backfill in the panels between stabilizing pillars.

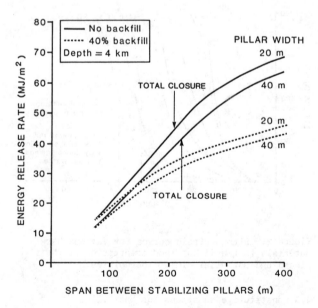

Figure 5. Relationship between ERR and pillar span for combinations of pillars and backfill at a depth of 4 km.

994

Figure 5 shows the relationship between ERR and
pillar span at a depth of 4 km, for pillars of width
20 m and 40 m, and with no backfill and backfill
covering 40 % of the area. The features of these
curves which should be noted are:
  (i) a greater reduction in ERR as a result of back-
      fill, with increasing span between the pillars·
 (ii) doubling the pillar width has little influence
      on the values of ERR; and
(iii) a smaller rate of increase in ERR with pillar
      span, once the span is sufficient for total
      closure to occur.
At the greatest span considered, namely, 400 m, the
ERR reduction as a result of backfilling is 33 % for a
depth of 4 km. Pillar spans of greater than 400 m
were not considered because the ERR values become
unacceptably high even with backfill and 40 m wide
pillars.
The second parameter which has to be considered when
designing stabilizing pillars is the average pillar
stress.

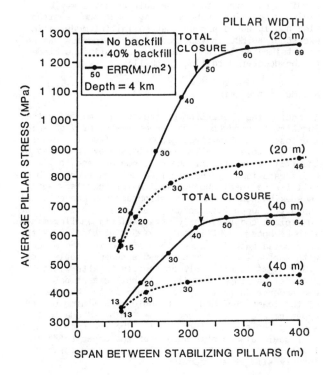

Figure 6. Relationship between APS and pillar span
for combinations of pillars and backfill at a depth of
4 km.

Figure 6 shows the variation in APS for 20 m and
40 m wide pillars at a depth of 4 km with increasing
pillar span, with no backfill and 40 % backfill. The
numbers adjacent to the curves are the ERR values for
the corresponding pillar span. The main features to
note are:
  (i) an increase in the effectiveness of backfill in
      reducing APS, as the span between pillars
      increases from 80 m to 250 m;
 (ii) a significant reduction in APS with increasing
      pillar width; and
(iii) almost constant values of APS at pillar spans of
      greater than 250 m.
At the greatest span considered, namely, 400 m, the
APS reduction as a result of backfilling is 32 % for a
depth of 4 km.

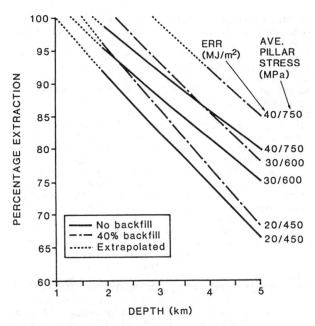

Figure 7. The influence of backfill on percentage
reef extraction with increasing depth for various
combinations of ERR and APS.

The ERR and APS results have been combined in
Figure 7 to illustrate the influence of backfill in a
stabilizing pillar layout on the quantity of additional reef that can be mined. Because the layout must be
designed to satisfy both the ERR and APS criteria, the
values have been paired. The benefit of backfill for
each combined criterion can be clearly seen. For
example, at a depth of 4 km, a stabilizing pillar layout which enables 86 % extraction of reef is required
to achieve an ERR of 40 MJ/m² and an APS of 750 MPa.
The amount of ore extracted can be increased to 92 %
if 40 % of the area is backfilled.
The theoretical studies described above and by de
Jongh (1986) show that backfill has the potential to
replace stabilizing pillars as a regional support
system over a wide, but finite, depth range. At
depths greater than 3,3 km stabilizing pillars should
be used for regional support but significant reductions in ERR and APS, or an increase in percentage
extraction, can be obtained by placing fair quality
backfill (b value = 0,3) in the panels between the
stabilizing pillars.

6 BACKFILL AS A LOCAL SUPPORT

6.1 Local stope support

Local support can be considered as that installed to
maintain the integrity of the hangingwall in the 10 m
wide working area along the stope faces, and along
access ways to the faces.
Conventional support in the face area consists of a
pattern of support units spaced between 1 m and 3 m
apart on strike and dip. They cover between 1 and
11 % of the exposed hangingwall and generate point
forces into the hangingwall. The effectiveness of the
system depends on the stiffness of the units, their
ability to yield without shedding load in response to
stope closure, the interaction between units through
fractured beams of rock and the ability of the system
to generate horizontal forces to clamp together the
fractured rock. This latter function is greatly
facilitated by the dilatation of fracturing rock ahead
of the stope face. However, the combination of shall-
owly inclined, mining-induced fractures or geological
planar weaknesses and bedding planes lead to unstable
key blocks (Jager and Turner, 1986). The dimensions
of these blocks are often less than the standard spacings between supports and are thus prone to fall out.

An initial minor fall of this nature causes a local reduction in the horizontal clamping forces which can then lead to the spread of the fall. Most falls of ground initiate either between the face and the first row of support, a distance which varies between 2,0 and 5,0 m, or between rows of support. The main advantage therefore of backfill is that it supplies total support coverage of the area in which it is placed and does not allow any fall out or reduction in clamping forces from this area. The problem of supporting the area between the face and the fill front is thus reduced but it should be emphasized that the need for face area support is not eliminated.

In non-rockburst prone areas the average support resistance generated by successful support systems is often less than 100 kN/m² or 0,1 MPa in the working area (Jager and Roberts, 1986). From this, the height of strata which could be supported as deadweight is found to be less than 4 m. This is substantiated by the fact that falls of ground seldom extend more than 1,5 m into the hangingwall. Also in areas where no back area support is used, the initial collapse is usually less than 1 m and the rock layers above this are remarkably stable. It is concluded that the support resistance required in the working area of stopes is generally low and that even under rockburst conditions the minimum support resistance can be as low as 200 kN/m² (Wagner, 1982). However this support resistance should be generated as close to the stope face as possible. Figure 8 shows the support resistance of 5 support systems required to maintain the stability of the immediate fractured hangingwall between the face and a distance of 10 m behind the face, assuming a stope closure rate of 15 mm/m of face advance.

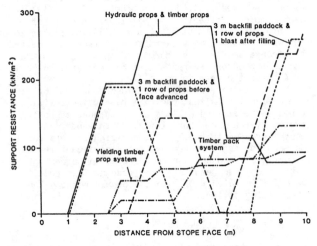

Figure 8. The variation in support resistance with distance from the face for 5 support systems. Backfill is supplemented by one row of face support.

This diagram shows that, because of low initial stiffness and some shrinkage, backfill cannot by itself supply enough support resistance in the face area and hence cannot prevent the fall out of key blocks even if it is kept within 5 m of the face. Moreover the curves depicting the support resistance of backfill combined with temporary face support show that most of the support in the first 10 m from the face is supplied by the temporary support which produces the initial build-up in support resistance. The difference in results from two mines, using different placement methods and amounts of conventional support in the area between the face and the fill, highlights the importance of the face area temporary support. In the first mine, where the fill front is allowed to lag between 3 m and 8 m from the face but where up to 3 rows of hydraulic props are used, the stoping width has been reduced by 10 %, despite above

average seismic activity in the area (Bruce and Klokow, 1986). In the second mine, where there is little seismic activity and the hangingwall is traversed by several weak parting planes, the fill is kept between 2 and 5 m from the face but only one row of props is installed adjacent to the fill. The initial results indicate that no significant overall decrease in stoping width has been attained.

There are thus three important requirements for backfill systems used for local support:
(i) the fill should be placed at as low a porosity as possible to reduce shrinkage and improve the initial stiffness;
(ii) the fill should be placed less than 8 m from the face to reduce the area from which fall-outs can occur and to develop support resistance and horizontal clamping forces as close to the face as possible; and
(iii) the area between the fill front and the face should be well supported by means of temporary support.

In some situations where very poor hangingwall conditions exist and where mining with conventional support is extremely difficult and costly, backfill with a much higher initial stiffness than is normally required may prove to be the only economic way of mining. One mine is backfilling under such a situation and it would appear that an initial modulus in excess of 40 MPa is required. Such a modulus can only be attained by the addition of a cementitious binder to the dewatered tailings backfill.

6.2 Multi-reef mining

An increasing proportion of the tonnage mined by the industry is produced from second reefs, many of which lie less than 60 m stratigraphically from the major reef. In this situation the main support function of the fill is to prevent the collapse of, and to limit bed separation in, the middling between stopes. Backfill is the ideal support medium for this task as it is work hardening and much stronger than conventional support. Support resistances of only 1 to 2 MN/m² are sufficient and are easily attainable in shallow mines with cemented backfill, after < 5 % strain, and with uncemented backfill after 15 % strain in deep mines with narrow reefs. By contrast a support resistance > 0,5 MN/m² is difficult and costly to attain with conventional support units. In both situations any bed separation that may take place in the hangingwall of the lower reef and which would make mining on the upper reef difficult, should be inhibited due to the support resistance of the fill. The presence of backfill could also enable mining of the upper reefs in low to moderate compressive field stresses instead of in a tensile stress zone.

6.3 High stoping widths

The main problems associated with support of wide stopes are:
(i) the reduction in stiffness of the longer support units leading to a lower support resistance close to the face;
(ii) the need for larger packs to maintain an adequate width to height ratio, and
(iii) difficulty in installation of tall support units particularly in stopes where the dip is greater than 20 degrees.

Both (i) and (ii) lead to the requirement of a greater support density and hence increased cost. These problems are inherent in most conventional support systems and little can be done to improve the situation. The stiffness of backfill can be improved in a number of ways, the simplest of which is the addition of a cementitious binder. It appears that backfill should prove to be the best option for support of wide stopes once the problem of shrinkage has been solved.

6.4 Rockbursts

Damage to the stopes caused by rockbursts consists of:
(i) falls of ground from the hangingwall. Smaller falls usually occur between rows of support whereas in major falls large areas of the stope collapse, breaking or dislodging support units. In the process large blocks of rock may fall or the rock can be highly fragmented;
(ii) sudden stope closure which may render access to the working area impossible. The dominant movement may be from either the footwall or the hangingwall, and
(iii) rock expelled from the stope face.

Backfilling obviously eliminates falls of ground from the areas where fill has been placed with resultant improvement in overall stability of the stope. Its influence on falls of ground in the area between the fill front and the face has not been fully ascertained even though several rockburst sites in backfilled areas have been investigated. The observations made at these sites indicate that face falls of ground occur if the fill front is more than 4 m from the stope face but do not occur if the fill front is closer than 4 m from the face. There was no indication that the energy of the rockburst was concentrated in the face area resulting in excessive damage as was hypothesized by Curtis (1984).

A marked benefit of backfilling has been the reduction in damage to stope gullies after a rockburst. Damage has been noted in these gullies only where particularly adverse geological conditions existed and the damage was less than would be expected with conventional support. However, those sections of gullies which are in advance of the backfill remain susceptible to rockburst damage.

The concern that partially or totally saturated backfill may liquefy or extrude into gullies and the face area as a result of rapid stope closure associated with rockbursts, has been allayed to a large extent. Several rockbursts causing stope closure of up to 300 mm have occurred in backfilled stopes and although minor damage to backfill barricades has been noted in some cases, only once has fill extruded into the gully. On this occasion the fill had been placed with a higher content of fines than was normal. (Klokow 1986).

The effectiveness of backfill in reducing rockburst induced stope closure in the working area is difficult to estimate mainly because of the fractured nature of the rock. However, the relative ability of backfill and other support systems to counteract the kinetic energy imparted to the rock by seismic ground motion can be calculated. Table 2 summarizes the results of such a calculation assuming a rapid stope closure of 300 mm for four of the support systems considered in Figure 8. The orders of magnitude superiority of backfill in absorbing seismically generated energy is a striking indication of the potential of backfill to reduce the effects of rockbursts.

It is therefore concluded that a major potential benefit of backfilling will be a reduction in damage caused by rockbursts and hence an overall improvement in safety and productivity.

Table 2. Average work done by four support systems between the stope face and 15 m back, during rapid convergence of 300 mm.

Support Type	Spacing	Work Done/m^2 kJ/m^2
Yielding Timber Prop	2m x 1,5m	15
1,1 m x 1,1 m Timber Pack	3m x 3m centres	43
Timber Props + Hydraulic Props	2m x 1,5m + 1m x 1,5m	39
Backfill	66 % coverage	19 000

7 AN INITIAL ASSESSMENT OF THE PERFORMANCE OF BACK-FILLING

Although the amount of continuous backfilling that has taken place in the gold mines is limited there are strong indications that significant strata control benefits will accrue when backfilling is introduced on a large scale. Improvements noted so far on various mines are summarized below:
(i) all mines using backfill have experienced an improvement in the conditions of access ways in stopes. (Joughin and Jager 1978, Bruce and Klokow 1986, Faure 1986);
(ii) two mines have recorded a reduction in stoping width, and hence in dilution of reef, in excess of 10 % (Bruce and Klokow 1986, Faure 1986);
(iii) measurements (up to 40 m behind the face) have shown that in certain instances the amount of inelastic deformation is reduced in backfilled stopes (Gay et al 1986):
(iv) a reduction in damage to the stope face area caused by rockbursts is evident where the backfill is kept close to the stope face;
(v) a decrease in accidents due to falls of ground of up to 50 % has been recorded since the implementation of backfilling (Faure 1986)·
(vi) at one site where approximately 200 m of backfill had been placed on strike an average reduction in ERR of 37 % compared to the non-backfilled situation was attained. A similar reduction in ERR could have been produced by leaving a stabilizing pillar which would have locked up 0,9 metric tons of gold valued at $11 million (Bruce and Klokow 1986)·
(vii) on some mines an increase in face advance per month and, hence, productivity has been attained (Bruce and Klokow 1986);
(viii) no deleterious effects have been experienced in off-reef tunnels passing above or below backfilled stopes.

8 CONCLUSIONS

From the above discussion, it is evident that backfill, if used effectively, has major advantages over currently used stope support methods. Backfill is a versatile support medium and can be used for several regional and local support applications. The choice, of which of the four backfill systems in use at present, is likely to be most suitable depends on the particular requirements of the support and on its cost effectiveness. Quantification of the benefits of the various systems is on-going and it will be some time before a realistic assessment of the cost effectiveness of the systems is obtained.

The option to fill only 40 % of an area to obtain a major portion of the regional support benefit seems only correct when the cost of the fill is high or the supply of backfill material is limited. In general, the significant additional local support benefits brought about by doubling the quantity of fill placed would justify the extra filling.

Rockbursts are a major hazard on some mines. Backfill can alleviate this problem by providing regional support, which would tend to reduce the number of significant seismic events, and by reducing the damage caused by rockbursts in the stope area. Moreover, it may be possible to relax the established ERR guidelines for regional support systems, based on the assessment of the likelihood of the occurrence of rockbursts, as a result of the improved conditions brought about by backfilling.

Mechanization of stoping is receiving increasing attention on the mines, in particular the transport of ore from the face area by LHD's. This implies that access ways have to be kept open and in good condition for longer time periods. A universal result of backfilling has been an improvement in gully conditions. Thus backfill seems to be the ideal support for this type of mining.

During this introductory phase of backfilling three important facts have emerged:
- (i) backfill should be placed at as low a porosity as possible;
- (ii) backfill should be placed as close to the face as possible and no more than 8 m from the face
- (iii) backfill should be integrated with a good temporary face support system and should be part of a standard mining cycle.

Future improvements in backfilling will come from further developments in these areas and in reducing costs of paddock construction, reducing the amount of conventional support used, and eliminating shrinkage. The cost effectiveness of using cementitious binders and tailings - aggregate mixtures also needs to be assessed.

Finally, it can be concluded that backfill has the potential for significantly improving the safety and productivity of deep gold mines.

ACKNOWLEDGEMENTS

The work described in this paper forms part of the research programme of the Chamber of Mines of South Africa. The co-operation of management and staff of the mines using backfill, and their permission to publish the paper, is gratefully acknowledged.

REFERENCES

Briggs, D.J. 1986. The load bearing properties of fill materials (uncemented fills). Proc. Backfilling Sym., Johannesburg, Assoc. Mine Managers S.Afr., in press.

Bruce, M.F.G. & J.W. Klokow 1986. A follow-up paper on the development of the West Driefontein Tailings backfill project. Proc. Backfilling Sym., Johannesburg, Assoc. Mine Managers, S.Afr., in press.

Cook, N.G.W., E. Hoek, J.P.G. Pretorius, W.D. Ortlepp, M.D.G. Salamon 1966. Rock mechanics applied to the study of rockbursts, Journal of the South African Institute of Mining and Metallurgy, Vol. 66, p.436-528.

Cook, N.G.W. & M.D.G. Salamon 1966. The use of pillars for stope support, Report to the Chamber of Mines of South Africa.

Curtis, J.F. 1984. Assoc. Mine Managers, S.Afr., Circular No. 3/84, Johannesburg.

de Jongh, C.L. 1986. The potential of backfill as a stope support in deep gold mines. GOLD 100 Proc. Int. Conf. on Gold, Vol. 1, Johannesburg, SAIMM, p.271-287.

Faure, M. 1986. Experimental backfilling at Harmony gold mine. Proc. Backfilling Sym., Johannesburg, Assoc. Mine Managers, S.Afr., in press.

Gay, N.C., A.J. Jager & P.S. Piper 1986. Quantitative evaluation of fill performance in South African gold mines. Proc. Backfilling Sym., Johannesburg, Assoc. Mine Managers, S.Afr., in press.

Hodgson, K. & N.C. Joughin 1967. The relationship between energy release rate, damage and seismicity in deep mines. Proc. of 8th Symposium on Rock Mechanics, University of Minnesota, C. Fairhurst. ed., American Institute of Mining Engineers, New York, p.194-209.

Jager, A.J. & M.K.C. Roberts 1986. Support systems in productive excavations. GOLD 100 Proc. Int. Conf. on Gold, Vol. 1, Johannesburg, p.289-300.

Jager, A.J. & P.A. Turner 1986. The influence of geological features and rock fracturing on mechanized mining systems in South African gold mines. GOLD 100 Proc. Int. Conf. on Gold, Vol. 1, Johannesburg, p.80-103.

Joughin, N.C. & A.J. Jager 1978. Solid waste packing as a support medium at depth. J.S. Afr. Inst. Min. Metall., Vol. 79, p.10-14.

Klokow, J.W. 1986. Personal Communication.

Napier, J.A.L. 1985. MINSIM-D User's Guide (Phase II), Unpublished Research Report, Chamber of Mines of South Africa.

Piper, P.S. & J.A. Ryder 1986. An assessment of backfill for regional support in deep mines. Proc. Backfilling Sym., Johannesburg, Assoc. Mine Managers S.Afr., in press.

Ryder, J.A. & H. Wagner 1978. 2D analysis of backfill as a means of reducing Energy Release Rates at Depth. Unpublished Research Report, Chamber of Mines of South Africa.

Salamon, M.D.G. & H. Wagner 1979. Role of stabilizing pillars in the alleviation of rockburst hazard in deep mines. Proc. 4th Int. Cong. I.S.R.M. Vol. 2, Montreux, p.561-566.

Salamon, M.D.G. 1983. Rockburst hazard and the fight for its alleviation in South African gold mines. Rockbursts : prediction and control. Institution of Mining and Metallurgy, London, p.11-36.

Stewart, J.M., I.H. Clark & A.N. Morris 1986. Assessment of fill quality as a basis for selecting and developing optimal backfill systems for South African gold mines. GOLD 100 Proc. Int. Conf. on Gold, Vol. 1, Johannesburg, p.255-270.

Wagner, H. 1986. Rock pressure problems in deep gold mines. GOLD 100 Proc. Int. Conf. on Gold, Vol. 1, Johannesburg, SAIMM, p.229-243.

Wagner, H. & E.H.R. Schumann 1971. The stamp-load bearing strength of rock - an experimental and theoretical investigation, Rock Mechanics, Vol. 3, p.186-207.

Wagner, H. 1982. Support requirements for rockburst conditions, Proc. 1st Int. Cong. on Rockbursts and Seismicity in Mines Ed. Gay, N.C. and Wainwright, E.H., Johannesburg, SAIMM, p.209-218.

Apports de la mécanique des roches à la maîtrise des phénomènes dynamiques dans les mines

Contribution of rock mechanics to the understanding of dynamic phenomena in mines
Der Beitrag der Felsmechanik zum Verständnis von dynamischen Erscheinungen in Bergwerken

J.P.JOSIEN, Centre d'études et recherches de charbonnages de France, BP
J.P.PIGUET, Laboratoire de mécanique des terrains, École des mines, Nancy, France
R.REVALOR, Centre d'études et recherches de charbonnages de France, École des mines, Nancy

ABSTRACT : Dynamic phenomena occur as the result of sudden and violent ruptures of the rock mass surrounding mine cavities. The management of these phenomena has been the subject of extensive studies in French coalfields over the past 10 years. This report summarizes the results of our research to date. The findings allow the definition of the natural and mining environments which are characteristic of the regions in which these phenomena occur. They also indicate the mechanisms of the ruptures which give rise to the phenomena. A knowledge of the environment and an understanding of these mechanisms are indispensable prerequisites for the deployment of appropriate forecasting and preventive measures.

RESUME : Les phénomènes dynamiques se manifestent par des ruptures soudaines et violentes du massif rocheux autour des cavités minières. Au cours des dix dernières années, des études importantes ont été conduites dans les mines de charbon françaises pour maîtriser ces phénomènes. L'article rassemble les résultats actuels de ces recherches. Ils permettent, d'une part de préciser l'environnement naturel et d'exploitation qui caracté rise les zones où ces phénomènes se manifestent. Ils débouchent d'autre part sur les mécanismes de rupture qui sont à l'origine de ces phénomènes. La connaissance de l'environnement et la compréhension des mécanismes sont un préalable indispensable à la mise en oeuvre au chantier des méthodes de prévision et de prévention adaptées.

ZUSAMMENFASSUNG : Die dynamischen Erscheinungen treten auf bei plötzlichem und heftigem Einbruch von Gestein rund um Bergwerks-Hohlräume. Im Verlauf der letzten 10 Jahre wurden in den französischen Steinkohlenbergwerken bedeutende Untersuchungen vorgenommen, um diesen Erscheinungen Herr zu werden. Der Aufsatz fasst die Ergeb nisse dieser Untersuchungen auf dem heutigen Stand zusammen. Diese erlauben, zum einen, die natürliche Umwelt und die Abbauumgebung zu bestimmen, die diese Zonen kennzeichnen, in welchen diese Erscheinungen auftreten ; zum anderen werden die Einbruchsvorgänge untersucht, die den Ursprung dieser Erscheinungen darstellen. Die Kenntnis der Umwelt und das Verstehen der Vorgänge sind eine unerlässliche Voraussetzung für den Einsatz im Abbaubetrieb von angemessenen Methoden zur Vorbeugung und Verhütung.

Le terme de "phénomènes dynamiques" s'applique à toute une série de manifestations, à caractère fondamentalement discontinu, et qui se matérialisent par des ruptures brutales du massif rocheux, localisées à proximité des cavités minières.

Dans les mines de charbon françaises, ces phénomènes, connus depuis de nombreuses années, ont subi une évolution caractéristique dans la dernière décennie : recrudescence et apparition dans des chantiers jusqu'alors préservés, mise en évidence de nouveaux types jusqu'alors inconnus.

Cette évolution, la violence des effets observés et donc l'incidence de ces manifestations sur la sécurité et le rendement des chantiers expliquent les efforts de recherche importants consentis pour les maîtriser.

Les études engagées dans ce domaine se sont fortement appuyées sur la Mécanique des Roches et cet article synthétise l'essentiel des résultats obtenus. Après une description des phénomènes observés, de leurs effets et de l'environnement dans lequel ils se sont placés, une analyse comparative des différents facteurs permet de préciser l'ensemble des conditions dont la conjonction favorise leur apparition.

Cette démarche débouche sur une présentation des mécanismes qui apparaissent aptes à déclencher ces phénomènes et à expliquer la brutalité des ruptures observées.

1 – LES PHENOMENES OBSERVES – LEUR ENVIRONNEMENT

Les phénomènes dynamiques observés au cours des dix dernières années dans les mines de charbon françaises, peuvent se regrouper en deux catégories :

* les coups de terrain ("coal bumps" de la terminologie anglosaxonne) qui affectent les couches de charbon au voisinage des chantiers exploités par longues tailles.

* les "surtirs" ("outbursts with rock and gas") qui se manifestent dans les creusements de galeries au rocher.

Les coups de terrain

Malgré la diversité des configurations dans lesquelles ils surviennent, de grandes analogies se retrouvent dans les descriptions (JOSIEN, 1980 et 1981). Les coups de terrain consistent en une expulsion brutale de la couche exploitée (coup de couche) accompagnée d'un ebranlement violent qui peut être ressenti jusqu'en surface (figure 1). Quelle que soit la lithologie de la veine de charbon, c'est souvent sa partie supérieure, immédiatement au contact du toit qui est expulsée. Le volume projeté est très variable, de quelques m^3 à 300 m^3 pour les plus importants. Le toit reste apparemment intact et sa descente est limitée à quelques centimètres. Sa surface est striée de marques brunâtres, témoins du frottement qui a eu lieu lors de l'expulsion. Le phénomène peut être accompagné par un soufflage du mur de la couche mais en aucun cas par un dégagement grisouteux important.

Les zones affectées se localisent en bordure des vides exploités au niveau des ouvrages "ouverts" dans la couche de charbon, fronts de taille ou parements de voies. Suivant l'environnement, deux situations se dessinent :

* localisation au niveau du carrefour taille-voie, sur une trentaine de mètres (figure 1)

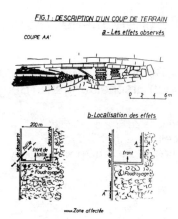

FIG.1 : DESCRIPTION D'UN COUP DE TERRAIN

a - Les effets observés

COUPE AA'

0 2 4 6m

b-Localisation des effets

200 m

front de taille

Foudroyage

voie de desserte

front

Foudroyage

~~~Zone affectée

* localisation dans les voies d'accompagnement, en avant ou en arrière du front de taille (figure 1) ou dans les voies de desserte du secteur (figure 6).

Dans ce cas, de grandes longueurs peuvent être affectées (supérieures à 100 m) et le soutènement porteur de la voie est chassé ou renversé et même écrasé si le déplacement n'est pas possible (figure 6).

Les veines de charbon concernées sont soit des "plateures" (pendage inférieur à 15°) exploitées par longues tailles foudroyées ou des "semi-dressants" (pendage compris entre 15 et 45°) exploitées par tailles remblayées. Dans ce cas, la veine est la première exploitée du secteur (veine de détente) et les effets se localisent en début ou en fin d'exploitation du panneau (figure 6).

Actuellement, la profondeur à laquelle ces phénomènes sont apparus varie de 600 à 1000 m.

Les surtirs

Les ouvrages concernés sont des galeries cintrées de 20 m^2 de section, creusées à l'explosif dans des grès et situées à plus de 1000 m de profondeur. Le phénomène se manifeste par des ruptures de la masse rocheuse en avant et autour du front de creusement, sur une dizaine de mètres de profondeur. Elles affectent un volume nettement supérieur à celui normalement abattu par le tir, d'où l'appellation "surtir" qui leur a été donnée (figure 2). Le massif est découpé en plaquettes de quelques centimètres d'épaisseur, sensiblement parallèles à la surface de dégagement. Le phénomène est accompagné par une émission significative de grisou. Pour les plus importants, les chiffres suivants peuvent être avancés (CHEMAOU, 1984) : fracturation de 1500 à 3000 m^3 de terrains dont 500 m^3 sont projetés dans la galerie, volume de gaz émis jusqu'à 8000 m^3.

Tous les phénomènes observés sont survenus immédiatement après un tir d'abattage et ont suivi précisément l'explosion du dernier retard comme l'a montré un enregistrement sismoacoustique réalisé lors d'un surtir.

FIG.2 DESCRIPTION SCHEMATIQUE D'UN SUR-TIR

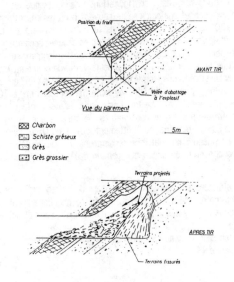

Position du front

AVANT TIR

Volée d'abattage à l'explosif

Vue du parement

▨ Charbon
▦ Schiste gréseux
▢ Grès
▨ Grès grossier

5m

Terrains projetés

APRES TIR

Terrains fissurés

2 - LES CONDITIONS A REUNIR

L'apparition de phénomènes dynamiques est généralement associée à la réunion de conditions particulières qui sont :
* l'état de contraintes in situ
* la nature des terrains (propriétés intrinsèques des roches)
* la lithologie et la tectonique.

Enfin, il faut ajouter la présence éventuelle, dans les terrains, de gaz et les modalités de cette présence.

Nous allons rassembler les observations et les résultats obtenus pour ces différents points.

2.1 - Les contraintes in situ

Des méthodes de mesures adaptées aux mines grisouteuses ont été développées pour évaluer le tenseur complet de contraintes originel. Il s'agit successivement des méthodes par vérin plat, par surcarottage et par fracturation hydraulique. Les difficultés de mise en oeuvre de ces méthodes et d'interprétation de leurs résultats ont fait l'objet de plusieurs publications (PIGUET et GEORGE, 1981 ; REVALOR, 1985 ; REVALOR et al, 1985 ; CHAMBON et REVALOR, 1985).

Elle sont actuellement bien maitrisées et leurs avantages et inconvénients respectifs ainsi que leur domaine d'emploi ont été correctement évalués (REVALOR et al, 1985).

Le tableau 1 réunit l'ensemble des campagnes de mesures réalisées au moyen des différentes méthodes dans les mines sujettes à phénomènes dynamiques. Pour ne pas alourdir la présentation, les directions principales des contraintes ne sont pas indiquées mais on retiendra qu'elles sont très généralement sub-verticales et sub-horizontales. Les directions horizontales sont très nettement liées à la structure géologique des gisements en particulier dans les zones plissées où on les trouve perpendiculaires et parallèles aux axes des plis (THOMAS et al, 1983 ; GAVIGLIO, 1985). Les traits structuraux peuvent également induire des écarts de 10 à 20° par rapport à la verticale de l'une des directions principales. Les valeurs reportées dans le tableau fournissent le rapport de la contrainte verticale σv sur la contrainte correspondant au poids du recouvrement pour chaque site (ρ g H, H étant la profondeur et ρ la masse volumique, prise ici constante et égale à 2500 kg/m^3). Les contraintes horizontales, quant à elles, sont ramenées à la valeur de σv. On voit de cette façon apparaître le résultat remarquable suivant : la contrainte principale majeure, dans tous les sites voisins de secteurs ayant donné lieu à phénomènes dynamiques, est horizontale et en moyenne deux fois plus importante que la contrainte verticale. Cette situation ne se rencontre que rarement et de toutes façons moins nettement dans les secteurs épargnés par les phénomènes dynamiques (surtout si on pondère les résultats du tableau par le fait que la méthode du vérin plat a donné des résultats beaucoup moins précis et fiables, en particulier lors des campagnes de mise au point sur les sites 1, 2, 3 et 8).

On remarque également sur ce même tableau que la contrainte verticale est dans la majorité des sites inférieure à la valeur ρ g H prévisible, plus particulièrement dans les secteurs sensibles aux phénomènes dynamiques.

2.2 - La nature des terrains

Une caractérisation géomécanique classique (mesure de résistance à la compression simple Rc, à la traction Rt, limite élastique Re, module d'Young E) a été conduite sur les roches ayant été le siège de phénomènes : charbons, grès, calcaires, ainsi que sur des roches de référence.

Elle a été complétée par la recherche d'indices particuliers supposés caractéristiques de l'aptitude des roches à donner lieu à ces phénomènes : coefficients représentatifs du bilan énergétique lors de cycles de chargement et déchargement d'éprouvettes - dit indice Wet - (KIDYBINSKI, 1981) ; coefficient représentatif du caractère fragile de la rupture comme par exemple le rapport des pentes des courbes effort-déformations après et avant rupture, etc... (HOMAND et al, 1986).

Les essais ont porté au total sur plus de 1000 éprouvettes de grès, environ 200 éprouvettes de charbon et plusieurs dizaines d'éprouvettes de roches de références comme le granite ou le calcaire (CHEMAOU, 1984 ; DA GAMA, 1986 ; GAVIGLIO, 1985).

N°	SITE	H en m	σv/pgH	σHmin/σv	σHmax/σv	METHODE	N°	SITE	H en m	σv/pgH	σHmin/σv	σHmax/σv	METHODE
	SECTEURS EXEMPTS DE PHENOMENES							**PROXIMITE DE SECTEURS AVEC PHENOMENES**					
1	Vouters	747	0,7	1,11	1,41	Vérin plat	8	Vouters	1250	0,58	1,1	1	Vérin plat
2	Vouters Nord	1036	1,35	0,54	0,57	Vérin plat	9	Vouters	1250	0,83	0,65	1,38	Surcar.
3	Vouters Sud	1036	0,68	0,88	1,53	Vérin plat	10	Vouters	1250	0,93	0,61	1,28	Fract. hydr.
4	Simon F1	850	1,08	0,28	0,63	Surcar.	11	Simon F1	1050	0,62	1,77	2,1	Fract. hydr.
5	Simon F1	850	1,13	0,34	0,62	Fract. hydr.	12	Etoile	700	0,39	0,23	2,38	Vérin plat
6	Ste Victoire	390	0,64	0,47	1	Vérin plat	13	Verdillon	550	0,71	0,50	1,94	Vérin plat
7	Ste Victoire	390	0,82	0,85	0,96	Vérin plat	14	Etoile	700	0,63	1,44	2,71	Fract. hydr.
							15	Etoile Sud	900	0,53	1,43	3,36	Fract. hydr.
MOYENNE			0,91	0,64	0,96		MOYENNE			0,65	0,97	2,02	

TABLEAU 1 : INFLUENCE DE L'ETAT DE CONTRAINTES NATUREL DANS LES TERRAINS

Enfin, des matériels et des procédures d'essais spécifiques ont été développés pour l'étude des roches sujettes à ces phénomènes : une cellule triaxiale permettant de réaliser un minicarottage dans des échantillons soumis à de très forts niveaux de contraintes (150 MPa), (MOUDAFI, 1986) ; une enceinte étanche contenant une éprouvette saturée par un gaz sous pression élevée (10 MPa) équipée pour déclencher une détente très rapide de ce gaz.

Ces appareillages ont permis de reproduire en laboratoire certains des comportements des roches observés in situ (disquage dans les sondages carottés ou réactivité de trous forés en destructif).

Pour résumer les résultats obtenus, on peut souligner les points suivants :

* les charbons des veines ayant donné lieu à des coups de terrains sont généralement plus résistants (Rc jusqu'à 46 MPa), plus raides (E jusqu'à 4000 MPa), plus fragiles. Certains des indices de susceptibilité, bien que statistiquement très dispersés, permettent de distinguer ces veines par des valeurs plus élevées. C'est le cas de l'indice Wet et du coefficient (Rc – Rt)/(Rc + Rt), représentatif de la pente de la courbe intrinsèque, au voisinage de l'origine (HOMAND et al, 1986)

* les grès ayant été le siège de phénomènes dynamiques (type surtirs) comparés à des grès de même faciès et de mêmes horizons stratigraphiques, apparaissent au contraire moins résistants (de l'ordre de 40 MPa au lieu de 110 à 140 MPa), moins raides (E = 20000 MPa au lieu de 25000 MPa), plus poreux (7 à 11 % au lieu de 1 à 2 %). En même temps, un indice tel que le Wet est nettement plus élevé (6 à 12 contre 2 à 5 pour les grès épargnés par les surtirs).

* les calcaires du recouvrement de certaines veines paraissent quant à eux plus résistants et moins poreux lorsqu'ils sont prélevés à proximité des grands accidents tectoniques, qui sont aussi les zones où se localisent fréquemment les coups de terrains (GAVIGLIO, 1985).

Ces particularités des propriétés sont généralement très accentuées par une forte hétérogénéité des terrains ou des veines.

De tous les essais est ressortie l'idée que les roches qui sont le siège de phénomènes dynamiques se singularisent systématiquement quand on les compare avec des roches de nature et de texture quasi identiques, prélevées dans des secteurs (souvent peu éloignés) exempts de phénomènes.

En outre, ces singularités ne sont pas les mêmes selon le type de phénomène considéré.

2.3 La lithologie et la tectonique

Tous les sites sujets à coups de terrain présentent des caractères lithologiques propres : le haut toit reste toujours massif et plutôt résistant tandis qu'un toit immédiat raide et résistant ou au contraire plus déformable influence la forme prise par les manifestations.

L'importance des contrastes de déformabilité dans le toit et les possibilités de déplacements au niveau des joints stratigraphiques ont été analysés dans un cas présenté plus loin.

Les cas de surtirs se sont toujours produits dans un contexte particulier de tectonique plissante (flexure importante et brutale dans un pli avec présence de micro plis d'entraînement, ou encore fermeture d'anticlinal).

Enfin, la tectonique cassante peut conduire à des configurations de failles découpant des polyèdres de dimensions importantes dans le haut toit sous cavés par les travaux d'exploitations et dont les déplacements en masse peuvent être à l'origine de certains phénomènes (figure 5).

3 – APPROCHE DES MECANISMES

Par rapport à l'ensemble des manifestations de ruptures observées autour des vides souterrains, une approche des mécanismes susceptibles d'initialiser "les phénomènes dynamiques" doit mettre l'accent sur l'aspect essentiel de ces phénomènes, à savoir le caractère brutal des ruptures observées.

Malgré des analogies concernant les conditions de formation (cf chapitre précédent), les deux grands types de phénomènes, les "surtirs" et les coups de terrain, sont envisagés séparément, compte-tenu des environnements d'exploitation radicalement différents dans lesquels ils se placent.

3.1 – Les surtirs

Les "surtirs" affectent des ouvrages isolés (galeries en creusement). Les conditions nécessaires à leur déclenchement sont réunies dans un volume limité autour de l'excavation.

Nous avons vu que les zones concernées par les surtirs sont soumises à grande échelle, à un tenseur de contrainte fortement anisotrope, avec des niveaux de contraintes élevés accentués par des anomalies micro tectoniques locales.

La galerie étant creusée à peu près parallèlement à la contrainte principale majeure, un calcul analytique simple (REVALOR et al, 1985) montre que des contraintes de traction se mettent en place juste en avant du front de creusement.

Elles disparaissent dès que la direction du creusement fait un angle de plus de 15° avec la contrainte principale majeure s1 (figure 3). L'existence de ces contraintes de traction appliquées à des roches localement moins résistantes (altérées) constitue un premier élément du mécanisme recherché.

Un deuxième élément a trait à la pression du gaz, estimée in situ à 4 MPa, et qui, dans ces roches poreuses et très imperméables sous contraintes, augmente encore les sollicitations en traction.

Il en résulte au moment de l'abattage un fort gradient de pression de gaz, situé juste en avant du front de creusement (JOSIEN et REVALOR, 1983).

1001

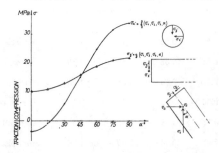

FIG. 3 - CONTRAINTES A FRONT DE LA GALERIE EN FONCTION DE
SON ORIENTATION PAR RAPPORT AU TENSEUR DE CONTRAINTES NATUREL

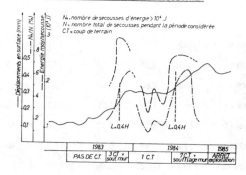

FIG. 4 - RELATION ENTRE LE COMPORTEMENT DU HAUT TOIT
ET LES COUPS DE TERRAIN AU VOISINAGE DU CHANTIER

Si la conjonction de ces différents éléments permet d'expliquer les ruptures observées, leur brutalité semble liée à deux conditions essentielles :

* l'existence de contraintes de traction met l'accent sur un type de ruptures qui se manifeste généralement violemment (ruptures des roches dans le domaine fragile)

* compte-tenu de l'abattage à l'explosif pratiqué dans ces chantiers, ces contraintes de traction sont elles-mêmes mises en place brutalement au niveau du front de creusement. Cette "brutalité" de l'abattage semble d'ailleurs constituer un élément prépondérant dans le déclenchement puisque les "surtirs" intempestifs (survenant en dehors d'un tir) sont très rares (ASTON et CAIN, 1986 ; JOSIEN et REVALOR, 1983).

3.2 - Les coups de terrains

Les coups de terrains observés affectent des ouvrages situés à proximité de chantiers exploités par longues tailles et, dans ce cas, l'analyse des mécanismes doit être conduite à une échelle radicalement différente.

En effet, compte-tenu de la profondeur et des dimensions des vides créés, ce type d'exploitation perturbe l'état d'équilibre des terrains dans un volume (le volume d'influence) très important puisque certains effets, les affaissements, sont ressentis jusqu'en surface. Dans ce cadre, tous les phénomènes qui surviennent au voisinage de ces chantiers, et en particulier les coups de terrain, doivent être intégrés dans un modèle global de comportement des terrains.

Les observations que nous avons pu conduire dans différents chantiers nous amènent à concevoir les effets observés (expulsion de la couche, soufflage du mur) comme une réaction à une rupture qui s'initialise dans le recouvrement des exploitations. Ces deux aspects du mécanisme justifient une analyse à deux niveaux.

a) Ruptures dans le recouvrement

L'ordre de grandeur des énergies libérées au cours de certains phénomènes permet déjà de suspecter la participation du recouvrement au déclenchement. En effet, ces énergies, évaluées à partir des magnitudes des évènements sismiques associés, sont beaucoup plus importantes (100 à 10000 fois) que celles que l'on peut déduire du volume de terrain affecté au niveau du chantier. En outre, cette participation a pu être mise en évidence sans ambiguité dans certaines configurations. La figure 4 en est un exemple. Les périodes à coups de terrain (fin 1983 et 1984) sont relativement bien corrélées avec des accélérations des déformations mesurées en surface qui traduisent une intensification des ruptures des bancs du haut toit. Ces ruptures, qui libèrent une forte énergie, sont repérées sur la chaine d'enregistrement sismoacoustique par une augmentation de l'énergie libérée par secousse.

Le comportement des terrains observé a pu être associé à une largeur exploitée de l'ordre de 0.4 H (H = profondeur) qui apparaît "critique" à travers la configuration tectonique du secteur (assemblage de failles délimitant "un coin auto desserrant" à l'aplomb du quartier exploité comme le montre la figure 5). Dans ce cas, le phénomène trouve son origine dans des ruptures de porte à faux dans le haut toit, soit par glissement le long des failles, soit par cisaillement à l'aplomb des bords fermes.

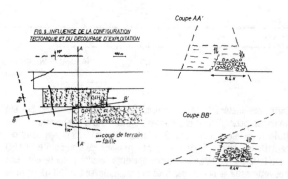

FIG. 5 - INFLUENCE DE LA CONFIGURATION
TECTONIQUE ET DU DECOUPAGE D'EXPLOITATION

Ce mécanisme de rupture n'est pas unique. Une autre conception a pu être formulée à partir d'une modélisation par éléments finis (*) de la configuration d'exploitation représentée sur la figure 6. Dans ce cas, la rupture des bancs du toit est la conséquence :

- des fortes contraintes normales aux épontes qui existent à l'aplomb du pilier ("effet pilier")
- des contraintes de traction qui apparaissent parallèlement aux épontes et qui sont liées à l'existence à ce niveau d'une hétérogénéité dans la lithologie du toit (veine de charbon enserrée par des bancs plus raides). La présence de cette hétérogénéité, donc des contraintes de traction induites, semble déterminante : en effet, des configurations d'exploitation analogues mais avec des toits beaucoup plus homogènes ont été traversées sans problème par le passé.

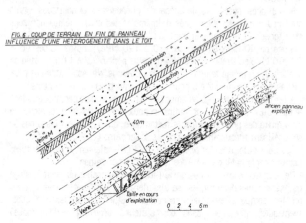

FIG. 6 - COUP DE TERRAIN EN FIN DE PANNEAU
INFLUENCE D'UNE HETEROGENEITE DANS LE TOIT

Ces deux mécanismes de rupture (cisaillement et traction-compression correspondent respectivement aux phénomènes type coup de charge et coup de toit, décrits par LABASSE (1973).

Ils conduisent à la libération d'une quantité d'énergie sous forme de déplacement d'ensemble vertical du haut toit dans le cas des coups de charge ou sous forme d'ondes sismiques dans le cas des coups de toit.

(*) Code de calcul ELFI3F développé par le Laboratoire de Mécanique des Terrains - Ecole des Mines de Nancy.

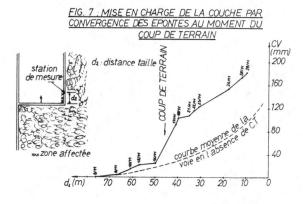

FIG. 7 : MISE EN CHARGE DE LA COUCHE PAR CONVERGENCE DES EPONTES AU MOMENT DU COUP DE TERRAIN

L'énergie libérée est d'autant plus importante :

- que le haut toit est massif, constitué de bancs épais, raides par nature et non détendus par une exploitation préalable, ce qui est une caractéristique commune relevée dans différents gisements affectés par ces phénomènes (JOSIEN et al, 1982).

- que les contraintes naturelles horizontales sont élevées ce qui a pour effet de maintenir fermées les fissures naturelles et permet à ces bancs d'atteindre des portées importantes sans se rompre.

b) Réaction de la couche

Quel que soit le mécanisme invoqué, les ruptures du haut toit provoquent une sollicitation dynamique de la couche de charbon. Dans le cas d'une configuration à coups de charge (figure 5), cette sollicitation se traduit par une mise en charge brutale de la couche provoquée par la convergence des épontes (figure 7). Dans le cas d'une configuration à coups de toit, nous ne disposons pas de mesures et observations analogues. Toutefois, on peut penser que l'énergie sismique libérée retentit sur la couche dont la mise en vibrations peut être amplifiée par l'effet d'ondes guidées.

Dans les deux cas, l'apport d'énergie est utilisé pour fracturer la couche, vaincre les frottements avec ses épontes et projeter les fragments résultants. Certaines conditions apparaissent nécessaires pour que ce mécanisme conduise par voie de conséquence à une expulsion violente de la couche.

La raideur et la résistance élevées du bas toit sont des éléments essentiels qui jouent à deux niveaux :

- transmission intégrale sans amortissement de la sollicitation imposée par le haut toit à la partie supérieure de la veine de charbon où se focalisent les effets

- existence à l'interface entre le toit et la veine d'un coefficient de frottement élevé qui freine l'expansion latérale de la veine et

l'empêche de se déformer de manière continue.

Dans le cas de veines hétérogènes constituées de bancs de caractéristiques différentes, ce mécanisme de blocage des déformations est amplifié par les frottements internes. La veine se déforme alors par saccades de manière caractéristique (figure 8).

L'existence au niveau de la zone affectée de contraintes très élevées et situées à proximité immédiate du parement a souvent été admise comme une nécessité. Elle est d'ailleurs à la base d'une méthode de détection universellement reconnue ("trou test"). Or, d'après les observations que nous avons pu conduire, cet élément, s'il existe effectivement pour certains coups de terrain, n'apparaît pas être systématique. La participation active de l'état de contraintes à l'expulsion de la veine peut donc dans certains cas être controversée.

Dans le cadre du mécanisme proposé, la figure 9 synthétise l'influence des facteurs repérés comme déterminants.

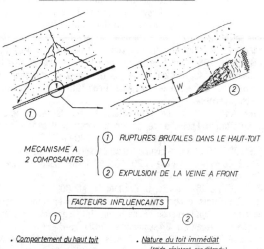

FIG.9 : MECANISME DES COUPS DE TERRAIN

MECANISME A 2 COMPOSANTES

① RUPTURES BRUTALES DANS LE HAUT-TOIT

② EXPULSION DE LA VEINE A FRONT

FACTEURS INFLUENCANTS

①

. Comportement du haut toit

- RAIDEUR NATURELLE (bancs résistants, épais)
- NON DETENDU

. Tectonique

- CONTRAINTES HORIZONTALES ELEVEES
- DISCONTINUITES DANS LE RECOUVREMENT

. Anciennes exploitations

- LIMITES, INFLUENCE LATERALE
- LARGEUR EXPLOITEE (0,4 H)

②

. Nature du toit immédiat (raide, résistant, non détendu)

. Nature de la couche

- Veine résistante
- Hétérogène (bancs à comportement différent)
- Veine non détendue (1er train)

4 - SYNTHESE - CONCLUSIONS

Depuis une dizaine d'années, la recrudescence des phénomènes dynamiques et le réel danger qu'ils constituent pour les exploitations minières expliquent les efforts importants consentis pour comprendre, prévoir et prévenir ces manifestations.

La comparaison de différents secteurs, affectés ou non par ces phénomènes, a permis de cerner les conditions qui favorisent leur formation : état de contraintes naturel, présence de gaz sous pression, nature des terrains, configurations d'exploitation.

Il apparaît que le déclenchement d'un phénomène en un point n'a généralement pas une cause unique mais résulte de la conjonction de ces différents facteurs.

Leur connaissance est importante. Elle permet d'une part de repérer dans le gisement des zones où de tels phénomènes sont susceptibles de se produire. Cette reconnaissance à grande échelle est un préalable indispensable à une prévision efficace de ces phénomènes.

Elle permet d'autre part de préciser les mécanismes de rupture qui sont à l'origine de ces manifestations. Au niveau de leur maîtrise, la compréhension de ces mécanismes est essentielle : elle constitue un guide pour la recherche et la mise en application au chantier de méthodes de prévision et de prévention adaptées.

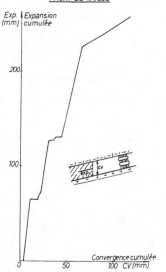

FIG.8 : EVOLUTION PAR SACCADES DE L'EXPANSION DE LA COUCHE LORS DE L'APPROCHE DU FRONT DE TAILLE

REFERENCES

ASTON T., P. CAIN, 1985. Gas and rock outbursts at N° 26 Colliery, Sydney Coalfield, Nova Scotia. A case history. Proc. 21 st Int. Conf. of Safety in Mines Research Inst., Sydney (21-25 octobre 1985) : 139-145.

CHAMBON C., R. REVALOR, 1986. Statistic Analysis Applied to Rock Stress Measurements. Proc. Int. Symp. "Rock Stress and Rock Stress Measurements", Stockholm (1-3 septembre 1986) : 397-410.

CHEMAOU O., 1984. Etude des phénomènes dynamiques dans les creusements au rocher et recherche de méthodes de prévision. Application au cas des surtirs de Merlebach (Houillères de Lorraine). Doctorat d'Ingénieur. Institut National Polytechnique de Lorraine. Ecole des Mines, Nancy (108 p.).

DA GAMA E., 1986. Comportement mécanique des charbons et des grès. Application aux phénomènes dynamiques. Doctorat d'Ingénieur. Institut National Polytechnique de Lorraine. Centre de Recherches en Mécanique et Hydraulique des Sols et des Roches, Nancy (193 p.).

GAVIGLIO P., 1985. La déformation cassante dans les calcaires fuveliens du bassin de l'Arc (Provence). Comportement des terrains et exploitation minière. Doctorat d'Etat. Université de Provence, Marseille (219 p)

HOMAND F., R. HOUPERT, E. DA GAMA, JP. PIGUET, R. REVALOR, O. CHEMAOU, 1986. Aptitude des roches à donner lieu à des phénomènes dynamiques. Application à l'étude des grès et du charbon des Houillères du Bassin de Lorraine. Proc. 9th Winter School Rock Mech., Wroclaw.

JOSIEN JP., 1980. Le comportement des terrains autour de l'exploitation. Les coups de terrain du bassin de Provence. Revue de l'Industrie Minérale. Suppl. Juin 1980 : 100-110.

JOSIEN JP., 1981. Lutte contre les coups de couche. Rapport final EUR 7869. Convention d'étude 7220-AC/307. Communauté Européenne du Charbon et de l'Acier (77 p.).

JOSIEN JP., J. BRENIAUX, C. DAUMALIN, M. DOLIGEZ, P. GEORGEL, 1982 : Effets dynamiques des pressions de terrains : coups de terrains. Compte-rendu 7ème Conf. Int. sur les Pressions de Terrains, Liège (20-24 septembre 1982) : 542-565.

JOSIEN JP., R. REVALOR, 1983. Prévision et maitrise des phénomènes dynamiques. Rapport final EUR 8795. Convention d'étude 7220-AC/309. Communauté Européenne du Charbon et de l'Acier (112 p).

KIDYBINSKI A., 1981. Bursting Liability Indices of Coal. Int. J. Rock Mech. Min. Sci. Geomech. Abstr. 18 : 295-304.

LABASSE H., 1973. Les pressions de terrains dans les carrières souterraines. Coups de toit et coups de charge. Revue de l'Industrie Minérale Mine. 3 : 141-160.

MOUDAFI S., 1986. Etude expérimentale de la réactivité des roches par foration sous contraintes. Application à la reconnaissance des zones sujettes aux phénomènes dynamiques (coups de terrain, surtirs). Doctorat d'Ingénieur. Institut National Polytechnique de Lorraine. Ecole des Mines, Nancy.

PIGUET JP., L. GEORGE, 1981. Influence de la profondeur et des facteurs naturels sur le comportement des ouvrages miniers. Rapport final EUR 7848. Convention d'étude 7220-AC/304. Communauté Européenne du Charbon et de l'Acier (50 p.).

REVALOR R., 1985. Maîtrise des phénomènes dynamiques. Rapport final EUR 10072. Convention d'étude 7220-AC/312. Communauté Européenne du Charbon et de l'Acier (91 p).

REVALOR R., J. ARCAMONE, JP. JOSIEN, JP. PIGUET, 1985. In situ Rock Stress Measurements in French Coal Mines. Relation between Virgin Stresses and Rock Bursts. Proc. 26 th U.S. Symp. Rock Mech., Rapid City (26-28 June 1985) 2 : 1103-1112.

THOMAS A., L. GEORGE, JP. PIGUET, 1983. Interprétation des mesures de contraintes en site minier à partir d'un modèle tectonique. Revue de l'Industrie Minérale - Les Techniques. 4-83 : 169-178.

The influence of geology and stresses on shaft excavations: A case study
L'influence de la géologie et fatigue sur l'excavation d'un puits: Une étude de cas
Einfluss der geologischen Eigenschaften und Spannungen auf Schächte: Ein Beispiel

R.W.O.KERSTEN, Anglovaal Limited, Republic of South Africa
H.J.GREEFF, Hartebeestfontein Gold Mining Co. Ltd, Republic of South Africa

ABSTRACT: There is an interactive relationship between the spatially varying Geological parameters and the changing stress conditions imposed by tectonics and mining activities for any particular area. These combination of factors influence the siting and support requirements of shafts and their service excavations. The evaluation of a typical case history could benefit similar projects in future.

RESUME: Il existe entre les paramètres geologiques à espace variable et les conditions de fatique changeantes rapport un rèciproque isposè par tectoniques et activitès de aines pour untel domaine. Ces considerations ci-dessus influencent le choix d'emplacement et les exigences de soutènement des puits et des excavations de service. L'évaluation d'une telle ètude des cas typique peut ainsi être avantageuse aux projets seablables á l'avenir.

ZUSAMMENFASSUNG: Es bestehen gewisse Verhältnisse zwischen geologischen Kennziffern, und tektonischen und abbau induzierten Spannungs veränderungen für bestimmte Gebiete. Die kombination dieser Faktoren beeinflusst die Position und Ausbau von Schächten und anderen Öffnungen. Eine typisher Fall wird hier besprochen und konnte vön nutzen sein für andere zukünftige Projekte.

1. INTRODUCTION

The stability of excavations, and the resulting support requirement, is a function of the geological environment and the acting stress levels.

The geological parameters varying in three dimensions are a function of the geological history of the area. The stress field is a function of the tectonic history, and the stresses induced by mining activity in the vicinity. Thus stability and support requirements can differ significantly over short distances.

The effect of the abovementioned factors was clearly illustrated when large service excavations were cut in a shaft pillar on a Witwatersrand Gold Mine at 2000 metres below surface.

Excessive movements occured in a cooling plant excavation, and additional support was installed in order to contain these.

An additional shaft was also planned for this area to increase the hoisting capacity in that part of the mine. With the experience of deformations on the service excavations, and deformations of the existing shaft it was possible to predict the support requirements in the new shaft, and sinking was completed according to schedule.

This paper deals with a detailed account of the deformation of important service excavations, and the support installed to stabilise these. Also, a detailed account is given of the support requirements of the new shaft in the light of variation of geological parameters, as well as the change in the stress field past, present and future.

2. HISTORY

Hartebeestfonein Gold Mine is situated in the Western Transvaal in the Republic of South Africa and is one of four mines exploiting the tabular Vaal reef of the Witwatersrand super group in the Klerksdorp area.

No 6 Shaft serves the western part of the mine and was sunk in 1972 when the area was still unmined. Mining commenced in earnest during 1973 with a resultant increase in the stress level in the vicinity of No 6 Shaft.

During 1980 it was decided to increase the cooling capacity and an underground cooling plant was designed for 71 level. Excavation started in 1981 and was completed in 1982, when the plant was commissioned.

It was during the development stage in 1981 that a sudden increase in shaft deformation, and tunnels in its immediate vicinity, was noticed. This increase in deformation was generally attributed to the increase in the number of excavations in close proximity to each other as well as the increase in the overall stress level due to mining on the western side of the shaft pillar.

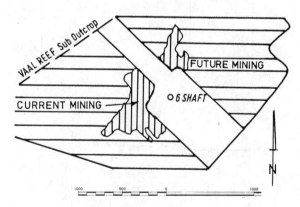

Fig 1. Plan of 6 Shaft pillar and extent of Mining

It was clear that a detailed analysis was required since at that stage planning of 6N Shaft was nearing finality, and it was obvious that problems could possibly be expected during sinking. Also, the effect of the increase in the overall stress level due to increased mining, to the end of life of the mine, had to be determined.

The investigation consisted of three parts.

1 Determination of the primitive stresses.
2 Calculation of induced and total stresses past, present and future.
3 Evaluation in rock properties for the various geological horizons.

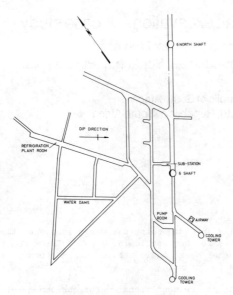

Figure 2. Location Plan of 71 level showing
ancilliary excavations

3 DEFORMATION OF EXCAVATIONS IN THE SHAFT PILLAR

Fig 2 is a plan giving details of all excavations on 71 level. Attention is drawn to
the fridge plant room and the pump chamber.
During excavation of the fridge plant room abnormal
amounts of sidewall/hangingwall/ footwall movements
were observed and measuring stations were installed.
The trend of measured displacements is shown in Fig 4.
The displacements appeared to carry on unabated and
the decision was made to increase the support density
as given in Fig 3.

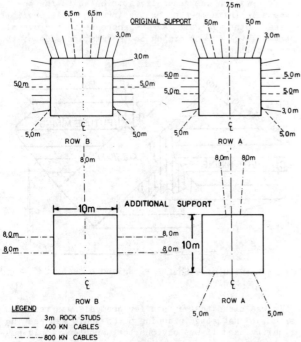

Figure 3. Support of fridge plant room

The displacements reduced after the support had been
installed, and the trend of displacement has settled
to a rate of 0,015 mm/day.
The pump chamber displacements in Fig 4 show a much
greater reduction in displacement after support installation and the continued displacement of 0,013mm/
day is similar to that in the fridge plant room.

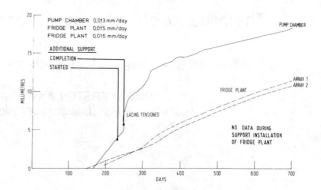

Figure 4. Displacement trends for the pump
and refrigeration chambers

The amount of failure around the pump chamber has
also been investigated by borescope and extensometers
and Fig 5 is a summary of all data to date.

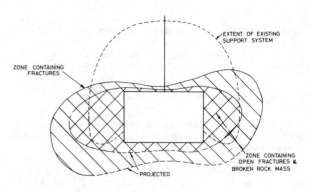

Figure 5. Zones of fracturing around pump chamber

4 THE PRIMITIVE STRESS

Two independant exercises of in-situ stress level determinations were made using the overcoring strain relief method employing the CSIRO hollow inclusion triaxial strain cell.
The results of the latest tests are tabulated below:

SHAFT		STRESS MPa	BEARING Degrees	DIP Degrees
6	σ_1	67,40	17,44	43,38
	σ_2	55,07	275,62	12,25
	σ_3	13,95	173,50	44,04

Noteworthy is the exceptionally high horizontal
stress, about 60% higher than normally expected.

5 TOTAL STRESSES AND THEIR COMPARISON WITH ROCK STRENGTHS

Fig 6 is a plot of the total stresses estimated by
numerical analysis of this area for various mining
geometries, incorporating the effect of the tectonic
stress, for past, present and future mining. Also
shown is the geological column and associated values
of the uniaxial compressive strengths.   There is a
sharp reduction in the strength value where as the
argillaceous quartzite horizons are intersected.

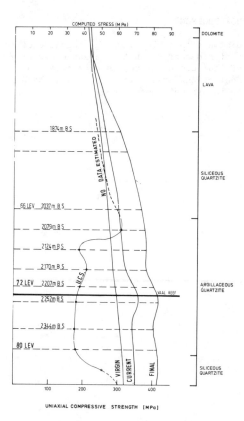

COMPUTED STRESS (MPa)

UNIAXIAL COMPRESSIVE STRENGTH [MPa]

Figure 6. Geological and rock strength data for 6
Shaft.

The stress increase along the shaft is only apparent
in the present and future stress fields and is highest
on the reef plane. This condition on the reef plane
on 72 level would under normal circumstances only be
found as deep as 80 level and below.

6   EVALUATION AND CONCLUSION

The abnormal displacements in the fridge plant room
and the pump chamber can be ascribed to the fact that
the critical stress level at which the strength of the
wall rock is exceeded, was attained on the reef plane
at 72 level and fracturing commenced.   This fracturing created an effectively bigger excavation, and
the process continued until a stable shape had been
attained - Fig 5.   Note that this is not the actual
geometry of the excavation but the interface between
fractured and intact rock.   The continued (0,015mm
/day) deformation is ascribed to the fact that the
total stress acting on the shaftpillar is increasing
gradually as mining progresses.
   A factor that exacerbated the condition is that the
induced stress level is highest on the reef plane,
creating conditions which can be expected on 80 level
and deeper.

   The knowledge gained from this experience is that:
1.Service chambers should not be placed on the plane
of the reef stoping excavation;
2. Where possible, select the stronger horizons above
or below the reef planes.
3. Sidewall and footwall support was underdesigned and
support of the hangingwall was overdesigned.
4. For conditions as described, service excavations
should be placed in smaller excavations spaced over
a greater volume of rock
5. The size of the excavation should be kept
to a minimum.
6. Location near the shaft pillar edges should be
avoided.

7   RECOMMENDATIONS WITH RESPECT TO   THE NEW 6 NORTH
SHAFT AND RESULTS

From the knowledge gained of rock mass behaviour it
was relatively easy to delineate the area where scaling would occur and hence extra support would be
required.
As seen from Fig 2, 6 North shaft was sunk within 170m
from 6 south shaft and intersected pre-developed excavations on all levels.   This added a further compli-
cation in that additional support was needed for all
these positions.
   Movement on the stations is also expected to occur
in future with possible damage to the concrete lining.
The following recommendations were implemented when
sinking 6 North shaft.

7.1 Support in the argillaceous quartzite.

Since the shaft was concrete lined, only the bottom
unlined portion of the sinking shaft required
temporary support.
   Conventional rockstuds and grouted smooth bar tendon
installation proved to be too time consuming for a
shaft sinking operation.
"Split-sets" were used extensively whilst the more
extreme conditions were countered with a left-hand
threaded type rebar which was grouted with fast set
cementitious grout capsules.   These rockbolts were installed with special adaptors fitted to jackhammers
which inserted and tensioned the rockbolt in one
operation.
   The pattern and density of installation varied
with the conditions encountered.

7.2 Support in the shaft haulage intersection.

Support consists primarily of grouted cable anchors
(Fig 7) which inhibit fracturing of the station
sidewall rock and thus control footwall and
hangingwall heave and sag respectively, which are
designed to yield in a controlled manner during
holing of the shaft.

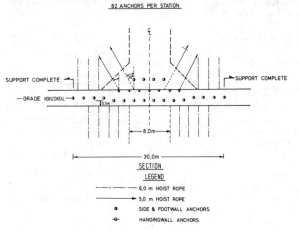

Figure 7. Station support layout

Fig 8 indicates the additional support in the form of
long grouted anchors consisting of 25 mm steel rope
installed in predetermined positions in order to effect holing with a supported brow.

7.3   Support on   the stations and the use of
"cushions".

Station support as described in 7.1 was supplemented
by meshing and lacing with 2,2m and 3m smooth bar full
column grouted tendons.
   A polystyrene cushion was designed to absorb movement of the rock towards the shaft excavation.   It
was installed between the station concrete lining and
the station floor on 74 level - Fig 9.

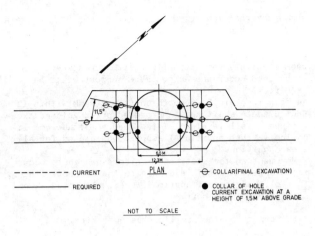

-------- CURRENT
———— REQUIRED

⊖ COLLAR(FINAL EXCAVATION)

● COLLAR OF HOLE
CURRENT EXCAVATION AT A
HEIGHT OF 1,5M ABOVE GRADE

PLAN

NOT TO SCALE

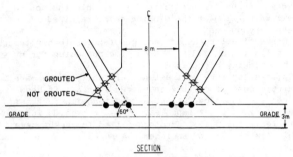

SECTION

Figure 8.   Layout of Holing support

8 ACKNOWLEDGEMENT

The assistance in the monitoring programme of the
Chamber of Mines of South Africa and especially P
Piper of that organisation, is acknowledged.   The
management of Anglovaal is also acknowledged for
allowing this paper to be published.

This large cushion of polystyrene sheets attached
to a steel bar frame was constructed on surface and
slung down the shaft. Installation took three days
to complete.
   This movement absorption concept is instrumented and
monitored at present to determine the degree of effectiveness. Fig 9 is a section of the station showing
the location of the cushion.

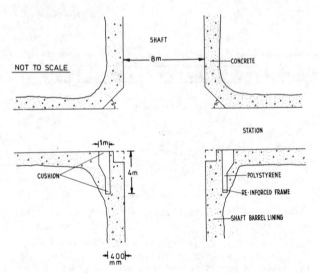

Figure 9.   Section of Shaft showing polystyrene
               Cushions

7.4 Shaft Sinking
Concrete lining should be carried close to the shaft
bottom, during sinking operations, even if this requirement results in reducing the number of blasts
per day.
   The shaft was completed within the forecast time
without any accidents due to fall of ground and is not
showing any signs of deterioration after one year.

Acoustic emission in the failure analysis of rocks

Analyse de la rupture de la roche à l'aide de l'émission acoustique
Die akustische Emission beim Bruch der Gesteine

P.KERTÉSZ, Budapest Technical University, Hungary
M.GÁLOS, Budapest Technical University, Hungary

ABSTRACT: For a correct evaluation of rock-mechanical rock properties and behaviour it is necessary to analyse the internal changing processes in them. On the basis of classical and fracture mechanics, using an elaborated petrophysical model, acoustic emission is analyzed as an indicator about failure process in the interior of rocks.

RESUME: Pour évaluation correcte des propriétés mécaniques des roches et de leur comportement c'est nécessaire de connaitre des procès de changement dans leur intérieures. Sur la base de la mécanique classique et de la fracture, employant un modèle de roche élaboré, l'émission acoustique est analysée comme un informateur des procès de rupture dans l'intérieur des roches.

ZUSAMMENFASSUNG: Zur genauen Kenntnis der mechanischen Eigenschaften von Gesteinen und ihres Verhaltens es ist notwendig, die internen Veränderungs-Prozesse zu klären. Aufgrund der klassischen und der Bruchmechanik, und ein speziell ausgearbeitetes Modell verwendend, wird die akustische Emission als Informator über die Bruchprozesse im Gesteinsintern analysiert.

1 INTRODUCTION

Adjoining the traditional testing and evaluation systems, the tests in fracture mechanics have been developed in the last years, they are suitable -- in connection with the analysis of acoustic emission-- for a theoretical approximation of mechanical-physical rock properties.

We do often use in our evaluation the condition, to consider the rocks as homogeneous and isotropic subtances, but this condition is largely dependent on the testing scale. The methods of classical mechanics are well applicable on the basis of these conditions: supposing that a failure occurs -- governed by theoretically determined failure criteria -- in form of a rupture.

The rock-mechanical evaluation of failure criteria has been assumed in Hungary by Asszonyi (1975), analysing the theories of
- limiting stress,
- limiting strains and
- rupture work,
following the rheological conditions of rock continua. Asszonyi renders the principle of the maximal dissipation work probable, as a suitable failure criterion, describing the failure according to the development and process of the phenomenon, starting from the principles of thermodynamics.

During a deformation process, depending upon the deformation-speed, the difference of the induced and of the accumulated energy must be continually dissipated. In this process, the required velocity of the dispersion may be determined by the velocity of energy-transfer, the properties of the system beeing fixed. If the level of the energy to dissipate cross a certain threshold, this energy can be ejected only by the mean of a failure, resulting a fracture in the rock.

The amount of accumulated energy in a rock is limited by the structural (chemical) bonds in it. The energy is storable until the displacements of the atoms (ions) in the bond are reversible. The part of the energy, resulting superior displacements, causes decomposition, disruption of the mentioned bonds, followed by the dissipation of the released energy in the form of thermal or acoustic vibration. From this results the fact, that the energy-transfer and the failure are strongly connected with the energy-content (energy-density) of the rock or of its elements.

The total (structural, cohesive) energy-content of a rocky system is equal to the summarized-quantums of structural bonds in it, for a crystal it is known as the lattice-energy. During the loading of a rock, the summarized energy-content decreases while an energy part is to dissipate. With the released and so dissipated energy the total (structural) energy-content of the rock is reduced. The failure -- similar to the weathering processes -- is a process, during which the energy-content is in diminution. Degree and temporal development of the diminution can be followed by registration of the acoustic emission too. Thus the role of the structural composition of rocks is doubtless, therefore analysis and evaluation of failure need tests dealing with the internal, petrophysical structure.

2 A PETROPHYSICAL MODEL

Mechanical criteria of failure-evaluation must be interpreted by the evaluation of the internal rupture-mechanism of the rock in question. Its basis is assured by a petrophysical model, outlined by Gálos and Kertész (1983). The application of this model opens for us a possibility to take in consideration the differences in the scales for the evaluation of a rocky environment of an engineering structure on the one hand and for the technical evaluation of a rock block, on the other hand.

In this model the rocky environment consists of continous rock blocks, corresponding to the three-phase, generalized rock system. In this three-phase system, the solid part corresponds to the minerals, the fluid and gaseous ones are represented by the porefilling substances. The joints (bonds) between rock-constituens (rock forming minerals) are modell-elements of the same importance as the minerals. Therefore the rock fabric, determined by minerals and joints is among other ones, the most important bearer of the mechanical rock properties.

Petrographically homogenous rock bodies are generally discontinous, the system of discontinuity in them is determining the mechanical properties including failure criteria.

Homogeneity and isotropy of rocks must be interpreted according to the geometrical extension of the concerned model-element. This virtual exactitude appears in the failure-evaluation, interpreting the material-parameters, determined by application of idealized mechanical (physical) conditions, through the existence or interaction of basic model-elements. Thus the characterizing model-elements are in the failure of a rock block the fabric-elements, in a rock body the system of discontinuity.

3 ANALYSIS IN FRACTURE MECHANICS

The theoretical and practical development of fracture mechanics succeeded due to the detailed analysis of the rupture mechanism. The hungarian tests in fracture mechanics have been executed in a common program at the Budapest Technical University (Department for mineralogy and geology, Institute for technical mechanics).

For the interpretation of the results received from this test series, Griffith's and Irwin's theoretical works gave the principles. The registration of acoustic emission was carried out in collaboration with the Central Research Station of Physics (KFKI) in Budapest, using different loading arrangements (Fig. 1).

Evaluation in testing technics of fracture mechanics has been elaborated by Czoboly, Gálos, Havas and Thamm (1986). For the determination of the parameter of critical stress-intensity (K_{Ic}) a three-point bending test on incised prismatic samples proved to be the most suitable, but experiments will be continued with cylindrical samples differently incised.

The mechanism of a failure process in rocks must be analyzed and geologically interpreted on two, subsequent model-elements in the rock mass: separately on the rock body and on the rock block. Following this way, our rock-mechanical way of thinking may be homogeneous and consequent.

The failure of a rock evolves following internal cracks, separeting or traversing minerals in such a way, that internal forces, assuring till now the integrity of a rock block disappear. The rupture-surface of a crack is not always a regular one, it may be composed by parts of different positions.

The internal energy of the elementary particles in a rock is equal to the amount of cinetic and potential energy, referred to the mass-center, thus it is a function of the internal, or thermodinamical state of the system. The energy-transfer, due to external loading alters the internal energy-distribution, causing the removal of atoms (ions) to critical position. Reaching the critical atomic distances it may result in the ceasing of the bonding forces. Its probability is higher at places of lattice deffects.

The different possibilities of energy-transfer are connected with internal phenomena in the rock. A conductive one can be supposed in the case, when the atoms or ions transferring the energy all remain in the structural bond: the energy-content doesn't change, the movements of the elements are reversible. Irreversible displacements, occur in the case of translation and disruption, the energy-transfer is a convective one. In a translation process, two parts of crystal lattices dislocate, but they doesn't leave the structural bond: the energy-content is unchanged, the displacements are in the order of crystal units. With the disruption, the elements remove irreversibly, a conductive energy-tranfer isn't more possible: with a convective one the energy-content in the rock is diminishing.

4 UTILIZATION OF ACOUSTIC EMISSION

Disintigration of bonds due to external loading (i.e. crack-formation and -propagation) can be observed by analizing acoustic emissions. In the phase of crack-formation, while loading increases unformly, an increasing number of acoustic events appears.

The further development of cracks to fissure and the formation of greater ruptures is followed by a considerable increase of acoustic events. For the evaluation of this phenomenon, the integral curves of acoustic events (ΣE) are plotted, where an event is considered as an acoustic vibration exceeding 35 dB.

The mentioned number of acoustic events summarizes the phenomena of the disruption of structural bonds. A repeated event means the further propagation of a former crack or the formation of a new one. The intensity of the acoustic emission is proportional to the cohesive energy of the disrupted joints and it may be represented by the number of oscillations.

The summarized number of oscillations (ΣO) during an acoustic event plotted versus loading force (Fig. 2) shows us the rapid increase of energy-dissipation in the failure-phase.

The registered oscillations of events may be negligible in the first phase of loading, they increase considerably in the crack-forming phase, it can be well seen in the distribution-curve of the peak-amplitudes of the results of a biaxial loading test.

With increasing cohesive energy of bonds the intensity of acoustic emission increases too. The disruption of intercrystalline joint shows us a smaller intensity than the intracrystalline rupture of a mineral without cleavage (f.i. quartz). The effect of this phenomenon is well visible on the plotted examples (Fig. 3.,4).

By the uniaxial loading of two rock types (Fig.3a) the formation of a great number of acoustic events occurs. The inelastic behaviour can be well seen, an intensive fracture-phase can be easily determined. The following equation has been established between the number of events (ΣE) and the loading force (F):

$$\log \Sigma E = c_1 \log F + c_2$$

c_1, c_2 : rock-dependent parameters

Integral curves plotted during a tensile test (Fig 3b) and fracture-mechanical test (Fig 4), according to similar loading conditions are similar in shape, both show us clear the active, cummulative phase of crack-propagation.

In triaxial state of stress, the opening of shear cracks is connected with a high number of events (Fig. 3c).

The so-called Kaiser's effect is a well known phenomenon in repeated loading. That means, that after unloading, a new acoustic emission occurs only in the case when a new state of stress surpasses the further maximum. This phenomenon, often characterized as a "memory" of rocks, is caused by the fact, that all joints must be disrupted already, where the energy to dissipate surpassed the cohesive energy, or the real atomic distances surpassed the critical ones. Under this state of stress only the phenomenon of fatigue is able to provoke new disruptions.

After surpassing this threshold, new joints are disrupted again and new acoustic events are observable, following the formation of new cracks.

From Fig. 5 we can see the above mentioned phenomenon, that the phase before the rupture can't be considered as an elastic one; using a constant load, acoustic events occur even during a constant load, it is a sign of a progressive crack-propagation.

As demonstrated, acoustic emission may be considered as an indicator of ruptures in rocks. With its registration the whole process of failure is more comprehensible and the different phases of the process may be unambiguously delimited. Therefore we shall continue our research work further in this direction.

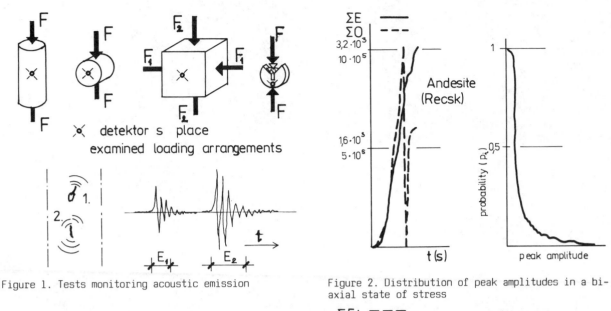

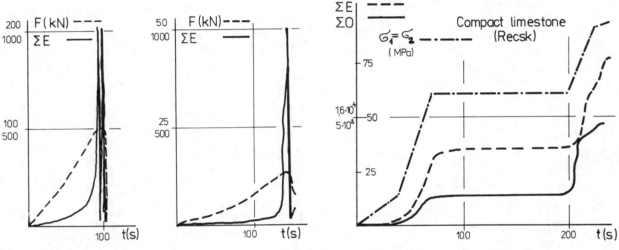

detektor s place
examined loading arrangements

Figure 1. Tests monitoring acoustic emission

Figure 2. Distribution of peak amplitudes in a bi-axial state of stress

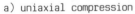

Figure 3. Acoustic emission during   a) uniaxial compression   b) traction   c) triaxial loading

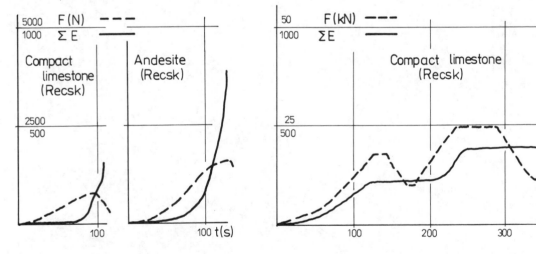

Figure 4. Acoustic emission during fract.-mech. test

Figure 5. Kaiser's effect in a tensile test

REFERENCES

Asszonyi, Cs. 1975. Kőzetkontinuumok reológiai elméle-
téről, Tatabánya: MTA.
Gálos, M. - Kertész, P. 1983. Results of rock mechanics in Hungary, An engineering model of rocks, 5nd
Int.Congr.ISRM. pp 259-283

Czoboly, E. - Gálos, M. - Havas. I. - Thamm, F.
1986. Appropriate fracture mechanics speciments
for testing rocks. Fract.Contr. of Eng.Struct.
Vol III. pp 2.lo5-2.114

On measuring gas pressure in coal seams

Mesure des pressions de gaz dans les couches de charbon
Bestimmung von Gasdruck in Kohleflözen

S.A.KHRISTIANOVICH, Member of the USSR Academy of Sciences, Head of Laboratory Institute for Problems in
Mechanics, the USSR Academy of Sciences, Moscow
R.L.SALGANIK, D.Sc., Professor, Institute for Problems in Mechanics, the USSR Academy of Sciences, Moscow
YU.F.KOVALENKO, Sc., Institute for Problems in Mechanics, the USSR Academy of Sciences, Moscow

ABSTRACT: The results of measuring gas pressure in sealed boreholes drilled in coal seams
are considered. Basing on hypothesis of complete absence of filtration ability in coal
seams unaffected by mining and of arising the filtration ability as a result of mining
induced deloading the coal from compressive stresses produced by lithostatic pressure,
it is given an explanation of all observed phenomena and substantiation of the hypothesis
that in initial state the pressure of free gas contained in pores of coal seam is close
to local lithostatic pressure. It is proved that the stationary gas pressure which is
achieved in test borehole in coal seam and which according to presently used instructions
is considered as equal to gas pressure in coal seam unaffected by mining is determined
actually by the stress state of coal around the borehole. An estimation of time of achieving
equilibrium states is given.

RESUME: On considére les résultats de la mesure de la pression du gaz contenus dans les
couches de charbon due sondage et de l'hermétisation suivante. Cette hypothèse sur l'absence de la capacité de filtrage dans les couches de charbon, qui ont de l'intégrité dans
les travaux miniers et sa formation du delestage de la tension sous la pression de couche
de charbon, quand on mène des travaux de mine, explique tous les phénomènes d'observation.
On prouve que la pression du gaz stabilisée dans les puits de mesure et qui est égal à la
pression du gaz dans l'integrité de la couche est determiné l'état tensif du charbon dans
le voisinage du puit. L'estimation du temps de stabilisation de l'état de l'équilibre
est donnée.

ZUSAMMENFASSUNG: In der Arbeit betrachtet man die Resultate der Druckmessung nach der
Bohrung und nachfolgender Abdichtung des Bohrloches. Die Erklärung der allen beobachteten
Erscheinungen wird auf Grund der Hypothese gegeben, dass Filtrationfähigkeit des vom
Tiefbau unberührten Flözes Null beträgt, und nur dann erscheint wenn die Bergdruckspannung
wegen des Tiefbaues abgenommen wird. Dabei wird auch eine Hypothese bestätigt, dass der
Druck des freien Gases in den Poren im Anfangszustand nah dem Lokalbergdruck ist. Es wird
bewiesen, dass der in den Messbohrungen setzende Gasdruck, der nach zur Zeit gültigen
Instruktionen dem im unberührten Floz gleich angenommen wird, in Wirklichkeit vom
Spanungszustand in der Nähe der Bohrung bestimmt ist. Die Auswertung der nötigen Zeit für
die Erreichung des Stabilzustandes wird gegeben.

The main problem in understanding the
mechanism of the formation of instantaneous
outbursts of gas and coal in coal seams
during their mining is the problem of initial state of gas (methane, carbon dioxide),
i.e. its state in a seam practically unaffected by mining. This problem, still deba-
table in spite of significant efforts to
solve it, is discussed in this paper.

Typical gas content in outburst-prone
coal seams is estimated as 15-25 (in some
cases 30-40) normal cubic metres per ton of
a coal. It is considered that a part of gas
in initial state is sorbed (dissolved) in
coal and the remaining part is in a free
state. When a coal seam contains water, the
gas can be partly dissolved in it.

Until recently, gas-containing coal seam,
even under very high lithostatic pressure,
was considered to possess some gas filtration ability (permeability), however small.
Conventional methods of gas pressure measurements in coal seams are based on this
hypothesis. According to it, coal seam gas
pressure achieved after sufficiently long
time in properly sealed borehole drilled
in the coal seam is adopted as the true
pressure.

Usually a bulk of crushed coal is taken
from boreholes during their drilling in coal
seams, the volume of this mass frequently
considerably exceeds the nominal volume of
the corresponding part of borehole. Sometimes, several days are needed to equip
such borehole for gas pressure measuring.
Considerable volume of gas emerges from this
open borehole. Nevertheless, the stationary
value of gas pressure eventually achieved
in sealed hole is not observed to be affected by this process which affects only the
formation time of this pressure. If the
borehole is opened after the stationary gas
pressure is achieved and closed again after
some time, the pressure value will be practically restored after a while in the
majority of cases.

Basing on the hypothesis of gas filtrating
ability of coal seams in initial state and
using measurements data on gas discharge
from open borehole and its time variation,
the permeability of seams are determined
using Darsy's law. Typical values of the
permeability are found to be in the range
10^{-2}-10^{-4}mD. This permeability is very
small compared to strata of gas and oil
deposits (10-100 mD). After the opening of

borehole, gas discharge sharply decreases with time and finally achieves very low values (several litres per hour or even per day). Gradually (after monthes), the discharge go on decreasing but does not vanish.

These facts seem to confirm the hypothesis of stable, though very small, initial permeability of coal seams in their initial state providing gas influx into the borehole from practically unbounded volume.

If this hypothesis is valid, only perfect sealing of boreholes and measurements during a period of time sufficient to achieve stationary gas pressure are needed to obtain reliable data on gas pressure in coal seams.

However, the bulk of observed phenomena is quite unexplainable by this hypothesis. First of all, there is a huge scatter of gas pressure measurements data obtained in different seams, in one seam and even in closely situated boreholes in the same seam. Presently there is a large number of data obtained in different coal deposits and outburst-prone coal seams.

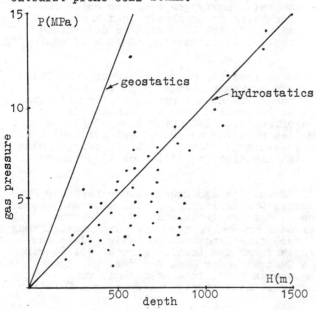

Figure 1. Data of measuring gas pressure in coal seams (Bobrov 1961, Airuni 1981, Kuznetsov and Krigman 1978).

Figure 1 shows some of the data published (Bobrov 1961, Airuni 1981, Kuznetsov and Krigman 1978). Other numerous data are similar. As can be seen, the data presented look as they were random. If the seam had stable initial permeability, the stationary gas pressure value in test boreholes drilled in the seam (at given depth) should have been the same. In this situation, if the coal seam is surrounded by water-saturated strata, gas pressure change with depth should be consistent with hydrostatic law. However, this is not the case.

There are numerous data on gas pressure variations in test boreholes in coal seams due to the approachment of working which were initially far enough from the holes. If coal seams possessed stable initial permeability, the formation of pressure difference and corresponding gas flux directed to the working should lead to some decrease of gas pressure in boreholes as workings approached them, anyway pressure should not increase. However, very frequently gas pressure in test borehole drilled in coal

seam is observed to increase as the working approaches it and sometimes reaches the level exceeding several times its initial value. Then gas pressure in the hole gradually decreases with further approaching of the working. As an example, tests can be mentioned, which were conducted in 1958 in an experimaental area in Derezovka coal seam (Bobrov 1961). It might be suppoused, as in (Bobrov 1961), that gas pressure increase in boreholes was caused by the compression of free gas contained in pores due to seam porosity decrease under increasing rock pressure as working approached the hole and by the decrease of borehole volume. However, such explanation does not look adequate. The data of measurement of seam roof and floor convergence show that this convergence would increase the gas pressure in the borehole only by several per cent.

On our part, we suggest that gas pressure increase in this case as well as in other similar cases, when drift or longwall face approached the holes can be explained only by the influx of gas high pressure from the neighbourhood of the hole as a result of the formation of filtration channels due to destressing towards the working face.

This also corroborates with fact that complete absence of permeability was proved in the seam far enough from working face in experiments with pumping helium in boreholes under pressure several times exceeding the stationary pressure of the seam gas determined in test boreholes (Bobrov 1961).

The analysis of above facts and some other known manifestations of gas action leads to the following two hypothesis.

1. In gas-containing, outburst-prone coal seams unaffected by mining the ability of gas filtration (permeability) is completely absent; it appears around openings and boreholes only in a region of considerable destressing from lithostatic pressure.

2. In coal seams free gas is contained in isolated pores, its initial pressure being close to lithostatic pressure at the given depth.

The initial state of gas-containing coal seams, corresponding to this hypothesis, is created due to the ability of coals in their natural state to flow slowly under stresses practically until the stresses vanish.

Basing on these two hypothesis, a large number of facts concerned with outbursts and related phenomena as well as with the action of antioutburst measures can be understood and explained (Khristianovich, Salganik 1983).

On the same basis, the facts concerned with the measurements of gas pressure and discharge in boreholes in coal seams can be understood and explained, including the ones which, at the first sight, seem to confirm the presence of initial permeability in such coal seams.

Consider a borehole drilled through rock strata over the hole thickness of coal seam perpendicularly to it. This situation is shown in Figure 2 presenting a vertical section in the plane of borehole axis. It is suppoused that conditions of plane strain take place.

It is natural to consider the state of stress in coal elastic far enough from the hole (zone V). Radial stresses ∂_r decrease and hoop stresses ∂_θ increase in zone V as the borehole is approached. Free gas is

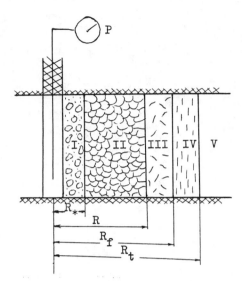

Figure 2. Zones around borehole in the coal
seam.
     I - "coal in gas" - zone
    II - zone of limit equilibrium
  III - elastic permeable zone
   IV - elastic zone (radial filtra-
        tion is absent)
    V - zone of the seam unaffected
        by mining

contained in isolated pores in this zone.
At its boundary, $r = R_t$, the diminishing of
δ_r causes coal rupture by gas contained
in pores and the crack growth from the pores
in hoop direction (perpendicular to radius)
so that a zone with a system of oriented
cracks is formed (zone IV). Gas pressure in
these cracks is less than the initial one
(which is close to lithostatic pressure,
according to the hypothesis 2), but it is
only slightly less as the calculations show
(Kovalenko 1980). According to the hypothesis 1, the permeability is completely absent
in zone V. Also no permeability arises in
the direction to the hole in zone IV, since
it's prevented by increased hoop compression.
    The inner boundary of zone IV ($r = R_f$)
is determined by the condition of the failure of crosspieces between cracks forming
a zone which is permeable in all directions
(zone III). The deformations of coal in
zone III can remain mainly elastic.
    Passing from zone IV to zone III, gas
pressure at the boundary $r = R_f$, decreases
stepwise to a value much lower than the
lithostatic pressure (to which the free gas
pressure is close in zone IV). This pressure
corresponds to stationary pressure measured
in the borehole. The equilibrium at the
boundary $r = R_f$ is maintained from inside
by the sum of gas pressure and the pressure
produced by coal itself. The decrease of
gas pressure in zone III causes advancing
boundary $r = R_f$. The stationary value of radius R_f achieved after sealing the borehole
is determined only by a mass of emerged gas.
    At the inner boundary of zone III ($r = R$)
shear stresses determined by the difference
between δ_θ and δ_r increase so that the
coal turns into the state of limit equilib-
rium (zone II). In zone II, just as in zone
III, the coal is permeable in all directions.
At the inner boundary of zone II ($r = R_*$)
coal is broken into separate pieces which
are not pressed to each other (the state of
coal in gas).

The considered above state of equilibrium
takes place when the borehole is sealed and
gas pressure in it is stationary. In this
case gas pressure being constant throughout
the total volume $r \leq R_*$ and acting from
inside on the zone of oriented cracks
(zone IV) plays an important role in maintaining the equilibrium.
    When the hole is opened, gas starts to
emerge from it, the equilibrium fails and
the boundaries of zones shown in Figure 2
expand into the seam. If than the hole is
closed again, the expansion of the boundaries into the seam will attenuate and stop
after some time, when a new equilibrium is
achieved. It is obvious that were the equilibrium maintained only by the gas pressure
the last one at the new state of equilibrium
would always restore to its initial value.
Analysis shows that similar situation occurs
when the equilibrium is maintained from
inside by both gas pressure and the pressure
produced by the coal.
    When the borehole opening is repeated suf-
ficiently many times or when the borehole
is opened for sufficiently long time, gas
pressure achieved in it after its sealing
reduces. This may be caused by several
factors. One of them is the fact that considerable expansion of the boundary of the
filtration zone leads to the filling of the
borehole by crushed coal which makes impossible further expansion of the boundary.
    Thus the results of the measurements of
gas pressure and discharge in a borehole
drilled in coal seam look as indicating
that the seam possesses stable permeability
throughout its total volume.
    The scutter of the data of measurements
of gas pressure in closely situated boreholes may be explained not only by a scutter
of seam characteristics but possibly by
difference in states of equilibrium.
    The calculation of time of achieving
stationary gas pressure in test borehole
shows that it is determined by a time scale
$t_o = (\mu m / K p_o r_o^2)$, where μ is viscosity of
gas, m is porosity of coal, p_o is initial
pressure of gas contained in pores of the
coal seam, K is the permeability in the
zone of filtration, r_o is an effective radius of the borehole. Typical values of t_o
(assuming real values of the other parameters) are about several decades of seconds.

REFERENCES

Airuni, A.T. 1981. Theory and practice of
    fight against mine gases at great depth.
    (in Russian). Moscow: Nedra.
Bobrov, I.V. 1961. Methods of safe conduc-
    ting of the preparative workings in out-
    burst-prone seams. (in Russian). Moscow:
    Gostechizdat.
Khristianivich, S.A. & R.L.Salganik 1983.
    Several basic aspects of the forming of
    sudden outbursts of coal (rock) and gas.
    Preprints of 5th Int. Congr. Rock Mech.
    Melbourn (Australia), E41-E50.
Kovalenko, Yu.F. 1980. Elementary act of
    the phenomenon of sudden outburst into
    borehole. (in Russian). Moscow: Inst.
    Probl. Mech. Acad. Sci. USSR, Preprint
    155.
Kuznetsov, S.V. & R.N.Krigman 1978. Natural
    permeability of coal seams and methods
    of its determination. (in Russian).
    Moscow: Nauka.

The development of mining-induced seismicity

Développement de la séismicité induite par l'exploitation minière
Die Entstehung von Erdstössen durch Bergbau

P.KONEČNÝ, Mining Institute of the Czechoslovak Academy of Sciences, Ostrava
J.KNEJZLIK, Mining Institute of the Czechoslovak Academy of Sciences, Ostrava
J.KOZÁK, Mining Institute of the Czechoslovak Academy of Sciences, Ostrava
M.VESELÝ, Mining Institute of the Czechoslovak Academy of Sciences, Ostrava

ABSTRACT: Data from a seismological network of 14 stations established in Ostrava-Karviná Coal Basin in 1982 have been used for the evaluation of mining-induced seismic events.

RESUME: Données d'un réseau séismique établi dans le bassin houiller d'Ostrava-Karviná en 1982 ont été utilisés pour l'évaluation des évenements séismiques causés par les travaux miniers.

ZUSAMMENFASSUNG: Die Daten von 14 seismologischen Stationen in Ostrava-Karviná Kohlen-revier, die im Jahre 1982 aufgebaut waren, wurden zur Auswertung der seismischen Erscheinungen, die durch Gewinnungsarbeiten verursacht wurden, ausgenützt.

1 INTRODUCTION

In the last time, the Ostrava-Karviná Coal Basin in Upper Silesia was characterized by a worsening of geological and mining conditions, a deepening of the mines /now-adays, the average depth is 665 m/ and by an ever increasing coal production per square kilometre of surface area /three times greater compared with that of the U.S.S.R., F.R.G. and Poland (Technical Yearbook 1986). In association with that, a number of rockbursts occurred. They were considered as mining-induced seismic events. They represent a serious danger not only for the safety of men working underground, but also for the whole operation of the mine. If the seismic activity is monitored continuously, it is possible to evaluate the relationship between the advance of mine workings and the seismicity, induced by the mining.

2 GEOLOGY

The Karviná part of the Basin where most rockbursts occurred is the Czechoslovak part of the Upper Silesian Coal Basin. The Upper Carboniferous /Namurian A, B, C, Westfalian A/ is divided into Ostrava and Karviná formations.

The Ostrava formation is composed of a great number of coal seams of a reduced thickness, predominantly with fine-grained marine and paralic sediments in the over-lying strata. The thickness of this formation reaches some 1400 m in the Karviná part of the Basin. For the time being, only a few seams deposited in the upper layers of the Ostrava formation have been extracted.

The younger Karviná formation of a thickness of 1200 m is composed exclusively of continental sediments. These are coarse-grained rocks, predominantly conglomerates and coarse grained sandstones as well as siltstones and claystones of a relatively high strength /their uniaxial compressive strength attains, as a rule, 80 -110 MPa/. The seams are medium thick and thick ones, so that the slicing method of mining had to be used in some instances. In the Karviná formation, 88 coal seams of an average thickness of 228 cm have been verified (Technical Yearbook 1986).

There is a Germano-type tectonic structure in the Karviná part of the Basin, with frequent occurrence of throws. Recent research proved that in some places, the tectonic structure is a complex one, with block-type of tectonics, some of the blocks having been rotated mutually (Dopita et al 1979). The deposition of the strata is subhorizontal.

In view of the exhaustion of coal reserves in the past, only 5 to 8 coal seams are being extracted in the Karviná part of the Basin at present. The development of these seams is variable due to the erosion of the seams at the time of their sedimentation /wash-outs and others/.

3 SEISMIC DATA OBTAINED

When evaluating the relationship between the development of the mining-induced seismicity and the advance of the mining in a particular zone, the different seismic events may be characterized in two ways:
- by locating their foci and
- by determining their energies.

In the Ostrava-Karviná Coal Basin, the seismological method has been used since 1982 when five seismic stations were established /Fig. 1/. The additional nine stations were equipped with recording apparatuses different from those used in the first five stations.

The five basic stations are provided with apparatuses for the classical photogalvanometric continuous recording of seismic events. Their frequency ranges from 1 to 10 Hz, the chart speed is approximately 118 mm/min. The time base is formed by minute marks

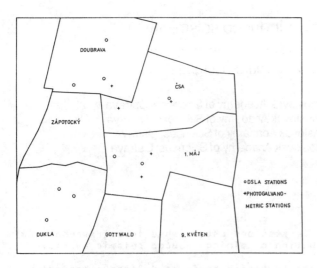

Figure 1. Distribution of seismological
stations in Ostrava-Karviná Coal Basin
photogalvanometric stations DSLA stations

generated by a timing source in-built into
the apparatuses. In view of the lack of
accuracy, it is impossible to locate the
foci using kinematic methods which require
very accurate determination of the times of
the first arrivals. That is why the location is deduced from time differences T_i
between the P-wave and the next pronounced
phase at the record from i station. Provided that the accuracy of the readings of
spacings between the arrivals of the investigated phases is 0,05 mm, the time accuracy has to be 0,025 s. It is evident that
such an accuracy cannot be obtained in all
cases.

The newly built nine stations were equipped with Czechoslovak made apparatuses type
DSLA. These are simple recording systems
with an accurate, radio-controlled time base. They serve for a quick automatic determination of the time of arrival of an event
to the station and for a subsequent approximate location of the focus of the event.
Since the first arrival is defined as an
excess to a pre-determined comparative level, the error in the determination of the
time of that arrival may reach the value of
50 ms and more (Veselý 1985).

The accuracy of the location of the foci
of events is affected, among others, by the
geology existent in a particular zone. From
a seismological point of view, the Ostrava-
Karviná Coal Basin is a region with a typical stratified structure. The velocity of
propagation of seismic waves is greater in
the deeper strata.

The seismological recording apparatuses
are situated more or less at the same plane
given by the actual depth of the mining,
i.e. at the depth of some 600 m. The strata
overlying these recorders have been disturbed by prolonged coal extraction, while the
floor is formed of undisturbed rock mass.
These facts influence the velocity of propagation of seismic waves in the following
way:

Waves from the immediate vicinity propagate, as a rule, through disturbed zones, i.
e. more slowly than the waves coming from
a greater distance and propagating through
the undisturbed rock mass. Recent investigations showed that the velocity of the P-

wave in horizontal direction is approximately 4800 m/s (Veselá 1985), while the
S-wave propagates at the rate of 2300 m/s.
The velocities of seismic waves in vertical direction are unknown up to now.

The records from the above described seismological network and the measurements
of the velocities of propagation of seismic
waves with the aim of relocating the sites
of known blasting works (Holub 1985) made
it possible to work out a methodology for
locating the foci of seismic events inside
the region covered by the stations with an
accuracy better than $\pm$ 130 m. Out of that
region, the accuracy decreases considerably (Knotek 1986).

The energetic parameters may be evaluated
using the following relationship:

$$E = K.A^2.f^2.D^2.t$$

where D is the distance of the stations
         from the focus,
      f is the predominant frequency on the
         record,
      A is the maximum displacement,
      t is the duration of the seismic
         event,
      K is the constant of the site which
         includes the effect of the density
         of the strata around the recorder,
         the velocity of the seismic wave,
         the sensitivity of the recorder
         and other influences (Knotek 1986).

Using the above formula, the energy of the
event may be calculated in units /j/, for
which the relation 1 j = 1 Joule applies.

In this connexion, it is important to
note that all calculations of the energy
are performed from Z components of the
seismometers only.

Using the above methodology, considerable
errors may be made both in the location of
the foci and in the calculation of energies
of single events. Nevertheless, in the average, the set of data obtained through the
recordings and subsequent calculations
yields valid results. As a consequence, an
analysis based on these assumptions will
have a general validity.

4 THE DEVELOPMENT OF MINING-INDUCED SEIS-
  MICITY

Records on the development of mining-induced seismicity /i.e. the location of the
foci and the calculations of the energies
of the events/ have been kept since 1982.
The number of seismic events, recorded in
the period of 1982 to 1985 and divided
according to their energies is shown in
Fig. 2.

The number of rockbursts and tremors,
i.e. those seismic events which caused damage to mine workings, is shown in Table 1
(Technical Yearbook 1986).

Table 1. The number of rockbursts and tremors in mines covered by the seismological
network

Year	1982	1983	1984	1985
No. of rock-bursts	4	5	3	0
No. of tremors	20	11	10	9
T o t a l	24	16	13	9

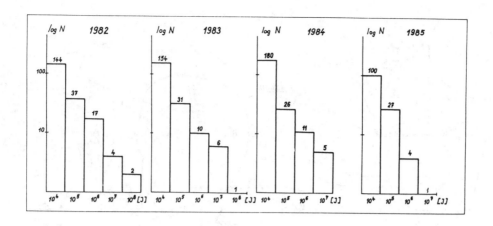

Figure 2. Frequency of seismic events divided according to their energies /years 1982-1985/

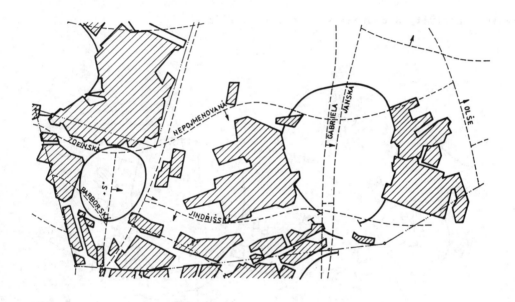

Fig. 3. Worked-out areas in the region by the end of 1981

Figures 3 - 7 give examples of the advance of coalfaces related to the verified seismic activity in the zone. Owing to the small scale, the individual seams were not differentiated in the figures. The figure 3 shows the state reached by the end of 1981 and the figs. 4 - 7 show further advance of coalfaces and the recorded seismic events.

It may be deduced from the above described seismic monitoring that:

- The number of rockbursts and tremors is much lower than the number of the recorded seismic events as shown in Fig. 2. This is a confirmation of the known fact that only a small number of seismic events causes damage to mine workings. Seismic events of a reduced energy do not manifest themselves necessarily as rockbursts - see also

(Salomon 1983) .

- It is evident that the seismicity was induced by the mining. The effect of the mining was a long-term effect /over months and even years/. An example of this is the zone A in Fig. 4 where seismic events in 1982 can be considered as a consequence of the mining carried out in the zone a year ago /see Figs. 3, 4/. In 1983, when the mining was started again in the A zone, a "concentration" of seismic events around the investigated coalface could be observed /Figs. 5, 6/. These events caused damage to mine workings and some of them were classified as rockbursts.

- Direct relationship between the seismicity and the mining is seen in the B zone. In 1983 when the mining was stopped in

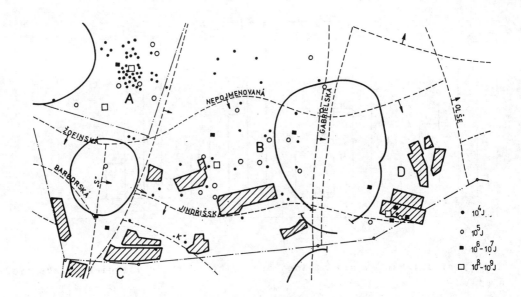

Fig. 4. Seismic activity and areas worked out in 1982

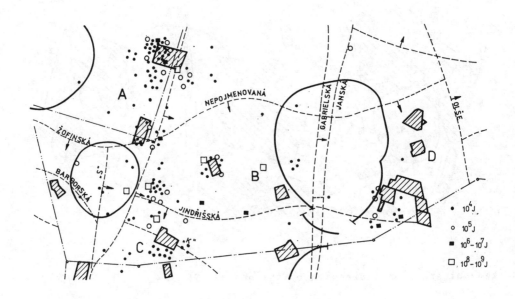

Fig. 5. Seismic activity and areas worked out in 1983

that zone, the number of seismic events fell to zero /compare the Figs. 3 - 7/.

- There are zones /e.g. the zone D/ where the mining does not induce seismic activity in the immediate vicinity of coalfaces but its effects are felt at a greater distance. This was the case of a tremor occurred at the boundaries of a shaft pillar in zone D /Figs. 3 - 7/.

- The zone C is an example of a relatively regular irradiation of seismic energy in dependence on the advance of the mining. Both the extraction of coal from the faces and the development of seismic activity exhibited a uniform trend /see Figs. 3 - 7/.

5 CONCLUSIONS

It may be deduced from the comparison of seismic activity and the advance of the mining in the investigated zones of the Ostrava-Karviná Coal Basin that - regardless of some uncertainty in the interpretation of seismic events -

- the mining may induce seismic activity even after a prolonged time /i.e. after several months, a year or more/ since it has been stopped;

- the seismic activity need not be induced only in the vicinity of the observed coal-

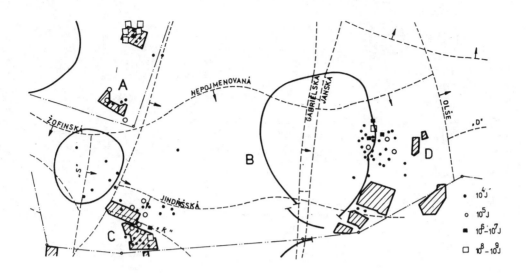

Fig. 6. Seismic activity and areas worked out in 1984

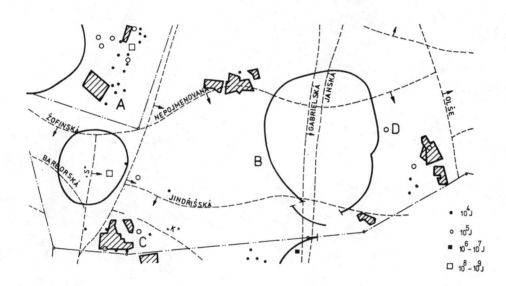

Fig. 7. Seismic activity and areas worked out in 1985

faces but also at a greater distance /of the order of hundreds of metres/, e.g. in natural and artificial discontinuities in the rock mass;

- only a small part of seismic events - usually those of the highest energy - causes damage to mine workings and manifest themselves as rockbursts or tremors.

REFERENCES

Dopita, M. et al 1979. New knowledge gained on block-type structure of Coal Measures in the Czechoslovak part of the Upper Silesian Basin. In National Geological Conference, Smolenec. /In Czech/
Holub, K. 1985. Results of investigations into the propagation of S-P waves in the O-K Basin. Acta Montana 71. /In Czech/
Knotek, S., Vajter, Z. 1986. Seismological activity in the Eastern part of O-K Basin Part IV 1985.In Bulletin geofyzikálni

služby dolu ČSA, Ostrava. /In Czech/
Salomon, M.D.G. 1983. Rockburst hazard and the fight for its alleviation in South African gold mines. In Rockbursts: prediction and control. The Institution of Mining and Metallurgy, London.
Technical Yearbook of the Ostrava-Karviná Coal Basin for the Year 1985, 1986. Ostravsko-karvinské doly, GŘ, Ostrava /In Czech/
Veselá, V. 1985. Analysis of velocity of propagation of seismic waves for the seismological network of DSLA apparatuses in Ostrava-Karviná Coal Basin. In Proceedings of 8 National Conference of Geophysicists, section S1, České Budějovice. /In Czech/
Veselý, M. 1986. Parametric measurements performed on seismological apparatuses in Ostrava-Karviná Coal Basin in 1985. Report of Mining Institute of Czechoslovak Academy of Sciences, Ostrava. /In Czech/

Investigation of deformation behaviour at a horseshoe-shaped deep cavern opening

Investigation des déformations d'une caverne profonde en forme de fer à cheval
Die Untersuchung der Deformation einer tiefen U-förmigen Kaverne

KEIGO KUDO, Tokyo Electric Power Co., Inc., Japan
KAZUTOSHI MATSUO, Tokyo Electric Power Co., Inc., Japan
YOUICHI MIMAKI, Kajima Institute of Construction Technology, Japan

ABSTRUCT: A considerable difference was observed in displacement of Breccia and other rocks surrounding a deep underground cavern, despite the symmetrical geometry of the cavern. To clarify the mechanism of deformation behavior, a triaxial compression test focused on lateral strain was conducted and unique dilatancy characteristics of Breccia were found.

RESUME: Une différenoe considérable a été observée dans le déplacement de la brèche et d'autres roches à la périphérie d'une caverne en terrain profond, malgré pression tectonique et des propriétés des matériaux identiques. Pour identifier ce phénomène, on a effectué du essai triaxial centrés sur la déformation latérale, et des caracteéristiques de dilatation uniques à la brèche sont apparues.

ZUSAMMENFASSUNG: Es wurde ein Verdrängungsunterschied zwischen Trümmergestein und anderem Gestein in der Umgebung einer tiefen unterirdischen Höhle festgestellt, obwohl der tektonische Druck und die Materialeigenschaften nahezu gleich waren. Um die Ursache zu identifizieren, wurde ein Druckversuch für die Seitendeformation durchgeführt, wobei spezifische Dilatanzeigenschaften von Trümmergestein ermittelt wurden.

1 INTRODUCTION

The Imaichi power station is now constructed by Tokyo Electric Power Co., Inc. in the cavern which is one of the largest ones in the world with a cross-sectional area of $1420m^2$. Due to the considerable depth of the overburden of 400m, a horseshoe-shaped cross-section was adopted and the cavern roof was not given a definitive lining. (Fig. 1)

Approximately 800 measuring instruments were used to monitor the behaviour of the surrounding rock during excavation of the cavern. According to measurment of multi-stage extensometers, marked asymmetrical deformation behaviour with respect to the center line of cross sections occurred in the rock surrounding the cavern. This report gives the results of triaxial test using stiff loading machine carried out on the rock core to obtain a quantitative understanding of the difference in deformation.

2 GEOLOGICAL OUTLINE OF THE UNDERGROUND CAVERN

The geology of the site is made up of three types of rock - paleozoic strata (alternate layers of sandstone and slate), siliceous sandstone and breccia. In this context the term breccia is used to mean 'vent breccia' formed by brecciation due to

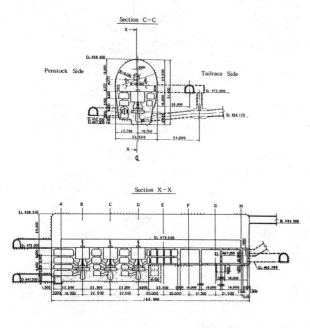

Figure 1. Imaichi Power Station

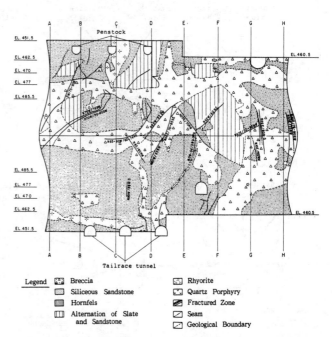

Figure 2. Developed diagram of geology of cavern walls

magma intrusion into the paleozoic strata at the end of the Mesozoic and Paleogene periods. (See Fig. 2)

The in-situ rock tests were conducted mainly on siliceous sandstone and breccia. In the jack tests and rock shear tests, very little difference was found between the two types of rock. In particular, the tangential elastic modulus at the time of unloading (average value for nine tests on both types of rock) was almost the same, that of breccia was 26.7 GPa, and that of siliceous sandstone was 26.6 GPa.

The longitudinal axis of the cavern was arranged almost parallel to the direction of the primary principal stress (12.3MPa) which was obtained by measurment.
As a result, ground pressure is symmetrically imposed to the rock with respect to the center line of the cross-section of the cavern.

3 RESULTS OBTAINED FROM MEASUREMENT OF ROCK BEHAVIOUR

Fig. 3 shows wall displacement after excavation obtained by multi-stage extensometers for each cross section.

The figure shows in general that displacement on the penstock side is larger than that on the tailrace side and this tendency was particularly marked at the machine hall (B, C, D sections) where the cross-section of the cavern is largest. Now concentrating on the wall of cross-sections (B, C, D), the displacement on the tailrace side is 10mm in maximum, whereas that on the penstock side is much larger ranging from 11 to 43 mm.

The measurements of the deformation behaviour mentioned previously were studied by a wide range of analyses including analysis of initial ground pressure, difference in the properties of the rock types involved and so on. However, in all cases, it was impossible to obtain satisfactory results. Therefore, the rock core tests described hereafter on two types of rock, breccia and siliceous sandstone, were carried out on the basis of the next reasons.

1. Making a comparison of the geology of the walls in cross-sections B to D, the tailrace side is composed mostly of siliceous sandstane and the penstock side is principally made up of breccia.

2. The results of conventional tests showed that there was practically no difference in physical properties of siliceous sandstone and breccia, nevertheless, a difference in deformation was demonstrated by measurements which cannot be explained by joint behaviour of rock in the vicinity of the walls.

4 LABORATORY TRIAXIAL TESTS USING STIFF LODING MACHINE

Fig. 4 shows a conceptual diagram of the stress working in the vicinity of the cavern walls. In the following laboratory tests, emphasis was placed on the confining pressure generated in the triaxial tests.

The equipment used in the tests was a triaxial testing apparatus for hard rock use with an overall stiffness of 1.5MN/mm.

Loading speed was set at 0.02%/min. Measurement of lateral strain was carried out on 50mm diameter test pieces. Two 120mm mono-axial strain gauges were wound around the circumference of the test piece and a simple average of the two measured values was taken.

4.1 Cyclic load releasing tests

In the first stage, load release of lateral pressure was conducted in a hydrostatic state, and the lateral strain was measured at this time.

Load was reimposed 10 times cyclically with the objective of resolving displacement into elastic and plastic components and axial deviation stress stood at 20MPa, 30MPa and at 40PMa. In the case of 20MPa, and 30MPa, one test sample of each type of rock was used. In the case of 40MPa, three samples of each type of rock were used. Fig. 5 shows the results, and conseqently the correlation between axial deviation stress and final strain is shown in

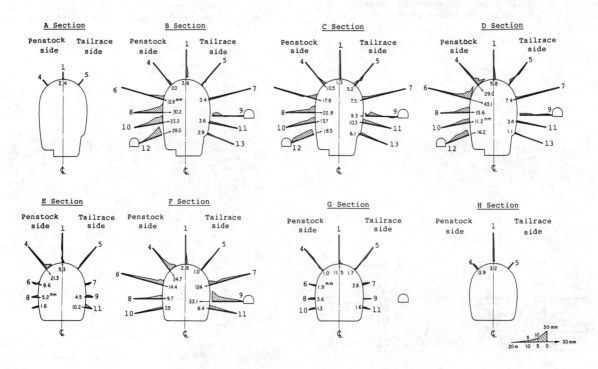

Figure 3. Displacement distribution of cavern walls

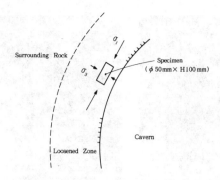

Figure 4. Conceptual diagram of the stress working in the vicinity of the cavern walls

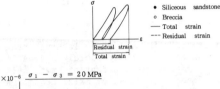

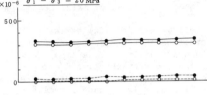

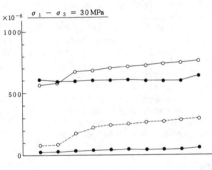

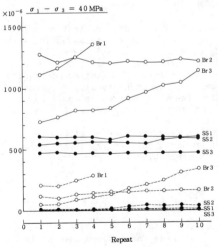

Figure 5. Strain history by cyclic load releasing

Fig. 6. As this figure clearly shows, at lower axial deviation stress there is almost no difference between samples. However, when axial deviation stress is increased to a higher level, the residual strain in the breccia builds up and at an axial stress level of 40MPa, total strain in breccia exhibits two times that of siliceous sandstone.

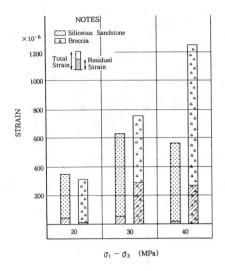

Figure 6. Final lateral strain

Table 1. Triaxial compression tests

	Confining pressure (M Pa)	Compressive strength (M Pa)	Modulus of elasticity (G Pa)	Poisson's ratio	Mode of failure
Siliceous Sandstone	1	109 (96−116)	34 (31−39)	0.11 (0.10−0.12)	A, A, B
	5	130 (110−147)	39 (19−50)	0.14 (0.11−0.16)	A, B', B
	10	149 (98−212)	42 (34−51)	0.11 (0.10−0.13)	B, B, A
Breccia	1	56 (54−57)	27 (25−32)	0.12 (0.10−0.13)	A', B, A'
	5	109 (99−121)	37 (33−43)	0.14 (0.12−0.16)	B, B', B
	10	122 (121−124)	43 (39−47)	0.18 (0.17−0.19)	A', B, B

Note : A : Vertical fracturing (strong crack directionality)
A' : Vertical fracturing (weak crack directionality)
B : Shear failure (single shear face)
B' : Shear failure (multiple shear faces)

4.2 Triaxial compression tests focussed on lateral strain near to the point of destruction

As seen in the previously described tests, there is a marked difference in the behaviour of breccia and siliceous sandstone at high levels of pressure. In the following tests of triaxial compression attention was focussed on lateral strain near to the point of destruction.

Tests were conducted on three core pieces for each breccia and siliceous sandstone at three levels of confining pressure, 1MPa, 5MPa and 10MPa (A total of 18 test pieces). Table 1 shows compressive strength, modulus of elasticity, poisson's ratio and mode of failure, and these values are more or less equivalent to values obtained by conventional tests.

Fig. 7 shows the correlation between axial deviation stress and axial, lateral strain at several confining pressures. In order to make a comparison between lateral strain in breccia and in siliceous sandstone, stress and strain are normalized by $(\sigma_1-\sigma_3)$max and $(\sigma_1-\sigma_3)$max/E at the time of destruction respectively. Based on the figure, there is practically no difference at a confining pressure of 1MPa, but as confining

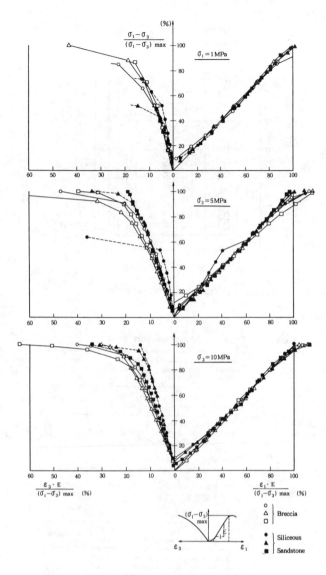

 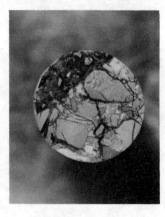

before   after

Photo 1. Core specimen before and after testing

pressure increases (5MPa, 10MPa), lateral strain of siliceous sandstone increases rapidly approaching the point of destruction but that of breccia tends to increase from a low level of axial deviation steress.

Ratio of the increase in axial stress ($\Delta\sigma_1$) to that in lateral strain ($\Delta\varepsilon_3$) is tentatively defined here as the modulus of lateral deformation $Dl(\Delta\sigma_1/\Delta\varepsilon_3)$, and the average value Dl at each stress is shown in Fig. 8. The figure shows that the ratio of $Dlss/DlBr$ is between 1 to 1.5 during low deviation stress but as the stress increases to some extent the ratio also increased.

With regard to rock core tests on breccia, Photo 1 shows a comparison of specimen before and after the test. In the photo taken after the test, a concentration of cracks in the boundary between the fragment and the matrix can be observed.

In confining hydrostatic pressure tests of breccia, pre-existing micro cracks were found out in the boundary between the fragment and the matrix. It is, therefore, assumed that as stress increases the cracks open wider due to dilatancy effect.

5 CLOSING REMARKS

As the asymmetrical wall displacement of the underground cavern during the construction was very pronounced, rock core tests and the analyses were carried out to determine the cause of the asymmetrical behaviour. The studies might give some informations for understanding deformation. In the cases like this cavern, it would be necessary to take into account the dilatancy of behaviour of the rock.

We wish to express our deep gratitude to Dr. Sugawara of Kumamoto University for his guidance and advice in connection with the compilation of this report.

REFERENCE

Mizukoshi T., & Y. Mimaki 1985. Deformation Behavior of a Large Underground Cavern. Rock Mechanics and Rock Engineering 18:227-251

Mimaki Y., & K. Matuo 1986. Investigations of Asymmetrical Deformation Behavior at the Horseshoe-shaped Large Cavern Opening. Large Rock Caverns:1337-1348

Figure 7. Correlation between axial deviation stress and axial, lateral strain

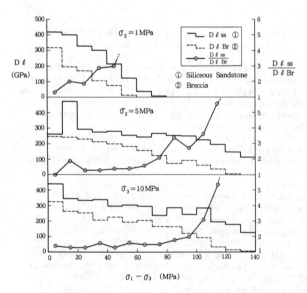

Figure 8. Modulus of lateral deformation

Integrated stability assessment of an underground cavern

Evaluation intégrée de la stabilité d'une caverne souterraine
Integrierte Stabilitätsbestimmung einer Untergrundkaverne

M.KUSABUKA, Hazama-Gumi Ltd, Tokyo, Japan
T.KAWAMOTO, Nagoya University, Japan
T.OHASHI, Hazami-Gumi Ltd, Tokyo, Japan
H.YOICHI, Hazama-Gumi Ltd, Tokyo, Japan

ABSTRACT: This paper describes stability assessment of an underground cavern for hydro-electric power plant in an integrated manner. The approaches covered in this study are fracture investigation, monitoring of deformation and properties of the rock mass and numerical stability analysis.

RESUME: Ce document décrit une manière intégrée d'evaluation de la stabilité d'une caverne souterraine pour une centrale hydro-électrique. Les méthodes que l'on a employé au cours de l'étude comprennent l'examen des fractures, la surveillance des déformations et des propriétés de la masse rocheuse et l'analyse numérique de la stabilité.

ZUSAMMENFASSUNG: In diesem Papier wird die Stabilitätsbestimmung einer Untergrund-Kaverne für ein Wasserkraftwerk in integrierter Weise beschrieben. Die in dieser Studie unternommene Vorgehensweisen sind Bruch-Untersuchungen, Überwachung von Deformationen, Eigenschaften von Felsmassen und numerische Stabilitätsanalyse.

1 INTRODUCTION

The structural stability of the underground cavern should be assessed in an integrated manner which combines all of the information available, to the full. This calls for close cooperation between the various engineers involved in the project. In this paper we describe how best to cope with the problem inherent in the the stability assessment of a certain underground cavern.

The cavern was excavated in the granite rock mass for the construction of an underground power plant. The cross section of the cavern is bullet shaped in appearance as illustrated in Figure 1. The rock making up the site is fairly good, as its uniaxial compression strength is approximately 1100 kgf/cm^2. But fractures (joint) were likely to affect the structural performance of the cavern. For excavation

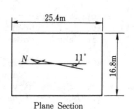

Plane Section

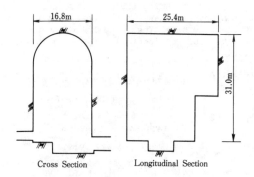

Cross Section   Longitudinal Section

Figure 1. Cavern dimensions.

the New Austrian Tunneling Method was employed. The task team on the contraction side was organized in order to perform safe and economical excavation work. This was made up of engineering geologists, rock mechanics engineers and numerical analysts, etc.

2 FRACTURE INVESTIGATION METHOD

Fracture characterization is one of the prerequisites in evaluating the cavern stability and in estimating the safety during excavation work. From a practical viewpoint, the fracture investigation method is required to possess the following capabilities; One is that it should be economical as well as practical. The other is that it should provide quantitative results that can be used directly as input data for numerical stability analysis.

In conventional methods, it seems hard to characterize the fracture distribution quantitatively, because firstly they mainly depend upon personal judgments of engineers and secondly limited number of sampling is inevitable.

In order to solve these problems, we have developed a new investigation and evaluation method that meets the above two requirements. We describe features of the method below.

In the investigation stage, we take photographs of excavated rock surfaces they are mutually perpendicular. Next, we pick out fractures by applying the image processing technique, and we can get fracture maps projected onto arbitrary three orthogonal rock surfaces.

In the evaluation stage, the fracture distribution is characterized quantitatively by employing stereology theory[1-3]. One of the features of this method is that a geometric relation between fractures and excavated surfaces is taken into consideration, and thus only fracture investigations on arbitrary three orthogonal surfaces are required. This also demonstrates the improved practical usefulness of the method.

Due to the lack of space, we have only described resultant equations. Detailed description of the method is presented by Kusabuka[4].

The distribution density $f(\theta_i)$ of fracture trace line with the direction θ_i is described as follows:

$$f(\theta_i) = \frac{C}{8\pi}\left\{1 - \sum_{n=2}^{\infty}{}'(n^2-1)(A_n \cos n\,\theta_i + B_n \sin n\,\theta_i)\right\}$$

$$C = \frac{2\pi}{M}\sum_{k=0}^{M=1}\overline{N}_k \,,\quad A_n = \frac{4\pi C}{M}\sum_{k=0}^{M-1}\overline{N}_k \cos\left(\frac{n\pi k}{M}\right) \qquad (1)$$

$$B_n = \frac{4\pi C}{M}\sum_{k=0}^{M-1}\overline{N}_k \sin\left(\frac{n\pi k}{M}\right)$$

where M indicates the total direction number of scanlines, which are generated in the fracture map. $\overline{N}_k$ is the average number that fractures intersect a scanline with k-th direction.

Here let us define the direction of fracture trace line as shown in Figure 2. Then, we get the following equations by considering the geometric relation between the fracture and the excavated surface.

$$f_1(\theta_1)/f_2(\theta_2) = |\sin\theta_j/\cos\theta_i| \simeq C_1$$

$$f_2(\theta_2)/f_3(\theta_3) = |\cos\theta_i/Q_{ij}| \simeq C_2 \qquad (2)$$

$$f_3(\theta_3)/f_1(\theta_1) = |Q_{ij}/\sin\theta_j| \simeq C_3$$

where $\quad Q_{ij} = \sqrt{\cos^2\theta_i \cos^2\theta_j + \sin^2\theta_i \sin^2\theta_j}$

Equation (2) means that one fracture set can be identified when an arbitrary triad f_1, f_2, f_3 satisfies the relation of equation (2).

Area density $f(\theta_i)$ of fractures, with directions θ, ϕ in the unit volume of the rock mass, is determined by:

$$F(\theta, \phi) = f_1(\theta_1)R_{ij}/|\sin\theta_j|$$

$$f_2(\theta_2)R_{ij}/|\cos\theta_i| \qquad (3)$$

$$f_3(\theta_3)R_{ij}/Q_{ij}$$

where $\quad R_{ij} = \sqrt{\cos^2\theta_i + \sin^2\theta_i \sin^2\theta_j}$

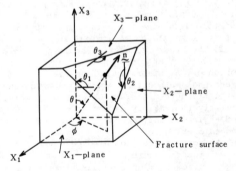

Figure 2. Definition of the direction of fracture.

3 MONITORING OF ROCK MASS PROPERTIES AND DEFORMATION DURING EXCAVATION

It seems crucial as well as useful to monitor changes in properties and deformation of the surrounding rock mass during the cavern excavation in evaluating the stability of the cavern. Measurements of seismic wave velocities, hydraulic conductivity and deformations of the surrounding rock mass were performed in every excavation stage as shown in Figure 3[5]. These measurement data are expected to provide us with useful information in the sense that they are measured before excavation as well as during excavation.

The features of each measurement system are;
 1. Seismic Wave Measurements : Using high frequency (10-5KHz) of wave in order to have accurate results.
 2. Lugeon Tests : Employing low injection pressures (0.5-2.0 kgf/cm^2) so that the side wall of the

cavern might not be damaged by the test itself. Also, measuring very low flow rate, as results of low pressure, by means of specially developed high accuracy of devices.
 3. Relative displacement Measurements : Installing the extensormeter from outside the cavern (the access tunnel) so that we could measure their initial values free of excavation effect.

In addition, some systematic deformation measurements were carried out[6], as also illustrated in Figure 3.

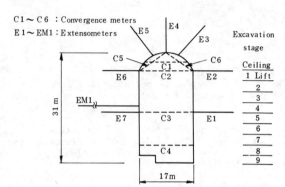

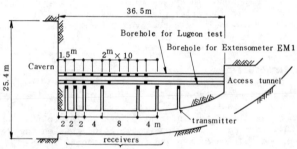

Figure 3. Measuring system for deformation and properties of the rock mass.

4 NUMERICAL MODELING OF THE ROCK MASS

4.1 Quantitative description of fracture systems

It is necessary to characterize fracture systems in the rock mass quantitatively, for the numerical stability analysis in which the effects of fractures are considered.

Fracture maps obtained through image processing are shown in Figure 4. By applying Eq.(1) and Eq.(2) to these fracture maps, we got the density functions, also shown in the figure. Interpretation of these density functions suggests that fracture trace lines in each map have two dominant directions. This also leads to the understanding that the fracture distribution in the rock mass is characterized by two dominant fracture sets. Then, the following two fracture sets could be identified by using Eq.(2) and Eq.(3).

Fracture set A
$$F = 0.782\,m^2/m^3, \quad \overline{\ell} = 0.89\,m$$
$$\underline{n} = (0.8466, -0.5269, 0.07513)$$

Fracture set B
$$F = 0.430\,m^2/m^3, \quad \overline{\ell} = 0.89\,m$$
$$\underline{n} = (0.2576, 0.3624, 0.8975)$$

where $\overline{\ell}$ and $\underline{n}$ indicate average spacing and unit normal vector of each fracture set, respectively.

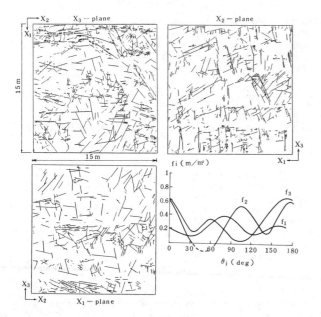

Figure 4. Fracture maps on X_1, X_2, X_3 planes and density functions f_1, f_2, f_3.

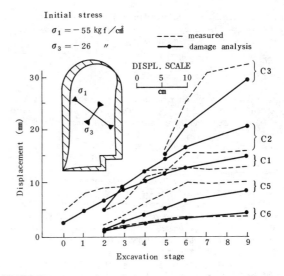

Figure 5. Comparison between measured convergences and computed results, the final deformation by the analysis and the initial stress state are illustrated.

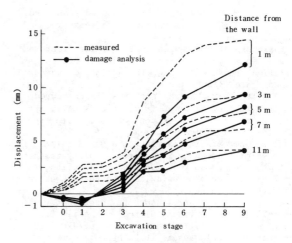

Figure 6. Comparison between measured relative displacements and computed results.

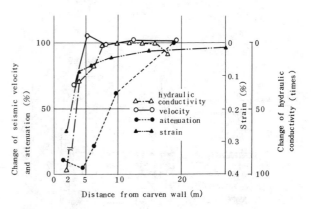

Figure 7. Changes in properties of the rock mass.

4.2 Numerical stability analysis

We used the damage model, developed by Kyoya et al, in order to consider the effect of fracture sets in the numerical stability analysis. The damage model has a merit that it can handle a great number of fractures efficiently. In this model, the geometry of fracture distribution in the rock mass is represented by the damage tensor.
From results of fracture characterization described above, the damage tensor for the rock mass was determined as follows:

$$\Omega = \sum_{i=1}^{m} \bar{\ell}_m F_m (\underset{\sim}{n}_m \otimes \underset{\sim}{n}_m)$$

$$= \Omega_A + \Omega_B = \begin{bmatrix} 0.5242 & -0.2748 & 0.1326 \\ & 0.2434 & 0.0966 \\ \text{Sym.} & & 0.3109 \end{bmatrix}$$

Here we could use the damage tensor directly as data for numerical analysis.
As detailed model descriptions are presented by Kyoya et al[7], it is noted here that the numerical analysis using the damage model requires not mechanical properties of the rock mass but those of the intact rock (rock specimens). Then, conventional triaxial tests were carried out for the intact rock specimens from the site. Young's modulus E, Poisson's ratio ν and a Mohr-Coulomb type of yield function are determined as follows:

$$E = 2.0 \times 10^5 \ \text{kgf}/\text{cm}^2, \quad \nu = 0.25$$

$$\tau = 210 + \sigma \tan 60° \ \text{kgf}/\text{cm}^2$$

We performed 2-dimensional finite element analysis under the initial stress states measured in situ[8].

5 RESULTS AND DISCUSSIONS

5.1 Deformation

Cavern Convergences measured between two points on the side wall are compared with computed results in Figure 5. The final deformation of the cavern estimated by the computation, is also presented in this figure. As it is not presented in this paper, the conventional FE analysis in which fracture ef-

fect was not taken into consideration was carried out as the preliminary study. But due to the presence of fractures, almost parallel to the longitudinal direction of the cavern, simulating larger horizontal deformation was not possible by the conventional FE analysis. On the other hand, the results using the damage model show fairly good agreement with measured deformation. As with the results obtained for convergences, relative displacements measured by extensometers are better simulated using the damage model(Figure 6).

5.2 Changes of rock properties

Figure 7 shows changes of such rock mass properties as hydraulic conductivity, seismic wave velocity and their amplitude. The results could be summarized as follows:
   1. As for the measured hydraulic conductivity, it increases remarkably around a range of 4m or so from the side wall.
   2. The seismic wave velocity decreases only within a rage of 2 - 4m from the side wall.
   3. Within a range of some 4m from the side wall, attenuation of seismic wave decreases remarkably. Though the stability assessment method that is based on rock properties has not been established yet, it could be said that the rock mass is damaged by excavations within a range of some 4m from the side wall of the cavern.

5.3 Fractured zones induced by excavation

Next, we focused the excavation induced fractured zones around the cavern through both measuring and analytical approach.
On the basis of the damage analysis, the fractured (yield) zone was estimated by using the fracture criterion presented before. The estimated fractured elements are illustrated in Figure 8.
The results show that the rock mass is fractured within a rage of 5m or so from the side wall. But, this is not sufficient enough for us to allow to make the final estimation.
Then, we show strain distributions based on measured relative displacements in Figure 7. These strain distributions could be calculated by assuming that the rock mass deforms uniaxially in the outward normal direction to the side wall.
Here, the critical strain concept, proposed by Sakurai[9], are very helpful tool for estimating fractured zones. As for the critical strain required for this, we could assume it to be some 0.1% from the triaxial tests mentioned before. Then, the rock mass was fractured within a range of 4m from the side wall in the sense that rock mass strains exceeded 0.1%.

Integrating the above two estimations with considerations for changes of rock properties, we finally estimated that the rock mass was fractured or damaged within a range of some 5m from the side wall.

6 CONCLUSIONS

In this paper, we took an integrated approach toward the stability assessment of an underground cavern. Through this study the following conclusions could be drawn;
   1. The fracture system in the rock mass can be best characterized quantitatively by using stereology theory.
   2. Changes in such rock properties as hydraulic conductivity, seismic wave velocity are useful indices that are sensitive to the effect of excavation.
   3. The damage model is capable of better simulating the anisotropic behavior of the discontinuous rock mass.
   4. The fractured zone in the vicinity of the excavated surfaces is assessed successfully by integrating monitoring results and the numerical analysis.

ACKNOWLEDGMENT

The authors wish to acknowledge Dr. Y.Ichikawa, Mr. T.Kyoya of Nagoya university for invariable suggestions and discussions.
We're also grateful to R.Yamasita of Hazama-Gumi, Ltd. for his kind assistance.

References

1. Weibel,E.R.1979. Stereological methods.,Practical methods for morphometry.vol.1,p.1-159.New York: Academic Press.
2. Weibel,E.R.1979. Stereological methods.,Theoretical foundations,vol.2.,p.264-308.New York:Academic Press.
3. Kanatani,K.1984. Stereological determination of structural anisotropy.,Int.J.Engng.Sci,vol.22.,5. p.531-546.
4. Kusabuka,M.1986. Quantitative description of discontinuities in the rock mass and a numerical analysis method for the rock structures.Doctor thesis of Tokyo university.(in Japanese)
5. Yoichi,H.et al.1986. Field measurements on the behavior of the rock surrounding an underground cavern during excavation.Proc.Int.Symp.Large Rock Cavern(Edited by K.H.O.Saari),vol.2,Helsinki, p.1099-1110.
6. Ohagi,T.et al.1985. Design and construction of Inagawa underground hyro-electric power station, Electric power civil engineering,vol.196. p.72-84.(in Japanese)
7. Kyoya,T.et al.1985. Damage mechanics theory for discontinuous rock mass.Proc.5th.Int.Conf.Num. Meth.in Geomech.(Edited by Kawamoto.T.),vol.2, Nagoya,p.469-480.
8. Kyoya,T.et al.1986. A damage mechanics analysis for underground excavation in jointed rock mass, Proc.Int.Symp.Engineering in Complex Rock Formations(Edited by Li Chengxiang),Beijing, p.506-513.
9. Sakurai,S.1981. Direct strain technique in construction of underground opening.Proc.22nd U.S. Rock.Mech.Symp,MIT,p.298-302.

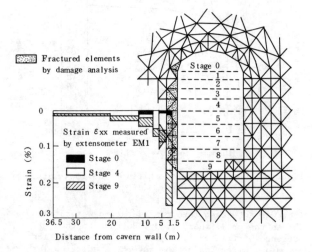

Figure 8. Estimated fractured zone by both the critical strain concept and the damage analysis.

A new linear criterion of brittle failure for rocks
Un nouveau critère linéaire de rupture fragile des roches
Ein neues lineares Bruchkriterium für Gesteine

M.KWAŚNIEWSKI, Silesian Technical University, Faculty of Mining, Gliwice, Poland

ABSTRACT: In the paper is presented a new linear strength criterion with the assumption that (i) decisive about the brittle failure of rock material are both the shearing stresses and the normal tensile stresses, (ii) in a clastic material, the tensile stresses may be evoked also when the applied stresses are the compressive ones, (iii) the relation between the shearing stresses and normal tensile ones is expressed by a linear function. In the new strength criterion $\sigma_1 = \sigma_C + d\,\sigma_{23}$, the coefficient d is a certain function of the so-called brittleness index of the rock $z=\sigma_C/\sigma_T$ and a parameter depending on the rock structure ν'. Taking as an example six different sandstones and granites, it has been shown that, contrary to Sondal-Botkin-Mirolubov criterion, Anderson criterion and modified Griffith criterion, the new linear criterion expresses at least well the ultimate strength of these rocks in the conditions of conventional triaxial compression, approximating the empirical data $\sigma_1=f(\sigma_3=\sigma_2)$ with the values of parameter ν' equal to 0.41÷0.45 for sandstones and 0.13÷0.21 for granites.

RESUME: Dans cet article on a présenté un critère nouveau de rupture fragile des roches en admettant que (i) ce sont des contraintes de cisaillement ainsi que des contraintes normales de traction qui décident de la rupture fragile des roches, (ii) qu'en matériau clastique des contraintes de traction peuvent apparaître aussi dans une situation où des contraintes appliquées ce sont des contraintes de compression, (iii) qu'on exprime la relation entre des contraintes des cisaillement et des contraintes normales de traction par la fonction linéaire. Dans le nouveau critère de rupture $\sigma_1 = \sigma_C + d\,\sigma_{23}$ le coefficient d c'est une fonction de l'indice de fragilité $z=\sigma_C/\sigma_T$ et du paramètre ν' dépendant de la structure d'une roche. On a montré sur six exemples des grès et des granites différents qu'au contraire du critère de Sondal-Botkin-Mirolubov, du critère d'Anderson et du critère modifié de Griffith, ce nouveau critère linéaire exprime au moins bien la résistance ultime de ces roches dans les conditions de la compression triaxiale de révolution, en approximant des données empiriques $\sigma_1=f(\sigma_3=\sigma_2)$ aux valeurs du paramètre ν' égales 0,41÷0,45 pour des grès et 0,13÷0,21 pour des granites.

ZUSAMMENFASSUNG: Im Artikel ist ein neues lineares Bruchkriterium angegeben worden, unter der Voraussetzung, daß (i) vom Sprödbruch des Gesteins sowohl Schubspannungen als auch normale Zugspannungen entscheiden, (ii) im klastischen Material können die Zugspannungen auch dann entstehen, wenn angelegte Spannungen Druckspannungen sind, (iii) der Zusammenhang zwischen Schubspannungen und Zugspannungen mit linearer Funktion ausgedrückt ist. Unter neuem Bruchkriterium $\sigma_1 = \sigma_C + d\,\sigma_{23}$ stellt der Koeffizient d eine Funktion des sogenannten Sprödigkeitswertes des Gesteins dar $z=\sigma_C/\sigma_T$ und des vom Gefüge abhängigen Parameters ν'. Am Beispiel von sechs unterschiedlichen Sandsteinen und Graniten ist gezeigt worden, daß im Gegensatz zum Sondal-Botkin-Mirolubov-Kriterium, Anderson-Kriterium und modifiziertem Griffith-Kriterium, das neue Linearkriterium mindestens gut die Grenzfestigkeit dieser Gesteine bei triaxialer Beanspruchung äußert, approximativierend empirische Angaben von konventionellen dreiachsigen Druckversuchen $\sigma_1=f(\sigma_3=\sigma_2)$ mit den Parameterwerten ν' gleich 0,41÷0,45 für Sandsteine und 0,13÷0,21 für Granite.

1 INTRODUCTION

The results of experimental and theoretical studies on the mechanism of brittle failure seem to indicate that in the case of the material subjected to a triaxial state of stress, the damage of its structure is not only the result of shearing stress or only normal stress; it is both of these stresses that are always decisive about the process of fracturing.

The shearing stresses loosen the material and prepare the fracture, however, the breaking of continuity (and macroscopic failure) of the material occurs as a result of the normal tensile stresses.

Such stresses may be evoked not only in the material subjected to tension (this case was analyzed by Pisarenko and Lebedev, 1969,

1976), but also when the so-called applied stresses are the compressive ones.

And it is just this case of a material subjected to triaxial compression in which tensile stresses occur as a result of compression, that will be examined in this paper with the purpose of deriving the criterion of brittle failure expressing the ultimate strength of such a material in the function of confining pressure.

2 THEORY OF BRITTLE FAILURE OF ROCKS UNDER TRIAXIAL COMPRESSION

On the basis of an analysis of the mechanism of deformation of a certain clastic model made up of discs with a definite diameter (representing atoms, molecules or grains) and connecting springs with a definite stiffness (representing atomic interaction forces, intermolecular forces or intergranular bonds), Trollope, 1966, 1968 (cf. also Brown and Trollope, 1967) has shown that if σ_1, σ_2 and σ_3 are applied stresses acting in three mutually perpendicular directions, then the corresponding effective stresses are defined by the following formulae:

$$\sigma_1' = \sigma_1 - \nu'(\sigma_2 + \sigma_3)$$

$$\sigma_2' = \sigma_2 - \nu'(\sigma_1 + \sigma_3) \qquad (1)$$

$$\sigma_3' = \sigma_3 - \nu'(\sigma_1 + \sigma_2)$$

where ν' is a certain parameter depending on the structure of the material (for ideally linear materials it may be equal to Poisson's ratio; in such case the effective stresses σ' may be expressed also by means of the strains corresponding to them ($\varepsilon' E$), and equations (1) take the form of the equations of the maximum elastic strain theory of strength).

From equations (1) it can clearly be seen that the stresses σ_1', σ_2' and σ_3' may take negative values (and so they may be tensile stresses) even when the "applied" stresses σ_1, σ_2 and σ_3 are positive, i.e. compressive ones.

In accordance with the assumption made at the beginning, that decisive about the failure of brittle material are the shearing stresses and the normal tensile stresses, the general strength criterion can thus be expressed by means of the function

$$F(\sigma_i, \sigma_3', p_j) = 0 \qquad (2)$$

where σ_i – stress intensity

$$(\sigma_i = \sqrt{2}\sqrt{(\sigma_1-\sigma_2)^2+(\sigma_2-\sigma_3)^2+(\sigma_3-\sigma_1)^2}/2 = 3\tau_{oct}/\sqrt{2}),$$

τ_{oct} – octahedral shearing stress,
σ_3' – minimum effective stress which for $\nu'(\sigma_1+\sigma_2) > \sigma_3$ is the tensile stress,
p_j – material constants.

Let us assume in the first approximation that the function (2) has a linear form

$$\sigma_i + p_1 \sigma_3' = p_2 \qquad (3)$$

From the condition that for $\sigma_3=\sigma_2=0$ (uniaxial compression), $\sigma_1=\sigma_C$, whereas for $\sigma_1=\sigma_2=0$ (uniaxial tension), $\sigma_3=-\sigma_T$, we get from (3) that

$$p_1 = \frac{\sigma_C - \sigma_T}{\nu'\sigma_C - \sigma_T} \qquad (4)$$

$$p_2 = \frac{(\nu' - 1)\sigma_C\sigma_T}{\nu'\sigma_C - \sigma_T} \qquad (5)$$

For the conventional triaxial compression and hence the case when the applied stresses satisfy the inequality $\sigma_1 > \sigma_2 = \sigma_3 (= \sigma_{23})$ the criterion (3) takes the form

$$(\sigma_1-\sigma_{23}) + p_1[\sigma_{23} - \nu'(\sigma_1+\sigma_{23})] = p_2 \qquad (6)$$

and, in the form of the function $\sigma_1 = f(\sigma_{23})$ –

$$\sigma_1 = \frac{p_2}{1 - p_1\nu'} + \frac{1 - p_1 + p_1\nu'}{1 - p_1\nu'}\sigma_{23} \qquad (7)$$

After substituting in (7) the expressions (4) and (5) we finally obtain the linear strength criterion in the form

$$\sigma_1 = \sigma_C + \frac{(1 - 2\nu')\sigma_C + \nu'\sigma_T}{(1 - \nu')\sigma_T}\sigma_{23} \qquad (8)$$

or

$$\sigma_1 = \sigma_C + \frac{\nu' + (1 - 2\nu')z}{1 - \nu'}\sigma_{23} \qquad (9)$$

where z – the so-called brittleness index, $z=\sigma_C/\sigma_T$.

Criterion (9) is valid for $0 \leq \nu' \leq 0.5$. For $\nu'=0$, it passes into the so-called Anderson criterion –

$$\sigma_1 = \sigma_C + z\sigma_3 \qquad (10)$$

whereas for $\nu'=0.5$ it takes the form of Huber-Mises-Hencky yield criterion –

$$\sigma_1 = \sigma_C + \sigma_{23} \qquad (11)$$

In Fig.1 are shown plots $\sigma_1/\sigma_C=f(\sigma_{23}/\sigma_C)$ expressing the general criterion (9) for the brittleness index z equal 12.0 and values of the parameter ν' varied over the range from 0.0 to 0.5.

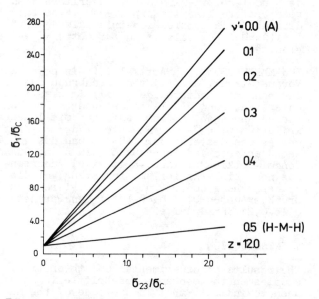

Fig.1. Graphical interpretation of the new, general linear strength criterion $\sigma_1=f(\sigma_{23},\nu',z)$ for different values of the structural constant ν' (with the brittleness index $z=\sigma_C/\sigma_T$ equal 12.0).

3 DISCUSSION

It should be noted that the new general criterion of brittle failure proposed here (9) belongs to a group of linear criteria of the type

$$\sigma_1 = \sigma_C + g(z)\sigma_3 \qquad (12)$$

just as the already-mentioned Anderson criterion (1942, 1951) -

$$\sigma_1 = \sigma_C + z\,\sigma_3 \qquad (13)$$

Sondal (1925) - Botkin (1940) - Mirolubov (1953) criterion -

$$\sigma_1 = \sigma_C + \frac{3z-1}{2}\,\sigma_{23} \qquad (14)$$

and modified Griffith criterion (McClintock and Walsh, 1962; Murrell, 1964) -

$$\sigma_1 = \sigma_C + \frac{1}{16}\,z^2\,\sigma_3 \qquad (15)$$

(if it is assumed that in the case of long, narrow microcracks, the stress σ_{cc} necessary to close them is equal to zero).

In Fig.2, all these three criteria as well as the new criterion

$$\frac{\sigma_1}{\sigma_C} = 1 + \frac{\nu' + (1-2\nu')z}{1-\nu'}\,\frac{\sigma_{23}}{\sigma_C} \qquad (16)$$

are illustrated by plots $\sigma_1 = f(\sigma_3)$ for $z=12.0$ (and $\nu'=0.3$):

S-B-M criterion -

$$\frac{\sigma_1}{\sigma_C} = 1 + 17.5\,\frac{\sigma_{23}}{\sigma_C} \qquad (17)$$

Anderson criterion -

$$\frac{\sigma_1}{\sigma_C} = 1 + 12\,\frac{\sigma_3}{\sigma_C} \qquad (18)$$

modified Griffith criterion -

$$\frac{\sigma_1}{\sigma_C} = 1 + 9\,\frac{\sigma_3}{\sigma_C} \qquad (19)$$

and the criterion proposed by the author of this paper -

$$\frac{\sigma_1}{\sigma_C} = 1 + 7.3\,\frac{\sigma_{23}}{\sigma_C} \qquad (20)$$

The new general linear criterion expresses better than those three criteria known so far, the actual triaxial strength of rocks without overestimating it as they do. (NOTE: However, for $\nu'<(z^2-16z)/(z^2-32z+16)$ the new criterion gives higher values of strength than those resulting from the modified Griffith criterion (15).)

This can be seen particularly well in Tables 1 and 2 (as well as Fig.3 and 4) in which are given the results of an examination to what extent the individual strength criteria correspond to the empirical data $\sigma_1 = f(\sigma_3 = \sigma_2)$ from the conventional triaxial compression tests of six different rocks.

Data for three granites were selected for the analysis: from Westerly (Rhode Island, USA), from Stripa (central part of Sweden) and from Strzelin (SW part of Poland), with

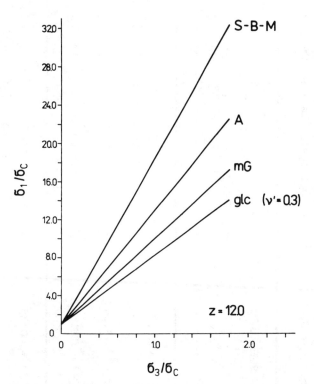

Fig.2. Comparison of the limiting major principal stresses σ_1 resulting from the new, general linear criterion of brittle failure (glc), from modified Griffith criterion (mG), from Anderson criterion (A), and from Sondal-Botkin-Mirolubov criterion (S-B-M).

the brittleness index $z=\sigma_C/\sigma_T$ equal respectively to 10.9, 13.9 and 15.0. They were tested under an axially symmetrical state of compresive stress with confining pressures up to 18 MPa (Strzelin-K granite), 30 MPa (Stripa granite) and 150 MPa (Westerly-B granite) and underwent brittle failure at these pressures.

Data for three sandstones were also selected: Kuzbass sandstone from the Kuznetsk Coal Basin (Asiatic part of the Russian Federal SSR), Izumi (BSH) sandstone from Shikoku island (Japan) and Anna sandstone from the Rybnik Coal District (S part of Poland), with the brittleness index equal respectively to 12.1, 16.5 and 9.9. The sandstones were tested at confining pressures up to 54 MPa - Anna sandstone, 94 MPa - Kuzbass sandstone and 150 MPa - Izumi (BSH) sandstone. At these pressures they were subject to brittle failure and only the Izumi (BSH) sandstone showed at pressures 100 and 150 MPa transitional behaviour (T) between the brittle (B) and ductile one (D).

Results of the analysis show, that in the case of all the three sandstones the strength criterion of Sondal-Botkin-Mirolubov, Anderson's criterion and modified Griffith's criterion markedly overestimate the triaxial strength, giving definitely poor, inadequate fit to the empirical data with the coefficient of accordance φ^2 (see Appendix) much higher than 0.5, equal at best to 2.005 (Anna sandstone, modified Griffith criterion).

The Sondal-Botkin-Mirolubov criterion describes inadequately also the triaxial strength of all the three granites. Only Anderson criterion sufficiently fits the empirical data in the case of Westerly-B and Strzelin-K granites, and satisfactorily in the case of Stripa granite, and the modified Griffith criterion sufficiently corresponds to the empirical results of triaxial tests of the

Table 1. Results of an assessment of the accordance of the four linear strength functions $\sigma_1 = f(\sigma_3)$ with the empirical data from the conventional triaxial compression tests of three different granites.

R O C K	ANDERSON CRITERION	MODIFIED GRIFFITH CRITERION	SONDAL-BOTKIN-MIROLUBOV CRITERION	NEW CRITERION OF BRITTLE FAILURE
	$\sigma_1 = \sigma_C + a\sigma_3$ $a = z$	$\sigma_1 = \sigma_C + b\sigma_3$ $b = z^2/16$	$\sigma_1 = \sigma_C + c\sigma_{23}$ $c = (3z-1)/2$	$\sigma_1 = \sigma_C + d\sigma_{23}$ $d = [\nu' + (1-2\nu')z]/(1-\nu')$
Westerly-B fine-grained granite (Brace, 1964) $\sigma_C = 229.0$ MPa, $\sigma_T = 21.0$ MPa, $z = 10.9$, $n = 0.9\%$ (cf. Wawersik and Brace, 1971) $\sigma_3 = \sigma_2 = p$ up to 150 MPa	$a = 10.9$ $\varphi^2 = 0.27345$ sufficient fit	$b = 7.43$ $\varphi^2 = 0.02869$ good fit	$c = 15.85$ $\varphi^2 = 2.248$ insufficient fit	$d = 8.26$ $(z = 10.9,\quad \nu' = 0.21)$ $\varphi^2 = 0.00161$ very good fit
Stripa (coarse-grained) granite (Swan, 1978) $\sigma_C = 207.6$ MPa, $\sigma_T = 15$ MPa, $z = 13.9$ $\sigma_3 = \sigma_2 = p$ up to 30 MPa	$a = 13.9$ $\varphi^2 = 0.16459$ satisfactory fit	$b = 12.08$ $\varphi^2 = 0.08480$ good fit	$c = 20.35$ $\varphi^2 = 1.607$ insufficient fit	$d = 11.98$ $(z = 13.9,\quad \nu' = 0.13)$ $\varphi^2 = 0.08461$ good fit
Strzelin-K fine-grained granite (Kwaśniewski et al., 1981, 1982) $\sigma_C = 202.6$ MPa, $\sigma_T = 13.5$ MPa, $z = 15.0$, $n = 0.92\%$ $\sigma_3 = \sigma_2 = p$ up to 18 MPa $\dot\sigma = 0.6 \div 1$ MPa/s	$a = 15.0$ $\varphi^2 = 0.43364$ sufficient fit	$b = 14.06$ $\varphi^2 = 0.25146$ sufficient fit	$c = 22.0$ $\varphi^2 = 3.585$ insufficient fit	$d = 11.52$ $(z = 15.0,\quad \nu' = 0.20)$ $\varphi^2 = 0.04266$ good fit

σ_C - uniaxial compressive strength, σ_T - uniaxial tensile strength, $z = \sigma_C/\sigma_T$ - brittleness index, n - porosity, $\dot\sigma$ - stress rate

Table 2. Results of an assessment of the accordance of the four linear strength functions $\sigma_1 = f(\sigma_3)$ with the empirical data from the conventional triaxial compression tests of three different sandstones.

R O C K	ANDERSON CRITERION	MODIFIED GRIFFITH CRITERION	SONDAL-BOTKIN-MIROLUBOV CRITERION	NEW CRITERION OF BRITTLE FAILURE
	$\sigma_1 = \sigma_c + a\,\sigma_3$	$\sigma_1 = \sigma_c + b\,\sigma_3$	$\sigma_1 = \sigma_c + c\,\sigma_{23}$	$\sigma_1 = \sigma_c + d\,\sigma_{23}$
	$a = z$	$b = z^2/16$	$c = (3z - 1)/2$	$d = [\nu' + (1-2\nu')z]/(1-\nu')$
Kuzbass sandstone (Kuntysh, 1964; Ilnitskaya et al., 1969)	$a = 12.1$	$b = 9.15$	$c = 17.65$	$d = 4.45$ ($z = 12.1$, $\nu' = 0.41$)
$\sigma_c = 115.2$ MPa, $\sigma_T = 9.5$ MPa, $z=12.1$ $\sigma_3 = \sigma_2 = p$ up to 94 MPa	$\varphi^2 = 7.618$	$\varphi^2 = 2.893$	$\varphi^2 = 22.624$	$\varphi^2 = 0.02558$
	insufficient fit	insufficient fit	insufficient fit	good fit
Izumi (BSH) mediumgrained sandstone (Hoshino and Mitsui, 1975)	$a = 16.5$	$b = 17.02$	$c = 24.25$	$d = 3.85$ ($z = 16.5$, $\nu' = 0.45$)
$\sigma_c = 187$ MPa, $z=16.5$ (cf. Otsuka and Kobayashi, 1982), $n=4.1$ % $\sigma_3 = \sigma_2 = p$ up to 150 MPa $\dot\epsilon = 10^{-4} \div 10^{-5}/s$	$\varphi^2 = 33.413$	$\varphi^2 = 36.191$	$\varphi^2 = 86.868$	$\varphi^2 = 0.01757$
	insufficient fit	insufficient fit	insufficient fit	good fit
Anna fine-mediumgrained sandstone (Borecki et al., 1982; Kwaśniewski et al., 1983)	$a = 9.9$	$b = 6.13$	$c = 14.35$	$d = 3.48$ ($z = 9.9$, $\nu' = 0.42$)
$\sigma_c = 115.4$ MPa, $z=(9.9)$, $n=4.6$ % $\sigma_3 = \sigma_2 = p$ up to 54 MPa $\dot\sigma = 0.4 \div 0.5$ MPa/s	$\varphi^2 = 11.675$	$\varphi^2 = 2.005$	$\varphi^2 = 33.414$	$\varphi^2 = 0.02421$
	insufficient fit	insufficient fit	insufficient fit	good fit

σ_c – uniaxial compressive strength, σ_T – uniaxial tensile strength, $z=\sigma_c/\sigma_T$ – brittleness index, n – porosity, $\dot\epsilon$ – strain rate, $\dot\sigma$ – stress rate

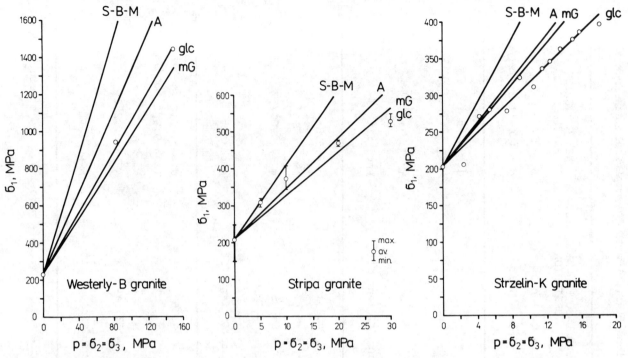

Fig.3. Empirical data from conventional triaxial compression tests of three different granites (circles), together with plots $\delta_1 = f(\delta_3)$ corresponding to Sondal-Botkin-Mirolubov criterion (S-B-M), to Anderson criterion (A), to modified Griffith criterion (mG), and to the new, general criterion given by the author (glc) (cf. also Table 1).

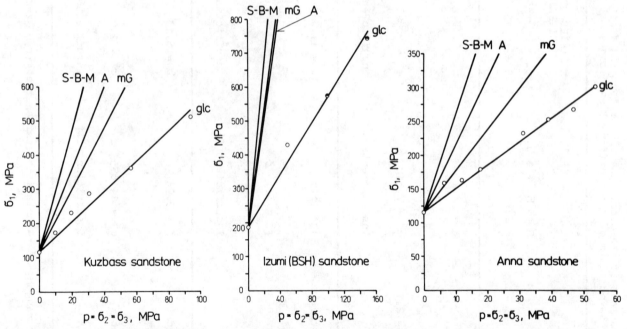

Fig.4. Empirical data from conventional triaxial compression tests of three different sandstones (circles), together with plots $\delta_1 = f(\delta_3)$ corresponding to Sondal-Botkin-Mirolubov criterion (S-B-M), to Anderson criterion (A), to modified Griffith criterion (mG), and to the new, general criterion given by the author (glc) (cf. also Table 2).

Strzelin-K granite and good to the results of tests of the Westerly-B and Stripa granites.

On the contrary, the new, general linear criterion proposed in this paper approximated in all the six cases of the rocks analyzed, the empirical data $\delta_1 = f(\delta_2 = \delta_3)$ at least good, giving once (for the Westerly-B granite) even very good fit ($\varphi^2 < 0.01$).

4 FINAL REMARKS

- A theory of brittle failure of rocks in the conditions of triaxial compression has been presented, the basis of which is the assumption that decisive about the failure of a material in a triaxial state of stress are shearing stresses as well as normal tensile stresses which may be induced in a material of clastic texture.

- The linear strength criterion resulting from this theory for the applied stresses $\sigma_1 > \sigma_2 = \sigma_3$ (an axially symmetrical state of compressive stresses) has the form

$$\sigma_1 = \sigma_C + d\,\sigma_{23} \qquad (21)$$

The coefficient d is a certain function of the so-called brittleness index of rock $z = \sigma_C/\sigma_T$ and a parameter depending on the rock fabric v'.

- It has been shown on the example of three granites and three sandstones that the new criterion describes good or even very good (as in the case of the Westerly-B granite), the triaxial strength of these rocks in the conditions of conventional triaxial compression. Using the least squares technique, the strength function (9) was fitted to the empirical data $\sigma_1 = f(\sigma_2 = \sigma_3)$. It has been found that the values of parameter v' for the granites analyzed (with the experimentally determined brittleness index z = 10.9 ÷ 15.0 and porosity lower than 1 %) are from 0.13 to 0.21, whereas for sandstones (with z = 9.9 ÷ 16.5 and porosity lower than 5 %)- from 0.41 to 0.45.

- It should be remembered that, as shown by Mogi (1974), the linear strength criterion of the type $\sigma_1 = \sigma_C + \alpha\sigma_3$ has, in principle, closely limited range of validity (applicability). It is not valid for high confining pressures $(p=\sigma_2=\sigma_3)$, at which rocks lose the brittle character and a transition occurs, from the brittle to the ductile state. In the case of rocks with appreciable crack porosity it is not valid also for low-pressure region characterized by the effect of high pressure -sensitivity of strength (cf. Fig.4).

APPENDIX

The coefficient of accordance

$$\varphi^2 = \frac{\sum(\sigma_{1_{exp}} - \sigma_{1_{cal}})^2}{\sum(\sigma_{1_{exp}} - \bar{\sigma}_{1_{exp}})^2} \qquad (A1)$$

where $\bar{\sigma}_{1_{exp}} = \frac{1}{m}\sum_{i=1}^{m}\sigma_{1_{exp}}$,

m – number of experimental data σ_{3_i}, σ_{1_i} (the sample size),

serves for estimating how well the particular strength functions $\sigma_1 = f(\sigma_3)$ correspond to the empirical data:

$\varphi^2 = 0$ – excellent (ideal) fit,
$0 < \varphi^2 \leqslant 0.01$ – very good fit,
$0.01 < \varphi^2 \leqslant 0.10$ – good fit,
$0.10 < \varphi^2 \leqslant 0.25$ – satisfactory fit,
$0.25 < \varphi^2 \leqslant 0.50$ – sufficient fit,
$\varphi^2 > 0.50$ – insufficient fit.

(Please note, that in this paper it was only the strength function (9) that was actually fitted, by means of a linear regression analysis, to the empirical data. In the case of strength criteria (13), (14) and (15) which are given in an explicit form, values of limiting maximum stress $\sigma_{1_{cal}}$ have been calculated from particular equations and, then, compared with experimentally determined

values $\sigma_{1_{exp}}$.)

REFERENCES

Anderson, E.M. 1942. The Dynamics of Faulting, London: Oliver and Boyd,

Anderson, E.M. 1951. The Dynamics of Faulting and Dyke Formation with Applications to Britain, 2nd edn. Edinburgh: Oliver and Boyd.

Borecki, M., M.Kwaśniewski, J.Pacha, S.Oleksy, Z.Berszakiewicz and J.Guzik 1982. Wytrzymałość trójosiowa dwu mineralogiczno/ -diagenetycznych odmian drobno-średnioziarnistych piaskowców karbońskich PNIÓWEK i ANNA badanych na ściskanie przy ciśnieniach okólnych do 60 MPa. Prace Instytutu Projektowania, Budowy Kopalń i Ochrony Powierzchni Politechniki Śląskiej, 119/2.

Botkin, A.I. 1940. O prochnosti sypuchikh i khrupkikh materialov. Izvestiya NII gidrotekhniki 26: 205-235.

Brace, W.F. 1964. Brittle fracture of rocks. In W.R.Judd (ed.), State of Stress in the Earth's Crust, p.110-178. New York: Elsevier.

Brown, E.T. and D.H.Trollope 1967. The failure of linear brittle materials under effective tensile stress. Rock Mechanics and Engineering Geology 5: 229-241.

Hoshino, K. and S.Mitsui 1975. Mechanical properties of palaeogene and cretaceous rocks in Shikoku under high pressure. Journal of the Japanese Association of Petroleum Technologists 40: 166-173 (in Japanese).

Ilnitskaya, E.I., R.I.Teder, E.S.Vatolin and M.F.Kuntysh 1969. Svoystva gornykh porod i metody ikh opredeleniya. Moskva: Nedra.

Kuntysh, M.F. 1964. Issledovanie metodov opredeleniya osnovnykh fiziko-mekhanicheskikh kharakteristik gornykh porod, ispolzuemykh pri reshenii zadach gornogo davleniya. Kandidatskaya dissertatsiya, Moskva.

Kwaśniewski, M., J.Pacha, Z.Berszakiewicz and S.Oleksy 1981. Odkształceniowe i wytrzymałościowe własności drobnoziarnistego granitu STRZELIN i trzech strukturalnych odmian piaskowców karbońskich PNIÓWEK i JASTRZĘBIE w warunkach konwencjonalnego trójosiowego ściskania przy ciśnieniach do 60 MPa. Prace Instytutu Projektowania, Budowy Kopalń i Ochrony Powierzchni Politechniki Śląskiej, 183/MR.I-16.

Kwaśniewski, M., J.Pacha and S.Oleksy 1982. Odkształceniowe i wytrzymałościowe własności średnioziarnistego granitu STRZEBLÓW i mikrytowego wapienia DĘBNIK w warunkach konwencjonalnego trójosiowego ściskania przy ciśnieniach do 48 MPa. Prace Instytutu Projektowania, Budowy Kopalń i Ochrony Powierzchni Politechniki Śląskiej, 183/MR. I-16.

Kwaśniewski, M., J.Pacha and S.Oleksy 1983. Wytrzymałość trójosiowa dwu mineralogiczno/ diagenetycznych odmian drobno-średnioziarnistych piaskowców karbońskich PNIÓWEK i ANNA. Zeszyty Naukowe Politechniki Śląskiej, Górnictwo 128: 265-287.

McClintock, F.A. and J.B.Walsh 1962. Friction on Griffith cracks under pressure. Proceedings of the 4th U.S. National Congress of Applied Mechanics, Berkeley, p.1015-1021.

Mirolubov, I.N. 1953. K voprosu ob obobshche-

nii teorii prochnosti oktaedricheskikh ka-
satelnykh napryazheniy na khrupkie materia-
ly. Trudy Leningradskogo tekhnologichesko-
go instituta 25: 42-52.

Mogi, K. 1974. On the pressure dependence of
strength of rocks and the Coulomb fracture
criterion. Tectonophysics 21: 273-285.

Murrell, S.A.F. 1964. The theory of the propagation of elliptical Griffith cracks under various conditions of plane strain or
plane stress: Parts II and III. Br.J.Appl.
Phys. 15: 1211-1223.

Otsuka, N. and R.Kobayashi 1982. Studies on
fracture toughness of various rocks. Journal of the Mining and Metallurgical Institute of Japan 98: 1-6 (in Japanese).

Pisarenko, G.S. and A.A.Lebedev 1969. Sopro-
tivlenie materialov deformirovaniyu i raz-
rusheniyu pri slozhnom napryazhennom sostoyanii. Kiev: Naukova Dumka.

Pisarenko, G.S. and A.A.Lebedev 1976. Defor-
mirovanie i prochnost' materialov pri slozhnom napryazhennom sostoyanii. Kiev: Naukova Dumka.

Sondal, 1925. Festigkeitsbedingungen. Leip-
zig.

Swan, G. 1978. The mechanical properties of
Stripa granite. Swedish-American Cooperative Program on Radioactive Waste Storage in
Mined Caverns in Crystalline Rock, Technical Project Report No.3. Berkeley: Lawrence Berkeley Laboratory, University of California.

Trollope, D.H. 1966. Discussion on Theme 3 -
"Properties of rock and rock masses". Pro-
ceedings of the 1st Congress of the ISRM,
Lisbon, Vol.III.

Trollope, D.H. 1968. The mechanics of discon-
tinua or clastic mechanics in rock problems. In K.G.Stagg and O.C.Zienkiewicz (eds),
Rock Mechanics in Engineering Practice, p.
275-320. London: John Wiley and Sons.

Wawersik, W.R. and W.F.Brace 1971. Post-
failure behaviour of a granite and diabase.
Rock Mechanics 3: 61-85.

Deformation and fracturing process of discontinuous rock masses and damage mechanics

Déformation et procédé de fracture d'une masse rocheuse discontinue et mécanique d'endommagement
Deformations- und Bruchprozesse von diskontinuierlichen Felsmassen und Schadensmechanik

T.KYOYA, Nagoya University, Japan
Y.ICHIKAWA, Nagoya University, Japan
T.KAWAMOTO, Nagoya University, Japan

ABSTRACT: A damage mechanics theory for discontinuous rock mass is proposed and applied to an excavation problems of underground cavern of three different shapes for a hydro-electric power station. Distributed joints in rock mass are characterized by a second-order symmetric tensor, called the damage tensor which is evaluated by a technique of three-dimensional photographic surveying with a micro-computer system. Mechanical effects of joints are represented by an additional force vector by introducing the net stress which is derived from the stress vector acting on the effective part of the material body. Results of the damage analysis are compared with those by a conventional analysis , and the influences of joints on the stability of each shaped cavern is checked

RESUMÉ: Une théorie de mécanique de dommage d'une mass rocheuse discontinue est proposée et appliquée à un problème d'excavation d'une caverne souterraine de trois formés différentes pour une centrale hydro-électrique. Les diaclases distribué dans la masse roche sont caractérisées par un tenseur symétrique de second-degré, designé comme le tenseur de dommage qui est ici mésuré par relevé photographie trois-dimensional assisté par ordinateur. Les effects de mécanique des diaclases peuvent etre représentés par un vecteur additionel en introduisant l'effect de contrainte réel dérivé du vecteur de contrainte qui agit sur le section effectif de corps matériau. Les résultats de l'analyse des dommages sont ensuite comparés avec ceux obtenus par analyse conventionnelle et l'influence des diaclase sur la stabilité de chaque forme de caverne est évaluée.

ZUSAMMENFASSLING: Eine Theorie für Schadensmechanik von diskontinuierlichen Felsmassen wird vorgeschlagen und auf Abtragungs-probleme bei drei unterschiedlich geformten Untergrundkavernen eines hydroelektrischen Kraftwerkes angewendet. Verteilte klüfte in Felsmassen werden durch einen symmetrischen Tensor zweiter Ordnung, den Schadenstensor, beschrieben, der über dreidimensionales fotografisches Vermassen mit Hilfe eines Mikro-Computers ausgewertet werden kann. Mechanische Effekte von klüften werden durich einen zusätzlichen Kraftvektor dargestellt, indem die resultierende Spannung eingeführt wird, die aus dem Spannungsvektor, der auf den effektiven Teil des Materials wirkt, abgeleitet wird. Resultate der Schadensanalyse werden mit denen einer konventionellen Analyse verglichen, und der Einfluss von klüften auf die Stabilität der unterschiedlich geformten kavernen wird untersucht.

1 INTRODUCTION

The stability of engineering structures constructed in/on rock mass is mainly concerned with the mechanical effects of discontinuities such as fault and joint. Several numerical models have been proposed to analyse the effects of discontinuities (e.g. Goodman, Taylor and Brekke 1968, Zienkiewicz, Best, Dullage and Stagg 1970, Ghaboussi, Wilson and Isenberg 1973), however, they are effective only for discontinuities of relatively large scale since they all attempt to model the geometrical and mechanical characteristics of distinct discontinuities. Distributed discontinuities, such as joints, are complicated in their geometry, so that it is difficult to model each of them explicitly. There is neither excellent theory nor powerful numerical method to analyse the effects of the distributed discontinuities

We have introduced the damage concept for the distributed discontinuities, and have proposed a theory to treat the the mechanical behaviours of discontinuous rock mass (Kyoya, Ichikawa and Kawamoto 1985a,b, 1986). In that theory, distributed discontinuities are characterized by the damage tensor which is determined from in-situ data of joint sets. Then, their mechanical effects are accounted by the net stress which is a transformation of the Cauchy stress by using the damage tensor. The components of the net stress have a meaning of the traction acting on the effective resisting part in the damaged material body. The constitutive law is given between the net stress and the strain, and it is determined only from the usual laboratory tests of intact rocks. By introducing the damage concept, mechanical behaviours of discontinuous rock mass are reasonably treated in a framework of continuum mechanics.

In the previous work (Kyoya, Ichikawa, Kusabuka and Kawamoto 1986), the damage mechanics theory was applied to several laboratory tests on cracked specimens and an excavation problem of a small sized underground cavern for a hydro-

electric power station. Numerical results gave good agreement with measured data, and the efficiency of the theory was confirmed.

In this paper, an excavation problem of an underground power station is solved for three different types of cavern, that is, the traditional mushroom shape, the horse-shoe shape (Wardech Ⅱ type) and the bullet shape. The damage tensor of the rock mass is evaluated by a three-dimensional photographic surveying with a micro-computer system proposed by Kondoh et al.(1986). Results of the damage analysis are compared with those by the conventional finite element analysis, and the influences of distributed joints on the mechanical behaviours of each shape of the caverns are discussed.

2 EXCAVATION ANALYSIS BY DAMAGE MECHANICS THEORY

The damage mechanics theory for the jointed rock mass is outlined here, and its numerical procedure for the excavation analysis is developed.

2.1 Damage tensor for rock mass

A rock mass is considered to consist of many foundamental block elements and joints which are distributed and may propagate along the boundary surfaces of the elements. Let the representative volume of the foundamental elements be v, which is related to the average spacing of joints l as $v=l^3$, then the damage tensor for the rock mass is defined by

$$\underset{\sim}{\Omega} = \frac{1}{V} \sum_{k=1}^{N} a^k (\underline{n}^k \otimes \underline{n}^k) \tag{1}$$

where V is the volume of a mass which includes N joints, a^k the area of the k-th joint surface, $\underline{n}^k$ the unit normal vector of the surface, and $\otimes$ denotes the tensor product. The damage tensor is associated with two physical meanings; the direction of the set of joints and its areal density.

It is impossible to measure a and n for all joints. However, we can fortunately observe some dominant directions and scales of joint sets in rock masses. Then, by in-situ observation, the damage tensor is specified for each set as

$$\underset{\sim}{\Omega} = \frac{1}{V} \, \bar{a} \, \bar{N} \, (\bar{\underset{\sim}{n}} \otimes \bar{\underset{\sim}{n}}) \tag{2}$$

where $\bar{a}$ is the average area of joint surfaces, $\bar{N}$ the number of joints involved in the mass volume V, and $\bar{\underset{\sim}{n}}$ the unit vector normal to the dominant direction of the joint set. Then, by summing up all of them, the damage tensor for the rock mass is obtained as

$$\underset{\sim}{\Omega} = \sum_{i} \underset{\sim}{\Omega}^{i} \tag{3}$$

where $\underset{\sim}{\Omega}^{i}$ is the damage tensor for i-th set of joints (Kyoya, Ichikawa and Kawamoto 1985a,b).

Kondoh et al.(1986) have proposed a photographic surveying system with a micro-computer for measuring discontinuity characteristics, called the Discontinuity Parameter Measuring System (DISPARMS).

On measuring by this system, some pairs of stereo photographs of outcrops where a reference frame is installed are used (See Fig.1). In the system, the relations between the coordinate systems as shown in Fig.2 are obtained by iterative calculation based on the Bundle Method with Self-Calibration, so that the actual position of any points on photographs can be evaluated with reference to the ground coodinates. Then, the strike and dip, the spacing and the trace length of joints can be measured. The strike and dip of a joint surface are obtained from the actual coordinates of three points on it, and trace length is calculated from the coodinates of the two end points of a joint. The spacing are evaluated from a scan line on the photographs.

Then, by using the system we can find some dominant joint sets and can evaluate the damage tensor for a rock mass by Eqn(2) and (3).

Fig.1 Stereo photographs in an exploration adit.

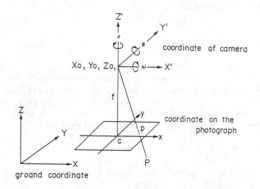

Fig.2 Coordinates of camera, photograph and ground.
(after Kondoh and Shinji 1986)

2.2 Net stress in rock mass

Distributed discontinuities reduce effective surface area in a material body of the rock mass. The stress tensor derived from force vectors acting on the reduced effective surface area is defined as the net stress (Murakami and Ohno 1980, Lemaitre 1985). If discontinuities are perfectly open and can not transmit any force, the net stress $\underset{\sim}{\sigma}^{*}$ is given by

$$\underset{\sim}{\sigma}^{*} = \underset{\sim}{\sigma}(\underset{\sim}{I} - \underset{\sim}{\Omega})^{-1} \tag{4}$$

where $\underset{\sim}{\sigma}$ is the Cauchy stress and $\underset{\sim}{I}$ the second-order identity tensor (Murakami and Ohno 1980). However, some of normal and shearing forces can be transmitted throughout joint surfaces. Then, the net stress in the rock mass must be modified as follows:

$$\underset{\sim}{\sigma}^{*} = \underset{\sim}{T}^{t}[\, \underset{\sim}{\sigma}_{n}'(\underset{\sim}{I} - C_{t}\underset{\sim}{\Omega}')^{-1} + \underset{\sim}{H}\langle\underset{\sim}{\sigma}_{n}'\rangle(\underset{\sim}{I} - \underset{\sim}{\Omega}')^{-1}$$
$$+ \underset{\sim}{H}\langle-\underset{\sim}{\sigma}_{n}'\rangle(\underset{\sim}{I} - C_{n}\underset{\sim}{\Omega}')^{-1} \,] \, \underset{\sim}{T} \tag{5}$$

where $\underset{\sim}{T}$ is the diagonalizer of $\underset{\sim}{\Omega}$:

$$\underset{\sim}{\Omega}' = \underset{\sim}{T} \, \underset{\sim}{\Omega} \, \underset{\sim}{T}^{t}$$

$$\underset{\sim}{\sigma}' = \underset{\sim}{T} \, \underset{\sim}{\sigma} \, \underset{\sim}{T}^{t}$$

The stress $\underset{\sim}{\sigma}'$ is separated into $\underset{\sim}{\sigma}_{n}'$ and $\underset{\sim}{\sigma}_{t}'$, which is written in a matrix form as

$$\underset{\sim}{\sigma}' = \underset{\sim}{\sigma}_{n}' + \underset{\sim}{\sigma}_{t}'$$

$$\begin{bmatrix} \sigma_{11}' & \sigma_{12}' & \sigma_{13}' \\ \sigma_{12}' & \sigma_{22}' & \sigma_{23}' \\ \sigma_{13}' & \sigma_{23}' & \sigma_{33}' \end{bmatrix} = \begin{bmatrix} \sigma_{11}' & 0 & 0 \\ 0 & \sigma_{22}' & 0 \\ 0 & 0 & \sigma_{33}' \end{bmatrix} + \begin{bmatrix} 0 & \sigma_{12}' & \sigma_{13}' \\ \sigma_{12}' & 0 & \sigma_{23}' \\ \sigma_{13}' & \sigma_{23}' & 0 \end{bmatrix}$$

$H\langle\cdot\rangle$ is a second-order tensor operator defined as

$$(\underset{\sim}{H}\langle\underset{\sim}{x}\rangle)_{ij} = \begin{cases} x_{ij} & \text{if } x_{ij} > 0 \\ 0 & \text{if } x_{ij} \le 0 \end{cases}$$

The coefficients Cn and Ct have values between 0 and 1, which characterize rate of transmission of compressive normal stress and shearing stress components through/along a joint surface, respectively.

2.3 Constitutive law and propagation equation of damage

In the damage mechanics theory, the Cauchy stress $\underset{\sim}{\sigma}$ plays a role of the statically admissible set. And the Cauchy stress is transformed to the net stress $\underset{\sim}{\sigma}^{*}$ due to the mechanical effects of the damage. A constitutive equation is introduced between the net stress and the kinematically admissible strain. It is generally given in the following form:

$$\underset{\sim}{\varepsilon} = \underset{\sim}{\Phi}(\underset{\sim}{\sigma}^{*}) \tag{6}$$

For the linear elastic case, it takes

$$\underset{\sim}{\sigma}^{*} = \underset{\sim}{D} \, \underset{\sim}{\varepsilon} \tag{7}$$

where $\underset{\sim}{D}$ is the Hookean stiffness tensor.

Since the size of rock specimens for laboratory tests is smaller than the size of the foundamental block element of the rock mass, there is no damage ($\underset{\sim}{\Omega}$=0) in it, so that the net stress $\underset{\sim}{\sigma}^{*}$ of the specimen coincides with the Cauchy stress. This means that the constitutive equation (6) can be determined from the usual laboratory tests for rock specimens.

If the fracturing process of the rock mass is considered, the damage should be a variable, and one more equation is needed for giving the state of the damage. Generally it is written as

$$d\underset{\sim}{\Omega} = \underset{\sim}{F}(\, \underset{\sim}{\sigma}, \, \underset{\sim}{\varepsilon}, \, \underset{\sim}{\Omega} \,) \tag{8}$$

This is called the propagation equation of damage. However, this should be refined if applied to macroscopic damages such as joints in the rock mass.

2.4 Damage analysis for excavation procedure

The virtual work equation for a damaged body is written as

$$\int \underset{\sim}{\sigma}^* \cdot \delta \underset{\sim}{\varepsilon} \ dv + \int \underset{\sim}{t}^0 \cdot \delta \underset{\sim}{u} \ ds_t + \int \underset{\sim}{f} \cdot \delta \underset{\sim}{u} \ dv + \int \underset{\sim}{\psi} \cdot \delta \underset{\sim}{\varepsilon} \ dv \qquad (9)$$

where $\delta \underset{\sim}{u}$ and $\delta \underset{\sim}{\varepsilon}$ are the variations of the displacement $\underset{\sim}{u}$ and the strain $\underset{\sim}{\varepsilon}$, respectively, $\underset{\sim}{t}^0$ the surface traction given on the boundary St, $\underset{\sim}{f}$ the body force vector, and $\underset{\sim}{\psi}$ a tensor given by

$$\underset{\sim}{\psi} = \underset{\sim}{\sigma}_t(\underset{\sim}{\phi}_t - \underset{\sim}{I}) + \underset{\sim}{T}^t \underset{\sim}{H} <\underset{\sim}{\sigma}'_n > \underset{\sim}{T} \ (\underset{\sim}{\phi} - \underset{\sim}{I}) + \underset{\sim}{T}^t \underset{\sim}{H} <-\underset{\sim}{\sigma}'_n > \underset{\sim}{T} \ (\underset{\sim}{\phi}_n - \underset{\sim}{I})$$

$$\underset{\sim}{\phi} = (\underset{\sim}{I} - \underset{\sim}{\Omega})^{-1} \ , \ \underset{\sim}{\phi}_t = (\underset{\sim}{I} - C_t \underset{\sim}{\Omega})^{-1} \ , \ \underset{\sim}{\phi}_n = (\underset{\sim}{I} - C_n \underset{\sim}{\Omega})^{-1} \qquad (10)$$

This $\underset{\sim}{\psi}$ represents mechanical effects of the damage (Kyoya, et al. 1985b).

The finite element discretization of Eqn(9) takes the following form

$$[K][U] = [F] + [F^*] \qquad (11)_1$$

$$[K] = \int [B]^t [D][B] dv \qquad (11)_2$$
$$(\ [D] \ ; \ \text{Hookean matrix} \ )$$

$$[F] = \int [N]^t [f] dv + \int [N]^t [t_0] ds_t \qquad (11)_3$$

$$[F^*] = \int [B]^t [\psi] dv \qquad (11)_4$$

where [U] is the nodal displacement vector, [N] the shape function matrix, [B] the strain displacement matrix, and $[\psi]$ a column vector $[\psi_{11} \ \psi_{22} \ \psi_{33} \ \psi_{12} \ \psi_{23} \ \psi_{31}]$ of which components are given by Eqn(10). In Eqn$(11)_1$, it should be noted that the mechanical effects of damage are represented by an additional force vector $[F^*]$.

Excavation analysis by the damage theory is performed in the following procedure: ① the equivalent nodal force vector [Fe] for each excavation stage is calculated from the initial Cauchy stress $[\sigma_i]$ by

$$[Fe] = \overset{M}{\Sigma} (\int [B]^t [\sigma_i] \ dv - \int [N]^t [f] \ dv) \qquad (12)$$

in which M is the number of excavated elements, and the integration is carried out on each of them and summed up; ② a simultaneous equation

$$[K][U'] = [Fe] \qquad (13)$$

is solved, then the displacement [U'] which corresponds to the non-damage state is obtained; ③ the Cauchy stress $[\sigma]$ is calculated from [U'] by the constitutive law

$$[\sigma] = [B][D][U'] \qquad (14)$$

which will be the initial stress for the next excavation stage; ④ by using $[\sigma]$ and the damage $[\Omega] = [\Omega_{11} \ \Omega_{22} \ \Omega_{33} \ \Omega_{12} \ \Omega_{23} \ \Omega_{31}]$, the net stress $[\sigma^*]$ is calculated by Eqn(4), and Eqn$(11)_4$ gives the additional force vector $[F^*]$; ⑤ an additional displacement vector [U''], which is due to the mechanical effects of the damage, is obtained by solving the equation

$$[K][U''] = [F^*] \qquad (15)$$

Finally, the displacement of the excavated rock mass is calculated as

$$[U] = [U'] + [U''] \qquad (16)$$

The sequential excavation is simulated by repeting the above procedure ① to ⑤.

3 EXCAVATION ANALYSES FOR VARIOUS SHAPED CAVERNS

In this section, results of excavation analyses for three different types of caverns for an underground hydroelectric power station are presented. Their shapes are shown in Fig.3.

The rock mass mainly consists of rhyolite. Young's modulus E and Poisson's ratio ν of the intact rock are obtained by laboratory tests as

$$E = 4.62 \times 10^5 \quad (\text{kg/cm}^2)$$
$$\nu = 0.22$$

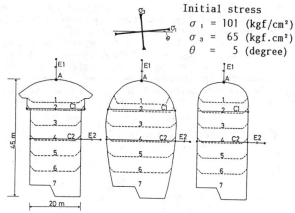

Initial stress
$\sigma_1 = 101 \ (\text{kgf/cm}^2)$
$\sigma_3 = 65 \ (\text{kgf.cm}^2)$
$\theta = 5 \ (\text{degree})$

(a) mushroom shape (b) horse-shoe shape (c) bullet shape

Fig.3 Shapes of cavern.

On the other hand, Young's modulus of the rock mass is estimated by results of in-situ plate loading tests:

$$E = 1.3 \times 10^5 \quad (\text{kg/cm}^2)$$

Three dominant sets of joints are found from the stereo photographs taken in exploration adits by using the formerly mentioned measurement system, and for each set the parameters of the damage tensor(3) are evaluated as followings:

set 1 ;
  unit normal vector $\bar{n}$ = (0.208, 0.122, 0.978)
  density of joints $\bar{N}/V$ = 18.04
  average area $\bar{a}$ = 0.203 m^2
set 2 ;
  unit normal vector $\bar{n}$ = (0.982, 0.105, 0.191)
  density of joints $\bar{N}/V$ = 21.95
  average area a = 0.175 m^2
set 3 ;
  unit normal vector $\bar{n}$ = (0.707, -0.602, 0.371)
  density of joints $\bar{N}/V$ = 11.24
  average area $\bar{a}$ = 0.15 m^2

The representative length l of the fundamental block element is specified as 0.2m which is the average spacing evaluated along scan lines. Thus, by Eqn(3) and (4) the damage tensor of rock mass is calculated as

$$\underset{\sim}{\Omega} = \begin{bmatrix} 0.954 & -0.045 & 0.383 \\ & 0.142 & 0.028 \\ \text{Sym.} & & 0.835 \end{bmatrix} \qquad (17)$$

Components Ω_{11} and Ω_{33} in the damage tensor(17) have large values. This implies that joints parallel to the side walls are dominant, especially in the horizontal direction.

The finite element meshes prepared for each cavern shape involve about 800 nodes and 500 triangular elements. The excavation procedure is simulated by seven stages of analyses (Fig.3), and we treat unsupported cases only. In the damage analyis, the contact conditions in Eqn(5) are assumed to be open, that is, Ct=Cn=1.0

Fig.4 shows the deformations of the cavern at the final stage. The maximum displacement in the horizontal direction is found at the center of the right wall for all cases, which is summarized in Table 1.

Table 1 Maximum horizontal displacement

cavern type analysis method	mushroom (cm)	horse-shoe (cm)	bullet (cm)
damage analysis	16.41	14.39	15.85
	(1.14)	(1.0)	(1.10)
conventional FEM	3.16	2.81	3.06
	(1.12)	(1.0)	(1.09)

It is observed that the displacements computed by the damage analysis are about five times larger than the conventional analysis. Thus, the damage effects of joints, which are dominant in the horizontal direction as mentioned previously, are clearly appeared. Figures written by parentheses in Table 1 denote the ratio of the displacement of each case to the horse-shoe shape one. For the bullet shape, the maximum displacement at the right wall is about ten percent greater than the horse-shoe one in any anysis. Furthermore, the displacement for the mushroom shape is two or four precent greater at the same place. Displacement distributions along the line E2 (Fig.3) at the final stage are shown in Fig.5.

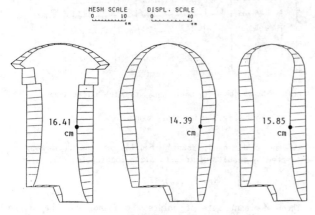

Fig.4 Deformations of cavern at final excavation stage.

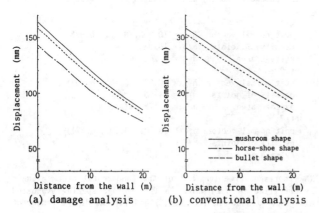

Fig.5 Displacement distributions along measuring line E2.

Settlements of the crown top (point A in Fig.3) at the final excavation stage are also summarized in Table 2 .

Table 2 Settlement at crown top

cavern type analysis method	mushroom (cm)	horse-shoe (cm)	bullet (cm)
damage analysis	1.08	0.29	0.23
conventional FEM	0.59	0.38	0.23

On the contrary to horizontal displacements at the side wall, the damage analysis gives mostly same settlements at the crown top as the conventional analysis except for the mushroom shape. This is because the damage effect of joints is dominant in the horizontal direction but not in the vertical direction. Thus, it is understood that the damage analysis can predict anisotropic behaviours of the caverns due to distributed joints.

Observing settlements at the crown top for each excavation stage (see Fig.6), a squeezing effect is clearly found due to the dominant horizontal stress and the joint set parallel to the side wall. That is, the crown top moves upwards. The effect differs depending on the cavern shape, and the bullet shape gives the smallest value. Fig.7 shows displacement distributions along the line E1 of Fig.3, and similarly we observe the squeezing effect at the final excavation stage.

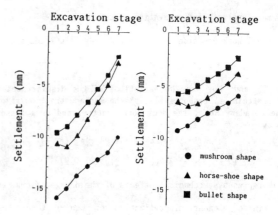

(a) damage analysis (b) conventional analysis

Fig.6 Accumulated settlements at crown top.

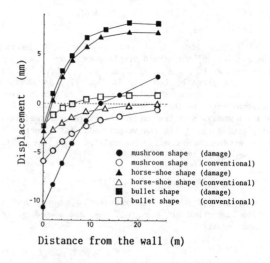

Fig.7 Displacement distributions along line E1.

CONCLUSIONS

We here presented a damage mechanics theory applied for the discontinuous rock mass, and its numerical analysis by finite elements was developed. An underground opening problem for the hydroelectric pumphouse was solved for several types of the cavern; the traditional mushroom shape, the horse-shoe shape (Waldeck II type), and the bullet shape. The damage tensor was determined by a microcomputer system acquiring the data from stereo photographs of the exploration adits. Since the joint sets parallel to the side walls of the cavern are dominant, we got much larger horizontal displacements by the damage analysis. On the other hand, the conventional finite element analysis using the data of in-situ rock mass tests gave relatively small displacements.

Some questions are still open in the damage mechanics theory, for example, the propagation equation of the damage. The geometrical data acquisition system in determining the damage tensor should also be refined.

REFERENCES

Kawamoto,T, Y.Ichikawa & T.Kyoya 1986. Computational Mechanics'86, Proc. ICCMM, Tokyo, vol.2, Springer, IX-27.
Kyoya,T., Y.Ichilawa & T.Kawamoto 1985a. Proc. JSCE, vol.358, III-3: 756-760 (in Japanese).
Kyoya,T., Y.Ichikawa & T.Kawamoto 1985b. Numerical Methods in Geomechanics Nagoya, Proc. 5th ICONMIG, Balkema,: 469-480.
Kyoya,T., Y.Ichikawa, M.Kusabuka & T.Kawamoto 1986. Proc. Symp. Eng. in Complex Rock Formations, Beijing, China: 506-513.
Kondoh,T., & M. Shinji 1986. Proc. Symp. Eng. in Complex Rock Formations, Beijing, China: 809-815.
Murakami,S. & N.Ohno 1980. Proc. 3rd IUTAM Symp. on Creep in Structures, Springer: 422-443.

The strength and deformation characteristics of Korean anthracite
Caractéristiques de résistance et de déformation de l'anthracite coréenne
Die Festigkeit und das Deformationsverhalten von koreanischem Anthrazit

KYUNG WON LEE, Korea Institute of Energy and Resources
HEE SOON SHIN, Korea Institute of Energy and Resources

ABSTRACT: Korean anthracite is generally so weak that its compressive strength only ranges from 0.28 MPa to 16 MPa. Furthermore the majority of them, approximately 70 %, reveal a compressive strength and modulus of deformation of less than 5 MPa and 150 MPa respectively. Their bearing capacity is from 0.07 MPa to 0.3 MPa excluding exceptional locations where it attains 5.6 MPa. Accordingly, it is obvious that the technology for developing deep seams should be renovated as early as possible in view of face support and efficient extraction.

RESUME: L'anthracite coréen est, en général, tellement fragile que sa résistance à la compression n'est autre que de 0,28 MPa à 16 MPa. Entre autre, la plupart des charbons, approximativement 70 %, révèlent une résistance à la compression et un module d'élasticité inférieurs à 5 MPa et a 150 MPa à 0,3 MPa sauf quelques cas exceptionnels où celle-ci atteint 5,6 MPa. Par conséquent, il est évident que la technologie pour développer des couches profondes devrait être améliorée le plus rapidement possible compte tenu du soutènement en taille et de l'abattage efficace.

ZUSAMMENFASSUNG: Koreanische Anthrazit ist gewöhnlich so weich, dass die Druckfestigkeit nur von 0.28 zu 16 MPa erreicht. Ausserdem ergibt sich dieser Anteil ungefähr 70 % der gesamten Kohle, und zwar kleiner als 5 MPa in Druckfestigkeit und 150 MPa in Verformungomodul. Die Tragfähigkeit der Kohle tiegt normalesweise zwischen 0.07 und 0.3 MPa, ausnahmsweise bis auf 5.6 MPa. Daher ist eine Weiterentwicklung der Bergbautechnik fur grosse Teufen in Anbetracht des Ausbaus und Ausbansystems erforderlich.

1 INTRODUCTION

Most of the Korean coal is classified as anthracite. It is used mostly for house heating and partly for power generation and industrial uses. Its chemical composition is fixed carbon(53.7%), ash(39.5%), H_2O (3.6%), volatile matter(3.2%) and sulphur(0.29%).

Coal production in Korea has increased from 5 million tons in 1960 to 22.5 million tons in 1985. Korea retains anthracite reserves of 1.63 billion tons. 90 percent of them was formed during Permaian period of Paleozoic era and the remaining 10 percent from the late Triassic to the early Jurassic period of Mesozoic era.

The coal bearing geologic strata are subjected to the earth crust disturbances in the early Tricassic period, followed by the orogenic action and strong metamorphism in the late Jurassic period. Korean coal seams therfore are in general very irregular, steep and discontinuous and comprise plenty of folding and faulting structure. The thickness of coal seams fluctuate from 0.6 m to 30 m. The thicker part of the seam occurs typically at the crest and trough of the fold. The predominant dipping of coal seam is from 30 to 90 degrees. The depth of working level at the present time averages 300 m to 400 m, the deepest faces 700 m. Regarding the annual rate of increase in the depth of operational levels, 25 m to 30 m per year, it will be expected to reach 800 m to 900 m from the adit level by the late 1990s.

It is therefore urgent to develop deep mining techniques in Korea concerning mechanized coal winning methods and roadway drivage and support systems.

The analysis of coal mine openings requires knowledge of mechanical behaviour of coal. Korean anthracite is extremely soft and friable so that coal specimens can not be machined and the laboratory tests on small coal specimens are not possible. Therefore it becomes important to investigate the accurate strength and deformation characteristics of in-situ coal.

This paper describes briefly the results of

- point load test
- impact test
- plate loading test
- borehole jack test

for determining the in-situ strength of coal.

2 MEASUREMENTS OF STRENGTH AND DERORMATION CHRACTERISTICS OF COAL

2.1 Point load test

With a point load tester (Terrametrics model T-1000), the test was carried out at coal faces of irregular lump coal samples ranging in size from 33 mm to 90 mm.

The range of uniaxial compressive strength(Cs) of anthracite was found out to be from 2.6 MPa to 11.2 MPa. The coal with high compressive strength of 28.5 MPa was rarely found at some places. The range of point load strength (Is(50)) are shown in table 1.

The bedding planes are well developed but cleat planes are less distinct.

2.2 Impact test

For determination of the compressive strength and

Table 1. The results of point load strength test.

Colliery	perpendicular to bedding planes		parallel to bedding planes	
	Is(50)*	Cs*	Is(50)*	Cs*
Dongwon	0.25	5.89	0.15	3.60
Kangwon	0.31	7.34	0.24	5.77
Kyungdong	0.26	6.24	0.20	4.91
Samtan	0.15	3.63	0.08	1.43
Euryong	0.29	7.09	0.18	4.33

* unit: MPa

strength index(f) of coal of Samtan, Protodyakonov's drop hammer test was carried out by using drop hammer apparatus, and compared with the result of plaster cast test. The compressive strength of the coal from Samtan colliery was estimated to be 1.5 to 1.9 MPa and the following formular was obtained from the plaster cast test.

$S_c = 0.8 \cdot (1 + 2.9 \ f \ )$, MPa (Kim, 1982).

This value is approximately same as the value obtained from the plate bearing test.

Table 2. The results of impact test.

Sample No.	Protodyakonov's coefficient(f)	Uniaxial compressive strength (MPa)
1	0.33	1.54
2	0.37	1.63
3	0.40	1.70
4	0.44	1.79
5	0.48	1.87

2.3 Plate loading test

The plate loading test was carried out   to predict the bearing capacity of coal floor subjected to load and the in-situ compressive strength, which is of great importance where proper support resistance is required. The results are shown in Table 3.

The bearing capacity which indicates the supporting ability of working face at minimum settlement of floor ranged from 0.45 to 2.76 MPa. Therefore a base plate for hydraulic prop should be sized about 700 cm^2 in area to maintain 100 KN load in case of Kangwon colliery.

The compressive strength of coal was estimated to be 0.3 to 1.7 MPa; it can be classified to very soft coal category.

Table 3. Summary of plate loading test results.

Colliery	Bearing capacity σf(MPa)	Ratio of load to settlement Bs(N/cm^3)	Compressive strength Cs(MPa)
Daesung	0.45-2.76	9-289	0.28-1.7
Kangwon	1.5	21	0.94
Samtan	0.82	11	1.63

The relationship between the values of the bearing capacity and those of the compressive strength is established in-situ and laboratory test. The ratio of bearing capacity to uniaxial compressive strength of coal is 0.5 - 2.0. A typical load-penetration curves for the plate loading are shown in figure 1 and figure 2.

In laboratory test, water reduced the bearing capacity and the compressive strength of coal by 18 %.

2.4 Borehole jack test

Goodman jack   tests were performed in 5 NX holes in coal seam at Samtan colliery to measure a deformation modulus of in-situ coal. A minimum of 8 tests were conducted in each hole, depending upon conditions of a hole. The model 52102 Goodman jack designed for soft rock was used in this test.

A special designed NX size bit was attached to longhole drill and with this 2 to 3.5 m holes were drilled in coal seam. A schematic view of borehole jack test is shown in figure 3.

The average modulus of deformation was found to be 99 MPa, with a range of 33.7 to 154 MPa. This value is less than one-tenth of those of coal in other countries.

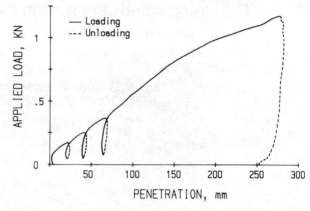

Figure 1. Load-penetration curve on coal in laboratory. (Samtan colliery)

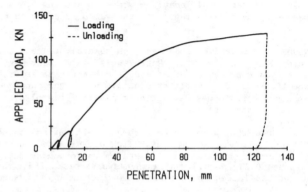

Figure 2. Load-penetration curve on in-situ coal. (Samtan colliery)

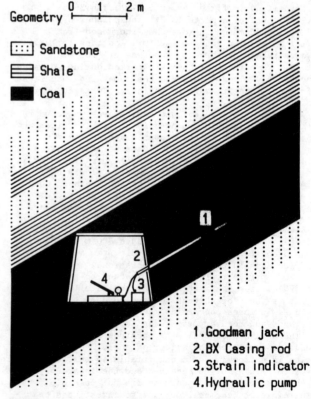

1.Goodman jack
2.BX Casing rod
3.Strain indicator
4.Hydraulic pump

Figure 3. Schematic view of borehole jack test in coal seam.

Table 4. Goodman jack test results(Poisson's ratio = 4.0).

Hole	Depth (m)	Applied pressure range(MPa)	Modulus of deformation(MPa)
A	2.4-3.0	to 3.20	81.8
B	2.0-3.2	2.10	33.7
C	1.8-2.8	2.90	154.0
D	2.0-3.2	3.10	122.6
E	2.0-3.5	3.40	102.0

The test results are given in table 4. Typical pressure-deformation curves of coal, such as that shown in figure 4, posses an initial non-linear portion which is associated with the closure of initial cracks, followed by a linearly elastic range. The slope of the curve becomes gradually steep until it becomes linear at full contact, giving the curve a "concave upward" appearance.

The modulus values obtained from Goodman jact tests are consistently smaller than those from plate loading tests. These values should be taken as the lower limit of in-situ coal.

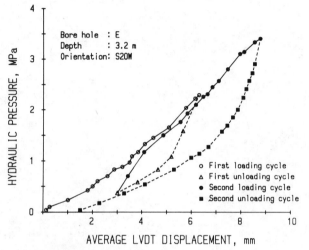

Figure 4. A pressure-displacement curve obtained from E hole.

3 CONCLUSION

Korean anthracite has a compressive strength of 0.28 MPa to 16 MPa and the range of a deformation modulus is 34 to 154 MPa, less than one-tenth of those of coal in other countries.
Although the compressive strength of the lump coal is ranging 1.4 to 7.3 MPa, the in-situ compressive strength is extremely low because of the crushed condition of coal seams. A plate loading test or other indirect testing methods can offer to estimate the strength in this case.

Furthermore, the mining techniques of foreign origin are not directly employable at Korean collieries, mainly because of the difference in geological conditions of coal bearing strata and coal characteristics.

REFERENCE

Shin, H.S. & K.W.Lee. 1984. Determination of coal strength by point load test on irregular lumps. J. of the Korean Institute of Mineral and Mining Engineers(KIME), Vol.21:38-43.
Kim, I.K. & C.Sunwoo. 1982. Measurement of coal strength by Protodyakonov's drop hammer test. J. of KIME, Vol.19:291-295.
Lee, K.W. & H.S.Shin. 1981. Bearing capacity of coal and its in-situ strength. J. of KIME. Vol.18: 115-121.
Kim, D.S. et al. 1982. A study of laboratory and in-situ plate bearing test at the new-mechanized face in Samchok coal field. J. of KIME. Vol.19:242-251.
Lee, K.W. 1985. Coal mining research for colliery mechanization. Proc. 2nd Korea-U.S.A. Joint Workshop on coal Utilization Technology, Seoul: KIER. p.23-37.
Ro, S.H. & K.W. Lee. 1986. Coal mining technology and analysis on the causes of mine accidents in Korea. 3rd U.S.A.-Korea Joint Workshop on coal Utilization Technology, Pittsburgh. p.10-22.
Shin, H.S. et al. 1987. In-situ determination of the modulus of deformation of coal. Seoul: KIER.

Measurement of rock displacements and its interpretation for the determination of a relaxed zone around a tunnel in a coal mine

Interprétation des mesures de déformations pour la détermination de la zone de relâchement en périphérie d'un tunnel dans une mine de charbon

Messung von Felsverformungen und die Bestimmung der Entspannungszone um einen Tunnel in einer Kohlegrube

KYUNG WON LEE, Korea Institute of Energy and Resources
HO YEONG KIM, Korea Institute of Energy and Resources
HI KEUN LEE, Seoul National University, Korea

ABSTRACT: When the amount of field measurement data is not enough to know the whole behavior of rocks around a tunnel, the computer analysis can be used to complement those informations. This paper presents an example which introduces the calculation method of the relaxed zone in rocks by BEM analysis using the data from field measurements. Taking into account the in-situ condition of rocks in BEM analysis, "an empirical failure criterion for rock" was used as follows; $\sigma_1/\sigma_c = \sigma_3/\sigma_c + \sqrt{mr \cdot \sigma_3/\sigma_c + sr}$, where mr and sr are constants which depend upon the in-situ properties of rocks. After determination of relaxed zone by measurement on certain locations, it is shown that the calculated relaxed zone by proper value of those constants coincides with the above result. Accordingly, the complete relaxed area of rocks around a tunnel can be determined.

RESUME: Lorsque le nombre de mesures sur le terrain n'est pas suffisant pour connaître, le comportement entier des roches autour d'une galerie, l'analyse par ordinateur peut être employée afin de compléter ces informations. Cet article présente un exemple qui introduit une méthode de calcul sur la zone détendue dans les roches par l'analyse des éléments frontiers(BEM) en utilisant les données saisies sur le terrain. Compte tenu de l'état des roches in-situ, en analyse des éléments frontiers, " un critère empirique de rupture pour la roche" est utilisé comme suit; $\sigma_1/\sigma_c = \sigma_3/\sigma_c + \sqrt{mr \cdot \sigma_3/\sigma_c + sr}$, où mr et sr sont les constantes qui dépendent des propriétés in-situ de roches. Après la détermination de la zone détendue par la mesure sur quelques endroits, il est démontré que la zone détendue calculée par les valeurs convenables de ces constantes coïncide avec le resultat ci-dessus. Par conséquent, la zone détendue entière autour d'une galerie peut être déterminée.

ZUSAMMENFASSUNG: Wenn die Messergebnisse für die Erklärung des Verhaltens der Gesteinsstrecke nicht vollständig sind, könnte die Bestimmung durch EDV die fehlende Information ergäzen. Im folgenden wird ein Beispiel dargestellt, wie das Rechenverfahren mit Feldmessungsdaten durch BEM Analyse die Relaxation des Gebirges auswertet. Berücksichtigt man die in-situ Bedingung des Gebirges im BEM Analyse, wird das folgende empirische Bruchkriterium benutzt, $\sigma_1/\sigma_c = \sigma_3/\sigma_c + \sqrt{mr \cdot \sigma_3/\sigma_c + sr}$, wo mr und sr die von der in-situ Eigenschaften des Gesteins abhängigen Konstante sind. Die Relaxation des Gebirges stimmt mit dem durch Einsatz von den ausgeglichenen Konstanten berechneten Resultat gut überein, das durch Messergebnisse punktweise festzustellen ist, somit der ganze Relaxationsumfang auswerten lässt.

1 INTRODUCTION

The main purpose of field measurements in rock mechanics is to know the behaviors of rocks around openings. Of various measuring techniques developed, monitoring the displacements has proven to be a simple and useful method to estimate the stability of a tunnel and the relaxed boundary in rocks. Especially the latter becomes an important parameter for designing the support systems.

After estimation of relaxed zone on certain locations around a tunnel by measurements, it is required to know the whole relaxed state in rocks, which can be established by computer analysis like the finite element method or the boundary element method.

This paper presents an example introducing the combining techniques of field measurements with computer analysis in calculation of the relaxed zone in rocks.

As a result, it is shown that even a small amount of data from measurements can offer useful informations to the computer analysis which is the next step of interpretations.

2 MEASUREMENT OF DISPLACEMENTS AND CONVERGENCES

2.1 Instrumentation

In monitoring the displacements of surrounding rocks and convergences of a tunnel, the single rod extensometers and the tape extensometer as shown in figure 1, were used; those are commercially available instruments.

Figure 2 shows the installation of single rod extensometers of which lengthes are 1 or 2 meters. By attching a hook on the cap of a rod, head of the rod became the measuring point for the tape extensometer.

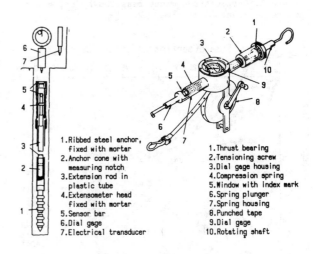

1. Ribbed steel anchor, fixed with mortar
2. Anchor cone with measuring notch
3. Extension rod in plastic tube
4. Extensometer head fixed with mortar
5. Sensor bar
6. Dial gage
7. Electrical transducer

1. Thrust bearing
2. Tensioning screw
3. Dial gage housing
4. Compression spring
5. Window with index mark
6. Spring plunger
7. Spring housing
8. Punched tape
9. Dial gage
10. Rotating shaft

Figure 1. Single rod extensometer and tape extensometer.

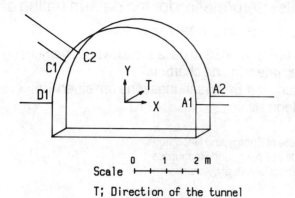

Scale 0 1 2 m

T; Direction of the tunnel

Figure 2. Installation of rod extensometers and measuring points for a tape extensometer.

2.2 Results and interpretations

The relative displacements between two fixed points - an end point is fixed in depth of rocks and a measuring point is fixed on the tunnel section, were measured for 68 days by rod extensometers. The results are shown in figure 3.

Considering those results, assumptions are possible as follows;

1) In view of the fact that the increasing rate of displacements of the rod A1 slowed down ahead of that of the rod A2, we can suppose two possible cases. Firstly, the end point of the rod A1 exists in the

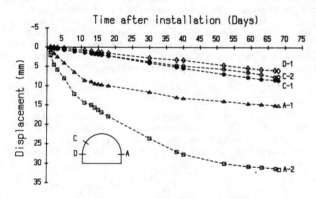

Figure 3. Rock displacements measured by rod extensometers in relation to time.

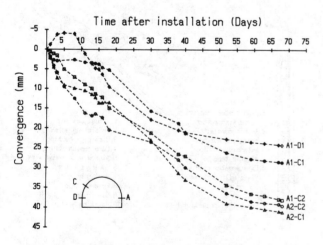

Figure 4. Convergences of the tunnel measured by a tape extensometer in relation to time.

relaxed zone, but the end point of the rod A2 still remains in the elastic region of rocks. Secondly, both of the end points of the rod A1 and A2 are sequentially embedded in relaxed zone.

2) For rod C1 and C2, it may be supposed that both of the end points remain in elastic region, or they are already included in relaxed zone.

3) For rod D1, the same cases as 2) can be considered.

Comparing the results with the measured data for convergences (see figure 4), we can make proper decisions. As given in table 1, the first two cases represent the large values of T-R, but the last two cases relatively small ones. It means that the end point of the rod A1 is existing in the relaxed zone, therefore, the resultant displacement is much smaller than the resultant convergence of the measuring point.

On the contrary, the resultant displacement of the end point of rod A2 having almost the same value of the resultant convergence of the measuring point, it may be provided that the end point of the rod A2 is still remaining in the elastic region.

For another rods, we have to put off making clear decisions.

Table 1. Comparison of the resultant displacements measured by a tape extensometer with those by rod extensometers.

Case	Line	Convergence (mm); T	Rod	Displacement (mm); R	T-R (mm)
1	A1-C1	28.6	A1	15.1	
			C1	8.2	
			summation	23.3	5.3
2	A1-C2	38.0	A1	15.1	
			C2	7.5	
			summation	22.6	15.4
3	A2-C1	41.1	A2	31.5	
			C1	8.2	
			summation	39.7	1.4
4	A2-C2	39.5	A2	31.5	
			C2	7.5	
			summation	39.0	0.5

3 CALCULATION OF RELAXED ZONE BY STRESS ANALYSIS USING THE BOUNDARY ELEMENT METHOD

3.1 Failure criterion

Since the theory on the failure of solids was reported by Griffith in 1921, many theories have been suggested about the failure criteria for brittle materials like rocks. It has been proved that the failure criterion including the effects of intermediate principal stress provides more exact calculations (Lee, 1977, 1978). In a general 2-dimensional stress analysis, however, a criterion without considering those effects may provide promising results.

In practical purpose, the latest useful theory is "an empirical failure criterion for rock" proposed by Hoek and Brown(1980) of the following form;

$$\sigma_1/\sigma_c = \sigma_3/\sigma_c + \sqrt{m \cdot \sigma_3/\sigma_c + s} \quad :(1)$$

where

σ_1 is the major principal stress at failure;
σ_3 is the minor principal stress;
σ_c is the uniaxial compressive strength of intact rock;
m and s are empirical constants which depend upon the properties of the rock.

The constants m and s are determined through tri-axial compression tests in laboratory and are reported as mi and si for intact rock. For in-situ rocks those are refered as mr and sr, of which values are much smaller than those of mi and si. In relation to the rock mass classifications a method for estimating mr and sr for rock masses has also provided (Bieniawski, 1984).

3.2 Properties of rocks at measuring site

The test tunnel locates in the coal bearing formation at a depth of about 800 meters from surface. The prevailing rocks of the strata is shale, of which mechanical properties are given in table 2.

Table 2. Some mechanical properties of shale at the test site.

Property	Tested result
Young's modulus (MPa)	36850
Poisson's ratio	0.272
Uniaxial compressive strength (MPa)	81.0
Tensile strength (MPa)	8.6

Presentation of the Mohr's failure envelope with the results of triaxial compression tests for shale is shown in figure 5.

3.3 Results of calculation

At the test site, initial rock stresses were measured(Lee et al, 1985), and secondary principal stress components in the perpendicular plane to the direction of the tunnel were calculated from that result.

Figure 6 shows the displacement and stress distributions calculated by 2-dimensional elastic analysis using the boundary element method. In calculation, Young's modulus of rock mass was used as reduced value of 20% from the vlaue of laboratory tests,

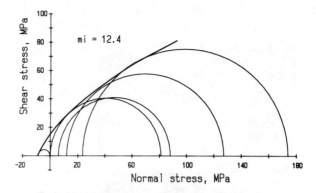

Figure 5. Mohr's failure envelope for shale.

which is based upon the suggested method by Bieniawski (1978) for estimating the rock mass deformability according to the R.Q.D.

Two constants mr and sr were determined in view of that the boundary of the relaxed zone lies between 1 m and 2 m points in the right part of the tunnel as determined in measurements; when mr equals to 0.992 - 8% of mi, and sr equals to 0.0001, the calculated relaxed zone coincides with those of measurement, as shown in figure 7.

At the test site, because the steel arches using GI-100 beam had been installed every 0.5 m, support pressure Pi was considered as 0.64 MPa.

As shown in figure 7, calculation result represents that the end points of the rod C1, C2, and D1 are existing in relaxed zone.

4 CONCLUSION

In the stress analysis for the example of this paper, only isotropic and elastic properties of rocks were considered. The other properties like anisotropy, rheological behaviors, and the effects of discontinuities in rocks, were not taken into account. It is assumed all those properties are comprised in the

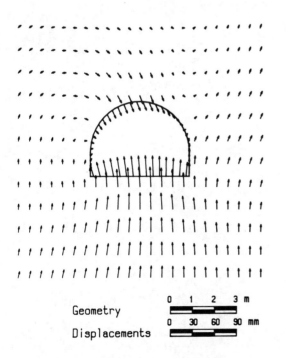

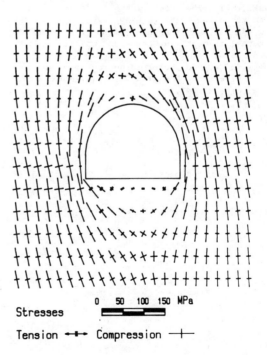

Figure 6. Displacement and stress distributions predicted by elastic analysis using the boundary element method.

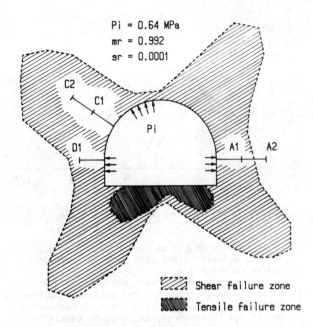

Pi = 0.64 MPa
mr = 0.992
sr = 0.0001

C2
C1
Pi
D1
A1   A2

▨ Shear failure zone
▓ Tensile failure zone

Figure 7. Calculated relaxed zone around the tunnel and comparison with the location of the rod extensometers.

values of mr and sr.

The predicted convergences by the computer analysis differ from the measured results, which is caused by the fact that the plastic deformation characteristics were not considered in the analysis.

REFERENCES

Lee, H.K. 1977 & 1978. Elasto-plastic analysis of structure considering characteristics of failed rock (I) & (II). J. KIMME. 14:300-316 & 15:120-132.
Hoek, E. & E.T. Brown. 1980 Empirical strength criterion for rock masses. J. Geotechnical Eng. Div. ASCE. 106:1013-1035.
Bieniawski, Z.T. 1984. Rock mechanics design in mining and tunneling. Rotterdam: Balkema.
Lee, K.W. et al. 1985. Design of support systems for rock tunnel in depth in Chang-sung colliery. Seoul: KIER.
Bieniawski, Z.T. 1978. Determining rock mass deformability - experience histories. Int. J. Rock Mech., Min. Sci. 15:237-247.

Seismic monitoring system using fiber-optic signal transmission
Réseau d'écoute séismique utilisant la fibre optique pour la transmission des signaux
Entwicklung eines seismischen Systems mit fiberoptischer Signalübermittlung

V.LABUC, Centre de Recherche Noranda, Pointe Claire, Canada
W.BAWDEN, Centre de Recherche Noranda, Pointe Claire, Canada
F.KITZINGER, Centre de Rechercher Noranda, Pointe Claire, Canada

ABSTRACT: A joint effort by the Noranda Research Centre and the Department of Energy, Mines and Resources Canada resulted in the construction and installation of a novel type of seismic monitoring system. This paper describes the development of a Canadian design that uses electronic, acceleration-sensitive geophones in a triaxial configuration coupled with multichannel fiber-optic signal transmission. A high speed computer digitizes and stores the complete waveform history of all channels as events are detected. A single optical fiber transmits triaxial information from each monitoring point.

RESUME: Un effort combiné du Centre de Recherche Noranda et du Département de l'Energie, des Mines et des Ressources du Canada s'est soldé par la construction et l'installation d'un noveau type de système de monitoring séismique. L'article décrit le développement d'un concept Canadien qui utilise des geophones électroniques sensibles à l'accélération, dans une configuration triaxiale connectée à une ligne de transmission par fiber optique à canaux multiples. Un ordinateur à haute vitesse digitalise et enregistre les ondes de tous les canaux à mesure que les événements sont détectés. Une fibre optique simple transmet l'information triaxiale de chaque point de monitoring.

ZUSAMMENFASSUNG: Noranda Research Centre, unterstützt durch Kanada's 'Department of Energy Mines and Resources', hat einen neuen Typ eines seismischen Systems entwickelt und eingesetzt. In diesem Bericht wird die Entwicklung des einzigartigen Entwurfs beshrieben, welches drei dreiachsig angeorndnete elektronische Beschleunigungsmesser pro Kanal verwendet, und welches die Signale eines jeden Kanals uber ein fiberoptisches Kabel übermittelt. Ein aussergewöhnlich schneller Komputer wird verwendet, welcher die Signale eines seismischen Ereignisses von jedem Kanal in seiner kompletten Form empfängt, die Signale in Zahlen umwandelt, und sie aufbewahrt für analytische Zwecke.

1 INTRODUCTION

The importance of microseismic monitoring in rockburst-prone mines is recognized as significant in terms of personnel safety and mine design. A survey of microseismic systems in South Africa (Bawden and Kitzinger 1986) showed a well-established seismic research technology necessitated by intensive production at deep levels involving large numbers of people. Until recently, little emphasis had been put on developing practical, low cost monitoring networks in Canada. Increased rockburst activity in several Canadian mining regions in the past 5-10 years has resulted in property damage and production delays, and in one case resulted in fatalities.

In an effort to promote the development of a Canadian microseismic monitor based on state-of-the-art technology, the Noranda Research Centre proposed a triaxial fiber-optic system to the Department of Energy, Mines and Resources, acting through the Mining Research Laboratory (CANMET), Elliot Lake, Ont. Noranda Inc. was subsequently contracted to design and construct a research instrument for installation and field trials at Elliot Lake, near the CANMET laboratory.

The system described consists of five triaxial sensors in an approximately 1 km² array mounted in boreholes from surface over a mine section that exhibits regular seismic activity. A mine-wide underground network is easily implemented by adding expansion modules to the system.

2 DESIGN CONSIDERATIONS

The early stages of system design resulted in several basic decisions concerning the fundamental structure of the microseismic network. Contract specifications required that events of Richter magnitude (M) 0 to 2.2 be picked up within the confines of the 1 km² array. These limits represent event magnitudes initiated by typical mine related seismic activity. Events smaller than M = 0 are generally studied using an underground Electrolab system equipped with wide band accelerometers. Magnitudes of M = 2.2 and higher are monitored by national seismic installations.

A design overlap margin was provided by targeting the dynamic range of the system to events between M = 0 at 1 km and M = 3.2 at 300 m. A derivation of average particle velocities (Campbell 1986) at the point of measurement showed that an overall dynamic range of approximately 60 dB (1000:1) would be required. The frequency band over which the network has to operate was determined by consulting available data which relates event magnitude to the frequency content of the generated wavefront (Spottiswoode 1984). Event magnitudes within the sensing range of the system contain little useful information above 300 Hz and the system was designed to sharply attenuate signals above this limit. Testing of the completed instrument showed that an overall signal to noise ratio of 75-80 dB over the range of 0.5 to 300 Hz had been achieved.

The choice of sensors was inspired by a promising development in geophone technology, as described by Klaassen & Van Peppen. An electronic feedback loop within a modified velocity geophone structure results in a servo-accelerometer type transducer with several advantages over standard velocity gauges. The feedback principle creates conditions where the vibration sensing coil remains virtually stationary with respect to the internal magnetic field. This allows for enhanced output linearity and an almost flat phase response over the frequency range of the sensor. The pronounced phase transition at resonance of standard passive geophones is eliminated. Further advantages are that the transducer provides an amplified low noise output signal readily suited for subsequent processing and, owing to the principle of a rigid sensing coil, the unit can be operated in any position. Separate vertical and horizontal sensors are not required.

Fiber-optic links between the sensor array and the central receiver were selected due to the immunity of optical transmission from electromagnetic interference problems. Electrical sources, such as radio transmitters, powerline cables, and atmospheric activity, do not affect the propagation of optical signals along glass fibers. Cost comparison between a suitable electrical cable and the fiber-optic equivalent in this case showed the fiber optics to be economically competitive with standard wiring. An environmentally suitable cable was designed and manufactured by CANSTAR Communications, a subsidiary of Noranda with extensive experience in the design and installation of fiber-optic systems.

Transmission of triaxial data over a single fiber was implemented using three multiplexed, high-frequency carrier signals modulated by the output of their associated geophones. Off-the-shelf optical transmitter and receiver modules, with a bandwidth of 10 Hz - 10 MHz, send and retrieve information through the optical link. Three carrier frequencies of 3.5, 4.5 and 5.5 MHz, corresponding to the X, Y and Z channels of the sensors, were chosen to lie within the passband of the system and to allow for future expansion. Temperature stability over an expected transmitter operating range of +40, -30°C was ensured by temperature compensation of the modulation circuits and automatic frequency tracking at the receiver.

Data analysis is performed by a MASSCOMP 5400 series computer. The unit consists of a 1 MHz multichannel data acquisition module, high speed processor with graphics, and a 71 megabyte hard disc storage unit with tape backup. This was regarded as a flexible system suitable for full waveform capture and recording in real time.

3 SYSTEM DESCRIPTION

A block diagram of the signal flow path in the microseismic monitor transmission system can be seen in Fig. 1. The outputs from a triaxial geophone assembly, consisting of three mutually orthogonal servo-accelerometer transducers, are fed, via short electrical cable, to an electronics package at the borehole collar. Internal frequency modulation circuits vary the frequency of three radio frequency carrier oscillators in proportion to input signal amplitude. These carrier frequencies are mixed together and subsequently modulate a light beam which propagates down the fiber-optic cable to the receiver.

The sensors have a current output scale factor of 1 mA/ms^{-2}, or approximately 100 µA/g. Current to voltage converters at the inputs condition the signals before frequency modulation, and provide a means of adjusting the scale factor for field calibration purposes. Sensitivities of 100 mV/g to 1 V/g, adjustable in four steps, were selected for the initial trials, and may be altered by replacing one resistor in each channel. Frequency modulation is performed by an off-the-shelf integrated circuit (Motorola MC 1376) designed for general purpose telemetry applications. Good oscillator stability, low input noise, wide frequency operating range, and ease of application were among the factors considered in choosing this particular device as the front end of the transmission network. An input signal of ≈350 mV rms from the current to voltage converters produces frequency deviations of ±100 kHz in the 3.5, 4.5 and 5.5 MHz carriers. This wide deviation capability significantly contributes to the low overall noise level of the system. The devices inherently exhibit a minimal temperature drift of ≈300 Hz/°C ambient. To counteract this effect, a solid-state temperature sensor was incorporated on each circuit card to provide a compensation signal acting to cancel thermally related frequency shifts.

The outputs from the three modulators are fed to a linear summing circuit which combines the signals into one waveform with minimum crosstalk interference between channels. At this point, the multiplexed representation of triaxial seismic data is in a form suitable for optical conversion and transmission. A commercially available, three-terminal module, designed for high bandwidth (10 Hz - 10 MHz) video telemetry, generates an amplitude modulated, visible (820 nm) light beam at a power output of ≈20 microwatts. This level is sufficient to operate the present system at distances up to 3 km without modification or repeater stages.

The complete electronics package is powered by a 12 VDC regulator in each unit. The servo-accelerometer sensors require an additional low current (5 mA) supply of -12 VDC derived from an inverter module. The total current consumption of the transmitter is less than 150 mA and is suited for remote battery operation. In the case of the present installation, however, power is supplied by two copper conductors built into each fiber-optic cable assembly. Lightning protection circuits are incorporated at each end of the supply lines to guard against damage from induced high voltage spikes. Individual power supplies at the receiver provide an isolated current supply to each transmitter. The driving supplies are unregulated and simply provide an energy source for accurate regulators in each electronics package.

Referring to Fig. 1, a fiber-optic receiver reconstructs the original waveform at the output of the multiplexing circuit by demodulating the optical signal on the fiber cable. The electrical output is fed simultaneously to a bank of three filters which separate out the combined triaxial carrier frequencies into individual channels. A simple multistage LC design is used which passes the desired carrier while suppressing adjacent channels. The subsequent demodulator circuits are sensitive to high levels of adjacent channel interference, therefore tuning of the filters, although uncomplicated, must be performed carefully to maximize the available signal to noise ratio.

Information on the frequency modulated carriers is recovered by utilizing an existing FM demodulator IC (Signetics CA 3089) designed for high-fidelity receiver applications. Specifications include an internal frequency tracking circuitry which follows the incoming carrier and nulls out any drifts that may be present at the transmitter.

Computer processing of analog data requires that the incoming signals be filtered sharply above a frequency of half the sampling rate of the digitizer. The demodulator outputs are therefore presented to a set of anti-aliasing filters which attenuate above

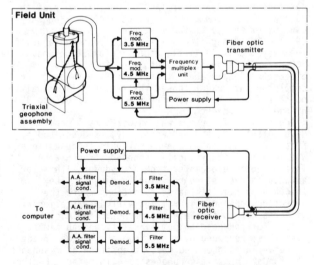

Fig 1. Block diagram of Microseismic Monitor transmission system.

300 Hz at a rate of 90 dB per octave. Further conditioning is provided by scaling circuits which set the full scale output of the receiver to the input requirements of the processing computer. The main bandwidth limiting components are the anti-aliasing filters at the outputs. The unfiltered frequency response of the instrument extends to 10 kHz (-3 dB), and is limited mainly by time constants in the demodulator section. High frequency piezoelectric accelerometers may be used in place of geophones by removing the filters. A dynamic range of 65 dB is attainable at the 10 kHz bandwidth.

3.1 Field installation

The fiber-optic network linking each geophone had to be laid over rough terrain, subject to environmental extremes. Resistance of the cable to damage by abrasion, crushing and animals was ensured by using the type of design depicted in Fig. 2.

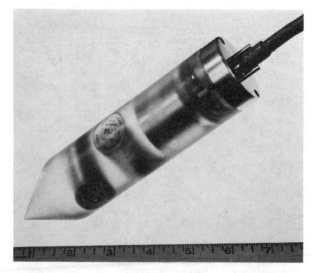

Fig. 3   Triaxial servo-accelerometer assembly.

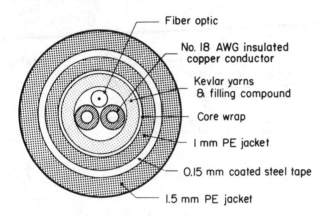

Diameter : 13 mm
Weight : 230 kg/km
Maximum installation tension : 2000 N

Fig. 2   Cross section of fiber-optic transmission cable.

A single 50 micron glass fiber in a protective jacket is bundled together with two 18 AWG insulated copper conductors in a grease filled wrap made up of a Kevlar strength member and a sealing layer. Mechanical protection is provided by two layers of tough polyethylene jacketing with an intermediate layer of hard steel tape. Field experience has shown that the design is highly suited for outdoor installation and direct burial.

Grouting of the triaxial geophone assembly into its associated borehole necessitated the use of a sealed and mechanically rigid casing to house the transducers. In addition, the assembly had to be orientable from surface to allow mutual alignment between stations. Since the triaxial sensors are directional in nature, much valuable data would be lost if random orientation were to be used. A photograph of the geophone assembly can be seen in Fig. 3.

Three mutually-orthogonal servo-accelerometer sensors are mounted in a machined polysulfone rod with a conical tip to facilitate pushing through borehole concrete. The transducers and sensor wires are sealed with epoxy and terminated in a potted metal cap at the top of the sensor. Alignment facility is provided by two short pins which protrude upwards next to the cable strain relief and mate with associated holes in the borehole insertion tool. Although other materials may be used, polysulfone was chosen for the casing because of its excellent

Fig. 4   Installation of triaxial sensor into surface borehole.

dimensional stability with temperature and high strength crack resisting mechanical properties. A five-conductor electrical cable, 15-20 ft in length, connects the sensor to the surface electronics.

During installation, tension on the cable holds the transducer casing against the tip of the insertion tool and maintains solid contact with the alignment pins. A photograph of a typical installation procedure can be seen in Fig. 4.

The insertion tool lowers the sensor and cable into a borehole (2 in dia.) loaded with about 2 ft of conventional sand mix concrete. As the sensor is pushed to the bottom of the borehole, a direction indicator, mounted at the top of the rod string and aligned with the sensor casing, is fixed to a predetermined compass orientation, e.g. north. Release of tension on the cable disengages the sensor and allows retrieval of the insertion assembly. Sand is poured to within 4 ft of surface and the rest of the borehole is filled with concrete into which the

Fig. 5   Electronics package installation at typical monitoring site.

Fig. 6   Fiber-optic receiver and data processing computer.

mounting bracket for the electronics package is cemented. The completed surface installation at one of the sites is shown in Fig. 5.

The sensor cable runs inside the mounting pipe which enters the box through a removable strain relief. Protection from weather is provided by the external enclosure which houses the main circuit package. All critical components are mounted in a waterproof package with an O-ring seal. Individual components or the entire surface box can be removed for laboratory problem diagnosis.

One of the installation sites was located immediately underneath a three-phase utility power line running on poles above surface. Testing of the associated sensor package showed 60 Hz interference components superimposed on the main signal. Analysis showed that the origin was from magnetic coupling between the sensors and the field around the power-lines. The transducers are manufactured in an aluminum case and, as such, have no inherent protection against auxiliary magnetic fields. The high gain servo-loop circuitry inside the device makes it sensitive to magnetic disturbance. Analog filtering was rejected because of the phasing problems introduced with respect to the other channels and, unless the component was pure sine wave, multiple stages would be needed to eliminate predominant harmonics.

An external sensing coil is used to generate a copy of the interfering signal being seen by the geophones. The coil's output is fed in inverse phase to a summing junction in the input circuits where it acts to cancel the power line noise while leaving the rest of the signal unaffected. In applications where interference signals are expected, the sensing coils may be incorporated as part of the transducer casing with a separate coil for each sensing axis.

3.2 Data analysis

Retrieval of seismic data for computer analysis is performed by the fiber-optic receiver located at a central control station. A photograph of the receiver, computer and graphic display terminal is shown in Fig. 6.

Fiber-optic lines from each location terminate at the rear of the receiver in optical connectors and power supply terminals. The front of the receiver contains access to individual triaxial channels via

BNC connectors. An audio output suitable for driving headphones and switchable between axes is provided for each channel to facilitate quick verification of system operation.

Power supply lines are individually fused and verification of correct current levels can be made via front panel monitor outputs. Data to the computer is sent via ribbon cables at the rear of the enclosure. The computer, a MASSCOMP 5400, can be seen directly beneath the receiver in Fig. 6 and a diagram of the software flow is outlined in Fig. 7.

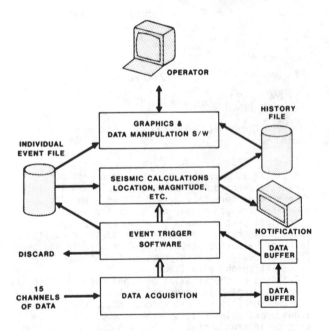

Fig. 7   Block diagram of processing computer software flow.

Fifteen channels of analog data are sampled at individual rates of 1 kHz by a 1 MHz data acquisition system. Events are recognized by software trigger routines which initiate the capture and storage of complete waveform data for all channels 200 ms before and 600 ms after a valid trigger. Waveform information is stored in labeled files on a 71 MB on-board hard disc. Event histories can be recalled when needed for computation. Manual processing, such as arrival time picking, can be done via an interactive graphic display that uses its own high speed CPU leaving the acquisition routines to a

separate host processor. Source location routines
and magnitude calculations are executed by software
in memory whose contents are protected by a tape
backup and an uninterruptable AC power supply.

4. CONCLUSIONS

Practical, low cost seismic monitoring networks using
fiber-optic transmission are feasible and can be
implemented using readily available components. The
use of multiplexed frequency modulation on an optical
channel allows for easy implementation of triaxial
sensor installations that are immune to transmission
line noise pickup without the expense of digitization
at source. Capture and storage of complete
multichannel event waveforms enables detailed
analysis and algorithm development.

ACKNOWLEDGEMENTS

The authors wish to thank D. Hedley and C. Graham for
their helpful cooperation, and the CANMET Mining
Research Laboratories for funding the project.
Special thanks are given to Harald Kanduth for
invaluable assistance during field installation.

REFERENCES

Bawden, W. and Kitzinger, F., 1986. Survey of South
  African Microseismic Systems, Noranda Research
  Centre.
Campbell, P., 1986. Dynamic Range Requirements for
  the CANMET Microseismic System, (internal
  memorandum, Noranda Research Centre).
Spottiswoode, S.M., 1984. Underground Seismic
  Networks and Safety, Symposium on Monitoring for
  Safety in Geotechnical Engineering, International
  Society for Rock Mechanics, S.A., National Group,
  Johannesburg.
Klaassen, K.B. and Van Peppen, J.C.L., 1983.
  Electronic Acceleration Sensitive Geophone for
  Seismic Prospecting, Geophysical Prospecting 31,457-
  480.

Application of the rigid block model to the study of underground cavern support
Modèle numérique des blocs rigides appliqué à l'étude du soutènement des cavernes souterraines
Anwendung des Models steifer Blöcke für die Bestimmung von Forderungen an den Ausbau in Untertagehohlräumen

R.LAIN HUERTA, ETS. Ing. Minas, Politechnic University of Madrid, Spain
P.RAMIREZ OYANGUREN, ETS. Ing. Minas, Politechnic University of Madrid, Spain

ABSTRACT : This paper deals with the Rigid Block Numerical Model applied to large underground caverns. An example of the application of the numerical model to estimate the most favorable orientation of rock caverns will be developed. In this example, a joint generation computer program is run. Two different orientations of the cavern have been analyzed: parallel and perpendicular to the dip of the joints.

The influence that the size and shape of the cavern exert upon support requirements have also been studied.

In order to demonstrate the utility of the numerical model, a brief study of the structural stability of a tunnel is carried out. In this example, pyramidal rock blocks in the roof and walls of the tunnel have been considered. The results obtained by the application of the Rigid Block Model to this problem are compared with the ones established by stereographic projection. It has been proved that the Rigid Block Model is valid for structural analysis of the support of underground opennings.

RÉSUMÉ : Cet article traite du modèle numérique des Blocs Rigides appliqué à de grandes cavernes souterraines. On montrera un exemple du modèle numérique pour trouver la plus favorable orientation de la caverne souterraine. Dans cet exemple, un modele numérique pour creer les diaclases sera utilisé. Deux orientations différentes de la caverne seront analysées : parallèle et perpendiculaire au plongement des diaclases.

Cet exemple montre aussi la posibilité d'utilisation du modèle des Blocs Rigides pour déterminer l'influence que produit dans les nécessités du soutenement la grandeur et la forme de la cavité souterraine.

Pour montrer l'utilité du modèle numérique, une brève étude de la stabilité structurelle d'un tunnel sera effectuée. Dans cet exemple, les blocs pyramidaux de roche du toit et des parois du tunnel seront considerèes. Les resultats obtenus de l'application du modèle numérique de Blocs Rigides seront comparés avec ceux de la projection stéréographique.

Il est démontré que le modèle des Blocs Rigides est valable pour l'analyse structurelle des massifs rocheux.

ZUSAMMENFASSUNG : Dieser Artikel handelt von Zhalenmodell der harten Blöcke, das auf die großen unterirdischen Hohlräume augewandt wird. Die Darstellung eines Beispieles des Zahlenmodelles soll die günstigste Orientierung der unterirdischen Kammern einschätzen. Bei diesem Beispiel benützt man ein Programm der Entstehungsweise der Blöcke.

Rau analysiert zwei verschiedene Orienterungen der Höhle : eine parallele und eine senkrechte hinsichtlich des Fallens der Verbindungstellen. Dieses Beispiel zeigt die Lebensfähigkeit des Zahlenmodelles der harten Blöcke zur Festlegung der besten Orientierung der Felsenhöhlen und des Einflusses der Größe und Form der Höhle in Bezug auf die Notwendigkeit einer Stützung.

Um die Lebensfähigkeit des Zahlenmodelles zu beweisen, verwirklicht man ein Studium über die Strukturstabilitat eines Tunnels. Rau hat erwogen ob Blöcke aus pyramidalen Felsen in der Dekke und den Giebeln des Tunnels das beste Wäre. Die Ergebnisse der Auwendung des Zahlenmodelles bei diesem Problem vergleicht man mit den bestimmten au Hand einer perpektivischen Zeichnung. Als Ergebnis dieser Vergleiche beweist man somit, daß das Zahlenmodell der Harten Blöcke in diesem Artikel vorgeschlagen für die Strukturanalyse eines Tunnels gilt.

1. INTRODUCTION. The growing use of the underground space for industrial, urban or military purposes, has encouraged the creation of larger underground structures and raised the necessity of new methods to analize their stability. The suggested method in this paper can be applied in jointed rock masses.

The largest difficulties for the modelization of a rock mass arise from the geological discontinuities (stratification, bedding, joints, faults, etc.) that are present in rock masses. The simulation of a discontinuous medium is a very difficult task. More-over, the engineers don't usually have the possibility to select the optimum orientation shape and size of large underground caverns. For example, if the problem consists on mining a mineral deposit, the orientation of the rooms is established by the location of the mineral. In other cases, the shape of the cavity is imposed by the equipment which is going to be installed inside the cavern, for example, turbines and alternating-current generators in a hydroelectric power plant.

What is intended to bring about in the design of large underground caverns is the optimum cavity orientation so that its stability is the best one and, therefore, the support element required are minima. Moreover, another fundamental parameters must be considered, for instance, the cavity shape and size, which are very important in the estimation of the cost of the support.

2. UTILITY OF THE NUMERICAL MODEL. The utility of the Rigid Block Numerical Model (Cundall, 1974) will be proved through an analisys of structurally controlled unstabilities in tunnels.

In the example proposed in this paragraph, a methodology is suggested in order to analyze the stability of shallow tunnels in jointed rock masses, in which most of displacements take place along the joints. The output of the program will be compared with the results obtained by conventional methods based upon the stereographic projection.

The example analized consists on the estimation of the bolt forces which stabilize the rock blocks on the roof and walls of a rectangular tunnel, built in a jointed rock mass with three joint sets.

Through the Rigid Block Method, the problem can be solved geometrically, by generation of twodimensional blocks produced by cross sections of the pyramidal rock block which is shaped by the three joint sets and the wall or roof of the tunnel.

A lot of sections would be necessary to cover the whole block; however, by extrapolation of a limited number of cross sections the accuracy of the results is enough, and usually greater than the precision obtained with the stereographic projection.

In the example below, the data analized are: Tunnel of 4 m high and 5 m width
Direction N-20°-W
Weight density of the rock 24 kN/m³
Safety factor 1,3

SET	DIP DIRECTION	DIP	FRICTION ANGLE
A	120°	50°	35°
B	220°	70°	30°
C	330°	60°	25°

Through the stereographic projection method, the results are:

    ROOF . . . . . Block weight 72 kN
                   Bolt force   93 kN
    RIGHT WALL . . Block weight 12 kN
                   Bolt force 3,2 kN

The Rigid Block Model has been applied on a group of twodimensional blocks generated by means of 15 cross sections of the pyramidal block, normal to the tunnel axis. These 15 sections cover a distance of 15 cm in theory, however by extrapolation, they cover the whole block, 301 and 247 cm along the tunnel axis for the roof and the right wall respectively.

On figure 1, the tunnel roof is modelled by a fixed block and the block which must be bolted in the 7th cross section of the pyramidal block is shown.

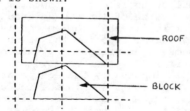

Figure 1. Rock block from the roof of
the tunnel

On figure 2, a rock block falling down from the tunnel wall is drawn. The block corresponds to the 5th cross section of the pyramidal block.

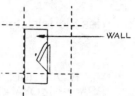

Figure 2. Rock block from the right wall
of the tunnel

The output of the program is resumed on Table 1.

TABLE 1
Bolt force in
Newtons

Section number	Roof	Right Wall
1	20,57	34,09
2	82,34	11,24
3	185,44	14,97
4	330,08	14,18
5	490,85	14,25
6	575,05	10,83
7	573,22	9,33
8	484,70	7,93
9	364,12	6,66
10	260,62	5,49
11	174,49	4,44
12	105,59	3,50
13	53,84	2,66
14	19,39	1,95
15	2,15	1,35
TOTAL	3722,53	110,39

Each 1 cm thick section represents actually a length of 20 cm, therefore the bolt force is :

        3722 x 20 = 74440 Newtons

On the right wall of the tunnel, the length represented by each section is 16,5 cm and the bolt force is :

        110 x 16,5 = 1815 Newtons

With a safety factor of 1,3 the results are

    Bolt force in the tunnel roof   97 kN
    Bolt force in the right wall   2,3 kN

As it can be observed, these values agree with those obtained by the stereographic projection method.

3. COMPUTER PROGRAM FOR THE GENERATION OF JOINTS AND BLOCKS. The joint and block generation program is recommended for the analysis of regularly jointed rock masses. In other cases, a digitalization system will be necessary.

Given a number of joint sets, the orientation of each set and the spacing, the program generates the planes which represent all the joints in the rock mass in three dimensions, by means of algebraic procedures. Afterwards, the lines resulting from the section of the joint planes and planes normal to the cavity axis are obtained. From these lines, the intersection points are mathematically deduced by the program.

The geometric problem limits must be specified and the outer points removed. At this stage, blocks can be generated. The block creation is carried out in two steps; in the first one primary blocks are generated through the intersection points of two joint sets. Secondary blocks are generated by intersections of the primary blocks with the lines representing the third joint set, allways on the cross section normal to the cavity axis. Then, small blocks are removed to avoid numerical unstabilities in the Rigid Block Model. Finally, given the cavity perimeter, the inner blocks are removed. At this moment, the block geometry is ready for input in the Rigid Block Program.

4. OPTIMUM ORIENTATION OF LARGE UNDERGROUND
CAVITIES. Once the utility of the suggested method for structurally controlled unstabilities in shallow tunnels has been proved, an example is carried out to determine the optimum orientation of large underground caverns and also the influence of the size and shape of the cavity upon the roof support requirements.

The geometry of the problem is given by the strike and dip of the joint sets, and also the joint spacing in each set. In this example, three joint sets will be considered.

The joint planes must be refered to a coordinate origin. It is necessary because the block generation computer program creates each cross section by the intersection of the joint planes with the plane of the section normal to the cavity axis. Moreover, in order to estimate the value of the external forces acting in the block system, the rock weight density and the depth of the cavern must be known.

The data analized in this example are:

Set	Dip Direction	Dip
R	130°	50°
S	310°	30°
T	325°	71°

The friction will be taken about 0,7

Two cavern orientations will be analyzed. The first one, whose direction is N-40°-E,is parallel to two of the joint sets, and the strike of the third one forms 15°with the tunnel axis. The second orientation analized is perpendicular to the first one as it can be observed in figure 3.

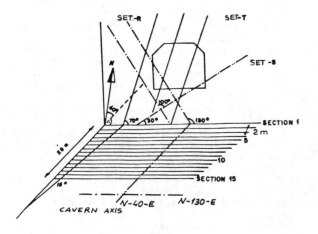

Figure 3. Schematic view of the 15 sections and the two orientations of the cavern

The roof of the cavern is located 100 m deep from the surface and the dimensions are drawn in figure 4.

The weight density of the rock is 26 kN/m^3
The joint set spacing,given in meters,is:

SET R (12 joints)
  20;10;8;7;10;10;7;8;10;9;11
SET S (11 joints)
  13;10;8;7;10;7;8;10;10;10

SET T (10 joints)
  9.7;13.6;8.7;6.8;7.8;11.7;7.8;6.8;6.8

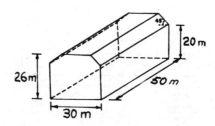

Figure 4. Dimensions of the cavern in m

With these data, the block generation program, briefly described in the former paragraph is run, and blocks are created. Later, the Rigid Block Model is applied to each cross section of the cavern. On figure 5, blocks and cavity perimeter are plotted.

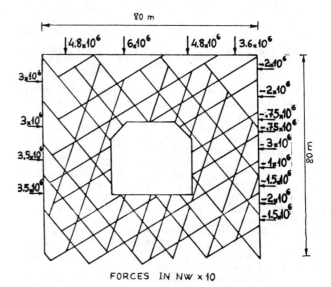

FORCES IN NW × 10

Figure 5. Section of the rock cavern normal to the axis, and boundary conditions

The first orientation to be analized, N-40°-E, is the most adverse, because the strike of two joint sets is parallel to the cavern axis and the third set is subparallel to it.

The second orientation analized is N-130-E , perpendicular to the first one. It must be the most favorable orientation, since the strike of the joint sets is perpendicular to the cavern axis.

On figure 6 a cross section of the cavern whose orientation is N-130°-E is shown.

For the two orientations, 15 different cross sections will be analized.The reason why 15 sections have been considered is because more sections will produce an increase in computer time without giving more accurate results and less than 15 sections can introduce some errors in the interpolation.

The length units are given in dm due to restrictions in the Rigid Block Model. The total length along the cavern axis covered by 15 sections of 1 dm thick is in theory 1,5 m.However as the joint spacing is about 10 m a length of cavity of at least 30 m must be covered in order to obtain significatives results, therefore each of the 15 sections will actually be representative of 2 m of cavern length and consequently the support forces obtained must be multiplied by 20.

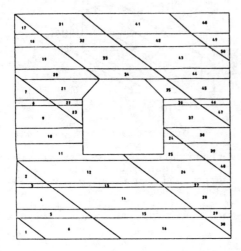

Figure 6. Cross section of the cavern. Orientation N-130°-E

The model extends over an area of 80 x 80 m^2 as shown in figure 5. In this figure the boundary conditions are also marked. These boundary conditions consist on a set of forces applied to the upper and side blocks on the limit of the model. These forces are produced by the natural stress field. On the other hand, some blocks will be fixed in the lower part of the model, to simulate the undisturbed area.

The forces to stabilize the system will be applied only on the unstable blocks of the perimeter of the cavity. These blocks have been identified after a first run of the Rigid Block Program, in which no forces have been introduced.

Relationships between a set of parameters and time will be obtained for the unstable blocks. In this way, forces that stabilize the system will be estimated. In a later run of the program, these forces will be introduced in the model and the stability of the system will be checked.

In Table 2, an output of the model for the two orientations analyzed is shown.

TABLE 2

Bolt force in kNewtons

Section number	Cavern N-40-E	Cavern N-130-E
1	1630	140
2	1720	112
3	1790	108
4	1870	93
5	1960	77
6	2050	62
7	2130	STABLE
8	2200	"
9	2280	"
10	2350	"
11	2180	"
12	2020	"
13	1850	"
14	1590	"
15	1460	"
TOTAL	29080	592

5. SIZE AND SHAPE EFFECT OF UNDERGROUND CAVITIES UPON THE SUPPORT REQUIREMENTS. In order to study the size effect upon the support requirements, an example with the same cavern mentioned in the former example whose dimensions have been divided by two, i.e., the cross section area divided by 4, has been analyzed. This cavern is also located at 100 m from the surface. The results are shown in Table 3.

The shape effect has been studied by comparing the results of the numerical model for two different cavities: the first one has 45 m width and 13 m hight and the second one 30 m width and 26 m hight, i.e. both cavities have the same cross section area, but different shape.

The output results are also presented in Table 3.

TABLE 3
Bolt force in kNewtons

Section number	Cavern 15x13 m^2	Cavern 45x17 m^2
1	380	2794
2	420	2843
3	480	2941
4	490	3052
5	490	3101
6	480	3089
7	520	3175
8	520	3065
9	570	3151
10	590	3249
11	560	3655
12	530	3754
13	440	3791
14	370	3840
15	360	3901
TOTAL	7200	49401

6. CONCLUSIONS. The bolt force neccesary to stabilize an underground cavity whose section is equal to 30 x 26 m^2 and its direction is N-40-E is 50 times greater than the bolt force for a cavern with the same cross section and direction N-130-E.

The preceedings results agree with the ones obtained in normal practice in civil and mining engineering. The best orientation of underground cavities is the normal to the maximum number of joint sets in the rock mass.

By comparison of the results obtained for the cavern of 30 x 26 m^2 and N-40-E, with another cavern with the same orientation and 15 x 13 m^2, about a quarter of the former cross section, it has been deduced that the bolt force necessary to stabilize the later one cavern is 4,04 times lower.

Finally, if the cavern of 30 x 26 m^2 and N-40-E is compared with a cavity of the same direction and a section of 45 x 17 m^2, i.e. the same cross area but wider, the bolt forces are 1.7 times greater in the last case.

The influence of the cavity shape in the volume of unstable rock is very important. If the cavity width increases and its height decreases, the resulting block dimensions in the roof are bigger, and therefore more support elements are necessary for ground control.

The cross section area is another factor whose influence in the support requirements is very relevant. When the tunnel size increases, greater blocks will fall more easily. The increase of the number of potentially unstable blocks is directly proportional to the increase of the cross section.

Finally, it has been shown that the Rigid Block Numerical Model, together with a joint generation program, is very useful to study the support of underground caverns in jointed rock masses subjected to low stresses, where deformations are produced principally along joints.

7. BIBLIOGRAPHY.

CUNDALL P. (1974) Rational Design of Tunnel Supports: A Computer Model for Rock Mass Behaviour Using Interactive Graphics for the Input and Output of Geometrical Data. Technical Report MRD-2-74, Missoury River Division,U.S. Army Corps Engineers

HOEK E. and BROWN E.T. (1980) Underground Excavations in Rock. Institution of Mining and Metallurgy. London.

GOODMAN R.E. and SHI G.S. (1985) Block Theory and its Application to Rock Engineering. Prentice-Hall, Inc. New Jersey

Sublevel stoping under high horizontal stress field at the Pyhäsalmi Mine

Chantiers d'abattage intermédiaires sous un champ de contraintes horizontales élevées à la mine Pyhäsalmi

Weitungsbau bei hohen horizontalen Druckspannungen in der Pyhäsalmi Grube

P.LAPPALAINEN, Outokumpu Oy Pyhäsalmi Mine, Finland
J.ANTIKAINEN, Helsinki University of Technology, Finland

ABSTRACT: Stability problems in sublevel stoping caused by excessive 30 - 70 MPa horizontal stresses, have been largely eliminated at the Outokumpu Oy Pyhäsalmi mine by pre-reinforcement with cable bolting, yielding pillars and stope shaping, rock destressing by sequential stoping technique and "slot stopes" and fast stoping from opening to backfilling. The basic idea in cable bolt pre-reinforcement is to help natural rock arching around stopes and thus prevent caving. Forty km of cable bolting is done annually mainly by a mechanized jumbo, Robolt H 690-5. "Hedgehog" bolting done from sublevel drifts has proved very effective. Parallel "mandolin" bolts along the ore contact are preferred to conventional "outside bolts" because progressive slabbing can be prevented by the use of mandolin bolts.

RESUME: Les problèmes de stabilité à l'exploitation en sous-étages causés par les contraintes horizontales de 30-70 MPa ont été éliminés dans une grande mesure à la mine de Pyhäsalmi de Outokumpu Oy au moyen du pré-renforcement par boulonnage au câble, des piliers coulissants et de la formation des abattages, du relachement des contraintes par l'abattage consécutif et les "gradins en fente" et de l'abattage rapide à partir de l'orifice jusqu'au remblayage. L'effet principal du boulonnage au câble est à aider la voussure naturelle de l a roche autour des gradins et empêcher l'éboulement. Le boulonnage au câble est fait 40 km par an, principalement par un jumbo mécanique Robolt H 690-5. Le boulonnage en "hérisson" exécuté par les galeries de tranches s'est montré très efficace. Les boulons parallèles en "mandoline" le long du contact de minerai sont préférés aux boulons conventionels parce que le décollement progressif de roche peut être empêché en utilisant les boulons en "mandoline".

ZUSAMMENFASSUNG: Stabilitätsprobleme im Weitungsbau verursacht von sehr hohen (30-70 MPa) Horizontalspannungen sind bei der Pyhäsalmi Grube der Outokumpu Ag im grossen eliminiert worden. Die wichtigsten Mitteln dazu waren Vorverstärkung mit Kabelankern, nachlässigen Pfeilern und Formengebung der Strossen, Felsentspannung mit sequentialem Abbautechnik und "Aufbruchstrossen", und schneller Abbau von Aufbruch bis Versatz. Der Grundidee der Vorverstärkung mit Kabelankern ist die natürliche Felsgewölbung um die Strossen zu hilfen und damit Einstürze zu verhindern. Kabelankerungen werden 40 km jährlich gemacht, hauptsächlich mit einem mechanizierten Bohrwagen, Robolt H 690-5. Sog. Igelankerung, aus Teilsohlen ausgeführt, ist sich als sehr effizient erwiesen worden. Sog. Mandolineankerung, Parallelankerung den Erzkontakt entlang, ist traditionaler Kabelankerung vorgezogen worden, weil die progressive Ablösung des Gebirges mit Mandolineankerung verhindert werden kann.

1 INTRODUCTION

The Pyhäsalmi mine is situated in the province of Oulu, about 4 km from the village of Pyhäsalmi.

The ore lens of Pyhäsalmi is 650 m long at the outcrop a 75 m wide, carrot-like formation. To date the ore is known to descend to 800 m. It is estimated that the mine will operate into the next century at its existing capacity of 900,000 tonnes of ore annually.

The mine produces annually over 500,000 tonnes of pyrite concentrate, 40,000 tonnes of zinc concentrate 25,000 tonnes of copper concentrate, and 10,000 tonnes of barite. Some silver and a little gold is recovered into concentrates.

The most important rock mechanical factors affecting the mining methods applied at the Pyhäsalmi mine are: the weak country rock enveloping the stiff, strong ore and high horizontal stress field. The maximum horizontal stresses below 400 m are between 40-70 MPa, which is about three times the vertical gravity load.

Since 1985 a new design for cable bolting has been used in three stopes at the Pyhäsalmi mine.The new design is a combination of "hedgehog" and "mandolin" cable bolting.

1. Hedgehog bolting: Cable bolts are installed in fan shape from sublevel drifts. The drifts are reinforced with wire mesh and the cable bolts are tied with bearing plates to the mesh. Finally the drifts are shotcreted. The cable bolts are installed with a fully mechanized electro-hydraulic cable bolting rig, Tamrock Cable Bolter H690-5, jointly developed by Tamrock and Outokumpu (Fig 1.)

2. Mandolin bolting: Cable bolts are installed sub-parallel to the stope walls (Fig.5). The spacing of the Ø 90 mm holes has been 1.0-1.5 m. Usually four Ø 15 mm steel strands are grouted into a hole, giving a total tensile strength of 100 tonnes. The downhand holes are drilled with a Solo 890 RF hydraulic longhole drill and the cables are installed manually.

2 EXPERIENCES ON THE NEW CABLE BOLTING DESIGN

2.1 Stope S71

The measured initial stress was moderate, horizontal maximum 32 MPa. The shape of the stope was dictated by the main fracture directons. BEM-calculations did not show any tensile stresses in the hanging wall, but the roof of the stope was loaded by excessive compressive stresses. In Figure 2 the roman numerals refer to excavation sequence. Reinforcement with cable bolting is presented schematically. The excavation sequence was:

I   Bottom opening
II   Opening raise and longitunal opening
III Footside expansion blasting
IV   Upper side expansion blasting
V   Hanging wall expansion blasting
VI   Final blasting at hanging wall, holes parallel
to the ore contact.
VII Final throwing blasting at footwall.

Figure 1. Cable bolting rig.

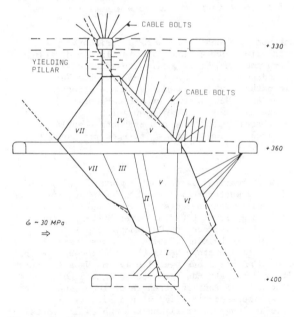

Figure 2. Stope S71. The excavation sequence and cable bolting, schematically.

The hanging wall, the draw points and the +330 sublevel were pre-reinforced with cement-grouted cable bolts, two steel strands per hole. One cable bolt per 10 m² of roof was installed, in de-stressing drift one cable bolt per 4 m² of roof. In the +330 sublevel drift cablebolting was combined with weldmesh and shotcreting to give the best possible reinforcement. The goal was to bind the weak hanging wall sericite quartzite to a thick beam and prevent the begining of rock failure in the upper corner of the stope.

The stope was opened longitudinally at the middle of the ore (Fig.2) and the stoping proceeded towards the ore contacts. In the final blasting the Ø 90 mm holes were parallel to the ore contacts. The holes

were loaded with ANFO and ignition made with NONEL.

Displacements were measured with extensometers. The hangingwall moved inwards 10-40 mm, which equals the elastic model. The horizontal pillar between the stope and the +330 sublevel (Fig.2) was broken in place due the high horizontal stress concentration caused by stoping. After that the stresses moved above the +330 sublevel. The roof of the +330 sublevel was broken, but thanks to the dense cablebolting and use of wire mesh with shotcreting, the sublevel remained safe. The horizontal pillar functioned as a yielding structure, and the stresses moved to less critical sites away from the top of the stope.

The main reasons for the good stability of S71 stope:
- initial horizontal stress (32 MPa) was moderate in Pyhäsalmi conditions
- the shape of the stope was favourable compared with the directions of stresses and fractures
- the hanging wall was pre-reinforced with cablebolts
- the bottom opening didn't break the hanging wall contact
- the longitudinal opening of the stope released the overstressed hanging wall gently
- a yielding horizontal pillar removed the excessive stresses far away from the upper corner of the stope
- when stoping near the ore contact, stoping holes was parallel to the contact

135,000 tonnes of ore were loaded from the stope, and waste rock dilution was only 5 %. Costs from cable bolting was USD 0.75/t. The reinforcement was profitable, because ore losses and waste rock dilution were low. The working profit from the stope was USD 2,000,000 (1985 metal prices).

2.2 Stope PK1

The stope is situated in an area where rock failures have occured in all the previous stopes, causing severe problems in waste rock dilution and ore losses.

The measured horizontal stresses above the stope were 60-70 MPa perpendicular to the ore. These are the highest in-situ stresses ever measured in Finland.

Although the compact S-Zn-Cu ore is stiff and strong, the country rocks are weak serisite quartzite with very weak mica rock veins. For rock quality estimation RQD-values were calculated and the in-situ rock mass module of deformation was measured with hammer seismograph. Laboratory tests were also made.

Rock was pre-reinforced with cable bolts, totalling 3800 m, mainly in the hanging wall. Most of the cable bolting was made from outside of the ore, one cable bolt per 8 m² of roof (Fig. 3). Part of the hanging wall was bolted from the stope. This reinforcement was not strong enough to prevent caving, but it gave enough time for loading, so waste rock dilution was only 15 %.

From the +425 sublevel drift the hanging wall contact was reinforced by cable bolting from inside of the ore, one cablebolt per 4 m² of roof (Fig.4). The drift was also reinforced by wire mesh and shotcrete. The goal was to make a strong, yielding beam in the most critical site in the stope.

According to the stress measurements stoping increased the stresses above the stope 20 MPa. The walls of the +425 sublevel drift were compressed in about 70 mm during three months. Rock above the stope was compressed and finally broken due to stoping. Effects of stress changes, such as rock loosening, were detected in the ramp, 70 m from the stope. In the upper part of the stope, the in-situ deformation modul of rock decreased, 11 % for the ore and 60 % for the country rock due the rock loosening. The roof of the stope moved downwards only 3-4 mm, but hanging wall displacements were 300-400 mm before caving begun.

In stope PK1 the cable bolting from outside the ore couldn't prevent caving of the hanging wall, because the rock broke to pieces and caved between the bolts. In the +425 sublevel drift the quite short, but dense cable bolting combined with wire mesh and shotcrete worked well and stopped the caving. The +425 sublevel drift remained safe, although 70 mm displacements was measured from the walls of the drift. Economically the stope was profitable due the succesful reinforcement. If the caving had started before loading was completed, the result would have been worse.

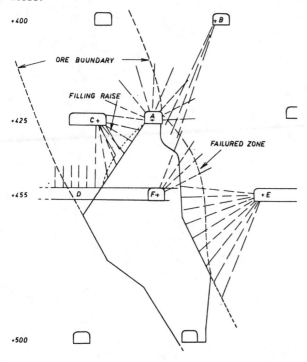

Figure 3. Stope PK1. Principle of cablebolting.

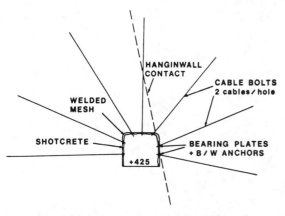

Figure 4. Cablebolting from the inside of the ore, "hedgehog bolting"; Ø 51 mm holes, two cement-grouted steel strands in each hole giving a tensile strength of 50 tonnes per bolt.

Cablebolting from inside of the ore, "hedgehog bolting" and the parallel cablebolting, "mandolin bolting" helps natural arching of the rock, so the rock supports itself and caving cannot begin. It's impossible to prevent rock loosening completely, when initial stresses are about 50-60 MPa as in Pyhäsalmi.

2.3 Stope X22

The parallel cable bolting with hole spacing from 1.0 m to 1.5 m, like in Pyhäsalmi mine, forms a kind of mesh or "skin support", which can tolerate large displacements. The both ends of "mandolin bolts" are grouted into "safe" rock, while almost the whole length of the bolts is situated in the "active zone" near the stope. The hanging wall began to cave in (Fig.5). That continued along the talc rock zone up to the "mandolin" bolts. The bolting did, however, stop the caving which could not penetrate the "bolt fence". The general stability of the X22 stope remained good and a waste rock dilution rate of only ten percent was experienced (Fig.5).

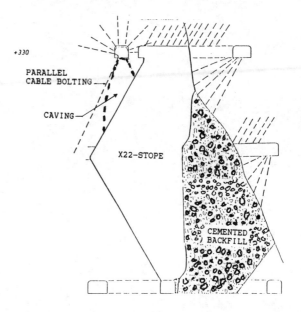

Figure 5. The reinforcement of stope X22. The spacing of parallel cable bolts is 1.0 m.

In subsequent stopes parallel or mandolin cable bolting has also been used successfully to prevent stope scale slabbing. In these cases "mandolin bolting" is nowadays in routine use.

Wirklichkeitsnahe Bemessung von oberflächennahen Tunnelbauwerken im Locker- und Festgestein

A realistic design of shallow tunnels in soils and rocks
Une conception réaliste des tunnels peu profonds dans les sols et dans les roches

G.LAUE, Stadtbahnbauamt Bochum, Bundesrepublik Deutschland
H.SCHULTZ, Zerna, Schultz und Partner, Ingenieurgesellschaft für Bautechnik, Bochum, Bundesrepublik Deutschland
H.KLÖNNE, Zerna, Schultz und Partner, Ingenieurgesellschaft für Bautechnik, Bochum, Bundesrepublik Deutschland

ABSTRACT: Since some years Bochum subway tunnels have been constructed due to the principles of the New Austrian Tunnelling Method (NATM), the final construction of which consists of a primary shotcrete shell and an additional in-situ concrete supplement and represents a compound structure. After analyzing the first results of a longterm-testing of a cross-sectional measurement and additional bench tests further effective rules of dimensioning had been developed for this economical construction.

RESUME: Depuis quelques années à Bochum les tunnels de métro sont construits selon les principes de la Nouvelle Méthode Autrichienne de construction des Tunnels (NATM). Leur construction définitive a une coupe transversale composite consistant en une coque primaire en béton armé projeté, suivie et complétée par une coque en béton armé coulé sur place. L'interprétation des résultats obtenus par un appareil de mesure dans le temps et par des analyses de laboratoire complémentaires a permis de développer des règles de dimensionnement plus réalistes pour cette méthode de construction économique.

ZUSAMMENFASSUNG: Seit einigen Jahren werden in Bochum U-Bahn-Tunnel nach den Prinzipien der Neuen Österreichischen Tunnelbaumethode (NATM) hergestellt, deren endgültige Ausbaukonstruktion ein aus primärer Spritzbetonsicherung und folgender Ortbetonergänzung bestehender Verbundquerschnitt ist. Nach Auswertung erster Ergebnisse eines für Langzeitmessungen ausgelegten Meßquerschnitts und ergänzender Laboruntersuchungen konnten für diese wirtschaftliche Bauweise wirklichkeitsnähere Bemessungsvorschriften entwickelt werden.

1. EINLEITUNG

Seit mehr als 15 Jahren werden in Bochum Tunnelanlagen der U-Bahn erstellt. Nach den Prinzipien der Neuen Österreichischen Tunnelbaumethode (NATM) wurden ein- und mehrgleisige Streckenröhren sowie Weichenanlagen und Bahnhofsröhren mit den zugehörigen Schrägstollen aufgefahren.

Zur Ausführung gelangten Querschnitte von 6 bis 21 m Breite. Die Firstüberdeckungen betrugen zwischen 3 und 16 m, in Sonderfällen - in Verbindung mit besonderen Abfangemaßnahmen - 0,0 m.

Der Baugrund in Bochum ist gekennzeichnet durch die typische Abfolge:
- Anschüttungen unterschiedlicher Mächtigkeit,
- sandige, schluffige, tonige Schichten des Quartär (Diluvium, Alluvium): Bachablagerungen, Löß-/Lößlehm, Geschiebemergel,
- Mergelschichten der Kreide (Cenoman und Turon) sowie deren schluffig-tonige Verwitterungsböden,
- Karbon (Sandstein, Tonschiefer, Kohlenflöze).

Die Vortriebe wurden in erster Linie in den Schichten des Quartär und der Oberkreide durchgeführt.

Kennzeichen der NATM ist, daß zunächst eine vorläufige Sicherung aus bewehrtem Spritzbeton (Tunnelaussenschale) eingebracht wird, die - in einem größeren zeitlichen Abstand folgend - durch eine Tunnelinnenschale aus Stahlbeton ergänzt wird (zweischalige Bauweise). Bei der Dimensionierung der Innenschale wird die Außenschale üblicherweise als nicht dauerstandfest und somit als nicht mittragend angesehen, obwohl sie wesentliche Tragfunktionen auch über den Auslegungszeitraum hinaus übernimmt. Durch gezielte Qualitätsverbesserung der Außenschale und intensive theoretische Vorarbeiten wurden die Voraussetzungen geschaffen, dieses Bauteil in den rechnerischen Nachweis der Dauerstandsicherheit für die Gesamtkonstruktion einzubeziehen.

Die folgend als "einschalig" bezeichnete Verbundbauweise ist dadurch gekennzeichnet, daß der äußere Schalenteil - die primäre Hohlraumsicherung - nachfolgend durch einen inneren Ortbetonschalenteil ergänzt wird, der weder durch Folien noch andere Trenn-

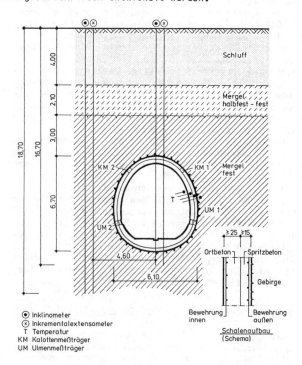

Abb. 1: Meßquerschnitt

schichten vom äußeren Schalenteil gelöst ist und so zusammen mit diesem zwangsläufig einen Verbundquerschnitt bildet.

Die geltenden Bemessungsregeln sehen folgende Vorgehensweise für Standsicherheits- und Gebrauchsfähigkeitsuntersuchungen vor:

Bis zum Einbau der Innenschale erfüllt die Außenschale -in Verbindung mit dem mittragenden Gebirge die Funktion der Sicherung des Hohlraumes. Der Beanspruchungszustand wird durch geometrische Gegebenhei-

ten (Querschnittsgröße und -form, Schalendicke, Oberdeckung usw.) und Materialeigenschaften (Beton, Gebirge) bestimmt. Die Dauer dieses Bauzustandes ist
zusätzlich zu berücksichtigen, wenn zeitabhängige
Spannungs-Verformungsbeziehungen des Gebirges und
des Betons eingerechnet werden müssen.

Während und nach Einbringen und Erhärten der inneren Schale wird zunächst aus zeitabhängigem Tempera-
turverhalten (Hydratationsprozeß) eine Wechselwirkung zwischen beiden Schalenteilen erzeugt. Danach
führt das Kriechen der vorbelasteten Außenschale zu
Lastumlagerungen auf den Innenteil. Die Größe dieser
Umlagerungen ist insbesondere abhängig von den bereits eingetretenen Kriechverformungen, d.h. von der
Dauer der Vorbelastung und somit vom Alter der vor-
laufenden Schale zum Zeitpunkt der Herstellung des
Innenteils und vom Kriechverhalten des Gebirges.

Nahezu zeitlich parallel zum Kriechvorgang verläuft der Schwindprozeß in der inneren Schale. Anders als bei Hochbaukonstruktionen laufen jedoch
Schwind- und Kriechprozesse bei Tunnelschalen wegen
der günstigeren Randbedingungen (Luftfeuchtigkeit
und Temperatur) langsamer und weniger ausgeprägt ab.

Eine weitere Belastungsphase wird durch spätere
Änderung der äußeren Belastung geprägt, wie sie
z. B. der sich aufbauende Wasserdruck und veränderte
Geländebelastungen erzeugen. Diese Belastungen bean-
spruchen die Gesamtschale nahezu wie eine monoli-
thisch hergestellte Konstruktion.

Als Kriterium für die Gebrauchsfähigkeit gilt im
allgemeinen die Einhaltung von zulässigen Beton-
druck- und Stahlzugspannungen im Zugbereich. Bei Ver-
bundsystemen müssen darüber hinaus die Verbundspannungen in der Arbeitsfuge nachgewiesen werden.

Die entscheidende Schwierigkeit eines realisti-
schen Nachweises ist die Behandlung der Verbundkräf-
te. In Anlehnung an die deutsche Stahlbetonnorm
DIN 1045 kann die Kontaktfuge zwischen beiden Schalenteilen als Arbeitsfuge betrachtet werden, in der
bei strenger Anwendung der Vorschrift eine Schubbewehrung anzuordnen ist.

Neben den verfahrenstechnischen Problemen, die beiden Tunnelschalen mit Schubbewehrung zu verbinden,
erhöhen vom äußeren zum inneren Schalenteil verlau-
fende Stäbe der Schubbewehrung die Wasserwegsamkeit
und stehen der Absicht, eine weitgehend wasserdichte
Konstruktion herzustellen, entgegen.

Unter Voraussetzungen, die bei dieser Bauweise gegeben und verfahrenstechnisch erzielt werden können,
werden - bei Verwendung der in Bochum ausgeführten
Betonqualitäten - begrenzte Schubspannungen in der
Verbundfuge ohne Schubbewehrung zugelassen.

Das Konzept der einschaligen Verbundkonstruktion
wurde in Grundsatzuntersuchungen in den Jahren 1980
bis 1986 verfolgt und konnte in Bochum bei der Ausführung von eingleisigen Tunnelröhren mit einer Gesamtlänge von ca. 3,8 km verwirklicht werden.

Zur Absicherung weiterer theoretischer Arbeiten
wurde ein für Langzeitmessungen ausgelegter Meßquerschnitt eingerichtet.

2. MESSQUERSCHNITT UND LABORVERSUCHE

Im Bereich einer eingleisigen U-Bahnröhre mit ca.
33 m² Ausbruchfläche wurde ein Meßquerschnitt eingerichtet (Abb. 1). Bei einer Oberdeckung von ca.
9,00 m wurde der Vortrieb in den festen Mergelschichten der Oberkreide, in die unregelmäßig Kalksand-
steinbänke eingelagert waren, durchgeführt. Der vor
Jahrzehnten in diesem Gebiet umgegangene Kohleberg-
bau in mehr als 200 m Teufe hat örtliche Störzonen
im Gebirge verursacht, die angefahren wurden.

Im Bereich des Meßquerschnitts wurde die Gelände-
oberfläche mit einem engen Raster von Oberflächenmeß-
punkten versehen.

Die Gebirgsverformungen wurden mit den bekannten
Inklinometern und Inkrementalextensometern gemessen.

Da der Meßquerschnitt Aufschlüsse über die Beanspruchung der Tunnelschalen in Bau- und Endzuständen
vermitteln sollte, wurden Meßträger System PH, mit

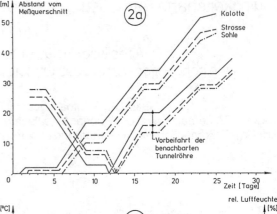

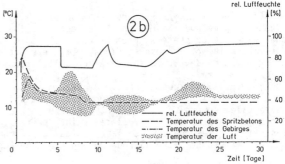

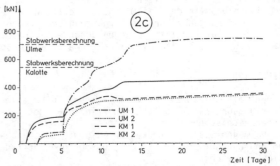

Abb. 2: Auswertung der Meßergebnisse
  a. Baufortschritt
  b. Temperatur- und Luftfeuchtigkeitsentwicklung
  c. Normalkräfte in den Meßträgern

denen der Verformungszustand über die Eigenfrequenz
der schwingenden Saite bestimmt wird, in die Schalen
eingebaut (Baumann, Th. 1985).

Zur Bestimmung von weiteren Parametern wurden Temperaturmessungen im Gebirge, in der Tunnelaußenscha-
le und in der Ortbetoninnenschale sowie Messungen
von Luftfeuchtigkeit und -temperatur im Tunnel durchgeführt.

Diese Messungen wurden ergänzt durch zeitliche
Aufnahmen des Arbeitsablaufs, der örtlichen geologischen Verhältnisse und aller sonstigen Randbedingungen wie z.B. Ausbruchkontur und Kluftwasseranfall.

Wie einleitend dargelegt, stellte bisher die Un-
sicherheit in der Frage nach den zulässigen Verbundspannungen in der Arbeitsfuge Spritzbeton-Ortbeton
eine entscheidende Schwierigkeit für rechnerische
Standsicherheitsuntersuchungen dar. Zur weitergehen-
den experimentellen Klärung dieses Komplexes - theoretisch schon in (Schmidt-Schleicher, H. und Lippert, 1982) und (Schmidt-Schleicher, H. 1986) behandelt - wurden Schubversuche an Probekörpern durchgeführt. Für diese Versuche konnte auf Arbeiten in
(Daschner, F. und Kupfer, H. 1982) und (Daschner, F.
1986) zurückgegriffen werden.

Zur Absicherung der aus Versuchen mit labormäßig
hergestellten Versuchskörpern erhaltenen Ergebnisse
wurden weitere Schubversuche an Bohrkernen, die fer-
tiggestellten Verbundkonstruktionen entnommen wurden, durchgeführt.

3. AUSWERTUNG

Folgend wird nur über die Auswertung der bis Oktober 1986 erhaltenen ca. 10 300 Werte der Meßträger in den Außenschalen berichtet. In Abb. 2 sind die Ergebnisse aus der besonders interessanten Phase im Zeitraum von 30 Tagen nach Durchfahren des Meßquerschnitts aufgetragen.

In Abb. 2a ist das Fortschreiten der Ortsbrust vom Meßquerschnitt dargestellt. Die unter dieser Darstellung befindliche Abb. 2b zeigt den gleichzeitigen Verlauf der Temperaturen im Gebirge, in der Spritzbetonschale und im Tunnel sowie den der Luftfeuchtigkeit. Während im unmittelbaren Ortsbrustbereich die täglichen Schwankungen signifikant sind, verhalten sich Temperaturen und Luftfeuchtigkeit ab einem Abstand von ca. 40 m hinter der Ortsbrust annähernd konstant und unabhängig von den baubetrieblichen Einflüssen. Auf Dauer betrugen dort die Luftfeuchtigkeit annähernd 95 % und die Temperaturen von Gebirge, Außenschale und Luft ca. 10^0 bis 12^0C.

In Abb. 2c sind die aus den Meßträgermessungen ermittelten Normalkräfte in der Außenschale aufgetragen.

Die dargestellten Mittelwerte für die Kalottenmeßträger 1 und 2 sowie den Ulmenmeßträger 2 (vgl. Bild 1) geben ein anschauliches Bild über den Zuwachs der Normalkraftbeanspruchung in der Außenschale in Abhängigkeit vom Abstand der Ortsbrust von der Meßstelle. Sie sind beispielhaft für einen Vortrieb im ungestörten Gebirge. In FE-Berechnungen gelang es nach wenigen Schritten, in denen die Streubreiten der Parameter variiert wurden, die hinreichende Übereinstimmung zwischen Messung und Berechnung zu erreichen.

Mit FE-Berechnungen rechnerisch nicht nachweisbar waren die aus den Meßwerten des Ulmenmeßträgers UM 1 abgeleiteten Normalkraftbeanspruchungen, die in Größenordnungen liegen, wie sie in der vor Baubeginn für das Genehmigungsverfahren durchgeführten Stabwerksberechnung zur Bemessung der Schale ermittelt wurden. Eine geologische Störung im Ulmenbereich hatte dort örtlich einen Mehrausbruch bis zu 90 cm verursacht.

Auf eine Darstellung der aus den Meßwerten ermittelten Momente wird verzichtet, da die Ergebnisse keine wesentlichen, neuen Erkenntnisse erbrachten. Weitere Meßträger wurden in die ergänzende Ortbetonschale eingebaut. Sie lieferten bereits Meßwerte, die in Verbindung mit den Temperaturmessungen Ermittlungen der Wechselbeanspruchungen in der Bauphase zulassen. Es ist zu erwarten, daß weitere Messungen bis zum Ende des Beobachtungszeitraumes 1991 das Tragverhalten der Gesamtkonstruktion bestätigen.

4. SCHUBVERBUND

Die in einer Arbeitsfuge zwischen Spritzbeton und Ortbeton übertragbaren Schubkräfte werden von einer Reihe von Parametern beeinflußt:
- Beschaffenheit der Kontaktfläche,
- Luftfeuchtigkeit und Temperaturen im Gebirge, in den Schalen und im Tunnel,
- Nachbehandlung des Ortbetonteils,
- Rezepturen und Festigkeiten beider Betone,
- Betonierpausen.

Auf der Grundlage der Messungen im Meßquerschnitt wurden die speziellen Verhältnisse im Tunnelbau durch folgende Vorgehensweise bei der Probenherstellung und -lagerung simuliert:
- Herstellung der Spritzbetonschalenteile in einem Tunnel unter Baustellenbedingungen.
- Lagerung der Spritzbetonprobenkörper im Tunnel bei ca. 95 % Luftfeuchtigkeit und 10^0 - 15^0 C Lufttemperatur bis zur weiteren Bearbeitung.
- Ergänzung der Spritzbetonprobenkörper durch Ortbeton im Labor.
- Lagerung der vervollständigten Probenkörper in einer Klimakammer bei 95 % Luftfeuchtigkeit und 15^0 C Raumtemperatur.

Tab. 1: Versuchsergebnisse von Abscherkörpern und Bohrkernen

	Probe	Fugenbeschaffenheit	τ_U	$\bar{\tau}_U$
	Nr.	-	N/mm²	N/mm²
Versuche Daschner,F. 1986	5 6 29 30 31	rüttelrauh	1.43 1.08 0.83 1.34 1.41	1.22
	74 75 76	mit Nagelrechen aufgerauht	1.94 1.82 1.33	1.70
Versuche Philipp Holzmann AG --- Zerna, Schultz und Partner 1986	2/4 2/5 2/6 5/14 5/15	Spritzbetonoberfläche	2.45 2.24 2.70 2.24 1.92	2.31
	10/28	(monolithisch)	2.49	-
	1 3 5 6 7	Spritzbetonoberfläche (Bohrkerne)	1.50 3.00 4.00 2.75 3.50	2.95

Für die Probenherstellung wurden Trockenmischungen des Betons aus dem laufenden Baubetrieb entnommen, die unter Einhaltung der Baustellenbedingungen weiterverarbeitet wurden.

Die Druckfestigkeitsprüfungen an normengemäßen Probekörpern nach 28 Tagen ergaben folgende Mittelwerte:

Spritzbeton β_{W28} = 36,0 N/mm²
Ortbeton β_{W28} = 46,0 N/mm²

Probenform, Versuchseinrichtung und Versuchsdurchführung wurden aus Referenzgründen in Anlehnung an die in (Daschner, F. 1986) beschriebenen Schubversuche mit Abscherkörpern 15 x 15 x 45 cm gewählt. Versuchsreihen an Probekörpern mit Scherflächen F = 10 cm x 10 cm sowie 20 cm x 20 cm ergänzten diese Versuche.

Bei Verschiebungen in der Arbeitsfuge in der Größenordnung von weniger als 1/10 mm trat der Bruch bei Schubspannungen von τ_U = 1,92 N/mm² bis τ_U = 2,70 N/mm² ein. Diese im Versuch nachgewiesenen Bruchschubspannungen betragen etwa das 1,9fache der bisher aus vergleichbaren Versuchen (Daschner, F. 1986) nachgewiesenen aufnehmbaren Schubspannungen bei rüttelrauhen Kontaktflächen und das 1,35fache bei künstlich aufgerauhten Kontaktflächen. Für monolithische Probekörper aus Ortbeton ergab sich eine vergleichbare Bruchschubspannung τ_U = 2,49 N/mm². Eine eingehendere Darstellung der Versuchsdurchführung und -ergebnisse erfolgt an anderer Stelle.

Die Auswertung aller durchgeführten Versuche an labormäßig hergestellten Abscherkörpern zeigte im Hinblick auf die in Bochum zugelassenen Schubspannungen folgendes Bild: Bei Annahme eines Sicherheitsbeiwertes von γ = 2,5 im Falle eines unangekündigten Bruches und unter Berücksichtigung eines Verhältnisses von Dauerstand - zu Kurzzeitfestigkeit von 75 % ergibt sich für die mittlere Bruchschubspannung τ_U = 2,27 N/mm² eine Gebrauchsspannung von zul. τ = 0,75 x 2,27/2,5 = 0,68 N/mm². In der Verbundfuge können demnach Schubspannungen τ = 0,25 N/mm² bis 0,50 N/mm² ohne Anordnung von Verbundbewehrung zugelassen werden. Die bisher in Bochum rechnerisch ermittelten Schubspannungen in der Verbundfuge liegen in dieser Größenordnung.

5. SICHERHEIT

Die Entwicklung schlüssiger und wirklichkeitsnaher Sicherheitskonzepte für Tunnelbauten in bergmännischer Bauweise stellt sich als außerordentlich schwierige Aufgabe des Bauingenieurwesens dar. Die Schwierigkeiten ergeben sich einerseits aus den variierenden Eigenschaften des Gebirges mit der Vielzahl der zur Beschreibung nötigen Parameter; andererseits ist eine zutreffende bruchmechanische Theorie der Flächentragwerke noch nicht entwickelt.

Gebirge sind als "Baustoffe" klüftige, geschichtete, anisotrope und inhomogene Körper, die überdies noch grobe örtliche Diskontinuitäten - Verwerfungen, fließfähige Einschlüsse, plastische oder extrem harte Einlagerungen - enthalten können und einem Primärspannungszustand unterliegen. Für die Berechnungen, gleichgültig ob konventionelle Stabwerksberechnungen oder hochentwickelte FE-Methoden, ist man immer gezwungen, Eigenschaften des Baustoffes Gebirge zu mitteln und mindestens bereichsweise Homogenität vorauszusetzen, wobei die Streuung der kennzeichnenden Werte in geeigneter Weise, z.B. durch Parameter-Variation zu berücksichtigen ist. Hinzu kommen noch die Fragen, die sich aus dem Bauablauf und durch die Festigkeitsentwicklung des Spritzbetons ergeben.

Allgemein erweist sich das Fehlen einer Bruchtheorie für Flächentragwerke oder Kontinua als besondere Schwierigkeit. Tunnelbauten sind hochgradig statisch unbestimmte Flächentragwerke. Die Idealisierung als Flächentragwerk ist notwendig, um das Problem des unterirdischen Hohlraumes der Berechnung zugänglich zu machen. Sie ist auch notwendig, um die durch die begrenzte Kapazität der Rechenanlagen vorgegebenen Bedingungen einhalten zu können. Diese Flächentragwerke sind in ihrem Tragverhalten wegen der sehr großen Möglichkeiten der Last- oder Beanspruchungsumlagerungen außerordentlich "gutmütige" Systeme. Wahrscheinliche bruchmechanische Modelle für Tunnel müssen Grenzzustände simulieren.

Als Elemente von Grenzzuständen werden gewöhnlich
- Biegeversagen,
- Scherbruch, insbesondere der Kalottenfüße sowie
- Undichtwerden des Tunnels

angenommen, während Verzweigungs- oder Durchschlagprobleme, die zum Stabilitätsversagen führen, bei den üblichen Abmessungen der Beton- oder Stahlbetonauskleidungen und den gewöhnlich starken Krümmungen der Tunnelwandungen ausgeschlossen werden können.

Bei Sicherheitsanalysen von Tunnelbauwerken ist immer zu beachten, daß die - wesentliche Lastspannungszustände erzeugenden - Erddrucklasten verformungsabhängig sind. Bei Verformungen des Tragwerkes infolge Überlastung werden diese Lasten im allgemeinen erheblich abgemindert und können auch zu Null werden. Der häufig ebenfalls zu berücksichtigende Wasserdruck ist ein nicht verformungsabhängiger Lastfall.

Bei den vorgenannten Grenzzuständen ist die Wasserdurchlässigkeit nur selten auf Risse infolge Biegeversagen zurückzuführen. In der Mehrzahl der Fälle werden hierfür Ausführungsfehler, Aggressivität des Grundwassers oder Risse infolge Schwinden oder ähnlicher nicht spannungsabhängiger Verformungen verantwortlich sein. Deshalb wird diese Versagensart hier nicht betrachtet. Es bleibt also letzlich als Bruchmodell für den Grenzzustand ein Biege- oder Scherversagen der Tunnelschale übrig, wobei die Kombination der beiden Versagensarten - auch nach den Bruchbildern zerstörter Tunnel - als das Wahrscheinlichste anzusehen ist.

Für Untersuchungen von Biege- und Schubbrüchen muß bei zweischaligen Bauweisen - zumindest bereichsweise - die Verbundwirkung beider Schalen mit erfaßt werden. Unsere Versuche haben gezeigt, daß die aufnehmbaren Verbundspannungen im Mittel über 90 % der für den monolithischen Werkstoff geltenden Werte erreichen. Schließt man die Verbundspannungen im Bereich der Sohle und der Firste aus, so können die für probabilistische Sicherheitsanalysen anzusetzenden Wahrscheinlichkeitsverteilungen gemäß der Abbildung 3 zugrunde gelegt werden.

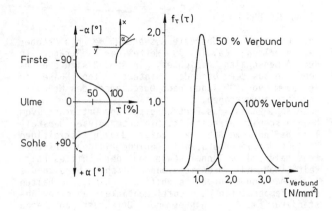

Abb. 3: a. Verbund über den Umfang
b. Wahrscheinlichkeitsdichten für Verbundspannungen

6. SCHLUSSBEMERKUNG

Welche Bedeutung haben die vorstehenden Überlegungen im Hinblick auf eine wirklichkeitsnahe Berechnung oberflächennaher Tunnelbauwerke in Verbundbauweise? Bei einer Vielzahl von innerstädtischen Tunneln treffen die folgenden grundlegenden Voraussetzungen zu: weitgehend "homogenes" Gebirge, ausreichend gekrümmte Konturen des Tunnelquerschnittes und Mindestdicke der Wandungen.

Zweischalig ausgeführte Tunnelbauwerke weisen neben dem Verbund Gebirge - Spritzbeton auch gute Verbundeigenschaften zwischen Spritzbetonaußenschale und der Ortbetoninnenschale auf.

Durch Laborversuche und Messungen an ausgeführten Bauwerken gelang es, zulässige Schubspannungen in der Arbeitsfuge festzulegen, die zu einer wirtschaftlichen und wirklichkeitsnäheren Bemessung der Tunnelschale führen. Zusätzliche Verbundmittel, die Schwachstellen darstellen, sind für den rechnerischen Nachweis des Verbundsystems nicht erforderlich. Es erscheint möglich, von dem ursprünglich von andersartigen Bauwerken übernommenen Sicherheitsniveau begründet abzuweichen und eine den wirklichen Verhältnissen näherkommende Sicherheitsphilosophie der Bemessung zugrunde zu legen.

Die Bereitschaft zur gezielten Weiterentwicklung und zum Einsatz innovativer Techniken bleiben Voraussetzung für eine Verbreitung dieser wirtschaftlichen Bauweise.

LITERATUR

Baumann, Th. 1985. Messung der Beanspruchung von Tunnelschalen, Bauingenieur 60: 449-454

Daschner, F. 1986. Versuche zur notwendigen Schubbewehrung zwischen Betonfertigteilen und Ortbeton, Deutscher Ausschuß für Stahlbeton, Heft 372

Daschner, F. und Kupfer, H. 1982. Schubtragverhalten, Deutscher Ausschuß für Stahlbeton, Heft 340

Sager, H.-J. 1985. Entwicklung für einschalige Bauverfahren mit Anwendungsbeispielen beim Stadtbahnbau in Bochum, Institut für Konstruktiven Ingenieurbau, Mitteilungen Nr. 85-6, Ruhr-Universität Bochum: 73-86

Schmidt-Schleicher, H. 1986. Bemessung von einschaligen Verkehrstunnelbauten aus Stahlbeton im oberflächennahen Bereich, Forschung und Praxis, Band 30: 153-157

Schmidt-Schleicher, H. und Lippert, D. 1982. Zur Berechnung und Konstruktion von einschaligen Verkehrstunnelbauten aus Stahlbeton im oberflächennahen Bereich, Konstruktiver Ingenieurbau, Berichte, Heft 40: 17-23

Schultz, H. 1982. Berechnung oberflächennaher Tunnel, Konstruktiver Ingenieurbau, Berichte, H. 40: 11-17

Fracturing and microseismicity ahead of a deep gold mine stope in the pre-remnant and remnant stages of mining

Formation des fissures et microséisme en avant d'un gradin d'une mine d'or en profondeur pendant les phases remnantes et pré-remnantes des opérations minières

Bruchbildung und Mikroseismizität im Vorfeld eines Abbauortes in einer tiefen Goldmine während der abschliessenden und letzten Abbauphasen

N.B.LEGGE, Rock Mechanics Laboratory, Chamber of Mines of South Africa, Research Organization, Johannesburg
S.M.SPOTTISWOODE, Rock Mechanics Laboratory, Chamber of Mines of South Africa, Research Organization, Johannesburg

ABSTRACT: The spatial, temporal and energy related aspects of fracturing ahead of a gold mine stope at a depth of 2 160 m are described with reference to microseismic activity monitored by a small-scale array of accelerometers. The indicated location, extent and orientation of fractures ahead of the advancing stope correlated well with a numerical analysis based on Excess Shear Stress. Changes in the nature of fracturing during the remnant stage of mining are described, and are discussed with reference to the increased amount of energy released by mining at this stage. A correlation between the Gutenberg and Richter b-values of micro- and macroseismic data and the fractal dimensions of the source region is confirmed, and provides useful information regarding the source dimensions and mechanisms of seismic activity.

RESUME: Les aspects spatiaux, temporaux et liés à l'energie de la formation des fissures en avant d'un gradin d'une mine d'or à une profondeur de 2 160 m sont décrits en tenant compte de l'activité microsismique controlée par un deploiement peu etendu d'accéléromètres. L'emplacement indiqué, l'étendu et l'orientation des fractures en avant du gradin avancant ont bien correspondu à une analyse numerique basée sur extrème contrainte de cisaillement. Les changements dans la nature de la formation des fissures pendant la phase remnante des operations minières sont décrits et sont discutés en tenant compte de la quantité augmentée de l'energie degagée au cours des operations minières à cette phase. Une correlation entre les b-valeurs de Gutenberg et de Richter des points microsismiques et macrosismiques et les dimensions 'fractal' de la region d'origine est confirmée, et a fourni d'information utile en ce qui concerne les dimensions d'origine et les mecanismes de l'activité sismique.

ZUSAMMENFASSUNG: Die räumlichen, zeitlichen und energiebezogenen Aspekte der Bruchbildung im Vorfeld eines 2 160 m tiefen Abbauortes in einer Goldmine werden beschrieben hinsichtlich der mikroseismischen Aktivität, die von einer beschränkten Anordnung von Beschleunigungsmessern überwacht wird. Die so bestimmten Werte für Lage, Umfang und Orientierung von Brüchen im Vorfeld eines vordringenden Abbauortes standen in guter Wechselbeziehung mit einer numerischen Analyse, welche die überschüssige Scherspannung zur Grundlage hatte. Veränderungen in der Art und Weise der Bruchbildung während der Schlussphase des Abbaus werden beschrieben, und werden in Bezug auf den Anstieg der durch den Abbau in dieser Phase freigesetzten Energie diskutiert. Eine Wechselbeziehung zwischen den von Gutenberg and Richter bestimmten b-Werten für mikro- und makroseismische Daten einerseits, und den Fraktaldimensionen des Ursprungsortes andererseits, findet Bestätigung, und lieferte nützliche Information über die Dimensionen des Ursprungsortes und über Mechanismen seismischer Aktivität.

1 INTRODUCTION

A complete understanding of the fracture and deformation processes occurring in the vicinity of the deep gold mine stopes of South Africa is essential for the safe and efficient extraction of these gold bearing strata. The behaviour of the rock mass in this region has important implications for stope layout and support design, for developing mechanized mining methods, for determining why rock ahead of the face fails in an unstable manner, occasionally resulting in rockbursts and, most importantly, for the safety of the many people who work at the stope face.

The pattern of fracturing occurring ahead of the stope face as the high mining induced stresses exceed the strength of the surrounding intact rock mass has been studied for a number of years (McGarr 1971, Roering 1978, Legge 1985 and Brummer and Rorke 1984). To gain a greater overall understanding of these processes of fracture and deformation, a cuboid microseismic network of 25 m side was installed symmetrically about the reef horizon ahead of an advancing stope face. As the stope face moved through the network many thousands of microseismic events were captured and recorded, yielding information concerning the temporal and spatial distribution of fracturing, and the energy content and frequency of occurrence of microseismic emissions accompanying the fracturing.

A great deal of seismic research has been conducted on the South African goldfields, and has involved the use of macro-, mini-, and microseismic networks, spanning seismic events with local magnitudes ranging from -4 to +5. Microseismic studies have included those by Brink and O'Connor (1983) on rockburst prediction, and by Pattrick (1984) who designed the system used in this study and applied it in an earlier experiment.

2 SITE DETAILS

The microseismic network was installed on a gold mine which is situated in the Klerksdorp district some 160 km from Johannesburg. The reef being mined at this site was the Vaal Reef which, stratigraphically, lies near the middle of the Central Rand Group quartzites. The Vaal Reef is a tabular, laterally extensive gold-bearing conglomerate, which in this region of the mine is about 1,1 m wide and dips at 8° to the west. Layers of quartzite varying from one to several metres in thickness, lie conformably above and below the Vaal Reef for many hundreds of metres; these strata have, on average, a uniaxial compressive strength of 200 MPa. Beds of shaly material varying from hairline thickness to bands several centimetres thick are present between the quartzite layers above and below the reef.

The stope where the experiment was carried out lies 2 160 m below surface. Figure 1 shows the mining configuration at the start and end of data recording.

The network was positioned such that the southward mining faces of the stope would move through the network as the portion of reef between this excavation and the stationary stope face was extracted. The initial distance between the two faces was about 45 m, and this reduced to 30 m during the two month duration of the experiment.

At the advancing stope face the initial average Energy Release Rate (ERR, a theoretically derived value used empirically in the gold mining industry to indicate the likely condition of stope faces - Cook et al 1966) was 15 MJ/m^2, a moderate to low value by industrywide standards. After 15 m of face advance the ERR had risen to about 30 MJ/m^2.

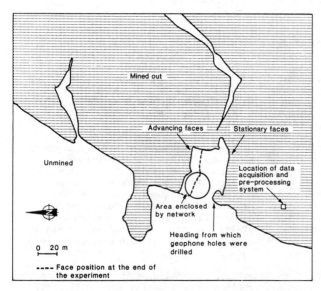

Figure 1. The mining layout in the area investigated and the location of the microseismic network.

3 SYSTEM DESCRIPTION

3.1 Data sensing

The acoustic signals were detected by high-sensitivity accelerometers. These were cast into resin 'boats' and grouted in fixed orientations at the end of percussion drilled boreholes. The boreholes used to position the accelerometers were drilled from a heading excavated 10 m ahead of the stationary stope face, towards which the stope to be monitored was advancing (Figure 1). This measure was employed, and proved highly successful, because of the difficulties associated with drilling boreholes from within the advancing stope where the highly fractured and discontinuous rock mass not only complicates drilling, but makes installing and maintaining contact with the accelerometers extremely difficult.

The array of accelerometers enclosed a volume of rock of some 25 x 25 x 25 m^3, situated symmetrically about the reef plane.

3.2 Data acquisition and pre-processing

Requirements for location accuracies of about 1 m and microseismic event rates in excess of one hundred per day lead Pattrick (1983) to design and develop the system illustrated in Figure 2.

Event detection was performed by a Trigger Intelligent Analogue Processor using the ratio of short-term to long-term averages of rectified signals. Simultaneously, three front-end processors sampled 12 channels, each at 30,75 kHz with a 12-bit resolution, and continuously stored 62 ms worth of data. When an event occurred, 19 ms of pre-trigger data and 43 ms of post-trigger data were 'frozen' in

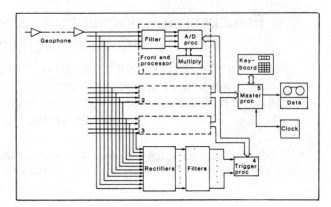

Figure 2. The data acquisition and pre-processing system.

memory and then processed to compute P-wave arrival times. In addition, each accelerogram was integrated to obtain ground velocity, and was squared and integrated again in preparation for calculating radiated energies. The arrival-time picking algorithm used by the system compared favourably with arrival times determined manually; system picks differed from the manually picked values with a standard deviation of less than 0,8 ms. This information, together with the time of day and certain diagnostic data, such as the signal-to-noise ratio which assisted in assessing the quality of the arrival times, was then passed to a digital data cartridge recorder. Total processing time for each event was 4 seconds, resulting in a maximum event rate recordable of 900 per hour. Magnetic tapes were changed each day of the experiment (which lasted about 2 months) and processed on surface using an HP 9845 desk-top computer. Full waveforms for at least one randomly chosen event were recorded each day; these were initially used to optimize automatic processing and then as a check on the system status. Detailed analysis of these waveforms has, to date, not been attempted.

The entire recording and pre-processing system was located underground in the stope. Although this reduced problems associated with signal telemetry, the harsh high temperature and humidity environment in the vicinity of the working faces necessitated special protection for the system circuitry.

3.3 Surface processing

Data stored on magnetic tape were retrieved and analysed on surface. Event hypocentres were located using the Salamon-Wiebols algorithm (Salamon and Wiebols, 1974). Hypocentral coordinates, event energy, and values representing a statistically determined location error range in metres, were then stored in files, each containing one hour's data. Locations were calculated using P-wave velocities of 5,24 km/s, determined from calibration blasts. Locations were considered acceptable only if they were based on five or more arrival times and had RMS errors of less than 2 m. As these located events consistently accounted for approximately 65 % of all triggers, they were taken as representative of the microseismic activity at the site.

Radiated energy E (in Joules) for each event was then calculated and converted to magnitudes using the Gutenberg - Richter relationship Log E = 1,5 M + 4,8, giving apparent magnitudes ranging from -4,8 to -2. Corrections were then made for the expected energy loss caused by the system's 10 kHz low-pass filter, amounting to as much as one magnitude unit for the smallest events. Saturation of the magnitude scale at about M = -2 was due to signal clipping and filtering limitations.

The data sensing, acquisition and pre-processing

systems operated satisfactorily for a 2 month period, during which time microseismicity associated with about 15 m of face advance was captured and recorded. At the end of this period the remaining reef area had been reduced by about one third and consequently average levels of vertical stress had increased by approximately 35 %. At this stage, quasi-stable ground movements around the heading shown in Figure 1 cut off signal lines, and the data collection programme was terminated. However, the information successfully gathered enabled a data set to be assembled concerning the spatial, temporal, and energy related aspects of stable fracturing ahead of an advancing stope face.

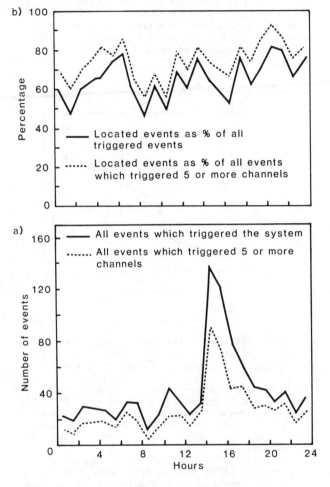

Figure 3. Microseismic data for a typical day, presented as (a) event frequency and (b) the percentage of events located.

Some data for a typical day are shown in Figure 3. In Figure 3a are plotted both the total number of events which triggered the system each hour, and those which triggered 5 or more channels and could, therefore, potentially be located by the data processing algorithm. The peak in the number of events occurring at about 15h00 corresponds to the time at which the face was advanced 1 m by blasting. Figure 3b represents for each hour, the number of located events as a percentage of all events which triggered the system, and as a percentage of all events which triggered on five or more channels. This information allowed the performance of the data sensing portion of the network to be assessed on a daily and hourly basis. It was found that about 65 % of all events which triggered the system were located continuously by the data sensing program; these events were therefore taken as an accurate portrayal of the overall microseismicity in the area.

4 DATA PRESENTATION

Data gathered during this experiment enabled the nature of fracturing ahead of a deep gold mine stope to be examined in 3-dimensions as the faces were advanced. As the remnant being mined became smaller, changes occurred in the pattern of microseismicity ahead of the advancing faces. In this paper the nature of fracturing ahead of the stope will be illustrated by data recorded on two typical days: February 18th 1985 in the 'pre-remnant' mining stage, and March 30th 1985 in the remnant mining stage. Approximately 25 m³ of rock was blasted out of the face advancing through the network on each of these days, and about 15 m of face advance occurred between these dates.

A plan and sectional view of the advancing stope face showing the location of microseismic activity on February 18 are presented in Figure 4. In plan these data reveal a concentration of events forming a lineation which follows the overall shape of the excavation, 5-10 m ahead of the stope. Seen in section two lobes of microseismic activity are apparent, symmetrically located above and below the plane of the reef and inclined away from the stope face at between 50° and 70° from the horizontal. Very little microseismic activity occurred ahead of the stationary face at this stage.

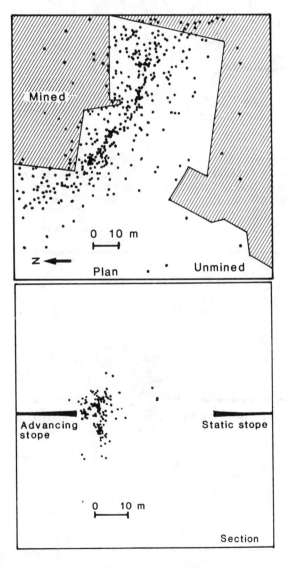

Figure 4. A plan and sectional view of micro-seismicity on February 18th 1985, pre-remnant stage of mining.

MINSIM-D, a three-dimensional, Boundary Element based, elastic, tabular mining stress analysis program, (Ryder and Napier, 1985), was used to examine the location, magnitude and orientation of regions of high shear stress ahead of the stope. The analysis predicted the formation of fractures ahead of and parallel to the stope face, inclined preferentially at 70° to the horizontal and symmetrically located above and below the reef plane. This prediction is in qualitative agreement with event locations shown in Figure 4. Figure 5 represents this information in terms of Excess Shear Stress (ESS), (Ryder 1987), which involves delineating those regions where the

prevailing shear stresses are in excess of the rock shear strength predicted by a Mohr-Coulomb failure criterion, assuming zero cohesion and an internal friction angle of 30°. The absolute values of ESS are shown above the reef plane, about which stresses are virtually symmetrical. Below the reef plane the change in ESS corresponding to a typical 1 m advance of the face is represented. This latter quantity represents the stress change giving rise to each day's post-blast microseismic activity.

The total amount of seismic energy radiated as the 25 m long face advanced 1 m was of the order of 2 kJ.

Figure 6 is a section through the advancing face, as at March 30th, and shows the microseismicity associated with a day during which blasting took place and at which time the width of the remnant had decreased from about 45 m to 30 m. Events defined steeply dipping planes less clearly, had a slightly greater vertical extent about the reef plane, and indicated a small amount of activity near to the stationary face.

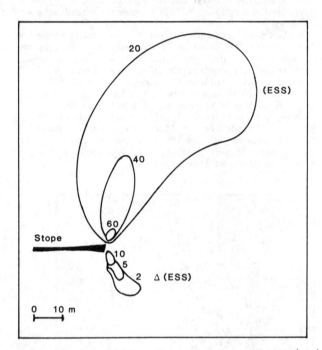

Figure 5. Excess Shear Stress and change in ESS (MPa) ahead of the advancing stope face for planes dipping at 70° and an internal friction angle of 30°.

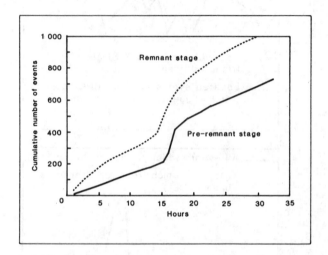

Figure 7. The cumulative number of events which triggered the system during a typical mining cycle in the pre-remnant and remnant stages of mining.

The cumulative number of events which triggered the system during a typical day's mining cycle in both the initial and remnant stage of the network are shown in Figure 7. The figure clearly shows the increased steady-state event rate during the remnant stage of the network.

The change in the nature of the microseismic frequency-magnitude relationship which was noted during the remnant stage is represented in Figure 8. Although both curves display a sensitivity cut-off at about M -3,7 and evidence of saturation as the event magnitudes approach M -2, the Gutenberg and Richter a- and b-values are clearly different. In the remnant stage of mining, the ESS analysis indicated coalescence of the shear stress lobes ahead of the approaching faces above and below the stope face. Microseismicity during the transition to the remnant stage confirmed this.

In Figure 9, averaged data from the microseismic network have been compared with data gathered by the Klerksdorp regional network by normalizing for area mined. The gradients of the lines are significantly different, with the b-value of the microseismic portion lying between 1 and 1,5, and that of the macroseismic portion being about 0,5. The table below summarizes changes in the b-values for these different sets of seismic data.

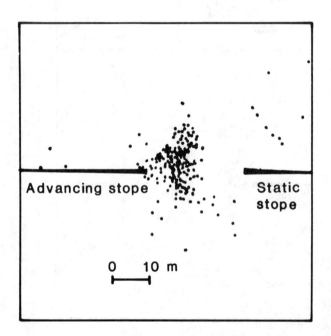

Figure 6. A sectional view of microseismic activity on March 30th 1985 during the remnant stage of mining.

	Macroseismic	Pre-remnant	Remnant
b-value	0,44	1,1	1,5

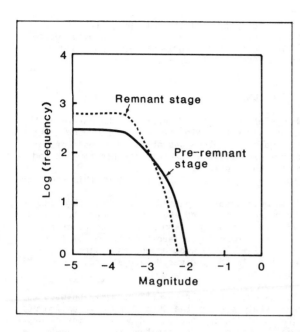

Figure 8. The Gutenberg and Richter event magnitude-frequency relationships obtained during the pre-remnant and remnant stages of mining.

5 DISCUSSION

Locations of microseismic events were in excellent agreement with previous findings concerning the nature of fracturing ahead of deep gold mine stopes based on in-stope observations, for instance those by Brummer and Rorke (1984). In plan, the region of primary fracturing was clearly delineated 5 m – 10 m ahead of the faces, and followed the general outline of the excavation (Figure 4). In section microseismic activity clustered on planes which extended 10 m – 15 m above and below the reef, were inclined away from the stope face at about 70° from the horizontal and were symmetrically orientated about the reef plane. The depth of fracturing ahead of the advancing stope decreased from the upper to the lower faces as shown in Figure 4; this is due to the higher stresses ahead of the upper face as a result of the geometry of the area being mined.

The location and orientation of the fractures developing ahead of the stope correlated well with numerical predictions based on Excess Shear Stress. In particular, the good correlation between microseismicity associated with a typical 1 m advance of the face and the change in Excess Shear Stress (Figure 5) indicated the applicability of a Mohr-Coulomb failure criterion to rock ahead of the stope, and strongly suggested a shearing mechanism for rock fracture at this location.

The change in Excess Shear Stress is a quantity which correlates well with the physical processes occurring ahead of the stope face. Before the face is advanced the rock mass around the stope is in equilibrium with the prevailing stresses, and, by definition, there is no Excess Shear Stress. After the face is advanced the changes in stress cause enlargement of the fractured zone of rock ahead of the stope, stresses are redistributed and equilibrium is again achieved. Furthermore, the good correlation between the numerical analysis and the field data support the use of an internal friction angle of 30° for the quartzite rock. It should be noted, however, that the elastic analysis used for calculating ESS did not allow for the presence of already fractured rock ahead of the stope face, and that, consequently, microseismic activity associated with the formation of new fractures was located some 5 m – 10 m ahead of the predicted positions (Figure 4). The depth of

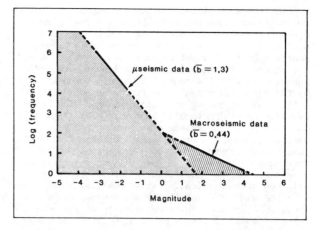

Figure 9. Normalized magnitude-frequency relationships for both micro- and macroseismic events.

fracturing ahead of the stope indicated by the microseismicity is in agreement with observations made in boreholes drilled ahead of stopes in similar locations.

The change in the nature of fracturing ahead of the stope after the transition to the remnant stage of mining was indicated by three independent variables: the spatial distribution of events (Figure 6), the change in event frequency with time (Figure 7), and changes in the a- and b-values obtained from the Gutenberg and Richter event magnitude-frequency relationship (Figure 8).

The more densely clustered pattern of microseismicity shown in Figure 6 indicates that in the remnant stage of the network, fracturing of a more intense nature was occurring ahead of the face. In addition, the changed pattern revealed that fracturing was no longer occurring along distinct planes, but that a volume of rock ahead of the stope, essentially the highly-stressed core of the remnant, was fracturing. As the distance between the approaching faces decreased, some microseismic activity was recorded near the stationary face. This indicated that relatively greater changes in stress were occurring ahead of this stationary face during the remnant stage of the network than prior to this stage, when most of the change in ESS and, therefore, microseismic activity occurred ahead of the advancing face. The increase in stress ahead of the stationary face at this stage was confirmed by the large quasi-stable rock mass deformations which sheared the boreholes in which the accelerometers were installed, and cut the cables to them.

Event frequency (Figure 7) in the pre-remnant and remnant stage of the network, clearly shows that as the distance between the approaching faces decreased and the rock to be mined became more highly stressed, the amount of microseismic activity increased during a typical mining cycle. This is indicative of the increased fracturing and deformation occurring ahead of the face in the remnant stage. Although the nominal ERR at the advancing faces had increased by about 100 % in the remnant stage, the number of microseismic events occurring each day increased by some 50 %.

Perhaps the most unexpected change in seismicity with time is the change in b-values, as deduced from Figure 8. In general the b-values typically varied by not more than 0,2 on consecutive days when blasting took place. Thus the observed increase in the average b-value from 1,1 in the pre-remnant stage to 1,5 in the remnant stage is highly significant. The physical implication of the change in b-value is that in the more highly stressed remnant stage of mining a greater number of smaller events occur. The smaller a-value indicates a decrease in the largest probable event magnitude. The Excess Shear Stress analysis for the remnant stage of mining indicated coalescence of the shear stress lobes above and below the reef plane, a fact confirmed by the microseismic data captured during the transition to the remnant stage. The decrease in the largest probable event magnitude and the increase in the number of small events occurring agrees well with the concept of events being limited by the surrounding region of fractured rock.

Although the amount of energy per unit area being released by mining doubled in the remnant stage, the quantity of energy released microseismically was similar in both the pre-remnant and remnant stages. That the increased amount of energy being put into the rock mass in the remnant stage did not result in more larger events, or significantly more energy being released microseismically, is in qualitative agreement with the work of Legge (1985) and Brummer (1986). These workers found that the majority of the energy released ahead of stopes could be entirely consumed in a stable manner by shear movements and frictional heating on fracture and parting planes.

The total amount of seismic energy radiated during 1 m of face advance was of the order of 2 kJ. This amount is a very small portion of the 375 MJ of energy which must, on average, be released by a 1 m advance of a 25 m panel, and also implies that most energy is released in a stable manner. This finding is also in qualitative agreement with the work of Salamon (1983), who found that energy dissipated during mining could be completely accounted for by quasi-stable aseismic processes.

Results from this study showed that mining reached a critical stage at which the fracture zone around the unmined reef began to coalesce when the faces were about 30 m apart. When this had happened microseismic event locations showed that the core of the remnant began to fracture; the rate of seismic energy radiation at this stage did not increase significantly and the indicated largest probable event magnitude decreased. These results concur with the work of Hill (1945), who studied 36 remnant extractions on the Central Witwatersrand goldfields and noted that the incidence of rockbursts increased when the remnant size was reduced to about 3 300 m^2, but decreased dramatically while the last 660 m^2 was mined - that is when the remnant was some 25 m wide. Spottiswoode (1986) reported on the extraction of two remnants at Blyvooruitzicht Gold Mine where, although the ERR nominally was 200 MJ/m^2, seismicity (M > 1,5) and rockburst occurrence were similar to areas mined at 30 MJ/m^2.

The b-value may be examined further in relation to Figure 9, in which are shown the magnitude-frequency relationships for micro- and macroseismic data. Aki (1981) has suggested that b-values are linked to the fractal dimensions of the seismic source region. A b-value of 1,0 would be associated with seismic events distributed uniformly in two dimensions, perhaps on a single plane. A b-value of 1,5 would correspond to a volume being filled up by many seismically active planes. At the other extreme, a b-value of 0,5 could account for seismic events occurring along a linear, essentially one-dimensional source.

The Klerksdorp mining district is broken up by major faults, dykes and other planar features which extend many kilometres in length. Mining of the tabular reefs is often extensive on dip, and the vertical extent of high mining-induced stresses ahead of the advancing faces rarely exceeds about 100 metres. The region of interaction between mining induced stress changes and the extensive planar discontinuities on which many of the larger seismic events occur is essentially one-dimensional; a b-value of 0,44 for seismic events with M = 1 to 4 is, therefore, entirely compatible with a fractal dimension of 1. Similarly, a b-value of around 1 indicating fracturing on distinct planes in the pre-remnant stage of mining, and the value of 1,5 associated wth the more intense uniform fracturing, as the core of the remnant failed, are both consistent with the hypothesis of Aki. Results from this study indicate, therefore, that

b-values obtained from the Gutenberg-Richter magnitude-frequency relationship may provide useful information regarding the source dimensions and mechanisms of seismic activity.

6 CONCLUSIONS

Much useful information concerning the nature of the fracture and deformation processes occurring ahead of a deep gold mine stope was obtained by monitoring microseismic activity, in both the pre-remnant and remnant stages of mining.

The principal conclusions drawn from this study are:

- Remnant extraction passed a critical stage when the fracture zone around the remnant began to coalesce and the core of the remnant began to fracture. When this happened a larger number of smaller microseismic events occurred, and there was a decrease in the indicated largest probable event magnitude.

- The Excess Shear Stress criterion may be used to adequately predict the extent and orientation of stable fracturing ahead of deep gold mine stopes. The change in ESS, rather than the absolute value, most realistically defines the extent of the fractured zone of rock ahead of the stope as the face is advanced by 1 m.

- The amount of energy released microseismically did not increase significantly in the remnant stage of mining, when the calculated energy release rate had doubled from its pre-remnant value. As the total amount of seismic energy released each time the stope advanced 1 m was a very small portion of the total energy released by mining, it may therefore be concluded that the majority of energy is dissipated by quasi-stable aseismic processes.

- The Gutenberg and Richter b-value was found to be related to the fractal dimension of the seismic source region, as suggested by Aki (1981), and, as such, may be used to assist in determining seismic source mechanisms.

ACKNOWLEDGEMENT

This paper describes work carried out as part of the research programme of the Research Organization of the Chamber of Mines of South Africa. Permission to publish the paper is gratefully acknowledged.

REFERENCES

Aki, K. 1981. A probabalistic synthesis of pre-cursory phenomena. In: D.W. Simpson and P.G. Richards (ed.). Earthquake Prediction - International Review, pp.566-574, Maurice Ewing Volume, Amer. Geophys. Union.
Brink, A.v.Z. & O'Connor, D.M. 1983. Research on the prediction of rockbursts at Western Deep Levels. Journal of the South African Institute of Mining and Metallurgy, 83, pp.1-10.
Brummer, R.K. & Rorke, A.J. 1984. Mining induced fracturing around deep gold mine stopes. Unpublished Research Report, Chamber of Mines of South Africa.
Brummer, R.K. 1986. Fracturing and deformation at the edges of tabular gold mining excavations and the development of a numerical model describing such phenomena. Ph.D. Thesis in preparation.
Cook, N.G.W., E. Hoek, J.P.G. Pretorius, W.D. Ortlepp & M.D.G. Salamon, 1966. Rock mechanics applied to the study of rockbursts. Journal of the South African Institute of Mining and Metallurgy, May 506-515.

Hill, F.G. 1945. A system of longwall stoping in a
  deep level mine, with special reference to its
  bearing on the pressure burst and ventilation
  problems. Association of Mine Managers Papers and
  Discussions, pp.259-265.
Legge, N.B. 1985. Rock deformation in the vicinity of
  deep gold mine longwall stopes and its relation to
  fracture. Unpublished Ph.D. Thesis, University of
  Wales, Cardiff.
McGarr, A. 1971. Stable deformation of rock near deep
  level tabular excavations. Journal of Geophysical
  Research, 76, pp.7088-7106.
Pattrick, K.W. 1983. The development of a data
  acquisition and pre-processing system for
  microseismic research. Unpublished M.Sc. Thesis,
  University of the Witwatersrand, Johannesburg.
Pattrick, K.W. 1984. The instrumentation of seismic
  networks at Doornfontein Gold Mine. In: N.C. Gay
  and E.H. Wainwright (ed.). Rockbursts and
  Seismicity in Mines, pp.337-340, Johannesburg :
  SAIMM.
Roering, C. 1978. In: The influence of geological
  structure on the instability of rock in and around a
  deep mechanized longwall stope. Rand Afrikaans
  University Press, Johannesburg.
Ryder, J.A. & Napier, J.A.L. 1985. Error analysis and
  design of a large-scale tabular stress analyser.
  5th International Conference : Numerical Methods in
  Geomechanics, Nagoya, Japan.
Ryder, J.A. 1987. Excess Shear Stress (ESS) : an
  engineering criterion for assessing unstable slip
  and associated rockburst hazards. 6th International
  Congress on Rock Mechanics, Montreal, Canada.
Salamon, M.G.D. & Wiebols, G.A. 1974. Digital
  location of seismic events by an underground network
  of seismometers using the arrival times of
  compressional waves. Rock Mechanics, 6, pp.141-166.
Salamon, M.G.D. 1983. The rockburst hazard and the
  fight for its alleviation in South African gold
  mines. IMM Conference : Rockbursts - prediction and
  control, pp.11-36.
Spottiswoode, S.M. 1986. Total seismicity and mine
  layouts : application of ERR and ESS. SAIMM
  Colloquium - Mining in the Vicinity of Geological
  and Hazardous Structures, Johannesburg.

Dynamic analysis of discontinua using the distinct element method

Analyse dynamique de milieux discontinus par la méthode des éléments distincts

Dynamische Analyse von Diskontinua durch die Einzelelementmethode

J.LEMOS, Itasca Consulting Group, Inc., Minneapolis, Minn., USA
R.HART, Itasca Consulting Group, Inc., Minneapolis, Minn., USA
L.LORIG, Itasca Consulting Group, Inc., Minneapolis, Minn., USA

ABSTRACT: Dynamic analysis of excavations in jointed rock is required for many extreme loading events such as explosive detonation, rockbursts, and earthquakes. A shortcoming of many analysis methods is the inability to model the effects of dynamic loading on rock media containing multiple, intersecting joint structures. The distinct element method is a discontinuum modeling approach which represents discontinuities as distinct features in the numerical formulation and employs an explicit time-marching algorithm. Consequently, large displacements and rotations, general non-linear constitutive behavior for both the rock mass and joints, and time domain calculations can be accommodated in a straightforward manner. This paper describes the basic features of the method and demonstrates several applications for both two- and three-dimensional problems.

RESUME: L'analyse dynamique des excavations dans les roches fracturées est nécessaire pour étudier de nombreux évenements particuliers tels que la détonation des expolsifs, les "rockbursts", et les tremblements de terre. Un défaut de nombreuses méthodes d'analyse est leur incapacité à modéliser les effets d'un chargement dynamique sur des roches contenant de multiples familles de joint sécantes. La méthode des éléments distincts consiste à modéliser un milieu discontinu en représentant ses discontinuités dans la formulation numérique comme des entités distinctes; elle utilise un algorithme explicite en temps réel. En conséquence, les grands déplacements et rotations, et des lois de comportement générales non linéaires pour la roche ou pour les joint peuvent être pris en compte d'une manière directe. Cet article décrit les principales caractéristiques de la méthode et en présente plusieurs applications, pour des problèmes en deux ou trois dimensions.

ZUSAMMENFASSUNG: Dynamische Analyse von Aüshuben in geklüftem Fels is für extreme Belastungsfelle, wie Detonationen, Gebirgsschläge order Erdbeben, oft verlangt. Viele der üblichen Methoden der mechanischen Analyse versagen bei der Modelierung des dynamischen Verhaltens von Felskörpern, welche mehere sich überschneidende Klüfte aufweisen. Die Einzelelementmethode (distinct element method) ist im wesentlichen ein numerisches Diskontinuum Verfahren, welches die Darstellung von Diskontinuiteten als Einzelstrukturen erlaubt und auf einem expliziten Zeitschrittalgorithmus sich stütz. Demzufolge können grosse Verformungen und Rotationen, und ein algemein nichtlineares Materialverhalten sowohl für den Fels als auch für die Klüfte berücksichtigt werden. In der vorliegenden Veröffentlichung wird die Einzelelementmethode kurtz beschrieben und anhand von zwei- und dreimensionalen Anwendungsbeispielen erläutert.

1 INTRODUCTION

Comprehensive design analysis of underground openings must take into account response due to excavation as well as any future dynamic loads which arise from sources such as earthquakes, rockbursts, or explosive detonations. Conventional analyses are often based on solving equivalent quasi-static problems. However, such analyses for jointed rock masses ignore differential acceleration of rock blocks, causing slip or separation at discontinuities.

The distinct element method presents a more rigorous approach by considering the rock mass to consist of a multiple degree-of-freedom system or assemblage of blocks with discontinuities regarded as boundary interactions between blocks. This method works equally well for dynamic and static computational analyses. In fact, the effect of the interaction of a dynamic stress wave with discontintuties can be evaluated as a function of the orientation and behavior of the discontinuities, the behavior of the rock mass, and the in-situ stress state.

This capability is demonstrated for several different applications of dynamic analysis. These examples illustrate the utility of this computational technique for studies of rockburst potential, seismic instability conditions, and vulnerability of tunnels subjected to dynamic pressure waves.

2 THE DISTINCT ELEMENT METHOD

The essential feature of the distinct element method is its ability to model arbitrary motion of each block with respect to any other. Large translations and rotations of blocks are thus allowed. The interaction forces between blocks are obtained by applying a joint constitutive relation, preferably in incremental form. Joint displacement is defined by the relative motion between the edges of adjacent blocks with the joint constitutive relation providing the corresponding contact stresses and forces.

The original distinct element codes assumed the blocks to be rigid. Later, simply-deformable (constant stress) blocks were formulated. The examples reported in this paper make use of a more general representation of block deformability. Each of these "fully-deformable" blocks is discretized into a mesh of triangular (2-D) or tetrahedral (3-D) finite-difference zones, as in standard continuum modeling. An automatic mesh generator provides any desired degree of refinement with the corresponding accuracy of the internal stress analysis. Within each zone, a state of constant strain (stress) is assumed. Therefore, block boundaries remain defined by straight piece-wise lines, allowing a simple determination of the relative displacements between adjacent blocks.

The method employs an explicit time-domain algorithm using a central difference scheme. For each timestep, two basic sets of calculations are executed: first, the application of the equations of motion allow the new kinetic quantities (accelerations, velocities, displacements) to be determined; then, the block and joint constitutive relations provide the new internal stresses and interaction forces. The computational steps within each time increment are as follows.

Gridpoint accelerations, $\ddot{u}_i$, are obtained from the equations of motion:

$$\ddot{u}_i = \frac{\int_S \sigma_{ij} n_j ds + F_i}{m} + g_i$$

where S is the Voronoi polygonal (2-D) or polyhedral (3-D) surface surrounding each gridpoint,

σ_{ij} is the stress tensor,

n_j are the components of the normal to S,

m is the mass lumped at each gridpoint, and

g_i are the gravitational accelerations.

The forces F_i include applied external loads and, for gridpoints on a block boundary, the contact forces.

Integration of the above accelerations provides the velocity and displacement increments. Therefore, new zone strains can be determined. The application of the block material constitutive relation gives the new internal stresses. In the following examples, both elastic and elasto-plastic (Mohr-Coulomb) relations are used.

The updated location of the block boundaries defines the new joint displacements; the new joint stresses (and forces) follow from the joint constitutive relations. Two joint models are available—an elastoplastic model with a Mohr-Coulomb friction law and a continuously yielding model with a non-linear stress-displacement relation, capable of simulating a peak-residual type of behavior. Once all new stresses and forces are known, a new calculation cycle can be initiated.

Viscous damping is included in the equations of motion. By making use of artificially high values for this damping, the distinct element method can be applied to obtain solutions for quasi-static problems. In dynamic analyses, both mass proportional and stiffness proportional damping are utilized, as in the well-known Rayleigh damping scheme.

A major problem in dynamic modeling with a finite numerical mesh can be caused by boundary reflections which do not allow the energy of outgoing waves to be properly radiated to the far field. To avoid this problem, non-reflecting boundaries are included in the distinct element method, following the viscous boundary formulation proposed by Lysmer and Kuhlemeyer (1969). Consequently, block edges lying on the outer boundary are subject to the viscous stresses

$$\sigma_s = \rho \, c_s \, \dot{u}_s$$

$$\sigma_n = \rho \, c_p \, \dot{u}_n$$

where σ_s and σ_m are the shear and normal stresses, respectively,

ρ is the mass density,

c_s, c_p are S- and P-wave velocities, and

$\dot{u}_s, \dot{u}_n$ are tangential and normal gridpoint velocities.

3 VERIFICATION ANALYSIS — A DYNAMIC LINE SOURCE PARALLEL TO A PLANAR DISCONTINUITY

An infinite elastic medium containing a planar discontinuity is subject to a dynamic line source of compression parallel to the discontinuity. The source time variation, S(t), is given by the following function:

$$S(t) = \begin{bmatrix} 0.5 \; (1-\cos\pi t/0.6), & t \leq 0.6 \\ 1.0 & , \; t \geq 0.6 \end{bmatrix}$$

The analytical solution can be derived as a special symmetry condition for the more general problem of interface slip due to a dynamic point source (Salvado and Minster, 1980). The analytical solution assumes the interface to have a viscous behavior in shear; the numerical simulations were carried out for the case of zero viscosity. The time history of slip at a point on the discontinuity is the basis of comparison between analytical and numerical solutions.

Figure 1 shows the region of study with the UDEC (Universal Distinct Element Code) [Cundall, 1980; Cundall and Hart 1985] model. The source is located at the origin of the coordinate axes; the discontinuity is located at y = -h. The y-axis is a line of symmetry and non-reflecting boundaries were used on the other sides of the model. The dynamic input was applied as a radial velocity history at the semi-circular boundary of radius 0.05h. This velocity was calculated from the solution for a line source in an infinite continuum. Slip movement was monitored at Point P on the discontinuity.

The continuous medium was modeled with two elastic, fully-deformable blocks discretized into a triangular finite-difference mesh. The discontinuity was modeled as a joint with zero shear stiffness and zero friction. A high-normal stiffness was prescribed, and no tensile failure was allowed in order to meet the assumptions implied by the analytic solution.

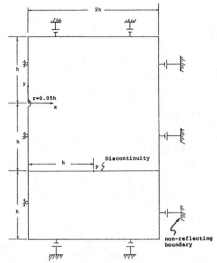

Figure 1. UDEC model

The results shown in Fig. 2 were obtained with a rather fine mesh (maximum zone length = 0.033h above discontinuity and 0.065k below discontinuity). The zone size above the discontinuity provides approximately 35 triangular zones per wavelength based on the dominant wavelength of the source time variations. Up to a dimensionless time of 2.0, the calculated slip is within 1% of the analytical solution. Results (not shown) from a much coarser mesh (maximum zone length = 0.13h) were up to 9% greater than the analytical solution. A spectral analysis of the slip response at point P produced significant frequencies up to 5.5 times higher than the dominant frequency of the input wave. This suggests that an appropriate relation for sizing the mesh is seven zones per wavelength of the highest frequency in the region of interest.

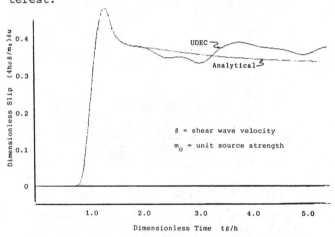

Figure 2. Comparison of UDEC results to analytical solution for slip at point P

In this problem, non-reflecting boundaries were located relatively close to the region of interest. The effects of boundary reflection can be observed after a dimensionless time of 2. One of the reasons for this is that the discontinuity at the far-field boundary violates the conditions assumed in formulating the viscous boundaries.

The analysis suggests that non-reflecting boundaries must be at least a factor of four times the distance from the dynamic source to the location of interest in order to reduce the effect of the discontinuity crossing the boundary.

4 DYNAMIC INSTABILITIES NEAR AN EXCAVATION IN JOINTED ROCK

Rock joints displaying a stress-displacement relation with post-peak softening can be a source for dynamic events. Previous studies by others (e.g., Rice, 1984) employing single degree-of-freedom systems have suggested that an instability may occur if the the post-peak slope of the joint stress-displacement curve is larger than the stiffness of the surrounding rock mass. In order to examine this assumption for a multiple degree of freedom system, numerical simulations were performed with UDEC using Cundall's continuously-yielding joint constitutive relation to describe joint behavior (Fig. 3) [Lemos, 1987]. The configuration and conditions of the numerical model are shown in Fig. 4. The model consists of two separate fully-deformable (elastic) blocks. A homogeneous in-situ state of stress with $\sigma_H/\sigma_V = 2$ was assumed.

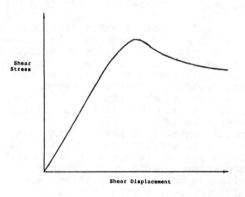

Figure 3. Continuously-yielding joint model

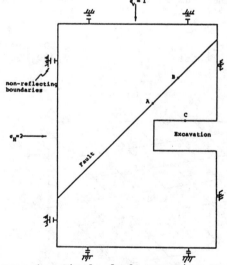

Figure 4. Single fault crossing an elastic rock mass near an excavation face

Excavation was simulated by slowly reducing the tractions applied on the excavation boundary. In this way, a quasi-static process takes place until a dynamic event is generated if unstable slip occurs in a segment of the fault. In the example shown, dynamic slip occurs along the fault between points A and B in Fig. 4. This slip results in a velocity "bump" in the roof (Point C) shown in Fig. 5.

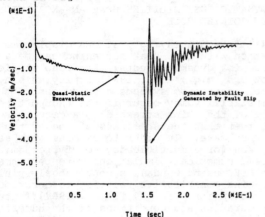

Figure 5. Vertical velocity recorded in roof (Point C)

5 ANALYSIS OF A SEISMIC-INDUCED ROOF FALL

If the dynamic event is sufficient to loosen a jointed rock mass surrounding an excavation, an unstable condition may result. The model shown in Fig. 6 is a two-dimensional, plane strain representation of a 5m high and 10m wide excavation. It was assumed that two continuous joint sets intersect the plane of analysis; one with an orientation of +45° and the other with an orientation of -9° from the horizontal. Both sets have a joint spacing of 5m. A near vertical "artificial" joint was also added to the block in the roof of the excavation to enhance the instability. The following material properties were assumed for the rock blocks and joints:

Block
density   3000 kg/m³
Young's Modulus   75,000 MPa
Poisson's Ratio   0.18

Joint
normal stiffness   20,000 MPa/m
shear stiffness   20,000 MPa/m
friction angle   30°
cohesion   0

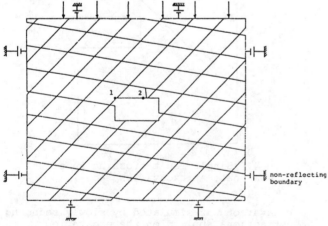

Fig. 6 Geometry and boundary conditions for excavation subject to dynamic loading

The in-situ stress state was assumed to be hydrostatic at 24 MPa at the excavation depth. The modeling sequence was performed as follows. First, the model, without the excavation, was consolidated under the in-situ stresses and gravitational acceleration. Next, the excavation was introduced and the model cycled to an equilibrium state. In the third stage, two dynamic loads with different peak velocities were evaluated.

Dynamic loads were represented by a sinusoidal y-directed stress wave applied at the top of the model for a duration of 0.0175 seconds. The applied stress wave was superimposed on the existing in-situ stresses. Mass proportional damping of less than 0.1% of critical damping was used in the motion calculations.

In the first simulation, a peak overstress of 0.625 MPa was applied. Displacements were monitored at the two points shown in Fig. 6. Displacement versus time plots for these points essentially showed an elastic response and stability of the roof block. In the second example, a stress wave with peak stress of 6.25 MPa was applied. Displacement versus time plots (Fig. 7) indicate that the roof block has loosened and is falling. As a matter of interest, the problem geometry and stress distribution at later times are presented in Fig. 8. It is interesting to note that the peak resultant boundary velocities for the two cases, 6 cm/sec (no roof fall) and 60 cm/sec (roof fall), are within the observed ranges for no damage and minor damage in underground excavations in rock subjected to earthquakes (Owen and Scholl, 1981).

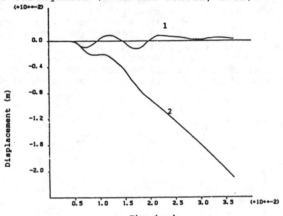

Fig. 7 Y-displacement histories For 2 points on excavation boundary
[applied stress = 6.25 cos(2π 100t)].

6 INTERPRETATION OF FAULT INTERACTION

The vulnerability of a tunnel to collapse due to a dynamic pressure wave is increased in a jointed rock mass. Explosion-induced motion along faults intersecting (or even in close proximity to) tunnels can produce substantial damage and total collapse. Therefore, predictions of tunnel vulnerability must account for the presence of major discontinuities.

The extent and magnitude of explosion-induced motion along faults is difficult to predict, however, in a multiple-faulted rock mass. Conventional analytical techniques can only account for the presence of single, nonintersecting discontinuities—the effect of the interaction of intersection faults cannot be analyzed. This interaction, though, can control the characteristics of the fault motion.

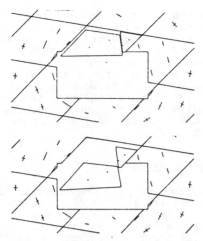

Fig. 8 Roof block displacement and principal stress distribution (at centroids of finite difference triangles) at various times

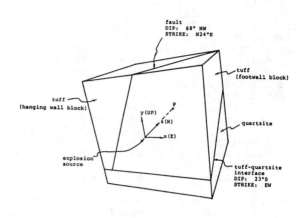

Fig. 9 Problem setting for fault interaction study

The influence of fault interaction is demonstrated with 3DEC (3-Dimensional Distinct Element Code) [Cundall et al, 1986]. This program presently is being used to provide a better understanding of the processes involved in explosion-induced fault motion in a study sponsored by the Defense Nuclear Agency. A demonstration of fault interaction is provided for a tuff/quartzite geologic setting. The simulation is representative of fault motion observations made in the Mighty Epic nuclear event at the Nevada Test Site.

Fault motion observed in one location of this test was opposite in nature to that predicted by the single-fault point estimate methods. It was surmised that this behavior was due to the interaction of the fault and a tuff-quartzite interface discontinuity in the vicinity of the test. The intention of the 3DEC analysis was to demonstrate whether such an interaction can produce motion in the direction observed.

The problem setting consists of two major discontinuities: a major north-trending fault along which motion was monitored and a south-plunging interface surface separating the tuff from underlying quartzite (Fig. 9). The fault motion was observed at a drift location denoted as position P(x=0, y=0, z=117m) in the figure. The movement was such that the footwall side of the fault moved up and to the north, relative to the hanging wall side (i.e., left lateral and normal fault motion).

Based on the orientation of the fault and the location of the fault/drift intersection relative to the explosive source, the normal fault motion is opposite in sign from that predicted by point estimate calculations. At this location, the direction of induced dip-slip would be estimated by point methods to be similar to reverse faulting.

For this investigation with 3DEC, a relatively coarse model was considered sufficient to provide some insight to the phenomena affecting the block motion. The study was taken as a qualitative validation of the model——i.e., the analysis attempted to reproduce the observed directions of motion on the fault and interface; an attempt was not made to match the magnitude of the measured motion.

The 3DEC model for the setting described in Fig. 9 consisted of three blocks—one for the hanging wall side of the fault, one for the footwall side, and one for the quartzite.

The blocks were fully-deformable, each block containing approximately 400 finite-difference tetrahedral zones. The deformable blocks were assumed to behave as non-dilatant Mohr-Coulomb elastoplastic materials. Representative values assumed for tuff, quartzite, and discontinuities are given in Table I.

Table I. Problem parameters for fault interaction study

Parameter	Tuff	Quartzite
Young's Modulus (GPa)	10.0	50.0
Poisson's Ratio	0.3	0.2
Cohesion (MPa)	9.4	14.0
Internal Angle of Friction	8°	30°

	Fault/Interface
Cohesion (MPa)	3.5
Friction Coefficient	0.2

The origin of the axis shown in Fig. 9 corresponds to the explosive source. For the 3DEC model, along the x-y plane at the origin, motion in the z-direction was prevented, which implies a symmetry condition on this face of the model. Along all other faces, viscous boundary conditions were imposed. An initial isotropic stress state of 6.3 MPa was applied to the model prior to the simulation of the blast loading.

A radial velocity history was prescribed on a hemispherical surface of radius 40m centered at the source. At this level of study, it was considered appropriate to choose a simplified input velocity which represents qualitatively the free-field velocity. The velocity input history was of the form

$$f(t) = 5 [1 - \cos(105t)] \text{ for } t \leq 0.2 \text{ sec}$$
$$= 0 \qquad\qquad \text{ for } t > 0.2 \text{ sec}$$

Dip-slip and strike-slip motion were monitored in the model along the planes representing the fault and the interface. Relative shear displacements on discontinuities can be calculated along any direction through the model. Directions were chosen to correspond to the dip direction of the interface and dip and strike of the fault.

Based on the coordinate system chosen for the 3DEC model, the projection of the blast source on the fault is at a point (x=-35.5m, y=15.6m, and z=15.7m). Using a single-fault point estimate calculation, the slip along the fault would be radial from this point— i.e., no slip would be calculated for points along the line y = 15.6m. Below this line (y < 15.6m), the sign of slip would be consistent with reverse fault motion; above this line (y > 15.6m), slip would resemble normal fault motion. However, as noted before, the observed slip at position P (y = 0) was in the direction of normal faulting.

The presence of the tuff/quartzite interface was considered to account for this discrepancy. In order to evaluate this effect, two cases were analyzed: one case with the quartzite block in the model and one case without. The difference in the resulting effects of these two cases on dip-slip along the fault is quite striking, as shown in Fig. 10.

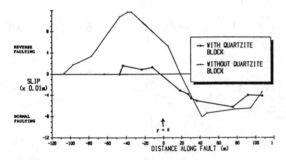

Figure 10. Dip slip along fault for section through point P

Without the quartzite block, the point of zero slip and directions of slip are consistent with those calculated by point estimates. However, with the quartzite block present, the point of zero slip is shifted downward along the fault. From a comparison of dip-slip shown in Fig. 10, it is seen that the sign of the dip-slip at y = 0 reverses due to the presence of the quartzite block. The block significantly inhibits the reverse faulting—thus extending the effect of normal faulting.

It is concluded that the model does indicate that the presence of the interface and the interaction of the fault may account for the sign of the measured slip. Even if the fault extends into the underlying quartzite, this conclusion is considered valid, because the interface motion precedes the fault motion, effectively locking the fault at the interface.

7 DISCUSSION

The distinct element method provides a useful tool for understanding a wide spectrum of dynamic problems in jointed rock masses, including potential sources of seismic excitation (i.e., earthquakes or rockbursts) as well as the effects of dynamic loading. With regard to the latter, the method has the following useful capabilities: (1) the ability to visualize displacements and stresses with real time; (2) the ability to model large displacements and post-peak behavior; and (3) the ability to model stress wave propagation. The method presently is being applied to various two- and three-dimensional problems to both validate the technique and provide insight to the behavior of jointed rock subject to dynamic loads.

8 ACKNOWLEDGEMENTS

Several of the examples described in this paper reflect the results of research performed by J. Lemos, under the guidance of P. A. Cundall, as partial fulfillment of the requirements for the Ph.D. degree at the University of Minnesota. A portion of this work was funded through a subcontract with Agbabian Associates as part of a block motion research study sponsored by the Defense Nuclear Agency. Additional funding was provided by Falconbridge Nickel Mines Ltd.

REFERENCES

Cundall, P. A. 1980. UDEC - a generalized distinct element program for modelling jointed rock," Report PCAR-1-80, Peter Cundall Associates; Contract DAJA37-79-C-0548, European Research Office, U.S. Army.

Cundall, Peter A. and Roger D. Hart. 1985. Development of generalized 2-D and 3-D distinct element programs for modeling jointed rock." Itasca Consulting Group; Misc. Paper SL-85-1, U.S. Army Corps of Engineers.

Cundall, Peter A., Roger D. Hart and J. Lemos. 1986. Program 3DEC (general 3-D distinct element code) user's manual. Itasca Consulting Group.

Lemos, J. 1987. Dynamic analysis of jointed rock masses using the distinct element method. Ph.D. thesis to be submitted to the University of Minnesota.

Lemos, José V., Roger D. Hart and Peter A. Cundall. 1985. A generalized distinct element program for modelling jointed rock mass: a keynote lecture, Proceedings of the International Symposium on Fundamentals of Rock Joints (Bjorkliden, 15-20 September 1985), pp. 335-343. Lulea, Sweden: Centek Publishers.

Lysmer, J. and R. Kuhlemeyer. 1969. Finite dynamic model for infinite media. J. Eng. Mcch. 95:EM4.

Owen, G. N. and R. E. Scholl. 1981. Underground structures intersecting active faults, FHWA/RD-80/195, U.S. DOT.

Rice, James R. 1984. Shear Instability in the Relation to the Constitutive Description of Fault Slip, Proceedings of the 1st International Congress on Rockbursts and Seismicity in Mines (Johannesburg, 1982).

Salvado, C. and J. B. Minster. 1980. Slipping interfaces: a possible source of S radiation from explosive sources. BSSA. 70:659-670.

A rheological model of the behaviour of rocks and rock masses
Un modèle rhéologique du comportement des roches et des massifs rocheux
Ein rheologisches Modell des Verhaltens von Gesteinen und Gebirgen

J.LOUREIRO-PINTO, Laboratório Nacional de Engenharia Civil, Lisbon, Portugal

ABSTRACT: This work presents rheological models of fissure closure body, creep body, yielding body and one body that are representative of rocks and rock masses, their rheologic equations, and their behaviour in creep and under constant load velocity. Lastly results of tests on two specimens of different rocks are given, and values of their rheologic constants are determined as well as deviation of unit strains calculated with reference to those measured.

RESUME: Dans ce travail on présente les modèles rhéologiques du corps de clôture des fissures, du corps de fluage, du corps de plasticité et du corps représentatif des roches et massifs rocheux, leurs équations rheo logiques et comportement en fluage et sous vitesse de chargement constante. À la fin du travail on présente les résultats des essais de deux éprouvettes de différentes roches, et on détermine les valeurs de leurs constantes rhéologiques aussi bien que les écarts des extensions calculées par égard aux extensions mesurées.

ZUSAMMENFASSUNG: In dieser Arbeit werden die rheologischen Modelle des Rißschließkörpers, des Kriechkörpers, des Nachgebkörpers und der Gesteine und Gebirge, sowie ihre rheologischen Gleichungen und ihr Verhalten beim Kriechen und bei konstanter Belastungsgeschwindigkeit vorgestellt. Als Abschluß der Arbeit werden die Ver - suchsergbnisse von zwei Proben verschiedener Gesteine vorgestellt und die Werte ihrer rheologischen Konstan- ten, sowie die Abweichungen der berechneten Dehnungen bezüglich der gemessenen Verformungen bestimmt.

1. INTRODUCTION

The study of the actual behaviour of rocks and rock masses has led to the preparation of several rheolo gical models intended to represent with considerable approximation the trend of the stress-stain curves obtained as a function of the type of load involved.

In fact the rheological models so far created ha ve not satisfactorily represented the very common phenomena of fissure closure and accelerated creep.

In this work two simple new rheological models are described, which aim at representing those two phenomena in approximate manner.

2. RHEOLOGICAL MODELS

2.1 Fissure closure body

The fissure closure body,fig.1, consists of two con stant-section cantilevers, of length 1 and curvatu re variable following a clothoid. They join at the nullcurvature ends and are acted by two forces (stres ses) σ at their extremities.

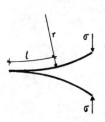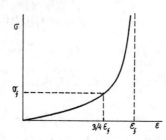

Figure 1. Fissure closure body

As a result of that spring-shape, cantilever span varies depending on the value of σ, and the higher is the stress the smaller will be the span, which beco mes null for $\sigma = \infty$, the maximum unit strain reaching

the value ε_f. In the limit the maximum bending mo - ment of the cantilever will attain the value m_f and, for the purpose of calculation, m_f can be considered equal to $\sigma_f 1$. On account of the hypotheses assumed, unit strains (deflections) and stresses will be given by:

$$\varepsilon = \varepsilon_f \left[1 - \left(1 + \frac{\sigma}{\sigma_f} \right)^{-2} \right] \; ; \; \sigma = \sigma_f \left[\left(1 - \frac{\varepsilon}{\varepsilon_f} \right)^{-1/2} - 1 \right] \qquad (1)$$

Curve $\sigma - \varepsilon$ is shown in fig. 1.

2.2 Creep body

The creep body is formed of a variable-section dashpot corresponding to a generalization of the Newton bo- dy,fig.2. This body is characterized by an initial zone of length ε_c which behaves like the Newton body:

$$\frac{d\varepsilon}{dt} = \frac{\sigma}{K_c} \qquad (2)$$

followed by other zone for which:

$$\frac{d\varepsilon}{dt} = \frac{\sigma}{K_c} \left(\frac{\varepsilon}{\varepsilon_c} \right)^n \qquad (3)$$

Exponent n defines the type of behaviour of the body as follows; if n<0 the body presents primary creep,if n=0 the creep is secondary,and if n>0 the body presents tertiary creep. Behaviour of this bo- dy depends on how the force acts. Below is presented a case of creep $\sigma = \sigma_0$, with initial unit strain ε_0.

$$- \varepsilon_0 < \varepsilon_c; \; 0 \leqslant t < t_c = \frac{K_c}{\sigma_0}(\varepsilon_c - \varepsilon_0);$$

$$\varepsilon = \frac{\sigma_0}{K_c}t + \varepsilon_0 \qquad (4)$$

$$- \varepsilon_0 < \varepsilon_c; \; t \geqslant t_c; \; n \neq 1;$$

$$\varepsilon = \varepsilon_c \left[1 + \frac{(1-n)\sigma_0}{K_c \varepsilon_c}(t - t_c) \right]^{1/(1-n)} \qquad (5)$$

$-\ \varepsilon_o < \varepsilon_c;\ t > t_c;\ n=1;$

$$\varepsilon = \varepsilon_c \exp\left[\frac{\sigma_o}{K_c \varepsilon_c}(t-t_c)\right] \tag{6}$$

$-\ \varepsilon_o \geqslant \varepsilon_c;\ n\neq1;$

$$\varepsilon = \varepsilon_o\left[1 + \frac{(1-n)\sigma_o}{K_c \varepsilon_c}\left(\frac{\varepsilon_o}{\varepsilon_c}\right)^n\right]^{1/(1-n)} \tag{7}$$

$-\ \varepsilon_o \geqslant \varepsilon_c;\ n=1;$

$$\varepsilon = \varepsilon_o \exp\left(\frac{\sigma_o}{K_c \varepsilon_c}\right) \tag{8}$$

Curves ε-t are presented in fig. 2

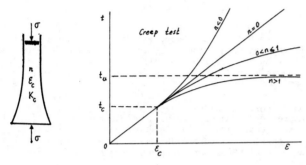

Figure 2. Creep body

Due to creep the body fails with infinite strain for infinite time, when $n\leqslant1$, and for finite time t_u when $n>1$. Expressions of t_u are given below:

$-\ \varepsilon_o > \varepsilon_c;\quad t_u = \frac{K_c \varepsilon_c}{(n-1)\sigma_o}\left(\frac{\varepsilon_c}{\varepsilon_o}\right)^{n-1} \tag{9}$

$-\ \varepsilon_o = \varepsilon_c;\quad t_u = \frac{K_c \varepsilon_c}{(n-1)\sigma_o} \tag{10}$

$-\ \varepsilon_o < \varepsilon_c;\quad t_u = \frac{K_c}{\sigma_o}\left(\frac{n\varepsilon_c}{n-1} - \varepsilon_o\right) \tag{11}$

2.3 Yielding body

The yielding body, presented in fig.3, consits of the association in parallel of a creep body and a St Venant body. The rheologic equations are as follows:

$-\ \sigma \leqslant \sigma_c;\qquad \frac{d\varepsilon}{dt} = 0 \tag{12}$

$-\ \sigma > \sigma_c;\ \varepsilon_{oy} < \varepsilon_c;\ \frac{d\varepsilon}{dt} = \frac{\sigma - \sigma_c}{K_c} \tag{13}$

$-\ \sigma > \sigma_c;\ \varepsilon_{oy} \geqslant \varepsilon_c;\ \frac{d\varepsilon}{dt} = \frac{\sigma - \sigma_c}{K_c}\left(\frac{\varepsilon}{\varepsilon_c}\right)^n \tag{14}$

The behaviour of this body depends on the type of loading, and irrecuperable deformations occur. When this body undergoes creep with initial unit strain ε_{oy} its behaviour is represented by expressions:

$-\ \sigma \leqslant \sigma_c;\qquad \varepsilon_y = \varepsilon_{oy} \tag{15}$

$-\ \sigma > \sigma_c;\ \varepsilon_{oy} \leqslant \varepsilon_c;\ t \leqslant t_y = \frac{\varepsilon_c - \varepsilon_{oy}}{\sigma_o - \sigma_c} K_c$

$$\varepsilon_y = \varepsilon_{oy} + \frac{\sigma_o - \sigma_c}{K_c} t \tag{16}$$

$-\ \sigma > \sigma_c;\ \varepsilon_{oy} \leqslant \varepsilon_c;\ t > t_y;\ n\neq1;$

$$\varepsilon_y = \varepsilon_c\left[1 - \frac{(n-1)(\sigma_o - \sigma_c)}{K_c \varepsilon_c}(t-t_y)\right]^{1/(1-n)} \tag{17}$$

$-\ \sigma > \sigma_c;\ \varepsilon_{oy} \leqslant \varepsilon_c;\ t > t_y;\ n=1;$

$$\varepsilon_y = \varepsilon_c \exp\left[\frac{\sigma_o - \sigma_c}{K_c \varepsilon_c}(t-t_y)\right] \tag{18}$$

When n>1 the body fails for the time t_u:

$$t_u = \frac{K_c}{(n-1)(\sigma_o - \sigma_c)}\left[\varepsilon_{oy} + n(\varepsilon_c - \varepsilon_{oy})\right] \tag{19}$$

$-\ \sigma > \sigma_c;\ \varepsilon_{oy} \geqslant \varepsilon_c;\ n\neq1;$

$$\varepsilon_y = \varepsilon_{oy}\left[1 - \frac{(n-1)(\sigma_o - \sigma_c)}{K_c \varepsilon_c}\left(\frac{\varepsilon_{oy}}{\varepsilon_c}\right)^{n-1} t\right]^{1/(1-n)} \tag{20}$$

$-\ \sigma > \sigma_c;\ \varepsilon_{oy} \geqslant \varepsilon_c;\ n=1;$

$$\varepsilon_y = \varepsilon_{oy}\exp\left(\frac{\sigma_o - \sigma_c}{K_c \varepsilon_c} t\right) \tag{21}$$

When n>1 the body fails for the time t_u:

$$t_u = \frac{K_c \varepsilon_c}{(n-1)(\sigma_o - \sigma_c)}\left(\frac{\varepsilon_c}{\varepsilon_{oy}}\right)^{n-1} \tag{22}$$

Curves ε-t are presented in fig. 3.

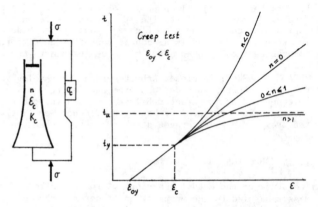

Figure 3. Yielding body

When this body undergoes constant rate of charge, corresponding to $\sigma = \sigma_o + at$, or $d\sigma/dt = a$, the behaviour is:

$-\ \sigma_o \leqslant \sigma_c;\ a < 0;\ \varepsilon_y = \varepsilon_{oy} \tag{23}$

$-\ \sigma_o \leqslant \sigma_c;\ a > 0;\ \varepsilon_{oy} < \varepsilon_c;\ t \leqslant t_c = \frac{\sigma_c - \sigma_o}{a};\ \varepsilon_y = \varepsilon_{oy} \tag{24}$

$-\ \sigma_o \leqslant \sigma_c;\ a > 0;\ \varepsilon_{oy} < \varepsilon_c;\ t_c \leqslant t \leqslant t_y = \frac{1}{a}\left[\sigma_c - \sigma_o + \sqrt{2aK_c(\varepsilon_c - \varepsilon_{oy})}\right];$

$$\varepsilon_y = \varepsilon_{oy} + \frac{a(t-t_c)^2}{2K_c} \tag{25}$$

$-\ \sigma_o \leqslant \sigma_c;\ a > 0;\ \varepsilon_{oy} < \varepsilon_c;\ t > t_y;\ n\neq1;$

$$\varepsilon_y = \varepsilon_c\left[1 - \frac{(n-1)}{K_c \varepsilon_c}\left[(\sigma_o - \sigma_c)(t-t_y) + \frac{a}{2}(t^2 - t_y^2)\right]\right]^{1/(1-n)} \tag{26}$$

$-\ \sigma_o \leqslant \sigma_c;\ a > 0;\ \varepsilon_{oy} < \varepsilon_c;\ t > t_y;\ n=1;$

$$\varepsilon_y = \varepsilon_c \exp\left[\frac{1}{K_c \varepsilon_c}\left[(\sigma_o - \sigma_c)(t-t_y) + \frac{a}{2}(t^2 - t_y^2)\right]\right] \tag{27}$$

When n>1 failure occurs for the value of t_u:

$$t_u = \frac{1}{a}\left[\sigma_c - \sigma_o + \sqrt{2aK_c\left(\frac{n}{n-1}\varepsilon_c - \varepsilon_{oy}\right)}\right] \tag{28}$$

$-\ \sigma_o \leqslant \sigma_c;\ a < 0;\ \varepsilon_{oy} \geqslant \varepsilon_c;\ t \leqslant t_y = \frac{\sigma_c - \sigma_o}{a};\ \varepsilon_y = \varepsilon_{oy} \tag{29}$

$-\ \sigma_o \leqslant \sigma_c;\ a > 0;\ \varepsilon_{oy} \geqslant \varepsilon_c;\ t > t_y;\ n\neq1;$

$$\varepsilon_y = \varepsilon_{oy}\left[1 - \frac{(n-1)}{K_c \varepsilon_c}\left(\frac{\varepsilon_{oy}}{\varepsilon_c}\right)^{n-1}\left[(\sigma_o - \sigma_c)(t-t_y) + \frac{a}{2}(t^2 - t_y^2)\right]\right]^{1/(1-n)} \tag{30}$$

$-\ \sigma_o \leqslant \sigma_c;\ a > 0;\ \varepsilon_{oy} \geqslant \varepsilon_c;\ t > t_y;\ n=1;$

$$\varepsilon_y = \varepsilon_{oy}\exp\left[\frac{1}{K_c \varepsilon_c}\left[(\sigma_o - \sigma_c)(t-t_y) + \frac{a}{2}(t^2 - t_y^2)\right]\right] \tag{31}$$

With n>1 the body fails for the value of t_u:

$$t_u = \frac{1}{a}\left[\sigma_c - \sigma_0 + \sqrt{\frac{2aK_c \varepsilon_c}{n-1}\left(\frac{\varepsilon_c}{\varepsilon_{oy}}\right)^{n-1}}\right] \qquad (32)$$

- $\sigma_0 \geqslant \sigma_c$; $a>0$; $\varepsilon_{oy} < \varepsilon_c$; $t \leqslant t_y = \frac{1}{a}\left[\sigma_c - \sigma_0 + \sqrt{(\sigma_0 - \sigma_c)^2 + 2aK_c(\varepsilon_c - \varepsilon_{oy})}\right]$;

$\varepsilon_y = \frac{1}{K_c}(\sigma_0 - \sigma_c)t + \frac{a}{2}t^2 + \varepsilon_{oy} \qquad (33)$

- $\sigma_0 \geqslant \sigma_c$; $a>0$; $\varepsilon_{oy} < \varepsilon_c$; $t > t_y$;

The body deforms according to the laws (26) or (27).
With n>1 failure occurs for the value of t_u:

$$t_u = \frac{1}{a}\left[\sigma_c - \sigma_0 + \sqrt{(\sigma_0 - \sigma_c)^2 + 2aK_c\left(\frac{n}{n-1}\varepsilon_c - \varepsilon_{oy}\right)}\right] \qquad (34)$$

- $\sigma_0 \geqslant \sigma_c$; $a>0$; $\varepsilon_{oy} \geqslant \varepsilon_c$; $n \neq 1$;

$\varepsilon_y = \varepsilon_{oy}\left\{1 - \frac{(n-1)}{K_c \varepsilon_c}\left(\frac{\varepsilon_{oy}}{\varepsilon_c}\right)^{n-1}\left[(\sigma_0 - \sigma_c)t + \frac{a}{2}t^2\right]\right\}^{1/(1-n)} \qquad (35)$

- $\sigma_0 \geqslant \sigma_c$; $a>0$; $\varepsilon_{oy} \geqslant \varepsilon_c$; $n=1$;

$\varepsilon_y = \varepsilon_{oy}\exp\left[\frac{1}{K_c \varepsilon_c}\left[(\sigma_0 - \sigma_c)t + \frac{a}{2}t^2\right]\right] \qquad (36)$

With n>1 failure occurs for the value of t_u:

$$t_u = \frac{1}{a}\left[\sigma_c - \sigma_0 + \sqrt{(\sigma_0 - \sigma_c)^2 + \frac{2aK_c \varepsilon_c}{n-1}\left(\frac{\varepsilon_c}{\varepsilon_{oy}}\right)^{n-1}}\right]$$

- $\sigma_0 \geqslant \sigma_c$; $\varepsilon_{oy} < \varepsilon_c$; $0>a<-\frac{(\sigma_0-\sigma_c)^2}{2K_c(\varepsilon_c - \varepsilon_{oy})}$; $t < t_0 = \frac{\sigma_c - \sigma_0}{a}$;

The body deforms according to the law (33).

- $\sigma_0 \geqslant \sigma_c$; $\varepsilon_{oy} < \varepsilon_c$; $0>a<-\frac{(\sigma_0-\sigma_c)^2}{2K_c(\varepsilon_c - \varepsilon_{oy})}$; $t > t_0$;

$\varepsilon_y = \varepsilon_{oy} - \frac{1}{K_c}\frac{(\sigma_0 - \sigma_c)^2}{2a} \qquad (37)$

- $\sigma_0 \geqslant \sigma_c$; $\varepsilon_{oy} < \varepsilon_c$; $0>a>-\frac{(\sigma_0-\sigma_c)^2}{2K_c(\varepsilon_c - \varepsilon_{oy})}$;

$\qquad t < t_y = \frac{1}{a}\left[\sigma_c - \sigma_0 - \sqrt{(\sigma_0-\sigma_c)^2 + 2aK_c(\varepsilon_c - \varepsilon_{oy})}\right]$;

The yielding body deforms according to the law (33).

- $\sigma_0 \geqslant \sigma_c$; $\varepsilon_{oy} < \varepsilon_c$; $0>a>-\frac{(\sigma_0-\sigma_c)^2}{2K_c(\varepsilon_c - \varepsilon_{oy})}$; $t > t_y$;

The body will deform following laws (26) or (27) until reaching time t_0. For $t>t_0$ the unit strain being maintained up to the end, but in the case of n>1, failure occurs for a time $t_u < t_0$ given by:

$$t_u = \frac{1}{a}\left[\sigma_c - \sigma_0 - \sqrt{(\sigma_0-\sigma_c)^2 + 2aK_c\left(\frac{n}{n-1}\varepsilon_c - \varepsilon_{oy}\right)}\right] \qquad (38)$$

provided

$$0 > a > \frac{-(\sigma_0-\sigma_c)^2}{2K_c\left(\frac{n}{n-1}\varepsilon_c - \varepsilon_{oy}\right)}$$

- $\sigma_0 \geqslant \sigma_c$; $\varepsilon_{oy} \geqslant \varepsilon_c$; $0>a$;

The body deforms following laws (35) or (36). Strains will be stationary for $t>t_0$, or in the case of:

$$0 > a > \frac{(1-n)(\sigma_0-\sigma_c)^2}{2K_c \varepsilon_c}\left(\frac{\varepsilon_{oy}}{\varepsilon_c}\right)^{n-1}$$

failure will occur for time:

$$t_u = \frac{1}{a}\left[\sigma_c - \sigma_0 - \sqrt{(\sigma_0-\sigma_c)^2 + \frac{2aK_c \varepsilon_c}{n-1}\left(\frac{\varepsilon_c}{\varepsilon_{oy}}\right)^{n-1}}\right] \qquad (39)$$

2.4 Rheological model of rocks and rock masses

On account of the actual rheologic behaviour of rocks and rock masses, the most suitable rheological model seems to be that of fig.4. This model consists of a series association of a Hooke body, with elasticity modulus E_h; a St Venant body, with ultimate strength σ_u; a fissure closure body, with constants ε_f and σ_f; a Kelvin body, with constants E_k and K_k and a yielding body, with constants n, ε_c, K_c and σ_c. Designating by:

ε_h - unit strain of Hooke body,

ε_a - unit strain of fissure closure body

ε_k - unit strain of Kelvin body,

ε_y - unit strain of yielding body,

ε_{ok} - unit strain of Kelvin body for t=0

ε_{oy} - unit strain of yielding body for t=0

ε - total unit strain of the rheological model,

the behaviour of the model presented, is given by:

$$\varepsilon = \varepsilon_h + \varepsilon_a + \varepsilon_k + \varepsilon_y \qquad (40)$$

$$\frac{d\varepsilon}{dt} = \frac{d\varepsilon_h}{dt} + \frac{d\varepsilon_a}{dt} + \frac{d\varepsilon_k}{dt} + \frac{d\varepsilon_y}{dt} \qquad (41)$$

and from values:

$$\varepsilon_h = \frac{\sigma}{E_h} \qquad ; \quad \frac{d\varepsilon_h}{dt} = \frac{1}{E_h}\frac{d\sigma}{dt} \qquad (42)$$

$$\varepsilon_a = \varepsilon_f\left[1 - \left(1+\frac{\sigma}{\sigma_f}\right)^{-2}\right] \quad ; \quad \frac{d\varepsilon_a}{dt} = \frac{2\varepsilon_f}{\sigma_f}\left(1+\frac{\sigma}{\sigma_f}\right)^{-3}\frac{d\sigma}{dt} \qquad (43)$$

$$\varepsilon_k = \frac{1}{E_k}\left(\sigma - K_k\frac{d\varepsilon_k}{dt}\right) \quad ; \quad \frac{d\varepsilon_k}{dt} = \frac{1}{K_k}(\sigma - \varepsilon_k E_k) \qquad (44)$$

$$- \sigma \leqslant \sigma_c; \quad \frac{d\varepsilon_y}{dt} = 0 \qquad (45)$$

$$- \sigma > \sigma_c; \quad \varepsilon_y \leqslant \varepsilon_c; \quad \frac{d\varepsilon_y}{dt} = \frac{\sigma - \sigma_c}{K_c} \qquad (46)$$

$$- \sigma > \sigma_c; \quad \varepsilon_y > \varepsilon_c; \quad \frac{d\varepsilon_y}{dt} = \frac{\sigma - \sigma_c}{K_c}\left(\frac{\varepsilon_y}{\varepsilon_c}\right)^n \qquad (47)$$

will be represented by equations:

$- \sigma \leqslant \sigma_c$;

$$\frac{d\varepsilon}{dt} = \left[\frac{1}{E_h} + \frac{2\varepsilon_f}{\sigma_f}\left(1+\frac{\sigma}{\sigma_f}\right)^{-3}\right]\frac{d\sigma}{dt} + \frac{1}{K_k}(\sigma - \varepsilon_k E_k) \qquad (48)$$

$- \sigma > \sigma_c$; $\varepsilon_y \leqslant \varepsilon_c$;

$$\frac{d\varepsilon}{dt} = \left[\frac{1}{E_h} + \frac{2\varepsilon_f}{\sigma_f}\left(1+\frac{\sigma}{\sigma_f}\right)^{-3}\right]\frac{d\sigma}{dt} + \frac{1}{K_k}(\sigma - \varepsilon_k E_k) + \frac{\sigma - \sigma_c}{K_c} \qquad (49)$$

$- \sigma > \sigma_c$; $\varepsilon_y > \varepsilon_c$;

$$\frac{d\varepsilon}{dt} = \left[\frac{1}{E_h} + \frac{2\varepsilon_f}{\sigma_f}\left(1+\frac{\sigma}{\sigma_f}\right)^{-3}\right]\frac{d\sigma}{dt} + \frac{1}{K_k}(\sigma - \varepsilon_k E_k) + \frac{\sigma - \sigma_c}{K_c}\left(\frac{\varepsilon_y}{\varepsilon_c}\right)^n \qquad (50)$$

The integration of this rheologic equations is simple as concerns creep and constant rate of load and much complex as regards relaxation and constant rate of deformation. In the latter cases it will be necessary to start from the analysis of curves (ε, t), (σ, t) and $(d\sigma/dt)$ in order to determine the constants of the rheological model through an harduous procedure.

2.4.1 Creep ($\sigma = \sigma_0$, $d\sigma/dt=0$)

In this case each body in the rheological model will deform regardless of the others, reason why it can be studied in separate. Thus, Hooke, fissure closure, and Kelvin bodies will present deformations as follows:

$$\varepsilon_h = \frac{\sigma_o}{E_h} \; ; \quad \varepsilon_a = \varepsilon_f\left[1-\left(1+\frac{\sigma_o}{\sigma_f}\right)^{-2}\right] \; ;$$

$$; \; \varepsilon_k = \frac{\sigma_o}{E_k} + \left(\varepsilon_{ok}-\frac{\sigma_o}{E_k}\right)\exp\left(-\frac{E_k}{K_k}t\right) \qquad (51)$$

The behaviour of the yielding body is presented by expressions (15) to (21).

Total deformation will be obtained through expression (40) by replacing the unit strains by its values.

2.4.2 Constant rate of load ($\sigma = \sigma_o + at$, $d\sigma/dt = a$)

As it occurs in creep, each body in the rheological model will deform regardless of the others, so deformations undergone by Hooke, fissure closure and Kelvin bodies will be:

$$\varepsilon_h = \frac{\sigma_o + at}{E_h} \; ; \quad \varepsilon_a = \varepsilon_f\left[1-\left(1+\frac{\sigma_o + at}{\sigma_f}\right)^{-2}\right] \; ;$$

$$; \; \varepsilon_k = \frac{\sigma_o + at}{E_k} - \frac{aK_k}{E_k^2} + \left(\varepsilon_{ok}-\frac{\sigma_o}{E_k}+\frac{aK_k}{E_k^2}\right)\exp\left(-\frac{E_k}{K_k}t\right) \qquad (52)$$

The yielding body behaves according to expressions (23) to (38).

Total deformation will be obtained through expression (40) by replacing the unit strains by its values.

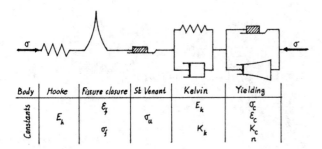

Figure 4. Rheologic model for rocks and rock masses

3. RESULTS OF TESTS

3.1 Creep test on granite

Table I presents values of unit strains, measured and calculated, for a creep test on granite, in which:

$\sigma_o = 152$ MPa; $\varepsilon_h + \varepsilon_a = 3690 \times 10^{-6}$; $E_k = 338$ GPa; $K_k = 3.87 \times 10^{16}$ Ns/m^2; $\varepsilon_{oy} = 1673 \times 10^{-6}$; n = 2.13337; $t_u = 92$ day; $\varepsilon = \varepsilon_h + \varepsilon_a + \varepsilon_k + (\varepsilon_y - \varepsilon_{oy})$;

$$\frac{(n-1)(\sigma_o-\sigma_c)}{K_c\varepsilon_c}\left(\frac{\varepsilon_{oy}}{\varepsilon_c}\right)^{n-1} = 1.25805 \times 10^{-7}/s;$$

$\varepsilon_y = 1673 \times 10^{-6}(1-0.01087t)^{-0.46874}$ (t in days)

Table I. Creep test on granite

t	Calculated strain (10^{-6})			Measured	δ	δ
(day)	ε_k	ε_y	ε_d	strain	ε_d-ε_m	%
0	0	0	3690	3690	0	0
5	439	33	4214	4232	-18	-0.43
10	449	179	4318	4315	3	0.07
15	449	285	4424	4402	22	0.50
20	449	404	4543	4548	- 5	-0.09
25	449	541	4680	4680	0	0
30	449	698	4837	4817	20	0.42
35	449	880	5019	4988	31	0.62
40	449	1096	5235	5286	-51	-0.96
45	449	1354	5493	5551	-58	-1.04
50	449	1670	5809	5823	-14	-0.24
55	449	2065	6204	6112	92	1.51
60	449	2576	6715	6715	0	0

3.2 Test on gneiss at constant load rate

Table II presents the values of unit strains measured ε_m, and unit strains calculated ε_d, in test under constant load rate, in which:

$\sigma = at = 0.25$ MPa/s; $\varepsilon_{ok} = \varepsilon_{oy} = 0$; $E_h = 95.8$ GPa; $\sigma_f = 9.25$ MPa; $\varepsilon_f = 924 \times 10^{-6}$; $E_k = 17.6$ GPa; $K_k = 6141$ GPa s; $\sigma_c = 18.2$ MPa; $\varepsilon_c = 301 \times 10^{-6}$; $K_c = 26087$ GPa s; n = 1.750; $t_c = 728$ s; $t_y = 1521$ s; $t_u = 1939$ s; $\varepsilon_h = 0.261 \times 10^{-6}t$; $\varepsilon_a = 924 \times 10^{-6}\left[1-(1+0.0027027t)^{-2}\right]$; $\varepsilon_k = 1.425 \times 10^{-6}t - 0.495626 \times 10^{-6}(1-\exp(-0.002866t))$; $\varepsilon_y = 0.4792 \times 10^{-6}(t-728)^2$ for $t_c < t < t_y$; and for $t_y < t < t_u$ $\varepsilon_y = 301 \times 10^{-6}\left[1-95.515 \times 10^{-6}(0.0125t^2-18.2t-1227.37)\right]^{-1.3333}$

Table II. Constant rate of charge on gneiss

t	σ	Calculated strain (10^{-6})					ε_m	δ	δ
(s)	(MPa)	ε_h	ε_a	ε_k	ε_y	ε_d		ε_d-ε_m	%
0	0	0	0	0	0	0	0	0	0
80	2	21	299	12	0	332	355	-23	-6.48
160	4	42	474	45	0	561	547	14	2.56
240	6	63	584	94	0	741	720	21	2.92
320	8	84	658	157	0	899	883	16	1.81
400	10	104	711	230	0	1045	1040	5	0.48
480	12	125	749	311	0	1185	1188	- 3	-0.25
560	14	146	778	399	0	1323	1330	- 7	-0.53
640	16	167	800	492	0	1459	1462	- 3	-0.21
720	18	188	818	589	0	1595	1592	3	0.19
800	20	209	832	689	2	1732	1730	2	0.12
880	22	230	843	792	11	1876	1878	- 2	-0.11
960	24	251	852	897	26	2026	2023	3	0.14
1040	26	271	860	1004	47	2182	2189	- 7	-0.32
1120	28	292	867	1112	74	2345	2354	- 9	-0.38
1200	30	313	873	1221	107	2514	2527	-13	-0.52
1280	32	334	878	1331	146	2689	2703	-14	-0.52
1360	34	355	882	1442	191	2870	2870	0	0
1440	36	376	885	1553	243	3057	3048	9	0.29
1520	38	397	889	1664	301	3251	3228	23	0.71
1600	40	418	891	1776	378	3463	3457	6	0.17
1680	42	438	894	1888	516	3736	3741	- 5	-0.13
1760	44	459	896	2001	804	4160	4197	-37	-0.89
1800	45	470	897	2057	1102	4526	4508	18	0.40

4. CONCLUSIONS

In the light of the results presented in the preceding chapter, the rheological model analysed is considered to fulfill the purpose it was conceived for. It is necessary to carry out several types of test on the same rock, in order to get a better knowledge of the model.

Due to great difficulty in obtaining rheologic constants, it seems that the most appropriate tests for determining those constants are creep tests and constant rate of load tests, though the latter are more difficult to perform than the constant rate of deformation tests and do not allows the analysis of behaviour after failure.

Creep tests by themselves do not permit to determine all rheologic constants; together with constant rate load tests, however, easier determination of those quantities may be achieved.

Values of t_u (failure time) are theoretical, since rock disaggregates and no longer behaves as a homogeneous body for comparatively values of ε_y.

REFERENCES

Goodman, R.E. 1980. Introduction to Rock Mechanics. New York, John Wiley & Sons.
Lama, R.D. & V.S. Vutukuri. Mechanical Properties of Rocks. Vol. III. Trans Tech Publications. Clausthal, Germany.

Rock failure near face IV of Chukha Power Tunnel

Coups de terrain près du front de taille no. IV du tunnel d'amenée de Chukha
Bergschlag an der Abbaufront des Chukha Zuleitungstunnels

K.MADHAVAN, Central Water Commission, New Delhi, India
M.P.PARASURAMAN, Central Water Commission, New Delhi, India
P.K.SOOD, Central Water Commission, New Delhi, India

ABSTRACT: In the Chukha tunnel large amounts of water were intersected in clayey and sandy materials which required a diversion of the tunnel. Water control measures and construction procedures are described.

RÉSUMÉ: Une partie d'un tunnel hydroélectrique de 6,5 km à Chukha devait passer très près de la surface dur roc avec des argiles sensibles susjacents. A près un sautage, 5 millions de mètres cubes de matériaux pénétrèlent dans le tunnel, ce qui obligea à contourner cette zone et à réaligner le tunnel.

ZUSAMMENFASSUNG: Im Chukha Tunnel wurde Wasser vermischt mit sandigem und tonigem Material angeschnitten. Der Einbruch von 5 x 10^6 m^3 Wasser in 24 Stunden machte eine neue Tunnelstrecke notwendig. Bauprobleme werden beschrieben.

1 GENERAL DESCRIPTION OF THE PROJECT

Chukha Hydro-electric project is a 336 MW run-of-the river hydro-electric scheme located in the Thimpu District of Bhutan - a north-easternly neighbour of India. The project features consist of a 40m high diversion dam constructed across Wangchu river for diverting a maximum of 120 cubic metre per second discharge into two semi-circular vertical intakes situated on the left bank of the tiver. Two de-silting chambers trap the silt particles down to 0.2 mm size and flush the same back to Wangchu river downstream of the dam via a free flowing steel-lined silt flushing tunnel consuming 24 cumecs of water in the process. Thus 96 cumecs of silt-free water is diverted into a 6.5 km long 4.9 m x 4.9 m D-shaped power tunnel which terminates into a 12.2m diameter surge shaft. An underground butterfly valve chamber is provided which houses two butterfly valves - one for each pressure shaft. Two inclined pressure shafts have been provided immediately downstream of the butterfly valve chamber. Each of the two pressure shafts bifurcates into two separate horizontal pressure shafts of 2.25m internal diameter to feed 4 Pelton Wheel turbines - each of 84 MW to generate 336 MW of peaking power under a head of 468 metres. The tail water is led back to Wangchu river via a 900 m long D-shaped free flowing tunnel. The schematic layout is presented in Fig.1.

2 GEOLOGY

Being primarily an underground project the geological and geophysical tests assumed great importance for evolving safe and economical designs for the underground works. The preliminery geological investigations of the project were carried out in April 63. As the works progressed, this information was supplemented with detailed geotechnical explorations providing useful data to the designers.

While most of the tunnel passes through competent granite gneisses a portion of the power tunnel is overlaid by a known weak zone - the Mebari Slide Zone made up of fluvio-glacial/Fluviatile terraces in the overburden. Seven drill holes were logged along the water conductor system out of which four are located along the tunnel alignment as shown in Fig.2. The bore logs are presented in Fig.3. These bore logs indicated the presence of quartz-schist-biotite-granatic gneiss bed rocks through which the tunnel was to be built. Water percolation tests in these holes indicated rapid loss of water pointing to the presence of open joints in the rock mass. Bore hole numbers 2 and 3 completely passed through loose talus material with large boulders of granite gneisses while in boreholes number 1 bed rock was met with at a depth of nearly 80m below the ground level. Geo-physical surveys carried out later established that the poor rock conditions of the Mebari Slide Zone did not extend down to the tunnel grade. A 24-Channel refraction seismograph was used for estimating the depth of the bed rock. It indicated the bedrock topography to be highly undulating and bedrock itself to be highly jointed. Two deep burried channels were indicated cutting across the tunnel alignment in the Mebari Slide Zone.

3 CHUKHA POWER TUNNEL

The 6.5 km long power tunnel was on the critical path for completion according to the construction schedule and hence an intermediate adit was provided to yield two additional faces for expediting its completion. Work was started simultaneously from Face I and Face IV at the upstream and downstream ends respectively of the tunnel in mid-79. The adit at Face IV also served as construction adit for the butterfly valve chamber and the surge shaft. The work on the power tunnel from this adit at Face IV was thus started in 1980 only. The principaal support elements consisted of shotcrete and rock bolts with wire mesh except in shear zone reaches where steel supports were provided.

3.1 Tunnel cross-Section

A 4.9m x 4.9m D-shaped cross-section has been provided as shown in Fig.4 for the power tunnel to suit the available construction machinery.

4. THE ACCIDENT

During the course of its excavation from Face IV which, incidently was located in the Mebari Slide Zone with minimum rock cover of 40-50m incompetent, saturated and blocky rock was encountered from chainage 565 onwards as shown in Fig.5. As the tunnel advanced further, quantity of seepage water flowing into the tunnel from its crown and right side wall increased and the tunnelling conditions worsened rapidly. The project execution engineers in consultation with resident geologist provided steel supports 150 x 150 I Sections at a close spacing. The intervening space was heavily lagged and backfilled with concrete. Advance probes were taken to determine the

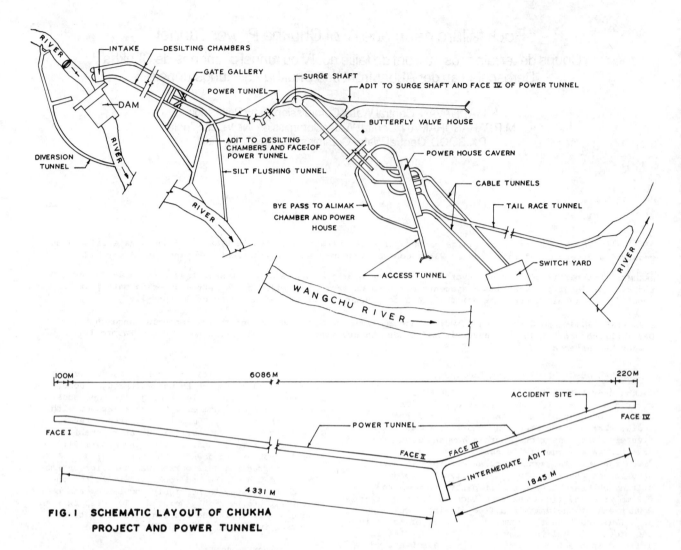

FIG.1 SCHEMATIC LAYOUT OF CHUKHA PROJECT AND POWER TUNNEL

rock conditions around the crown of the tunnel. These probes indicated the presence of highly saturated clayey material present above the tunnel crown and cutting across the right side wall. In January 81, at a distance of 636m measured from Face IV as soon as the blast was taken water mixed with sandy and clayey material gushed out with a tremendous force. Steel supports erected near the heading were twisted, collapsed and thrown out from Face IV. A thick, concentrated and continuous stream of water followed the debris and dropped nearly 500m down in the valley below Face IV. The flow of water which was estimated to be of the order of 1.5 cumecs in the beginning reduced to about 0.025 cumecs in approximately 24 hours. An estimated 5 million cubic metre of perched water escaped through the tunnel within this period. An inspection of the collapsed heading revealed chimney formation in the roof and right side wall of the tunnel.

5 THE PROBABLE CAUSE

A visual inspection of the site of accident and the loose much thrown out and further probes taken around the collapsed portion of the tunnel evidenced the presence of a shear zone in the crown portion and cutting across the right side wall of the tunnel. The shear zone measuring nearly 80 cm thick and mainly composed of highly saturated clayey material was acting as a barrier against the perched water stored in the mountains as depicted in Fig.5. As soon as the shear zone was ruptured, the entire perched water got an opportunity to be released and escaped through the tunnel heading bringing along with it tons of loose much which resulted in a chimney formation over the tunnel crown.

6 REMEDIAL MEASURES

After the discharge from the tunnel had reduced considerably, attempts were made to proceed further. All the twisted steel supports and clayey muck which had filled the tunnel near the heading were carefully removed - first from the sides and then from the middle after erection of temporary supports. Advance probes indicated extremely difficult tunnelling conditions. Yet the work proceeded; though at a snail's pace. The stand-up time of the excavated heading reduced to only about a few minutes. Nearly six months of valuable construction time was lost in tackling this reach In the meantime, probes taken on the right side indicated fair tunnelling conditions and in view of the drainage provided by the excavated heading a decision to realign the tunnel as shown in Fig.6 was taken in August 81. Controlled blasting techniques coupled with pre-drainage and pre-grouting of the rock mass were adopted while tunnelling through the re-aligned route and the tunnel was finally brought back to its original alignment in Feb. 82.

Since a lot of seepage water was still flowing into the collapsed and abandoned portion of the tunnel, it was decided to completely backfill the abandoned portion with lean concrete and masonry. However, in order to minimise the danger of a build up of water pressure which might act on the re-aligned tunnel, a perforated drain-pipe with a graded filter around it was embedded in the backfill and the perocating water discharged into the power tunnel.

Being a hydraulic tunnel, the entire power tunnel is lined with 225 mm thick plain cement concrete lining of M:20 grade concrete. However, as a pre-

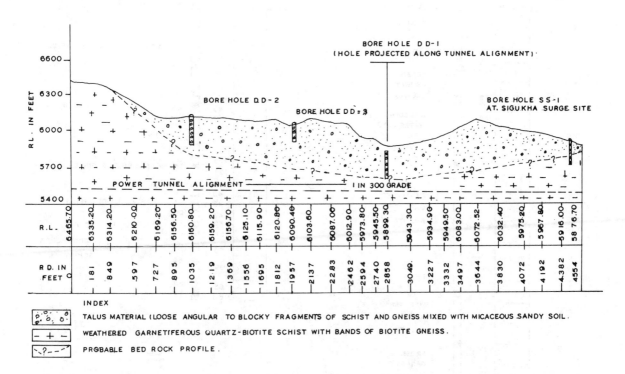

FIG.- 2 GEOLOGICAL SECTION ALONG CHUKHA POWER TUNNEL

INDEX

TALUS MATERIAL (LOOSE ANGULAR TO BLOCKY FRAGMENTS OF SCHIST AND GNEISS MIXED WITH MICACEOUS SANDY SOIL.

WEATHERED GARNETIFEROUS QUARTZ-BIOTITE SCHIST WITH BANDS OF BIOTITE GNEISS.

PROBABLE BED ROCK PROFILE.

cautionery measure in the low cover reaches, in the vicinity of the re-aligned route and in the re-aligned route itself, the tunnel has been lined with reinforced cement concrete lining of 250 mm thickness in a length of approximately 500 metres.

7 THE POST-CONSTRUCTION PROBLEMS

The Chukha power tunnel was completed in January 1986. During its final cleaning, a crack was observed in the concrete lined invert of the tunnel. The location of the crack was close to the collapsed portion mentioned above. The crack had a maximum width of 3mm and was located more or less in the centre of the invert and continued for a length of 30 metres. Water was continuously oozing from this crack.

In order to assess the quantum of pressure built up behind the concrete lining, the concrete in the invert was chipped to the rock level. A spring of water rising in height by about a metre was observed. The spring disappeared with the gradual dissipation of the accumulated water pressure. While contact and consolidation grouting had been done at the tunnel crown and on both the sides, no grouting had been done for the invert portion. This had rendered the rock mass fairly water tight on both its sides and the crown whereas the rock mass remained porous and shattered below the invert floor of the tunnel. A shear seam controlled the built up of ground water as in the case of the accident reported earlier in the paper. Water thus found its way down to the tunnel invert causing an uplift of the tunnel floor resulting in the development of cracks.

8 REMEDIAL MEASURES

The remedial measures required the treatment of the invert floor of the tunnel by grouting under flowing water conditions, provision of drainages and sealing of the cracks to restore the concrete lining to its original conditions. Cement grouting under flowing water conditions would not have yielded the desired

results. Epoxy grouting by either of the two suitable products AQUAGEL-9 or POLY COAT-EP was recommended. For the water pressure relief, 75 mm diameter pressure relief holes 3m deep and spaced at 6m centres along the centre-line of the invert were provided. These holes were blinded with graded filter for the lower half and with porous concrete for the upper half of their depth and finished flush with the invert concrete. The cracks were then sealed with AQUA GEL-9 epoxy mortar and the tunnel was finally charged with water in April 1986.

9. CONCLUSIONS

It is of interest to note that a similar accident had occured at the Blue Mountain Tunnel near Carlisle where a wide seam filled with sand and water was encountered. The recommendations of the Underground Research Council set up by the American Society of Civil Engineers and National Science Academy of the U.S.A. are of particular interest in this connection. The suggestions contained therein may be adhered to as far as the investigations and instrumentation of underground structures is concerned particularly where new construction techniques are contemplated. Detailed Geological and geophysical investigations - more so in the low cover reaches could perhaps reveal the hidden geological surprises and adequate measures taken before hand to deal with such problems.

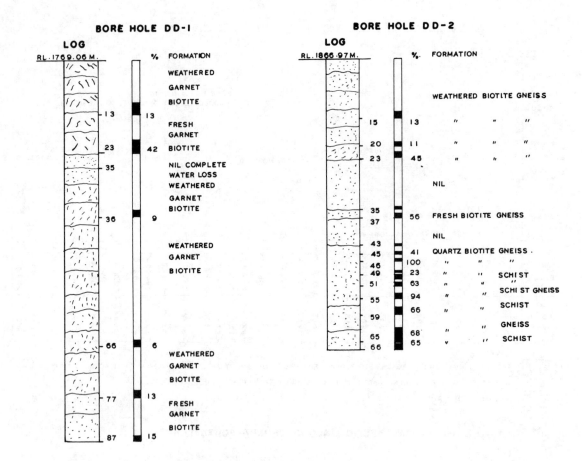

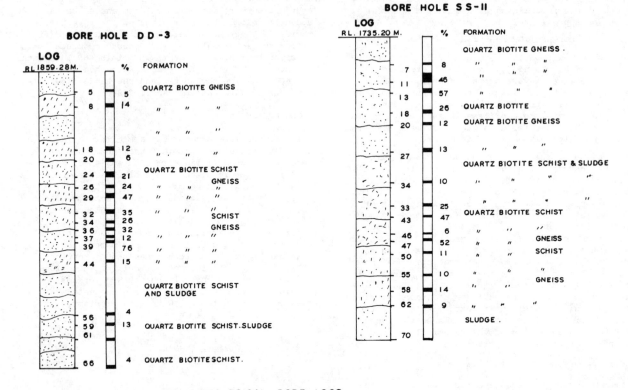

FIG. 3 GEOLOGICAL BORE LOGS.

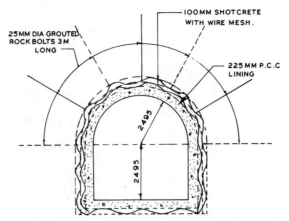

FIG.4 TYPICAL CROSS-SECTION OF POWER TUNNEL

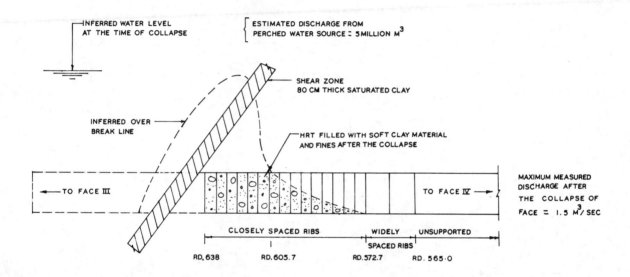

FIG. 5   HEAD RACE TUNNEL COLLAPSE AT FACE NO. IV CHUKHA H.E. PROJECT (BHUTAN)

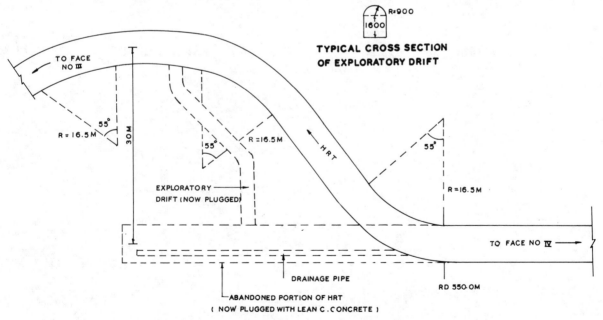

FIG · 6   RE-ALIGNMENT OF HRT AT FACE IV CHUKHA PROJECT (BHUTAN)

Stability of hanging walls at the Viscaria Copper Mine

Stabilité des toits à la mine Viscaria Copper
Standfestigkeit des Hangenden in der Viscaria Kupfer Grube

ILPO MÄKINEN, M.Sc., Rock Mechanics Engineer, LKAB, Department of Research and Development, Sweden
TAISTE PAGANUS, Chief Rock Mechanics Engineer, LKAB, Department of Research and Development, Sweden

ABSTRACT: The Viscaria Mine employs large scale mining methods with open stopes, namely raise mining and sublevel open stoping. In the first 20 stopes mined, the behaviour of the hanging wall has varied from stable to failing. The back analysis showed that the flattening dip of the ore is the dominant parameter when considering hanging wall stability where a graphite schist exists in the vicinity of the hanging wall. Our experiences tell us that the critical dip lies between 60°-70°. Stopes with dips 60° or less, are considered hazardous, and stope height must be decreased. Stopes dipping 70° or steeper are sufficiently stable to be exploited without extra reinforcements. Intermediate stopes are to be excavated either by employing cable bolts, or by special designs, i.e. changes of mining sequences, extra careful blasting, etc.

RESUME: L'exploitation de la mine souterraine de Viscaria se fait par grande chambres ouvertes. Une cheminee est tracé dans chaque chambre, suivi par l'abattage de minerai en tranche montantes. Au cours de l'abattage des 20 premiers chambres le comportement des toits a varié entre la stabilité et l'affaissement. Une analyse a montré que l'angle d'inclinaison du minerai est le paramètre le plus important pour la stabilité des toits. D'après nos expériences, la présence d'un schiste graphitique à proximité d'un toit entraîne un angle d'inclinaison critique entre 60° et 70°. Une chambre avec une inclinaison du minerai de 60° ou moins est considéré comme risqué et entraine le besoin de reduire l'hauteur de la chambre. Des gradins, avec une inclinaison du mineral a 70° ou plus sont assez stables pour être exploitees sans renforcements supplémentaires. Les chambres intermédiaires seront creusés, après stabilisation soit par des boulons, soit par l'emploi d'autres méthodes, telles que changement des séquences d'exploitation, choix très soigneux du mode de tir etc.

ZUSAMMENFASSUNG: In der Viscaria Grube wird Etagenbau mit grossem Abbauraum in aufwärtsgerichteten Bruch im Anschlus an einen ansteigenden Ort betrieben. In zwanzig gebrochenen Etagebauen variierte die Standfestigkeit des Hangenden zwischen stabil und unstabil. Eine Untersuchung ergab, dass der Stand des Hangenden nicht nur durch die Neigung des Erzkörpers sondern vor allem durch eine Grafitschiefereinlagerung im Hangenden beeinflusst wird. Bei Auftreten des Grafitschiefers lag der kritische Neigungswinkel für Instabilität zwischen 60 und 70 Graden. Ein Erzgang von weniger als 60 Grade Neigung wird als Abbauschwierig eingestuft und die Abbauhöhe wird vermindert. Bei einer Neigung von über 70 Graden wird der Erzgang ohne Verstärkung des Hangenden abgebaut. Erzgänge die zwischen 60 und 70 Graden Neigung liegen müssen entweder vor dem Abbau durch Kabelverstärkung des Hangenden stabilisiert werden oder auch müssen andere Abbaumethoden angewendet werden, wie zum Beispiel, Änderung der Abbaufolge, vorsichtiges Sprengen oder anderes.

1. INTRODUCTION

The purpose of this paper is to discuss problems which have arisen in connection with large scale mining methods used at the Viscaria Mine, and the efforts made to solve these problems. It has not been possible to report the outcome of most of the recommended measures included in this paper, since they had not been finished at the time this paper was delivered.

Because of the low copper contents (2.5%), the future of the Viscaria Mine depends greatly on our ability to compensate for the high mining costs owing to the expensive labour, the Arctic environment, and the high safety standards, by an effective mining operation. Viscaria has met this challenge with large scale mining methods using the latest mechanization.

Sublevel open stoping and raise mining methods, developed by LKAB's Viscaria Group, were chosen, using stopes 55m high and 55m wide. A detailed description of the mining methods has been prepared by Aaro in 1986.

Following the selection of the mining methods, a new problem arose in the form of graphite schists in the hanging wall. Due to low cohesion of the graphite schists, the hanging wall between a stope and a graphite schist can be considered as an unconstrained beam. This report discusses recommendations made and the measures implemented to tackle the problems posed by the graphite schists and large scale mining.

2. VISCARIA COPPER MINE

The story of Viscaria started with a flower. Viscaria Copper Mine derives its name from a particular flower; alpine campion, Viscaria Alpina. This flower is one of a few that thrives in copper-rich soil.

This fact is well-known to geologists. One day in the early 1970's the LKAB geologist Paul Forsell noticed a colony of alpine campion whilst walking near Kiruna. His observation resulted in test drilling, and it shortly became clear that a valuable copper deposit had been discovered.

The mine was opened by the Swedish State Owned Company LKAB(Luossavaara-Kiirunavaara AB) and is situated 5 km to the west of the city of Kiruna in Northern Sweden.

Some important figures concerning the Viscaria Mine are presented in table 2.1.

Table 2.1 Important figures of Viscaria

Annual ore production:	1.3 Mt
Percentage of copper:	2.5 %
Total number of employees including all contractors:	250
Productivity of one employee of the mining department (incl. contractors and officials):	40.1t/sh
Investments (1982):	US$ 55 M
Proven ore reserves:	14 Mt

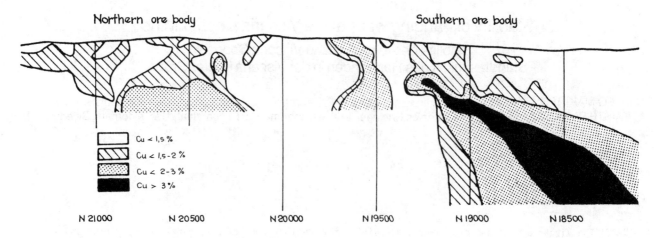

Figure 3.1 Long section of Viscaria ore bodies

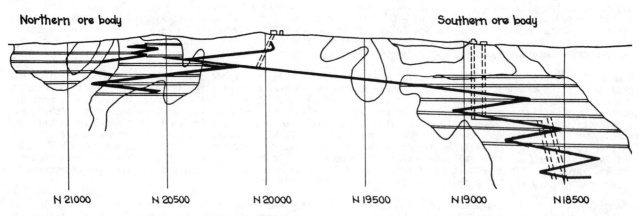

Figure 3.2 The general mining layout of Viscaria Mine

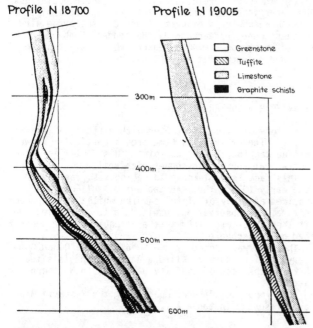

Figure 3.3 Two cross-sections through the southern deposit of Viscaria

3. THE VISCARIA DEPOSIT

Only a brief summary of the Viscaria Mine is presented here. A detailed description has been provided by Kisiel and Godin in 1985.

Mineralization of the Viscaria deposit occurs over a distance of 3.5km. A 500m stretch of low-grade mineral-ization near the center seperates the high-grade north-ern and southern areas. From surface to a depth of 200m, the southern deposit dips 70°- 80° to the east. Below 200m , the dip declines to 45 . The dip in the northern area is 80°-90° to the east. In the Viscaria area, the sedimentary sequence has been classified to form four major horizons, labeled A, B, C and D. Present economic ore reserves are all located in horizon A. In horizon B 8Mt of currently uneconomic mineralization has been indicated.

The A zone sediments have a mean width of 30m and consist of inter-bedded layers of tuffite, albite fels, copper-rich graphite, and limestone being the main components. The limestone is host to magnetite and copper in grades varying from 3% to 6%. The average width of the deposit is 10m and ranges from 5 to 30m (Fig. 3.1, 3.2 and 3.3).

Chalcopyrite is the only economic mineral in the Viscaria ores. Gold grades at less than 0.05ppm, and silver varies from 2 to 5ppm. Zinc exists in the form of sphalerite, the content varying considerably 0 to 5%.

4. MINING METHODS

For profitability, the Viscaria Mine should be non-labour intensive with large scale mining methods. Thus, sublevel open stoping (Fig. 4.1) and raise mining (Fig. 4.2) using the latest mechanization were selected. Since mining problems have arisen, new variation of these methods have been proven

The concept of the new mining method, namely raise mining, was introduced by Marklund, but the machinery and equipment to meet the requirements did not exist. Raise mining means that relatively narrow ore is mined in stopes that are 55m long and 50-65m high. For a discussion on raise mining, see Marklund (1982).

The other principal mining method is sublevel open

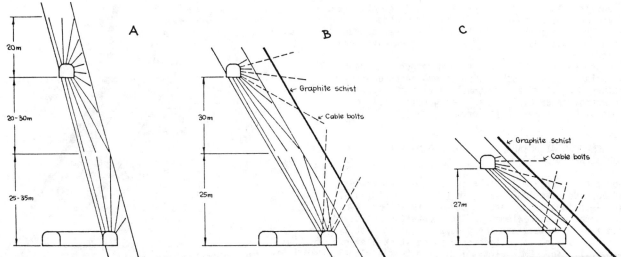

Figure 4.1 A) Sublevel III method for favourable ground conditions; B) Sublevel II method is used for the typical mining conditions at the Viscaria Mine; C) Sublevel I method is used in flat and hazardous stopes which dip less than 60 .

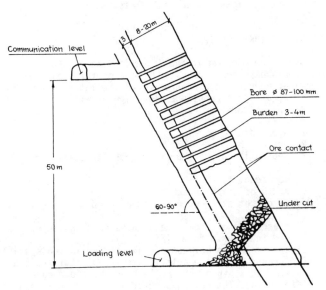

Figure 4.2 Principle sketch of raise-mining

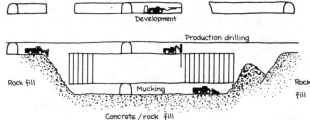

Figure 4.3 Sublevel stoping with rock/concrete fill

5. ROCK MECHANICS AND MINE PLANNING

The latest long term mining plan was completed in July 1986. A risk factor ranging from 1 (unfavourable ground conditions) to 5 (favourable ground conditions) was given to every mining stope. The purpose of the risk factor is to find an alternative to the vague adjectives formerly used to describe the ground condition. Mine managers prefer information in numerical form. The risk factor was estimated (hopefully further work will enable the calculation of these factors) using the following parameters, in order of importance:

1. Presence of graphite schist in the hanging wall
2. Dip of the stope
3. Space between the ore and the graphite schists
4. Rock quality designation
5. Mining methods employed
6. Water conditions

6. ROCK MECHANICS AND MINING OPERATION

The successful practical application of the rock mechanics know-how depends greatly on the communication between the rock mechanics department and mining production department . If these two do not co-operate, the benefits arising from the rock mechanics departments are slight, e.g. what is the sense of cable bolting, if the hanging wall is simultaneously undercut by a mining malpractice. The quality of long hole drilling, charging and blasting are the subjects of several debates in many mines, and unfortunately the Viscaria Mine is not an exception.

The control of the hole deviation and blasting is the responsibility of production drift foremen. Rock mechanics engineers perform random controls. The blasting operations are monitored by vibration measurements.

Tessen and Wennberg (1981) hit the nail on the head

stoping. With varying ground conditions, three variations of sublevel open stoping are used. The sublevel III method is used in favourable ground conditions, i.e. where no graphite schist occurs and the dip is greater than 70° (Fig. 4.1A). Sublevel II is the normal method for stopes, with a dip of 60°- 70° and where the graphite schist is located in the hanging wall (Fig. 4.1B). For stopes, which are classified as hazardous the sublevel I method is recommended.

For a stope with a poor copper content, a longitudinal mucking layout is recommended instead of the usual transverse mucking layout.

Sublevel stoping with the rock/concrete fill method was designed for the higher-copper areas with a Cu-content of more than 4.5%. The principal difference between sublevel I and this method is that the stopes are between during ore extraction. The type of filling depends on the subsequent mining sequence (Fig. 4.3). The preliminary mining plan, which advised leaving horizontal rock pillars between every mining stage, has been altered. Thus, the copper-rich limestone (4%) in the middle of a crown pillar is to be exploited by a cut and fill method. Concrete/rock filled pillars are reinforced by cable bolts.

by saying:" An acceptable result can only be obtained, if you can ensure that a predetermined procedure is followed. Unless this control continues, various un- authorized methods will be gradually tried until you again have a problem."

To prevent possible unauthorized methods at the Viscaria Mine, the rock mechanics engineers present lectures on practical and theoretical rock mechanics to the operation crew. This has seen great success amoungst miners and shift foremen.

7. ROCK MECHANICAL ANALYSES AND DEFORMATION MEASUREMENTS

When the Viscaria Mine first opened, two FEM analyses had been done by JOBFEM-program (Stille et al 1982). The dimensions of the second model were 1000m*1500m, and contained 3438 elements. The excavation was divided into twelve stages, and the simulation of the cable bolting was done for stages 6 -8 using stick elements (Fig. 7.1). The FEM analysis indicated that a global stability would not be jeopardized if the crown pillars were left intact. However, exploitation of the crown pillars would remove the global stability.

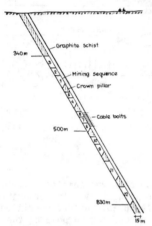

Figure 7.1 The basic model of the second FEM analysis

Table 7.1 Rock mechanical values of the A zone

Rock type	RMS class	Compress. strenght Mpa	Tensile strength Mpa	E-mod. GPa	Ø	Cm Mpa
Greenstone	II	225	19	50	35	0.4
Tuff,massive	II	210	17	42	35	0.4
Tuff,banded	III	180	15	20	30	0.3
Graphite schist in ore	III	110	12	5.0	22	0.8
Graphite schist in hw	III	100	10	1.3	18	0.1

At the Viscaria mine, deformations are measured with rod extensometers. A comparison between measured displacements and calculated deformation values does not clarify the usefulness of the finite element analysis in heterogeneous, low stressed rock (Fig. 7.2). However, the following comments are given:

- in heterogenous rock environments, 2-D FEM analyses are inadequate for predicting deformations, as the deformations have four degrees of freedom
- an extensometer hole should be core drilled in order to document local geology

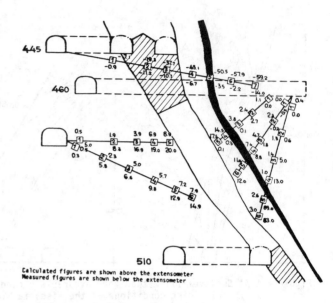

Calculated figures are shown above the extensometer
Measured figures are shown below the extensometer

Figure 7.2 Comparison between the displacements measured and calculated

8. MONITORED DESIGN

Since the opening of the Viscaria Mine in 1983, 20 stopes have been excavated in the southern part of the mine. Before the commencement of the latest long term plan, all 20 stopes were investigated with regards to six parameters:

1. Stope dip
2. Ore width
3. Seperation between ore and graphite schist in hanging wall
4. Rock quality
5. Width of graphite schist
6. Height of stope when caving occured

To preclude a verbal inexactness each parameter was listed in diagram form (Fig. 8.1).

The correlation between dip and caving is indisputable The ground condition deteriorate the flatter the ore body. With a dip of 55 or less, every stope caved (Fig. 8.1A). The caving occured after mining in several stopes, with dips between 60° and 70°. The flatter the stope, the earlier the caving occured (Fig. 8.1F). The other parameters stated have not yet revealed clear correlations.

9. BEHAVIOUR OF CAVING AT THE VISCARIA MINE

Two alternative explanations to the hanging wall failure have been presented . The first emphasizes the importance of horizontal stresses; which, according to this explanation release strains from a tuff beam.

The other explanation is based on toppling and inadequate cohesion of graphite schists. After unloading a stope, the hanging wall begins to topple to the empty opening (Fig. 9.1A). A tuff beam between the empty stope and the caved graphite schist is left without vertical adhesion, and therefore it begins to fall in the form of boulders (Fig. 9.1C). The horizontal stresses increase in the crown pillar as the failure advances upwards. There must exist at some point an equilibrium where the stresses have increased sufficiently to press the graphite schist tight and stop the caving (Fig. 9.1D).

In spite of the fact that there are more detailed factors to consider, such as the rotation of crown pillars owing to stress directions, the latter theory has been accepted.

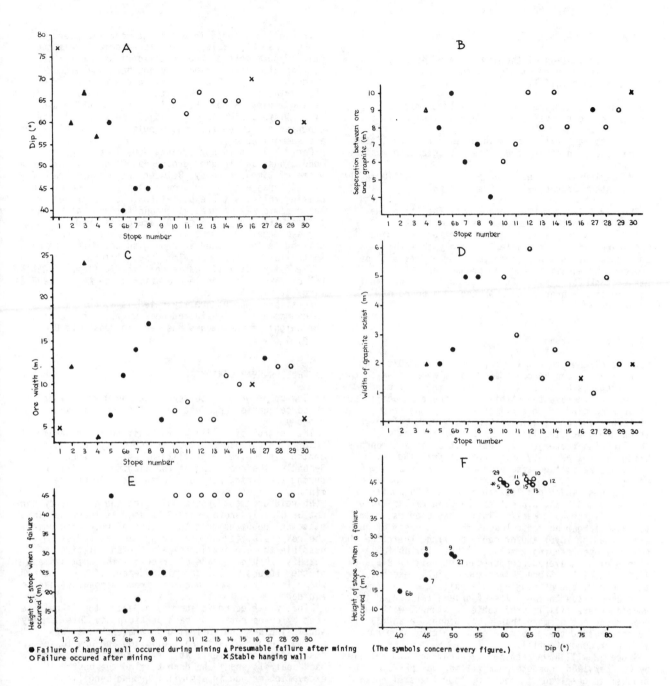

Figure 8.1 A) Stope dip; B) Seperation between ore and graphite schist in hanging wall; C) Ore width; D) Width of graphite schist; E) Height of stope when a failure occured; F) Comparison between dip and height of stope when a failure occured.

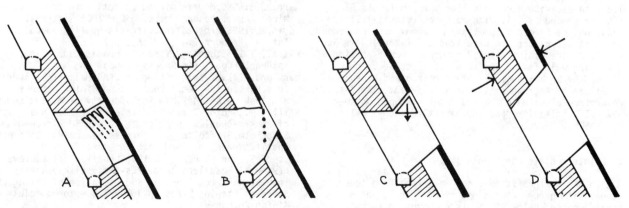

Figure 9.1 A) Toppling of tuff beam; B) Caving of graphite schist; C) Falling of boulders; D) Horizontal stresses press a crown pillar.

10. METHODS SUGGESTED TO PREVENT CAVING

After determination of the mechanism of hanging wall failure, suitable solutions had to be found. An ideal solution would fulfill the demands listed below:

1. Economic criterium: The costs allowed must be related to the value of a stope and the risk of a failure.
2. Operational criterium: Methods must require little manpower, in other words, the mechanization must be high.
3. Valuableness: It should be possible to estimate the advantage of the methods unambiguously.

The first two demands are obvious. The third one is important when the production department hits difficulties and wants to give up the predetermined procedure. If one is unable to clearly show the advantages of ones procedure ,ones arguments will be overcome by the importance of production. (It is an easier task for a mine manager to explain a ground failure than a production shortage to a board of directors.)

10.1 Unstable stopes

Because of the present low copper price (£930/t), no reinforcement can be considered for stopes of which the Cu-content is below 2.0%. Therefore, the only conceivable way to improve the stability of these stopes is to leave some kind of internal support. By the term "internal support", we mean temporary rock bridges, which are good stabilizers of the hanging wall.
The limiting factor for the utilization of a temporary bridge is the accuracy of long hole drilling. Plans for stope 22 include attempting to leave temporary pillars.
In stopes, which dip less than 60° and Cu-content between 2 and 3.5%, a predicted hanging wall failure with ore losses and waste dilution has been estimated to cause larger economic losses than the costs of an extra loading level and/or cable bolting. Therefore the sublevel open stoping I method is used in unstable stopes with an average copper-content between 2.5 and 4.0% (Fig. 4.1). Stope 25 was successfully excavated by this method, and no hanging wall failure was observed.
In comparison to the price of ground failure, using rock/concrete fill is profitable in stopes, with a copper content of more than 4.0%. Stopes 32 and 33 will be mined using the method we call sublevel stoping I with rock/concrete fill.
Stope 32 will have been partly excavated and refilled by October 1986. Conrete/rock filling was performed by trucks; to one load of concrete, four loads of rock fill were poured into the stope. The greatest problems to date have been delays, caused primarily by a lack of experience in refilling, and secondly by an incorrect choice of mining sequence. Upwards mining was planned, i.e., to excavate a crown pillar between levels 580 - 596 first, and fill it with rock/concrete fill. This phase was performed without any problems. In the second phase it was planned to mine stopes between levels 552 - 580. Before mining started attempted cable bolting of a hanging wall failed due to drill holes becoming blocked by broken rock. For this reason and because of a failure of a crown pillar of an adjacent stope, the determined plans were altered to mine stopes between 527 - 552 initially, and subsequently stopes between 552 - 580.

10.2 Intermediate and stable stopes

Due to the low profit that can be gained from those stopes with ground conditions rated 3 or 4, and a copper content below 2%, it is recommended to increase the control of the drilling and blasting work, but not to use any extra reinforcement.

In similar stopes, in which the copper-content is higher than 2%, cable bolting is recommended as the main support method. Previous experiences of cable bolting at Viscaria Mine have not been unambiguously described. Stopes 8 and 15 were cable bolted and mined out during 1985. Stope 8, with a dip of only 45°, caved during mining. Stope 15, having a dip of 65°, caved after mining, similarly the adjoining stope 14. The experiences gained from cable bolting at the Viscaria are summarized below.
Cable bolting is not advantageous in poor ground conditions, and in good ground conditions, it may be a waste of time and money. Between these two there exists a range of other ground conditions, where cable bolting fulfills the expectations and has proven to be an economic way to prevent ground failure. At Viscaria, stopes with a dip of 60°- 70° are considered suitable for cable reinforcement. Testing of parallel cable bolting is recommended according to the Mount Isa Mine, Australia, and the Pyhäsalmi Mine, Finland.
The fully automatic cable bolting rig (Tamrock Cable Bolter H690) has been in operation since August 1986 at Viscaria Mine.
In stopes, ground quality rated as 5, no reinforcement is considered necessary. To the contrary, the height of the stopes has been increased to 85m (Fig, 4.1).

11. FOLLOW-UP PROGRAMS

Follow-up programs recommended can be diveded into three categories: production, reinforcement, and economy.
The control of drilling and blasting is of paramount importance to ensure a good ground condition. This control is the responsibility of shift foremen. Secondly, the recording of the hanging wall failure during the mining period is also the duty of the miners.
No reinforcement system, of which the efficiency can not be clearly proven should be used at the mine. This statement requires that the method of recording output, the responsibility of monitoring, and as far as possible the explication of all the speculative outputs, should be made clear before the commencement of the project. Thus, the misjudgement of a new method, which is very common concerning rock mechanics, can be avoided.
Finally, an economic summary of each stope should be done by co-operation of each participant. This summary should include the following parameters: cost of mining, incomes, waste rock dilution and ore losses, costs of reinforcement. At the base of this summary, a back analysis should be done to find the best excavation method in a similar ground conditions.

12.REFERENCES

Aaro, L-E.1986.Modern mining operation in the Viscaria Mine: Annual general meeting CIM 1986,Montreal.
Godin, L.1985.Prospecting.Viscaria quality education, Internal report (in Swedish), p17.
Kisiel, T.1985.Geology.Viscaria quality education, Internal report (in Swedish), p36.
Marklund, I.1982.Vein mining at LKAB,Malmberget,Sweden. In W.Hustrulids (ed.) Underground Mining Methods. Handbook, p.441-442,New York:Society of Mining Eng.
Stille, H.,Groth, T.and Frediksson, A.1982.FEM-analysis of rock mechanical problems with JOBFEM (In Swedish), Stockholm:Stiftelsen Bergteknisk Forskning, BeFo och Institutionen för jord- och bergmekanik, KTH.
Tessen, S. and Wennberg, S.1981.Longitudinal sublevel caving at Fosdalens Bergverks-Aktieselskab,Norway. In D.Stewart (ed.) Design and Operation of Caving and Sublevel Stoping Mines, p.365-371,New York:Society of Mining Engineers.

Geomechanical criteria of prediction and technology of prevention of rock bursts during excavation in tectonically stressed rocks

Critères géomécaniques des prévisions et mesures technologiques de la prévention des coups de pression pendant l'exécution des excavations dans les roches de contrainte tectonique
Geomechanische Kriterien der Bergschlagsvorhersage und technologische Massnahmen zur Bergschlagsverhinderung bei den Vortriebsarbeiten in tektonisch beanspruchten Gesteinen

G.A.MARKOV, All-Union Extramural Polytechnic Institute, Moscow, USSR
A.N.VOROBIEV, All-Union Extramural Polytechnic Institute, Moscow, USSR
V.S.URALOV, All-Union Extramural Polytechnic Institute, Moscow, USSR

ABSTRACT: On the basis of theoretical and experimental studies made by the authors the paper presents: 1) mathematical model of distribution of gravitational tectonic stresses in an untouched rock mass (before excavation); 2) criteria of rock failure after excavation in a tectonically stressed rock mass obtained from the results of field investigations. Behaviour of the rock around excavations is considered not only in the elastic state but in the past peak strength one as well, where the stresses on the outline are 2-4 times higher than the ultimate strength of the rock.

RESUME: Compte tenu des résultats des études théoriques et des expériments réalisés par les auteurs, dans le rapport sont envisagés: 1) modèle mathématique de la répartition des contraintes gravitationnelles et tectoniques du massif rocheux sain (avant l'exécution des excavations); 2) critères de la manifestation des ruptures des roches après l'exécution des excavations dans le massif de contrainte tectonique, obtenus sur la base des études in situ. En même temps, on examine l'état et le comportement des roches autour des excavations non seulement dans le domaine élastique, mais, aussi, dans l'état postlimite quand les contraintes agissant sur le contour dépassent la limite de la résistance des roches 2 à 4 fois.

ZUSAMMENFASSUNG: Auf der Grundlage eines Komplexes theoretischer und experimenteller Arbeiten, die von den Autoren durchgeführt wurden, werden im Vortrag behandelt: 1) mathematisches Modell der Verteilung von schwerkraftbedingten und tektonischen Spannungen des anstehenden (vor dem Ausbruch) Felsgesteins; 2) aufgrund der Großversuche gewonnene Kriterien des Gesteinsbruchs mach dem Ausbruch des Hohlraums im tektonisch beanspruchten Felsmassiv; dabei wird das Verhalten des Gesteins um den Hohlraum herum nicht nur im elastischen Bereich, sondern auch in dem Bereich, wo die an der Ausbruchkante wirksame Beanspruchungen die Bruchfestigkeit des Gesteins zwei- bis vierfach überschveiten.

1 NATURAL STRESSED STATE OF ROCKS

In 2-3 recent decades it has been determined that high stress state of rocks occurs not only at large depths, but at comparatively small ones as well. Essential lack of correspondence between calculated geomechanical prediction and actually observed rock pressure manifestation has been indicated under such conditions. Dynamic rock pressure manifestation, for example, is observed at depths several times less than those predicted on the basis of gravitation. According to statistics such discrepancy is found at 60% of projects in igneous rocks and 20% of those in sedimentary metamorphic rock complexes. The cause, or additional stimulus of high natural (tectonic) stress state in rocks is geotectonic processes in the Earth's crust. They are of various extent, duration and intensity. With active tectonic forces horizontal (or inclined) components of stress tensors may be much higher than vertical ones calculated from gravitation. In this case the principal stresses σ_1, σ_2, σ_3 are determined as follows:

$$\sigma_1 = \rho gH + \psi_1 T$$
$$\sigma_2 = \xi \rho gH + \psi_2 T \qquad (1)$$
$$\sigma_3 = \xi \rho gH + T$$

In the formulas (1) the components of the tectonic stress field are characterized by the vector modulus T, azimuth direction of the vector α_T, slope β_T and two coefficients of tectonic stress state anisotropy: vertical ψ_1 and horizontal ψ_2 (taking into account the boundary conditions on the surface, $\psi_1 \approx 0$; $\beta_T \approx 0$) should be considered the most general values of the indexes ψ_1 and β_T. The components of the gravitation stress field are characterized by the moduluses ρgH (vertical) and $\xi \rho gH$ (horizontal). ξ is lateral outward pressure coefficient, ρ - rock density; g - gravity acceleration.

The analysis of the formulas (1) shows essential difference in natural stress state between rocks subject to tectonic forces ($T \neq 0$) (2) and those unaffected by them ($T = 0$) (2a)

$$/\sigma_1/ < /\sigma_2/ < /\sigma_3/ \qquad (2)$$

$$/\sigma_1/ > /\sigma_2/ > /\sigma_3/ \qquad (2a)$$

The laws of natural tectonic stress variation depending on various natural factors (the Earth's crust movement, tectonic block structure, hilliness of the relief, variability of rock properties) are discussed in a special paper (Markov 1982).

Tectonic stress field components are determined from experimental data. A model of uniform pieces including volumes V, W is applied to describe the initial field of tectonically stressed rock masses. The results of stress state determination in the volume of V-type (the volume of V-type is commensurable with gauge length, or with characteristic dimension of a solid rock sample) contain two parts of information. The first depends on the stress values at the point of observation and is, therefore, the function of the space coordinates; the second is related to incidental vibrations. The nature of those constituent parts is different: while the first is connected with the geometry of stressed rock mass location, the second is of probabilistic character. Besides the volume V, W- and Ω- types are introduced, with the following relation of their characteristic dimensions:

$$\Omega \gg W \gg V \qquad (3)$$

The stresses in the volumes V and W are expressed by the formulas:

$$\bar{\sigma}_v = \frac{1}{n} \sum \sigma_i K_i \qquad (4)$$

$$\sigma_w = \bar{\sigma}_v \pm \frac{t_i \Delta \sigma}{\sqrt{n}} \qquad (5)$$

$$\Delta \sigma = \sqrt{\frac{\sum (\sigma_{vi} - \bar{\sigma}_v)^2 \cdot K_i}{n-1}} , \qquad (6)$$

where n is the number of stress measurements in the uniformly stressed volume W consisting of subvolumes of V-type; K is frequency of the measured stress values; t_c - Student's coefficient determined by the number of measurements with the given reliability; $\Delta \sigma$ - standard deviation. By uniformly stressed macrovolume of W-type is meant one where the difference between the stresses in each its subvolume and the stress σ_w does not exceed the incidental deviation. The quasi-uniformity criterion is:

$$\frac{t_c \cdot \Delta \sigma}{\sqrt{n} \cdot \sigma_w} \ll 1 \qquad (7)$$

The stress state in the volume of Ω-type may be represented by the set of tensors the significant difference of which in the function of space coordinates is characterized by the gradient:

$$/\text{grad } \sigma / = \nabla \sigma = \frac{\sigma_{w1} - \sigma_{w2}}{L_{1,2}} , \qquad (8)$$

where $L_{1,2}$ is difference of coordinates between the rock masses W_1 and W_2.

2 GEOMECHANICAL CRITERIA OF ROCK PRESSURE MANIFESTATION AROUND EXCAVATIONS IN TECTONICALLY STRESSED ROCK MASSES

Long-term observations (for more than 10 years) of the nature of failures, effective stresses and strength of rocks have been carried out in situ to determine the criteria of dangerous stress state around excavations. The observations were conducted at special stations in excavations with different cross-sections and different stress states. The stresses acting around the outline of the opening were ascertained by instrumental and analytical methods in the elastic zone, and by the analytical method in the past peak strength zone. The initial rock strength was determined on rock samples in a laboratory.

The general analysis of failures shows that rock failure close to the outline of the opening in tectonically stressed rock masses is on the whole brittle and depends on the two compressive stresses: tangential stress σ_θ, acting tangentially to the cross-section outline, and longitudinal stress σ_L, acting close to the outline and parallel to the longitudinal axis of the opening. The process of failure may proceed with attenuated, uniform, or accelerated speed depending on the correlation of values of the effective stresses and rock compressive strength index $\sigma_{compr.}$, that is $K_s = \frac{\sigma_\theta + \sigma_L}{\sigma_{compr.}}$. Statistical analysis of the experimental data has made it possible to ascertain such important characteristics of failure as statistical coefficient K_s and statistical (frequency) probability of failures V_s.

The coefficient K_s is a certain interval of the ratio of the sum of effective stresses $\sigma_\theta + \sigma_L$ to the rock compressive strength:

$$K_s = \frac{\sigma_\theta + \sigma_L}{\sigma_{compr.}} .$$

The probability V_s is the ratio of the number n of rock failures which occurred at the given interval of values K_s to the general number N of the determinations within the same interval K_s, i.e.:

$$V_s = \frac{n}{N}$$

According to the variation of $\frac{\sigma_\theta + \sigma_L}{\sigma_{compr.}}$ the probability V_s of brittle rock failure on the outline of the opening varies from 0 to 1. This relationship is shown in Fig. 1.

The following statistically grounded criteria are adopted for engineering prediction:

$$\left.\begin{array}{l} /\sigma_\theta + \sigma_L / < 0.3/ \sigma_{compr.} / \\ V_s = 0 \end{array}\right\} \qquad (9)$$

$$\left.\begin{array}{l} /\sigma_\theta + \sigma_L / > 0.8/ \sigma_{compr.} / \\ V_s = 1 \end{array}\right\} \qquad (10)$$

$$\left.\begin{array}{l} 0.3/ \sigma_{compr.} / < /\sigma_\theta + \sigma_L / < 0.8/ \sigma_{compr.} / \\ 0 < V_s < 1 \end{array}\right\} \qquad (11)$$

The condition (9) satisfied, the probability of brittle failure of rock bursting type in openings is statistically close to zero, and the opening is stable with conditional (statistical) reliability $P_v \approx 100\%$. With the condition (11) satisfied the statistical probability V_s of failure ranges from 0 to 1. The conditional (statistical) reliability of the opening stability varies respectively from 0 to 100%.

The instrumental studies carried out in mines show that without technological effects of explosions the openings remain stable for a long time if the condition (9) is satisfied. The investigations in tectonically stressed rock masses did not indicate

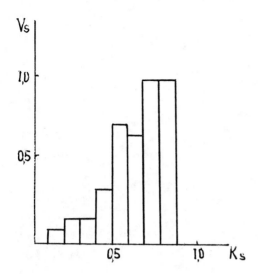

Figure 1. Bar graph of frequency of failures V_s on the outline of the opening at different values of the ratio K_s of the stresses to the rock strength .

any statistically significant dependence of the obtained criteria on the opening cross-section size. The investigations were performed in development workings of about 10 m² in cross-section, tunnels with the cross-section of about 40 m², slot-like stopes with that of about 70 m². With the condition (10) satisfied brittle failure was observed even in the boreholes of 50-70 mm in diameter. As results from the field observations of bursts, the role of rock structure and scale factor in the process of brittle failure is insignificant. Numerical values of the obtained criteria for brittle failure risk are confirmed by the investigations in many ore deposits. Naturally, with the change of forces due to technological operations experimental-analytical curves may have different statistically specified values.

Original results have been obtained by the authors on brittle failure intensity increase in time in the zone of past peak strength rock deformation, i.e. on condition that $/\sigma_\theta + \sigma_L/ > \sigma_{compr}$. Such observations were performed both in vertical and horizontal openings where under the technological conditions of their usage the effective stresses $/\sigma_\theta + \sigma_L/$ exceeded the rock ultimate strength σ_{compr} by 1.5-2.5 fold. Under such conditions accelerated failure takes place. Thus with $K_s = \dfrac{\sigma_\theta + \sigma_L}{\sigma_{compr}} = 1$ the rate of failure and the opening cross-section increase in a certain direction is about 0.3 m/year, and with $K_s = \dfrac{\sigma_\theta + \sigma_L}{\sigma_{compr}} > 2$ the rate of the process is more than 2 m/year (Fig. 2).

With technological effects in the past peak stressed state the intensity of failure increases sharply. Field observations have shown that due to ore chuting failure rate of the ore chute stressed walls rises by 2 and more fold, as shown in Fig. 2. The observations were carried out in open ore chutes 400-500 m deep (Markov 1982). The rate of ore release through the ore chute ran into 10 mtons per year.

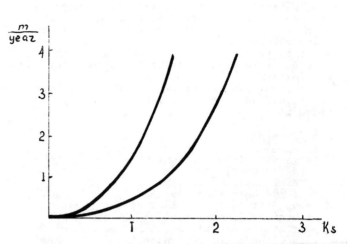

Figure 2. Failure rate change of ore chute vertical shaft walls: 1 - with the increase of the ratio of the effective stresses $\sigma_\theta + \sigma_L$ to the rock compressive strength σ_{compr}.; 2 - with an additional action of loads due to ore chuting.

3 PREVENTION OF OPENING FAILURE

Detailed theoretical and experimental analyses make it possible to find great reserves for increasing safety, strength and durability of underground excavations at the expense of high natural strength of tectonically stressed rock masses. The criterial relationships found are necessary for solving the problems of the use of these reserves. Detailed calculations for the stresses around excavations allowing for the peculiarities of natural stressed state of rocks are required for comprehensive study. Such calculations and substantiations are presented in the authors' special work (Markov & Savchenko 1984). As proved by industrial experiments, the effective measures for opening stability control in tectonically stressed rock masses are the following:

a) change of opening direction orienting the longitudinal axis mainly in the direction of the highest compression in the rock mass;

b) arrangement of parallel openings in the plane of the highest tectonic compressive stress;

c) use of horizontal openings with pyramidal cross-section;

d) use of vertical openings with elliptic cross-section, the long axis of the ellipse oriented in the direction of the highest tectonic compression;

e) drilling and blasting according to special diagrams taking into account tectonic stresses: blasting around the outline, camouflet outside the opening, formation of unloading slots;

f) special supporting with spray concrete lining and rock reinforcement with rasin injection.

The tests made verified the two-threefold increase of strength and durability of excavations due to the above listed measures. The experimental data show that rock burst probability decreases sharply, if the stresses on the opening outline are reduced even by a small value (Fig. 1). There are data

available showing that the change of the opening direction or its cross-section shape may
reduce the outline stresses by a factor of
1.5-2 and more. It follows from the relationships obtained that the probability of dynamic rock pressure manifestation may be thus
reduced by several times up to its complete
elimination.
  The geomechanical criteria obtained enable
one to evaluate the efficiency of the applied measures.

BIBLIOGRAPHY

Markov, G.A, 1986. Estestvennoye napriazhen-
  noye sostoyanie gornykh porod. In: Gornaya
  entsiklopediya, v. 2, p. 299. Moscow: So-
  vetskaya entsiklopediya Publishing House.
Markov, G.A. 1982. The laws of distribution
  of high horizontal stresses in the upper
  part of the Earth's crust and peculiarities of underground construction. In: Proc.
  IV Congr. Intern. Association of Engineering Geology. 5. India.
Markov, G.A. 1982 . Stability of vertical
  longlife shafts in tectonically induced
  massif. In: Proc. ISRM Symposium "Rock Mechanics: Caverns and Pressure Shafts",
  p. 1261-1264, Rotterdam: Balkema.
Markov, G.A. & S.N. Savchenko 1984. Napria-
  zhennoye sostoyanie porod i gornoe davle-
  nie v strukturakh goristogo reliefa. Leningrad: Nauka Publishing House.

Rock bursting and related phenomena in some Swedish water tunnels

Coups de terrain et phénomènes connexes dans certains tunnels d'amenée d'eau suédois
Gebirgsschläge und verwandte Erscheinungen in einigen schwedischen Wassertunnel

J.MARTNA, Chief Engineering Geologist, Swedish State Power Board, Stockholm
L.HANSEN, Consulting Geologist, Trias Geologi AB, Uppsala, Sweden

ABSTRACT: Overstressed rock has now and then occurred in Swedish water tunnels. When competent rock is excavated, high stresses may manifest themselves by some kind of induced instability, such as rock bursting, spalling or flaking off and fracturing as well as various sound phenomena. Some types of excavation-induced fractures and their relationship to the stress field and geological structures are discussed. The stress-induced phenomena may appear and disappear abruptly at successive sections of a proceeding excavation without any obvious reason. Measured stresses vary independently of the overburden, and initial stresses up to 70 MPa have been measured a few hundred metres below the surface. In some instances faults have demonstrably influenced the magnitudes and the directions of the stresses. It is suggested that the overstressed rock discussed here is caused by present-day tectonics and occurs in limited zones, which sometimes also feature faults or fractured rock.

RESUME: Il y a eu, de temps à autre, des massifs rocheux sous grandes contraintes dans des galeries suédoises. Lorsqu'on creuse dans un terrain compétent, de fortes contraintes peuvent se manifester par une sorte d'instabilité induite, telle que coup de terrain, éclatement ou effeuillement et fracturation, aussi bien que divers phénomènes acoustiques. Certains types de fracturations induites par le creusement, et leur relation au domaine de la contrainte et aux structures géologiques y sont traités. Les phénomènes induits par la contrainte peuvent apparaître et disparaître abruptement aux tronçons successifs d'un creusement en cours sans aucune raison évidente. Les contraintes mesurées varient indépendamment de la hauteur de recouvrement, et des contraintes initiales allant jusqu'à 70 MPa ont été mesurées à quelques centaines de mètres en-dessous de la surface. Dans quelques cas, il peut être prouvé que des failles ont influencé l'ampleur et la direction des contraintes. Il est suggéré que les massifs rocheux sous grandes contraintes traités ici sont causés par les tectoniques actuelles et existent dans des zones limitées, où il y a également quelquefois des failles ou des terrains fracturés.

ZUSAMMENFASSUNG: Beim Bau von Wasserstollen in Schweden ist man dann und wann auf überbeanspruchtes Gebirg gestoßen. Wenn kompetentes Gebirge ausgebrochen wird, können hohe Spannungen auftreten, die sich in einer Art von induzierter Instabilität wie etwa in Gebirgsschlag, im Abplatzen oder im Abbröckeln oder in der Zerklüftung des Gesteins, aber auch in verschiedenen akustischen Phänomenen äußern können. Einige Typen von ausbruchsbedingten Klüften und ihr Verhältnis zu dem Spannungsfeld und den geologischen Strukturen werden hier behandelt. Beim Vorantreiben des Stollens können die spannungsinduzierten Phänomene ohne ersichtlichen Grund auf der einen Strecke plötzlich auftreten und auf der darauffolgenden Strecke ebenso plötzlich wieder verschwinden. Die gemessenen Spannungen variieren unabhängig von dem überlagernden Gebirge, Initialspannungen von bis zu 70 MPa sind einige hundert Meter unter der Erdoberfläche gemessen worden. In einigen Fällen haben Verwerfungen bewiesenermaßen die Magnituden und die Richtung der Spannungen beeinflußt. Wir sind der Meinung, daß die hier behandelte Überbeanspruchung des Gebirges durch die rezente Tektonik verursacht wird und in begrenzten Streckenabschnitten auftritt, in denen manchmal auch Verwerfungen oder zerklüftetes Gestein vorkommen.

1 INTRODUCTION

The occurrence of overstressed rock has been known for some time in Swedish tunnelling experience (Martna 1970, 1972; Hiltscher et al. 1979; Carlsson & Olsson 1982b; Bergman & Johnsson 1985; Martna & Hansen 1986c). It is in practice usually called "banging rock" or "cracking rock" and expresses itself by a variety of cracking or banging sounds, usually combined with a more or less violent form of rock instability, such as bursting or slabbing. The occurrence of these phenomena is rather erratic and unpredictable. In this paper some cases from the experience of the Swedish State Power Board are treated.

2 VIETAS

In some restricted stretches of the Headrace Tunnels 2 and 3 at the Vietas hydroelectric power station, manifestations of high stress such as rock bursting, fracturing, rock instability and various banging and cracking sounds did occur during the excavation. A large number of stress measurements were carried out in both tunnels (Hiltscher 1969, 1972; Hiltscher & Ingevald 1971; Martna 1970, 1972; Hallbjörn 1986; Martna & Hansen 1986a, 1986c).

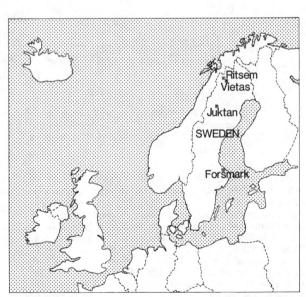

Fig. 1. Locations of the power stations discussed in this paper.

Fig. 2. Vietas hydropower station, Headrace Tunnel 3, chainage 5+290 in Cambrian siltstone. Steep, excavation induced, tensile fractures caused by high rock stresses can be seen in the tunnel face and in the wall to the left in the photograph. The horizontal roof has been caused by the latent bedding planes of the siltstone, re-opened during excavation by high horizontal stress.

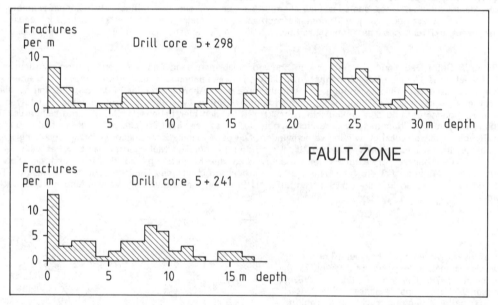

Fig. 3. Vietas hydropower station, Headrace Tunnel 3. Fracture frequency in drill cores drilled at chainages 5+241 and 5+298 respectively. The bars represent both horizontal, oblique and vertical fractures striking in various directions.

Thanks to a detailed geological mapping (Martna & Hansen 1986a), the geology of the Vietas headrace tunnels is well documented. The tunnels pass through metamorphic rocks of granitic and sedimentary origin belonging to the Caledonian mountain chain as well as autochthonous Cambrian sediments (siltstone) and Precambrian basement rocks (quartzite).

In both tunnels, the high rock stresses have caused fracturing of the rock. Most of the fractures are more or less parallel to the excavation face, and thus perpendicular to the tunnel walls.

In Headrace Tunnel 2, where the overstressed area is entirely situated in the Precambrian basement, the most frequent fractures strike NNE-SSW (athwart the tunnel),

and dip 60 to 85 degrees to the west, i.e. in the direction of the tilted bedding planes of the basement quartzite and metaarkose as well as the fault planes (Martna 1970, 1972). Although the bedding planes are clearly visible, they do normally not constitute fractures in intact rock.

Tunnel 3 was partly excavated through Cambrian, horizontally bedded siltstone, overlying the basement. In the siltstone two fracture sets are common in the tunnel. One set consists of horizontal fractures along the bedding planes. Core drillings have shown that these fractures are common, but mostly healed with quartz, calcite and chlorite. These latent fractures re-opened during the excavation, and caused planar overbreaks in the roof (Fig. 2).

Fig. 4. Vietas hydropower station, Headrace Tunnel 3. Overbreak caused by tensile fractures in the roof, coinciding with poor cohesion in a horizontal thrust fault. Vertical rock stresses have been determined to 35-70 MPa near this section. Note the frequent bolting of the roof.

The other set strikes parallel to the tunnel face and is vertical or dips steeply to the east. In the tunnel, this direction is by far the most frequent one in overstressed siltstone (Fig. 2). The spacing varies between 15 and 60 cm but can occasionally be less than 5 cm. The fracture surfaces are quite free from coating minerals and they often feature so-called hackle structures. These fractures have been induced by the excavation and are much less frequent in the drill cores (Fig. 3).

Usually, overstressed rocks were found in the autochthonous rocks where the main principal stress is more or less horizontal and up to about 50 MPa in magnitude (Martna & Hansen 1986c). An exception was a mylonite in the thrust rocks found in tunnel 3. This rock is extremely brittle and minor rock falls as well as cracking sounds were common during the excavation. High vertical stress (70 MPa) was measured. The two other principal stresses are nearly horizontal and almost equal. Long-extended water-bearing fractures with a strike at a low angle to the tunnel axis occurred parallel to the plane constituted by the vertical stress and one of the horizontal stress directions.

Due to tension perpendicular to these fractures caused by high vertical stress and the poor cohesion of the crushed rock in a horizontal thrust fault, overbreaks developed within a long stretch of the tunnel. Fig. 4 shows such an overbreak with comprehensive bolting in the roof.

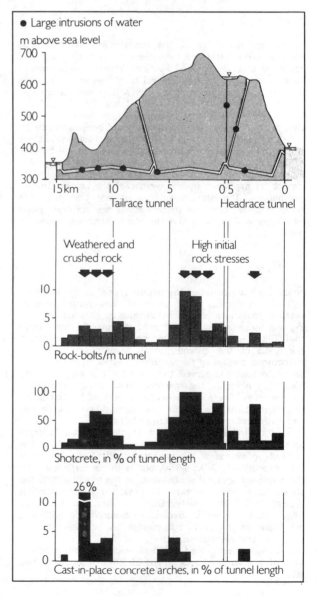

Fig. 5. Longitudinal section of the tunnels at the Juktan pumped storage power station. Total rock reinforcement measures in relation to high initial stresses, weathered rock and large intrusions of water. Modified from Carlsson et al (1982).

Fig. 6. Ritsem hydro power station, access tunnel to machine hall. Tension fracturing of the roof and vertical slabbing of the wall due to high vertical compressive stress.

3 JUKTAN

Within certain areas of the tunnels at Juktan hydro-electric power station, rock bursting and slabbing has occurred in granite (Fig. 5). Rock stress measurements have been carried out at several localities in the tunnels (Olsson 1979; Carlsson et al. 1982). The results showed horizontal rock stresses between 2 and 10 MPa in the access tunnel near the powerhouse. In magnitude, although not in direction, the stress could be in accordance with the overburden of up to 400 metres. However, at the restricted chainages of the headrace and tailrace tunnels where rock bursting occurred, the maximum horizontal stresses exceeded 50 MPa. Stresses of such a magnitude cannot be explained by the weight of the rather limited overburden. The present authors suggest that the high stresses encountered in these tunnels are chiefly caused by tectonic forces, active in the restricted zones in question.

4 RITSEM

When the excavation of the access tunnel to the machine hall of the Ritsem hydro power station passed a near-vertical fault in a brittle mylonite, a continuous fracturing and loosening of rock started abruptly in the tunnel roof (Fig. 6) and a pressure-induced slabbing took place in the walls of the curved tunnel. (Figs. 6 and 7). The phenomena continued for about 30 metres from the fault. Since the tunnel is curved, the same fault was likely to be met with again in the machine hall, about 150 metres further away. There, however, some sheared joints possibly indicated the position of the fault and no trouble was met with during the excavation or later on.

At the first-named place, a uniaxial, vertical, compressive initial stress of 20 MPa was measured. This value is much in excess of the expected load of the 130 m of rock overburden (3-3.5 MPa), but is in concordance with the observed physical phenomena. In the machine hall the stress measurement showed the presence of almost equal biaxial initial stresses (vertical 12 MPa, horizontal 10 MPa). Also this stress level is in concordance with the behaviour of the rock, but cannot be explained by the weight of the overburden.

The above-mentioned incident was the only notable case of overstressed rock in the 18 km of tunnels at this site.

Fig. 7. Ritsem hydro power station, access tunnel to machine hall. Detail of vertical slabbing parallel to the curved wall, caused by high vertical stress. The slabbing proceeded about 0.5 m per month until stopped with rock bolting.

5 FORSMARK

Thanks to extensive excavation and construction work, the geology and rock stresses in the Forsmark area are well known (Carlsson 1979; Carlsson & Christiansson 1986). Also, two research boreholes 250 and 500 metres deep have been carried out here by the State Power Board for the purpose of development of the Hiltscher probe (Hiltscher et al. 1979) and investigation of deep seated rock masses (Carlsson & Olsson 1982a; Martna et al. 1983). The results of these investigations show high horizontal compressive stresses, highly variable around a median of about 17 MPa near the rock surface in this area with a practically horizontal bedrock surface. The highest principal stresses are in a general way more or less parallel to the main strike direction of the foliation (NNW-SSE to NW-SE) of the bedrock gneiss, and also to the major tectonic structure in the area, the Singö Fault. Some measurements close to this fault constitute, however, exceptions to the rule and show deviations from the above-mentioned direction. These deviations are in concordance with the measurements near faults in the Vietas tunnel.

The only case of overstressed rock in this well studied area has been described by Carlsson & Olsson (1982b). The incidence occurred at shallow levels (5-15 m below the rock surface) in a short section of a cooling water tunnel at Forsmark 3 nuclear power plant up to four months after excavation and caused no serious problems. It expressed itself mainly by banging sounds and by minor rock pieces being expelled.

6 DISCUSSION AND CONCLUSIONS

The effect of overstressed rock depends not only on the magnitude and anisotropy of the stresses but largely also on the mutual directions of the stress field, geological structures and the tunnel. Of practical interest are not only the immediate and sometimes violent stress manifestations upon excavation, but also stress-induced fracturing and the occasionally substantial overbreaks.

Various types of overbreak, caused by overstressed rock have been observed in the discussed tunnels. Rounded (Martna 1972), angular (Hiltscher et al. 1979) and planar (Martna & Hansen 1986b, 1986c) types have, for instance, been described from the Vietas tunnels.

In all the case histories discussed, supporting and protective measures have consisted of rock bolting combined with either steel netting or steel-fibre reinforced shotcrete. In general, this has worked out well. In some cases, however, falls of rock and shotcrete have occurred, presumably at places with locally exceptionally high stresses. Cases of ruptured bolts are also known. Whereas shotcrete is the more permanent support of the two, netting or multiple netting seems to offer better security during the excavation in areas with high stresses. Shotcrete without reinforcement will do as support when the stresses are low, while steel-fibre reinforced shotcrete offers a good support at moderate stresses.

A phenomenon of special interest is the fracturing of the face and the adjacent rock during excavation, observed in the Vietas tunnels. The fracturing is parallel to the face and may cause bursting or spalling of the face but otherwise little overbreak, except when horizontal latent fractures are present.

The mechanism is thought to be as follows: by each round, the supporting rock is removed from the face. Thus, the stress in the direction of the tunnel will be released and tension in that direction will occur due to the two-dimensional stress field athwart the tunnel. The details of this mechanism have been described by Fairhurst & Cook (1966).

In general, the phenomenon has been treated by Joughin & Jager (1983) and by Hanssen & Myrvang (1986). The former authors describe three types of fractures developing near the tunnel face: 1) planar, steeply inclined fractures with no displacement, 2) faults with a displacement of up to 150 mm dipping 60-75 degrees, and 3) short,

planar fractures with no displacement and with a gentle to moderate dip.

It is probable that the vertical and subvertical fracturing of the face in Vietas corresponds to Type 1 fractures of Joughin & Jager (1983). In the Precambrian metaarkose, the direction of these fractures is not vertical but follows the direction of the steeply inclined latent bedding.

Type 2 fractures have not been recognized in the Vietas tunnels. Fractures somewhat resembling Type 3 have, however, been observed in Tunnel 3. At these localities, approximately horizontal latent fractures, such as mineral-coated tectonic ones in some fault zones or along the bedding planes of the siltstone, have been re-opened by excavation. They occur as horizontal planar overbreaks in the tunnel roof (Fig. 2). As shown by core drillings, all these induced fractures extend only a metre or two from the tunnel walls.

Some features in common emerge from the case histories discussed in this paper:

o The measured stress magnitudes as well as the stress manifestations vary independently of the thickness of the overburden and the stress levels are far in excess of the values which could be caused by the weight of the overburden.

o High stresses and their manifestations (rock bursting, slabbing, fracturing and sound effects) have normally appeared and disappeared rather suddenly in successive sections of a proceeding tunnel. Without any obvious reason, they have only occurred in some tunnels and only in a minor part of these tunnels. Their appearance is not predictable without comprehensive rock stress measurements.

o High stresses and stress manifestations have in some cases appeared abruptly in connection with faults or fractured zones of rock. Measured stress fields have changed magnitude and direction adjacent to faults.

The authors suggest, that the occurrence of overstressed rock is limited to certain restricted zones of the bedrock. The high stresses responsible for the overstressing are chiefly caused by tectonic forces active at present in these zones. The change of magnitude and direction of stresses observed in some cases in the vicinity of faults indicates that tectonic forces may also be active in certain faults in the Swedish bedrock.

ACKNOWLEDGEMENTS

The rock stress measurements have been performed by K. Ingevald, T. Öhman, L. Strindell and M. Andersson, all of the Swedish State Power Board. With Dr. R. Hiltscher the authors have had fruitful discussions concerning the interpretations of the results.

REFERENCES

Bergman, S. G. A. & Johnsson, N-E., 1985. Rock burst problems in ÖEF 600 m^2 tunnels in Brofjorden, Sweden. Swedish Rock Engineering. Ed. M. Finkel, 111-123. Sw. Rock Eng. Res. Found - BeFo, Stockholm.

Carlsson, A., 1979. Characteristic features of a superficial rock mass in southern central Sweden. Striae, vol 11, 1-79. Dept. of Quarternary Geology, Uppsala.

Carlsson, A. & Christiansson, R., 1986. Rock stresses and geological structures in the Forsmark area. Int. Symp. Rock Stress, 457-465. Stockholm.

Carlsson, A. & Olsson, T., 1982a. Characterization of deep seated rock masses by means of borehole investigations, Final Report, 1-155. Swedish State Power Board, Stockholm.

Carlsson, A. & Olsson, T., 1982b. Rock bursting phenomena in a superficial rock mass in southern central Sweden. Rock Mechanics 15, 99-110. Springer, Wien.

Carlsson, A., Fredriksson, S. & Olsson, T., 1982. Tunnelling and rock support at Juktan hydro power plant, Sweden. Tunnelling -82, 65-72. Inst. Mining and Metallurgy, London.

Fairhurst, C. & Cook, N. G. W., 1966. The phenomenon of splitting parallel to the direction of maximum compression in the neighbourhood of a surface. Proc. 1st Int. ISRM-Congr. Rock Mechanics, vol 1, 687-692. Lisboa.

Hallbjörn, L., 1986. Rock stress measurements performed by the Swedish State Power Board. Int. Symp. Rock Stress, 197-205. Stockholm.

Hanssen, T. H. & Myrvang, A., 1986. Rock stresses and rock stress effects in the Kobbelv area, northern Norway. Int. Symp. Rock Stress, 625-634. Stockholm.

Hiltscher, R., 1969. Bergtryck och spänningsfördelning omkring tilloppstunneln Suorva-Vietas. Bergmekaniskt diskussionsmöte, IVA-rapport nr 18, 229-34. Stockholm.

Hiltscher, R., 1972. Anwendung der Gebirgsspannungs-messung bei der Schwedischen Staatlichen Kraftwerks-verwaltung. Int. Symp. Untertagbau, 555-560. Luzern.

Hiltscher, R. & Ingevald, K., 1971. Bergmekaniska undersökningar beträffande smällbergszonens inverkan vid pallsprängning i tilloppstunneln Suorva-Vietas. (Summary: Rock-mechanical investigations concerning the influence of a rock-burst zone on the bench excavation in the headrace tunnel at Suorva-Vietas hydro power plant.) IVA-rapport nr 38, 153-164. Ingenjörsvetenskaps-akademin, Stockholm.

Hiltscher, R., Martna, J. & Strindell, L., 1979. The measurement of triaxial rock stresses in deep boreholes and the use of rock stress measurements in the design and construction of rock openings. Proc. 4. Int. ISRM-Congr. Rock Mechanics, vol 2, 227-234. Montreux.

Joughin, N. C. & Jager, A. J., 1983. Fracture of rock at stope faces in South African gold mines. Symp. on Rockbursts: Prediction and Control. Inst. of Mining and Metallurgy, London.

Martna, J., 1970. Rock bursting in the Suorva-Vietas headrace tunnel. 1st Int. Congr. of the International Association of Engineering Geology, vol 2, 1134-1139. Paris.

Martna, J., 1972. Selective overbreak in the Suorva-Vietas tunnel caused by rock pressure. Int. Symposium on Underground Openings, 141-145. Luzern.

Martna, J. & Hansen, L., 1986a. Vietas hydro power station, Headrace Tunnel 3. Documentation of geology and excavation. 1-90. Swedish State Power Board, Stockholm.

Martna J. & Hansen, L., 1986b. The influence of rock structure on the shape and the supports of a large headrace tunnel. Int. ITA Congr. Large Underground Openings, vol 2, 260-269. Firenze.

Martna, J. & Hansen, L., 1986c. Initial rock stresses around the Vietas Headrace Tunnels nos. 2 and 3, Sweden. Int. Symp. Rock Stress, 605-613. Stockholm.

Martna, J., Hiltscher, R. & Ingevald, K., 1983. Geology and rock stresses in deep boreholes at Forsmark in Sweden. Proc. 5. Int. ISRM-Congr. Rock Mechanics, F 111-116. Melbourne.

Olsson, T., 1979. Hydraulic properties and groundwater balance in a soil-rock aquifer system in the Juktan area, northern Sweden. Striae, vol 12, 1-72. Dept. of Quarternary Geology, Uppsala.

Discontinuity mapping – A comparison between line and area mapping
Cartographie des discontinuités – Comparaison entre une cartographie suivant une ligne et une autre selon une surface déterminée
Kluftkartographie – Ein Vergleich zwischen linearer Kartographiertechnik und einem Flächenkartographiesystem

JAMES I.MATHIS, Sweden

ABSTRACT: This study was conducted in order to determine the accuracy of an area discontinuity mapping system when compared to the more conventional line mapping techniques. An area mapping technique is generally faster and covers larger areas than line mapping. This is to be preferred in underground excavations where both time and access are limited. For this study, a mapping technique was to be developed which involved a minimum of data collection, was faster than scanline mapping, and gave results which were equivalent in accuracy to those obtained from scanline mapping. A comparison of the mapping results shows this goal was achieved and that area mapping (cell mapping) is a viable system for discontinuity data collection.

RESUME: Cette étude a été menée dans le but de déterminer l'exactitude d'un système discontinu de levés régionaux et de le comparer à des techniques conventionnelles de levés cartographiques. Une technique de cartographie régionale se déroule rapidement et couvre des régions plus grandes que le levé cartographique conventionnel. Celle-ci est à préconiser dans les excavations souterraines puisque le temps et l'accessiblité sont limités. Dans cette étude, la technique cartographique devait être développé et inclure un minimum de recueil de données, et s'est montrée plus rapide que la cartographie de l'exploration puisqu'elle a abouti à des résultats d'exactitude équivalents à ceux obtenus avec la cartographie de l'exploration. Une comparaison des résultats cartographiques a démontré que le but était atteint et que la cartographie régionale (cell mapping) est un système fiable pour recueillir les données discontinues.

ZUSAMMENFASSUNG: Diese Studie wurde durchgeführt, um die Genauigkeit eines Flächen-Kartographiersystems für Diskontinuitäten in Vergleich zur herkömmlichen linearen Kartographiertechnik zu ermitteln. Flächenkartographie ist im allgemeinen schneller und deckt größere Bereiche als herkömmliches lineares Kartographieren. Dieses Verfahren ist in unterirdischen Ausschachtungen vorzuziehen, da dort die verfügbare Zeit unde die Zugänglichkeit beschränkt sind. Für diese Studie wurde eine Kartographiertechnik angestrebt, die ein Mindestmaß an Datenerfassung beinhaltet, schneller als das Kartographieren mit Abtastlinien ist und Ergebnisse zeitigt, die ihrer Genauigkeit nach dem Abtastlinien-Verfahren gleichwertig waren. Ein Vergleich der Kartographierergebnisse zeigt, daß dieses Ziel erreicht wurde und daß die Flächenkartographie (Zellenkartographie) ein zuverlässiges System für das Erfassen von Diskontinuitätsdaten ist.

1 INTRODUCTION

The study described in this article is a part of a study being conducted at the University of Luleå in connection with the development of a statistical 3-dimensional rock discontinuity model. In order to accurately model discontinuities (joints), statistical values which accurately reflect the discontiniuities geometrical properties must be collected. In an underground environment this can be very difficult. There are several limiting factors: time, access, lighting, and exposure. This means that the mapping method must be rapid in order to map as much of an area that is accessable during a fixed time. It must also be robust, being relatively insensitive to errors introduced by poor lighting and limited exposure. These demands exclude scanline mapping in most cases. Line mapping is time demanding as every structure which intersects a baseline is measured as to length, orientation, and position. It was thus decided to proceed with the development of an area mapping technique which should be faster than, and hopefully give as good as, or better, results than scanline mapping.

2 GEOLOGY

Mapping was conducted in various tunnels in Forsmark SFR, an underground nuclear waste storage facility being constructed under the Baltic Sea near Forsmark, Sweden. The rocks, which are highly metamorphosed, consist of gneisses, a coarse grained granite (pegmatite), a fine grained granite (bordering on a

gneiss), and greenstones. Three rock types were mapped: coarse grained granite, fine grained granite, and gneiss. Difficulty was encountered in differentiating between fine grained granite and gneiss. Pegmatite stringers often were found dispersed throughout the rock mass. Folding was common.

3 FIELD MAPPING

The scanline mapping technique used was developed after Priest's (Priest 1981) method. A baseline was established approximately 1 meter above, and parallel with, the tunnel floor (Figure 1). The portion of the joints which extended above this line (semi-trace length) and which were longer than the truncation limit (15cm) were included in the sample. Properties measured were; semi-trace length, orientation, and intersection point with the line. Vertical lines were also sampled in order to collect data concerning the flat lying joints.

Cell mapping was conducted by defining (painting) a 4m by 4m square window on the tunnel wall (Figure 2). Joint sets were then mapped within this sample window. By this it is meant that joint sets were visually defined by the person mapping and the mean orientation estimated. Three to four discontinuities which belonged to each set but deviated markedly from the mean orientation were also recorded in order to estimate the orientation deviation parameters for the joint set. The number of joints per joint set, the number of joints with one end within the window, and the number of joints with no ends within the window were also recorded in order to estimate length

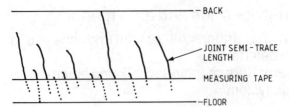

Figure 1. Scanline mapping - Forsmark SFR

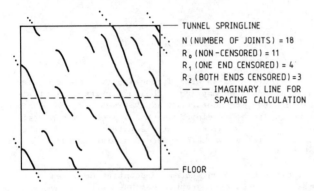

— TUNNEL SPRINGLINE
N (NUMBER OF JOINTS) = 18
R_0 (NON-CENSORED) = 11
R_1 (ONE END CENSORED) = 4
R_2 (BOTH ENDS CENSORED) = 3
— — — IMAGINARY LINE FOR
            SPACING CALCULATION

— FLOOR

Figure 2. Cell mapping - Forsmark SFR

according to Kulitalake's (Kulatilake 1984) method.
An apparent spacing was also measured my counting the
number of fractures which intersected an imaginary
line within the cell.

Mapping was done in a recurrent manner at Forsmark.
That is, an area was mapped with cell mapping and
then re-mapped with line mapping.

4 DATA REDUCTION THEORY

4.1 Orientation

Discontinuity sets were determined by orientation
using a method described by Grossman (Grossman 1983).
This method is based on the assumption that if the
discontinuity orientations are Poisson distributed in
space, a limiting angle can be calculated between the
discontinuities. If this limiting angle is violated,
the discontinuities are not independent and belong to
a discontinuity set. The equation given by Grossman
is:

$$\cos \alpha_L \geq 1 - \frac{1.67835}{N} \qquad (1)$$

where: N = number of discontinuities

This method, has, of course, problems with overlapping joint sets. When this problem was encountered, an interactive computer routine was used to
define the discontinuity set boundaries.

The orientation distributions were described using
Grossman's (Grossman 1985) bivariate normal distribution on a tangent plane to the Schmidt hemisphere. No corrections were made for non-uniform
sampling.

Scanline mapping used the above described methods
without modification. Cell mapping necessitated a
conversion of the discontinuity set orientation mean
and maximum deviation members to parameters on the
tangent plane to the sphere. The discontinuity sets
for each cell were thus described by a bivariate
normal distribution on this plane. After this was
done, cell discontinuity sets could be combined in
the same manner as scanline discontinuities to give
global discontinuity sets.

4.2 Length

Discontinuity mean lengths were calculated by
distribution free methods. The method utilized for
scanline data reduction is based upon a method
proposed by Priest (Priest 1985). This method can be
described as follows:

$$h(l) = (1/\mu) * \int_{l}^{\infty} f(m)dm \qquad (2)$$

where: $h(l)$ = probability density function of the
                semi-trace lengths
       μ = mean trace length

Therefore:

$$h(l) = (1-F(l))/\mu \qquad (3)$$

where: $F(l)$ = cumulative distribution function of true
                discontinuity lengths

Thus:

$$H(l) = (1/\mu) * \int_{0}^{l} (1-F(m))dm \qquad (4)$$

where: $H(l)$ = cumulative distribution function of
                semi-trace lengths

It can then be shown that equation (4) becomes:

$$H(l) = 1/\mu \quad \text{as} \quad l \to 0 \qquad (5)$$

Thus, the expected value of the joint length distribution can be obtained at the zero point of the
semi-trace length cumulative distribution. This
value cannot be measured. It can, however, be estimated in the following manner:

Let c equal any arbitrary censoring level,
equivalent to l; let N equal the total number of
fractures intersecting the scanline; let r be the
number of fractures which have length (l<c). If
equation (4) is analyzed as l→0, it can be shown
that:

$$\mu = Nc/r; \quad \text{as} \quad l \to 0 \qquad (6)$$

This solution is presented in Figure 3.

Cell mapping mean lengths were solved for using a
method proposed by Kulitalake (Kulatilake 1984). This
method is based on the theory that joint centers are
randomly distributed in space. The number of joints
that intersect a rectangular window can be described
as:

$$N = \lambda(wh + w\mu B + h\mu A) \qquad (7)$$

where: N = expected number of discontinuities
               intersecting the window
       λ = joint center density
       μ = joint mean length
       w = window width
       h = window height
       A = $E(\cos \varphi)$
       B = $E(\sin \varphi)$
       φ = structure apparent dip

Further details can be found in Kulatilake's paper.
The important fact is that the discontinuity mean
length can be estimated without measuring the length
of a single structure as long as the window dimensions, the discontinuity orientation distribution, the
total number of discontinuities per set, and the
number of singly and doubly truncated discontinuities
are known.

Mean lengths for scanline and cell mapping were not
corrected for truncation bias (truncation length 15
cm). This was because that this correction requires
the introduction of a distribution in a previously
distribution free calculation and because that the

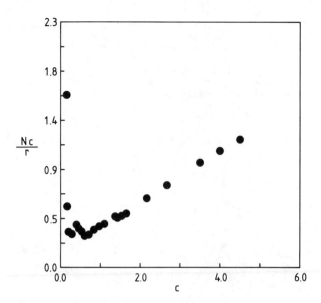

Figure 3. Scanline joint mean length estimation

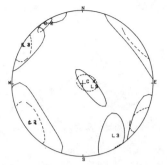

SCHMIDT NET
LOWER HEMISPHERE
MEAN ORIENTATION VECTORS
w/95% CONFIDENCE INTERVALS

C (---) = CELL MAPPING
L (——) = SCANLINE MAPPING

Figure 4. Comparison of cell and scanline mapping joint set orientations

mean lengths were considered long enough that truncation bias was small.

4.3 Spacing

Mean spacings for the scanline data were calculated according to simple geometry. All discontinuities within a joint set were considered to have the mean orientation of the set. The true spacing was then calculated from the apparent line spacing.

Spacing for the cell mapping method was calculated by two methods. The first method was the reduction of the apparent spacing data collected for every cell joint set. This procedure was the same as for scanline mapping. The second method involved the calculation of lengths using the discontinuity length values obtained from Kulatilake's method. It can easily be proven from equation (7) that if the values of λ, μ, A, and B are known, a width w is assumed and h is allowed to go to zero, the number of discontinuities (N) along the resulting line can be calculated. The true spacing can be calculated from this value in the same manner as the spacing was calculated for scanline data.

5 RESULTS

Results can be discussed in two stages; one where Grossman's set selection method was used without modification for the individual areas (Table 1) and the other where engineering judgment was used to separate overlapping joint sets for overall design joint sets (Table 2).

In both cases, the comparison scheme is the same. A Schmidt pole plot of the joint set mean vectors with associated 95% confidence intervals was plotted as an overlapping plot for both line and cell mapping (Figure 4). The joint set means from line mapping that fell within the 95% confidence interval for a cell set mean (or vice-versa) were compared as to differences in mean orientation (measured as the dihedral angle between vectors), mean length, and mean spacing. Comparisons were conducted without a weighting factor, all joint sets having equal weight, and with a weighting factor, where consideration was taken to the number of line and cell discontinuities which described a joint set.

As is obvious from the tables, the mean orientation values do not differ significantly between the two mapping methods. Mean length estimations are also

Table 1. Mean differences between line and cell mapping for individual areas.

| | | Non-weighted | | Weighted * | |
	No.	Mean	Std. Dev.	Mean	Std. Dev.
Dihedral angle	20.	6.499	3.815	6.245	4.094
Mean length	20.	.014	.357	.006	.344
Mean spacing	15.	-.245	.800	-.305	.530

* Weighting factor = 5.840
** Differences are (line value - cell value);
*** Dihedral angle measured in degrees, length and spacing in meters

Table 2. Mean differences between line and cell mapping for design areas (design sets)

| | | Non-weighted | | Weighted * | |
	No.	Mean	Std. Dev.	Mean	Std. Dev.
Dihedral angle	19.	6.981	3.456	6.725	3.341
Mean length	19.	.028	.630	.053	485
Mean spacing	14.	-.254	.719	-.281	.541

* Weighting factor = 5.840

consistent with an average difference fluctuating around zero though the standard deviation is rather large. Mean spacing is the only value which differs significantly between the two mapping methods with cell spacing being considerably wider than line spacing. This, however, can be explained by a variety of means.

The main problem associated with line spacing is that the spacing is calculated under the assumption that the spacings were, in fact, measured along a line. A line, has, of course, no thickness, only length. This is not the case encountered when mapping blasted rock. The line assumption is violated by projecting rock and joint surfaces, overbreak, etc. due to blasting damage. Instead of a line a half-cylinder of space is mapped with a diameter depending upon rock damage. This gives a tighter spacing than a true line would give. A deviation from a true line of 30cm was estimated for the scanlines studied in Forsmark. This deviation was then included in the calculations for cell spacing using equation (7). For the area studied, the difference in spacing between cell and line mapping was reduced to zero for both the weighted and non-weighted case with an associated reduction in standard deviation.

Another reason for the difference between spacing results is that cells were rarely composed entirely of one rock type. Thus, joint sets that were present

1113

in one rock type, but only sparsely represented in the second rock type would have a relatively wide resulting cell spacing.

6 COMMENTS AND CONCLUSIONS

The cell mapping technique proposed within this paper appears to be a viable method of collecting statistical data on discontinuities. It has several advantages when compared to standard scanline mapping.

The method is relatively fast. Once the window is delineated on the rock wall, which takes only slightly longer than setting up a line for scanline mapping, no further boundary definitions are need. Joint sets are defined by the mapper for an area of the rock mass, i.e. the window defined by the cell boundaries. This eliminates, in large part, the problem associated with overlapping joint sets on the Schmidt net. The joint sets can always be combined later statistically if desired. Only the necessary orientations needed to describe the joint set statistically are recorded. This involves between 4-5 orientation measurements per joint set per cell which is only a fraction of the total number of discontinuities included in the set. Joint mean length estimation procedures all also simplified as only the number of discontinuities, as previously described, are recorded. No lengths are measured. The mean joint set spacing is then derived from this mean length calculation, thus no spacings are measured.

Additional advantages are that sparse sets which are not readily detected with line mapping are generally detected by cell mapping. Thus local variations in properties are more easily detected.

Another comparison which can be made, which is possibly unfair, is that the results obtained from cell mapping are based on a larger number of discontinuities than scanline mapping. This advantage is in some part, perhaps entirely, neutralized by the greater amount of precision data collected with scanline mapping. However, as this study shows, cell mapping can give equivalent results when compared with scanline mapping.

Disadvantages of the method are that the person mapping must be schooled in three-dimensional thought processes and be able to differentiate between joint sets. The mapper must also be observant in that the number of joints per joint set within each cell must be accurately counted.

The next stage of this project is to predict the discontinuity statistics for the rock silo at Forsmark. The present statistics mapped on a 4m high wall will then be compared to the those obtained from mapping on an area measuring 60m tall by approximately 50m in width (the silo's circumference).

ACKNOWLEDGEMENT

This study is supported by the Swedish Nuclear Power Inspectorate.

REFERENCES

Grossman, N.F. The bivariate normal distribution on the tanget plane at the mean attitude, Proceedings of the Int. Symp. on Fundamentals of Rock Joints, Björkliden, Sweden, Sept. 1985

Grossman, N.F. A numerical method for the definition of discontinuity sets, ISRM Meeting, Melbourne, Australia, 1983

Kulatilake, P.H.S.W, Wu, T.H. Estimation of mean trace length of discontinuities, Rock Mechanics and Rock Engineering, No. 17, 1984, pp 215-232

Priest, S.D. Workshop on statistical methods in rock mechanics, Swedish Rock Engineering and Research Foundation - BeFo 134:S/85 Stockholm, 1985

Priest, S.D., Hudson, J.A. Estimation for discontinuity spacing and trace length using scanline surveys. Int. J. Rock Mech. Min. Sci & Geomech. Abstr., Vol. 18, 1981, pp 183-197

Deformational behavior and its prediction in longwall gate roadways under pillars and ribs

L'influence de piliers du reste sur la déformation des galeries du panel de la longue taille sousjacente

Der Einfluss der Restfesten und Abbaukanten auf die Konvergenz von darunter liegenden Abbaustrecken

K.MATSUI, Kyushu University, Fukuoka, Japan
M.ICHINOSE, Kyushu University, Fukuoka, Japan
K.UCHINO, Kyushu University, Fukuoka, Japan

ABSTRACT: The effect of remnant pillars and ribs on the deformation of gate roadways in underlying longwall panels has been clarified by means of both in situ measurements and quasi-three dimensional stress analysis. On the basis of these results an equation for predicting the deformation of gate roadways is proposed which enables to be taken the influence of the previously worked face into account.

RESUME: L'influence de pilliers du reste sur la déformation des galeries du panel de la longue taille sous-jacente est clarifié au moyen des mesures in situ et l'analyse des contraintes quasi-trois dimensionales. L'équation proposée permet de predire les déformations des galeries en consideration de l'influence de la taille déjà exploitée.

ZUSAMMENFASSUNG: Der Einfluß der Restfesten und Abbaukanten auf die Konvergenz von darunter liegenden Abbaustrecken wird durch die ortlichen Vermessungen der Querschnittsverlusten geklärt. Der Zusammenhang zwischen dem Gebirgsdruck und der Konvergenz von Abbaustrecken wird auch mittels quasi-dreidimensionaler Gebirgsdruck-analyse eingehend untersucht. Auf Grund dieser Ergebnisse von Untersuchungen wird ein Verfahren zur Voraus-berechnung der Abbaustreckenkonvergenz vorgeschlagen.

1. INTRODUCTION

In 1984, Japanese coal mines produced 16.83×10^6 t of coal. Underground longwall workings accounted for 71.6% of this output, and 90% of longwall workings used the retreat system.

The formation of longwall workings is characterized by the creation of high stress zones in the solid around the extraction zone. Gate roads, the access roadways to longwall faces are predominantly located along the solid ribsides of the face. Therefore, gate roadways are subjected to high stresses and significant strata movements, especially behind the longwall working faces. Many gate roadways are deformed to the extent of needing maintenance, generally dinting, but in some cases requiring ripping.

The deformation of a gate roadway is a function of several factors including extracted seam height, geological properties of seam roof and floor, panel width and length, type and spacing of support and so on. In multi- seam mining, remnant pillars and ribs left in the worked out seams cause additional and significant deformation on gate roadways in current longwall workings.

In this paper, measured results regarding deformational behavior under remnant pillars and ribs are presented. Comparisons are made with computed results derived from quasi-three dimensional stress analysis. A method of predicting gate roadway deformation is proposed.

2. DESCRIPTION OF SITE AND MEASURING TECHNIQUE

Fig.1 shows the Hikishima area of Ikeshima colliery where this investigation into gate roadway deformation was carried out.

The geological structure in the investigation area is comparatively simple and stable. There are three minable coal seams with a gentle dip of 1-5. Each coal seam is overlain and underlain by shales.

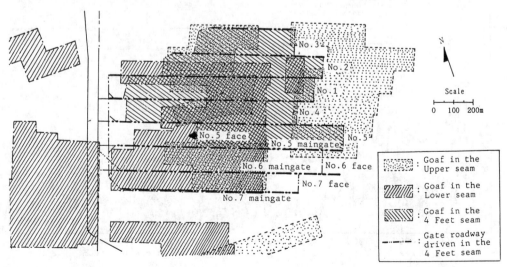

Fig.1 Layouts of No.5 face

These shales are weak and their mechanical properties deteriorate greatly in the presence of water(Ihara et al.1985). The Lower and Upper seams, situated at distances of about 20m and 40m respectively above the 4 Feet seam had been partially extracted, and remnant pillars were left in the Upper seam(Fig.1).

The gate roadways were formed by drilling and blasting, the standard method used in Japanese collieries. Support in the gate roadway was set by using 2.8m x 4.8m three piece steel arches (cross-section 0.115m x0.095m) as shown in Fig.2 and spaced at 0.65m intervals. The supports were tied and lagged fully with wood.

No.5 retreating face was 110m long and situated under the old worked areas in the Lower and Upper seams. The face was supported by self-advancing powered shield supports, and the coal was extracted by a double ranging drum shearer. The roof in the extracted area was allowed to cave fully. Behind the face line, face side packing with fly-ash and cement slurry was used in the gate roadway(Fig.2).

Measurements were carried out on the maingate of No.5 panel. Marks for measurement were placed on the inside flanges of the arch girders at about 5.5m intervals. The roof mark was at the center of the arch crown. Roadway closure was determined by using a telescopic scale (accurate to 1cm) and measuring the height between the roof mark and the true floor. Measurements were taken along the entire length of gate roadways during development and mining.

3.MEASURING RESULTS

In presenting gate roadway deformation survey, four measurements were taken along the gate roadway, as follows:

1st. Measurement - Development of No.5 maingate
2nd. Measurement - Before starting of No.5 maingate working
3rd. Measurement - Retreating of No.5 face working
4th. Measurement - Retreating of the adjacent No.6 face working

Closure in arched profile gate roadways may be defined in terms of vertical and lateral components. However, in this survey only the vertical closure was taken, by measuring the height of the roadway. Fig.3 illustrates the vertical closure during the period of study. Although there is a fairly wide scatter in the values plotted, since there are several contributory factors such as interaction effects from the worked out area in the overlying seams, the curves are characteristic and indicate

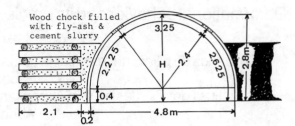

Fig.2 Support system in gate roadway

the general influence of the remnant pillar and the rib edge on the overall roadway stability.

Referring to Fig.3, vertical closure during the period between 1st. and 2nd. measurement was observed to be small under the goafs. The mean values of vertical closure of 1st. and 2nd. measurement under the Lower seam were 29cm and 34cm, respectively. Under the goafs in the Lower and the Upper seams, the mean values were 17cm and 19cm, respectively. The relatively small time-dependent closures which are represented as the rate of closure per day, were 0.29mm/day under the Lower goaf and 0.15mm/day under both the Lower and the Upper goafs. However, localized and excessive deformation in the gate roadway was experienced under the rib edge or remnant pillar of the seams above. The mean value of deformation under a remnant pillar was 38cm at 2nd. measurement. Under the rib edge the values of the 1st. and 2nd. measurement were 53cm and 103cm, respectively. From these values the time-dependent closure was calculated as 2.94mm/day. The length of gate roadway affected by the rib edge required a large amount of dinting owing to the great time-dependent closure.

If the standby period was longer under the remnant pillar, the vertical closure would become extremely large.

During the retreating stage of No.5 face ,the vertical closure was extremely large under the remnant pillar, and the greatest vertical closure was about 170cm. Closure under the pillar edges was greater than at the center of the pillar. It was observed that closure under the pillar occurred rapidly behind the face line, and most supports distorted significantly and some failed. To maintain the gate roadway profile and limit closure, dinting was carried out and supplementary supports were introduced. The mean amount of dinting was 100cm. However, these measures could not restrain the closure and ripping and new support setting were done finally. After setting new arch supports, the

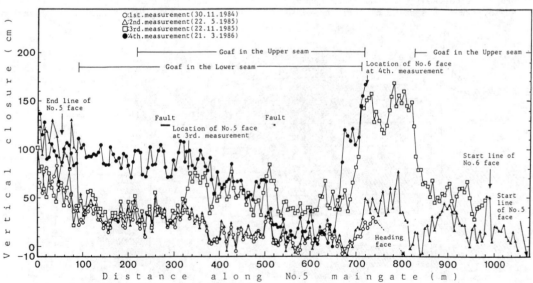

Fig.3 Measured vertical closure along No.5 maingate

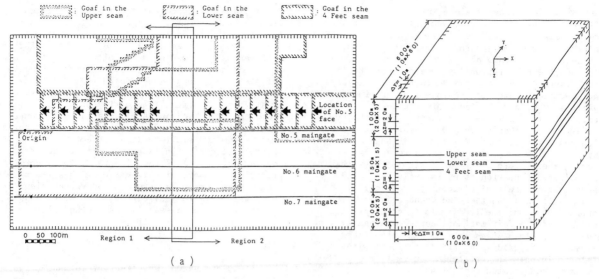

Fig.4 Analytical model (a) plain view of model (b) size of each region

measured time-dependent closure was 3.45mm/day.

Similar deformational behavior was also observed under the rib edge.

Even under the goafs, the gate roadway deformed after longwall mining but the closure that occurred behind the face line reached a level of 80cm under the goaf in the Lower seam and less under both the goafs in the Lower and Upper seams. The roadway, however, kept the original profile and little repair work was needed for re-use as a tailgate of the adjacent No.6 face working.

4.STRESS ANALYSIS

The quantitative discussion of gate roadway stability requires an estimation of the amount and trend of the redistributed stresses and displacements caused by longwall mining. The distribution of stresses and displacements around the longwall workings depends on the behavior of the surrounding strata, which generally consist of blocks of strata divided by discontinuities such as joints and faults. It is not practical to attempt to model the detailed strata conditions, and some simplifying assumptions must be made. It seems to be reasonable to assume that the strata behave in a linearly elastic fashion.

Even with the assumption of linearly elastic strata, there still exists the problem of computing stresses and displacements caused by longwall mining. Three dimensional finite element analysis on any practical scale is economically beyond the limits of current capabilities. To cope with this

problem , a quasi-three dimensional analysis established by Everling et al.(1972) for mining layouts was used. This analysis is a kind of finite difference method, and the special feature is that for analytical simplification the major displacement caused by seam mining is normal to the plane of the coal seam and the two minor displacements parallel to the seam are neglected.

Fig.4(a) shows the plain view of the analytical model considering the measured site. This analytical region is divided into regions 1 and 2 in order to evaluate the stresses along the entire length of the gate roadway. As shown in Fig.4(b), the size of each region is 600m x 600m x 350m and divided into rectangular solid elements whose sizes are generally 10m x 10m x 10m and partly 10m x 10m x 20m.

The depths of each seam are 600, 620 and 640m below surface and the extracted heights are 1.5, 2.4 and 1.8m for the Upper, the Lower and the 4 Feet seam, respectively.

The elastic constants of the rock mass and seams are taken as $E = 1.0 \times 10^3$ MPa and $\nu = 0.25$.

The applied loads are the overburden load on the upper boundary of the model and the body forces. The unit weight of surrounding strata is taken as $25kN/m^3$.

The computed stress distributions along the gate roadway are shown in Fig.5. The dotted line is the primitive vertical stress(Po = 16MPa) without longwall workings in both the current and above seams. The small differences in the stresses are shown in the length where regions 1 and 2 are superimposed, but these differences do not seem to seriously affect results in the following

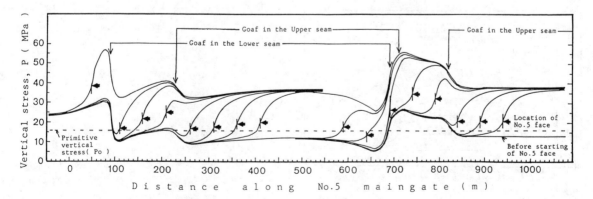

Fig.5 Stress distribution along No.5 maingate

discussion. The stresses along the roadway before starting of retreating No.5 face are much higher under the remnant pillar and rib edge. Under the goafs the stresses are smaller and relatively uniform, except near the rib edge of the Upper seam.

The advance of a retreating face changes the stress pattern. The stresses begin to increase at about 100m in advance of the face line and increase rapidly at 20m from the face. After passing by the face, the stresses increase rapidly to a peak level. The maximum value of stress level achieved was about 3.5Po under the remnant pillar and rib edge. However, the values under the goafs were less, reaching about 2.5Po. These indicate the fact that the roadway deformation is excessively large and supports deteriorate significantly under pillar or rib edge. It is shown that the trend of the stress distribution along the roadway is in good agreement with that of the deformational curve obtained by measurements. This seems to mean that the stress analysis can be used as an aid to rational planning of mining layouts and maintenance of gate roadways under more complicated conditions.

5.PREDICTION OF GATE ROADWAY DEFORMATION

The complexity of the factors involved in gate roadway deformation makes prediction of the amount of deformation particulary difficult. An empirical approach has been used to predict roadway deformation in West Germany(Götze et al.1976). This approach involves a method of estimating the vertical closure of gate roadways based on observations of arched roadways. The method can be used to estimate vertical closure of roadways on single unit longwall faces within very high probability limits. But this method has a drawback in that a great deal of data including many factors is needed to produce a practical empirical equation, especially in multi-seam longwall mining.

To cope with this problem, the following analytical method can be applied. The trend of the stress distribution along the gate roadway evaluated by the quasi-three dimensional stress analysis is in good agreement with that of the measured vertical closure, as mentioned in section 4. In single seam longwall mining, the vertical closure caused by mining can be predicted by using the following equation(Ihara et al.1985).

$$VC = K \cdot \triangle P$$

where VC is the vertical closure in cm, $\triangle P$ is the stress increase in MPa derived from the stress analysis and K is a closure factor in cm/MPa which depends on the mechanical properties of surrounding strata, support system, the nature of packs, etc.

Fig.6 illustrates the vertical closure along No.5 maingate using each closure factor. In the figure,

the solid marks represent the measured vertical closure caused by No.5 longwall mining. It is noted that the closure factors under the remnant pillar are relatively large and small under the goafs. The differences in closure factor along the gate roadway result from the differences in magnitude of acting stresses along roadway before current No.5 longwall mining. The mechanical properties of surrounding strata tend to deteriorate easily under high stress. Therefore, the subsequent closure factor becomes greater. The small closure factors under the goafs result from the smaller extent of deterioration of strata. It is of significant importance to estimate the correct values of the closure factor in order to predict the amount of deformation with a reasonable degree of accuracy by using the above equation. It is necessary to collect a great deal of data in order to ascertain this analytical method.

6.CONCLUSIONS

This study has shown the effect of remnant pillars or ribs left in other seams on deformation of the current gate roadways. The existence of a pillar or rib edge increase roadway closure excessively. The advance of retreating longwall working promotes gate roadway deformation even under the goafs but much less than under remnant pillar. Under remnant pillar or rib edge, time-dependent closure is significant. These deformational behaviors can be analyzed by stress analysis. The stress analysis shows that the amount and the trend of redistributed stresses along gate roadway are major contributory factors to deformation of gate roadways. Moreover,it is shown that using the stress increases derived from stress analysis and closure factors, the gate roadway deformation caused by longwall mining can be predicted well. As more experience is gained in multi-seam mining, the analytical prediction method presented in this study becomes more useful in planning and layout of longwall mining.

ACKNOWLEDGMENTS

The authors wish to thank the management and personnel at Ikeshima Colliery.

REFERENCES

Everling,G. & A.G.Meyer(1972). Ein Gebirgsdruck-Rechnenmodel als Planungshilfe, Glückauf Forschungshefte, 3:81-88
Götze,W. & W.Kammer(1976). Die Auswirkungen von Streckenführung und Ausbautechnik auf die Querschnittsverminderung von Abaustrecken, Glückauf, 112:846-853
Ihara,M., K.Matsui, Y.Ichikawa & M.Ichinose(1985). Gate Roadway Deformation with Advancing Longwall Coal Face, J.J.M.M., 101:409-414

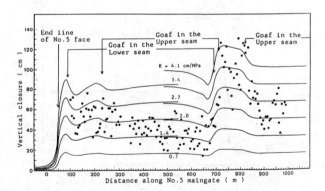

Fig.6 Comparisons between measured and calculated vertical closure

Observations, recherches et résultats récents sur les mécanismes de ruptures autour de galeries isolées*

Observations, researches and recent results about failure mechanisms around single galleries
Beobachtungen, Untersuchungen und Neuergebnisse über Bruchmechanismen isolierter Stollen

V.MAURY, Président de la Commission, rapporteur Elf-Aquitaine, Pau, France

SUMMARY : This report summarizes the conclusions drawn from observations of failures occurred around single galleries submitted to simple loadings. Although the liaison between members of the Commission are somewhat difficult, the exchanges were very profitable. This report leads to the inability of Mohr-Coulomb's type criteria to predict quantitatively the appearance of failure ; but these criteria might give the modes of failures occurring at the wall (shapes of failures lines). This report insists upon the contradiction induced by plastic or elastic theories when applied to the wall of the gallery ; the failure initiation is probably located inside the wall, but more precise observations are necessary to decide if and when the propagation is in true shear mode or extension ; this last one being probably more frequent than thought in the past. This report summarizes recent results of researches undertaken following the first recommendations of some national working groups. Many other cases of failures were collected by members of the Commission ; the analyses and interpretation of these cases must be the following task of this Commission.

RESUME : Ce rapport présente les conclusions tirées d'observations de ruptures survenues autour de galeries isolées en conditions simples de chargement et de comportement. Malgré des difficultés de liaison, les échanges entre experts de la Commission ont été très fructueux. Ce rapport aboutit à démontrer l'inaptitude quantitative des critères type Mohr-Coulomb à prédire les modes de ruptures capables d'apparaître. Il signale les paradoxes auxquels conduisent les théories conventionnelles élastiques et plastiques ; l'amorce de la rupture semble se faire à l'intérieur du parement, mais il est recommandé de procéder à des observations plus fines qui sont nécessaires pour savoir comment la fracture se propage (véritable cisaillement ou extension, condition qui paraît maintenant plus fréquente qu'on le pensait). Il résume des résultats de recherches récentes, obtenus à la suite de premières recommandations des groupes de travail. De nombreux autres cas de ruptures ont été collectés grâce aux groupes nationaux ; leur analyse et interprétation devront faire l'objet de la poursuite des travaux de la Commission.

ZUSAMMENFASSUNG : Der vorliegende Bericht enthält Schlussfolgerungen aus Beobachtungen betreffs um einzeln liegende Stollen in einfachen Belastungs- und Verhaltensverhältnissen eintretende Brüche. Trotz Verbindungs-schwierigkeiten erwies sich der Erfahrungsaustausch unter den Fachleuten des Ausschusses als äusserst ergiebig. Der Bericht stellt dar, warum die Mohr-Coulomb-Kriterien zu einer mengenmässigen Voraussage der möglichen Brucharten ungeeignet sind. Er deckt Widersprüche auf, zu den die herkömmlichen Elastizitäts- bzw. Plastizitäts-theorien führen. Der Bruch setzt anscheinend im Stoss an, jedoch ist es empfehlenswert, die Sache einer genaueren Betrachtung zu unterziehen, um zu erfahren wie sich die Kluft verbreitet (echte Scherung oder Ausdehnung ? Letzteres scheint häufiger vorzukommen als bisher angenommen wurde). Neuliche Forschungsergebnisse werden zusammengefasst, die den ersten Vorschlägen der Arbeitsgruppen entsprechen. Dank den nationalen Gruppen wurden viele andere Bruchfälle gesammelt ; ihre Analyse und Auswertung soll Ziel weiterer Tätigkeit des Ausschusses sein.

1. INTRODUCTION - PRESENTATION

En 1983 à Melbourne, il était décidé de poursuivre l'action d'une Commission sur la rupture d'ouvrages souterrains, confiée au Professeur ROCHA peu avant sa disparition. L'expérience d'une Commission anté-rieure avait montré un besoin de bien définir les objectifs et les limiter, une grande difficulté pour rassembler des données sur des cas de ruptures et la nécessité de s'appuyer sur des groupes nationaux pour apporter une contribution relative à des ouvrages très divers. Les objectifs de la Commis-sion, discutés, proposés et retenus au cours des réunions tenues à AIX-LA-CHAPELLE (1982) et MELBOURNE (1983) consistent en un rassemblement, une classification préliminaire de cas de ruptures pour aboutir aux principaux mécanismes déterminant la rupture d'ouvrages souterrains. A partir de là, les experts réunis au sein de la Commission se proposent de tirer des conclusions dans deux directions : celle des Universités et Centres de Recherches en dégageant les voies où les recherches fondamentales sont le plus nécessaires pour faire progresser la mécanique des roches dans les travaux souterrains, celle des projeteurs et concepteurs d'ouvrages souterrains ensuite, de façon à ce qu'ils profitent des outils disponibles en mécanique des roches, en étant conscients de leurs limites, sans croire à l'existence de marges ou de coefficients de sécurité, toujours évalués par rapport à un seul mécanisme alors que ceux-ci peuvent être multiples dans le cas des cavités souterraines. La première étape est donc de cerner les principaux mécanismes à l'origine des ruptures de massifs rocheux autour d'ouvrages souterrains.

Sous le terme "rupture", on a considéré deux types de manifestations : la rupture de la roche autour de l'excavation avec apparition de discontinuités macroscopiques, ou rejeu important le long de discon-tinuités existantes ; on a englobé également le cas de déformations permanentes importantes même continues bien qu'il ne s'agisse pas à proprement parler de ruptures. On a également considéré sous le

---

* Rapport de la Commission S.I.M.R. sur les Mécanismes de Ruptures autour d'Ouvrages Souterrains

terme "rupture", le résultat d'un comportement inha-
bituel et imprévu de la roche aboutissant à une
interruption de construction ou de fonctionnement
d'un ouvrage, pour ne pas laisser échapper de
ruptures d'importance pratique considérable, dont le
comportement du massif est responsable même s'il n'a
été que peu apparent ou peu détecté.

De façon à couvrir un large éventail de spécialités
et conditions géologiques, la Commission a été
constituée d'un groupe d'Experts provenant chacun
d'un pays où il animerait si possible un Groupe de
Travail national permettant de démultiplier la
tâche.

Actuellement, quelques groupes nationaux ont été
constitués et sont actifs, sur des sujets précis ou
généraux à propos desquels ils peuvent apporter une
contribution utile. Ce sont :

- le groupe australien, sous la direction de
B. BAMFORD : collection de cas de ruptures en
ouvrages miniers et génie civil, en roches dures et
tendres,

- le groupe belge (P. BRYCH), qui a déjà réalisé un
travail important également sur deux séries de
ruptures (voir conclusion dans le rapport de la
journée du 30.05.85 du groupement belge), a décidé
un ambitieux programme de laboratoire in-situ,

- le groupe japonais, (N. SAKURAI), particulièrement
orienté vers des ruptures survenant en ouvrages de
génie civil et en roches tendres,

- le groupe américain (E. CORDING), a entrepris un
intéressant classement de ruptures selon des styles
régionaux observés aux Etats-Unis,

- Le groupe norvégien (N. BARTON) est orienté sur
des cas typiques de ruptures survenues en roches
dures mais dues à des remplissages de matériaux
mous,

- le groupe sud-africain (D. STACEY), travaillant
sur des cas de ruptures de roches fragiles sous très
fortes contraintes. Les travaux antérieurs exécutés
en Afrique du Sud constituent un apport fondamental
à toute recherche et progrès sur la rupture,

- des contacts très fructueux ont été maintenus avec
les experts néerlandais (GRAMBERG, ROEST, SPIERS des
universités de DELFT et UTRECHT), et finlandais
(P. SARKKA),

- Le groupe français (V. MAURY), grâce à un
démarrage plus précoce sous l'impulsion du Président
TINCELIN en 1980, a pu dégager ses premières
conclusions et recommandations en 1983 ; certaines
d'entre elles ont été suivies d'actions, de premiers
résultats échangés avec les membres de la Commission
(Rapport du Groupe de Travail Français 1986).

Ce rapport présente un résumé des travaux de la
Commission concernant des observations et des
conclusions tirées sur les mécanismes de ruptures
autour de cavités isolées en conditions simples de
chargement (pression interne, température) ; il
s'agit donc du problème de base de l'apparition et
de la propagation de la rupture autour de galeries
uniques de formes géométriques simples, ou
courantes. Suite aux premières conclusions tirées
notamment par le Groupe de Travail français en 1983,
des recherches ont été entreprises à l'échelle
mondiale ; ce sont les résultats de ces recherches
acquises lors des trois dernières années, qui ont
été résumés et les conséquences en sont tirées dans
les conclusions. L'éclairage nouveau apporté par les
recherches entreprises dans d'autres domaines que le
génie civil et minier, le domaine pétrolier
notamment, sont un des résultats les plus payants et
un des retours les plus positifs de la multi-
disciplinarité de la Commission.

La tâche entreprise est parfois une gageure : mettre
la ou les théories en face de réalités bien observées
est parfois un exercice décevant. L'analyse de cas
est souvent rendue difficile car peu nombreux sont
ceux où l'on dispose de toutes les observations et
mesures nécessaires ; on a ainsi conservé certains
cas de ruptures très intéressants pour leurs morphologies mais insuffisamment documentés, ils permettent
cependant des rapprochements entre des cas apparemment très dissemblables.

Autre difficulté inattendue : lorsque la rupture
devient trop habituelle et que "l'on vit avec", comme
c'est le cas dans les mines sédimentaires, il peut
être difficile de faire saisir aux mineurs l'intérêt
que présentent des amorces de ruptures qu'ils sont
trop habitués à cotoyer. Une des conséquences posi-
tives de ce rapport serait de leur faire bien appréhender le niveau des questions qui se posent dès ce
moment.

Enfin, le rapport actuel n'est basé que sur un nombre
limité d'observations ; il en existe beaucoup
d'autres, peut-être plus illustratives, didactiques
ou documentées que les lecteurs sont invités à nous
signaler, de façon à améliorer ce point des connaissances, nécessairement évolutif.

2. RESUME DES CONCLUSIONS D'OBSERVATION DE CAS VECUS DE RUPTURES

Pour faciliter l'analyse des cas vécus de ruptures,
on avait distingué les cavités isolées non revêtues
en matériau homogène à formes courbes ou voisines
(fer à cheval), des galeries très élancées à
parements verticaux, situations que l'on peut
retrouver même dans des mines en chambres et piliers
où les taux d'extraction sont faibles : en dehors des
croisements, les galeries sont pratiquement en
conditions de galeries isolées.

Les ruptures autour de galeries à formes courbes,
soit parfaitement circulaires car taillées à la
machine, soit à formes voisines de formes circulaires
abattues à l'explosif, montrent un tracé incurvé ;
elles se localisent aux extrémités d'un diamètre, pas
toujours horizontal, effet le plus probablement
attribuable à une dissymétrie du champ de contrainte
latéral (perpendiculaire à l'axe de la galerie) ;
initialement considérées comme résultant de
mouvements de cisaillement, ces ruptures observées en
calcaires, grès, marnes se manifestent soit instantanément après le passage du front, soit d'un façon
différée (Fig. 1).

De telles ruptures peuvent également s'observer à
plus petite échelle autour des coins arrondis de
galeries carrées ou rectangulaires (mines de sel en
Allemagne Fédérale).

En matériaux plus résistants, granites, grès durs,
quartzites, les lignes de ruptures qui évoquent des
ruptures en cisaillements conjugués deviennent
beaucoup moins inclinées, presque "tangentes" à la
périphérie de la galerie (cf. photo fig. 116, p. 200
- HOEK et BROWN, 1980) ; ceci serait normal si l'on
admet que les lignes de ruptures font un angle de :
$\Pi/4 \pm \emptyset/2$ avec les isostatiques, et que l'angle de
frottement de ces matériaux est très élevé. A
l'examen plus détaillé de ces ruptures, il est moins
certain qu'il s'agisse de véritables cisaillements,
car il est difficile d'apercevoir sur des écailles
des traces de friction ; les mouvements relatifs
semblent plus résulter de la seule possibilité cinématique restante que d'un véritable cisaillement ; on
les appellera ici "pseudo-cisaillements". Les observations sur les ruptures survenues en matériaux
tendres, semblent permettre d'y appliquer cette
remarque.

En matériaux très résistants, grès quartzites ou roches cristallines en grande profondeur, de telles ruptures observées en AFRIQUE du SUD sont considérées comme résultant d'une extension radiale et un critère en extension a été proposé (par extension radiale, on entend un état de déformation simple parallèle à la contrainte mineure, qui est radiale).

A l'opposé, les ruptures survenant en parements droits (verticaux) de grandes cavités (à élancement vertical) ressortent plus sûrement d'un mécanisme en extension ; elles sont observables en tous types de matériaux : craie (carrières à très faible profondeur), sel (LOUISIANE, ALLEMAGNE FEDERALE), roches cristallines (usines souterraines - ALPES - cavités de stockages à parements verticaux - PAYS SCANDINAVES -). Dans ces cas, la rupture aboutit à des écailles en dalles successives parallélépipédiques d'épaisseur décimétrique à métrique parfois gigantesques ; les plans de rupture suivent le réseau d'isostatiques, et évoquent la rupture en colonnettes d'une (demi) éprouvette en compression simple. On rejoint la fracturation par clivage axial (GRAMBERG 1965 et 1977, GRAMBERG et ROEST 1984, FAIRHURST et COOK 1966), formalisée et appliquée par STACEY et de JONGH (1977) et STACEY (1981).

Quelques particularités sont à signaler à propos de ces ruptures :

- Les cisaillements "vrais" ou "pseudos" observés autour de galeries à formes courbes présentent une contradiction par rapport à une théorie de distribution de contraintes élastiques classique : dans celle-ci, la concentration de contraintes maximales est sur le diamètre perpendiculaire à la contrainte majeure et en parement (en A, Fig. 4 colonne de gauche). Or les ruptures observées débouchent en parement, mais pas en ce point ; elles débouchent même symétriquement par rapport à celui-ci en C et C' ; la concentration de contraintes maximales surviendrait donc à l'intérieur du parement en B, comme dans une distribution plastique ; toutefois, le caractère quasiment intact des écailles rend cette théorie peu applicable. Il y a là un paradoxe qu'une théorie acceptable de la rupture devra expliquer.

Ce point a été constaté par HOEK et BROWN (1980) en essai au laboratoire sur une plaque percée d'un trou dans un matériau très dur (chert) ; après l'apparition de fentes de traction sur le diamètre parallèle à la contrainte majeure, la fissuration est apparue à l'intérieur du parement, avant celle classique en cisaillement (théorique) sur le diamètre perpendiculaire à la contrainte majeure.

- Lorsque le mode d'excavation est fait par une séquence de tirs, les formes successives de la galerie ont pu subir des ruptures de modes différents. De telles observations ont été signalées à propos d'excavations abattues à l'explosif en plusieurs phases (galeries de tête et stross) et sont importantes pour les zones d'ancrages de boulons (MORE O'FERRAL et BRINCH, 1983) ; on peut même se demander si les aspects en dents de scie des fractures, signalées par ces auteurs en parements, ne viennent pas d'une découpe de zones déjà fracturées en extension par une séquence de tirs ultérieure à l'échelle même du micro-retard dans une même séquence.

Enfin, on constate parfois lors du creusement d'un tunnel, l'alternance le long du tracé, de zones relativement intactes non fracturées, et de zones fracturées ; les zones intactes peuvent être le siège de ruptures en parement ultérieures, ruptures qui n'apparaissent pas dans les zones déjà fracturées. Il est d'ailleurs parfois difficile de savoir s'il s'agit de zones fracturées "naturel-

lement" à l'échelle géologique, ou s'il s'agit de zones rompues en avant du front comme cela a été mentionné en AFRIQUE du SUD, et évoqué à propos du discage (STACEY), mais dont la périodicité s'expliquerait par la réflexion de phénomènes dynamiques sur des accidents géologiques.

En conclusion de ces observations, les ruptures survenant en grands parements verticaux ressortent clairement d'un mécanisme en extension ; sur les parements courbes, les matériaux très résistants sous fortes contraintes montrent des ruptures qui sont peut-être d'un mécanisme analogue ; il est à l'heure actuelle plus difficile de dire si en matériaux plus tendres sous plus faibles contraintes, il s'agit de véritables cisaillements ou "pseudo" cisaillements.

A propos de l'influence du temps, certaines de ces ruptures ont été quasi immédiates, d'autres différées à l'échelle du mois ou de l'année. Lorsque l'on cherche à décrire ces phénomènes au moyen de lois visqueuses, on peut être amené à prendre des viscosités en place différentes de celle du laboratoire ; ceci paraît normal car il s'agit de progression de ruptures et non de phénomènes visqueux véritables. Précisons toutefois que ceci ne s'applique pas au sel.

En outre, les ouvrages souterrains subissent plus souvent qu'on ne le croit des chargements différés tels que thermiques, dus aux écoulements, et à des transformations minéralogiques telles que dissolutions, recristallisations, altérations ; de nombreux comportements différés leurs sont imputables, et pourraient être prévus, sans invoquer nécessairement des phénomènes visqueux.

3. CAS DE STABILITES ANORMALES IN-SITU

Bien que l'objectif de ce travail soit l'évaluation des mécanismes de rupture affectant les cavités souterraines, il a semblé utile de noter quelques cas de stabilités anormales d'ouvrages souterrains (galeries, puits, forages). Il s'agit de stabilité "anormale" par rapport à une distribution élastique classique de contrainte, et des critères types Mohr-Coulomb ou courbe intrinsèque. On ne dispose que de peu d'exemples certains, mais deux d'entre eux méritent d'être signalés : dans les galeries du stockage du VEXIN en FRANCE dans la craie Turonienne (MAURY, 1977), dont les parois travaillaient, en théorie, au niveau de la résistance en compression simple (confirmé par mesures au vérin plat) et la clé beaucoup plus, des sondages ont été faits et les parois de ces sondages, soigneusement observées n'ont montré aucun signe de rupture, bien que soumises à 2 fois le niveau de contraintes correspondant à la résistance en compression simple Q, d'après la théorie de l'élasticité conventionnelle.

D'assez nombreuses galeries d'adduction d'eau ont été construites à faible profondeur dans les silstones calcaires sous la ville de CHARLESTON (CAROLINE du Sud, U.S.A.) ; par rapport à un état de contraintes déduit du poids des terrains, les parois de ces excavations auraient été en théorie soumises à 2 à 3 fois Q.

Dans le cas de la craie, des examens très précis et des mesures (vitesses du son, sclérométrie, densité) ont été faits pensant qu'il s'agissait d'une adaptation plastique, envisageable dans le cas de matériaux à fortes porosités. En fait, il n'a pas été possible de mettre en évidence quelque variation significative de propriétés du matériau justifiant l'explication de ces stabilités par une adaptation plastique.

Il existe probablement beaucoup d'autres cas de stabilités "anormales" passant inaperçus du fait de la méconnaissance des propriétés du matériau laissé en place par l'excavation et de l'état de contraintes in-situ ; un certain nombre de forages pétroliers en milieux peu cohérents à forte profondeur (> 2 000 m) sont probablement dans ce cas.

Si la prévision des ruptures présente un intérêt pratique considérable, il en est de même de ces mécanismes de stabilité dont on voudrait pouvoir profiter, en ajustant les soutènements (ou les pressions internes dans le cas des forages) au strict nécessaire pour rester en conditions stables.

De façon à progresser dans cette voie, des recommandations avaient été faites en 1981-83 concernant une meilleure connaissance de ces phénomènes de stabilité anormale.

CAS DE STABILITE ANORMALE EN LABORATOIRE

D'assez nombreux essais ont été réalisés sur tubes épais avec des dimensions extérieures suffisantes pour représenter le problème du trou en milieu infini (diamètre du trou de l'ordre du tiers du diamètre extérieur). Les contraintes tangentielles σ_t en paroi ont été estimées par la théorie élastique courante à partir de la valeur du chargement extérieur correspondant à l'apparition de la rupture à la paroi. En appelant Q la résistance en compression monoaxiale, ou triaxiale, selon que la pression interne est nulle ou non, tous les auteurs (OBERT-STEPHANSON, BEREST-BERGUES-NGUYEN, HAIMSON, GUENOT, SIMONYANTS, GEERTSMA, DARLEY, GAY, voir références complètes dans GUENOT, 1987) ont noté des rapports anormalement élevés de σ_t/Q, c'est-à-dire des contraintes théoriques en paroi excédant largement, de 2 à 8 fois, la résistance "conventionnelle". Sur la plupart des matériaux testés, grès, calcaires, craies sèches, marbres, granite, la plupart des auteurs ont remarqué qu'il n'y avait pas de zones annulaires travaillant en grandes déformations continues évoquant une déformation plastique, mais rupture avec apparition de surface de discontinuités et déplacements éventuels de morceaux (écailles) dont la matrice rocheuse reste apparemment intacte.

4. MOYENS DISPONIBLES ACTUELS POUR L'EVALUATION DE LA STABILITE D'UNE CAVITE SOUTERRAINE (unique, en matériau homogène)

A ce jour, un certain nombre de modèles théoriques ont été proposés, mais aussi pris en défaut ; il fallut les compléter par des règles empiriques qui n'ont jamais été formulées explicitement.

4.1 Modèles théoriques analytiques et autres (pour la bibliographie complète des références citées dans ce paragraphe voir BROWN et al., 1983)

Depuis KIRSCH (1898), de très nombreux modèles ont tenté d'appliquer la théorie de la plasticité au problème du comportement du bord du trou. Les premiers (FENNER, KASTNER, LABASSE, MORRISON-COATES, HOBBS, BRAY, DIEST) ont envisagé des critères types MOHR-COULOMB sans hypothèse sur les déformations "plastiques". HOBBS a supposé qu'à la valeur de contrainte correspondant au "pic", la rupture survenait brutalement déterminant une zone à propriétés (E et ν) affaiblies. Par la suite, on s'est préoccupé de la loi de comportement : SALENCON a traité le cas de critères de TRESCA et MOHR-COULOMB, avec loi de comportement associée, et variation de volume plastique. DAEMEN et FAIRHURST, LOMBARDI, HENDRON et AYER, LADANYI, EGGER, PANET, KORBIN, KENNEDY et LINDBERG, ont introduit des améliorations : critères bilinéaires, de pic et

résiduels, écrouissage négatif dit "radoucissement", diverses lois de comportement, influence du passage du front (PANET). FLORENCE et SCHWER, NGUYEN MINH et BEREST ont introduit l'influence d'une pression radiale ; SCHWARTZ et EINSTEIN, HOEK et BROWN, KAISER, ont essayé d'adopter des critères empiriques en pic et en résiduel, avec diverses lois de comportements. RISNES, BRATLI et HORSRUD ont traité le cas complet des diverses zones en plasticité selon la contrainte longitudinale.

A l'heure actuelle, ces modèles donnent une évaluation théorique des niveaux de contraintes en paroi. En ce qui concerne la prédiction de l'apparition de la rupture, il faut garder à l'esprit que :

a - Ces modèles supposent les déplacements et donc les déformations continues ; or, même si, comme le supposent les modèles radoucissants, la zone rompue a certainement des propriétés affaiblies, elle n'est plus continue.

b - De ce fait, la démarche consistant à prendre un critère de rupture comme potentiel plastique (lois de comportement associées) est pratique pour le traitement analytique du problème mais masque la vraie nature physique du problème, l'apparition de discontinuités, qui est extrêmement différente.

c - Ces modèles essaient de décrire le comportement de la paroi du trou à partir d'essais conventionnels de laboratoires interprétés avec les hypothèses classiques auxquelles on peut faire deux critiques ; la première est celle de l'indépendance du critère vis-à-vis de la contrainte intermédiaire et du chemin de contrainte suivi ; les essais triaxiaux classiques - contraction longitudinale et extensions radiales, ou contractions radiales et extension longitudinale - seraient en principe équivalents, or on sait que les essais en extension longitudinale, correspondant à un état de contrainte plus proche de celui en paroi de galerie (à pression atmosphérique) donnent souvent des résultats inférieurs à ceux en contraction simple ; la seconde est que l'état de contrainte en paroi est rapidement variable (décroissance de la contrainte tangentielle en $1/r^2$) alors que l'on cherche en essais conventionnels à réaliser des états de contraintes aussi homogènes que possible.

d - Le caractère quasi intact de la matrice rocheuse constituant les écailles ne milite pas en faveur de la validité de l'interprétation en plasticité. Par ailleurs, comme on l'a vu, les distributions élastiques de contraintes amènent à une contradiction sur la concentration de contraintes, flagrante dans le cas de chargement anisotrope.

Pour pallier ces insuffisances extrêmement gênantes dans la conception d'ouvrages souterrains non revêtus, et tenir compte qu'au laboratoire, les résistances à long terme -toujours en essais conventionnels de fluage par exemple- sont inférieures à celles à court terme, on admet parfois des règles empiriques.

4.2 Règles empiriques

Sans entrer dans le détail des règles de renforcements proposées d'après les classifications géotechniques de massifs rocheux, HOEK et BROWN (1980) ont repris l'analyse des contraintes autour des formes de galeries usuelles et les concentrations maximales correspondantes (p. 222, op. cit.) ; ils donnent également une échelle d'intensité de désordres autour des galeries à profil carré en fonction du rapport contrainte in-situ verticale/ résistance en compression simple. Pour ces auteurs, les ruptures survenues en coins de galeries carrées seraient en cisaillement ; on peut se demander si les ruptures en parement ne sont pas elles franchement en extension (photo Fig. 117, op. cit.) ainsi que pour la rupture en coin (Fig. 116) de galerie carrée, qui rappelle ce que l'on voit au moyen de vernis craquelants. Il est vrai qu'il s'agit là d'un matériau extrêmement rigide et fragile.

Bien que ceci n'ait jamais été formulé expressément, on a parfois recommandé de s'arranger dans le projet d'excavations souterraines à ne tolérer en paroi qu'une contrainte tangentielle égale à une fraction de la résistance en compression simple. A partir de là, on a tenté de proposer des chiffres plus ou moins inspirés de valeurs déduites d'essais en fluage conduisant à la ruine : 70 % pour certains sels, 50 % pour la craie humide, 20 à 30 % pour la potasse (MORLIER 1964) ; mais il n'existe rien pour les matériaux très résistants.

Ces règles, tout à fait informelles, plutôt "tuyaux" ou règles du pouce que l'on échange entre projeteurs ont été trop souvent prises en défaut pour être encore considérées : stabilités "anormales" dans la craie, fluages et peut-être ruptures, et effondrements de cavités lessivées dans le sel, soumis à des contraintes soit excessives soit insuffisantes en principe pour provoquer de tels comportements. Ces insuffisances reflètent un besoin impératif pour le projeteur d'avoir des guides même très empiriques ; ceux-ci ne parviennent toutefois pas à intégrer des observations disparates et contradictoires.

4.3 Critères en extension

Depuis longtemps, il a été reconnu que la forme de la rupture en colonettes dans l'essai de compression simple résultait d'une déformation latérale en extension excessive, et que ce même phénomène pouvait affecter les parois de galeries à profil carré ou rectangulaire (FAIRHURST et COOK, 1970). La déflection de lignes de ruptures au voisinage des extrémités de l'échantillon pouvant être due soit à une déviation des trajectoires des isostatiques par l'effet des plateaux (NEHMAT-NASSER et FAIRHURST) soit le dégagement vers l'extérieur des colonnettes de hauteurs décroissantes vers l'intérieur de l'échantillon. On rejoint dans ce cas la théorie de la fracturation par clivage axial de GRAMBERG.

Suite à des analyses de cas (STACEY et de JONGH, 1977), STACEY (1981) a proposé un critère en extension simple, considérant que l'initiation de la fracture se fait lorsque la déformation dépasse un taux d'extension limite ("e_c"), mesurée en essais conventionnels très soigneux, et valables pour des classes de roches à définir ; STACEY insiste sur le fait que le critère concerne surtout les faibles confinements ; il attribue à ce genre de phénomènes l'initiation du discage (STACEY, 1982). Ceci semble particulièrement intéressant car dans ce cas, les extensions maximales se situeraient en avant du front de taille, relâchant les contraintes jusqu'à ce que le carottier soit en position pour recréer une distribution de contraintes apte à provoquer de nouvelles extensions, et ainsi de suite. L'évaluation des zones en extensions potentielles successives est très riche en enseignement pratique (MORE O' FERRALL, BRINCH). L'évolution des formes et écailles (ORTLEPP et GAY) mérite d'être reconsidérée en tenant compte très fidèlement de la séquence de tir dans le cas du creusement à l'explosif.

5. RESULTATS DE RECHERCHES RECENTES ET EN COURS

Les résultats des premières observations et conclusions obtenues (Groupe de Travail Français, discussions et confrontations avec les Groupes de Travail sud-africain, belge, U.S., norvégien et spécialistes néerlandais) ont incité au développement de recherches avec une optique plus large d'application à d'autres domaines, pétrolier notamment, dont les résultats sont en cours de publications et dont on donne un résumé ci-dessous.

5.1 Résultats acquis dans le domaine pétrolier

Dans le but d'élucider les mécanismes de tenue de forages pétroliers, et les cas, extrêmement onéreux, où ils sont en défaut, l'analyse des contraintes a été reprise pour des forages verticaux ou inclinés dans un champ de contrainte in-situ quelconque, soumis à des écoulements, et à des chargements thermiques (GUENOT, 1987) ; le régime thermique des forages a été analysé en toutes phases de creusement (CORRE, EYMARD et GUENOT, 1984). Les différences essentielles des forages pétroliers avec les galeries et mines sont, outre leur diamètre, leur orientation et inclinaison variables, leur forme géométrique parfaite au départ, leur pression interne établie grâce au poids de la colonne de boue dont on peut choisir la densité, les variations thermiques liées à l'approfondissement et l'écoulement de la boue et les écoulements radiaux divergents (éventuellement en cours de forage), ou convergents (en cours de production) qu'ils subissent.

CONTRAINTES EN PAROIS

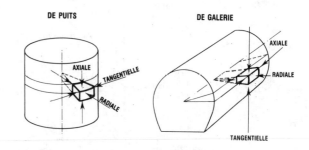

RUPTURE ASSOCIÉE

SI TANGENTIELLE > AXIALE > RADIALE → RUPTURE TYPE A   INTERSECTION DES RUPTURES ∥ AXIALE,
(d. boue faible ou forage à l'air)   CONTRAINTE INTERMEDIAIRE

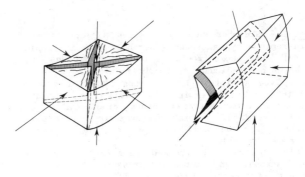

FIGURE 1

Fig. 1 - Sketch showing conventionnal stresses around a borehole (top left), a tunnel (top right) and a type of wall rupture (bottom left and right) : axial stress is intermediate : type A rupture (surfaces intersecting parallel to axial stress).

1123

Il ressort de cette recherche (Fig. 2 et 3), dans l'hypothèse où il n'y a pas d'écoulement, qu'en appelant σ_r, σ_t, σ_l respectivement les contraintes radiale, tangentielle et axiale (ou longitudinale),

- en cas de pression interne faible ou nulle -cas des galeries de génie civil et mines- on a une disposition de contraintes du type : $\sigma_r < \sigma_l < \sigma_t$, dite "α",

- si la pression interne augmente, à contraintes latérales in-situ constantes (ou en résumé que le rapport des deux croisse) on arrive à une disposition $\sigma_r < \sigma_t < \sigma_l$ dite "β",

- si ce rapport croît encore, on a la disposition $\sigma_t < \sigma_r < \sigma_l$ dite "γ".

Si maintenant on admet que la rupture se fait en cisaillement selon un critère type courbe-intrinsèque (MOHR-COULOMB à cohésion nulle et frottement interne de 20°) (Fig. 3 d'après GUENOT 1987), la disposition α poussée à l'excès (tangentielle excessive, radiale insuffisante) conduira à une rupture dite en "mode A" (Fig. 1) avec lignes de ruptures conjuguées dont l'intersection se fait selon la direction de la contrainte intermédiaire (axiale), les écailles auront la forme de pelures d'oignons, isotropes ou anisotropes selon le champ latéral.

MODES DE RUPTURES EN PAROIS

CONTRAINTES PERPENDICULAIRES A L'AXE

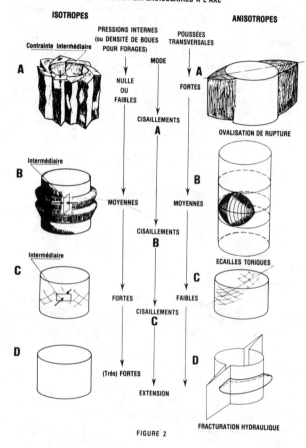

FIGURE 2

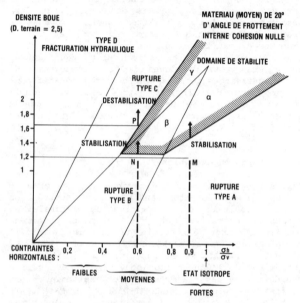

FIGURE 3

Fig. 2 - Sketches showing detailed aspects of the main rupture types around a borehole with isotropic (left) and anisotropic (right) conditions of stresses perpendicular to borehole axis.
Top : Type A. Intermediate stress is axial stress and internal pressure is nil or very low ; rupture appears in shear surfaces intersecting parallel to axial stress. If lateral stresses are not equal, borehole section will have an elliptical shape (borehole caving).
Middle top : type B. Intermediate stress is tangential stress, and internal pressure is medium. Rupture fragments have toroîdal shape (like "O"-ring).
Middle bottom : Type C. Intermediate stress is radial stress, internal pressure is high. Borehole wall rupture appears like multishear surfaces intersecting parallel to radius.
Bottom : Type D : Internal pressure is very high : hydraulic fracturation : isolated fracture in extension.

Fig. 3 - Normalized nomogram for the analysis of stability conditions of a vertical borehole. The vertical axis is normalized in mud weight (density) units (for a rock density of 2.5) and the horizontal axis describes the state of stresses with the ratio Ko, between horizontal and vertical stress around the borehole.
Example is given for a rock with medium to low mechanical characteristics (internal friction Ø = 20° and cohesion, So = 0).
Right : for a high state of stress (M point, Ko = 0.9), Type A failure appears when drilling with a too low mud density (1.2) : stability need to use mud density higher than 1.4.
Left : if the horizontal stresses are lower (N Point, Ko = 0.6), type B failure appears for mud density lower than 1.25 ; stability needs using mud density higher than 1.25 but not than 1.6 (P point) because instability type C can appear from this excessive value of internal pressure.

De même, la disposition β, excessive, conduira à un type de rupture différent, dit en "mode B", donnant des écailles toriques sur la périphérie entière du trou (en champ latéral isotrope) ou sur un secteur seulement (anisotrope), l'intersection des lignes de ruptures se faisant selon la contrainte intermédiaire maintenant tangentielle.

Ensuite, la disposition γ devrait aboutir à une mosaïque de lignes de cisaillement en parement, rupture dite "en mode C", la contrainte intermédiaire étant alors radiale (JAEGER and COOK, 1976, p. 161).

Enfin, en champ latéral isotrope, on arrive à la génération de contraintes de traction en paroi et rupture, dite en "mode D", en extension ; en champ anisotrope ces ruptures peuvent apparaître beaucoup plus tôt. Leur analyse complète est beaucoup plus complexe.

Il est intéressant de noter que si des écailles type pelures d'oignons ont été vues en essais de laboratoires et en galeries, des écailles toriques ont été obtenues récemment en laboratoire et même dans les "retombées" de forage pétrolier, morceaux écaillés provenant de ruptures des parois du forage ; celles-ci sont très nettes dans le cas d'argilites anisotropes. Les lignes de cisaillement correspondant au mode C auraient été vues (sous réserve de confirmation) en essais types pressiométriques, et sont peut-être le prélude à la fracturation des formations peu perméables dans l'essai de puits pétrolier dit d'injectivité ou "Leak-off Test".

Sans préjuger de la validité et des limites d'application de critères types MOHR-COULOMB ni des anomalies de tenue par rapport aux valeurs numériques qu'ils amènent à prévoir, ces critères fournissent peut-être lorsqu'ils s'appliquent (cas de véritables ruptures en cisaillements) l'allure suivant laquelle ces lignes de ruptures seront défléchies au voisinage des parois ; la forme des écailles produites serait un précieux indicateur de l'état de contraintes et du classement des contraintes les unes par rapport aux autres.

V. MAURY et J.M. SAUZAY (1987) ont signalé des cas précis dans le domaine pétrolier où ces phénomènes ont été observés et traités en conséquence.

De façon à confirmer leur existence au laboratoire, des essais ont été entrepris à BERKELEY (Dpt of Geotechnical Engineering), LONDRES (Imperial College) et LILLE (Laboratoire de Mécanique) où un appareil triaxial de fortes dimensions (cubes de 60 cm de côté chargés jusqu'à 60 MPa) est en cours de construction.

Les essais réalisés à BERKELEY sur un matériau modèle de faible résistance ont montré des amorces de ruptures (photo dans GUENOT 87), évoquant plutôt dans leur phase initiale, un mécanisme par extension en chargement latéral isotrope, dégénérant en pseudo-cisaillements localisés probablement par suite de défauts d'isotropie du chargement ou d'homogénéité du matériau (Fig. 4 colonne centrale).

Des résultats analogues ont été obtenus par MASTIN (1984) et ZOBACK et al. (1985) qui ont considéré l'angle d'ouverture de l'écaille comme représentatif de l'anisotropie de contraintes latérales.

De façon à préciser et mettre en évidence les ruptures d'origine thermique, une recherche a été entreprise à LONDRES (Imperial College). Devant la difficulté de mesures des véritables modules à introduire pour évaluer les valeurs réelles des contraintes thermiques, SANTARELLI (1986, 1987) a repris la distribution des contraintes également en élasticité pour tenir compte des augmentations de modules constatés et mesurés pour de nombreuses roches ; il prend en compte l'effet d'augmentation du module E en fonction de σ_r lorsqu'on s'éloigne de la paroi. La distribution de σ_t et σ_r a donc été recalculée en prenant $E(\sigma_r)$ et $\nu(\sigma_r)$ et appliquée au cas d'un grès carbonifère britannique, en prenant des modules "raidis" linéairement et bilinéairement, permettant dans ce cas d'approcher la phase "pré-pic".

Cette façon de voir apporte des résultats intéressants :

a - La concentration de contrainte tangentielle peut se faire cette fois à l'intérieur du parement, à une distance qui est de quelques % du rayon, et non plus en parement. Ceci expliquerait la naissance des ruptures à l'intérieur et lève le paradoxe lié aux distributions élastiques classiques en champ anisotrope (Fig. 4 colonne centrale).

b - Du fait de cette diminution de la contrainte tangentielle en paroi, comme en distribution élastoplastique mais selon un mécanisme très différent, il faut charger beaucoup plus les tubes épais que ne le laissent prévoir les distributions de contraintes élastiques classiques pour obtenir une concentration de contraintes à l'intérieur qui provoque la rupture. On expliquerait ainsi les cas de tenues anormales qui ont été signalés.

c - Enfin, ce mode de distribution aboutit à des déformations tangentielles plus importantes que celles prévisibles et réputées tolérables par la théorie classique par rapport à un module moyen correspondant à des tests triaxiaux, qui ont été notées par de nombreux observateurs. Cette distribution explique également que la rupture, initiée à l'intérieur du parement et se propageant soit en vrai cisaillement ou en extension selon le comportement du matériau et le réseau des isostatiques, puisse produire des écailles de matériau quasi intact, puisque celui-ci n'a subi que des déformations correspondant à une phase "pré-pic" (le pic ne serait alors plus une caractéristique intrinsèque du comportement du matériau).

d - En conséquence de cette approche, le comportement du matériau laissé par le passage du front, c'est-à-dire sur une distance d'un rayon environ, apparaît comme absolument fondamentale. Ce point était connu, il peut maintenant être appréhendé par cette approche d'une façon plus cohérente avec les observations que la plasticité. Pour les projets d'excavation à l'explosif, on serait tenté de "repenser" les méthodes de prédécoupage pour obtenir ce genre de comportement et distribution de contraintes associées, permettant de ne pas avoir de ruptures. Pour le projet d'excavation à la machine, ou les forages pétroliers, la forme du front, la rotation du champ de contraintes au raccordement du front avec les parois méritent la plus grande attention selon que l'on peut ou que l'on veut faire apparaître une telle zone en paroi, créant par ce procédé une zone quasi-intacte, mais légèrement affaiblie permettant une redistribution de contraintes insuffisantes pour faire apparaître la rupture.

e - Lorsque les valeurs de modules restent indépendantes du confinement, on retomberait sur les distributions classiques élastiques. Cette éventualité paraît rare avec les matériaux rocheux ; ceci expliquerait que des observations correspondantes de ruptures de tubes épais survenant précisément sur le diamètre au point de concentration maximale (A, Fig. 4) aient été exceptionnelles.

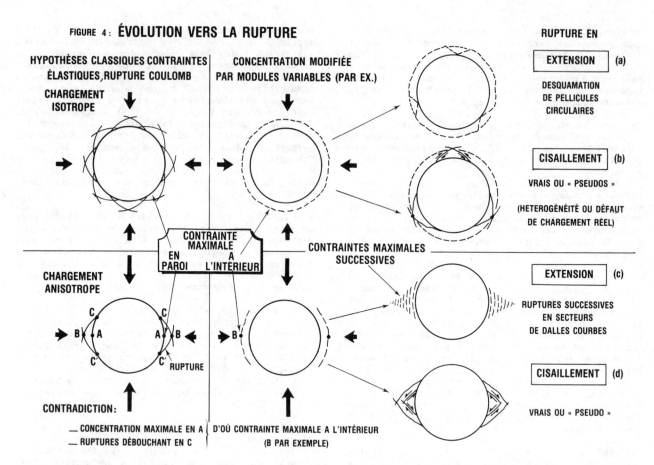

FIGURE 4 : **ÉVOLUTION VERS LA RUPTURE**

HYPOTHÈSES CLASSIQUES CONTRAINTES ÉLASTIQUES, RUPTURE COULOMB

CONCENTRATION MODIFIÉE PAR MODULES VARIABLES (PAR EX.)

RUPTURE EN

CHARGEMENT ISOTROPE

CONTRAINTE MAXIMALE EN PAROI A L'INTÉRIEUR

CONTRAINTES MAXIMALES SUCCESSIVES

EXTENSION (a)

DESQUAMATION DE PELLICULES CIRCULAIRES

CISAILLEMENT (b)

VRAIS OU « PSEUDOS »

(HETEROGÉNÉITÉ OU DÉFAUT DE CHARGEMENT RÉEL)

CHARGEMENT ANISOTROPE

EXTENSION (c)

RUPTURES SUCCESSIVES EN SECTEURS DE DALLES COURBES

CISAILLEMENT (d)

VRAIS OU « PSEUDO »

CONTRADICTION :

— CONCENTRATION MAXIMALE EN A
— RUPTURES DÉBOUCHANT EN C

D'OÙ CONTRAINTE MAXIMALE A L'INTÉRIEUR (B PAR EXEMPLE)

Fig. 4 - Sketches showing the new concepts for the understanding of rock ruptures around a hole
Top : isotropic lateral stress - Bottom : anisotropic
Left : Conventional hypothesis : elasticity + Coulomb's type criteria
Question : How the rupture can appear at the wall borehole at point C, since stress concentration are maximum at point A ?
Middle right : New hypothesis ex. : elastic modulus is changing with confining pressure (radial stress here)
Answer : the maximal (tangential) stress is now inside the borehole wall (B point)
Right : from this point, the failure can appear, depending on various rock-type behaviours, as "extensions" -(a) or (c) cases, or "shears" ((b) or (d) cases-.

f - Enfin, une nouvelle approche de la rupture inspirée de la mécanique des sols (VARDOULAKIS, 1984 et 1986) consiste à considérer la localisation de la déformation comme un cas de bifurcation, constituant un autre phénomène physique, justiciable d'autre formulation théorique, un peu comme le flambage eulérien. De maniement mathématique complexe, cette théorie fait prévoir deux modes de localisation des déformations possibles, soit par instabilité de surface, soit par bande de cisaillement, susceptibles d'apparaître l'un avant ou après l'autre. On manque encore de recul pour juger de l'apport de cette théorie qui paraît prometteuse, évoquant déjà les deux types d'instabilités notés sur des cas vécus et au laboratoire.

g - La mécanique linéaire de la rupture semble indiquer, elle aussi, des possibilités de démarrage de la rupture à l'intérieur du parement à partir de défauts existants ou apparaissant à des distances du parement de quelques pourcents du rayon. Des travaux sur ce sujet sont en cours à l'Université de Lille (France).

6 CONCLUSION

Des séries d'observation de cas vécus de ruptures en galeries isolées et en mines amènent aux conclusions suivantes :

a - Il semble que dans la grande majorité des cas, la rupture s'initie à l'intérieur des parements (point B, Fig. 4).

b - La propagation de la rupture peut se faire soit en extension suivant l'isostatique de contrainte majeure (Fig. 4, a et c), soit en cisaillement (Fig. 4 b et d). Selon les matériaux et le mode de chargement, le mode en extension apparaît comme un mécanisme dominant, susceptible de produire une desquamation en dalles courbes, voire aussi des lignes de ruptures courbes à apparence de cisaillements (pseudo-cisaillement analogue à la déflexion des colonnettes près des plateaux de presse, dans les éprouvettes de laboratoire).

c - Les critères types MOHR-COULOMB sont inapplicables au moins quantitativement à la prévision de la stabilité des ouvertures souterraines. Ils fournissent cependant peut-être une prévision de divers modes et morphologies de rupture que l'on obtient en parement.

d - La question-clé est maintenant de savoir si dès l'initiation de la rupture (en B), celle-ci se propage en extension, voire avec une trajectoire courbe guidée par le réseau des isostatiques, ou bien si elle se propage en vrai cisaillement avec maintien du contact entre les deux surfaces de la discontinuité, et alors quel critère permet de le prévoir ? Les observations en place devront s'efforcer de préciser ce point.

e - Il existe une insuffisance du côté des théories classiques : l'élasticité classique donne des concentrations de contraintes excessives et une contradiction entre les observations et la théorie bien visible en champ anisotrope ; la plasticité ne peut pas représenter des phénomènes dont l'essentiel est l'apparition de discontinuités de déplacements.

f - Il conviendra de bien distinguer à l'avenir, comme l'a proposé GRAMBERG le type de déformations permanentes auxquelles le matériau peut se prêter ou qu'il peut subir : adaptation microgranulaire, peut-être justiciable de théories plastiques à lois de déformations contractantes, plasticité cristalline, rare, et apparition de microruptures (micro-cataclase) dont la localisation constitue un autre phénomène.

Si la trajectoire des lignes de ruptures est bien guidée par l'ordre des contraintes en paroi et sa position par rapport aux critères classiques (MOHR-COULOMB ou courbes intrinsèques), des recherches récentes montrent que l'on peut aboutir à 3 modes de ruptures en cisaillements lorsque le rapport pression interne/contraintes latérales croît, avant d'atteindre un régime de fractures en extension (cas de la fracturation hydraulique). Un modèle de distribution élastique avec module dépendant de la contrainte radiale a été proposé, qui permet d'expliquer une concentration maximale de contrainte à l'intérieur du parement, point à partir duquel peuvent s'initier les fractures, se propageant soit par un mécanisme d'extension, soit par un mécanisme de cisaillement selon le comportement des roches, les modes de chargement et les contraintes résultantes. Elle permet d'expliquer également les cas de stabilité anormale par rapport aux critères conventionnels, les déformations tangentielles excessives que l'on constate dans de nombreuses expériences sur tubes épais, et le fait que les écailles tombées soient constituées de matériau quasi intact.

Plus important encore, le point crucial sur lequel butent toutes les théories, à savoir la discrétisation pseudo-périodique des lignes de rupture que l'on voit lorsque les essais sont bien faits, peut être appréhendé par cette théorie, en imaginant des reports successifs de concentration de contrainte à l'intérieur de la paroi, et dont la forme se modifie progressivement avec celle du contour de l'excavation en matériau intact, quelle que soit la cadence à laquelle cette modification se produit.

Ces schémas en cours de confirmation doivent faire l'objet de recherches théoriques et en laboratoire pour être rendus pratiquement plus utilisables.

A l'échelle des projets, il apparaît souhaitable :

a - De bien séparer les types de comportements auxquels on a affaire, et comme GRAMBERG, les types très différents de déformations permanentes auxquelles ils peuvent donner lieu : vraie déformation plastique cristalline du sel et des évaporites, ou dissolution-recristallisation, "plasticité" granulaire des matériaux poreux, à laquelle on serait tenté d'ajouter "élasticité variable" en fonction de σ_r, microcataclase pour reprendre le terme de GRAMBERG, en roches compactes.

b - Essayer d'approcher au mieux le comportement réel de la roche laissée en place par le mode d'abattage et la forme du front que l'on se propose de réaliser. Les laboratoires de terrains sont là un outil irremplaçable.

c - Commencer à "ouvrir l'oeil" lorsque le niveau théorique de contraintes se rapproche de la résistance conventionnelle en compression simple, sans condamner pour autant le projet, qui peut tenir "anormalement" par rapport à ces critères.

d - Surveiller scrupuleusement toutes les possibilités de chargement différés : mise en place des écoulements, variation du régime de ces écoulements, contraintes thermiques, expansions volumiques d'origines diverses, effets minéralogiques, altérations, etc.

e - En cas de doute, tout mettre en oeuvre pour détecter des ruptures dont le présent travail évoque l'initiation possible à l'intérieur du parement, de façon à savoir si elles correspondent à un mécanisme autostabilisant, ou si au contraire, elles risquent d'évoluer vers une interruption de la construction ou du fonctionnement de l'ouvrage.

f - En cas de doute, ne pas hésiter à conseiller de recourir à des galeries d'essais, seul moyen d'intégrer le rôle de paramètres très difficiles à séparer et à mesurer, comme les contraintes in-situ, et le comportement de la roche.

g - Ne pas omettre, compte tenu des conclusions de ce rapport, de faire la transposition entre essais en place en galerie expérimentale et ouvrage réel, si les modes d'excavation ne sont pas identiques : cette différence est susceptible de changer radicalement le comportement futur de la paroi.

REFERENCES (résumé)

L'article de BROWN et al (1983) contient la bibliographie complète des modèles élastoplastiques proposés.
L'article de GUENOT (1987) contient celle sur les cas de stabilité anormales.

C.R. des conseils de la S.I.M.R. du Symposium d'AIX-LA-CHAPELLE - 1982

C.R. des conseils de la S.I.M.R. du Congrès de MELBOURNE - 1983

Groupement belge de Mécanique des Roches - Journées d'Etudes "Mécanismes de rupture autour d'ouvrages souterrains", Louvain-La-Neuve, 30.05.1985

Rapport du Groupe de Travail du Comité Français de Mécanique des Roches sur l'analyse de la rupture, (1986). Disponible sur demande auprès du Secrétariat du Comité Français. (c/o B.R.G.M., B.P. 6009, 45018 ORLEANS - FRANCE)

BRATLI, R.K. and RISNES, R., "Stability and failure of sand arches", SPE Journal, April 1981, 236-248.

BROWN, E.T., BRAY, J.W., LADANYI, B., HOEK, E., "Ground response curves for rock tunnels", Journ. of Geot. Eng., Vol. 109 n° 1 Janvier 1983.

CORRE, B., EYMARD, R., GUENOT, A., "Numerical computation of temperature distribution in a wellbore while drilling", SPE 13208, HOUSTON Sept. 1984.

FAIRHURST, C. et COOK, N.G.W., "The phenomenon of rock splitting parallel to the direction of maximum compression in the neighbourhood of a surface", C.R. 1er congrès Int. Mec. Roches, LISBONNE, Vol. 1 - p. 687, 692 (1966).

GRAMBERG, J., "The ellipse with notch theory to explain axial cleavage fracturing of rocks (a natural extension to the first Griffith theory)", Int. J. of Rock Mechanics and Mining Sc., Vol. 7 n° 5 - p. 537-559.

GRAMBERG, J., and ROEST, J.P.A., "Cataclastic effects in rock salt laboratory and in-situ measurements", Comm. of Eur. Com., Nuclear Science and Technology, Rapport EUR 9258 EN. 1984.

GUENOT, A., "Contraintes et ruptures autour de forages pétroliers", C.R. Cong. Int. S.I.M.R., Montréal, 1987.

HOEK, E., and BROWN, E.T., "Underground excavation in Rock", Institution of Mining and Metallurgy, London, 1980.

JAEGER, J.C., and COOK, N.G.W., "Fundamentals of Rock Mechanics, 2nd Edition, Chapman and Hall, Londres, 1976.

MASTIN L., "Developpement of borehole breakouts in sandstone" M.S. thesis - Stanford University Ca, 1984.

MAURY, V., "An example of underground storage in soft rock : the chalk", Rockstore, Stockholm 1977.

MAURY, V., FOURMAINTRAUX, D., et SAUZAY, J.M., "Mécanique des roches pétrolières : application pratique à des cas de forage difficiles", C.R. Congr. Int., S.I.M.R., Montréal 1987.

MAURY, V. et SAUZAY, J.M., "Borehole instability : case histories, rock mechanics approach and results", SPE/IADC n° 16051, March 1987.

MORLIER, P., "Etude expérimentale de la déformation des roches" - Rev. I.F.P. Vol. XIX N° 10 et 11, Oct. et Nov. 1964

MORE O' FERRAL and BRINCH, G.H., "An approach to the design of tunnels for ultra-deep mining in the Klerksdorp district", SANGORM Symposium on Rock Mechanics on the design of tunnels, 1983.

ORTLEPP, W.D., and GAY, N.C., "Performance of an experimental tunnel subjected to stresses ranging from 50 MPa to 230 MPa", C.R. Symposium S.I.M.R., Cambridge, 1984.

RISNES, R., BRATLI, R.K., and HORSRUD, P., "Sand stresses around a wellbore", SPE 9650, paper presented at the Middle East O.T. Conf., Bahrain, March 9-14, 1981.

STACEY, T.R. and de JONGH, C.L., "Stress fracturing around a deep-level Board Tunnel", J.S. Afr. Inst. Min. Met., Vol. 78 n° 5, p. 124, p. 133, 1977

STACEY, T.R., "A simple extension strain criterion for fracture of brittle rock", Int. Journ. Rock. Mech. Min. Sci., Vol. 18, p. 469 to 474, 1981.

SANTARELLI, F.J., BROWN E.T., MAURY V., "Analysis of borehole stress using pressure-dependent linear elasticity", Technical Note, Int. Journ. of Rock Mec. and Min. Sciences, à paraître Décembre 1986.

SANTARELLI, F.J., BROWN E.T., "Performance of deep wellbores in rock having a confining pressure-dependent elastic-stiffness", C.R. Int. Cong. I.S.R.M., Montréal, 1987.

VARDOULAKIS, I., "Rock bursting as a surface instability phenomenon". Int. Journ. of Rock Mechanics Min. Sci., 21 - p. 137-144, 1984.

VARDOULAKIS, I., SULEM J., GUENOT A., "Stability analysis of the borehole in deep rock layer", à paraître 1986.

ZOBACK, M.D., MOOS, D., MASTIN, L., and ANDERSON, R.N., "Wellbore breakouts and in-situ stress", Journ. Geophys. Res., 90 (5523 - 5530), 1985.

Alternatives to shaft pillars for the protection of deep vertical shaft systems
Alternatives aux piliers de puits pour la protection des systèmes de puits profonds verticaux
Alternative zu Schachtpfeilern zur Sicherung tiefer senkrechter Schachtsysteme

S.D.McKINNON, Rock Mechanics Laboratory, Chamber of Mines of South Africa, Research Organization, Johannesburg

ABSTRACT: In the gold mines of South Africa, conventional shaft pillars do not provide adequate protection to the shaft and associated service excavations at depths exceeding approximately 3 km. Alternative methods of protection being considered generally involve extraction of the reef around the shaft at the start of mining operations, which solves the problem of high stress build-up, but permits large deformations of the rock mass to occur. These deformations, particulary the reef-level displacement discontinuity, could cause severe problems to the shaft steelwork and lining. Numerical modelling has shown that deformation can be controlled to a degree by the use of backfill, and suitable mining layouts. Delaying shaft steelwork and lining installation until residual deformations are tolerable also alleviates the problem of excessive deformations.

RESUMÉ: Dan les mines d'or d'Afrique du Sud, des piliers des puits conventionnels ne fournissent pas une protection suffisante au puits et aux puits de service auxilliaires à l'excès de 3 km environ. Des techniques de protection en consideration engagent, en général, l'extraction du filon autour du puits au commencement des opérations minières ce qui rèsolvent le problème du générateur de contrainte mais qui laissent la deformation non restreinte à la masse rocheuse. Ces deformations, particulièrement le deplacement discontinu au niveau du filon provoqueraient au structure en acier et au soutènement du puits. Modelage numerique a indiqué que la deformation peut être controllée à un certain degré par l'usage de terre de remblayage et la planification appropriée des mines. Le retardement de l'installation du structure en acier et du soutenement du puits jusqu' aux deformations residuelles sont tolérables allège le problème des deformations excessives.

ZUSAMMENFASSUNG: Die konventionellen Schachtpfeiler in den Südafrikanischen Goldminen bieten keinen ausreichenden Schutz für den Schacht und die verbundenen Hilfsausschachtungen in Tiefen von mehr als ca. 3 km. Alternative Schutzmethoden, die erwogen werden, befassen sich durchweg mit dem Abbau des Erzganges rings um den Schacht bei Beginn des Abbaus; dies löst zwar das Problem von hohem Spannungsaufbau, lässt aber eine vollkommen freie Verformung der Gesteinsmasse zu. Diese Verformungen, vor allem die Verschiebungsdiskontinuität auf der Ebene des Erzganges, könnte ernste Probleme für die Stahlkonstruktion des Schachts und seine Auskleidung verursachen. Numerisches Modellieren hat gezeigt, dass Verformungen zu einem bestimmten Grade unter Kontrolle gebracht werden können durch Hinterfüllung und durch geeignete Abbauplanung. Das Problem übermässiger Verformungen kann auch dadurch verringert werden, indem man die Stahlkonstruktion im Schacht und das Anbrigen der Auskleidung verzögert, bis verbleibende Verformungen annehmbar sind.

1 INTRODUCTION

The conventional method of protecting vertical shafts and their associated service excavations from the effects of mining of tabular orebodies, is to leave a sufficiently large block of unmined reef (shaft pillar) such that mining induced plus virgin stresses are kept below damaging levels. In the gold mines of South Africa, depths of shaft-reef intersections are beginning to exceed 3 km. At these great depths, virgin stresses are so high that shaft pillar sizes required to reduce induced stresses to acceptable levels become impractically large (Wagner and Salamon 1973). Also,depending on the strength of the particular rock mass, virgin stresses alone may be above damaging levels. Shaft pillars therefore no longer provide an adequate method of protection at great depth.

Since any unmined reef in the vicinity of the shaft will become highly stressed at great depth, the principle of protection for all alternatives being considered is to either carefully engineer the location of any unmined reef (in the form of crush pillars or satellite pillars), or to completely mine the reef around the shaft, the 'shaft-reef', prior to commencement of regular stoping activities.

Complete removal of the shaft-reef reduces stress levels but permits unrestrained deformation of the rock mass to occur prior to total closure. These deformations may be acceptable from a rock mechanics point of view, but they assume particular importance when considering the stability and design of shaft steelwork and linings.

The purpose of this paper is to review the possible alternatives to use of deep shaft pillars, and to quantify certain practical methods that are available to reduce or virtually eliminate the undersirable effects of strata deformation.

2 MODIFICATIONS AND ALTERNATIVES TO SHAFT PILLARS

Various alternatives to shaft pillars have been discussed in the past (More O'Ferrall 1983) but little experience has been accumulated by the Industry in their planned application. A review of some of the more important methods of shaft system protection being considered will highlight some of their common features, and will help illustrate their differences when compared to the use of conventional shaft pillars.

2.1 Satellite pillars

A logical progression from a solid shaft pillar experiencing high stress levels would be to overstope service excavations and thereby leave a segmented pillar, or a series of satellites. A slight variation of this method would be to extend a pattern of stabilizing pillars into the vicinity of the shaft. By moving the satellites away from the shaft, the zone of stress relief, in which the service excavations may be placed, increases. At the same time, however, convergence takes place in the unsupported span of the central area allowing deformation of the shaft.

Such convergence can be limited by the use of backfill, and in general, displacements in the shaft will be less than in a scheme involving total extraction of the shaft-reef area.

2.2 Differential stoping width

As a means of controlling deformations, backfill has the effect of reducing the effective stoping width and has been the traditional method of limiting convergence. In the case of wide reefs or closely spaced multireef geometries, however, the concept of differential stoping width may be used more advantageously. Rather than mining the full width of the reef everywhere, a narrower section could be mined in specific areas, concentrating on the highest grade zones. Around the shaft, larger stoping widths would ensure that when convergence takes place, closure first occurs in the surrounding zones of lower stoping width, effectively creating "satellite pillars". In this manner, most of the reef can be recovered, convergence is reduced, and shaft deformations can be controlled. Additionally, stresses beneath those zones mined at full width will be reduced.

2.3 No Shaft pillar

The most extreme alternative to a shaft pillar is to mine the shaft-reef as extensively as possible from the outset of mining. In layouts involving satellite pillars or differential stoping width, shaft-reef extraction would be limited in order to reduce the elastic convergence and associated rock mass deformations, while still providing a destressed umbrella of overstoping for nearby service excavations. The alternative of complete extraction would be optimum from a reef recovery viewpoint, but would result in increased deformations. As depth is increased, moreover, ground control considerations lead to a reduction in free spans which can be mined without the introduction of stabilizing pillars or backfill.

2.4 Comments on alternatives

The alternative methods of shaft protection described above are variations on a common theme, but all are radical departures from the use of conventional shaft pillars. Since the stability of shafts and their associated excavations are vital to the operation of a mine, any proposed alternatives must be thoroughly studied prior to their use. Major differences between the use of shaft pillars and alternative protection schemes, and methods of reducing problems associated with their application will be addressed in the following sections.

3 MAJOR DIFFERENCES BETWEEN SHAFT PILLARS AND ALTERNATIVE PROTECTION SCHEMES

The optimum layout for a given situation will depend on many factors, and variations of the alternatives described previously may be adopted. These particular alternatives all involve extraction of the shaft-reef which, from a design viewpoint, leads to a number of differences in rock mass behaviour when compared to leaving a shaft pillar. Figure 1 illustrates the shaft axis and the essential differences in rock mass behaviour between conventional and alternative protection schemes, while Table 1 summarizes how these differences apply to the shaft system.

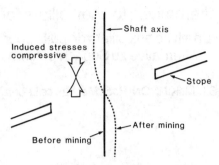

(a) Conventional shaft pillar.

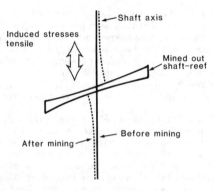

(b) Extraction of shaft-reef.

Figure 1. Sketch of section through shaft axis showing major differences in rock mass behaviour between (a) conventional shaft pillar and (b) extraction of shaft-reef.

Table 1. Major differences between conventional and alternative methods of shaft system protection

	Shaft pillar	Shaft-reef extraction
Reef-level stress	high	low
Reef-level deformation/ dislocation	small/absent	large
Induced stress, strain along shaft axis	compressive	tensile
Shaft axis tilt	larger	smaller

While the problem of the high stress build-up associated with shaft pillars is solved by shaft-reef extraction, new problems arise due to the differences in rock mass behaviour and its effect on the shaft steelwork and lining.

Firstly, reef-level deformations are increased, and in particular, a discontinuity in displacement is created in the shaft axis between the hangingwall and footwall of the mined-out reef as convergence of the shaft-reef area takes place. This type of dislocation can be, and has been, accommodated by novel shaft steelwork and lining designs, such as those used during shaft pillar extraction (Van Emmenis and More O'Ferrall 1970). However, these modifications add a further dimension of complexity and cost to shaft equipping.

1130

The second major difference associated with shaft-reef extraction alternatives is that the latter generate induced tensile stresses and strains along the shaft axis. Rock mass behaviour in compression is reasonably well understood, but in tension it is far less predictable.

In the immediate hangingwall and footwall of the mined out shaft-reef, there will be a surrounding envelope of mining induced fractures such as is seen around a stope (Jager and Roberts 1986). Formed by high compressive and shear stress concentrations near the stope face, this envelope is generally limited in extent to some tens of metres from the plane of the reef. It is this near-reef zone, however, which is subjected to the largest induced tensile stress after the passage of the stope face. Also, bedding planes or other reef parallel planes of weakness may open up under the influence of induced tensile stresses. The resulting inelastic behaviour of the rock mass, and the uncertainty of its magnitude and spatial extent complicate shaft steelwork and lining designs.

These differences were examined in detail using numerical modelling with the objective of identifying ways in which shaft deformations and consequent steelwork design difficulties could be reduced as much as possible.

4 NUMERICAL MODELLING OF SHAFT-REEF EXTRACTION

4.1 Description of model and layout used

In order to quantify the effects of progressive extraction of the shaft-reef, a numerical study was carried out using the MINSIM-D program (Ryder and Napier 1985). This is a 3-D tabular mining stress analyser, based on the boundary element method, developed at the Chamber of Mines. Figure 2 shows a plan of the selected mining layout, at an assumed depth of 3 000 m. The shaft is located in the centre of a completely mined out area which in turn is surrounded by stoped ground protected by stabilizing pillars. The stoping width is 1,0 m. This configuration, while idealized, nevertheless embodies current concepts of practical mining and rock mechanics for layouts at great depth. Variations of this basic layout were used at other depths.

Progressive extraction of the shaft-reef was modelled by symmetrically mining out squares of increasing size around the shaft up to a maximum size of 450x450 m. More extensive mining, incorporating the layout shown in Figure 2, was modelled up to a total mined out area with a span of 3,8 km.

The most rapid and significant mining-induced effects occurred within the area modelled, with further mining being remote enough that only minor changes were induced at the shaft location. Other typical model parameters are listed in Table 2.

Table 2. Typical values of parameters used in MINSIM-D models

Parameter	Symbol	Value
Young's modulus	E	70 GPa
Poisson's ratio	ν	0,2
Stoping width	Sm	1,0 m
Vertical stress gradient		0,027 MPa/m
Horizontal to vertical virgin stress ratio	k	0,5
Backfill placement height		0,9 m
Good quality backfill	a	5,0 MPa
	b	0,3
Poor quality backfill	a	3,0 MPa
	b	0,5

The sign convention for stress is such that compression is positive.

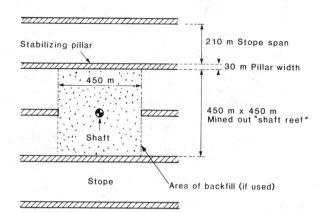

Figure 2. Plan of model layout for 3 000 m depth.

Due to physical constraints, backfill must be placed a number of metres behind the stope face. At the great depths being considered, a certain amount of elastic (and inelastic) convergence will already have taken place within this short distance from the stope face such that backfill cannot be placed at the original stoping width. For modelling purposes, it has been assumed that 0,1 m of prior convergence will have taken place. Backfill placement height is, therefore, 0,9 m as detailed in Table 2. This reduces the effective backfill stiffness as a means of controlling convergence. Methods of reducing this effect are currently being investigated.

Two types of backfill were considered in the modelling to bound the range of support effect, these being 'good' quality (stiff) and 'poor' quality (more compressible) backfill. As backfill stiffness increases, so does its ability to resist convergence and, therefore, to decrease the energy release rate (ERR) at the stope face. Backfill load-deformation behaviour was modelled using the following hyperbolic reaction curve (Ryder and Wagner 1978)

$$\sigma_f = \frac{a\varepsilon}{b-\varepsilon}$$

where 'a' and 'b' are the characteristic stress and strain parameters respectively. Values corresponding to 'good' and 'poor' quality backfills are shown in Table 2; these are representative of qualities currently attainable. Higher quality behaviour may be achieved, at higher cost, by use of appropriate grading/blending/placement techniques.

The benefit of backfill is largely dependent on the 'b' value which represents the limiting strain asymptote of the material. Good quality backfill is therefore more effective in reducing ERR and reef-level displacements.

4.2 Criteria used to evaluate results of modelling

Certain parameters relevant to the design of stoping layouts and shaft system stability were examined in detail. At this conceptual stage of modelling, relatively simple criteria were used which, without exception, were developed empirically for conditions prevailing in South African gold mines. These parameters and typical design criteria included:
- Energy Release Rate (ERR). In unfaulted ground, ERR is used as an indication of stoping conditions (Hodgson and Joughin 1967). For this study, a limiting value of 30 MJ/m² was chosen.
- Reef-level stress and displacement. The major principal stress should not exceed approximately 100 MPa to ensure the stability of near-reef service excavations. Movement across the

reef-level displacement discontinuity should be as small as possible although no firm criteria are available.
- Induced stress and strain along the shaft axis. Induced stress change should be less than 18 MPa (Wilson 1971) to avoid damaging the concrete lining, and induced tensile stresses should be avoided. Steelwork damage may occur if induced strains exceed 0,0004 in either compression or tension (Wilson 1971, van Emmenis and More O'Ferrall 1971).
- Tilt of shaft axis. No firm criteria exist. It would appear however, that tilts, unless provided for, should be limited to 0,001 measured as an out of plumb strain (horizontal offset over a vertical drop).

4.3 Potential seismic risk

Consideration of any alternative mining concepts in deep-level mines would not be complete without addressing the potential seismic risks, and specifically, the possibility of rockbursts. Presently, rockburst mechanisms are imperfectly understood, but on a qualitative basis, the seismic risks of leaving a shaft pillar can be compared to those of the proposed alternative of pre-mining.

The relative seismic activity of these two situations can be assessed by examining three fundamental requirements for a seismic event (Wagner 1985). Firstly, there must be a situation of unstable equilibrium in the rock mass (such as a fault with incipient shear failure). Secondly, there must be a source of energy for the event. Finally, the source of energy must be sufficiently large that it provides a sustained driving force as opposed to only momentary stimulation.

Shaft pillars are designed to provide a low stress (protected) interior, but, as the extent of surrounding mining increases, the pillar edges become highly stressed, providing a large store of strain energy. At the end of the life of the shaft, pillar recovery results in extraction of the low stress interior, further increasing the store of strain energy in the final remnants (satellite pillars). Shaft-reef pre-mining, on the other hand, is carried out in a lower stress environment in the absence of induced stresses from far-field mining and causes stress relief of the immediate rock mass around the shaft. It can be concluded that the store of energy is lower in the case of shaft-reef pre-mining, and this is therefore expected to be in general a seismically more favourable scenario than that of leaving a shaft pillar.

5 DISCUSSION OF RESULTS

The mining sequence, support type, and layout at reef-level are the only means of controlling convergence and, therefore, the effects of mining on conditions at reef-level as well as along the axis of the shaft. Numerical modelling allowed the behaviour of these components to be compared, which aided in developing the solution strategy.

5.1 Effects of mining on conditions at reef-level

On the basis of ERR considerations, free spans of 450 m could not be mined around the shaft at depths greater than 3 km without exceeding the limiting value of 30 MJ/m^2. Complete backfilling of the shaft-reef area was considered the most effective means of reducing ERR for two reasons. Firstly, complete support of the hangingwall would be provided, thereby ensuring the integrity of the rock mass immediately around the shaft (in particular, elimination of bed separation). Secondly, backfill could provide uniform distribution of load transfer between hangingwall and

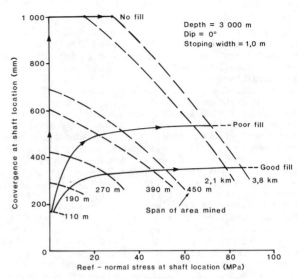

Figure 3. Convergence versus reef-normal stress at shaft/reef intersection.

footwall, giving flexibility in the siting of major service excavations. Stabilizing pillars are more efficient in reducing ERR but may cause unacceptable stress concentrations in the rock mass.

At reef level, changes in stress and deformation as a function of the extent of mining can be conveniently portrayed in the form shown in Figure 3. As mining progresses outwards from the shaft, indicated by the dashed lines, convergence takes place in the stope and reef-normal stresses increase. Stress levels are mainly of concern for the stability of service excavations, whereas the displacement across reef must be accommodated by the shaft steelwork and lining.

In the models it was assumed that mining took place symmetrically around the shaft, and as a result, convergence is maximum at the shaft location. For the model layout used in producing Figure 3, convergence results in only vertical displacements between the hangingwall and footwall due to the reef being horizontal.

The stress build-up is also determined at the shaft/reef intersection. If no backfill is used, stress build-up only occurs after total closure has taken place. With complete backfilling of the shaft-reef area, it was found that reasonably uniform stress build-up occurred throughout the backfill, without peripheral stress concentrations characteristic of shaft pillars. This behaviour provides increased flexibility in locating near-reef service excavations.

The particular combination of stress build-up and displacement is dependent on the type of backfill placed in the shaft-reef area, if any. For example, Figure 3 shows that if a good quality backfill is placed, and an area of 270 m x 270 m has been mined around the shaft, backfill stress build-up would be 26 MPa and the vertical displacement (convergence) would be 326 mm. If a poor quality backfill had been used instead, the stress build-up would only have been 6 MPa, and a larger vertical displacement of 401 mm would have occurred. For the case of no backfill, the stress and displacement path follows the displacement axis until complete closure occurs. Similarly, if the stress-displacement path for a shaft pillar were to be plotted, it would closely follow the stress axis, starting at the virgin stress level, and building up to high stresses for large mined areas.

Important features shown by reef-level stress-displacement diagrams such as Figure 3 are:
- the compromise between stress build-up and the reduction in convergence which can be engineered to some extent by the use of backfill,
- the asymptotic nature of convergence when backfill is used, and

- the rapid approach of the stress-displacement path to the displacement asymptote at relatively small spans.

Figure 3 also shows that after large areas have been mined, relatively high stresses are re-introduced due to backfill compression. In certain layouts at great depth, consideration must therefore be given to decreasing the stiffness of backfill support by decreasing the quality of the material directly, or indirectly. This can be achieved by delaying placement behind the stope face, or by partial backfilling, in order to avoid the build-up of damaging stresses generated by backfill compression. By the same token, specially placed backfills with very low 'b' values (as low as 0,1), currently being investigated for use in stopes, would be far too stiff for use in the shaft-reef area as they could result in the build-up of stresses higher than virgin levels.

For dipping strata, convergence of the hangingwall and footwall results in horizontal as well as vertical components of displacement discontinuity at the shaft/reef intersection, which must be negotiated by shaft steelwork. This effect is shown in Figure 4, which depicts the horizontal and vertical components of the reef-level discontinuity in the shaft, for the case of a reef dipping at 30° with a shaft/reef intersection at a depth of 4 000 m. In this example, the displacement asymptotes are closely approached at spans of between 300 m and 400 m, depending on the type of backfill used. The beneficial effect of backfill on controlling deformations is again clearly illustrated.

5.2 Effects of mining on conditions along the shaft axis

The effects of mining, and backfilling, are also felt along the axis of the shaft. For any proposed mining layout, it is therefore important to be able to predict the types and magnitudes of rock mass deformations along the axis of the shaft in order to assess stability of the shaft steelwork and linings.

Criteria for assessing damage to the shaft steelwork and lining are expressed in terms of induced stress, strain and tilt. Figure 5 shows the induced vertical strain profiles along the shaft axis, at various stages of mining, for the case of a 30° dipping reef with the shaft/reef intersection at 3 000 m depth. A poor quality backfill was used in the shaft-reef area.

Some of the more important trends brought out by the modelling were:
- initial large near-reef induced tensile strains followed by compressive strain changes as backfill compressed,
- magnitudes of strain tending to limit values at large spans, and
- the existence of a maximum strain envelope.

Referring to Figure 5, when the mined area is small (110 m x 110 m profile), large near-reef vertical strains are induced, but these decay rapidly with distance away from reef level. As the area of mining increases, so does the extent of its influence into the hangingwall and footwall; but, at the same time, compression of the backfill reduces the near-reef induced vertical tensile strain such that subsequent near-reef strain changes are compressive. This behaviour is particulary important when considering unpredictable inelastic effects, such as bed separation, which are more likely to occur in a tensile regime than one in which strain changes are compressive.

During the re-introduction of compressive stresses by backfill compression, previously open parting planes would close, and further inelastic effects would be unlikely to occur. Shaft equipping after the closure of at least those parting planes close to the reef plane would clearly be more desirable than in one where strain changes were tensile.

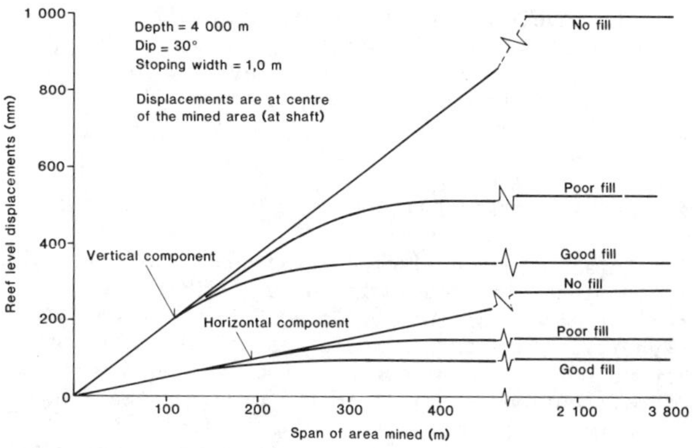

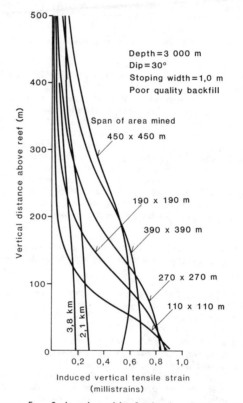

Figure 4. Reef-level displacements versus span of mined area, 4 000 m depth, 30° dip, with and without backfill in shaft-reef area.

Figure 5. Induced vertical strain along axis of shaft, 3 000 m depth, 30° dip, poor quality backfill in shaft-reef area.

After a sufficiently large area has been mined, further mining has a negligible effect on the rock mass at the shaft location. Strains in this case tend towards a limiting profile, which in Figure 5, is closely approximated by the profile for a mined area of 3,8 km x 3,8 km. The position of the limiting profile is dependent on the relative stiffness of the mining layout outside the shaft-reef area compared to that of the backfilled shaft-reef area itself, and also on the size of the shaft-reef area. These are factors which can be controlled to a certain extent by suitable layouts.

Families of induced vertical strain profiles, such as those shown in Figure 5, sweep out a distinct area such that an envelope of maximum induced elastic vertical strain experienced over the life of the shaft can be drawn. This envelope is useful in bounding the effects of mining on the shaft steelwork and lining.

Tilt profiles along the shaft axis for various layouts involving pre-mining of the shaft-reef were also examined. In all cases, the maximum envelope was well below the empirical damage level, indicating that this mode of damage plays only a minor role in the overall problem.

6 APPLICATION OF MODELLING RESULTS

A natural method of implementing shaft-reef pre-mining would be to sink and equip the shaft, followed by shaft-reef extraction utilizing full shaft hoisting capacity. This option would require steelwork and lining designs to accommodate the types and magnitudes of deformation described in the previous section. These deformations are not only difficult to accommodate, but would require frequent adjustments of steelwork, which, cost-wise, can be considered to be a form of 'damage'.

Figure 3 shows that it is not possible to minimize simultaneously both reef-level stress build-up and convergence. Since stress levels re-introduced by backfill compression are in most cases acceptable, the main problem is to further reduce the reef-level displacements and induced strains along the shaft axis.

Two methods of reducing the deformations are being considered, both based on the use of backfill in the shaft-reef area. In Figures 3 to 5 it was noted that as the extent of mining increases around the shaft, induced strains and displacements tend towards asymptotic values. After a certain extent of mining, therefore, a proportion of the total deformation is dissipated, and the remaining deformation up to the asymptotic value is reduced. For convenience, this remaining balance of deformation will be referred to as the 'residual'. Acceptable residual deformations, i.e. deformations which are either below damaging levels or which would require only minor steelwork and lining re-design, can be achieved by two methods:

(i) mine out a portion of the shaft-reef concurrently with shaft sinking and before shaft equipping, or

(ii) halt sinking and mine out the shaft-reef, then complete sinking and commence steelwork installation.

Until the shaft has been equipped, hoisting can be carried out using shaft sinking equipment, or an independent mid-shaft loading system (Tregoning 1983). In order to minimize any delays in production build-up, the area of pre-mined shaft-reef must be as small as possible, subject to the requirement of protecting service excavations. The extent of pre-mining, however, depends on the tolerable deformations that the steelwork and lining can safely withstand. Residual deformations must, therefore, be less than the tolerable deformations by some safety margin but, once this is determined, it is a simple matter to use the results of modelling (such as Figures 3 to 5) to obtain the area which should be pre-mined to yield the desired residual. The solution is somewhat gradational in that many combinations of pre-mined area and steelwork designs could work. However, from all points of view, there are clearly many advantages to pre-mining at least that relatively small area of shaft-reef for which deformation dissipation is large. Analysis of the modelling results shows that if areas with spans of between 300 m and 400 m are pre-mined, maximum dissipation of displacements is achieved, and residuals are reduced to easily-accommodated levels. Futhermore, complications due to inelastic effects, which are more likely to occur at small spans prior to backfill compression, will be avoided.

7 CONCLUSIONS

Shaft pillars do not provide an adequate means of protecting shaft systems from the damaging effects of stress at depths in excess of approximately 3 km. Alternatives being considered generally involve mining out the shaft-reef (reef around the shaft) to reduce these stresses, but in turn, lead to large rock mass deformations. Backfilling of the shaft-reef area is shown to be necessary in terms of reducing ERR and improving stability conditions around the shaft. Compressive stresses re-introduced by backfill compression are in most cases below damaging levels. Deformations, both at reef-level and along the shaft axis, approach limiting values at relatively small spans.

By delaying shaft equipping until sufficient mining and filling of the shaft-reef has taken place, a large proportion of the total deformation can be dissipated. The residual deformations to be accommodated by the shaft lining and steelwork are thereby significantly reduced.

The proposed methods of implementing this alternative to shaft pillars provide solutions to the problem of high stress build-up in pillars, and to the problem of large deformations associated with mining out the shaft-reef.

8 ACKNOWLEDGEMENT

This paper describes work carried out as part of the research programme of the Research Organization of the Chamber of Mines of South Africa. Permission to publish this paper is gratefully acknowledged.

REFERENCES

Hodgson, K. & N.C. Joughin 1967. The relationship between energy release rate, damage and seismicity in deep mines. In C. Fairhurst (ed.). Failure and breakage of rock. Proceedings 8th Symposium on Rock Mechanics, University of Minnesota.

Jager, A.J. & M.K.C. Roberts 1986. Support systems in productive excavations. Proceedings of the International Conference on Gold, volume 1, Johannesburg : SAIMM.

More O'Ferrall, R.C. 1983. Rock Mechanics aspects of the extraction of shaft pillars. In S. Budavari (ed.). Rock mechanics in mining practice, Johannesburg : SAIMM.

Ryder, J.A. & J.A.L. Napier 1985. Error analysis and design of a large-scale tabular mining stress analyser. In Procs. 5th Int. Conf. on Numerical Methods in Geomechs., Nagoya.

Ryder, J.A. & H. Wagner 1978. Analysis of backfill as a means of reducing energy release rates at depth. Unpublished material.

Tregoning, G.W. 1983. Mid-shaft loading development at Cooke 3 shaft. Association of Mine Managers of South Africa, Papers and Discussions 1982-1983. Johannesburg : Chamber of Mines of South Africa.

Van Emmenis, R.J. & R.C. More O'Ferrall 1970. The extraction of the Toni shaft pillar. Association of Mine Managers of South Africa, Papers and Discussions 1970-1971. Johannesburg : Chamber of Mines of South Africa.

Wagner, H. 1985. Unpublished lecture notes.

Wagner, H. & M.D.G. Salamon 1973. Strata control techniques in shafts and large excavations. Association of Mine Managers of South Africa, Papers and Discussions 1972-1973. Johannesburg : Chamber of Mines of South Africa.

Wilson, J.W. 1971. The design and support of underground excavations in deep-level, hard-rock mines. PhD thesis, University of the Witwatersrand Johannesburg.

The phenomena, prediction and control of rockbursts in China
Les phénomènes, la prédiction et le contrôle des coups de terrain en Chine
Das Auftreten, die Vorhersage und die Kontrolle der Gebirgsschläge in China

MEI JIANYUN, Institute of Geophysics, Academia Sinica, Beijing, People's Republic of China
LU JIAYOU, Institute of Water Conservancy and Hydroelectric Power Research, Beijing, People's Republic of China

ABSTRACT: This paper summarized some typical phenomena and effective methods for rockburst prediction and control in China. The cases cited here include 6 coal mines, 2 metal mines, 2 largescale hydropower stations and 1 railway tunnel.

RÉSUMÉ: Le papier decrit des phénomèmes typiques de roche-éclat en China et les methodes de prevision et contrôle. Ces génie-ouvrages comprendent six mines de Charbon, duex mines metalliques, duex usines hydrauliques à grande echelle et un tunnel de chemin de fer.

ZUSAMMENFASSUNG: Dieser Artikel beschreibt einige typische Erscheinungen der Felsbersten in China und die gültige Methoden für den Bericht und die Kontrolle. Diese Felsbaus enthält sechs Kohlenminen, zwei Metallminen, zwei grossen Wasserkraftwerke und eine Eisenbahntunnel.

1 INTRODUCTION

Rockburst is one of the main disasters in mines and underground works especially in coal mines. In China, the earliest record of rockburst can be traced back to 1933 and an incomplete statistic shows that about 2,000 rockbursts have occurred in 32 coal mines since 1949. It can be expected that the occurrence of rockburst will tend to increase in number with increasing coal production and mining depth. Rockbursts of various intensity have also been encountered in metal mines, hydropower stations and railway tunnels. In the past decade, numerous studies on the prediction and control of rockburst were carried out in China and encouraging results in eliminating and reducing the catestrophic consequence of disasters were achieved. In this paper, the term "rockburst" refers to coal burst and large-area roof falling as well.

2 PHENOMENA OF ROCKBURST

Rockburst mostly occurs in underground works built in hard rocks. Some typical case histories are given below.

1. The railway tunnel at Guanchunba. This tunnel is situated in a mountain and-canyon region in Southwestern China. Its buried depth is 600-1500 m. The surrounding rock is a limestone hard and intact with a uniaxial compressive strength of 95-128 MPa. During its construction, 7 larger rockbursts occurred in the period from January to March, 1976. They have two types of manifestation. In one type, smaller rock fragments were projected out with a rather loud pop. The area of fragments is generally a few square centimeters and the largest volume is 0.5X0.37 X0.08 m^3. In the other type, the lumpiness of the rock block thrown out is larger and is generally 2.5-4.2 m long, 1.0-3.0 m wide and 0.1-0.3 m thick. The largest volume amounts to 3.5X1.5X0.3 m^3. In contrast, the accompanying pop is less loud. Rockbursts mostly occurred within 2-3 hours after shoot. Some of them even delayed for several days and thus led to some troubles to the construction.

2. The Tunnel of Tianshengqiao Hydropower Station. The branch tunnel No. 2 mentioned here is 10 m in diameter and 1.3 km in length. It was excavated by TBM. In a section with a buried depth of 200-500 m,

a number of rockbursts occurred. Five among them were of larger scale. The bursts mostly occurred at the arch-top. The rock is a limestone fresh and kard with a uniaxial compressive strength of 60-80 MPa. The failure caused by rockburst was in the form of cleavage spalling accompanied by sounding. The spalled body was generally 7-10 cm wide, 3-5 cm thick and was thicker at the center than on the margin. The largest dimension of the spalled zone was about 2 m in length, 1.5 m in width and 0.6-0.7 m in depth.

3. Ertan Hydropower Station. The host rock there is a hard and intact syenite with a uniaxial compressive strength of 200 MPa, elastic modulus of 70 GPa and Poisson's ratio of 0.22. Among the 203 boreholes in the dam area, disc-shaped cores occurred in 84 of them . In 40 of the 48 boreholes in the river bed zone with high stress concentration, core discing was observed. The disc thickness is 1-3 cm and the major principal stress measured in the borehole is 65 MPa. Besides, when preparing rock specimens 40X40 cm^2 in area for shear tests in an adit with a 400 m thick overburden, the specimen threw rock fragments out or fractured along the interface of specimen and host rock with a pop whenever the cutting depth around it attained to 10 cm or so. When continued to cut, the same phenomenon occurred again and thus obstructed the preparation of specimen. It is suspected that rockburst is likely to occur in the future when excavating the underground power house.

4. Pangushan Tungsten Mine. The rock surrounding the gallery is a quartzite which is hard and intact with a uniaxial compressive strength of 200 MPa, elastic modulus of 60-70 GPa, Poisson's ratio of 0.18 and a longitudinal wave velocity of 5 km/s. Disc-shaped cores with a disc thickness of 0.4-1.5 cm were found in boreholes. The measured geostress is 20-35 MPa and the stress concentration factor of the gallery is 3-6. 13 rockbursts happened when mining at the depth of 400-600 m. When bursting, the rock projected powder or fragments out or outburst by a length of up to 10 m. Consequently, pillars were crushed.

5. Jinchuan Nickel Mine. The uniaxial compressive strength of the ore body is 75 MPa. Disced cores were found in boreholes. At the mining depth of 400-500 m, rockburst occurred in the gallery. The principal stresses are horizontal in direction with a

major one of 30 MPa and increase with depth.

According to imcomplete statistics, coal bursts have occurred in about 32 coal mines up to 1985. Among them, the destructive ones amount to 2,000 in times. The maximum seismic magnitude ranges from 2.5 to 3.8 in Richter scale. The total length of destroyed gallery attained to 17 km. Nowadays, the depth of coal mining in China extends 20-30 m downward every year and thus leads to an increasing potential menace of rockburst.

Some information about the six typical coal mines where rockbursts occurred is given in Table 1.

Table 1. A compendium of the statistics of typical coal bursts in China.

Name of mine	Coal seam		Roof			Physico-mechanical properties of coal						Mining depth (m)	Depth of coal burst (m)	Statistic of coalburst		Richter magnitude	Characteristics of coalburst
	Thickness (m)	Dip angle (°)	Rock type	Thickness (m)	Compressive strength (MPa)	Porosity (%)	Water content (%)	Compressive strength (MPa)	Elastic modulus (GPa)	Poisson's ratio (γ)	Bursting liability (W_{Et})			time	number		
Hentougou	0.7-3.5	3-15	Sandstone	5-30	130-190	8.1	3.00	19-30	6.6-9.0	0.21-0.32	7.00	600-900	200-500	1949-1982	288	2.4-3.8	Typical case of coalburst occurs most frequently and with highest intensity in China
Tangshan	2.0-10.0	5-20	Sandstone	10-30	150	-	1.67	17-40	4.3	0.23	7.40	700	600-700	1964-1984	46	3.4	Typical case of coalburst in depth in China
Tianchi	0.7-4.0	40-70	Limestone	220	120	-	0.19-0.71	9-10	2.4	-	3.84	240-700	400-700	1958-1985	49	-	Typical case of tectonic type coalburst in China
Taozhuang	2.0-5.0	10-25	Sandstone	20-40	90-150	-	1.00-2.50	10-20	5.6	0.23-0.41	-	960	480-960	1976-1982	146	1.5-2.7	Typical case of serial coalburst in China
Longfeng	50.0-60.0	0-60	Shale	100-200	200	5.7-8.0	1.73-2.30	10-14	1.9-3.3	0.25-0.30	3.31	730	600-700	1976-1983	675	2.5	Typical case of thick-layered coalburst in China
Datong	1.7-4.6	3	Sandstone	4-11	100-800	-	-	25-50	0.29	-	-	200-300	50-100	1960-1970	26	3.2	Typical case of large-area caving in China

The characteristics of rockbursts in China can be summarized as follows.

1. Rockbursts mostly occur in rocks which are hard and intact and of high geostress. Disc-shaped cores are often found in boreholes. According to the statistics of several hydraulic tunnels in China where rockbursts have taken place, the maximum peripheral stress at the entrance of the tunnel is less than or close to one half of the ultimate compressive strength of the rock. It indicate that the condition of rockburst occurrence is the fracture under low stress of a brittle material with high strength.

2. The coal seams in which coal bursts occur are usually 0.7-10.0 m in thickness with a variety of dip angles. The roof rock is hard and intact with a uniaxial compressive strength of 100-200 MPa. The coal has a water content less than 30%, uniaxial compressive strength 10-50 MPa, elastic modulus 2-9 GPa, Poisson's ratio 0.2-0.3, bursting liability index 3-7 and Richter magnitude 1.5-3.8.

3. The occurrence of rockburst is generally at the depth of 200-700 m and increases in number with increasing depth. Rockbursts mostly occur in coal pillars and advancing faces in highly stressed zones. Some of them may also occur in zones having subjected to intense tectonic movements or showing local geological anomalies.

4. For a hard sandstone roof overlying the coal seam, large area integral falling often happens to result in disaster.

3 PREDICTION OF ROCKBURST

The methods commonly used for rockburst prediction in China are as follows.

3.1 Identification of rockburst liability

Strain energy storage index W_{ET} (A. Kidybinski, 1981) is rather widely used for this purpose. It is defined as:

$$W_{Et} = \phi_{sp}/\phi_{st} = \text{elastic strain energy/dissipated strain energy}$$

where ϕ_{sp} and ϕ_{st} are determined from the areas under the load-deformation curves for loading up to and unloading from 80-90% of the ultimate strength respectively. It is specified that

$W_{ET} \geqslant 5.0$, high bursting liability,

$2.0 \leqslant W_{ET} < 5.0$, low bursting liability,

$W_{ET} < 2.0$, no bursting liability.

This specification has been verified by experiments and agrees well with the field situations in China. However, it often fails owing to the difficulties encountered in estimating the ultimate strength accurately beforehand. Furthermore, it does not reflect the failure and post-failure mechanical behavior of the specimen. In view of this, attempts have been made to establish a bursting liability index by making use of the complete load-deformation curve. The two approaches proposed are to take the ratio between the areas under the load-deformation curve in the pre-failure and post-failure stages and to take the time elapsed from peak load to ultimate failure, respectively, as the index of bursting liability (Wang, 1985). However, the author of this paper considers that how to take the effect of deformation rate into consideration in using the complete load-deformation curve is still a problem for further study since both the shape of this curve and the time elapsed from peak load to ultimate failure are sensitive to deformation rate and different deformation rates usually lead to curves quite different from each other (see Fig. 1).

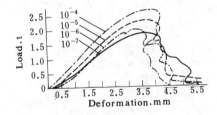

Fig. 1. Complete load-deformation curves at various deformation rates (after Zhou, 1986).

3.2 Drilling yield testing

This method is one of those which are ripe and very widely used for rockburst prediction in the world.

1136

When this method is used, the distance from the working face to the location of peak stress, the depth of possible rockburst and the dynamic effect as indicated by drill-jamming are also measured along with the drilling yield.

The following indices are adopted in coal mines of China.

1. Tangshan Mine
The index for judging the bursting liability used in this mine is given in Table 2.

Table 2. Class of bursting liability

Class of bursting liability	Drilling yield ratio		
	H/M < 1.5	H/M=1.5-3.0	H/M > 3.0
I	$\chi \leqslant 2.0$	$\chi \leqslant 2.5$	$\chi \leqslant 3.5$
II	$\chi > 2.0$	$\chi > 2.5$	$\chi > 3.5$

where H-depth of borehole (m); M-thickness of coal seam (m); χ-drilling yield ratio.

$$\chi = \frac{\text{actual drilling yield per meter (kg/m)}}{\text{normal drilling yield per meter (kg/m)}}$$

2. Tianchi Mine
The index adopted in this mine is as shown in Fig. 2. In the figure, class I refers to high bursting liability, class II erfers to medium bursting liability and class III refers to low bursting liability.

3. Mentougou Mine
Fiven in Table 3 is the index for this mine.

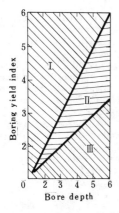

Fig. 2. A nomograph for predicting the class of rockburst

Table 3. Indices for judging bursting liability from drilling yield

Testing depth (multiple of mining height)	1-1.5	2	2.5	3
Critical drilling yield (kg/m)	4-5	8	16	25
Dynamic effect	drill-jamming			

4. Longfeng Mine
The classification of rockburst and testing standard of this mine are listed in Table 4.

Table 4. Class of rockburst and testing standard

Case of Rockburst	Maximum drilling yield (kg/m)			Constituent percentage of drill chips larger than 3 mm (%)	Dynamic Phenomena	Estimated supporting pressure (MPa)
	L<4 m	L=5-6 m	L>6 m			
I	>5.0	>5.5		>30	drill-jamming, impulsive sound, jerking of drill rod	>26
II	3.5-5.0	4.0-5.5	>6.0	>30	pop drill-jamming	15-26
III	<3.5	<4.0	<5.5	<30	none	about 15

In this table, class III indicates no bursting liability, class II indicates bursting liability which requires careful obervation, strengthened supporting and necessary depressurizing treatment and class I indicates serious bursting liability which requires depressurizing measures to be adopted immediately and mining activities to be stopped until the bursting liability has already been eliminated as confirmed by drilling yield testing.

3.3 Accoustic emission and microseismic monitoring

Since the drilling yield testing is laborous and time-consuming and can hardly be carried out continuously, accoustic and microseismic methods are used as important complementary means.

The rock releases strain energy in the form of elastic pulse and thus radiates stress wave when it fractures. This phenomenon is called acoustic emission (AE). To monitor the AE signal caused by rock or coal fracture enables the prediction of rock or coal burst to be made. In addition, the rock or coal releases elastic energy of low frequency and high intensity in the form of microseismic activity. To monitor the microseismic activities in rock or coal makes the prediction of rockburst in a larger scale possible. The frequency of microseismic activity is lower than that of acoustic emission by 4 orders of magnitude. The former is a manifestation of macrofracturing, while the latter is that of microfracturing. The following experiences in applying AE and microseismic techniques to rockburst monitoring have been accumulated in China.

1. The Research Institute of Safety Technology of the Ministry of Metallurgical Industry and Pangushan Tungsten Mine have systematically monitored the failure of hard rockmasses using AE technique and established the following criterion (Li, 1980):

$N < 10$, $R < 10\%$, the rockmass is in a relatively stable condition;
$10 < N < 20$, $R < 10\%$, the rockmass is in the stage of slow fracturing;
$20 < N < 30$, $R < 60\%$, the rockmass is in the stage of accelerate fracturing;
$N > 30$, $R > 60\%$, the rockmass is in the stage of violent fracturing;

where N is the accumulated count of sounding and R is the percentage ratio of the count of loud and ultra-loud sounding in the total count.

This standard has been used to monitor the rockburst and rockmass failure in Pangushan Tungsten Mine, Kuimeishan Tungsten Mine, Bali Tin Mine and Daye Iron Mine and successful forecasts were made. However, it should be pointed out that the values of N and R may be different for various rocks and should be modified in practice depending on the experience of the user. For this reason, long-term monitoring and data accumulation should be upheld in order to determine the location and extent of rockmass failure accurately.

2. AE technique has been used in Langfeng Mine to study the relationship between the energy rate of AE and rockburst occurence (see Fig. 3), to determine the regions of stress concentration and to select the locations for drilling yield testing. By adopting such a comprehensive means in rockburst liability prediction, the labor and time spent on testing were largely reduced and the density of testing increased.

3. Microseismie monitoring has been successfully used in Taozhuang Mine to determine the amount of energy released by coal burst (in Richter scale), as shown in Fig. 4, to circle the region of stress concentration and the extent and time of coalburst occurrence approximately, and to indicate the precursory sign of coal burst.

4. Large-scale microseismic monitoring together with AE sourse location were used successfully in Datong Mine to predict the regularity of roof movement. By doing so, the harm to production brought about by large-area roof caving has been largely reduced.

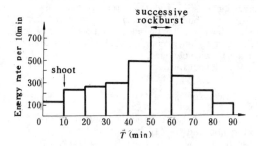

Fig. 3. Variation of energy rate after shoot

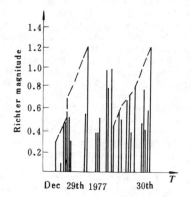

Fig. 4. Gradual increase of microseismic intensity leads to coal burst

3.4. Indices of core discing

This method has been used to evaluate the bursting liability in metal mines in many countries because core discing is caused by high stress concentration. However, it does not apply to coal seams since drilling in highly stress concentrated coal seams will make the coal powdered. Therefore, it is the drilling yield of coal that has been used for evaluating coalburst liability.

Disc-shaped cores were found in boreholes in all the above-mentioned places where rockbursts occurred, viz. , the Ertan Hydropower Station, Pangushan Tungsten Mine and Jinchuan Nickel Mine. It is especially obvious in Ertan Hydropower Station. With the aid of in-situ testing, 3-D photoelastic analysis and 3-D finite element calculation, the cause of rockburst occurrence has been analysed, formulae of the initial and critical stresses for core-discing

have been derived, the relation between the disc thickness and the stress in surrounding rock has been obtained and a scheme of rockburst classification shown in Fig. 5 has been proposed (Hou and Jia, 1986). It is considered by the author of this paper that this classification does not necessarily apply to other places. Nevertheless, it is still of some referential value.

3.5 Preliminary estimation of the bursting location by calculation according to the theory of rock dilatation.

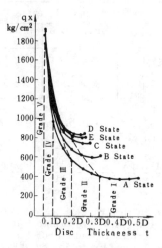

Fig. 5. Relation between disc thickness and surrounding-rock stress for syenite in Ertan Hydropower Station

By this method, the region of microfracturing can be determined from the criterion of $\sigma_{oct} > f_s'$ and the potential energy and volume of this region can be evaluated. Therefore, the amount of work for predictive monitoring can be reduced (Tan, 1986). In addition, there are also the methods of measuring the geostress, crustal deformation, surrounding rock deformation, etc., which will not be discussed here owing to the limited space.

It should be pointed out that rockburst prediction is an extremely complicated problem which requires the coordination of various methods and comprehensive analysis in order to achieve satisfactory results. The goal of rockburst prediction can hardly be attained by any individual method.

4 CONTROL OF ROCKBURST

There are two types of measures for rockburst control. One is the measure which has an overall and fundamental feature. It is to make efforts to eliminate the condition of rockburst occurrence. This type of measures includes controlling the stress within an acceptable level, using reasonable stope arrangement and mining method, and giving none-bursting treatment to the rock or coal body. The other type is the measure which has a local and temporary feature, i.e., to give the burst prone region some liability releasing treatment. It includes high-pressure water-injection, unloading blast, borehole unloading, liability releasing treatment on pillars, etc.. Owing to the complexity of geological conditions and the difficulties encountered in judging the condition of rockburst occurrence, it seems that both types of measures are necessary.

Measures now adopted for rockburst control in China are as follows.

4.1 Stress control

The aim of stress control is to avoid the formation of highly stress-concentrated region or to change the shape of stress concentration zone by improving the stope arrangement and mining procedure. It should be noted that coalbursts are mostly related to the choice of mining procedure. In other words, they mostly occur when excavation is carried out in highly stress-concentrated zone or partly through a region under high stress. Therefore, once the cause, location and conditions for transfer of stress concentration are found out and reasonable mining procedure is adopted, it is quite possible to reduce the frequency of rockburst occurrence and the bursting intensity.

The measures for stress control commonly adopted in coal mines of China can be stated as follows:

1. Mining in faces towards each other and excavating in isolated coal body should be avoided.

2. Excavation in highly stressed region can be avoided by extracting and excavating altermately in different sections.

3. In regard to the mining procedure, to cut multiple extraction galleries in coal seams in a same section is advised. The spacing between parallel galleries should generally be greater than 10-20 m and two galleries should intersect, when necessary, at right angle.

4. The arrangement of pillars should be improved.

4.2 High-pressure water-injection

Water injection method has been successfully used in many countries and is also widely used in China. Its role is to improve the texture of coal body so as to cause a variation of the existing stress concentration region, and to change the physico-mechanical properties of coal so that high stress concentration can be avoided.

In Mentougou Mine, it is found that water-injection can soften the coal and reduce its elastic energy as well as the value of W_{ET}, as shown in Fig. 6. By doing so, rockbursts of intensity higher than 2.3 have been reduced by 79% in number.

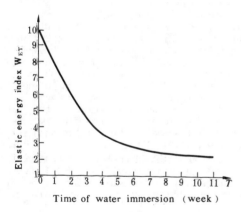

Fig. 6. Elastic energy storage index as a function of the time of water immersion

In Longfeng Mine, before injecting water to the coal body, the width of stress concentration zone is 15-20 m, the width of peak stress zone is 2-3 m the distance from peak stress zone to the working face is 3-7 m and the peak stress value is 30-50 MPa, respectively. After water injection, the peak stress zone extends by 2 m towards the depth, the peak stress reduces by 20%, the value of W_{ET} reduces by 40-60% on average, and the maximum drilling yield reduces by 20%. It is shown by experiments that the bursting liability can

substantially be eliminated when the water content in coal exceeds 3.8%.

Testing in Taozhuang Mine shows that water injection can reduce the energy released by a coal body when it fails and lengthen the time of dynamic failure. As a result, the energy release decreases significantly and tends to be stabilized. Fig. 7 gives the curves of time of dynamic failure versus stress for coal specimens under natural and water saturated conditions.

The purpose of injecting water to roof is to propagate the cracks in roof rock, to increase the water content, to reduce the strength and to increase the possibility of caving.

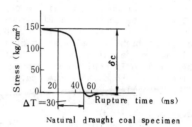

Natural draught coal specimen

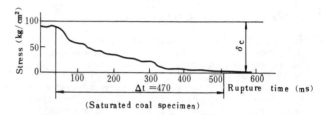

(Saturated coal specimen)

Fig. 7 Curves showing the process of dynamic failure of coal specimen.

The sudden large-area caving of thick-layered graywacke roof in Datong Mine has once caused serious harm to coal production. After studying the microscopic feature of rock and the mechanism of water-injection softening, special measures of water-injection have been adopted and good results achieved. The strength of rock reduces by 26-49% and the number of cracks increases after water-injection. Accordingly, measures have been taken to control the roof caving manually. By making the roof to cave part by part and layer by layer instead of to cave as a whole, the cost of roof ripping has been reduced by 85.7% and the cost of equipment exhaustion by 73.7% (Pan et al., 1983).

In Longfeng Mine, the roof subsidence rate before water-injection is 0.167 mm/day whilst it increases to 4.16 mm/day during the period of water-injection which is 25 times of the former. This fact shows that water-injection can reduce the roof rigidity and thus increase its deformability.

It must be noted, however, that the coal seam for water-injection should have a water-weakening feature and its porosity should generally be no less than 4%. Furthermore, the higher the rock stress, the higher the resistance against water-injection and the lower the efficiency of water-injection will be. In view of this, it is inadvisable to drill holes for water-injection in highly stressed zones.

4.3 Unloading blast Method

This method is to drill holes and shoot in stress concentration zone and to release elastic strain energy by blast in order to change the shape of stress concentration zone and the value of stress, so that the bursting liability can be eliminated by unloading. The position of shoot should be as close as possible to the location of prak stress. In using this method, hard roof is generally desired.

In Mentougou Mine, once a maximum drilling yield was detected at a distance of 5.54 m from the coal face. After blast unloading, the maximum energy rate monitored by geophone reduced to one third of the critical energy rate and the bursting liability was eliminated.

In Longfeng Mine, the location of peak stress concentration is generally 3 m apart from the coal face. By arranging boreholes for blasting 4-6 m in depth and 3-5 m in spacing along vertical working face, the drilling yield is reduced after blast, the stress concentration is mitigated and hence unloading effect is achieved.

In Tangshan Mine, the drilling yield is reduced by 7.5-17.3% and the location of peak stress concentration shifts towards the depth after blast unloading.

In many mines, bursts often occur in coal pillars. Therefore, bursting liability mitigation by lossening blast on coal pillars has also been used. For instance, the coal extraction in Tianchi Mine has been increased up to 84% using this method. Nevertheless, drilling large boreholes in pillars is inadvisable since this will menace the safety. This method has also been used in metal mines limitedly.

Using the method of blast unloading requires safety, reliability and practical economic efficiency.

4.4 Borehole unloading method

This method is to drill holes of large diameter (ϕ 100 mm) so as to unload the ore body or lower the peak stress and to increase its distance to the working face. The higher the stress concentration, the larger the range affected by borehole unloading and the better the unloading effect will be. The selection of diameter, spacing and depth of boreholes should be made in accordance with actualities and complemented by drilling yield and microseismic testings.

4.5 Excavation of protective seams

By doing so, the surrounding rock and coal body are unloaded beforehand so that stress concentration can be avoided and the bursting liability eliminated. This method applies to multiseam working and requires suitable protective seams to exist for excavation. Generally, the protective seam of the roof, coal seams of lowest bursting liability and those of smallest thickness are excavated first.

4.6 Strengthening supporting

This method is mostly used in metal mines and underground excavations in hard rocks. In the railway tunnel of Guancunba, the time for the rock surface to expose freely in air was shortened as for as possible by lining immediately after excavation. Safe operation can be assured by using anchor bolts together with steel wire gauze. In addition, some effects were also achieved by mist spray and water sprinkling in the tunnel, controlling the amount of explosive and excavating parallel leading adit, etc..

The tunnel of Tianshengqiao Hydropower Station is now being in construction. Steel web supporting, shotcrete and steel wire gauze are applied immediately after excavation. Other more effective measures are also under investigation.

In the variety of engineering works mentioned above, especially in coal mines, significant safety and economic effects have been achieved by using the measures for rockburst control described above.

In recent years, no casualty due to rockburst happened in Mentougou Mine and Longfeng Mine. 1.225 million tons of coal were extracted safety from burst prone regions in 1981-1985. The average increase in annual output value in the four years amounts to 1.6 million of RMB yuan.

In Tianchi Mine, no rockburst casualty has happened since 1976. Merely in the year of 1984, output value of 0.82 million RMB yuan was created by extracting coal from burst prone regions.

In Tao zhuang Mine and Tangshan Mine, casualty has been reduced and production promoted as well by taking measures to control rockburst.

5 CONCLUSION

Rockburst is an extremely complex dyanmic phenomenon. In China, some effective experiences in rockburst control have been accumulated through laboratory experiments and in-situ tests: Studies on the mechanism of rockburst are being strengthened. However, further improvement based on practice, communication and cooperation between scholars and engineers from different countries are still needed. Though the problem is very complicated, these still exists the possibility of eliminating the menace of rockburst according to the regularities of its occurrence.

ACKNOWLEDGEMENTS

The author wishes to express his thanks to the Academy of Coalmining Sciences of the Ministry of Coal Industry for offering him the collected paper, "Research on the mechanism of shock ground pressures and experiences in their prevention", from which many data concerning coal mines have been cited. He also achnowledges the help from Mr. Chen Chengzong and Mr. Ding Jiayu who have sent him the data of Guancunba and Pangushan.

REFERENCES

Ganzhou Research Institut of Non-ferrous Metallurgy, Jiangxi Research Institute of Metallurgy and Pangushan Tungsten Mine, 1985, A research report of the study on the ground pressure activity in the lower-middle section of Pangushan Tunsten Mine and methods for its control.
Kidybinski, A., 1981. Bursting liability indices of coal. Int. J. Rock Min. Sci. No.4, V.18, 295-304.
Hou Faliang, Jia Yuru 1986. The relations between rockburst and surrounding rock stress in underground chambers-with a tentative gradation of rockburst intensity Proc. Int. Symp. on Engineering in Complex Rock Formation. Beijing, 297-505.
Li Dianwen, 1980. Study and application of the technique of acoustic emission detection in rockmasses, Proc. 1st Annual Meeting on Mining Sciences Chinese Society of Metal Sciences.
Pan Qinglian, Xing Yumei, Wang Shukun, Niu Xizhuo, 1983. An application to research on rock behavior and its microstructure for coal mining engineering, Chinese Journal of Rock Mechanics and Engineering, V.2, No.1, 77-83
Scientific Documentation Center for Mining Pressure of the Ministry of Coal and shoock Ground Pressure Station, 1985. Research on the mechanism of shock ground pressures and experiences in their preventions.
Tan Tjong Kie, 1986. Rockburst, case record, theory and control. Special lecture in Proc. Int. Symp. ECRF'86, Beijing.
Wang Shukun, 1985. An experimental study on the bursting liability of coal.
Zhou Simeng, Wu Yushan, Lin Zhuoying, 1986. Post-failure behavior of rocks under uniaxial compression, Proc. Int. Symp. on Engineering in Complex Rock Formation, Beijing, 253-261.

Rock mass anisotropy modelling by inversion of mine tremor data
Modèle de l'anisotropie des masses rocheuses determiné par l'inversion des données de secousses séismiques souterraines
Die Entwicklung eines Modells für anisotropische Gesteinsmassen durch Inversion von Messwerten von Bergwerksbeben

A.J.MENDECKI, Anglo American Corporation of South Africa, Rock Mechanics Department, Welkom

ABSTRACT: A generalized, adaptively damped inversion technique with partitioned matrices was applied to determine the locations of mine tremors as well as the parameters of an assumed velocity model. This method is applied to two datasets of arrival times in order to determine the P-wave velocity anisotropy within underground seismic networks of radius 15m and 2000m respectively.

RESUME: On s'est servi d'une methode généralisée d'inversion adaptable amortie avec des matrices cloisonnées pour determiner l'emplacement des secousses sismiques souterraines aussi que les paramètres d'un modèle à vitesse supposée. Cette méthode est appliquée à deux ensembles de données d'heure d'arrivée afin de determiner l'anisotropie de la vitesse de l'onde P des réseaux sismiques souterraines d'un rayon de 15m et de 2000m respectivement.

ZUSAMMENFASSUNG: Dargestellt wird die generalisierende und kontrollierbar gedämpfte Inversionstechnik zur Ortung von Bergwerksbeben wie auch zur Ermittlung der Parameter eines angenommenen Geshwindigkeitmodells. Diese Methode wird angewandt auf zwei Gruppen von Ankunftszeitmeszwerten, um das anisotropische Charakter der P-Wellengeschwindigkeit in unterirdischen seismischen Bezirken, im Radius von jeweils 15 und 2000 Metern, zu ermitteln.

1 INTRODUCTION

The quantification of the velocity structure, in order to determine accurate locations of mine tremors, requires a number of calibration blasts that in the case of an extensive mine area, covered by one seismic network, can be prohibitively expensive.

This paper describes a method that uses the P arrival times (generalization for P+S or S-P is straightforward) from a group of mine tremors to evaluate simultaneously: the parameters of an assumed velocity structure (in this case the average values of velocities of seismic waves from the region of seismic activity to the particular stations) and the co-ordinates of hypocentres. An efficient algorithm of partitioned matrices (Spencer, 1985) was used, where unknown parameters are separated into two distinct sets on the ground of physical differences and for computational reasons. This enables joint hypocentre and velocity inversion to be carried out on a personal computer and at only marginally greater cost (in terms of computer memory and processing time) than that of a velocity inversion alone. Additional procedures improving the numerical properties of the specific inverse problem have been used, namely weighting (Inman, 1975), centering (Draper and Smith, 1981; Lee and Lahr, 1972; Lienert et al., 1986), scaling (Lee and Lahr, 1972; Inman, 1975; Smith, 1976; Lienert et al., 1986) and damping (Marquardt, 1970; Crosson, 1976; Aki and Lee, 1976; Lienert et al., 1986). These techniques are often used in geophysical inverse theory and are not fully described here.

The algorithm, on which a Fortran program for an IBM AT personal computer is based, is given in a general form below. This is followed by a description of the practical application of the technique to P-wave arrival time data sets from two seismic networks in South African gold mines, one of 15m radius and the other of 2000m radius.

2. GENERAL DESCRIPTION OF ALGORITHM

Consider the rock mass as a statistically slightly disturbed anisotropic and inhomogeneous medium, in which v_j is the average value of P-wave velocity for the rock mass containing the ray paths between the sources of NE events and the j-th seismic station. Thus, the arrival time, t_{ij} of the direct P-wave from the source of the i-th mine tremor, with co-ordinates x_{qi} (q=1,2,3,4), where x_{4i} is origin time, to the j-th seismic station, xs_{qj} (q=1,2,3) can be given as follow

$$t_{ij} = x_{4i} + h_{ij}/v_j, \quad i=i,\ldots,NE; \quad j=1,\ldots,NS, \quad (1)$$

where h_{ij}=SQRT (SUM $[(x_{qi}-xs_{qj})^2]$), q=1,2,3.
  (q)

After the approximation of equations (1) by a first order Taylor series and after separation of unknowns, which are refinements of the co-ordinates of the tremors and the velocities to the initial model, into two distinct sets (Spencer and Gubbins, 1980; Spencer, 1985), one can write a residual vector as

$$R = Z - X*Dx - V*Dv \quad (2)$$

where Dx and Dv represents 4*NE and NS-dimensional vectors of correction parameters respectively, X is an (NE*NS) x (4*NE) matrix of partial derivatives of the hypocentres, V is an (NE*NS) x NS matrix of partial derivatives of velocities, and Z is an NE*NS-dimensional vector of arrival time residuals.

Differentiating $|R|^2$ with respect to Dv and Dx, equating to zero and applying simple transformations, gives a solution which, in the full rank case, has a form

$$Dv = (C*V)^{-1}*C*Z, \quad (3)$$

$$Dx = (X^T*X)^{-1}*(X^T*Z - X^T*V*Dv), \quad (4)$$

where $C = V^T - V^T*X*(X^T*X)^{-1}*X^T$.

Equations (3) and (4) require the inversion of two matrices. Because all elements of the matrix (X^T*X) which are outside of those 4 x 4 submatrices on the diagonal are zero, its inverse is a matrix in which the inverses of the submatrices $(X_i^T*X_i)$ lie along the diagonal, with all other elements zero. Thus the entire system of equations of order 4*NE+NS is solved with the inversion of an NS x NS matrix C*V and NE inversions of a 4 x 4 matrices $(X_i^T*X_i)$. Applying a centering procedure to the simple location problem, (Dv=0), the diagonal submatrixes $(X_i^T*X_i)$ are 3 x 3, because then the origin time is defined as the mean arrival time minus the mean travel time (Lienert, et al., 1986).

The generalized inverse solution of equations (3) and (4) omits or suppresses (by damping factors dl_i and d2) those eigenvectors corresponding to zero or near to zero eigenvalues from singular value decomposition (SVD) of the matrices X_i and C*V. Thus

$$(X_i^T*X_i)^{-1} = Q1_i*diag\left(\frac{1}{sl_{iq}^2 + dl_i}\right)*Q1_i^T, \quad q=1,\ldots,4. \quad (5)$$

and

$$(C*V)^{-1} = Q2*diag\left(\frac{1}{s2_q + d2}\right)*Q2^T, \quad q=1,\ldots,NS \quad (6)$$

because the matrix C*V is, in this case, symetric, where $Q1_i$ and Q2 are the orthogonal matrices whose columns are the eigenvectors and sl_{iq} $s2_q$ are eigenvalues of X_i and C*V respectively.

The equations (3) to (6) give the generalized inverse of the given problem, in terms of the inverses of partitions, with four limits of the solution:

1. dl=0, i.e. no damping of the hypocentre part of the problem. Resolution of co-ordinates of tremors will be perfect.

2. d2=0, i.e. no damping of velocity part of the problem. Resolution of velocities will be perfect.

3. dl=∞, refinements of co-ordinates of tremors will be zero, i.e. equivalent to a simple velocity inversion.

4. d2=∞, refinements of velocities will be zero, i.e. a simple location problem.

Generally, it is desirable to choose damping factors as small as possible to achieve maximum resolution but large enough to stabilize the solution. In an adaptive damping technique (Lienert et al., 1986), the damping factor depends on changes of the Euclidean norm of the residual vector R. If ∥R∥ increases in any iteration, the procedure is to return to the previous solution, increase the damping factor and try again.

3 APPLICATIONS

3.1 A very small Micro-Network

In 1985, a 15m radius seismic network of 8 geophones was installed by the SA Chamber of Mines Research Organisation in a pillar 2160m below surface in a gold mine in the Klerksdorp district. The main purpose was to investigate the fracture process ahead of the stope in the pre-remnant and remnant stages of mining by analysis of micro-seismicity (Legge and Spottiswoode, 1987). In order to obtain reliable locations of seismic activity, 8 calibration blasts were performed and average values of P-wave velocities to the particular geophones were determined. Arrival times of micro-seismic events were accumulated over a period of two months and a sample of the data was kindly made available for numerical analysis.

Seven of the eight calibration blasts form a relatively tight spatial group and seismic events

were selected which cluster in the same source region. Fig.1 shows the mining situation, positions of the calibration blasts, the relocated positions of seismic events and contours of expected location errors (Mendecki and van Aswegen, 1986). The contour values were calculated assuming maximum errors for the P-wave arrival and the P-wave velocity of 0,05 ms and 480 m/s respectively.

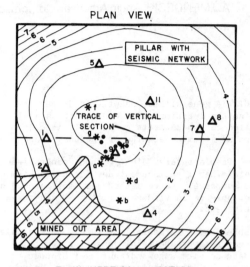

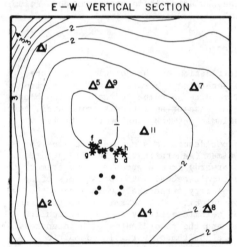

Fig.1 Plan and section of the very small micro-seismic network showing the locations of the calibration blasts and of those seismic events used in the analysis.

The events shown in Fig.1 yielded a set of 54 arrival times, which, after inversion using the procedure described above, gave small corrections to the initial positions as determined by the Chamber of Mines.

Starting from an initial isotropic velocity model, $v_j=5200$ m/s, the P-wave velocities calculated simultaneously with the hypocentres, are compared with velocities determined from the calibration blasts in the following table:

GEOPHONE	INVERSION PROCEDURE	CALIBRATION BLASTS
1	5109	5070
2	4814	4830
4	5149	5270
5	4739	4840
7	5589	5710
8	5650	5780
9	5245	5440
11	5280	5230
MEAN	5197 m/s	5271 m/s.

The results show that seismic wave velocities as calculated by the given inversion technique are in good agreement with those determined by the calibration blasts. The wide range in velocities, from 4739 to 5650 m/s indicates a relatively strong velocity anistropy within the pillar.

3.2 Orange Free State Seismic Network

A part of the Orange Free State Goldfields is covered by a network of underground seismic stations, feeding data to a central micro-computer from which arrival time data is transferred to a mini-computer for analysis (Lawrence, 1984). The P-wave arrival times and calculated source locations of approximately 40 000 seismic events have been recorded since 1980. The main purpose of the present exercise was to refine the P-wave velocity model around a mine shaft area which was particularly prone to seismicity.

Data was selected from an area, 3km by 3km square, around the No.3 Shaft of Western Holdings Mine for the time interval 1 April 1980 to 28 March 1981. The data set consists of the arrival times of 52 tremors at 6 geophone sites. Not all geophone sites were triggered by all of the events, and thus only 268 arrival times were obtained. The locations from the system database together with an isotropic velocity structure (P-wave velocity = 5500m/s) were used as an initial model.
Fig.2 shows the relocated positions of the mine tremors in relation to geology, seismic stations and contours of expected location errors. In this case, the contour values were calculated assuming the maximum errors for the P-wave arrival time and P-wave velocity as 2,5 ms and 200 m/s respectively. The maximum difference between the new locations and the original ones is about 180m, and, in general, the differences are mainly in the vertical. The calculated average P-wave velocities (m/s) between the region defined by selected tremors and particular seismic stations are: WH1-5528, WH2-5544, WH4-5406, WE2-5470, WE3-5609, PB3-5420 (see Fig.2 for locations of seismic stations). Calculations performed after separating the tremors into two groups (A and B, Fig.2) yielded very similar velocities, indicating that velocity structure does not vary over the area of interest. These values for the P-wave velocities are currently used in the routine location procedure for the area around Western Holdings N°3 Shaft.

4   CONCLUSIONS

Advantages of the described technique can be considered in terms of immediate practical application and with reference to possible future research:

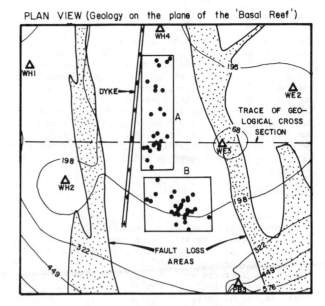

PLAN VIEW (Geology on the plane of the 'Basal Reef')

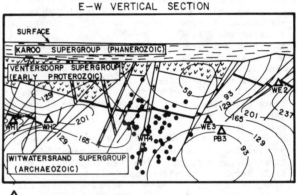

E—W VERTICAL SECTION

WH1   POSITION AND NUMBER OF SEISMIC STATION
•   CALCULATED LOCATION OF SEISMIC EVENT
✕   BASAL REEF, DISPLACED BY A FAULT
⌲⌲   DYKE (MOSTLY DOLERITE)
−58−   CONTOUR OF EXPECTED MAXIMUM LOCATION ERROR (metres)

0   500   1000   1500 m
HORIZONTAL & VERTICAL SCALE

Fig.2 Plan and section of part of the OFS Goldfield seismic network showing the distribution of selected mine tremors and simplified geology.

1. Having accurate arrival times, recorded by a well configured network, the parameters of a seismic wave velocity model for a seismically active area can be determined by the inversion technique described in this paper. This technique can also be applied to more complex velocity models and seismic ray tracing for minimum time path (Lee and Stewart, 1981, 76-104) This should lead to more accurate locations of mine tremors and a reduction in the need for expensive calibration blasts.

2. Applying this procedure to a region of high seismic activity for some period of time one can expect to detect spatial and temporal variation in the velocity structure caused by changes in the density, distribution and saturation of cracks. A combination of such changes is a probable driving mechanis for most precursors of earthquakes and mine tremors (Crampin et al., 1984).

ACKNOWLEDGEMENTS

The cooperation of Dr N.B. Legge, Dr S.N.
Spottiswoode and Mr K. Pattrick of the Chamber of
Mines Research Organisation, who provided the data
and preliminary locations for the small micro-network
case study is gratefully acknowledged, as is the
permission granted by the Research Organisation to
use the data. I would like to give special thanks to
G. van Aswegen (AAC of SA) for selecting data from
the OFS Network and for his suggestions for the
improvement of the manuscript. I. de Lange typed the
manuscript and S. Gruenbaum did the draughting.

REFERENCES

Aki, K. and W.H.K. Lee 1976. Determination of
    three-dimensional velocity anomalies under a
    seismic array using first P arrival times from
    local earthquakes. 1. A homogeneous initial model.
    J. geophys. Res. 81: 4381-4399.
Crampin S., R. Evans and B.K. Atkinson 1984.
    Earthquake prediction: a new physical basis.
    Geophys. J.R. astr. Soc. 76: 147-156.
Crosson, R.S. 1976. Crustal structure modelling of
    earthquake data. 1. Simultaneous least squares
    estimation of hypocentre and velocity parameters.
    J. geophys. Res. 81: 3036-3046.
Draper N.R. and H. Smith 1981. Applied Regression
    Analysis. New York. John Wiley & Sons Inc..
Inman, J.R. 1980. Resistivity inversion with ridge
    regression. Geophysics 40, 798-817:
Lawrence, D. 1984. Seismicity in the Orange Free
    State gold-mining district. Proceedings of the 1st
    Int. Congress on Rockbursts and Seismicity in
    Mines, Johannesburg 1982: 121-130.
Lee, W.H.K. and J.C. Lahr 1972. HYPO71: a computer
    program for determining hypocenter, magnitude and
    first motion pattern of local earthquakes. U.S.
    Geol. Surv. Open-File Rept.: 75-311.
Lee, W.H.K. and S.W. Stewart 1981. Principles and
    applications at microearthquake Network.
    New York. Academic Press.
Legge, N.B. and S.M. Spottiswoode 1987. Fracture and
    microseismicity ahead of a deep gold mine stope in
    the pre-remnant and remnant stages of mining, 6-th
    ISRM Congress. Montreal 1987.
Lienert, B.R., E. Berg and L.N. Frazer 1986.
    Hypocenter: an earthquake location method using
    centered, scaled, and adaptively damped least
    squares. Bull. Seism. Soc. Am. 76: 771-783.
Marquardt, D.W. 1970. Generalized inverse, ridge
    regression, biased linear estimation, and nonlinear
    estimation, Technometrics 12: 591-611.
Mendecki, A.J. and G. van Aswegen 1986. A method for
    the optimal designing of mine seismic networks in
    respect to location errors, and its applications.
    Anglo American Corp. SA, RMD. Welkom. SP 7/86.
Smith, E.G.C. 1976. Scaling the equations of
    condition to improve conditioning. Bull. Seism.
    Soc. Am. 66: 2075-2081.
Spencer, C. and D. Gubbins 1980. Travel-time
    inversion for simultaneous earthquake location and
    velocity structure determination in laterally
    varying media, Geophys. J.R. astr. Soc. 63, 95-116.
Spencer, C. 1985. The use of partitioned matrices in
    geophysical inverse problem, Geophys. J.R. astr.
    Soc. 80, 619-629.

The geomechanics of Panasqueira Mine, Portugal
Le géomécanique de la mine de Panasqueira, Portugal
Die Geomechanik des Bergwerks Panasqueira, Portugal

F.MELLO MENDES, Professor, Technical University of Lisbon, Consultant, Hidromineira, Portugal
A.CORREIA DE SÁ, General Manager, Beralt Tin & Wolfram Portugal, SARL
J.FERREIRA E SILVA, Head of Planning, Beralt Tin & Wolfram Portugal, SARL

ABSTRACT: At Panasqueira mine the mining method for the subhorizontal quartz veins has changed from longwall stopes with partial filling to room and pillar stopes. To obtain with the last referred, under safe conditions, the maximum ore recovery, it became necessary to design different stope patterns based on the knowledge of the geomechanical implications of mining. The paper describes the fundamental part of this knowledge which data collection began more than 25 years ago. The geological and geomechanical characteristics of the rock mass, where low deformable schists are dominant, are referred. The behaviours showed by both the longwall and the room and pillar stopes, as well as its interpretations, are described. The subsidence and also the whole mine stability are analysed on the basis of the actual knowledge and considering the modifications to be implemented on the stoping method.

RÉSUMÉ: A la mine de Panasqueira, la méthode d'exploitation des filons de quartz sub-horizontaux a evolué de la longue taille avec remblayage partiel jusqu'aux chambres et piliers. Pour obtenir, au moyen de ces dernières méthodes, en de bonnes conditions de securité, un maximum de récupération du gisement, il fut nécéssaire d'envisager des variantes basées sur la connaissance des implications géomécaniques de l'exploitation. Dans ce travail on présente l'essentiel de cette connaissance dont l'acquisition se fait depuis 25 années. On mentionne les charactéristiques géologiques et géomécaniques du massif, où prévalent des roches schisteuses très raides. On référe les comportements observés dans les chantiers tabulaires de longue taille et de chambres et piliers, tout comme les interprétations correspondantes. La subsidence et la stabilité générale de la mine sont analysés, d'après les connaissances obtenues jusqu'à present, et aussi face aux variantes d'exploitation qu'on envisage d'introduire.

ZUSAMMENFASSUNG: In Bergwerk Panasqueira hat sich das Abbauverfahren der subhorizontalen Quarzgaenge von Strebbau mit Teilversatz in Kammerpfeilerbau geaendert. Um mit diesem letzteren Verfahren, und unter guten Sicherheitsbedingungen, ein Maximum an Lagerstaettengewinnungsgrad zu erreichen, wurde as noetig, auf Grund der Kenntnis der geomechanischen Auswirkungen des Abbaus, Varianten zu entwerfen. In der Arbeit wird das Wesentliche jener geomechanischen Kenntnis vorgestellt, mit deren Erwerbung vor mehr als 25 Jahren begonnen wurde. Die geologischen und geomechanischen Merkmale des Gebirges, in dem sehr gering Schiefergesteine, werden angegeben. Danash werden die in den tafeligen Streb-und Kammerpfeilerbauen beobachteten Verhalten erwaehnt, sowie der darueber gemachten Auslegungen. Abschliessend, werden Betrachtungen aufgestellt ueber die im Licht der bis jetzt gewonnenen Erkenntnisse untersuchten Gebirgssenkung und allgemeinen Bergwerksstandfestigkeit, angesichts der gegenwaertig eingefuehrt werdenden Varianten des Abbauverfahrens.

1 INTRODUCTION

At Panasqueira mine, the greatest underground wolframite production unit in the world, subhorizontal quartz veins with some tens of cms in thickness are mined, being the distance between them usually of some metres; in a more rigorous description we can say that geological units, which for exploitation purposes are named veins, are in fact formed by series of quartz lens. The referred veins can be regularly mined by subhorizontal tabular stopes.

As shown in fig.1, the orebody as a whole has a tendency to dip to SE. The exploitation of a multiple vein system necessitated the development of a complex access and haulage network, the actual extent of which is illustrated in fig.2.

The constant increase in production costs forced to perform an evolution on the stoping method, along the years; the follow up of this evolution, partially assisted by the progress on the mining equipment, only can be carried out within acceptable safety levels if an improvement on the existing geomechanical knowledge of the mine is achieved.

We can say that, from the start of the mine exploitation, towards the beginning of this century until the sixties, the several stoping methods used were not very different from the conventional longwall, sometimes filling partially the voids using blocks of surrounding waste rock to build packwalls or using some other support systems which were expected to

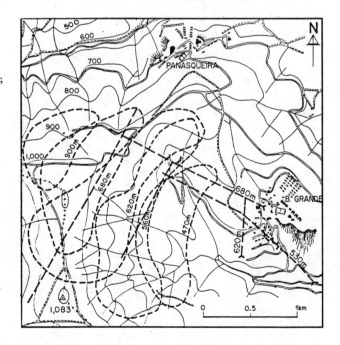

Figure 1. Mine location and orebody contours.

perform the same role of the packwalls. After the sixties and for economical reasons related with the
unbearable costs of erecting packwalls, longwall methods were replaced by room and pillars in which mechanization, introduced at pre-programmed steps, is now
a fact. However, a final satisfactory solution for the
main problem of this method which is the attainment
of an acceptable level on the orebody recovery, has
since then being under investigation.

One attempt that during some time seemed adequated
consisted in total removal, with the help of timber-
packs, of pillars which were previously left; but due
to the cost increases both on timber and labour and
the low value of wolfram, that solution soon became
impracticable.

Actually, it is intended to avoid totally the use
of timber, leaving definitively pillars sufficiently
small to allow a good global recovery of the orebody
but simultaneously strong enough to keep a reasonable
roof control. Hence the design of this alternative for
the stope method required a more deep geomechanical
knowledge of the mine and lead to the necessity of
subdividing the mine using control pillars, if the
areas to be stoped are too wide.

The existing geomechanical knowledge of Panasqueira
mine is already based on considerable data, which
begun to be collected 25 years ago. From these, the
most relevant elements, which are indispensable for
the understanding of the geomechanical restrictions
of the mine project will be presented. This project,
being one of the main concerns of Beralt Planning
department, determined a global evaluation of all
geomechanical data available in Panasqueira, which is
the purpose of the present paper.

2 GEOLOGICAL AND GEOMECHANICAL CHARACTERISTICS OF THE ROCK MASS

The published geological papers about Panasqueira can
nowadays be considered numerous, having the most of
them been issued after the first synthesis paper about
the subject (Thadeu 1951). However, for the geomechanical rock mass characterization, referred later for
the first time (Mello Mendes 1960), the bibliography
is much less available.

Panasqueira rock mass is basically formed by schists
which present different levels of metamorphism generated by an underground granite intrusion and for the
mineralized quartz veins. During the fifties the underground works and several drill holes reached one granite cupola; however in the actual works no granite is
present. For the metamorphized sedimentary formations
is particularly important, due to its abundancy, the
lithological type "spoted schist", which gradually
turns into "unspoted schist". The rock mass shows also
local differentiations as those related with siliceous
impregnated zones or bands in veins vicinity where
silicified grauwacke rock rich in tourmaline appears,
which can be important for the rock mass workability.

Geomechanically and at the exploitation scale, only
under particular cases the rock mass can be considered
continuous; as a whole, it must be viewed as an
intensely jointed rock mass. Between the discontinuity
systems which contributed for their jointing we must
separate: a) quartz veins; b) doleritic dykes; c) big
fractures and faults with carbonated and argillaceous
fillings; d) joints. All these systems have already
been studied under geological point of view (Thadeu
1951; Mello Mendes 1960), but their geomechanical
studies are only now receiving the required attention.

The big fractures and faults are vertical or quite
dipped, and seem to be grouped in two different systems: one that strikes between N-S and NW-SE, mainly
with argillaceous fillings; the other that strikes
between E-W and NE-SW, usually filled by carbonates
(Thadeu 1951; Mello Mendes 1960). The big faults of
these two systems intersect themselves under angles
between 90° and 45° and seem to determine blocks with
more or less independent geomechanical behaviours
(Mello Mendes 1985).

In what joints are concerned their attitudes are
widely spread and it is not easy to correctly determine joint-systems (Mello Mendes 1960). The main joint
system is practically parallel to the vein system,
which can indicate a common origin for these joints
and for the fractures occupied by the mineralized
veins; this system has a great practical importance
because it generates individual slabs on the stope
roofs which sometimes need local support. It seems
also possible to define two different inclined joint
systems (Thadeu 1951; Mello Mendes 1960). On the stope
roofs, the cohexistance of joints belonging to the
two last referred systems with subhorizontal system
joints and more, eventually, with faults and filled
subvertical fractures allows the definition of polyhed-
ric rock blocks likely to fall only by load action
when the horizontal confinement is sufficiently
released. In order to foresee the roofs behaviour is
also important to know the position of doleritic
dykes, which great alterability leads to spheroidal
exfoliation.

Besides the jointing, Panasqueira rock mass is
affected by marked subvertical schistosity, which
reflects in a rather characteristic way in the mechanical behaviour of the rock material (Mello Mendes
1960). This schistosity seems to be independent of the
original stratification, which layers, strongly affected by a complex tectonics are more hard to define
than schistosity.

The schistosity of Panasqueira rock material has
already originated various papers, especially based
in laboratorial experimentations (Mello Mendes 1960).
The results of this experimentation allowed to a
proposal for a rheological model which reasonably fits
the real behaviours (Mello Mendes 1960; Mello Mendes
1966).

The more relevant aspects related with schistosity
are the mechanical anisotropies generated by that,
which grow in a very clear way with the increase of
rock alteration; this one takes place preferably
through the schistosity, helped by the destressing of
the surrounding ground (Mello Mendes 1966). Near surface, where the alteration is greater, the anisotropy
is intensely marked and the rock loses continuity
becoming divided in "leaves"; underground the
anisotropy does not seem to be important, although we
may admit that it has some influence on the rock mass
destressing, and thus on the establishment of the
decompressed zones appearing near the stoped areas
(Mello Mendes 1985).

The mechanical influence of schistosity is also well
seen in the behaviour of the core drillings; in fact,
even with a vertical starting, they show a natural
tendency to follow the schistosity strike (Mello Mendes 1960).

It must be noted that no important joint system parallel to the schistosity has been described; besides,
in decompressed and greatly alterated zones the rock
starts "opening" along the schistosity plans, forming
then a very important joint system (Mello Mendes 1985).

For the geomechanical characterization of Panasqueira rock material, spoted and unspoted schists were
already submitted to laboratorial tests (Mello Mendes
1960). Uniaxial and triaxial compression tests (maximum lateral pressure 120 bar) were carried out, as
well as tension tests by diametral compression and
also bending tests. The anisotropy caused by schistosity was well shown by all these tests, in which drill
cores, collected in the sixties, at the same depth of
stoping levels, were used.

The spoted and unspoted schists showed deformability
moduli in compression, according to the test conditions, ranging from $2x10^4$MPa to 10^5MPa, the lower values corresponding to the direction perpendicular to
the schistosity and the highest to the direction parallel to the schistosity. For any of these kinds of
rock the Poisson ratio ranged frequently between 0.2
and 0.3. The spoted schists revealed deformability
moduli in tension between 10^4MPa and $5x10^4$MPa, parallel
to the schistosity, lower, but in the same range of
the corresponding deformability moduli in compression.

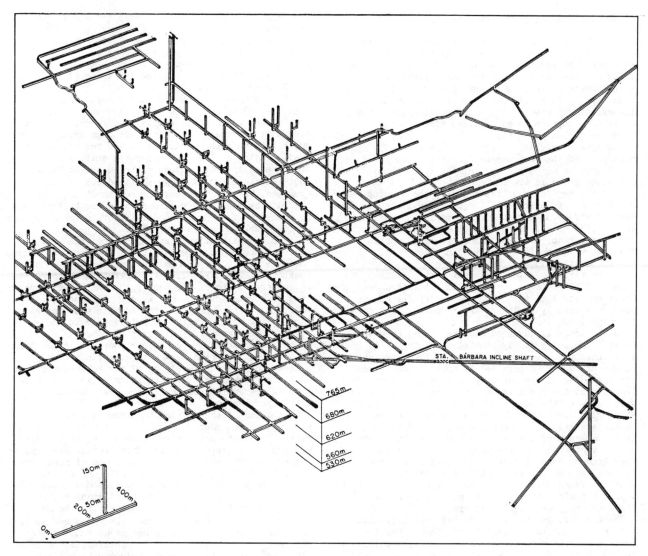

Figure 2. Main development.

The schists exhibited strengths in uniaxial compression parallel to schistosity from 100 to 150MPa. Superposing lateral pressures these resistances were strongly increased: 1.5 to 2.0 times with lateral pressures of 40 bar, and 2.5 to 3.0 times with lateral pressures of 120 bar.

The spoted schists were tested in uniaxial compression along several different directions regarding schistosity; they showed that strengths were directly dependent of these orientations. The greater values, corresponding to compression along schistosity plans, ranged from 75 to 125MPa, and the lowest, between 50 and 70MPa, as a result of the compression along an axis forming a 40° angle with schistosity. Later, after a detailed analysis of these results, it became apparent that the failure features along schistosity are markedly different from those along anyother direction, i.e. physically affecting the rock material: while for those the internal friction angle is much lower, about 38°, corresponding to a sliding along such plans, for the others an internal friction angle of 57° would be found (Mello Mendes 1971a). On the other side it was concluded that only when the schistosity plans are not confined, and is physically possible sliding along them, the orientation of the failure surfaces can be affected; if this is not the case, the orientation of failure surfaces seems to follow the original Griffith theory (i.e. without closure of microfissures before failure starts).

A good explanation for the anisotropy of the uniaxial compression strength that the schists revealed consists in the growing of microfissures and its stopping when schistosity plans are reached (Mello Mendes 1971a).

As a conclusion of the research results here presented we can assume that Panasqueira rock material is highly resistant and has a very low deformability. The same occurs for the whole rock mass if any local destressing does not disturbe it, allowing joint plans to act; if this occurs the behaviour of the rock mass will change from one characteristic of a continuous ground to another typical of a jointed ground.

Unfortunately there is not yet any results of "in situ" measurements of the underground stress field which will enable the characterization of the stress conditions before mining exploitation. The intense ground jointing, that prevents the rock to act as a continuous rock mass, blocked all the attempts to evaluate the proportion between horizontal and vertical stresses through the analysis of the behaviour of mine voids with different sections and directions (Mello Mendes 1960). However, the evolution of the dome that covers the destressed ground superimposed to the tabular subhorizontal mined zone, that has already been studied (Mello Mendes 1960), showed a form in which the relation height/width was about 1/3 and, furthermore, the maintenance of large unsupported spans in longwall tabular stopes, strongly indicate the existance of stress states in virgin rock mass which have greater horizontal than vertical components. Actually there are no arguments against the hypothesis that takes the vertical component as a result, at any depth, of the overlying ground load.

Panasqueira rock mass has such characteristics that usually there are no problems in opening linear voids even when the sections are widely big. As a rule the roofs of those voids stand without support, sometimes with occasional bolting or any other kind of discontinuous support where the jointing seems unfavourable; the exceptions are the important faulted zones where continuous support based on lined timber sets or concrete arches are used.

The main geomechanical problems are related with the behaviour of the rock surrounding the stope cavities, which are, as already said, tabular and subhorizontal and wide spread in area.

Till the beginning of the sixties, the stoping methods have always been some kind of longwall, advancing very slowly (0.3m/day average) sometimes parallel to its initial orientation, some others rotating around one of its ends; for both cases, we could consider that the void sections were rectangular greatly elongated with 1.3 to 1.6m height.

In the first solutions for roof control, the voids were partially filled by packwalls but later, near the faces and parallel to them, some stockpiles of waste were additionally built for the retention of wolframite fines.

As the longwall advanced and the stoped area increased the roofs slowly started deforming until critical spans were reached (40 to 50m); then, near the advancing faces and parallel to them, in the zones corresponding to the extremities of the referred critical spans, the roofs showed important shear fractures, which grew to the center of the spans under less than 45° angles. Until the fractures started the roofs had small sags even in the center, what could be detected by the absence of compressions showed by the packwalls, as the roofs behaved like very stiff plates, eventually submitted to high horizontal stresses that prevented its collapse by jointing; only after the fractures near the faces the roofs started lowering on the packwalls, pressing them strongly and thus decreasing significantly the height of stopes.

These roofs, due to its small deformability, had little sags and though seldom experienced vertical tension fractures in the middle of spans; usually before that, when critical spans were reached, the existing shear tension which values were evidently higher near the faces where the roofs were fixed, compelled the roofs rupture in the vicinity of these faces. In these zones, where the fractures were also parallel to the faces or to the fixation lines of the roofs, the highest compressions were installed perpendicularly to those lines because of the existing negative bending of the roofs. From the faces to the center of the spans these strongly compressed zones successively changed into less stressed and soon after the inflection points of the roof bending to tensional ones; the last ones were never very important not only due to the rock stiffness but also rather probably as a consequence of the high horizontal tensions previously existing in the rock mass.

The evolution of such roof bendings in longwall stopes could reasonably be followed by roof-floor convergence measurements. Not only the convergences revealed the roof sags but also the variations with time of the convergence rates allowed to foresee where on the roofs the shear stresses tended to reach values capable of generating ruptures. Even the location of these zones became possible throughout the mapping of convergence rates; in 1960, based on the already acquired knowledge, the convergence rate of 0.1mm/hour was considered a real danger of future failure, and more, the rate of 0.7mm/hour was admitted as leading to inevitable failure (Mello Mendes 1960).

Later, during 1970, the failure near the faces was interpreted through the Bieniawsky mechanism for brittle fracture considering that increments of convergence rate were due to Poisson effect when on the roofs near and perpendicularly to the faces the stresses suddenly increased, meaning this, as mentioned before,

that the shear stresses would also quickly increase in the same zones; the rate of 0.1mm/hour could then mean that the compressive stress level correspondent to the rupture beginning had already been passed and the rate of 0.7mm/hour that the rupture becomes instable, irreversible and leading to complete failure (Mello Mendes 1970).

The fractures near the faces occurred simultaneously with big noises and violent air blow-ups during some tens of hours and could really be considered as rockbursts. Its mechanical effects although important were not highly ruinous, certainly due to the relatively small bending of the roofs over the critical spans, in spite of the high values of those spans: in consequence the elastic energy envolved was not exaggerated. The low tendency for rockbursts of Panasqueira rock mass was in 1970 pointed out by Potts when he commented the low deformability of the country rock (Potts, Szeki & Patchet 1970/73).

After a rockburst and after rebuilted the affected face the stope could restart until its progress reached the same critical span and then most probably a new rockburst could take place. In what this new progress is concerned, the fallen material resulting from previous fractures would become a rear abutment for the roof; as it was much more compressible than the stope face no new shear fractures could appear in the correspondent area but only near the face.

The measurements of roof-floor convergences in longwall stopes supported by waste packwalls not only allowed to a better knowledge of the behaviour of the rock mass regarding mining but also established a good warning system for roof collapses, very useful for safety reasons.

By the end of the fifties a change in longwall stoping was introduced to replace the very expensive support by waste packwalls; it consisted on building columns of concrete disks slightly ironed. When erecting the columns edges or any other wood elements were put near the floor and the roof. Although the close grid of this kind of support, the system did not satisfy. Really, for identical critical spans rockbursts occurred near the faces very similarly to the ones occurring in the previous method. On the other hand, when the roofs moved down by weight action, those concrete pillars, with low deformability, could not stand and were fully smashed, reducing the stope heights, in a few hours, from 1.4m to only some tens of cms. This alternative method was much more dangerous than the previous one, once the roofs could not be controlled in the same way after face ruptures, conducing in a very short time to the almost complete closure of the stope voids.

Much better results were obtained in the same stopes when the control of the roofs was sistematically done by timber packs as tighten as possible. Although the reactions against the roofs were also weak and the critical spans did not change significantly, the fact is that, when the roofs failed near the faces, as timber packs have good deformability characteristics, either before pick loads or after its occurrence, they may provide a readily flexible system to control the movement of the roof, as there is a good load distribution over each of the packs. However, due to economical reasons this option was not feasible as it envolved big quantities of timber and was a labour intensive one; on the other hand, its mechanization seemed rather difficult. Consequently, the system had to be abandonned and room and pillar commenced to replace the other stope methods.

4 GEOMECHANICS OF ROOM AND PILLAR STOPES

As it is well known, in a subhorizontal tabular room and pillar stope, the acting vertical load only affects in a small extent the adjacent virgin area; its main action is concentrated on pillars, with an increased intensity as the pillars become closer to the center of the stope area; there, and for a given room and pillar pattern, that intensity approaches the theoretical one, established regarding tributary area, as much as the stoped area increases. This is the reason why, when evaluating the loads acting on the pillars, we may generally assume that a safety factor has been already provided, which increases with the distance between pillars and the center of the area (Salamon & Oravecz 1976).

The concentration of greater loads near the center of stoped area can also be looked as the result of the bending of a roof supported by independent resistant elements, like described before; but now these elements are abandoned pillars of rock and hence less deformable and much more resistant than the artificial supports.

As the roof bends, in spite of the resistance of the pillars, we must consider the possibility of the failure in that roof, as a result of either tension fractures near central area or shear fractures near the fixation line on the surrounding virgin rock; if any of these failure modes occur, taking into account the type of support, it will only happen in connection with critical spans significantly wider than those related with the longwall stopes. However, in the usual circumstances where pillar sections are not too large, the pillars start collapsing before any roof failure also in the dependence of a critical span.

As the load is redistributed gradually, in connection with the collapse of the more stressed pillars, there is a tendency to open a void near the central zone of the stope; the limits of the related span are not anymore installed on the virgin area but are laying on a much more deformable abutment - the remaining pillars. Consequently, wave bending of the roof appears, and some divergences may occur; in this case, if important shear fracturing of the roof starts, it may be displaced from the virgin ground towards the stopes to the places where the divergences are noticeable.

In the room and pillar stopes the collapse of the pillars and the collapse of the roof must both be taken into account as distinct and sometimes independent happenings. Once the discontinuities generated by failure start acting we are really far from the behaviour correspondent to elastic material and we can not expect the roof and the pillars to interact normally. The consequences of the two kinds of failure can be completely different, those of the pillars being usually confined and revealing a small tendency to generate rockbursts, and those of the roof, since they are related with big spans, and consequently being able to engage and store great amount of elastic energy, much more likely to produce rockbursts.

On the other hand, among all the technical and economical criteria directing mining the optimization of the recovery of the ore deposit is particularly important, namely for room and pillar stoping methods. For safety reasons and also for the interest in grade control techniques of gradually increasing the knowledge of the deposit, the maximum recovery is seldom aimed at only one stage of stoping: it starts at a low level by leaving unnecessary big pillars, and only further it increases either by splitting those pillars or by reducing their action, increasing the size of the rooms; the reducing of pillar sections is only adviseable if the roofs are exceptionally self-supporting, as the width of the initial chambers, which is determined by economy envolved in its opening, is usually the maximum possible without the necessity of additional support. The final recovery if not total must be as big as safety allows it, corresponding this to the definitive abandonement of a minimum part of the deposit under the forms of residual or control pillars. In both cases the compromise between recovery and safety is always present.

When, from a certain phase of recovery, the existing pillars are removed, the roof of the void, usually limited by pillars which at the beginning were resistant enough, has usually an intermediate behaviour between those referred for the open longwall stope and that of the void resulting from the progressive collapsing of the pillars in a stope that initially had subdimensioned pillars.

At Panasqueira mine the experience pointed out the convenience of limiting the room widths to 5m (Potts, Szeki & Patchet 1970/73; Streets 1968/69), although an increase of the widths by adopting additional techniques is not yet disregarded. Concerning pillars we can say that, assuming a regular grid with rooms 5m wide and square pillars (there was no arguments in favour of any other geometry), the dimensions of the last evoluted in the dependence of the acquired knowledge as its behaviour and according to specific punctual circumstances.

When mechanical room stopes began to be tested in the early sixties, some delays in layout and development of the mine with consequent difficulties in accessing the veins that, under a correct mining sequence, should be mined, in a first step, forced such sequence to be abandoned and stoping started on lower level veins under unstoped mineable areas; hence, the pillars to be left had to be undoubtly stable to assure the workability of the last referred. Under this constraint the option was to have 15mx15m pillars and 5m wide rooms; those pillars were splitted in 5mx5m pillars, in a further stage and in the descending sequence. Some preliminar tests with 6mx6m pillars, and 6m or 5m wide rooms, were also performed (Potts, Szeki & Patchet 1970/73).

It is interesting to note that although there was no previous experience about room and pillar design at Panasqueira mine it is possible to find a basis to the establishment of the referred 5mx5m pillars and the related rooms; it is enough, for example, to extend the criteria used for the selection of pillar dimensions in South Africa coal mines (Mello Mendes 1985).

For those mines, it was admitted that patterns with square pillars with width W and height H and rooms wide B (m) shall be stable, without any restrictions both in how much time such stability will be assured or regarding the total area of that stope, if at least the safety factor S equalls 1.6; S being the ratio between the pillars resistance, calculated by the formula $\sigma_r = 7.3W^{0.46}H^{-0.66}$ (MPa), and the average load on the pillars, that one, considering the tributary area and assuming a pre-existing average vertical stress only due to the overlying rock weight, calculated from $\sigma_p = 10^{-3}(W+B)^2W^{-2}\gamma h$ (MPa), where γ and h are respectively the average specific rock weight (kN.m^{-3}) and the height (m) of the rock mass lying over the referred mining works (Salamon & Oravecz 1976).

If the rock mass is different and no more data is available it is acceptable to admit in a first step the maintenance of the influence of the section of the pillars and that of its ratio width/height (which is the same as maintaining exponents of W and H in the expression established for σ_r) and the correction of the numerical constant in σ_r expression (which represents compression strength over a rock cube with 1m side), proportionally to the uniaxial compression strengths, obtained in laboratorial tests carried on coal and rock material (Mello Mendes 1985). If this approach is used we can conclude that for the evaluation of the resistance (MPa) of the pillars in Panasqueira mine the formula $\sigma_r = 62W^{0.46}H^{-0.66}$ is valid.

Being W=5m and H=2.3m (average height of the pillars) this last expression gives $\sigma_r = 75.12$MPa and, for B=5m, $\gamma = 27.5$kN.m^{-3} (average specific weight of the rock) and h=300m (overlying rock height) the tributary area allows to calculate, by the above referred expression, $\sigma_p = 33$MPa. For these, S=2.28, value that is much superior than the minimum indicated S=1.6, which guarantees for a long time the stability of the remaining pillars in horizontal stopes, independently of stoping areas.

The practice completely proved that the above justification for the dimensions of remaining pillars in Panasqueira was valid under the referred conditions, since in the mine there are already several room and pillar stopes of that kind which connect each other through very extensive areas and stand during several years without any failure that can not be explained by preferential influence of other kind of circumstances (e.g. superimposed stopes). The unexistence, in these cases, of collapses both for pillars and stope roofs induces the possibility of the non-existence in Panasqueira mine of a critical span limiting the lateral progress of stopes where the pillars are left within this geometry.

But this geometry is related with a vein recovery of only $R=1-W^2/(W+B)^2=0.75$, which for wolframite mining is rather low and must be increased. For that, a second phase of recovery in this kind of stope was tried by removing completely the left pillars in the reachest areas of the veins. When the 5mx5m pillars, also separated by 5m wide rooms were removed, timber packs of 1.8mx1.8m were installed in such a way that after the removal of rock pillars an orthogonal grid of artificial pillars with center distances of 5m was placed.

In one of these stopes after the generation of an open span ranging 120m, the stope roof collapsed by shear fractures in a place with timber packs, where the pillars had already been stoped, in a zone near the pillar removing face. Between the extremity of the open span and the stope border, some small roof-floor divergences have been noticed. Before the roof starts breaking, strong spaced noises, which seemed to be coming from the roof, were heard, during some tens of hours. The first fractures were followed by either shear or tension failures with big subsidence of the roof, ranging some tens of cms and significantly ruining the unremoved closer pillars; all this happened in a gradual way having no clear rockbursts characteristics. Undoubtly a good control of this event had been reached, probably due to the important delaying action of the timber packs.

It must be pointed out that in this case just before the roof starts failing, the highest convergence rates were observed not in the places where greater compressive stresses could be expected in its roof but around a point located near the face of complete pillars removal, already in a zone only supported by timber packs. We believe that during a phase when important load is being transferred from highly ruined pillars to timber packs the last ones temporarily gave bigger resistance than the first ones; on the other hand, in the zones where the roof surface present the greatest compressive stresses, the divergences should not allow the rate of convergence to increase due to the growing of Poisson ratio during pre-rupture phase (Mello Mendes 1986). This event (about which we do not want to make here a detailed analysis) is evidenciated by the markedly different configuration of isometric rate of convergence lines in relation to those appearing in longwall stoping just before failures, that ones with rockbursts characteristics; there, the isometric rate of convergence lines were open against the advancing face. On another hand, for this pillar recovery where the failure of the roof was better controlled, convergence rates reaching 1.2mm/h were determined greater if compared with 0.7mm/hour of the longwalls.

The good control reached in pillars recovery both in the collapse of pillars (with the consequent pillar load transferences to other pillars and to timber packs), and in what roof failures are concerned gave very good perspectives for the evolution of stoping method design, which is aimed to be the more economical and also the more safe. Really it was shown that the utilization of timber packs at least allowed, for load transmission between pillars, a good utilization of their residual resistances, preventing the failure of pillar groups to reach the rockburst characteristics. On the other side, regarding roof failure, it was noticed that if the control of its bending was maintained by the timber packs reaction and also by highly fractured pillars, it was not only possible to greatly increase the critical spans related with those failures but also to avoid these failures to reach rockburst characteristics, like those in

longwall stopes; this fact is probably the most significant for the improvement of stoping method, possibly due to the greater deformability showed by the roof abutments (which are now formed by remaining pillars instead of virgin rock mass, as in longwall stope cases).

Regarding the importance of an eventual chain-type failure of pillars, or roof collapse correspondent to a great span where the intermediate support conditions are suddenly altered, it has been developed at Panasqueira mine a big effort to quantitatively control the load transferences between pillars or timber packs, as qualitatively was described; such quantitative control logically requires data collection for deformation and load (or at least load changes) acting on the pillars.

Regarding deformations, the possibility of its evaluation by measuring roof-floor convergences was readily proved. For this purpose, mechanical rod convergencemeters were used; when the necessity of controlling strong deformations of pillars during recoveries arose, like in the previous example, it was indeed possible to obtain the convergences by measuring with a tape the height variations of timber packs.

For load or load change determinations, great difficulties have been found. The use of extensometric cells fixed with special glue in holes done on the pillars and also on the stope borders was not good enough. Another trial, started to collect data about load-deformation states of the pillars, consisted in using laboratorial tests with a rigid press on small dum-bell shaped models of pillars, as described elsewhere (Mello Mendes 1971b; Mello Mendes & Dinis da Gama 1973), in order to find the complete load-deformation curves; although not yet convinceable, these tests are encourageable (Barroso & Mello Mendes 1985).

Meanwhile, the use of timber packs to fully remove the remaining pillars became unpracticable for economical reasons. Taking in consideration the pillars behaviour, it was therefore decided to eliminate completely the timber, changing to the definitive abandonment of pillars smaller than those previously adopted.

In the stopes where this decision was taken a regular grid of 11mx11m pillars separated by 5m wide rooms was used, with the aim of improving the knowledge of vein values; this corresponds to an initial recovery $R=1-W^2/(W+B)^2=0.53$. In a second phase, these pillars are splitted by 5m wide rooms in a way to obtain, from each, 4 pillars of 3mx3m; those will be definitively abandoned, once the overall recovery reaches $R=1-W^2/(W+B)^2=0.86$, this value being difficult to be exceeded in practice with a room and pillar stope even if the complete removal of pillars is minded.

Regarding the stabilities offered by 11mx11m and 3mx3m grids, the first show oversized pillars; using the expressions mentioned before, this is effectively confirmed as the load over them, if they are 300m deep, is about $\nabla_p=18MPa$, being its resistance $\nabla_r=108MPa$; the resulting safety factor comes S=6, which is much high. Yet for the 3mx3m grid, the same formula gives $\nabla_p=58.9MPa$ and $\nabla_r=59.4MPa$ and the safety factor becomes S=1.01; this value is not only the practical minimum allowed for any safety factor but it also is far from the stated S=1.6 recommended.

Owing to the followed methodology, the factor S=1.01 by no means corresponds to imminent failure, but it must be viewed as it is less than S=1.6, as a result of a grid for which the stability shall not be maintained whatever the size of stoped area is; in other words, a critical span corresponding to an already splitted area must be considered, above which the 3mx3m pillars under unfavourable conditions from a load point of view are expected to start failing.

According to the actual practice in this kind of stopes the failure of these pillars start in stoped areas corresponding to spans greater than 120m. The transference of load between pillars as its failure advances, seems a gradual process, revealing advantageous profitability of pillar residual strengths. In a particular case, having roof-floor convergences, in the zone where the pillars were more affected, of already some tens of mm, the roof as a whole was good, showing no signs of failure. Owing to all the acquired experience, there seems that no motives exist to fear a rapid succession of 3mx3m pillar collapses resembling a rockburst; however the possibility of roof failures must at the actual stage be kept in mind. These roof failures, considering the big spans involved, the relative big stifness of its abutments (now 11mx 11m pillars, compared with 5mx5m pillars in the total removal operations) and also the absence of timber packs (which effect in previous analysed cases, although it could not be completely explained, was probably relevant) can eventually undertake rockburst characteristics with undesirable dynamic effect.

In the present position of geomechanical knowledge on Panasqueira mine, it was though as prudent to limit the importance of any similar event; for the purpose, the abandonment of control pillars was foreseen, with the objective of limiting the spans generated by the development of the 3mx3m pillar areas. It is already acceptable that since a rigorous mining disciplin is undertaken, namely in what downwards sequence for the several vein stoping is concerned, the control pillars to be left in a vein may partially be recovered by the end of this vein stoping.

The different aspects considered altogether for the establishment of the division of the stope levels by control pillars, its placing and its sizing were the following: i) the advantage of splitting later the control pillars abandoned in the first phase of mining, which determines its 11x11 sizing and also its insertion in the general grid; ii) the interest, if possible, of using the control pillars as protection pillars for upper main galleries to be kept till one vein is entirely mined out or after that; iii) the need to respect the geometry and location of the mine layout altogether with the respective 100m side grid; iv) the advantage of profiting, if possible, from low grade or waste zones at the vein levels to be used as control pillars.

All this leaded to the settlement of control pillars sized 11mx11m at each vein level which should form barriers to last at least during the stope life. These control pillars also at each vein level, should in principle be placed along double lines being the axes of these barriers maximum distant 200m; between these barriers, the normal sequence of stoping can be done.

If we admit, for the stope depths we are considering, that during a final phase all the 3mx3m pillars, which lay between two of the double control pillar lines, fully collapsed, then each control pillar placed on both sides of that area must support a maximum load related with an average compressive stress of $\bar{r}_p=(0.001x27.5x300)x(11+5)x200/2x11x11=109MPa$; as the correspondent resistance is $\bar{r}_r=108MPa$, as previously referred, then their safety ratio becomes approximately S=1. However, the complete failure of 3mx3m pillars places between the two adjacent control barriers seems to be a rather irrealistic situation resulting from a too pessimistic analysis; in fact, that could only occur if the roof was fractured, simultaneously, at both sides of the area, along vertical shear planes which if not reaching surface should at least create a very high pressure arch; the load acting in the 11mx11m pillars would then be much lower, resulting in a safety ratio much more confortable. Since it is not expectable that the 3mx3m pillars will simultaneously fail at both sides of the same barrier and that each barrier has two lines of 11mx11m pillars, each line can be considered, at least temporarily as an extra safety condition to the other.

Therefore, under a descendent stope sequence, it is excessive to predict that the 200m stepped barriers, left in a vein, having each two lines of 11mx11m pillars, as above described, shall in what static is

concerned, guarantee enough safety, not allowing the progress of stope collapses through them. However, regarding dynamics we must always admit the possibility of rockbursting as a result from the failure of bended roof zones over large stope areas already mined in 3mx3m, which may generate shockwaves, being its consequences not yet known at all. But on the other hand, the acquired knowledge about stope behaviour, in what load transference between pillars and the concomitant progressive bending of roofs are concerned, and also if we mind the roof failures, after advanced pillar failures makes the occurrence of rockbursts as rather improbable; even so, if eventually it happens, the control pillars certainly will constitute themselves a barrier preventing the spreading of its effects. To allow the standing of upper development works, the lines of control pillars to be left in the successive vein levels should be superposed and its axes should lay in the same vertical planes of the mentioned works. To have in such way the whole of control pillars which have to be considered for each vein is nevertheless irrealistic because the valuable areas to stope in each vein are not superposed; so, keeping all the barriers superposed for the protection of a few development works and on the other hand leaving some other natural barriers in the vein levels due to the existing low grade zones or else vein absences (natural barriers that merely will superpose), at the end, for the great majority of the veins and also for the whole mine, the recovery would become unacceptable as too low. The division of the mine throughout the control pillars must be viewed vein by vein without any superimposing constraints. The greatest objections to these procedures will hopefully soon be reduced, as from a safety point of view it becomes feasible the pillar splitting in a certain vein before the stope of vein just below starts; then, as much as possible, the abnormal situations of local load concentration over the underneath stopes would be eliminated, although a detailed evaluation case by case must be considered. The development works to be protected must be reduced to the lowest possible number and some temporarily solutions for its protection must be found.

The evaluation of each vein partition will most probably and for the majority of the cases reduce the number, or even spare, the need of control pillars, as low grade or waste areas which must be left behind can obviously meet the same purpose. In practice, this means an increase in the recovery of the deposit.

5 SUBSIDENCE AND OVERALL MINE STABILITY

The voids resulting from tabular stopes in Panasqueira mine develop connecting each other at different vein levels and if the right stope sequence is followed they advance under those that correspond to the already mined upper levels. The state of the rock mass around each mined area which locally show a certain trend to generate unstressed zones under the rock compressed domes which are lying over the cavities, degenerate (in a progressive way capable of reaching the surface as a result of its sequential accumulation) in a situation that consists on having an extensive single roof where a strong pressure dome is installed, affecting now a wide spread zone of the rock mass which already is in some extent disturbed and subsided. It is evident that the progression of general subsidence always becomes easier when it reaches and joins overlying subsided areas from previous mining.

The form of the generated pressure dome, depending on the importance of previous horizontal stress field, is scarcely investigated. At the stope scale, the utilization of the tape extensometers to control the differential deformations in roofs, used during the seventies (Potts, Szeki & Patchet 1970/73) and reintroduced a few years ago, did not give conclusive information. Much better but reduced information could be obtained for the load dome superimposed to the whole mined area; according to the investigation in 1960 (Mello Mendes 1960) this dome had height/width rates ranging 1/3, what could indicate that the virgin rock mass had higher than vertical stresses and thus the already low deformability of the ground should accentuate.

Undoubtly, the existence of significant horizontal stresses is a major contribut for the general stability in the mine but it is not yet completely studied. The horizontal stresses not only exert an opposite influence on the bending of stope roofs but also opposes to the sliding along the faults or vertical and subvertical joint fractures, which at different scales divide the rock mass.

The generation of destressed zones over the roofs at a stope scale, no doubt, turns effective the action of the weight of the blocks defined in the roofs by jointing; as these destressed zones increase in a time dependable way, it is of great interest that, to avoid unnecessary problems with roof behaviours, the different phases of room an pillar stoping take place one soon after the other, with as less time as possible between them. At the mine scale, the destressing surely facilitates the action of the big existing faults and not only makes the related big block to act differently in what general subsidence is concerned but also lets the load conditions on stope areas lying below to be much distinct; under this scale, the close possibility that such destressing action takes place is sufficient reason to condemn any mining plan which foresees the deposit recovery through vertical panels rather than following a descendant sequence where slices are mined throughout the deposit.

All these complementary features must be studied and are chiefly important in what overall mine stability is concerned. For the optimization of both stope method and its sequence this stability must be carefully taken into consideration; the partial recovery of control pillars is one of the possibilities for which the same stability must be regarded as most important. This partial recovery of control pillars is certainly far from being the last improvement to be introduced in stoping methods by increasing the geomechanical knowledge of the mine: really, even now the possibility to increase recovery by totally removing 3mx3m pillars in specific stope zones where they have been left and more over to fully and systematically stope these pillars by a method consisting in a controlled large scale caving under the protection of 11mx11m pillars, is not considered anymore as unreasonable or else to be likely to endanger the mine safety.

REFERENCES

Barroso, M.J.G. & F.Mello Mendes 1985. Estudo do comportamento de pilares da mina da Panasqueira. Lisboa: Laboratório Nacional de Engenharia Civil (unpublished report).

Mello Mendes, F. 1960. Comportamento mecânico de rochas xistosas; contribuição para o estudo da influência da pressão do terreno na exploração do campo filoneano de Panasqueira. Porto: Imprensa Portuguesa.

Mello Mendes, F. 1966. About the anisotropy of schistose rocks. Proc.1st Cong.Int.Soc.Rock Mechanics, I, 607-611. Lisboa: Laboratório Nacional de Engenharia Civil.

Mello Mendes, F. 1970. Application of the Bieniawsky brittle fracture mechanism to the interpretation of rockbursts in tabular stopes. Proc.2nd Cong.Int. Soc.Rock Mechanics, 2, 3-44. Beograd.

Mello Mendes, F. 1971a. About the anisotropy of uniaxial compressive strength in schistose rocks. Rock Feature, Proc.Int.Symp. Rock Mechanics, I, II-13, Nancy.

Mello Mendes, F. 1971b. Acerca da previsão da resistência de pilares de mina. 1º Cong.Hispano-Luso-Americano Geol.Econ., 5:307-319. Madrid-Lisboa.

Mello Mendes, F. 1985. A geomecânica das minas da Panasqueira, estado dos conhecimentos em 31 de Dezembro de 1985 (unpublished report).

Mello Mendes, F. 1986. Acompanhamento geomecânico da exploração das minas da Panasqueira, 1º semestre de 1986 (Anexo). Lisboa: Hidromineira - Consultores de Minas e Geomecânica, Lda. (unpublished report).

Mello Mendes, F. & C. Dinis da Gama 1973. Laboratory simulation of mine pillar behaviour. New Horizons in Rock Mechanics, Proc.14th Symp. Rock Mechanics (Pennsylvania, 1972) ed. by H.R.Hardy & R.Stefanko. New York.

Potts, E.L.J., A. Szeki & S.J. Patchet 1970-1973. Several reports on geomechanics of Panasqueira mine (unpublished).

Salamon, M.D.G. & K.I.Oravecz 1976. Rock Mechanics in Coal Mining. Johannesburg: Chamber of Mines of South Africa.

Streets, C.G. 1968-1969. Several reports on geomechanics of Panasqueira mine (unpublished).

Thadeu, D. 1951. Geologia do Couto Mineiro da Panasqueira. Com.Serv.Geol.Portugal, XXXII, 5-64. Lisboa.

Rôle des effets à court terme sur le comportement différé des galeries souterraines
Role of short term effects on the delayed behaviour of tunnels
Der Einfluss der Bauphase auf das Dauerverhalten der Untertagebauten

D.N.MINH, Laboratoire de Mécanique des Solides, Ecole Polytechnique, Palaiseau, France
G.ROUSSET, Laboratoire de Mécanique des Solides, Ecole Polytechnique, Palaiseau, France

ABSTRACT : We study the influence of initial critical phases (drifting, support setting, ...) on a drift's final equilibrium, when the surrounding rockmass presents a delayed behaviour.

For this purpose, we propose a model with viscoplastic behaviour taking into account instantaneous failure ; the influence of the loading rate on failure growth around a circular gallery is then exposed.

RESUME : On étudie l'influence des phases critiques initiales (creusement, pose du soutènement, ...) sur l'équilibre final d'une galerie, lorsque le massif encaissant présente des effets différés.

Pour cela, on propose un modèle de comportement viscoplastique avec rupture instantanée ; on examine ensuite pour une galerie circulaire l'influence de la vitesse de sollicitation sur le développement de la rupture.

ZUSAMMENFASSUNG : Wir studieren den Einfluss der anfänglichen kritischen Phasen (Förderung und Verkleidungslegung) auf das endliche Gleichgewicht einer Strecke, wenn das umliegende Gebirge ein Zeitverhalten vorstellt.

Anf diesem Grunde wird ein viskoplastisches Modell mit Rücksicht auf den augenblicklichen Bruch proponiert ; der Einfluss der Verladungsgeschwindigkeit auf das Bruchswachsen um eine kreisformige Strecke herum wird dann angestellt.

1 INTRODUCTION

Pour analyser les phénomènes différés affectant la plupart des galeries souterraines profondes, divers modèles rhéologiques de milieux continus ont été proposés dans la littérature ; ces modèles pourraient être classés suivant leur aptitude à décrire le comportement à court terme ou à long terme de ces ouvrages (Nguyen Minh, 1986.....).

Ainsi, les modèles viscoplastiques de type Bingham, avec prise en compte du radoucissement pour interpréter la rupture différée, semblent convenir pour le long terme (Borsetto et al., 1978 ; Nguyen Minh et al., 1984 ; Rousset, 1985).

Ces modèles sont cependant insuffisants pour décrire les phases critiques initiales (creusement et pose du soutènement), durant lesquelles les grandes vitesses de déformation et les variations importantes des contraintes peuvent entraîner la rupture prématurée dans le massif.

En pratique, l'utilisation de soutènements provisoires (boulonnage, treillis, béton projeté, ...), la mise en oeuvre de cintres coulissants dans les ouvrages miniers, servent en partie à pallier ces inconvénients. La méthode d'exécution par section divisée pourrait relever d'un principe analogue.

Dans cet article, nous nous proposons d'étudier l'influence des paramètres qui décrivent ces évènements à court terme, sur l'équilibre final de la galerie.

Plus précisément, nous décrivons un modèle rhéologique de nature viscoplastique, avec plasticité instantanée pour tenir compte de la résistance à court terme.

Il s'agit d'une étude essentiellement qualitative qui a pour thème l'influence de la vitesse de sollicitation sur le comportement différé d'une galerie circulaire ; le modèle retenu est donc assez schématique.

2 CHOIX DU MODELE DE COMPORTEMENT

Le comportement unidimensionnel du modèle est représenté par le schéma rhéologique de la figure 1.

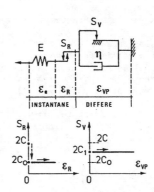

Figue 1 : Modèle rhéologique.

La déformation différée ε_{vp} vient du groupement viscoplastique de Bingham, de seuil égal à $2\,C_1$.

La déformation instantanée inclut une déformation de rupture fragile ε_R produite par un patin plastique particulier dont le seuil initial, égal à $2\,C$, chute brusquement à un niveau résiduel égal à $2\,C_0$.

La réponse du modèle à un chargement de vitesse de déformation constante est donnée sur la figure 2 ; il existe une vitesse de déformation critique $\dot{\varepsilon}_C = 1/\eta\;(C/C_1 - 1)$, au-delà de laquelle la rupture a lieu.

De façon qualitative, on retrouve ainsi certains résultats d'essais de laboratoire sur des matériaux fragiles (Bieniawski, 1970) et sur des matériaux plus visqueux (Nguyen Minh et al., 1984).

Il est important d'insister sur l'ordre des valeurs des cohésions :
$$C_0 < C_1 < C\;.$$

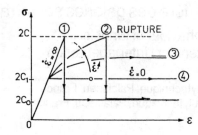

Figure 2 : Réponse du modèle à un chargement $\dot{\varepsilon}$ = Cte.

En particulier, il est logique d'associer la rupture au minimum de résistance du matériau ($2\,C_0$). Il en résulte un gel de la déformation différée, dès que la rupture se produit ; c'est sur ce point essentiel que le modèle proposé se distingue d'autres modèles destinés à étudier des phénomènes analogues : celui de Lombardi (1979) comporte deux éléments de Bingham en série, le rôle d'élément de rupture étant dévolu à celui dont le seuil est le plus élevé et la viscosité la plus faible ; celui de Sakuraï (1970) comporte un patin plastique en série avec un élément de Kelvin.

Dans le cadre de cette étude, nous adopterons pour le calcul les hypothèses simplificatrices suivantes : représentation tridimensionnelle des patins par des critères de Tresca, incompressibilité du matériau. Remarquons que les deux mécanismes de déformation irréversible du modèle ne peuvent être activés simultanément ($\dot{\varepsilon}_R\,\dot{\varepsilon}_{VP} = 0$) ; ce qui nous permet d'utiliser des solutions classiques à l'intérieur des zones correspondantes.

3 CONTRACTION D'UNE GALERIE CIRCULAIRE EN DEFORMATION PLANE

La figure 3 décrit la géométrie du problème traité ; les zones visqueuses et plastiques sont supposées en charge continue.
$\sigma_\infty = -\,P_\infty$ est la contrainte géostatique, exercée à l'infini.
$\sigma_i(t) = -\,P_i(t)$ est la contrainte en paroi, variable au cours du temps t à partir de la valeur initiale σ_∞. La variation $\sigma_i(t)$ en fonction de l'avancement du front, dans le cadre de la méthode de convergence-confinement, est largement discutée par ailleurs (Panet et al., 1982).
Enfin, pour simplifier l'écriture, on utilise le temps adimensionnel $T = t\,E'/\eta$, par rapport auquel la dérivée est notée par un point ($\overset{\cdot}{\frown}$).

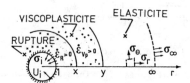

Figure 3 : Géométrie du problème.

3.1 Equations générales

L'incompressibilité du matériau et la continuité du déplacement radial u à la frontière viscoplastique donnent :

$$\frac{u}{r} = -\,\frac{2\,C_1}{E'}\,\frac{y^2}{r^2} \tag{1}$$

$E' = \dfrac{4}{3}\,E$ Module d'Young 'équivalent'.

La partition de la vitesse de déformation donne :

$$\frac{2\,C_1}{E'}\,\frac{\dot{y^2}}{r^2} = \frac{\overset{\cdot}{\overparen{(\sigma_r - \sigma_\theta)}}}{E'} + \dot{\varepsilon}_{ir} \tag{2}$$

$\dot{\varepsilon}_{ir} = \dot{\varepsilon}_R + \dot{\varepsilon}_{VP}$ Vitesse de déformation irréversible radiale ($\dot{\varepsilon}_{VP}\,\dot{\varepsilon}_R = 0$).

Les solutions à l'intérieur de chacune des zones de déformations irréversibles sont les suivantes :

1. Zone viscoplastique ($r \in [x,y]$) :

$$\frac{\partial}{\partial T}\,(\sigma_r - \sigma_\theta) + (\sigma_r - \sigma_\theta) = 2\,C_1\left(\frac{\dot{y^2}}{r^2} + 1\right) \tag{3}$$

$$\frac{\partial\sigma_r}{\partial T} + (\sigma_r - \sigma_\infty) = C_1\left(1 + \mathrm{Log}\,\frac{y^2}{r^2} + \frac{\dot{y^2}}{r^2}\right) \tag{4}$$

en particulier, lorsqu'il n'y a pas rupture, on a pour r = 1 :

$$\dot{\sigma_i} + (\sigma_i - \sigma_\infty) = C_1\,(1 + \mathrm{Log}\,y^2 + \dot{y^2}) \quad. \tag{5}$$

2. Zone de rupture ($r \in [1,x]$, s'il y a lieu) :

$$\sigma_r = \sigma_i - C_0\,\mathrm{Log}\,r^2 \quad. \tag{6}$$

A la frontière x, on a continuité de $\sigma_r(x)$, alors que $(\sigma_r - \sigma_\theta)$ est en général discontinu ; si la fracture se propage, on a $\dot{x} > 0$ avec $\sigma_r - \sigma_\theta = 2\,C$ en avant du front (x^+). C'est ce qui se passe dans un des cas que nous examinons ; dans un autre cas, la zone de rupture reste fixe.

3.2 Cas d'une évolution progressive de la zone de rupture

L'intégration de (3) dans la zone viscoplastique donne :

$$\frac{\sigma_r - \sigma_\theta}{2C_1} - 1 = \frac{1}{r^2}\int_{T_0(r^2)}^{T}\dot{y^2}\,\exp\,(\tau - T)\,d\tau \quad. \tag{7}$$

$T_0(r^2)$ est l'instant où le rayon de la zone viscoplastique a pour valeur r.

Avant rupture, $\sigma_r - \sigma_\theta$ est maximal en paroi ; la rupture a lieu si cette valeur atteint 2C et se développe par la suite à partir de la paroi ; $(\sigma_r - \sigma_\theta)(x) = 2C$ décrit alors l'évolution du front de rupture ; il est plus intéressant, pour la solution, de dériver cette expression par rapport au temps pour obtenir l'équation différentielle suivante :

$$\overset{\cdot}{\overparen{x^2}}/x^2 = \frac{\dot{y^2}/x^2 - C/C_1 + 1}{C/C_1 - 1 + \exp(T_0(x^2) - T)} \quad. \tag{8}$$

Une deuxième équation différentielle s'obtient par (4) et (6) en exprimant la continuité de $\sigma_r(x)$ et $\dot{\sigma_r}(x)$, en remarquant que :

$$\frac{\partial\sigma_r}{\partial T}(x) = \overset{\cdot}{\overparen{\sigma_r}}(x) + C\,\frac{\overset{\cdot}{\overparen{x^2}}}{x^2} \quad,$$

d'où :

$$\frac{\overset{\cdot}{\overparen{y^2}}}{x^2} = -\,(1 + \mathrm{Log}\,\frac{y^2}{x^2}) + \frac{\overset{\cdot}{\overparen{x^2}}}{x^2}\left(\frac{C - C_0}{C_1}\right) - \frac{C_0}{C_1}\,\mathrm{Log}\,x^2 + \frac{\dot{\sigma_i} + (\sigma_i - \sigma_\infty)}{C_1} \quad. \tag{9}$$

(8) et (9) forment un système d'équations différentielles pour x et y. On remarquera la similitude des équations obtenues avec celles relatives au modèle viscoplastique radoucissant (Nguyen Minh, 1986).

4 CONVERGENCE D'UNE GALERIE A VITESSE $\dot{U}_i$ CONSTANTE, JUSQU'A UNE VALEUR PRESCRITE U_i^0

D'après (1), la convergence U_i et le rayon y de la zone viscoplastique sont proportionnels. Le cas de charge étudié s'écrit donc :

$$\begin{cases} y^2 = A_0 \, T & ; \quad A_0 = \text{constante positive} \\ y^2 < y_1^2 & \text{valeur prescrite,} \end{cases}$$

la rupture une fois atteinte se développe suivant le scénario précédent ; (7) s'exprime simplement ici :

$$\frac{C}{C_1} - 1 = \frac{A_0}{x^2} \left(1 - \exp\left(\frac{A_0}{x^2} - T \right) \right) \qquad (10)$$

d'où l'on déduit :

- la rupture n'apparaît qu'à partir d'une certaine vitesse $A_0 = A_C$, avec $A_C = C/C_1 - 1$;
- même si la convergence n'est pas limitée, la zone de rupture tend vers une valeur limite x_∞, fonction croissante de la vitesse A_0 :

$$x_\infty^2 = A_0/A_C \quad .$$

Enfin, une fois la valeur y_1^2 atteinte, y^2 (et x^2 éventuellement) se figent ; la pression de confinement P_i croît alors asymptotiquement avec le temps.

Sur la figure 4, on pourra noter les différences de comportement entre le modèle étudié et le modèle classique de Bingham.

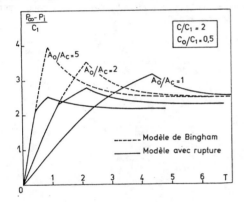

Figure 4 : Evolution du chargement pour différentes vitesses de convergence de la galerie.

5 APPLICATION AU PROBLEME DU SOUTENEMENT D'UNE GALERIE

On s'intéresse maintenant au problème plus spécifique de l'avancement de la galerie et de la pose de son soutènement.

Il est commode de se rapporter au diagramme classique de convergence-confinement (figure 5). Pour bien voir l'influence de la vitesse d'avancement, considérons deux cas extrêmes :
- dans le cas d'un avancement infiniment lent, il n'y a pas de rupture ; le massif converge suivant une courbe de réponse élastoplastique (C_1) qui est la réponse à long terme du modèle viscoplastique classique :

$$(C_1) = \begin{cases} P_i = P_\infty - C_1 (1 + \text{Log } y^2) \\ U_i = \dfrac{2C_1}{E'} y^2 \end{cases} \qquad (11)$$

- Dans le cas d'un avancement infiniment rapide, le massif converge d'abord suivant la courbe de réponse "instantanée" (C_0) qui est la réponse "élastofragile" du modèle :

$$(C_0) = \begin{cases} P_i = P_\infty - C - C_0 \text{ Log } x^2 \\ U_i = -\dfrac{2C}{E'} x^2 \end{cases} \qquad (12)$$

Par "instantanée", on entend que les durées des phases initiales du creusement et de la pose du soutènement sont très petites devant le temps caractéristique de la réponse différée du massif, $\tau = \eta/E'$ ou $T = 1$. La courbe (C_0) définit alors l'ensemble des états initiaux possible de la galerie, $M_0 = (P_i^0, U_i^0)$, à partir duquel les phénomènes différés traduisant l'interaction entre le massif et le revêtement vont conduire à l'état d'équilibre final $M_\infty = (P_i^\infty, U_i^\infty)$.
On trouvera en annexe les détails de cette évolution dans le cas où le revêtement est supposé élastique, de rigidité K :

$$P_i - P_i^0 = -K (U_i - U_i^0) \qquad (13)$$

Il est intéressant de noter que la courbe de convergence à long terme du massif rompu n'est pas unique ; on obtient en fait une famille de courbes ordonnées suivant la valeur de la rigidité K du revêtement (figure 5).

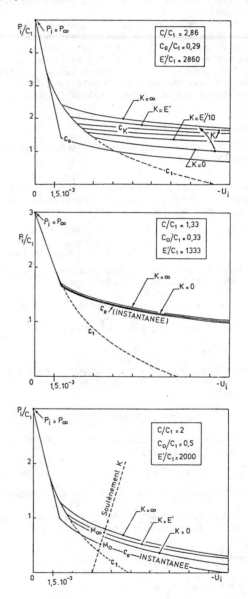

Figure 5 : Courbes de Convergence-Confinement à long terme.

La courbe ($C_{K=\infty}$), par exemple, se déduit de (C_0) par une translation suivant l'axe $O P_i$, de module :

$$P_i - P_i^0 = C - C_1 - C_1 \text{ Log } C/C_1 \quad .$$

Alors que ($C_{K=0}$) se déduit de (C_0) par une affinité de même axe, de rapport :

$$U_i / U_i^0 = C_1 / C \ \exp(C/C_1 - 1) \ .$$

Ce dernier cas pourrait simuler, par exemple, l'action d'un soutènement formé de cintres coulissants.

Ces résultats se démarquent tout à fait des résultats classiques obtenus avec les modèles de type "Bingham", ainsi qu'avec des modèles tenant compte des ruptures à court terme mentionnés plus haut ; pour tous ces modèles, en effet, la courbe de réponse à long terme est unique et ne porte pas trace des évènements initiaux, ruptures en particulier.

6 CONCLUSION

La simplicité du modèle de comportement utilisé n'ôte en rien à la portée générale des résultats obtenus ; grâce à la modélisation de la rupture adoptée ici, on a pu mettre en évidence l'incidence du mode d'exécution sur le comportement à long terme de l'ouvrage.

Si le modèle est fondé sur des considérations assez naturelles, il peut être amélioré sur certains points (prise en compte de l'effet de la pression moyenne, du radoucissement, ...) ; il n'en reste pas moins que c'est l'expérience en laboratoire et in-situ qui permettra, sur chaque type de matériau, de préciser l'importance des phénomènes qualitatifs mis en évidence.

ANNEXE :

EQUILIBRE D'UNE GALERIE SOUTENUE, APRES CREUSEMENT INFINIMENT RAPIDE

Soit $M_0 = (P_i^0, U_i^0)$ représentant l'état initial de la galerie (figure 6). L'évolution ultérieure $M(P_i, U_i)$ est décrite d'une part par (13), qui s'écrit aussi :

$$P_i(T) = P_i^0 + \alpha \ C_1 \ (y^2 - y_0^2) \quad \text{où } \alpha = 2K/E' \quad , \qquad (14)$$
$$\text{et } y_0 = \text{rayon initial de la zone viscoplastique,}$$

et, d'autre part, par une des deux équations (15) ou (17). On distingue en effet deux cas :

1- Il n'y a pas de rupture initiale ($P_i^0 > P_\infty - C$) :

$$P_i/C_1 = P_\infty/C_1 - (1 + \text{Log } y^2) - \overset{\Large\frown}{\dot{y}^2} \ (1 + \alpha) \quad , \qquad (15)$$

par ailleurs, $\overset{\frown}{y^2} = P_\infty - P_i^0 / C_1$;

à l'équilibre $(\overset{\frown}{\dot{y}_2} = 0)$, on reconnaît l'équation de la courbe de convergence du massif sain (modèle de "Bingham").

2- Il y a rupture initiale ($P_i^0 < P_\infty - C$) :

Il se développe instantanément une zone de rupture de rayon x_0 (voir (12)) :

$$P_i^0 = P_\infty - C - C_0 \ \text{Log } x_0^2 \qquad (16)$$

On vérifie que cette zone ne se propage pas par la suite.

L'évolution de y^2 est donnée par (5), dans laquelle $x^2 = x_0^2$, soit, en tenant compte de (14) :

$$\frac{P_i}{C_1} = \frac{P_i^0}{C_1} + \frac{C}{C_1} - \left(1 + \text{Log } \frac{y^2}{x_0^2}\right) - \overset{\Large\frown}{\dot{y}^2}\left(\frac{1}{x_0^2} + \alpha\right) \qquad (17)$$

Dans ce cas, $y_0^2 = C/C_1 \ x_0^2$.

A l'équilibre $(\overset{\frown}{\dot{y}^2} = 0)$, cette équation définit une famille de courbes paramètrées en M_0. En éliminant P_i^0 et x_0^2 entre (14), (16) et (17), on détermine, pour une valeur donnée de la raideur K du soutènement, la courbe de convergence à long terme.

BIBLIOGRAPHIE

Berest, P., Nguyen Minh, D. (1983), "Time-dependent behaviour of lined tunnels in soft rocks", Eurotunnel 83 Conf., Basle : 57-62.

Bieniawki, Z.T. (1970), "Time-dependent behaviour of fractured rock", Rock Mechanics, 2 : 123-137.

Borsetto, M., Ribacchi, R. (1978), "Influenza degli effetti reologici sull interazione tra terreno envestimento di une galleria", XIII Convegno Nazionale di Geotecnica, Merano.

Lombardi, G., Amberg, W. (1979), "L'influence de la Méthode de construction sur l'équilibre final d'un tunnel", Proc. 4ème Cong. SIMR, Montreux, 1 : 475-484.

Nguyen Minh, D. (1986), "Modèles rhéologiques pour l'analyse du comportement différé des galeries profondes", Proc. Cong. Int. Grands Ouvrages en Souterrain, ITA/AITES, Florence : 659-665.

Nguyen Minh, D., Habib, P., Guerpillon, Y. (1982), "Time-dependent behaviour of a pilot tunnel driven in hard marls. Design and performance of underground excavation", ISRM/BGS, Cambridge : 453-459.

Panet, M., Guenot, A. (1982), "Analysis of convergence behind the face of a tunnel", Proc. Tunnelling 82, Brighton :

Rousset, G. (1985), "La faisabilité d'un enfouissement de déchets radioactifs dans une formation argileuse profonde. Aspects mécaniques et géotechniques", Ann. Ponts & Chaus., 4ème trim. : 14-39.

Sakurai, S. (1970), "Stability of tunnel in visco-elastic-plastic medium", Proc. 2nd Cong. ISRM, Belgrade : 4-20.

Stability of deep underground excavations in strongly cohesive rock
Stabilité des excavations souterraines profondes dans des roches très résistantes
Stabilität von Untertagebauten im stark kohäsiven Gebirge mit grosser Zugfestigkeit

H.-B.MÜHLHAUS, Institute of Soil and Rock Mechanics, University of Karlsruhe, FRG

ABSTRACT: Conditions which are decisive for the stability of deep underground excavations are investigated in a numerical study. First a nonassociated elasto-plastic constitutive relation for brittle materials is derived. It is shown that an originally load controlled boundary value problem is equivalent to a displacement controlled boundary value problem. As an application circular - and rectangular cavities in an isotropic stress field are investigated.

RESUME: Les conditions décicives pour la stabilité des cavités très profondes ont été étudiées par les études numériques. D'abord on décrive une loi des conditions elasto-plastiques non associatives pour des matériaux cassants. On démontre comment un problème aux conditions aux limites originalement contrôllé par la charge est équivalent à un problème aux conditions aux limites contrôllé par le déplacement. Comme application on étude des cavités circulaires et rectangulaires dans un champ de tension isotropique.

ZUSAMMENFASSUNG: Anhand numerischer Studien wird überprüft, unter welchen Bedingungen tiefliegende Hohlräume in sprödem Gestein versagen können. Zunächst wird ein nichtassoziiertes elasto-plastisches Stoffgesetz für sprödes Gestein hergeleitet. Es wird gezeigt, wie ursprünglich lastparametergeregelte Randwertprobleme in äquivalenter Weise verschiebungsgeregelt formuliert werden können. Als Anwendungen werden kreisförmige und rechteckige Untergrundhohlräume in einem isotropen Spannungsfeld behandelt.

1. INTRODUCTION:

One is frequently confronted with the problem of stability of deep underground excavations in brittle rock in connection with geophysical explorations or energy production.

In this paper constitutive relationships are derived for the desription of the mechanical behaviour of strongly coherent rocks such as granite, sandstone etc. On the basis of this constitutive model the stability of circular and rectangular cavities is studied by means of the finite element method.

It was pointed out by EGGER (1973) in an analytical study that the most important aspect of the material behaviour with respect to the stability of cavities is the law that describes how the material looses coherence. In the present paper this property will be considered in an elasto-plastic constitutive relation which allows the representation of frictional hardening and simultaneous cohesion softening. In finite element solutions in softening material behaviour one is confronted with two major problems:

1. Load control in the incremental solution is not possible because the load-displacement relation is not monotoneous in general. This means in a tunnel problem that the intensity of the support pressure cannot be used as a control parameter in a finite element solution because in softening material the support pressure first decreases to a certain threshold level and has to be increased to maintain stability then. By applying the so called arc-length algorithm for solution of nonlinear systems of equations this problem can be overcome (RIKS, 1979; CRISFIELD, 1985; de BORST, VERMEER, 1984). In arc-length algorithm the increments in externally prescribed loads or displacements do not have to be definded by the user. An increase or decrease of the loads is determined by the algorithm in such a way that e.g. a certain norm of the incremental displacements do not exceed a specified value. Alternatively, as will be done here, the surface deformation which is conjugate in energy to the surface pressure can be used as a control parameter (provided this surface deformation is monotoneously increasing). In an tunnel problem this means that the volume change of the tunnel is prescribed instead of the support pressure.

2. The finite element solution becomes strongly mesh dependent. One of the reasons for this is the tendency of the displacement gradients to localize in thin shear bands. However, if the classical continuum theory is underlain the thickness of the shear bands is undetermined. To remedy this desease, it has been proposed to make the softening modulus dependent on the element size (PIETRUSCZAK, MROZ, 1981; WILLAM et al. 1984, 1986). A physically more evident approach is to underlay a generalized continuum theory as e.g. the Cosserat theory. In such a continuum the thickness of shear bands is defined by certain material parameters depending on the dimension of length (Grain diameter etc.) (BESDO 1985; MÜHLHAUS 1986a, b; MÜHLHAUS, VARDOULAKIS, 1986).

In this article we content with the classical continuum description. It will be shown that in the case of the boundary value problems considered here the essential properties of the solution are mesh independent. In the following section a constitutive relation is defined. Subsequently it is shown that an originally load controlled boundary value problem can be made displacement controlled. In the fourth section the stability of the surface of a rectangular cavity in an infinite plate is considered. The in-plane principle stresses at infinity are assumed to be equal. To get an idea of the influence of the shape of the cavity on the stability, the case of a circular cavity is also considered.

2. CONSTITUTIVE EQUATIONS

In this section a simple isotropic hardening/softening model of elasto-plasticity is desribed. The existence of a range in stress space where the response of the material is essentially linear elastic with Young's modulus E and Poisson number ν is assumed. The range of elastic deformation is bounded by a yield criterion of extended v. Mises type (e.g.: RUDNICKI, RICE, 1975):

$$f = \tau + \mu_M p - k_M = 0, \tag{1}$$

where

$$\tau = \left(\frac{3}{2} \sigma'_{ij} \sigma'_{ij} \right)^{\frac{1}{2}}, \quad p = \frac{1}{3} \sigma_{kk}, \tag{2}$$

μ_M is a mobilized friction factor and k_M is a mobilized cohesion factor. In (2) σ_{ij} denotes the stress tensor. Here and in the following the superimposed stroke marks the deviators of the respective tensors.

Under triaxial compression test condition (1) coincides with the Mohr-Coulomb criterion if

$$\mu_M = \frac{6 \sin \phi_M}{3 - \sin \phi_M}, \quad k_M = \frac{6 c_M \cos \phi_M}{3 - \sin \phi_M}, \tag{3}$$

where ϕ_M is the mobilized angle of friction and c_M is the mobilized cohesion. The direction of plastic strain rates $\dot{\varepsilon}_i^P$ is defined by the gradients of the plastic potential

$$g = \tau + \beta_M p, \quad \dot{\varepsilon}_{ij}^P \sim \frac{\partial g}{\partial \sigma_{ij}}, \tag{4}$$

$$\beta_M = \frac{6 \sin \psi_M}{3 - \sin \psi_M}, \tag{5}$$

where β_M is the mobilized dilatancy factor. The definition (5) of the mobilized angle of dilatancy ψ_M implies that in triaxial test conditions

$$\frac{\dot{\varepsilon}_{ii}^P}{\dot{\varepsilon}_1^P} = - \frac{2 \sin \psi_M}{1 - \sin \psi_M}, \tag{6}$$

where $\dot{\varepsilon}_1^P$ is the irreversible part of the axial strain rate.

It is assumed that the mobilized angle of friction and the mobilized cohesion depend on a single hardening parameter γ_p, where

$$\dot{\gamma}^P = \left(\frac{2}{3} \dot{\varepsilon}'^P_{ij} \dot{\varepsilon}'^P_{ij} \right)^{\frac{1}{2}}. \tag{7}$$

As a hardening law we assume a slightly modified version of the relations (fig. 1) proposed by VERMEER and de BORST (1984), de BORST (1986):

$$\sin \phi_M = \begin{cases} \sin \phi_0 + 2 \dfrac{\gamma^P \gamma_f}{\gamma^P + \gamma_f} (\sin \phi - \sin \phi_0), & \gamma^P \leqslant \gamma_f \\[2mm] \sin \phi, & \gamma^P > \gamma_f \end{cases} \tag{8}$$

and

$$c_M = c \exp \left(- \left(\frac{\gamma^P}{\gamma_c} \right)^2 \right), \tag{9}$$

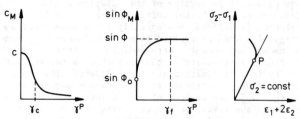

Fig. 1: History dependency of strength parameters for the hardening/softening model

The physical meaning of the parameters $\sin \phi_0$, $\sin \phi$, γ_c, γ_f and c is shown in fig. 1. The cohesion c and the angle of friction ϕ_0 are the strength parameters at initial yield. Initial yielding occurs as soon as inelastic volume change is detected (fig. 1c) in trixial test (BRACE et al., 1968). The results of BRACE et al. indicate that in the case of Westerly Granite, ϕ_0 is strongly dependent on the loading velocity v. It seems that $\phi_0 \to 0$ in the limit for $v \to 0$.

For definition of the development of plastic volume change ROWE's (1972) stress-dilatancy relation is assumed which states that

$$\frac{\dot{\varepsilon}_{ii}^P}{\dot{\varepsilon}_1^P} = 1 - \frac{K_M}{K_C}, \tag{10}$$

where

$$K_M = \frac{1 + \sin \phi_M}{1 - \sin \phi_M}, \tag{11}$$

$$K_C = \frac{1 + \sin \phi_C}{1 - \sin \phi_C}, \tag{12}$$

In (12) ϕ_C is that angle of friction where $\dot{\varepsilon}_{ii} = 0$.

The physical model from which (10) originally has been deduced does only apply for granular materials. However, the general trend of the $\dot{\varepsilon}_{ii}^P - \dot{\varepsilon}_1^P$ relation, predicted by Rowe's theory, can be observed in most pressure sensitive materials (fig. 2).

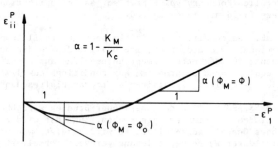

Fig. 2: Assumed relation between inelastic axial strain and inelastic volume change

Eliminating $\dot{\varepsilon}_{ii}^P / \dot{\varepsilon}_1^P$ from (6) and (10) one obtains:

$$\sin \psi_M = \frac{\sin \phi_M - \sin \phi_C}{1 - \sin \phi_M \sin \phi_C}. \tag{13}$$

In some cases it is more convenient to use the residual angle of dilatancy ψ ($: = \psi_M (\phi_M = \phi)$) as a material constant instead of ϕ_C. It follows:

$$\sin \phi_C = \frac{\sin \phi - \sin \psi}{1 - \sin \phi \sin \psi}. \tag{14}$$

If the normality rule is assumed, then $\psi_M = \phi_M$ and from (14) it follows that $\phi_C = 0$. If $c_M \to 0$ or in the limit as $c_M \to 0$ (fig. 1a) it follows from the requirement that the dissipation power density has to be positive, that $\phi_M \leqslant \phi_M$. According to (13, 14) the latter is the case if

$$\psi \leqslant \phi. \tag{15}$$

In section four two types of materials will be considered: In one case the cohesion is decreasing slowly (solid line in fig. 3) and in the other case the cohesion is decreasing rapidly (dotted line). Fig. 3 shows evaluations of the constitutive equations described in this section.

In the formulation of the material law a number of properties of brittle materials have been neclected (yielding in isotropic stress states, Bauschinger effect, decrease of unloading moduli due to cracking). In nonproportional stress path these simplifications could result in a too stiff material response. It will be assumed here that these simplifications do not alter the basic mechanical character of the stability problems considered in the following section.

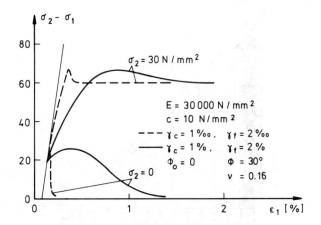

Fig. 3: Numerical simulation of triaxial tests by the hardening/softening model. σ_1: Axial stress, ε_1: Axial strain

3. BOUNDARY CONDITIONS

Consider a cavity of arbitrary shape (fig. 4). The components of the unit normals of the boundary of the cavity are denoted as n_i. The support pressure p is

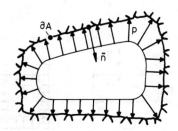

Fig. 4: Geometry and boundary conditions for a cavity

assumed to be uniform. At the boundary ∂A (fig. 4) the stresses have so satisfy the boundary condition

$$\sigma_{ij} n_j = -p\, n_j \ . \tag{16}$$

The rate of work of $-pn_i$ in the displacements on ∂A is

$$\dot{W} = p\dot{A} \ , \tag{17}$$

$$\dot{A} = -\int_{\partial A} \dot{u}_i n_i\, dA, \tag{18}$$

where $\dot{A}$ is the rate of volume change of the cavity.
   In the solution of the boundary value problem where the relationship between p and $\dot{A}$ has to be determined, one has two possibilities:
1. The support pressure is used as a control parameter of the incremental solution process.
2. The volume change of the cavity is used as a control parameter, i.e. values are prescribed for ΔA. In this case, p is a Lagrange multiplier, i.e. the value of the support pressure is part of the solution of the boundary value problem.
   In softening materials it is more convenient to apply the latter formulation. In infinitesimal deformations the corresponding boundary condition is:

$$-\int_{\partial A} u_i n_i\, dA = \Delta A \ , \quad \Delta A \text{ given.} \tag{19}$$

In finite element method (19) becomes an algebraic equation:

$$\sum_{J=1}^{M} a_J (n_i u_i^J) = \Delta A \ , \tag{20}$$

where M is the number of nodal points at the boundary of the cavity and u_i^J is the nodal point displacement vector. The coefficients a_J are obtained by integration of the displacement shape functions.

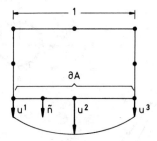

Fig. 5: Determination of displacement weight factors a_J

In the case of a single 8-noded element, the left member of (20) reduces to

$$-(\tfrac{1}{6}u^1 + \tfrac{2}{3}u^2 + \tfrac{1}{6}u^3), \tag{21}$$

i.e. in this case the coefficients a_J are equal to the nodal point contributions of an uniform surface load. Constraint equations of type (20) are available in a number of finite element codes (eg Nastran, Adina). Originally these constraint equations were provided for representation of rigid structural members (rigid foundations etc.).

4. RESULTS FOR CIRCULAR AND RECTANGULAR CAVITIES

First a circular cavity under uniform radial stress at infinity is considered. We use the notations $p_\infty := -\sigma_r(r \to \infty)$; $p = -\sigma_r(r = R)$; R is the radius of the cavity. Because of axial symmetry, the constraint (19) reduces to $-2\pi R u_r(r = R) = \Delta A$. The constitutive relationship from section two is used in connection with the parameters shown in fig. 3. The initial conditions of the finite element calculations are $\sigma_r \equiv -p_\infty$, $u_r \equiv 0$. Subsequently $\Delta A/A = -2u_r(R)/R$ is monotoneously decreased. The BFGS algorithm (MATTHIES, STRANG, 1979) was applied for solution of nonlinear equations. If convergency was not achieved in a certain configuration, the corresponding increment of the displacement control parameter ($= u_r(R)$) was successively subdivided until convergency (with respect to incremental energy norm) was achieved.

Finite element calculations have been performed for the case of the material with rapidly decreasing cohesion ($\gamma_c = 1$ ‰) and for the case of the material with slowly decreasing cohesion ($\gamma_c = 1$ %; fig. 3). Fig. 6 shows evaluations of these calculations for the case of mesh (b) in fig. 7. The support pressure p is defined as the average value of the radial stress at the integrations points of the element adjacent to the cavity. Independent of the angle of dilatancy the cavity is stable for $\gamma_c = 1$ %. Failure does not occur even if significant tensile stresses are acting at the boundary of the cavity. The structure becomes stiffer with decreasing angle of dilatancy. The situation is reverse in most of the limit load problems associated with close surface structures (e.g. de BORST, VERMEER, 1984).
   In case of the rapidly softening material ($\gamma_c = 1$ ‰) p/p_∞ has to be increased when the material in the vicinity of the boundary of the cavity is in the softening regime. When $c_M \sim 0$ the support pressure decreases again; the material behaves like a cohesionless soil then. In this case, for $p/p_\infty \to 0$, $|\Delta A/A| \to \infty$.

1159

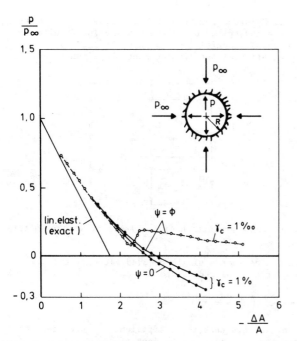

Fig. 6: Results of F.E. calculations
$p_\infty/E = 0.767 \cdot 10^{-3}$

To get an idea of the mesh dependency of the results, calculations have been performed with different meshes (fig. 7). Although the general trend of the p/p_∞ –$\Delta A/A$ relation is similar, quantitative differences are present. The differences result from the definition of p in part: p was defined as the average value of the radial stress at two integration points. However, in both meshes the amount of $\Delta A/A$ increases indefinitely as $p/p_\infty \rightarrow 0$.

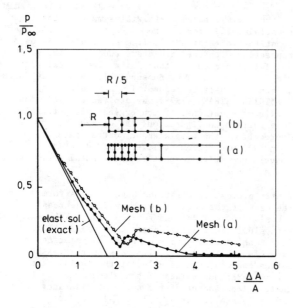

Fig. 7: Comparison of the result for different meshes

In fig. 9 results for a quadrilateral cavity are represented. In this case the general form of the constraint equation (20) had to be applied. The p – ΔA curves are surprisingly similar to those for the case of a circular cavity (fig. 6). Note that p_∞/E in fig. 9 is smaller than p_∞/E in fig. 6. This explains the slightly negative values of p in the vicinity of the jump in the p/p_∞ – $\Delta A/A$ ($\gamma_c = 1$ %o) relation.

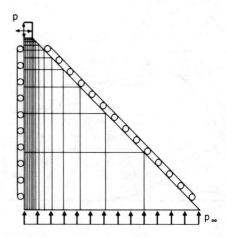

Fig. 8: F.E. mesh for quadrilateral cavity

EGGER (1973) investigated the stability of circular cavities in an analytical study. In the p – ΔA curves found by EGGER even for rapidly decreasing cohesion, the support pressure is monotoneously decreasing to zero as $|\Delta A| \rightarrow \infty$. In the results found here this is not the case (fig. 6, 7, 9). The reason for this

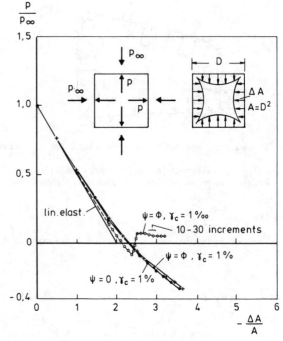

Fig. 9: Results of F.E. calculations.
$p_\infty/E = 0.633 \cdot 10^{-3}$

qualitative difference is that EGGER assumed an (nonisotropic) softening rule which is not consistent with the theory of plasticity assumed here. The results of the present analysis lead to similar conclusions despite of the differences in the formulation: The cavity is instable for $p \rightarrow 0$ if γ_c (fig. 1a) is smaller than a certain threshold value, provided p is sufficiently large. The precise value of the angle of dilatancy is of minor importance in this context.

REFERENCES

Besdo, D.: Inelastic behaviour of plane frictionless block-systems described as Cosserat media. Arch. Mech. 37, 6, 603-619, WARSZAWA, 1985.

de Borst, R: Non-linear analysis of frictional
    materials. Dissertation; Delft University of
    Technology, Delft, 1986.

de Borst, R.; Vermeer, R.A.: Possibilities and
    limitations of finite elements for limit analysis.
    Géotechnique 34, 2, 199-210, 1984.

Brace, W.F., Paulding; B.W.; Scholz, C.: Dilatancy
    in fracture of crystalline rocks. J. of geoph.
    res. 71, No. 16, 3939-3953, 1966.

Crisfield, M.A.: A fast incremental/iterative
    procedure that handles snap-through. Comp. struct.,
    13, 55-62, 1981.

Egger, P.: Einfluß des Post-Failure-Verhaltens von
    Fels auf den Tunnelausbau. Veröffentlichungen des
    Institutes für Bodenmechanik und Felsmechanik der
    Universität Karlsruhe, Heft 57, 1973.

Matthies, H., Strang, G.: The solution of nonlinear
    finite element equations. Int. J. Num. Meth. Eng.
    14, 1613-1626, 1979.

Mühlhaus, H.-B.: Surface instability of a half space
    with bending stiffness (in German). Ing. Arch. 55,
    388-400, 1985.

Mühlhaus, H.-B.; Vardoulakis, I.: Axially-symmetric
    buckling of the surface of a laminated half-space
    with bending stiffness. Mechanics of materials 5,
    109-120, 1986.

Mühlhaus, H.-B.: Consideration of material
    inhomogeneities in a continuum theory (in German).
    Habilitationsschrift: Submitted to the Faculty of
    Civ. Eng. Univ. Karlsruhe, 1986.

Pietruszcak, St.; Mroz, Z.: Finite element analysis
    of deformation of strain softening materials. Int.
    J.Num. Meth. Eng., 17, 327-334, 1981.

Riks, E.: An incremental approach to the solution of
    snapping and buckling problems. Int. J. Solids and
    Struct. 15, 529-551, 1979.

Rowe, P.W.: Theoretical meaning and observed values of
    deformation parameters for soil. In: Proc. Roscoe
    Mem. Symp. on stress-strain behaviour of soils.
    Foulis, Henley-on-Thames, 143-194, 1982.

Rudnicki, J.W., Rice, J.R.: Conditions of the
    lokalization of the deformation in pressure
    sensitive materials. J. mech. phys. solids 23, 371-
    394, 1975.

Vermeer, P.A.; de Borst, R.: Non-associated
    plasticity for soils, concrete and rock, Heron, 29,
    No. 3, 1-64, 1984.

Willam, K.J., Bicanic, N; Sture, S.: Constitutive and
    computational aspects of strain-softening and
    localization in solids. In: Winter annual meeting
    of ASME, New Orleans, Louisiana, 1984.

Willam, K.J.; Pramono, E.; Sture, S.: Stability of
    strain-softening computations. In: Finite element
    methods for nonlinear problems. Europe-US-
    Symposium. Trondheim, Norway 1985. Springer,
    Berlin, Heidelberg, 1986.

Yielding precast reinforced concrete tunnel lining in Oligocene clay
Revêtement de tunnel souple préfabriqué en béton armé pour des roches argileuses de l'Oligocène
Ein nachgiebiger Tunnelausbau aus armiertem Fertigbeton im Oligozän-Ton

M.MÜLLER, Technical University Budapest, Hungary

ABSTRACT: To utilize Hungarian energy resources, several ignite beds are under development. Entry tunnels are driven in weak Oligocene clay /$\emptyset$ = 30°, S_o = 500 kPa, Q = 1732 kPa/ with a diameter D = 5 m. In case of such a low rock strength, if the support resistance is low, very large tunnel wall displacements are needed to stabilize the surrounding rock after the cut and to lower geostatic pressures. If a yielding support is applied together with a significantly higher support resistance than for normal forces and moments, then significantly smaller tunnel wall displacements suffice to stabilize the surrounding rocks.

Outlines are given of results of theoretical investigations, and of structural solutions.

RESUME: Pour utiliser les ressources d'énergie hongroises, plusieurs gisements de lignite viennent d'etre aménagés, dont les roulages, au diamétre D = 5 m, sont poussés dans de l'argile loigocéne peu résistant /$\emptyset$ = 30°, S_o = 500 kpa, Q = 1732 kpa/. Pour des résistances de soutément réduites dans des roches peu résistantes, des déplacements importants des bords de vacité sont nécessaires pour stabiliser la roche environnante aprés la construction des roulages, pour réduire les poussées géostatiques importantes. L'application d'un esclimbage compressible d'une résistance á l'effort normal et au moment augmentée du souténement permet une réduction importante du déplacement de bords de cavité nécessaire á la stabilisation de la roche environnante.

Une déscription est donnée sur les résultats de l'examen théorique de probléme, ainsi que sur la solution structurelle.

ZUSAMMENFASSUNG: Um die Energiequellen in Ungarn besser ausschöpfen zu können, wurden mehrere Kohlgruben geöffnet. Die Förderstrecken dieser Gruben werden mit einem Durchmesser von 5 m in einem Oligozän-Ton kleiner Festigkeit ausgebaut /$\emptyset$ = 30°, S_o = 500 Kpa, Q = 1732 Kpa/. Sind der Ausbauwiderstand und die Felsfestigkeit gering, so sind sehr grosse Deformationen am Streckenrand nötig, um Felsumgebung nach der Bauarbeit zu stabilisieren, und den grossen Gebirgsdruck zu vermindern. Ist ein biegsamer und nachgiebiger Ausbau mit viel grösserem Widerstand eingebaut, sind bedeutend wenigere Streckenranddeformationen genug, um die Felsenumgebung zu stabilisieren.

In dieser Vortrag werden die theoretischen Ergebnisse dieser Frage, die Konstruktion und die Festigkeitsprüfungen beschrieben.

Cross section of large-diameter entry tunnels built in Oligocene clay, supported on TH vault and lining boards soon much decreased, and in its surrounding, extensive elastically failed plastic rock zones arose.

This is due to the low rock strength compared to its depth, low load capacity of the roadway supports /Fig. 1./ as well as to the increased possibility of rock loosening between the irregular void surface of excavation by blasting and the support.

In the following, a roadway support of increased load bearing capacity will be presented, permitting a rock deformation and loosening not greater than needed to reduce the initial rock stresses in the void edge surrounding to match the wall support resistance.

To solve this problem, on one hand, structural behavior of the rock in elastic and post-failure condition has been analyzed; on the other hand, a roadway support material and structure likely to provide for a continuous, and - alongside with rock deformations - nearly constant supporting force to the excavated irregular rock surface was chosen.

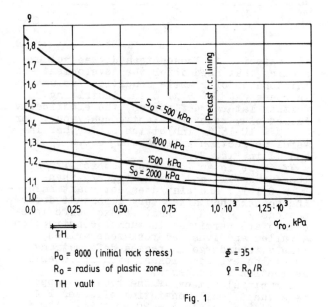

p_o = 8000 (initial rock stress) Φ = 35°

R_o = radius of plastic zone ϱ = R_q/R

TH vault

Fig. 1

Strength testing of the rock was made in a triaxial cell. Elastic /E_o/ and residual /E_r/ material characteristics:

E_o = 800 Mpa ν = 0,12 $\varnothing$ = $30°$
S_o = 3,0 Mpa E_r = 130 Mpa

Our tests on loosening after elastic failure /Figs 2 and 3/ show that, letting

Effect of loosening possibility to the deformation diagram of clay

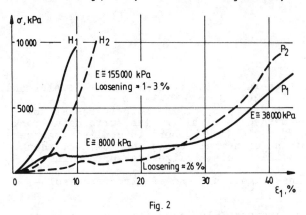

Fig. 2

Effect of hindered loosening possibility to the deformation diagram of clay

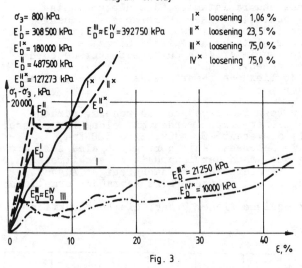

σ_3= 800 kPa $\quad$ I$^×$ loosening 1,06 %
E_D^I = 308 500 kPa $\quad E_D^{III} \approx E_D^{IV}$ = 392750 kPa $\quad$ II$^×$ loosening 23,5 %
E_D^{Ix} = 180 000 kPa $\quad$ III$^×$ loosening 75,0 %
E_D^{II} = 487500 kPa $\quad$ IV$^×$ loosening 75,0 %
E_D^{IIx} = 127273 kPa

Fig. 3

the material to loosen to a degree not exceeding that needed for the development and maintenance of plastic condition /volume increase V % $\leqq$ 5 %/ then, inhibiting further lateral deflection, the elastically failed clay recovers elastic behaviour. For V % $\leqq$ 10-15 %, inhibition of further lateral deflection will make E_r = 0.6 E_o /I and II in Fig. 2./

If there is an increased possibility of loosening /III and IV in Fig. 2/, then E_r = 0,1 E_o, and clay does not start hardening sooner than after very important deformations /$\xi \approx$30 %/. Thus, if excessive loosenings are permitted in tunnel construction, a failed and loosened rock zone arises, likely to undergo increased deformations under the expected loads, even if lateral deflection is hampered.

Structural member of the roadway support is a tunnel wall consisting of seven B 400 precast r.c. units d = 20 cm thick, 1 m in length along the tunnel axis, assembled to cyrcular lining in situ. Wall units within

the ring are hinged, and yield in annular direction due to special rubber joints /with characteristic curves seen in Fig.4/.

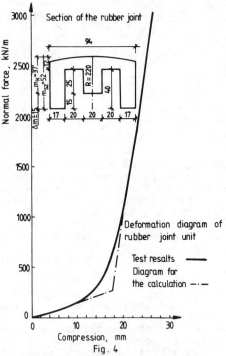

Fig. 4

There is no structural connection between rings 1 m wide. The back void between tunnel wall and rough-surfaced rock is grouted by a cement mortar with synthetic aggregates and special admixtures, such as not to transmit load from the rock to the wall exceeding its load capacity but gets compressed to a smaller volume. The mortar actually under development is able to be compressed by 30 to 35 % of its thickness without increasing the transmitted load /Fig. 5/.

Deformation diagram of polystyrene added cement mortar

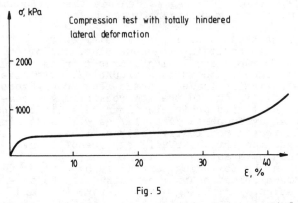

Fig. 5

Radial yielding of the wall, hence radial displacement of the excavated rock edge, that is, decrease of the initial, excessive loads to values not exceeding the wall load capacity is due to the interaction of rubber joints and back void grouting. Wall flexibility is provided for by the hinged connection between units.

Variation of the stress/strain tensor space of this roadway support in interaction with the surrounding rock, calculation of the necessary displacement, deformations and stresses in the lining has been done

with finite element computer program making
use of material laws corresponding test results gained both with grouting material
and rock. Elastic failure has been reckoned
with in terms of the Mohr-Coulomb failure
condition.

At the elastic failure of a rock finite
element, plastic yield starts at a level
below ultimate stress Q /Q^X = nQ/, and after
failure, the rock behaves as a quasi-elastic
material, its stress is obtained from:

$$p = nQ + /\bar{p} - Q/\frac{E_r}{E_0}$$

Stress $\bar{p}$ is a physically impossible auxiliary magnitude of calculation /Fig. 6/.

Constitution law of rock

Reduced stress after failure ($\bar{p}'$)

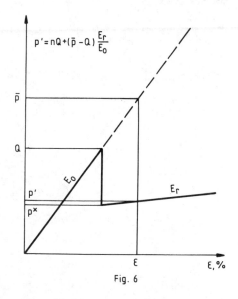

Fig. 6

Back void grouting has been simulated, in
conformity with its characteristic curve
obtained in the tests, by a material law
involving a linear-elastic section described
by constants E_0 and $0 < V < 0.5$, then, after
elastic failure, a quasi-elastic section
where $E_r = \frac{E_0}{100}$ and $V = 0$. Thus, it is a
material with no lateral displacement in
compression after failure. Elastic failure
is where the deviator stress /$\sigma_1 - \sigma_3$/ exceeds the typical ultimate stress of the
material, actually 600 kpa.

Computation results have led, among others
to the conclusions:
- Maximum compressibility of the grout 200
mm thick is 65 mm, less than 0.3 . 200 = 70
mm admissible without hardening.
- Mean value of the stress transmitted to
the lining is 130 kPa, and its maximum is
at the wall hinge displacing outwards
/Fig. 7/.
- Maximum normal force in the lining amounts
to 606 kPa. Ultimate load capacity of the
wall, with a safety factor of 1.5, is
2500 kpa /Fig. 8/.
- Bending moments are unimportant, with 55
to 65 kNm/m as maxima.
- Evaluation of stresses in the rock surrounding the tunnel showed the development
of an arching effect, that is, a rock ring
of increased stress around the wall /Fig.
9/, while plastic zones develop only there,
and to a degree as needed for the development of this arching effect.

Stresses transmitted to the lining

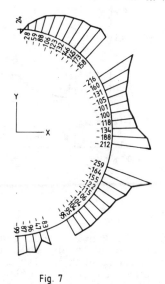

Fig. 7

Diagram of normal forces

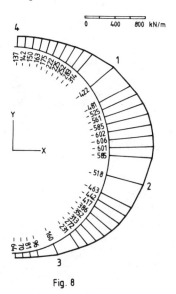

Fig. 8

The experimental structure is being constructed, to the subject to measuring
characteristic wall deformations, compression of the back void grout, the load acting on the wall.

The presented roadway support is advantageous in that, once completed, an equilibrium condition free from displacements
will develop in the void surroundings,
where loads from the void trimming are
supported jointly by the wall and the
rock zone.

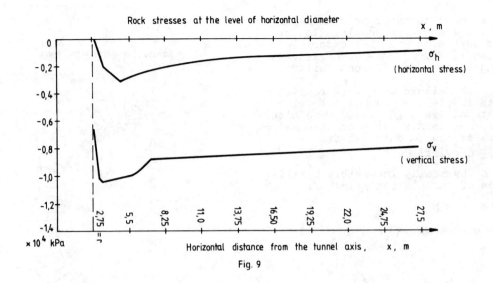

Fig. 9

REFERENCES

EGGER, P. 1973. Einfluss des Post-Failure
    Verhaltens von Fels auf den Tunnelaus-
    bau. Veröffentlichungen des Institutes
    für Bodenmechanik und Grundbau der Tech-
    nischen Hochschule Fridericana in Karls-
    ruhe.
BALLA, Á. 1960. Stress Concitions in Tri-
    axial Compression. ASCE Proceedings
    Vol. 86. No.5M6. Dec. New York.

Analytical calculation of the strength behaviour of an interbedded rock mass
Calcul analytique du comportement de résistance d'un massif rocheux à couches alternées
Analytische Berechnung des Festigkeitsverhaltens von Wechsellagerungen

TH.O.MUTSCHLER, Lehrstuhl für Felsmechanik, Universität Karlsruhe, Bundesrepublik Deutschland
B.O.FRÖHLICH, Lehrstuhl für Felsmechanik, Universität Karlsruhe, Bundesrepublik Deutschland

SUMMARY: The strength behaviour of mixed-layers under general triaxial states of stress is analytically calculated in a simplified model. The modell is based on test results and observations from large scale triaxial tests on representative samples. Both shear and tensile failure of the layers is considered. The equations include anisotrpic strength behaviour depending on the stress level and being different for compression and extension conditions. The closed form solution allows a fast variation of parameters.

RESUME: Le comportement de résistance d'un massif rocheux à couches alternées soumis à un état de contraintes triaxial est calculé par un modèle analytique simplifié. Les bases du modèle sont des observations et des résultats de mesures d'essais triaxiaux à grande échelle. A c té de la rupture au cisaillement on considère aussi la rupture à l'extension. Les équations montrent que le comportement de résistance est anisotrope, dépend du niveau des contraintes et est différent pour les conditions de compression et d'extension. La solution directe permet une variation rapide des paramètres.

ZUSAMMENFASSUNG: Die Festigkeitseigenschaften von Wechsellagerungen unter einem allgemeinen räumlichen Spannungszustand werden an einem einfachen, analytisch berechenbaren Modell abgeleitet. Grundlage für das Modell sind Beobachtungen und Messergebnisse aus Triaxialversuchen an repräsentativen Großproben. Neben dem Schubversagen der einzelnen Schichten wird auch deren Zugversagen infolge innerer Zwängspannungen erfaßt. Die Lösungen zeigen, daß das Festigkeitsverhalten anisotrop ist, vom Spannungsniveau abhängt und unterschiedlich für Kompressions- und Extensionsbedingungen ist. Die geschlossene Lösung gestattet eine schnelle Parametervariation.

1. INTRODUCTION

Most tunnelling is done in low geological depth where the sedimentary rocks which cover about 75% of the surface of the continents are the most crossed rock types. Due to often changing conditions during sedimentation the rocks are layered and textured. Mixed-layers of e.g. sandstone and claystone or marl and limestone are found very often. Their strength and deformation behaviour is not enough researched yet.

Stratified rocks with nearly horizontal bedding often show loosening of the rock mass caused by slop forces, creeping behaviour, high transversal extension of the softer layers and water pressure in the joints of the stiffer layers. The horizontal stresses are changing and are transmitted in the softer layers only. The interaction of the different layers has a dominant importence for these penomena.

In the following the strength behaviour of mixed-layers is investigated. It depends on the deformation and strength properties of the single layers, geometrical parameters like thickness and orientation of the layers and the interaction between them. PINTO, L. (1966), SALAMON, M. D. (1968), GERRARD, C. M. (1982)) gave mathematical models of mixed layers on the basis of a composite material. To get the total deformation they calculated the deformation of each layer according to the inhomogenious stress field and added them. REIK and HESSELMANN (1981) used an empirical model to estimate the strength of mixed-layers.

A model of a two-component-material is used in this paper too. The anisotropic state of stress is calculated and compared to the boundary condition for each layer. Failure is assumed if the boundary condition is violated by the stress field in one layer at least.

2. MECHANICAL MODEL ON BASIS OF OBSERVATIONS

The first step of every calculation must be a model of the problem. The model should describe the problem as far as intended and must be able to be solved. The expense and complexity increase with the number of details which shall be modeled.

Figure 2.1 shows a simplified model of a mixed-layer in form of a two-component-material . The different layers always have the same thickness and alternate regularly. The material behaviour must be given by a deformation law and a boundary condition. Generally this model allows the most complex behaviour but in order to limit the expense of calculation and keep it clear and in order to use material parameters from standard rock testing a restriction to an isotropic, linear elastic behaviour with a boundary condition of circular cone type with a tension cut-off is assumed. The material law shall be independent of the stress level. Thus it can be determined by the deformation parameters Young's modulus E and Poissons ratio ν and by the strength parameters friction angle ϕ, cohesion c and uniaxial tensile strength σ_t. The parameters of each layer can be determined in uniaxial and triaxial compression tests, tension tests and direkt shear tests.

Further assumptions have to be made about the state of deformation and stress and the interaction between the layers. Fig. 2.2 shows two extremes of behaviour of a cylindrical sample under triaxial conditions. They differ in the assumption of the interaction of the two layers.

a) Figure 2.2 a shows the case that the two layers can deform independently without influencing each other. Thus the state of stress in each layer is homogenious and equal to the global state of stress. The deformations are homogenious in every single layer but different from one to the other. This model conforms to the static boundary conditions. It is obvious that the strength calculated with this model is equal to the strength of the weakest layer.

b) Figure 2.2 b shows the case that the two layers are completely shear stiff connected at their planes of contact. The deformations parallel to the bedding planes are equal in each layer. This causes the state of stress to be different in every layer because of the constraints. Under certain parameter combinations even tensile stresses may occur. This model conforms to the kinematic boundary conditions. The global strength can be related to the shear failure or to the tensile failure of one of the layers.

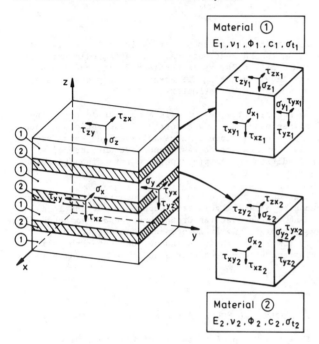

Fig 2.1: Model of a two-component material

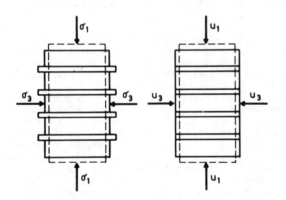

Fig 2.2: a) Static boundary conditions fullfilled
b) Kinematic boundary conditions fullfilled.

The quality of a model must be prouved by observations on prototypes. In this case we look at large scale triaxial tests with a diameter of 60 cm and a height of 120 cm. The large scale triaxial testing procedure is described in a former publication (NATAU, O. et al., 1983). Fig. 2.3 shows a specimen of a mixed-layer of limestone and marl and one of sandstone and claystone. Two facts support the assumption of kinematic boundary conditions.

1.) Although the global stress conditions in the test were purely compressive, tension cracks which run through the intact material can be seen in the

stiffer limestone . This indicates that the interior stress field must be different from the outer one and partially tensile (fig 2.3 a).

2.) Although the thickness of each layer of sandstone and claystone are more than five centimeters there are no differences in the lateral deformations at the surface of the specimen (Fig. 2.3 b).

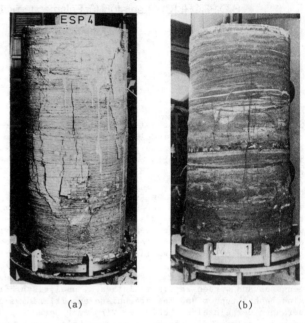

    (a)   (b)

Fig. 2.3: 60-cm-diameter specimen of limestone-marl(a) and sandstone-claystone (b)

3. BASIC EQUATIONS

This chapter gives the equations for calculation of the interior stresses in a two-component material assuming that the deformations parallel to the contact planes are equal in each layer (Fig 2.2 b). The obtained stresses are to be compared with the boundary conditions of the material of the layers and thus give the global strength of the material.

Figure 2.1 shows a cube of the composite material. Material 1 has a thickness d_1 and material 2 one of d_2. The parameters in the frames are explained in chapter 2.

A right-hand coordinate system x,y,z is given so that the z-axis is normal to the bedding planes. The outer, global stress tensor $\underset{\sim}{T}$ is:

$$\underset{\sim}{T} = \begin{bmatrix} \sigma_x & \tau_{xy} & \tau_{xz} \\ \tau_{yx} & \sigma_y & \tau_{yz} \\ \tau_{zx} & \tau_{zy} & \sigma_z \end{bmatrix}$$

Because of the constraints a internal, local state of stress $\underset{\sim}{S}_i$ is induced in each layer which causes a local state of deformation $\underset{\sim}{E}_i$:

$$\underset{\sim}{S}_i = \begin{bmatrix} \sigma_{x_i} & \tau_{xy_i} & \tau_{xz_i} \\ \tau_{yx_i} & \sigma_{y_i} & \tau_{yz_i} \\ \tau_{zx_i} & \tau_{zy_i} & \sigma_{z_i} \end{bmatrix} \quad ; \quad \underset{\sim}{E}_i = \begin{bmatrix} \sigma_{x_i} & \tau_{xy_i} & \tau_{xz_i} \\ \tau_{yx_i} & \sigma_{y_i} & \tau_{yz_i} \\ \tau_{zx_i} & \tau_{zy_i} & \sigma_{z_i} \end{bmatrix}$$

$i = 1, 2$

The relation between $\underset{\sim}{S}_i$ and $\underset{\sim}{E}_i$ is given by the material law $\underset{\sim}{C}_i$ of the layer:

$$\underset{\sim}{S}_i = \underset{\sim}{C}_i \cdot \underset{\sim}{E}_i \quad \text{or inverse:} \quad \underset{\sim}{E}_i = \underset{\sim}{C}_i^{-1} \cdot \underset{\sim}{S}_i$$

The tensors $\underset{\sim}{S}_i$ and $\underset{\sim}{E}_i$ are symmetrical, this means:

$\tau_{lm} = \tau_{ml}$ and $\gamma_{lm} = \gamma_{ml}$ $l,m = x,y,z$; $i=1, 2$

The assumed interaction causes the relations:

$- \sigma_{z_1} = \sigma_{z_2} = \sigma_z$

$- \tau_{yz_1} = \tau_{yz_2} = \tau_{yz}$; $\tau_{zx_1} = \tau_{zx_2} = \tau_{zx}$

$- \varepsilon_{x_1} = \varepsilon_{x_2} = \varepsilon_x$; $\varepsilon_{y_1} = \varepsilon_{y_2} = \varepsilon_y$

$- \gamma_{xy_1} = \gamma_{xy_2} = \gamma_{xy}$

$- \sigma_x = \dfrac{d_1 \sigma_{x_1} + d_2 \sigma_{x_2}}{d_1 + d_2}$; $\sigma_y = \dfrac{d_1 \sigma_{y_1} + d_2 \sigma_{y_2}}{d_1 + d_2}$

$- \tau_{xy} = \dfrac{d_1 \tau_{xy_1} + d_2 \tau_{xy_2}}{d_1 + d_2}$

$- \varepsilon_z = \dfrac{d_1 \varepsilon_{z_1} + d_2 \varepsilon_{z_2}}{d_1 + d_2}$

$- \gamma_{yz} = \dfrac{d_1 \gamma_{yz_1} + d_2 \gamma_{yz_2}}{d_1 + d_2}$; $\gamma_{zx} = \dfrac{d_1 \gamma_{zx_1} + d_2 \gamma_{zx_2}}{d_1 + d_2}$

The relation between the deformations and stresses in each layer can be expressed by:

$$\begin{bmatrix} \varepsilon_{x_i} \\ \varepsilon_{y_i} \\ \varepsilon_{z_i} \\ \gamma_{xy_i} \\ \gamma_{yz_i} \\ \gamma_{zx_i} \end{bmatrix} = \begin{bmatrix} \frac{1}{E_i} & -\frac{\nu_i}{E_i} & -\frac{\nu_i}{E_i} & 0 & 0 & 0 \\ -\frac{\nu_i}{E_i} & \frac{1}{E_i} & -\frac{\nu_i}{E_i} & 0 & 0 & 0 \\ -\frac{\nu_i}{E_i} & -\frac{\nu_i}{E_i} & \frac{1}{E_i} & 0 & 0 & 0 \\ 0 & 0 & 0 & \frac{1+\nu_i}{E_i} & 0 & 0 \\ 0 & 0 & 0 & 0 & \frac{1+\nu_i}{E_i} & 0 \\ 0 & 0 & 0 & 0 & 0 & \frac{1+\nu_i}{E_i} \end{bmatrix} \cdot \begin{bmatrix} \sigma_{x_i} \\ \sigma_{y_i} \\ \sigma_{z_i} \\ \tau_{xy_i} \\ \tau_{yz_i} \\ \tau_{zx_i} \end{bmatrix}$$

The 18 equations which describe the interaction of the layers and the two times 6 equations which describe the stress-strain relation in particular layers are sufficient to get the local stress and strain tensors $\underset{\sim}{S}_i$ and $\underset{\sim}{E}_i$ and the global strain tensor $\underset{\sim}{D}$.

The solution of the equations system shall not be demonstrated in this paper. The local stresses can be written as:

$\sigma_{x_i} = A_i \cdot \sigma_z + B_i \cdot \sigma_y + C_i \cdot \sigma_x$

$\sigma_{y_i} = A_i \cdot \sigma_z + C_i \cdot \sigma_y + B_i \cdot \sigma_x$

$\sigma_{z_i} = \sigma_z$

$\tau_{xy_i} = D_i \cdot \tau_{xy}$

$\tau_{yz_i} = \tau_{yz}$

$\tau_{zx_i} = \tau_{zx}$

The coefficients are:

Material 1

$A_1 = \dfrac{(\nu_1 - e\,\nu_2)(1 + \nu_1) + (1 + \nu_2)\,ea}{(1 + ea)^2 - (\nu_1 + ea\nu_2)^2}$

$B_1 = \dfrac{e(a + 1)(\nu_1 - \nu_2)}{(1 + ea)^2 - (\nu_1 + ea\nu_2)^2}$

$C_1 = \dfrac{e(a + 1)(1 - \nu_1\nu_2) + (1 - \nu_2^2)ea}{(1 + ea)^2 - (\nu_1 + ea\nu_2)^2}$

$D_1 = \dfrac{e(a + 1)(1 + \nu_2)}{ea(1 + \nu_2) + (1 + \nu_1)}$

Material 2:

$A_2 = -a\,\dfrac{(\nu_1 - e\,\nu_2)(1 + \nu_1) + (1 + \nu_2)\,ea}{(1 + ea)^2 - (\nu_1 + ea\nu_2)^2} = -a\,A_1$

$B_2 = -a\,\dfrac{e(a + 1)(\nu_1 - \nu_2)}{(1 + ea)^2 - (\nu_1 + ea\nu_2)^2} = -a\,B_1$

$C_2 = \dfrac{(a + 1)(1 - \nu_1^2) + (1 - \nu_1\nu_2)ea}{(1 + ea)^2 - (\nu_1 + ea\nu_2)^2} = (a + 1) - a\,C_1$

$D_2 = \dfrac{(a + 1)(1 + \nu_1)}{ea(1 + \nu_2) + (1 + \nu_1)}$

The coefficients e and a mean:

$e = E_1 / E_2$ ratio of moduli
$a = d_1 / d_2$ ratio of layer thicknesses

The local stresses are composed of the global stresses which are influenced by the factors A_i, B_i, C_i and D_i. These factors depend on the deformation parameters and the thicknesses of the layers only.

3.2 Simulation of a triaxial test

The global strength of a mixed layer is determined by simulating a triaxial test. Failure is assumed if in one of the layers the boundary condition for either shear failure or tensile failure is reached.

Fig 3.1 shows a triaxial test on a specimen with an inclination α of the z-axis. A new material orientated coordinate system x',y',z' is defined according to fig. 3.1.

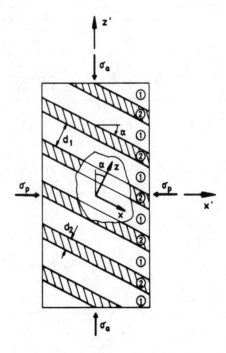

Fig. 3.1: Global and local coordinate system

1169

The global state of stress of triaxial test is:

$$[\underset{\sim}{T}]_{x',y',z'} = \begin{bmatrix} \sigma_p & 0 & 0 \\ 0 & \sigma_p & 0 \\ 0 & 0 & \sigma_a \end{bmatrix}$$

Transformation to the global coordinate system x,y,z:

$$[\underset{\sim}{T}]_{x,y,z} = \begin{bmatrix} \sigma_a \sin^2\alpha + \sigma_p \cos^2\alpha & 0 & (\sigma_p - \sigma_a)\sin\alpha\cos\alpha \\ 0 & \sigma_p & 0 \\ (\sigma_p - \sigma_a)\sin\alpha\cos\alpha & 0 & \sigma_a\cos^2\alpha + \sigma_p\sin^2\alpha \end{bmatrix}$$

The local stess tensor $S_{\sim i}$ is calculated in layers:

$$\sigma_{x_i} = A_i(\sigma_a\cos^2\alpha + \sigma_p\sin^2\alpha) + B_i\sigma_p + C_i(\sigma_a\sin^2\alpha + \sigma_p\cos^2\alpha)$$

$$\sigma_{y_i} = A_i(\sigma_a\cos^2\alpha + \sigma_p\sin^2\alpha) + C_i\sigma_p + B_i(\sigma_a\sin^2\alpha + \sigma_p\cos^2\alpha)$$

$$\sigma_{z_i} = \sigma_a\cos^2\alpha + \sigma_p\sin^2\alpha$$

$$\tau_{xy} = 0 \; ; \quad \tau_{yz} = 0 \; ; \quad \tau_{zx} = (\sigma_p - \sigma_a)\cos\alpha\sin\alpha$$

The local stresses can be inserted into the boundary condition which can be generally expressed as:

$$F_i = q_i - M_i\,p_i - N_i = 0$$

$$p_i = \sigma_{oct_i} = (\sigma_{x_i} + \sigma_{y_i} + \sigma_{z_i})/3$$

$$q_i = 3\,\tau_{oct_i}/\sqrt{2} =$$

$$= \{[(\sigma_{x_i} - \sigma_{y_i})^2 + (\sigma_{y_i} - \sigma_{z_i})^2 + (\sigma_{z_i} - \sigma_{x_i})^2 + 6(\tau_{xy_i}^2 + \tau_{yz_i}^2 + \tau_{zx_i}^2)]/2\}^{1/2}$$

The following table gives some examples of strength parameters M_i and N_i:

Criterion	M	N
Axial compression cone	$\dfrac{6\sin\phi}{3 - \sin\phi}$	$\dfrac{6c\cos\phi}{3 - \sin\phi}$
Axial extension cone	$\dfrac{6\sin\phi}{3 + \sin\phi}$	$\dfrac{6c\cos\phi}{3 + \sin\phi}$
Drucker-Prager	$\dfrac{3\sin\phi}{\sqrt{3} + \sin^2\phi}$	$\dfrac{3c\cos\phi}{\sqrt{3} + \sin^2\phi}$
Tension cut-off	3	$3\sigma_t$

The boundary condition is squared and after putting in the local stresses $S_{\sim i}$ a quatratic equation for the global axial limit stress σ_a is obtained:

$$a_i\,\sigma_a^2 + b_i\,\sigma_a + c_i = 0$$

with the coefficients:

$$a_i = a_{0_i} = [g_{2_i}^2 c^2 + g_{3_i} S^2 + (g_{1_i} g_{2_i} + 3)SC] - M_i^2[g_{5_i}C + g_{1_i}S]^2/9$$

$$b_i = b_{0_i} + b_{1_i}\sigma_p$$

$$b_{0_i} = -2M_i N_i[g_{5_i}C + g_{1_i}S]/3$$

$$b_{1_i} = [2(g_{2_i}^2 + g_{3_i} - 3)SC + g_{4_i}S + g_{1_i}g_{2_i}(S^2 + C^2 + C)] - 2M_i^2[g_{5_i}^2 SC + g_{1_i}g_{5_i}(S^2 + C^2 + C) + g_{1_i}^2(SC+S)]/9$$

$$c_i = c_{0_i} + c_{1_i}\sigma_p + c_{2_i}\sigma_p^2$$

$$c_{0_i} = -N_i^2$$

$$c_{1_i} = -2M_i N_i[g_{5_i}S + g_{1_i}(C+1)]/3$$

$$c_{2_i} = [g_{5_i}^2 S^2 + g_{2_i}(C^2+1) + g_{4_i}C + g_{1_i}g_{2_i}(SC+S)+3SC] - M_i^2[g_{5_i}S + g_{1_i}(C+1)]^2/9$$

The following abreviations are introduced:

$$S = \sin^2\alpha \; ; \quad C = \cos^2\alpha$$

$$g_{1_i} = B_i + C_i \; ; \qquad g_{2_i} = A_i - 1 \; ;$$

$$g_{3_i} = B_i^2 - B_i C_i + C_i^2 \; ; \quad g_{4_i} = 4B_i C_i - B_i^2 - C_i^2$$

$$g_{5_i} = 2A_i + 1$$

The general solution of the quadratic equation is:

$$\sigma_{a1/2_i} = \frac{-(b_{0_i} + b_{1_i}\sigma_p)}{2a_0} \pm \frac{\sqrt{(b_{0_i}+b_{1_i}\sigma_p)^2 - 4a_{0_i}(c_{0_i}+c_{1_i}\sigma_p+c_{2_i}\sigma_p^2)}}{2a_{0_i}}$$

The multiple reduction of coefficients hides the parameters which determine the solution for σ_a

$$_{a1/2_i} = f(\sigma_p, \nu_1, \nu_2, E_1, E_2, d_1, d_2, \alpha, F_i)$$

4. STRENGTH PROPERTIES OF MIXED-LAYERS

This chapter discusses the results of the calculations in general. The equations are able to describe basic phenomena of the strength behaviour of mixed-layers. These phenomena are demonstated in numerical examples:

- Dependence on the stress level

The dependence of the strength of a mixed-layer on the stress level is shown in fig. 4.1. The global boundary condition for a given combination of parameters is crestfallen twice. Part I represents tensile failure in material 1, part II shear failure in material 1 and part III shear failure in material 2.

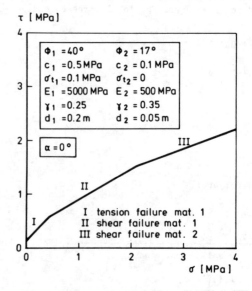

Fig 4.1: Dependence of the system strength on the stress level

- Anisotropy

The anisotropy can be seen obviously from the equations which depend on the orientation α of the bedding planes to the stress axises.

- Influence of the joint distance in a rock mass

The influence of the joint distance on the global strength is investigated by an example of a rock mass with filled joints. The rock substance corresponds to material 1 the joint filling to material 2. Fig. 4.2 shows the development of the axial failure stress σ_a^{RM} of the rock mass related to the failure stress σ_a^{RS} of the rock substance over the joint distance a_j related to the joint thickness d_j. For ratios $a_j/d_j \gtrless 50$ the strength rock mass reaches more than 90% of the strength of the rock substance.

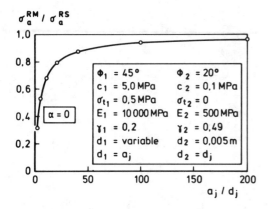

Fig. 4.2: Influence of joint distance in rock mass

- Difference between compression and extension

The method is able to model both compression and extension load paths. Fig. 4.3 shows a diagramm $q = \sigma_a - \sigma_p$ over $p = (\sigma_a + 2\sigma_p)/3$.

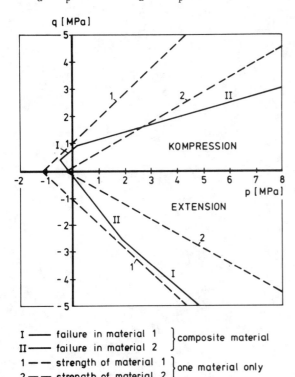

I —— failure in material 1	} composite material
II —— failure in material 2	
1 — — strength of material 1	} one material only
2 — — strength of material 2	

Fig. 4.3: Difference between compression and extension stress paths

The broken lines indicate the strength cones of the two matarials themselves. The full lines give the strength of the copmosite material whose layers are normal to the axial stress. The strength for extension conditions is greater than the one for compression. This turns to the contrary if the orientation of the layers is parallel to the axial stress. In every case the boundary condition of axial cone type is changed.

5. COMPARISON WITH LARGE SCALE TRIAXIAL TESTS

A series of large scale triaxial tests of the limestone and marl (fig. 2.3 a) resulted in friction angles of more than 60° and cohesion values of less than 1 MPa in a confining pressure range from 0.1 to 1.8 MPa. The seperate tests of small specimens of limestone gave friction angles about 45° which is less than the one of the mixed-layer. The marl of which no samples could be kept can be assumed to be weaker. A parameter variation of an analytical model explained the high fricton angle and the small cohesion gained in the large scale tests. They resulted from tensile failure of the large specimen.

6. CONCLUSIONS

The described method is able to explain basic phenomena of mixed-layers and rock masses with filled joint systems. Although the parameters of the material components are assumed most simplified the composite material shows non-linear boundary conditions including tensile failure under pure compressive states of stress, anisotropy and difference between compression and extension stress paths. The influence of the joint distance of a rock mass is pointed out too.

The closed form solution of the method enables fast parameter variations. The posibility to introduce more detailed behaviour of the material components enlarges the application.

Test results of large scale triaxial tests confirm the model and the obtained solutions.

REFERENCES

Gerrard, C.M. 1982. Equivalent elastic moduli of a rock mass consisting of orthorombic layers. Int. J. Rock Mech. Min. Sci. & Geomech. Abstr. Vol. 19, 9 - 14.

Natau, O.P.; Fröhlich, B.O. and Mutschler, Th.O. 1983. Recent developments of the large scale triaxial test. 5th Int. Congr. of ISRM, Melbourne, 1983.

Pinto, J.L. 1966. Stresses and strains in an anisotropic-orthotropic body. 1st Int. Congr. of ISRM, Lissabon 1983.

Reik, G. and Hesselmann, F. 1981. Verfahren zur Ermittlung der Gebirgsfestigkeit von Sedimentgesteinen. Rock Mechanics Suppl. 11, 59-71

Salamon, M.D.G. 1968. Elastic moduli of a stratified rock mass. Int. J. Rock Mech. Min. Sci. & Geomech. Abstr. Vol. 5, 519 - 527.

Underground observation of high propagation-rate, extension fractures

L'observation souterraine des fractures qui se propagent à haute vitesse
Untertagebeobachtung ultraschneller Ausbreitung von Zugbrüchen

W.D.ORTLEPP, Anglo American Corporation Ltd, Johannesburg, Republic of South Africa
M.A.MOORE, Fulmer Research Laboratories Ltd, Slough, UK

SYNOPSIS

The photographic record of unusual fracture surfaces revealed by a roof collapse in a
shallow coal mine was analysed in terms of fracture mechanics principles.
Certain textural features suggested strongly that the collapse had occurred violently.
The most conspicuous of these features were segments of circular mirror-zones which
indicated that the propagation velocity of the fracture front had reached terminal velocity
of nearly 2000 m/sec. An estimate of the stress perpendicular to the plane of the fracture
was made from the dimensions of the mirror-zone. This showed that the direct gravity
effect was totally inadequate to cause the unstable failure. Buckling of a thinly-
laminated plate seemed the most likely mechanism for the observed collapse.
Although their traces are very rarely seen, fractures extending at terminal velocity are
probably essential accompaniments of seismic spalling in deep gold mines.

RESUME

Le toit d'une mine de charbon peu profonde s'est éffondré, et des fractures inhabituelles
sont apparues. Ces surfaces fracturéés ont ete analysée dans le contexte des lois de
l'étude de fractures. Certaines caracteristiques de texture semblent indiquer que
l'éffondrement s'est déclenche violamment. Des segments de "zones-mirroires" indiquant que
la vitesse frontale avait atteint présque 2000 m/sec, constituent la caracteristique la
plus frappante de ce phenomene. On à calcule la contrainte perpendiculaire d'aprés les
dimensions de la zone-mirroire, et on a pû voir que l'éffet de gravite directe n'avait pas
été suffisamment prononce pour provoquer l'éffondrement. Le déjettement d'une tôle laminée
mince semblerait être la cause la plus plausible de l'éffondrement. Bienque ceci n'ait
rarement été observe, les fractures en prolongement, qui ont lieu a une telle vitesse
frontale, sont nomralement associées au écaillage sismique dans les mines d'or profondes.

ZUSAMMEMFASSUNG

Photographien von ungewöhnlichen Bruchflächen die sich während eines Dacheinsturzes in
einer flachen Kohlengrube offenbarten wurden analysiert.
    Gewisse Gefügezüge deuten stark an dass der Einsturz gewaltsam passierte. Das
auffallendeste dieser Züge waren kreisförmige Spiegelzonen die andeuteten dass Ausbreitungs
geschwindigkeit der Vorderseite des Bruches eine Spitzengeschwindigkeit von beinahe
zweitdusend Metern pro Sekunde erreicht hatte.
    Eine Schätzung des senkrechten Druckes auf die Bruchfläche wurde vom Ausmas der
Spiegelzonen gemacht. Dies zeigte dass die direkte Schwerpunktwirkung vollkommen
unzulänglich war den um den instabilenBruch zu erzugen. Die Beulung einer dünn
laminierten Platte scheint der wahrscheinliche Mechanismus des wahrgenommenen Falles zu
sein.
    Obwohl man ihre Spuren kaum sehen kann sind Brücherweiterungen mit Grenzgeschwindigkeiten
an wahrscheinlich wichtige Begleiter der seismischen Abblätterung im tiefen Goldbergbau,
der Bruchkanle.

INTRODUCTION

The most widely exploited seam in the
Witbank coalfield of South Africa is known
as the No. 2 seam. The roof of the seam is
typically formed by a very fine-grained
carbonaceous shale which although massive
and homogeneous in appearance, is actually
fissile and strongly anisotropic.
    After a primary bord and pillar
extraction at 2,5 to 3 m working height
with 6,5 m bords at 13 m centres, a
secondary top-coaling operation was carried
out in a section of the colliery where the
seam roof was only 22 m below surface.
    At one intersection a collapse of roof
occurred whilst the rock-stud support was
being installed by drillers standing on the
pile of top-coal which had been blasted
down shortly before.
    The most striking aspect of the fall was
the strongly textured conchoidal nature of
the shallow saucer-shaped depression which
formed the new roof. No stratification was
apparent and the sparse jointing did not

control the overall fracture surface. This indicated convincingly that the fall was not simply a gravity-induced collapse but that some other failure mechanism had prevailed.

Because the textural features were so unusual detailed photographs were taken and, subsequently, 70mm diameter cores were drilled from block samples of the roof strata. The compressive strength and Brazilian tensile strength of the shale were determined from these specimens.

Some years later the significance of some of the textural features recorded in the photographs was recognised. This led to the analysis and interpretation which form the substance of this paper and which enable a plausible, albeit conjectural, account to be advanced for the mechanism of the collapse.

OBSERVATIONS

The most important overall impressions gained at the scene of the collapse were the shallowness of the cavity in the roof - Fig 1 - and its strongly textured appearance. The slenderness of the shale slabs which lay on top of the pile of blasted coal confirmed that only a thin layer of the shale roof was involved in the failure - Fig 2.

Figure 3 is an enlargement of the upper central portion of Fig 2 showing a cluster of radial traces extended from circular arcs. These formed the most conspicuous of the textural features imprinted on the roof. Fig 7 shows clearly how the smooth enclosed surface or "mirror-zone" breaks down beyond the arc into "hackled" radial traces extending in the direction of propagation of the fracture front.

The location of this arc and two other intriguing but unexplained features detailed in Figures 5 and 6 are indicated in the general view of Fig 4 and in the plan of Fig 1.

The difficulties of preparing specimens from the weak shale roof limited the number of tests to one uniaxial compressive, and nine tensile values - Table 1(a).

Mechanical properties obtained from the same stratigraphical horizon on an adjacent colliery are shown in Table 1(b).

TABLE 1(a)

	STRENGTH	MPa
	Perpendicular to Strata	Parallel to Strata
U.C.S.	64,2	
Brazilian Tensile Strength	1,68	-
	1,39	-
	2,59	-
	2,63	-
	1,46	-
	-	3,34
	-	3,60
	-	5,35
	-	4,53

TABLE 1(b)

U.C.S. (MPa)	YOUNG'S MODULUS (MPa)	POISSONS RATIO
87	12 600	0,17
120	18 700	0,10

INTERPRETATION

The circular mirror-zones of Figs 3 and 7 were recognised as the typical pattern for a high velocity, high energy release rate failure in type I (tensile opening mode) fracture of brittle materials.

The clearly discernible markings on these features permit both qualitative and quantitative interpretation.

The failure process commences with a pre-existing flaw (or a small fracture that developed relatively slowly in a stable manner) which is subjected to a normal tensile stress.

At some critical dimension, which is a function of the strength properties of the material and the normal stress, the edge of the discontinuity begins to extend rapidly in all directions and the failure becomes catastrophic.

The fracture front accelerates rapidly to a limiting or terminal velocity of about 0,6 times the elastic shear wave velocity of the material (Bansal, 1977). This limit is indicated clearly by the circular boundary separating the smooth mirror zone from the radial hackled region - Fig 7.

At this stage the increasing rate at which new surface area is being created by the extension of a single surface is curtailed by the inability of the fracture front to propagate any faster. The energy released by the main failure process can now no longer be balanced by the energy demand (in the form of surface energy) of the expanding fracture surface and the phase of crack branching commences. This phenomenon is expressed by the radial features of the hackled region which are the traces of intersection of planes symmetrically inclined about the plane of the initial fracture.

The higher the uniform applied stress the more rapidly the crack accelerates and the shorter is the distance travelled before branching occurs. There is a well established experimental relationship between the size of the mirror region and the applied stress (Congleton and Petch, 1967; Kirchner and Gruver, 1974 and Shetty et al, 1983) of the form :

$$\sigma_f \cdot r_m^{\frac{1}{2}} = \text{constant} \quad \dots (1)$$

This permits a quantitative estimate to be made of the stress acting across the fracture surface during the failure process. For the case of a circular flaw in a semi-infinite solid, as in the observed mirror-zone :

$$K_I = \frac{2}{\pi^{\frac{1}{2}}} \sigma_f a^{\frac{1}{2}} \quad \dots (2)$$

Where K_I is the type I stress intensity factor of the crack, σ_f is the fracture stress and "a" is the crack radius.

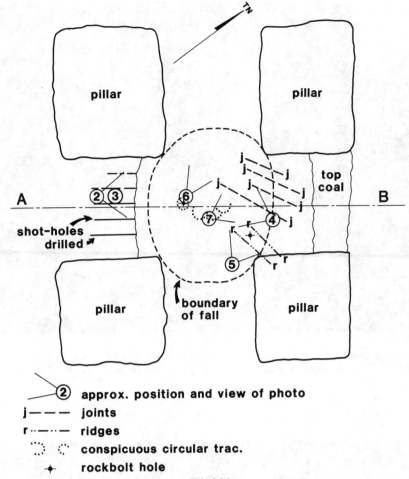

shot-holes drilled

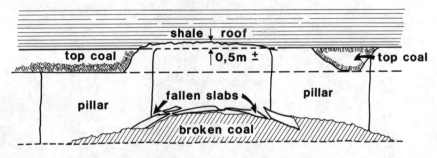

② approx. position and view of photo

j——— joints

r·—·—· ridges

⌒ ⌒ conspicuous circular trac.

✦ rockbolt hole

PLAN

Fig 1 - Plan and Section of Fall

At the limiting stage when crack branching commences, the relationship can be written as:

$$K_{IB} = \frac{2}{\pi^{\frac{1}{2}}} \sigma_f r_m^{\frac{1}{2}} \quad \dots (3)$$

where K_{IB} is the dynamic stress intensity factor at crack branching and r_m is the radius of the mirror zone.

K_{IB} values are not available for many materials but there is an empirical relationship between K_{IB} and Young's modulus E (Kirchner, 1978) whereby

$$K_{IB} = E \times (3,3 \times 10^{-5} m^{\frac{1}{2}}) \quad \dots (4)$$

From tests carried out on samples of carbonaceous shale of the No. 2 seam roof, from the Young's modulus was taken as

E = 20 000 MPa - Table 1(b).

Thus from equation (4)

$$K_{IB} = 0,66 \text{ MNm}^{-3/2}$$

From the mirror zone which was photographed close-up in Fig 7, r_m was estimated at 0,28 m Putting these values for K_{IB} and r_m into equation (3) provides a value for σ_f of 1,1 MPa. This is the applied tensile stress normal to the crack

1175

Fig. 2 - General View from floor elevation towards north showing
slenderness of fallen shale slabs and shallowness of roof cavity

Fig. 3 - Enlargement of upper central portion of Fig. 2 showing
an array of mirror-zones and hackled surfaces.
Dotted line encloses area shown in detail on Fig. 7

Fig. 4 - General view towards south-west showing relative positions
of main features detailed in Figs. 5, 6 and 7

Fig. 5 View of ridge 'scarps' with 'wave-reflection' traces
on the nearest scarp

Fig. 6 Detail of arcuate 'gouges' with small
adhering particles on right, possibly
electrostatically charged

plane at the time the crack front reached
the mirror-zone boundary.

It is considered to be significant and
not coincidental that the stress initiating
and driving the fracture was of similar
value to the measured static tensile
strength of the rock. The significance
becomes apparent when the paradox
surrounding the mechanism of the failure
process is considered.

MECHANICS OF THE FAILURE PROCESS

In the great majority of cases a rock fall
is caused simply by gravity acting on blocks
of rock which are bounded by pre-existing
discontinuities such as joints, bedding
planes or shear surfaces. In this case
there was no evidence of any discontinuity
and it was clear that the fracture had
traversed pristine rock through its entire
path.

Although the strength tests had shown a
marked anistropy - Table 1(a), the range of
values was not unusually large and even the
lowest (1 397 kPa) indicated a cohesion
substantially above zero. The mass of rock
which became detached was only some 0,5m
thick at maximum, thus, due to gravity, was
able to generate only 0,012 MPa tensile
stress across the fracture. This is two
orders of magnitude smaller than the
measured tensile strength so the failure
could not have occurred under simple
gravity loading - Fig 8 (b).

If the value assumed for K_{IB} was
realistic and the fracture stress
calculated from equation (3) consequently a
good estimate, then a tensile stress of
about 1,1 MPa was necessary to cause
unstable collapse.

To understand the phenomenon then,
requires that a mechanism be found that
could generate a vertical tensile stress of
this magnitude 0,5 m above the immediate
roof at the centre of the bord
intersection.

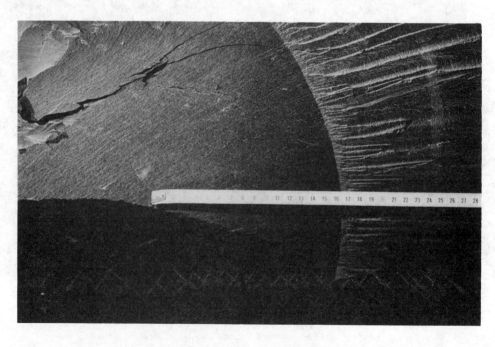

Fig. 7 Detail of mirror-zone and hackled region

1178

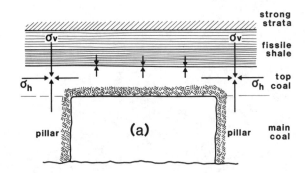

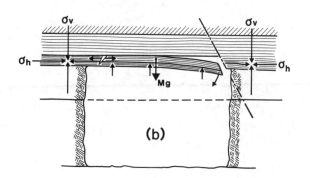

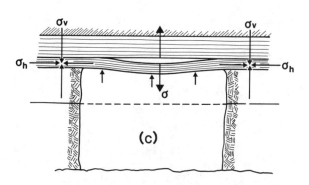

Fig. 8 Section through collapse showing
a) the estimated stress situation
b) gravity failure mechanism
c) plate-buckling failure mechanism

At a depth of 25m the vertical primitive stress would be about 0,6 MPa compressive. At this shallow depth the horizontal component could be as much as two or three times this value. With a percentage extraction of 75% the average vertical stress in the pillars would be about four times the primitive stress. The Poisson effect would increase the horizontal component to a similar total value of about 2,5 MPa.

The compressive strength of the shale was measured at 62 MPa which is high relative to the very low 1,5 MPa tensile strength normal to the bedding. When subjected to the horizontal compressive stresses of about 2,5 MPa estimated above, The shale layer across the intersection would tend to buckle like an arrangement of thin plates stacked on a flat unyielding surface and subjected to compressive edge-loading parallel to the surface - Fig 8(c).

Unstable equilibrium is maintained while the intact top coal provides some confinement - Fig 8(a).

Shortly after the top coal is blasted down small flaws grow in a quasi-stable

Fig 9 - Mirror-zone in Tunnel

fashion to critical dimensions near the centre of the intersection.

There is elastic strain energy stored in the compressed rock in and above the pillars. Buckling is an essentially unstable mode of failure and the system response is "soft" in the horizontal sense. The stored strain energy thus becomes available to drive the failure into the catastrophic regime and to imprint the characteristic textural features of violent failure on the fracture surface. The most conspicuous mirror- zones are clustered fairly close to the centre of the collapse, suggesting that the failure process started near the centre of the intersection where one would expect the buckling tendency to be a maximum. The fact that there appeared to be several initiating flaws is difficult to explain since one would expect a brittle fracture to have one unique starting point. Even if the process leading to instability was as simple as the buckling mechanism suggested in Fig 8(c), the violent part of the collapse must be extremely complex. The intriguing features of Figs. 5 and 6 are evidence of this complexity.

CONCLUSIONS

Although their importance .was not appreciated initially the subsequent recognition of circular mirror-zones as definitive indications of cracks reaching terminal velocity, confirmed that the collapse was a violent event.

Two important practical implications follow from this interpretation :

1) The existence of a mirror-zone on the rock walls of any excavation is clear evidence that violent failure occurred in which gravity did not play a direct role.

2) It is possible to estimate the transient stress which acted perpendicular to the crack surface during the failure.

In deep hard rock mining this could provide valuable insight into the dynamics of the damage process which accompanies a rock burst. Knowledge of the velocity and energy content of a violently expelled slab is necessary for the design of a support tendon that would then yield rather than fail and so prevent disruption of the rock. This would, ultimately, make possible the design of a total support system that could control the affects of even very large rock bursts.

Although rock bursts are not rare in South African gold mines no previous observation or analysis of mirror-zones has been made. With this new awareness of their existence in mind, a search through some hundreds of photographs of rock burst damage extending back more than 20 years has revealed only one clear instance of mirror-zones and a few indications of the radial traces of the associated hackled region. It is probable that a fine-grained rock is necessary to register these textural imprints and most gold mines are dominated by relatively coarse-grained quartzite.

However, a good example of a mirror-zone was recently found in a tunnel close to the located focus of a large seismic event which occurred in October 1986 at a depth of 2400m - Fig 9

Although the imprinted evidence appears to be very rare, the phenomenon of violent extension fracturing is, nevertheless, considered to play an important role in the mechanics of rock burst damage and consequently deserves continued study.

ACKNOWLEDGEMENT

This study was pursued some years after the undergound investigations were carried out by W D Ortlepp while in the employ of Rand Mines Limited.

The interpretation of the significance of the observed features is largely due to Dr D M Moore of Fulmer Research Ltd., England. Conclusions do not necessarily represent the convictions of Rand Mines Limited or of Fulmer. Their permission to publish the results is assumed and gratefully acknowledged.

REFERENCES

Bansai, G.K., 1977. Phil.Mag. 35 : 953
Congleton, J., & Petch, N.J., 1967.
  Phil.Mag. 16 : 749
Kirchner, H.P., 1978. Engng. Fracture
  Mechanics 10 : 283
Kirchner, H.P. & Gruver, R.M., 1974 :
  Fracture Mechanics of Ceramics, ed. R.C.
  Brandt et al, Plenum Press, New York :
  309
Shetty, D.K., Rosenfield, A.R. &
  Duckworth, W.H., 1983. Commun. American
  Ceram. Soc: 66 : C-10

A presentation of the ISRM suggested methods for determining fracture toughness of rock material

Présentation d'une méthode suggérée par la SIMR pour déterminer la résistance à l'extension d'une fracture dans le rocher

Eine Darstellung der von der IGFM vorgeschlagenen Methoden zur Bestimmung der Bruchzähigkeit von Gesteinen

FINN OUCHTERLONY, ISRM working group coordinator, Swedish Detonic Research Foundation, SveDeFo, Stockholm

ABSTRACT: The new ISRM Suggested Methods for determining fracture toughness of rock material are presented. The motivation behind and background of their introduction are given followed by test specimen specifications and an outline of the specified test and evaluation procedures. Finally, some fracture toughness test results are discussed.

RESUME: Les nouvelles méthodes sugérées par la SIMR pour la détermination de la ténacité de rupture des matières de roche sont présentées ainsi que les causes de leur introduction. La présentation est complétée par des spécifications des éprouvettes des roches et une description de la méthode d'essai et du procédé d'evaluation. En conclusion quelques résultats d'essai à la ténacité de rupture sont discutés.

ZUSAMMENFASSUNG: Die neu vorgeschlagenen Verfahrungen der IGFM zur Bestimmung der Bruchzähigkeit von Gesteinen werden präsentiert. Die Gründe ihrer Einführung werden gegeben wie auch die Prüfstabspezifikationen und die Hauptzüge der Prüf - und Berechnungsverfahrungen. Am Ende werden einige Bruchzähigkeitswerte diskutiert.

1 INTRODUCTION

Fracture mechanics is an engineering discipline which primarily is used to prevent and predict catastrophic failure of structures of man made materials such as metals, plastics and ceramics (Kanninen and Popelar 1985). Its application to the cracking of concrete is becoming important too (Wittman 1983, Shah 1985).

Simultaneously fracture mechanics concepts are receiving an increased interest in rock engineering. In most of these applications, though not all, the cracking is considered to be beneficial such as in fragmentation processes.

Historically fracture mechanics is a development of the strength of materials approach. In present day elastic fracture mechanics, the governing parameter is the stress intensity factor K_I, at least in the linear case. It is, on one hand, a measure of the strength of the stress field at a loaded crack tip and, on the other, intimately related to the available energy release rate. K_I is usually determined by analysis and its dimensions are stress·(crack length)$^{0.5}$, i.e. Pa·(m)$^{0.5}$ or N/m$^{1.5}$.

The basic relation equates K_I to a critical value, which is often taken as a material property, and called the plane strain fracture toughness K_{Ic}. When K_I reaches K_{Ic} catastrophic crack growth is assumed to occur. Thus, a structure can be designed to be safe if K_I is kept below K_{Ic} and failure or fragmentation could be achieved if K_{Ic} is exceeded.

The relation between fracture toughness and (effective) surface energy is simply

$$K_{Ic} = (2\gamma_{eff}E')^{0.5}, \qquad (1)$$

where the factor 2 connects two crack faces with each crack tip, the index "eff" implies that all dissipative effects at the crack tip and in its immediate surroundings are included in γ, and E' is an appropriate modulus of the material.

Obviously available and dissipated energy rates must also balance during crack growth. Thus, the fracture toughness of a material expresses its resistance to (catastrophic) crack propagation, or the fracture energy consumption rate required to create new surfaces. Some applications of such values for rock are as:

(i) A parameter for classification of rock.

(ii) An index of fragmentation processes such as tunnel boring etc.

(iii) A material property in the modelling of rock fracturing processes like hydraulic fracturing for stress field measurements and hot dry rock geothermal energy extraction, gas driven fracturing, explosive stimulation of gas wells, radial explosive fracturing and crater blasting, rock cutting as well as in stability analysis and in the interpretation of geological features.

Articles covering such applications may be found in monographs by Rossmanith (1983), Fourney et al. (1985), Bazant (1985) and Atkinson (1986) or in recent proceedings from the yearly US Symposium on Rock Mechanics e.g.

2 BACKGROUND ON WORKING GROUP

A comprehensive review (Ouchterlony 1982) showed that a wide variety of specimen types and evaluation methods had been used in the fracture toughness testing of rock. The resulting values were generally not comparable, implying that the fracture toughness values thus measured usually did not represent a material property. Testing methods which consistently yield accurate and precise values are however important if the use of fracture mechanics in rock engineering is to proceed from qualitative estimates to quantitative predictions.

In response to this situation and based on further work (Ouchterlony 1983, Ouchterlony and Sun 1983) a working group (WG) within the ISRM Commission on Testing Methods was formed at Swedish initiative in 1984. Its purpose was to deliver "Suggested Methods for Determining Fracture Toughness of Rock Material", a testing standard, in time for the ISRM Congress in Montreal 1987.

The WG has members from Canada, Chile, People's Republic of China, Germany, Great Britain, Japan, Sweden, USA and the USSR. The work has proceeded to a point where a final draft of the Suggested Methods (SM:s) has been approved by both WG and Commission members and sent to the International Journal of Rock Mechanics and Mining Sciences and Geomechanics Abstracts for publication.

Apart from the drafting of the SM:s, some essential supporting work has been carried out by WG members:

(i) Experimental calibration of test specimens (Ouchterlony 1986a, Takahashi et al. 1986).
(ii) Numerical verification of specimen calibration (Gerstle 1985, Ouchterlony 1986b).
(iii) Derivation of evaluation formulas and test requirements (Ouchterlony 1986c).
(iv) Testing according to SM intentions (Meredith 1983, Müller and Rummel 1984, Sun and Ouchterlony 1986, Takahashi et al. 1986).

These SM:s will hopefully provide fracture toughness values for rock which are material properties, i.e. are reasonably independent of specimen size, specimen type, loading rate etc. Even for other rocks, where a material property doesn't result after testing, the number has a practical value if it can be used as an index. In such cases standard testing methods with well defined results still let a user profit by the experience of others more easily.

3 OUTLINE OF SUGGESTED METHODS

The SM:s specify the use of two core based test specimens or methods, the Chevron Bend specimen (Ouchterlony 1980) and the Short Rod (Barker 1977). Since rock material is usually available in the form of cores, this implies ease of machining and easier acceptance by prospective users than rectangular specimens.

The Chevron Bend specimen has a notch cut perpendicular to the core axis, see Figure 1, and the notch in the Short Rod is cut parallel to the core axis, see Figure 2. In both cases the ligaments of the notched section have the form of a V or chevron. This shape results in a relatively long period of stable crack growth under increasing load before the point at which fracture toughness is evaluated. The advantages of using chevron notched specimens are also considered important in metals and ceramics testing (Underwood et al. 1984).

The use of two specimens with different crack orientations permits an assessment of fracture anisotropy from one core while saving core material, since the bend specimen is long enough for the halves remaining after testing to be used for Short Rod testing.

Two levels of testing are offered in each method, giving the prospective user a choice between accuracy and cost. Level I requires only the registration of maximum load to be made during the test. Level II requires continuous load and displacement measurements to be made during the test.

While level I testing is simple and cheap, yielding index like results, level II testing is more complicated. Since level I testing of normal core sizes tends to yield fracture toughness values which are lower than values obtained from larger specimens, but level II testing in some cases yields values that are independent of core size and that correlate well with results from other test methods, level II testing is justified in some cases.

The SM:s follow other ISRM Suggested Methods closely, the general contents are:
GENERAL INTRODUCTION
TECHNICAL INTRODUCTION
METHOD 1: Suggested Method for Determining Fracture Toughness Using Chevron Bend Specimens.
METHOD 2: Suggested Method for Determining Fracture Toughness Using Short Rod Specimens.
NOTES
BIBLIOGRAPHY

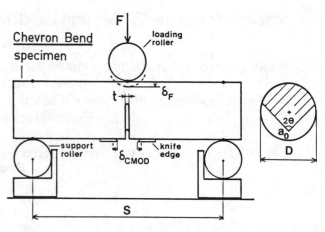

Notation:

D = diameter of specimen
S = distance between support points
2θ = chevron angle
a_0 = chevron tip distance from specimen surface
t = notch width
δ_F = deflection of load point relative to support points, LPD
δ_{CMOD} = relative opening of knifes edges, CMOD

Figure 1. Geometry of Chevron Bend (CB) specimen and related notation.

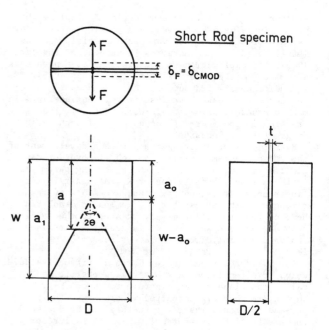

Notation:

D = diameter of short rod specimen
w = length of specimen
2θ = chevron angle
a_0 = chevron tip distance from load line
a_1 = maximum depth of chevron flanks
t = notch width
a = crack length
F = load on specimen
δ_F = load line displacement, LPD
δ_{CMOD} = crack mouth opening displacement, CMOD $\approx$ LPD

Figure 2. Geometry of Short Rod (SR) specimen and related notation.

Each SM contains detailed specifications of testing, as e.g. in
METHOD 1 (Chevron Bend Testing),
Scope:
Specimen Description:
Apparatus: Specimen Preparation Equipment, Testing Machine and Load Fixtures, Specimen Alignment Aids, Displacement Measuring Equipment and Recording.
Procedure: Specimen Selection and Preparation, Calibration, Setting Up and Testing.
Validity: Crack Path Geometry, Specimen Size and Type and Use of Fracture Toughness Values.
Calculations: Calculation of Loading Rate, Calculation of Slope Values, Calculation of Fracture Toughness, Non-linearity Corrected Fracture Toughness, Calculation of Additional Quantities (Young's modulus, Energy Rate Resistance and Specific Work of Fracture).
Reporting of Results: General Data, Test Sample Data and Specimen Data.

The specifications concerning METHOD 2 (Short Rod Testing) are of course as identical as possible.

In what follows the parts most relevant to a fracture toughness determination are sketched.

4 SPECIMENS AND LEVEL I FORMULAS

The specified proportions of the Chevron Bend specimen in Method 1 are, including tolerances,

$$S = 3.33D \pm 0.02D \text{ or } |\Delta S| \leq 0.02D,$$
$$a_o = 0.15D \pm 0.10D, \qquad (2\text{-CB})$$
$$2\theta = 90.0° \pm 1.0°, \text{ and}$$
$$t \leq 0.03D \text{ or } 1 \text{ mm, whichever is greater.}$$

See Figure 1 for notation. The specified proportions of the Short Rod Specimen in Method 2 are

$$w = 1.45D \pm 0.02D \text{ or } |\Delta w| \leq 0.02D,$$
$$a_o = 0.48D \pm 0.02D,$$
$$a_1 - a_o = 0.97D \pm 0.02D, \qquad (2\text{-SR})$$
$$2\theta = 54.6° \pm 1.0°, \text{ and}$$
$$t \leq 0.03D \text{ or } 1 \text{ mm, whichever is greater.}$$

See Figure 2 for notation. The tolerances are a compromise between ease of machining and the available calibration data.

The stress intensity factor may be written

$$K_I = AF/D^{1.5} \qquad (3)$$

where the non-dimensional factor A depends on the geometry and mainly on crack length a. The predetermined evaluation point of fracture toughness is where A(a) has a minimum, $A = A_{min}$ at $a = a_m$ say. For the specimens conforming to Eqns. 2

$$A_{min} = [1.835 + 7.85(a_o/D) + 9.15(a_o/D)^2] \cdot S/D \quad (4\text{-CB})$$

$$A_{min} = 24 \cdot [1 - 0.6\Delta w/D + 1.4\Delta a_o/D - 0.01\Delta(2\theta)], \quad (4\text{-SR})$$

(Ouchterlony 1986c). The level I fracture toughness is then calculated from

$$K_{CB} \text{ or } K_{SR} = A_{min}F_{max}/D^{1.5}. \qquad (5)$$

For a brittle material the failure load F_{max} would coincide with $a = a_m$.

5 LEVEL II TESTING FORMULAS

Level II testing takes non-ideal material behavior into account and specifies the evaluation of a non-linearity corrected fracture toughness value along the lines of Barker's plasticity correction (1979), which requires that load point displacement (LPD) measurements are made.

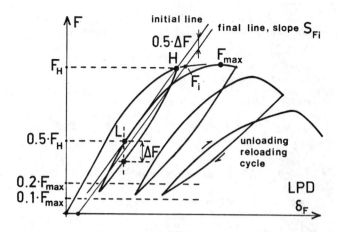

Figure 3. Construction of linearized unloading lines from smoothed F vs. LPD curve.

Assuming that the specimen compliance is a good measure of crack length there are three steps involved. The first step is a specification of how the compliance λ_F or stiffness $S_F = 1/\lambda_F$ is to be determined at a number of points on the F vs. LPD curve. Barker's (1980) linearization procedure is specified, see Figure 3. It assumes that a partial unloading cycle down to 0.1-0.2 F_{max} is made for each point.

The final line of cycle i has slope S_{Fi}, its intersection with the smoothed F-LPD curve defines a point F_i and its intersection with the LPD axis defines the residual LPD, δ_{Fi}^r.

The second step is finding the desired evaluation point F_m corresponding to $a = a_m$, for which $S_{Fm} = 1/\lambda_F(a_m)$. The specified way is depicted in Figure 4. First find the two points with slope values that span S_{Fm}, F_{i-1} and F_i say. Then find the point F_ℓ on the unloading line through F_i. It shall have the same amount of displacement recovery as F_{i-1}, i.e.

$$\ell = \delta_{Fi-1} - \delta_{Fi-1}^r = F_{i-1}/S_{Fi-1} = F_\ell/S_{Fi}. \quad (6)$$

Then draw a straight line through F_{i-1} and F_ℓ and find the interpolation point whose load is

$$F_e = S_{Fm} \cdot \ell = F_{i-1} \cdot (S_{Fm}/S_{Fi-1}). \qquad (7)$$

The corresponding displacement is δ_{Fe}. Next find the point $(\delta_{Fe} - \ell, 0)$ on the LPD axis. Draw a straight line through it and F_e. It will automatically have the slope S_{Fm}. Let its intersection with a F-LPD curve, which has been smoothed to remove

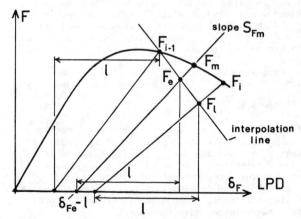

Figure 4. Construction of evaluation point F_m using an interpolation procedure with constant displacement recovery ℓ.

any dips due to the unloading reloading cycles, define the evaluation point F_m with coordinates (δ_{Fm}, F_m).

This procedure is a linear interpolation in terms of slope and contains a second order error term. In practice S_{Fm} is determined as $S_{Fm} = S_{Fo} \cdot s_m$, the slope value of the initial tangent times the slope ratio, either

$$s_m = 0.85 \cdot (1 - 1.04 \Delta a_o/D) \text{ or} \qquad (8\text{-CB})$$

$$s_m = 0.5. \qquad (8\text{-SR})$$

If $F_m \approx F_{max}$, then F_{max} is used instead.

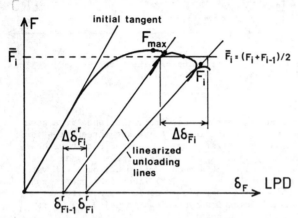

Figure 5. Construction of non-linearity correction factor p.

The third step concerns the non-linearity correction. To this end define first the average load $\bar{F}_i = (F_i + F_{i-1})/2$, see Figure 5. Next let this load define two matching displacements along the unloading lines through F_{i-1} and F_i. Their difference

$$\Delta\delta_{\bar{F}i} = \delta_{\bar{F}i} - \delta_{\bar{F}i-1} \quad \text{and} \quad \Delta\delta^r_{Fi} = \delta^r_{Fi} - \delta^r_{Fi-1} \quad (9)$$

are used to define p,

$$p = \Delta\delta^r_{Fi}/\Delta\delta_{\bar{F}i}. \qquad (10)$$

cf. Barker (1979).

The non-linearity corrected fracture toughness is then defined as

$$K^c_{CB} = \left(\frac{1+p}{1-p}\right)^{0.5} K_{CB} \text{ or } \left(\frac{1+p}{1-p}\right)^{0.5} A_{min} F_m/D^{1.5} \quad (11\text{-CB})$$

if $F_m \neq F_{max}$. An identical relation yields K^c_{SR} from K_{SR}. If the LPD measurements are calibrated then the SM:s also contain expressions for the non-dimensional compliance of the uncracked specimen, g_o, with which the Young's modulus may be determined from $E = S_{Fo} g_o/D$ and then fracture toughness in K-terms can be converted to energy rate resistance.

The loading rate in LPD controlled testing and crack speed around $a = a_m$ are related through

$$\dot{LPD} = 6.4 \, K_{CB}/(E\sqrt{D}) \cdot \dot{a}_m \qquad (12\text{-CB})$$

for the Chevron Bend specimen. For the Short Rod the numerical prefactor is 17.0. It is specified that $\dot{a}_m \geq 0.001$ m/s during testing so as to minimize the influence of subcritical crack growth on the fracture toughness measurements.

The Short Rod is sensitive to transverse tensile failure of the specimen arms. In order to avoid this the specimen size should fulfil

$$D > 1.15(K_{SR}/\sigma_t)^2. \qquad (13\text{-SR})$$

The Chevron Bend specimen is much less sensitive to this type of failure.

6 SOME FRACTURE TOUGHNESS VALUES

Table 1 contains a list of toughness values obtained from core bend specimens. Müller and Rummel's (1984) tests were made on 30 mm diameter cores. Their data has been recalculated using the current formulas. Meredith's (1983) data refers to flat-jack testing of 25 mm diameter cores. The tests of Takahashi et al. (1986) were made on 52 mm cores. The other data is taken from Ouchterlony and Sun (1983) and Sun and Ouchterlony (1986) and recent tests at SveDeFo and at the Luleå University of Technology (Kråkemåla granite), mostly on 42 mm diameter cores.

The level II toughness values of Meredith (1983) were evaluated using the method specified in the SM:s. They showed an excellent overall agreement with fracture toughness values obtained from double torsion tests on the same rocks. The level I toughness values fail in this respect since they are substantially lower.

Costin's (1981) Short Rod testing of Anvil Points oil shale gave similar results. His non-linearity corrected (essentially level II) fracture toughness values agree reasonably well with three-point bend results. His level I values are significantly lower.

For Stripa granite (Sun and Ouchterlony 1986) the same conclusions hold, K^c_{SR} for specimens of different sizes were effectively identical and agreed well with fracture toughness values from both round compact tension specimens and bend specimens with a straight notch. The corresponding K_{SR} values were significantly lower and depended on specimen size.

The evidence is thus that the LPD measurements associated with level II testing hardly can be avoided for the small specimens used here, if a correct toughness value is desired. Since the present p-values range up to 0.5 it would seem that much larger specimens usually are required to make level I and level II toughness values agree. Further Chevron Bend testing at Tohoku University in Japan and Short Rod testing at Luleå University in Sweden are underway to investigate this aspect.

The agreement between toughness data from Chevron Bend testing and from Short Rod data is probably insufficiently investigated. The Kråkemåla granite results agree well, the Iidate granite ones are significantly different. The matter is not simple to clarify in that anisotropy should be accounted for both in specimen orientation and evaluation formulas.

7 CONCLUDING REMARKS

The ISRM Suggested Methods outlined here provide an international standard for the fracture toughness testing of rock. For them to gain widespread acceptance however, they will have to give accurate results at acceptable costs. Even if these methods have the qualifications to provide accurate results, more testing is needed to determine how well they perform in practice.

To meet this need an interlaboratory testing program is planned by the working group. In a first round a single block of well characterized rock will be used to evaluate first of all interlaboratory variations and secondly the influence of specimen size and type on the fracture toughness values measured according to these SM:s.

Table 1. Fracture toughness values obtained from core specimens, given as $\text{mean}_{\text{no.tests}} \pm \text{std.dev}$ in $MN/m^{1.5}$.

Rocks:	Chevron Bend		Short Rod	
	K_{CB}	K_{CB}^c	K_{SR}	K_{SR}^c
Whin Sill dolerite /1/	–	–	2.86_6 ±0.12	3.26_6 ±0.09
Bohus granite	1.46_5 ±0.07	–	1.8_3	2.4_3
Cornwall granite /2/	1.32_4 ±0.10	–	–	–
Epprechtstein granite /2/	1.74_8 ±0.18	–	–	–
Falkenberg granite A /2/	0.65_4 ±0.14	–	–	–
B /2/	–	1.52_5 ±0.20	–	–
Iidate granite /2/	1.09_5 ±0.13	1.73_5 ±0.21	–	–
Iidate granite /3/	1.37_3 ±0.13	2.26_3 ±0.65	1.01_4 ±0.18	1.12_4 ±0.35
Kråkemåla granite	1.64_3 ±0.04	2.16_3 ±0.23	1.69_6 ±0.17	2.22_5 ±0.24
Merrivale granite /1/	–	–	1.50_{24} ±0.10	1.80_{24} ±0.13
Pink granite /1/	–	–	1.58_4 ±0.04	2.03_4 ±0.08
Strath Halladale granite /1/	–	–	1.80_{11} ±0.10	2.19_{11} ±0.11
Stripa granite /4/	–	–	2.01_5 ±0.14	2.36_{11} ±0.13
Stripa granite	–	–	2.37_9 ±0.15	2.70_9 ±0.27
Westerly granite /1/	–	–	1.64_9 ±0.03	1.82_9 ±0.07
Westerly granite	–	–	2.04_4 ±0.05	2.27_4 ±0.03
Grey norite /1/	–	–	2.23_{11} ±0.11	2.69_6 ±0.16
Pennant sandstone /1/	–	–	1.98_6 ±0.06	2.56_6 ±0.07
Ruhr sandstone /2/	1.03_{10} ±0.04	–	–	–
Klinthagen limestone	1.31_2	–	1.41_8 ±0.19	1.87_8 ±0.25
Shelly limestone	–	–	1.40_5 ±0.03	1.44_5 ±0.04
Carrara marble /2/	1.26_5 ±0.08	1.38_5 ±0.09	–	–
Ekeberg marble	1.89_{16} ±0.12	–	1.83_9 ±0.35	2.25_9 ±0.36
Ekeberg marble -ST	–	–	1.48_3 ±0.16	1.82_3 ±0.10
-D	–	–	2.28_2 ±0.01	2.62_2 ±0.05
Treuchtlingen marble /2/	1.26_6 ±0.07	1.70_6 ±0.09	–	–

Notes: Numbers within / / refer to data from Meredith (1983) /1/, Müller and Rummel (1984) /2/, Takahashi
et al. (1986) /3/, and Sun and Ouchterlony (1986) /4/.
ST and D mean short transverse and divider orientations of crack with respect to discernible
structure in rock, see Ouchterlony (1983) e.g.

ACKNOWLEDGEMENTS

The author expresses his gratitude to the following colleagues in the WG, without whose active
support these Suggested Methods would never have
materialized: Dr John A Franklin (Canada); Prof.
Sun Zongqi (China); Drs Barry K Atkinson and
Philip Meredith (England); Dr Wolfram Müller
(Germany); Prof. Hideaki Takahashi and his colleagues at Tohoku University and Prof. Yuichi
Nishimatsu at University of Tokyo (Japan); Prof.
Ove Stephansson (Sweden); Dr Laurence S Costin
and Prof. Anthony R Ingraffea (USA).

REFERENCES

Atkinson, B.K. (ed.) 1986. Rock Fracture Mechanics.
London: Academic Press, to appear.
Barker, L.M. 1977. A Simplified Method for Measur-
ing Plane Strain Fracture Toughness. Engng.
Fract. Mechs. 9:361-369.
Barker, L.M. 1979. Theory for Determining K_{Ic}
from Small Non-LEFM specimens, Supported by
Experiments on Aluminum. Int. J. Fracture 15:
515-536.
Barker, L.M. 1980. Data Analysis Methods for Short
Rod and Short Bar Fracture Toughness Tests of
Metallic Materials. Report TR 80-12. Salt Lake
City UT: Terra Tek.
Bazant, Z.P. (ed.) 1985. Mechanics of Geomaterials;
Rocks, Concretes, Soils. Chichester: John Wiley.
Costin, L.S. 1980. Static and Dynamic Fracture
Behaviour of Oil Shale. In S.W. Freiman & E.R.
Fuller Jr (eds.), Fracture Mechanics for
Ceramics, Rocks, and Concrete. ASTM STP 745,
p.581-590. Philadelphia PA: ASTM.
Fourney, W.L., R.R. Boade & L.S. Costin (eds.) 1985.
Fragmentation by Blasting. Brookfield Center
CT:SEM.
Gerstle, W.H. 1985. Finite and Boundary Element
Modeling of Crack Propagation in Two- and Three-
dimensions Using Interactive Computer Graphics.
Dept. Structural Engng., PhD Thesis, 220p.
Ithaca NY: Cornell Univ.
Kanninen, M.F. & C.H. Popelar 1985.
Advanced Fracture Mechanics. Oxford Engineering
Science Series 15. New York: Oxford University
Press.
Meredith, P.G. 1983. A Fracture Mechanics Study of
Experimentally Deformed Crustal Rocks. PhD Thesis,
350p. London: Univ. of London.

Müller, W. & F. Rummel 1984. Bruchzähigkeits-messungen an Gesteinen. Bericht zu den BMFT-FE-Vorhaben 03E-3068-B, 133p. Bochum FRG: Ruhr University.

Ouchterlony, F. 1980. A New Core Specimen for the Fracture Toughness Testing of Rock. Report DS 1980:17. Stockholm: Swedish Detonic Research Foundation.

Ouchterlony, F. 1982. Review of Fracture Toughness Testing of Rock. SM Archives 7:131-211.

Ouchterlony, F. 1983. Fracture Toughness Testing of Rock. In H.P. Rossmanith, Rock Fracture Mechanics, CISM Courses and Lectures No. 275, p.69-150. Wien: Springer.

Ouchterlony, F. 1986a. A Core Bend Specimen with Chevron Notch for Fracture Toughness Measurements. In H.L. Hartman (ed.), Proc. 27th US Symp. on Rock Mechanics, p.177-184. Littleton CO: SME.

Ouchterlony, F. 1986b. An FEM Calibration of a Core Bend Specimen with Chevron Notch for Fracture Toughness Measurements. Report manuscript. Stockholm: Swedish Detonic Research Foundation.

Ouchterlony, F. 1986c. Evaluation Formulas for Rock Fracture Toughness Testing with Standard Core Specimens. In Proc. 1986 SEM Spring Conference on Experimental Mechanics, p.115-124, Bethel CT: SEM.

Ouchterlony, F & Z. Sun 1983. New Methods of Measuring Fracture Toughness on Rock Cores. In R. Holmberg & A. Rustan (eds.), Proc First Intnl. Symp. on Rock Fragmentation by Blasting 1, p.199-223. Luleå Sweden: Luleå Univ. Techn.

Rossmanith, H.P. (ed.) 1983. Rock Fracture Mechanics. CISM Courses and Lectures No. 275. Wien: Springer.

Shah, S.P. (ed.) 1985. Application of Fracture Mechanics to Cementitious Composites. NATO ASI Series E: Applied Sciences No. 94. Dordrecht: Martinus Nijhoff.

Sun, Z. & F. Ouchterlony 1986. Fracture Toughness of Round Specimens of Stripa Granite. Int. J. Rock Mech. Min. Sciences & Geomech. Abstracts, to appear.

Takahashi, H., T. Hashida & T. Fukazawa 1986. Fracture Toughness Tests by Use of Core Based Specimens. GEEE Research Report T-002-86. Sendai Japan: Tohoku Univ. Faculty of Engineering.

Underwood, J.H., S.W. Freiman & F.I. Baratta (eds.) 1984. Chevron-Notched Specimens: Testing and Stress Analysis. ASTM STP 855, p.152-166. Philadelphia PA: ASTM.

Winter, R.B. 1983. Bruchmechanische Gesteinsunter-suchungen mit dem Bezug zu hydraulischen Frac-Versuchen in Tiefbohrungen. Institut für Geo-physik, PhD Thesis 179p. Bochum FRG: Ruhr University.

Wittman, F.H.(ed.) 1983. Fracture Mechanics of Concrete. Development in Civil Engineering 7, Amsterdam: Elsevier Science Publishers.

Vorbeugende Sicherungsmassnahmen zur Beherrschung charakteristischer Verbruchsituationen im Tunnelbau

Preventive supporting measures to control typical collapse situations in tunnelling
Mesures préventives pour maîtriser des situations caractéristiques d'éboulements dans la construction de tunnels

F.PACHER, Dr.-Ing., Ingenieurgemeinschaft für Geotechnik und Tunnelbau, Salzburg, Österreich
G.M.VAVROVSKY, Dipl.Ing., Ingenieurgemeinschaft für Geotechnik und Tunnelbau, Salzburg, Österreich

SUMMARY: Collapses which do not occur directly at the face often appear to be caused by an overloading of the shotcrete lining. In almost all cases examined the rupture of the lining was only an initial factor. The reason for the failure was not so much in the shotcrete lining itself, but rather in the loss of strength and load carrying capacity of the rock mass. The tunneler's profession is the ability to distinguish between such critical situations and the less harmful redistribution of stress. This can be achieved by a thorough examination and precise evaluation of the continual measurements.

RESUME: En cas d'éboulements qui ne se produisent pas directement au front de taille, on a souvent l'impression que la surcharge de la coque en béton armée est la cause principale de l'éboulement. Toutefois, dans presque tous les cas examinés, la rupture de la coque n'a provoqué que le déclenchement. La véritable cause de l'éboulement ne réside pas tant dans la coque en béton armée que dans la perte de rigidité et de capacité d'auto-portance de la roche. Tout l'art du constructeur de tunnel consiste à pouvoir faire la différence entre ces situations critiques et des phénomènes de déplacement non dangereux. Il n'y arrive qu'en s'occupant du problème de façon plus intensive et qu'en exploitant les mesures de façon bien précise.

ZUSAMMENFASSUNG: Bei Verbrüchen, welche sich nicht unmittelbar an der Ortsbrust ereignen, entsteht häufig der Eindruck, daß die Überlastung der Spritzbetonschale die maßgebende Ursache für einen Verbruch ist. In fast allen untersuchten Fällen war der Bruch der Schale aber nur das auslösende Moment. Die wahre Versagensursache lag nicht so sehr bei der Spritzbetonschale, als vielmehr im Verlust an Festigkeit bzw. an Eigentragvermögen des Gebirges. Die Kunst des Tunnelbauers liegt darin, derartige kritische Situationen von ungefährlichen Umlagerungsvorgängen unterscheiden zu können. Dies gelingt ihm nur durch noch intensivere Befassung und eine besondere, gezielte Auswertung der laufenden Messungen.

1. ALLGEMEINER TEIL

Der wesentlichste Unterschied des Tunnelbaues gegenüber anderen Bauingenieurfächern wie Stahlbau, Betonbau, Holzbau besteht darin, daß der uns von der Natur aus vorgegebene Baustoff das Gebirge ist, das Gebirge so wie es ansteht mit allen seinen Strukturen, Defekten aus der Entstehung, der tektonischen Entwicklung und Abtragung, mit all seinen Vorbelastungen (Eigenspannungszustand) aus Gewicht, geometrischer Form, aber auch rezenten Gebirgsbewegungen, ungünstig beeinflußbar noch durch die Anwesenheit von Bergwasser, besonders wenn das Gebirge darauf empfindlich eingestellt ist. Der Berg ist ein Riese und reagiert als Masse. Demgegenüber nehmen sich unsere Maßnahmen, das Öffnen und Sichern eines Hohlraumes, die "machbaren" Tragfähigkeiten des Ausbaus, wie das Werk von Zwergen aus. Dies begreift man sofort, wenn man eine Tunnelinnenschale unter 1000 m Gebirgsüberlagerung bemessen soll oder muß.

Das Problem besteht dann nicht darin den entstehenden Gebirgsdruck zu errechnen, sondern darin, Mittel und Wege zu finden - durch partielles Ausbrechen und Anwendung eines nachgiebigen Ausbaues - die Deformations-Entspannung in einem Ausmaß herbeizuführen, die für das Bauwerk und die Mannschaft noch gefahrlos ist und die es uns ermöglicht, mit unseren schwachen Kräften die Hohlraumwandungen zu sichern.

Dabei kommt es auf die Wechselwirkung zwischen Gebirge und Auskleidung (Schale) an, ein Vorgang,
- welcher in und über einen gewissen Zeitraum abläuft, eigentlich so lange bis der Tragring geschlossen ist und die plastischen Gebirgsverformungen ausgeklungen sind,
- bei welchem sich der Spannungszustand im Gebirge, aber auch in der Schale umlagert,
- bei welchem sich Gebirge wie Schale verformen und
- die Kontaktspannungen zwischen Gebirge und Schale ständig wechseln (können). Man denke nur an den mehrstufigen Ausbruch von Kalotte, Strosse und Sohle.

Nicht nur bei großer Überlagerungshöhe entstehen Schwierigkeiten, sondern auch bei geringer Überlagerung, nämlich dann, wenn die Verspannung über dem Hohlraum und in dem dazugehörigen Gebirgsverband nicht ausreicht, um eine ausreichende und dauernde Tragwirkung zu erzielen.

2. WO UND WORIN LIEGT DIE GEFÄHRDUNG?

Die Schwächestellen sind allen Tunnelbauern wohl bewußt, sie werden daher nur symptomatisch aufgezählt. Verbruchgefahr besteht in den Bereichen
- der Brust, wenn freiwerdende Energie der geöffneten Querschnittsfläche und Gebirgsfestigkeit sich nicht die Waage halten etc., Auswirkungen bis zum Tagbruche;
- der Firste durch Senkung der Kalottenfüße, unter Umständen Schlotbruch bis zur Oberfläche;
- der Ulmen durch Überbeanspruchung der seitlichen Wände, die zum Scherbruch führen. Bei gutmütigem Gebirge bewirkt ein solcher eine zeitlich begrenzte Entspannung, die für die Sanierung genützt werden kann, bei ungünstigen Gebirgsformationen unter Umständen aber Zusammenbruch der Tunnelschale;
- letztlich der Sohle. Alle kennen die ungünstigen Auswirkungen einer überbeanspruchten Sohle aus vertikaler Überbelastung der Kalotten-oder Ulmenfüße (Grundbruch).

Eine gegensinnige Wirkung resultiert aus Quell- oder Schwelldruck bei Anwesenheit von dafür sensitivem Gebirgsmaterial in der Sohle. Solche Situationen können zu einem Auswölben oder Bruch der Sohle führen.

Aber auch wenn es nicht zum Verbruch kommt, sondern nur sehr starke und unterschiedliche Senkungen oder Hebungen eintreten, besteht die Gefahr, daß die Schale örtlich bricht und es kommt zu zeit- und kostenaufwendigen Überfirstungs- bzw. Nacharbeiten.

3. WORIN LIEGEN DIE URSACHEN?

Die Ursachen, die zu einer eventuellen Verbruchsituation führen können, sind mannigfaltig. Sie können einzeln oder in Kombinationen auftreten, meist ist letzteres der Fall. Sie liegen
- erstens im Nichterkennen oder in einer Fehleinschätzung der geologischen Verhältnisse und der daraus abzuleitenden Materialkennwerte,
- zweitens im Eintritt unerkannter Materialveränderungen im Lauf der Zeit, z.B. infolge längerer Stehzeit, durch Wasserzutritt o.ä.,
- drittens in unerkannten Spannungsumlagerungen im Gebirge - hervorgerufen durch ungleiche oder übergroße Setzungen - die zu einer bedeutenden Änderung der Verhältnisse führen können.

Die Gefahr bezieht sich aber nicht auf Vorgänge im Querschnitt, sondern wegen der räumlichen Lastabtragung - wie in besonders gefährdetem Vortriebsbereich - auch auf solche in Längsrichtung. In Abschnitt 4 wird darauf ausführlicher eingegangen.

4. CHARAKTERISTISCHE VERBRUCHSITUATIONEN BEIM VORTRIEB VON KALOTTE UND STROSSE

Bei all jenen Verbrüchen, welche sich nicht unmittelbar an der Ortsbrust ereignen, entsteht häufig der Eindruck, daß die Überlastung der Spritzbetonschale die maßgebende Ursache für einen Verbruch war. In fast allen Fällen war der Bruch der Schale aber nur das auslösende Moment. Die wahre Versagensursache lag nicht nur so sehr bei der Spritzbetonschale, als vielmehr im Verlust an Festigkeit bzw. Eigentragvermögen des Gebirges. Ursache der überhöhten Belastung, welche in der Folge zum Verbruch führte, war fast immer der Aufbau von Porenwasserdruck, eine tiefgreifende Auflockerung des Gebirges, eine starke voreilende Entspannung oder eine rasche Entfestigung des Gebirges.

Solange das Gesamtsystem noch fähig ist Spannungen umzulagern, bleibt in der Regel genügend Zeit, um gezielte Verstärkungsmaßnahmen setzen zu können. Wirkliche Verbruchgefahr besteht immer erst dann, wenn das Umlagerungsvermögen im Gebirge als auch im Ausbau erschöpft ist.

Die Kunst des Tunnelbauers liegt darin, derartige kritische Situationen von ungefährlichen Umlagerungen und Überlastungsvorgängen unterscheiden zu können und rechtzeitige Warnsignale dort zu geben, wo sie vom Gebirge nicht mehr zu erwarten sind. Eine treffsichere Diagnose setzt jedoch ein hohes Maß an praktischen Erfahrungen voraus. Erkenntnisse aus Analysen früherer Schadensereignisse sollten daher nicht hinter Schloß und Riegel verwahrt, sondern möglichst vielen interessierten Fachkollegen zugänglich gemacht werden.

Aufgrund der bisherigen Erfahrungen können Verbrüche von Kalottenvortrieben unter geringer Überlagerung häufig auf einen der nachstehend genannten Versagensmechanismen
- Durchbrechen der Kalottenfirste
- Überlastung bzw. Entspannung des Gebirges in den Kämpferbereichen des räumlichen Schutzgewölbes
- Überlastung des Kalottenauflagers im Bereich des Strossenabbaus
- Absacken des gesamten Kalottengewölbes im Tunnel-Eingangsbereich
zurückgeführt werden.

4.1 Durchbrechen der Kalottenfirste

Dieser Verbruchtyp kann bei Kalottenvortrieben in schwach kohäsivem Gebirge speziell in Situationen auftreten, in denen
- die nachdrängende Firstbelastung zufolge einer hohen Entfestigungs-Sensitivität des Gebirges auch bei relativ kleinen Setzungen rasch ansteigt,

- das Gebirge im unverritzten Zustand nur wenig verspannt ist,
- die Verspannung über der Firste wegen einer geringen Überlagerung rasch verloren geht,
- sich keine entsprechende Selbstverspannung des Gebirges einstellen kann,
bzw. auch
- der Aufbau von Porenwasserdruck möglich ist und damit der Effekt einer raschen Festigkeitsminderung im Gebirgstragring gegeben ist.

Voraussetzung ist stets, daß die Kalottenschale im Fußbereich ein relativ widerstandsfähiges Auflager vorfindet, während die Bettung und die Verzahnung mit dem Gebirge eher weich und nachgiebig ist.

Trotz geringer Firstsetzungen können in den Kämpfern und in der Firste zufolge einer hohen, nachdrängenden Vertikalbelastung Biegemomente entstehen, welche unter Umständen einen Biegedruckbruch in Kämpfer und Kalotte und damit ein Durchbrechen der Firste zur Folge haben können.

4.2 Überlastung bzw. Entspannung des Gebirges in den Kämpferbereichen des räumlichen Schutzgewölbes

Hat sich als Folge eines Vortriebsstillstandes, einer geologischen Störzone, starker Anfangssetzungen, einer starken Nachgiebigkeit der Ortsbrust oder ähnlichem, über der Firste eine quer zur Vortriebsrichtung liegende Zone mit stark aufgelockertem Gefüge ausgebildet, so kann dies zu Schwierigkeiten führen, welche erst erheblich später offenbar werden.

a) Abgleiten des räumlichen Gewölbes an einer Schwächezone:

Im Zuge des weiteren Vortriebes wird die geschwächte bzw. entspannte Zone unterfahren, die Scherbewegung konzentriert sich daher nicht mehr auf diesen, sondern auf den jeweils weiter vorne liegenden Ortsbrustbereich.

Mit zunehmendem Abstand von der Ortsbrust gelangt die entfestigte Zone aber in den portalseitigen Kämpferbereich des räumlichen Gewölbes, in welchem sie erheblichen Schubspannungen ausgesetzt wird. Dies kann zur Folge haben, daß das räumliche Gewölbe an der bestehenden Schwächezone "abgleitet" (Abb. 1) und eine starke Beanspruchung der ortsbrustseitig gelegenen Nachbarfelder (etwa 4 - 5 m) nach sich zieht. Das Abreißen von Ankerplatten im unteren Kämpferbereich ist hiefür ein sichtbares Zeichen.

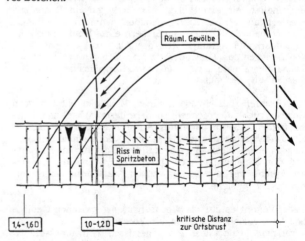

Fig. 1: Diagram of a spatial lining "sliding" at an already existing shearing zone.

Fig. 1: Représentation schématique d'une voute dans l'espace qui glisse à une zone de cisaillemnent déjà existante.

Abb. 1: Schematische Darstellung eines räumlichen Gewölbes, welches an einer schon bestehenden Scherzone "abgleitet".

b) Verlust der räumlichen Gewölbeverspannung durch einen größeren Nachbruch an der Ortsbrust

Kommt es aufgrund geologischer Randbedingungen zu einer raschen Entwicklung der Ausbaubelastung, so gewinnt das räumliche Gewölbe, welches den im Ortsbrustbereich noch nachgiebig reagierenden Ausbau schützend überspannt, entscheidende Bedeutung für die Standsicherheit des Hohlraums. Gerät nun die Ortsbrust unvermittelt in eine Störzone mit sehr geringen Festigkeitseigenschaften, so kann dies unter Umständen einen örtlichen Nachbruch über der Firste zur Folge haben. Durch die rasch um sich greifende Entspannung geht dabei aber auch der Kämpferwiderstand verloren, den das räumliche Gebirgsgewölbe bis dahin schräg über der Ortsbrust vorfand.

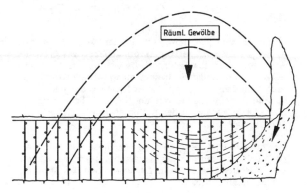

Fig. 2: Loss of the spatial rock arch in case of rock fall at the face.
Fig. 2: Perte de la tension de voûte dans l'espace par suite d'un réglage sur le front de taille.
Abb. 2: Verlust der räumlichen Gewölbeverspannung durch Nachbruch an der Ortsbrust.

4.3 Überlastung des Kalottenauflagers im Bereich des Strossenabbaus

Grundsätzlich kann dieser Versagenstyp sowohl mit, als auch ohne Kalottenzwischengewölbe auftreten. Wurde beim Kalottenvortrieb auf den Einbau eines Sohlgewölbes verzichtet, so kündigen sich Überlastungserscheinungen in der Regel schon frühzeitig durch erneute Zunahme der Setzungen an. Da rechtzeitig reagiert werden kann, ist diese Situation durch entsprechende Maßnahmen meist problemlos beherrschbar. Kritischer wird es hingegen, wenn eine eingebaute Zwischensohle annähernd vorwarnungslos versagt und dem Gebirge in kürzester Zeit ein nicht unbeträchtlicher Ausbaustützdruck entzogen wird.

Hier sind zwei Versagensvarianten bekannt, welche sich dadurch unterscheiden, daß die Zwischensohle im einen Fall erst zufolge des Umlagerungsgeschehens beim Strossenabbau zu Bruch geht (Variante I), während sie im anderen Fall zufolge einer hohen Setzungsgeschwindigkeit schon in Ortsbrustnähe abgeschert wird (Variante II).

In der Variante I stellt sich in der tonnenförmigen Spritzbetonschale eine Längstragwirkung ein, welche den vor der Strossenbrust liegenden Bereich mit deutlichen Zusatzlasten beansprucht. Wegen der nachgiebigen Verzahnung zwischen Ausbau und Gebirge und der steifen Auflagerung der Schale zieht der Kalottenfuß diese Lasten, welche aus einem Durchhängen des Kalottengewölbes über der offenen Ulme resultieren, in einem hohen Maße an sich.

Schert nun das relativ steife Zwischengewölbe aufgrund der Lasterhöhung im Gewölbeanschluß ab, oder kommt es, sofern mit Mittelschlitz gearbeitet wird, zu einer neuerlichen Nachgiebigkeit des Kalottenfußbereiches (Einstanzen oder Grundbruch), so führt dies in beiden Fällen zu einer raschen und starken Reduktion der Tragfähigkeit der Spritzbetonschale.

Wenn das Zwischengewölbe hingegen schon im Zuge der Kalottenstabilisierung abgeschert wurde (Variante II), beteiligt es sich nur mehr mit Hilfe einer durch die Gebirgsentspannung aufgezwungenden Normalkraft (Druck) über Reibung und Verzahnung an der Ableitung von Vertikalkräften.

Bedingt durch die relativ hohe Verzahnungssteifigkeit der Rißufer im abgescherten Spritzbetongewölbe kann der kritische Zustand beim Herannahen des Strossenabbaus durch Verformungsmessungen allein nicht rechtzeitig erkannt werden. Meßgeräte zur Beobachtung der Spritzbetonbeanspruchung können jedoch nur begrenzt eingesetzt werden und sind daher an der entsprechenden Stelle oft nicht vorhanden. Die Folge ist unter Umständen, daß eine auf objektivierbaren Wahrnehmungen aufbauende Beurteilung der Situation das dem Strossenabbau vorauseilende Aufschlitzen des Zwischengewölbes zur Herstellung einer Auffahrsrampe in die Kalotte als vertretbar erscheinen läßt.

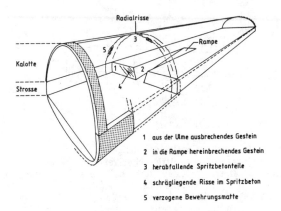

1 aus der Ulme ausbrechendes Gestein
2 in die Rampe hereinbrechendes Gestein
3 herabfallende Spritzbetonteile
4 schrägliegende Risse im Spritzbeton
5 verzogene Bewehrungsmatte

Fig. 3: Diagram of an evident indication of a slide during bench excavation with platform.
Fig. 3: Pepésentation schématique des signes visibles d'un début d'éboulement lors de la taille de la stross avec rampe.
Abb. 3: Schematische Darstellung der sichtbaren Anzeichen eines beginnendes Verbruches beim Strossenabbau mit Rampe.

Mit dem Aufschlitzen verliert das Zwischengewölbe seine aussteifende Funktion. Der von der Spritzbetonschale mobilisierte Ausbaustützdruck kann somit relativ rasch und stark abfallen, wenn das Gebirge unter den Kalottenfüßen auf die zusätzlichen Vertikalkräfte nachgiebig reagiert.

Ist in dieser Situation nicht mit Hilfe einer massiven Kämpferankerung für eine entsprechende Tragreserve des Gebirgstraggewölbes vorgesorgt, besteht die Gefahr, daß das Gebirge zwischen den hochwachsenden Scherzonen "durchzurutschen" beginnt. Die zunehmenden Setzungen führen in der Folge zum weiteren Anstieg der nachdrängenden Lasten. Ist die Spritzbetonschale diesen Belastungen gewachsen, wird sich die Kalotte lediglich mehr oder weniger stark in die Sohle einstanzen. Wird der Ausbau jedoch überlastet, wird die Spritzbetonschale, wie in Kap. 4.1 beschrieben, in der Firstkappe durchschlagen.

4.4 Absacken des Kalottengewölbes im Tunnel-Portalbereich

Die Situation zeigt einen Kalottenvortrieb in der Portalzone bzw. im Bereich mit einer Überlagerung zwischen $H_ü$ = 15 - 20 m. Das Überlagerungsmaterial ist kohäsionsarm bzw. stark klüftig und wenig verspannt.

Die auf den Ausbau wirkenden Kräfte sind zufolge der geringen Verspannung relativ hoch. Ein räumliches Schutzgewölbe kann sich aufgrund des fehlenden portalseitigen Widerlagers noch nicht ausbilden.

Kommt es in dieser Situation zu einer Überlastung der Scherfestigkeiten im Gebirge, welche durch Ausrieseln von Kluftfüllungen, Verlust der Verspannung im Ortsbrustbereich, Setzungen, Wassereinfluß etc. ausgelöst werden kann, erhöht sich die Belastung der Schale und damit der Kalottenfüße vor allem dann sehr rasch, wenn die Überlastung eine Entfestigung der Kohäsion mit sich bringt.

Da die Spritzbetonschale den bei diesen Überlagerungen auftretenden Belastungen in der Regel noch gewachsen ist, kommt es vorwiegend auf die Tragfähigkeit der Kalottenfüße an, ob die nachdrängenden Gewichtskräfte abgetragen werden können. Gibt das Gebirge unter den Kalottenfüßen nach, so ist meistens ein Absacken der mehr oder minder intakten Kalottenschale die Folge (Abb. 4).

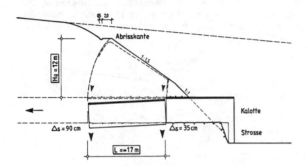

Fig. 4: Failure of calotte during heading through loose material with little cohesiveness.

Fig. 4: Calotte afaissée lors du creusement dans un matériau meuble à faible cohésion.

Abb. 4: Abgesackte Kalotte beim Vortrieb in kohäsionsarmem Lockermaterial.

Der geschilderte Verbruchmechanismus ist relativ häufig, zumal aufgrund der noch geringen Überlagerungen meist nicht mit dessen Auftreten gerechnet wird.

5. WIE KANN MAN GEFAHREN IM VORAUS ERKENNEN?

Eine Gefährdung auf die man gefaßt ist, ist nur mehr halb so schlimm. Tritt sie unerwartet ein, weil unerkannt ein, fehlen meist Zeit zur Untersuchung und Mittel zur Stabilisierung der Situation. Oft bleibt nur der Ausweg, den Hohlraum bis zum nächsten sicheren Stützring zu räumen.

Man spricht gerne von Stabilität. Das Wort Stabilität kann sich beziehen
- auf einzelne Bauglieder,
- auf einen bestimmten Zustand,
- auf einen kontinuierlichen Vorgang, d.h. solange der Vorgang läuft bleiben die Verhältnisse gleich, man darf aber in bestimmten Formationen nicht stehen bleiben, z.B. weil sich sonst das Gebirge erweicht.

Zur Beurteilung eines Zustandes, der Gefahr bzw. der Standsicherheit benützt man verschiedene Kriterien:
- erstens das Auftreten von Rissen im Spritzbeton oder das Reißen von Ankern in bestimmten Bereichen. Solche Anzeichen gelten bereits als grobe Warnsignale.
- zweitens die Spannungen. Diese werden indirekt oder direkt über Druckmeßdosen gemessen, sie geben uns Hinweise über die Beanspruchung des Ausbaues, in besonders gut instrumentierten Fällen auch über die des Gebirges;
- drittens: die wichtigsten und am leichtesten meßbaren Werte sind die Formänderungen. Hier gibt es viele Möglichkeiten der Beobachtung, um das Gebirgs- und Ausbauverhalten zu studieren, zu analysieren und zu kontrollieren. Um nur einige Möglichkeiten zu nennen seien angeführt:

= Die Absolutwerte an Setzung, Konvergenz etc., d.h. Mindestmaße, die vom jeweiligen Gebirge und dem Umfeld abhängen und die nicht überschritten werden sollten.
= Die Tendenz, zunehmende Verformung signalisiert Gefahr.
= Die Verformungsgeschwindigkeit. Mit diesem Hilfmittel der Steuerung hat man schon am Arlberg begonnen und sich eine Grenze gesetzt, ab welcher Zusatzmaßnahmen gesetzt werden mußten.
= Nicht übersehen werden sollten auffallende Unstetigkeiten in den Kurvenverläufen, wenn sie einwandfrei nicht auf Meßfehler zurückzuführen sind. Unstetigkeiten sind immer ein Zeichen, daß etwas nicht in Ordnung ist.
= Darüberhinaus geht man neuerdings den Weg einer weiteren Verfeinerung in der Auswertung einer systematischen Analyse, in der Zusammenschau wie im Detail.

Darunter sind Möglichkeiten in der Auftragung der Meßergebnisse zu verstehen, um frühzeitig wertvolle Aussagen über das Gebirgsverhalten bzw. sich anbahnende Veränderung in diesem zu erhalten. Dies kann man zum Beispiel durch die Auftragung der der Brust voreilenden Setzungen an der Oberfläche erreichen oder durch Vergleich der Linien stationsgleicher oder zeitgleicher Zustände.

Ganz wesentlich erscheint ferner die Rückkoppelung des Beobachteten mit der Rechnung, das Heranziehen von Modellen und Modellerfahrungen und das Studium von Bruchabläufen.

6. WELCHE FOLGERUNGEN KANN MAN LETZTLICH AUS EINER SOLCHEN ZUSAMMENSCHAU ZIEHEN?

Für den Tunnelbauingenieur stellt sich in einer solchen Situation klarerweise die Frage - stehenbleiben oder weiterfahren.

Für den Fall "Stehenbleiben" sprechen Gründe, sobald die Situation unklar und unüberschaubar ist, d.h. z.B. man steht vor einer Störung, deren Ausdehnung, deren Materialverhalten erst erkundet werden muß. Und in diesen Fällen ist es durchaus ratsam lieber an Bauzeit zu verlieren, als weiter vorzutreiben und einen Verbruch zu riskieren.

Wurden die Gebirgsverhältnisse ausreichend erkundet, Unregelmäßigkeiten der Meßansatz oder von Verhaltensanomalien zufriedenstellend geklärt und der vorhandene bzw. erwartbare Sicherheitsspielraum ausgelotet, so liegt damit die erforderliche Gesamtinformation vor und es können die entsprechenden Entschlüsse gefaßt und die weiteren Schritte eingeleitet werden. Welche, das kommt ganz auf die Umstände an.

Zum Beispiel kann das Umstellen auf eine andere Bauweise bzw. auf eine weitere Unterteilung des Querschnittes, wie Ulmenstollen sinnvoll sein, oder eine voreilende Verfestigung des Materials in der zu durchörternden Strecke etc.

Entschließt man sich zur Weiterführung der Arbeiten, muß man auf folgende Dinge achten:
a) die meßtechnische Überwachung muß so intensiv sein, daß man die leiseste Verschlechterung sofort erkennen kann,
b) man muß in der Lage sein, die Ausbaumaßnahmen sofort zu verstärken bzw. ausfallende Stützmittel in ausreichendem Ausmaß ersetzen zu können,
c) man muß den Sohlschluß so nahe an den Kalottenausbruch heranbringen, daß kurzzeitig der Ringschluß erfolgen kann.

Literaturverzeichnis:
Vavrovsky, G.-M. 1987. Entspannung, Belastungsentwicklung und Versagensmechanismen bei Tunnelvortrieben mit geringer Überlagerung. Dissertation an der Montanuniversität Leoben.

An empirical approach to open stope design

Approche empirique à la conception des chantiers à ciel ouvert
Ein empirischer Ansatz zur Technik des Kammerbaus

R.C.T.PAKALNIS, Ph.D., University of British Columbia, Vancouver, Canada
H.D.S.MILLER, Ph.D., University of British Columbia, Vancouver, Canada
S.VONGPAISAL, Ph.D., CANMET, Mining Research Laboratories, Ottawa, Canada
T.MADILL, HBSc., Sherritt Gordon Mines Ltd, Leaf Rapids, Manitoba, Canada

ABSTRACT: Studies have been conducted by the authors on the development of a rational approach to open stope design for the Ruttan Mine of Sherritt Gordon Mines Ltd. The underground mine has been in operation since 1979. This has resulted in a large data base of information which yielded forty-three stopes at various stages of extraction. An empirical formulation has been developed based upon the Ruttan data base whereby dilution was quantified in terms of the rock quality, the area exposed and the rate at which the hanging wall is mined. This relationship was found to accurately predict the resultant dilution that one can expect from a particular stope.

RÉSUMÉ: Les auteurs ont effectué des études en vue de développer une méthode empirique s'appliquant à la conception des chantiers ouverts à la mine Ruttan de la Sherritt Gordon Mines Ltd. La mine souterraine est exploitée depuis 1979, et par conséquent, un fichier informatisé important concernant quarante chantiers d'abattage à différentes étapes d'exploitation est déjà établi. Les auteurs ont élaboré une formule empirique, à partir du fichier Ruttan, voulant que la dilution soit quantifiée en fonction de la qualité des roches, de la surface exposée et de la vitesse à laquelle la lèvre supérieure de la faille est exploitée. Le rapport entre ces éléments permet de prédire de façon exacte le coefficient de dilution d'un chantier d'abattage donné.

ZUSAMMENFASSUNG: Die Verfasser stellten Untersuchungen zur Entwicklung eines rationellen Verfahrens für die bergbautechnische Planung des kammerbaus in der Ruttan-Grube der Sherritt Gordon Mines Ltd. an. Dieses Unternehmen ist bereits seit 1979 in Betrieb und lieferte umfangreiches Datenmaterial, dem Angaben über 43 Stösse in verschiedenen Stufen des Abbaus entnommen werden konnten. Auf der Grundlage der von der Ruttan-Grube gewonnenen Datenbank wurde eine empirische Formulierung entwickelt, in der der Bergeanteil Abhängigkeit von der Gesteinsbeschaffenheit, der freigelegten Fläche und des Firstenabbaufortschritts quantitativ bestimmt wird. Mit Hilfe dieser Beziehung gelang es, den an einem bestimmten Stoss zu erwartenden Bergeanteil mit Genauigkeit vorherzusagen.

1 INTRODUCTION

Approximately 51% of all ore production by underground metal mines in Canada is derived directly from open stoping operations (Pakalnis, 1986). This method of mining requires that large excavations remain open until all the ore is extracted under a minimal acceptable dilution. There exists no accepted design methods for predicting stable stope dimensions in jointed materials. Beam theories, numerical models and empirical criteria have been employed in the past to some degree of success. The classical design approaches are limited to simple structural con-figurations whereas numerical methods can be employed for complex structures. Empirical design methods relate practical experience gained on previous projects to the conditions anticipated at the proposed site. It was decided to incorporate an empirical approach at this mine whereby historic observations are employed as a predictive tool for future stope design.

This project is a joint effort involving industry, university and government. It initially commenced in 1982 as a doctoral research project at the University of British Columbia, and is presently being conducted under the auspices of the Canada Centre for Mineral and Energy Technology as part of the Canada/Manitoba Mineral Development Agreement. The ultimate objective is to develop design guidelines for predicting stable stope dimensions for echelon type deposits. This paper will present empirically derived stope design guidelines as they have been developed at the Ruttan operation. Test stopes are being monitored and evaluated in response to the design criteria set forth in this paper.

2 BACKGROUND

The Ruttan Mine of Sherritt Gordon Mines Ltd., a 6000 tpd underground operation is located in northern Manitoba, Canada. The Ruttan copper-zinc orebody is a multi-lensed, steeply dipping (70°) en echelon deposit, Fig. 1. Individual lenses have a maximum strike length of 350 m with widths varying from 7 to 61 m. The mining method is open stoping with delayed backfilling and extends from surface to 860 m below the surface. Stope spans at Ruttan were designed initially employing classical beam theory. Typical stope and pillar dimensions are shown in Fig. 2. Failures at the mine were found to be controlled primarily by structure, and could not be explained by a previous homogeneous model. In 1982, a study was undertaken to re-evaluate the method of stope span determination at Ruttan.

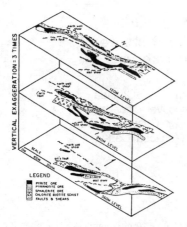

Figure 1. Isometric diagram -Ruttan deposit

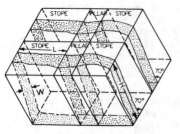

TYPICAL STOPE
DIMENSIONS

STOPE HEIGHT (H) = 40 − 100m
STOPE LENGTH (L) = 40 − 50m
STOPE WIDTH (W) = 6 − 30m
RIB PILLAR LENGTH = 6m

Figure 2. Stope isometric

3 APPROACH TO DEVELOPMENT

3.1 Stress

Prior to the acceptance that failure at Ruttan is
entirely structurally controlled, one has to
investigate the influence that stress has on the
stability of the hanging wall. The contributing
stress factors that would affect the stability of an
individual block are gravitational, hydrostatic and
confining stresses. It has been determined through
modelling and instrumentation that the hanging wall
and footwall of stopes at Ruttan are in a state of
relaxation. This dictates that the induced stresses
acting on a particular wall segment are tensile.

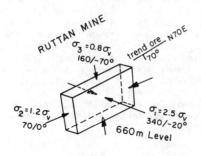

Figure 3. In situ stress at Ruttan 660 m level

Consequently that particular segment is not confined
and coupled with a jointed material would result in
the block to slip out under its' own weight.
  In situ stress determinations were carried out as
part of this study on the 620, 660, and 790 m levels
below surface. Generally it was concluded that the
major principal stresses are as shown in Fig. 3. In
addition, a geological and historical interpretation
verified the previously recorded stress directions
and magnitudes. Due to the geometrical complexity of
the Ruttan orebody, Fig. 1, individual stopes were
classified as having an isolated, rib or echelon
configuration, Fig. 4. Parametric studies were
performed on the individual stope configurations.
The variance of axial to diametrical (span/width)
dimensions were evaluated in terms of the effect on
the state of induced stresses. In addition the
variance in stress ratio with depth was assessed in
terms of its effect on the tensile zone. Generally
it was found that the hanging and footwall at Ruttan

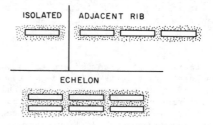

Figure 4. Plan view of stope geometries analyzed

will always be in a state of relaxation and the
extent of this relaxed zone will always envelope the
actual observed zone of slough. Two dimensional
boundary element methods using simple homogeneous,
isotropic and elastic models were employed as part of
the parametric study. A three-dimensional boundary
element model was employed to verify that the hanging
wall was in a state of relaxation, Fig. 5.
Instrumentation placed since the commencement of
mining in terms of extensometers and stress meters
has indicated that the hanging wall was in a state of
relaxation. A test stope is now completed that was
instrumented by extensometers and stress meters
specifically to verify the above parametric
observations. It showed similar results and is
presently being modelled and analysed. It was
concluded that the relaxed zone encompasses all
isolated, rib and echelon stopes at Ruttan. The
resultant dilutions are well within the zones of
relaxation and consequently are not considered to be
greatly effected by shearing stresses.

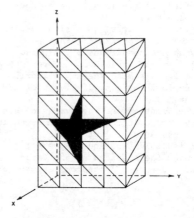

Figure 5. Relaxed zone XZ and XY plane

3.2 Rock Mass Characterization

A detailed structural investigation of the Ruttan operation was conducted. This was derived from underground mapping and through core logs. Fig. 6 is a stope characterization procedure employed for all stopes at Ruttan. A stope is characterized by two drill holes per section and two sections per stope. In essence, four holes are analysed which are representative of the footwall, ore, and hanging wall for a particular stope. Three meters of the immediate hanging wall and footwall were evaluated in addition to the full length of the ore for a particular drill hole. However, the rock quality of the drill core within two stope diameters was studied, but in less detail. Major structures are recorded in addition to the geometry of the proposed stope and its configuration at the time of mining. Through visual estimates and historic observation, it is generally accepted that wall slough at Ruttan is primarily confined to the hanging wall. This has been determined primarily through observation at the drawpoints, and the drill levels. It is for this reason, that the rock quality in the vicinity of the hanging wall will generally be employed in characterizing an individual stope. In certain instances, however, it was found that an assessment of the footwall was more critical when:
- a major fault intercepts the footwall
- the rock mass rating of the footwall was much lower than that of the hanging wall (one class or more difference)

A further practice was reducing the rock mass rating by a single class when a major structure intercepted either of the wall contacts. The RMR System classifies the mean rock mass parameters. A major structure must be treated separately from adjacent bounding units. This modification was shown to empirically be a good estimator of dilution. It is realized that kinematically each major structure should be analyzed in greater detail. However, for purposes of this investigation, it was found that by reducing the RMR value by a class, resulted in statistically higher correlations between the modified RMR and dilution. The orientation of the individual structures was not incorporated into the classification, since structures at Ruttan are generally oriented parallel to the stope contact. The major difference is found in the frequency and strength of the individual jointing. Groundwater was not incorporated into the characterization. This is primarily due to the absence of water as evidenced by dip tests conducted on the 430, 370, and 260m level. The close diamond drill pattern generally ensures that any trapped groundwater has an access to drain. The rock units at Ruttan can be generalized as follows:

Hanging Wall and Footwall Units
Chlorite Talc Schist
Quartzite
Acid Sediments
Basic Dyke
Massive Sulphide
Ore Units
Semi-Massive Sulphide with Basic Dyke
Semi-Massive Sulphide with Quartzite
Semi-Massive Sulphide with Chlorite Talc Schist
Semi-Massive Sulphide with Chlorite Schist
Massive Sulphide

The average RMR rating for the individual wall contacts are as follows:

Location	RMR (%)	
Footwall (54)	58	18
Ore (54)	63	17
Hanging Wall (54)	60	19

() refers to number of observations

The dominant joint sets at Ruttan are shown in Fig. 7. Stereonets and a kinematic analysis indicated that structures exist that would generally yield failure in the hanging wall as shown in Fig. 8. This method of toppling was observed as the most common mode of failure at Ruttan. Parting is generally along set 1 and sliding along set 2. It was concluded that failure due to structural instability is the most likely failure mechanism prevalent at Ruttan.

Figure 6. Stope characterization

The rock mass rating (RMR) developed by Bieniawski was further grouped into the following classes:

RMR Value	Class
81 - 100%	A
61 - 80%	B
41 - 60%	C
21 - 40%	D
0 - 20%	E

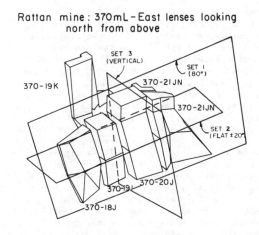

Rattan mine: 370mL-East lenses looking north from above

Figure 7. Design sets

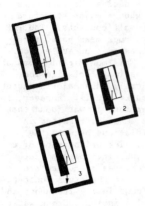

Figure 8. Base friction model showing typical failure mode obeserved at Ruttan (section showing sequence 1-3)

3.3 Dilution

This parameter is a measure of the quality of the stope design. Various definitions exist, however, it is a parameter recorded by most open stope operators. Dilution will be considered as the dependent control variable. The significance of the individual characteristics will be evaluated empirically in terms of dilution. Dilution figures from forty-three stopes will be analyzed at various stages of extraction yielding 432 dilution values. These values are subsequently averaged into 133 observations.

Dilution at Ruttan is visually assessed on a daily basis for purposes of this data base. A visual interpretation of the volume percentage of massive sulphide, semi-massive sulphide, disseminated material and dilution based on iron content is made

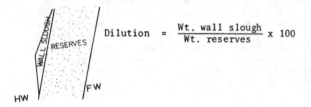

$$\text{Dilution} = \frac{\text{Wt. wall slough}}{\text{Wt. reserves}} \times 100$$

Figure 9. Definition of dilution

for each producing drawpoint. Dilution is classified as anything with less than 10 % iron or less than one percent copper. It can be classified as either internal or external to the ore reserve. Based on prior knowledge of the rock types inherent to a particular stope, the size of muck and/or the presence of blast holes within the muck, the geologist can estimate what percentage of the dilution is internal or external. These visual estimations are compiled at the end of each month and a total average dilution figure for internal and external dilution is derived. The external dilution is evaluated as shown in Fig. 9. It is important to comment on the accuracy of the procedure. Dilution is an observed parameter that has been recorded in most instances on a daily basis for each drawpoint. The frequency of observation and the consistency of measurement has produced a figure that is a reliable measure of the waste tonnage that comes from a particular stope. In addition, grab samples are taken which are representative of the degree of waste and ore for a particular stope. These samples

are assayed and consequently the sample taken reflects the degree of dilution observed. The observed grades were found to be historically within 3% of milled grade. It is assumed that the accuracy of grade observation extends to that of dilution estimation since one has an affect on the other. However, the dilution approach at Ruttan is valid for Ruttan and the absolute values must be calibrated for other operations intending to employ this methodology towards stope design.

3.4 Statistical Analysis

The observed dilution is a measure of the external waste for a particular stope. Dilution measurements are recorded daily and tabulated monthly. It is converted into a given volume of stope excavation exhibiting a particular level of dilution. The volume excavated is determined through recording the trammed tonnes for a particular stope and by the exposed surface area. Table 1 shows distribution of the data comprising of the Ruttan data base. Forty-three stopes were assessed from a total of forty-six since the commencement of mining. It was intended to incorporate all stopes at Ruttan and not to bias the results by a selection procedure. An observation consists of the following:

(1) No. of observation
(2) Stope name
(3) Rock mass rating (RMR(%)) - for the critical wall contact
(4) Stope height (H(m)) - refer to Fig. 2
(5) Stope width (W(m)) - refer to Fig. 2
(6) Stope volume (m^3) - measured at various stages of extraction
(7) Dilution (%) - amount of waste associated with volume of stope mucked
(8) Span (L(m)) - refer to Fig. 2
(9) Area exposed (m^2) - refer to Fig. 10
(10) Hydraulic radius (m) - parameter as defined by Laubscher (1976)
(11) Exposure rate (m^2/mth) - refers to rate that the hanging wall is mined
(12) Blast correction factor (BCF(%)) - dilution observed during slot mining

Generally, zero dilution should be observed during excavation of the slot since minimal area is exposed under the most confined situation. Slot dilution is generally attributed to blasting. This dilution is then assumed to be constant over the remainder of the stope life, since the blast practice does not change. The BCF factor would also tend to compensate for observation error in estimating dilution, since this value should also remain constant for a given stope.

The exposure rate is calculated as the rate of excavation divided by the true stope width (W) and is recorded as m^2/mth.

Multivariate analysis on the individual parameters were evaluated in terms of their effect on dilution.

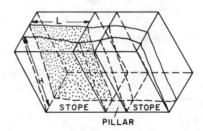

Figure 10. Exposed surface area shaded

Table 1. Data base - statistical summary

PARAMETER	TOTAL	ISOLATED	ECHELON	RIB
No. OF STOPES	43	22	12	28
No. OF OBSERVATIONS	133	61	44	28
Rock Mass Rating (%)	55±19 (56±20)	64±19 (59±22)	47±14 (45±15)	48±16 (49±17)
EXP. RATE (1000m²/mth)	.18±.09 (.18±.08)	.18±.09 (.17±.07)	.15±.07 (.15±.08)	.22±.09 (.18±.09)
HYD. RADIUS (m)	7±4 (11±3)	7±4 (11±3)	7±4 (11±2)	8±5 (10±4)
SPAN (m)	20±14 (31±13)	21±14 (32±13)	19±14 (34±8)	21±15
SURFACE AREA (m²)	1450±1120 (2250±1120)	1420±1020 (2254±970)	1360±1085 (2360±924)	1705±1545 (2090±1550)
SPAN/WIDTH RATIO	1.6±1.3 (3.4±1.7)	1.6±1.5 (3.2±1.8)		
STOPE HEIGHT (m)	71±21 (68±20)	68±15 (68±16)	72±23 (68±21)	78±28 (70±25)
BCF (%)	3±6	3±9	3±3	3±7
DILUTION (%)	7±6 (10±6)	6±5 (10±6)	6±4 (10±4)	10±7 (12±7)
STOPE DEPTH (m)	368±54 (360±48)	356±65 (346±55)	385±36 (382±40)	369±47 (369±40)
STOPE INCLINATION (°)		67±9 (68±9)		
STOPE WIDTH (m)		18±9 (15±8)		
LONG SPAN/SHORT SPAN		5.1±5 (2.1±1.1)		
( ) Refers to Final Stope Configuration				

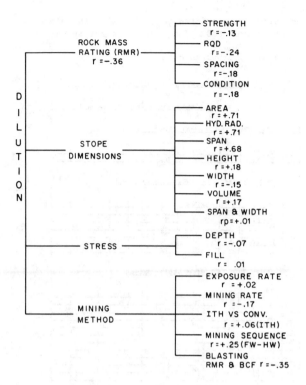

Figure 11. Mutual correlations between dilution and an individual parameter

The dilution sensitive parameters were evaluated with respect to the isolated data base, thereby ensuring that stope configuration would not bias the resulting correlations. Mutual correlations (r) between dilution and individual parameters is shown in Fig. 11. The parameters having the most significant effect on dilution were combined into a best fit multi-linear equation. A multi-quadratic equation was attempted with only minor improvements in correlation. The value of r generally varies between ±1 and 0. The higher the absolute value, the better the predictive equation, in fitting a mathematical expression to the observed dilution observations.

A sensitivity analysis performed on the parameters yielding the best correlation are shown in Fig. 12. The exposed surface area and the rockmass rating are the most dilution sensitive parameters, whereas the exposure rate is the least. The exposure rate was included in the governing equation since it significantly improved the overall correlation of the predictive equation. In addition, a single equation relating observed to predicted dilution for the entire data base was found to be a significantly poorer estimate than if a single equation was employed for each of the individual stope configurations. The resultant governing equations are shown in Fig. 13. The average absolute difference in estimation is ± 2.4% (absolute) for the isolated, ±1.8% for the echelon, and +2.7% for the rib configurations. These errors of estimation are determined for the sample and for the population estimate. Fig. 14 graphically depicts the predicted versus observed dilution for the individual stope data base. The upper and lower bounding lines represent a 68% (one standard deviation) confidence that the observed value will fall within the ranges delineated, or one is 84% confident that the observed dilution would be less than the mean estimated dilution plus one standard deviation.

4. APPLICATION

The governing equations, Fig. 13, were employed to estimate dilutions for eight stopes that were mined subsequent to the study. the results are summarized in Table 2. An error of estimate of ±2.4% was recorded (absolute dilution). The values were tabulated by Ruttan mine personnel and the BCF values were recorded as dilution attributed to the slot removal. The Ruttan approach of empirical was

compared to the empirical methods of Mathews et al (1981), Voussoir Arch by Beer et al (1982), Laubscher (1976), Barton (1974) and Bieniawski (1984). Good correlations existed among the individual methods, however, for purposes of brevity, only Fig. 14 is reproduced. Mathews method of stope design, Fig. 14, employs a stability number which accounts for the rock quality, stress, stope and structure orientation and relates this value to the exposed surface geometry. The shape factor is defined as the surface area divided by the perimeter (hydraulic radius). The stability number is plotted versus the shape factor to yield zones of stable, potentially unstable and potentially caving regions. These zones have been empirically calibrated. The dilutions observed at Ruttan generally conform closely to the categories defined by Mathews.

It must be noted that the dilutions predicted and observed have been corrected for blasting. This was necessary due to lower correlations achieved when individual observations were not adjusted for the initial dilution observed as the slot was excavated. Therefore at Ruttan the total dilution for a stope is

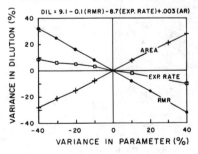

Figure 12. Sensitivity analysis - data base = all stopes

actually the predicted (Fig. 13) plus the slot dilution expected. The average at Ruttan is 3%, however, through better control blasting this value is being reduced significantly.

Table 2. Calibrated data base

No.	Stope	RMR(%)	E.R.(m²/mth)	Area(m²)	Res. Dil(%)	Pred.Dil(%)	Diff.(%)
	CALIBRATED DATA BASE - STOPES MINED SUBSEQUENT TO STUDY (8 Stopes)						
	ISOLATED STOPE: DIL(%)=8.6 - .09(RMR) - 13.2(E.R.) + .0038(AREA)						
1	320 0 I	90	0.210	750	4	1	3
2	320 0 I	30	0.240	1625	5	4	1
	ECHELON STOPES: DIL(%)=10.3 - .13(RMR) - 14.8(E.R.) + .003(AREA)						
3	490 1 D	49	0.120	225	3	3	0
4	490 1 D	49	0.130	450	3	3	-0
5	490 1 D	49	0.140	900	3	5	-2
6	490 1 D	49	0.160	1350	5	6	-1
7	490 1 D	49	0.160	1800	7	7	0
8	370 14 CW	47	0.150	750	5	4	1
9	370 14 CW	47	0.150	1250	5	6	-1
	RIB STOPES: DIL(%)=15.8 - .18(RMR) - 7.7(E.R.) + .0026(AREA)						
10	260 17 J	53	0.210	375	1	6	-5
11	260 17 J	53	0.250	760	2	6	-4
12	260 17 J	53	0.310	1560	3	8	-5
13	260 17 J	53	0.330	2400	8	10	-2
14	260 17 J	53	0.330	3000	12	12	0
15	260 4 /SCROP	67	0.260	375	3	3	0
16	260 4 /SCROP	67	0.300	750	3	3	-0
17	260 4 /SCROP	67	0.330	1575	4	5	-1
18	260 4 /SCROP	67	0.350	2250	5	7	-2
19	260 4 /SCROP	67	0.370	3075	5	9	-4
20	260 4 /SCROP	67	0.360	3750	6	11	-5
21	260 15 H	78	0.570	2496	3	4	-1
22	340 9 C	26	0.350	840	9	11	-2
23	340 9 C	26	0.250	1400	14	12	2
24	420 20 J	71	0.110	288	1	3	-2
25	430 20 J	71	0.130	576	3	4	-1
26	430 20 J	71	0.190	1200	4	5	-1
27	430 20 J	71	0.200	1776	6	6	-0
28	430 20 J	71	0.210	1920	8	6	2

ISOLATED STOPES (61 OBS)
$DIL(\%)=8.6-0.09(RMR)-13.2(EXP.\,RATE)+0.0038(AREA)$
$r=\pm0.79$ $\hat{s}=\pm3\%$
ECHELON STOPES (44 OBS)
$DIL(\%)=10.3-0.13(RMR)-14.8(EXP.\,RATE)+0.0026(AREA)$
$r=\pm0.83$ $\hat{s}=\pm2\%$
RIB STOPES (28 OBS)
$DIL(\%)=15.8-0.18(RMR)-7.7(EXP.\,RATE)+0.0026(AREA)$
$r=\pm0.8$ $\hat{s}=\pm4\%$
where:
DIL(%) — DILUTION (%)
RMR — ROCK MASS RATING (%)
EXP.RATE — EXPOSURE RATE (100 m²/mth)
AREA — EXPOSED SURFACE AREA (m²)

Figure 13. Governing equations

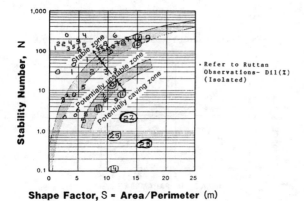

Figure 14. Calibration - Mathew's method (1981)

5 CONCLUSIONS

Presently the Ruttan operation is employing the predictive equations to determine the degree of wall slough that can be expected given the rock quality, the areas exposed, and the exposure rate. It is also interpreted in terms of what stope dimensions can one live with given a maximum acceptable level of dilution. It is our firm conviction, that historic data should be used where possible to shape predictive tools. It is generally concluded by the Ruttan operation that the work has already had a significant impact on the mines intuitive feeling about stope layouts, and is allowing Ruttan to layout stopes that give lower dilutions than would have been the case in the past. The governing equations are applicable for the Ruttan operation, however the methodology is valid for all open stoping operations. The absolute magnitudes of stope design are expressed in terms of dilution. This parameter is difficult to assess accurately, however once calibrated, it is a useful indicator of instability for any mine operation. It is suggested that the empirical methods of design outlined in this paper be attempted/calibrated for other operations where similar failure mechanisms prevail. It should be noted that the characteristics unique to the Ruttan operation be assessed prior to extending the study to other operations.

ACKNOWLEDGEMENTS

The authors gratefully acknowledge the research effort that is being presently conducted by Sherritt Gordon Mines Ltd. and CANMET as well as by the University of British Columbia in the past.

BIBLIOGRAPHY

Barton, N., Lien, R. & Lunde, J. 1974. Engineering classification of rock masses for the design of tunnel support. Rock Mechanics. 6, no. 4:183-236.

Beer, G. & Meek, J.L. 1982. Design curves for roofs and hanging walls in bedded rock based on voussoir beam and plate solutions. Trans.Instn.Min.Metall. January: A18-22.

Bieniawski, Z.T. 1984. Rock mechanic design in mining and tunneling. Rotterdam: Balkema.

Laubscher, D.H. & Tayler, H.W. 1976. The importance of geomechanics classification of jointed rock masses in mining operations. Rotterdam:Balkema.

Mathews, K.E., Hoek, E., Wylie, D.C. & Stewart, S.B. 1981.Prediction of stable excavation spans for mining at depths below 1000 meters in hard rock. CANMET, Energy, Mines and Resources, Canada.

Pakalnis, R.C.T. 1986. Empirical stope design at the Ruttan mine, Sherritt Gordon Mines Ltd. Vancouver: Ph.D thesis, University of British Columbia.

CANMET 1985. Review of rock mass classification systems relevant to underground mine prepared for CANMET by R.C.T. Pakalnis and Ruttan operation of Sherritt Gordon Mines Ltd. CANMET project no. 4-9147-1. Scientific Authority:
Dr. S. Vongpaisal. Energy, Mines and Resources Canada.

Effect of three-dimensional state of stress on increasing the stability of underground excavations

Effet d'un état tridimensionnel de contraintes sur l'accroissement de la stabilité d'excavations souterraines

Der Einfluss des dreidimensionalen Druckspannungszustandes auf die Standfestigkeitserhöhung der Untertagebauten

R.D.PARASHKEVOV, Higher Institute of Mining and Geology, Sofia, Bulgaria

ABSTRACT: Possibilities for increasing stability of underground constructions by creating three-dimensional state of stress in the surrounding rock massif have been estimated on the basis of a new energetic criterion. By analyzing boundary variations of the stressed state of rocks in proximity of the walls of underground constructions, advisability and necessity of using technical means and technologies for increasing the own resistant capacity of the rock massif by accelerated formation of three-dimensional stressed state have been proved.

RESUME: A la base d'un critère énergétique nouveau de la résistance des roches, il est effectué une estimation des possibilités d'augmentation de la stabilité des travaux souterrains avec création d'une contrainte volumique dans le massif environant. Par analyse des changements limitatifs de l'état de contrainte des roches au voisinage des parois des travaux souterrains, on a prouvé la nécessité et l'utilité de l'application des moyens techniques et des technologies pour l'augmentation du niveau de la résistivité propre du massif par formation accélerée de l'état de contrainte volumique.

ZUSAMMENFASSUNG: Auf Grund des neuen energetischen Festigkeitskriteriums der Gesteine ist eine Beurteilung für Standfestigkeitserhöhung der Untertagebauen bei der Formierung des Raumspannungszustandes in dem Umgebungsmassives gemacht. Durch Analisierung der Grenzveräuderungen des Spannungszustandes der Gesteine in der Nähe der Untertagebauenstöße wird die Notwendigkeit von Anwendung der technischen Mittel und Verfahren zur Standerhöhung des Eigenwiderstandes von Raumspannungszustand nachgewiesen.

Investigations of the possibilities and advisability for more complete utilization of own resistance of rocks and massifs, aiming at increasing stability of underground mine openings, have been carried out in connection with the project for development of Dobrudja coal deposit, located at a depth of 1600-1850 m.

Analysis and assessment of strength-deformation parameters of rocks and a massif, serving as a basis for predicting the character and intensity of geomechanical processes around underground openings and structures, are of great importance for designing and application of technical and technological decisions in driving mine openings at great depths and assuring their stability.

The state of natural factors in an undisturbed massif at great depths has been poorly studied so far. It can be accepted with a relative validity that the resistance of the rocks in a massif is comparatively high, because under the action of the gravity force porosity and water content decrease, packing and closing of cracks are observed and the strength increases.

On the other hand, in conditions of a high level of the stressed state at great depths, the construction of underground openings leads to more intensive change of the energy balance, rise of the absolute value of the stresses, intensification of the process of loosening, destruction, plastic flow or rheological creeping. The possibility for brittle-elastic rock burst increases. This imposes higher requirements to the bearing capacity of support constructions and to the technological processes in driving, supporting and maintenance of underground openings.

It is necessary to make a quantitative prognosis and assessment of the more complete utilization of own resistance of rocks and a massif which they possess in an undisturbed massif, so that well-grounded technical and technological decisions can be reached. An attempt for such an assessment is made by the author further down on the basis of a new energetic criterion for the rock strength in triaxial non-uniform compression stress state.

For prognosticating geomechanical processes and grounding the possibility for making a quantitative assessment of the effect of the own rock resistance, results obtained by analytical and experimental studies, based on a new energetic criterion have been used. It is assumed that rock failure in triaxial non-uniform compression stress state ($\sigma_1 > \sigma_2 = \sigma_3$) is determined by the energy necessary for changing the shape, conditioned by the difference between maximum and minimum principal stresses ($\sigma_1 - \sigma_3$) and the mean hydrostatic pressure ($\sigma_2 = \sigma_3 = p$) in the form

$$\delta\sigma = (\sigma_1 - \sigma_3) - p = \sigma_1 - 2\sigma_3 = A = const \quad (1)$$

when

$$0 < n = \frac{\sigma_3}{\sigma_1} = \frac{\sigma_3}{2\sigma_3 + A} < 0,5$$

At a depth $H > H_s = \dfrac{A}{2\gamma(1-2n)}$ (2)

a disturbed zone with dimension

$$R = a\sqrt{k + \frac{k\gamma H}{A}}, \quad (3)$$

will be formed around a circular opening with radius a where

γH - natural pressure at a depth of H ,

A - triaxial strength of rocks,

k - coefficient of stress concentration.

It is clearly seen from formula (3) that dimensions of the disturbed zone diminish with increasing rock strength or keeping it constant.

At a more accurate solution from mechanical pont of view, the dimension of the disturbed zone around an unsupported opening can be found by the term

$$R = \frac{2}{3} \cdot \frac{(\gamma H + A)a}{A} , \qquad (4)$$

and for a supported opening, respectively

$$R^* = \frac{2}{3} \cdot \frac{(\gamma H + A)a}{p^* + A} . \qquad (5)$$

Expression (5) shows also the quantitative effect of increasing support resistance for acfieving a quick stabilization of the opening - should diminish.

When destabilization process has to be completely prevented, support must possess maximum resistance of a value

$$p_{max}^* = \frac{2}{3} \cdot \gamma H - \frac{1}{3} A \qquad (6)$$

Equation (6) clearly shows the role of preservation, maintaining or increasing the own strength of the rocks during the process of construction and supporting of openings at great depths. With increasing triaxial strength of an massif, the maximum necessary support resistance diminishes.

Support resistance could be of smaller value, if prior to reaching an equilibrium state of the rock-support system, certain uniform shrinkage of the cross section is allowed, i.e. some radial displacements of rocks occur

$$p^* = \frac{2}{3} \cdot \frac{\gamma H + A}{1 + \frac{2u^*}{a(k_v - 1)}} - A \qquad (7)$$

where $k_v > 1$ is the volumetric expansion during loosening and destruction of rocks.

However it is not advisable to allow for great depths significant displacements u^* and loosening of rocks, as they spread all over large areas in the massif around the opening in reatively short time. When observing the rquirement for the principal normal tangential stress to be equal or higher than the pressure in the undisturbed massif

($\sigma_\theta^R \geqslant \gamma H$) an active loosening zone R reaching size of

$$R_1 = \frac{3}{4} R , \qquad (8)$$

is formed within the disturbed zone with radius R_1, which indicates the great activity of loosening processes at great depths. Furthermore intensity of loosening σ_3^R reaches its greatest values at low values of the radial support resistance (conditioned by the support reaction and that of the broken rock material) when the positive energy balance increases at diminishing the ratio

$$n = \frac{\sigma_3}{\sigma_1}$$

$$W = \frac{k^2 \sigma_1^2}{8G} (1 - n)^2 , \qquad (9)$$

where: $0 < n = \frac{\sigma_3}{\sigma_1} < 0,5$

k - coefficient of stress concentration ($k > 1$),

$G = \frac{E}{2} (1 + \mu)$,

E - Young's modulus,

μ - Poisson's ratio.

At one and the same value of the maximum principal stress ($\sigma_1 = k\gamma H$) and equal deformation properties of the massif the energy balance increases more than three times when changing

$$n = \frac{\sigma_3}{\sigma_1}$$

from 0,1 to 0,45. This means that in case of delay of supporting or limiting application of packing-strengthening operations (n decreases), the energy saturation will increase and loosening processes will become very intensive.

In triaxial compression strength of rocks A is expressed, although approximately (on the basis of experimental data) by uniaxial compression strength σ_s

$$A = k_0 \sigma_s \qquad (10)$$

where $k_0 \geqslant 1$, an approximate quantitative assessment of the influence of ratio between maximum and minimum (σ_1 and σ_3) principal stresses for maintaining or decreasing the own resistance capacity of the massif, can be made:

(a) at $\sigma_3 = 0$, unsupported oprning, n = 0 $K_0 = 1$,

(b) at weak reaction of the support $K_0 = 1,05 \div 1,1$; $0 < n < 0,15$;

(c) with application of strengthening-consolidating technological action on the massif and high response of the support

$K_0 = 1,2 \div 1,6$ at $0,15$ $n = \frac{\sigma_3}{\sigma_1} < 0,5$.

From the point of view of practical application of geomechanical prognoses, particular attention should be given to those technologies that preserve or increase the own resistance capacity of a rock massif, prior to disturbing its equilibrium state as a result of driving underground openings. For instance, setting expansion frictional pipe anchors in a massif in front of the faces of openings in heavy conditions at great depths should result in:
- preliminary additional stressing of a massif and diminishing the difference ($\sigma_1 - \sigma_3$) between principal stresses when shaping the cross section of the opening;
- limitimg intensity and extent of massif disturbance around a mine opening;
- restricting the possibilities for loosening, giving rise to additional internal friction, limiting rock displacements, diminishing shrinkage of the useful cross section of the opening;

1198

-reduction of the energy balance, dimini-
nishing support loading, a total increase of
opening stability in the period of construction and exploitation etc.

On the basis of the above mentioned considerations and conclusions, the following recommendations for increasing stability of
underground openings in deep mines can be
made:

1) Developing and application of technical
and technological decisions aiming at preserving or increasing the own resistance capacity of the rocks in a massif, some of
them being applied prior to excavation of
rocks in construction of mine openings and
 structures (preliminary plugging up,
packing, anchoring etc.);

2) Prognosis of the expected geomechanical
processes around underground openings at
great depths requires an analysis and
assessment of a united system, comprising the
rock massif, application of measures for
foregoing consolidation and strengthening,
technology of rock excavation, technology of
support construction and quick realization
of a close contact between support and rock
massif, fast reaching the working bearing
capacity of the support etc. A quantitative
assessment of the contribution of different
technical and technological decisions for
preserving or increasing resistance capacity
of rocks in a massif, diminishing energy balance and reducing intensity of the processes of loosening, destruction, plastic flow
rheological creeping or brittle-elastic rock
burst, has to be made.

An alternative solution for the in situ stress state inferred from borehole stress relief data

Une solution alternative pour déterminer l'état de contraintes in situ à partir de données de surcarottage

Eine Alternativlösung für die Bestimmung von Druckspannungen im Fels aus Verformungsmessungen in Bohrlöchern

W.G.PARISEAU, University of Utah, Salt Lake City, USA

ABSTRACT: The conventional analysis of stress about a long cylindrical opening in linearly elastic rock that has been the basis for inferring the in situ stress state from overcored deformation gages is shown to be in error. The correct solution is given. Luckily, a second error compensates for the first and minimizes the practical consequences, at least in isotropic ground.

RESUME: Il a ete demontre que l'analyse traditionnelle de gauges de deformation dans les sondages qui sert de base afin de deduire la tension in situ d'une longue perforation cylindrique dans de la roche elastique lineaire est fausse. Heureusement, une erreur secondaire annule l'erreur primaire, ainsi minimisant les consequences pratique, du moins dans de la roche isotrope. Une version de l'analyse corrigee est donnee.

ZUSAMMENFASSLING: Gezeigt wird, dass die herkommliche Analyse von Deformationsmessungen von vergebohrten land-zylindrischen Bohrlochern in Bohrkernen als Methode zur Berechnung von in situ Sapnnungsverhaltnissen im elastischem Gestein schon an der Basis fehlerhaft ist. Gezeigt wird auch, wie aber ein sekundarer Berechnungsfehler der fehlerhaften Grundannahme entgegenwirkt und somit die praktischen Konsequenzen der fehlerhaften Analylse, jedenfalls auf isotropisches Gestein bezogen, auf ein Minimum reduziert. Schliesslich wird eine berichtete Version der Analyse gegeben.

1 INTRODUCTION

The stress relief overcore method is a frequently used technique for obtaining data from which the in situ state of stress is inferred. Gages used in a pilot hole respond to changes in displacement and strain at the borehole wall during overcore drilling. The U. S. Bureau of Mines (USBM) borehole deformation gage is an example of a gage that responds to displacements only. The CSIRO gage is an example of a gage that responds to strains. A nominal 150 mm (6 in.) diameter overcore hole provides the requisite stress relief for these gages. The USBM gage requires measurements in three non-parallel holes while the CSIRO gage requires only one hole for the determination of the three-dimensional stress state in situ.

These gages and similar devices depend on the same theoretical solution to the problem of determining the state of stress about a long cylindrical hole drilled in initially stressed elastic rock. The initial stress state is, of course, the in situ stress to be determined. The solution to this problem, and therefore the data reduction formula for isotropic rock, was first obtained by Hiramatsu and Oka (1962). Subsequent equipment development and theoretical refinements have extended the technology to solid and hollow inclusion gages and to anisotropic rock (Fairhurst, 1964, 1968; Panek, 1966; Leeman, 1964, 1967, 1968; Merrill, 1967; Moody, 1968; Hirmatasu and Oka, 1968; Bonnechere, 1969; Oka and Bain, 1970; Niwa and Hirashima, 1971; Rocha and Silverio, 1969, 1974; Martinetti and Ribacchi, 1974; Bonnechere and Cornet, 1977; Ribacchi, 1977; Herget, Miles and Zawadski, 1977; Hirashima and Koga, 1977; Duncan Fama and Pender, 1980; Morgan, 1982; van Heerden, 1983; Amadei, 1983; Borsetto, Martinetti and Ribacchi, 1984; Rahn, 1984; Kanagawa and others, 1986).

One of the early theoretical refinements was the recognition that the present or post-hole stress about a borehole is composed of contributions from (i) the pre-hole or initial stress state σ^0 and (ii) the stress change σ' caused by introduction of the hole. In the elastic case $\sigma = \sigma^0 + \sigma'$. The stress change σ' is computed on the basis of a plane analysis in the sense that derivatives in the direction of the hole axis are zero. The initial stress is arbitrary. Strains ε^0 and ε' are associated with σ^0 and σ' through the stress strain relations (Hooke's law). Displacements u^0 and u' are obtained by integration of the strain displacement equations. Overcoring of the gage relieves the stress σ; a gage thus responds to total strains and displacements ε and u. The strains $\varepsilon = \varepsilon^0 + \varepsilon'$ and are generally not plane in the sense that derivatives in the direction of the hole axis are not zero.

However, the original solution by Hiramatsu and Oka (1962) is based on the plane assumption. They solved the problem of determining the stress distribution about an infinitely long circular hole loaded by a set of prescribed stresses at infinity. The medium is isotropic and linearly elastic. Independence of all quantities with respect to the direction of the hole axis is explicitly assumed. Their solution therefore tacitly implies that the displacements and strains associated with the pre-hole stresses also do not vary in the direction of the hole. Alternatively, they view the loads as being applied only after the pilot hole is drilled. Subsequent refinements including the more recent extension to anisotropic rock by Amadei (1983), for example, are also based on the plane assumption in whole or in part. In this regard, although Fairhurst (1964, 1968) explicitly recognizes the problem, the actual treatment of shear stress in the direction of the hole is, in effect, plane as is the treatment by Panek (1966). The plane assumption for the post-hole state is a special case at best and is obviously inconsistent where more than one borehole is used.

This paper presents an alternative solution for interpreting borehole stress gage data that is free of any plane assumption regarding the pre-hole and post-hole states. However, the changes caused by introduction of the pilot hole are considered plane in agreement with previous investigators.

Comparison with some existing data reduction schemes are then made and the consequences for in situ stress measurements summarized.

2 PRELIMINARIES

Both rectangular (xyz) and cylindrical (rθz) coordinates are useful in viewing the problem. Under the coordinate transformation from cylindrical to rectangular coordinates

$$\begin{Bmatrix} x \\ y \\ z \end{Bmatrix} = \begin{bmatrix} c & -s & 0 \\ s & c & 0 \\ 0 & 0 & 1 \end{bmatrix} \begin{Bmatrix} r \\ - \\ z \end{Bmatrix} \qquad (1)$$

where $c = \cos(\theta)$, $s = \sin(\theta)$ and (-) means zero as shown in Figure 1, the displacements transform according to

$$\begin{Bmatrix} u_x \\ v_y \\ w_z \end{Bmatrix} = \begin{bmatrix} c & -s & 0 \\ s & c & 0 \\ 0 & 0 & 1 \end{bmatrix} \begin{Bmatrix} u_r \\ v_\theta \\ w_z \end{Bmatrix} \qquad (2)$$

where the displacements on the right are relative to (rθz); those on the left are relative to (xyz). The stresses transform according to

$$\begin{aligned}
\left.\begin{matrix} \sigma_r \\ \sigma_\theta \end{matrix}\right\} &= \left(\frac{\sigma_x + \sigma_y}{2}\right) \pm \left(\frac{\sigma_x - \sigma_y}{2}\right)\cos(2\theta) \pm \tau_{xy}\sin(2\theta) \\
\tau_{r\theta} &= -\left(\frac{\sigma_x - \sigma_y}{2}\right)\sin(2\theta) + \tau_{xy}\cos(2\theta) \\
\sigma_z &= \sigma_z \qquad (3) \\
\tau_{rz} &= \tau_{xz}\cos(\theta) + \tau_{yz}\sin(\theta) \\
\tau_{\theta z} &= -\tau_{xz}\sin(\theta) + \tau_{yz}\cos(\theta)
\end{aligned}$$

3 PILOT HOLE CHANGES

The stress, strain and displacements caused by introduction of the pilot hole are the "changes" and are indicated by a prime. The stress and displacements changes are recorded here for definitiveness. They are obtained analytically by superposition of solutions to a plane and an anti-plane problem, both of which are independent of the z-direction chosen to coincide with the axis of the pilot hole. These results are in agreement with previous investigators. The stress changes in cylindrical coordinates are

$$\sigma'_r = -\left(\frac{\sigma_r^0 + \sigma_\theta^0}{2}\right)\left(\frac{a^2}{r^2}\right) + \left(\frac{\sigma_r^0 - \sigma_\theta^0}{2}\right)\left(\frac{3a^4}{r^4} - \frac{4a^2}{r^2}\right)$$

$$\sigma'_\theta = \left(\frac{\sigma_r^0 + \sigma_\theta^0}{2}\right)\left(\frac{a^2}{r^2}\right) - \left(\frac{\sigma_r^0 - \sigma_\theta^0}{2}\right)\left(\frac{3a^4}{r^4}\right)$$

$$\sigma'_z = -(\nu)\left(\frac{\sigma_r^0 - \sigma_\theta^0}{2}\right)\left(\frac{4a^2}{r^2}\right) \qquad (4)$$

$$\tau'_{r\theta} = \tau_{r\theta}^0\left(-\frac{3a^4}{r^4} + \frac{2a^2}{r^2}\right)$$

$$\tau'_{rz} = \tau_{rz}^0\left(-\frac{a^2}{r^2}\right)$$

$$\tau'_{\theta z} = \tau_{\theta z}^0\left(\frac{a^2}{r^2}\right)$$

where the superscript ()° refers to the pre-hole condition. These stresses satisfy boundary conditions at the hole wall (r = a) and at infinity. All changes approach zero with large r. Tractions on the hole wall are equal but opposite in sign to the corresponding pre-hole stresses.

The displacement changes relative to (rθz) are

$$u'_r = \left(\frac{1}{2G}\right)\Big\{\left(\frac{\sigma_r^0 + \sigma_\theta^0}{2}\right)\left(\frac{a^2}{r^2}\right) + \left(\frac{\sigma_r^0 - \sigma_\theta^0}{2}\right)\Big[(1-\nu)\left(\frac{4a^2}{r}\right) - \left(\frac{a^4}{r^3}\right)\Big]\Big\}$$

$$v'_\theta = \left(\frac{1}{2G}\right)\Big\{(\tau_{r\theta}^0)\Big[(1-2\nu)\left(\frac{2a^2}{r}\right) + \left(\frac{a^4}{r^3}\right)\Big]\Big\} \qquad (5)$$

$$w'_z = \left(\frac{1}{G}\right)\Big\{(\tau_{rz}^0)\left(\frac{a^2}{r}\right)\Big\}$$

where G = shear modulus, and ν = Poisson's ratio.

4 ALTERNATIVE SOLUTION

In the writer's view, the total or post-hole displacements obtained by adding the changes to the pre-hole displacements are given by

$$u_r = \left(\frac{1}{2G}\right)\Big\{\left(\frac{\sigma_r^0 + \sigma_\theta^0}{2}\right)\Big[\left(\frac{a^2}{r}\right) + r\left(\frac{1-\nu}{1+\nu}\right)\Big] + \left(\frac{\sigma_r^0 - \sigma_\theta^0}{2}\right)\Big[r + (1-\nu)\left(\frac{4a^2}{r}\right) - \left(\frac{a^4}{r^3}\right)\Big] - \left(\frac{\nu}{1+\nu}\right)r\sigma_z^0 + [z\tau_{rz}^0]\Big\}$$

$$v_\theta = \left(\frac{1}{2G}\right)\Big\{(\tau_{r\theta}^0)\Big[r + (1-2\nu)\left(\frac{2a^2}{r}\right) + \left(\frac{a^4}{r^3}\right)\Big] + [z\tau_{\theta z}^0]\Big\} \qquad (6)$$

$$w_z = \left(\frac{1}{2G}\right)\Big\{2\tau_{rz}^0\left(r + \frac{a^2}{r}\right) + \left(\frac{z}{1+\nu}\right)\Big[\sigma_z^0 - \nu(\sigma_x^0 + \sigma_y^0)\Big] - [r\tau_{rz}^0]\Big\}$$

These results differ from the conventional relationships for the total or post-hole displacements in the shear stress terms in brackets on the right hand side of Eqs (6). These terms are missing in the conventional formulas. The expressions (6) indicate that the radial and circumferential displacements depend on z as well as on r and θ , unlike the conventional formulas. The expression for the axial displacement differs from the conventional expression in magnitude only.

5. ALTERNATIVE INITIAL DISPLACEMENTS

The reason for the difference between Eqs (6) and those given by, say, Oka and Bain (1970) must lie in the displacement field associated with pre-hole or in situ stress state. The writer's in situ displacement field relative to (xyz) is

$$
\begin{Bmatrix} u^o_x \\ v^o_y \\ w^o_z \end{Bmatrix} = \begin{bmatrix} S^o_x/E & \tau^o_{xy}/2G & \tau^o_{xz}/2G \\ \tau^o_{xy}/2G & S^o_y/E & \tau^o_{yz}/2G \\ \tau^o_{xz}/2G & \tau^o_{yz}/2G & S^o_z/E \end{bmatrix} \begin{Bmatrix} x \\ y \\ z \end{Bmatrix} \quad (7)
$$

where
$$S^o_x = \sigma^o_x - \nu(\sigma^o_y + \sigma^o_z)$$
$$S^o_y = \sigma^o_y - \nu(\sigma^o_z + \sigma^o_x)$$
$$S^o_z = \sigma^o_z - \nu(\sigma^o_y + \sigma^o_x)$$

and E = Young's modulus. In cylindrical coordinates, one has, instead of Eqs (7)

$$
\begin{Bmatrix} u^o_r \\ v^o_\theta \\ w^o_z \end{Bmatrix} = \begin{bmatrix} S^o_r/E & \tau^o_{r\theta}/2G & \tau^o_{rz}/2G \\ \tau^o_{r\theta}/2G & S^o_\theta/E & \tau^o_{\theta z}/2G \\ \tau^o_{rz}/2G & \tau^o_{\theta z}/2G & S^o_z/E \end{bmatrix} \begin{Bmatrix} r \\ - \\ z \end{Bmatrix} \quad (8)
$$

where (-) means zero. The rigid body motions have been removed from Eqs (7), (8), and (9) following.

6 CONVENTIONAL INITIAL DISPLACEMENTS

The initial displacements implied by the conventional relationships may be determined by subtracting the displacement changes from the total displacements as given, for example, by Oka and Bain (1970). The results are

$$
\begin{Bmatrix} u^o_x \\ v^o_y \\ w^o_z \end{Bmatrix} = \begin{bmatrix} S^o_x/E & \tau^o_{xy}/2G & 0 \\ \tau^o_{xy}/2G & S^o_y/E & 0 \\ \tau^o_{xz}/G & \tau^o_{yz}/G & S^o_z/E \end{bmatrix} \begin{Bmatrix} x \\ y \\ z \end{Bmatrix} \quad (9)
$$

which clearly reveal the tacit plane assumption on the initial or in situ displacements with respect to the z-direction. Since the z-direction is arbitrary, such an assumption is generally not tenable.

7 EXPERIMENTAL TEST

One is naturally reluctant to challenge 25 years of conventional wisdom with no more than an alternative analysis. An experimental test was therefore sought. A physical test of the competing hypotheses represented by Eqs (7) and (9) is desirable, but difficult (Larson, 1987). However, it is a simple matter to do a test on the computer with the aid of a finite element program. In this regard, it is not necessary to determine the total or posthole displacements by simulating the overcoring operation because the difference is in the pre-hole displacements, not in the changes. One need therefore only load a block of undrilled ground in some arbitrary manner and then compare the experimental results with the two sets of competing calculations.

The computer experiment is ideal and produces exact results to within numerical accuracy which is quite sufficient for the test. If on loading a cube with the origin fixed at the center and rigid body rotation prevented, one specifies all stresses to be zero except the z-direction shears as shown in Figure 1, then according to conventional wisdom, Eqs (9), the x- and y-direction displacements should be zero everywhere. But according to Eqs (7), they should vary linearly with z. Also, according to Eqs (7), the z-direction displacement should vary linearly with x and y but only at one-half the rate given by conventional wisdom, Eqs (9). Thus, according to conventional wisdom

$$
\begin{Bmatrix} u^o_x \\ v^o_y \\ w^o_z \end{Bmatrix} = \begin{bmatrix} 0 & 0 & 0 \\ 0 & 0 & 0 \\ \tau^o_{xz}/G & \tau^o_{yz}/G & 0 \end{bmatrix} \begin{Bmatrix} x \\ y \\ z \end{Bmatrix} \quad (9')
$$

According to the alternative solution

$$
\begin{Bmatrix} u^o_x \\ v^o_y \\ w^o_z \end{Bmatrix} = \begin{bmatrix} 0 & 0 & \tau^o_{xz}/2G \\ 0 & 0 & \tau^o_{yz}/2G \\ \tau^o_{xz}/2G & \tau^o_{yz}/2G & 0 \end{bmatrix} \begin{Bmatrix} x \\ y \\ z \end{Bmatrix} \quad (7')
$$

With the finite element results indentified as experimental and the values from Eqs (7) and (9) as calculated, one can ascertain from inspection of the computer output that the calculated results from the alternative solution are almost in exact agreement with experiment, while those calculated from conventional wisdom show no agreement whatsoever. This may also be seen in regression data of calculated on experimental displacements in Tables 1 and 2.

Table 1. Regression data--alternative solution[*]

Displacement	Intercept	Slope	Correlation Coefficient
u^o	-0.000 020	0.970 468	0.999 998
v^o	-0.000 020	0.971 381	0.999 998
w^o	-0.000 016	1.030 854	1.000 000

Table 2. Regression data--conventional solution[*]

Displacement	Intercept	Slope	Correlation Coefficient
u^o	0	0	0
v^o	0	0	0
w^o	-0.000 032	2.061 708	1.000 000

[*]Based on 729 points

Table 1 shows the regression data obtained from the alternative analysis. The fit of the alternative solution is not quite exact to six decimal places because of a slight rigid body rotation that appears in the computer output. In order to prevent rigid body rotation under entirely traction boundary conditions, one needs to specify conditions on displacements derivatives which cannot be done directly with the finite element code used (UTAH-III). Although one could extract the rigid body parameters from the data and therefore the rigid body rotation displacements, they are several orders of magnitude less than the deformation displacements. The case for the alternative solution over the conventional solution is so clear that such a refinement seems quite unnecessary. Although the z-direction displacement from either solution correlates well with the experimental data, the slope of the regression line using the conventional solution is twice what it should be, as seen in Table 2.

8 CONSEQUENCES

The erroneous conventional solution for the displacements relieved by overcoring the pilot hole appears to have consequences for displacement gages because the radial displacement is, in fact, influenced by the z-direction shear stresses. This may be seen by evaluating the radial displacement at the hole wall. Thus, the radial displacement at the hole wall is

$$
u_a = \left(\frac{a}{E}\right)\left\{\left(\sigma_x^0 + \sigma_y^0\right) + 2(1 - \nu)\left[\left(\sigma_x^0 - \sigma_y^0\right)\cos(2\theta)\right.\right.
$$
$$
\left.\left. + 2\tau_{xy}^0\sin(2\theta)\right] - \nu\sigma_z^0\right\} \tag{10}
$$
$$
+ \left(\frac{z}{2G}\right)\left[\tau_{xz}^0\cos(\theta) + \tau_{yz}^0\sin(\theta)\right]
$$

In this regard, the usual practice of computing the change in pilot hole diameter by simply doubling the radial displacement is not generally valid. The proper method is to compute the relative displacement between opposite points on the hole wall. As luck would have it, the net contribution of the z-direction shear stresses to the diameter change still vanishes. Thus, the alternative solution has no consequences for diametral displacement gages such as the USBM borehole deformation gage. Fortunately, the error in the original solution and the error in computing diameter changes nullify each other.

Interpretation of data from gages that respond to axial displacements is in error using the conventional solution. The correct axial displacement of the borehole wall is

$$
w_a = \left[\frac{3a(1 + \nu)}{E}\right]\left[\tau_{xz}^0\cos(\theta) + \tau_{yz}^0\sin(\theta)\right]
$$
$$
+ \left(\frac{z}{E}\right)\left[\sigma_z^0 - \nu\left(\sigma_x^0 + \sigma_y^0\right)\right] \tag{11}
$$

The conventional solution overestimates the shear contribution to the axial displacements by 33 percent. A coefficient of 4 instead of 3 appears in the conventional solution.

9 CONCLUSION

The analytical basis for interpreting in situ stress measurement data has been in error for 25 years. Fortunately, compensating errors have rendered the practical consequences nil for most stress measurement devices used in overcored pilot holes. These conclusions are based on a detailed

investigation of the isotropic case. Although the consequences for the anisotropic case remain to be quantified, the procedure is essentially the same.

REFERENCES

Amadei, B. 1983. Rock anisotropy and the theory of stress measurements. Berlin: Springer Verlag.

Borsetto, M., Martinetti, S. and Ribacchi, R. 1984. Interpretation of in situ stress measurements in anisotropic rock with the doorstopper method. Rock Mechanics and Rock Engineering, Vol. 17, pp. 167-182.

Bonnechere, F. 1969. La cellule "Universite de Liege" de mesure des deformations d'un forage. International Symposium on the Determination of Stresses in Rock Masses, Lisbon, pp. 300-306.

Bonnechere, F. J. and Cornet, F. H. 1977. In situ stress measurements with a borehole deformation cell. International Symposium on Field Measurements in Rock Mechanics, Zurich, pp. 151-159.

Duncan Fama, M. E. and Pender M. J. 1980. Analysis of the hollow inclusion technique for measuring in situ rock stress. International Journal of Rock Mechanics, Mining Sciences, & Geomechanics Abstracts, Vol. 17, pp. 137-146.

Fairhurst, C. 1964. Measurement of in situ rock stresses, with particular reference to hydraulic fracturing. Felsmechanik und Ingenieurgeologie, Vol. II. pp. 129-147.

Fairhurst, C. 1968. Method of determining in-situ rock stresses at great depths. Missouri River Division, Corps of Engineers, Technical Report No. 1-68, Omaha, Nebraska.

Herget, G., Miles, P. and Zawadaski, W. 1977. Equipment and procedures to determine ground stresses in a single drill hole. Canmet Report 78-11.

Hiramatsu, Y. and Oka, Y. 1962. Stress around a shaft or level excavated in ground with a three-dimensional stress state. Memoirs of the Faculty of Engineering, Kyoto University, Vol. 24, Part I, pp. 56-76.

Hiramatsu Y. and Oka, Y. 1968. Determination of the stress in rock unaffected by boreholes or drifts, from measured strains or deformations. International Journal of Rock Mechanics and Mining Sciences, Vol. 5, pp. 337-353.

Hiroshima, K. and Koga, N. 1977. Determination of stresses in anisotropic elastic medium unaffected by boreholes from measured strains or deformations. International Symposium on Field Measurements in Rock Masses. Zurich. pp. 173-182.

Kanagawa, T., Hibino, S., Ishida, T. Hayashi, M., and Kitahara, Y. 1986. In situ stress measurements in the Japanese Islands: over-coring results from a multi-element gauge used at 23 sites. International Journal of Rock Mechanics, Mining Sciences, & Geomechanics Abstracts, Vol. 23, No. 1, pp. 29-39.

Larson, M. K. 1987. In situ stress measurement in Utah coal mines. M.S. Thesis, University of Utah.

Leeman, E. R. 1964. The measurement of stress in rock--Part I., II, III. Journal of the South African Institute of Mining and Metallurgy, Vol.

65, pp. 48-81, 82-114, 254-284.

Leeman, E. R. 1967. The borehole deformation type of rock stress measuring instrument. International Journal of Rock Mechanics and Mining Sciences, Vol. 4. pp. 23-44.

Leeman, E. R. 1968. The determination of the complete state of stress in a single borehole--laboratory and underground measurements. International Journal of Rock Mechanics and Mining Sciences, Vol. 5, pp. 31-56.

Martinetti, S. and Ribacchi, R. 1974. Result of state-of-stress measurements in different types of rock masses. Proc. 3rd Congress of the International Society for Rock Mechanics, Vol. II-A, Denver, pp. 458-463.

Morgan, H. S. 1982. Analysis of borehole inclusion stress measurement concepts proposed for use in the waste isolation pilot plant in situ testing program. Sandia Report Sand 83-1192.

Merrill, R. H. 1967. Three-component borehole deformation gage for determining the stress in rock. U.S. Bureau of Mines R.I. 7015.

Moody, W. T. 1968. Contribution to the discussion of E. R. Leeman's paper International Journal of Rock Mechanics and Mining Sciences. Vol. 5, 00. 123-125.

Niwa, Y. and Hirashima, K. 1971. The theory of the determination of stress in an anistropic elastic medium using an instrumented cylindrical inclusion. Memoirs Faculty of the Faculty of Engineering. Kyoto University, Vol. 33, pp. 221-232.

Oka, Y. and Bain, I. 1970. A means of determining the complete state of stress in a single borehole. International Journal of Rock Mechanics and Mining Sciences, Vol. 7, pp. 503-515.

Panek, L. A. 1966. Calculation of the average ground stress components from measurements of the diametral deformation of a drill hole. U.S. Bureau of Mines R.I. 6732.

Rahn, W. 1984. Stress concentration factors for the interpretation of "doorstopper" stress measurements in anisotropic rocks. International Journal of Rock Mechanics, Mining Sciences, & Geomechanics Abstracts, Vol. 21, No. 6, pp. 313-326.

Ribacchi, R. 1977. Rock stress measurements in anisotropic rock masses. International Symposium on Field Measurements in Rock Masses, Zurich, pp. 183-196.

Rocha, M. and Silverio, A. 1969. A new method for the complete determination of the state of stress in rock masses. Geotechnique, vol. 19, No. 1, pp. 116-132.

Rocha, M., Silverio, A., Pedro, J., and Delgado, J. 1974. A new development of the LNEC stress tensor gauge. Proc. 3rd Congress of the International Society of Rock Mechanics, Denver, Vol. 2A, pp. 464-467.

Van Heerden, W. L. 1983. Stress strain relations applicable to overcoring techniques in transversely isotropic rock. International Journal of Rock Mechanics and Mining Sciences and Geomechanics Abstracts, Vol. 20, pp. 277-282.

ACKNOWLEDGEMENT

The financial support of the Utah Mineral Leasing Fund and the McKinnon Chair is gratefully acknowledged.

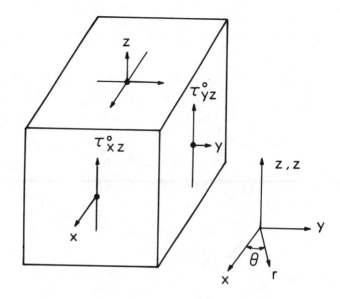

Figure 1. Coordinates and block loaded in shear.

Forecasting and combating rockbursts: Recent developments

Prédiction des coups de toit et leur prévention: Développements récents
Vorhersage von, und Kampf gegen Gebirgsschläge: Neue Entwicklungen

I.M.PETUKHOV, Prof.Dr., Techn. Sc., VNIMI, Leningrad, USSR

ABSTRACT: A summary is given of recent work carried out by members of the Working Group on rockbursts of the International Bureau for Strata Mechanics.

RESUME: On fait un résumé du travail sur les coups de terrain que les membres du groupe de travail du Bureau International de Mécanique des Terrains ont récemment exécuté.

ZUSAMMENFASSUNG: Dieser Bericht ist eine Zusammenfassung der Arbeit die die Arbeitsgemeinschaft des Internationalen Büros für Gebirgsmechanik über Gebirgschläge neulich gemacht hat.

The problem of rock burst prevention in mineral mining in many countries exists already about two hundred years. It is to be supposed,that unusual rock pressure manifestation,just as rock outbursts,thrusts etc.had occurred ever since mineral deposit mining has been initiated,especially in mountainous regions,where,as we know from present-day practice,high tectonic stresses and rock bursts or their signs have been noted.In some cases such dynamic phenomena manifest at mining depth of 50 m - 100 m. The necessity of solving the rock burst problem became especially acute in the past fifties due to increasing the depth of mining. Long-term practice of deposit mining is rich in heavy rock failure, even catastrophic failure in some ore- and coal mines with complete or partial ceasing the mining operations.

Wide experience in mining the hazadous deposits has been gained in ore- and coal mines in Belgium,Great Britain,GDR,India, China,Poland,Czechoslovakia,France,FRG, South Africa and in other countries.

In the USSR the first rock bursts had occurred in coal deposits in the fourties, in ore deposits - in the sixties of the XX-th century.

Many countries pay great attention to solving the problem of rock burst prevention,because it has not only scientific and economical importance,but also a major social significance.

A special interest in this problem arose, in particular,since 1977,when in International Bureau of Strata Mechanics the Working Group on rock bursts has been formed, the members of which are representatives from GDR,India,Poland,USSR,Czechoslovakia, France and FRG. Due to its activities the Group succeeded in formulating the uniform principles in the problem of rock burst prevention,in putting in good order collection of information and particulars on rock burst occurrence by filling in a proper rock burst record form,as well it managed to make up "Annotated Bibliography on rock bursts (1900-1979)"/4/(will be published by Balkema Publishers) and to set up the catalogue of equipment for rock burst prediction (is being published in Czechos-

lovakia).

A wide exchange of experience in solving the problem of rock burst prevention in various countries takes place at the meetings of the Group.Summing-up the ten-years activities of the Group on rock bursts,one can say,that it was rather useful. Prof.,Dr.techn.sc.A.Kidybinski and Dr.Ing. M.Kwasnewsky,as the competent leaders of International Bureau of Strata Mechanics, have made a plenty of work.

Let us consider up-to-date state of solving the problem of rockburst prevention in ore- and coal mines in the USSR. Comprehensive mine,laboratory and analytical investigations,as well extensive experimental works,carried-out during the past thirty-five years,permitted practically to reduce hazard from rock burst manifestation in coal mines. In ore deposits,however,where investigations are being carried-out only for recent ten years,complete solving the problem of rock burst prevention is still remote.

At the beginning of the fifties the author submitted the working hypothesis upon nature of rock bursts, according to which all the system "coal seam - surrounding rocks" takes part in initiation and manifestation of rock bursts.

Rock burst occurs due to exceeding the rate of load pressure over the rate of stress relaxation in coal seam.The energy of rock burst consists of energy of elastic compression of faulted coal seam and strain energy of surrounding rocks /1/. The later long-term investigations allowed to confirm this hypothesis and on its basis to develop modern theory of rock bursts /1,2/.Without detailed consideration of aspects of the rock burst theory it is necessary to note only interrelation of nature of rock bursts and thrust-type deformations of rock mass /1,3/. Block system of rock mass in the earth's crust is in a constant interaction and movement along boundaries of its blocks. Rock movements occur like thrusts along relaxation surfaces and on initiated fractured zones.

Thrust-type deformations of block rock mass naturally occur in the earth's crust, as well under strain conditions,induced by

mining operations,and manifest as thrusts, tectonic shocks,outbursts etc.
Where and why do thrust-type movements occur?

Thrusts occur ahead of stoping face and in hard rocks of roof due to movements along contact rocks and bending fracture surfaces. Heavy thrusts occur in rock mass in the process of mining operations,often at a considerable extent from mining site. Here, as an example,one can make mention of coal mining in Tkibuli-Shaorsk coal field,where thrusts are regarded as weak earthquakes. Heavy thrusts more often occur in ore deposit mining. Stress relief of rock mass in the direction of mined-out space is accompanied by thrust-type movements along contact boundaries of blocks.

Seismic energy of thrusts recorded with special microseismologic stations was

within the range of $10^2 - 10^9$ J. Thrusts with seismic energy below 10^2 J more often occur, but they can not be recorded with these stations. Due to thrusts uneven deformation of rock mass occurs induced by worked-out space,especially in ore deposits under strain conditions,when horizontal stresses are main components of principal stresses. Initiation of numerous thrusts is due to that complicated process. Thrust-type movements induce the initiation of movements in other parts. All this confirms,that block rock mass constantly changes its state of stress,adapting itself to a new stress state and trying to make it constant. Effects of waves induced by thrusts may manifest in two ways:
- roof caving and wall failure due to unsupported roof or with roof and wall bolting to be poor;
- rock burst occurence,when hazadous conditions had already existed: rock burst would occur,but a little later.

Factors mentioned above allow to explain the nature of rock bursts as follows.
R o c k   b u r s t is the result of rock failure under strain conditions induced by mine workings,when the rate of stress change exceeds the rate of stress relaxation. Energy balance consists of energy of elastic compression of coal seam and strain energy of surrounding rocks.

Rock burst has some phases in its occurrence,i.e.
- phase,preceding the rock burst,when block rock mass is stress-strain state;
- phase of strength loss,when released energy of surrounding rocks exceeds energy consumption under strain conditions;
- phase of wave propagation of dynamic pressure - spontaneous increasing the rock failure zone and its decreasing.

On the basis of theoretical knowledge of nature of rock bursts their classification was formulated. According to this classification rock bursts are subdevided into five categories.

R o c k   o u t b u r s t - brittle fracture of rocks (ore,coal) on exposed surface in the form of violent torning off the lens like plates with strong sound. The volume of rock failure,as a rule,does not exceed $0,5$ m^3.

M i c r o b u r s t - brittle fracture of edges of pillar or coal seam at the depth of 1 m below exposed surface in the form of

throwing out rocks (ore,coal) into mine workings with total volume of 2 - 3 m^3 without harmful aftereffects in mining operations. Microburst is accompanied by sharp sound,slight shake of rock mass and dusting.

S e l f-b u r s t (or destructive rock burst) - brittle fracture of pillar, its part or part of rock mass at the depth of above 1,0 m below exposed surface with throwing out rocks up to 3 m^3 and with harmful failure aftereffects,interrupting the mining operations. Rock burst is accompanied by strong sound,heavy shaking, a large amount of dust and air wave. Energy level of rock failure is more than 100 J.

T h r u s t - brittle fracture inside rock mass without rock failure in the vicinity of mine workings. Thrust is accompanied by rock shaking and strong sound. It is possible dusting,fracturing the concrete supports and rock caving in poor supported areas.

T e c t o n i c   r o c k b u r s t - brittle fracture inside rock mass in the form of thrusts,inducing rock failure in pillars and seam edges abbutting the mine workings.

Special commission is engaged in investigating only self-bursts and tectonic rock bursts. Other categories of rock bursts characterize mainly hazardous state of rock mass.Hence it is necessary to systematize rockbursts and not less than once a month to analyze them with the view of taking measures for their prevention.

We'll have more clear idea of nature of rockbursts and their energy level,if one would regard rock mass,from one hand, as a single system having a certain strain energy,variation of which is induced by mining operations,and from other hand, as block system,stress-strain state of each block depends on its location in rock mass and its interaction with adjoining blocks.

Theoretical and practical basis for prediction and prevention of rock bursts is geodynamic zoning of mineral deposits (mine fields),including study of block structure of rock mass,its initial stress state,tectonic stresses and use this information in designing,building and exploitation of mining ventures.This new branch of mining science has been developed recently in the USSR; it concerns first of all ore deposits /3/.

We distinguish three types of forecasting the rock bursts /1,2/:
- determination of hazadous deposits, coal seams;
- regional forecast of degree of rockburst hazard within the deposit (mine field);
- prediction of degree of rockburst hazard in operative stoping faces and development workings.

Timely definition of hazadous deposit (coal seam) allows in good time to provide safety measures with low operating costs. Tendency of a deposit to be prone to rock bursts depends on stress state of undisturbed rock mass and brittle fracture of rocks induced by mine workings. Up to now in the USSR about 70 ore deposits and 847 coal seams emerged to be prone to rockbursts.

Regional prediction of rock bursts is carried-out by means of seismologic stations Coordinates of potential thrusts and rock burst shocks,their seismic energy and hazadous rock within the deposit(mine field) are defined.

Prediction of degree of rock burst hazard in certain sections is performed only in dangerous zones,defined by regional pre - diction. Systems for continuous automated control of rock burst hazard are being developed.

In the process of deposit mining the service of rock burst prediction is en - gaged in determination of degree of rock burst hazard in some zones of rock mass in the vicinity of mine workings.The matter of the service of rock burst prediction is to find hazadous rock. Degree of rockburst hazard is evaluated by defining the dis- tance up to zone of virgin (initial) pressure at a given point and by high load pressure.Intensity of load pressure is de- termined by values of mineral fines out- crop,power resistance,core disc diameter etc.

Hence,the service of rock burst prediction in operative coal-and ore mines reveals the place of potential rock burst manifes- tation,defines the efficiency of methods for rock burst prevention and thereby ensures safe and no-failure mining opera- tions.

This approach to rock burst prediction has been successfully applied in coal mines over thirty years and in ore mines - over ten years in the USSR. Now the specialists in VNIMI are engaged in developing the new methods for rock burst prediction and equipment with recording facilities of high frequency acoustic emission,both initial and induced,and natural impulse electro- magnetic radiation.

Magnificent basis for safe mining of a de- posit prone to rock bursts is as follows:

1. Foundation of the service for predic- tion and prevention of rock bursts,which ensures:
   - timely reveal of hazadous rock, coal seams,ore bodies;
   - .prediction of degree of rock burst hazard in definite sections;
   - taking the preventive measures to reduce hazardous state;
   - evaluation of efficiency of safety measures to be taken;
   - well-grounded selection of methods for safe mining operations.

2. Relief of gas- and rock pressure in rock mass by means of:
   - deviding a deposit into mine fields, adopting the calendar plan of their mining,that ensures regular mining operations and prevents initiation of high stress concentrations;
   - necessary use of advance mining of protecting layers;
   - mining without pillar leaving;
   - reducing a number of mine workings ahead of stoping face;
   - elimination of oncoming and overtaking fronts of room works;
   - mine workings would be driven in the di- rection of high stress concentrations.

3. Minimizing the ability of coal seam edges to store strain energy by blasting, water injection,discharge boreholes, relief slots etc.

4. Lessening the rock burst hazard by appropriate selection of safe and effi- cient width of entrap of mining machines and their velocity in mining operations.

5. Protection of men and mine workings against rockburst aftereffects by:
   - limiting rock burst intensity by means of protective shields,metal supports,

drilling and blasting operations,combined machines etc.
   - complete ceasing the drivage of hazar- dous mine workings.

6. Use of rock pressure energy of rock mass to facilitate mine working drivage by:
   - remote control of mining machines with permissible rock burst manifestation without interrupting mining operations;
   - cutting of separate blocks and subse- quent self-failure of rocks;
   - remote rock failure by resonance vibra- tions,rope saws etc;
   - laying-out the mine workings,taking into account inherent stress field in undis- turbed rock mass and degree of rock burst hazard.

7. Continuous improvement a set of measu- res for prevention of rock bursts and out- bursts,timely and efficient application of them by:
   - organizing the permanent commissions of representatives from mining plants, scientific institutions,Mine Technical Inspection,who will solve the problems of mining in severe conditions;
   - coordination of work of designers and scientists,engaged in solving the prob- lem of rock burst prevention.

All these measures for safe mining are used in dependance on certain geological and mining conditions,nature and manifes- tation of dynamic phenomena /1,2/.

In ore-and coal mines,where heavy thrusts may occur,it is necessary to take safety measures to avoid their harmful effects on mine workings.

One of obligatory requirements for safe and efficient operation of mine workings is that they must be reliably supported. Permanent tightening or metal safety nets should be provided within the supports to keep rock pieces from falling-out.

When it is predicted that the whole working or its parts are hazadous,all the preventive measures must be timely taken. Requirements for supports in such workings are quite the same as for supports at all.

Let us refer to the experience,that coal miners have acquired in Tkibuli-Shaorsk coal field. Coal seam under the name of "Thick",consisting of four heavy beds,was being regionally wetted through boreholes from water entry. During past ten years there was no failure of rocks,when weak earthquakes had occurred,although every month from 6 to 20 weak earthquakes occur with mean seismic energy within the range of $3 \cdot 10^5 J$ to $5 \cdot 10^8 J$.

The practice of successful prevention of rock bursts in Tkibuli-Shaorsk coal field, use of proper arched and iron-ring supports to protect mine workings against harmful effects of thrusts,attracts great attention.

Thus,timely prediction of degree of rock burst hazard in mine working,induced by thrusts,proper supports in all workings within the mine field,prevention of rock peaces from falling out during thrust occurrence - all these measures provide safe and no-failure mining operations.

In conclusion the author would like to draw attention to the necessity of wide use of geodynamic zoning of mineral deposits. This new branch of mining science may be of great importance for many other pur - poses,e.g.for building the hydrotechnical underground and surface constructions,for townbuilding, for exploration and mining the liquid and gaseous deposits.

Bibliography

Petukhov,I.M. 1972. Rock bursts in coal
    mines. Moscow, Nedra.
Petukhov,I.M. & Yegorov,P.V. & Vinokur,B.Sh.
    1984. Prevention of rock bursts in ore
    mines. Moscow, Nedra.
Recommendations for rock burst prevention
    with due account of geodynamic phenomena
    of deposits. 1983. Leningrad,VNIMI.
Annotated Bibliography on rock bursts
    (1900-1979). 1982,1983. Leningrad,VNIMI.

Excess Shear Stress (ESS): An engineering criterion for assessing unstable slip and associated rockburst hazards

Contrainte de cisaillement en excès (ESS): Un critère technique pour évaluer les glissements instables et risques associés aux coups de charge
Die überschüssige Scherspannung: Ein technischer Kennwert, um labilen Schlupf und die damit verbundene Gebirgsschlaggefahr abzuschätzen

J.A.RYDER, Rock Mechanics Laboratory, Chamber of Mines of South Africa, Research Organization, Johannesburg

ABSTRACT: Excess Shear Stress, a paraphrase of the widely-used Seismological concept of 'stress-drop', appears to have direct control over the occurrence and magnitude of large shear-driven seismic events in mining. It can readily be calculated by standard stress modelling programs. Used as an engineering criterion, it could ultimately assist in the design of safer mining layouts at depth.

RESUME: Contrainte de cisaillement en exces, une paraphrase du concept très usité en sismologie de 'dimunition de contrainte', paraît avoir un control direct sur l' occurence et la grandeur de grands secousses sismiques entrainès par cisaillement dans les operations minieres. On peut la calculer facilement par des programmes actuels de modelage de contrainte. Employé comme un critère technique, en fin de compte on pourrait en profiter dans la planification plus sûre des mines à profondeur.

ZUSAMMENFASSUNG: Die überschüssige Scherspannung, eine Umschreibung des gängigen seismologischen Begriffes 'Spannungsabfall', scheint direkten Einfluss auf Vorkommen und Grösse seismischer Ereignisse zu haben, die durch grosse Scherkräfte bewirkt werden. Die überschüssige Scherspannung kann leicht mit gewöhnlichen Programmen zur Spannungsmodellierung berechnet werden. Als technischer Kennwert könnte sie in der Zukunft dazu beitragen, erhöhte Sicherheit in der Planung tiefer Minen zu erlangen.

1 INTRODUCTION

Rockbursts are a premier hazard in deep hard-rock mining. In the South African goldfields, the concept of Energy Release Rate (ERR) (Cook et al 1966) has become an established and popular measure of the severity of general mining conditions. Certain shortcomings of this elegantly simple concept have, however, started to emerge. ERR does not directly 'see' the hazard-amplifying effects of planes of weakness (faults/joints), inhomogeneities (dykes/sills) or abnormal tectonic stress fields. Yet seismic events associated with such structures seem to account for a majority of damaging rockbursts in geologically-disturbed districts such as the far west and southern portions of the Witwatersrand goldfields. Urgently needed then, is a more physical view of the rupture mechanisms underlying seismic events to supplement the ERR approach in the regional design of deep mines. By drawing on established seismological theory, this paper attempts to construct one such simple model and at the same time introduces the 'new' engineering criterion of Excess Shear Stress (ESS).

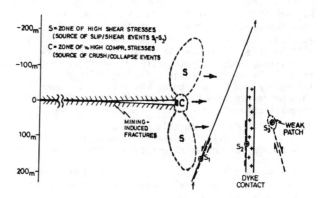

Figure 1. Origin of Crush and Shear-type events near face of advancing isolated stope at depth.

2 CLASSES OF SEISMIC EVENTS

Mining seismologists (e.g. Stiller et al 1983, Rorke & Roering 1984; van der Heever 1984) have proposed that mine seismicity falls into at least two distinct classes: one associated with crushing of highly-stressed volumes of rock nearby a mined-out void, the other with unstable slip or rupture along planes in the rockmass. Figure 1 sketches the possible nature of these classes for the case of an advancing tabular excavation at depth. In front of the immediate face is a zone C characterized by high compressive stress concentrations; the resulting intense fracturing is a (harmless) manifestation of ERR. Occasionally though, a volume of rock may collapse violently as a 'strainburst' or 'outburst' in which severe localized damage may occur. Such Crush-type events were likened by Cook (1965) to the behaviour of brittle rock in a 'soft' testing machine.

Above and below the face in Figure 1 are also large lobes S of shear stress concentrations. As the face advances, these lobes sweep through volumes of intact rock normally leaving no evidence of their passage behind them. Occasionally however, a lobe may impinge on a pre-existing plane of weakness which permits a Shear-type event to occur: sudden and violent slip along the plane (events S1 and S2 in Figure 1). Similarly, if a lobe encounters and activates a suitable patch of relatively weak rock, a damaging rupture event can be triggered creating a new 'fault' in the previously intact rockmass (event S3 in Figure 1). Evidence for the dominating importance of these shear-driven events in the South African goldfields is partly seismological (Spottiswoode 1984) and partly observational (Ortlepp 1978); it is such events which form the topic of this paper.

3 FRICTION AND EXCESS SHEAR STRESS

The level of shear stress required to initiate slip on a pair of contacting rock surfaces is described by the 'static friction' or 'strength' τ_s. This increases linearly (Byerlee 1977) with confining stress σ_N according to the well-known friction law

$$\tau_s \simeq S_0 + \mu_s \sigma_N \qquad (1)$$

where S_0 is a cohesion (which will vary radically along an in-situ fault plane due to local variations in infilling or asperities), and $\mu_s = \tan\phi_s$ is the coefficient of static friction (typical values for quartzite being 0,6; with friction angle $\phi_s = 30°$). Most researchers in earthquake theory agree on the concept of 'dynamic friction' τ_d: a reduced level of frictional resistance which pertains once slip has been mobilized and cohesion-generating asperities have been sheared off:

$$\tau_d = \mu\sigma_N = (\tan\phi)\ \sigma_N \qquad (2)$$

in which cohesion is negligible and a typical dynamic friction angle ϕ is 30°. The average 'stress-drop' between eqs.(1) and (2), illustrated in Figures 2 and 3, has been estimated from mining seismological data (Spottiswoode & McGarr 1975) to range around 0,1 – 10 MPa. Laboratory stress-drops of 5 – 10 % of the τ_s level have been reported, and after slip has ceased the surfaces are said to 'heal', i.e. slowly regain their static strengths (Dieterich 1977). Stress drops are a necessary condition underlying mining shear-type events; just as in Cook's (1965) model of Crush events, violent collapse is impossible without brittle failure and rapid post-failure load-shedding. The peak stress-drop potential $\hat{\tau}_e$ (Figure 3) of a plane of weakness in the rockmass will be affected by the nature of contacting rock asperities, degree of planarity of the fault, and presence of water or gouge. Pending further laboratory and field studies the following values for this parameter have been suggested (Ryder 1986) for quartzitic rock environments at depths to 4 km:

$$
\begin{aligned}
\hat{\tau}_e &\simeq 5 - 10 \text{ MPa for unstable movement on a}\\
&\qquad \text{pre-existing plane of weakness;}\\
&\simeq 20 \text{ MPa for unstable rupture of intact rock} \qquad (3)
\end{aligned}
$$

Excess Shear Stress (ESS = τ_e) represents the nett force available to power a shear-type seismic event once rupture has commenced:

ESS = Shear stress prior to slip – Dynamic friction

$$\text{or} \quad \tau_e = \left|\tau\right| - \mu\sigma_N \qquad (4)$$

in which τ and σ_N are the stress components acting prior to slip, and $\mu = \tan\phi$ is the dynamic coefficient of friction.

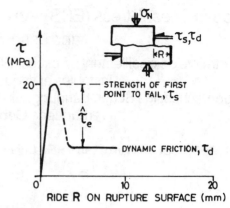

Figure 3. Stress-drop once sufficient stable ride has occurred to precipitate rupture (failure) at a point on a fault surface.

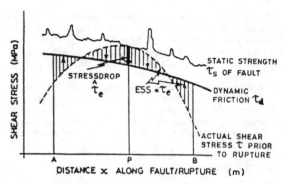

Figure 4. Static and dynamic friction, and driving shear stress on plane of weakness just prior to rupture.

Consider Figure 4 which depicts possible conditions along a typical rupture plane. The static 'strength' τ_s is sketched as the light irregular line; according to eq.(1) this varies smoothly with the level of σ_N but also erratically due to effects of asperities and irregularities of the contacting surfaces. Conversely, the heavy line representing dynamic strength τ_d after fault mobilization, is sketched in Figure 4 as a smooth curve varying only with σ_N acting along the fault. This assumption, of constancy of dynamic coefficient of friction along a rupture plane, is a crucial one for the practical applicability of the ESS concept to mining problems. In this sense, ESS analysis gives a pessimistic upper bound of Shear-type activity: event magnitudes would be significantly reduced if regions of high frictional resistance were to persist under dynamic conditions.

The dashed curve in Figure 4 represents a typical profile of prevailing shear stress τ; i.e. virgin stress plus stresses induced by a nearby slowly advancing mining excavation. If this stress reaches a value equal to the static strength τ_s at some point P along the fault, slip will initiate at this point. By assumption, dynamic friction now pertains and there will be a stress drop, denoted τ_e in Figure 4. This stress drop is unbalanced and therefore forces ride to occur along a small segment of the plane centred at P. This ride in turn generates very large ('infinite') stresses at the tips of the broken segment: stresses sufficient to cause the rupture to grow into regions where the shear stress was previously less than the static strength, even to the extent of rupturing through high-strength asperities on the plane. Eventually, rupture comes to a halt at points A and B where the ESS (prior to slip) is sufficiently negative to inhibit further movement. This process is sketched in Figure 5 at a series of time instants during the genesis of the slip event.

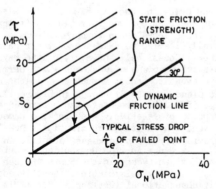

Figure 2. Friction as function of normal stress σ_N.

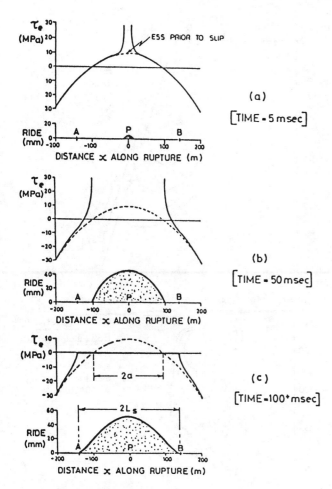

Figure 5. Schematic ESS and Ride distributions along plane during rupture event, calculated quasi-statically. Final distribution (c) represents possible static equilibrium if effects of dynamic overshoot are small.

The final ride distribution illustrated in Figure 5 was calculated purely from knowledge of the shaded ESS distribution in Figure 4 by applying the fact that for static equilibrium, high stresses cannot persist at the tips of the ruptured part of the plane. This implies low or, for convenience, zero (Nur 1977) rates of change of ride at these tips. Figures 5 and 6 were calculated in terms of this static equilibrium precept.

An underlying assumption here is that forces and energies needed to propagate the rupture are small in relation to other forces and energies involved (Dmowska & Rice 1983). Dilatation effects during movement on the fault have also been ignored - under reasonable assumptions, these can be shown to reduce ride by 10 % at most. A more profound simplification is the neglect of dynamic overshoot effects, which have been shown by advanced numerical modelling to enhance rides by about 20 % only (Aki & Richards 1979), and which in any event are partially cancelled by the first two simplifications listed above.

It follows then that ESS is truly the force powering, controlling and bringing to a halt, dynamic slip events. It can readily be calculated in a mining context by standard stress modelling programs. The complexities of static strength variations along a plane (light irregular line in Figure 4) which would be impossible to quantify in practice, fortunately become irrelevant once rupture has initiated and assumed constant-ϕ dynamic conditions pertain. According to the techniques outlined below, knowledge of an ESS distribution suffices to determine extent, potential magnitude, and somewhat less precisely, probability of seismic shear activity.

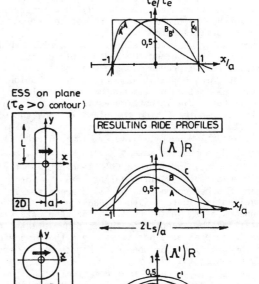

$$\Lambda = G/2\,(1-\nu)\hat{\tau}_e a;\ \Lambda' = (1-\tfrac{\nu}{2})\Lambda;\ \nu = \text{POISSON's RATIO}$$

2D	$M_0/(\hat{\tau}_e a^2 L)$	$(\Lambda)\hat{R}$	$\bar{\tau}_e/(\hat{\tau}_e)$	$L_s/(a)$
A	3,97	0,75	0,60	1,49
B	5,03	0,94	0,67	1,41
C	5,03	1,00	1,00	1,00

$\bigcirc^r$	$M_0/(\hat{\tau}_e a^3)$	$(\Lambda')\hat{R}$	$\bar{\tau}_e/(\hat{\tau}_e)$	$L_s/(a)$
A'	1,28	–	–	1,23
B'	1,74	0,52	0,57	1,23
C'	2,37	0,64	1,00	1,00

Figure 6. Ride profiles and key seismic parameters for three possible profiles of ESS, where the contour of positive ESS on a given rupture plane approximates a 2D (corresponding to profiles A, B, C) or circular (profiles A', B', C') outline.

4 ESTIMATION OF SEISMIC EVENT PARAMETERS

A robust measure of magnitude of a shear-type seismic event is its 'seismic moment' M_0, defined by

$$M_0 = GV \qquad \text{(MN–m)}$$

$$V = \int \text{Ride.dA} \qquad \text{(m}^3\text{)} \qquad (5)$$

where G is the shear modulus (MPa), and the "volume of ride" $V=\bar{R}A$ where $\bar{R}$ is mean ride (m) and A is the area of the rupture surface (m²). The following widely-used formula (Hanks & Kanamori 1979) relates M_0 to Richter magnitude M:

$$1,5\,M = \log_{10} M_0 - 3,1 \quad (M_0 \text{ in MN–m}) \qquad (6)$$

In principle these quantities, as well as the actual size of the rupture area, can be estimated by numerical modelling of a particular mining scenario if explicit slip (with friction parameters $S_0=0$ and $\mu=\tan\phi$) is permitted to take place along the rupture plane and the ride distribution R noted. Boundary-element models such as MINSIM-D (Ryder & Napier 1985), NFOLD, or BESOL (Crouch 1986) are particularly suitable for this purpose.

A somewhat simpler procedure, which avoids explicit modelling of slip along the rupture plane, is available if the proposed plane does not intersect any mined-out voids and if the modelled distribution of positive ESS prior to slip on the rupture plane approximates a simple outline as in Figure 6. In particular, if the ESS on the plane is constant within a circular outline, the popular Brune model (e.g. Spottiswoode & McGarr 1975, c.f. Figure 6 case C') can be invoked to predict moment M_0, peak ride $\hat{R}$, mean stress-drop τ_e, and rupture half-length L_s of the resulting seismic event. Analytic results for a number of other common ESS profiles are also summarized in Figure 6 (Ryder 1986); note the strong dependence of M_0 on the positive ESS lobe size (a^2 or a^3); and the expectation that rupture dimension L_s is in general somewhat larger than the dimension a of the exciting ESS lobe (c.f. Figure 5(c) where a=100 m and L_s=141 m).

As an example, suppose that numerical modelling of ESS on a plane prior to slip gives a 2D profile similar to ESS profile A (Figure 6) with $\hat{\tau}_e$ = 10 MPa, L = 200 m and a = 50 m. Elastic constants are G = 29 200 MPa and υ = 0,2. Then $M_0 = 3,97 \times \hat{\tau}_e a^2 L = 2 \times 10^7$ MN-m (equivalent to a Richter magnitude M = 2,8); peak ride R = 0,021 m; mean stress drop = 6,0 MPa; rupture half-length L_s = 75 m.

More generally, if a modelled ESS distribution has a complex profile or outline, or if mining transects the rupture plane, the explicit-slip modelling step already mentioned can be used. A 'mined' plane having zero thickness and dynamic friction characteristics is introduced on the rupture plane and explicit slip is permitted to occur. The modelled ride profile will automatically have zero slope at the tips of the rupture, and numerical evaluation of eqs.(5) and (6) then permits the magnitude of the seismic event to be estimated. This, coupled with information on distance to the nearest workings and the quality of support installed, in principle allows an assessment of potential rockburst damage (Wagner 1984) to be carried out.

5 ESS ANALYSES OF SELECTED TABULAR MINING SITUATIONS

A number of mining situations have recently been analysed to estimate the size of mining-induced ESS lobes and the magnitude of associated seismic activity. The simple examples in Figure 7 were evaluated using Salamon's (1968) analytic solutions for a partially closed stope and for an isolated remnant. ESS values on planes at varying inclinations were evaluated analytically, and event magnitudes for given maximum stress drops estimated from the theory underlying Figure 6 case A. The following salient points emerged from a detailed suite of parameter sensitivity runs (Ryder 1986).
- Rupture planes dipping at angles of around $70°$ or $130°$ are most favourable for generating energetic shear activity; this correlates well with field observations of many actual events.
- Reductions in the effective mined stoping width (accomplished for example by extensive placement of stiff backfill) are very beneficial in reducing lobe sizes and event magnitudes. For example, a reduction in stoping width from 1 m to 0,5 m reduces the moment and energy associated with potential seismic events 8-fold.
- Conversely, low dynamic friction angles or small horizontal virgin stress levels lead to similarly massive increases in lobe size and associated event magnitudes.

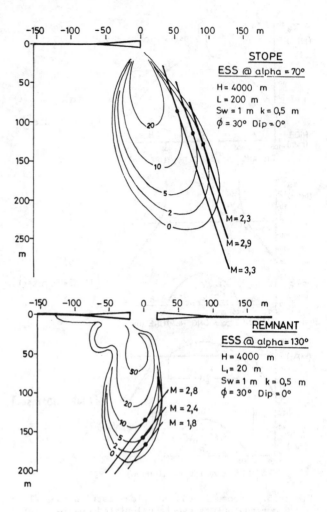

Figure 7. Analytically calculated contours of ESS associated with an isolated 200 m half-span stope (for $70°$ dipping rupture planes), and a 20 m half-width remnant (for $130°$ dips), at 4 000 m depth. Rupture planes and event magnitudes (assuming $L \approx 2L_s$) are illustrated for three possible peak stress drops of 2, 5 and 10 MPa.

Napier (1986) used the MINSIM-D tabular stress program to analyse a number of idealized scattered-mining scenarios involving two reef horizons separated by a loss of ground due to faulting. Briefly, he found that:
- ESS was strongly affected by the strike length of mining on one or both reef horizons, and by areas of intact reef left near the fault boundaries as 'bracket' pillars;
- backfill and dip stabilizing pillars were relatively ineffective in improving ESS stability in this scattered mining situation; and
- sequences of seismic events could occur in certain situations, and that these could be modelled.

Spottiswoode (1986) has correlated general seismicity in a geologically undisturbed mining environment with both ERR and a simplified ESS measure, and found strikingly better agreements with the latter parameter. Other workers (More O'Ferrall 1986) have attempted back-analyses of major events and found, in general, good correlations between areas of positive ESS and event occurrence. An example is given in Figure 8, in which a magnitude 4,5 event on a major fault caused measurable ride dislocations in numerous tunnels transecting the fault as well as widespread damage. These observations correlated quite well with ESS and explicit slip analyses (O'Hare 1986, Ryder 1986).

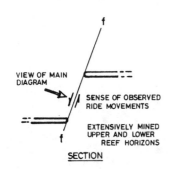

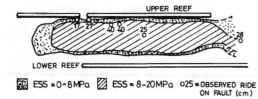

Figure 8. ESS back-analysis of a large (M=4,5) seismic event after extensive mining on either side of a large fault loss.

6 DISCUSSION AND CONCLUSIONS

Further work is required before ESS techniques can confidently be applied to the design of safer mining layouts. It is true that back-analyses seem invariably to reveal zones of mining-induced positive ESS levels in the inferred slip area of actual events. Conversely though, high ESS levels exist in many areas where seismic events have not yet occurred, and where significant seismic activity may never occur. Part of the reason for this is ignorance of the numerical values of certain key parameters: friction properties of specific faults, and detailed in-situ tectonic stress conditions. However, pending the results of further numerical back-analyses, field geotectonic studies and laboratory friction measurements, the following conclusions can reasonably be drawn to describe the present state of the art:
- ESS analyses, given realistic values of fault friction and stress conditions, permit the estimation of an upper bound of the magnitude of potential seismic activity in a given locale. The hazard-amplifying effects of geological weaknesses or tectonic anomalies (e.g. low horizontal virgin stresses) can in principle be evaluated on a more quantitative basis.
- In this light, safer mining layouts and sequences can be evaluated numerically. For example, the merits of leaving dip or strike stabilizing pillars, the staggering of mining sequences, or the use of backfill, can in principle be assessed quantitatively.
- The possibilities of carrying out control measures such as destressing by blasting or the induction of controlled slip on known planes of weakness, can more rationally be explored by prior ESS analyses as well as by careful subsequent seismic and in-situ monitoring. Such techniques hold out hope for a more active role of the rock mechanics engineer in rendering safer and more productive environments in ultra-deep level mining situations.

ACKNOWLEDGEMENT

The work described was carried out as part of the research program of the Chamber of Mines of South Africa. Permission to publish this paper is gratefully acknowledged.

REFERENCES

Aki, K. & P.G. Richards 1979. Quantitative seismology, p.880. San Francisco: W.H. Freeman.
Byerlee, J. 1977. Friction of rocks. In Proc. 2nd conference on experimental studies of rock friction with application to earthquake prediction, U.S.D.I., Office of Earthquake Studies, Menlo Park, p.55-79.
Cook, N.G.W. 1965. A note on rockbursts considered as a problem of stability. J.S. Afr. Inst. Min. Metall. March: 515-523.
Cook, N.G.W., E. Hoek, J.P.G. Pretorius, W.D. Ortlepp & M.D.G. Salamon, 1966. Rock mechanics applied to the study of rockbursts. J.S. Afr. Inst. Min. Metall. May 506-515.
Crouch, S.L. 1986. BESOL proprietary documentation.
Dieterich, J.H. 1977. Time-dependent friction and the mechanics of stick-slip. In Proc. 2nd conference on experimental studies of rock friction with application to earthquake prediction, U.S.D.I., Office of Earthquake Studies, Menlo Park, p.81-115.
Dmowska, R. & J.R. Rice 1983. Fracture theory and its seismological applications. In R. Teisseyre (ed.) Continuum theories in solid earth physics. Poland: Elsevier.
Hanks, T.C. & H. Kanamori 1979. A moment magnitude scale. J. Geophys. Res. 84: 2348-2350.
More O'Ferrall, R. 1986. Procedures for mining in the vicinity of faults and dykes in the Klerksdorp area. In Mining in the vicinity of geological and hazardous structures, S. Afr. Inst. Min. Metall., Mintek, Transvaal.
Napier, J.A.L. 1986. Application of ESS concepts to the design of mine layouts. In Mining in the vicinity of geological and hazardous structures, S. Afr. Inst. Min. Metall., Mintek, Transvaal.
Nur, A. 1977. Nonuniform friction as a physical basis for earthquake mechanics: a review. In Proc. 2nd conference on experimental studies of rock friction with application to earthquake prediction, U.S.D.I., Office of Earthquake Studies, Menlo Park, p.262-264.
O'Hare, M. 1986. Unpublished material.
Ortlepp, W.D. 1978. The mechanism of a rockburst. In Proc. 19th U.S. Symposium on Rock Mechanics, U. of Nevada Reno, p.476.
Rorke, A.J. & C. Roering 1984. Source mechanism studies of mining-induced seismic events in a deep-level gold mine. In Rockbursts and Seismicity in Mines, S. Afr. Inst. Min. Metall., Johannesburg, p.51-56.
Ryder, J.A. 1986. ESS assessment of geologically hazardous situations. In Mining in the vicinity of geological & hazardous structures, S. Afr. Inst. Min. Metall, Mintek, Transvaal.
Ryder, J.A. & J.A.L. Napier 1985. Error analysis and design of a large-scale tabular mining stress analyser. In Proc. 5th Int. Conf. on Numerical Methods in Geomechs., Nagoya, p.1549-1555.
Salamon, M.D.G. 1968. Two dimensional treatment of problems arising from mining tabular deposits in isotropic or transversely isotropic ground. Int. J. Rock Mech. Min. Sci. 5: 159-185.
Spottiswoode, S.M. & A. McGarr 1975. Source parameters of tremors in a deep-level gold mine. Bull. Seism. Soc. Am. 65: 93-112.
Spottiswoode, S.M. 1984. Source mechanisms of mine tremors at Blyvooruitzicht Gold Mine. In Rockbursts and Seismicity in Mines, S. Afr. Inst. Min. Metall., Johannesburg, p.29-38.
Spottiswoode, S.M. 1986. Total seismicity & mine layouts: application of ERR and ESS. In Mining in the vicinity of geological and hazardous structures, S. Afr. Inst. Min. Metall., Mintek, Transvaal.
Stiller, H., E. Hurtig, H. Grosser and P. Knoll 1983. On the nature of mining tremors. J. Earthq. Predict. Res., Tokyo. 2:61.
Van der Heever, P.K. 1984. Some technical and research aspects of the Klerksdorp seismic network. In Rockbursts and Seismicity in Mines, S. Afr. Inst. Min. Metall., Johannesburg, p.349-350.
Wagner, H. 1984. Support requirements for rockburst conditions. In Rockbursts and Seismicity in Mines, S. Afr. Inst. Min. Metall., Johannesburg, p.209-218.

Performance of deep wellbores in rock with a confining pressure-dependent elastic modulus

Comportement des puits profonds dans des roches dont le module élastique dépend de la pression de confinement
Das Verhalten von tiefen Bohrlöchern in Felsgesteinen mit einem Elastizitätsmodul, der vom Seitendruck abhängt

F.J.SANTARELLI, Société Elf Aquitaine (Production), Pau, France (on secondment to Imperial College, London, UK)
E.T.BROWN, Imperial College, London, UK

ABSTRACT: A complete closed form solution is presented for the stresses and displacements induced around an axi-symmetric wellbore whose elastic modulus is a function of the minimum principal stress or confining pressure. The theory which is a generalization of classic linear, constant modulus elastic theory, provides satisfactory explanations of the displacements and apparent strength enhancement recorded at the walls of model wellbores. An important ramification of the theory is the prediction of a failure pattern in which fractures parallel to the borehole wall may initiate some distance inside the surrounding rock.

RESUME: Une solution analytique complète pour les contraintes et déformations est présentée dans le cas d'un puits axisymétrique dans une roche dont le module élastic est fonction de la contrainte principale mineure ou pression de confinement. La théorie qui est une généralisation de l'élasticité linéaire classique permet d'expliquer convenablement les déplacements et résistance anormaux enregistrés a la paroi de modèles réduits de puits. Un aspect important de la théorie est la prédiction d'un mode de rupture ou la fracture peut s'initier à l'intérieur du rocher et se propager parallelement à la paroi du puits.

ZUSAMMENFASSUNG: Es wird eine analytische Auslösung presentiert für Verformungen und Drücken um einem achsen-symmetrishen Bohrloch das in einem Felsgestein gebohrt ist wessen Elastizitätsmodul mit der kleinsten Haupt-spannung oder Seitendruck variiert. Diese Lösung die die klassische lineare Elastizität verallgemeinert, kann die seltsamen Festigkeit und Verformungen erläutern die an den Wänden von kleinmasstäblichen Bohrlochmodelver - suchen vermessen werden. Ein wichtiger Anblick is die Aussage für eine Bruchfläche die in der Felsgestein-innenseite beginnen kann und parallel mit der Lochwand ausgebaut wird.

INTRODUCTION

Wellbore instability is a continuing problem in the oil industry adding substantially to drilling and operating costs. This has led to the establishment of a number of large research programs aimed at im-proving understanding of the initiation and develop-ment of rock failure around wellbores. The study described here forms part of a major industrial research effort being undertaken by Elf Aquitaine (Production) and outlined by Guenot (1987). It is concerned with hole enlargement due to brittle fracture. This is but one of the several modes of wellbore instability identified by Bradley (1979); its occurrence in the field has been widely reported (e.g. Bell & Gough, 1979; Blüming, Fuchs & Schneider, 1983; Teufel, 1985; Klein and Barr, 1986). The principal concern of this paper is with the accurate prediction of the conditions of brittle fracture initiation.

Classical studies of this and related problems usually assume that the rock behaves as a linear elastic material with failure being predicted by com-paring the stresses at the borehole wall, calculated using elastic theory, with the peak strength of the rock measured in laboratory compression tests (Bradley, 1979; Zoback et al, 1985). More advanced approaches use elasto-plasticity in order to attempt to take into account zones of fractured rock (Brown & Bray, 1982) or yield zones (Kaiser, 1981; Kaiser, Guenot & Morgenstern, 1985). Despite the fact that these methods are widely used in the design of under-ground structures in rock, carefully monitored lab-oratory experiments show that they are rarely able to explain satisfactorily the observed behaviour of even idealized underground excavations. For example, Daemen and Fairhurst (1971) found that there were no indications of fracturing or loosening of the material around the borehole when the external hydrostatic pressures applied to thick-walled hollow cylinders

of Indiana limestone and concrete reached levels at which linear elastic theory gave tangential stresses at the borehole wall of at least four times the measured uniaxial compressive strength of the material. Final collapse occurred at even higher pressures. Similar apparent strength enhancement has been recorded in hollow cylinder tests by Hobbs (1966), Hoskins (1969) and Haimson and Edl (1971), among others. In a detailed study of the behaviour of model tunnels in plates of coal loaded biaxially, Kaiser, Guenot and Morgenstern (1985) found that the tunnel convergence near failure could not be pre-dicted adequately using linear elasticity.

In an attempt to explain these experimental obser-vations, Santarelli, Brown and Maury (1986) used a numerical model incorporating the well documented phenomenon of an increase in the stiffnesses of porous and clastic rocks with increasing confining pressure (King, 1970; Kulhawy, 1975). They gener-alized Hooke's law of linear elasticity for plane strain by replacing the constant Young's modulus, E, with a value $E(\sigma_r)$ which varies as a power law function of the radial stress, σ_r, in the axisymmetric wellbore problem (Fig.1). This method provided a satisfactory explanation of the convergence measured and the apparent strength enhancement recorded in model wellbore tests on a Carboniferous sandstone. The stress distributions around wellbores computed using confining pressure dependent elasticity were found to differ markedly from those given by constant modulus linear elasticity.

This paper presents a generalization of the approach used by Santarelli, Brown and Maury (1986). The method developed by Biot (1974) is used to establish a closed form solution for the axisymmetric plane strain stresses and strains around a deep wellbore in rock having a confining pressure dependent elastic modulus. The mathematical model used to link con-fining pressure and elastic modulus is based on the extensive amount of data analysed by Kulhawy (1975). However, its form is such that $E \rightarrow 0$ as $\sigma_r \rightarrow 0$ and

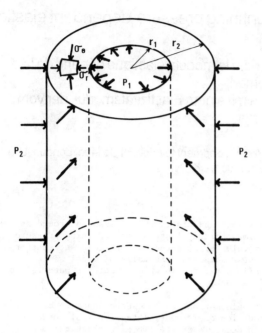

Figure 1. Geometry and boundary conditions of the axisymmetric wellbore problem.

so its use is restricted to cases such as the deep wellbore problem in which the borehole pressure does not approach zero.

DERIVATION OF THE CLOSED FORM SOLUTION

Biot (1974) showed that when the constitutive law of the material is defined by $\varepsilon_r = \varepsilon_r(\sigma_r, \sigma_\Theta)$ and $\varepsilon_\Theta = \varepsilon_\Theta(\sigma_r, \sigma_\Theta)$, the stresses in an axisymmetric hollow cylinder (Fig.1) satisfy the differential equation

$$\frac{d\sigma_r}{d\sigma_\Theta} = \frac{-(\partial\varepsilon_\Theta/\partial\sigma_\Theta)}{\dfrac{\varepsilon_r - \varepsilon_\Theta}{\sigma_r - \sigma_\Theta} + \dfrac{\partial\varepsilon_\Theta}{\partial\sigma_r}} \tag{1}$$

where σ_r, σ_Θ and $\varepsilon_r, \varepsilon_\Theta$ are the radial and tangential stresses and strains, respectively.

Assume that, for plane strain,

$$\varepsilon_r = \frac{1}{E(\sigma_r)} [(1-\nu)\sigma_r - \nu\sigma_\Theta] \tag{2}$$

$$\varepsilon_\Theta = \frac{1}{E(\sigma_r)} [-\nu\sigma_r + (1-\nu)\sigma_\Theta] \tag{3}$$

where $E(\sigma_r)$ is the confining pressure dependent Young's modulus and ν is the constant Poisson's ratio. In this case, equation (1) becomes

$$\frac{d\sigma_r}{d\sigma_\Theta} = \frac{-(1-\nu)}{\left[1 - \dfrac{dE(\sigma_r)/d\sigma_r}{E(\sigma_r)}\{-\nu\sigma_r + (1-\nu)\sigma_\Theta\} + \nu\right]} \tag{4}$$

By rearrangement, equation (4) may be written

$$\frac{d\sigma_\Theta}{d\sigma_r} = \frac{[dE(\sigma_r)/d\sigma_r]\sigma_\Theta}{E(\sigma_r)} - \frac{\nu[dE(\sigma_r)/d\sigma_r]\sigma_r}{(1-\nu)E(\sigma_r)} - 1 \tag{5}$$

This first order linear differential equation of the function σ_Θ and the variable σ_r may be solved by the method of variation of the constant using the observation that $E(\sigma_r)$ is an evident solution of the associated homogeneous equation.

The solution to equation (5) can be written as

$$\sigma_\Theta = C(\sigma_r)\, E(\sigma_r) \tag{6}$$

where $C(\sigma_r)$ is given by

$$C(\sigma_r) = -\int\frac{d(\sigma_r)}{E(\sigma_r)} - \frac{\nu}{1-\nu}\int\frac{dE(\sigma_r)}{d\sigma_r}\cdot\frac{\sigma_r}{[E(\sigma_r)]^2}\,d\sigma_r \tag{7}$$

in which the symbol $\int$ indicates the undefined primitive. Equation (7) can be transformed by integrating its second term by parts. Substitution back into equation (6) and rearranging gives the solution to equation (5) as

$$\sigma_\Theta = \frac{\nu}{1-\nu}\sigma_r - \frac{E(\sigma_r)}{1-\nu}\int\frac{d\sigma_r}{E(\sigma_r)} \tag{8}$$

At this stage, an empirical relation describing the variation of Young's modulus with confining pressure must be introduced. This relation should provide a well documented fit to experimental data for a wide range of rocks and must be mathematically well suited, i.e. the primitive of its invert must be easily expressible. After examining a wide range of data, Kulhawy (1975) proposed an expression of the form

$$E(\sigma_3) = E_0\, \sigma_3^{\alpha} \tag{9}$$

where σ_3 is the minor principal stress. Values of E_0 and α for a given rock type may be determined by the least squares method from $\log(E)$ versus $\log(\sigma_3)$ plots. The modulus exponent α varies between 0 and 1 in almost all cases, and the constant E_0 may be interpreted as the value of Young's modulus measured in uniaxial compression tests. Other empirical formulae have been proposed, but none have been supported by a data base as extensive as that assembled by Kulhawy (1975).

Substitution of equation (9) into equation (8) gives

$$\sigma_\Theta = \left[\frac{\nu(1-\alpha) - 1}{(1-\nu)(1-\alpha)}\right]\sigma_r - \frac{CE_0}{1-\nu}\sigma_r^{\alpha} \tag{10}$$

where C is a constant of integration. The boundary condition $\sigma_r = \sigma_\Theta = P_2$ at $r = \infty$ gives the value of C in the case of an axisymmetric wellbore as

$$C = \frac{(2\nu-1)(1-\alpha) - 1}{(1-\alpha)E_0}P_2^{1-\alpha} \tag{11}$$

Substituting for C in equation (10) and dividing throughout by P_2 gives

$$\sigma_{\Theta n} = K_1\sigma_{rn} - K_2\sigma_{rn}^{\alpha} \tag{12}$$

where σ_{rn}, $\sigma_{\Theta n}$ are the radial and tangential stresses normalized with respect to the external pressure P_2 and K_1, K_2 are functions of α and ν,

$$K_1 = \frac{\nu(1-\alpha) - 1}{(1-\nu)(1-\alpha)} \tag{13}$$

$$K_2 = \frac{(2\nu-1)(1-\alpha) - 1}{(1-\nu)(1-\alpha)} \tag{14}$$

Biot (1974) showed that it is possible to establish the complete closed form solution from an equation similar to (12) by solving the following equation in σ_{rn}:

$$\int_{P_1/P_2}^{\sigma_{rn}} \frac{dt}{K_1 t - K_2 t^{\alpha} - t} = \log(r/r_1) \tag{15}$$

It can be seen from equations (13) and (14) that $K_2 = K_1 - 1$. Therefore, equation (15) may be written as

$$\int_{P_1/P_2}^{\sigma_{rn}} \frac{t^{-\alpha}dt}{t^{1-\alpha} - 1} = K_2\log(r/r_1) \tag{16}$$

Integrating gives the equation in σ_{rn}

$$\log\left[\frac{\sigma_{rn}^{1-\alpha} - 1}{(P_1/P_2)^{1-\alpha} - 1}\right] = K_2(1-\alpha)\log(r/r_1) \tag{17}$$

the solution to which is

$$\sigma_{rn} = \{[(P_1/P_2)^{1-\alpha} - 1](r/r_1)^{K_3} + 1\}^{\frac{1}{1-\alpha}} \tag{18}$$

where $K_3 = (1-\alpha)K_2 \tag{19}$

Equations (2), (3), (12) and (18) provide a complete

closed form solution to the plane strain stresses and strains induced around an axisymmetric wellbore subject to internal and far-field pressures of P_1 and P_2 (Fig.1).

STRESS AND STRAIN DISTRIBUTIONS

It is instructive to examine the influence that the confining pressure dependency of E has on the stresses and strains induced around the wellbore. The modulus exponent α provides a measure of this dependency; α is zero in the case of pressure independent or constant modulus linear elasticity.

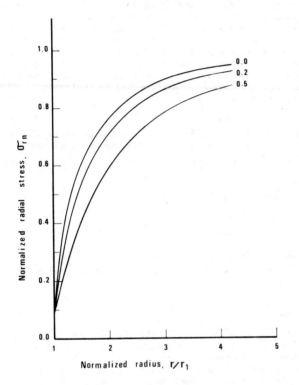

Figure 2. Distribution of normalized radial stress with distance from the borehole wall for $P_2 = 100$ MPa and $P_1 = 10$ MPa. The numbers on the curves are values of the modulus exponent α.

Figure 2 shows the distribution of normalized radial stress around a wellbore for $P_2 = 100$ MPa, $P_1 = 10$ MPa and $\nu = 0.20$ calculated from equation (18). These results show that as the pressure dependency of the modulus of the rock increases (i.e. as α increases), the larger is the radius of influence of the wellbore on the radial stress and the smaller is the radial stress gradient at the wellbore wall. Figure 2 also shows that σ_{rn} increases monotonically with r/r_1 for all three values of α. In fact, differentiation of equation (18) gives

$$\frac{d\sigma_{rn}}{d(r/r_1)} = [(P_1/P_2)^{1-\alpha}-1] \; K_3 \frac{(r/r_1)^{K_3-1}}{1-\alpha} \{[(P_1/P_2)^{1-\alpha}-1]$$
$$(r/r_1)^{K_3} +1\}^{\frac{\alpha}{1-\alpha}} \qquad (20)$$

If $0<\alpha<1$ and $0<\nu<0.5$, both K_3 and $(P_1/P_2)^{1-\alpha}-1$ are strictly negative and, therefore, $d\sigma_{rn}/d(r/r_1)$ is strictly positive in which case σ_{rn} always increases monotonically with r/r_1. This result is important because it means that there is an unique relation between radius and radial stress. Therefore, the normalized radial stress can be used as a space variable in which case equation (12) is sufficient to describe the stress distribution around the wellbore.

By substituting from equations (12) and (18) into equation (3), the tangential strain at the wellbore wall is given as

$$\epsilon_\theta = \frac{(1+\nu) \; P_2}{E_0}^{1-\alpha} [\frac{(P_1/P_2)^{1-\alpha}}{1-\alpha} - (1-\nu)K_2] \qquad (21)$$

This shows that, except in the case in which $\alpha = 0$, ϵ_θ will not vary linearly with either P_1 or P_2.

The tangential strain has sometimes been measured in model tests on underground excavations. Figure 3

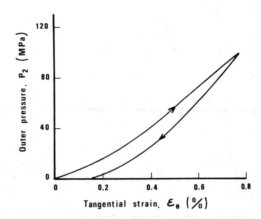

Figure 3. Tangential strain at the inner wall as a function of the external pressure for a hydrostatic test on a hollow cylinder of Carboniferous sandstone.

shows a typical plot of the variation with external pressure of the tangential strain at the borehole wall measured by Santarelli (1987) in tests on hollow cylinders of dry Carboniferous sandstone. Similar results have been reported by Kaiser and Morgenstern (1982) and by Santarelli, Brown and Maury (1986). These curves display a concave upwards curvature which cannot be explained satisfactorily by assuming elasto-plastic material behaviour. Equation (21) leads to

$$\frac{\partial^2\epsilon_\theta}{\partial P_2^2} = \frac{(1-\nu^2)}{E_0} K_2 \; \alpha(1-\alpha)P_2^{-(1+\alpha)} \qquad (22)$$

Thus, $\partial^2\epsilon_\theta/\partial P_2^2$ is always negative if $0<\alpha<1$ and $0<\nu<0.5$ which shows that equation (21) predicts correctly the form of the observed curvature of experimental plots of P_2 against ϵ_θ. Figure 4 shows such curves calculated from equation (21) for $P_1 = 1.0$ MPa, $E_0 = 10$ GPa, $\nu = 0.20$ and three different values of α. These results show that the value of α influences both the shape of the curve and the value of ϵ_θ at the wellbore wall for a given value of P_2. This may provide an explanation of the apparently abnormal strains measured by Kaiser, Guenot and Morgenstern (1985) in their model tests.

The distribution of tangential stress within the rock surrounding the borehole is also of interest. Because the normalized radial stress can be used as a space parameter, equation (12) will be used to study this distribution. Figure 5 shows the calculated relation between σ_{rn} and $\sigma_{\theta n}$ for $P_2 = 100$ MPa, $\nu = 0.20$ and three values of the modulus exponent α. When $\alpha \neq 0$, the maximum tangential stress concentration occurs within the rock and not at the borehole wall. Differentiation of equation (12) gives

$$\frac{d\sigma_{\theta n}}{d\sigma_{rn}} = K_1 - \alpha K_2 \; \sigma_{rn}^{\alpha-1} \qquad (23)$$

which indicates that the maximum tangential stress occurs when the normalized radial stress takes the critical value

$$P_c = \left(\frac{K_1}{\alpha K_2}\right)^{\frac{1}{\alpha-1}} \qquad (24)$$

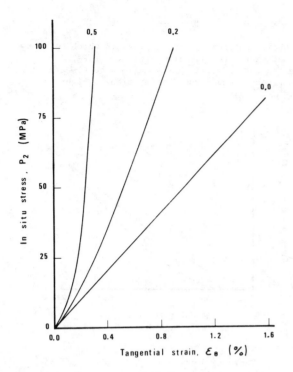

Figure 4. Tangential strain at the wall of a borehole as a function of the in situ hydrostatic stress. The borehole pressure is $P_1 = 1$ MPa, and the modulus number of the rock is $E_0 = 10$ GPa. The numbers on the curves are values of the modulus exponent α.

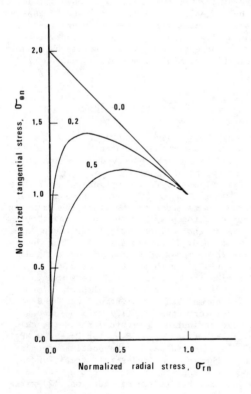

Figure 5. Variation of normalised tangential stress with the normalised radial stress for $P_2 = 100$ MPa. The numbers on the curves are values of the modulus exponent α.

In the case of a wellbore, the borehole pressure, P_1 is the support pressure applied by the mud. In deep wellbores this value will always be greater than zero, so only parts of the curves shown in Fig.5 will apply. The maximum tangential stress will occur within the rock only if P_1/P_2 is less than the value of P_c given by equation (24).

WELLBORE STABILITY

Wellbore stability analysis is taken to be the determination of the conditions under which fracture will initiate in the rock surrounding the wellbore and the location at which this happens. Clearly, the results obtained in the previous section concerning the value and location of the maximum stress concentration will have a major influence on such analyses.

As an illustration of the influence of a confining pressure dependent modulus on wellbore stability analysis, consider the case in which the initiation of rock fracture occurs according to Coulomb's criterion when

$$\sigma_\Theta > \sigma_c + M \sigma_r \qquad (25)$$

where σ_c is the uniaxial compressive strength of the rock and M is a constant. This may be rewritten as

$$F(\sigma_{rn}) > 0 \qquad (26)$$

where F is the failure function. From equation (12) it follows that, in the limiting case,

$$\frac{dF}{d\sigma_{rn}} = (K_1 - M) - K_2 \alpha \sigma_{rn}^{\alpha-1} \qquad (27)$$

The sign of $dF/d\sigma_{rn}$ determined from equation (25) indicates that F will increase as σ_{rn} increases from zero to a critical value

$$\psi_c = \left(\frac{K_1 - M}{\alpha K_2}\right)^{\frac{1}{\alpha-1}} \qquad (28)$$

and that it will decrease when σ_{rn} increases from ψ_c to 1. Since M is positive, P_c is always greater than ψ_c. As noted above, σ_{rn} will vary between P_1/P_2 and 1 for a real wellbore, and so a study of wellbore stability must be made by comparing P_1/P_2 with the critical value ψ_c.

If $P_1/P_2 > \psi_c$, F is maximum at the wellbore wall and stability depends on the sign of $F(P_1/P_2)$. In this case, fracture will initiate at the wall and the shape of the fracture surface will be similar to that derived from classical linear elasticity (King, 1912).

If $P_1/P_2 < \psi_c$, F takes a maximum in the rock some distance from the wellbore wall and stability depends on the sign of $F(\psi_c)$. In this case two situations may arise:

(i) If $F(P_1/P_2)$ is also positive, the wellbore will be highly unstable. However, a study of F alone cannot determine the full extent of fracturing.

(ii) If $F(P_1/P_2)$ is negative, failure will initiate some distance from the wellbore wall. The rock adjacent to the wellbore will remain intact and a ring of fractured rock will develop some distance from, and concentric with, the wellbore wall. Failures of this type have been observed in real and in model underground excavations (Maury, 1987).

EXPERIMENTAL STUDIES

In order to test the applicability of the theory developed above, a laboratory study was undertaken in which air dried specimens of a pale brown, fine grained Carboniferous sandstone were subjected to triaxial compression using the apparatus and techniques described by Elliott and Brown (1986). Further details are given by Santarelli (1987). For values of confining pressure of up to about 10 MPa the peak strength data were well represented by the linear relation

$$\sigma_1 = 96.15 + 7.944 \, \sigma_3 \quad (MPa)$$

in which $\sigma_c = 96.15$ MPa, $M = 7.944$ and σ_3 is the confining pressure.

The triaxial tests also showed that the tangent modulus increases with confining pressure. Linear regression gave the best fit of an empirical power law to the plot of tangent modulus at 50% peak strength against confining pressure (Fig.6) as

$$E = 15.08 \sigma_3^{0.195} \quad (GPa)$$

so that $E_o = 15.08$ GPa and $\alpha = 0.195$

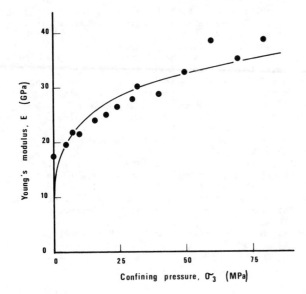

Figure 6. Values of the tangent modulus at 50% peak strength measured in triaxial compression tests on dry Carboniferous sandstone.

Hollow cylinders of the sandstone, 204mm long and with internal and external diameters of 25 and 102mm, were then loaded externally with hydrostatic pressures. No axial loads were applied. The tangential strains developed at the inner wall were measured with high yield electric resistance strain gauges. Typical test results are shown in Figs. 3 and 7. The experimental $P_2 - \varepsilon_\theta$ curves were concave upwards to about $P_2 = 80$ MPa after which the sense of the curvature reversed. Visual inspection of the tested samples indicated that fracture at the borehole wall initiated at an external pressure of about 130 MPa. Classical constant modulus linear elastic theory is quite unable to explain these two experimental observations. It predicts that failure at the inner wall should have occurred at an external pressure of 45 MPa and that the $P_2 - \varepsilon_\theta$ curves should be straight lines.

The solution developed above cannot be used directly to interpret these experimental results because
(i) the radius at which the external pressure was applied is not infinite. However, Fig.2 shows that a reduction of the normalized outer radius to 4 makes only a small difference to the radial stress distribution and therefore also to those of tangential stress and strain;
(ii) as $P_1 \rightarrow 0$, the value of E given by equation (9) also approaches zero in which case the solution may not be applied. This does not represent a constraint to applications of the theory to deep wellbores for which $P_1 \gg 0$.
Because P_2 reaches large values in the experimental case, the tests may be simulated reasonably well by an analysis in which $P_1 = 1.0$ MPa. Figure 7 shows that values of tangential strain at the inner wall predicted by equation (21) with $P_1 = 1.0$ MPa agree

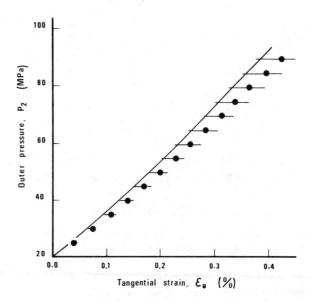

Figure 7. Mean tangential strains measured at the inner walls of hollow cylinders of Carboniferous sandstone subjected to increasing outer pressure. The curve is predicted by pressure dependent elasticity theory. The error bars indicate the ranges of measurement.

reasonably well with the experimental data for external pressures above the bedding down value of 20 MPa. The difference between the predicted and the measured values increases for high values of P_2, i.e. when the $P_2 - \varepsilon_\theta$ curve starts to curve downwards.

A stability analysis carried out using equations (26) to (28) shows that, for $P_1 = 1.0$ MPa, the rock around the hole should fracture when P_2 reaches 108 MPa and that fracture should initiate 0.2mm from the borehole wall. The latter value is small compared with the 0.5 - 2.0mm thicknesses of the chips detached from the walls of the sandstone samples when they were tested to failure. This small predicted value may well correspond to a very early stage of the failure process in which Daemen and Fairhurst (1971) found that fine dust particles were loosened from the borehole walls. Larger slabs or chips of rock become detached when the external pressures are increased above those at which this initial boundary loosening occurs.

CONCLUSIONS

The closed form solution derived for the stresses and strains developed around an axisymmetric wellbore in elastic rock having a confining pressure dependent modulus shows that
(i) the tangential stresses at and near the wellbore wall may be much lower than those predicted by constant modulus linear elasticity;
(ii) the maximum tangential stress may be induced some distance from the wellbore wall. Accordingly, fracture may initiate away from the wall in certain cases;
(iii) the strains at the wellbore wall will vary non-linearly with the normalized far field stress;
(iv) better predictions were given of the behaviour of model wellbores in a Carboniferous sandstone than those obtained using constant modulus linear elasticity.

The limitations of the solution presented are:
(i) it takes no account of pre-peak yield and so should still over estimate the value of the tangential stress at the wellbore wall and under estimate the strains and convergence for large values of the pressure difference $P_2 - P_1$. A numerical method able to account for both pre-peak yield and the variation of Young's modulus with confining pressure has now been developed by Santarelli (1987);
(ii) it cannot be applied to unsupported excavations

1221

because equation (9) does not model satisfactorily the values of Young's modulus at very low values of confining pressure. In this case, alternative representations of the confining pressure dependency of the elastic modulus must be used. Some of these formulations do not permit closed form solutions to be obtained and so numerical methods must be used to solve the problem (e.g. Santarelli, Brown and Maury (1986). McLean (1987) used a linear relation. Another possibility is the alternative form of the power law

$$E = E_o(1+\sigma_3)^\alpha$$

which leads to the solution

$$\sigma_\Theta = K_1\sigma_r - (K_2P_2 - K_4)\left(\frac{1+\sigma_r}{1+P_2}\right)^\alpha - K_4$$

where

$$K_4 = \frac{1}{(1-\alpha)(1-\nu)}$$

In cases such as this, studies similar to that reported in this paper can be carried out, but the equations involved become much more cumbersome.

ACKNOWLEDGEMENTS: The work described herein was sponsored by Elf UK plc as part of a larger research programme sponsored jointly with BP International. The authors are grateful to Drs. A. Guenot and V. Maury of Elf Aquitaine (Production) for their support and encouragement, to Dr. M.R. McLean of BP for his contributions to many useful discussions and to Mr.J.W. Dennis for his assistance with the laboratory work.

REFERENCES

Bell, J.S. & D.I. Gough 1979. Northeast-southwest compressive stress in Alberta: evidence from oil wells. Earth Planet. Sci. Letters 45: 475-482.
Biot, M. 1974. Exact simplified non-linear stress and fracture analysis around cavities in rock. Int. J. Rock Mech. Min. Sci. 11: 261-266.
Blüming, P., K. Fuchs & T. Schneider 1983. Orientation of the stress field from breakouts in a crystalline well in a seismic active area. Phys. Earth Planet. Interiors 33: 250-254.
Bradley, W.B. 1979. Failure of inclined boreholes. J. Energy Resources Technol., Trans Am. Soc. Mech. Engrs 101: 232-239.
Brown, E.T. & J.W. Bray 1982. Rock-support interaction calculations for pressure tunnels. In W. Wittke (ed.), Rock mechanics: caverns and pressure shafts 2: 555-565. Rotterdam: Balkema.
Daemen, J.J.K. & C. Fairhurst 1971. Influence of failed rock properties on tunnel stability. In G.B. Clark (ed.), Dynamic rock mechanics, Proc. 12th Symp. Rock Mech., p. 855-875. New York: AIME.
Elliott, G.M. & E.T. Brown 1986. Laboratory measurement of the thermo-hydro-mechanical properties of rock. Quart. J. Engng. Geol. 19:
Guenot, A. 1987. Contraintes et ruptures autour de forages pétroliers. Proc. 6th Congr., Int.Soc.Rock Mech., Montreal (this volume)
Haimson, B.C. & J.N. Edl 1972. Hydraulic fracturing of deep wells. Soc.Petrol.Engrs Paper No.4061.
Hobbs, D.W. 1966. The strength of coal under biaxial compression. Colliery Engng 39: 285-290.
Hoskins, E.R. 1969. The failure of thick walled hollow cylinders of isotropic rock. Int. J. Rock Mech.Min.Sci. 6: 99-125.
Kaiser, P.K. 1981. A new concept to evaluate tunnel performance-influence of excavation procedure. In H.H. Einstein (ed.), Proc. 22nd U.S. Symp. Rock Mech. p.284-291. Cambridge, MIT.
Kaiser, P.K., A. Guenot & N.R. Morgenstern 1985. Deformation of small tunnels - IV Behaviour during failure. Int.J.Rock Mech.Min.Sci. & Geomech.Abstr. 22: 141-152.
Kaiser, P.K. & N.R. Morgenstern 1982. Time independent and time dependent deformation of small tunnels - III Pre failure behaviour. Int.J.Rock Mech.Min.Sci. & Geomech.Abstr. 19: 307-137.

King, L.V. 1912. On the limiting strength of rocks under conditions of stress existing in the earth's interior. J.Geol. 20: 119-137.
King, M.S. 1970. Static and dynamic moduli of rocks under pressure. In W.H. Somerton (ed.), Rock mechanics - theory and practice, Proc. 11th Symp. Rock Mech. p.329-351. New York: AIME.
Klein, R.J. & M.V. Barr 1986. Regional state of stress in Western Europe. In O. Stephansson (ed.), Proc. Int.Symp.Rock Stress and Rock Stress Measurements, p.33-44. Luleå: Centek Publishers.
Kulhawy, F.H. 1975. Stress deformation properties of rock and rock discontinuities. Engng Geol. 9: 327-350.
Maury, V. 1987. Observation, recherche et résultats récents sur les mécanismes de rupture autour de galeries isolées. Rapport de la commission de la SIMR sur les mécanismes de rupture autour d'ouvrages souterrains. Proc. 6th Congr., Int.Soc.Rock Mech., Montreal (this volume).
McLean, M.R. 1987. Wellbore stability analysis. Ph.D. thesis, Univ. of London (in preparation).
Santarelli, F.J. 1987. Theoretical and experimental investigation of the stability of the axisymmetric wellbore. Ph.D. thesis, Univ.of London(in preparation).
Santarelli, F.J., E.T. Brown and V. Maury 1986. Analysis of borehole stresses using pressure-dependent, linear elasticity. Int.J.Rock Mech.Min. Sci. & Geomech.Abstr. 23, 445-449.
Teufel, L.W. 1985. Insights into the relationship between wellbore breakouts, natural fractures and in situ stress. In E. Ashworth (ed.), Research and engineering applications in rock masses, Proc. 26th U.S. Symp.Rock Mech. 2: 1199-1206. Rotterdam: Balkema.
Zoback, M.D., D. Moos, L. Hastin & R.N. Anderson 1985. Wellbore breakouts and in situ stress. J. Geophys. Res. 90: 5523-5530.

Stabilité mécanique du chenal de liaison dans la gazéification souterraine profonde du charbon

Mechanical stability of the linkage in deep underground coal gasification
Mechanische Festigkeit der Verbindung für Kohlevergasung in grosser Tiefe

N.SCHMITT, Laboratoire de Mécanique des Solides, Ecole Polytechnique, Palaiseau, France
D.NGUYEN MINH, Laboratoire de Mécanique des Solides, Ecole Polytechnique, Palaiseau, France

ABSTRACT : We study the mechanical stability of the linkage in deep underground coal gasification (U.C.G.). The coal under study was a non swelling anthracite. The main thermomechanical characteristics were determined, namely, by developping an original methodology with regard to the high temperatures involved (1000 K) and the heterogeneity of the material. Then a physical coal model of the channel has been set up, and different thermomechanical loadings examined. The results were interpreted by a mathematical model and a stability criterion was proposed.

RESUME : On étudie la stabilité mécanique du chenal de liaison dans le procédé de gazéification profonde du charbon ; nous avons considéré ici le cas d'un charbon non gonflant. Les principales caractéristiques thermomécaniques du charbon ont d'abord été déterminées, en développant notamment une procédure expérimentale originale pour tenir compte de fortes températures mises en jeu (≈ 1000 K) et de l'hétérogénéité du matériau. Ensuite une étude sur maquette du chenal a été réalisée et différents trajets de chargement thermomécanique ont été examinés. On interprète les résultats par un modèle théorique et on propose un critère de stabilité pour la liaison.

ZUSAMMENFASSUNG : Wir studieren die mechanische Festigkeit der Verbindung der tief unterirdischen Vergasung der Kohle. Die hier geprüfte Kohle ist eine nicht schwellende Kohle. Zuerst wurden die wichtigsten thermomechanischen Kennzeichen der Kohle geprüft. Ein besonderes experimentales Verfahren wurde entwichelt, um die hier in Betracht genomenen hohen Temperature (≈ 1000 K) und die Ungleichartigkeit des Materials zu berücksichtigen. Danach wurde ein Modellstudie der Verbindung des Kanals durchgeführt ; verschiedenen Verladungswege des Modells wurden durch einen theoretischen Modell erklärt und einen stabilitätskriterium für die Verbindung vorgeschlagen.

1 INTRODUCTION

En Europe, on cherche à démontrer la faisabilité de la gazéification à grande profondeur (600 à 1000 m), (Pottier et al., 1978). La stabilité mécanique du gazogène souterrain (figure 1) constitue un des problèmes essentiels de ce procédé, du fait des fortes températures mises en jeu et des pressions lithostatiques élevées. Dans le gazogène, la partie terminale de la liaison est l'une des zones les plus exposées vis à vis de la rupture car elle est encore soumise à la circulation des gaz chauds (800 à 1 000 K), alors que les réactions chimiques tendant à élargir l'ouverture sont arrêtées.

L'objectif de cette étude est de mettre au point un modèle théorique simple permettant de prédire le comportement du chenal de gazéification en condition de fonds.

Ce modèle s'appuie d'une part, sur les essais de caractérisation des propriétés thermomécaniques du charbon, et, d'autre part, sur l'interprétation d'essais sur maquette du chenal, dont l'échelle de réduction géométrique est modérée (1/5) ; les essais sur modèles permettent de corriger les éventuels effets d'échelles sur les valeurs des caractéristiques thermomécaniques, de définir d'autres paramètres non mesurables par les essais plus classiques, et enfin, de mettre en évidence un critère de stabilité pour le chenal.

2 CARACTERISTIQUES THERMOMECANIQUES D'UN ANTHRACITE

Le charbon testé est un anthracite provenant du bassin houiller d'Estevelles (Nord de la France). Il a été prélevé dans la veine Pérus à 860 m de profondeur, près du site expérimental de la Haute Deule où ont été réalisés les derniers essais de gazéification in situ (Gadelle et al., 1986).

L'anthracite Pérus a une structure anisotrope, marquée par une forte hétérogénéité en raison notamment de la fissuration induite par l'ouverture de la mine. Il en résulte une telle dispersion qu'il serait nécessaire de recourir à un grand nombre d'essais pour évaluer le comportement mécanique du matériau (Vutukuri et al., 1974) sans toutefois être certains d'arriver à des résultats vraiment représentatifs de l'état de massif vierge.

Pour tenter d'obtenir des résultats significatifs et mettre en évidence l'effet de la température, nous avons préféré procéder à une suite de sélections : prélèvement des blocs dans la mine selon des critères d'homogénéité ; sélection due au carrotage des éprouvettes (≈ 20 % de récupération) ; élimination des essais douteux (détachement d'un coin suivant une fracture préexistante...)

2.1 Caractéristiques mécaniques à la température ambiante

Dans les autres essais, on étudie les caractéristiques dans la direction perpendiculaire à la stratification
1. L'anisotropie a été évaluée sur des essais de compression simple.
2. Le module d'Young E et la résistance à la compression simple Rc sont les plus élevés dans la direction perpendiculaire aux strates. Ils varient dans un rapport de 1 à 2,4 environ avec l'inclinaison de la stratification, mais E/R_c reste pratiquement constant (≈ 200).

Cependant l'anisotropie du charbon est fortement masquée par la dispersion. De plus la direction des strates in situ est très variable, aussi avons nous préféré considérer le matériau comme globalement isotrope. Pour l'interprétation des essais sur maquette, l'incertitude sera levée par la détermination de la déformabilité élastique de la maquette.
3. Le charbon présente une rupture à caractère fragile, même sous des confinements de l'ordre de 20 à 30 MPa. La courbe intrinsèque de rupture, obtenue par l'essai triaxial, est très ouverte pour les faibles pressions où elle peut être décrite par un critère de

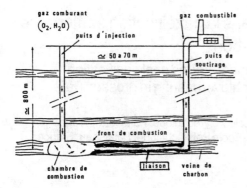

Figure 1. Principe de la gazéification souterraine profonde du charbon.

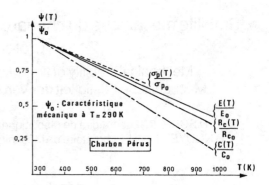

Figure 3. Evolution des caractéristiques mécaniques avec la température.

Coulomb (Cohésion C = 2 MPa, angle de frottement interne φ = 52°). Ces valeurs doivent être corrigées pour ces confinements supérieurs à 10 MPa.

2.2 Effet de la température

Une connaissance précise du comportement thermomécanique du charbon n'étant pas indispensable, l'essai triaxial sous haute température est remplacé par une procédure expérimentale plus simple. Pour évaluer la courbe intrinsèque de rupture, nous avons réalisé trois types d'essais : l'essai de compression simple, l'essai de poinçonnement et l'essai de cisaillement direct qui donnent respectivement les valeurs de Rc, σ_p (résistance au poinçonnement) et C. Moyennant une interprétation théorique, il est possible de déterminer trois points de la courbe intrinsèque de rupture et de définir deux critères de Coulomb différents selon que la pression de confinement est faible ou moyenne. Une discussion plus complète est présentée dans Schmitt et al. 1986.

Le même dispositif expérimental est utilisé pour les trois essais (figure 2). L'éprouvette est chauffée par palier de 100 K/heure. Lorsqu'il y a risque de combustion du charbon à l'air libre (environ 720 K ici), elle est plongée dans un bain de plomb fondu. Le dispositif a permis de réaliser des essais pour des températures T allant jusqu'à 1000 K.

Les principaux résultats sont les suivants :

1. On remarque que pour des températures élevées, le matériau conserve son caractère fragile ; aucune viscosité particulière n'a été décelée.

2. Les résultats de ces essais sont reportés sur la figure 3. Les caractéristiques E, Rc, C et σ_p diminuent à peu près linéairement avec la température. Il est intéressant de constater que les rapports $\frac{Rc}{C}$ et $\frac{\sigma_p}{Rc}$ restent pratiquement constants. Il en résulte notamment que φ ne dépend pas de la température (φ ≃ 55° pour les faibles confinements, φ ≃ 28° pour les confinements moyens)

3. Des mesures des caractéristiques thermiques ont été faites par ailleurs (Schmitt et al., 1986) ; on remarque que le coefficient de dilatation thermique est élevé pour une roche courante, mais qu'aucun gonflement n'est apparu jusqu'à 970 K.

2.3 Modélisation du comportement de l'anthracite Pérus

A priori, un modèle élasto-plastique avec radoucissement semble convenir pour schématiser le comportement de ce charbon. Ce modèle a fait ses preuves (Berest et al., 1979) et a déjà été appliqué pour étudier le comportement du gazogène (Pater, 1985). Ici, nous adopterons une version simplifiée avec un radoucissement brutal pour limiter le nombre de paramètres caractérisant la rupture.

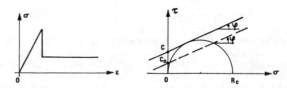

Figure 4. Modèle simplifié du comportement mécanique de l'anthracite Pérus

3. REALISATION DES ESSAIS SUR MAQUETTES

Le modèle physique est un tube dont les dimensions ont été choisies de façon à simuler la liaison percée dans un milieu infini.

Le chargement thermomécanique du modèle est limité à :

1. Une pression extérieure uniforme Pe exercée par des vérins plats spéciaux.

2. Un échauffement radial par une résistance chauffante placée dans l'axe du chenal T(r,t).

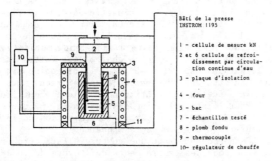

Figure 2. Dispositif expérimental pour les essais sous température.

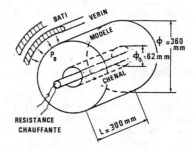

Figure 5. Principe de l'essai sur maquette.

On mesure la température en divers points de la maquette par des thermocouples, ainsi que les convergences du chenal dans deux directions perpendiculaires, par des capteurs de déplacement munis de palpeurs en silice.

Les faces latérales du modèle sont bloquées par des flasques rigides pour éviter les expansions longitudinales lors du chargement.

3.1 Procédure expérimentale et résultats de deux essais.

On a considéré deux chargements standards dans lesquels la pression et la température sont élevées séparément pour faciliter l'interprétation.

Premier essai : chargement thermique suivi d'un chargement mécanique.

La maquette est d'abord chauffée jusqu'à ce que la température dans la masse soit stabilisée. Ensuite, à partir de cet état, on charge mécaniquement la maquette jusqu'à la rupture. Comme il est plus aisé de contrôler la pression Pe que la température $T(r,t)$, un critère de stabilité du chenal sera facilement mis en évidence.

Dans cet essai, les plans de stratification du bloc sont pratiquement parallèles à l'axe du chenal.

La répartition radiale de température dans le modèle au moment du chargement mécanique a été ajustée par une courbe d'équation $T(r) = 290 + 360 \exp(-0,85(r-1)^{0,8})$; la température en paroi est estimée à 650 K. Une légère dilatation du chenal a été observée durant la période de chauffe (10 heures environ) ; on peut l'attribuer à un microécaillage en paroi, ou à des transformations physico-chimiques

Les capteurs de déplacement sont inclinés à 45° par rapport à la stratification et donnent des valeur moyennes de la convergence. Les courbes pression extérieure-convergence du chenal sont présentées sur la figure 8. De cette figure, on peut définir un critère de stabilité de la galerie basée sur la valeur limite de la pression ($\simeq$ 7 MPa) au delà de laquelle une convergence excessive conduirait à l'effondrement du chenal. De façon analogue, on peut définir une limite de convergence ($\simeq$ 4 %).

Second essai : chargement mécanique suivi d'un chargement thermique.

Dans cet essai, le chenal est percé perpendiculairement à la stratification. Ici, on soumet une maquette préalablement chargée mécaniquement à un échauffement conduisant à la rupture éventuelle du chenal. Toutefois, comme la température doit être limitée en paroi pour éviter la combustion il est nécessaire que le niveau de chargement initial soit suffisamment élevé. Ce niveau sera choisi comme une limite "élastique" au delà duquel, un faible accroissement de la température provoque une convergence du chenal. Dans la pratique, ce seuil a été recherché en augmentant alternativement par petits paliers la pression et la température de la résistance chauffante Tr. L'expérience s'est déroulée sur trois journées.

Durant les deux premiers jours, pour Pe $\leqslant$ 5,8 MPa et Tr $<$ 630 K, aucun effet de la température n'a été mis en évidence et aucune convergence significative due au fluage n'a été décelée. L'influence de la température apparaît au cours de la troisième journée pour des chargements thermomécaniques un peu plus élevés ; l'historique du trajet de chargement de cette journée est présentée sur la figure 6. Différents états de chargement remarquables (I, II, ...,V) ont été considérés en vue de l'interprétation.

La convergence du chenal au cours du temps est présentée figure 7. Seule la courbe du capteur 2 sera retenue pour l'interprétation, l'autre étant suspecte par suite de l'apparition d'une écaille sous le point de mesure. On pourra remarquer l'accélération de la convergence au cours du dernier chargement thermique ; le chenal est alors sérieusement endommagé et est proche de la rupture.

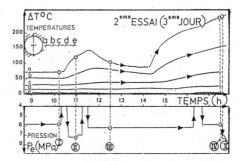

Figure 6. Second test : historique du chargement thermomécanique du 3ème jour.

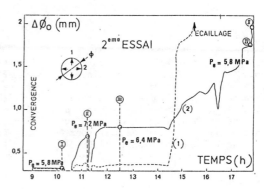

Figure 7. Second essai : convergence au cours du temps

3.2 Interprétation

Pour modéliser le comportement mécanique du chenal, un calcul semi explicite a été développé (Schmitt, 1986). Dans ce calcul, on considère un chenal circulaire creusé dans un milieu infini et chargé uniformément par une pression extérieure Pe à l'infini égale à la pression de soutènement Pi et une température $T(r,t)$ simulant l'échauffement du massif par les gaz chauds. On suppose que le milieu a un comportement élastofragile. Les détails de l'interprétation sont donnés dans (Nguyen Minh et Schmitt, 1986).

L'ajustement des courbes de convergence du premier essai permet de déterminer les paramètres de rupture du modèle $(C/Co, \beta)$, (fig 8) ; le paramètre de dilatation thermique α semble être soumis à un effet d'échelle car sa valeur a due être divisée par 4 par rapport à la mesure directe.

Le modèle ainsi calibré simule de façon très satisfaisante les résultats correspondant aux états de chargement I...V du second essai (fig 10). Toutefois, la zone endommagée prédite est deux fois plus grande que celle qui a pu être observée sur la maquette ($\simeq$ 5 mm d'épaisseur) et suggère qu'un modèle prenant

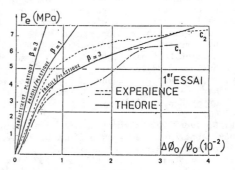

Figure 8. Premier essai : comparaison entre la théorie et l'expérience.

en compte un radoucissement progressif donne des résultats encore meilleurs.

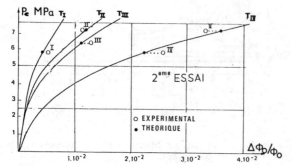

Figure 9. Second essai : comparaison entre la théorie et l'expérience.

4 CONCLUSION

Le comportement mécanique du chenal de gazéification soumis à un chargement plus réaliste peut être prédit par le modèle théorique précédent. On retrouve cette notion de pression limite qui caractérise le critère de stabilité ; une pression minimale de soutènement, dépendant de la profondeur et de la durée d'exploitation d'un doublet est nécessaire pour éviter la fermeture de la liaison (exemple figure 10).

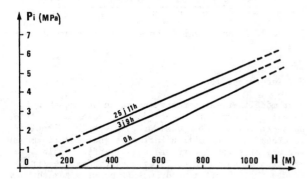

Figure 10. Pression intérieure minimale en fonction de la profondeur de la liaison.

La procédure que nous avons mise au point peut être adaptée pour étudier d'autres types de charbon, tels les charbons gonflants.

Il convient de remarquer que l'essai sur maquette, au delà de la représentation du modèle réduit de chenal, est aussi un essai de laboratoire évolué, participant entièrement à la mise au point d'un modèle de comportement du matériau.

REFERENCES

Berest, P., Bergues, J. & Nguyen Minh, D. 1979. Comportement des roches au cours de la rupture. Application à l'interprétation d'essais sur des tubes épais. Rev. Française de Géotechnique n°9, : 5-12.
Gadelle, C. Marrast, J. & Raffoux, J.F. 1986. An attractive Pilot test site for UCG. 12th Symp. UCG Sarrebrück, : 110-117.
Schmitt, N., Bergues, J., Luong, M.P. & Nguyen Minh, D. 1986. Characterization of the thermomechanical behaviour of Coals. 12th Symp. UCG Sarrebrück, : 399-407.
Schmitt, N. 1986. Comportement thermomécanique du charbon. Application à la gazéification souterraine. Thèse Université Paris 6 à paraitre.
Vutukuri, V.S., Lama , R.D. & Saluja, S.S. 1974. Handbook on mechanical properties of Rocks volume 1 (1er Ed.). Trans. Tech. Publications.

Nguyen Minh, D. & Schmitt, N. 1986. Physical Modelling for stability analysis of the linkage in deep inderground gazeification. 12th Symp. UCG. Sarrebrück, : 278-286.
de Pater, C.J. 1985. Thermal and Mechanical simulation of the gasification cavity. 11th Symp. UCG Denver.
Pottier, M., Chaumet, P. & Lechevin, L. 1978. Etudes Préliminaires à la gazeification souterraine profonde des charbons. Perspectives et problèmes. Rev. Ind. Pr. Petr. Vol 33 n° 5 : 279-746.

Prediction of closures and rock loads for tunnels in squeezing grounds
Prévision de la convergence et des pressions de roche pour des tunnels dans un terrain plastique
Konvergenz und Gebirgsdruckvorhersage für Tunnel im druckhaften Gebirge

V.M.SHARMA, Central Soil and Materials Research Station, New Delhi, India
T.RAMAMURTHY, Indian Institute of Technology, New Delhi, India
C.SUDHINDRA, Central Soil and Materials Research Station, New Delhi, India

ABSTRACT: The Hoek and Brown yield criterion was adopted for the prediction of convergence in a number of tunnels in the Himalaya region and compared with measurements in squeezing ground for a hydro-electric power development

RÉSUMÉ: Le critère de rupture sous contrainte maximale de Hoek et Brown a été vérifié dans un tunnel hydroélectrique de l'Himalaya et une relation simplifiée a été sélectionnée pour estimer la convergence des terrains fortement contraints

ZUSAMMENFASSUNG: Das Hoek und Brown Kriterium wurde für die Vorhersage von Untertagekonvergenzen benutzt und mit Geländemessungen in verschiedenen Himalaja-Tunneln verglichen.

1 INTRODUCTION

In so far as assumptions of forces in tunnel design are conserned, the remark of Terzaghi in 1942 that there was little doubt in his mind that such forces which acted on the tunnel are very much smaller than those assumed by the designers holds good even today. Inadequacy of knowledge regarding the real intensity of stress distribution makes one to lean on customary method of design. This opens up large scope for understanding of the same and every contribution, however small, in this field is likely to lead towards rationalization and economy in design.

The state of the art regarding design practice of tunnels virtually being at stand-still, most of the designs methods in use today are based on empirical formulations. It is heartening that emphasis is presently being given to the observational approach adopting 'build as you go' technique. Improvements in the design methods is the need of the day. This evidently calls for an increased awareness of the mechanism and modes of behaviour of the system composed of tunnel and surrounding medium. Such awareness can best be obtained by a combination of theoretical considerations and analysis of data of the observed behaviour of tunnels in the field.

The Himalayas form the north-western boundary of the Indo-Australian plate. This is a continent to continent collision boundary, the deliniation of which runs along the axis of the Himalayas. The region remained under water for the greater part of its history. This continental collision has resulted in some of the highest mountain ranges and deepest valleys surpassing those on any of the other continental plate boundaries.

The Himalayan region has a potential in abundance for the development of hydroelectric power by virtue of the available water and enormous potential heads.

Construction of tunnels in such a terrain has been extremely difficult. The ground squeezes enormously and continues to do so for long periods. It is seen that the convergence confinement method of rock-support interaction is more applicable in these conditions than the empirical formulations.

2 CONVERGENCE CONFINEMENT METHOD

The technique of employing two characteristic curves - one for the ground and the other for the support - has been described in detail by Lombardi(1970), Daeman and Fairhurst(1970), Ladanyi(1974)and Lombardi (1977). It is considered as one of the most promising methods for understanding the mechanics of tunnel deformation and development of rock loads. The evaluation of the characteristics calls for assumptions that do simplify the theoretical solution obtained for the complex practical problem, but the analytical model can be made sufficiently comprehensive to cover a wide range of the practical situations.

The ground convergence characteristic is the relation between the inward radial displacement at the tunnel periphery and the radial pressure or support pressure applied at the periphery. It is assumed that the sequence of face advance and excavation can be represented by a gradual continuous increase of the tunnel convergence and the ground characteristic is then obtained by calculating the support pressure that would be required to maintain equilibrium at each point of this convergence curve. The support characteristic is the load deflection curve of the support. The inter-section between the two characteristic curves refers to the equilibrium point between the ground movement and the support reaction, and thus the load mobilised by the support to maintain the stability of the tunnel at the corresponding convergence(fig.1). A complete evaluation of the two characteristic curves requires knowledge of the virgin stresses and the strength and deformation characteristics of the rock mass and the support.

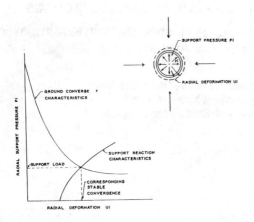

Figure 1. Convergence confinement method.

2.1 Yield criterion

Most of the earlier research workers such as
Fenner(1938), and Morrison and Coates(1955)
adopted Mohr-Coulomb yield criterion for
calculating the stresses. Hobbs(1966) calculated the stress distribution around a tunnel using a non-linear failure criterion.
Kennedy and Lindberg(1978) used a step-wise
linear Coulomb approximation of non-linear
Mohr envelope. The yield function was so
chosen that the internal friction decreased
with increasing normal stresses. Hoek and
Brown(1980)evolved a non-linear peak and
residual strength criteria and gave examples
of step by step calculations for covering
complete ground-support-interaction in
elastic-brittle-plastic ground. Brown et al
(1983) gave a step by step method for the
calculation of ground convergence characteristics in elastic-strain softening-plastic
ground.

In the analysis presented in this paper,
the non-linear failure criterion of Hoek and
Brown has been used. Accordingly,

$$\sigma_1 = \sigma_3 + (m\sigma_c\sigma_3 + s\sigma_c^2)^{\frac{1}{2}} \qquad (1)$$

where

σ_1 = major principal stress
σ_3 = minor principal stress
σ_c = uniaxial compressive strength of rock
      material
m,s= constants which depend upon the nature
      of the rock mass and the extent to
      which it is broken, before being
      subjected to stress σ_1 and σ_3

Similarly, the residual strength of the rock
mass is given by

$$\sigma_1 = \sigma_3 + (m_r \sigma_c\sigma_3 + s_r \sigma_c^2)^{\frac{1}{2}} \qquad (2)$$

where m_r and s_r are parameters for the broken rock mass corresponding to the parameters m and s of the intact rock mass.

2.2 Volume changes

In an attempt for making a more realistic
assessment of the volume changes in rock
during failure, results of tests conducted
on stiff testing machines(Crouch 1970) as
well as other laboratory techniques(Ladanyi

and Don 1971) have been used in the study.
It is observed that as the rock fails, the
rate of volume increase is initially high
but the same reduces when the axial strains
increase . A similar relationship between
radial and tangential strains of yielding
rock around the circular tunnel could be
anticipated. Correspondingly, if the initial
slope of the ε_r - ε_θ curve was h_1 and if the
secant slope was reduced by an amount h_2 as
the major principal strain was increased
from $\varepsilon_{\theta e}$ to infinity, the secant slope h of
ε_r - ε_θ curve, corresponding to any value of
strain ε_θ can be represented by

$h = h_1 - h_2(1 - \dfrac{\varepsilon_{\theta e}}{\varepsilon_\theta})$ where h is such that

$\varepsilon_r = -h \varepsilon_\theta$. ε_θ and ε_r are two principal
strains and $\varepsilon_{\theta e}$ is the principal strain
corresponding to the yielding point.

2.3 Ground behaviour model

The ground behaviour assumed for the studies
is shown in fig.2.

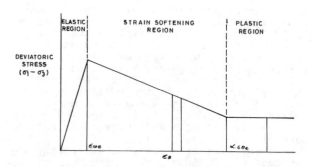

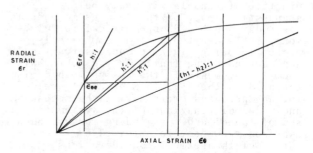

Figure 2. Assumed ground behaviour model.

$\varepsilon_{\theta e}$ is the major principal(elastic) strain
at which the rock yields and strain-soften-
ing starts. $\alpha\varepsilon_{\theta e}$ is the major principal
strain at which the strain softening is over
and the rock becomes completely plastic. The
rock mass is assumed to be linearly elastic
with Young's modulus E and Poisson's ratio
μ, until the initial strength for the appropriate value of σ_3 is reached. Thereafter,
the strength reduces gradually as shown in
fig.2.

Strength reduction from the peak and continued deformation at residual strength are
accompanied by plastic dilation. Some elastic volume increase is also possible when
the stresses are reduced. These aspects are
covered while dealing with the appropriate-

ness of α, h_1 and h_2.

3 PART INTEGRATION - PART NUMERICAL TECHNIQUE

The entire rock mass surrounding the tunnel is divided into a number of thin concentric annular rings. If the radius, stresses and strains at one surface of any ring are known, radius, stresses and strains at the other surface of the ring can be calculated using some approximate method. Brown et al(1983) have used the finite difference method. It is, however, possible to use an exact integration of the differential equation within the thin annulus and obtain the required parameters. However, the integration is justifiable for thin annulus only. As the strength and strains in the rock mass are changing continuously, an approach of part integration and part numerical has to be used. To start with, the radius of the broken zone and the corresponding stresses and strains are calculated on the basis of closed form solution for the elastic-brittle-plastic ground conditions.

The tangential strains corresponding to the two radii of the first annulus are taken as $\varepsilon_{\theta e} (= \varepsilon_{\theta_1})$ and $\varepsilon_{\theta e} + d\varepsilon_\theta$ $(=\varepsilon_{\theta_2})$. A suitable value of $d\varepsilon_\theta$ is assumed for this purpose. The values of radial strains corresponding to the two tangential strains are calculated in accordance with the relationship assumed in fig.2. The radius of the starting annulus is $r_e (=r_1)$. Using the values of ε_{θ_1}, ε_{r_1}, r_1, ε_{θ_2} and ε_{r_2}; the radius r_2 is calculated by the following equation

$$r_2 = \frac{r_1}{\{\frac{(h+1)}{2\varepsilon_{\theta_1}}\varepsilon_{\theta_2} - \frac{(h-1)}{2}\}^{1/(h+1)}}$$

Once the other radius of the annular ring is known, the radial stresses can be calculated. This process of calculating r_2 and σ_{r_2} is continued, till we reach a radius r_2 for which σ_r equals p_i or the internal pressure. At this stage the radius r_2 (so calculated) is equated to the radius tunnel r_1 and all other radii adjusted in the same proportion.

It is seen that for the particular cases for which closed-form solutions are available, the use of the modified procedure results in stress distribution much closer to the theoretical one. Also the iterative cycles converge faster, resulting in a saving of time.

4 PARAMETRIC STUDY

A computer programme has been developed to compute the ground convergence curve for the tunnel through the ground described by the model above. The influence of different parameters on ground convergence has been evaluated by varying one parameter at a time, keeping all others constant. For each case the ground convergence curve was computed and plotted. It is established from this sensitivity analysis that the peak strength of the rock mass, the modulus of deformation and strain softening characteristics form the more important parameters.

5 CORRELATION

Utilizing the data generated by running the computer programme mentioned above, ground convergence has been correlated with three dimensionless factors, enumerated below.

i) $(\sigma_1-\sigma_3)/p_0$ where $(\sigma_1-\sigma_3)$ is the peak yield strength of the rock mass and p_0 is the in situ stress.

ii) p_i/p_0 where p_i is the internal pressure provided by the support system.

iii) nE_r/K_n where n is the joint frequency i.e., the no. of joints per metre, E_r is the modulus of elasticity of rock material and K_n is the average joint stiffness i.e. compressive stress required to produce unit deformation.

The relationship between ground convergence and the factors described above is given by

$$\frac{u_i}{r_i} \times 10^3 = 1 + \frac{.0525 \ J.B.}{(p_i/p_0)^C} , \text{ where}$$

u_i = radial deformation

r_i = radius of the tunnel

$J = n E_r/K_n$

$B = \{1/1_n(\frac{\sigma_1-\sigma_3}{p_0})\}^{0.5}$

$C = \{1/1_n(\frac{\sigma_1-\sigma_3}{p_0})\}^{0.15}$

6 PRACTICAL UTILITY OF THE CORRELATION

For working out the ground convergence and the corresponding support pressure, using the convergence confinement method, it is required that two characteristic curves be drawn. As far as the support characteristics are concerned, the calculations are comparatively straight forward and equations given by Daeman(1975) or Hoek(1981) may be used. For the purpose of ground convergence the parameters required for utilizing the correlation suggested in this paper can be estimated from observations in the field and laboratory tests, as follows :

1. Rock samples collected from the field can be tested in the laboratory for the uniaxial unconfined compressive strength σ_c.

2. Field studies are required to obtain information regarding the rock quality designation(RQD), the spacing of discontinuities, condition of discontinuities, the orientation of discontinuities and ground water conditions. Utilizing the information collected from the field, the rock mass rating(RMR) can be decided. The parameters m and s can be estimated on the basis of suggestions made by Hoek(1983). The amount of overburden can be used to get an idea of the in situ stress p_0. Using Hoek and Brown's equations, the peak strength of the rock mass can be estimated and knowing p_0, the ratio $(\sigma_1-\sigma_3)/p_0$ can be calculated.

3. The modulus of elasticity of rock material E_r can be determined in the laboratory. The joint frequency can be obtained from field studies. Tests on representative samples of rock joints can give the value of stiffness across the joints K_n. Combining these parameters, the other dimensionless

factor nE_r/K_n can be calculated. Alternatively, the modulus of rock mass can be determined in the field using plate jacking test and borehole extensometers. The ratio of the two moduli can be used to obtain the factor nE_r/K_n from the relation

$$E_m/E_r = 1/(1 + nE_r/K_n)$$

In case, no in situ tests have been conducted, the modulus of deformation of rock mass can also be estimated from the RQD (Deere 1967, Kulhawy 1978) or the RMR (Bieniawski 1984).

Having got the dimensionless factors, it is possible to use the equation and calculate the convergence u_i/r_i for different assumed values of p_i/p_o. These values can be plotted to obtain the ground convergence curve.

7 FIELD DATA AND ANALYSIS

Field data from four projects have been collected, and analysed in the light of rock support interaction to arrive at the rock load and tunnel convergence. Due to limitations of space, example from Maneri Bhali project is given below to illustrate the methodology.

7.1 Maneri Bhali project

7.1.1 General features

The stage I of this scheme harnesses the power potential of the river Bhagirathi flowing down between Maneri and Uttar Kashi. The project work included the construction of an 8.63 km. long 4.75 m. diameter circular concrete lined tunnel, besides a concrete dam, an intake-cum-sedimentation chamber and a powerhouse. The tunnel passes through complex geological formations under rock covers of 30m. to 900m.

7.1.2 Engineering behaviour of rock material and rock mass

A number of plate jacking and flat jack tests have been conducted inside drifts excavated for geological exploration. The average modulus of rock mass (secant value) is of the order of $0.5 \times 10^5 kg/cm^2$ and Elastic modulus of rock material is $3.6 \times 10^5 kg/cm^2$. Compressive strength of the rock material is 310 kg/cm² and triaxial shear tests on rock material have shown c=75 kg/cm² and $\phi = 43.5$.

7.2 Ground behaviour on tunneling

The tunnel has been constructed by conventional drilling and blasting technique. Steel supports of 150m x 150m H sections have been provided at a spacing of 0.3m, 0.5m and 1.0m depending upon the judgement of the engineer at site. Pre-cast concrete laggings have been provided in between and the remaining space filled back with lean concrete. The system started showing distress in course of time when the steel supports started twisting and buckling. As the foundation provided for the steel girders was inadequate, not much resistance was offered to converging movements. At some places, a horizontal strut was welded to increase its stiffness. These girders got sheared off in course of time.

Tunnel closures and rock loads have been measured at a number of locations. A reach in metabasics, which is located between chainages 1000 - 1200 from Heena adit is considered for the purpose of illustration.

7.3 Typical calculation of rock loads and closures

i) Terzaghi's rock load theory

For a classification of rock as completely crushed but chemically intact rock, the rock load work out as 2.64 kg/cm².

ii) Rock Mass Quality Index

Considering a RQD of 60%, two joint sets, with joint softening and clay filling, Q works out as 1.26 kg/cm².

iii) The proposed method of interaction analysis

Estimating Hoek and Brown's parameters on the basis of description given above as m = 0.30, s = 0.0001 and using the other available data from the project. The ground convergence is given by

$$\frac{u_i}{r_i} \times 10^3 = 1 + \frac{6 \times .0525 \times 5.0487}{(p_i/p_o)^{1.625}}$$

Fig.3 shows the ground convergence characteristics

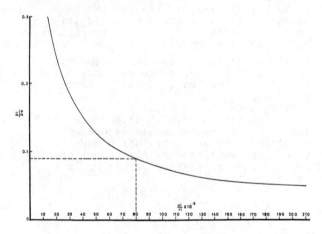

Fig.3. Ground convergence curve.

From fig.3, it can be seen that for a measured convergence of 380mm, the radial deformation corresponds to 190mm, the ratio $u_i/r_i \times 10^3$ would equal 190/2.4 = 79.16. For this ratio, the graph shows a value of $p_i/p_o = 0.09$, which means a support pressure of 7.87 kg/cm². The actual observations are plotted in fig.4. It is seen, the recorded rock pressures are close to those calculated. Similar calculations for other reaches of the tunnel have shown that prediction of closures and rock loads match well with the deserved values.

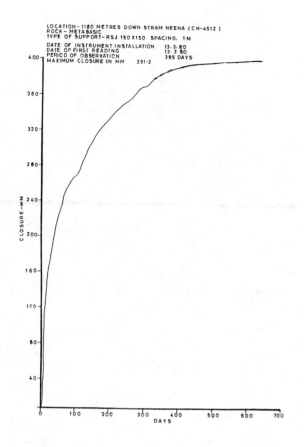

LOCATION-1180 METRES DOWN STRAM HEENA (CH-4512)
ROCK - METABASIC
TYPE OF SUPPORT-RSJ 150 X150 SPACING. 1M
DATE OF INSTRUMENT INSTALLATION   13.3.80
DATE OF FIRST READING   13.3 80
PERIOD OF OBSERVATION   365 DAYS
MAXIMUM CLOSURE IN MM   391.2

Fig.4. Tunnel convergence as actually observed.

8 CONCLUSIONS

Convergence confinement method offers a better choice in the estimation of the radial closures and rock loads for the tunnels through squeezing grounds. With appropriate modifications, the existing methods of evaluation of the ground convergence characteristics can be made more effective.

9 REFERENCES

Bieniawski, Z.T. 1984. Rock mechanics design in mining and tunneling. A.A.Balkema.
Brown,E.T., Bray, J.W., Ladanyi, B. & Hoek, E. 1983. Ground response curves for rock tunnels. Journal of the geotechnical engineering division, ASCE, vol.109, no.1, pp.15-39.
Crouch 1970. Experimental determination of volumetric strains in failing rock. Int. journal of rock mechanics & mining sciences, vol.7, no.6, pp.589-603.
Daeman 1975. Tunnel support loading caused by rock failure. Technical report MRD-3-75, Missouri river division, U.S.corps of engineers, Omaba, Neb.
Daeman & Fairhurst 1970. Influence of failed rock properties on tunnel stability. Twelfth symposium on rock mechanics,Rolla,

Missouri, pp. 855-75.
Deere, D.U.1964. Technical description of rock cores for engineering purposes. Rock mechanics and engineering geology vol.1, no.1, pp.17-22.
Fenner 1938. Untersuchungen zur erkenntnis des gebirgsdruckes. Glucka, vol.74,Essen, Germany, pp.681-695 and 705-715.
Hobbs 1966. A study of the behaviour of broken rock under triaxial compression and its application to mine roadways. Int. journal of rock mechanics and mining sciences, vol.3, pp 11-43.
Hoek 1983. Strength of jointed rock masses. Geotechnique vol. 33, no.3, pp.187-223.
Hoek & Brown 1980. Underground excavations in rock. The institution of mining and metallurgy, London, England.
Kennedy & Lindberg 1978. Tunnel closure for nonlinear Mohr Coulomb function. Journal of the engineering mechanics division, ASCE, vol. 104, no.EM6, proc. paper 14245 pp.1316-1326.
Kulhawy, F.H. 1978. Geomechanical model for rock foundation settlement. J.Geotech Engr.div. ASCE vol.4, pp 211-227.
Ladanyi, B & Nguyen, Don 1971. Study of strains in rock associated with brittle failure. Proceedings of the sixth Canadian rock mechanics symposium, Dept. of Energy, Mines and Resources, Ottawa, Canada, pp 49-64.
Ladanyi 1974. Use of the long-term strength concept in the determination of ground pressure on tunnel linings. Advances in rock mechanics, proceedings of the third congress of the international society for rock mechanics, vol.2, part B, National Academy of Sciences, Washington,DC, pp.1150-1156.
Lombardi 1970. Influences of rock characteristics on the stability of rock cavities. Tunnels and tunneling, vol.2,no.1, London, England, pp.19-22, vol.2, no.2, pp.104-109.
Lombardi 1977. Long-term measurements in underground openings and their interpretation with special consideration to the rheological behaviour of rock. Field measurements in rock mechanics, K.Kovari, ed., vol.2, A.A.Balkema, Rotterdam, Holland, pp.835-858.
Morrison & Coates 1955. Soil mechanics applied to rock failure in mines. The canadian mining and metallurgical bulletin, vol.48, no.523, Montreal, Canada, pp.701-711.

Rock shock rise by seismic action

L'origine des chocs de roche sous l'influence séismique

Die Schockwellenentstehung bei der Erdbebentätigkeit

A.P.SINITSYN, 'Hydroproject' Institute, Moscow, USSR
V.V.SAMARIN, 'Hydroproject' Institute, Moscow, USSR
A.D.BORULEV, 'Hydroproject' Institute, Moscow, USSR

ABSTRACT: Instability of seismic wave propagation in soil and rock massif was investigated by a theoretically designed scheme the so-called cap-model and by using the finite element method. As a result it was obtained that the cause of rock shocks is the plastic dilatancy, investigated by a cap-model.

RESUME: La stabilité des ondes séismiques propagées dans la terre et le massif roché sont étudiès par le schema nommé "cap"-model. Des calculations sont faites par le méthode des elements finis. Des résultats de ces études montre que les chocs en roche sont evoqués par la dilatance plastiquee obtenu par "cap"-model.

ZUSAMMENFASSUNG: Die Stabilitat der Wellen, die das Erdbeben durch die Erde oder das Felsstein verbreiten, wird mit dem theoretischen Schema und so genaunten Modell "cap" studiert. Die Berechnungen waren mit der Methode der endlichen Elemente ausgeführt. Das Ergebnis zeigt dass die Ursache der Felsstosse die plastische Dilatancy ist, die mit dem Modell "cap" beschrieben ist.

INTRODUCTION

As it is know strong earthquakes may give rise to shock waves which have been investigated insufficiently. For example during the earthquake in 1985 in Mexico the underground seismic wave entered the town from the south and brought no destruction to that part of city. The wave propagated to the city centre and as if "exploded" there with great shock and went away to the north. And the areas almost completely destroyed neighbour with the blocks of the city absolutely untouched.

Analogous natural phenomenon arise when seismic waves propagate the rock massif on great depths where the rock structures are situated. As it was stated in scientific work of Ivanzov (1986) the large destructions in rock massif are ground conductors for seismic waves, and give rise to rock shocks. These local effects and other similar effects observed in experiments have up-to-date no suffient theoretical explanation.

In this connection it should be mentioned, that in the last ten years the solution was obtained for nonlinear waves which propagate without changing the form and energy loss (solitons) (Lonngren, 1978). The equations having solutions as solitons (equation of Sine-Gordon, the equation of Korteweg-de Vries, nonlinear equation of Shredinger etc.) have been very well investigated as general equations for solution of any physical problem (Eilenberger 1981, Tahtagian and Fadeev 1986).

But in order to apply these equations to soil and rock materials it is necessary to take into consideration the peculiarity of the problem, which does not permit to restrict it to the investigation of fully integrated systems. For example in rock shocks we have no solitons but solitary waves with complex properties. In this case it is necessary to investigate the problems which cannot be described by the analitic equations mentioned above. To these natural phenomena besids a rock shock belong many other cases of instability of waves propagation in soils and rock massifs which are followed by local magnification of velocities and stresses.

Modern theory of deformation of the solids show that this process depends on complex laws. Depending on material properties, dimensions and shape of solids, and conditions of loading these laws are determined by a number of factors. For soils it is mostly the compression deformation and the ability of energy dissipation.

The scientific work of Nikolaevskiy (1975) shows that the general way of development in soil mechanics is connected with investigations of general problemes by using the elasto-plastic dilatation model. The principal question is which elasto-plastic model must be chosen for automatic taking into consideration the influence of strengthening of material dilatancy, effect of relaxation and energy dissipation.

MATEMATICAL ANALOG OF PHYSICAL MODEL

According to scientific work of Nowogilov (1984) the plastic deformation is a process of transition of multiatom system from one state of stability equilibrium into another by overcoming the energy barriers which separate the two states of equilibrium.

The same is the modern law of dry friction by sliding one solid against another. The deformations in plastic material according to finite displacements of solids caused by small demuniation of stresses cannot occur because the plastic material and the loads which cause plastic deformations cannot do even small work. But it is not pos-

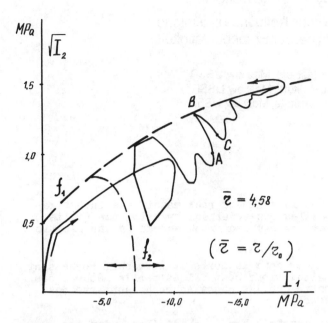

Figure 1. Fluidity surface (f_1), loading surface (f_2) and the stress path in a point of rock massif at a distance $\bar{z}$ = 4.58 from the centre of rock structure (z_0 - radius of a rock structure).

sible to determine the stability of system with friction by any finite work realizing infinite small indignation.

But in spite of the similarity there is a difference between the friction and the plasticity in modeles. In system with friction the finite displacements are possible by infinite small decreasing of pressing load which is equal to instability of system. For the full analogy between plastic model and instability processes in friction, there must be such a plastic model which can describe the mechanism of physical instability at certain conditions of generation.

Mathematical model which takes into consideration this mechanism is a space singular dilatation model of strengthening elastoplastic solids (DiMaggio, 1971) in which the plastic surface is obtained from three dimension nonisothermic experiments and consists of motionless stationary fluidity surface (f_1) and mobile loading surface (f_2) (Fig. 1). The fluidity surface is a modified surface of Drucker-Prager. The mobile loading surface has a shape of ellipse and its displacements are connected with the volumetric deformation of the strengthening law.

Now there is a volumetric elasto-plastic model which does not depend on the velocity of deformation and includes ideal plasticity and deformation strengthening. Both parts of plastic surface together with the laws of deformation and fluidity satisfy Drucker postulate of stability. On the basis of this principle law of active loading the function f_2 is equivalent to dissipative function, which evokes the nonreturned losses of energy on the strengthening part of the surface. These losses are similar to those which arise in a vescous-plastic model of solids.

In the intersection point of curves (Fig.

1) in space of stresses there is a rib where during plastic loading the vector of velocity of volumetric plastic deformation has a direction which is limited by two normales to the crossing surfaces. The movement of ellips is connected with on increase or decrease of volumetric plastic deformation and the strengthening in this model is reversive. This may happen if plastic compression is followed by elastic deformations and the point of stresses is placed on the curve of plastic fluidity surface below the singularity point. In this point the plastic dilatancy occurs and the velocity vector of volumetric plastic deformation which was on loading surface now passes to fluidity surface. Thus if the average stresses decrease slightly the velocity vector rotates unti-clockwise to the finite angle of that of dry friction modelled mathematically.

The occured plastic dilatancy causes the decrease of the plastic surface (ellips). This decreasing of the ellips lasts until the ellips reaches the above-mentioned point and the latter becomes a new singularity point. Now the velocity vector of volumetric plastic deformation returns to the vertical position; the decreasing of ellips automatically stops and the plastic dilatancy disappears.

If the loading point is in the space invariants ($I_1 - \sqrt{I_2}$) and is situated on immobile fluidity surface the increment of stresses is determinated by a following formul

$$d\sigma_{ij} = K\delta_{ij}\left[d\varepsilon_{KK} - 3d\lambda\gamma\beta\sqrt{I_2}\,e^{\beta I_1}/(\alpha - \gamma e^{\beta I_1})^2\right. +$$
$$\left. + 2G\left\{de_{ij} - d\lambda S_{ij}/[2\sqrt{I_2}(\alpha - \gamma e^{\beta I_1})]\right\},\right.$$

where

$$d\lambda = \frac{3\gamma\beta\sqrt{I_2}\,e^{\beta I_1}K d\varepsilon_{KK} + G S_{z,s} de_{z,s}/f_1}{9\gamma^2\beta^2 f_1^2\, e^{2\beta I_1} K + G};$$

$$de_{ij} = d\varepsilon_{ij} - \frac{1}{3}d\varepsilon_{KK}\delta_{ij}; \quad S_{ij} = \sigma_{ij} - \frac{1}{3}I_1\delta_{ij}$$

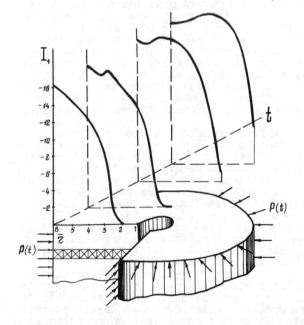

Figure 2. Fragment of stress change depending on time and distance from the centre of rock structure.

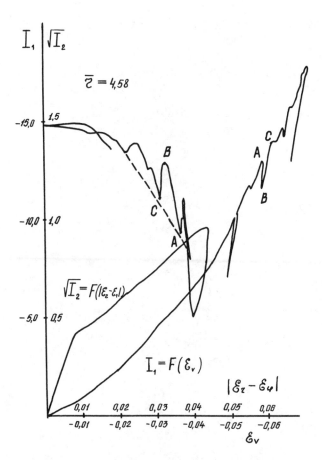

Figure 3. Diagramm of stress-strain and stress intencity-difference of strain at a distance of $\bar{z}$ = 4.58.

α, β, γ - constants of materials; K and G - bulk modulus and shear mudulus.

If the loading point is on a moving surface, which is described by a family of ellipses the increment of stresses is evaluated by formul

$$d\sigma_{ij} = K\delta_{ij}\left[d\varepsilon_{\kappa\kappa} - 3d\lambda \frac{2(I_1 - C)}{R^2 b^2}\right] + 2G\left[de_{ij} - d\lambda \frac{S_{ij}}{b^2}\right],$$

where

$$d\lambda = \frac{6(I_1 - C)Kd\varepsilon_{\kappa\kappa}/(R^2 b^2) + 2GS_{z,s}de_{z,s}/b^2}{36K(I_1 - C)^2/(R^4 b^4) + 4GI_2/b^4 + A};$$

$$A = \frac{6(I_1 - C)e^{-D(x-z)}\left[2(I_1 - C)/(R^2 b^2) + 2f_z\beta(1 - \frac{\alpha}{b})\right]}{R^2 b^2 DW\left[1 + \beta(\alpha R + X - C)\right]};$$

$$X = \left[\ln(1 + \varepsilon_v^p/W)\right]/D + z.$$

ε_v^p - volumetric plastic deformation; W, D, z - constants of material. Parameter C is determined by an equation as a function of X

$$C = X + R\alpha - R\gamma e^{\beta C}.$$

The problems of seismic wave propagation in rock are described by finite element method (FEM). The equation of rock motion is obtained from the law of energy conservation by a formul (Oden, 1972)

$$M_{FD}\Delta\ddot{u}_i^D + \Delta K_{F_i} + \mathring{H}_{FD}\Delta u_i^D + \Delta H_{FD}\mathring{u}_i^D +$$

$$+ \Delta H_{FD}\Delta u_i^D = P_{F_i} - \left(M_{FD}\mathring{\ddot{u}}_i^D + \mathring{K}_{F_i} + \mathring{H}_{FD}\mathring{u}_i^D\right). (1)$$

Equation of motion (1) is valid for any material and gradients of deformation and change of loading. If deformations are large the equation (1) should be nonlinear. But the seismic load increases to maximum value in a limited area near to heart of earthquake. The dimensions of this area are considerably smaller than of the area of investigations. In this case a linearised form of equation (1) may be used. The last component of the left part of the equation omitted. Such form of equation is usefull especially for dynamical problems with elastoplastic materials. The solution is obtained by a system of nonlinear algebraic equations (Morosov, 1980).

In FEM three iteration methodes are mostly used. But the method of initial stresses is the best as it is suitable for any connection between stresses and deformation and for ideal plastic material too. This general preferance of the method of initial stresses permits to apply it widely in practice as the decrease of load can also be taken automatically into consideration. This is extremely necessary for the investigation of seismic wave propagation.

RESULTS OF CULCULATIONS AND THEIR ANALYSIS

Detailed algoritmus for solving the problems of seismic wave propagation in conjunction with the above-mentioned model and the finite element method as well as preliminary resultes were published in proceedings of the 5-th International Conference on numerical methods in geomechanics (Sinitsyn and Samarin, 1985). This report is a further study of the problem of seismic wave propagation by nonlinear theory of local physical instability in elaborated mathematical model.

For example the investigation of stresses around a cylindrical rock structure at a symmetric seismic load exerted on it within a distance of $\bar{z}$ = 4.58 shows that a certain moment there arises a new solitary wave (Fig. 2) which propagates from this point in all directions with the velocity of a volumetric elastic wave. This wave turns around the general wave and magnifies its amplitude.

The appearance of this local wave can not be explained by mathematical instability of problem solution or by a mistake in calculations. First of all, there are confirmations about the conservation of stability when a stable system is placed on a new linear bond as in this problem. Secondly the analysis of calculations show that the rise of local wave in rock massif exactly coincides with the time moment when the path of stress changes touches the fluidity surface, which is the cause of the instability process, as is the case with the model of the real physical instability due to rock shock.

The path of stress changes is described by a complex curve (Fig.1), which periodically touches the fluidity surface and each touching gives rise to a new wave. From the

diagramm ($I_1 - \mathcal{E}_v$) (Fig. 3) it is possible
to make a conclusion that each touching coincides with unloading and after this a
new loading, which results in magnified stresses. The results obtained theoretically
from such evaluations when rock shocks followed one another very well confirm experimental observations.

Thus the existing opinion regarding the
conditions of rock shock rising during seismic wave propagation must be defined more
precisely in the sense that the plasticity
and arising plastical dilatation are the
cause of a rock shock and other instability effects.

REFERENCES

Ivanzov, N.M. 1986. Modern problems of rock
mechanics in energy construction. M.:
Energoizdat.
Lonngren K., Scott A. 1978. Solitons in action. New York: Academic Press.
Eilenberger G. 1981. Solitons. Mathematical method for physicists. Berlin: Springer.
Tahtagian L.A., Fadeev L.D. 1986. Method of
Gamilton in theory of solitons. M.: Nauka.
Nikolaevskiy V.M. 1975. Modern problemes in
soil mechanics. In "General laws of soil
mechanics. M.: Mir, p. 210-229.
Novogilov V.V. 1984. Nonlinear models in
problems of mechanics of solids. M.: Nauka.
DiMaggio F.L., Sandler I.S. 1971. Material
model for granular soils. In J.eng.mech.
div., EM3., p. 935-949.
Oden J.T. 1972. Finite elements of nonlinear continua. New York: McGRaw-Hill book company.
Morosov E.M., Nikishcov G.N. 1980. Method
of finite elements in destruction mechanics. M.: Nauka.
Sinitsyn A.P., Samarin V.V. 1985. Numerical
investigations of stresses in rock structures with deformations beyond the elastic limit due to seismic waves. In proc.
V-th Int.conf.on Numer. methods in geomech. (Nagoya), p. 1329-1335.

Statistical analysis of the joint physical field measurements during the process of deformation and failure of large-scale rock blocks

Analyse statistique des mesures des champs physiques des fissures au cours de la déformation et de la destruction des blocs rocheux de grandes dimensions
Statistische Auswertung der Messergebnisse der physikalischen Felder der Klüfte während der Deformationen und des Bruches von grossen Felsblöcken

G.A.SOBOLEV, Institute of the Physics of the Earth, USSR Academy of Sciences, Moscow
K.BADDARI, Moscow Geological Prospecting Institute, USSR
A.D.FROLOV, Moscow Geological Prospecting Institute, USSR

ABSTRACT: The possibility to distinguish the early stages of microfailures and to predict the local macrofailures in a stress-strained rock by means of complex statistical parameters which generalized a few physical precursors are shown.

RESUME: On a montré la possibilité de distinguer les stades précoces des microruptures et de prévoir les macroruptures locales dans une roche se déformant sous l'action d'une tension, grace à des parametres statistiques complexes généralisant quelques précurseurs physiques.

ZUSAMMENFASSUNG: Hier ist die Möglichkeit der Aussonderung der früheren Statien von Mikrozerstorungen und der Prognose von lokalen MakrozerriBen in deformierten Spannungsdestein mit Hilfe der komplexen statistischen Parametern, gezeigt, die Abmessungen der Reihe von physikalischen Vorboten zusammenfassen.

INTRODUCTION

The most complicated problems of rock mechanics are those of revealing the early stages of fracturing and predicting catastrophic phenomena (earthquakes, rock bursts, quarry collapses etc.) in a stress-strain rock masses. In such cases loss of strength occur under the stresses far less than the yielding point due to brittle destruction prevailing.

Any rock is always characterized by a definite space crystallizing structure (Rebinder 1979) with some strength distributions of grain contacts, grains themselves and intergrain zones. The process of deformation and failure of such a structure under loading is accompanied by regular changes of these distributions caused by distruction of some of the less strong contacts and grains and formation of microfractures. In real rock masses these changes are not uniform in space and time and sometimes anisotropic, which makes the deformed rock mass state control much more difficult.

The principal physical subsystems (Frolov 1986) associated with the spatial crystallizing structure of any rock as a complex microsystem and conditioning the response to external mechanical influence are: ionic, dipolar and electronic holes. The changes of spatial structure are accompanied by the corresponding responses of these subsystems and therefore by certain changes of mechanical. electric and some other properties of the medium (emanation, permeability, etc.), most of which can be recorded by remote control methods. Physical precursors of rock mass destruction have been studied for a long time and certain progress has been achieved (Sobolev 1982; Scheidegger 1975; Yamshchikov 1982; etc.).

However, the multifactor connection of individual physical properties with micro- and macrodestruction of spatial structure of rock makes it difficult to interpret the measurements of individual physical parameters and to predict fracturing, especially on its early stages.

SAMPLES AND METHODS

Unique experiments on laboratory modelling of rocks and concrete failure consisting in uniaxial loading of large (0.5-1 m³) blocks with 50000 tons press have been carried out in the Institute of Physics of the Earth, USSR Academy of Sciences. The rate of loading was 10^{-3} MPa/s. Experiments were made under step-increasing and cyclic loading until the failure of the blocks. Simultaneously a number of physical fields have been measured on many points of every block.

Five physical parameters were considered as principal ones: facet plane strain tensor (I), elastic longitudinal wave velocity (V), acoustic emission intensity (N), electric resistivity and electric potential changes (and U respectively).

Results of the statistical analysis of the experimental data during deformation and failure processes of a granite block (700x x700x700 mm) and a dolerite block (600x600x x900 mm) are discussed in the paper. Experiment was stopped after microcracking of the block. Maximum loading (P_r) is 140 MPa for granite and 75 MPa for dolerite.

For analysis and transformation of raw data the method of natural ortogonal functions was chosen, its basic system of functions is not pre-determined, but is determined by correlation matrix of normalized raw data X_{ik}. Close interrelation between the registered physical parameters was supposed, since all of them reflect the responses of the same physical subsystems in the spatial structure of the rock. That was confined by the high values of the pair correlation coefficients (0.6÷0.94).

Besides, our calculations show that for all analyzed cases 95-98% of information were concentrated in the first ( ) and second ( ) components of the eigenvector ( ). The following complex parameter was chosen as the common characteristic for the total complex of physical fields:

$$S_{nk} = \quad _{hi}X_{ik} \qquad (1)$$

and the first S_{1k} and the second S_{2k} components of S_{hk} were used for the detailed analysis of rock deformation process. For them concrete relations like (1) were found, and $S_1( )$ and $S_2(P)$ were calculated both differencially, i.e. for the parts of the blocks with different strength (destroyed by tensile cracks and placed far from the blocks) and integrally, i.e. on the average for each block.

RESULTS AND DISCUSSION

The results obtained show the universality of the description of the process of deformation and failure of granular medium with crystallizing spatial structure using complex parameters $S_1(P)$ and $S_2(P)$. For example, for the parts destroyed by tensile cracks the equations of these parameters are:

$$S_{1k} = -0.22V-0.18I-0.20N+0.19\ U+0.19\ S$$

$$S_{2k} = 0.01V+0.25I-0.19N+0.21\ U+0.22\ S$$

for the divine dolerite block and

$$S_{1k} = 0.15V-0.10I+0.23N+0.24\ U+0.23\ S$$

$$S_{2k} = 0.27V+0.34I-0.04N-0.005\ U-0.003\ S$$

for the granite block.

By typical changes of S_2 (Fig. 1) one can see with confidence 3 stages of stress-strain states of rocks.

The first stage is observed at little loading and is characterized by increasing negative values of S_2 up to S_2 0 (Fig. 1, curves 1), when the load P (0.5÷0.6)P_F. This stage is called quasielastic, meaning practically reversible deformation with predominance of surface reduction at the side of the block. In the parts of the blocks with hight solidity disposed far from macrofailures the quasielastic stage is observed at loading P (0.8÷0.9)P_F.

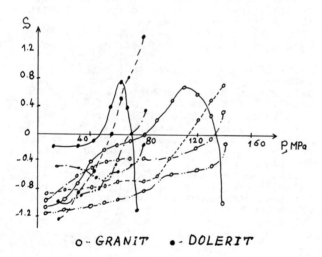

o -- GRANIT   • - DOLERIT

Figure 1. The variation of complex parameters $S_2(P)$ and $S_1(P)$: 1 ——— S_2 and 2----S_1 for macrofailure formation zones; 3 -.-.-S_2 and 4 -..-..-S_1 for stable deformation parts.

The second stage is characterized by the transition of S_2 to the positive values. It is called quasiplastic stage corresponding to the onset of irreversible microfailures. It can be observed at the loading (0.7-0.8)P_F, when S_2 reaches the maximal value and S_1 0 (curves 2). In the parts of blocks with[1] stable deformation, the failure process ends at this stage and S_2 (Fgi. 1, curves 3) does not reach the maximal value and S_1 (curves 4) is always negative.

The third stage called progressing macro-failure and formation of tensile local zones is characterized by the transition of S_2 through maximum value and sharp decrease up to the negative value. The complex parameter S_1 becomes positive and quickly increases (curves 2).

Such interrelation between S_2 and S_1 corresponds to predominance of dilatancy[1] in the zones of future macrofailures. In this case the junction of microfissure system is observed, which leads to splitting of the rock block. For the parts of blocks with a stable deformation, this stage is not observed.

The reaching by S_2 the maximum followed by sharp decrease combined with increase of S_1 are the sure criteria - predictors of macro-failure in definite parts of the rock mass.

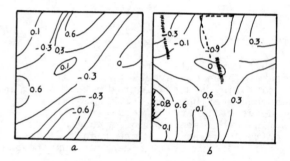

Figure 2. Spatial distribution of S_2 for a side of a granite block: a) at P (0.7-0.8)P_F b) at P P_F.

Maps of isolines of parameter S_2 for the side of a granite block are shown[2] in Fig. 2. These maps illustrate the space variability of S_2 in connection with non-homogenity of rock failure. The parts of the block that are most dangerous in relation to macrofailure are clearly separated by the maximum values of S_2 at loads P = (0.7-0.8)P_F (in Fig. 2a) it is a small part of the left side of the block and its upper section. In all other sections in accordance with the abovementioned criteria there is no danger of approaching failure. The lower and middle sections of the block are still in the stage of quasielastic deformation. This prognosis is fully confirmed by Fig. 2b, according to the stress-strain state of the rock at loading P P_F. Ruptures in the upper parts and a peeling of a small part from the left side of the sample (after the maximum S_2 sharply decreased up till negative values were observed) within the forecasted local parts while the block itself did not undergo macro-failure. In accordance with the values of S_2 (Fig. 2b), deformation of the majority of the block parts ended at the quasiplastic stage. This is fully confirmed by $S_2(P)$ and $S_1(P)$, calculated integrally (in average for the granite block).

At the same time integral parameters $S_2(P)$ and $S_1(P)$ for the block of olivine dolerite which at P=P_F was cut by several major

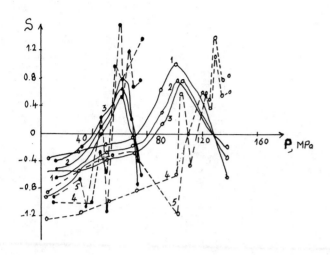

Figure 3. Examples of variation of $S_2(P)$ and $S_1(P)$ with the diminution of raw physical parameters: 1 - $S_2(V,N,U,S)$; 2 - $S_2(V,I,N)$; 3 - $S_2(V,N,U)$; 4 - $S_1(U,S)$; 5 - $S_1(V,N)$.

ruptures all over volume (1) show characteristic changes according to the criteria - predictors of the above-mentioned macrofailure. This bears witness to the fact that in more homogeneous and fine dolerite the failure occurred more uniformily over the block volume than in the granite.

The minimisation of raw data is interesting in practical use of multiparametric physical information. Using many combinations from 4 to 2 of physical registered parameters $S_2(P)$ and $S_1(P)$ were calculated. On curves 1-3 (Fig. 3) are well visible all characteristic changes (though with some drift on axis P for granite) allowing one to separate three stages of deformed state of the rock and to predict the approaching macrofailure.

On decreasing the number of physical parameters in use to 2, the criteria connected with characteristic changes of $S_2(P)$ practically disappear and the prediction of approaching of macrofailure becomes insufficient and is possible only through $S_1(P)$ (Fig. 3, curves 3, 4).

CONCLUSIONS

1 - Complex statistical parameters S_2 and S_1 effectively generalize information of a series of physical (mechanical and electrical) failure precursors of a stress-strained rock. Parameter S_2 gives a lot of details of changes depending on stress-strain state of rock and is basic for diagnosis of the state of a rock block and prediction of its failure.

2 - The $S_2(P)$ changes indicate that the rock stress-strain states could be devided into three stages:

a - quasielastic stage - $S_2(P)$ has negative values and at the end of this stage tends to zero at $P = (0.5-0.6)\ P_F$;

b - quasiplastic (the onset of irreversible microfailures) - $S_2(P)$ has a tendency to increase, reaching the maximal value at $P=0.7$--0.8 at failure load;

c - progressing macrofailure stage - $S_2(P)$ after reaching maximum decreases quickly, changing again towards negative values at load P P_F.

3 - The first component S_1 of the complex parameter does not allow the exact prediction of earlier stages of failure but at the same time it can be used to confirm the transition of the block to the progressing-failure stage.

4 - The transition of $S_2(P)$ over the maximum and quickly increasing of $S_1(P)$ are the predictors of approaching brittle macrofailure.

5 - The informativity of S_2 and S_1 is always sufficient to detect all three stages using four or three acoustic and electric parameters. That allows one to make a sure acoustic-electric system (including automatic definition of complex parameters S_2 and S_1) in order to control the state of the rock mass and predict catastrophic phenomena on different parts of it.

6 - Space-time variability of values of parameters S_2 allows the explanation of the non-homogeneity and anisotropic deformation and failure of rock blocks prediction of macrofailures in local parts of stress-deformed rock masses.

REFERENCES

Baddary, K., G.A. Sobolev & A.D. Frolov 1986. Analyse deformaszionnikh predvestnykov kroopnogo obrasza doleryta. Izv.Acad.Nauk USSR, Phys.Zem.:10.

Rebinder, P.A. 1979. Poverkhnostnye yavlenia ve disperssyonnykh systemakh y physico-khimicheskaya mekhanikha. Moscow: Nauka.

Sobolev, G.A. 1982. Predvestnyky razruchenia bolchogo obrassa guornoi porody. Izv.Acad. Nauk USSR, Phys.Zem.: 8.

Frolov, A.D. 1986. Guornaya poroda kak slojnaya makrosystema. II Vsesoyouzn. Conf. Systemny podkhod ve gueologui. Tezisy doklada, Ch. 1: Moscow.

Scheidegger, A.E. 1975. Physical aspects of natural catastrophes. Elsevier SPG. Amsterdam-Oxford-New York.

Yamshchikov, V.S. 1982. Metody y sredstva islodovaniya y kontrolia guornykh porod y processov. Moscow: Nedra.

Recent advances in the interpretation of the small flat jack method
Progrès récents dans l'interprétation de la méthode des vérins plats
Ein neues Auswertungsmodel der Methode der kleinen Druckkissen

C. SOUZA MARTINS, Promon Engenharia, São Paulo, Brasil
L. RIBEIRO E SOUSA, Research Officer at Laboratório Nacional de Engenharia Civil (LNEC), Lisbon, Portugal

ABSTRACT: LNEC has been developing the small flat jack (SFJ) method for the in situ measurements of the state of stress based on stress relief by opening slots, and also to evaluate the deformability of the rock mass. A new interpretation model for this method was developed, using a 3-D finite element model.

RESUMÉ: LNEC a amélioré la méthode des vérins plats pour mesurer l'état de contrainte sur place, qui est basée sur la libération des contraintes par l'ouverture de saignées, laquelle sert aussi à évaluer le déformabilité du massif rocheux. On a préparé un nouveau modèle d'interprétation pour cette méthode, utilisant un modèle tri dimensionnel d'éléments finis.

ZUSAMMENFASSUNG: Bein LNEC ist bislang die Methode der kleinen Druckkissen (SFJ = Small Flat Jack) entwickelt worden. Diese Methode dient der "in situ" durchzuführenden Messungen im Gebirge, zur Ermittlung der Spannungen und der Verformbarkeiten, aufgrund der erhaltenden Entspannungen, die als Foge von eingesägten Schlitzen im Gebirge entstehen. Es ist ein neues Auswertungsmodell dieser Methode aufgrund eines "3-D finite element" Modelles.

1 INTRODUCTION

In the analysis of underground structures in rock masses it is necessary to predict actions on these structures, particularly those regarding initial stress relief along the boundaries created by the excavations. These actions depend on the initial state of stress prior to the construction of the work and also on the adopted construction sequence. Accurate forecasts of the state of stress must be made by means of in situ measurements. Among the various methods used for the purpose, emphasis is laid on those based on stress relief. LNEC has been developing its own test methods in this domain, particularly the SFJ (small flat jack) method and the STT (stress tensor tube) method.

The SFJ method is based on partial stress relief by the opening of slots.

For this method, a new interpretation model was developed, using a finite element tridimensional model. There the rock mass is assumed with mechanical characteristics of isotropy or such as an isotropic-transverse body. One also considers the hypothesis of a crack being generated in the plane of the slot for the usual test depths. The results obtained with the finite element model are compared with the values obtained by the previous experimental model used and with an analytical solution.

2 SFJ METHOD

2.1 Test technology

The flat jack method consists of making a slot in an exposed surface and restoring measurement points to their initial position by applying pressures through flat jacks. Pressures required are in correspondence with the normal stresses acting on the slot surface. This method uses flat jacks, which consist of two steel blades with a given shape, welded one another along their contour, a certain pressure being transmitted to the flat jacks through the oil that is inside them.

As regards techniques developed for this method, reference should be made to the LNEC technology using small flat jacks (Rocha et al. 1966). It may be briefly described as follows (Fig. 1): i) on a surface to be

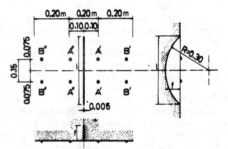

Figure 1. SFJ system-geometry of a slot

studied apply pairs of measuring bases between which distances are measured; ii) with a diamond disk sawing machine, open a slot between the bases, so relieving normal stress on its plane and consequently distance between measuring bases will vary; iii) insert a flat jack to fill up the slot and apply pressurized oil in the jack, measuring the distance between bases until the initial position is restored and so, pressure introduced in the flat jack to restore to the initial position (cancelling pressure) corresponds to the normal stress.

It is assumed that the cancelling pressures is equal to the normal stress acting on the surface of the slot less a correction factor that essentially depends on the shape of the slot segments. LNEC has been using circular segment shapes with slot depths (f) of 10.5, 17 and 24 cm, the corresponding correction factors being 9%, 8% and 7%.

The innovation brought about by the LNEC technique with respect to others alread in use lies in the way of opening the slot to accomodate the flat jack. In this case a diamond disk saw is used, whereas conventional slots have been obtained by means of circular parallel holes, partly overlapping. With the latter, the cutting is very lengthy, it disturbs the rock mass and makes it necessary to use cement group to pack the jack into the slot, with the further drawback that the elastic characteristics of cement are markedly different from those of the surrounding rock mass.

The SFJ method also permit to evaluate the deformability of the rock mass in the test site. The modulus of deformability has been obtained through experimental curves plotted from results of reduced model tests (Rocha et al. 1966).

2.2 Difficulties in the application of the method

The SFJ technique presents some advantages, namely simplicity and the fact that mechanical characteristics of the rock mass need not to be known beforehand. These tests ave been applied by LNEC, assuming that the rock mass is a homogeneous, isotropic and elastic medium and that the cancelling pressures correspond to the normal stress acting on the slot. In fact there are some factors that make its application difficult, particularly the fact that rock mass behaviour differs from the interpretation model, due to the loading system, and to the measuring technique used.

Thus experience shows that reversible behaviour of the loading-unloading curve which is inherent to the assumption of homogeneous, elastic and isotropic rock mass, does not occur. Sometimes a hysteresis effect and a residual deformation are detected, which are more marked in the first cycle. The real behaviour may be justified on account of the influence of several factors, such as heterogeneity and discontinuities of the rock mass. Another difficulty in the interpretation of the test arises owing to the fact that the zone tested is decompressed and, sometimes, fractured because of excavation, and this must be taken into account in the specific interpretation of tests. It is therefore recommended that they should be done in sound zones, so as to, avoid as far as possible the influence of decompression and fracturing. Problems also occur as regards the hypotheses assumed for the behaviour of the rock masses, namely when there are failures and plasticization in the rock mass zone involved or when there is marked aniso-tropic behaviour of the rock mass (Faiella et al. 1983).

Other difficulties in the interpretation of these tests are related to the load applying procedure. As regards the loading system, it should be noted that the cancelling pressure depends on the measurement points, that is, it is not possible with the same cancelling pressure restoring all measurement points to their original position. Among other reasons this may be due to the normal stress not being uniform along the slot surface, and also to the fact that pressure exerted by the flat jacks is not uniform over all slot area. Other difficulty may arise from the existence of shear stresses, whose influence on displacements resulting from slot opening are difficult to evaluate. As thickness of the slot is small (5 mm) with the SFJ system, that effect can be neglected in such case, as verified by Rocha et al.(1966). Due to stress concentration at the edges of the slot during application of pressure by a flat jack, existing fractures may open or cracks occur. As their depth is not known a priori, such events may hamper interpretation of results.

With reference to the technology used is test, experience has shown that the technique is comparatively easy, quick and unexpensive, and the error does not exceed 20% in the determination of the state of stress. Nevertheless some improvement in the test technique is felt necessary, namely as regards location of measurement points and of the displacement measuring system.

3 CALCULATION MODEL FOR TEST INTERPRETATION

As an attempt to improve the model that interprets the results of the SFJ method, a tridimensional finite element model was built up. A parametric study was made for different mechanical characteristics of the rock mass, by assuming the hypothesis of a crack being generated in the plane of the slot during application of pressure by the flat jack for the two

slot depths (17 and 24 cm) more used by LNEC (Martins, 1985).

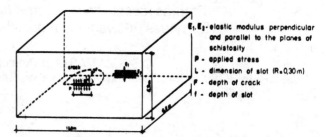

Figure 2. Conceptual model for test interpretation

In Fig. 2 shows the model adopted. The face where the slot is cut corresponds to the boundary condition of null state of stress; normal displacement is assumed to have been prevented in the remaining faces.The rock mass is assumed a homogeneous and elastic medium supposed to be either isotropic or anisotropic. For anisotropy, a transverse isotropic body is considered, with the isotropy planes (strata) parallel ($\theta=0^{o}$) or normal to the slot ($\theta=90^{o}$), the modulus of elasticity perpendicular to the stratum E_1 being equal to 1, and that parallel to stratum E_2 being variable, and the Poisson's ratios (ν) being equal to 0.2 or 0.1. Anisotropy ratios E_2/E_1 equal to 4 and 8 are considered and, in isotropic conditions with a slot depth of 24 cm, several crack lengths in the plane of the slot were considered.

A 3-D finite element model was built to one fourth of the block, due to some simplifications introduced, symmetry planes were assumed. The mesh has 2091 nodal points.

In order to simulate the pressure applied by flat jacks to the slot surfaces, uniform unit pressure (σ_z^o) was assumed for any of the different hypotheses of crack length. Consideration of the initial state of stress assumes the existence of other components, such as those in the plane of the slot surface (σ_x^o and τ_{xz}^o), whereas the ramaining can be neglected since the depth of the slot is small.

By assuming the unit stress component σ_x^o, calculation was made for a situation of isotropy with slot depth equal to 17 cm, and without crack, that component being found not to affect the test results. As regards stresses τ_{xz}^o, their influence on test results could not be checked since the model adopted has two planes of symmetry.

4 ANALYSIS OF THE STUDY

4.1 Isotropic medium

The first analyses carried out refer to the situation of isotropy and slot opening with f equal to 17 and 24 cm. Fig. 3 was then plotted, showing displacements between pairs of measurement surface points at several distances of the slot plane, along two axes perpendicular to the slot, one through the center (section AA'), the other to the distance of 7.5 cm (section BB'). Results are referred to the hypothesis of $\nu=0.2$. For the case of f = 24 cm, further calculation was made with $\nu=0.1$. The values were found to be practically coincident, with variations that do not exceed 2% as a rule. Values of displacements obtained through an analytical approach are also presented for an elliptically — shaped slot with infinite length under stresses (Martins, 1985).

Fig. 3, which referes to the diagrams of displacements calculated for points at 7.5 cm from the axes of the slot, presents the corresponding values as determined by laboratory tests on a prismatic model, 40x50x100 cm in size, made of a mixture of plaster, diatomite and water, submitted to uniaxial compression (Rocha et al. 1966).

Studies carried out made it possible to obtain a

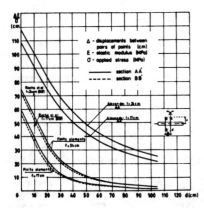

Figure 3. Curves of displacements as a function of the distance between pairs of points.

complete picture of the fields of stresses and displacements produced by opening of the slot and application of the cancelling pressure, and allowed us to better understand phenomena involved in the use of the method. Besides comparison could be made with results obtained through other approaches, namely experimental and analytical approaches. Thus if diagrams of displacements with slot depths 17 and 24 cm are analysed, a similar evolution is found, as well as about 20% attenuation points 30 cm or 40 cm distant from the slot respectively for slot depths of 17 and 24 cm. As can be seen, for distances of 30 cm which correspond to the most distant pins used in SFJ tests, the values of displacements are fairly reduced as compared to maximum displacements measured at the slot surface.

For pins at 10 cm whose displacements are ordinarily used for the determination of the state of stress and deformability, displacements are 66% and 74% on the maximum displacement value, respectively for f = 17 and 24 cm. By considering the two slots, displacements obtained are abot 40% larger for a depth 24 cm and a distance of 10 cm, and they are much larger for the distance of 30 cm, reason why it should be recommended to locate the pins at smaller distances when slots are less deep.

Comparison between displacements given by numerical solutions and those calculated through the analytical (Alexander) solution provided a first checking of results obtained in calculations the latter solution pointing to an upper limit since hypotheses therein assumed involve a slot that extends indefinitely.

By comparing values of displacements obtained in the experimental model mentioned and those given by the numerical solution (Fig. 4) one will find out that the latter values for pairs of points at 10 cm from the slot and at 7.5 cm of the axis (Δ_1) are about 75% of the former values in case of a slot depth of 17 cm, which becomes 73% in case of f=24 cm. Nevertheless contrary to the above statement, for relative displacements between points 10 cm distant and points 30 cm distant (Δ_2) displacements calculated through the experimental model are found to be much smaller than those given by numerical models, in 54%

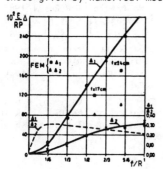

Figure 4. Curves $E\Delta/RP$ versus f/R

for f = 17 cm and 53% for f = 24 cm, respectively. Results given by finite elements show that displacements Δ_2 are considerable in absolute value when compared with Δ_1, reaching values of about 44% of the maximum displacement.

These differences in results can be justified for the following reasons: i) in the experimental model slots opened concern a model in the geometric scale of 1/5 and so slots have a radius of 6 cm and thickness of 1 mm; ii) as free surfaces were considered, the block tested was submitted to uniaxial compression, thus displacements having been induced which are likely to exceed slightly those in actual conditions; iii) the numerical solutions in finite elements as they use a formulation in displacements are more rigid solutions as regards the exact solution of the problem, reason why the maximum error by default is estimated not to exceed 5 % on account of the discretization used.

Based on results obtained with the calculation model, curves E versus Δ/RP were plotted, Δ being the displacement of pairs of points at the same distance to the slot of radius R, and P the cancelling pressure. Fig. 5 shows these curves for slot depth of 24 cm and points in planes perpendicular to the slot, at 7.5 cm from the center.

The modulus of deformability E can be obtained by means of charts, or else by the expression

$$E = C_i \frac{P}{\Delta_i} \tag{1}$$

where Δ_i is the displacement between pairs of points at the distance Z_i from the slot, and C_i is the coefficient obtained with the calculation model for these points i. E and P are expressed in MPa, C_i and Δ_i in cm.

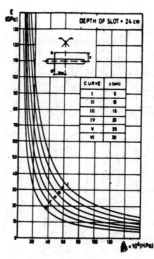

Figure 5. Curves E versus Δ/RP for f = 24 cm.

As said, it was found on the one hand that in the test zone the rock mass is sometimes decompressed and fractured due to the influence of excavation and surface levelling, on the other hand owing to stress concentration in the edges of the slot and to slot opening methods it may be possible a priori to produce opening of the existing fractures and/or cause new fractures to appear, whose depth is unknown. These conclusions obviously make the analysis of test results difficult, and thus an attempt was made to analyse the influence of a crack, for simplicity assumed to be in the plane of the slot. Cracks with F/f length equal to 0; 1/8; 2/3; 2; 4.5 and infinite were considered, for the case of 24 cm depth and isotropic medium with elastic linear behaviour, being F the length of the crack.

Fig. 6 shows displacements obtained at pairs of superficial equidistant points, along two planes perpendicular to the slot at distances x = 0; 7.5 cm for the different crack lengths, including cases of null length and of the analytical solution. The values obtained for the analytical solution are higher, even if compared with the infinite crack solution.

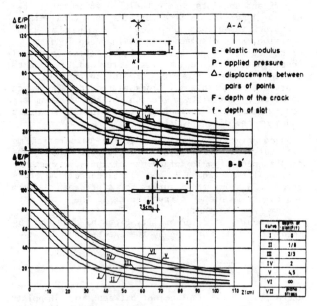

Figure 6. $\Delta E/P$ relations as a function of distance to slot in case of cracks

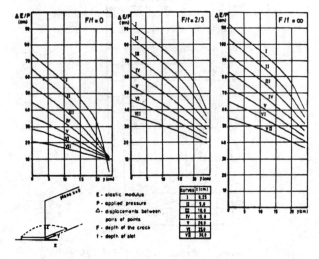

Figure 7. $\Delta E/P$ curves for inner points (x = 0)

Fig. 7 shows charts that relate parameter $\Delta E/P$ to the distance y from the surface, for measurement points at distance z from the slot, with some slot lengths defined, and in a plane perpendicular to the plane of slot at zero distance to the axis.

The modulus of deformability as a function of the measurement point and crack length can be expressed as in (1), changing the parameter C_i by K_i. K_i being a parameter that represents displacement of point i for E and P equal to 1. The values of K_i are obtained by charts (Martins, 1985).

4.2 Anisotropic medium

Let us also refer to the analyses carried out for the hypothesis of anisotropy, in which a transverse isotropic material was assumed, with direction normal to the plane of isotropy parallel or perpendicular to the plane of slot, considering f = 24 cm and no cracks existing.

Fig. 8 shows values of $\Delta E_N/P$ versus E_N/E_P, E_N being the modulus of elasticity perpendicular to the plane of slot and E_P the modulus parallel to that plane. Displacements obtained in the test are found to be influenced both by the modulus of elasticity normal to the

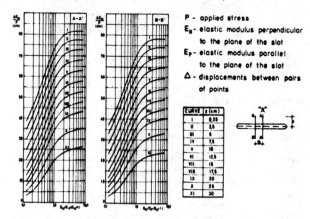

Figure 8. Variation $\Delta E_N/P$ in function of E_N/E_P

plane of slot and by the ratio between moduli. When the modulus of elasticity parallel to the slot is 4 or 8 times higher than the perpendicular modulus, displacements between pairs of points, 10 cm distant from the slots and along an axis 7.5 cm distant from the center, are found to reduce to 56% and 37% respectively, when compared with the hypothesis of isotropy with modulus equal to E_N. On the other hand, when testing is carried out perpendicularly to strata, displacements between those pairs of points are higher than 15% and 16% respectively for ratios of the moduli equal to 4 or to 8.

For these reasons, when these values are used for the determination of the moduli of deformability in directions normal or parallel to the plane of stratification, it is not always acceptable to adopt the hypothesis of isotropic and equivalent continuous medium. Hence on account of values of the moduli determined in both directions for the hypothesis of equivalent continuous medium, one should recommended that these values should be corrected on basis of the ratio of the moduli already calculated and using charts presented in Fig. 8. An iterative sequence can thus be established; for points near the slot two or three iterations may be enough, whereas the number of necessary iterations augments for more distant points.

REFERENCES

Faiella, D., G.Manfredini & P.Rossi 1983. In situ flat jack tests: analysis of results and critical assessment. Int Symp. In Situ Testing, Paris.

Martins, S. 1985. Contribution to the study of underground structures associated with hydroelectric development (in Portuguese). LNEC Research Officer Thesis, Lisbon.

Rocha, M., B.Lopes & N.Silva 1966. A new technique for applying the method of the flat jack in the determination of stresses inside rock masses. 1st Cong. of ISRM, Lisbon.

Mechanisms of backfill support in deep level tabular stopes

Mécanismes de soutènement par remblayage dans des chambres de minage tabulaires et profondes
Stützwirkung durch Versatz im Kammerbau für tafelförmige Erzkörper

T.R.STACEY, Steffen, Robertson and Kirsten, Johannesburg, Republic of South Africa
H.A.D.KIRSTEN, Steffen, Robertson and Kirsten, Johannesburg, Republic of South Africa

ABSTRACT : In deep level gold mining operations, the beneficial effect of backfill has traditionally been assessed in the light of its ability to reduce the amount of energy available to cause damaging rockbursts. The elastic methods usually used to quantify the benefits of this effect indicate that, owing to the high compressibility of the fill, the support provided in areas of small convergence will be negligible. This is contrary to the beneficial effects observed in practice.

The results of analyses of the hangingwall as an elastically supported beam, and the stope surrounded by a jointed rock medium, are presented. These indicate that, contrary to the approach of conventional elastic analyses, the non-elastic effects are very significant. Backfill is shown to be very effective in reducing the non-elastic deformations and the jointed hangingwall shows a strong tendency to act as a beam. It is shown that, for stopes which are not subject to large elastic closures, the support effect of the backfill is not sensitive to the quality of the backfill.

RESUMÉ : En ce qui concerne les operations de mines d'or à grandes profondeurs, on a évolué l'effet avantageux du "backfilling" (remplissage) tenant compte de sa capabilité de réduire la quantité d'énérgie disponible à produire des "coups de toits" violents. Les méthodes elastiques employées d'habitude pour quantifier les avantages de cet effet montrent que à cause de la haute compressibilité du remplissage, le support pourvu dans les endroits à petites convergences sera négligeable. Ceci est en opposition avec les effets avantageux observés en pratique.

Les résults des analyses du mur de suspension en forme de poutre supporté elastiquement et du chantier entourrée d'un milieu de roches jointes sont présentés. Ceux-ci indiquent que, contrairement à la facon conventionelle d'aborder les analyses elastiques, les effets non-elastiques sont très importants. Ils montrent que le "backfill" est très efficace pour réduire les déformations non-elastiques et le mur de suspension joint montre une forte tendance à agir comme une poutre. Il est indiqué que, pour les chantiers qui ne sont pas sujets à de grandes fermetures élastiques, l'effet de support du "backfill" n' est pas sensible à la qualite de celui-ci.

ZUSAMMENFASSUNG : Der vorteilhafte Effekt des Wiedereinfüllens bei Bauarbeiten in tief liegenden Goldminen, ist bis jetzt an der fähigkeit, um die vorhandene Energie für einen zerstörenden Gebirgsschlag zu verringern, gemessen worden. Die elastischen Methoden, die gewöhnlich gebraucht werden, um die Vorteile dieses Effektes zu quantifizieren, zeigen, dass wegen der hohen Zusammendrückbarkeit der Füllung, die Stützen in Gegenden kleiner Konvergenzen unbedeutend sind. In der Praxis wurde aber das Gegenteil wahrgenommen.

Die Resultate von der Analysierung der liegenden Gesteinsmassen, gemessen als ein elastischer Stützungsträger und des Abbaus, umgeben von einen kluftigen Felsformation, werden besprochen und vorgelegt. Diese Resultate zeigen widersprechende, der conventionellen elastischen Analyse, dass die nicht-elastischen Effekte sehr bedeutend sind. Es ist gezeigt worden, dass die Wiedereinfüllung sehr effektiv ist, um die nicht-elastischen deformationen zu reduzieren, dass die kluftige liegende Gesteinmasse eine starke Tendenz zeigte, um sich wie ein Träger zu verhalten. Es wird gezeigt, dass für einen Abbau, der nicht grosse elastische konvengenzen enhält, der Stützungs-effekt der Füllung nicht von der Qualität des Füllungsmaterials abhängig ist.

1 INTRODUCTION

Gold is mined in South Africa at deep levels in narrow tabular stopes. At these levels, stress magnitudes are very high, and significant convergence between hangingwall and footwall occurs as soon as these surfaces are exposed. The elastic energy stored as a result of the mining is also high and the release of this energy occurs frequently in the form of seismic events, which may result in rockburst damage. It has been shown (Ortlepp 1983a) that the energy available to cause a rockburst is proportional to the volume of elastic convergence that occurs in the stopes. The reduction of this volumetric convergence has consequently been seen as a means of reducing the incidence of rockbursts. It is in this light that backfill has been evaluated for its ability to reduce the volumetric closure. While this regional support effect is real and very important, the methods of analysis commonly

used to quantify it indicate that, owing to the high compressibility of the fill, the support that it provides in areas of small convergence will be negligible. However, conditions observed in practice indicate the contrary, namely, that even very "weak" fill does provide considerable beneficial effect (Close and Klokow 1986). Therefore, in addition to the regional support effect, mechanisms of local support must be active which, cumulatively, in smaller span stopes, can be more important than the regional effect.

2 CHARACTERISTICS OF BACKFILL SUPPORT BEHAVIOUR

The use of backfill is common practice in many mass mining operations, where the benefits of weak, often uncemented fill materials are well appreciated. In these situations, the purpose of the fill is not usually to resist high stresses, but, for example, to provide a working surface, to provide some confinement to pillars, and to fill large stopes for later extraction of pillars. These operations tend to involve large scale filling operations, often through large diameter boreholes with no difficulty of accessibility. The deep level gold mining operations, on the contrary, usually involve extraction of gold bearing reef from shallow dipping narrow tabular stopes of the order of 1 m in stoping height (MacFarlane 1983). The areal extent of stoping is great, access difficult and volumes of extraction per area relatively small. Consequently placement of fill is a complex management operation. Once in place the fill is called upon to provide three forms of support :

.   regional support and reduction of volumetric closure

.   local support to provide better conditions in the working areas adjacent to the stope face and in gullies

.   support under the dynamic loading conditions imposed by seismic events.

In this environment, two broad support category models have been identified. These are :

i)   reduction in energy release (RER) support model - if the potential energy release associated with rock bursts is to be reduced, the fill should develop support resistance as rapidly as possible to inhibit **elastic** stope closure

ii) inhibition of block movement (IBM) support model - to preserve competent hanging- and footwalls, and to resist the adverse effects of a rockburst, the backfill should minimise the fall out or ejection of potentially loose material.

The IBM and RER support models contribute simultaneously to the stability of the hanging-wall. The contribution of the IBM model is constant and does not depend upon the depth, span and width of the stope, nor on the stiffness of the fill.

The contribution of the RER model to the stability of the hangingwall depends on all of these parameters.

2.1 Characteristics of Hangingwall Behaviour

The hangingwall can be divided into two domains, the detached beam which lies immediately above the stope and the mass of rock constituting the super-hangingwall. The detached beam is intersected by natural joints and mining-induced stress fractures, and depends for its integrity on the extent to which the freedom of movement of individual blocks is inhibited. Its deformation behaviour, unlike the predominantly elastic super-hangingwall, will be distinctly non-linear, being controlled by the rock structure, geometry and movement of the constituent blocks rather than by the regional stresses.

The behaviour of the detached hangingwall will be influenced significantly by the type of support system. Under conditions of considerable closure and isolated point support, a state of unstable equilibrium will pertain and a small disturbance may cause a run of collapse. The presence of a continuous support will convert this condition to a state of neutral equilibrium. If, in addition, this continuous support is stiff, a condition of stable equilibrium will be obtained in the back area.

2.2 Support Mechanisms of Backfill in Deep Level Tabular Stopes

It has been shown theoretically that backfill can be very effective in reducing the rate of energy release associated with deep level stopes (Ryder and Wagner 1978). For a 1 m stope at a depth of 3 000 m, 100% filling with no lag behind the face of hydraulically placed backfill reduces the energy release rate by 45%. With packed waste rock this improvement increases to 65%, and with a specially tailored fill, the improvement can be up to 85%. These effects, however, can only be realised where the stope spans are such that the potential for elastic closure is high. Close to the stope face, and in scattered mining environments where spans are relatively short, elastic convergence is small and therefore support mechanisms other than the generated fill support reaction must be responsible for the stabilising effect of the fill. Whereas the individual effect of any of these mechanisms may be small, their cumulative contribution is expected to affect the stability of the hangingwall significantly. The secondary support mechanisms which have been identified may be summarised as follows:

Retention of keystones

The joints and fractures dividing the hangingwall beam into blocks or slabs, will define a number of critical blocks or "keystones" which have a greater potential for movement than other blocks. Depending upon the lateral freedom of movement, the tendency for dilation on the joints demarcating a keystone will promote stability owing to the roughness and waviness of the joints and fractures. The effect of this stabilising mechanism will increase with an increase in the depth of the keystone. As a result, detached hangingwalls of lesser thickness will be less stable, but more likely to be supported effectively by relatively soft fills.

The most important function of the fill is to provide a link between the keystone and the surrounding strata. It will retard and ultimately arrest any tendency of the keystone to be dislodged and, in so doing, enhance the self-supporting capacity of the hangingwall. In the absence of the fill, potentially loose keystones are free to fall away from the hangingwall, thus giving rise to a run of collapse. The position of the keystone will be further ensured by the fines in the fill which will find their way into open joints and cracks around the keystones.

Reduction of Unsupported Exposed Hangingwall Surface Area

Since backfill "covers" the majority of the hangingwall and footwall surfaces, the probability of occurrence of exposed blocks with fallout potential is greatly reduced compared with the conventional support situation. The probability of maintaining continuous hangingwall and footwall beams is therefore substantially increased.

Transmission of seismic waves

A significant link function which the fill may provide, will be manifest under seismic conditions (Ortlepp 1983b). In the absence of fill the compressive shock waves caused by a seismic event will be reflected at the rock-air interface provided by the hangingwall and footwall surfaces. On reflection they will become tensile and will tend to eject the detached hangingwall (or footwall). The presence of the fill in contact with the rock will enable a fraction of the shock waves to travel through the stope with partial reflection at the rock-fill interface. The "throwing-off" action will therefore be much reduced and fill in the stope will serve to inhibit dislodgement of keystones under these conditions. In addition, by shortening the length of unsupported hangingwall beam, the resonant frequency of the beam is increased (Spottiswoode, 1986). At the higher frequencies, less energy is available from seismic events to induce vibration, and backfill also provides a much more effective damping action than conventional support. Soft fill is also likely to behave as a fill of stiffer quality under dynamic short duration loads.

Delimitation of curvature reversals

The detached hangingwall is subjected to several curvature reversals from the time that it is first exposed at the face until it comes to rest in the back area. These curvature reversals fatigue the joints and fractures tending to eject relatively free blocks progressively by the successive clamping and declamping action, and associated rotation and shearing to and fro on the joints and parting planes. Since the presence of fill will result in a significant reduction in the non-elastic displacement of the detached hangingwall beam, it will reduce these adverse effects. Consequently the deflection profile will be smoothed and greater shearing strength will be promoted on the parting planes.

Preservation and augmentation of joint and fracture infill

Upon placement of the fill the fines will enter the cracks and fissures in the hanging- and footwall rocks. In addition, the proximity of the fill to the hangingwall will prevent any of the natural fines which occur on the joints and fractures from running out during periodic curvature reversals of the strata. This will enhance the dilatant capability of the joints and fractures and hence the confinement of keystones, thus promoting competence of the hangingwall beam.

3 STRESS AND DEFORMATION ANALYSES

The prediction of the effectiveness of backfill in a mining environment has commonly been made using two- or three-dimensional mining simulation analyses which are elastic methods (Ryder and Wagner 1978). In a scattered mining environment spans are quite small and elastic closures are limited. The application of the elastic mining simulation techniques in this situation indicates that benefit of only a few per cent can be expected from backfill. This is contrary to the fact that significant benefits are observed, and therefore it is considered that it is the reduction of non-elastic behaviour by backfill which produces the observable benefits. To investigate this, two sets of analyses were carried out:

. beam analyses to assess the potential behaviour of the hangingwall as a beam on an elastic foundation

. non-linear finite element stress analyses, with the rock mass being modelled as a jointed medium.

3.1 Hangingwall beam on an elastic foundation

These analyses were carried out since it was known that the bending behaviour of a beam on an elastic foundation is governed by a damped sine wave form (Hetenyi, 1971) and that this might have a contributary effect to the form of fracturing ahead of the stope face. The results of the analyses did indeed

illustrate this type of behaviour as shown in Fig 1, in which it can be seen that the bending moment shows a series of sign changes ahead of the stope face. Corresponding with this are relative deflection reversals and a cyclic variation of stress maxima and minima ahead of the face. Apart from the first cycle, the effects are of a secondary nature which would have only a minor contributary effect on distortion of the hangingwall (and footwall), opening and closing of joints and fractures, and shearing on parting planes. The first cycle however, would result in an increased vertical stress at the face, and a reduced vertical stress a short distance ahead of the face, which may contribute to the discrete fracturing that has been observed to occur ahead of the stope face (Adams and Jager 1980).

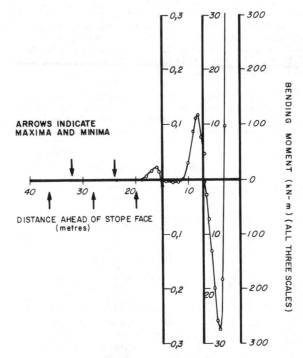

Figure 1 : Hangingwall beam bending moment sign reversals

3.2 Jointed rock analyses

Stress analyses of a stope surrounded by jointed and fractured rock were carried out using a finite element program (Stacey et al 1981). Deformational behaviour associated with the redistribution of excessive tensile and shear stresses acting on joints and fractures was handled by an iterative procedure. Since it is not practical to model a large number of joints and bedding planes discretely, a continuum approach was used in which the joints were included as sets. The orientation, continuity and spacing of the sets are specified as well as their cohesive and frictional strength, and dilational properties. The continuum approach has the disadvantage that it is limited to relatively small strains and does not cater for movements of specific blocks. It is therefore not suitable for examining collapse behaviour, but, with a large number of iterations, can model quite large deformations and give an indication of instability.

The stope modelled had a width of 1 m and a span of 80 m at a depth of 2 000 m. The rock mass contained continuous horizontal parting planes and cross joints with a 50% continuity. Most analyses were carried out with vertical cross joints, but two analysis took into account orientations 10° either side of vertical. In addition, a localised fall of hanging was modelled.

Single stage cycloned tailings backfill was modelled, being placed 4 m from the face and allowing 7% stope closure before placement.

3.2.1 Results

The presence of joints results in a major departure from elastic behaviour. Shearing occurs on parting planes and joints, and deformations increase as a result of redistribution of tensile and excessive shear stresses. The greater the dilation that occurs on joints, the more likely it is for the deformations to stabilise, indicating that joint infill/rock flour should be preserved in place if possible.

Deformations

Numerical convergence was not achieved for the stope without backfill, implying the onset of collapse. Maximum stope closure calculated for this case was in excess of 400 mm. In the presence of backfill, the maximum closure calculated was approximately 200 mm.

Profiles of hangingwall deflections adjacent to the face are shown in Fig 2. As can be expected, the deflections for jointed rock are substantially greater than the corresponding elastic values.

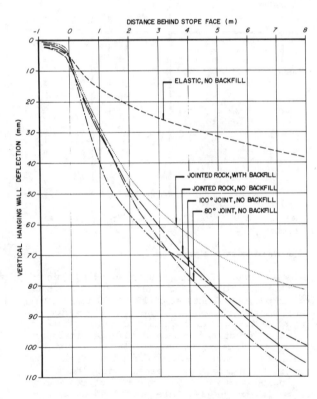

Figure 2 : Vertical deflection of hangingwall

The introduction of backfill results in a reduction in the deflection in the working area of approximately 10% adjacent to the backfill and a 100% reduction 40 m behind the face. The reason for the small reduction in the working area is that much of the movement occurs before the backfill is placed, and also since the non-linear backfill characteristic does not allow support resistance to develop rapidly. It is of interest to note that, with backfill support, it is calculated that some 70% of the ultimate closure in the stope has already occurred 5 m from the face.

For low convergence stopes there are significant implications which arise from these results:

• backfill has a substantial effect in stabilising the hangingwall deflection and in reducing the deflection far behind the face, but has a much smaller effect on deflections in the working area

• the deflection behaviour in both areas is not likely to be much affected by different backfill qualities, unless the quality is improved by an order of magnitude, and shrinkage is eliminated completely.

Stresses

The elastic stress distribution shows very large compressive stresses at the stope face and a very large zone of tension in the stope hangingwall. Failure on the joints and bedding planes results in the re-distribution of the excessive stresses - the stope face stress is reduced considerably and the tensile zone obviated almost completely. There appears to be a very strong tendency for a clamped beam action to develop in the immediate hangingwall, probably promoted by dilation on the joints. This corresponds with the behaviour described by Pender (1985).

The distributions of vertical and horizontal stresses ahead of the stope face are shown in Figs 3 and 4. The vertical stresses show a similar trend to the elastic form, but with much reduced magnitudes. The placement of backfill results in a reduction of these stresses, but this is not a major effect. Horizontal stresses ahead of the face for jointed rock are substantially reduced compared with the elastic values.

The effect of alternative inclinations of joints is also shown in Figs 3 and 4. The 80° inclination of joints has a very significant effect on the form of the stress distribution ahead of the stope face and the corresponding horizontal stresses are also substantially lower than for the other analyses.

The occurrence of a fall of hanging reduces the horizontal stresses to very low values for several metres ahead of the face. Unlike the other cases in which confinement of the rock by the horizontal stress field is high, the low horizontal stress conditions in the case of 80° joints, and resulting from a fall of hanging, are much more conducive to the formation of fractures ahead of the face.

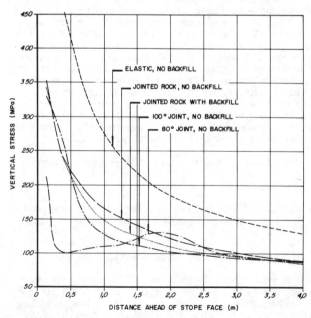

Figure 3 : Distribution of vertical stress ahead of the stope face.

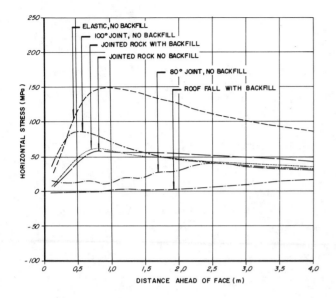

Figure 4 : Distribution of horizontal stress ahead of the stope face.

6 CONCLUSIONS

Several mechanisms of local support by backfill in low convergence stopes have been described, and non-linear stress analyses of a stope have been carried out. From these studies the following main conclusions can be drawn:

(i) conventional elastic stress analysis methods are unrealistic when applied to local behaviour around stopes.

(ii) the stress and deformation analyses show that curvature reversals occur with associated shearing on parting planes and joints. Backfill provides some reduction in these curvature reversals, but its main function will be to prevent loss of joint infill thus promoting dilation, and to prevent loss of keystones by falls of hanging, with consequent loss of competence of the hangingwall beam. The non-linear stress analyses therefore confirm the postulated local support mechanisms of backfill.

(iii) backfill has a major effect in reducing the overall closure of the stope. In the 80 m span stope modelled, backfill halved the total closure. In the working area there is not a major difference between the closures with and without backfill. In the presence of backfill 70% of the ultimate total closure occurs within 5 m of the face.

The inclusion of the effects of jointing results in a substantial reduction in the magnitudes of stresses at the stope face. The placement of backfill does not have a further major effect on the stress distribution.

The implication of these closure and stress findings is that, for low convergence stopes, the behaviour is unlikely to be sensitive to the quality of the backfill, unless the quality is improved by an order of magnitude. Consequently, small changes in backfill quality are likely to be of much less significance than the effectiveness of placement procedures.

(iv) in the elastic stress distribution, the major arching action develops immediately above the face producing very high stresses at the face and a large tensile zone in the stope hangingwall. In the jointed rock situation the major arching occurs at a much higher level, and the tensile stresses are redistributed. A clamped beam action develops in the immediate hangingwall, above which is a zone of low arching stresses. A fall of hanging destroys this clamped beam action in the immediate hangingwall, shifting it to a higher level, and providing the opportunity for unstable blocks to fall out. It also causes a major reduction in the horizontal stresses ahead of the face, resulting in conditions more conducive to fracturing.

(v) small variations in the orientations of joints and fractures could have a significant influence on the stress distribution at the face as well as the deformation behaviour ahead of the face.

Acknowledgements

The permission of Vaal Reefs Exploration and Mining Company Limited to publish this paper is gratefully acknowledged.

REFERENCES

Adams, G. and A.J. Jager 1980. Petroscope observations of rock fracturing ahead of stope faces in deep level gold mines, J. S. Afr. Inst. Min. Metall. 80, pp 204-209.

Close, A.J and J.W. Klokow 1986. The development of the West Driefontein tailings backfill project, S.A Mining World, Vol. 5, No. 4, pp 49-67.

Hetenyi, M. 1971. Beams on elastic foundation, Ann Arbor: The University of Michigan Press.

Macfarlane, A.S 1983. Development of a Hydraulic Backfilling System, Ass. of Mine Managers of S.Afr., Papers and Discussions, 1982-1983, pp 109-141.

Ortlepp, W.D. 1983a. Mechanism and control of rockbursts, in S.Budavari (ed.), Rock Mechanics in Mining Practice, pp 257-281, The S.Afr. Inst. Min. Metall.

Ortlepp, W.D. 1983b. Discussion, Ass. of Mine Managers of S. Afr. Papers and Discussions 1982-1983, pp 146-151.

Pender, M.J. 1985. Prefailure joint dilatancy and the behaviour of a beam with vertical joints, Rock Mech. and Rock Engineering, Vol. 18, pp 253-266.

Ryder, J.A. and H. Wagner, 1978. 2D analysis of backfill as a means of reducing energy release rates at depth, Chamber of Mines of S.Afr. Research Report No 47/78.

Spottiswoode, S.M. and J.M. Churcher 1986. The effect of backfill on the transmission of seismic energy, Proc. Back Filling Symposium, Ass. Mine Managers of S.Afr. and S.Afr. Inst. Min. Metall, 17 pp.

Prediction of coal outburst
Prévision des affleurements de charbon
Vorhersage von Gebirgsschlägen

KATSUHIKO SUGAWARA, Department of Mining Engineering, Kumamoto University, Japan
KATSUHIKO KANEKO, Department of Mining Engineering, Kumamoto University, Japan
YUZO OBARA, Department of Mining Engineering, Kumamoto University, Japan
TOSHIRO AOKI, Kumamoto University, Japan

ABSTRACT: This paper presents a practical procedure to predict the hazard of coal outburst in the longwall mining. The mechanism of coal outburst is discussed, and an use of AE monitoring and test drilling in combination is presented to investigate the de-stressed area in front of the longwall face.

RESUME: Cette étude présente un moyen pratique pour prévoir les affleurements de charbon dans une exploitation par longues tailles. Le mécanisme des affleurements est étudié, et un emploi combiné de moniteur AE et de trous de sondage est présenté pour étudier la zône non soumise aux contraintes face aux longues tailles.

ZUSAMMENFASSUNG: Diese Abhandlung gibt ein praktisches Verfahren für die Vorhersage der Gefährdung durch Gebirgsschlag bei Strebbruchbau. Der Mechanismus des Gebirgsschlags wird diskutiert, und gemeinsame Verwendung von AE-Überwachung und Prüfbohren wird präsentiert, um den entspannten Bereich vor dem Strebstoß zu untersuchen.

1. Introduction

Many technical problems of underground mining arise with increasing the working depth. Among them, coal outburst which is discussed in the present paper is a difficult problem of long standing. At the deep longwall face in Miike coal mine, Japan, coal outbursts have frequently occurred, and the prediction of coal outburst has become an urgent problem as well as its preventives. They are pure coal outbursts with no gas and belong to a transversal type of rock burst. A series of field measurements and analyses has been performed to predict the hazard of coal outburst, and various types of preventives have been applied. Fortunately, the serious damages have been avoided by making effective use of de-stress drilling, de-stress blasting and water injection in parallel. However, they are not fundamentally settled, even today.

In the present paper, the features of coal outburst in Miike coal mine will be described, and its mechanism will be presented and discussed by analyzing the stress state and AE activity near the longwall face. Then, to investigate the extent of the de-stressed zone in front of the face, the use of AE monitoring and test drilling in combination will be presented as a promising means, and it will be shown how the hazard of coal outburst can be predicted.

2. Features of coal outburst

In the present field, there are two workable coal seams dipping about 5 degrees, namely in descending order, the upper coal seam, from 1.7m to 2.5m thick, and the main coal seam, from 5.0m to 7.0m thick. The vertical distance between them is about 60m, and the working depth of the upper coal seam is ranging from 500m to 650m from the sea level. Firstly, the upper coal seam has been mined by the usual retreating method. After that, under the goaf of the upper coal seam, the main coal seam has been mined with dividing into two slices, in the order from the upper slice.

The coal outbursts of the upper coal seam have been frequently occurred in the district characterized by (1) the working depth more than 550m, (2) the immediate roof of rigid sandstone, 20m or more in thickness, (3) a small convergence of gate road and (4) a frequent occurrence of audible rock noise. The coal outbursts have mostly occurred at a certain part of the face just after cutting or blasting. As shown in Fig.1(a), the floor has been covered by the scattered fragments, while no fracture of sandstone roof has been observed. The coal outburst has occurred anywhere in the face, but more frequently and severely on the tailgate side in the face as shown in Fig.2, (Kimura and co-workers, 1982). In the severest case, the length of damaged face has reached about 40m, and a lenticular opening of about 6m in depth has appeared between the sandstone roof and the fractured coal seam. On the surface of the roof in the opening, the streaks resulted by a shearing fracture has been observed.

The coal outbursts of the main coal seam have been occurred on the face of upper slice, equipped with a remote-control mining system. As shown in Fig.1(b), the floor coal has been violently heaved in an instance. In the severest case, the length of damaged floor has reached about 60m. Such a damage has been caused by a shearing fracture along the bottom

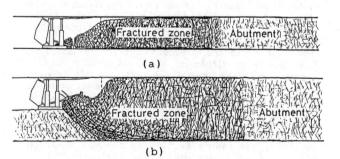

Fig.1. Schematic sections of coal outburst:
(a) the upper coal seam;
(b) the main coal seam.

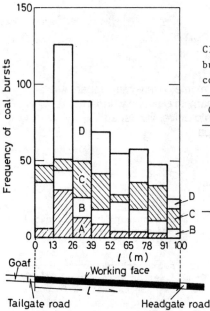

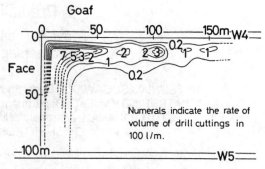

Classification of coal
bursts by the length of
collapsed face ; $L*$.

Class	$L*$ (m)
A	>26.1
B	13.1~26.0
C	6.6~13.0
D	0.1~6.5
E	Only rock noises

Fig.3. Distribution of the value of drill
cuttings obtained by the test drilling.

Fig.2. Classification of coal outbursts in the
upper coal seam and histogram of coal outbursts
in relation with each class and position in the
face.

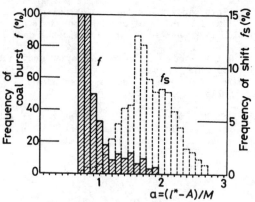

Fig.4. Relation between the frequency of
coal outburst and the value of α evaluated
from the actual results.

of coal seam. The width of fractured zone in front
of the face has been about two times of the thickness of the main coal seam.

It is no doubt that the present coal outbursts are
connected with the large-scale shearing fracture
between the sandstone roof and the coal seam and/or
between the floor sandstone and the coal seam. Such
a shearing fracture can initiate in a highly
stressed area, which is called the abutment. Usually
the abutment is at a some distance from the face,
and there is always a zone of de-stressed coal seam
between the abutment and the face, like the fractured zone in Fig.1. In the present paper, such a
zone is named the protecting zone, because it controls the fracture propagation. There is no problem
if the fracture propagates slowly in a stable manner, and the damage can decrease with increasing the
width of protecting zone even if a large-scale fracture occurs in the abutment. Therefore, it is concluded that the hazard of coal outburst can be
predicted from the extent of the protecting zone.

3. In-seam test drilling

In order to investigate the extent of the protecting
zone, the two types of test drilling have been
successfully performed. One is the large diameter
drilling having a diameter of 10cm and the volume of
cuttings is measured for each unit drill-length. The
other is the small diameter drilling with a handy
auger, and the drillable length is surveyed. The
latter method is based on the fact that jumming is
sure to occur in the hazardous district and drilling
is unable to do within the abutment.

From the former type of test drilling, it is made
clear that the abutment in the upper coal seam is
about 6m distant from the tailgate and the face
respectively, and that the width of the abutment is
ranging from 15m to 30m, as shown in Fig.3.

The latter type of test drilling, carried out from
the face, has been effective to control the face
advance in the upper coal seam. The rate of face
advance, A, for each working shift has been determined by eq.(1).

$$A = l* - \alpha M \qquad (1)$$

where $l*$ is the drillable length, M is the thickness
of coal seam and α is a coefficient. In eq.(1), $l*$
should be measured for each working shift prior to
the face operation. In this time, $l*$ is the width of
the initial protecting zone, and αM is considered to
represent the protecting zone sufficient to prevent
the coal outburst.

The coefficient α can be statistically evaluated
from the actual results (Sugawara and co-workers,
1981). For this purpose, the value of α, obtained
by substituting the actual values of A, $l*$ and M
into eq.(1), must be assigned for each working
shift, and the frequency of coal outburst must be
computed for each group of the working shift assigned the same value of α, as shown in Fig.4. After
that the minimum of the coefficient α should be
estimated under the condition of no coal outburst,
e.g. α=2.0.

4. AE monitoring

For the thin coal seam, the test drilling has been a
promising method to investigate the extent of de-
stressed zone in the coal seam. However, its application to the main coal seam needs a drilling of 10m
or more in length. Therefore, the investigation only
by the test drilling is not desirable in the thick
coal seam. This disadvantage will be overcome by the
combination with AE monitoring, as discussed latter.
The main coal seam has sometimes additional problems, that mainly come from the irregular shape of
the upper-lying goaf (Sugawara and Kaneko, 1986).
The ununiformity of the initial field stress has
been analyzed by FDM, and the stress concentration
along the face and the gate road, see Fig.5, and the
strain energy release rate due to the face advance
have been also analyzed. From these analyses, it has
been concluded that the working within the painted
area in Fig.8 needs a strict watch over the coal
outburst.

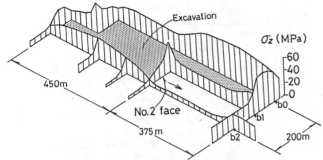

Fig.5. An example of the vertical stress distribution along the gate road and the face of the main coal seam, analyzed by FDM (Sugawara and Kaneko, 1986).

The source location system has been installed to monitor AE with low frequency ranging from 10Hz to 60Hz. Twelve seismographs are used for the present source location and the evaluation of the seismic energy radiated. But the latter is given by a relative value, E, that is computed by the empirical equation between the maximum amplitude and the distance of propagation (Sato and co-workers, 1986).

Fig.6 shows all of AE observed during the working of the main coal seam. Each circle indicates one event, and the area of which is proportional to the

seismic energy radiated. The broken line shows the outline of the upper-lying goaf, and the solid lines indicate the longwall panels in the main coal seam. In each panel, the face advanced from the right to the left, and the working moved to the adjacent dip-side panel. In order to examine the AE activity neighboring the face, the AE events in Fig.6 have been classified into the two types, from the geometrical relationship between the epicenter and the face. One is called N-type, that is located in the square region, involved the face within, which is ranging from 40m just in front of the face to 10m in rear of the face within the longwall panel. The others are called O-type.

The distribution of the O-type events has expressed the existence of some linear structure, as shown with line C-C and line D-D in Fig.7. Among total six lines found out in the same manner, C-C and F-F are confirmed to coincide with two faults dipping 80-90 degrees respectively. Then, the others have been presumed to correspond to similar discontinuities respectively, but of small displacement. As shown in Fig.8, they divide the hazardous area painted into the five regions, i.e. I, II,...,V.

It is noticeable that the N-type events concentrates in the region I, III, and V, as shown in Fig.9. This suggests an important role of structural discontinuities. If the roof strata are divided by them into several roof plates, the abutment pressure can increase with decreasing the area of the unmined

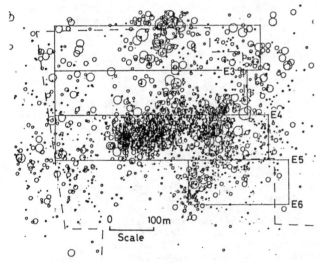

Fig.6. Source locations of AE on the main coal seam.

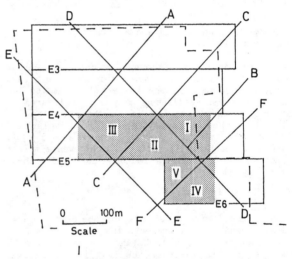

Fig.8. Geometrical relations between the six structural discontinuities, A-A~F-F, and the hazardous area forecasted by the analysis of the strain energy release rate, painted.

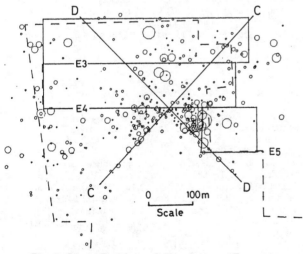

Fig.7. Distribution of the O type AE events observed during the working of 100 days and the structural discontinuities, C-C and D-D.

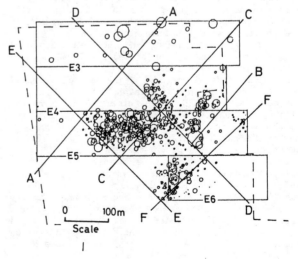

Fig.9. Distribution of the N type events.

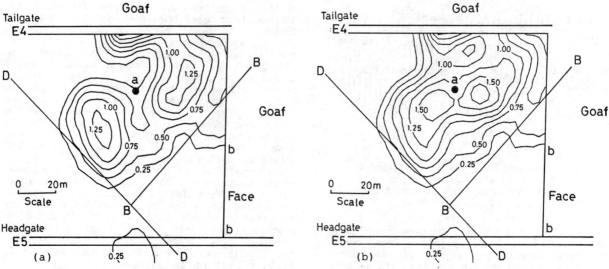

Fig.10. Change of rate of de-stress (RDS) in the main coal seam resulted from a large scale AE event located at the point a : (a) contour map of RDS just before the occurrence of the event; (b) contour map of RDS after the event.

coal seam which should independently support the over-lying roof plate. This assumption can give an explanation to the concentration of AE events in the region I, III and V.

5. Analysis of energy release density and rate of de-stress

A new parameter named the energy release density (ERD) is introduced to analyze the rate of de-stress in the coal seam from the AE activity. ERD is defined as the released energy for every unit area on the coal seam plane, and it can be given by the following scheme. Firstly, the coal seam plane is divided into 1m meshes, and the seismic energy, E, radiated by each event, is distributed evenly to the meshes within a circle of radius R computed according to eq.(2).

$$E=KS=K\pi R^2 \qquad (2)$$

where K is the energy release for the unit area by one event and $S=\pi R^2$ represents the de-stressed area. After such an iteration, ERD can be given for each mesh by the summation of the distributed energy. Therefore, ERD is integral multiples of K.

If the energy release in the abutment is always accompanied with AE event observed and its magnitude is proportional to the seismic energy radiated, the de-stressed zone will be expected to be found out as a zone of high value of ERD. In the following analysis, K is assumed to be a constant and its value has been back-analyzed from the seismic energy, E, observed for a coal outburst, and the area S assessed by the field investigation of the fractured zone in front of the damaged face.

The rate of de-stress (RDS) in the coal seam has been represented in a dimensionless form, that is the mean value of ERD, in the circle having a radius of 10m, divided by K. Such a value of RDS is given to the center point of the circle and the contour map of RDS has been drawn as shown in Fig.10. Fig.10 demonstrates the change of the contour of RDS with one large scale AE event. The epicenter of this event is indicated by point a, which is located in a valley between two peaks in the contour map. Since the large scale AE events can occur in the vicinity of the area of RDS>1 and on the face in the vacant area, this example maintains that the hazard exists along b-b in Fig.10 and that the de-stress operation and test drilling should be reinforced in the head-gate side of the face.

However, it is resulted that the face has advanced

with no problem. The geometrical condition that b-b is located in the headgate side of the face and in the outside of the region I is considered to be a main reason of this result. Therefore, it has been concluded that the vacant area in the tailgate side of the region I, III and V is mostly dangerous. Since the limitation of the present analysis of ERD and RDS is conditioned by the reliability of the source location and the seismic energy evaluation, it must be pointed out that the re-examination of the de-stressed zone by means of the test drilling is indispensable for the face advance in the hazardous region. Namely, the AE monitoring and the test drilling should be made up for defects of each other.

6. Conclusion

It has been concluded that the hazard of the coal outburst on the longwall face can be predicted from the extent of the de-stressed zone in front of the longwall face. For the investigation of the extent of the de-stressed zone and the rate of de-stress in the coal seam, the test drilling and AE monitoring have been presented and discussed. The practical analysis of the test drilling data to determine the rate of face advance for each working shift and the practical scheme to evaluate the energy release density and the rate of de-stress from the AE activity have been presented. Additionally, the limitation and disadvantages of the two methods have been made clear, and the combination of them has been concluded to be mostly effective at the present time.

References

Kimura, O., K.Sugawara and K.Kaneko: Study on the controlling of coal burst in Miike mine, Proc. of 7th Int. Strata Control Conf., Liege(Belgium), pp.431-448, 1982.

Sugawara, K. and K.Kaneko: Rock pressure and Acoustic Emission in the longwall coal mine, J. Min. Metall. Inst. Jpn, 102, 1177, pp.143-148, 1986.

Sugawara, K., H.Okamura, K.Kaneko and O.Kimura: Mechanical behavior of coal seam sandwiched by rigid strata, Proc. of the Int. Symp. on Weak rock, Tokyo(Japan), pp.561-566, 1981.

Sato, K., T.Isobe, N.Mori and T.Goto: Microseismic activity associated with hydraulic mine, Int. J. Rock Mech. Min. Sci. & Geomech. Abstr., 23, 1, pp.85-94, 1986.

The quality control of rock bolts
Contrôle de la qualité des boulons de roc
Qualitätskontrolle für Gesteinsanker

S.SUNDHOLM, Embassy of Finland, Tokyo, Japan

ABSTRACT: A two-year series of tests on the long-term load bearing capacity of rock bolts has been carried out. The bolt types tested were cement grouted rebar, resin grouted rebar, expansion shell anchor, Split-set and Swellex. Five different mines were used as test sites to find out the effect of different environments. The work was done with both pull-out tests and overcoring and inspection of bolts. Overcoring gives the most reliable estimate of the quality of bolt installation, even if pull-out tests may be used as first estimate. In certain environments even correctly installed bolts may loose 20 % of their load carrying capacity per year.

RESUME: Une série des essais de deux ans concernant la puissance de levage de longue durée des boulons a été exécutée. Les types des boulons essayés étaient la cheville cimentée, la cheville résineuse, le boulon d'ancrage à coquille d'expansion, Split-set et Swellex. Afin d'observer l'effet des environnements différents, cinq mines ont été utilisées pour les terrains d'essai. L'essais d'attraction et le surforage ainsi que l'inspection des boulons ont été exécutés. C'est le surforage, qui le mieux estimes l'état de l'installation des boulons, bien que les résultats des essais d'attraction puissent être tenus pour les premières estimations. En certains environnements même un boulon correctement installé peut perdre 20 % de sa puissance de levage par an.

ZUSAMMENFASSUNG: Eine Testserie von zwei Jahren ist über die langwierige Belastungstragfähigkeit von Gesteinsankern ausgeführt geworden. Die geprüften Ankerarten waren mit Zement eingegossener Kammstahl, mit Kunststoff eingegossener Kammstahl, Spreizhülsenanker, Split-set und Swellex. Fünf verschiedene Gruben dienten als Teststellen, die Einwirkung der verschiedenen Umgebungen auszufinden. Die Arbeit wurde sowohl mit Auszugtesten als auch mit Freibohrungen und Untersuchungen von Ankern durchgeführt. Die Freibohrung gibt die zuverlässigste Bewertung über die Qualität der Ankereinrichtung, obgleich die Auszugtesten als erste Bewertung angewandt können werden. In gewissen Umgebungen können sogar richtig eingerichteten Ankern 20 % von ihrem Tragkraft pro Jahr verlieren.

1 INTRODUCTION

In Finland the only type of rock bolts used for long-term purposes is a cement grouted rebar. It has shown to be reasonably well protected against corrosion. However, a relatively long curing time has to be accepted when cement grout is used. This is not in accordance with the requirements of modern mining techniques.

Need to achieve higher work efficiency has led to the development of several new types of rock bolts for supporting elements that are simpler, cheaper and faster, like resin grouted rebar, split set friction stabilizer and Swellex (Hoek & Brown 1980). Some manufacturers have tried to improve the characteristics of a cement grouted rebar by packing the grout into cartridges, adding chemicals to shorten the curing time, and mechanizing the whole installation process.

The Laboratory of Mining Engineering, Helsinki University of Technology, has carried out research work on long-term performance of rock bolts from the year 1980 (Sundholm 1984). The studies began with pull tests to find out the actual need for support quality control. These tests led to further questions: Which other methods are available? Is the information obtained reliable and sufficient?

The need for accurate control device was evident but the intermediate target was to learn to predict and prevent the possible defects that can cause the rock bolts to loose their load-bearing capacity. The future aim is to gain sufficient knowledge of factors effecting the long-term performance of different types of rock bolts in different environments.

The project consisted of two phases: pull tests made on experimental rock bolts, and overcoring of in-situ rock bolts. On one hand information about the condition of the existing "old" rock bolts was needed, on the other hand the recent development had to be followed and data on the behaviour and quality of new support methods and rock bolts had to be gathered.

The purpose of the work was to find out if there are any damages or defects in the rock bolts that require quality control which covers the original components, installation procedure and environmental problems affecting both immediate and long-term behaviour of the bolts.

2 TEST PROGRAM

2.1 General

The test program began with pull tests because these have been a widely used control method when checking that bolts have been properly installed. Complementary to the pull tests it was decided to try to have a more detailed look at different rock bolts in-situ. The basic requirements for this were:
- the rock bolt itself should not be damaged.
- the contact between the bolt and the surrounding rock should remain intact.

Overcoring seemed to be a suitable method to meet these requirements. It was decided to use a Ø 150 mm diamond drill bit and overcore each bolt in one piece.

2.2 Bolt types

Five different types of rock bolts were chosen to be investigated:
- cement grouted rebar, because it is the most common and the only bolt type accepted for permanent installations,
- resin grouted rebar, because it gives immediate support, and there was little experience about the bolt,
- expansion shell anchor as a representative of old bolt types,
- split set friction stabilizer, because it provides flexible support, and no previous experience was available of it, and
- swellex tube for same reasons as split set.

2.3 Test sites

Pull tests were carried out in five separate mines (Figure 1) in order to find out the possible effects of different environments:
- the Pyhäsalmi mine of Outokumpu Oy has massive sulphide ore, and a rather heavy water flow creates a very corrosive surrounding for steel bolts.
- the Keretti mine of Outokumpu Oy was the first mine in Finland to use resin grouted rebars. It also had the only mechanized rock bolting unit for these bolts.
- the Hammaslahti mine of Outokumpu Oy had difficulties with nearly vertical schistosity and slabbing of drift walls.
- the Otanmäki mine of Rautaruukki Oy was a hard rock iron ore mine with quite a neutral surrounding.
- the Tytyri mine of Lohja Oy is an underground limestone mine.

Figure 1. Location of test sites.

2.4 Pull tests

Four types of rock bolts were tested: cement (Ø 20 mm) and resin (Ø 16 mm) grouted rebars, expansion shell anchor and Split-set (Ø 39 mm), see Figure 2. Ten pieces of test bolts of each type were installed in the research sites. All the bolts were of standard length (in average 2.4 m).

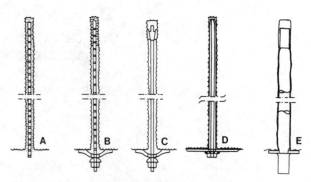

Figure 2. The types of rock bolts under investigation. A = cement and B = resin grouted rebar, C = expansion shell anchor, D = Split-set and E = Swellex.

The pull tests were made 1 day, 1 week, 2 months and 6 months after installation. The bolts were loaded with a hydraulic jack and the results were simultaneously recorded on a xy-recorder as a continuous load-displacement curve.

2.5 Overcoring

This part of project began in November 1982 at the Pyhäsalmi mine of Outokumpu Oy. Later samples were drilled also in the Keretti and Hammaslahti mines of Outokumpu Oy. Altogether 27 separate rock bolts (5 cement grouted rebars, 5 resin grouted rebars, 10 split-set and 7 swellex tubes) were overcored, and their functioning evaluated at the Laboratory of Mining Engineering, Helsinki University of Technology.

The bolts were overcored in one piece, i.e. in 2.4 m long cores with a Ø 150 mm diamond drill. After the samples were brought to the laboratory they were cut into shorter pieces, split in halves and photographed.

Split-set and swellex tubes were cleaned, the corrosion products were removed from the surfaces and the specimen wall thickness was measured with a micrometer. Surface areas were determined from several crosscuts of split set and swellex specimens in order to estimate the possible loss in area and thus in load-bearing capacity, too.

The original aim was to overcore the oldest possible bolts in the worst possible conditions. Overcoring itself was found to be extremely time-consuming and expensive work. The core size, the location of the bolts, the hardness and fractures of the rock, and finally the straightness of the hole were but few of the difficulties met during the work. In some places even the advance of stoping caused the bolts to be out of reach without heavy platforms for the drilling equipment.

3 PULL TESTS

The load-displacement curve describes well the behaviour of different types of rock bolts under tensile load (Figure 3), even if it still has limitations.

The maximum force on the curves means for grouted rebars maximum load they can sustain without breaking (Table 1). The maximum loads applied on expansion shell anchors and Split-sets mean their initial or maximum residual load-bearing capacity. If the load-bearing capacity decreases during the pull test, initial load-bearing capacity is given, else maximum (Table 1).

3.1 Grouted rebars

For both cement and resin grouted rebars there is no significant displacement before the rebar breaks (Figures 4 and 5).

Table 1. Maximum loads from the pull tests.

MINE/bolt type	LOAD (kN)			
	1 day	1 week	2 months	6 months
TYTYRI				
cement grouted		87.3 93.2 71.6	72.6 78.5	76.5 71.6
resin grouted		147.2 142.2	107.9	139.3 136.4
split set	77.5	81.4 39.2	86.3 84.4	89.3 120.7 110.9
exp. shell	94.2 97.1		108.9 35.3	114.8 101
HAMMASLAHTI				
cement grouted		71.6 91.2	61.8 76.5	73.6 83.4
resin grouted	163.8 171.7	129.5 107.91*	109.9 96.1	
split set	64.7 113.8	46.1 44.1•	54 76.5	84.4 101 78.5•
exp. shell	98.1 104	61.8 63.8	60.8 56.9	
KERETTI				
cement grouted		82.4 85.3 91.2	74.6 77.5 91.2	79.5 82.3
resin grouted	122.6 129.5	117.7 110.9	96.1* 94.1	121.6 107.9
split set	45.1 54.9	77.4 45.1 54.9	66.7 53 89.3	91.2 90.3
		78.5		
exp. shell		69.7 85.3	109.9 98.1	107.9 107.9
PYHÄSALMI				
cement grouted		75.5 76.5	93.2 88.3 86.3	71.6
resin grouted		145.2 150.1	142.2 137.3 138.3	127.5 (107.9)
split set			86.3 87.3 67.7	89.3 68.7 95.2
			74.6	81.3
exp. shell	57.9	66.7 65.7	28.4 76.5	39.2 48.1
OTANMÄKI				
cement grouted		110.9* 108.9	78.5 76.5	78.5 83.4
resin grouted	141.3 147.2	137.3 106	132.4 133.4	110.9 122.6
split set	(68.7 54.9)	(88.3 23.5)	(80.4 46.1)	(58.9 65.7)
exp. shell	126.5 68.7	75.5 107.9	93.2 92.2	113.8 105

* broken at threads, • jack slipping

If grouting is well done and sufficient, the shape of the curve is determined by the length of the free end of the rebar, the quality of the threading and the material characteristics of the rebar (Poitsalo 1983).

However, in the Otanmäki and Keretti mines one out of eight (12.5 %) pull-out tested resin grouted rebars showed to be of poor quality (Figure 6). These curves emphasize that intensive care and inspecta- tion should be done during installation.

In the Keretti mine an automatic rock bolting jumbo was used to install the bolts. The two bolts could take a reasonable load but not as much as expected. When they were pretensioned to about 40 kN it did not appear that they were ineffective.

3.2 Expansion shell anchor

In Figure 7 it can be seen that expansion shell anchor is most effective in hard rock. The pull-out curves in hard rock remind those of friction stabilizers. In softer rock the anchors show slight signs of loosing load-bearing capacity with time (Figure 8).

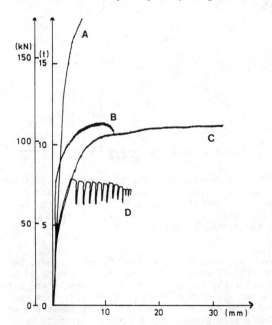

Figure 3. Load-displacement curves for different types of rock bolts. A = cement grouted rebar, B = resin grouted rebar, C = expansion shell anchor, D = split set.

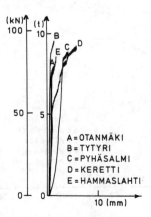

A = OTANMÄKI
B = TYTYRI
C = PYHÄSALMI
D = KERETTI
E = HAMMASLAHTI

Figure 4. Load-displacement curves for cement grouted rebars.

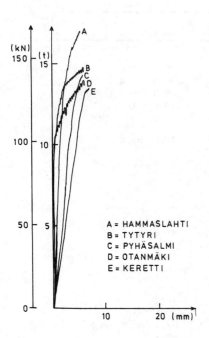

Figure 5. Load-displacement curves for resin grouted rebars.

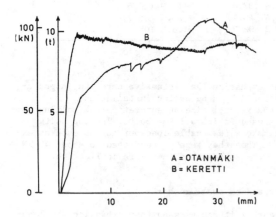

Figure 6. Poor resin grouting. The rebars should be able to take higher load than 100 kN without remarkable displacements.

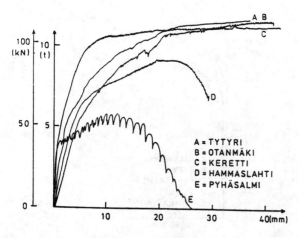

Figure 7. Expansion shell anchor pull-out tests in hard (A, B, C) and soft (D, E) rocks.

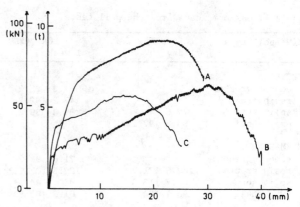

Figure 8. The load-displacement curves for expansion shell anchors with time. A = 1 day, B = 1 week, C = 2 months after installation in the Hammaslahti mine.

3.3 Split set friction stabilizer

The load-displacement curve for split set bolt is quite different from the curves of conventional grouted rebars. The effect of friction is clearly seen from the curves (Figures 9 and 10).

The obtained results were very promising although no earlier test experience of the stabilizers was available. Generally the maximum loads for a 2.4 m long bolt varied from 45 kN to 100 kN.

When the initial peak load (Figure 9, point A) of the split set bolt is reached, the bolt begins to slide. The movement stops in point B (Figure 9) which is the residual load bearing capacity of the split set. The curves show that the initial load-bearing capacity increases with time.

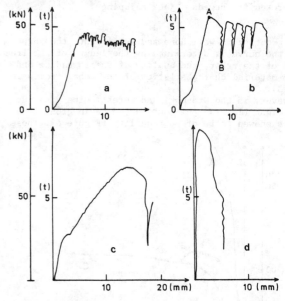

Figure 9. The load-displacement curves for split set bolts. a = 1 day, b = 1 week, c = 2 months and d = 6 months after installation in the Keretti mine. The peak load (A) seems to increase with time in contrary to the residual load (B).

The residual load-bearing capacity decreases to 80 % of the initial peak load after one day, 50 to 60 % after one week, 30 to 60 % after two months and 5 to 20 % after six months (Figures 9 and 10). This can be explained partly by corrosion which smooths down the surface roughness. The corrosion products also diminish the sliding friction between the bolt and the borehole wall.

The contact between the bolt and the borehole varies greatly. Sometimes it may take place only through some points on the tube surface (Figure 11). That causes large variation to the load-bearing capacity (Figure 12).

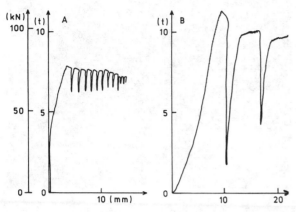

Figure 10. The load-displacement curves for split set bolts (A) 1 day and (B) 6 months after installation in Tytyri mine.

Figure 11. Cross-cut of split set bolt no 3 from the Pyhäsalmi mine at the depth of 135 cm. The bolt is only partly in contact with the borehole wall.

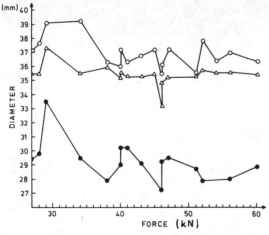

Figure 12. Initial pull-out loads for split set bolts as a function of the smallest measured hole diameter (Δ), the average hole diameter (o) and the average split set diameter (•).

4 OVERCORING

4.1 Cement grouted rebar

The core samples did not reveal any unexpected results. Cement grouting seems to be a very good shelter against environmental effects.

The age of the samples varied from 1.5 to 7 years. Although there were fractures and empty spaces in the grouting (Figures 13, 14, 15, 16 and 17) the rebars were mostly clean. The lowest quality grouting was found in the beginning or at the bottom of the hole.

Even if it is difficult for air and water to penetrate the fractures in the rock these fractures always create a risk of corrosion. The most dangerous are fractures which serve as routes for continuous waterflow. On the other hand cement together with water scapes alcalic compounds around the rebar. In such an environment the rebar is passive and does not corrode.

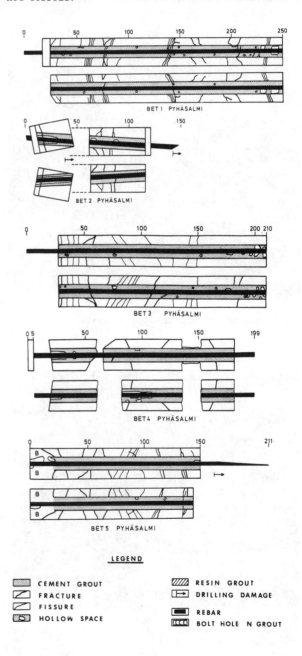

Figure 13. Schematic drawing of the cement grouted rebar bolts Bet1, Bet2, Bet3, Bet4 and Bet5 from the Pyhäsalmi mine.

Figure 14. Bet1/Pyhäsalmi. A major crack is cutting the grouting but the rebar is still quite clear.

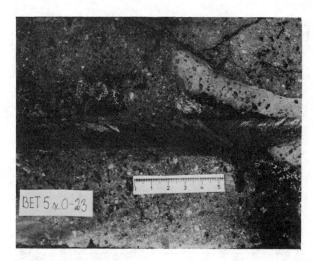

Figure 17. Damages due to corrosion in the rebar near the contact between the rock and the shotcrete.

Figure 15. Bet3/Pyhäsalmi. Hollow spaces in the grouting.

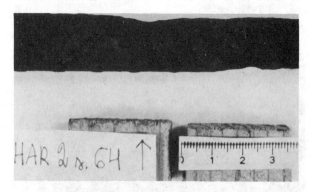

Figure 18. Har2/Keretti. The free end of the rebar was badly corroded.

Figure 16. Bet4/Pyhäsalmi. Grouting not complete in the beginning of the hole.

Figure 19. Har1/Keretti. Pieces of the plastic cover were found to prevent the contact between the rock and the grouting.

4.2 Resin grouted rebar

The results obtained from the overcored resin grouted rebars show that several things can affect the quality of the grouting.

Two of the five bolts were not properly grouted: there was only 20 cm resin around the bolt marked Har1 and no resin was found around the bolt Har2. The free parts of the rebars had been bended in rock deformation. Strong corrosion also occurred all along (Figure 18). Plastic cover pieces of the resin cartridges were found in grouting at the border against the borehole wall (Figure 19).

Figure 20. Har3/Keretti. The resin grouting was fractured.

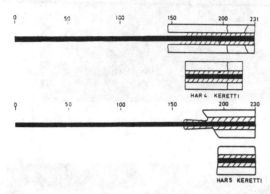

Figure 21. Schematic drawing of the resin grouted rebar bolts Har4 and Har5 from the Keretti mine. Legend, see Figure 13.

In the case of the bolt Har3 the bolt was partly in fibrous rock. The resin did not seem to adhere this type of rock. The grouting was also fractured (Figure 20).

The last bolts marked Har4 and Har5 (Figure 21) gave an impression of proper grouting. Still there could be found empty spaces and fractures in the resin. The ungrouted part of the rebar had been damaged to some extent by corrosion but the grouted parts were clean.

4.3 Split-set

All the Split-sets had been corroded to some extent as could be expected. The bolts were covered with corrosion products over their entire length (Figure 22). Heavy point corrosion was the general feature (Figure 23).

Figure 22. Sample from an overcored Split-set.

Figure 23. Heavy point corrosion on a cleaned Split-set bolt.

The thickness of the tube wall was measured with a micrometer from several points. An average value for the specimen wall thickness was determined. In Table 2 the age of the Split-sets and the wall thicknesses are presented.

The wall thickness of a new tube ought to be 2.3 mm. The measured values differ remarkably from this. In the cases the tube was thicker the maximum value measured was used as the original wall thickness.

Table 2. Wall thickness measurements of Split-sets.   P = Pyhäsalmi, H = Hammaslahti, G = Galvanized

Bolt place and type	P1	P2	P3	P4	P5	P6	P7G	P8G	H1	H2
Age (years)	1.5	1	1	1	1	1.5	1	1	2	2
Smallest wall thickness measured (mm)	1.44	2.12	2.12	1.76	0.96	1.06	2.18	2.33	1.40	1.20
% of original	62.6	92.2	92.2	76.5	41.7	46.1	87.2	93.2	59.6	51.1
Average wall thickness on evenly corroded surface (mm)	2.19	2.24	2.26	2.21	2.27	2.16	2.48	2.47	2.28	2.25
% of original	95.2	97.6	98.2	96.0	98.7	93.9	99.2	98.8	97.0	95.7
Average wall thickness in corrosion craters (mm)	1.87		2.17	2.05	1.98	1.68	2.30	2.40	1.92	1.84
% of original	81.3		94.1	89.0	86.1	73.2	92.0	96.0	81.7	79.8

Table 3. Cross-section areas of Split-sets.

P1 (cm^2)	P1 (%)	P2 (cm^2)	P2 (%)	P3 (cm^2)	P3 (%)	P4 (cm^2)	P4 (%)	P5 (cm^2)	P5 (%)
1.88	84.0	2.01	89.7	2.14	95.5	2.02	90.6	2.05	84.4
2.07	92.4	2.03	90.6	1.99	87.7	1.97	81.1	2.02	89.7
1.86	83.0			2.07	91.2	2.13	87.7	2.08	89.7

P6 (cm^2)	P6 (%)	P7G (cm^2)	P7G (%)	P8G (cm^2)	P8G (%)	H1 (cm^2)	H1 (%)	H2 (cm^2)	H2 (%)
2.01	90.1	2.33	95.9	2.33	95.9	1.96	86.0	2.03	89.0
1.93	86.6	2.29	94.2	2.38	97.9	2.16	94.7	2.14	93.9
2.09	93.3	2.31	95.1			2.05	89.9	2.07	98.1

P = Pyhäsalmi, H = Hammaslahti, G = galvanized

Figure 24. A crosscut used for surface area measurement.

Figure 25. Sample from an overcored Swellex.

Figure 26. Cleaned Swellex samples show heavy corrosion.

Figure 27. Some Swellex tubes were partly broken.

The surface area of some cross-sections was determined (Figure 24). It showed that the cross-section of one year old Split-sets had diminished by 5 to 19 %, in the average 11 % (Table 3). On the other hand, the minimum wall thickness measured was only 42 % which indicates that corrosion rate could be even higher.

If the tensile strength of the bolt is allowed to decrease to the expected holding force of the installed bolt (80 - 90 kN) the allowed loss of material could be about 25 %. In this specific case and place it means an effective lifetime of 2 to 3 years.

Galvanization of bolts reduces the corrosion rate remarkably (Table 3). The minimum wall thickness

measured for a galvanized Split-set was 87.2 % from the original. The reduction in cross-section was only 2 to 6 %.

4.4 Swellex

The swellex tubes are exposed to same kind of corrosion problems as the Split-sets (Figure 25). Although the swellex tube is thinner the corrosion rate is lower than with Split-sets because the Swellex is a closed tube and deforms more accurately to the borehole. The water used in the installation often remains inside the tube and prevents the corrosion process from the beginning.

The principles of corrosion presented under previous headline apply also to Swellex (Figure 26). According to the measurements the minimum wall thickness measured was 55.1 % (Table 4). The cross-sections of the bolts had diminished by 3 to 17 % in 1.5 years (Table 5). On this ground a loss of 7 to 8 % in load-bearing capacity can be expected in one year. This does not mean that the corrosion rate would be constant all over the surface. It is more probable that there are some weakness points or zones but even one weak point in a critical area can be a controlling factor.

Table 4. Swellex wall thickness measurements

Bolt place and type	H1	H2	H3	H4	H5	K1	K2
Age (years)	1.5	1.5	1.5	1.5	1.5	2	2
Smallest wall thickness measured (mm)	1.13	1.40	1.21	1.51	1.42	1.76	1.92
% of original	55.1	66.7	60.5	73.7	69.3	87.1	89.3
Average wall thickness on evenly corroded surface (mm)	2	1.99	1.86	1.95	1.95	1.97	2.09
% of original	97.6	94.8	95.4	95.1	95.1	97.5	97.2
Average wall thickness in corrosion craters (mm)	1.67	1.67	1.53	1.77	1.65	1.88	1.98
% of original	80	79.5	78.7	86.3	80.5	93.1	92.1

Table 5. Cross-section areas of swellex bolts.

H1		H2		H3		H4	
(cm²)	(%)	(cm²)	(%)	(cm²)	(%)	(cm²)	(%)
2.36	96.3	2.17	88.6	2.09	85.3	2.30	93.9
2.30	93.9	2.19	89.4	2.11	86.1	2.38	97.1

H5		K1		K2	
(cm²)	(%)	(cm²)	(%)	(cm²)	(%)
2.19	89.4	2.29	91.2	2.08	79.4
2.03	82.9	2.25	89.6	2.29	87.4

H = Hammaslahti, K = Keretti

Two of the five bolts overcored from the Hammaslahti mine were nearly broken (Figure 27). The tube bolts follow the borehole very tightly. This makes them less flexible to tolerate horizontal movements, so they act like fully grouted rock bolts.

5 DISCUSSION

The project proved to be a success in many respects. The knowledge obtained exceeded even the fondest hopes. For the first time it has been possible to see how different types of rock bolts are in contact with the rock to be supported and what happens to them with time.

It could be shown that overcoring, although a complicated phase of work based completely on the use of very skilled and careful personnel, is for the moment the only way to get undisturbed test samples. With pull tests and overcored samples the usability of the two methods, and also information supplied by them could be compared. The pull tests which were recorded as continuous load-displacement curves allowed the study of the behaviour of the bolts under increasing tensile load. These field tests give a sound basis for future research work on the long-time performance of different types of rock bolts.

The results convince that the environmental properties like the quality of the rock, the amount, quality and flow of water, rock mechanics characteristics and the stability of the rooms can be controlling factors in choosing proper type of support.

Most of the bolts tested, even the newer types, proved to work better than was expected. Many of the defects resulted from using a wrong type of bolt or from incomplete installation. Only careful installation leads to effective and continuously resistant support.

In both cement and resin groutings cracks and empty spaces, even major lack of material was found. Also major pieces of the resin cartridge cover were seen between the grouting and the borehole wall. All these factors can diminish the holding force of the bolt depending on the size and location of the defect.

Corrosion is undoubtedly one of the major problems in long run. It determines which bolts can be used as permanent support. Sofar the cement grouted rebar seems to be best protected against corrosion, because even if water is present alkalic and thus non-corrosive surroundings are created. Damage due to corrosion occured only on the free end of the rebar or near the beginning of the borehole where the grouting often was very poor. Cracks or empty spaces deeper in the grouting had not led to any corrosion. The critical size of cracks or fractures and the rate of waterflow remains to be solved which could favour corrosion processes. In the case of resin grouted rebars the free parts of the rebars were very badly corroded only in two years. The diameter of the original Ø 20 mm rebar was found to be diminished to 16 mm.

The split set and swellex tubes were corroded all over their entire length. The point corrosion was strong, the minimum measured value of wall thickness being for Split-set 41.7 % one year after installation and for Swellex 55.1 % 1.5 year after installation. The surface area of the bolt crosscuts had been reduced 11 % for Split-set and 7.5 % for Swellex in the average. Galvanization reduced the corrosion rate of the Split-sets to about a half.

For the moment special concern should be paid on developing a reliable non-destructive testing method and devices for rock bolts. These devices should be fit for continuous use at work site. They ought to be able to discover possible defects arising at time of installation as well as to follow up the influence of corrosion and rock movements. At this time the best results can be achieved by overcoring. Although pull tests and overcoring are usable methods they are too expensive and too slow. They are more research than control methods.

6 SUMMARY

Sofar there has not been any reliable method available to get information about rock bolts in situ ie. how effectively their support is utilized. At the moment only overcoring can provide us with the most

comprehensive details. Pull tests may be used as a
kind of first aid in quality control but if there is
any doubt of major defects overcoring is advisable.

Conventional cement grouted rebar carefully installed is an effective support for long term use.
Grouting protects it against corrosion, even if fractured. Its load-bearing capacity can be defined accurately enough from the material characteristics.

A resin grouted rebar provides a high load bearing
capacity. The possibility to poor grouting is greater than for cement grouted rebars. If resin grouted
rebars are used as point anchored bolts, the free
end of the rebar can be severely corroded in certain
circumstances.

Split-set and Swellex are quick and easy to install. They support immediately after installation.
The corrosion damages are obvious but their lifetime is still long enough for temporary use even under difficult conditions. The load bearing capacity
of split sets increases with time. The residual holding force strongly decreases to 80 % just after installation and to 5 to 20 % six months later. If the
load bearing capacity of a swellex is completely utilized and the bolt is tight installed, it acts like
a grouted rebar and can survive only relatively
small deformations.

ACKNOWLEDGEMENTS

The author appreciates the financial help from the
Finnish Fund for Labour Protection Research which
was fundamental for the successful completion of the
project.

The co-operation of Finnish mining industry and
its personnel, especially in Outokumpu Oy, Rautaruukki Oy and Lohja Oy, is gratefully acknowledged.

My thanks are also due to the project leader,
prof. Raimo Matikainen, and the personnel of the Laboratory of Mining Engineering for many inspiring
discussions and other help during the work.

REFERENCES

Hoek, E. and Brown, E.T. (1980). Underground excava-
    tions in rock. Institute of Mining and Metallurgy,
    London. 527 p.
Poitsalo, S. (1983). The strengthening efficiency of
    different rock bolts. Rock Bolting, pp. 459-464.
    A.A.Balkema, Rotterdam.
Sundholm, S. (1984). Pultituksen valvonnan tarpeelli-
    suus ja valvontamenetelmät (The quality control of
    rock bolts). Lic.Techn.Thesis, Helsinki University
    of Technology, Department of Mining and Metallurgy.
    97 p (in Finnish).

The NATM studied from the viewpoint of rheology and geodynamics

La NATM envisagée du point de vue de rhéologie et géodynamique
Die NATM untersucht vom Gesichtspunkt der Rheologie und Geodynamik

TAN TJONG KIE, Professor and Director, Institute of Geophysics, Academia Sinica, Beijing, People's Republic of China

ABSTRACT: In order to get a better understanding of the interrelated factors leading to the successes and failures with the NATM the author has carried out a basic study from rheological principles. The geological condition, creep and dilatancy and longterm strength properties of rocks are considered important factors. Based on his constitutive equations the author presents a F.E. Analysis; especially the regions of dilatancy-creep and the influence of tectonic stress are studied. Various suggestions for the strengthening of tunnels unstable with the time are recommended.

RÉSUMÉ: Afin d'obtenir une comprehension plus profonde des facteurs de corelation qui avaint menés aux succes et fiascos de l'ATMM l'auteur a fait une étude fondée sur les principes rheologiques. La condition géologique, le fluage et dilatancy et puissance a long duration de roches sont considerés des facteurs importants. Au fond de ses equations constitutives, l'auteur présente un Analyse d'Elements Finis et especiallement les regions du fluage dilatants et l'influence des tractions tectonics sont étudiés. Plusieurs suggestions pour la fortification des tunnels instabiles avec le temps sont presentées.

ZUSAMMENFASSUNG: Um ein besser Begriff über die zusammenhängende Faktoren welche nach   Erfolge or Katasprophen mit NATM, geleitet haben hat der Verfasser ein grundlegendes Studium gemacht auf Grund von rheologische Grundsätzen. Die geologische Bedingung, das Kriechen, die Dilatanz und langzeitliche Festigkeit werden als wichtige Faktoren studiert. Mit Hilfe seiner Material gleichungen hat er eine F.E. Analyse ausgeführt. Insbesonders die Mächtigkeit der Kriech Dilatant Gebiete und der Einflusz der tectonische Spannungen sind analysiert. Verschiedene Vorschläge für die Verstärkung von Tunnels instabiel mit dem Zeit sind empholen.

INTRODUCTION

The New Austrian Tunneling Method has been succesfully applied in many tunnels all over the world, in strong tectonic areas, at depths of 1000m under the influence of blast waves from mining operations, as well at low and medium overburdens in weak soils as well as in strong and hard rocks. The Method has been applied with success to squeezing tunnels in Japan (Otsuka et al 1981). In our country the NATM has also found application in the Jin Jiayan tunnel (Chengdu Railways Bureau 1984), the Nan Ling tunnel (Nanling tunnel experimental station, 1984) at shallow depth, the instable galleries of the Qing Juan Mine (Tan 1981) and many others.

Despite its many successes however the physics underlying the NATM still needs further clarification. The shotcrete, steel web, bolts and anchors technique is frequently considered as an omnipotent remedy in complex tunneling. In some cases successes have been achieved, without a pre-analysis based on basic principles. However in many other cases there was no clear definite result and some failures-technical as well as economical-clearly indicate that the basic philosophy of the method was not fully grasped. Therefore Müller (1978) correctly stressed that the NATM is based on the commentment to certain principles, which have been substantiased by successful use in practice. Thus the engineer must utilize the load bearing capacity of the rock mass to an extent that guarantees an optimum on safety and economy.

In this paper the author will emphasise the importance of geodynamics and rheology in tunneling as it is not yet adequately considered in the NATM.

BASIC FACTORS IN TUNNELING

The author suggests that the following basic factors must be considered:

1. tectonics, tectonic stress history;

2. geology of the region (basic state of the rock mass, usually jointed, fissured and cut by weak intercalations);

3. rheological properties of the rock mass; many complicated fissuring, bulging and failure phenomena occur not earlier than after weeks, months, years and decades;

4. proper choice of direction of tunnel axis and most feasible tunneling crossection; in order to utilise the load bearing capacity of the surrounding rock mass efficiently, a finite element analysis based on appropriate geological descriptions and constitutive equations must be carried out;

5. on the basis of the above 4 points to design the strengthening method;

6. back-analysis+field convergence measurements; improved measures.

GEODYNAMICAL CONSIDERATIONS

Usually a rock-engineering site may cover some square kilometers, but viewed on the scale of the lithosphere it is only a point on the skin of the upper crust. During the tectonic history of hundreds of millions of years the crust was strained, in different manners. It is known that the earthcrust is strained continuously at a rate of 10^{-14}-10^{-16}/sec. In definite parts the straining may be larger than 10^{-14}/sec and then earthquakes are frequently generated. The present imprints of all these tectonic sequences can be found in:

1. families of big faults and dikes as studied in structural geology;

2. many families of small faults and joints filled with weaker intercalations as studied in engineering geology;

3. folding;

4. presence of a tectonic stress, which usually acts tangentially to the crustal surface;

5. presence of internal stresses which are locked in the rock mass during rock genesis and tectonic straining and can be liberated gradually by the formation of cracks or dynamically by the generation of rockbursts (Tan 1986).

TECTONIC STRESS σ_T

The fact that the crust is in the strained state indicates that crustal stresses must be working. They   may be due to the collision of plates and mantle convection or regional topography. On the basis of a geodynamic analysis of the past and present seismicity of the region the direction of the present maximal normal stress, can be estimated from fault-plane solutions. This is very important as it gives the stress directions in the large region.

In special cases the dip of the big faults is nearly vertical and then the directions of main normal stresses coincide with the bisectrix of the angle formed by the big faults.

Furthermore stress measurement which can be made by the overcoring and hydraulic fracturing method are indicated.

The tectonic stress σ_T which is a governing factor in geodynamics must not be overlooked. As we will discuss further on in this paper it is of great influence on the stability of the bottom and roof of underground cavities.

SOME CASE HISTORIES

The time dependent deformation and long term strength properties of the rock mass are fundamental features which must be considered. In this respect the author wishes to present two typical cases from his practice:

In 1973 he was invited to an underground machine hall. The rock was slate, with plates slanting at 45° to the horizon. Thin sheets of weak intercalations, silt and clays were embedded between the plates. It was possible to pull the plates out of the rockwall by hand. The overburden was 130-200 m, the width and height of the machine hall were 20 m and 40 m respectively. Signs of rock falls from the roof during excavations were obvious. The question was whether a further deepening of the floor with 2m was possible. According to the author's finding the machine hall was unstable, in the stage of tertiary creep and could collapse every moment, especially after some raining; therefore an immediate strengthening of the walls with bolts, steelweb and concrete was imperative, furthermore convergence monitoring was necessary.

After two years the author came back on the site and was informed that the underground hall had completely collapsed one year before... no strengthening measures were undertaken. It is evident that the load bearing capacity of the surrounding rockmass was decreasing with the time, as it will be discussed further on.

In 1983 the author was invited to a railway tunnel between Sydney and Wolonggong. Although the tunnel was completed in igneous rock more than 100 years ago many repairs were necessary when the concrete lining showed too much cracking in the course of a few years. Similar problems the author has met many times in China. The method of repair was always the same i.e. the cracked parts were removed and a new lining was applied. As the crucial problem i.e. instability of the force bearing surrounding rock was not fundamentally solved, of course cracking repeated after many years etc etc.

The essential problem with these cracking tunnels is that the deviatoric stresses in the surrounding rocks exceed an upper stress limit:(the upper yield value called f_3 by the author in previous papers (Tan 1980); when this limit is exceeded then cracking increases in intensity which leads to loosening. The process of formation of new cracks and opening and propagation of existing cracks-whereby porespace is created under deviatoric stresses-is known as dilatancy.

This dilatancy is a time process which gradually loosen and destrengthen the rock mass. It is obvious that merely the application of a surface layer of concrete will not be effective to subdue the process of loosening in the rock's interior. The potential remedy must be searched in the strengthening of the surrounding rocks which can be achieved by bolts, anchors in combination with an iron web shotcrete layers, and grouting. Therefore it is very plausible in the NATM that it is based on the fundamental concepts that:

1. The major load bearing element of a tunnel is the surrounding rock mass;

2. The original rock mass strength must be preserved as much as possible and dilatation (dilatancy) should be kept a minimum as its lowers strength.

In order to grasp the basic principles stated above and many working methods adopted in the NATM, it is necessary to have a clear understanding of the rheology of rocks. The time dependent behaviour is inherent to rocks and should always be considered. As it is described above tunnel linings and tunnels may fail after a lapse of months and years i.e. the load bearing capacity is decreasing with the time (see form (7)) and the convergence increasing.

RHEOLOGICAL PROPERTIES OF ROCKS

In order to have a better insight into the longterm stability of tunnels it is necessary to have an idea of the time dependent properties of rocks: From many tests on igneous and sedimentary rocks the stress strain time relationships can be represented in Fig. 1. Here the relationship between the deviatoric stress $\sigma_1 - \sigma_3$ is plotted against the vertical strain e_1, volumetric strain e and horizontal strain e_3.

The following features may be described:

1. The deviatoric stress strain relationships are nearly linear in the beginning; in this lower range the deviatoric stress-volume strain relations are also linear. The rock assumes a creep deformation increasing with log t (Fig. 2A);

2. As soon as the stress difference exceeds a certain limit $f^*(=\sqrt{3}f_3)$ nonlinear deformations due to micro-fracturing and volume expansion (dilatancy) occurs. This stage is also characterised by the onset of accoustic emissions and the increased lowering of the porewater pressure in undrained tests (Tan 1986 Fig. 7A); So f_3 is the limiting shearing stress marking the boundary between stable logarithmic creep and unstable creep leading to failure (Fig. 2A, B);

3. The radial strain e_3 grows faster than the axial strain e_1 in uni-and triaxial tests;

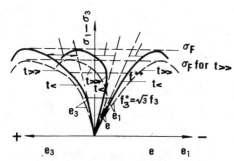

Fig. 1. Creep dilatant behavior of rocks deviatoric stress $\sigma_d = \sigma_1 - \sigma_3$, plotted against axial strain e_1 lateral strain e_3 and volumetric strain e. Dashed curves show stress strain relations for large values of the time.

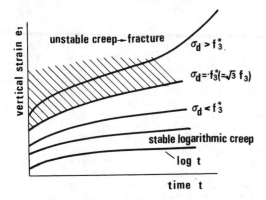

Fig. 2A. Creep curves for increasing values of $\sigma_d = \sigma_1 - \sigma_3$.

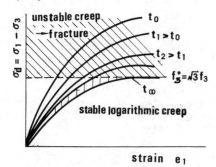

Fig. 2B. Stress strain isochrones derived from Fig. 2A (not to be confused with the stress strain curves obtained from tests at different constant rates).

4. With higher confining pressures cracks are closed and crack propagation is anticipated by the friction of the crack surface. Hence dilatancy decreases with the increase of the confining pressure;

5. Dilatant volume deformations, the axial and radial deformations show an instantaneous part followed by a time dependent part increasing with the time t (Fig. 2A); the time dependent stress strain relations are shown by dashed curves (Fig. 1);

6. At low confining pressures shear softening occurs after the stress difference has attended the maximum strength σ_F. But the dilatancy continues to increase steadily.

In addition to the above basic features the following three assumptions will be made:

1. The rock is a continuum; tensile stresses are positive;

2. The behaviour after the stresses exceed σ_F is

associated with cataclastic deformations, where the continuum is transformed into a discrete medium; this final stage is beyond the scope of our modelling;

3. Complex heterogeneous and layered rock masses can be modelled by ascribing different values of the parameters to the discrete elements.

CONSTITUTIVE EQUATIONS

The following constitutive equations are suggested (Tan 1986 , 1980).

$$e_x(t) = \frac{s_x}{2G} + \frac{1-2\nu}{E}p + \int_{-T}^{t} \psi(t-\theta)\, d\frac{s_x(\theta)}{d\theta}d\theta +$$
$$\int_{-T}^{t}\phi(t-\theta)\frac{dp(\theta)}{d\theta}d\theta + D*\left\langle F_1\left(\frac{\sigma_{oct}}{f_3}\right)\right\rangle +$$
$$C\left\langle F_2\left(\frac{\sigma_{oct}}{f_3}\right)\right\rangle \frac{s_x}{\sigma_{oct}} +$$
$$\zeta\int_{-T}^{t}\left[D*\left\langle F_1\left(\frac{\sigma_{oct}}{f_3}\right)\right\rangle + C\left\langle F_2\left(\frac{\sigma_{oct}}{f_3}\right)\right\rangle \frac{s_x}{\sigma_{oct}}\right]dt \quad (1)$$

$$\tfrac{1}{2}\gamma_{xy}(t) = e_{xy}(t) = \frac{\tau_{xy}}{2G} + \int_{-T}^{t}\psi(t-\theta)d\frac{\tau_{xy}(\theta)}{d\theta}d\theta +$$
$$C\left\langle F_2\left(\frac{\sigma_{oct}}{f_3}\right)\right\rangle \frac{\tau_{xy}}{\sigma_{oct}} +$$
$$+ C\zeta\int_{-T}^{t}\left\langle F_2\left(\frac{\sigma_{oct}}{f_3}\right)\right\rangle \frac{\tau_{xy}}{\sigma_{oct}}dt \quad (2)$$

and similar expressions for $e_y(t)$, $e_z(t)$, $\gamma_{yz}(t)$, $\gamma_{zx}(t)$.

$$e(t) = 3\frac{1-2\nu}{E}p + 3\int_{-T}^{t}\phi(t-\theta)\frac{dp(\theta)}{d\theta}d\theta +$$
$$3D*\left\langle F_1\left(\frac{\sigma_{oct}}{f_3}\right)\right\rangle + 3D*\zeta\int_{-T}^{t}\left\langle F_1\left(\frac{\sigma_{oct}}{f_3}\right)\right\rangle \quad dt \quad (3)$$

whereby: $S_x = \sigma_x - p$; $p = (\sigma_1 + \sigma_2 + \sigma_3)/3$;

$$e(t) = e_x(t) + e_y(t) + e_z(t) = \text{volume strain};$$

$$\sigma_{oct}^2 = J_2 = 1/6\left[(\sigma_x - \sigma_y)^2 + (\sigma_y - \sigma_z)^2 + (\sigma_z - \sigma_x)^2\right]$$
$$+ \tau_{xy}^2 + \tau_{yz}^2 + \tau_{zx}^2 \quad (3a)$$

The first two terms in (1) and the first terms in (2) and (3) represent the elastic part of the deformation. The creep deformation is represented by the third and fourth terms in (1) and the second terms in (2) and (3);

$$\psi(t) = \beta_s \log(1+\alpha t) = \text{creep compliance for unit stress difference;}$$

$$\phi(t) = \beta_v(1 - e^{-t/r}) = \text{volume creep compliance for unit hydrostatic stress;}$$

whereby β_s, β_v, α and r are constants. The instantaneous dilatant deformations e_{dji} are represented by the fifth and sixth terms in (1) and the third terms in (2) and (3). The time dependent dilatant deformation is described by the seventh term in (1) and the fourth term in (2) and (3).

It is understood that the onset of dilatancy occurs for $\sigma_{oct} > f_3$, i.e.

$$\langle F_1(\tfrac{\sigma_{oct}}{f_3})\rangle = (\tfrac{\sigma_{oct}}{f_3})^n - 1$$
$$= 0$$
$$\langle F_2(\tfrac{\sigma_{oct}}{f_3})\rangle = (\tfrac{\sigma_{oct}}{f_3})^{n*} - 1$$
$$= 0$$

$$\text{For } \tfrac{\sigma_{oct}}{f_3} \begin{cases} > 1 \\ \leq 1 \end{cases} \qquad (4)$$

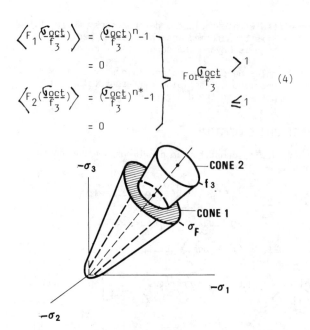

Fig. 3. Yield f_3 and failure cones σ_F.

In this paper a cone is assumed for the yield surface; a coaxial cone is assumed for the failure surface. A set of 2 coaxial hexagonal pogramides is another alternative but the formulas become a little more complicated. In σ_1 σ_2 σ_3 space (Fig. 3) the cones are described by:

$$J_2^{\frac{1}{2}} = f_3 = f_{3o} - mp,$$
$$J_2^{\frac{1}{2}} = \sigma_F = \sigma_{Fo} - m'p; \qquad (5)$$

The space between these cones is the zone of dilatancy, microcracking and accoustic emissions. In the above theory we try to model the processes prior to failure. The above theoretical approach is designed for creep dilatant rocks. In weak plastic rocks intense creep is observed when $\sigma_{oct}/f_3 > 1$. In hard brittle rocks cracking is accompanied by accoustic emissions for $\sigma_{oct}/f_3 > 1$. So the computations can be used as well for plastic bulging as for rockbursts (Tan 1986).

TIME OF RUPTURE

A rough estimation of the ultimate breakdown time of the tunnel after a complex sequence of time dependent displacements is very helpful for the construction engineer as he has to know the time span in which he can place the lining most effectively. For the longterm failure of samples the following formula has been suggested (Tan 1981).

$$(\tfrac{t_R}{t_{Ro}})^\gamma = \text{Const}/(\sigma_{oct}/f_3 - 1)^n \qquad (6)$$

whereby f_3 is given in (5)

t_{Ro} = Rupture time for short term quick tests;

t_R = Rupture time of longterm failure;

σ_{oct} = given in (3A) is $> f_3$

n = 4 to 5 for granites and 6 to 7 for weak rocks.

The constant in formula (6) can readily be estimated. For axi-symmetrical loading $\sigma_1 > \sigma_3$; $\sigma_2 = \sigma_3$ we get:

$$(\tfrac{t_R}{t_{Ro}})^\gamma = \text{Const}/[(\sigma_1-\sigma_3)/f_3^* - 1]^n: \text{ From many}$$
experiments we know $\gamma \sim 1$ and $\dfrac{\sigma_1-\sigma_3}{f_3^*} = \dfrac{\sigma_F}{f_3^*} = 2 \to 3$

So we get: Const $= \tfrac{1}{2} \to 1$; note $f_3^* = \sqrt{3} f_3$

The average value of $\overline{\sigma}_{oct}$ can be estimated from F.E. computations, and then t_R for tunnels can be roughly predicted: days, weeks, months or years.

THE ZONE OF LOOSENING, INFLUENCE OF TECTONIC STRESS σ_T, F.E. ANALYSIS

The tunneling engineer is much concerned with the site and depth of the dilatant zone, as it is an unstable region of loosening. The depth is necessary to design the length and mutual distance of bolts and anchors. Unnecessary large lengths are time consuming and uneconomical. Whereas short lengths will not be efficient to solve the bulging problems. Many factors as the overburden, tectonic stress, rheological properties of rocks, geological variations and homogeneity are fixed and are governing factors leading to the stability or instability of tunnels; on the other hand their exist subjective factors, which are at the disposal of the engineer such as excavation and strengthening techniques. The crucial point is to construct a tunnel which is stable on the long term.
As we can see from the formulas (4) and (5) the decisive factor is the ratio:

$$\eta = \frac{\left[1/6 \left((\sigma_1-\sigma_2)^2 + (\sigma_2-\sigma_3)^2 + (\sigma_3-\sigma_1)^2\right)\right]^{1/2}}{f_{3o} - m(\sigma_1+\sigma_2+\sigma_3)/3} \qquad (7)$$

For simplicity we consider σ_1 and σ_3 main normal stresses acting in planes porpendicular to the tunnel axis, whereas σ_2 acts parallely to it. Then we have the main normal stresses:

$$\sigma_1 = \gamma h + \sigma_{T1}$$
$$\sigma_2 = \tfrac{\nu}{1-\nu} \gamma h + \sigma_{T2}$$
$$\sigma_3 = \tfrac{\nu}{1-\nu} \gamma h + \sigma_{T3}$$

$\sigma_{T1}, \sigma_{T2}, \sigma_{T3}$, are the components of the tectonic stress σ_T in the directions 1, 2, 3.

(an approximate semi-infinite mass is assumed)
Most favourable is when $\sigma_{T1} = \sigma_{T3} = 0$; $\sigma_{T2} = \sigma_T$ i.e. the tectonic stress acts in the direction of the tunnel. It contributes to $p = (\sigma_1 + \sigma_2 + \sigma_3)/3$ thus increases f_3 and decreases the danger of loosening (Form 7).
Most unfavourable is when σ_T acts perpendicular to the tunnel axis in a horizontal direction.

F.E. Analysis

In order to study the effects of the tectonic stress on loosening we have programmed the equations (1)......(5) for finite element analysis. The following inputs have been choosen;
overburden H=5000 m, tunnel width 6m, height 9m; $E = 10^4$ kg/cm^2; $\nu = 0.3$; $\gamma = 2.7$ tons/m^3; $f_{3o}=30$ kg/cm^2; m=0.1; n=5; C=9.22×10^{-5}; D*=9.87×10^{-5};

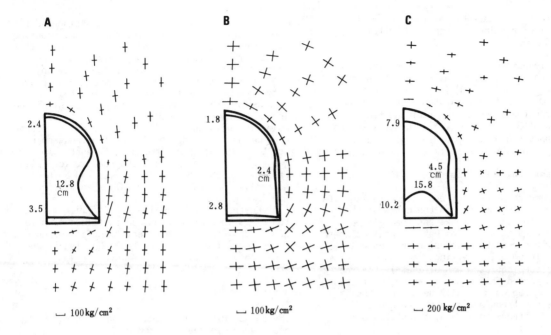

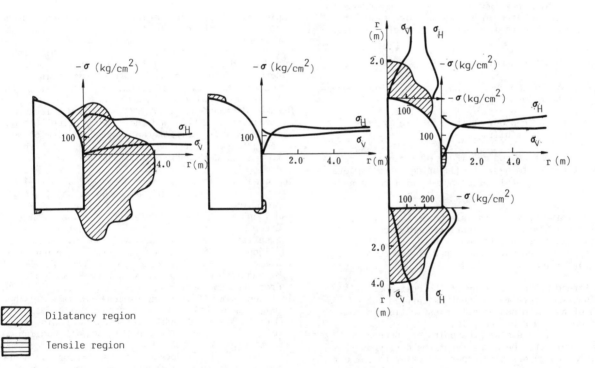

▨	Dilatancy region
▤	Tensile region

Fig. 4. Stress distributions, displacements (cm); Dilatancy and tensile regions from F.E. Analysis.

σ_T = 0, 80, 150 kg/cm^2 thus σ_1/σ_3 = 2.3, 1.0, 0.65 or $\sigma_1 - \sigma_3$ = +77, 0, -73 kg/cm^2.

In this preliminary analysis only the instantaneous stress distributions, displacements and loosening zones are considered. The results are shown in the figures 5A, B and C. The following analysis can be presented.

The site and extent of loosening zone depend on the stress difference $\sigma_1 - \sigma_3 = \sigma_d$ between vertical overburden stress σ_1 and horizontal stress σ_3: but not on the individual magnitudes

of σ_1 or σ_3 themselves:

A. For $\sigma_d \approx$ +77 kg/cm^2 the dilatancy zones appear at the side-walls. Some small tensional regions are found at the bottom and roof. The wall displacements amount to 12.8 cm;

B. For $\sigma_d \approx$ 0 kg/cm^2, the dilatancy zones are very small and restricted to the roof and corners;

C. For $\sigma_d \approx$ -73 kg/cm^2 the dilatancy zones appear at the bottom and roof. The influence of the large tectonic stress is remarkable at roof and bottom in the form of roof sagging (7.9 cm) and bottom up-heaval (15.8 cm).

The above results can be expected from the formula (7): increases of the hydrostatic pressure p leads to a decrease in the ratio η; for instance an increase of σ_T suppresses dilatancy in the side walls but promotes loosening at roof and top; further small values of σ_d gives only η values less than 1 and dilatancy can only appear in regions of high concentrations of stresses (corners) where $\eta > 1$.

Time Dependent Weakening-Prognosis

The time dependent process of desintegration and its consequences can be analysed on the basis of the rheological properties of rocks and the related constitutive equations. In the hatched regions where the deviatoric stress are exceeding the upper limit f_3 and $\eta > 1$ (formula 7) cracking keeps on con-tinuously; cracks coalesce and ultimately the rock-mass in this region fails. As the strength of this desintegrating region gradually decreases, the stresses are now transferred to the neighbouring regions. For instance in case A, the stresses will be transfered to:
  1) the roof and bottom, accompanied by roof falls and excessive bottom upheaval;
  2) to the more remote parts in the interior of the surrounding rock, which can lead to an expansion of the loosening zone in horizontal direction. This is accompanied by an increased bulging of the wall.

CONCLUSIONS AND SUGGESTIONS

On the basis of the above study the following sugges-tions can be presented:

1. A geodynamic study of the tectonic history of the region is helpfull;

2. The direction and magnitude of tectonic stress σ_T must be determined. It will bring extra explora-tion and experimental costs, but it will be reward-ings;

3. A Finite Element analysis based on a proper rheological basis and appropriate geological des-cription is necessary to know roughly the sites and depths of loosening zones and influence of tectonic stress;

4. Strengthening by means of the bolt-anchor-steelweb shotcrete method must be carried out on the basis of a minute geological investigation in com-bination with a F.E. Analysis.
   It is usual to emplace the bolts or anchors perpendicular to the boundary of the cavity. So the bolts (anchors) are placed perpendicular to the cir-cumferential stress i.e. it can only have little contribution in the taking up of stresses acting in the loosening and tensile zones such that the high tensile strength of the steel bolts can not be utilised efficiently. Hence it is recommended that in these crucial regions bolts must be installed obliquely to the rock surface and must mutually be crossed.
   In this manner the loosening and tensile zones can be reinforced and expectedly the engineer can fulfill with shorter bolts in dense formation as a reinforced ring around the tunnel can be created in this way;

5. In cases of important loosening zones and es-pecially at large tectonic stresses a firm bottom element is indicated and the load bearing ring around the tunnel must be installed as properly and soon as possible;

6. In order to decrease the propagation and fur-ther deterioration of the loosened zone, grouting with cement and or chemicals is recommended. It in-creases the resistance between crack walls and decreases the risk of further disintegration;

7. Bulging always starts in small regions and expand with the time in extension and intensity.
   Therefore for regions of loosening as predicted in the hatched portions it is wise to take strength-ening measures as quickly as possible. The prognosis depends on the method and time of strengthening. May be short bolts of appropriate lengths in a dense web may rescue the problem supposed it is applied rapidly enough, whereas long bolts and anchors will not solve the problem adequately when applied at too late a period;

8. Further closure of the bearing ring around the tunnel is indicated when important loosening re-gions, desintegrations stress transfer and consider-able time effects are expected, the closure must take place as rapidly as possible.

ACKNOWLEDGEMENT

The author wishes to thank Mr. Wu Haiqing for his assistance in the F.E. computations.

REFERENCES

Engineering Department of the Chengdu Railway Bureau 1984, Application of bolt shotcrete method in the railway tunnel of Jin Jiayan (in Chinese).
Müller Salzburg, L. and Fecker, E. 1978. Philosophy and basic principles of the New Austrian Tunneling Method in Grundlagen und Anwendung der Fels Mechanik Kolloquim Karlsruha, Frabs Tech: Publica-tions Clausthal 1978, p. 247-262.
Otsuka, M. and Kondoh, T. 1981. On the displacement forecasting methods and their application to tunnel-ing by NATM. Proc. International Symposium on weak rock, Tokyo. in Akai, K., Hayashi, M. and Nishimatsu, Y. (eds.), Weak Rock: Soft, Fractured and Weathered Rock vol. 2, p. 945-950.
Southwest Institute of the Academy of Railways 1984. Tunneling technique in weak rocks at shallow depth (in Chinese).
Tan Tjong Kie and Kang Wenfa 1980. Locked in stresses, creep and dilatancy of rocks, and constitutive equations, Rock Mechanics, vol. 13, p. 5-22.
Tan Tjong Kie 1981. Discussion. Proc. International Symposium on weak rock, Tokyo. in Akai, K., Hayashi, M. and Nishimatsu, Y. (eds.), Weak Rock: Soft, Fractured and Weathered Rock vol. 3, p. 1373-1375.
Tan Tjong Kie 1982. The mechanical problems for the long-term stability of underground galleries. Chinese Journal of Rock Mechanics and Engineering, vol. 1, No. 1, p. 1-20 (in Chinese with English Summary).
Tan Tjong Kie 1986. Rockbursts, Case Records, Theory and Control. Proc. International Symposium on Engineering in Complex Rock Formations, Beijing, China, p. 32-47.

Relationship between deformation and support pressure in tunnelling through overstressed rock

Relation entre la déformation et la pression sur le soutènement en creusant un tunnel au travers de roches trop contraintes
Die Beziehung zwischen Verformung und Auflastdruck durch Tunnelvortrieb in überspanntem Fels

CHIKAOSA TANIMOTO, Kyoto University, Japan
SHOJIRO HATA, Kyoto University, Japan
TOSHIO FUJIWARA, Ohbayashi Corporation, Tokyo, Japan
HISAYA YOSHIOKA, Ohbayashi Corporation, Tokyo, Japan
KAZUTOSHI MICHIHIRO, Setsunan University, Osaka, Japan

ABSTRACT: Using experience and data from recent tunnel construction in Japan, the authors propose a method to find the appropriate range of maximum allowable convergence measurements. Five types of geologic conditions are discussed: tertiary mudstone, tuff breccia, tuff, limestone and alternative layers of slate and chert. The behaviour of these rock types in response to tunnelling are discussed in terms of the stress-strain behaviour, actual deformations, maximum allowable convergence (MAC), practical design and monitoring. The emphasis is placed on the importance of the stress-strain behaviour for the determination of the MAC for weak rocks. Based on the analysis of the relationship between deformation and support pressure, the classification of support loads for overstressed rocks is proposed, and it is suggested that allowable limit of deformation in tunnelling must not be beyond 2% of an original diameter.

RÉSUMÉ: En usant de l'experience et données dans les travaux souterrains en Japon, les auteurs proposent une méthode pour trouver la limite appropriée de la convergence maximale et permissive pour les roches en conformité avec la mesure des convergences. Cinq types geologiques des roches se discutent, glaises tertiares, ardoises, grèss, calcaires, les alternances des couches d'ardoises et roches sedimentaires de quartz micro-crystalin. Leur comportement en réponse au creusement d'un tunnel est discutée au sujet de la relation entre la contrainte et la deformation, de la deformation actuelle, de la convergence maximale et permissive, du dessin pratique d'un tunnel et sa surveillance. Nous appuyon sur la importance de la relation entre la contrainte et la deformation pour la quantité de la convergence maximale et permissive des rouches faibles de resistance. En conformité avec la analyse de la relation entre la deformation et les charges sur le soutènement, la classiflcation de charge sur le soutènement dan les roches trop contraintes est proposée, et il est suggere qu'il ne faut pas que la limite de la deformation permissive est plus de deux pour-cent de la largeur du tunnel originaire en radier.

ZUSAMMENFASSUNG: Unter Benutzung der Ergebnisse und Daten vergangener Tunnelbauprojekte in Japan, versuchen die Verfasser eine Methode aufzuzeigen, den erlaubten Bereich der Konvergenzen von Fels, basierend auf die Tunnelkonvergenzmessungen, zu bestimmen. Fuenf verschiedene geologische Zustaende werden untersucht: tertiaerer Ton, Schiefer, Sandstein, Kalkstein, sowie wechselnde Schichten von Schiefer und Kieselschiefer. Das Verhalten dieser Gesteinstypen wird in Bezug auf die Spannungs-Dehnungs-Beziehung, auftertende Verformungen, maximal erlaubte Konvergenzen (MAC), Bemessung und Schadensbilder diskutiert. Mit Nachdruck wird auf die Wichtigkeit der Spannungs-Dehnungs-Beziehung zwischen Verformung und Auflast auf die Bestimmung der maximal erlaubten Konvergenz (MAC) von bruechigem Fels hingewiesen. Auf Grund der Analyse der Beziehung zwischen Verformung und Auflast, wird eine Einteilung der Auflasten fuer uberspannen Fels vorgeschlagen. Gleichzeitig wird darauf hingewiesen, dass Verformungen im Tunnelbau 2% des geplanten Durchmessers nicht ueberschreiten sollen.

1. INTRODUCTION

260 tunnels of 300 kilometers long which have been constructed in Japan since 1975 have been driven by using rockbolts and shotcrete as the main supporting elements. The reasons for the high deformation rates observed during construction are as follows: 1) low competence factor; 2) allowance of too much deformation to establish the minimum support load; 3) delayed ring closure and loss of the potential bearing capacity of rock; 4) incapability of the stiffness of the support elements relative to the stiffness of the ground; 5) too slow feed back of convergence measurement for checking of the support installation timing.

The convergence measurement is the most popular and practical measurement in tunnelling. When the ground behaves as an elastic body with no time dependency, the result of convergence reaches a constant within the range of 2*D (D: tunnel diameter) in driving direction. On the other hand, when the ground is subject to the non-elastic behaviour such as the strain-softening and/or plastic-flow, deformation does not reach a constant level within 2*D distance, and converges in 3-10*D distance. As the ground arch formation based on the stress distribution is mobilized along the elastic/non-elastic boundary around the tunnel opening, it has been verified through several case histories that the convergence curve suggests the development of the non-elastic zone.

From the observations, it is pointed out that the allowable deformation accepted in tunnelling is often too large, resulting in support loads larger than necessary (Tanimoto et al., 1983). The authors analyzed the rasults of measurements obtained from several tunnel projects in detail, and proposed that support load classification (Tanimoto et al., 1983; 1986) which enables to estimate the magnitude of final displacement and available support pressure by means of an initial deformation rate obtained from a first 3 days driving.

In the construction of Orizume Highway Tunnel, which was subject to heavy support pressure and non-elastic behaviour, it was verified that the support load classification mentioned above is appropriate and that the relationship between the minimum support load and the corresponding magnitude of displacement has been clarified for several rock types in the state of being overstressed.

2. OBSERVED DEFORMATIONS IN TUNNELLING

In most cases the results of convergence measurements are plotted on a deformation-time graph, which gives the apparent deformation rate. However, because of the dependence of deformatin on the position of the mining face, it is more informative to plot the results of the convergence survey relative to the distance to the face and tunnel walls.

There are many uncertainties involved in the interpretation of geologic conditions for tunnelling from exploratory borings. Terzaghi (1946) and Peck (1969) proposed the use of an "observational construction method" for tunnelling in difficult ground. Under such conditions, a convergence survey may serve the purpose of providing information on the deformation rate at the early stages of excavation, from which it is possible to estimate the approximate support load and timing for the installation of the support elements.

The optimum goal in tunnelling is to advance the face in such a manner as to maintain the original stress state of the ground. However, the stress distribution is altered by the removal of the rock from within the opening and the displacement of the surrounding rock with the loss of inner confining pressure. It has been found that by allowing some displacement of the rock within the non-elastic zone around the opening that the shear strength of the rock is mobilized so that it is capable of supporting some of its own load by the ground arch effect.

The largest deformation occurs in the vicinity of the advancing face and is subject to the advance rate and stress-strain behaviour of the rock. As the face advances, artificial supports take the place of the half-dome action of the face. To fully mobilize the bearing capacity of the rock mass, the supports must be installed within a distance of one tunnel diameter from the face (Terzaghi, 1946; Mueller, 1977; AFTES,1978; Tanimoto, 1980; Tanimoto et al.,1981a,b). Experience and observations from convergence surveys have shown that the deformation rate is most influenced by the position of the face relative to the installation of the support elements and that time-dependent behaviour of the rock is of little consequence (Tanimoto et al., 1980). In the case of some tunnels through tertiary mudstone which exhibit strong time-dependent behaviour, the deformation rate decrease to almost zero with a halt in the advance over an extended holiday.

3. MEANING OF DEVELOPING NON-ELASTIC ZONE

Based on many observations, it has been suggested that actual deformation caused by tunnelling does not converge within 2*D distance and does in the distance of 3-10*D. The execution of multiple headings is one of the reasons for it. Even if this influence is taken account into further analysis, it should be concluded that ground shows rather remarkable non-elastic behaviour in many cases and convergence is strongly influenced by the development of non-elastic zone around a tunnel opening.

Fig. 1 shows the change of non-elastic zone in association with the decrease of inner pressure given by a support system for the case in which ground has yhe competence factor of 0.5 and is in the state of being overstressed (Tanimoto et al., 1983). It can be seen that a small change of supporting effect (as an inner pressure) such as 0.1 MPa has a large influence on the extent of non-elastic zone, and the state of stresses on the elastic/non-elastic boundary does not vary (always keeps constant) and only the thickness of non-elastic zone around an opening changes in correspondence with the magnitude of inner pressure. It means that the ground around a tunnel trends to equilibrate for itself by developing a non-elastic zone which enables to provide a constant inner pressure toward the outer ring in the state of elastic. This can be obtained by allowing a

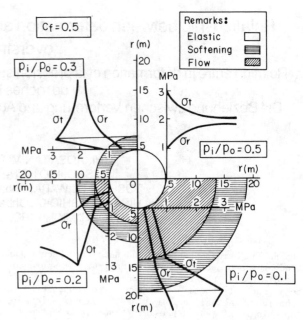

Fig. 1 Non-elastic ring and ground arch

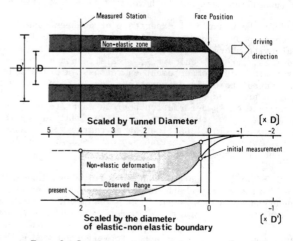

Fig. 2 Convergence curve in case of causing non-elastic deformation

certain deformation, and a magnitude of allowable deformation is limited in practice.

Although many discussions on minimum support load have been developed, there are few papers in which the relationship between the minimum support load and the corresponding magnitude of displacement was clearly defined. Using experience and data from recent tunnel construction in Japan and the United States, Tanimoto et al. (1983) proposed a method to find the approriate range of maximum allowable convergence of rock based on tunnel convergence measurement, and also Tanimoto et al. (1985) showed that, when a change (ΔD) of the original tunnel diameter or height (D) is over 1.0% (as $\Delta D/D$), a very heavy support pressure such as 0.6-1.0 MPa has been observed. Then, to reconsider the tendency of convergence in several cases in which the effect of supporting system was well confirmed, it has been clarified that the outer diameter of non-elastic zone (hereafter, it is denoted as D') corresponds with the tendency of convergence. Namely, the distance of 2*D' agrees with the point which reaches a constant displacement. It is the same as if a tunnel with the diameter of 2*D'is driven instead of a tunnel with 2*D. It was clearly observed in the construction of Orizume Highway Tunnel, in which a deformation reached constant within 4-5*D (D=10m) and the observed non-elastic zone was 5-6 m wide around an opening. 2*D' was equal to 40-45 m as shown in Fig.2.

4. ESTIMATED SUPPORT PRESSURE

Regardless of the type of support/lining elements employed, it is possible to estimate an equivalent support pressure, the change in the half-dome action and timing for the installation of the support elements from the strain-softening analysis (Tanimoto et al., 1983). The analysis procedure for fully bonded rockbolts is outlined by Tanimoto et al. (1981a,b).

The basic concept for the support pressure has been described in the publications, e.g. Hoek et al. (1980), Tanimoto et al. (1982a,b; 1983). The problem on the estimation of support pressure comes from the difference between the theoretical values and observed ones. An assumed cross-section in design is apt to vary very much from a realistic one, which is rather irregular-shaped at excavation in most cases, and also it is very difficult to evaluate the realistic stiffness of a support system in-situ after installation. The authors consider that variation in installation timing and skillfulness in construction give the wide range of support/lining stiffness in reality with the different magnitude of 1/2 to 1/20 less than the theoretical value. For instance, a thin shotcrete layer such as 5 cm thick, which is applied onto the ideally circular opening, shows very high stiffness and bearing capacity, but many measurements aiming at monitoring the magnitude of stresses in a shotcrete lining by means of pressure cells indicate quite low values compared with a theoretical ones. The same tendency can be seen on rockbolt system either.

Since it does not make any sense to compare many data obtained under different conditions on a same level, the authors tried to estimate the approximate values for various support systems by employing the data which have been obtained directly by themselves and are considered reliable with the proper measurement. They are plotted onto the Hoek's diagram as shown in Fig. 3. Circled A to H with thin curves are given by Hoek, et al. (1980) and circled alphabets with asterisks show the observed ranges for the case D=10 m by Tanimoto et al. (1983). Support systems are: Ⓐ shotcrete, 5 cm thick, 1 day strength is 14 MPa; Ⓑ shotcrete, 10 cm thick, 28 days strength is 35 MPa; Ⓒ concrete lining, 30 cm thick, 28 days strength is 35 MPa; Ⓓ concrete lining, 50 cm thick, same strength as Ⓒ; Ⓔ steel ribs, H125, 1m pitch; Ⓕ steel ribs, H175, 1.5 m pitch; Ⓖ steel ribs, H250, 1.5 m pitch; Ⓗ medium bolting, 25 mm in dia., 0.5 bolt/sq.meters; Ⓘ dense bolting, 25 mm dia., 2 bolts/sq.meters. These were obtained theoretically. Ⓑ*, Ⓕ*, and Ⓘ* are realistic ranges based on the observations at sites by the authors (1986), and correspond with Ⓑ, Ⓕ and Ⓘ respectively.

5. RELATIONSHIP BETWEEN SUPPORT LOAD AND DEFORMATION

In order to estimate the approximate support load and to control the magnitude of deformation during construction, the authors propose the rock classification showing the relationships between support load, initial deformation rate and deformation as shown in Table.1, which summarizes their past studies to be listed at the end.

Using Fig.3 and Table.1, let us discuss the case of Orizume Tunnel which has been driven through the highly fractured and overstressed tertiary formations. The precise measurements were carried out at 8 sites, consisting of five rock types, i.e. 1) tertiary mudstone, fractured, Vp= 1.7 km/sec (where Vp:primary wave velocity); 2) tuff breccia, Vp= 2.4 km/sec; 3) tuff, containing high amount of montmorillonite, high lateral pressure was observed; 4) limestone, Vp= 4.5 km/sec; 5) alternative layers of slate and chert, Vp= 2 km/sec. As the deformation behaviour of ground associated with tunnelling is highly influenced by strength of surrounding rock, primary stress field and reaction of support system,

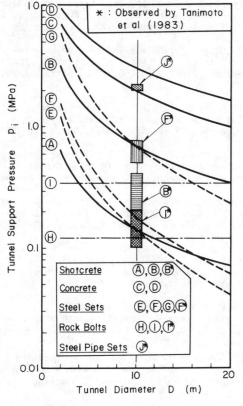

Fig. 3 Support pressure for various support elements (after Hoek et al.,1980; Tanimoto et al.,1983)

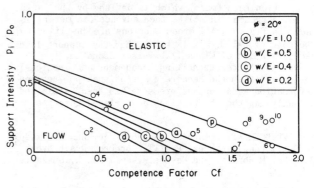

Fig. 4 Support intensity and competence factor

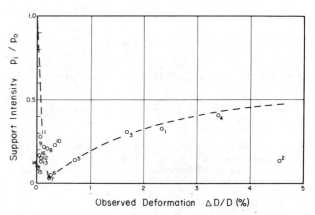

Fig. 5 Support intensity and deformation

Table. 1 Relationship between support load, initial deformation rate and deformation (for D = 10 m)

Class	Support Load	Initial Deformation Rate (mm/m)	Observed Deformation $\Delta D/D$ (%)	Estimated Support Pressure Pi (MPa)	cf. Terzaghi's Rock Class & Load $\gamma \cdot H$ (MPa)
I	Slight	less than 0.3	less than 0.05	less than 0.1	1 - 3 ; 0 - 0.1
II	Medium	0.3 - 2	0.05 - 0.3	0.1 - 0.3	4 - 5 ; 0.1 - 0.3
III	Heavy	2 - 7	0.3 - 1	0.3 - 0.6	6 - 7 ; 0.3 - 0.7
IV	Very Heavy	7 - 15	1 - 2	0.6 - 1	8 - 9 ; 0.7 - 1.0
V	Extremely Heavy	over 15	over 2	over 1	10 ; 1.0 - 2.2

the concept of "support intensity" which is defined by the ratio of support pressure (pi) versus major primary stress (po), namely pi/po, was conveniently introduced into the analysis. Fig.4 shows the plot onto the relation between support intensity and competence factor. In the figure, w/E is a ratio of negative gradient of softening to modulus of elasticity, and a high value indicates rapid drop of peak strength. The line expressed by the circled p corresponds with the boundary of elastic and non-elastic behaviours. The farther left a plot approaches, the easier plastic flow occurs.

The site of circled 2 is plotted entirely in the plastic flow state, and it required to reexcavate for the improvement of a cross section due to too high deformation. The introduction of the support system with high support intensity was necessary. Fig.5 suggests the relationship between minimum support load and deformation. According to Table.1, when high deformation rate is observed at the initial stage of excavation, stiffer support system and earlier timing of installation are compulsory.

In order to reconfirm the verification of Table.1, the results of the observation in Orizume Tunnel were plotted onto the diagram of deformation ($\Delta D/D$) vs. deformation rate (mm/m) which is defined by the ratio of convergence (in mm) to face advance (in meter) as shown in Fig. 6. All cross-sections are justified to the case of D=10m. It is known that the support load classification of Class I to V is reasonable.

For the better tunnelling management and rational design work, the concrete relationship between deformation and support pressure has been explained by analyzing the observations and related theories.

ACKNOWLEDGEMENT: Grateful acknowledgements are made to Drs.T.Lang, E.Hoek, Prof.T.L.Brekke, Drs.B.Myers-Boehlke, M.Horita, and Messrs.H.Azuma and H.Yamazaki.

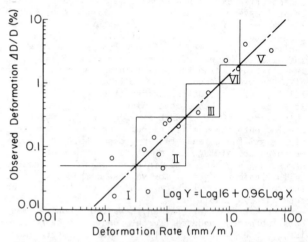

Fig. 6 Deformation rate and total deformation

REFERENCES:

AFTES(1978): General Report,Conference on Convergence Confinement Method, Paris, Oct.26; J. of Tunnels et Ouvrages Souterrains,No.32(1979)(in French); Underground Space,4-4, pp.225-232(1980) (in English).

Hoek,E.,E.T.Brown(1980): Underground Excavations in Rock,The Institute of Mining and Metalblurgy, London, Ch.8.

Mueller,L.(1977): The Use of Deformation Measurements in Dimensioning the Lining of Subway Tunnels, Proc. Int.Sympo. on Field Measurements in Rock Mechanics: edited by Kovari,Balkema(1979), pp.451-472.

Peck,R.B.(1969): Deep Excavation and Tunnelling in Soft Ground, State-of-the-Art Report, 7th ICSMFE, Mexico, State of the Art Volume, pp.225-290.

Tanimoto,C.,S.Hata(1980): Fundamental Concept of Designing Tunnel Supports in Consideration of Elasto-plastic and Strain Softening Behaviour of Rock, Memoirs of the Faculty of Engineering, Kyoto University, Vol.42, pp.349-376.

Tanimoto,C.,et al.(1981a): Interaction between Fully Bonded Bolts and Strain Softening Rock in Tunnelling, Proc. 22nd U.S.Sympo. on Rock Mech.,pp. 347-352.

Tanimoto,C.,et al.(1981b): Interaction between Rockbolts and Weak Rock in Tunnelling, Proc. Int. Sympo. on Weak Rock, Tokyo,Sept.,Vol.3, pp.157-162.

Tanimoto,C.,et al.(1982a): Interpretation of Characteristic Line for Tunnel Stability, 14th Sympo. on Rock Mech., JSCE, pp.86-90 (in Japanese).

Tanimoto,C.,S.Hata(1982b): Fundamental Concept of Tunnel Supports to Be Placed in the Vicinity of Mining Face, Proc. of JSCE, No.325, pp.93-106. (in Japanese).

Tanimoto,C.,et al.(1983): Allowable Limit of Convergence in Tunnelling, 24th U.S.Sympo. on Rock Mech., pp.251-263.

Tanimoto,C.et al.(1986): Allowable Limit of Deformation Rates in Tunnelling,18th Sympo. on Rock Mech., JSCE, pp.431-435 (in Japanese).

Terzaghi,K.v.(1946): Rock Defects and Loads on Tunnel Supports,Publ. in"Rock Tunnels with Steel Supports" by Proctor and White, Commercial Shearing, Inc.

Innovations in rock reinforcement technology in the Australian mining industry
Innovations en ce qui concerne la technologie du renforcement des roches dans l'industrie minière australienne
Neuerungen in der Felsbewehrungstechnik in der australischen Bergbauindustrie

A.G.THOMPSON, CSIRO Division of Geomechanics, Melbourne, Australia
S.M.MATTHEWS, CSIRO Division of Geomechanics, Melbourne, Australia
C.R.WINDSOR, CSIRO Division of Geomechanics, Melbourne, Australia
S.BYWATER, Mount Isa Mines Ltd, Australia
V.H.TILLMANN, The Zinc Corporation Ltd, Broken Hill, Australia

ABSTRACT: Recently, the Australian underground mining industry has seen a number of innovations in rock reinforcment technology. These changes have brought about improvements in safety and productivity during development, cut and fill mining and long hole open stope extraction. Field trials of the innovations have been complemented by the measurement of rock mass movements and induced rock reinforcement loads with prototype instruments. The field trials have demonstrated that reinforcement characteristics should be varied according to site specific structural geology and stress regimes.

RESUME: L'industrie minière australienne a récemment assisté à une série d'innovations en ce qui concerne la technologie du renforcement des roches. Ces changements ont abouti à une amélioration de la sécurité et de la productivité lors de la reconnaissance, de l'exploitation minière à déblai-remblai et de l'extraction en gradin ouvert à cavitié longue. On a complété les essais pratiques de ces innovations en mesurant au moyen d'instruments prototype les mouvements de la masse rocheuse ainsi que les charges induites du renforcement de la roche. Les essais sur place ont démontré qu'on doit faire varier les caractéristiques du renforcement selon les régimes de la géologie structurelle et des contraintes spécifiques au terrain.

ZUSAMMENFASSUNG: In den letzten Jahren hat man in der australischen Bergbauindustrie die Einführung mehrer Neuerungen im Bereich der Felsbewehrungstechnik gesehen. Diese Veränderungen haben bei der Aufschliessung, beim Firstenstossbau mit Versatz, sowie beim Langlochabbauverfahren mit offenem Abbauraum zu Sicherheitsund Leistungsverbesserungen geführt. Die Testergebnisse dieser Neuerungen von Versuchen im Bergbau werden mit Messungen von Gebirgskörperbewegungen und Felsbewehrungsbelastungen mit Hilfe neuer Messinstrumente ergänzt. Die Felsversuche haben gezeight, dass die Bewehrungseigenschaften verändert werden müssen um den Verhältnissen der Baustelle im Hinblick der geologischen Struktur und der Gebirgsspannung gerecht zu werden.

1 INTRODUCTION

In recent years, the Australian underground metalliferous mining industry has seen the adoption of innovative technology in rock reinforcement practice. These new technologies have developed largely as a result of collaboration between mining companies and the Commonwealth Scientific and Industrial Research Organisation (CSIRO) Division of Geomechanics in projects sponsored by the Australian Mineral Industries Research Association Limited (AMIRA).

The changes in practice have been facilitated by the willingness of the mining companies to trial different reinforcing systems in their production areas. Prototype instrumentation was developed by CSIRO Division of Geomechanics personnel. This instrumentation provided quantitative information on reinforcement performance related to measured and visual observations of rock mass behaviour.

Technology changes were introduced to the mining industry to address perceived deficiencies in existing reinforcing systems. The modifications and their applications in a variety of mining environments are presented in the following sections.

2 CEMENT GROUTED CABLE REINFORCEMENT

The reinforcement systems are based mainly on cement grouted high tensile strength steel cables adopted from the prestressed concrete industry. The introduction of these standard 15.2 mm diameter seven wire cables replaced the use of 7 * 7 mm diameter high strength steel wire dowels in cut and fill mining. Subsequently, the use of cable reinforcement was extended to crown pillar and wall support around long hole open stopes. A review by Fabjanczyk (1982) found that in many cases there was poor performance of these cables demonstrated by the lack of broken cables. These visual observations suggested that load was not being transferred from the grout to the standard cables. Therefore, measures were taken to improve the load transfer characteristics of the cables by the addition of anchors and by modification of the cable profile.

2.1 Supplementary anchors

The early applications of untensioned dowels simply used factory supplied steel cable. The load transfer could be improved locally by the use of anchors. Three types of internal anchors (Fig. 1) have been used at various mine sites. Tests by O'Bryan (1982) with both the barrel and wedge and swaged steel anchors showed in hard rock, with percussion drilled holes, approximately 250 mm of embedment length was sufficient to transfer the design load of the cables (25 tonne). Bywater and Fuller (1983) showed that the swaged aluminium anchor yielded at approximately 20 tonne.

These anchors are added singly or at regular intervals along the embedment length and also at the collar of the borehole to retain surface hardware such as mesh, straps and bearing plates. Currently, the steel swage and the barrel and wedge anchor are used extensively as internal anchors. A single barrel and wedge anchor is used exclusively for securing surface hardware.

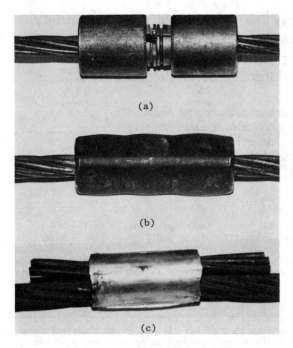

Figure 1. Internal anchors used to improve load transfer. a) barrel and wedge, b) swaged steel and c) swaged aluminium.

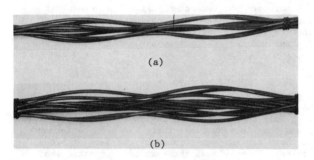

Figure 2. Birdcage pattern cables. a) seven-wire and b) combination with standard cable.

2.2 Birdcage cable

In cut and fill mining with preplaced supports up to 20 m long, it was not possible to accurately determine where intermittent anchors at a nominal spacing will be located relative to the advancing back. Therefore, a cable with enhanced load transfer characteristics for the entire length was developed (Hutchins et al, 1986). In its simplest form shown in Fig. 2a, a very stiff reinforcement results due to the mechanical interaction of the non uniform cross-section and the increased steel/grout interface area compared with a standard cable. Less than 1 m of embedment length has been found necessary to cause rupture of birdcage cables in pull tests. For standard cables, the corresponding embedment length may be in excess of 1.5 m.

The manufacturing process of birdcage cable allows for a wide variety of configurations and capacities up to 50 tonne. Combination of a standard cable and a birdcage pattern (Fig. 2b) allows for both improved load transfer and the means for securing surface hardware.

Figure 3. Jacking tool for installation of surface anchors.

2.3 Surface anchor installation

The increased use of surface anchors created the need for a more efficient method of installation and cable tensioning. Two methods were used initially to install surface anchors. In the first case, a known tension was applied to the cable by jacking against the rock face or bearing plate and the wedges were installed in the barrel. The jacking load was then released, leaving a significantly lower and unknown amount of pretension in the cable due to pull in of the wedges. The second method of installation involved pushing directly on the wedges which protruded from the barrel as shown in Fig. 1a. This method had two undesirable characteristics. Damage to the outer wires of the cable was caused by the teeth of the hardened wedges and again, an unknown pretension was produced in the cable behind the faceplate due to the significant friction between the cable and the wedges.

An installation tool was developed in which the load was applied to the barrel directly. The wedges were pushed into place by a stiff spring contained in the tool. Whilst this device caused no damage to the cable during installation, significant pretension loss still occurred after releasing the jacking load. A compromise was to redesign the barrel taper so that the wedges and face of the barrel were aligned during pushing. In this way, damage to the cable is reduced and the relaxation after release of jacking load is minimised. The tool shown in Fig. 3 only requires a short length of protruding cable because gripping wedges are incorporated into the front of the jacking device.

3 APPLICATIONS OF CEMENT GROUTED CABLES

3.1 Cut and fill mining reinforcement

The development of quick and efficient surface anchor installation devices enabled the minimisation of secondary support installation in cut and fill mining with preplaced support. It is now possible to incorporate the temporary support with the preplaced support as shown schematically in Fig. 4. Previously, short bolts were installed with mesh or strapping to support closely jointed rock in stope backs. Improvements to productivity and safety have been demonstrated using the existing reinforcement.

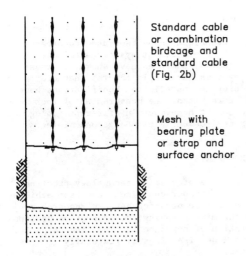

Figure 4. Use of preplaced cables with mesh and strap support in cut and fill stope.

3.2 Crown pillar reinforcement

In this application, a number of crown pillars were created between cut and fill mined areas above and open stopes below. Plain cables were used in the first three of these crown pillars in an attempt to stabilise the rock mass. In all cases extensive pillar overbreak occurred due to high stress (measured up to 100 MPa) accompanied by large vertical dilations. Typically, long lengths of cable were observed hanging from the back, stripped of grout and rock, with the capacity of the cables not utilised.

Two factors were apparently required to be satisfied by the reinforcing system:
1. It was necessary to increase the load transfer in order to take full advantage of the strength of the cables.
2. A low stiffness reinforcement system was required, to be compatible with the pillar dilation. These apparently conflicting requirements were satisfied by the use of internal anchors along the cables and artificial debonding between the anchors as described in Matthews, Worotnicki and Tillmann (1983).

The reinforcing system shown schematically in Fig. 5 was used in two crown pillars at the same horizon as the previous three failures. Although yielding of the pillars occurred, in one case violently as evidenced by vibrations recorded at the surface, overbreak was minimal and confined to unreinforced areas of the pillars and walls.

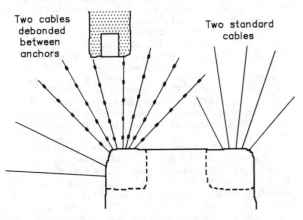

Figure 5 Reinforcement of crown pillar with cables with regularly spaced internal anchors and artificial debonding.

3.3 Hangingwall reinforcement

In this application, large scale overbreak occurred from the laminated hangingwalls of large open stopes despite the use of arrays of standard cables. To utilise the tensile strength of the cable, surface strapping and anchors were used. The dowels were artificially debonded for 2m from the surface to the internal anchor as shown in Fig. 6. and more fully described by Windsor, Bywater and Worotnicki (1984) and Greenelsh (1985). This permitted a small installation prestress (2 tonne) to be used and allowed for significant displacements (measured up to 50mm) to occur in the laminated surface rock. This reinforced section of hangingwall reduces the down dip unsupported span. Overbreak has been significantly reduced by the application of this reinforcing technique. In some cases where overbreak has occurred below the cable array, the support has prevented failure progressing upwards.

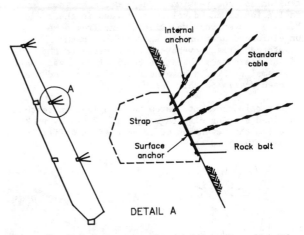

Figure 6. Reinforcement of a high hangingwall with cables and surface strapping.

4 EVALUATION OF CABLE REINFORCING SYSTEMS

To quantitatively evaluate reinforcement performance, it was necessary to measure loads developed in the various reinforcing systems and to relate the results to measured rock mass behaviour and visual observations. Because of the lack of suitable commercial instruments to measure the loads in cables, prototype instruments were developed. These instruments are based on strain gauge technology and have been described in detail by Windsor and Worotnicki (1986). These writers have also demonstrated how the loads in grouted cables are non uniform and vary according to the rock structure in the vicinity of the measuring point.

The monitoring of rock mass movements and cable loads was in many cases an important factor in the conduct of mining operations and also was able to provide information governing further improvements in mining and reinforcement procedures.

5 HIGH SHEAR CAPACITY DOWEL REINFORCEMENT

In this novel reinforcing scheme, it was necessary to stabilise the walls of a longhole open stope in blocky rock. Mining in adjacent areas had caused destressing of the region both horizontally and vertically as described by Cutjar, Gooding and Matthews (1985) and Matthews et. al. (1986). In particular, the stope was intersected and undercut by an extensive tension crack up to 500mm wide. Analysis of the structural geology in the region indicated a potential for large numbers of blocks to fall out from the walls of the stope.

With the above information, limited access from adjacent development and the knowledge that cable arrays had not proved successful in blocky rock (Fuller, 1983), a high strength, high density reinforcement system was proposed. The reinforcement consisted of a string of socketted 114.3mm diameter steel pipes installed in boreholes drilled nearly parallel with the stope walls as shown in Fig. 7. The purpose of the pipes was to increase shear resistance across discontinuities. In addition, three high tensile strength steel cables were installed untensioned down the centre of the pipe string to provide tensile strength across the joins in the pipes. The entire assembly was cement grouted. The tops of the cables were also grouted into the wall above the dowel. The dowels reinforced approximately one-third of the wall area as shown in Fig. 8. The installed cost of the wall dowels was A$0.57 per tonne.

Instrumentation of the dowels confirmed loading in tension and compression. There was also some indication of the dowel being loaded in shear and bending. Overbreak was significant from some unreinforced sections of the walls. In the reinforced areas, overbreak was limited to the upper 10m of the dowels where the cover of rock was small. Some smaller blocks were held in place behind the exposed dowels in this region.

This reinforcement system has application to thin, non load bearing diaphragm pillars of blocky rock used for the retention of uncemented sand fill in previously mined stopes.

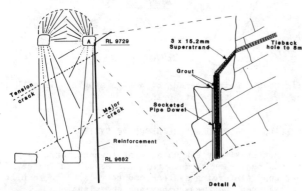

Figure 7. Cross section of stope showing high shear strength dowel reinforcement.

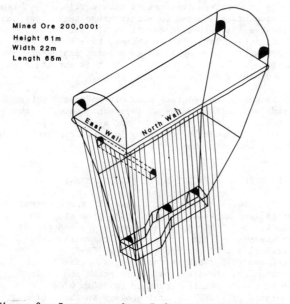

Figure 8. Isometric view of the stope showing areas of reinforced wall.

6 CONCLUSIONS

A number of techniques have been used to expand the range of rock reinforcement systems available to design engineers. Examples have shown how these different systems have exploited the characteristics of the reinforcing materials. Modifications have evolved as a result of the measurement of rock mass and reinforcement behaviour in different geological and stress environments in underground mining.

7 ACKNOWLEDGEMENTS

The changes in reinforcement technology practice outlined in this paper could not have been achieved without the full cooperation between research project teams and the mine planning and production personnel. The cooperation of personnel at The Zinc Corporation, Limited and Mount Isa Mines is gratefully acknowledged.

The authors acknowledge the assistance of Mr. W. Hutchins, Rock Engineering Pty. Ltd., Melbourne in the development of the anchor installation tool and in the modifications to surface anchor geometry and in the development of birdcage cable configurations.

The technical assistance of Mr. G. Cadby and Mr. M. Cassetta in the development and manufacture of prototype instruments for testing and field monitoring is also gratefully acknowledged.

REFERENCES

Bywater, S. and P.G. Fuller 1983. Cable support of lead open stope hangingwalls at Mount Isa Mines Limited. Proc. of the International Symposium on Rock Bolting, Abisko, Sweden, 539-555.

Cutjar, L.J., J.E. Gooding and S.M. Matthews 1985. Longhole open stoping in The ZC "Crack" Zone. Underground Operators Conference, Aus.I.M.M., Kalgoorlie Branch, 123-130.

Fabjanczyk, M.W. 1982. Review of ground support practice in Australian underground metalliferous mines. Proc. Aus.I.M.M. Conference, Melbourne, Australia, 337-349.

Fuller, P.G. 1983. Cable support in mining - A keynote lecture. Proc. of the International Symposium on Rock Bolting, Abisko, Sweden, 511-522.

Greenelsh, R.W. 1985. The N663 stope experiment at Mount Isa Mine. International Journal of Mining Engineering, 3, 183-194.

Hutchins, W.R., S. Bywater, A.G. Thompson and C.R. Windsor 1986. A versatile grouted cable dowel reinforcing system for rock. Submitted to Australasian Institute of Mining and Metallurgy Proceedings.

Matthews, S.M., A.G. Thompson, C.R. Windsor and P.R. O'Bryan 1986. A novel reinforcing system for large rock caverns in blocky rock masses. Large Rock Caverns: Proceedings of the International Symposium, Helsinki, Finland, V2, 1541-1552.

Matthews, S.M., G. Worotnicki and V.H. Tillmann 1983. A modified cable bolt system for the support of underground openings. Proc. Aus.I.M.M. Annual Conference, Broken Hill, Australia, 243-255.

O'Bryan, P.R. 1982. A preliminary assessment of cable bolt anchorage. Unpublished B.E. (Mining) thesis, University of New South Wales, Australia, pp131.

Windsor, C.R., S. Bywater and G. Worotnicki 1984. Instrumentation and observed behaviour of N663 trial stope, Racecourse Area, Mount Isa Mines. Confidential Report No. 23. CSIRO Geomechanics of Underground Metalliferous Mines Project.

Windsor, C.R. and G.Worotnicki 1986. Monitoring reinforced rock mass performance. Large Rock Caverns: Proceedings of the International Symposium, Helsinki, Finland, V2, 1087-1098.

Zones de traction au cours d'essais de compression
Tensile zones during compressive tests
Zugspannungszonen während Kompressionsversuchen

S.M.TIJANI, Centre de Mécanique des Roches de l'Ecole Nationale Supérieure des Mines de Paris, France

ABSTRACT: The laws governing the mechanical behavior of rocks are generally deduced from the results of classical tests performed in laboratory on samples with simple geometric shapes.
Particularly to obtain the failure criterion of a rock, the tests are often performed on cylindrical samples subjected to an axial pressure (applied by the rigid platens of a press) and to a lateral pressure (confinement).
The aim of this note is to draw attention to the difficulty of the interpretation of such tests during which tensile zones may take place inside the sample.

RESUME: Les lois de comportement mécanique des roches sont généralement dégagées à partir des résultats d'essais classiques réalisés au laboratoire sur des éprouvettes de formes géométriques simples.
En particulier, pour obtenir le critère de rupture d'une roche, le plus souvent, les essais sont effectués sur des éprouvettes cylindriques que l'on soumet à une pression axiale (appliquée par les plateaux rigides d'une presse) et à une pression latérale (confinement).
Le but de cette communication est d'attirer l'attention sur la difficulté de l'interprétation de tels essais au cours desquels des zones de traction peuvent prendre place à l'intérieur de l'échantillon.

ZUSAMMENFASSUNG: Die Gesetze des mechanischen Verhaltens von Gesteinen ergibt sich im allgemeinen aus Klassischen Laborversuchen, die an Proben mit einfachen geometrischen Formen durchgeführt werden.
Insbesondere, um das Kriterium für den Bruch eines Gesteins zu erhalten, werden die Versuche meist an zylindrischen Proben getätigt, die man einem axialen Druck (durch die festen Klammern einer Presse hervorgerufen) und einem seitlichen Druck (Manteldruck) beansprucht.
Das Ziel dieses Aufsatzes ist es, die Aufmerksamkeit auf die Schwierigkeit der Interpretation solcher Versuche zu lenken, bei denen Zugspannungszonen innerhalb der Proben entstehen können.

1. LES ESSAIS CLASSIQUES DE COMPRESSION

Pour les essais de compression réalisés entre les plateaux de la presse on utilise généralement des éprouvettes cylindriques. L'élancement (rapport de la hauteur par le diamètre) est choisi suffisamment petit pour éviter le flambage et assez grand pour pouvoir négliger l'effet des contacts entre l'échantillon et les plateaux. Les mesures sont alors analysées comme si le contact était parfait (hypothèse du glissement sans frottement).

Le choix de l'élancement a fait l'objet de nombreuses études expérimentales, analytiques (théorie de l'élasticité linéaire) et numériques (comportement non linéaire de la roche étudiée et/ou des contacts).
Les divers auteurs de ces études aboutissent à la même conclusion : "dés que l'élancement est assez grand, les perturbations qui ont lieu aux extrêmités de l'éprouvette n'ont pratiquement aucune influence sur les relations entre les pressions appliquées et les déformations mesurées (jauges électriques collées au milieu de l'éprouvette ou rapprochement des plateaux)".

Ce résultat, valable pour l'étude du comportement ductile des roches (déformabilité), est utilisé aussi pour interpréter les essais de rupture (cercles de Mohr courbe intrinsèque).

Cette communication est consacrée à un contre-exemple qui illustre la nécessité de la prise en compte des effets de bord lors de l'analyse des essais de rupture.

2. ANALYSE "REALISTE" D'UN ESSAI DE COMPRESSION

2.1. Présentation

Prenons une éprouvette cylindrique de rayon 1 (unité de longueur) et de demi-hauteur 2 (élancement fréquent). Supposons qu'elle soit constituée d'une roche dont le comportement ductile suit la loi de J.Lemaître (fluage

en loi puissance du temps et du déviateur); ce qui est le cas des roches salines par exemple.

t : temps (seconde)

$\tilde{\sigma}$: tenseur de contrainte (Mega Pascal)

$\tilde{\varepsilon}$: tenseur de déformation totale

$\tilde{\varepsilon}^{vp}$: tenseur de déformation viscoplastique

ξ : variable scalaire d'ecrouissage

Etat initial : $t = 0$, $\tilde{\sigma} = \tilde{\sigma}_0$, $\tilde{\varepsilon} = \tilde{0}$, $\tilde{\varepsilon}^{vp} = \tilde{0}$, $\xi = 0$

$\operatorname{tr} \tilde{\sigma} = \sigma_{11} + \sigma_{22} + \sigma_{33}$: trace de $\tilde{\sigma}$

$\tilde{\sigma}' = \tilde{\sigma} - (\operatorname{tr} \tilde{\sigma} / 3)\, \tilde{1}$: déviateur de $\tilde{\sigma}$

$\sigma_{eq} = \sqrt{(3/2) \sum\limits_{i=1}^{3} \sum\limits_{j=1}^{3} \sigma'^2_{ij}}$: contrainte équivalente

E, v : module d'Young et coefficient de Poisson

α, β, K : paramètres rhéologiques

$\tilde{\varepsilon} = [(1 + v)/E]\,(\tilde{\sigma} - \tilde{\sigma}_0) - (v/E)\,\operatorname{tr}(\tilde{\sigma} - \tilde{\sigma}_0)\,\tilde{1} + \tilde{\varepsilon}^{vp}$

$d\tilde{\varepsilon}^{vp} = A\,[d(\xi^{\alpha})/\sigma_{eq}]\,\tilde{\sigma}'$

$d\xi = (\sigma_{eq}/K)^{\beta/\alpha}\,dt$

$E = 20000 \qquad v = 0,25$

$\alpha = 0,5 \qquad \beta = 5 \qquad K = 10 \qquad A = (3/2)10^{-6}$

Faisons subir à cette éprouvette un essai de compression simple avec adhérence totale entre les faces de l'échantillon et les plateaux de la presse.
Ces derniers appliquent sur chaque face des efforts dont la résultante est une force parallèle à l'axe de l'éprouvette, dirigée vers son centre O et d'intensité qS où S est l'aire de la face (q : pression axiale).
Même pour un historique simple de q, on ne dispose pas de solution analytique pour un tel problème.

2.2. Compression simple à vitesse constante

Le problème sera donc abordé numériquement (Méthode des Elements Finis) en étudiant le 1/4 d'une section méridienne Orz : $0 \leqslant r \leqslant 1$ et $0 \leqslant z \leqslant 2$.
Et on considérera le cas d'une pression axiale évoluant à vitesse constante : $q = Vt$.
La solution du problème lorsque $V=V^*=0,01$ MPas^{-1} sera repérée par une astérisque.
La figure 1 montre le 1/4 de la section méridienne de de l'éprouvette à l'instant $t = 0$ (trait continu) et à l'instant $t = q^*/V^*$ lorsque $q^* = 50$ MPa (trait pointillé). Sur la figure 2, on a porté en abscisses la diminution relative de la hauteur et en ordonnées la pression axiale q^*. On remarque que les points $(-\Delta h/h, q^*)$ sont pratiquement sur la courbe effort-déformation théorique correspondant au cas du glissement parfait.

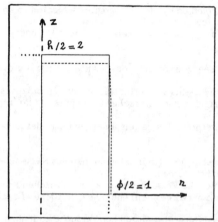

Figure 1. Le quart de la section méridienne. Etat initial (trait continu) et sous 50 MPa.

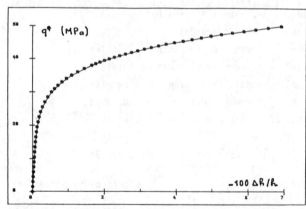

Figure 2. Courbe effort-déformation. Comparaison entre le cas sans frottement (courbe) et le cas de l'adhérence (points).

L'élancement 2 est donc suffisant pour négliger les effets de bord lors de l'étude du comportement ductile de cette roche. Ce résultat important était déjà connu par d'autres auteurs.
Cependant, dans le cas du glissement sans frottement le champ des contraintes est homogène dans l'éprouvette et ses composantes, dans le repère cylindrique (r,θ,z), sont toutes nulles sauf σ^*_z qui est égale à $-q^*$.
En revanche, dans le cas de l'adhérence totale, non seulement le champ des contraintes n'est pas homogène mais de plus le maximum dans l'éprouvette de la contrainte principale majeure est une traction.
Ce maximum, noté s^* est atteint le long de deux cercles de rayons 1 (donc à la surface latérale) et de cotes : $\pm h/4$ (donc loin des deux faces de l'échantillon).

Sur ces deux cercles, la contrainte principale majeure coïncide avec la contrainte tangentielle σ^*_θ.

Par ailleurs, s^* est une fonction croissante de q^* dont la courbe représentative (figure 3) présente une asymptote passant par l'origine : $s^* \sim lq^*$ $l \approx 0,114$ Cette valeur est à comparer à la pente à l'origine qui est $m \approx 0,0095$.

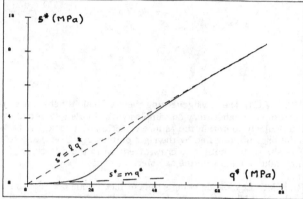

Figure 3. Maximun dans l'éprouvette de la contrainte principale majeure fonction de la pression axiale.

Si la roche étudiée a une résistance à la traction T alors dès que la pression axiale q^* atteint une valeur Q' telle que $s^*(Q')=T$, une fissuration par traction tangentielle apparaît aux cotes $\pm h/4$.
Par conséquent, au cours d'un essai de compression simple, si la rupture a lieu pour une pression axiale $q^* = Q$, on a nécessairement $Q \leqslant Q'$ et de plus si $Q=Q'$ la rupture est dûe quasicertainement à la traction.
Ainsi Q' n'est que la résistance à la compression avec adhérence. Ce n'est donc pas un paramétre intrinsèque.

2.3. Influence de la vitesse de mise en charge

Il est aisé de montrer que l'on passe des résultats du calcul numérique de l'essai à la vitesse $V^* = 0,01$ MPa à ceux relatifs à une vitesse V par une transformation simple utilisant le coefficient :

$$A = (V/V^*)^{\alpha/(\alpha+\beta-1)}$$

Le vecteur de déplacement et le tenseur de contrainte sont des fonctions de $q = Vt$ de la forme $f(q)=Af^*(q/A)$. En particulier le maximum de la contrainte principale majeure est $As^*(q/A)$ et si la résistance à la traction est $T = 2,5$ MPa (sel gemme), alors la pression axiale de rupture sera fonction de la vitesse V (figure 4) :

$$T/l \leqslant Q' = As^{*^{-1}}(T/A) \leqslant T/m \qquad 21,9 \leqslant Q' \leqslant 263,1$$

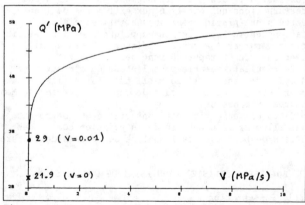

Figure 4. Résistance apparente à la rupture en compression simple fonction de la vitesse de chargement

2.4 Influence du confinement

Si l'éprouvette est soumise à un confinement p constant et à une pression axiale q = p + Vt avec adhérence, les déplacements et les déformations sont les mêmes qu'en compression simple (p=0) sous la pression axiale q - p. Et on passe des contraintes de l'essai simple à celles de l'essai triaxial en retranchant le tenseur p1.

Donc au cours d'un essai triaxial classique la contrainte principale majeure est maximale le long des 2 cercles de cotes ± h/4. Le maximum est une fonction croissante de la pression axiale q, décroissante de la vitesse de mise en charge V et du confinement p :

$$s(q,p,V) = As*((q-p)/A) - p$$

Pour un confinement et une vitesse donnés, s finit par atteindre la valeur T de la résistance à la traction quand la pression axiale q atteint :

$$Q' = As*^{-1}((T+p)/A) + p$$

La figure 5 représente Q' en fonction de p pour V=V*.

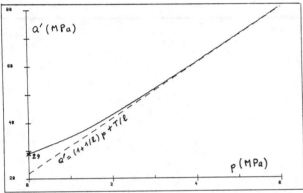

Figure 5. Pression axiale de rupture en fonction du confinement pour une vitesse de 0.01 MPa/s.

Si l'essai est réalisé à faible vitesse, alors :
$$Q' \approx (1+1/l)p + T/l$$
De plus, cette approximation est valable, pour toute vitesse V, lorsque le confinement est assez élevé.

Ainsi, lorsque l'on néglige l'effet de l'adhérence dans l'analyse des essais triaxiaux réalisés à faible vitesse (pratiquement : V < 0,001 MPa/s) ou sous des confinements élevés (pratiquement : p > 2,5 MPa), on a l'impression que la roche admet un critére de rupture par 'cisaillement' et que ce critère est une propriété intrinsèque obéissant à la loi de Coulomb avec une cohésion C ≈ 3,5 MPa et un angle de frottement ≈ 55 deg.

Alors qu'en réalité la courbe intrinsèque déduite des essais triaxiaux classiques réalisés sur des roches telles que le sel gemme, dépend non seulement de la vitesse de mise en charge, mais aussi de l'élancement des éprouvettes et des conditions aux limites (contact avec les plateaux de la presse). Seule la résistance à la traction est une carctéristique de la roche.

3. CONCLUSION

Cette analyse de la rupture des roches ductiles est encore récente. De plus elle n'a comme base expérimentale que les résultats d'une cinquantaine d'essais triaxiaux classiques réalisés sur des éprouvettes de sel d'élancement 2. Par conséquent, il convient de prendre la nouvelle théorie avec précaution en attendant les résultats de nouveaux essais : étude en cours en faisant varier l'élancement et les conditions de contact avec les plateaux de la presse.

Cependant, malgré sa fragilité, la méthode présentée permet d'expliquer :
- l'influence de la vitesse de mise en charge sur la rupture au cours d'essais de compression
- la rupture différée (au cours d'essais de fluage)
- la fissuration des piliers de mine (sel gemme) avec apparition de discontinuités dans des plans méridiens (excès de contrainte tangentielle).

REFERENCES

J. Lemaitre, 1971. Sur la détermination des lois de comportement des matériaux élasto-visco-plastiques. Paris. Thèse. Office National d'Etudes Aérospatiales.
H.R. Hardy & M.Langer, 1984. The mechanical behavior of salt. U.S.A. Trans Tech Publications. Particulièrement, page 421 : Creep Testing of Salt. par W.R. Wawersik & D.S. Preece.
A. Templier & J. Fine, 1969. L'effet de la forme et de l'hétérogénéité des échantillons sur leur résistance en compression. Revue de l'Industrie Minérale Vol.51, n°4, page 365.

Rockburst research in Canada – 1987

Recherche sur les coups de toit au Canada – 1987
Gebirgsschlagsforschung in Kanada – 1987

JOHN E.UDD, Director, Mining Research Laboratories, CANMET, EMR, Government of Canada
D.G.F.HEDLEY, Research Scientist, Mining Research Laboratories, CANMET, EMR, Government of Canada

ABSTRACT: Rockbursts have occurred in the mines of the Province of Ontario for nearly sixty years. After nearly two decades of diminished rockbursting, however, the problem has once again attained major importance. As a result, research efforts have been intensified in several directions.

In this paper, the authors present an overview of the rockburst research which is in progress in Canada.

RÉSUMÉ: Depuis près de soixante ans des coups de toit se produisent dans les mines de l'Ontario. Toutefois, la fréquence des coups de toit avait diminué depuis deux décennies, mais le problème a de nouveau atteint une importance considérable. Par conséquent, les efforts consacrés à la recherche ont été intensifiés dans plusieurs directions.

Dans la présente communication, les auteurs donnent un aperçu de la recherche qui se déroule actuellement au Canada sur les coups de toit.

ZUSAMMENFASSUNG: Gebirgsschläge treten in der Provinz Ontario seit etwa 60 Jahren auf. Nach etwa 20 Jahren mit veringerter Gebirgsschlagshäufigkeit hat diese Problematik jetzt wieder zugenommen, und der Forschungseinsatz wurde in verschiedenen Richtungen verstärkt. In dieser Arbeit geben die Authoren einen Überblick über die Forschungsvorhaben, die sich mit der Problematik der Gebirgsschläge in Kanada befassen.

INTRODUCTION

Wherever an underground opening is made in a stressed rock mass the pre-existing stresses are redistributed and concentrated.The speed with which these adjustments to the presence of an opening take place, and the physical results, are dependent upon the strength and deformational properties of the geological materials. In strong, hard, elastic, brittle rocks, it is not uncommon that an instantaneous concentration of stresses beyond the strength of the rock results in an explosive failure. Such events are known to hard rock miners as "rockbursts".

Rockbursts in underground mines result from the practices of extracting ores. The causes include excessively high stress concentrations on the boundaries of openings, sudden failures of supporting pillars, and movements along faults and other weaknesses. In all instances substantial energy may be released. Fault-slip bursts, however, are the most dangerous and damaging because of the larger masses of rock and changes in potential energy which can be involved. At the other end of the scale, the smallest bursts may take the form of "spitting" or "popping" as wall rocks dissipate excessive stored strain energy.

Pillar bursts occur when either increased loading or decreased strength causes a sudden failure. Both strain energy and potential energy are released; the latter being associated with rapid convergence between the hanging wall and footwall.

Of the 217 rockbursts (of magnitude 1.5 to 4.0 on the Richter Scale) which were recorded in Ontario mines during 1984-1985; 5% were classified as strain energy bursts; 81% as pillar bursts; and 14% as fault-slip bursts (Brehaut/Hedley 1986).

A BRIEF HISTORY OF ROCKBURSTING IN ONTARIO MINES

Rockbursts have occurred in Ontario mines since 1929, and possibly earlier (Morrison 1942). It was only in the mid 1930s, though, that the increasing numbers of such events began to cause concern in the mining industry. The Ontario Mining Association, in response to the needs of its member companies, formed a Rockburst Committee and, in 1940, engaged the late Professor R.G.K. Morrison (then Superintendent, Nundydroog Mines Ltd., Ooregum, Mysore Mines Ltd., Ooregum, Mysore State, South India) to study and report on the situation. His report (Morrison 1942) which is equally relevant in many respects to present-day practices, is a classic in the field. Known as the father of rock mechanics in Canada, he is credited with introducing the concepts of "doming" and "sequential mining" into Canadian practice.

At the time that Morrison's report was written, severe and frequent rockbursts were being experienced in the gold mines of Kirkland Lake, Little Long Lac, and the Porcupine District, as well as in the nickel-copper mines of the Sudbury basin. The underlying major causes were given as being: the sizes of mining excavations; the depths of mining; and the local rock types. The same list, with a few additions (to include: the shapes of openings; the pre-mining conditions of stress; and, the rates of mining), would apply today.

During the following four decades, roughly from 1941 to 1981, the occurrences of rockbursts in Ontario mines decreased greatly (Pakalnis 1984). The reasons certainly included improvements in mining practices and ground control measures, but it is also very significant that many of the rockburst-prone mines were closed for economic reasons during this period.

A notable exception in this long, relatively tranquil period, was the severe rockburst of May 5, 1964, which resulted in the closure of the Wright Hargreaves gold mine at Kirkland Lake, Ontario. The same event was also said to have been indirectly responsible for the closure of the adjacent Lake Shore Mine.

With seismic events having become much more frequent again in the early 1980's, however, the rockbursting problem has once again attained serious proportions. The principal reason is that a number of deposits are now in the final, or pillar recovery, stages of extraction. Additionally, however, mining production openings have become larger in order to benefit from economies of scale, and mining has progressed to greater depths. These factors have combined to create, once again, conditions which are favourable to rockbursting.

During the past four years, serious rockbursting has taken place in Ontario in the Elliot Lake, Sudbury, and Balmertown areas. The economic consequences have been very severe as one mine (Falconbridge) and major parts of others have been closed to production. The economic consequences, in Ontario, are estimated by us to be about $200 million per annum.

The most serious recent events have been:

1) Elliot Lake

In March, 1982, a major series of rockbursts occurred at the Quirke Mine, of Rio Algom Ltd., as the result of violent pillar failure. The mine is a room-and-pillar uranium operation in a gently-dipping strong and brittle quartz pebble conglomerate.

After a period of over two years, in which little further bursting took place, seismic activity resumed. Between September, 1984 and April, 1985 over 150 rockbursts were recorded in an area of the mine measuring 1100 m by 600 m. The hangingwall above the ore zone has now become fractured up to the surface - a distance of some 500 m. One of the evidences of this was the disappearance of a beaver pond during the spring of 1986. Most of the rockbursts in the Quirke Mine have been of the pillar burst type.

2) Sudbury

In mid-1984, two major series of rockbursts took place in important nickel-copper mines of

the Sudbury area. In the first of these, in June, four miners were killed at the Falconbridge Mine when the mat above them collapsed as the result of seismic disturbances caused by slippage along an important fault. The mine, which was in the tertiary stage of mining, through recovery of the shaft pillar by undercut-and-fill methods, was immediately closed.

Just one month later, in July, a similar series of seismic events took place in the Number 5 shaft area of INCO's Creighton Mine. As the operations were in a period of vacation shut-down there were, fortunately, no injuries. Damage to the area of the bursts was substantial, however, and this section of the mine was closed to further production.

Subsequent to 1984, rockbursting has become a problem in other Sudbury-area mines, notably, INCO's North Mine and Falconbridge's Strathcona Mine.

3) Balmertown

In December, 1983, a major series of rockbursts took place in an area of sill pillars at the Campbell Red Lake gold mine. Damage was substantial and the mining of a complete ore zone was suspended.

ROCKBURSTING IN OTHER CANADIAN MINES

Much of Canada's hard-rock underground mining production is derived from the mines of the Province of Ontario. For this reason, and also conditions of local geology and mining methods, the rockbursting problem has, for the most part, been confined to that Province.

Nonetheless, occasional rockbursts occur at the lead-zinc operations of Brunswick Mining and Smelting, in New Brunswick, and in some of the potash mines in Saskatchewan. Isolated rockbursts have also been reported in the Val D'Or gold mining area of northwestern Quebec.

A BRIEF HISTORY OF ROCKBURST RESEARCH IN CANADA

Rockburst research in North America probably commenced with the pioneering work of Obert in the Ahmeek amygdaloidal copper mine of Michigan's Keweenaw penninsula (Obert 1941). In the following year, he and his colleague, Duvall, studied the problem: at the Sunshine Mine, in Idaho; at INCO's Frood Mine, in Sudbury; and at the Lake Shore Mine, in Kirkland Lake (Obert/Duvall 1942). Their work led to the development of test equipment, and testing and analytical procedures (Obert/Duvall 1945 a,b).

On the Canadian side of the Canada-U.S.A. border, the pioneering work was done by Dr. E.A. Hodgson, of the Dominion Observatory, at the Lake Shore mine (Hodgson 1958). For a period of about six years, considerable attention was given to the work.

In both the American and the Canadian efforts, attempts were made to apply geophysical methods of seismic monitoring to the rockburst problem. It was hoped that a means might be found of predicting the occurrences of rockbursts. While much progress was, and has been, made in identifying rockburst-prone areas in mines, the goals of prediction remain elusive today.

In fact, because of the speed of failure, prediction in time may never be feasible in the practical sense of being able to give a reliable warning. Prediction in space, however, through the identification of potentially high-risk areas, is thought to be attainable.

In the mid-1940's, the occurrences of seismic events in Canadian mines had decreased substantially. Concurrently, little success had been obtained in the use of geophysical methods for predictive purposes. For both of these reasons, there was little further development for several years.

In the mid-1960's, however, the United States Bureau of Mines undertook to improve the microseismic monitoring technique. The system which was subsequently developed by Blake (Blake 1982) and others represents the state-of-the-art of the present technology (Leighton 1983). Systems based on the American design have, as the result of the problems of the past four years, now been installed in 13 mines in Ontario as well as one mine in Saskatchewan and one in New Brunswick. The units are manufactured by Electrolab, a firm located in Spokane, Washington.

A basic multi-channel microseismic monitoring system consists of a number of sensors (or, geophones) connected to an automatic monitoring system. This may be pre-set to calculate, through an algorithm, the location of the source of an assumed seismic event once a prescribed minimum number of geophones have received waves of first arrivals within a selected "time window". Once the time window is reopened the system is ready to record the next event.

Geophones, which may either be velocity gauges or accelerometers, are located around the volume to be monitored. The time interval which is usually selected for the "window" is that which would be required for a wave to pass diagonally through the array of monitoring sensors. This is to help ensure that only events which occur within the volume being monitored will cause a calculation sequence to be commenced. An algorithm which is commonly used causes the system to go into a calculation mode once 5 or more geophones have detected first arrival waves within 100 milliseconds. The pre-set values may, and are, adjusted to suit local conditions of installations.

No data goes into a buffer. One of the disadvantages of this type of system is that although the complete waveform signal from each sensor is transmitted to the processing unit only the arrival time of the P-wave is recorded. The rest of the information, which would be valuable for mechanism evaluation, is discarded.

Very large events, which can consist of a succession of several large tremors can cause such systems to become so swamped with data that all information is lost after a certain point. This has happened in 1984 during some of the large events in the Sudbury area. Unfortunately, when this does happen, it frustrates one's efforts to obtain the locations of the sources. From a mine stability monitoring point of view, it is essential that locations of tremors should be established as quickly as possible. Part of the solution to this problem lies in the development of a more powerful and intelligent system with real-time monitoring capabilities.

In the meantime, the large events are normally detected and recorded at at least some of the stations of the Eastern Canada Seismic Network, which is maintained for earthquake monitoring purposes by the Geological Survey of Canada. Data, which is recorded continuously on the drum recorders of the short-period seismographs installed at all such stations, are transmitted regularly to Ottawa for processing. Ultimately, the times and locations of major events occurring in mining localities can be established from these data. Unfortunately, because of the distances between the field stations, the accuracies of locations are not sufficiently precise for mine-monitoring purposes. There is also some delay in obtaining the data.

PRESENT THRUSTS IN ROCKBURST RESEARCH

Government

As the result of the seismic events which took place in Ontario mines, commencing in 1982, it became evident that an intensification of rockburst research was necessary. In May of 1984, at a consultative meeting between CANMET, industry, and representatives of the Ontario Government, this was identified as the highest priority.

The events which took place, in Sudbury, within the following two months emphasized the urgency of the needs.

During the following few months, a large number of meetings were held with representatives of the Government of Ontario and of the companies which had experienced severe rockbursting. The result was a proposal for a major tri-partite research project, in which each of the federal and provincial governments, and the industry would contribute funds and/or services in the amount of $1.4 million (Udd 1984).

A Memorandum of Understanding was subsequently signed in September, 1985. Under it, CANMET will provide a team of 5 persons dedicated to the project for 5 years, together with operating funds. The Government of the Province of Ontario, will contribute up to $1.4 million over the five-year period for the purchase of capital equipment and services. The Industry of the Province of Ontario was requested to provide a matching contribution, with a value of up to $1.4 million, through the provision of monies, goods, and services, to the project.

Even at this early stage in the project, it is clear that more will be committed to the research than was originally visualized.

Given the needs of the industry for more rapid and precise mine monitoring, and the technological short-comings which have been identified, the research is proceeding along three lines:

1) To enhance the seismic monitoring capabilities in all mining camps. A very high priority is to develop a seismic monitoring system that will capture wave forms (as compared with triggered first arrivals) and provide information on first motion, peak particle velocity and seismic energy.

Ideally, an "intelligent" real-time system should be provided, with the software to permit automatic differentiation between signals originating from seismic sources and those being generated by other sources such as blasting, drilling, rock tumbling down ore and waste passes, equipment operating, and so on.

Another disadvantage of present technology is that waves arising from any kind of source can arm and activate a microseismic monitoring system. An ability to separate real events from "noise" would improve the operating efficiencies of systems and system operators enormously. We refer to this as an ability to recognize "footprint" signals.

There are many opportunities for research into the characteristics of signals arising from different sources, and in the development of intelligent sensors.

2) Even with the above, however, it is likely that local mine monitoring systems will be saturated by the signals coming from large rockbursts of magnitude 3.0 (Local Richter scale) or greater.

To alleviate that problem, it is planned to provide additional coverage to the Eastern Canada Seismic Network through the installation of seismograph stations in the major mining camps. Through the generosity of Denison Mines Ltd., a short-period seismograph has been installed in the Mining Research Laboratory of CANMET, at Elliot Lake. This is used to monitor events occurring both in Elliot Lake, and as far away as Sudbury.

In Sudbury, two additional stations are being installed in order to provide greater accuracy through finer scale triangulation. (Figure 1). Data from these stations will be transmitted via dedicated telephone lines to a computer located at Science North, in Sudbury, and also to the Geophysics Division of the Geological Survey of Canada, in Ottawa. Provision will be made to enable both of the companies operating in the Sudbury basin, INCO and Falconbridge, to gain access to the data through data ports. Seismograph stations will be installed at Red Lake and Kirkland Lake to provide coverage to these mining camps.

3) Between the seismic stations of the Eastern Canada grid at one end of the scale, and the local mine microseismic monitoring systems at the other, there is a need for an intermediate out-of-mine system having the capability of being able to record the complete waveforms of large seismic events. For such systems a small number of strong motion triaxial sensors are used as the geophones.

At present, macroseismic systems of this kind are being installed; at Falconbridge's Strathcona Mine and INCO's Creighton Mine, both in the Sudbury basin; at Rio Algoma's Quirke Mine, at Elliot Lake; at Campbell Red Lake Mines, at Balmertown; and at the Macassa Mine, in Kirkland Lake. Waveforms from large local events will be stored on computers at these sites and down-loaded daily to CANMET's Elliot Lake Laboratory via telephone.

The objectives of the Canada/Ontario/Industry rockburst research project are to add to our knowledge of the causes, origins, effects, energy sources, and mechanisms of rockbursts.

The information to be derived from the three levels of monitoring systems mentioned will permit much greater accuracy in locating the origins of mining-induced seismic events. Macroseismic systems, in particular, will be very valuable when events occur outside of existing microseismic sensor arrays.

Likewise, the recording of waveforms will permit determinations of peak particle velocities and seismic energies liberated. This will add considerably to our knowledge of the driving forces and the mechanisms which may be involved.

The Mining Industry

During the past two years there has been a great increase in the number of microseismic monitoring systems installed in Canadian mines. At the time that the Canada/Ontario/Industry project was proposed, in late 1984, there were 6 systems operating and 3 others planned. At the time of this writing there are 15 systems in use: 13 in Ontario, and 1 each in New Brunswick and Saskatchewan. INCO and Falconbridge, in the Sudbury basin, now have systems installed at several of their mines.

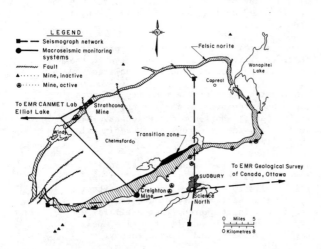

Figure 1. The seismograph networks which are being established for rockburst monitoring in the Sudbury Basin.

The primary function of a mine microseismic monitoring system is to provide the locations and relative magnitudes of seismic events in real-time. After a rockburst has occurred the immediate concern of management is to determine the location and if there has been injury and/or damage. From an operator's viewpoint, understandably, research needs are secondary.

All of the mine operators are involved to greater or lesser degrees in the development of software and the addition of hardware which will improve the accuracy of calculations and facilitate the graphical portrayal of data. In the future, perhaps using CAD (Computer Assisted Design) technology, the location and magnitude of an event will be shown on monitor images of plans and sections (or isometrics) shortly after it occurs. For the present, however, because of the difficulties with the algorithms that have been mentioned, there is much checking of data and plotting to be done. Most mine operators are making efforts to improve the operating efficiencies of their systems.

The Canada/Ontario/Industry Rockburst Project, which has been described, is having a significant effect on this through providing an on-going forum by which systems operators can meet regularly and exchange information on both problems and solutions. The Project is guided by a Technical Advisory Committee, which includes as members, representatives of the federal and Ontario governments and all Ontario companies operating mine microseismic monitoring systems. Meetings are held quarterly, and provide an invaluable opportunity for operators to compare their data and approaches to interpretations.

With a number of algorithms available, and numerous possibilities as regards computing systems and software, it can be appreciated that, without some standardization there would be little ability to compare results. The Technical Committee provides an opportunity to discuss matters of common concern and to obtain a consensus. In addition to CANMET/MRL and the Ontario Ministries of Labour and Northern Development and Mines, the mine operators participating in Ontario at present are: Campbell Red Lake Mine Ltd., Denison Mines Ltd., Falconbridge Ltd., INCO Ltd., Lac Minerals Ltd., and Rio Algom Ltd.

An important objective of the research is to relate rockburst activity to: mine design, the mining methods used, the sequencing of extraction, the local rock types, the depth of operation, and other factors.

Through understanding the causes of rockbursts the industry will be able to develop strategies which will minimize risks.

Equipment Designers and Manufacturers

Two Canadian organizations are presently involved in the design of improved systems for mine monitoring. These are completely complementary inasmuch as one system is intended for macroseismic monitoring applications, while the other is intended for use in underground mine microseismic monitoring. Both involve improved sensors and recent advances in communications, such as fibre-optics technology.

The first system mentioned, for out-of-mine local macroseismic monitoring is being developed by the Noranda Research Centre of Noranda Mines Ltd. An installation of the newly-developed system will be field tested at the Quirke Mine in Elliot Lake. The system will permit the recording of complete waveforms of large seismic events.

The second system, for in-mine monitoring, is being developed by Instantel Inc., of Kanata, Ontario. Involving tri-axial sensors with local microprocessors, fibre optic data transmission, more intelligent triggering algorithms, and dedicated computers, the system is intended to be the next generation of monitoring equipment. The installation of a prototype unit at an Ontario mine should take place in the not-too-distant future.

Both systems are being developed with financial assistance from the Government of Canada - the former through the Unsolicited Proposals Program of the Department of Supply and Services; the latter through the Projects for Industry/Laboratory Participation (PILP) program of the National Research Council. The first-mentioned author is the Scientific Authority on the Instantel project, while his co-author author is Scientific Authority for the Noranda work.

Academia

Research into some of the various aspects of rockbursting is underway at two Canadian universities; Queen's at Kingston, Ontario, and the University of Saskatchewan, at Saskatoon. At the former, there are projects in each of the Departments of Mining Engineering and Geological Sciences.

In the Queen's Mining Engineering Department, a team is studying the waveforms emitted from rock subjected to increasing loads. The intent is to determine, both in the laboratory and in the field, if the waveform characteristics can be used to establish load levels. Field work, to date, has consisted of monitoring at different locations which were known to have been subjected to different stresses. The initial results have been encouraging and point towards a different approach to monitoring.

In the Department of Geological Sciences at Queen's an attempt is being made to apply the principles of tomography to determining the integrity of a rock structure, such as a supporting pillar. If successful in detecting and permitting the mapping of fractures in large in-situ structures, the method might permit periodic rapid assessments of rock mass integrity. This would permit one to study the degradation of a rock mass, and to relate this to other factors such as loading, changes in conditions, and rockbursting.

At the University of Saskatchewan, a team is involved in the development of better monitoring systems for use in the local potash mines. Improvements include triaxial sensors, and approaches to the design of monitoring systems and analysis of waveform records.

Summary

In this paper, a review has been made on the research into rockbursts which is presently in progress in Canada. The listing is impressive, including work by governments, the mining industry, the manufacturing industry, and academia. It is clear that the state of technological development has increased rapidly in a very short period, and that further very substantial gains can be anticipated. Much of the research in progress is the "world-class" level and is directed at developing the next generation of equipment and approaches.

The work is essential, for Canadian mines are progressing to greater depths. The economics of scale and mass production are dictating that openings must be larger and productions (or sites of mining advance) greater. All of these factors are forces towards rockburst-prone conditions.

Knowing this, some of the fundamental concerns must be addressed now.

Acknowledgements

The authors are grateful for the cooperation and assistance which they have received from their colleagues in industry, government, and academia during the evolution of new thrusts in rockburst research in Canada. In particular, the mining industry and the Government of the Province of Ontario must be thanked for their enthusiastic support of, and collaboration in, the major tripartite project which has been mentioned.

REFERENCES

Blake, W. 1982. "Microseismic Application for Mining - A Practical Guide", Final Report, U.S. Bureau of Mines, Contract No. JO215002.

Brehaut, C.H. and Hedley, D.G.F. 1986. "1985-1986 Annual Report of the

Canada-Ontario-Industry Rockburst Project", CANMET Special Publication SP86-3E, ISBN 0-662-14984-X, Minister of Supply and Services, Canada.

Hodgson, E.A. 1958. "Dominion Observatory Rock Burst Research 1938-1945", Vol. XX, No. 1, Publication of the Dominion Observatory, Ottawa.

Leighton, F. 1983. "Growth and Development of Microseismics Applied to Ground Control and Mine Safety", Mining Engineering, pp. 1157-1162.

Morrison, R.G.K. 1942. "Report on the Rockburst Situation in Ontario Mines", Transactions CIM, Volume XLV, pp. 225-272.

Obert, L. 1941. "Use of Subaudible Noises for Prediction of Rock Bursts", U.S Bureau of Mines, R.I. 3555.

Obert, L. and Duvall, W. 1942. "Use of Subaudible Noises for the Prediction of Rock Bursts, Part II", U.S. Bureau of Mines, R.I. 3654.

Obert, L. and Duvall, W. 1945 - [a]. "Microseismic Method of Predicting Rock Failure in Underground Mining, Part I, General Method", U.S. Bureau of Mines, R.I. 3797.

Obert, L. and Duvall, W. 1945 - [b]. "The Microseismic Method of Predicting Rock Failure in Underground Mining, Part II, Laboratory Experiments", U.S. Bureau of Mines, R.I. 3803.

Pakalnis, V. 1981. "Strengths and Limitations of Microseismic Monitoring for Rockburst Control in Ontario Mines", Proc. 3rd Conf. on Acoustic Emission/Microseismic Activity in Geologic Structures and Materials, the Pennsylvania State University, Trans Tech. Publications, Clausthal, Germany, 1984, pp. 549-558.

Udd, J.E. 1984. "A Proposal for a Major Research Project on Rockbursts", CANMET Division Report MRP/MRL 84-84(TR), 18 pages.

Comparative stress measurement by two overcoring methods in deep holes at Siilinjärvi Mine, Finland

Comparaison de deux méthodes de surcarottage dans des forages profonds pour la détermination des contraintes à la mine finlandaise de Siilinjärvi

Vergleich von Druckspannungswerten von zwei Überbohrmethoden in tiefen Bohrlöchern der Siilinjärvi Grube, Finland

A.VÄÄTÄINEN, Helsinki University of Technology, Espoo, Finland
P.SÄRKKÄ, Helsinki University of Technology, Espoo, Finland

ABSTRACT: Triaxial measurements of the in situ state of stress were conducted by two overcoring techniques in order to compare their field applicability and likelihood of stress results. The techniques to be tested were that of the Swedish State Power Board with the Leeman-Hiltscher gauge and that of Interfels, both based on the Leeman calculation method. The measuring depth was 70 - 90 metres in four adjacent holes. These were drilled into an apatite orebody to a dip of 40 - 45°. The results, error sources, restrictions and field applicability of the techniques are discussed and compared.

RESUME: Des mesures de l'état des contraintes tridimensionel in situ étaient exécutées par deux téchniques de surcarottage afin de comparer leur aptitude en pratique et les résultats obtenus. Les deux téchniques à comparer étaient la technique de la Direction nationale de l'énergie éléctrique de Suède et cela d'Interfels, toutes les deux étant fondées sur la méthode de calcul de Leeman. Les mesures étaient faites à la profondeur de 70 - 90 mètres dans quatre trous de forage peu éloignés inclinés a 40 - 45°, forés dans le minerai d'apatite. Les résultats, sources d'erreur, restrictions et l'utilité des deux téchniques à la roche précambrienne sont discutés et comparés.

ZUSAMMENFASSUNG: Messungen des dreiaxialen in-situ Spannungszustandes wurden mit zwei Überbohrungsmethoden ausgeführt, ihre Feldanwendbarkeit und die Ähnlichkeit der Spannungsresultaten zu vergleichen. Die geprüften Methoden waren die Methode von staatlicher schwedischer Kraftwerksverwaltung (Vattenfall) und die Methode der Interfels GmbH, Deutschland. Diese beide ausnützen das Rechnungsverfahren von Leeman. Die Messungen nahmen Platz in vier nahezu Bohrlöchern mit einer Neigung von 40° bis 45° in einem Apatiterzkörper. Die Messungstiefe war zwischen 70 und 90 m. Die Resultaten, Fehlerquellen, Begrenzungen und Feldanwendbarkeit werden diskutiert und miteinander vergleicht.

1 INTRODUCTION

The state of stress is one of the main factors effecting the stability of rock caverns and tunnels. Accordingly it is of great importance during development work to be able to determine the magnitude and orientations of stresses in the rock mass in order to choose the location, shape and orientation of rock chambers successfully.

For the mining industry, stress determinations ought to be done in water filled boreholes down to 300 metres. So far that has been possible by local investigation facilities in dry 25 metre boreholes.

The Finnish mining industry together with the Laboratory of Mining Engineering, Helsinki University of Technology, and the Ministry of Trade and Industry started a two and a half year project in 1985 in order to procure or construct stress measurement equipment for deep boreholes, applicable in Finnish Precambrian bedrock.

Part of the project were test measurements carried out in 1986 at the Siilinjärvi open pit, Finland. The techniques tested were chosen on the basis of a literature study taking also into consideration the availability of equipment.

2 STRESS MEASUREMENT METHODS FOR DEEP BOREHOLES

Various techniques for stress determination have been presented in the literature. By considering only tested methods four practical in-situ stress determination techniques for deep boreholes were found applicable in Finnish bedrock. They were
- hydraulic fracturing,
- differential strain curve analysis (DSCA),
- overcoring technique of the Swedish State Power Board, and
- overcoring technique of Interfels.

A closer study of these techniques gave the impetus for a comparative measurement programme using the two overcoring methods.

2.1 Hydraulic fracturing

Hydraulic fracturing measurements have been conducted in various geological environments in boreholes down to thousands of metres. The first measurements in Scandinavia were conducted in Sweden in 1981 to a depth of 370 metres (Doe 1983).

In 1985 a test measurement programme was carried out in Finland by a research group from the Luleå University of Technology, Sweden, in a 500 metre test hole (Bjarnason 1986, Johansson & al. 1986).

2.1.1 Measurement procedure of hydraulic fracturing

In hydraulic fracturing, an intact region of a borehole is isolated by packers and subjected to increasing radial fluid pressure until a fracture develops in the borehole wall.

It is assumed that the fracture initiates when the internal fluid pressure is sufficient to exceed both the tangential stress on the borehole wall and the tensile strength of the rock. The hydrofracture is assumed to propagate from the borehole in a plane normal to the minimum principal stress.

The calculation of stresses from hydrofracturing is based on a two dimensional analytical solution of stresses around a circular opening in an infinite plane. The rock is assumed to be isotropic, linear elastic and impermeable.

The solution only considers the stresses in a plane perpendicular to the borehole axis. It is further assumed that the borehole axis is parallel to one of the principal stresses.

2.1.2 Comments on hydraulic fracturing

According to common experience hydraulic fracturing is a usable method especially when defining the state of stress in large scale in deep boreholes because it is not sensitive to differential stresses of the rock.

Practical difficulties initially arise from the need for a completely intact rock region of about 10 x hole diameter in the borehole. Furthermore the assumption concerning the parallelism of the borehole axis and one principal stress places a considerable restriction on the technique.

The hydrofractures in test measurements in Finland (Bjarnason 1986, Johansson & al. 1986) often developed horizontally, against the demands of the calculation method. This was possibly caused by the principal stress orientations not fulfilling the conditions of hydrofracturing theory.

The primary cause for rejecting the hydraulic fracturing technique when choosing the measurement system for mine use was it's ability to determine only the two-dimensional state of stress.

A method for establishing the triaxial state of stress by hydraulic pressure tests using preexisting fractures has been developed in France (Cornet & Valette 1984) but had not been tested in practice in Precambrian rocks before the measuring technique was selected. However, a successful test of the method was performed in Sweden by the Luleå University of Technology at the end of 1986 (Ljunggren & Raillard 1986).

2.2 Differential stress curve analysis

Based on the theoretical work of Walsh (1965), Simmons & al. (1974) developed an experimental method, differential strain curve analysis (DSCA), for measuring total microcrack volumes and orientations in rock samples in the laboratory.

In the USA the technique has been successfully used in various studies in deep rock masses, especially in petroleum exploration and production (Dey & Brown 1986, Ren & Rogiers 1983).

2.2.1 Measurement procedure of DSCA

In the DSCA stress measuring technique, core samples taken from the measuring point in the borehole are prepared with strain gauges attached in appropriate orientations. Then the specimen is loaded hydrostatically to a pressure sufficient to close essentially all crack porosity (about 200 MPa).

By measuring the crack closure with strain gauges in different orientations a strain tensor with three principal crack strains and their directions can be determined. Under suitable conditions, this crack strain tensor can be interpreted to give the in situ stress state.

A necessary condition is that the majority of microcracks are a result of the relief of the in situ stress in coring and cutting the sample. A further assumption is that microcrack porosity oriented in any given direction is produced in proportion to the magnitude of the effective compressive stress that was relieved in that direction (Dey et Brown 1986).

2.2.2 Comments on DSCA

In spite of numerous successful measurements in the USA, the DSCA method is still in the experimental stage and was rejected in this project. Generally, the advantage of the DSCA method is that all the measuring work can be carried out in a laboratory.

2.3 Overcoring method of the Swedish State Power Board

The Swedish State Power Board's (SSPB) equipment for three-dimensional rock stress measurements in vertical boreholes was developed about ten years ago. A detailed description of the measuring system has been published (Hiltscher & al. 1979).

The system is based on strain gauges and overcoring. Measurements have been performed in water-filled vertical boreholes down to a depth of 500 metres (Hallbjörn 1986). In 1986 the equipment was tested in 45° inclined holes in Finland. The greatest measuring depth was 90 metres.

2.3.1 Measurement procedure of SSPB's overcoring technique

The technique is based on the well-known calculation method of Leeman (Leeman & Hayes 1966). The measurement procedure is shortly illustrated in Figure 1 and the arrangement of the foil strain gages in the pilot borehole is presented in Figure 2.

The strain gauge assembly has three plastic tongues with a butyl-rubber tape pad, strain-gauge rosette and a layer of polyurethane foam. The tongues are 120° apart and each rosette consists of three strain gauges (Figure 2).

The purpose of the polyurethane foam is to hold a sufficient amount of glue immediately before pressing the plastic tongues with the strain gauge rosette against the borehole wall.

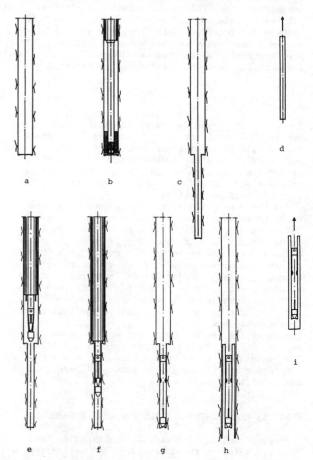

Figure 1. Procedure for cementing and measuring: a) Ø 76 mm borehole, b) centering Ø 36 mm bit, c) Ø 36 mm 40 cm long pilot borehole, d) inspection of small core, e) probe before releasing the mechanism, f) cementing gauges to the pilot borehole wall under pressure from cone, first measurement, g) release of the carrier and hoisting of the probe, h) overcoring, i) second measurement.

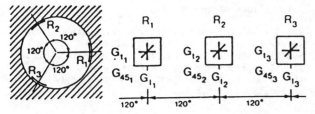

Figure 2. Setting of strain gauges in pilot borehole.

The glue is an acryl-resin, metamethylacrylate mono-mer, with catalyst, bensoylperoxide, and a coupling agent. The glue must have a certain viscosity and an average pot life of about 20 minutes.

This can be regulated by proper measures i.e. storing cold in a freezer, sometimes increasing the viscosity by heat-treatment, mixing young and old resin and test mixing before measurements. Hardening time of the glue is about two hours (Hallbjörn 1986).

The rock cores are test loaded afterwards. In a biaxial loading cell hydraulic pressure acts radially on the outer surface of the hollow core. Strain gauge readings are taken at different load levels.

From these it can be judged whether the strain gauges are properly cemented to the rock or not. From the readings, Young's modulus and Poisson's ratio of the core are calculated (Hallbjörn 1986).

The probe is shown in Figure 3. The probe hangs in a combined carrying and measuring cable. The cable contains a carrying part of kevlar and 12 conductors. The strain gauges are submerged in glue in a glue pot at the bottom end of the probe.

When the pinpoints touch the bottom of the large borehole the mechanism is released and the cone is pushed downwards by the weight including the upper part of the probe. The glue pot sinks to the bottom of the small borehole and the strain gauges are pressed to the borehole wall by the cone.

During the glue's hardening time a pre-heated liquid around the compass-needle solidifies by cooling down and gives the orientation of the probe. The liquid mixed must be modified depending on the actual rock temperature (Hallbjörn 1986).

2.3.2 Comments on overcoring method of SSPB

During the ten years since 1976 three-dimensional rock stress measurements with the described SSPB-Hiltscher system have been succesfully performed in Sweden and other countries. As a three-dimensional measuring method, long used in Precambrian rocks in Sweden, the SSPB's overcoring system was chosen to be tested in Finland in 1986.

2.4 Overcoring method of Interfels

The stress measuring equipment of the German Interfels company is a combination of the Australian CSIRO measuring device and the original Leeman measuring device planned for deeper boreholes than any of these. The Interfels' system has been used mainly in Central Europe but even on other continents.

The first measurements in hard Precambrian rocks were carried out in 1986 in Finland. The deepest measurement hole was about 70 metres dipping 45°. In vertical, water filled boreholes the state of stress has been measured down to 120 metres with Interfels' equipment. According to the manufacturer it is possible to measure at least down to 200 metres.

2.4.1 Measurement procedure of Interfels' method

Interfels' system, which is also based on Leeman's calculation method is theoretically similar to that of SSPB (2.3.1). Instead of three strain foils in every strain gauge rosette there are four of them in Interfels' probe. The setting of the rosettes on the circle of the pilot borehole is analogous to SSPB's system.

The measuring procedure is essentially similar to that of SSPB (2.3.1). The main differences are a bigger hole diameter in overcoring (116 mm), a longer pilot hole (min 120 cm) and the method of gluing.

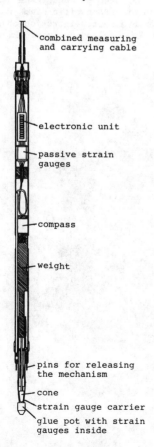

Figure 3. Probe for rock stress measurements in deep vertical boreholes.

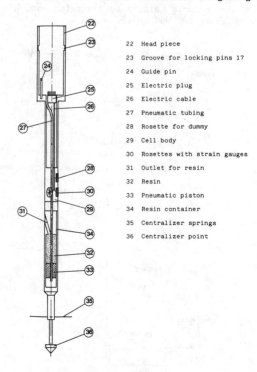

22	Head piece
23	Groove for locking pins 17
24	Guide pin
25	Electric plug
26	Electric cable
27	Pneumatic tubing
28	Rosette for dummy
29	Cell body
30	Rosettes with strain gauges
31	Outlet for resin
32	Resin
33	Pneumatic piston
34	Resin container
35	Centralizer springs
36	Centralizer point

Figure 4. Measuring probe of Interfels.

The strain foil rosettes are situated slightly inside of an epoxy strain gauge body. When gluing, the three rosettes are pressed pneumo-mechanically from the central body and pushed onto the borehole wall while glue runs out from a glue cell situated at the end of the probe.

After gluing, the probe is surrounded by glue and the strain gauge rosettes are connected with the rock by means of a thin glue film. The composition ratio of hardener and bonder of the two-component glue is constant and not dependent on the temperature. The hardening time of the glue is about ten hours.

The measuring probe is presented in Figure 4. By means of a string of setting rods, also intended for orientating of the probe in the pilot hole and a cable, the probe is led into the pilot borehole by means of a conical centralizing part (36). The glue is squeezed out of the reservoir (32) through a distributor by a piston (33) powered by compressed air.

To determine Young's modulus and Poisson's ratio needed in calculation of stresses, a rock specimen with strain gauges cemented on it is tested in a laboratory press. The rock mass moduli can also be determined on site by means of a borehole deformation probe (dilatometer).

2.4.3 Comments on overcoring method of Interfels

As a widely used, commercially available measuring equipment for deep water filled holes Interfels' equipment was chosen for testing at a Finnish mine.

The main test objects were its usability in hard rocks and in inclined holes as well as the orientation system in deep boreholes. The possibility to use a smaller hole diameter in overcoring instead of the normal 116 mm was tested and the possibility to increase the measuring depth to 200 metres in future was examined.

3 TEST SITE

The stress measurements were carried out at the Siilinjärvi apatite mine of Kemira Oy in Finland in 1986, Figures 5 and 7. Production capacity of the mine is about 500 000 mt apatite concentrates a year and as by-products calcite concentrates and phlogopite-mica about 100 000 mt/yr.

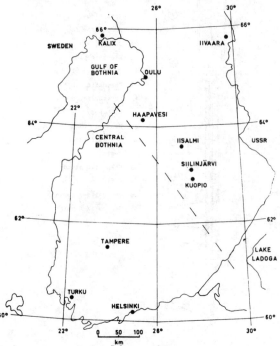

Figure 5. Situation of test site.

3.1 Outlines of geology

The Siilinjärvi carbonatite complex, with minable reserves totaling more than 500 million mt, forms a roughly tabular, vertical igneous body as an intrusion in surrounding granite gneiss, comprising (from the oldest to the youngest) glimmerite, syenite and carbonatite, Figure 6.

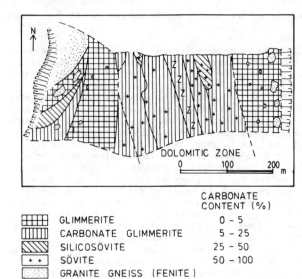

	CARBONATE CONTENT (%)
GLIMMERITE	0 – 5
CARBONATE GLIMMERITE	5 – 25
SILICOSÖVITE	25 – 50
SÖVITE	50 – 100
GRANITE GNEISS (FENITE)	
Z Z SHEAR ZONE	

Figure 6. A geological cross-section of the formation.

The widespread formation of mixed rocks between glimmerite and carbonatite is a result of the emplacement in the 'deep plutonic zone' where the upward propagation of carbonatitic magma was only able to take place in a narrow intrusion channel. The complex is 16 km long and up to 0.6 km wide (Puustinen 1971).

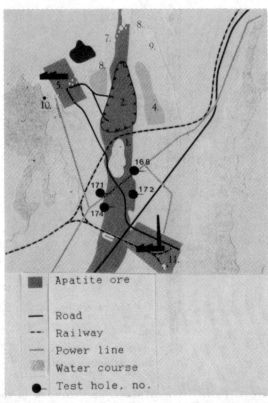

Figure 7. Test holes at the Siilinjärvi mine area.

3.2 Mine

The location of the open pit in the southern part of the ore formation is presented in Figure 7. The deposit is composed of nearly vertical, N-W oriented ore lenses.

The most abundant minerals of the ore are apatite 10 %, calcite 16 %, dolomite 3%, mica 65 % and other silicates 5 %. Apatite is distributed fairly evenly through the deposit but carbonatite and mica vary significantly; so the ratio of apatite to carbonates ranges between 1:1 and 1:4.

In the current pit area, selected because of it's low percentage of waste rock inclusions in the ore, there are six main stringers of diabase dikes. Dike thicknesses vary from 10 centimetres to 20 metres.

There is also a shear zone in which the rock condition prohibits processing. Waste rock in situ is about 9 % of the rock total.

Current planning at Siilinjärvi calls for an excavation that will reach 1250 m in length, 200-650 m in width and a depth of 215 m. Bench heights are usually 12 m but may range from 9 m to 14 m depending on the dip. The pit slopes range from 41° to 46°.

Preparations have been started to open a new pit south of the railway. These stress measurements are a part of that expansion programme.

4 TEST MEASUREMENTS OF STRESS STATE

The overcoring methods of the Swedish State Power Board and Interfels were tested and compared with each other in the field at Siilinjärvi in April - November 1986.

4.1 Measurements of Swedish State Power Board

The Swedish State Power Board with assistance of a Finnish drilling crew carried out seven measurements five of which were technically successful. The measurement holes are presented in Table 1 and marked in Figure 7.

Measurements were taken at a depth of 74 metres in the hole no. 168 (slope 45° E), at 71 metres in hole no. 171 (slope 41° E) and at 92 metres in hole no. 174 (slope 45° E).

In borehole no. 168 the rock was so jointed that small pieces of rock fell from the borehole wall and blocked the pilot hole. The hole had to be grouted before measurements. The grouting was successful. No siltation appeared afterwards.

However the second measurement was erroneous due to a break in one of the small cables in the hoisting cable. This made it impossible to connect the gauges correctly when the probe was down in the hole. The result had to be excluded.

Borehole no. 171 was noted to have obviously passed a fracture zone because in spite of a thorough rinsing of the hole silt was found both in the tubular core and in the probe.

It appeared during the calibration that the cementing of the rosettes was not completely successful on the lower side of the sloping hole because of siltation. The result of the third determination in hole 171 was rejected.

In the present Siilinjärvi case the measuring equipment was modified due to the sloping holes. In ordinary vertical holes the weight of the probe is sufficient to lead the gauge into the pilot hole and result in proper cementing of the rosettes. Now the weight had to be increased by about 50 kgs to compensate friction forces and the lower probe gravitation force component in the hole direction.

Young's modulus (E) and Poissons's ratio (ν) of the rock cores needed for the calculation of stress values were determined by calibration in a laboratory. The cores with the measuring gauge cemented inside them were radially loaded with hydrostatic pressure of 5 MPa and 10 MPa and the strains were measured.

4.2 Interfels' measurements

Two measurements with Interfels' system by Interfels with assistance of a Finnish drilling crew were carried out. The measurements were taken in hole no. 172 (Table 1, Figure 7) at a depth of 72 metres. The slope of the hole was 45° E.

The first measurement failed due to an unsuccessful positioning of the gauge. Because of a fracture zone at the depth of 30-40 metres the hole had to be grouted before measurements.

In setting the gauge cement pieces fallen from the hole walls partly filled the opening of the pilot hole obstructing the entry of the probe. Resistance of the cement fragments against the centralizer point at the bottom of the big hole gave the impression that the centralizer springs were squeezing into the pilot hole and cementing was started. Not before the overcoring was the failure exposed.

In the present case the general Interfels' laboratory determination for the elasticity modulus and Poisson's ratio was not done but the results were calculated on the basis of E- and ν-values determined in Finland.

Table 1. Boreholes in measurements of state of stress at Siilinjärvi in 1986.

Hole	Rock type	E (MPa)	ν	Comments
168	Apatite, mica			Grouted hole
SSPB 1		58	0,35	
2				Rejected meas.
171	Granodiorite			Siltation
SSPB 1		58	0,20	
2		74	0,19	
3				Rejected meas.
174	Apatite, mica, calcite			
SSPB 1		68	0,38	
2		62	0,33	
172	Apatite, mica			Grouted hole
Int. 1				Failed meas.
2		50	0,25	

5. RESULTS

The calculated values of principal stresses and vertical stresses from the measurements are presented in Table 2 and Figure 8.

Table 2. Results of stress measurements.

Point	$\sigma 1$	$\sigma 2$	$\sigma 3$	Slope/bear. $\sigma 1$	σvert.
168/1	18,6	6,1	-0,2	23/62	2,2
171/1	11,9	1,7	0,6	3/40	1,5
/2	8,2	4,1	-0,3	14/53	1,2
174/1	29,5	12,8	5,0	18/94	9,3
/2	28,9	16,7	8,3	23/20	11,9
172/1	4,0	2,6	1,0	8/92	1,9

stresses in MPa, directions in degrees

Applying the statistical theory of Fisher (1953) for the orientations of the principal stresses and the ordinary Gauss distribution for their magnitudes the stress values are concluded to have a significant mutual deviation for all other determination points except the two points in hole no 174, where the stress values and directions clearly present the same state of stress.

The orientation and magnitude of the state of stress determined in hole no 172 by Interfels is inside the mutual deviation of the states of stress in the other holes investigated by SSPB. So the difference of the stress results is probably caused by geological factors, not by different measuring techniques.

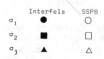

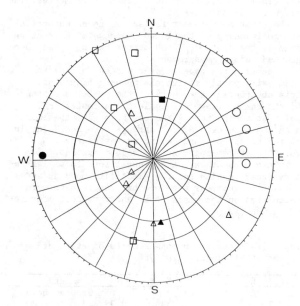

Figure 8. Orientations of principal stresses in the Schmidt stereographic projection.

With respect to the rock types in the measuring points no clear mutual difference between the stress-values in apatite ore (holes 168, 174 and 172) and country rock (hole 171, granodiorite) can be noticed.

Vertical stresses calculated from the principal stresses agree with the stress due to overburden except in hole no 174 where the weight of overburden is noticeably exceeded and the state of stress is exceptionally high.

The very low values of principal stresses in hole 172 are possibly due to a strong fracture zone found at a depth of 30-40 metres in the hole.

6 CONCLUSIONS

Successful drilling as well as grouting of a hole, when necessary, is a main condition for a successful measurement of state of stress by overcoring systems.

Difficulties in the first measurement in the grouted hole no 172 proved the usability of an experimental probe used by SSPB in the other test holes. With the experimental probe the openness and central location of the pilot hole is confirmed before the measuring probe is led into it.

Gluing in waterfilled boreholes succeeded for the most part (80%). An essential difference with influence on costs between the techniques is the different hardening time of the glue.

Lower drilling costs in SSPB's system due to a smaller hole diameter and a shorter pilot hole are obvious. According to the manufacturer the hole diameter for Interfels' system could be decreased to the same diameter as SSPB's in hard rocks.

In present tests the orientation of probes of both measuring techniques proved to work well. However, the usability of setting rods in orientation, especially in deeper inclined holes, should be ensured.

With regard to the confidence of the stress results obtained by the two measuring techniques tested at Siilinjärvi both of them fulfill the requirements of the project. When selecting a technique emphasis will be laid on field applicability and costs of the technique as well as its commercial availability.

ACKNOWLEDGEMENTS

The authors wish to thank Kemira Oy which allowed the test results be used for this report and Suomen Malmi Oy which was responsible for general arrangements and drilling work during all the measurements. The Swedish State Power Board and Interfels GmbH are acknowledged for giving abundant material and information concerning their measuring methods.

REFERENCES

Bjarnason B., 1986. Hydrofracturing Rock Stress Measurements in the Baltic Shield. Licentiate Thesis 1986:012 L, Division of Rock Mechanics, Luleå University of Technology, Luleå, Sweden.

Cornet F. and Valette B., 1984. In Situ Stress Determination From Hydraulic Injection Test Data. J. Geoph. Res. 89, B13, 11.527-11.537.

Dey, T.N., Brown, D.W. 1986. Stress measurements in a deep granitic rock mass using hydraulic fracturing and differential strain curve analysis. Int. Symp. on Rock Stress and Rock Stress Measurements 1986, Stockholm.

Doe, T.W., Ingevald, K., Strindell, L., Leijon, B., Hustrulid, E., Majer, E. and Carlsson, H. 1983. In Situ Stress Measurements at the Stripa Mine, Sweden. Report LBL-15009, SAC44. Swedish-American Cooperative Program on Radioactive Waste Storage in Mined Caverns in Crystalline Rock.

Fisher R.A., 1953. Dispersion of a sphere. Proc. Roy. Soc. London 217, pp. 295-305.

Hallbjörn L., 1986. Rock stress measurements performed by Swedish State Power Board. Int Symp. on Rock Stress and Rock Stress Measurements 1986, Stockholm.

Hiltscher R., Martna J., Strindell L. 1979. The measurement of triaxial rock stresses in deep bore holes. Proc. 4th Int. Congr. on Rock Mechanics, Montreux 1979.

Johansson E., Väätäinen A., Särkkä P., 1986. The Interpretation of Rock Stress Measurements Using a Hydraulic Fracturing Method in the Lavia Test Bore hole, Finland. Report YJT-86-07 Part II. Nuclear Waste Commission of Finnish Power Companies.

Leeman E.R., Hayes D.I., A technique for determining the complete state of stress in rock using a single borehole. Proc. 1st Congr. ISRM 2, 17-24. Lisboa 1966.

Ljunggren C., Raillard G., 1986. In-situ stress determination by hydraulic tests on preexisting fractures at Gideå test site, Sweden. Report. Luleå University of Technology, Div. of Rock Mechanics. Luleå, Sweden, November 1986.

Puustinen K. 1971. Geology of the Siilinjärvi Carbonatite Complex, Eastern Finland, Bull. Geol. Comm. Finland 249. Helsinki, Finland 1971.

Ren, N.K., Roegiers J.C., 1983. Differential strain curve analysis - A new method for determining the pre-existing in situ stress state from rock core measurements. Proc. 5th Int. Conf. ISRM, Melbourne, Australia: F117-F128.

Simmons G., Siegfried R.W., Feves M. 1974. Differential strain analysis: A new method for examining cracks in rocks. J. Geoph. Res. 79, pp. 4383-4385.

Walsh J.B. 1965. The effect of cracks on the compressibility of rock. J. Geoph. Res. 70, pp. 381-389.

Determination of deformational characteristics of rock using wave velocity methods

Détermination des caractéristiques de déformation de la roche en employant des méthodes de vitesse des ondes
Bestimmung von Deformierungscharakteristiken im Gestein mittels Schallgeschwindigkeitsmethoden

W.L. VAN HEERDEN, Senior Specialist Researcher, National Mechanical Engineering Research Institute, CSIR, Pretoria, Republic of South Africa

ABSTRACT: Sonic velocity tests were carried out on six different rock materials in the laboratory. The results show that the dynamic modulus of elasticity is higher than the static modulus and that the static modulus can be predicted from a knowledge of the dynamic modulus at different stress levels.

RÉSUMÉ: Des essais de laboratoire de vitesse somique ont été et écutés sursit éspece de roche différents. Les résultats obtenus ont moutré que le module d'élasticité dynamique est plus haut que le module statique et que ce module peut être prédit en sachant le module dynamique à des niveaut de contrainte différents.

ZUSAMMENFASSUNG: Schallgeschwindigtkeitsversuche wurden im Laboratorium an sechs verschiedenen Gesteinsarten durchgeführt. Die Ergebnisse zeigen, dass der dynamische Elastizitätsmodul höher ist als der statische Modul, welcher vorausgesagt werden kann, sofern der dynamische Modul bei verschiedenen Spannungsstufen bekannt ist.

1. INTRODUCTION

Wave velocity methods such as the sonic velocity technique have been used for many years [1,2,3] to determine the so-called dynamic elastic moduli of rock. The technique is popular as tests are quick, simple and inexpensive compared to static tests where strain gauges are used to measure strain on the specimens.

One disadvantage of the technique is that for most rock materials the dynamic moduli differ from the normal static moduli. In general most of the results published to date [3,4,5,6] indicate that the dynamic moduli are greater than the static moduli. For design purposes it is essential to know the static moduli of the rock, and although work aimed at finding general correlations between static and dynamic moduli of rock has been reported [7,8], it is still impossible to predict the static moduli of rock from a knowledge of the dynamic moduli.

As part of a long-term research project, sonic velocity and static loading tests were conducted simultaneously on six different rock types. The methods used and results obtained are reported in this paper.

2. EXPERIMENTAL PROCEDURES

The rock samples used in the programme were received in the form of NX-sized core pieces ranging in diameter from 54 to 56 mm. From these samples, cylindrical specimens approximately 80 mm long were cut on a diamond saw and their flat surfaces ground parallel to 0,03 mm on a modified milling machine. After preparation the specimens were air-dried for a period of 48 hours after which their masses and dimensions were measured. Strain gauge rosettes consisting of an axial and a lateral gauge were attached to the circumference of each specimen at two directly opposite points.

Discs with a thickness of 10 mm were also prepared from each rock material for the purpose of determining the porosities of the different rock types. Details of all specimens are given in Table 1.

For testing, a specimen was installed in a loading frame between two sonic velocity transducer heads so that dynamic and static measurements could be carried out at the same stress increments and without having to unload the specimen. The pulse transmission method using combined compressional (P) and shear (S) wave transducers described by King [9] was used for the dynamic measurements. Lead foil discs approximately 0,05 mm thick were employed to provide acoustic coupling between the rock and the transducer holders.

Before doing actual tests on rock, a series of tests were carried out on specimens of different lengths of steel, aluminium and brass. It was found that measured velocities agreed with theoretical velocities to within 2 per cent.

The procedure adopted was first to load the specimen to an axial stress of 2 MPa to bed-in the lead foil. Thereafter the load was reduced to zero and P and S-wave transmission times measured. Zero readings of static strain were also recorded. Static and dynamic measurements were repeated at stress increments up to a maximum stress of 40 MPa for the stronger materials.

3. RESULTS

Before the results are presented it should be mentioned that both tangent and secant static moduli were calculated. Unless otherwise indicated the results presented refer to the tangent static modulus at the relevant stress level.

Typical results in the form of curves showing variation in static and dynamic moduli (modulus and Poisson's ratio) with stress level is shown in Figs 1 and 2 for specimen 9 (yellow sandstone) and in Figs 3 and 4 for specimen 11 (grey sandstone).

It can be seen that the dynamic modulus is greater than the static modulus for both materials and at all stress levels, while the dynamic Poisson's ratio is smaller than the static Poisson's ratio at all stress levels. This was found to be the case for all specimens tested.

4. DISCUSSION

While examining the shape of the curves of dynamic modulus versus stress level it was noticed that for materials where there was a small change in dynamic

Table 1. Details of rock specimens

Specimen	Material	Diameter (mm)	Length (mm)	Density (kg/m³)	Porosity (%)	Strength (MPa)
1	Fine-grained	54,9	79,0	2920	0,14	341
2	Norite	55,0	78,3	2920		349
3	Coarse-grained	55,1	78,5	2926	0,17	284
4	Norite	55,0	78,4	2926		292
5	Pink Sandstone	55,0	78,5	2278	12,3	100
6	(fine)	55,0	78,7	2276		96
7	Red Sandstone	54,9	78,4	1930	25,3	-
8	(coarse)	54,9	77,9	1922		24
9	Yellow Sandstone	55,1	78,5	2194	15,4	81
10	(fine)	55,1	78,9	2207		84
11	Grey Sandstone	53,6	79,0	2464	8,1	155
12	(fine)	53,5	78,7	2468		158

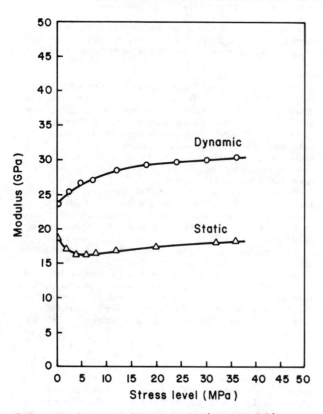

FIG. I Modulus v Stress level (Yellow SS)

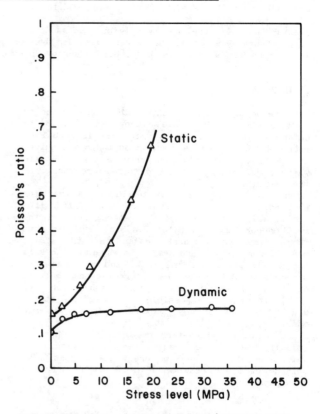

FIG. 2 Poisson's ratio v Stress (Yellow SS)

modulus with stress there was also a small difference between the static and dynamic moduli. For materials where there was a large change in dynamic modulus with stress there was also a large difference between the static and dynamic moduli.

It therefore appeared possible to estimate the static modulus of these materials from the dynamic data alone.

For this purpose a dynamic index was defined as the dynamic modulus at any stress level divided by the dynamic modulus at zero stress.

Fig. 5 shows a plot of the static modulus on a logarithmic scale versus the dynamic index for various stress levels. It can be seen that there is a definite relationship between the dynamic index and the static modulus of the rocks tested and that all the materials tested conform to the relationship. The same applies to the modulus of deformation or secant modulus as shown in Fig. 6.

However, it should be noted that these conclusions are based on a limited number of tests and that further tests are necessary to substantiate the relationships before they can be used with confidence.

With regard to Poisson's ratio the only conclusion that can be drawn is that the dynamic value is smaller than the static value. For uniaxial loading tests it is believed that for certain rocks the lateral to axial strain ratio has very little meaning at high stress levels. A typical example is shown in Fig. 2 where values greater than 0,7 are indicated.

It is also interesting to note that one of the materials (grey sandstone) yielded negative dynamic Poisson's ratio values at low stress levels (see Fig. 4). No relationship between the static and dynamic Poisson's ratio values could be established from the results.

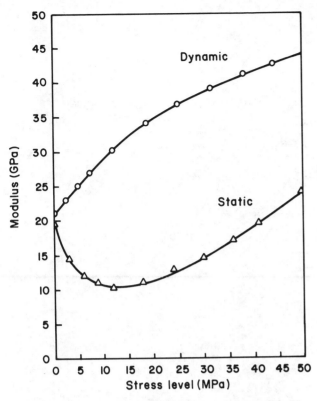

FIG.3 Modulus v Stress (Grey sandstone)

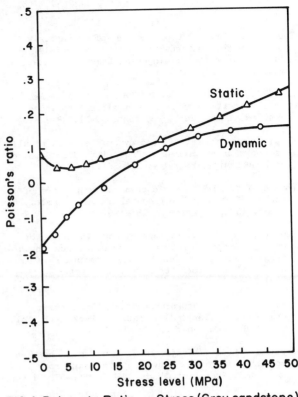

FIG.4 Poisson's Ratio v Stress (Grey sandstone)

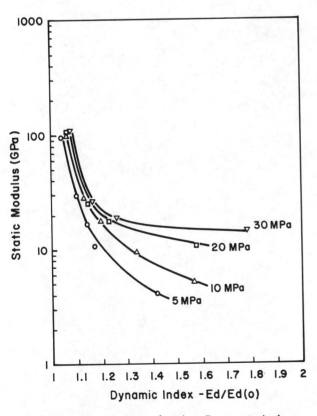

FIG.5 Static Modulus (tan) v Dynamic Index

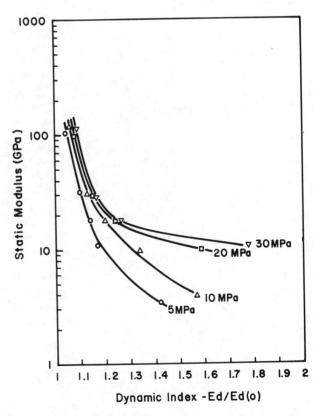

FIG.6 Static Modulus (sec) v Dynamic Index

REFERENCES

Shook, W.B. 1983. Critical survey of mechanical
property test methods for brittle materials. Ohio
State University, Engineering Experiment Station,
Columbus, Ohio.

Lama, R.D, & V.S. Vutukuri 1978. Handbook on mechanical properties of rocks - testing techniques and
results. Trans. Tech. Publication, 11:218-223.

King, M.S. 1970. Static and dynamic elastic moduli
of rocks under pressure. Proc. 11th U.S. Symp. on
Rock Mechanics, Berkeley, 329-351.

Youash, Y.Y. 1970. Dynamic physical properties of
rocks: Part II, Experimental results. Proc. 2nd
Congr. ISRM, Belgrade. 1:185-195.

Howarth, D.F. 1985. Development and evaluation of
ultrasonic piezoelectric transducers for the determination of dynamic Young's modulus of triaxially loaded rock cores Geotech. Test. J.,
8(2):59-65

Ramana, Y.V. & B. Venkatanarayana 1974. Laboratory
studies on kolar rocks. Int. J. Rock Mech. Min.
Sci. & Geomech. Abstr., 10:465-489.

Kujundzic, B, & N. Grujic 1966. Correlation between
static and dynamic investigations of rock mass
(in-situ). Proc. 1st Congr. ISRM, 1:565-570 Lisbon

Savich, A.I. 1984. Generalized relations between
static and dynamic indices of rock deformability.
Translated from Gidroekhnicheskoe Stroitel'stvo,
(8):50-54.

King, M.S. 1983. Static and dynamic elastic properties of rocks from the Canadian Shield. Int. J.
Rock Mech. Min. Sci. & Geomech. Abstr., 20(5):
237-241.

Flexible lining for underground openings in a block caving mine
Revêtement souple pour des chambres souterraines dans une mine par foudroyage
Biegsamer Ausbau für Untertagestrecken in einer Grube mit Blockbruchbau

M.L. VAN SINT JAN, Universidad Católica de Chile, Associate Geotéchnica Consultores
L. VALENZUELA, Geotéchnica Consultores, Chile
R. MORALES, El Salvador mine, Codelco, Chile

ABSTRACT: A case history of ground support in a large block caving mine located in altered andesite is presented. Rock support traditionally used in the underground openings consists of steel frames and yielding steel arches with wood lagging, sometimes complemented with grouted bolts and locally in heavily loaded areas, with reinforced concrete. Based on field observations of ground conditions, general recommendations from rock mechanics classification systems and on the results of simple analytical models, a new support system has been adopted for certain areas of El Salvador mine in Chile, a flexible lining obtained with rock bolts and mesh-reinforced shotcrete applied in stages. The satisfactory behavior of this support in the LHD zone of the mine is described and the associate savings are commented.

RESUME: Présentation d'un cas historique de soutènement de roche dans une grande mine de "block caving". Le soutènement de roche utilisé traditionnellement dans le travaux souterrains consiste de cadres d'acier et de cintres d'acier glissants avec du bois entre le cadre et la roche, complétés parfois de boulons injectés, aussi bien que de béton armé dans des localisations très sollicitées. D'accord aux conditions du terrain, aux recommendations générales issues des systèmes de classification géotechnique des roches et aux résultats de modèles analytiques simples, on a adopté un nouveau système de soutènement pour certaines zones de la mine El Salvador au Chili. Ce système consiste d'un revêtement flexible obtenu au moyen de boulons d'ancrage et de béton projeté, renforcés de grillages d'acier et appliqués par étapes. On décrit le comportement satisfaisant de ce soutènement dans la zone LHD de la mine, en ajoutant un commentaire sur les économies associées.

ZUSAMMENFASSUNG: Vorgestellt wird ein geschichtlicher Fall ueber einen Felsstuetzsystem innerhalb eines grossen Bergwerkes im ungeschischtetem Andesit. Das ueblich benutzte Felsstuetzsystem in diesen unterirdischen Bauwerk besteht aus Stahlrahmen, Stahlboegen die mit Holzkeile angepasst werden sowie auch mit Felsbloecke und injekzierte Schrauben. Je nach den Felsanforderungen wird auch armierter Beton benutzt. Aufgrund von Beobachtungen der Gelaendebedingungen, allgemeine Empfehlungen die aus den geotechnischen Felsklassifizierungen gegeben sind, und aus den Ergebnissen der einfachen analytischen Modellen, ist ein neues Stuetzsystem angenommen worden fuer einige Zonen des Bergwerkes in El Salvador, Chile. Dieses biegsame Stuetzsystem wird aus de Verbindung von Stahlgitterverkleidung mit Ankerschrauben und Spritzzement erreicht. Das ausgezeichnete Verhalten dieser Stuetzart in der LHD-Zone des Bergwerkes wird hier beschrieben sowie die verbundenen materiellen Vorteilen.

1 INTRODUCTION

El Salvador copper mine is exploited by block caving in an isolated mountain such that the average depth from the ground surface to the multiple excavations of the undercut and haulage levels ranges from 260 to 700 m. The main portion of the underground workings covers an approximately rectangular area 1300 m long and 500 to 600 m wide. Figure 1 shows a simplified view of the underground openings. Several undercut galleries 2.2 m-wide and 2.5 m-high, are excavated at 14.3 m on centers; parallel to such galleries, and 14.6 m below, the 4.0 m-wide by 3.4 m-high production galleries are excavated. The broken ore descends through the ore passes and is removed with mechanical equipment (LHD) at the lateral draw points, connected to the production drifts.

The exploitation system described above results in multiple underground openings at different orientations, leaving rock pillars of complicated geometry, usually an undesirable situation from the rock mechanics point of a view. Moreover, as the mining proceeds, both the areal extent of the broken ore and the boundaries between broken and intact rock change with time. Thus, an extremely complicated stress distribution develops on the rock pillars and around the underground openings. Field observations of ground and lining behavior carried out at the mine are in agreement with the results observed elsewhere under similar circumstances where large stress concentrations occur as the limit between broken and intact ore approaches and passes over a particular observation point (Bolmer 1965, Legast 1981, Björnfot & Stephansson 1984).

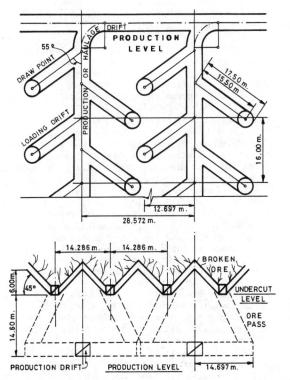

Figure 1. General layout (plan view and cross section) of the LHD zone at El Salvador mine, Chile.

The steel lining support traditionally used in the mine has been designed based on precedent; in this paper emphazis is placed on describing the selection and behavior of a flexible lining consisting of rock bolts and mesh-reinforced gunite. Such support system, very common in civil engineering projects, have not been so widely used in mining engineering, particularly when the block caving method is used, where dramatic changes in ground stresses are induced. Thus, there is a need to document cases where a flexible support, mainly based on shotcrete and rock bolts, has been used, whether successfully or not, under the difficult environment found in caving mines.

2 GEOLOGY

The mine is located within a body of granodioritic porphyries which intruded andesites and rhyolites. Although several joint orientations are recognized, three sets of discontinuities are prominent at the site, with the following average orientations (strike and dip) :
- Set 1 : N45W/vertical
- Set 2 : N45E/vertical
- Set 3 : N30E/40NW

Most joints are tight, planar and sligthly rough, many have coatings of soft minerals or quartz. Even though faults parallel to the second set are common, the most relevant structural features at the mine are steeply dipping faults and dykes parallel to the first set. Faults have widths ranging from a few centimeters to 1 m, and consist of fractured rock and soft gauge seams with kaolin montmorillonite. The dykes have widths ranging from 1 cm to 2 m and they are filled with clastic material. Both the dykes and faults are major discontinuities which are usualy surrounded by more heavily fractured rock. During excavation of the underground workings such zones of more heavily broken rock would tend to result in larger than usual overbreak.

The major faults and dykes of set 1 also control design of the undercut retreat direction because large stresses are concentrated over the pillars between such major discontinuities and the undercut face when the said face is parallel to the discontinuities.

3 BEHAVIOR OF STEEL LININGS

The typical lining used at the production level consisted of 44 kg/m yielding steel arches at 0.75 m on centers, as shown on fig. 2, lagged with wood and rock cribbage. Heavy steel frames (10 WF 112 of 167 kg/m) were used at the intersections of loading drifts next to draw points with the production drifts.

Although the largest overbreak occurred in the zones of more heavily broken and weathered rock, in places where the strength of the intact rock was high, failure and overbreak also were present, controlled by the steeply dipping discontinuities. Delay in placing the steel support resulted in large roof cavities which stopped where an important horizontal discontinuity was encountered. Thus it was difficult to obtain a good and continuous stiff blocking of the lining to the rock. Scaling of loose blocks would tend to increase the overbreak and the concentrated loads on the lining which at the same time was being given a more heterogeneous support, with several soft spots.

The excavation of the intersections presented particularly difficult support problems because it was not possible to excavate them with a geometry suitable for standard yielding steel arches; there was a tendency to support them with square frames made of heavy steel beams, being difficult to obtain a good blocking of the set to the rock. In some cases the excavation and final support of the intersections were completed as the face of the production drift was near. In other cases the excavation of the intersections is completed only after the production or haulage drift had been supported. Removal of the steel

arches to replace the support with square sets, would tend to increase the overbrak. In addition to the gravity loading from loose blocks, the lining was subjected to a complicated system of overpressure loadings due to the mining activity. The decrease in radial lining stiffness due to inadequate blocking plus the increase in concentrated gravity and overpressure loading resulted in large bending moments and deformations of the linings. Fig. 3 and 4 show deformed steel arches and steel frames after the undercut retreat face passed over those areas.

Figure 2. Tunnel lining with yielding steel arches and excesive wood lagging.

Figure 3. Failure of yielding steel arches support in altered andesite.

4 NEW ROCK SUPPORT APPROACH

The steel support described above had been installed based primarily on past precedent. The amount of support was preliminary estimated based on the lithology (andesite, rhyolite, etc.) and the type and dimensions of the opening (haulage drift, undercut drift, etc.). However, support dimensions were not specifically related to a design procedure which would include consideration of rock mass strength properties, failure mechanism, in situ state of stress or ground-support interaction.

In general, the design approach used led to high costs due to overdimensioned support elements but also to frequent support failures caused by local underestimation of rock mass quality and by the construction problems previously described. Moreover, it was apparent that although the yieldable steel arches provided a convenient means to maintain safety and opening stability, they did not take full advantage of the compression and bending capacity of the steel section.

Figure 4. Typical failure of steel frame support in intersection areas.

A comprehensive analysis of the geotechnical charac teristics of the rock and of the behavior of the different types of support that was carried through out the mine during 1982 led to the adoption of a new approach to rock support design considering the following aspects :
1. Recognition of relatively high horizontal in situ stresses.
2. Identification of the important influence of the main geological structures (N45W/vertical) on the general behavior of the support and particularly its relationship with the orientation of the undercut retreat face.
3. Set up of a rock quality characterization program based in an universal rock mechanics classification system, as a basis for support design and selection.
4. Preferential adoption of active support systems based on relatively flexible elements (rock bolting and shotcrete) instead of the used passive system (steel frames, concrete lining).
5. Improvement of construction techniques in order to preserve the integrity of the surrounding rock mass.
The first program of in-situ stress measurements confirmed the general regional trend with horizontal stresses higher than vertical stresses and vertical stresses corresponding approximately to rock burden, but the limited data available showed an important scatter, some computed values being suspect. Probably the measurements were influenced to varying degrees by the nearby block-caving activity. In 1982 a program of rock mass characterization using Laubscher's (1977) classification method was started,which was applied to all new underground workings (Pesce 1983). After some training it was found that relatively little additional effort was required from the field geologists mapping rock cores or tunnel walls in order to include the additional data needed to perform such classification. Further, it was found that such excercise would force field geologists to pay attention to the major factors which would control ground behavior. Moreover, using a systematic rock mass classification provided a quantitative description of the ground quality to engineers in charge of support design, in such a way that ground behavior could be predicted more accurately and support design changed accordingly.

It was found that discontinuities and planar or linear structures, such as dykes, faults and associated zones of broken rock, would control ground behavior to a larger degree than intact rock strength or average RQD. Thus, a zone consisting of hard rock with major discontinuities or zones of broken rock, would have more support problems than one having softer rock but no major discontinuities, even though the rock mass may have been strongly fractured. The right application of a rock mass classification system should show this characteristic. This correct calibration of the classification method was the aspect that needed more attention from the geologists and engineers involved in the field program and in the support selection.
Typical values of Laubscher rock quality range from class III-A to IV-B (approximately Barton's Q=5 to 0.5) with some zones classifying as II-B (approximately Q=20).

5 FLEXIBLE SUPPORT DESIGN

Following the design philosophy which has generally been associated to the New Austrian Tunneling Method (Rabcewicz, 1969) it was decided to try a new lining such that it could allow controlled ground displacements and its capacity could be increased or repaired easily. A preliminary design was developed which included a combination of rock bolts and mesh-reinforced shotcrete, applied in stages, assuming a circular tunnel in a hydrostatic stress field and following the procedure described by Brown et al (1983) for an axisymmetric tunnel problem. Although in was recogniced at the outset that such model was a drastic simplification of the actual situation, it was also clear that not enough data was available to validate a more comprehensive analysis. The scope of the preliminary design was to assess the sensitivity of ground and lining response to changes in major parameters, to provide a simple reference frame to interpret field observations of such response and to establish the feasibility of providing the proposed lining support under the field conditions described.

The procedure incorporates the semi-empirical failure criteria proposed by Hoek (1983) trough the strength envelope parameters "m" and "s" which are dependent on intact rock properties and rock mass classification. Typical values used for the parameters needed in the analysis are shown in Table 1.

Table 1. Typical ground parameters for the analysis

σ_c= uniaxial compressive strength of intact rock	=	52.5 MPa
E = Young's modulus of original rock mass	=	15,395 MPa
ν = Poisson's ratio of original rock mass	=	0.24
m = material constant for original rock mass	=	0.34
s = material constant for original rock mass	=	0.0001
m_r = material constant for post failure condition	=	0.09
s_r = material constant for post failure condition	=	0.00001
p_o = in situ hydrostatic field stress	=	13.9 MPa
r_i = internal tunnel radius	=	2 m

The analysis included the weight of the loosened rock over the crown of the support during post-peak behavior. The reaction curve for the mesh-reinforced shotcrete lining was developed considering a ring under hydrostatic compression and the reaction curve for fully grouted rock bolts was obtained from the bolts' strain computed assuming the bolts displaced the same amount that the ground around them.

A typical result of this type of analysis is shown in fig. 5. The results of this fig.5 suggest that for the case where fair ground (type III) would exist under in situ hydrostatic stresses of 13,9 MPa (equivalent to 600 m of rock burden) a support system consisting of rock bolts alone (298.9 KN yield load, at 0.50 m x 0.50 m spacing) would allow large rock displacements before full stabilization is achieved. If such primary support is complemented with a 0.10 m - thick reinforced shotcrete lining after approximately 30 mm of wall convergence has been allowed, final ground displacements would be reduced to about 33 mm. In order to include the influence of non-hydrostatic stresses in the analysis, the resulting bending moments can be estimated assuming a certain lining distortion.

Based on similar analysis it was decided to try such lining in actual field conditions. Most of the tunnels which have so far been supported with rock bolts and mesh-reinforced shotcrete have been excavated in class III ground, although some openings in zones with bad quality rock (class IV-A) have also been recently treated in the same way with satisfactory results in some of the drifts, but some difficulties related to excessive convergency and wear problems in draw points areas still need to be solved.

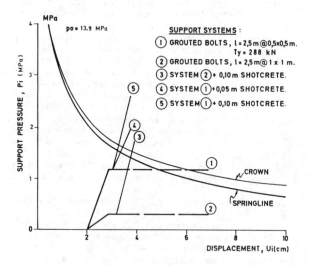

Figure 5. Ground support interaction curve and support design.

6 BEHAVIOR OF THE NEW SUPPORT SYSTEM

A typical lining currently used for support of the production drift in type III ground in the LHD mine consists of a primary support made of 1.8 m-long steel friction rock bolts at an average spacing of 1 m x 1 m. This support allows advancing the face some 50 m before the next stage is placed. For the next stage a steel mesh with 0.15 m-wide openings and 6 mm-diameter steel bars is placed with 25 mm-diameter and 2.4 m-long rock bolts at a spacing of 1 m x 1 m and covered with 0.10 m of gunite. The rock bolts are made of concrete reinforcing steel bars, threaded at the outer end, provided with steel plate and nut, fully grouted in place with lean cement grout; the plate is simply pressed against the steel mesh and the rock with a wrench.

Figure 6 shows the appearence of the intersection between a production drift and a loading drift next to a draw point, which has been supported with the reinforced lining described above.

A program of field measurements has recently been initiated in order to control lining displacements and to compare how do the different linings already

Figure 6. Production drift and draw point (right, background) supported with rock bolts and mesh-reinforced gunite.

described behave in different areas, but the results are not available yet. However, it is clear from the observations that the new lining has performed satisfactorily to date. In the case of the production or haulage drifts in the LHD zone, the application of the new support approach (system B in table 2) has produced savings in rock support of the order of US $ 1000 per meter of tunnel when compared with the previous system (system A in table 2), with the exceptions of local zones of very poor quality rock where additional reinforcement has been necessary in order to avoid frequent repairs in critical areas (system C in table 2). It should be noticed that at present more than 30% of the total mine production of 36.000 tons of mineral per day corresponds to the LHD zone.

Table 2. Average rock support costs in U$/m

| Drift | Rock Support System | | | |
	A	B	C	D
Haulage	1.405	390	1.795	--
Loading	710	200	--	570

The costs indicated in table 2 (Valenzuela 1986) correspond to the following support systems :

System A: grouted bolts, steel mesh, steel flange beams and yielding steel arches.

System B: friction rock bolts, grouted bolts and mesh-reinforced gunite.

System C: idem as system B plus reinforced concrete in zones of poor rock.

System D: idem as system B plus yielding steel arches in zones of poor rock.

Although the cost of system C used in zones of poor quality rock is higher than B, only one repair is necessary, while for the same period of time three repairs were necessary when system A was used. The construction crews have adapted rapidly to the new procedures and a shotcrete crew of 3 can perform all the job required in a particular area of the mine.

A readily apparent advantage has been obtained both for the excavation and the support of the intersections. The site of the future drawpoint is usually lightly supported, with plain gunite and with few or no bolts. The perimeter of the future excavation, however, is pressuported with a line of rock bolts spaced at some 0.40 m. The result, as shown in figure 6, is a very neat intersection even though there is a complicated geometry. So far, some 4.000 m of production tunnels and loading drifts have been supported in the way decribed. Behavior during operation

of the loading drifts and draw points and as the
block caving face proceeds over a particular point
has been quite satisfactory, with only occasional
fallouts. The simple fact that several production
tunnels and their draw points have been successfully
supported in the same area where the steel supports
had shown poor performances is a clear demonstration
of the advantages of this lining support.

7 CONCLUSIONS

The case history presented in this paper illustrates
not only the feasibility of the use of a flexible
lining consisting of rock bolts and mesh-reinforced
shotcrete for support of the underground openings at
a deep block caving mine in weak rock where large
dynamic stresses develop around such opening, but
also the technical and economical advantages associate to this type of rock support.

8 ACKNOWLEDGMENTS

Part of the data summarized in this paper was obtained by the authors for the development of design recommendations of rock support for El Salvador mine.
Other data has been furnished by the Mining Engineering Departament of El Salvador. The authors are
gratefull to Codelco Chile and particularly to Mr.
Fernando Riveri, G. Muller, J. R. Pesce and C. Valenzuela for providing access to the required information.
   Thanks are also due to the Pontificia Universidad
Católica de Chile for financial support provided
through grant DIUC 30/85, for the preparation of
this paper.

9 REFERENCES

Björnfot, F. & Stephansson, O. 1984. Interaction of
   grouted rock bolts and hard rock masses at variable loading in a test drift of the Kiirunavaara
   mine, Sweden. In O. Stephansson (ed.). Rock bolting,
   theory and applicationin mining and underground
   construction, p.377-395. Rotterdam : Balkema.

Bolmer, R. L. 1965. Stresses induced around mine
   development workings by undercutting and caving
   Climax molybdenum mine, Colorado. US/B. Mines RI
   6666.

Brown, E. T., Bray, J. W., Ladanyi, B. & Hoek, E.
   1983. Ground response curves for rock tunnels. Geotechnical Engineering ASCE 109, N° 1 : 15-39.

Hoek, E. 1983. Strength of jointed rock masses. Géotechnique 33. N° 3 : 187-223.

Laubscher, D. H. 1977. Geomechanics classification
   of jointed rock masses - mining applications. Trans.
   Inst. Metall, 86, A1-8.

Legast, P. 1981. Dissertation on block caving. PhD
   thesis McGill University, Montreal, Canada.

Pesce, J. R. 1983. Clasificación geomecánica de rocas
   en mina El Salvador. Dissertation Universidad de
   Chile, Santiago, Chile. In spanish.

Rabcewicz, L. V. 1964. The new austrian tunneling
   method. Water Power, Nov, Dec, Jan 1985.

Valenzuela, C. 1986. Fortificación en minería metáli-
   ca. IV encuentro de seguridad y productividad en
   la minería. Universidad de Santiago y Sernageomin,
   Santiago, Chile. In spanish.

Considerations on the properties and modeling of the deformational behaviour of weak rock masses

Considérations sur les propriétés et la modélisation de la déformabilité des roches faibles
Einige Überlegungen über die Verformungseigenschaften und die entsprechenden Modellidealisierungen weicher Felsen

E.VARGAS JR., Department of Civil Engineering, Catholic University, Rio de Janeiro, Brazil
L.DOBEREINER, Enge-Rio Engenharia e Consultoria S.A., Rio de Janeiro, Brazil
J.P.MILLAN, Sondotechnica, S.A., Rio de Janeiro, Brazil

ABSTRACT: This work makes some considerations on the present knowledge of the deformability properties of weak rock masses together with comments on the adequate modeling of engineering problems in these materials. An analysis of a hypothetical foundation consisting of a fractured weak rock loaded on the surface was carried out by various techniques and the results are presented.

RESUMÉ: Ce travail présente des considerations sur l'état actuel de connaissance des propriétés de deformabilité des roches faibles et sur la modelization numérique des problèmes du génie des ces matériaux. On présente encore les resultats de l'analyse numérique d'une fondation hypothétique que consiste de roche faible fracturée, chargée à la surface. Cette analyse a été effectuée en utilisant plusieurs techniques numériques.

ZUSAMENFASSUNG: In diesem Beitrag werden einige Überlegungen gemacht über den stand der Forschung betreffend der Verformungseigenschaften weicher Felsen und die entsprechenden Modell-idealisierungen für die Anwendung auf dem Gebiet der Ingenieurwissenchaft. Ein hypothetischer, aus weichem Felsen in gerissenen Zustand bestehenden Grundbau wird an der Oberfläche geladen und nach unterschiedlichen Verfahren untersucht.

1. INTRODUCTION

Recently, a growing number of engineering works are being built or designed in weak rocks. Such undertakings require a proper knowledge of the properties of the rock mass comprising the intact rock itself and possible existing features such as faults, fractures, bedding planes, etc. Sections 2 and 3 of this paper present some available data concerning the deformability of some weak rocks together with a discussion on influencing factors. In section 4 comments are made on the problems associated with the numerical modeling of fractured rocks with relevance to fractured weak rocks and results are presented for loads on a hypothetical foundation. In this analysis, various analytical techniques modeling the fractured medium were used either by means of an equivalent continuum or a medium composed of discrete blocks.

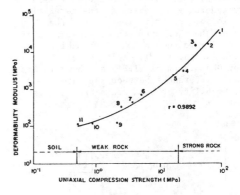

Fig. 1 - Deformability Modulus of Weak Sandstones (Dobereiner and de Freitas, 1986)

2. DEFORMABILITY PROPERTIES OF INTACT WEAK ROCKS

Special relevance is made in this section on the properties of weak sandstones. Typical results of the deformability moduli of weak sandstones are shown in Fig. 1 (Dobereiner and de Freitas, 1986). Each value represents the average of at least five tests and they were obtained on saturated samples tested in uniaxial compression. The range of the tangent modulus of deformability (E) measured before the onset of dilatancy varied between 100 and 3000 MPa. On average, the value obtained for the correlation between the deformability modulus and the saturated uniaxial compressive strength (σ_c) was $E = 140 \, \sigma_c$. This value is lower than the 200 proposed for weak rocks in general by Rocha(1975). It also produces a lower modulus ratio than the one defined for weak rocks by Deere and Miller (1966). The Poisson's ratio obtained also for weak sandstones is in the range .27 - .40, calculated for stress levels below the onset of dilatancy.

The lateral, axial and volumetric strain typical of the weak sandstones tested in uniaxial compression is illustrated in Fig. 2. The plot of volumetric strain records an initial decrease in

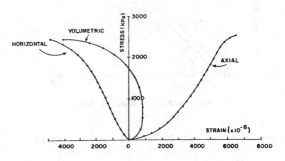

Fig. 2 - Volumetric, axial and lateral strain of weak sandstones (Dobereiner and de Freitas, 1986)

volume followed by an increase. This increase in the volume of the sample starts at low levels of stress, often at about one third of the peak compressive strength. If those stress-strain relations are compared with those for strong rocks as described by Goodman (1980) and Jaeger and Cook (1976), some discrepances can be observed. There is no marked linear relationship between stress and strain even in the pre-peak stage. Typically an inelastic concave upwards stress-strain section is observed, and also that the rate of lateral strain begins to increase, relative to the rate of axial strain, at very low stresses.

The dilatancy observed in the volumetric strain curve of Fig. 2, is probably caused by the beginning of the formation of microcracks inside the most critically stressed portions of the specimens.

It is clear that at stress levels above dilatancy (sometimes at less than one third of peak strength), the strains are inelastic and non recoverable. Therefore it is advisable to use volumetric strain curves as a standard procedure indeformability tests, and to determine the deformability modulus at stress levels below dilatancy (Dobereiner and O- liveira, 1986 or Dobereiner and Dyke, 1986). At least the deformability modulus should not be obtained at high stress levels as there is the certainty of non-recoverable and inelastic strains. At stress levels below the onset of dilatancy the strains are probably a combination of elastic and non-elastic strains due to the closing of pores and fissures, but they should at least be partially recoverable.

3. DEFORMABILITY OF WEAK ROCK MASSES.

The influence of discontinuities on the rock mass deformability generally decreases as the material gets weaker. In some cases in weak rocks, the deformability of the rock mass can be considered of the same order of magnitude as the deformability of the intact rock (Rocha, 1975). This is well illustrated by Okamoto et al (1981) as shown in Fig. 3. These small differences between laboratory and "in-situ" deformability of weak rocks are the result

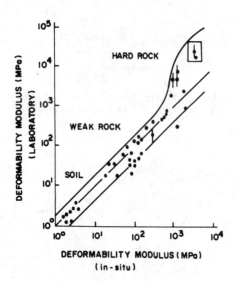

Fig. 3 - Deformability Modulus of Weak Rocks; in situ and in the laboratory (Okamoto et al, 1981)

of minor facturing, as observed in some weak rock masses (Dobereiner, 1984), or due to the plasticity of these materials and/or the fact that the fractures are infilled with material which has a similar strength to the intact rock. It has even mentioned by Rocha (1975) that in certain cases the characterization of the deformability in weak rock masses can be made based on laboratory tests.

Recently the study of hard rock joint deformation has shown remarkable developments, as described by

Kulhawy (1978), Bandis et al (1981), Bandis et al (1983), Barton et al (1985), Leichnitz (1985), Herdocia (1985), Maki (1985), Sun et al (1985) and Barton (1986). However, very little has been done on the deformability of joints in weak rocks. Yoshinaka and Yamabe (1986) for example have studied the joint deformation behaviour of a weak welded-tuff (Ohaya-Stone, Japan).

In relation to the anisotropy in joint deformability, Bandis et al (1983), testing weathered joints in sandstone and dolerite obtained Kn/Ks relations with lower values than those for fresh surfaces. By contrast, weathered limestone joints showed higher Kn/Ks ratios than for fresh surfaces. These examples show that the detailed knowledge of the characteristics of deformation of single joints in weak rocks is still not well known and only when these are studied more fully can a better understanding of weak rock masses be achieved.

4. MODELING OF WEAK ROCK MASS

As described in sections 2 and 3, the deformability of intact weak rocks can be high. Although very litle data concerning the deformability of joints in weak rocks is available today, it if reasonable to assume that the deformability of the joints can be comparable to the one of the intact rock. Therefore, the proper modeling of the deformational behaviour of weak rock masses should take into consideration the basic characteristics of the intact rock and the joints. This can be achieved in two ways either by means of a so called "equivalent", elastic, continuum (Amadei, 1983 among others) or by means a discrete representation of the discontinuities present in the rock mass. In this section, some results concerning idealized problems on a blocky type foundation are analysed either by means of the above mentioned "equivalent" continuum analysis or by means of discrete type modeling. Furthermore, the parameters used in the analysis related to the deformability of the intact rock and the joints, are in the range of the available data for weak rock masses. In the present study, only linear stress-strain (displacements) were analysed, and further analysis should incorporate more representative relations as reported in the literature. (Bandis et al, 1985).

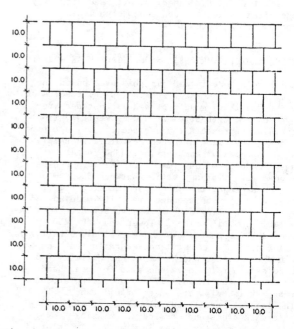

Fig. 4 - Geometry analysed

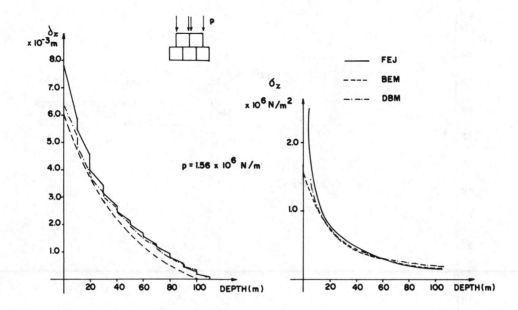

Fig. 5 - Distributed loads Kn/Ks = 1; FEJ, BEM and DBM

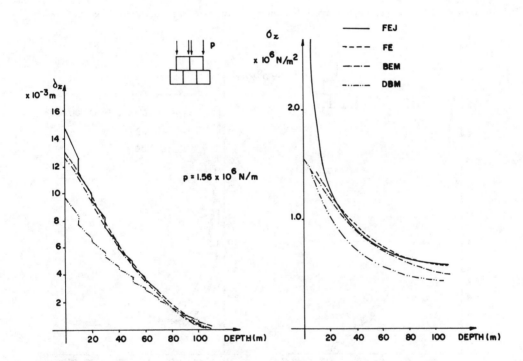

Fig. 6 - Distributed loads Kn/Ks = 100; FEJ, FE; BEM and DBM

The discontinuities present in the rock mass may introduce deformation mechanisms such as separation, slip and large rotations which the standard methods based on the theory of elasticity of small strains are not able to simulate (Bray, 1967). Hence, apart from representing a fractured rock mass by an equivalent continuum, special numerical techniques have been developed in order to properly model mechanisms such as slip, separations and large rotations. Previously, results with the discrete modeling of the fractured rock mass were obtained solely by means of the Discrete Block Model (DBM) (Vargas, 1982 and Vargas, 1985). In the present work, the

same geometry was analysed by means of the Finite Element Method incorporating joint elements (FEJ) (Hittinger and Goodman, 1978 and Millan, 1987). The DBM formulation used incorporates both the deformability of the intact rock and the deformability of the joints. Only one type of geometry has been considered (Fig. 4), one containing one persistent and one non-persistent joint sets. The basic properties for the joint sets and the intact rock are:

modulus of elasticity of intact rock (E)=10^4MPa
Poisson's modulus of intact rock = .2
joint normal and shear stiffness(Kn and Ks)=10^{10}N/m

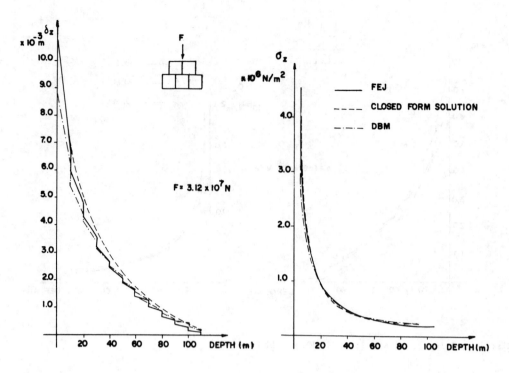

Fig. 7 - Line load Kn/Ks = 1; FEJ, Closed Form Solution (Gerrard and Wardle, 1980) and DBM

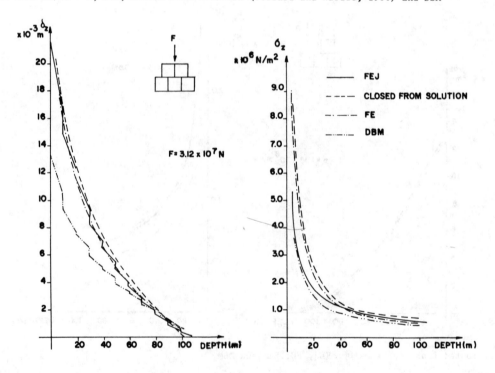

Fig. 8 - Line load Kn/Ks = 100; FEJ, Closed Form Solution (Gerrard and Wardle, 1980), FE and DBM

Normal and shear stiffness are point contact stiffness as required by the DBM (see Vargas, 1982).

The above properties can be used to establish the parameters of the "equivalent" orthotropic continuum. (Singh, 1973) Several methods were be used to analyse continuum condition for the anisotropic media, Finite Elements (FE), Boundary Elements (BEM) and closed form solutions. (Gerrard and Wardle, 1980).

The results obtained in the present paper concern linear and distributed lods on the surface of a stack of blocks as shown in Fig. 4. Figs. 5 - 9 show comparisons between the above described methods of

analysis using DBM and FEJ and solutions for the respective "equivalent" continuum. The presentation of these results refers to:
- vertical displacements (δ_z) along the vertical centerline of the model.
- vertical stresses (σ_z) along vertical centerline of the model.

Here, ratios of Kn/Ks = 1 and 100 are considered. If may be noticed that as the ratio Kn/Ks grows, also grows the discrepancy both in displacements and stresses between the various continuum models and the solutions by DBM and FEJ. This difference appears to be more pronounced for the case of linear loads than for distributed loads. The solutions obtained from FEJ appear to be closer to the continuum solutions than the ones obtained with the DEM. This may be due to factors such as different discretization patterns or to the non allowance for rotation in the FEJ. These factors should be further clarified. Fig. 9 shows the distorced shapes of the two load bearing blocks (line loads and Kn/Ks = 100) both in continuum and discontinuum modeling. Fig. 9 also shows vertical displacements also for Kn/Ks = 100 and line loads along horizontal surfaces at different depths. Discrepancies between continuum and discontinuum arise mainly in the region next to centerline and the surface.

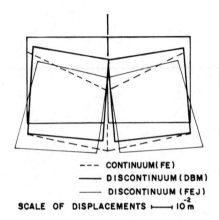

--- CONTINUUM (FE)
— DISCONTINUUM (DBM)
— DISCONTINUUM (FEJ)
SCALE OF DISPLACEMENTS ⊢———⊣ 10^{-2} m

Fig. 9 - Distorted Geometry of Load Bearing Blocks

5. CONCLUSIONS

Various considerations have been made in this paper concerning the modeling of the deformational behaviour of weak rock masses. From the experimental standpoint there is a strong need to establish parameters especially concerning joints in weak rocks and the applicability of models such as the one described by Bandis et al (1985) could be verified for weak rocks. From the numerical analysis standpoint, such models should be implemented and also studies should be made in order to establish the most appropriate techniques to properly represent mechanisms that arise during deformation of the rock mass.

REFERENCES

Amadei, B. 1984. Rock Anisotropy and the Theory of Stress Measurement. Springer Verlag, Berlin.
Bandis, S.; Lumsden, A.C.; Barton, N.R. 1981. Experimental Studies of Scale Effects on the Shear Behaviour of Rock Joints". Int. Jour. Rock Mech. Min. Sci. Vol. 18 pp. 1-21.
Bandis, S.C.; Lumsden, A.C.; Barton, N.R. 1983. Fundamentals of Rock Joint Deformation. Int. Jour. Rock Mech. Min. Sci. Vol. 20, pp. 249-268.
Bandis, S; Barton, N. & Christianson, M. 1985. Application of a New Numerical Model of Joint Behaviour to Rock Mechanics Problems. Proc. Symp. (ISRM) on Fundamentals of Rock Joints, Bjorklidden, Sweden.
Barton, N. 1986. Deformation phenomena in jointed rock. Géotechnique, vol. 36, No. 2, pp. 147-167.
Barton, N.R.; Bandis, S.C.; Bakhtar, K.; 1985. Strength, Deformation and Conductivity Coupling of Rock Joints. Int. Jour. Rock. Mech. Min. Sci, Vol. 22, pp. 121-140.
Bray, J. 1967. A Study of Jointed and Fractured Rock. Part 1: Fracture Patterns and their Failure Characteristics. Rock Mechanics and Engineering Geology. 5(4): 117-136.

Deere, D.V. and Miller, R.P. 1966. Engineering Classification and Index Propertia of Intact Rock. Technical report NAFNL-TR-65-116, US AIR FORCE, Weapons Lab, Kirtland Air Force Base, New Mexico.
Dobereiner, L. 1984. Engineering Geology of Weak Sandstones. Ph.D. Thesis, University of London, 471 p.
Dobereiner, L. and Dyke, C. 1986. Características de deformabilidade de arenitos em função da variação de umidade na rocha. II South American Rock Mechanics Symp. Vol. 2. pp. 57-66, Porto Alegre.
Dobereiner, L. and de Freitas, M.H. 1986. Geotechnical Properties of Weak Sandstones. Geotechnique, Vol. 36 - N1. pp. 79-94.
Dobereiner, L. and Oliveira, R. 1986. Site investigations on Weak Sandstone. 5th International IAEG Congress, Vol. 2, pp. 411-422, Buenos Aires.
Gerrard, G. & Wardle, L. 1980. Solutions for Linear Loads and Generalized Strip Loads. Applied Within Orthorrombic Media. Publ. 31. Div. of Applied Geomechanics, CSIRO. Australia.
Goodman. 1980. Introduction to Rock Mechanics Publ., New York; John Wiley and Sons, pp. 1-478.
Herdocia, A. 1985. Direct shear tests of artificial joints. Proc. Int. Symp. on Fundamentals of Rock Joints, pp. 123-132, Sweden.
Hittinger, M. and Goodman, R. 1978. JTROCK a Computer Program for Stress Analysis of Two-Dimensional, Discontinuous Rock Masses. Res. Report No. UCB/GT/78-04. University of California, Berkeley.
Jaeger, J.C. and Cook, N.C.W. 1976. Fundamentals of Rock Mechanics, 3rd edn, pp-1-593 - London: Chapman and Hall.
Kulhawy, F. 1978. Geomechanical Model for Rock. J. Geot. Eng. Div. ASCE, (111) n7.
Leichnitz, W. 1985. Mechanical properties of Rock joints. Int. Jour. Rock. Mech. Min. Sci. Vol. 22, pp. 313-321.
Maki, K. 1985. Shear Strength and Stiffness of Weakness planes by controlled facturing of intact specimens. Proc. Int. Symp. on Fundamentals of Rock joints, pp. 133-142, Sweden.
Millan, J.P. 1987. Numerical Analysis of Fractured Rock Masses by Means of the Finite Element Method (in Portuguese). MSc. thesis, Catholic Univ. Rio de Janeiro.
Okamoto, R., Kojima, K.; Yoshiaka. 1981. Distribution and Engineering Properties of Weak Rocks in Japan. Proc. Int. Symp. on Weak Rock, Vol. 5, pp. 89-103.
Rocha, M. 1975. Alguns fatores relativos a mecânica das rochas dos materiais de baixa resistência. Proc. 5th Pan Am. Congr. Soil Mech. Fd. Engng, Buenos Aires.
Singh, B. 1973. Continuum Characterization of Jointed Rock Masses. Part 1: The Constitutive Equations; Part 2: Significance of Low Shear Modulus. Int. J. Rock. Mech. Min. Sci., 10: 319-349.
Sun, Z.; Gerrard, C.; Stephansson, O. 1985. Rock Joint Compliance tests for compression on shear Loads. Int. Jour. Rock. Mech. Min. Sci., Vol. 22, pp. 197-213.
Vargas, E. 1982. Development and Application of Numerical Models to Simulate the Behaviour of Fractured Rock Masses. Ph.D. Thesis, University of London.
Vargas, E. 1985. Continuum and Discontinuum Modeling of Some Blocky Type Foundation Problems. Proc. Symp. (ISRM) on Fundamentals of Rock Joints, Bjorklidden, Sweden.
Yoshinaga, R.; Yamabe, T. 1986. Joint Stiffness and the Deformation Behaviour of Discontinuous Rock. Int. Jour. Rock Mech. Min. Sci. Vol. 23. pp. 19-28.

Verification by the finite element method of the influences on the roof conditions in longwall faces

Vérification par la méthode des éléments finis de l'influence du soutènement marchant sur les conditions du toit dans les exploitations 'longwall'
Die Bestätigung der Einflüsse auf den Zustand des Hangenden im Streb durch eine Finite-Element-Berechnung

A.VERVOORT, Candidate for Ph.D. degree, Faculté Polytechnique de Mons, Belgium
J.-F.THIMUS, Senior Assistant, Université Catholique de Louvain, Louvain-la-Neuve, Belgium
J.BRYCH, Professor, Katholieke Universiteit Leuven, Belgium
O.DE CROMBRUGGHE, Professor, Katholieke Universiteit Leuven, Belgium
E.LOUSBERG, Professor, Université Catholique de Louvain, Louvain-la-Neuve, Belgium

ABSTRACT: By observations in longwall faces, different parameters, related to the powered roof support and determining the occurrence of roof cavities or fall outs, above and before the powered support, are defined. Some of these influences are now also verified by a finite element model. So, we affirm that the probability to observe roof cavities increase with an increasing unsupported distance, an increasing distance between the face and the first legs, and with a lower vertical force of the support : all these parameters have a direct influence on the extent of a zone of insufficient pressure in the first beds of the roof above and before the support.

RESUME: En nous basant sur des observations en taille, nous avons constaté l'influence de différents paramètres, liés au soutènement marchant, sur les chutes de toit. Ces influences sont également vérifiées par un modèle d'éléments finis. Ainsi, nous confirmons que la fréquence d'observation d'une chute au-dessus ou en avant d'un élément de soutènement marchant augmente pour une plus grande distance non-soutenue, pour une plus grande distance front-première ligne d'étançons et pour une force plus faible du soutènement : tous ces paramètres ont une influence directe sur l'étendue d'une zone détendue (compression insuffisante) dans les premiers bancs du toit au-dessus et en avant du soutènement marchant.

ZUSAMMENFASSUNG: Durch betriebliche Strebbeobachtungen, hat man verschiedene Einflussgrösse, verbunden mit dem Ausbau, auf die Anwesenheit von Ausbrüche festgestellt. Diese Einflüsse wurden weiter bestätigt durch ein endiges Elementenmodell. Wir bestätigen also dass die Warscheinligkeit um eine Ausbrüche fest zustellen zunimmt mit einem grösseren nicht-unterstützten Abstand, einem grösseren Abstand Abbaustoss-erste Stempelreihe und mit einer kleineren Ausbaukraft : all diese Parameter haben einen direkten Einfluss auf die Ausgestrecktheit von einer Zone von unzureichendem Druck in das Hangende über und vor dem Ausbau.

1. INTRODUCTION

In the roof behaviour of longwall faces, there are two contradictory aspects :

- on the one hand, due to productivity and security reasons, the roof in the longwall has to be kept as intact as possible; so the opening of fractures, the forming of steps, roof cavities,... above and before the support have to be limited;

- on the other hand, behind the roof support, the caving of the roof has to happen immediately.

The longwall process already induces ahead of the face a lot of different fractures which are very advantageous for the caving of the roof behind the support, but not at all for having good roof conditions above and before the support. One of the principal tasks of the powered roof support is to establish a satisfactory stress distribution in the roof (exposed by the abutment), so that no fractures are opened and consequently no roof cavities are formed.

The presence of roof cavities is still an important problem in Belgian coal mines, where the nature of the roof is mainly shale and sandstone : generally, above or before 10 to 50 % of all the support elements, a cavity is present.

To learn more about the influences of the roof support on the presence of roof cavities, a lot of detailed statistical examinations of underground observations were done : these examinations have shown a significant influence of the unsupported distance, of the distance face - first legs and of the total vertical force of the powered roof support, on the presence of roof cavities and on their

height [1, 2]. Apart of these main influence parameters, related to the roof support, there are also a lot of indirect parameters, as the distance between the face and the roofbar tip and as the pressure of the support on the roof [1, 2].

Simultaneously with these observations, a finite element model was established to simulate the abutment process. Although, an accurate simulation of the roof behaviour in longwall faces and the forming of cavities is a very complicated problem, we think that by making the right hypothesis and by simplifying the problem in a correct way, we have succeeded - with an economically justified model - in verifying the relations between roof support and roof conditions; also, this simulation teaches us more about the process of forming roof cavities.

2. FINITE ELEMENT MODEL

We simulate the longwall process by considering only a small part of the whole massif (height : 10 meters, length : 26 meters & width : 1 meter), but we apply on it the strata loading pressure, due to the abutment process (see figure 1). The model is composed of two main parts : firstly, the massif with the roof, the coal seam and the floor, and secondly the powered roof support [3].

We consider the longwall and the rock massif around it, at a certain moment : we neglect the influence of the time factor and of dynamic phenomenons, which are really present in coal mines. Also, we suppose that all the fractures, already formed ahead of the face, are closed : so, the beds of the roof, the coal seam and the floor can be considered as continuous layers. Further, we also suppose that these

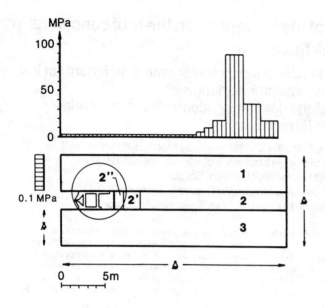

: complete fixation

Number	Material	E (MPa)	γ	δ (t/m³)
1	roof	6000	0.2	2.70
2	coal	3000	0.4	1.40
2'	fractured coal	2250	0.45	1.30
2"	fractured coal	1500	0.45	1.20
3	floor	5000	0.3	2.40
4	steel	210000	0.3	7.85
4'	spring	1500	0.01	0.00
4"	loose rock	400	0.45	0.00

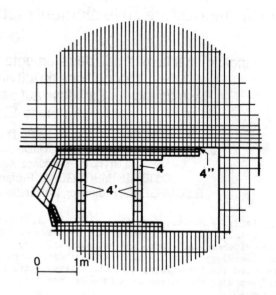

Figure 1 : Finite Element Model for the longwall and the rock massif around it.

beds are homogeneous. Based on these hypothesis, we do not consider a visco-elastic behaviour nor a visco-elastoplastic (influence of the time) nor a discontinuous one (presence of fractures). Finally, we have - mainly for economic and practical reasons- chosen for a linearly elastic model instead of a plastic one. [4]

In a first step, we calculate the stresses in the elements and the displacements of the nodes for the whole model, in which the elements of the roof in the first two meters from the face are squares of 0.25 x 0.25 meter. In a second step, we consider - mainly for economic reasons - a reduced model (length : 8 meters & height : 6 meters), on which we impose the displacements, calculated by the whole model. We now calculate the stresses, for the first two meters of the roof from the face, in elements of 0.10 x 0.10 meter.

2.1 The massif (roof, seam and floor) around the longwall

As already mentioned, we consider only a small part of the whole rock mass; the tickness of the roof and also of the floor is 4 meters, and that of the seam 2 meters (see figure 1). The total length of the model is 26 meters. The length of the roof in the longwall is 6 meters. Notice, that in fact all these dimensions are relative. We consider for the roof and the floor one homogeneous bed; we do not decrease their modulus of elasticity due to the presence of fractures. In the coal seam, we take one half of the modulus of elasticity for the first meter from the face, and for the next two meters three quarters : the main reason is that the coal is more fractured than the roof and the floor; a supplementary reason is that doing so, the contact roof-seam and seam-floor would not be too firm at the face.

2.2 The powered roof support

The support modelised is a modern shield support with two rows of vertical legs, an extensible roofbar, ... It is composed of 3 materials (see figure 1) :

✶ Steel : most of the elements are composed of this material.
✶ Spring : some explorative calculations have shown clearly (and confirm in this way underground examinations) that the effect of a support element is more than the introduction of a force in the roof. Of equal importance is the resistance of the support to limit the movement of the roof. Under an overload, the support element is deformed elastically as long as the pressure is less than the yield pressure. Therefore, we introduce 8 elements (spring-effect) in the model of the legs, so that the whole support model is compressing vertically in the same way as a real support element.
✶ Rock : we very often observe in coal faces a layer (normally ± 0.1 meter) of loose rock on the roof-bars. In our model, there is - on the whole length of the roofbar - a layer of 0.10 meter.

We apply the settling force of the support, which is the result of the original hydraulic pressure in the legs, in 6 nodes (6 up and 6 down). Notice that apart from this force, a possible elastic deformation of the legs, caused by an overload, results in a supplementary pressure on the roof (effect of the resistance of the support). The total force is limited by the yielding pressure in the legs. We apply the force of the extension bar in two nodes.

We will calculate the stresses in the roof, for 3 values of the three main influence parameters considered (see figure 2) :

* unsupported distance : 0.5, 1.0 and 1.5 meters. By suppressing elements of the rock layer, we increase this parameter;
* distance face - first row of legs : 1.5, 2.0 and 2.5 meters. We only modify the position of the first legs;
* total vertical force : 540, 1500 and 2400 kN. We increase or decrease proportionally the forces in all the different nodes.

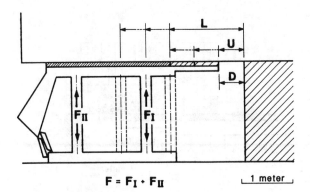

$$F = F_I + F_{II}$$

U : unsupported distance
L : distance face - first row of legs
D : distance face - roofbar tip
F_I : force in the first row of legs
F_{II} : force in the second row of legs
F : total force

Figure 2 : The different main influence parameters.

Notice that we can modify these 3 parameters independently of the others; we can also make a combination of values of these parameters, which rarely occur in the underground or we can extrapolate them out of the area of today's values.

2.3 Boundary conditions

For this two-dimensional model, a hypothesis of plane strain is valid. As already mentioned, we apply on the upper side of the model the strata loading pressure [5, 6], distributed by the abutment process (see figure 1) : far before the face, the initial pressure (18 MPa at more than 17 meters before the face) is present; shorter to the face, the pressure increases and reaches a maximum of 5 times the initial pressure. Still shorter to the face, the pressure decreases and above the face the pressure is about 2 MPa. Notice again, that all the dimensions are in fact relative.

On the side of the massif (at right) and on the left side under the caving zone, we do not apply horizontal stresses, but we fix the nodes completely in the space : under the elastic hypothesis, the instantaneous displacements are equal to zero, but the horizontal stresses are not so exactly known as these displacements. Also, we fix the nodes of the lower side of the model. We simulate the "silo" effect of the caving zone by application of an horizontal pressure (0.1 MPa) [7].

3. VERIFICATION OF THE 3 MAIN PARAMETERS OF INFLUENCE

We are going to determine the extension of the relaxed zones (zones of insufficient pressure) in the roof; the extent of these zones will be the base of comparison between the influences of different values for the support parameters on the occurrence of roof cavities. We are not going to determine the zones in the roof and the conditions in which fractures can be formed; but we will determine the zones

where existing fractures (for example already formed ahead of the face) can be opened and cavities will be formed. As a criterion of the relaxed zones, we will consider an insufficient value of the smallest principal stress in any element. This smallest principal stress must be a sufficient pressure and in any case no traction, so that existing or potential fissures cannot be opened [3].

3.1 Influence of the unsupported distance

By increasing the unsupported distance (from 0.5 to 1.5 meter) the extent of the relaxed zone increases clearly (see figure 3). Although the smallest principal stresses act on nearly horizontal planes ($\pm$ 20 to 30 degrees), there is also a clear decrease of the stresses acting on vertical and inclined planes. Notice, that this parameter has an influence as well as on the height as on the length of the relaxed zone (for example with a principal stress less than 0.3 MPa) :

Unsupported distance (meter)	Height (meter)	Length (meter)
0.5	0.2	0.6
1.0	0.3	0.8
1.5	0.5	1.3

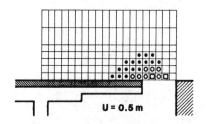

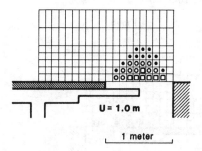

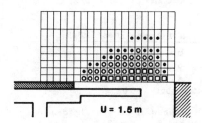

Smallest principal stress (MPa)

□	<0.1
○	0.1...0.3
•	0.3...0.5
no symbol	>0.5

Figure 3 : Influence of the unsupported distance U (L = 2.0 m; F = 2400 kN).

The relaxed zone is clearly situated above the unsupported roof. Roughly, we can consider the relaxed zone as the half of a circle with a diameter equal to the unsupported distance (we will deal with this item later).

3.2 Influence of the total vertical force

For a small unsupported distance (0.5 meter), we modify the total vertical force of the support between 540 and 2400 kN (see figure 4). The influence of this modification is a lot less important than that of the unsupported distance, established in previous paragraph; the relaxed zone is clearly influenced by the small unsupported distance. By increasing the vertical force from 540 to 2400 kN, the height and the length of the relaxed zone are (with for example a principal stress less than 0.5 MPa) decreased only by 0.1 meter. In 3.4, we will deal with the influence of the vertical force for more unfavourable values of the other parameters.

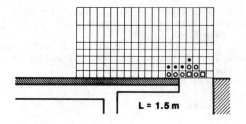

L = 1.5 m

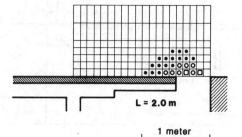

L = 2.0 m

1 meter

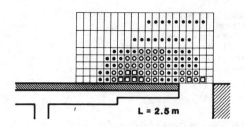

L = 2.5 m

Smallest principal stress (MPa)

□	< 0.1
o	0.1...0.3
•	0.3...0.5
no symbol	> 0.5

Figure 5 : Influence of the distance face-first row of legs L (U = 0.5 m; F = 2400 kN).

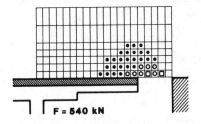

F = 540 kN

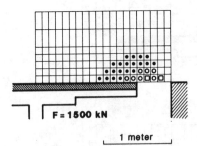

F = 1500 kN

1 meter

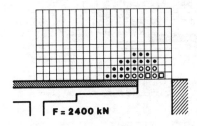

F = 2400 kN

Smallest principal stress (MPa)

□	< 0.1
o	0.1...0.3
•	0.3...0.5
no symbol	> 0.5

Figure 4 : Influence of the total force F (U = 0.5 m; L = 2.0 m).

3.3 Influence of the distance from the first row of legs to the face

At the moment that the distance between the roof-bar tip and the first legs becomes very important, the stress distribution in the roof is strongly influenced by this parameter (see figure 5).
An increase of the distance face-first legs from 1.5 to 2.0 meters has a less influence on the extent

of the relaxed zone. Probably, when the distance between the first legs and the bartip is large the rigidity of the bar gets a big influence (maybe, the bar in this model is weaker than in reality).

3.4 Influence of the 3 main parameters

The comparison of 18 different combinations (3 values for the unsupported distance and the vertical force and 2 values for the distance face-first legs) confirm the conclusions already established. In figure 6, we indicate the extent of the relaxed zone, with in each element of it a principal stress less than 0.3 MPa :

* The relaxed zone is mainly influenced by the unsupported distance or the distance between the face and the first contact point "roof-bar" (direct or indirect). The length of this relaxed zone is approximately the unsupported distance and the height is nearly half of it. So we can conclude that if we want to limit the probability of having roof cavities higher than h meter (for example 0.3 meter), we have to decrease the unsupported distance to minimal 2 h (for example 0.6 meter). Or in other words, if the first contact point is normally at 2 h from the face, we can accept a lot of roof cavities, with a height equal to h.

* An increase of the vertical force has little or no influence on the length of the relaxed zone, but

F = 540 kN F = 1500 kN F = 2400 kN

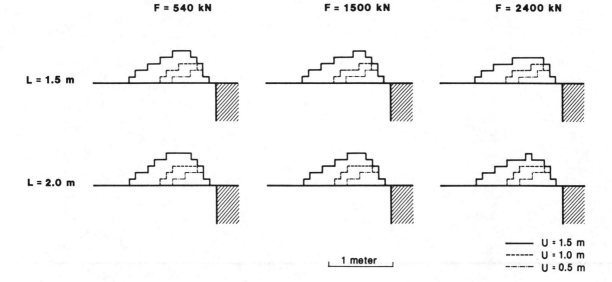

L = 1.5 m

L = 2.0 m

1 meter

——— U = 1.5 m
------ U = 1.0 m
—·—· U = 0.5 m

Figure 6 : Influence of the three main parameters on the extension of the relaxed zone
with a principal stress less than 0.3 MPa.

has still an influence on the height of it.
* An increase of the distance face - first legs from
1.5 to 2.0 meters has a small influence on the
height of the relaxed zone.

We conclude that to obtain a satisfactory stress distribution in the roof of a longwall face and consequently to reduce the probability of observing roof
cavities above and before the powered support, we
have mainly to limit the unsupported distance. We
can realise this by advancing the roof support as
near as possible to the face or by making the roofbar longer, and by coinciding the contact point with
the roofbar tip. We also have (but this has less
influence) to achieve that the first legs are close
to the face and that the settling force of the support is high.

4. CONCLUSIONS

Notwithstanding all the difficulties occuring in
simulating the complex abutment process in longwall
faces, we have succeeded by making realistic suppositions and simplifications, in realising it. We
have not only verified the relations, determined by
observations in coal faces, between the support parameters and the occurrence of roof cavities (fall
outs), but we have also proposed a possible explanation of the process of producing roof cavities :
when there is an insufficient pressure (a relaxed
zone) in the first beds of the roof above and before
the powered support, the fractures which are already
formed ahead of the face, are opened and roof cavities are produced. In these circumstances, the
relaxed zone has a bigger extension with a greater
unsupported distance, a greater distance face -
first legs and a less vertical force of the support.

It is the opinion of the authors, that by this finite element model still a lot of problems can be
researched and resolved; of course, a short relationship with underground observations will always
be necessary. We hope in the future not only to
verify afterwards the determined relations in the
underground, but also to predict the roof conditions, for example of other coal basins and for new
roof support types. So we hope that such a model
will eventually become useful in designing roof
supports and in determining the right handling of
them.

Notice that probably for researching other roof
conditions than the presence of roof cavities, other
models or suppositions are more useful : for instance in researching the forming of roof steps, discontinuous models have to be chosen, and in researching
convergence and other dynamic phenomenons, visco-
elastic and visco-elastoplastic models are more
useful.

ACKNOWLEDGEMENTS

The authors acknowledge greatly ir J.P. Rossion of
Laboratoire du Génie Civil at Louvain-la-Neuve for
his consulting to use Finite Element program.

REFERENCES

[1] O. Jacobi : Praxis der Gebirgsbeherrschung.
Verlag Glückauf GmBH/Essen/1981

[2] A. Vervoort : Interprétation systématique du
comportement du toit dans la longue taille
(objectifs et méthode de travail).
Journée d'études (30 mai 1985) : Mécanismes
de rupture autour des ouvrages souterrains -
GBMR-BVRM - Louvain-la-Neuve

[3] A. Vervoort : Le comportement du toit dans la
longue taille : Vérification des influences de
paramètres, liés au soutènement marchant, par
la méthode des éléments finis.
Publication n° 227 du Centre de Formation Post-
Universitaire en Mécanique de roches (F.P.Ms./
Mons/Prof. J. Brych/15-10-86)

[4] Programme "Elfinis-Plan".
Université Catholique de Louvain; Laboratoire
du Génie Civil

[5] C. Jeger & M. Lory : Mécanique du front de taille.
Charbonnages de France/Documents Techniques/ N 4-
1970

[6] G. Everling : Prediction and estimation of the
strata loading pressure in coal mines.
Fifth International Conference on the strata
pressure/1972/London

[7] E. Tincelin : Le choix des méthodes d'exploitation dans les couches en plateure.
Annales des Mines/Avril 1983.

Seismic response analysis of rock masses by the Finite Element Method
Analyse de la réponse séismique de la masse rocheuse par la méthode d'éléments finis
Erdbebenreaktionsanalyse des Gesteins mittels der Finite Element Methode

WANG QIZHENG, Tsinghua University, Beijing, People's Republic of China
ZENG ZHAOYANG, Tsinghua University, Beijing, People's Republic of China
LI WEIXIAN, Tsinghua University, People's Republic of China

ABSTRACT: The seismic responses of abutment rock mass of two arch dams were analysed by F.E.M. The influence of boundary conditions of the calculation model on the dynamic response of the abutment were studied. The results agreed quite well with those of in-situ measurements and model tests.

RESUME: La réponse sismique de la foudation d'aboutement en deux barrages en voûte est analysée par la méthode d'element fini. L'influence des conditions frontières du calcul modelé sur les propriétés dynamiques de la masse de rocke a été étudiés. Les résultats du mesurage de sur place correspond bien à celui de l'essai modelé.

ZUSAMMENFASSUNG: Die Erdbebenreaktion der seitlichen Gesteinsauflager in zwei Bogenstaumauern wird durch F.E.M untersucht. Der Einfluss der Randbedingungen des numerischen Modells auf die dynamische Eigenschaft wird bewertet. Die Resultate stimmen mit der Feldsvermessungen und der Modellversuchen ueberein.

INTRODUCTION

Engineering structures such as dams, bridges, mines, etc., located in areas of high seismic risk must be such designed as to safely resist dynamic excitation by earthquakes. The amplitude of ground motion at the top of a canyon could be 1-3 times greater than that at the toe. Obviously, this amplification behavior is unfavorable for the aseismic stability of the slope of rock mass. Therefore, to study the dynamic behavior and seismic response of rock mass is important and to find out a method to evaluate them is imperative.

This paper conducted a study on the dynamic behavior of rock mass by finite element method (F.E.M), and calculated the seismic responses of abutment foundations for two arch dams. The results were compared with those obtained from in-situ measurements of ambient vibrations and geomechanic model tests.

ANALYSIS METHOD

With certain restraint conditions on the boundary of rock mass given, the canyon can be idealized as an assemblage of finite elements as shown in Fig.1. The applicability of this model depends on the simulation of the boundary, topographic and geological conditions, and on the degree of absorption of the reflected waves. The effects of the boundary on the dynamic behavior of rock mass will be described later.

The equations of motion of a discretized system can be written as

$$M\ddot{V} + C\dot{V} + KV = -Mr\ddot{V}_g$$

where M, C, K are the mass, damping, and stiffness matrices of the system respectively; V is the vector of relative displacement of the nodal point; the vector $\ddot{V}_g$ includes 3 components of earthquake acceleration imposed upon the rigid base, and r is an influence coefficient matrix.

Two typical methods were used to solve the equations. One is the "response spectrum" method. It was assumed that only a limited number of lower modes of the system evaluated by the subspace iteration method was excited and only the maximum value of each component was estimated on the basis of the definition of standard spectral accelerations.

Another procedure for solving the equations is the response history method. The complete history of displacement and acceleration of each degree of freedom can be obtained by a step-by-step direct integration scheme. By selecting the time step Δt properly, the higher modes of the system can be considered.

8-node 3-D solid elements were used. The finite element models are shown in Fig.1 and Fig.2. Model I (Fig.1) is the model of the abutment of Longyangxia arch dam located on the Yellow river in Northwest China. The dam is 178m high and the canyon wall is of nearly the same height as the dam. The main fault F7 and the variation of elastic modulus of the rock with

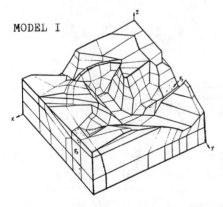

figure 1. Discretization of Longyangxia abutment.

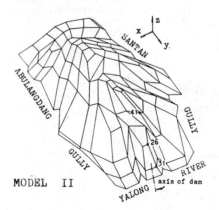

Figure 2. Discretization of left abutment of Ertan dam.

elevation were simulated in the model.Model II (Fig.2) is the model of the left abutment of the Ertan arch dam located in Southwest China. The dam is 240m high and the canyon is much higher than it.

The calculation were performed by using the program SAP5. The details of the formulation are described elsewhere (Bathe & Wilson 1976) and need not be repeated here.

In order to check the results obtained by F.E.M, a comparison with those obtained from in-situ measurements of ambient tests in the sites of Longyangxia and Ertan (Zeng & Li 1985) and from geomechanic model tests. The excitation of the tests were analysed by a stochastic procedure. The transfer function of each point of rock mass was obtained by in-situ measurements or by model tests in which excitation was effected by random or impulsive loading. The power spectral density function such as standard white noise and filtered white noise with predominant frequency (fg) of 2.5,5, and 10Hz were input. Then the seismic response of rock mass can be estimated in the frequency domain.

SEISMIC RESPONSE OF ROCK MASS

Analyses of the two models shown in Fig.1 and 2 show that the natural vibration behavior of rock mass is different from that of usual structures. It has both global and local vibrations. The local vibrations depend on local topographic and geologic conditions of the local site, and, generally, have higher frequencies. While the frequencies of global vibrations are lower. This character causes the closely spaced natural periods and leads to a considerable expense of the computational effort to evaluate higher modes by solving the eigenproblems. According to the evaluations the maximum difference between two adjacent natural periods is about 0.1 seconds, and the minimum one is only 0.001 seconds or so. Furthermore,the influence of the closely spaced modes on the dynamic response of rock mass must be taken into account when the mode superposition is carried out.

For the model I the response spectrum method was used to calculate the seismic responses of the abutments. A standard spectum shown in Fig.3 was used as the input for the x (transversal) and y (longitudinal) components of the ground motion. The first 30 modes of the system were calculated. In view of the closely spaced natural frequencies the complete quadratic combination technique (Wilson et al.,1981) was used for combining modal maxima. The amplification factor of the maximum superposed acceleration along the elevation are plotted in Fig.4. The analysis shows that the first 10 lower modes play an important role in the overall seismic force. It means that because the canyon wall of the dam site is only as high as the dam, the safety of the dam depends mainly on the global stability of the abutment. Thus the global vibration of lower frequencies should be considered.

For the model II, both the response spectrum method and the direct integration method were used. The time step Δt was taken as 0.01 seconds to consider higher modes. The seismic records of Qian'an (NS,1976,China) and El Centro (NS,1940)earthquakes were used. The response spectra of both records for 5 per cent of critical damping are given in Fig.3. The Qian'an

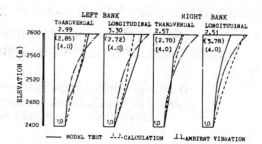

Figure 4. Acceleration amplification factor of the abutments of Longyangxia arch dam

record is shown in Fig.5, the maximum acceleration is 0.1805g. The acceleration histories of node 26,31 and the bed are shown in Fig.6. The response accelerations against elevations are also shown in Fig.6 including the transient distributions and the maximum envelopes. Because the effects of higher modes on the acceleration response were not taken into account, the maximum accelerations calculated by the response spectrum method were much less than those obtained by the direct integration method, though 30 modes were taken. It means that for the topography of Ertan dam site, the abutment is only a local portion in comparison with the entire ridge, even though the dam is 240m high. It seems necessary in this case to take the local vibration into account.

The comparisons with the results obtained by the in-situ ambient vibration and the model tests are shown in Fig.4 and Fig.7. The agreement among the results by all three kinds of methods is fairly satisfactory.

On the basis of the results obtained above. the behavior of seismic response of the rock mass can be summarized as follows:

1. The topography of rock mass has significant effects of amplification on the ground motion. Generally, the higher the elevation is, the greater the response acceleration would be. The magnitude of ampli-

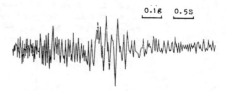

Figure 5. Qian'an record (NS,9th Aug.1976)

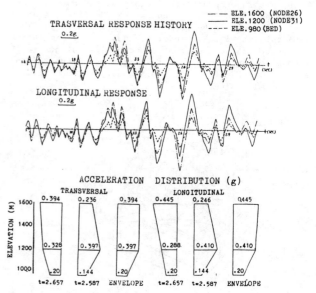

Figure 6. Response history of left abutment of Ertan dam to Qian'an record.

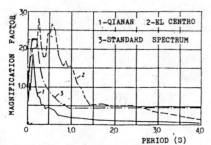

Figure 3. Acceleration response spectrum

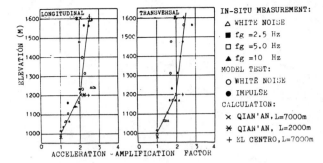

Figure 7. Comparison between the amplification factors obtained by three kinds of method.

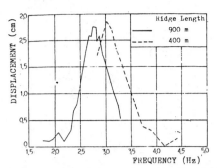

Figure 9. Comparison between the displacement responses for defferent extents of model I

fication considerably depends on the local topographic conditions.For the two calculated dam sites, the maximum response accelerations at the top of the abutments could be 2-4 times as large as that at the bottom.

2. It can be seen clearly from the transient distributions of the response acceleration shown in Fig.6 that the maximum responses of different points do not appear simultaneously. It means that the actual sliding force acting on the rock mass during earthquake would be less than that calculated according to the envelope of the maximum response. Obviously, this would be on the safe side for the aseismic stability of rock mass. On the other hand, the stability of local rock mass should be paid extra attention.

EFFECTS OF BOUNDARY ON DYNAMIC RESPONSE OF ROCK MASS

The extent of rock mass and the boundary conditions in the calculation model must be selected reasonably to reduce the effects of the reflected waves on the dynamic response of rock mass to an acceptable level. 2-D and 3-D models were used to study the effects of the boundary. Different lengths of the ridge were taken and three kinds of restraint conditions were used, i.e., the nodes on the boundary of the model are assumed to be free, sliding(free in the horizontal direction but fixed in the vertical) and fixed.

The natural frequencies of aforementioned cases for the 2-D model shown in Fig.8, which is the section along the left ridge route of the model I, are listed in table 1. It shows that with the extending of the ridge, the natural frequencies for different restraint conditions decrease and get close to each other. When the length of the ridge equals 1900m, they tend to be constant. Thus the effects of the boundary on the dynamic behavior of rock mass can be ignored. A compar-

ison between the displacement responses for different ridge lengths at the top of the left abutment of 3-D model I is shown in Fig.9.It shows that the maximum response and peak frequency increase slightly when the length of the ridge reduces from 900m to 400m. Similarly, for model II, a comparison between the responses when the length of the ridge reduces from 7000m to 2000m is given in Fig.7. The conclusion is much the same.

Therefore, so long as the boundary of the model is far enough from the canyon wall, the dynamic response of rock mass is not seriously affected by the boundary conditions.Thus, for saving the computational effort, the size of the model can be reduced under the prerequisite of satisfying the accuracy of aseismic design.

CONCLUSION

The seismic responses of abutments for two arch dams were analysed. The results agreed quite well with in-situ measurements of ambient vibration and geomechanic model tests. It means that the seismic response of rock mass can be analysed successfully by F.E.M so long as the size and the boundary conditions of the calculation model are reasonably selected.

The dynamic behavior of rock mass is different from that of usual structures. There are not only lower modes of global vibrations but also higher modes of local vibrations, and their natural periods are closely spaced. The slope of rock mass has significant effects of amplification on the ground motion, but the maximum responses of different points are out of phase. These characteristics must be taken into account when carrying out the dynamic analysis of rock mass.

REFERENCES

Bathe,K.J.& E.L.Wilson 1976, Numerical methods in Finite Element Analysis. New Jersey, Prentice-Hall, Inc..
Zeng,Z.Y.& W.X.Li 1985. The application of ambient vibration test to the analysis of the dynamic behavior of mountain valleys. China J. Hydraulic Engineering, 10:54-59.
Wilson,E.L., A.D.Kiureghian & E.P.Bayo 1981. A replacement for the SRSS method in seismic analysis. Earthquake Engineering and Structure Dynamics. 9:187-194.

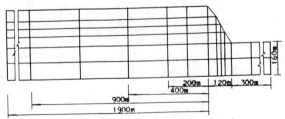

Figure 8. 2-D model of left abutment of model I

Table 1. The natural frequencies (Hz) of rock mass for different boundary conditions

EXTENT OF RIDGE	1900m			900m			400m			200m
RESTRAINT CONDITION	FREE	SLIDE	FIXED	FREE	SLIDE	FIXED	FREE	SLIDE	FIXED	SLIDE
MODE 1	2.10	2.11	2.10	1.90	2.09	2.32	1.90	2.18	2.88	2.29
MODE 2	2.39	2.45	2.50	2.44	2.69	3.12	2.96	3.33	3.60	3.89
MODE 3	2.56	2.57	2.59	2.75	2.80	3.19	3.45	3.78	4.21	4.64

The bolted rockmass as an anisotropic continuum – Material behaviour and design suggestion for rock cavities

La roche armée comme une continuum anisotrope – Comportement du succédané et proposition de dimensionnement pour les excavations

Das geankerte Gebirge als anisotroper Stoff – Materialverhalten und Dimensionierungsvorschlag für den Felshohlraumbau

D.WULLSCHLÄGER, Lehrstuhl für Felsmechanik, Universität Karlsruhe, FRG
O.NATAU, Lehrstuhl für Felsmechanik, Universität Karlsruhe, FRG

ABSTRACT: This paper describes a novel method for the calculation of a systematic bolting around an underground opening. Bolts and rock mass are combined to an anisotropic substitute material. Its parameters are determined in plane model tests and numerical studies. The most important results of these studies concerning the increase of strength and stiffness, anisotropic cohesion etc. due to the bolting are explained. The determined parameters of the substitute are fitted to the geometrical boundary conditions in the surrounding of the opening. At a tunnel configuration it is shown that the hypothesis of the substitute and the application of discreet bolts leads to nearly the same results while dimensioning a bolted carrying ring in the rock mass.

RESUME: Dans cet article on décrit une neuve hypothèse pour le dimensionnement d'un support systématique avec boulons pour les excavations. Le boulonnage et la roche sont considerés comme milieu continué anisotrope, dont les paramètres du succédané sont déterminés dans des essais sur modèles réduits plans et dans des études numériques. Les résultats, les plus importants, des variations des parametres, concernant l'augmentation de la resistance, de l'augmentation de la contrainte, de la cohésion anisotrope et autres comme effet du boulonnage sont expliqués. Les paramètres du succédané sont adaptés aux conditions aux limite géométriques de l'environ de la cavité. On montre dans un example qu'on obtient presque des résultats identiques pour le dimensionnement avec l'hypothèse du succédané qu'avec des boulons discrètes.

ZUSAMMENFASSUNG: In vorliegendem Beitrag wird ein Ansatz zur Dimensionierung einer Systemankerung im Hohlraumbau beschrieben. Ankerung und Gebirge werden zu einem anisotropen Ersatzstoffkontinuum zusammengefaßt. Dessen Parameter werden in ebenen Modellversuchen und numerischen Studien bestimmt. Die wichtigsten Ergebnisse der Parameterstudien, betreffend Festigkeitserhöhung, Steifigkeitszunahme, anisotroper Kohäsion u.a. als Folge der Ankerung werden erläutert. Die ermittelten Ersatzstoffparameter werden an die geometrischen Randbedingungen in der Hohlraumumgebung angepaßt. An einem Beispiel wird gezeigt, daß Ersatzstoffansatz und diskrete Anker bei der Dimensionierung des geankerten Gebirgstragrings zu nahezu identischen Resultaten führen.

1. INTRODUCTION

Rockbolts belong to the most frequently used support elements for underground openings in rock masses. Because of its simple handling and its positive influence to the convergence the fully bonded rockbolt, which is installed untensioned and gets its strain from the deformations of the rock mass, succeeds more and more. Meanwhile, support by bolts after the principle of the carrying ring in the rock mass is drawn up for constructions in squeezing rock and large depths, too. Mostly, the design is done on an empirical point of view. There is no question that a sufficiently dense bolting with an exact dimensioned bolt length is an active support with statical reserves.

In research and industry a lot of work is done on the complete description of the mode of action of a systematic bolting and on the development of a suitable calculation method for some time. Bolt density, bolt length and bolt diameter must be tuned one upon another to contribute most favourably to the increase of the limit load of the carrying ring. Furthermore it is important to develop simple analytical and numerical calculation methods for the use in practice.

For some years a novel hypothesis is followed up by the authors, which is based upon the idea of a substitute material, composed by bolts and rock mass. In this paper, the most important own research results are described and a suggestion for a dimensioning method is submitted which is explained at an example.

2. CONCEPTION FOR THE DIMENSIONING OF A SYSTEMATIC BOLTING

The fundamental conception of the dimensioning method is presented in detail in former papers of the authors (Natau/Leichnitz (1978), Wullschläger/Natau (1984), Natau/Wullschläger (1986)). The bolted rockmass is no longer interpreted in his mechanical properties through its discreet components bolt, mortar and rock mass. Rather it is considered as an anisotropic substitute material in the sense of a continuum, whose smeared material parameters are implemented in an analytical and/or numerical calculation method. Inhomogeneities (joints, schistosity) are considered by efficiently diminished rock mass parameters. The mechanical properties of the substitute material are determined in model tests and by finite element calculations.

While transfering to the opening situation rotationally symmetric geometries and load cases are chosen first. An uniaxial state of stress can be approximated in the immediate surrounding of the opening ($\sigma_t \gg \sigma_r$). Considering a sufficiently high bolt density resp. a great tunnel radius the bolt's deviation from parallelity is small (see fig. 1).

To determine the parameters of the substitute material a cubic element with bolts, transversal arranged to the direction of stress (σ_t resp. σ_z), is used. In the model test arrangement resp. in the numerical calculation the bolted test specimen is hindered in its transverse extension transverse to the bolt direction, corresponding to the situation at the surrounding of a tunnel (plane strain).

Some of the most important results of the model tests and the finite element calculations are described in the following. The stress strain relations of the unbolted and of the bolted rock mass matrix, determined in model tests, show, that the modulus of deformation in load direction is hardly influenced by bolting (figure 2). The increase of the modulus in bolt direction however, is important (not figured here). The transition into the overproportional (plastic) phase takes place relatively smoothly. A strain

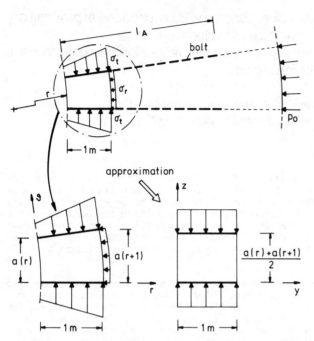

Fig. 1 Idealization of a carrying ring element

hardening effect can be observed, which is more distinct with an increasing bolt density f_b. A further mark is the increase of the ultimate strain by bolting, e.g. the working capacity of the bolted rock mass is considerably higher. The higher limit load σ_{SYST}, caused by the bolting, is evident. In figure 2, σ_{SYST} is the limit load of the bolted matrix, σ_u^* is the correspondent strength of the model material, σ_z is the compression of the test cubes in load direction and f_b is the related bolt density $f_b = A_b/(a_a \cdot a_t)$ with A_b = bolt cross section, a_a = axial bolt spacing, a_t = tangential bolt spacing. In this example a_a is constant.

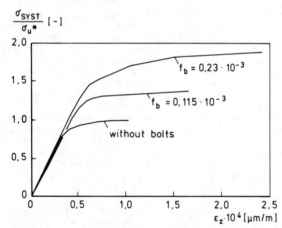

Fig. 2: Example for the stress strain relations in dependence of the bolt density f_b (Natau/Wullschläger, 1986)

The tear of the bolts was the reason of the system failure for densely bolted test cubes. Shear planes in the matrix could be observed just before the failure of the bolts. A sufficient bolt density hinders the early grow of shear planes which lead to the break down of the system. The failure planes were exclusively oriented transverse to the unhindered direction under a constant degree of failure, not in-

fluenced by the bolting. From this it can be assumed that the angle of friction φ is hardly influenced by the bolting under the given boundary conditions.

The central point of the hypothesis of the substitute material is the description of the discreet bolts by higher strength parameters. In several numerical parameter studies with element meshes, which were loaded up to the limit load with defined outer stress like a numerical triaxial test, the qualitative relation, given by Egger (1978), could be confirmed quantitavely (figure 3): The substitute material consisting of bolts and rock mass has the same angle of friction as the unbolted rock mass (see above). Its strength is controlled by the additional cohesion c_b, which is a function of the related bolt density f_b. The hypothesis of the anisotropic cohesion by bolting according to Bjurström (1974) is thereby confirmed.

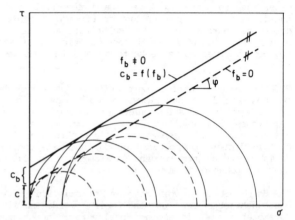

Fig. 3: Change of the rock mass rupture curve as a function of bolting

The system limit load of the bolted rock mass is determined at the named basis element. In the model test the criterion for the limit load is the ultimate tensile stress of the bolts. In the finite element calculation the yield limit of the bolts, their ultimate stress or some intermediate stress can be chosen as the limit load criterion. With the assumption, that the angle of friction of the substitute material is not influenced by the bolting, the additional cohesion c_b as a function of the bolt density follows directly from the limit load σ_{SYST}. The increase of stiffness in load direction and in bolt direction is recorded and is defined as fictitous modulus. The additional cohesion and the fictitous modules are parameters of the bolted rock mass (the substitute material) for the analytical resp. numerical calculation in dependence of the bolt length l_b.

The material behaviour of the substitute resp. of the rock mass is described first without considering any loosening of strength by an elastoplastic material law just as DRUCKER/PRAGER or MOHR/COULOMB. The anisotropic stiffness induced by the bolts can be considered by a suitable anisotropic material law. If there is no such material law available, an isotropic adjustment of the Young´s modulus in bolt direction leads to useful approximations. This could be proven by comparisions with discreet bolts.

Through the odds of a maximum allowed convergence the criterion for the design is given. Calculations by variation with different parameters of the substitute material lead to the dimensioning of the bolt support. With this method substantially simplier structured element meshes are possible in the numerical calculation and the calculation time in the computer is shortened considerably.

3. DIMENSIONING OF A BOLTED CARRYING RING

The above described proceeding was applied to an exemplary numerical calculation. Given a tunnel with circular cross section (r = 5 m) in large depth (t = 800 m, situated in an isotropic stress field (rotational symmetry). The primary stress is $p_0 = \gamma_{rm} \cdot t = 20$ MN/m^2 with $\gamma_{rm} = 0,025$ MN/m^3 (unit weight of the rock mass). The rock mass is assumed to be homogenious and isotropic and follows the DRUCKER/PRAGER material law. Its parameters are:
- Young´s modulus   E = 3.000 MN/m^2
- Poisson´s ratio ν = 0,35
- angle of friction φ = 20°
- cohesion   c = 2,75 MN/m^2
- ultimate strength σ_u = 7,885 MN/m^2

While simulating the fully bonded rockbolts an explicite modelling of the mortar was not taken into consideration. It was postulated, that the mortar strength is equal to the rock mass strength. There are no shotcrete shells or bolt plates, because only the pure bolt effect is of interest in this example. The properties and the dimensions of the bolts orient themselves on those bolts which are used in the German coal mines. Their specifications are:
- Young´s modulus   E = 210.000 MN/m^2
- yield strength σ_y = 400 MN/m^2
- ultimate strenth σ_{ub} = 700 MN/m^2
- ultimate strain ε_u = 14,5%
- diameter d_b = 25 mm
- cross section A_b = 491 mm^2

For this example a bolt length l_A = 5 m $\hat{=}$ radius of the tunnel is chosen. It was the aim of the calculation, to keep a given convergence measure with the aid of the bolts, in this case K = 0,01 · r = 5 cm. For the stabilization of the opening a very dense bolting was expected, considering the relatively low cross sections of the bolts and the low steel qualities. Moreover the relatively high convergences must lead to a straining of the bolts over their yield limit. Therefore a design of the bolt support must follow the limit load principle, i.e. plastic bolt deformations have to be allowed.

The finite element calculations were carried out with a program which allows the activating resp. inactivating of elements and groups of elements at any time step. By this means the simulating of the excavation and the installing of the bolts are possible. Realistic stress and strain fields can be generated. Up to their activating the bolt elements are free of stress. The plane continuum elements for the rock mass are 8-knots-elements with a parabolic displacement formula. The area of excavation is inactivated step by step after the generating of the primary stress field.

Tunnels in large depth and squeezing rock mass require to allow a part of the convergences before installing of the bolts. Otherwise the bolts will be overstressed and they loose their capacity. It is assumed in the exemplary calculation, that at the time of installing 50% of the convergences have passed. The allowable convergence ratio is related to the radius at the time of installing the bolts.

The calculations were made for plane strain conditions. Three variations are described:
a) Opening without support.
b) Systematic bolting with fully bonded bolts, simulated by discreet truss elements, bolt distance at the boundary of the opening a_i = 0,4 m.

The trusses are connected with the rock mass at their element knots in a space of 0,25 m. For variations a and b sectors are choosen as element meshes, which are subdivided in radial direction into 20 (excavation area), 14 (bolted area) and 5 (outer area) elements. In tangential direction 12 elements were arranged. A somewhat more wide-meshed configuration had influenced the results only in a negligible way.
c) Systematic bolting with fully bonded bolts after the hypothesis of the substitute material, bolt distance at the opening boundary a_i = 0,4 m.

The bolts were simulated by superposed 2-D-elements

with differential additional cohesion c_b and differential additional Young´s modulus as a function of the bolt length resp. bolt distance. In a finite element calculation the additional cohesion c_b, caused by the bolts, was determined from the compound´s limit load σ_{SYST} after MOHR/COULOMB:

$$c_b = 0,5 \cdot \sigma_{SYST} \cdot \tan(45^o - \varphi/2) - c_{rm}$$

In this example the criterion for the limit load σ_{SYST} is the ultimate strength σ_{ub} of the bolts. c_b is an exponential function of the tangential bolt spacing which increases with the distance from the opening boundary. It was determined from further calculations at cubic elements and subsequent interpolation. An analytical formula for this relation will be developed in near future. For this type of variation the structure of the element mesh was only a tenth as fine as in variation a and b. This leads to a shortening of the calculation time about more than 80%.

The results of these variations are described in figure 4 (distribution of stress, radius of the plastified zone) and figure 5 (distribution of strain).

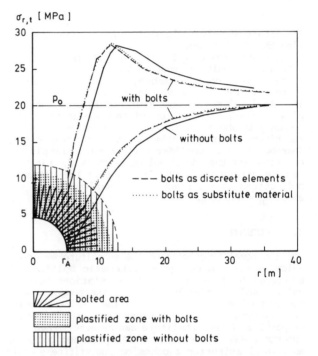

Fig. 4 : Distribution of radial and tangential stresses, radius of the plastified zone

From figure 4 the following results as a consequence of the bolting are derivable:
- The distinct increase of the radial support stresses up to areas behind the bolted zone;
- the increase of the tangential stress on the boundary of the opening up to the value of the limit load σ_{SYST};
- a small decrease of the thickness of the plastified zone about 10%;
- a slight increase of the peak of the tangential stress;
- a smaller deviatoric stress in the elastic region. The curves of the radial and tangential stresses show the excellent conformity of the hypothesis of the substitute material to the conservative modelling of bolts with discreet truss elements.

In figure 5 the radial displacements are layed out which arise after 50% of the convergence resp. after the installing of the bolts. The elastic predeformations of the element mesh before the excavation of the opening are substracted.

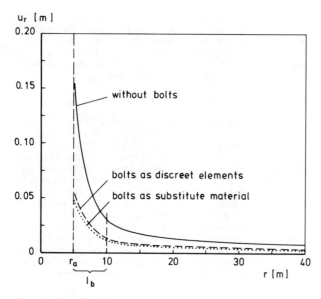

Fig. 5 Distribution of the radial displacements

The effect of bolting is still more evident concerning the displacements:
- The displacement at the boundary of the opening is reduced by bolting about 67,5%.
- The prevention of loosening by dense bolting reaches far into the outer region.

The displacements of both bolt models show a harmonical course, the maximum difference at the inner boundary is about 15% at which the hypothesis of the substitute material yields to somewhat lower values. Thereby it is to consider that in this example the condition for the additional cohesion is the ultimate tensile stress of the bolts. In fact the ultimate bolt stress is not reached. Its maximum value was about σ_b = 440 MPa.

4. CONCLUSION

For the novel hypothesis of the substitute material, which is used to design a systematic bolting, it succeeded to determine a lot of relations between bolting and material behaviour of the compound with the aid of model tests and plane finite element calculations. It could be proven that the material properties of the substitute can be determined on a numerical way. The effect of fully bonded bolts as additional anisotropic cohesion and stiffness could be confirmed also quantitavely. This is described in detail in former papers of the authors and is the basis of the presented calculation. The calculation confirmes the ability of the substitute hypothesis for tunnelling problems. In the moment of composing this paper some work is done on the generalization of the relations relative to the material properties. At the same time the dimensioning method will be extended to not rotational symmetric states of stress and geometries. In addition to this, the shear modulus of the bolted matrix has to be determined in useful tests and calculations. Then the substitute material can be described completely.

REFERENCES

Bjurström, S. (1974): Shear strength of hard rock joints reinforced by grouted untensioned bolts. Proc.III.Int.Congr.Rock Mechanics, Denver 1974, 1194-1199

Egger, P. (1978): Neuere Gesichtspunkte bei Tunnelankerung. Felsmechanik Kolloquium Karlsruhe 1978, Trans Tech, Clausthal, 263-276

Natau, O., Leichnitz, W. (1978): Die Verbundwirkung Systemankerung – Gebirge. Felsmechanik Kolloquium Karlsruhe 1978, Trans Tech, Clausthal, 297-317

Natau, O., Wullschläger, D. (1986): Ansatz zur Berücksichtigung des geankerten Tragringes bei der statischen Bemessung im Tunnelbau. Proc.7.Nat. Felsmechanik-Symposium, Aachen 1986

Wullschläger, D., Natau, O. (1984): Studies on the composite system of rock mass and non-prestressed grouted rockbolts. Proc.Int.Symp.Rock Bolting, Abisko 1983, 75-85

Seismic behavior of a rock tunnel
Comportement séismique d'un tunnel de roche
Seismische Beobachtungen im Felstunnel

YASUKI YAMAGUCHI, Technical Research Institute of Hazama-Gumi Ltd, Yono, Saitama, Japan
MITSURU TSUJITA, Technical Research Institute of Hazama-Gumi Ltd, Yono, Saitama, Japan
KAZUSHI WAKITA, Technical Research Institute of Hazama-Gumi Ltd, Yono, Saitama, Japan

ABTRACT: In order to investigate the seismic behavior of a cavern and the surrounding rock, earthquake observations have been carried out in the rock tunnel started July, 1983. The characteristics of the earthquake motion in rock, deforming behavior of the cavern, and the relation between the strain of the cavern and the particle velocity of the surrounding rock were clarified.

RESUME: Afin d'étudier le comportement séismique d'une caverne et des roches avoisinantes, des observations de secousses telluriques ont été effectuées à l'intérieur d'un tunnel de roche commencé en juillet 1983. Les caractéristiques des déplacements telluriques à l'intérieur des roches, le comportement de déformation de la caverne et le rapport entre les contraintes subies par la caverne et la vitesse de particule des roches avoisinantes ont été mesurés.

ZUSAMMENFASSUNG: Zur Untersuchung des seismischen Verhaltens der Kaverne und des umgebenden Felsens wurden seit Juli 1983 Erdbeben-Beobachtungen im Felstunnel durchgeführt. Die Charakteristika der Erdbebenbewegung im Felsen, das Verformungsverhalten der Kaverne und die Beziehung zwischen der Beanspruchung der Kaverne und der Partikelgeschwindigkeit des umgebenden Felsens wurden geklärt.

1 INTRODUCTION

In recent years, new type of structures such as underground nuclear power stations, storage caverns for high level radioactive nuclear waste disposal, and oil storage caverns have been planned.

In Japan, for these structures which are to be constructed in seismic region, reliable seismic design should be achieved based on the observed data. However, a quantitative method of evaluating seismic stability is yet to be established, and from this viewpoint, the seismic designs providing high reliability will be required in the future utilization of rock cavern.

Therefore, in order to clarify the seismic behavior of a cavern and surrounding rock based on the observed data, we, in cooperation with JNR, carried out earthquake observations in Shin-Usami Tunnel of JNR's Itoh Line started July, 1983.

2 OUTLINE OF EARTHQUAKE OBSERVATION

Shin-Usami Tunnel is a single tracked railway tunnel having a 3000 m of overall length. Its internal cross section has a circular form with inner diameter of 6 m, and the lining concrete is 30 cm in thickness. The observation section is a 100 m section located at approx. 1500 m from each entrance, and the depth from top of the mountain is approx. 260 m. The observation section is composed mainly of Alternated Basalt, and the velocity of the S wave Vs is 1.1 to 1.6 km/sec.

The earthquake observaiton is carried out using 8 accelerometers including one at the Entrance, 10 strain gauges set on the lining concrete, and 6 strain gauges set in the rock. Fig.1 and Fig.2 show the layout of measuring instruments for the earthquake observation. The accelerometers used are the servo type with measurable range of frequency 0.1 to 30 Hz and the minimum resolution 0.01 gal. The strain gauges are the differential trans type having the measurable range of frequency 0.1 to 30 Hz and the minimum resolution 0.03×10^{-6}. The measuring unit is set so as to be triggered when anyone of the 3 components of the

A-6 accelerometer installed 40 m below the tunnel's bottom receives 0.3 gal.

Fig.3 shows the position of epicenters of the earthquakes observed. As the examples of the observation records, the acceleration waveforms in rock are shown in Fig.4.

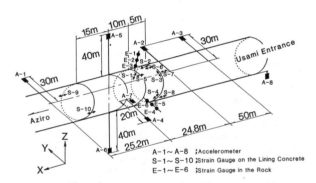

Fig.1 Layout of Measuring Instruments.

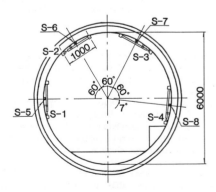

Fig.2 Layout of Strain Gauges on the Lining Concrete (View from Aziro).

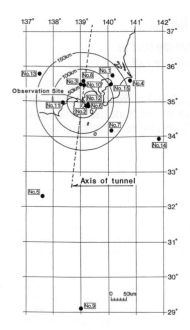

Fig.3 Epicenters of Observed Earthquakes

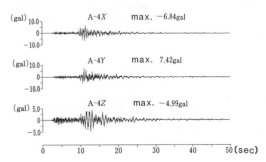

Fig.4 Acceleration in Rock (Earthquake No.3)

3 PRINCIPAL AXIS OF EARTHQUAKE AND WAVE PROPAGATION

3.1 Calculation method for principal axis of earthquake

To clarify the properties of earthquake motion in rock, the principal axes of observed earthquakes were studied. In calculating the principal axes of earthquake motion: ① principal axis I ••• the principal axes varying in time domain (t), where the cross power spectrum at the time (t) becomes the maximum, medium and minimum, and ② principal axes II ••• the fixed principal axes, along which the total energy becomes the maximum, medium and minimum, are used out of the methods using the cross power spectrum proposed by Hoshiya[1].

3.2 Direction of fixed principal axes

Fig.5 shows the direction in horizontal plane (X-Y plane) and the direction in vertical plane (Z-X, Y plane) of the fixed principal axes of earthquake motion in rock A-4.

Conventionally, there have been reports on the direction of the fixed principal axes in horizontal plane of earthquake motion, one insisting that the maximum principal axis and the direction of epicenter show a favorable correspondence[2], and ones insisting that the intermediate principal axis and the direction of epicenter show a favorable correspondence[1],[3]. According to this study performed in a deep rock, however, both the maximum and intermediate principal axes can be considered to have the

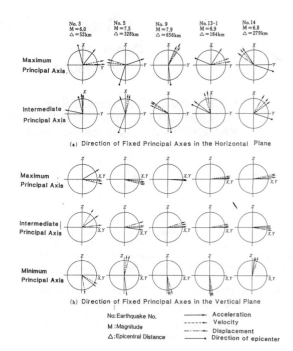

Fig.5 Direction of Fixed Principal Axes in Rock (A-4).

Table 1 Ratio of Total Vibrational Energy in the Direction of Fixed Principal Axis.

No	Acc.		Vel.		Disp.	
	λ_2/λ_1	λ_3/λ_1	λ_2/λ_1	λ_3/λ_1	λ_2/λ_1	λ_3/λ_1
3	0.76	0.63	0.70	0.45	0.26	0.17
5	0.93	0.86	0.61	0.40	0.43	0.23
9	0.43	0.24	0.40	0.20	0.55	0.26
13-1	0.65	0.31	0.62	0.21	0.33	0.09
14	0.90	0.60	0.68	0.35	0.40	0.16

No: Earthquake No

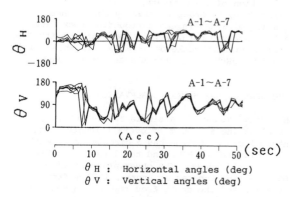

θ H : Horizontal angles (deg)
θ V : Vertical angles (deg)

Fig.6 Variation of Maximum Principal Axis in Time Domain (A-1 ~ A-7), Earthquake No.3.

possibility to become relatively near the direction of epicenter.

Also, the degree of correspondence of these principal axes with the direction of epicenter cannot be said to be so good as was said conventionally, as a whole. As the reason for this, it may be pointed out that the dominant vibration direction in earthquake motion is influenced not only by the direction of epicenter, but also by the geology along a path through which th earthquake motion propagates to reach the observation point, topographical

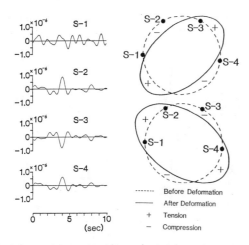

Fig.7 Filtered Strain of the Lining and Deformation Mode of the Cavern.

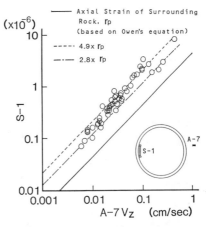

Fig.8 Relation between the Strain of Cavern and the Particle Velocity of Rock (S-1 and A-7Vz).

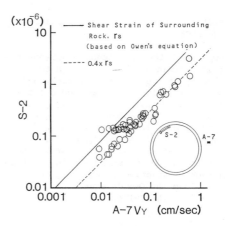

Fig.9 Relation between the Strain of Cavern and the Particle Velocity of Rock (S-2 and A-7Vy).

variation, or reflection and refraction of wave motion at the observation point.

As for the direction of fixed maximum principal axis in a vertical plane, it is considered to be near the horizontal direction, as was reported conventionally. Consequently, it can be assumed that, near the observation point, the earthquake motion propagates vertically in any earthquake.

3.3 Distribution of observed values around the fixed maximum principal axis

The degree of dispersion of observed values around the fixed maximum principal axis is studied using the ratio of total vibrational energy λ_2 in the direction of intermediate principal axis and that λ_3 in the direction of minimum principal axis to the total vibrational energy λ_1 in the direction of maximum principal axis, namely λ_2/λ_1 and λ_3/λ_1.

The dispersion is considered large when λ_2/λ_1 or λ_3/λ_1 is closed to 1 (one), or it is considered small when both of these ratios is near 0 (zero).

Table 1 shows the λ_2/λ_1 and λ_3/λ_1 obtained from acceleraiton, velocity and displacement in rock A-4.

From Table 1, it is considered that, in rock, there is a large dispersion around the maximum principal axis. Therefore the degree that the three directional components of observation (vertical and 2 horizontal components) are affected by the predominant vibrational direction of earthquake motion, is considered small, and three components of earthquake motion can be considered to be independent of each other.

3.4 Maximum principal axis varying in time domain

Fig.6 shows the variation in time domain of the horizontal and vertical angles of the maximum principal axis of acceleration in rock A-1 ~ A-7 for earthquakes No.3.

From Fig.6, the principal axes of earthquake motion cannot be said to be fixed throughout the duration time of an earthquake motion, and particularly during and after the main motion, they are considered to vary largely.

Also, the fact that the vertical angle of maximum principal axis is near $0°$ or $180°$ during the initial motion and it varies around $90°$ during and after the main motion, is presumably indicating that, during the initial motion, the primary wave propagating vertically becomes eminent, and during and after the main motion, the shear wave progapating vertically becomes eminent.

4 DEFORMATION MODE OF A CAVERN

In the case of this observation site, Vs is 1 to 1.6km/sec, and from the observed results, the predominant period of the spectrum of earthquake motion is 1 to 2 sec for earthquakes in remote places and 0.3 to 0.5 sec for those in near places. Therefore, the wavelength passing the cavern is estimated as 1000 to 4000 m at the longest, and 300 to 1000 m at the shortest. Since the tunnel currently observed has a circular cross section of 6 m in inner diameter, the ratio of the diameter of the tunnel to wavelength of incident wave is 1/170 to 1/670 for long wavelengths and 1/50 to 1/170 for short wavelengths, and it is considered that a first deforming mode is dominant in the cavern in each case.

Fig.7 relates to the earthquake No.5 and shows the strain waveforms S-1, S-2, S-3, and S-4 in the circumferential direction of tunnel lining during the main motion for 10 sec. In order to distinguish the correlation of waveforms, these waveforms are applied with 0.2-1.2 Hz narrowband pass filter where the power spectrum of strain is eminent.

From Fig.7, by comparing the time histroy of the main motions among S-1, S-2, S-3 and S-4, the phase of S-1, S-2 and S-4 is nearly the same, and that of S-3 is reverse of those.

From the waveform of circumferential strains, the tunnel in th lateral direction, is considered to be governed by a first deforming mode as shown in the right half of Fig.7.

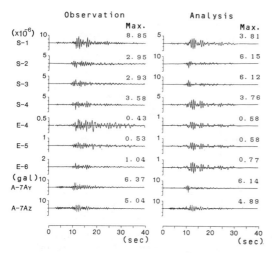

Fig.10 Comparison between Observation and Analysis (Waveform), Earthquake No.3

5 STRAIN OF CAVERN AND THE PARTICLE VELOCITY OF SURROUNDING ROCK

Fig.8 and Fig.9 show the relationship between the circumferential strain of the cavern and the particle velocity of the surrounding rock, A-7.

From these figures, it can be pointed out that the circumferential strains S-1 and S-2 are in proportion to A-7 Vz, the Z component of the particle velocity of A-7, and A-7Vy, the Y component of the particle velocity of A-7, respectively. S-1 and S-2 are in proportion to the axial strain γp and the shear strain γs of the surrounding rock, respectively, where γp and γs are calculated from the Owen's equation[4] as $\gamma p = (A-7Vz)Vp$ and $\gamma s = (A-7Vy)Vs$.

6 DYNAMIC RESPONSE ANALYSIS USING THE BOUNDARTY ELEMENT METHOD

To simulate the seismic behavior of the cavern and the surrounding rock, the boundary element method was employed for dynamic response analysis.

The analytical model was set as two-dimentional model and the boundary between the cavern and the surrounding rock was divided by 30 elements. Horizontal boundary was set 260 m above the cavern.

As the input earthquake motion, the observed acceleration of earthquake No.3 at A-6, 40 m below the cavern, was employed. In the dynamic response analysis, the horizontal component of the acceleration was employed as the shear wave propagating vertically, and the vertical component of the acceleration was employed as the primary wave propagating vertically.

Fig.10 shows the comparison between the observed waveforms and those obtained from the analysis. Fig.11 shows the comparison between the fourier spectrums of the observed waveform and those obtained from the analysis.

Analytical results show the good accordance with the observed records both in the waveforms and in the shapes of the spectrums.

7 CONCLUSION

From the analysis of the observed data, the following effective results on the seismic design of the underground structures were obtained.

1 In the rock around the cavern, the primary wave of the initial motion and the shear wave of main motion propagate in the upward direction nearly vertically.

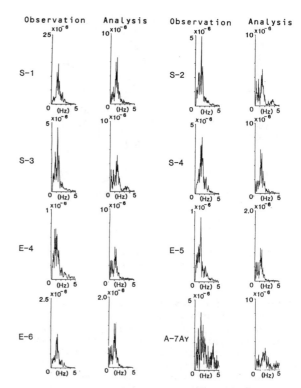

Fig.11 Comparison between Observation and Analysis (Fourier Spectrum), Earthquake No.3

2 The three components of earthquake motion can be considered to be independent of each other.
3 The cavern is governed by a first deforming mode accompanying the propagation in the upward direction of the primary wave and the shear wave, the wavelength of which is very long.
4 There is the strong correlation between the strain of the cavern and the particle velocity of the surrounding rock.

Additionally, the boundary element method was shown to be available for simulating the seismic behavior of a cavern.

8 ACKNOWLEDGEMENTS

As aforementioned, this observation was conducted in cooperation with JNR, and in this connection, we would like to express our sincere thanks to JNR's Tokyo Second Construction Bureau and those concerned to Railway Technical Research Institute for kindly permitting the publication of the observed results.

REFERENCES

1. Hoshiya, M. et al.: Principal Axes and Wave Characteristics of Ground Motion, Proceedings of the JSCE, No.268, pp.33~46, Dec., 1977.
2. Watabe, M.: Simulation of 3-Dimensional Earthquake Ground Motions, Annual Reports of the Building Research Institute, pp.201-206, 1974.
3. Hamada, M. et al.: Observation and Study on Dynamic Behavior of Rock Cavern During Earthquake, Proceedings of the JSCE, No.341, pp.187-196, Jan., 1984.
4. Owen, G.N., Scholl, R.E., Brekke, T.L.: Earthquake engineering of tunnels, RETC Proceedings, pp.709-721, 1979.

Experimental study on the rock bolt reinforcement in discontinuous rocks
Etude expérimentale sur le renforcement des boulons à roche dans la roche discontinue
Die experimentelle Forschung der Verfestigung durch Steinschrauben im diskontinuierlichen Fels

R.YOSHINAKA, Saitama University, Japan
S.SAKAGUCHI, Saitama University, Japan
T.SHIMIZU, Kumagai Gumi Co. Ltd, Tokyo, Japan
H.ARAI, Kumagai Gumi Co. Ltd, Tokyo, Japan
E.KATO, Kumagai Gumi Co. Ltd, Tokyo, Japan

ABSTRACT: This paper presents the result of experimental study using model joints, and the numerical study to propose a mechanical/practical model to express reinforcing effect of rock bolt in jointed rock. Numerical model for bolt effect derived from the experiments involves such terms as joint roughness, mechanical properties of rock, normal stress, stiffnesses of bolt and its installation angle, etc. Stress-displacement behaviors are well expressed by hyperbolic function.

RÉSUMÉ: Ce rapport présente le résultat d'une étude expérimentale avec des joints modèles et d'une étude numérique pour proposer un modèle mécanique pratique pour présenter les effects des renforcements des blutoirs dans les joints. Le modèle numérique des effects des blutoirs provenant des expériences implique des terms come la regosité du joint, les propriétés mecaniques des roches, les contraintes normales, la dureté du blutoir et son angle d'installation. Les déplacements de contraites sont bien exprimés par les fonctions hyperboliques.

ZUSAMMENFASSUNG: Diese Abhandlung präsentiert die Ergebnisse der experimentellen Forschung unter Verwendung eines Knotenmodells und die Forschungsergebnisse für die numerische Analyse für die Wahl eines praktischen mechanischen Modells für Ausdruck der Verstärkungswirkung von Steinschrauben gegenüber verbundenem Fels. Das experimentell eingeführte numerische Analysemodell enthält Ausdrücke für Grobheit der Verbindung, mechanische Eingeschaften des Muttergesteins, Normalspannung, Schraubensteifheit, Installierungswinkel usw. Das Spannungsversetzungsverhalten wird durch Hyperbelfunktionen gut ausgedrückt.

1 INTRODUCTION

Rock bolt plays an important role as one of major support members for underground excavation in rock and rock slope. Some method has been proposed to estimate the reinforcing effect, but the mechanism and effect of bolt action in jointed rock mass are complicated and not enough clarified. So we excute a series of laboratory shear tests to explain the support mechanism of rock bolt in jointed rock mass subject to shear deformation.

Test specimens made of mortar with single or layered joint, as jointed rock model, are 80x40x20cm in size. Regular asperities (or teeth) with dilation angles of 0°/10°/20° and Barton's JRC roughness profiles are used as roughness of joint. Rock bolts, 10∿25mm in diameter, are installed with angles of 45°/90°/135° and 30°∿90°. Shear tests are performed under plain strain condition with normal stress of 0∿60Kgf/cm^2 (0∿5.88MPa) on joint surface. Some portion of test method and results are presented in our paper (1986).

2 JOINT STRENGTH

2.1 Strength of regular teeth joint

From the test result it is found that there are two phases in the behavior of regular teeth joint, that is : joint surface slides along the teeth when the relative normal stress σ_n/σ_c is small, and teeth are sheared off before sliding occur if σ_n/σ_c become greater. In the former cases, shear strength can be expressed by Patton's equation. As for the joint strength of the latter case, it can be estimated by appling limit equilibrium theory to local stress state of teeth.

Mortar strength is usually expressed by Fairhurst's equation calculated from the result of splitting tensile test and uniaxial compression test, however, the strength from the results of triaxial compression test were represented by straight line. Fig.1 shows experimental and calculated results of the joint strength, and the strength curves of mortar written above are

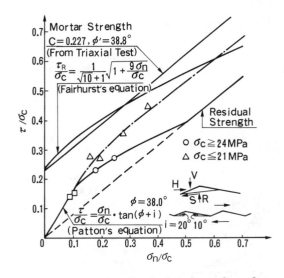

Fig. 1 Experimental and calculated results of τ vs. σc, and mortar strength curves in regular teeth joint

also included.

The joint strength of the latter case seems to be connected with deformation behavior of teeth and to be divided into two groups. It is asummed for one group that failure of teeth occures under biaxial stress state and little restraining pressure act on back face of teeth, because relatively small displacement of joint arise at failure of teeth. It is also asummed for the other that failure of teeth occures under triaxial stress state and some restraining pressure act on back face of teeth, because relatively large displacement of joint arise at failure of teeth. Fairhurst's equation may be applied for the former and results of triaxial test for the latter. When we calculate the latter, 38% of normal stress is distributed as restraining pressure on back face of a tooth.

2.2 Strength of irregular teeth joint

Comparisons are made in Fig.2 between Barton's equation and test results excuted by using two dimensional joint surface based on Barton's JRC roughness profiles. In cases uniaxial strength of mortar σc >23MPa, test results agree with Barton's equation fairly well when JRC=20, but test results are 15∿20% greater than values from the equation when JRC=10. In cases σc<13MPa, test results are 25∿30% and 20∿25% greater than values from the equation when JRC=10 and JRC=20 respectively. It is similar with the cases of regular teeth joint that relative strength of low σc is greater than that of high σc.

If equation of Ladanyi & Archambault may be used, As, portion of joint area sheared through asperities, and $\dot{v}$ or dv/du, dilatancy speed when peak shear stress arise, may be as shown in Fig.3.

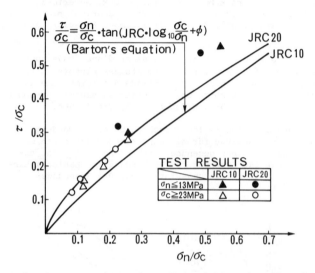

Fig. 2 τ vs. σn, test results and Barton's strength
curve in irregular teeth joint

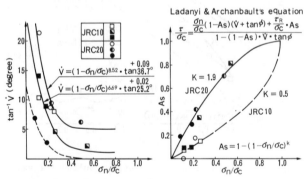

Fig. 3 Estimated As and $\dot{V}$ from test results in
irregular teeth joint

3 EXPERIMENTAL RESULTS OF ROCK BOLT MODEL

Fig.4 shows an example of shear stress-shear displacement relation curves of flat joint model with bolts (2x D16) by direct shear test. The less bolt setting angle is, the bigger bolt effect is mobilized in an early stage. The best bolt angle is φ in theory, where φ is friction angle of the joint. φ=36°∿40° in the test. It is known by the previous tests that normal stress and dilation angle give little affection to bolt effect on shear strength except when asperities of joint are sheared off before bolt effect is fully mobilized.

Fig.5 shows the relation between shear stress and bolt setting angle at each stage of shear displacement.

It is clearly shown how the bolt effect is changed by bolt setting angle and shear displacement.

Fig.6 shows an example of bolt bending strain-shear displacement relation curves of the model. The bigger bolt setting angle is the bigger bending strain arise in the bolt. Fig.7 shows an example of axial bolt stress-shear displacement relation curves of the model. It is also shown evidently that the bolt effect well correspond to axial bolt stress.

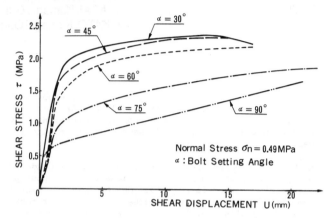

Fig. 4 Relation between shear stress and shear
displacement

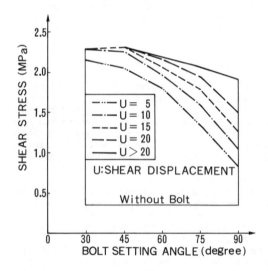

Fig. 5 Relation between bolt setting angle and shear
stress at each level of shear displacement

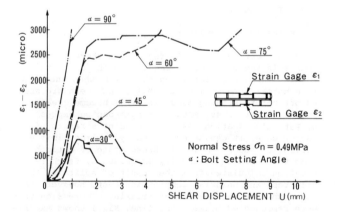

Fig. 6 Relation between bending strain and shear
displacement

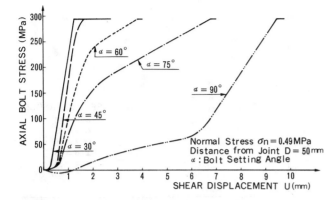

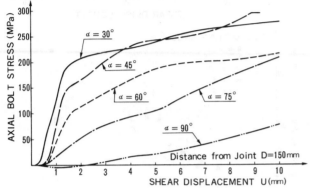

Fig. 7 Relation between axial bolt stress and shear displacement

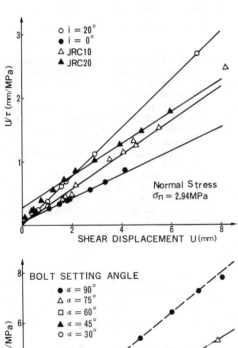

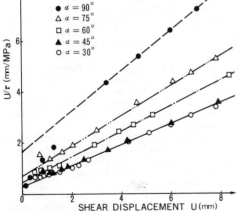

Fig. 8 Relation between U/τ and U

4 DEFORMATION BEHAVIOR

If the test results are plotted in the form of (shear displacement/shear stress)-shear displacement, we get the relations as shown in Fig.8. As good linear property is shown in every case, hyperbolic approximation can be made by following equation.

$$\tau = \tau_{ult} \cdot K_{si} \cdot U / (\tau_{ult} + K_{si} \cdot U) \qquad (1)$$

where τ: shear stress, τ_{ult} : limit value of τ, K_{si}: initial shear stiffness, U: shear displacement

We now consider the cases without bolt. τ_{max} written in Sec.2 may be used approximately as τ_{ult}. As for K_{si} of joint, following facts are found from the tests.

1) K_{si} increase according to increment of σ_n in planar joint.
2) In regular teeth joint with high dilation angle such as 20°, K_{si} decrease according to increment of σ_n, because of the large deformation of the teeth.
3) In irregular teeth joint, K_{si} decrease according to increment of σ_n or JRC, as shown in Fig.9.
4) Fitness of teeth or asperities much affect initial stiffness and K_{si} decrease if fitness become worse.

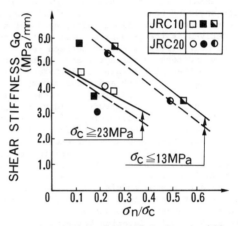

Fig. 9 Relation between initial shear stiffness and relative normal stress

5 NUMERICAL CONSIDERATION ON REINFORCING EFFECT

5.1 Effect on initial shear stiffness

Illustration of bolt at initial stage of loading is shown in Fig.10a. Reinforcing effect of bolt on initial shear stiffness by tention is given as follows:

$$K_{si}^{N} = N/U \cdot \{ \sin\alpha \tan(\phi + i_0) + \cos\alpha \}/A \qquad (2)$$

$$N/U = Ab \cdot C \cos(\alpha - i_0)/\cos i_0 \qquad (3)$$

where N,U: axial bolt force and shear displacement at an early stage respectively, α: bolt setting angle, io : initial dilation angle; C, from the paper (Saito), can be derived as follows in the terms of Young's modulus of bolt E_b, bolt radius R, adhesion spring coefficient of pull out test C_o.

$$C = \sqrt{2 E_b C_o / R} \qquad (4)$$

Additional initial shear stiffness by dowel action is given as follows (Chang's equation):

$$K_{si}^{S} = 2 E_b I / A \cdot \{ K_H B / (4 E_b I) \}^{3/4} \cdot \sin(\alpha - i_0)/\cos i_0 \qquad (5)$$

where I: moment of inertia of area of bolt, K_H: lateral coefficient of rock reaction, B: bolt diameter; $K_H B$ is given as follows (Vesic's equation):

$$K_H B = 0.65 E/(1 - \nu^2) \{ E B^4/(E_b I) \}^{-1/12} \qquad (6)$$

where E, ν are the elastic parameter of rock material.

5.2 Effect on peak shear strength

Tipical illustration is shown in Fig.10b. Reinforcing

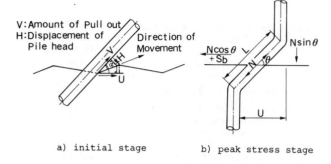

a) initial stage   b) peak stress stage

Fig.10   Sketch illustrating a rock bolt situation

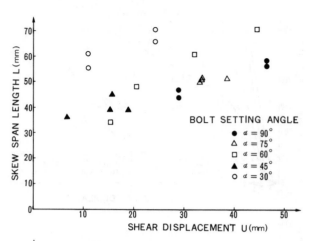

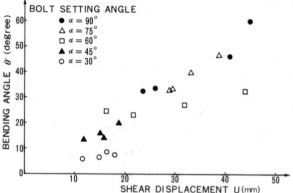

Fig.11   Experimental results of a) skew span length and
b) bending angle vs. shear displacement

effect of bolt on peak shear strength is given as follows:

$$\tau_{ult}^{b} = N/A\{\sin\theta \tan(\phi + i) + \cos\theta\} + S_b/A \qquad (7)$$

where N: axial bolt force, θ: bolt angle at joint, ϕ: friction angle of contact surface, i: dilation angle S_b: shear force of bolt along joint, A: joint area

N is derived as the product of bolt section area A_b and axial bolt stress σ_b. σ_b is between tension yielding stress and tension strength, and can be derived according to the axial strain level. S_b are derived as the product of A_b and $\sin\theta$ and shear stress of bolt τ_b. τ_b may be given as 0.3~0.4 times the shear yielding stress, because the bolt already yield by tension. θ, in connection with the bolt deformation behavior, may be affected many terms such as stiffness of bolt and rock material, shear displacement, etc.. From measurements of bolt shape after the test written in Sec.3, relations between skew length L and shear displacement U are shown in Fig.11a. Although L has been considered to be constant in some models, it is clearly shown in the figure that L tend to increase according to the increment of U under the condition of rigid fixation at both ends of bolt. Relation between bending angle θ' ($=\alpha-\theta$) and U are shown in Fig.11b.

Stress-displacement relation of rock joint with bolt can be derived by summing τ_{ult} and K_{si} of rock joint and bolt effect respectively. This numerical model can be easily introduced to joint element in FEM analysis.

6 APPLICATION TO UNDERGROUND OPENINGS

We think about underground openings supported by rock bolts. The most effective bolt angle is 35°~55° against joint, except when we expect the bolt effect to be kept even after large shear displacement. As bolt effect are mobilized at an early stage when α=35°~55°, rock deformation is restrained and range of displacement to be considered in each joint may be relatively small. In pattern bolting system, the dangerous joint angle for stability of underground openings is considered to be 35°~85° against the direction transverse to the opening surface. If the major joint direction are in range of 60°~85°, it is more favorable that the bolt are installed in the angle of 5%~30° against the direction transverse to the opening surface. Bolt effect increase 70%~90% if the angle change 85° to 55°.

As rock bolts must keep their stress for a long time in the openings without lining concrete, it must be noted that rock bolts are in a vulnerable state against corrosion if displacement of the opening is relatively large and bolts yield at joint. Bolt design considering corrosion may be needed in that cases.

7 CONCLUSIONS

Experimental and numerical study has been made to explain the reinforcing effect of rock bolt in jointed rock mass subject to shear deformation. Main results in this study are concluded as follows:

1) Shear strength of regular teeth joint can be estimated by appling limit equilibrium theory to local stress state of tooth, when the teeth are sheared off before sliding occur.
2) Relative shear strength of low material strength is greater than that of high material strength in regular or irregular teeth joint, in connection with deformation behavior of the teeth.
3) The most favorable bolt angle is assumed to be 35°~55° against joint.
4) In the practical range, shear stress-shear displacement relations of joint with or without bolt may be expressed by hyperbolic function. This numerical model can be easily introduced to joint element in FEM analysis.

REFERENCES

Yoshinaka,R. & Sakaguchi,S. & Shimizu,T. & Arai,H. & Kato,E. 1986. Reinforcing effect of rock bolt in rock joint model, Proc. Int. Symp. ECRF, p922-928
Barton, N. & Choubey,V. 1977. The shear strength of rock joint in theory and practice, Rock Mechanics, 10, p1-54
Ladanyi,B. & Archambault,G. 1970. Simulation of shear behaviour of a jointed rock mass, Proc. 11th Symp. on Rock Mechanics, AIME, p105-125
Saito,T. & Amano,S. 1982. Fundamental consideration on support design of rock bolt, Proc. 14th Symp. on Rock Mechanics, JSCE, p76-80
Patton,F.D. 1966. Multiple modes of shear failure in rock, Proc. 1st Cong. ISRM, Vol.1, p509-513
Fairhurst,C. 1964. On the Validity of the Brazilian test for brittle material, Int. J. Rock Mech. & Min. Sci., 7, p125-148
Chan,Y.L. 1937. Discussion on "Lateral pile-loading tests" by Feagin, Trans., Vol.102, ASCE, p272-278
Vesić,A.B. 1961. Bending beams resting on isotropic elastic solid, Proc. ASCE, Vol.87, No.EM2, p35-53

Concurrent seismic tomographic imaging and acoustic emission techniques: A new approach to rockburst

Imagerie par tomographie et émission acoustique: Une combinaison pour une nouvelle technique séismique destinée à l'investigation des éclatements de roches

Tomographische Darstellung und akustisches Emissionsverfahren: Eine neue seismische Technik zur Untersuchung von Felsspannungen

R.P.YOUNG, D.A.HUTCHINS, W.J.McGAUGHEY, T.URBANCIC, S.FALLS & J.TOWERS, Rock Physics Laboratory, Departments of Geological Sciences and Physics, Queen's University, Kingston, Canada

ABSTRACT: Seismic experiments have been undertaken in underground hard rock pillars and the results have been used to obtain geotomographic images of anomalous ground conditions. The aim is to identify and locate areas of high stress concentrations for the application of destress blasting techniques. In addition, acoustic/microseismic emission data from sensors sensitive to motion in three directions, show that shear waves are polarised after travelling through the rock mass. These effects are discussed and their importance in minimising rockburst hazards explained.

RESUME: Des expériences de sismique ont été entreprises sur des colonnes de roches dures, et les résultats ont été utilisés pour obtenir des images tomographiques de roches comportant des anomalies. Le but est de déceler ces anomalies en présence de fortes contraintes, afin de les réduire dans les mines. En outre, des données recueillies après émission acoustique sur des capteurs sensibles au déplacement triaxial font apparaître une polarisation des ondes de cisaillement après leur passage à travers la roche. On présente ces effets et on explique leur importance pour la réduction des éclatements de roches.

ZUSAMMENFASSUNG: An Pfeilern aus hartem Gestein wurden seismische Esperimente ausgefuhrt. Die Messresultate abnormaler Stellen wurden zu tomographischen Bildern ausgewertet. Ziel der Untersuchung war die Lokalisierung hoher Spannungen im Fels von Bergwerken in Hinblick auf Behebung solcher Gefahrenzonen und Verhinderung von Grubenzusammenbruchen. Anhand akustischer Emissionsmessungen mit dreidimensional empfindlichen Bewegungssensoren wurde eine Polarisierung von Transveralwellen im Fels festgestellt. Diese Effekte wurden untersucht, sowie deren Wichtigkeit in Zusammenhang mit Verhinderung von Felsbersten aufgedeckt.

1 INTRODUCTION

Rockbursts are catastrophic explosive events which occur in deep hard rock mines with, in many instances, severe and sometimes very tragic socio-economic consequences. Although some significant progress has been made in rockburst research, little is known about the mechanisms of these events and even less about their prediction and prevention. Fundamental research into the problem has often been hindered by the very real and pressing need for short term solutions which can be put into effect immediately, unfortunately with only limited success. It was for these reasons that a major research initiative was started in 1986 at Queen's University, supported by NSERC and the Canadian mining industry, to attempt to delineate the physical processes involved and evaluate the geophysical methods which could best be used to study mining induced seismicity and rockburst phenomena.

In recent years, rockburst research has been characterised by the aim of prediction. This search for predictive techniques stems from the fact that many mines have, at present, a serious rockburst problem. An alternative viewpoint is that research efforts should also be directed at developing methods for the prevention, or at least minimisation of the risk of rockbursts. Seismic tomographic imaging from one mine level to a deeper proposed level could be used to produce stress/integrity maps of the rock mass insitu and this in turn could be used at the planning stage to identify anomalous "rockburst prone" ground ahead of mining. Methods have been developed for the destressing of rock masses in mines and these have been shown to be successful at minimising the rockburst problem in already known problem areas (Blake 1984). A combination of geotomographic imaging and destress blasting techniques would therefore be a useful method for recognising, locating and subsequently controlling highly stressed rock masses. Seismic tomographic imaging techniques could also be used before and after destress blasting to determine the efficiency of the destressing operation (Blake 1982, Young and Hill 1985).

2 PREVIOUS INVESTIGATIONS: AN OVERVIEW

Investigations of rockburst phenomena in mines are usually undertaken with a multichannel microseismic network. Numerous transducers, positioned throughout the mine, are used to detect seismic events, which are designated as such when a preset amplitude trigger is exceeded. The arrival times of these events at the various locations are calculated and used to compute the source location. These data are then used to plot location maps as a function of time and also to produce histograms of the number of events per unit of time. This approach has been well documented in the literature by many authors, notably McDonald (1983), Leighton (1984) and several South African researchers, notably Brink and O'Conner (1983). The philosophy behind this approach is that a build up or pattern of microseismic activity in a selected area can be used to give prior warning of a major event. However, it has been shown by selected studies (Blake 1982) that only in approximately 30% of cases did some form of precursor phenomena precede a major rockburst, while 60% of recognisable precursor patterns were not followed by a rockburst. It is possible that certain rock mass conditions will permit a moderate level of microseismic activity, to prevent a large build up of strain energy, which could in turn cause a major rockburst. Conversely, it may be the case that areas which exhibit a low level of microseismic activity, or no significant change in the rate of activity during stress build up, are more likely to produce major microseismic events. Also, the Kaiser effect (the absence of detectable acoustic emission until previously applied stress levels are exceeded) could well be a major factor controlling pre-burst acoustic emission (Hardy in press). Thus if stress concentrations could be identified, the condition of the rock mass characterised and its microseismic activity monitored, it may be possible to explain the statistics described by Blake.

3 NEW APPROACH: PRELIMINARY RESULTS

The primary objective of the field research programme is the development of techniques, by which geophysically anomalous volumes of rock in underground mines may be identified as rockburst prone, enabling preventative action (blast destressing or avoidance) to be taken. The field programme is separated into two phases, the distinction being primarily one of scale. The small scale phase involves evaluation of the potential use of seismic tomographic imaging methods, for identifying and delineating rockburst prone ground. This will be accomplished by initially carrying out concurrent geotomographic imaging and microseismic monitoring of mine rock masses as a function of time in a dynamic stress regime. The large scale phase involves rock mass investigation on the scale of the mine itself, and is concerned with the enhancement of presently employed microseismic monitoring systems and the development of signal processing techniques for the study of rock burst source mechanisms. These two aspects and the preliminary results of the field research programme are now described separately.

3.1 Seismic tomography experiments

Seismic tomography is a data processing technique which allows seismic data, collected from multiple source/receiver positions around the rock, to be analysed so as to provide images of the rock between boreholes (Wong et al 1984) or inside structures (similar to CAT scanning of the human body using x-rays), see figure 1. These images can then be related directly to the condition/quality of the rock mass through which the waves travelled. Many of the problems of applying tomography to rocks stem from the fact that if rocks are inhomogeneous and have a velocity structure, seismic waves will not travel in straight lines, but travel minimum time paths. The

technique developed as part of this project extends the work of Wong et al and offers a solution for correcting for ray bending, to obtain a better estimate of the velocity structure. Three stages are involved in the new analysis technique to produce the corrected high resolution image. Firstly, a traditional convolution or filtered back projection technique, similar to that used in x-ray tomography, is used to produce a crude image based on straight ray theory. At the second stage, the algebraic reconstruction technique (ART) which is iterative and also based on a straight ray approximation, is applied to refine the image. The second stage image is then used to compute theoretically the ray behaviour, if the rays had propagated through this image, and then the image is corrected for ray bending and refraction in a third stage iteration of the ART algorithm.

In order to quantify the resolution of this technique, models and rocks of known properties were imaged. The models showed that a combination of imaging techniques could be used to increase resolution and correct for ray bending effects. Figure 2 shows a schematic of the laboratory instrumentation system used to collect data for model studies. A modified optic bench and transducer mounting system allow compressional and shear wave signals to be transmitted and received through rock samples, with reproducible travel times and amplitudes. A panametrics pulse generation system provides the electronics for excitation of the transmitter and a Lecroy 3500, 32MHz transient recorder is used to amplify, capture and digitise the received signals for subsequent digital processing. Figure 3 shows a geotomographic velocity image of a laboratory sample of rock buried within a cube. The image was produced from data collected from 24 source/receiver locations around the sample. The imaged position and velocities are in good agreement with the known location and velocities of the two media (rock salt P wave velocity of 4,600 m/s embedded within a 6cm plexiglass cube with P wave velocity of 2,700 m/s).

Preliminary field trials were carried out in early 1986 in order to identify potential practical problems, in the instrumenting of rock masses in underground mines for the collection of geotomographic data. The experiment was conducted on an approximately 100m x 200m pillar at a Sudbury hard rock mine, 760m below ground surface. The pillar consisted primarily of metamorphic crystalline rock containing a large pocket of backfill where mining had previously occurred, see figure 4a. An array of 36 single component ADR 711 accelerometers were rigidly fixed to the pillar wall so that they surrounded the pillar, see figure 4a. Seismic sources, consisting of blasting caps and boosters, were positioned down shallow holes, fired at each location and recorded for all 36 locations. Recording was carried out using an OYO multichannel engineering seismograph. The data showed clear P and S waves at bandwidths up to 2.5KHz, which is very promising in terms of the ultimate resolution of any internally

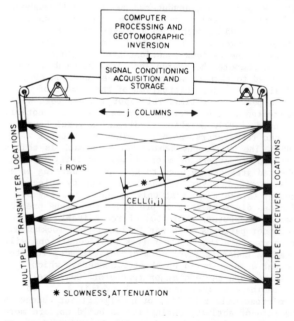

Figure 1: Schematic of geotomographic imaging system. Inversion process considers the effect of slowness or attenuation that each cell has on the rays passing through it.

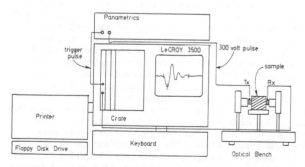

Figure 2: Schematic of modified optic bench and Lecroy ultrasonic transient acquisition system

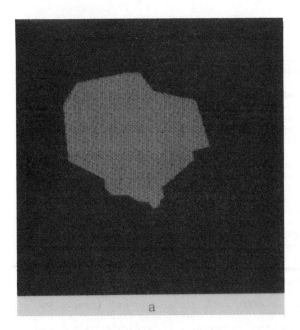

Figure 3: Geotomographic reconstruction of laboratory sample; a) actual plan outline of rock salt specimen (4,600 m/s) embedded within a 6cm plexiglass cube (2,700 m/s); b) reconstructed image of rock fragment within cube from 6 ultrasonic source and receiver locations per side. Maximum imaged velocity is 5,300 m/s (marked X on figure) and minimum imaged velocity is 2,300 m/s (marked Y on figure. The contour interval is 500 m/s.

imaged features. The fact that both P and S wave arrivals could be identified for selected source/receiver geometries from small shallow blasts, is also promising in terms of the calculation of internal rock mass elastic moduli.

Direct P wave arrival times were picked for over 1,200 records (36 seismic sources and 36 receiver locations) and used as input to the geotomographic imaging program outlined above. The image shown in figure 4b indicates the presence of a low velocity zone, approximately 2,500 m/s, increasing in the backfilled and adjacent destressed portion of the pillar (marked A on figure 4b). The majority of the pillar's central region (marked B on figure 4b) is characterised by a high velocity area, approximately 6,500 m/s. The edges of the pillar are destressed and this has been confirmed by a seismic refraction interpretation giving a velocity of 5,500 m/s. To date, only P wave velocity has been used to image

internal rock structure. However, during the main phase of the project, triaxial accelerometers will be used at each source/receiver location. This will enable more accurate detection of S wave arrivals using digital polarisation filtering, which will facilitate the computation of S wave velocity tomograms. Attenuation spectral ratio studies are also planned to enable images to be computed on the basis of attenuation spectral ratios of seismic energy at selected frequencies. This technique has been shown to be extremely sensitive to determining the integrity/quality and anisotropy of insitu rock masses (Young and Hill 1986). Studies carried out to date indicate that utilising P and S wave velocity in conjunction with attenuation spectral data for geotomographic imaging, should provide the necessary resolution to identify highly stressed ground. It is expected that velocity and attenuation anomalies of 5-10% or greater will be resolved with the refined

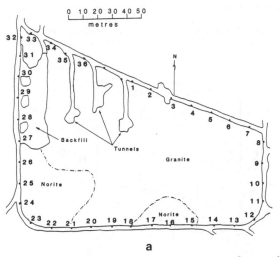
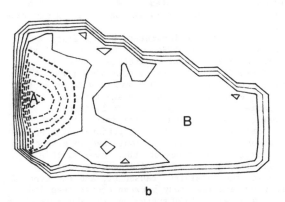

Figure 4: Geotomographic reconstruction of an underground hard rock pillar; a) actual plan of pillar showing marked source/receiver positions and the location of a backfilled area; b) image shows a clear low velocity anomaly in the location of the backfill (2,500 m/s at the position marked A in the figure) and the associated destressed area. The imaged velocity in the centre of the pillar is 6,500 m/s (marked B in the figure).

technique. These encouraging results indicate that the geotomographic reconstruction software which is still at an early stage of development, can be applied to a field situation to distinguish and locate large scale (several metres) velocity anomalies, and verify the integrity of rock masses.

3.2 Microseismic monitoring experiments

Although many researchers have investigated mining induced microseismicity in terms of locating events and monitoring the time relationship between events, very little work has been carried out in the analysis of whole waveform multichannel signals. Whole waveform research to date has been limited to studies carried out in South Africa, where whole waveform triaxial data are used to pick P and S wave arrivals for determining source locations (Brink and O'Conner 1983), although a few studies have investigated rockburst source mechanisms (notably Spottiswoode 1984, Hanks 1984). It is the aim of this research to extend the work of the South African researchers and evaluate the usefulness of whole waveform data in characterising microseismically active rock masses.

In order to achieve this goal, the development of an effective monitoring system and a familiarity with the type of signals (microseismic events) expected is a required prerequisite. Preliminary field trials were carried out in early 1986 in order to identify potential practical problems in the instrumenting of rock masses in underground mines, for the collection of whole waveform digital microseismic data. The experiment was conducted 760m below ground surface in the same mine as that used for the geotomographic experiments and incorporated both analogue and digital data acquisition systems. The analogue system involved recording 14 channels of microseismic data into a Honeywell 101 FM tape recorder. The main disadvantage of this system is that its dynamic range is limited. However, the continual recording of data did allow for studies of mine noise, and background signal levels were evaluated. A 16 channel digital acquisition system based on an IBM PC/AT was used in parallel to the analogue system. The major component of the system is manufactured by RC Electronics and consists of a 16 channel A/D board, an external instrument interface and acquisition software capable of acquiring data at an aggregate sampling rate up to 1MHz. Digital conversion is with 12 bit accuracy (72dB) over an input range of +/-10v. The software allows for data acquisition with user selected channels, waveform manipulation, cursor measurement and display, trigger control, pre-trigger memory, memory buffer and high speed transfer to hard disk storage, and eventual archival storage to inexpensive VCR tapes. Although many more expensive systems are available, this approach allows a multichannel whole waveform acquisition system to be interfaced to a traditional MP250 microseismic monitor for a moderate cost.

Bruel and Kjaer accelerometers with a flat frequency response to 5KHz, were used in a triaxial configuration and signal conditioned, prior to digitisation using the system described above. A low profile direct fixture method was used in order to overcome resonance effects from any triaxial accelerometer housing. Spectral analysis of microseismic signals with higher resonant frequency accelerometers, showed that the predominant frequency content of events is usually contained in a spectrum with a maximum frequency of 5KHz. However, this is strongly dependent upon the distance between the source and the receiving transducer, and the type of rock. A comparative study between events recorded using a mine wide MP250 system and the whole waveform data acquisition confirmed suspected problems. The MP250 system triggers from several single component detectors and many significant events were detected from S wave arrivals because of the orientation of uniaxial detectors for that particular event. This

resulted in erroneous source locations using arrival times computed by the MP250 system. The use of three component accelerometers, although more expensive, clearly did not suffer from this major disadvantage. In addition, by examining three component records, it is possible to determine the direction of propagation and the distance of travel in that direction, using the relative arrival times of the P and S waves. This technique can then be used in conjunction with traditional methods to refine source location estimates.

During this study, a number of events clearly showed shear wave splitting, indicative of seismic propagation in anisotropic rock. Anisotropy can manifest itself in many forms including bedding, parallel fractures, parallel alignment of crystals and stress anisotropy. Conventional analysis of seismic compressional wave arrival times is not a sensitive indicator of anisotropy in the earth. Some measure of the anisotropy can be made by recording P wave arrivals at various directions with respect to the anisotropy, but this is often not possible or is inconvenient to do. Shear waves, however, have been found to be much more sensitive to anisotropy in rocks (Crampin 1985, Young in press). Crampin and others have shown theoretically that if P and S waves are recorded on mutually perpendicular triaxial detectors, very good estimates can be made of the direction and degree of anisotropy in the ground through which the waves have passed. This theory has not been fully tested experimentally, although in recent years much research has been carried out into the triaxial recording of signals generated by earthquakes, in order to detect anisotropy in the upper crust.

When a shear wave enters a zone of anisotropy, it splits into two waves known as quasi S waves and the P wave is then called the quasi P wave. These three quasi seismic waves each propagate at different velocities with mutually perpendicular directions of vibration. The velocity and direction of vibration (polarisation) will vary with respect to the orientation of anisotropy. The shear waves can no longer be called horizontal or vertical as the polarisation of each wave could be at any angle to these directions. The difference in the velocity of propagation of these two waves results in the phenomena known as "shear wave birefringence" or "shear wave splitting" and is similar in principle to optical birefringence in minerals, see figure 5. One shear wave travels through the zone of anisotropy at a higher velocity than the other and when the shear wave emerges, it has split into two distinctively polarised waves, one travelling ahead of the other. If the wave then enters an isotropic medium, the splitting is maintained as the waves are now travelling at a common velocity, The degree of splitting or delay between the two shear waves should be directly related to the

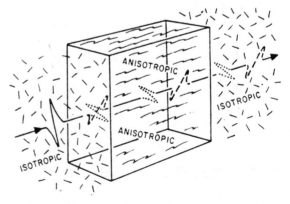

Figure 5: Diagrammatic representation of shear wave splitting. Inside the zone of anisotropy two quasi polarised shear waves exist, one travelling at a higher velocity than the other (Crampin 1985).

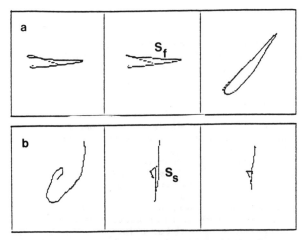

Figure 6: Typical particle motion diagrams for segments of a microseismic waveform containing shear wave motion; a) shows the early polarisation of the fast shear wave (S_f); b) shows the orthogonal motion of the later arriving slow shear wave (S_s).

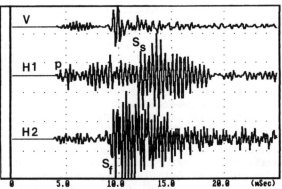

Figure 7: Example microseismic event showing shear wave splitting recorded using a triaxial cluster of accelerometers (V = vertical component and H1 & H2 = horizontal components). P marks the arrival of the compressional wave, S_f indicates the arrival of the fast shear wave and S_s indicates the slow shear wave polarised orthogonal to S_f.

degree of anisotropy (magnitude of the velocity difference between the two S waves) and the distance the waves travelled within the zone of anisotropy. The polarisation of the waves will depend on the orientation of the fractures/stress with respect to the direction of propagation of the shear waves. It is very difficult to measure the delay between the two quasi shear waves experimentally, because the arrival of the second wave is usually masked by the noise and the wave train of the first wave, although digital polarisation filters can be used to enhance arrival recognition. Particle motion diagrams are often useful in identifying the polarisation of the shear waves which should indicate the orientation of the anisotropy. Essentially, particle motion diagrams map out the motion of the ground in three mutually perpendicular planes. This is achieved by plotting 3 graphs using pairs of the three signals from a triaxial sensor as values for the x and y axes. The orientation of the anisotropy is usually indicated by the polarisation of the first motion of the fast shear wave which is polarised parallel to the strike of the anisotropy. Figure 6 shows typical particle motion plots for segments of a microseismic waveform containing S wave motion. Figure 6a shows the polarisation of the fast shear wave, while the orthogonal motion of the slow shear wave can be seen in figure 6b.

Figure 7 shows an example of a microseismic event recorded during the preliminary trials at the mine under study, where the horizontal stresses are 2–4 times the vertical overburden stress. The upper trace is the vertical component and the lower two traces the horizontal components. The horizontal components clearly indicate the presence of two quasi shear waves travelling at different propagation velocities. The fast direction indicated on the lower trace and partly on the top trace, coincides with a NW/SE orientation which could be a result of pre-aligned structure or stress anisotropy. Calculations show that the shear wave birefringence for several similar events varies between 400 and 800 m/s.

A study of several hundred microseismic signals showed that only 39% of the signals recorded were true microseismic events. Of this 39%, different seismic event signatures can be identified. Events, which when source located are associated with major structures such as lithologic boundaries, dykes and faults, contain considerable shear wave energy probably associated with a shear slip mechanism. Events associated with pillar bursting appear to have a more compressional source mechanism, although the signals still contain shear energy, but to a lesser

degree (probably mainly due to P to S wave mode conversion). Considerable work is still required in this area, but the application of pattern recognition techniques may prove useful in this regard (Young and Hill 1982). The remaining 61% of recorded events has been attributed to mining related noise (for example ore-chutes, drilling and blasting). During the main phase of the project when microseismic data will be collected around mine pillars and on a mine wide scale, complex triggering will be used to minimise the storage of non-microseismic events.

4 Main experimental phase

Field techniques and processing software have been developed to allow geotomographic imaging of underground rock masses and the whole waveform acquisition and processing of microseismic data. The next phase of the work will involve the concurrent instrumentation of a volume of rock mass in an area of current mining activity, where local stress levels will change as a result of mining. The instrumentation will consist of triaxial seismic transducers, a series of multiple conponent extensometers and insitu stress cells. Over a period of several months, a multichannel seismic system will continuously monitor microseismic activity within the pillar, detecting and recording acoustic/microseismic emission data originating from within the rock mass volume under study. These microseismic data will be processed using both P and S wave data to determine source location. Inferences about rock quality/state of stress through which the seismic waves have propagated, as well as the characterisation of the source mechanism to classify the mode of failure, will be made on the basis of statistical analysis of the recorded waveforms. In addition, the pillar will be periodically imaged geotomographically in order to resolve the state of stress in a complete two dimensional section across it, as well as to locate any geotomographically identifiable anomalous zones which may be due to properties of the rock mass other than stress, for example lithologic boundaries, crack/joint density, local pillar geometry.

Numerical models will be generated using insitu stress measurements and compared to geotomographic interpretations. The geotomographic imaging, acoustic emission and static stress and strain measurement data will be analysed together, in order to determine relationships between the changes in geotomographic images with time and the actual crack propagation within the pillar causing the microseismic activity. The objective of the study will be to develop

sophisticated techniques for obtaining geotomographic
data and reconstructing images, and to state criteria
for the interpretation of geotomographic and
microseismic data in terms of rockburst prone ground.
These studies will be carried out in conjunction with
scaled laboratory experiments where rock samples will
be geotomographically imaged and acoustic emission
data will be collected at different stages during a
simulated deformation process.

In addition to the localised study described above,
whole waveform microseismic data will be collected on
a mine wide scale. This will be achieved by
interfacing the microseismic system described in
subsection 3.2 to a MP250 currently in operation at
the mine site under study. Two systems will be
utilised to provide high and low gain recording, in
order to cover the dynamic range of microseismic
signals. Five triaxial accelerometer systems will be
used in order to further study shear wave
polarisation phenomena and microseismic source
mechanisms.

5 Conclusions

The preliminary studies from a major Canadian
research initiative for investigating mining induced
seismicity and rockbursts are now complete. The
research project is divided into three phases:
concurrent geotomographic imaging and microseismic
monitoring of mine rock masses, mine wide whole
waveform microseismic monitoring, and laboratory
experiments and modelling. To date, preliminary field
and laboratory experiments have been carried out, and
the detailed design of the final experiments is
nearing completion. The insitu pillar experiments
have demonstrated that large underground rock mass
volumes of anomalous velocity may be readily
identified by geotomographic imaging. This work will
be extended for increased resolution using shear
waves and attenuation spectral ratios in order to
study changing rock mass conditions in the vicinity
of an advancing mine front. The data obtained from
this technique will also be used to provide input and
verification for numerical analysis programs.

The representative microseismic data so far
collected, have indicated that insitu rock masses at
the mine under study are anisotropic. Shear wave
splitting has been observed and further studies are
planned to see if these observations can be used to
map the changing principle stress directions and/or
rock texture in the vicinity of the mine and major
geologic structures. The implications of seismic ray
bending to the location of microseismic events will
be investigated and development of source location
techniques in anisotropic rock masses will need to be
studied. In the final stages of this research
project, it is hoped that a technique of
acoustic/microseismic emission geotomography can be
developed. If this is possible, the multiple sources
of mine microseismic activity, sometimes several
hundred events a day, could be used to update the
velocity structure of the mine and to map changing
stress and fracture patterns in the vicinity of the
mine workings.

The ultimate aim of this work is to provide
fundamental scientific information about rockburst
phenomena. The practical objective of the research is
to provide techniques which can be used to identify
anomalous ground conditions and rockburst prone
areas, prior to instability, so that they can be
destressed or taken into consideration in the design
of subsequent mining operations. It is our hope that
by reducing the risk of rockbursts, less emphasis
will need to be placed on the prediction of major
events. Even if prediction is eventually possible,
the consequences of a major event which results in
loss of life and/or total closure of the mine
workings would be unacceptable.

Acknowledgements

The authors would like to thank the Natural Sciences
and Engineering Research Council (NSERC) of Canada,
Queen's University, Falconbridge Mines and Noranda
Research who are jointly sponsoring the project.

References

Blake, W. 1982. Microseismic applications for
mining. U.S. Bureau of Mines report, contract
number J0215002.
Blake, W. 1984. Rock preconditioning as a seismic
control measure in mines. Proc. 1st Int. Con. on
rockbursts and seismicity in mines, S.A.I.M.M.,
Johannesburg: 225-230.
Brink, A.B.Z. & O'Conner, D. 1983. Research on the
prediction of rockbursts at Western Deep Levels
Limited. J. S.A.I.M.M. 83: 1-10.
Crampin, S. 1985. Evaluation of anisotropy by shear
wave splitting. Geophysics 50: 142-152.
Hanks, T.C. 1984. A rms and seismic source studies.
Proc. 1st Int. Con. on rockbursts and seismicity in
mines, S.A.I.M.M., Johannesburg: 39-44.
Hardy, H.R., Zhang, D. & Zelanko, J. in press. Recent
studies relative to the Kaiser effect in geologic
materials. Proc. 4th conference on acoustic
emission/microseismic activity in geologic
structures and materials, Pennsylvania State
University: Trans. Tech. publications.
Leighton, F.W. & Steblay, B.J. 1977. Applications of
microseismics in coal mines. Proc. 1st conference
on acoustic emission/microseismic activity in
geologic structures and materials, Pennsylvania
State University: Trans. Tech. publications: 205-
229.
MacDonald, P. & Muppalaneni, S.N. 1983.
Microseismic monitoring in a uranium mine. In
rockbursts: prediction and control, p. 141-145.
London: I.M.M.
Spottiswoode, S.M. 1984. Source mechanisms of mine
tremors at Blyvooruitzicht gold mine. Proc. 1st
Int. Con. on rockbursts and seismicity in mines,
S.A.I.M.M., Johannesburg: 29-38.
Wong, J., Hurley, P., & West, G.F. 1984. Cross-hole
audio-frequency seismology in granitic rocks
using piezoelectric transducers as sources and
detectors. Geoexploration 22: 261-279.
Young, R.P. & Hill, J.J. 1982. Statistical analysis
of seismic spectral signatures for rock quality
assessment. Geoexploration 20: 75-91.
Young, R.P. & Hill, J.J. 1985. Seismic
characterisation of rock masses before and after
mine blasting. Proc. 26th U.S. symp. rock
mechanics, Rapid City: 1151-1158.
Young, R.P. & Hill, J.J. 1986. Seismic attenuation
spectra in rock mass characterisation: a case study
in openpit mining. Geophysics 51: 302-323.
Young, R.P. in press. Seismic propagation in rock
masses: implications for acoustic
emission/microseismic activity. Proc. 4th
conference on acoustic emission/microseismic
activity in geologic structures and materials,
Pennsylvania State University: Trans. Tech.
publications.

Rock-support interaction study of a TBM driven tunnel at the Donkin Mine, Nova Scotia

Etude de l'interaction massif-supports pour un tunnel foncé au moyen d'un tunnellier à la mine Donkin, Nouvelle Ecosse

Verhalten des Tunnelausbaus während des Vortriebs mit einer Lovat Vortriebsmaschine im Donkin Bergwerk, Nova Scotia

C.M.K.YUEN, Golder Associates (Eastern Canada) Ltd
J.M.BOYD, Golder Associates (Eastern Canada) Ltd
T.R.C.ASTON, Cape Breton Coal Research Laboratory, Energy, Mines and Resources Canada, Sydney, Nova Scotia

ABSTRACT: The initial phase of the Donkin-Morien Development Project in Sydney, Nova Scotia comprises the excavation of two 7.6 m diameter access tunnels to intersect the "Harbour" coal seam at approximately 3.5 km offshore. The first tunnel was driven using a shielded Lovat TBM. The second tunnel was partially excavated by the drill-and-blast method and the remaining length is being driven using the TBM. An extensive geotechnical instrumentation program was undertaken during the TBM drivage to monitor the stability of the tunnel and to study the aspect of rock support interaction. This paper describes the tunnel support system, the geotechnical instrumentation program implemented, and the method of rock support interaction analysis. A comparison of the measured data and the results of the rock support interaction analysis was presented.

RESUMÉ: L' étape initiale du projet de développement Donkin-Morien, situé près de Sydney, N.E., comprend l'excavation de deux tunnels d'accès, de 7.6 m de diamètre, pour rejoindre le lit de charbon Harbour, à une distance de près de 3.5 km au-delà de la côte de l'Ile du Cap Breton. Le premier tunnel fut excavé au moyen d'un tunnelier Lovat. Le deuxième tunnel fut partiellement excavé au moyen de la méthode de forage et sautage puis complété au moyen du tunnelier. Un programme complet d'instrumentation géotechnique fut établi durant l'avance du tunnelier pour recueillir l'information nécessaire à l'évaluation de la stabilité du tunnel, mais surtout pour étudier l'interaction entre le massif et les supports. Cette contribution présente le système de support du tunnel, le programme de surveillance mis en place et la méthode d'analyse d'interaction massif-supports. Une comparaison entre les données obtenues et les resultats d'analyse d'interaction massif-supports est aussi présentée.

ZUSAMMENFASSUNG: Als erste phase im Donkin-Morin Projekt, Sydney, Nova Scotia, wurden zwei Tunnelröhren mit einem Durchmesser von 7.6 m aufgefahren, um das Harbour Flöz etwa 3.5 km vor der Atlantik küste zu erreichen. Der erste Tunnel wurde mit einer Lovat Vortriebsmaschine mit Schild vorgetrieben. Der zweite Tunnel wurde zumteil im konventionellen Bohr/Sprengbetrieb vorgetrieben und der Rest mit der Vortriebsmaschine. Während des Vortriebs wurden die Stabilität und die Verformung von Tunnel und Ausbau gemessen. Diese Arbeit beschreibt den Tunnelausbau, die geotechnischen Messungen und die Analyse der Wechselwirkung von Ausbau und Tunnelwänden.

1, INTRODUCTION

The Sydney Coalfield, owned and operated by the Cape Breton Development Corporation (CBDC), is located in a 2200 m sequence of strata known as the Pictou Group of Upper Carboniferous Age. Exploration carried out in the late 1970's in the vicinity of Cape Percé, had identified a significant resource block containing approximately two billion tonnes of coal. The Donkin-Morien Development Project was set up to develop this resource in the early 1980's.

The initial phase of the project comprised the excavation of two 7.6 m diameter tunnels to intersect the Harbour Coal Seam (the upper-most seam of the resource block) at a depth of 200 m below sea bed, 3.5 km offshore. The location of the Donkin-Morien site is shown in Figure 1.

A rock tunnel boring machine built by Lovat was used to develop the tunnel. The experimental and background activities leading to the use of a Lovat TBM was discussed by Hunter and Lovat (1983). Excavation of the first tunnel (Tunnel No. 2) commenced in January, 1984 and was completed in December, 1984. During drivage of this tunnel, a research contract funded by Canada Centre for Mineral and Energy Technology (CANMET), of Energy, Mines and Resources Canada, was awarded to Golder Associates Ltd. to undertake a joint geotechnical research program with CANMET and CBDC to study the performance of the tunnel and the aspect of rock support interaction.

This paper describes the rock support system, the geotechnical instruments used, and a comparison of the field measurements and the results of rock-support interaction analysis.

2, TUNNEL GEOLOGY, LAYOUT AND SUPPORT SYSTEM

The tunnel was driven northward through an interbedded sequence of sandstone, siltstone, mudstones and coal seams which dip approximately 7 to 9 degrees offshore. The tunnel followed a thick sandstone unit for the first 1,000 m after which it passed stratigraphically upward through the coal measures succession. The tunnel was at a -20% grade for the first 900 m, followed by a vertical curve to approximately chainage 1100 m and a final straight segment (Figure 2) to chainage 3580 m at a grade of -1.027%.

The Lovat TBM used for the drivage is of the trailing shield variety which permits the erection of the support inside the periphery of a trailing, full circumference steel shield. The components in the support system were required to provide reactions to the TBM thrust and to accommodate a variety of geological conditions anticipated along the tunnel alignment. Figure 3 shows the layout of the support system.

The major joint sets encountered in the tunnel were generally subvertical and striking in the east-west orientation. No major faults had been intersected in the tunnel and the seepage into the tunnel was less than 450 litres/min. The rock encountered was grouped into seven types (I to VI A) and their mechanical properties are shown in Table 1.

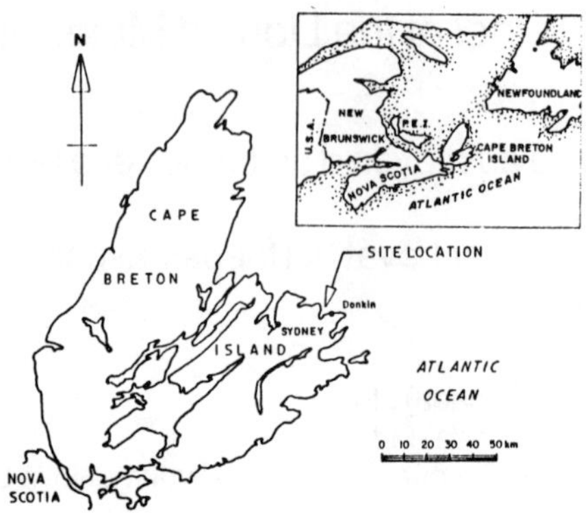

Fig. 1. Site location plan

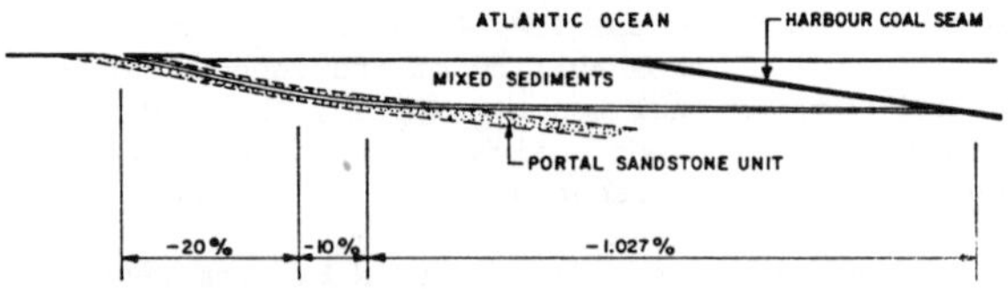

Fig. 2. Tunnel alignment

Table 1. Rock types and rock properties

ROCK TYPE		ROCK MASS PARAMETER				
		E (MPa)	ν	σc (MPa)	m	s
I	Sandstone	24000	0.25	92	5	0.1
II	Interbedded Sandstone - Siltstone	17600	0.25	121	5	0.1
III	Siltstone	11000	0.25	53	5	0.1
IV	Interbedded Siltstone - Mudstone	9000	0.25	36	5	0.1
V	Mudstone	5400	0.25	36	5	0.1
VI	Carbonaceous Mudstone	3450	0.1	16.6	0.05	0.00001
VIA	Coal Seam	3450	0.1	16.6	0.05	0.00001

E = deformation modulus ν = Poisson's ratio

σc = Uniaxial compressive strength of intact rock

m, s = strength parameters for fractured rock (Hoek and Brown, 1980)

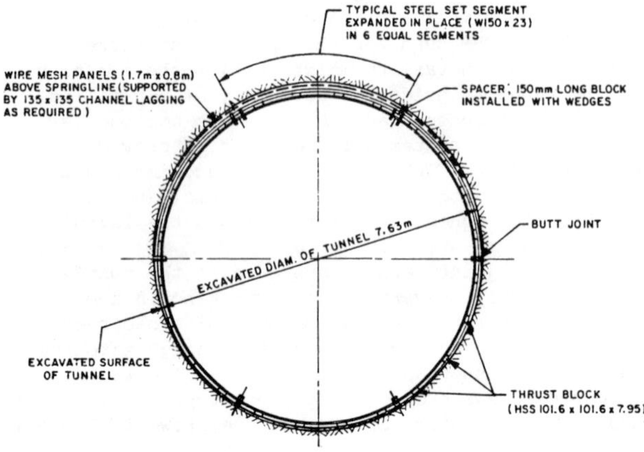

Fig. 3. Tunnel support system

3, INSTRUMENTATION

A total of four sets of ring beam in the tunnel, at chainages 984 m, 1431 m, 1831 m and 3353 m were instrumented to study rock-support interaction. In addition, multiple point extensometers were installed to monitor rock displacements at the crown of the tunnel at various locations along the tunnel to cover a range of geological conditions. The types of instruments used and their functions are summarized in Table 2.

Table 2. Summary of instrumentation

INSTRUMENT (FUNCTIONS)	DESCRIPTION
MPBX's (Extensometer) (to measure rock displacements)	Two types of MPBX's were used: a. Irad Gage Sonic probe (rigid probe) multiple-point extensometers generally with 3 anchors with deepest one at 21m. Resolution ± 0.002 in. (0.05 mm). Readout unit MB-7 (intrinsically safe for use in coal mines). b. Irad Gage Sonic probe (flexible probe) multiple-point extensometers generally, with 10 anchors with deepest one at 7.6 m. Resolution ± 0.002 in. (0.05mm). Readout unit MB-7.
STRAIN GAUGES (to measure strains in ring beams and thrust blocks)	Irad Gage Model SM2W weldable vibrating wire strain gauges welded on steel set with spot welder. Sensor for gauge either spot welded or epoxied. Gauges read with MB-6LU readout unit (intrinsically safe for use in coal mines). Maximum strain range of gauge is 2500 microstrain units.
PRESSURE CELLS (to measure axial pressures at butt joints and radial pressures on circumference)	Two types of Irad Gage vibrating wire pressure cells were used: a. Curved pressure cell model EPC-500: max. pressure range 500 psi (3.5 MPa), sensitivity 0.5 psi (3.5 kPa), custom made to 0.9 m x 0.15 m x 6.3 mm thick to fit profile of ring beam. b. Square pressure cell model EPC-1000: max. pressure range 1,000 psi (7 MPa), and 2,000 psi (14 MPa), sensitivity 1 psi (7 kPa) custom made to 0.15 m x 0.15 m x 6.3 mm thick to fit between butt joints of ring beam. Pressure cells read with MB-6LU readout unit.

4, ROCK SUPPORT INTERACTION STUDY

The rock support interaction study for the Donkin-Morien tunnel has two primary objectives. The first is to monitor the stability of the tunnel with the aid of the field measurements. The second is to study the response of the tunnel support to the rock loads under various geological conditions and to assess whether a simple analytical procedure could be of use in providing a reasonable basis for support design.

In the following sections, the axial stresses (with respect to the ring beam section) sustained by the support system and the radial pressures applied on the ring beams are presented and compared with the results of analysis.

4.1, Support Installation Sequence

In order to provide a basis for the discussion of the data from field measurements, an outline of the support installation and rock-support interaction sequence during the TBM drivage is presented below.

The ring beam segments for tunnel support were initially assembled within the TBM shield. When the TBM started to advance, the ring beam was slowly pushed off the shield and the machine thrust of up to 9000 kN was transmitted to the thrust blocks connecting the ring beams (see Figure 3). Even though the pressures in the 24 hydraulic rams of the TBM were uniform, there were considerable variations in the thrust loads sustained by the support system (Figure 4). In general, the thrust loads were concentrated at the lower half of the tunnel.

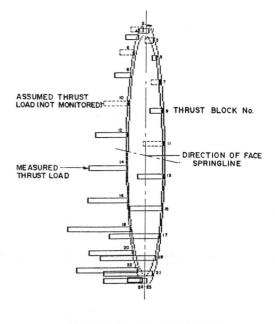

ASSUMED THRUST
LOAD (NOT MONITORED)

● THRUST BLOCK No.

DIRECTION OF FACE
SPRINGLINE

MEASURED
THRUST LOAD

● LOADING RAMS USED FOR THRUSTING
1,2,10,13,14,15,16,17,18,19,20,
21,22,23,24.

SCALE 200   0   200   400 kN.

Fig. 4. Distribution of thrust load

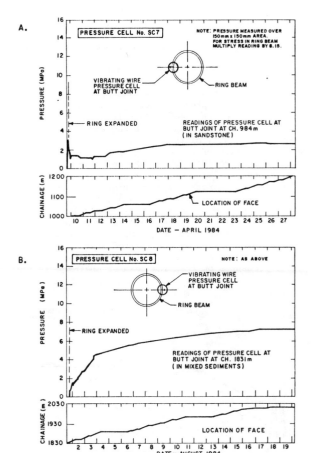

Fig. 5. Distribution of axial stresses at
chainage 984 m and 1831 m

Table 3. Summary of axial pressure

Chainage (m)	Rock Type	Instrument No. [1]	Measured *[2] Axial Stresses (MPa)	Special Features of Support System	Estimate depth of loosened rock based on extensometer data [3] (m)
984	Sandstone	SC7 SC8	0.82 7.6 (Av. = 4.2)	No movement took place along sliding joing at crown segment	0.1 - 0.2
1,423	Mixed sediment (Type II)	SC7 SC8	79.2 59.7 (Av. = 69.5)	Sliding joint allowed to move by 150 mm	2 - 2.5
1,831	Mixed sediment (Type IV and V)	SC7 SC8	- 62.3	Sliding joint movement limited to 50 mm	1.8 - 2.0
3,353	Mixed sediment (Type V and IV)	SC7 SC8	79.3 104.5 (Av. = 92)	Sliding joint not allowed to move	2.0 - 2.5

Note

1. SC7 and SC8 are pressure cells on left hand and
   right hand sides of springline, respectively.

2. Corrected for area factor.

3. Refer to Yuen et al (1985).

When the ring beam was extruded out of the shield,
it was then expanded against the tunnel wall by means
of a hydraulic jacking system known as a "rib
expander". To facilitate expansion of the ring
beam, several different details (see Table 3) were
used at the butt joints connecting the top and
adjacent segments (Yuen et al, 1985). Because of
this machine thrusting methodology, the process of
rock-support interaction did not occur until the
ring beam was expanded in place. At this point,
the face would be about 8 m ahead.

4.2, Results of Field Measurements

Axial stresses in ring beam

Figures 5A and B illustrate the axial pressures
measured at the springline butt joints of two ring
beams, one in the sandstone and one in the mixed
sediments. The pressures shown were measured over
an area of 150 mm x 150 mm. To evaluate the actual
stresses in the ring beam section, a multiplying
factor of 8.15 is required. The results generally
indicate that the ring beams were subjected to
substantial changes in axial stress prior to the
expansion of the ring beam. , At the moment when the
ring beam was expanded, the axial pressure increased.
It then decreased slightly as the "rib expander"
was released due to relaxation of the system. As
the rock load was transferred to the ring beam, the
axial stresses gradually increased and stabilized
when equilibrium conditions were reached.

A summary of the measured axial stresses in the
ring beam and the estimated depth of rock loosening
at the crown of the tunnel (Yuen et al, 1985) is
presented in Table 3 below.

In general, the axial pressures were not
symmetrical due probably to slight eccentricity in
the applied load. It may also be noted that the
loading from the more competent sandstone is
substantially lower than that from the mixed
sediments.

1341

Radial Pressures on Rock-Ring Beam Interface

The radial pressures measured by the curved pressure cells at the rock-ring beam interface were generally very low at the crown segment with the exception of the one at chainage 3353. In general, the rock-ring beam contact at the crown was not as uniform as those at other pressure cell positions on the circumference of the beam. This is primarily due to the fracturing (spalling) of the rock in the immediate vicinity of the crown. As the pressure cell on the circumference of the ring beam is a contact pressure cell which can only sense pressure applied over the area of the cell, the interlocking of the fractured rock fragments might have prevented direct contact. On the other hand, if a rock fragment happened to be positioned directly on the pressure cell, the measured data would reflect the localized pressure which might be very high because the load was applied over a limited area.

Figure 6 shows a typical distribution of the radial pressures on the ring beam at chainage 1831 m.

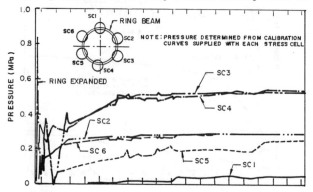

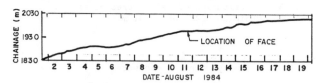

Fig. 6. Distribution of radial pressures at chainage 1831 m

In a manner similar to the response of the axial pressure cell, there were fluctuations in the radial pressure cell readings prior to and soon after expansion of the ring beam. Thereafter, the pressure gradually increased to a stabilized value.

A summary of the radial pressures measured at the instrumented ring beam locations is presented in Table 4. Within the sandstone, the radial pressures on the ring beam are substantially lower.

Table 4, Summary of measured radial pressures

Measured Radial Pressures (kPa)

Instrument No.	Ch. 984	Ch. 1431	Ch. 1831	Ch. 3353
SC1	0	40	110	470
SC2	0	400	240	190
SC3	230[2]	380	300	250
SC4	10	280	470	820
SC5	-[3]	580	230	230
SC6	20	170	280	360

Notes:
1. SC1 to SC6 are curved pressure cells. SC1 and SC4 are at crown and invert, respectively. Instruments are numbered clockwise.
2. Possibly point contact.
3. Pressure cell not functioning.

Within the mixed sediments, the crown segment of the instrumented ring beams at chainage 1420, 1831 and 3353 m were in ascending order of rigidity because of the different rigidity at the crown butt joints (see descriptions in Table 3). At chainage 3353, there was little allowance for displacement to occur at the butt joint whereas at chainage 1420 and 1831 m, the butt joints were allowed to displace up to 210 and 50 mm, respectively. Consequently, the rate of increase in radial stresses was faster at chainage 3353 m and the axial stresses were higher.

4.3, Rock-Support Interaction Analysis

On the basis of the observations made in the tunnel, the measured displacement pattern, and the results of stress analyses (Yuen et al, 1985), failure around the tunnel occurred only in the crown and invert of the tunnel. Hence, the assumption of a uniform plastic zone around the tunnel in the rock-support interaction models (e.g. Hoek and Brown, 1980 and Ladanyi, 1974) did not appear to be appropriate in this case.

Rock movements as measured by the extensometers were most prevalent at the crown where rock loosening occurred (Boyd et al, 1986). The ring beam was therefore primarily subjected to external loading from the weight of loosened rock. This loading configuration is consistent with the field observation and is considered to be realistic in regards to the analysis of rock-support interaction.

In order to study the effect of rock-support interaction under applied loading over a limited area, numerical modelling was considered to be most appropriate. The numerical model (Figure 7) used for the analysis was a two-dimensional finite element model which comprised 24 beam elements which represented the tunnel liner (ring beam). Each of the nodal points on the liner was connected to a "spring element" oriented in the radial direction representing the rock surrounding the tunnel. The modulus of the rock was represented by the stiffness of the spring. The bar elements representing the liner could transmit axial load, moment and shear while the spring elements representing the rock could only transmit normal load.

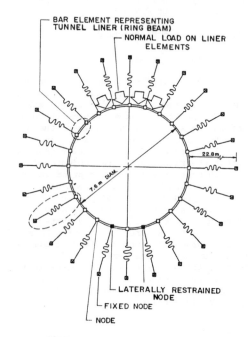

• NODES ON TUNNEL LINER ARE EQUALLY SPACED

• NOT DRAWN TO SCALE

Fig. 7. Numerical model

The rock load on the support system was assumed to be applied over the four beam elements representing the crown segment of the ring beam.

The applied loads on the beam elements varied between different ring beam arrays. The depth of the loosened zone at the position of the instrumented ring beam was estimated on the basis of the measured displacement distribution recorded in the near-by extensometers (Boyd et al, 1986). Based on field observations, it was anticipated that the loosened zone was thickest at the crown line and then gradually reduced in thickness towards the edge of the area of loosening. Two loading configurations as shown below were studied.

Load Pattern A: uniform loading across the two central beam elements. The loading then gradually reduced to 10 per cent of the maximum loading over the length of the adjacent elements.

Load Pattern B: uniform loading across the four beam elements.

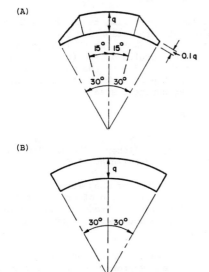

Note: q = Applied crown surcharge

Fig. 8. Loading pattern on crown segment

4.4, Comparison of Measured Data with Results of Analyses

In this section, the results of numerical analyses are compared with the measured data. The intent was to assess whether a simplified numerical model would have been useful as a design tool to evaluate the adequacy of the TBM excavation support system for the rock conditions at Donkin-Morien prior to construction.

Axial Stresses in the Ring Beam

The calculated axial stresses in the ring beam based on the two loading cases are shown in Figure 9 with the average axial ring stresses plotted against the maximum extent of crown surcharge (q). The maximum extent of the loosened rock (above the crown) corresponding to the maximum crown surcharge pressure is also shown as a separate horizontal axis in the plot.

It may be noted that all the measured data falls within the range of axial pressures computed by the numerical model. On the basis of the results shown on the plot, the measured data generally indicates that loading pattern A is the more likely load distribution from the loosened rock. This is

consistent with the assessment based on field observations.

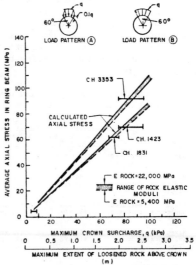

Fig. 9. Comparison of measured and calculated axial stresses

The above comparisons provide a good calibration of the numerical model in that the axial stress data is least affected by factors related to the propulsion of the TBM and the contact at the interface of the ring beam and the rock. The data suggests that the graphs shown on Figure 9 provided a reasonable relationship between the axial stress in the ring beam and the anticipated rock load. Similar graphs can be developed for other beam sections and other loading patterns.

Radial Stresses on the Ring Beam

On the basis of the data from the axial stress measurement, it is considered that for the purpose of the rock-support interaction analysis, graph A in Figure 9 was reasonably representative of the loading condition and the measured rock response. On this basis, the input crown surcharges for analyzing the instrumented arrays are summarized below:

Chainage (m)	Input Crown Surcharge for Rock-Support Interaction Analysis (kPa)	
984	6	)
1431	79	) Based on Loading
1831	73	) Pattern A
3353	104	)

In summary, the measured and computed radial stresses at the ring beam rock interface are of the same order of magnitude (Table 5). The sum of the measured radial stresses at positions SC2, SC3, SC5 and SC6 are generally symmetrical with respect to the vertical centreline of the tunnel. The contact pressure at the crown segment depends on whether there is direct contact between the loosened material and the ring beam.

It is of interest to note that the butt joints at the crown segment in each of the instrumented arrays in the mixed sediments have slightly different design features (see Table 3). The ones at chainage 3533 m had the least clearance for changes in the circumferential dimension of the ring beam whereas the sliding joint at chainage 1423 m allowed circumferential movements (closures) of up to 210 mm when the sliding joint closed. The clearance for circumferential movement at chainage 1831 m was limited by the spacer blocks and was intermediate (50 mm) between the other two. Based on the magnitudes of the radial stresses measured at the crown segment at these three stations, it appears that a relatively more "rigid" support (at chainage 3533 m) resulted in better rock-ring beam contact at the crown segment.

Table 5, Comparison of measured and computed radial pressure on ring beams

Chainage (m)	Rock Type		Sum of radial stress at SC5 and SC6 (Left of ℄)	Sum of radial stress at SC2 and SC3 (right of ℄)	Average radial stress at positions SC2 to SC6
984	Sandstone	Measured	(1)	(1)	(1)
		Computed	60	60	30
1,423	Measured sediments (Type III)	Measured	750 kPa	780 kPa	350 kPa
		Computed	710 kPa	710 kPa	350 kPa
1,841	Mixed Sediments (Type IV)	Measured	540 kPa	510 kPa	304 kPa
		Computed	655 kPa	655 kPa	323 kPa
3,353	Mixed Sediments (Type II and III)	Measured	950 kPa	440 kPa	370 kPa
		Computed	936 kPa	936 kPa	462 kPa

(1) Insufficient data

In the case where the circumferential dimension of the ring beam was allowed to change by a significant amount (as in the case at chainage 1431 m) the decrease in the circumferential dimension of the beam would be translated into vertical displacements. Such beam displacement would allow the loosened rock fragments at the crown to re-orientate and interlock. The wire mesh supporting the loosened rock would likely sag by a larger amount because the ends of the mesh panel were not tightly restrained by the ring beam. In turn, the ends of the mesh might tilt slightly and prevent good contact between the rock and the beam. The total rock load would still be transmitted to the ring beam but the radial stress at the crown segment would be low.

This loading mechanism is reflected in the results at chainage 1431 and 3353 m. While the average measured radial stresses at positions SC2 to SC6 at these two arrays are approximately the same (see Table 5) there is substantial difference between the radial stresses measured at the crown segments at these two sections. The simplified numerical model used for the analysis is limited in taking this loading mechanism into account.

Both the measured and computed results indicated that the loading on the support system was substantially less in the sandstone than in the mixed sediments. The agreement between measured and computed radial stresses at individual pressure cell positions was only fair due to non-uniformity of contact at the interface. However, there is reasonable agreement between the overall average of the measured and computed radial pressures at the positions of pressure cells SC2 to SC6.

5, CONCLUSIONS

Rock-support interaction was analyzed using a simplified two-dimensional finite element model with beam elements representing the ring beam and spring elements representing the rock. The rock loading configuration was estimated on the basis of the measured displacement pattern and field observations.

A comparison of the measured data and results of analysis indicate that: -

o The model provided a reasonable assessment of the axial stresses in the ring beam. Both the measured and computed results indicate that the rock load on the support system was substantially less in the sandstone than in the mixed sediments.

o The model indicated a fairly uniform distribution of radial stresses at the contact of the ring beam and the rock below the crown segment. The measured radial pressures were generally non-uniform and highly dependent on the contact condition. The rigidity of the butt joints at the crown segment affected the ring beam-rock contact and hence the resulting radial pressure at the crown. The sum of the measured radial stresses were generally symmetrical with respect to the vertical centreline of the tunnel. The overall average of the measured and computed radial stresses agreed reasonably well.

In the overall context, the numerical model used provided a reasonable assessment of the performance of the ring beam but it was limited in simulating the fine details of the loading mechanism such as that resulting from the rigidity of the butt joints at the crown segment and the effects of out of plane bending of the ring beam.

ACKNOWLEDGEMENTS

The authors would like to thank the Cape Breton Development Corporation and the Coal Research Laboratory of CANMET for their permission to publish the results of the tunnel research study. Thanks are also due to the CANMET Research Program office and the Department of Supply and Services, Science Procurement Branch.

REFERENCES

Hoek, E. and Brown, E.T. "Underground Excavation in Rock", publisher: The Institution of Mining and Metallurgy, 1980.

Ladanyi, B. "Use of Long Term Strength Concept in the Determination of Ground Pressure on Tunnel Linings". Proceedings, 3rd International Congress on Rock Mech., Denver, Vol. 2, Part B, 1974.

Boyd, J.M., Yuen, C.M.K. and Marsh, J.C. "Design and Performance of Donkin-Morien Tunnels",Proceedings 6th Canadian Tunnelling Conference, 1986.

Yuen, C.M.K., Gilby, J.L. and Boyd, J.M. "Measurement and Analysis of Rock Deformations and Support System Response in the Drill and Blast and Bored Access Drivages at the Donkin-Morien Project"(UP-G198), Vol. 1-4, 1985.CANMET Contract Report No.26SQ.23440-2-9159.

Implementation of finite element model of heterogeneous anisotropic rock mass for the Tkibuli-Shaor coal deposit conditions

Application de la méthode des éléments finis au gisement houiller de Tkibuli-Shaor situé dans des roches anisotropiques et hétérogènes

Die Anwendung der Finite-Element-Methode auf die anisotropen und heterogenen Gesteine der Tkibuli-Shaor Kohlenlagerstätte

S.A.YUFIN, Moscow Academy of Civil Engineering, USSR
I.R.SHVACHKO, Moscow Academy of Civil Engineering, USSR
A.S.MOROZOV, 'Hydroproject' Institute, Moscow, USSR
T.L.BERDZENISHVILI, Geophysics Institute, Tbilisi, USSR
G.M.GELASHVILI, VNIMI, Tbilisi, USSR
Z.A.GORDEZIANI, Gruzugol, Kutaisi, USSR

ABSTRACT: Numerical modelling of complex geologic and mining conditions of Tkibuli-Shaor coal deposit region using the finite element method is performed and proven to be effective as the means for repetitive operative predictions of burst-free mining sequence.

RESUME: On présente la méthodologie efficace basée sur la méthode des éléments finis afin de prévoir les coups de toit lors de l'exploitation du gisement de houille de Tkibuli-Shaor dans les conditions géologiques complexes au moyen de la simulation numérique.

ZUSAMMENFASSUNG: Die auf der FEM enturckelte effektive Methodik der operativen Bestimmung des Bergschlagungefährdeten Abbauverhältnisse des Kohlevorkommens Tkibuli-Schaor in erschwerten geologischen Bedingungen durch das Sequentialrechenverfahren wird erklärt.

The Tkibuli-Shaor coal field is one of the most complicated coal deposits of the country whose exploitation presents considerable difficulties stipulated by the construction geology. Thick coal strata occurs at a large depth and is blast-prone due to gas and coal dust and is self-igniting. Frequent dynamic phenomena caused by high elastic properties of the coal, the presence of tectonic field exceeding the gravitational one and seismic activity of the deposit can be observed at large depths (Geology... 1963; Geology... 1964).

The coal field region with a total area of 25 thou.sq.m is a rugged terrain. Its major orographic feature is the Nakeral ridge confining the Tkibuli depression from three sides.

As far as the geology is concerned, the coal field region is composed of the rocks of Jurassic, Cretaceous and Quarternary age. The stratigraphic section base of the region is represented by alternation of porphytes, sandstone tuffs and breccia tuffs of total thickness of 2500 m called a porphytic Bajoss series. The Bajoss formation is overlaid by arkose sandstones with tuff sandstone and clayey shale interbeds, 200 m thick, alternating with a series of paper shales referred conventionally to Bathonian and represented by argillites, clayey shales with interlayers of sandstones, marls and microlaminated aleurolites, 150 m thick.

A coal-bearing formation overlies carboniferous shales in which three layers are distinguished and they are as follows: quartz and felspar "lower sandstones" with interbeds of argillites and aleurolites, 150 m thick, a coal strata itself, of the average thickness of 50 m and "upper sandstones" of quartz-feldspar composition, 100 m thick.

The coal strata within the occurrence is represented by two coal bedding divided by a coal free intermediate section with traces of ancient scouring in the lower sandstones roof. At some places the coal strata comes out to the surface or is underlain with Jurassic and carbonaceous rocks of Cretaceous period of a thickness from 600 to 1700 m.

The coal strata is of a complex structure consisting of numerous, not persistent in thickness and alternating benches of humic and liptobiolithic coals, coaly and coaly clay shales, argillites, aleurolites and sandstones of the thickness from one centimeter to 6-8 m. Its total thickness varies from 0 to 80 m averaging 50 m. And the effective coal mass share makes up 50-60% of the total thickness in the areas being mined at present.

The upper Jurassic deposits locally called a "multicoloured series" are envountered within the "upper sandstone" strata crowning a section of a coal-bearing series and the Middle Jurassic series.

It is overlapped by Creataceous deposits composed of heavily karsted organogenic and quartzous limestones averaging 500 m in thickness. The benches of mountain ridges are composed of these limestones. The Quarternary system is represented mainly by talus deposits draping over the slopes of the Nakeral ridge and local uplands by a mantle of variable thickness (from 0 to 85 m). The tectonics of the deposit is rather complicated. Both large and fine folds and dislocations with a break in continuity are found here. The dip angle of fold limbs reaches 40°-50° at coal strata outcrops and decreases rapidly to 20°-11° with depth manifesting even a reverse dip.

Tectonic discontinuities are represented mainly by the Sabelasur and Moharaul faults. The Sabelasur fault, almost of a latitudinal strike is the largest dislocation of Pre-Cretaceous age; the fault amplitude is 400-500 m; dip angle is 80-85°. The Moharaul fault, a tremendous dislocation of the Post-Cretaceous age crosses the Shaor coal field from South-East to North-West. The fault amplitude varies from 750 m to 350 m, dip angle is 50-70°. Apart from large tectonic dislocations in the underground excavations numerous small discontinuities are observed with amplitudes up to 15 m and dip angle of 60-80°.

1345

Table 1. Physical and mechanical characteristics of rock mass of Tkibuli-Shaor coal field.

Sl. Nos	Rock description	Bed thickness, m	Young's modulus E, GPa	Poisson's ratio	Volumetric weight, tf/m³	Strength compr., MPa
1	Limestone	500	24	0.22	2.5	120
2	Sandstone of multi-coloured series	700	30	0.25	2.43	70
3	Heaving clay	100	20	0.45	1.88	20
4	"Upper" sandstone	100	32	0.2	2.6	100
5	Coal strata	50	2	0.3	1.36	10
6	"Lower" sandstone	150	33	0.2	2.6	100
7	Paper shale	150	15	0.25	2.5	50
8	Arkose sandstone	200	35	0.15	2.6	120
9	Porphyric series	2500	40	0.12	2.65	130

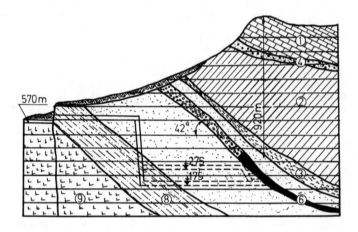

Figure 1.

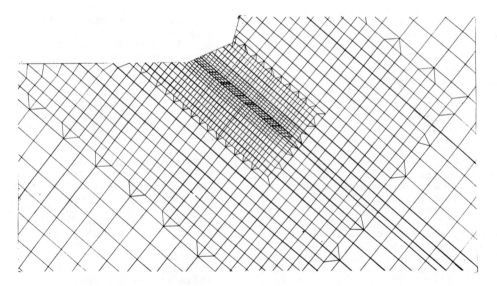

Figure 2.

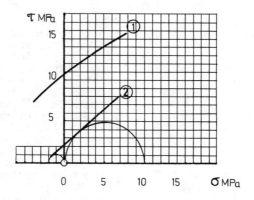

Figure 3. Strength diagram of tuff sandstone: 1 - intact rock; 2 - shear test along cracks.

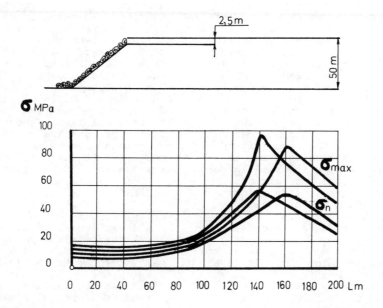

Figure 4.

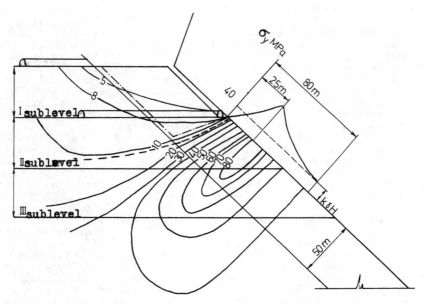

Figure 5.

1347

The seismic studies conducted in the area of the Tkibuli-Shaor coal field showed that horizontal stresses are 1.3-1.8 times more than vertical stresses and this has been confirmed by the results of mine survey and geophysical studies.

The coal stratum to be mined out is located at the depths from 600 to 1100 m. A tectonic fault is encountered in a roof at an angle of 70° to the horizon. The fault amplitude makes up 200 m. The natural stresses in the mass are conditioned by tectonic manifestations and at the depth of 1000 m the average values of horizontal stresses σ_x = = 37.5 MPa, that of vertical stresses σ_y = = 25.0 MPa. A coal seam, 50 m thick, is inclined at an angle of 40° to the horizon and is about 700 m long at the stretch under consideration.

It is established that a coal seam is characterized by intensive jointing. The major systems of joints divide the seam into sufficiently isometric and monolithic blocks to a certain extent (joints are not visible) and due to this a coal strata is characterized by a capability to pseudo-plastic deformation at high elasticity and accumulation of potential energy of elastic compression in monolithic blocks.

The rock mass in question consists of nine beds. Each bed possesses the properties of a respective rock featuring the geological structure of the Tkibuli-Shaor coal field. The numerous data on the confined and unconfined tests, the data on separate large-scale tests and the known close analogues were used for choosing the design deformation and strength parameters of enclosing rock and coal. The physical and mechanical characteristics of rock mass are presented in Table 1.

The purpose of the present paper is to analyse the optimum sequence of coal mining at a certain coal deposit and to formulate the software needed for speedy study of the similar conditions in other areas.

Mathematical modelling has been based on a large programme complex "STATAS", using the finite element method. The programme has been worked out in the Moscow Institute of Civil Engineering named after V. Kuibyshev (Yufin et al. 1985; Morozov et al. 1985).

The selected rock mass sections of the coal seam under consideration are represented in the FEM models by the second-order isoparametric quadrilateral elements, the size of the area was about 1100x650 m. The rest portion of the enclosing rock is represented by quadrilateral and triangular elements of the first order. The size of the analysis area (13500 m x 6500 m) has been selected so that the exterior boundaries are removed from the zone in question, i.e. a coal strata and from the adjoining enclosing rocks. The FEM mesh used in the analysis consists of 3663 nodes and 2409 elements out of which 674 elements are of the second order.

The problem is solved to meet the plane deformation conditions.

Cross-section of the geological region and part of the FEM mesh are shown respectively on Fig. 1 and Fig. 2. Deformation models were formed for each type of rock as based on the in-situ and laboratory triaxial tests performed on cubic samples mathematical model is in general hardening/softening elastic-visco-plastic model of solid formed with particular influence of papers (Hughes

& Taylor 1978; Lade & Nelson 1984), verified during mathematical modelling of testing conditions and paramtrically formulated for specific properties of given rock type. An example of strength curve for one of the rock types is presented on Fig. 3; rather detailed description of such curves has been given in (Morozov et al. 1985; Gaziev et al. 1984).

Research performed during last several years permitted the prediction of burst-prone regions in the rock mass during mining process and evaluation of rational sequence of excavation and optimal layout of adits. Illustrative results are given on Fig. 4 and 5. Thus on Fig. 4 distribution of maximal main stresses (I) and normal to layer (II) stresses obtained as elastic (1) and elasto-plastic (2) solution for 2.5 m thick cut in a 50 m coal layer is shown. Fig. 5 shows distribution of normal to layer stresses and boarders of burst-protected zones after excavation of heading on the first sublevel.

The important conclusion during this research has been the recommendation to divide every 100 m of coal layer in three sublevels and to place all adits in the surrounding rock mass not closer to the coal layer than 20 m. Analysis of variations in excavation sequences and sizes of different cuts permits to recommend for given mining situation the best solution. Essential is a good coincidence of numerical and in-situ data obtained in abundance at the Tkibuli-Shaor coal deposits by "Gruzugol" (Georgian Coal) Mining Organization and Georgian branch of VNIMI.

REFERENCES

Geology of USSR deposits of coal and flammable shales. 1963. Vol. 1. Moscow: Gostekhizdat (in Russian).
Geology of the USSR. 1964. Vol. X. "Georgian". Moscow: Nedra (in Russian).
Blake, W. 1971. Rock bursts research at the Galena mine, Wallace, Idaho. U.S. Bureau of Mines Technical Progress Report - 39, Denver, Colorado, August 1971.
Fadeev, A.B. & E.K. Abdyldaev 1979. Elasto-plastic analysis of stresses in coal pillars by finite element method. Rock Mechanics, 11: 243-251.
Yufin, S.A., Postolskaya, O.K., Shvachko, I.R. & V.I. Titkov 1985. Some aspects of underground structure mechanics in the finite element method analysis. Fifth International Conference on Numerical Methods in Geomechanics (Nagoya): 1093-1100.
Hughes, T.I.R. & R.L. Taylor 1978. Unconditionally stable algorithms for quasi-static elasto/visco-plastic finite element analysis. Computers and Structures, 8: 169-173.
Morozov, A.S., Postolskaya, O.K. & S.A. Yufin 1985. Structure sensitivity to different geologic conditions as the base to optimization of rock mechanics research setup. International Symposium on the Role of Rock Mechanics in Excavations for Mining and Civil Works. Zacatecas, Mexico, 1985, 2B.12.
Lade, P.V. & R.B. Nelson 1984. Incrementalization procedure for elasto-plastic constitutive model with multiple, intersecting yield surfaces. Int.J. for Num.Anal.Meth. in Geomech., 8: 311-323.
Gaziev, E., Morozov, A. & V. Chaganian 1984. Comportement expérimentale des roches sous contraintes et peformations triaxiales. Revue Francaise de Géotechnique, 29: 43-48.

Modelling design of the Caiyuan Tunnel in Chongqing
Modèle de conception du tunnel de Caiyuan à Chongqing
Entwurf auf Grund von Modelluntersuchungen des Caiyuan Tunnels in Chongqing

ZHU JINGMIN, Chongqing Institute of Architecture and Engineering, People's Republic of China
ZHU KESHAN, Chongqing Institute of Architecture and Engineering, People's Republic of China
GU JINCAI, Chongqing Institute of Architecture and Engineering, People's Republic of China
WANG LIWEI, Chongqing Institute of Architecture and Engineering, People's Republic of China
WANG JIANTING, Chongqing Designing Institute of Buildings and Surveying, People's Republic of China

ABSTRACT: A modelling design approach for the preliminary dimensioning of underground opening supports is presented in this paper. The method is based on the model test results under the condition of plane deformation. It is to be used for the analysis and design of the primary support measures and the proper thickness of secondary linings as well as appropriate construction techniques adapted to the condition of the Caiyuan Tunnel in Chongqing.

RESUME: Dans ce texte sera présentée une méthode de la conception similaire d'avant-projet sur le support de l'ouverture souterraine. Cette méthode resulte des épreuves d'un modèle sous la condition de deformation de la plane. L'utilisation de cette méthode consiste à analyser et concevoir la mesure des supports de première étape et l'epaisseur propre au linge secondaire sinsi que la technique de construire le Tunnrl Caiyuan, convenable aux conditions de Chongqing.

ZUSAMMENFASSUNG: Es wird ein Verfahren zur Vordimensionierung der Tragkonstruktion von unterirdischen Räumen auf Grund von Modelluntersuchung im ebenen Deformationszustand aufgestellt. Mit Hilfe von diesem Verfahren werden die Bemessungen der zuerst erstellten (primären) Aussteifungskonstruktion, der danach erstellten Innenausfütterungen und die Auswahl des günstigen Bauverfahrens durchgeführt.

1 INTRODUCTION

Chongqing, situated at the confluence of the Jialing and Yangtze (Chang) rivers, is the economic center and communication hub of the upper Yangtze. It is a mountainous city in the southeastern part of the Sichuan Basin, some 2000 km upstream the Yangtze from Shanghai. Railways, airliners and highways take passengers and cargoes to and from the city at an ever-increasing speed. As the metropolitan center is located in a narrow peninsula with a building density as high as 85%, brosd expansion plan and improvement of the existing traffic transportations has been under way. Five highway and railway bridges have been built across the Yangtze and Jialing rivers. Another highway bridge across the Jialing is under construction. The most important item for traffic improvement is the construction of quite a number of roadway and highway tunnels in recent years. Among them, the Xiangyang Twin Tunnel near the Chongqing railway station, the Eling Tunnel under the Eling Park, the Lanping Tunnel on the south bank of the Chongqing Changjiang Bridge, the Geleshan Tunnel near the Chongqing Airport, etc. To aese the traffic tension in the Daping region, the Caiyuan Tunnel connecting the Chongqing railway station with the higher level Yuajiagang is now in the stage of preliminary design. The design technique of tunnel linings presented in this paper is based on the findings from model tests which will be used as a guide in selecting and dimensioning the linings, forecasting the optimum convergence and time of lining as well as back analyzing the field measurements to improve the current design practice.

2 GEOLOGICAL CONDITION

The site of the Caiyuan Tunnel, the No.2 tunnel on the Caiyuan Road, is located in a hilly stretch along the norhtern bank of the Yangtze. The terrian elevation rises from 235m to 303m with slopes of 10 to 45 degrees and even ateeper locally near the east portal, which rises at an elevation of 276. The terrian near the west portal is open and rather flat with an elevation of 289m. The tunnel pierces the southwest wing of King Dragon Cave Anticline. Geological exploration by 23 drill holes with a total length of 502m and 15 auger borings with a total length of 46m showed that the strata belong to the Shaxmiao formation of the middle Jurassic series. They mainly consist of thin to thickly bedded silty mudstone with intercalated layers of thickly and very thickly bedded, medium to fine-grained felspathic sandstone as well as thickly to very thickly bedded, fine-grained argillaceous sandstone. Most strata appear as lenticular. Apparently, no fault has been explored. Outcrops can be seen near the east portal area. Other portions are covered with fills of 1 to 3.7m. The weathering depth of rock is 1 to 4m with a maximum of 7m. Since the strata are not much influenced by the tectonic movements, shear fractures or tension fractures are rare. Most of them appear as blind joints. The aperture of tension fractures is 0.2 to 0.1 cm with limited length. The dip direction of the strata is evaluated to be 255 to 265 degrees; they dip 5 to 9 degrees.

3 TUNNEL LAYOUT AND ROCK MASS CHARACTERIZATION

The Caiyuan Tunnel is a twin double-lane roadway tunnel. Its plan layout, cross section and longitudinal cross section are shown in Figs.1.2. and 3. The linear length of the tunnel is 405m with 10.13m difference in elevation between the portals.

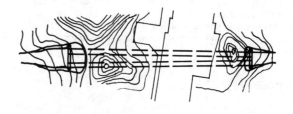

Figure 1. Plane layout of the Caiyuan Tunnel

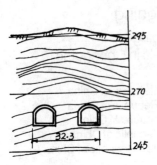

Figure 2. Cross section and associated geological condition.

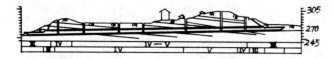

Figure 3. Longitudinal cross section and associated geological condition.

For rock mass classification, specimens taken from three boreholes have been used for laboratory studies. They include physical properties, uniaxial compression test, split tension test and triaxial test. A short summary is listed in Table 1:

Table 1. Laboratory test results

Rock	Uniaxial compre. strength	Tan Ø/c	E	Mu	F_k
	MPa	/MPa	GPa		
	Borehole ZK 2				
Ms	8.5/10.1		7.0	.22	3-4
Ss	67.0/70.4	1.09/8.3	92.0	.28	3-5
As	13.6/	0.99/2.9	59.8	.20	3.5-4
	Borehole ZK 9				
Ms	24.0/33.4	0.94/3.4	52.8	.27	3-4.5
Ss	48.0/51.8		131.3	.30	5-6
As					3.5-4.5
	Borehole ZK 18				
Ms	5.8/8.5	0.89/2.0	9.7	.22	3-4
As	9.2/13.4		15.5	.29	3.5-4

Wherein Ms stands for mudstone, Ss for sandstone, As for argillaceous sandstone; the numerator in the column of uniaxial strength stands for saturated specimens, the denominator for dry specimens.

It may be seen that in Fig. 1 the tunnel axis is nearly parrallel to the dip direction of the strata. In Figure 3, there lists rank of rockmass classification for railway tunnels and those chosen for preliminary design.

4 TUNNEL LININGS

As the geological conditions at the Caiyuan Tunnel are very similar to those at the Xiangyang Tunnel, the same design procedures were followed for preliminary dimensioning of the linings except that both rock-

bolts and shotcrete were chosen for temporary supports.
Rock bolts of 22mm in diameter at a spacing of 50cm by 100cm are used on the roof while the length is 3m for rank III and IV rock and 2.5m for V.Wire mesh is also specified for rank III and IV rock. Shotcrete on the roof is 10cm thick and that on the side wall 7cm. Formed concrete was chosen as primary support, see Table 2.

Table 2. Formed concrete lining.

Rank	Roof arch		side wall
	Crown	Springing	
III	80cm	120cm	115cm
IV	70cm	110cm	105cm
V	60cm	100cm	95cm

The successful applcation of shotcrete as primary support in jointed hard rock hyraulic tunnels (Zhu and Li, 1983) and especially the successful application of the compound lining of shotcrete, rock bolt and formed concrete at the Dayaoshan Tunnel (Li, 1983) have prompted reconsideration of the design procedure. Thus, a model test program has been developed to evaluate the effectiveness of the compound lining.

5 MODEL TEST PROGRAM

Quite a number of model tests have been done to elucidate the failure mechanism of the rock mass surrouding the excavated cavern including time effects (Zheng et al, 1986) and the strengthening effects of shotcrete and rock bolts (Zhu et al, 1986) in homogeneous rocks. To simplify the geological conditions at Caiyuan Tunnel, alternating layers of mudstone and sandstone are assumed. As a further simplification, plane deformation models are adopted. A third assumption is that the virgin rock stresses are uniform: σ_v =0.87 MPa. σ_h =0.29 MPa. Overloading technique is justifiable for such drastic simplifications.

By dint of the experiences gained at the Dayaoshan Tunnel, it was decided to decrease the thickness of formed concrete lining to 40cm at crown and 60cm wall. The width of the rock pillar is reduced to 12m. For comparison, three scaled models 70cm wide, 50cm high and 20cm thick,are used to simulate the rock tunnel as excavated, with shotcrete and rock bolts, and also with formed concrete, for Model I II III respectively. All the dimensions of shotcrete, rock bolts and formed concrete are duely scaled to a length ratio of 1:100. The strength and deformational behaviour of the modelling materials are conformal or nearly conformal to similarity requirements, see Table 3.

Table 3. Model materials test result

Material	Uniaial comp. str MPa	E GPa	Mu	C MPa	Ø Deg
Mudstone	1.1	2.9	.15	.3	45
Sandstone	3.6	4.9	.20	.8	47
Shotcrete	1.8	4.7	.17		
Concrete	2.0		.21		
Al wire	-140	28.4			

Model rocks and concrete were made by varying proportions of plaster of Paris, sand and water. Modelling material of shotcrete was a mixture of sand and a cementing material. Aluminum wires were used to model the rock bolts. 10cm cubes were used to investigate the strength properties of model mudstone compared with those strengthened by aluminum wires at a spacing 2cm by 2cm either parallel or normal to the principal stress in triaxial tests. At the peak stresses, the cohesion is quite close: .2, .19, .21 MPa while the friction angle is different: 44, 54 and 52 degrees.

Test were carried out on the PYD-501 Geomechanics Model Testing Equipment as shown in Fig. 4.

Figure 4. PYD-501 Geomechanics Model Testing Equipment

Each 20cm thick model was made up of two 10cm thick plates bonded together by cementing plaster. Electric resistance strain gages were embedded in the bonding surface. When the model was loaded to the virgin rock stress state, openings were excavated under stressed state by maintaining axial plane deformation and virgin rock loading around outer boundaries. The full depth of the openings was excavated in four steps. Rock bolts and shotcreting dry sand and cementing materials were done after each step for Model II and III. After shot-creting Model III prefabricated inner lining was put in and all the cracks left between them were carefully filled up.

Creep tests were performed after excavation and lining under virgin rock loading until strain and displacement readings showed that their time increments approaching zero. Then, overloading tests were run to failure occurred. Finally, after unloading, surface cracks were observed and surveyed and the cracked portions were carefully removed for further investigation.

Though the rheological properties of weak rock are well known, it is not practical to measure them in laboratory. Creep test results of 5cm Ø specimens of model rocks under uniaxial compression are shown in Figs. 5 and 6. It is to be noted that the model mudstone enters the tertiary stage at 70% uniaxial compressive strength, while model sandstone at less than 50%. To evaluate the rheological parameters, a rheological model consisting of a spring, a Kelvin element and a Bingham element in series was adopted.

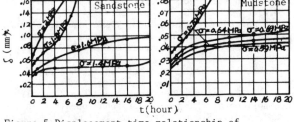

Figure 5 Displacement-time relationship of modelling material of rock.

6 MODEL TEST RESULTS

6.1 Creep test

Displacements at sidewall and roof crown are shown in Fig. 7. Figs. 8 and 9 show the time dependent strain on the midheight of the sidewall, and on the lining

surface of Model III.

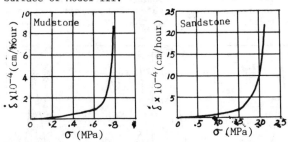

Figure 6.Displacements rate-stress relationship of modelling material of rock.

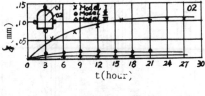

Figure 7.Displacement-time relationship of opening

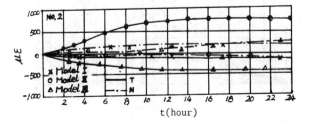

Figure 8. Strain-time curves of No.2

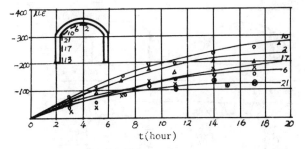

Figure 9. Strain-time curves of secondary lining

It can be seen that in all cases the strains and displacements stabilized within 20 hours.

6.2 Overloading test

Fig. 10 shows strains near the contour of the opening in Model I. It can be seen that normal stains are essentially tensile in the sidewalls with increasing load while those at the arch crown little change occurs. The tangential strains are slmost always compressive. Similar paterns are observed both in Models II and III.

Convergence of all the models are shown in Fig. 11. The convergence rate increases with increasing load and it is always higher for the sidewalls.

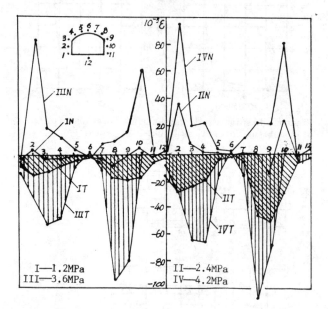

Figure 10. Strain distribution around the opening of Model I.

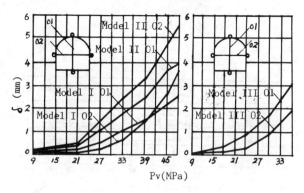

Figure 11. Convergence displacement-load curves

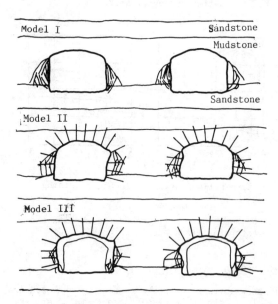

Figure 12. Type of failure of models

The failure mode of all the three models was very much similar by cracking into wedges at sidewalls. Yet, the sidewalls of Model I cracked at a load 10/3 times the virgin load and failed at 17/3 times. No spalling occurred in Models II and III and cracks at linings first appeared at 14/3 times the virgin load. Model II and Model III failed at 62/9 and 59/9 times respectively.

Fig. 12 shows the cracking paterns in the three models.

Dense cracking exists in all the model pillars but no cracking exist in the overlying model sandstone.

7 DISCUSSIONS AND CONCLUSIONS

The strength and deformational properties of layered rock and tunnel linings have been studied with the aid of models. Though strict similarity laws could not be applied to derive the behaviour of the prototype, findings from model tests can serve as a guide. The strengthening effects of shotcrete and rock bolts are evident from Fig. 7. No spalling has been observed and the cracking load was increased 40%. The rock pillar may be quite safe even the pillar width was reduced by 30%. The ineffectiveness of the formed concrete may be exaggarated due to the poor workmanship in filling the gap between the prefabricated lining and the shotcreted surface. Yet, it does in someway reflect the current practice.

FEM analysis (Wang, 1986; Sun and Li, 1983) showed that there were stress concentration at the springing and wall corners and tensile stresses occurred near the crown and in the middle span of the floor. Thus curved walls and inverts are preferred. Strains in the pillar agreed quite well with the model test results.

From the foregoing discussions, we might conclude that with qualified workmanship, shotcrete and rock bolts can be used as permanent support and formed concrete, as thin as practical, can be used only to improve smoothness for better ventilation and appearance.

ACKNOWLEDGEMENT

This work has been supported by Mr. Xiang Changren, chief engineer of the Chongqing Company of Urban Construction Development. The authors also wish to acknowledge the efforts made by Mr. Wu Yingshi in the preparation of the manucript.

REFERENCES

Li Zifan 1983. Design of Dayishan Tunnel. Underground Engineering. No. 11.
Sun J. and Y. S. Lee 1985. A viscous elastoplastic numerical analysis of the underground structure interacted with familied of multilaminate rock mass using FEM.Proc. 5th ICNMG, V2, p. 1121-1134.
Sun Jun and Zhang Yusheng 1983. Visco-elastic plastic finite element analysis of underground structure with large cross-sections. Journal of Tongji University. No.2, p. 10-25.
Wang Liwei 1986. Model study of underground openings in layered rock.
Zheng Li, Zhu Keshan, Zhu Jingmin and Fan Zebao 1986. A experimental study of underground openings in isotropic media, Symposium of numerical methods and model tests, CSRME.
Zhu Jingmin, Zhu Keshan, Gu Jingcai and Wang Lin 1986. Model study of the failure characteristics of underground openings with shotcrete linings and rockbolts. Symposium of tunnel and underground engineering, CSCE.
Zhu Keshan and X. Li 1983. Pressure tests in rock chambers. Proc. 5th ICRM, preprints, D267-9.

Some practical cases of back analysis of underground opening deformability concerning time and space effects

Cas pratiques de rétro-analyse de la déformabilité de cavités souterraines en fonction du temps et du milieu

Einige praktische Fälle der Rückanalyse von Verschiebungen in Untertagebauten mit Berücksichtigung von Zeit und Raum

ZHU WEISHEN, Institute of Rock and Soil Mechanics, Academia Sinica, Wuhan, People's Republic of China
LIN SHISHENG, Institute of Rock and Soil Mechanics, Academia Sinica, Wuhan, People's Republic of China
ZHU JIAQIAO, Institute of Rock and Soil Mechanics, Academia Sinica, Wuhan, People's Republic of China
DAI GUANYI, Institute of Rock and Soil Mechanics, Academia Sinica, Wuhan, People's Republic of China
ZHAN CAIZHAO, Institute of Rock and Soil Mechanics, Academia Sinica, Wuhan, People's Republic of China

ABSTRACT: A description is given to the in-situ deformation monitoring results and some back analysis results concerning threecases of underground galleries or tunnels. These underground openings were excavated in three different rocks, i.e., hard, weak and clayey rocks.Of these three openings, two are deeply burried and the other is a shallow one. In displacement analysis, the effects of the advance of working faces and of reheological properties of the surrounding rocks on the deformability of the rocks were taken into consideration. The mechanical parameters of rheology and the geostress field of in-situ rock masses are also derived by the back analyses method.

RESUME: On present quelque resultats de l'observation des deplacemants in situ et de la contre analyse concernant trois cas de galeries ou tunnels. Ils sont les roches de trois defferentes sortes: la roche dure, la rock faible et largile. les deux tunnels sont a la couverture profonde er lautre est a la couverture non-profonde. Quqnd on fair lanalyse des deplacements,on prend en consideration linfluence de lavancement du travail et laquelle des caracteres rheologiques de la roche encaissante.Les parameters mecaniques de rheologie et le champ des contraintes geologiques concerant la roche massive sur le terrain sont derives de la contre analyse.

ZSUAMMENFASSUNG: In vorliegenden Beittrag werden die Ergebnisse von der in-situ-Messungen der verschiebungen in 3 unterirdisechen stollen or Tunnels und der Ruck-analyse erlautert. Es wird sowohl hart und weich Stein als auch letten Gesteinsmental betrachtet. Davon sind die 2 Tunnels tief liegt, und der andere untief. In der Verschiebung Berechnungen werden der Einfluss des Verhicbes und die Wirkung der Rheologisch-Eigenschaft des Nebengesteines berucksichtigt. Die Rheologisch-Mechanik Parameter und ursprungliche Belastungszustand des in-situ Gebirge werden mit der Ruckrechnung Methode berechnet.

1 INTRODUCTION

In recent years , there has been an increasing trend of using back analysis method to assess the deformability of rocks surrounding underground openings, to judge their stability and to derive the rocks' mechanical parameters or geostress field.

The purpose of this paper is to describe our research work of using this method to primarily assess the stability of several practical engineering cases according to in-situ measured data. In analyses, the influences of the working face advance and the rocks'rheological properties are fully taken into consideration.

2 JUNDUSHAN TUNNEL

2.1 Monitoring

A new railway line in construction, the Datong-Qinhuangdao line,will pass through a 9km long shallow tunnel which is located in the northern suburb of Beijing and 65km away from the city.The tunnel,named Jundushan Tunnel, has a cross sectional area of 11.86×11.32 (m×m). The overburden with a thickness of about 12 m consists of loess. In order to understand the stability of the Tunnel during construction,the deformation mornitoring by using extensometers and distometers has been carried out in the testing section of the tunnel. Most boreholes for monitoring were predrilled downwards from the ground surface in front of the tunnel to monitor the whole process of deformation.The NATM was employed to excavation the testing tunnel with new loess overlying it.The upper part of the stratum is sandy clay with rudaceous lenses,and the lower is a fine sand layer, The specific weight the cohesion and internal friction angle are γ=1.97t/m and C=0.05-0.08 MPa and ϕ=25° respectively.The WRM-3 mechanical extensometer developed by our institute,having a sensitivity of 0.01 mm,was used to measure the displacements in the boreholes.The arrangment of measuring points is shown in Figs 1 and 2. Here after,The measuring results and analyses will be described.The monitoring work during

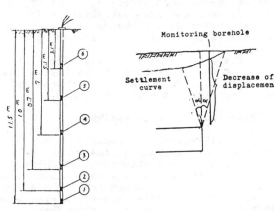

Fig.1 Layout of measuring points

Fig.2 Place of monitoring borehole

the excavation of the upper part of the test-ing tunnel lasted about 65 days,from Nov. 27 of 1985 to Jan. 30 of 1986.

In Fig. 3 shown is the deformation—time relationship. For measuring point 1 (the deepest one),the maximun value of deformation was U=15.29 mm.The deformation process can be derived into three stages.

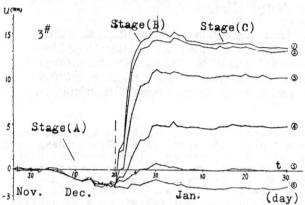

Fig. 3 Relation between displacement and time

(A) Predeformation stage (compression). Starting from the commencement of excavation and ending at the time when the working face advanced to the measuring hole, the surroun-ding rocks has a largest compressive defor-mation of -1.91 mm, which lasted 24 days, i.e.,deformation rate is e=-0.08 mm per day. All displacements measured at 6 points with different depths were negative. At the be-gining of measurement, the results were so-mewhat unstable 10 days later,however,the negative displacements became larger and the curves clustered which showed that the dis-placements were close to each other. All this indicated that the deformation of surround-ing rock was caused by the subsidence of ground surface.

(B) Abrupt deformation stage Soon after the excavation had crossed the measuring hole, the measured displacement became positive and increased rapidly. The largest one was up to 15.29 mm. This stage lasted about 10 days and the deformation rate,e=1.72 mm per day.

(C) Gradual stabilization stage When the tunnel was excavated 14 m away be-yond the measuring point, the displacements stopped increasing, the measured curves were slightly fluctuating, but tended to be stable The deformation rate is e=0.06 per day.This stage lasted about one month, except that during the first day, the largest displace-ment at the deepest measuring points (point 1 and 2) were somewhat decreasing,which show-ed that consolidation phenomenon may be cau-sed at this place by the existence of sandy layers.

For the later excavation,when the excava-vation of the lower part of the tunnel was finished, the displacements increased 1-5 mm The deformation rate is e=0.14 mm.

2.2 Visco-elastic back analyses

(A) Axially symmetric model analyses The numerical analysis of the tunnel shoud belong to the three dimensional problem, in order to avoid too large amount calculation work, however, the axially symmetric analy-sis was firstly carried out for approximate estimation. At the begining of excavation, the excavation height and width of the upper part of the tunnel are 4m and 10m respecti-

vily. For the purpose of approximate simula-tion, the equivalent diameter of the tunnel was chosen as the average of its height and span, i.e., a=7m.

(a) The model for approximate analysis is shown in Fig. 4a
(b) The rheology model of the medium is as-sumed as in Fig. 4b

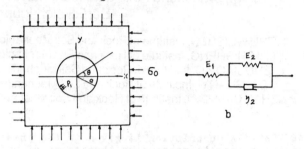

Fig.4 Analytic model and viscoelastic model

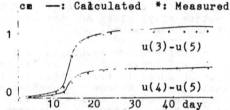

Fig.5 Comparison between monitored and calculated displacements

(B) Imaginary supporting force calculation Considering spacial effects, the authors chose the equivalent supporting force as $P=(1-\lambda)\sigma_0$ to simulate the real one,where $\lambda=0.5+\frac{1}{\pi}arctg(t-t_0-2)$, t_0 is the time necesa-ry to excavate to the measuring points (Ges-ta,1986).

(C) Displacement determination Assuming that the volume strain is elastic and the poisson's ratio is const., the phy-sical equation of the deviator parts follows the rheology model of elements (Fig.4b).From the visco-elastic correspondence law, one can obtain

$$U=\frac{1+\mu}{E_0}\sigma_0\cdot\frac{R^2}{r}\{\frac{1}{2}+\frac{1}{\pi}arctg(t-t_0-2)\}\{1+\frac{E_1}{E_2}-\frac{E_1}{E_2}e^{-\frac{E_2}{\eta_2}(t-t_0)}\} \quad (1)$$

According to the in-situ measured curves and the eq.(1), E_1,E_2 and η_2 can be calculated by back analysis method as follows.

E_1=64MPa η_2=5504.3MPa.day E_2=193Mpa $\quad$ (2)

Substituting E_1,E_2 and η_2 of eq.(2) into eq. (1), a curve Fig.5 which is fairly coincident with the in-situ measured results can be ob-tained through numerical calculation.

Note that it is in the tenth day (Fig.5) when the working face was advanced to the place just below the measuring points.

3 UNDERGROUND STORAGE OF YICHANG WINE MILL

3.1 Monitoring

The underground wine storage of the Yichang wine mill is located at the foot of Duijing shan mountain in Yichang city.The overburden above the cellar is only 30m in thickness. The storage has a length of about 300m and a cross section of 6.7m (width)×7.5m(height). The overlying rocks are subhorizontal inter-bed conglomerate and sandstone layers with a dip of 4-7°. Each rock layer is generally 3-5 m in thickness. Joints and cracks do not develop and the density of joints is 1.7/m². Underground water do not develop either. The sandstones with lower strength are divided into three kinds according to their grain sizes, namely, coarse, medium and fine ones. The fine sandstones are purple lutaceous siltstone whose compressive strength is 10-20 MPa and Young's modulus is 1500 MPa. In a hydraulic project several km away from the storehouse, the major principal geostress measured in-situ is 3.0 MPa with a direction of NE 34°. A 10 cm thick shotcreting was used for supporting the storehouse.

According to the plan of experiment research, the testing section I was arranged at a place 117m away from the east mouth of the storehouse . It includes a vertical hole on the surface and an inclined hole of 45° dip in the house. Extensometers were used for monitoring inclined hole and inclinometer was used for vertical holes. Besides, a convergence measuring section was arranged in the storehouse. The testing section was arranged 78 m away from the west mouth. One vertical hole and one horizontal hole were drilled in the storehouse for monitoring the deformation of surrounding rocks at different depths. Meanwhile, a convergence section with five measuring points was arranged to monitoring the convergence of the storehouse (Fig.6).

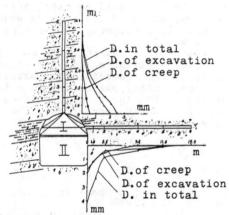

Fig.6 Monitoring of section II

3.2 Back analyses

Because the overburden depth, joint configuration and rock layer distribution are different for various sections, calculations were performed for three different sections.

Since the overburden is shallow, only elastic analysis of the tunnel excavation was performed and only were considered gravitational and small lateral geostresses. The calculation was carried out in two steps to simulate actual excavation operation.

As mentioned above section I is 117 m away from the east mouth. In calculation 321 elements (including 35 joint ones) and 337 nodes were used.

Section is 75 m away from the west mouth. In calculation, the calculated area was divided into 289 elements (including 25 joint ones) with 337 nodes altogether.

The mechanical indexes of rock masses and joints are listed in Table 1.

The elastic back analysis results based on the values of section II measured in excavation are shown in Fig. 7 which indicates that there exists a good agreement between the measured and calculated results. The

Table 1 Mechanical indexes of rock masses and joints

Para-meters \ Rocks	Conglo-merate	Sand-stone	Silt-stone	Joint
Young's moduls E(MPa)	1.5×10⁴	4.8×10³	1.6×10³	20/cm 3/cm
Poisson's ratio μ	0.25	0.28	0.35	
Friction angle φ	33,43 50.9,53	33,43 45,53	44 46	18.26
Cohesion C(MPa)	062,141 14,0.82	112,14 052,06	025,03	0.04
Tensile Strength	0.5	0.4	0.2	

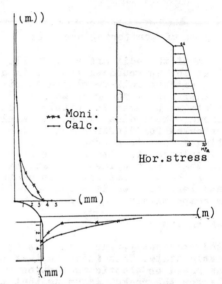

Fig.7 Comparison between back analyses and measurement

mechanical parameters of surrounding rock mass derived from back analysis are much smaller than those anticipated before. They are given in Table 2.

Table 2 Change of Young's Modulus

	Original value	Computed value
Sandstone	48000	9600
Conglo-merate	150000	30000
Siltstone	16000	3200

4 TESTING ADIT FOR ERTAN HYDROPOWER STATION

4.1 Monitoring

The research work concerning the underground chambers of the Ertan Hydropower Station has been introduced by the authors else (Zhu, 1986). A 30 meters long testing adit was excavated at the place where the main underground chambers will be built in order to investigate the actual deformability of chambers with high side walls and in a highly stressed region (Fig.8). Three monitorring sections were arranged in it, each having 5-6 deep drilling holes. The results of elastic wave measurements show that the changes of wave velocities before and after

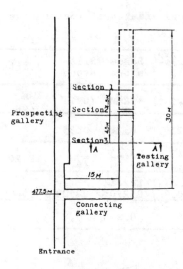

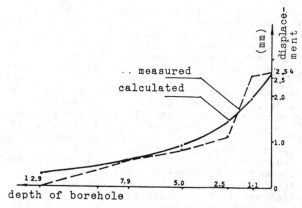

Fig.9 Comparison between measured and calculated results

Fig.8 Layout of monitoring sections

excavation of the adit are very small. It means that the surrounding rock is in a good condition with no well-developed joints in it. In other words, the rock mass is in its elastic state. Because the displacements were missed, three dimensional FEM analyses were performed for different situations of excavating stages to correct the above mentioned displacements. It is known from calculation that the displacements at the mouthes of the two drilling holes on the right and left side walls shoud be 2.54 and 2.45 mm respectively.

4.2 Back analyses

As stated previously, the rock mass is in its elastic state. This indicate that back analyses based on elastic assumption are suitable for the rocks. Assuming that the initial stress field arround the testing adit does not chande seriously and is uniform, the stress field can, when Young's modulus and poisson's ratio are known, be described from the following eqs. with back analysis method (Sakurai, 1983).

$$\{\sigma_o\} = \{\{A\}^T \cdot \{A\}\}^{-1} \cdot \{A\}^T \cdot \{u\}$$

In computation, Young's modulus E is taken to be 4.0×10^4 MPa and poisson's ratio $\mu = 0.15$. The initial principal stresses σ_1 and σ_2 obtained by such a method are 23.7 MPa and 18.7 MPa, while the angle between σ_1 and the horizontal plane is 31.5°. A comparison between measured data and calculated results from back analyses of the left wall is shown in Fig.9, in which one can find the fitting degree is fairly good.

The above statement indicates that the elastic analysis for the problem under consideration is feasible. The calculated stress field is very close the abundant results measured in recent years.

5 DISCUSSION AND CONCLUSION

1.Through the in-situ deformation monitoring of three engineering cases mentioned above, it can be seen that for rock masses with obvious elasticity, displacements basically become stabilized when the working face advances a distance of one height; but for weak rocks, the first and second cases

for example, the displacements will not be stabilized untill the working face advances a distance of three height.

2. Either in the first or in the second case, the rocks surrounding tunnels all exhibit a certain rheological deformation along beddings makes up about 50% of the total.

3. For weak rocks as those in the first case, it seems feasible to introduce a 3-element rheology model for simulating surrounding rocks. It should be noted that in the same case, there exists a sandy bedding with consolidation property in the upper part of the vault.

4. For case A, the mechanical properties fitted by back analysis method are close to those obtained from geotechnical tests;for case B,the mechanical parameters so obtained are much lower than those obtained from Lab tests; and for case C (hard rock), the back analysis results are quite coincident with Lab tests

REFERENCES

Gesta,P. 1986.Tunnel support and lining. Tunnels. N° 73,P18-38
Sakurai,S. 1983. Displacement measurements associated with the design of underground openings. Proc. of Intern. Symp. on Field Measurements in Geom., Zurich, Sept..
Zhu, W. and others. 1986. Three-dimensional FEM analyses and back analyses for deformation monitoring of Ertan Hydropower station chambers. Proc. of Intern. Symp. on ECRF, Beijing, Nov.

A proposed mechanism of rock failure and rockbursting
Le mécanisme fondamental de rupture violente de la roche
Angenommener Mechanismus von Gesteinsbruch und Gebirgsschlag

ZOU DAIHUA, Department of Mining and Mineral Process Engineering, The University of British Columbia, Vancouver, Canada

HAMISH D.S.MILLER, Department of Mining and Mineral Process Engineering, The University of British Columbia, Vancouver, Canada

ABSTRACT: Sudden rock failure in the form of rockbursting has long been a problem in underground mines. The basic mechanism of this phenomenon is still unresolved. This paper describes the research work conducted at the University of British Columbia to study the basic mechanism of violent rock failure, and to identify reliable precursive behaviour. Acoustic emissions were tested from rock specimens, the rock failure mechanism postulated and the experimental results obtained are in agreement with measurements made in situ in a deep level South African mine. The shear failure mechanism proposed has been modelled and this numerical model allows rock tests and their associated acoustic emissions to be realistically simulated.

RESUME: Les affaissements violents de roche ont été étudié de façon à identifier un comportement precurseur fiable. Un mécanisme de fracture de la roche a été postulé d'après les emissions accoustiques d'échantillons de roche. Le mécanisme de cisaillement a été modelisé et il permet une simulation des essais de roche et des emissions acoustiques associées.

ZUSAMMENFASSUNG: Um eine zuverlaessige Voraussage ueber Gesteinsbruch zu erhalten wurden im Test Bruchgeraeusche gemessen. Der Mechanismus des Scherbruches ist ebenfalls im Modell untersucht worden und dies erlaubt eine realistische Simulation von Gesteinstests und ihre begleitenden Bruchgeraeusche.

INTRODUCTION

The phenomenon of rockbursts has long been a problem in underground mines. It has been associated with mining excavations throughout the world in all rock types and at all depths. As mining depth continues to increase, this problem is becoming more serious. Despite the work of many researchers, the basic mechanism of this phenomenon is still unresolved. With the development of microseismic monitoring, warning about of impending rockbursts has been greatly improved. However, the low reliability of this technique as a predictive tool has largely limited it to event location. The commonly faced problem is the failure without anomaly or an anomaly with no accompanying failure.

Generally, a rockburst can be described as a sudden release of strain energy stored in the rock mass, sometimes resulting in catastrophic failure and extensive damage to underground openings and mining facilities. It is characterized by expulsion of rock in varying quantities from the surface of an opening. This phenomenon usually happens instantly and without any visual warning. Once it happens it threatens miners' lives and gives rise to considerable operational problems in the mine.

The rock mass is highly jointed and anistropic. Before excavation takes place, a stress field exists in the rock mass, known as the virgin stress. As an opening is excavated, the virgin stress field is disturbed, resulting in stress redistribution around the opening. Close to the boundary of the excavation a zone of stress concentration is formed. If the maximum stress is less than the strength of the rock mass, the structure will be stable. Otherwise, failure occurs either by yielding or cracking. At the same time, the stored strain energy which is proportional to the square of the stress, is released. If the energy release happens suddenly, the failure will be violent. However, until today, little is known about the way this energy release occurs.

This research work is intended to study the basic mechanism of rock failure and rockbursting. We are trying to approach this problem from the very beginning, studying both the conditions under which a burst is likely to occur, and the release of acoustic energy prior to rock failure.

1. FAILURE MECHANISM OF ROCK MASS

1.1 Failure principle

As a geological material, rock mass is generally jointed and anistropic. The failure process of rock mass is also complex. An examination of the failures of rock mass, such as of a pillar, or of the remnants of a rock specimen, shows that failure usually takes place along a surface. This surface makes an acute angle with the major loading direction. For an isotropic material, this surface can be determined on Mohr's circle, Figure 1[1]. However, this kind of ideal material rarely exists. The failure of rock mass will occur along the weakest surface, which may be a major fault, joint, or any other weakness when the shear stress on that surface reaches the corresponding shear strength. As can be seen, the failure of jointed rock mass will be controlled by the shear process.

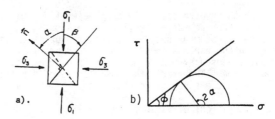

Fig. 1 Schematic showing shear failure plane.

For intact rock, this principle of shear failure may not apply until a failure surface is formed. In this case, the failure is believed to initiate from local microfracturing, since even intact rock always contains microcracks. When a force is applied to a rock specimen, these microcracks initially close. Then the rock mass exhibits perfect elastic deformation under further load. As the stress reaches some level, these cracks start to develop, or micro-fracturing initiates. This results in acoustic emissions that occur due to the vibration of rock particles in the immediate vicinity of the fracturing. These vibrations are extremely small. As the load continues to increase, the number of microfractures increases and fracturing propagates further. At first the fracture propagation is stable and fracturing can be halted by maintaining a constant load. When the fracture propagation reaches the unstable stage, the fracture process is self-sustaining because the energy required to maintain the crack propagation decreases. Even if the load is held constant, fractures will develop. Any increase of the load will accelerate the fracture propagation. Apparently the seismic event rate increases due to the intensive fracturing, but the increase of acoustic energy is not significant because the vibrations of rock particles remains low. Meanwhile, the released energy from crack extension increases with the crack length. When additional energy is available, the crack tends to fork in the weakest direction. The onset of forking represents a transition within the process of the unstable stage. Once this transition has taken place, successive forking will lead to co-alescence of many fractures, forming macrofractures. Thus, the acoustic energy is expected to increase dramatically. These macrofractures will join together to form a surface on which the final failure takes place. From now on the shear principle controls the failure process.

1.2 Stick-slip in shear failure.

During the shear process, once the shear stress reaches the shear strength, slip begins. However, the behaviour of slipping varies with the loading conditions and the surface properties. In shear experiments, either stable sliding or stick-slip is observed. Generally, the phenomenon of stick-slip is expected to occur if the shear surfaces are very smooth or if the normal stress is very high. [2,3]

The stick-slip is of significance. During stick time, shear stress and potential energy gradually build up. When slip occurs, the shear force drops and the potential energy is released. If the slip takes place suddenly, the energy will be released very quickly, resulting in violent failure. The phenomenon of stick-slip has drawn great attention from seismologists and is considered as a mechanism of shallow earthquakes [4], especially for those occurring along geological faults. The high stress field in the earth's crust tends to initiate relative movement along the fault. Once the potential energy exceeds the shear strength of the fault, stored energy is released by a sudden slip in the crust.

A natural earthquake and a rock burst are extremely similar in terms of seismic events. The only difference between the two is a matter of scale. They both involve the violent release of seismic energy. For an earthquake, the stress build-up is the result of many decades or even centuries of movement in the earth's crust. For a rockburst, it is usually caused by mining activity in a relatively short time. Therefore, the shear failure is significant to the study of rockbursting.

1.3 Conditions of violent failure.

In this research, the shear behaviour is analyzed on a numerical model. Figure 2 schematically illustrates the model. The whole system can be described by an equilibrium equation.

$$mx = F(t,x) - f(u,P,x) \ldots \ldots \ldots (1)$$

where m is the mass of the particle, P the normal pressure, u the frictional coefficient, f the total resistance, F the shear force, x, $\dot{x}$ and $\ddot{x}$ the slip distance, velocity and acceleration respectively.

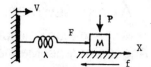

Fig. 2 Simple shear model.

The shear force F which is a function of time and slip distance, is modelled by a spring which represents the elasticity of rock mass. This spring is supported by a moving base. The moving speed V simulates the loading speed. The effects of slip velocity and seismic radiation are also considered in this model. Scholz and Engelder (1976) observed from experiments that the frictional coefficient is inversely proportional to the logarithm of slip velocity [5]. The seismic radiation is complex. One method in which radiation effects can be simulated without making the model unduly complicated is to attach a semi-infinite string to the mass particle in such a way that the motion of the mass particle excites an elastic wave which propagates along the string. The derived force exerted by the string on the mass is linearly proportional to the slip velocity. Thus, the term f includes the frictional resistance and the seismic dissipative force, both of which are functions of slip velocity.

The stick time, which is the peace time between adjacent slips, obviously varies with conditions. If stable sliding is considered as a special case of stick-slip, in which the stick time is zero, a transition condition between the stable sliding and stick-slip exists. In this model, stable sliding is considered to occur when the stick time is equal to or less than 10^{-5} seconds instead of zero, mainly due to the numerical approximation and computing costs. Figure 3 shows some typical numerical results. This figure shows the transition chart. As expected, the driving speed, normal pressure and elasticity of the rock mass have a significant effect, although the effect from frictional coefficient is negligible. For a given material, the elastic modulus is specified. If the loading conditions of loading speed and normal pressure fall within the lower part of this chart, the shear behaviour will show stick-slip. The upper part represents the stable sliding. This chart can also be used to determine the maximum loading speed or minimum normal pressure for stick-slip for a given type of rock mass. It should be pointed out that this chart only shows the basic principle. The transition condition for a particular material should be obtained from experiments.

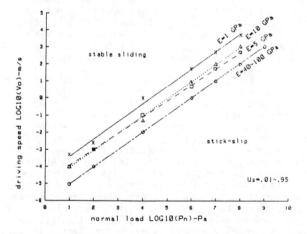

Fig. 3 Transition between stable sliding and stick-slip.

From the physical conditions of the shear process and the transition chart, violence failure is expected to occur during shear in the following three cases:

MODE I. Violence is the stick-slip under very high normal pressure because of the large amount of energy released at each slip.

MODE II. Violence comes from the transition from stick-slip to stable sliding. If a shear system which shows stick-slip behaviour suddenly changes into stable sliding for some reason, such as sudden reduction of normal pressure or quick increase of loading speed, extra energy is available. This energy has to be released quickly in order to keep up with the sudden change of loading condition.

MODE III. Violence occurs under sudden loading. Whether the shear behaviour is stable sliding or stick-slip, violent failure is bound to happen if a shear force much higher than the shear strength is applied to the system suddenly, because extra potential energy is always available. The failure of intact rock, such as a rock specimen under conventional compressive test, belongs to the Mode III violence.

If the shear stress starts from zero, which is usually the case, failure happens only when the shear stress reaches or exceeds the shear strength. Under this condition, there are only two modes of violent failure, namely Modes I and II.

2. ACOUSTIC SIGNALS FROM ROCK SPECIMENS

During this research, laboratory experiments on rock specimens were carried out to study the acoustic activity prior to rock failure. They were designed to examine the acoustic emissions pattern from similar rock types loaded under compression and shear. The testing results are very encouraging and agree well with the mechanism postulated previously. Details of the test program and results have been previously published [6].

During compressive tests, generally few acoustic signals were observed before stress had reached some level, say 75% or 80% of the compressive strength, Figure 4. This stress level corresponds to the fracture initiation. As the fracture propagates further under continuous loading, the acoustic activity becomes more intense. In the period between the fracture initiation and the final failure, the acoustic emission is most active. The most significant phenomenon during this period is the fact that the event rate increases rapidly initially and then dies down immediately preceding the specimen failure. At the same time, the energy release rate increases. When failure is approached, the energy rate shows a peak value. The drop of event rate and the peak value of the energy may indicate the coalescence of microfractures. These events are in perfect agreement with the failure principle discussed above.

In the shear tests, both sawcut surfaces and natural breakage surfaces were loaded to failure. The surface roughness seems to have little effect on acoustic emission pattern, although the magnitude of the acoustic signal from the breakage surface is higher for the sawcut surface. In general, the acoustic activity is low before the slip and remains somewhat unchanged as sliding continues.

The effect of normal pressure on acoustic emission seems significant. As normal pressure increases, the acoustic activity increases throughout the shear process, characterized by higher magnitude, but the acoustic emission pattern changes little, Figure 5. As previously described, stick-slip phenomena were observed under high normal pressure. Following each slip, the acoustic emission shows a sharp increase. The acoustic activity remains at a low level as the stress builds up again.

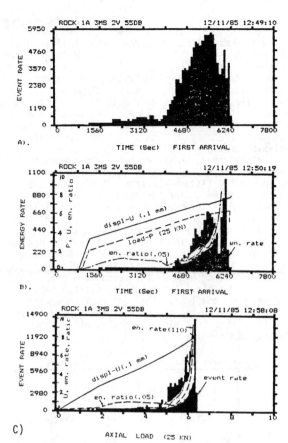

Fig. 4 Acoustic emission vs. axial load for specimen #1.

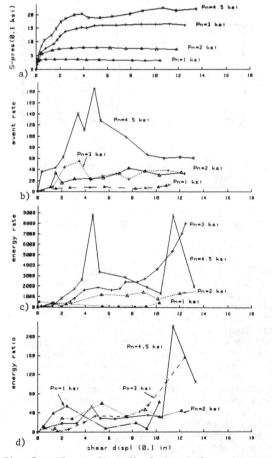

Fig. 5. AE vs. shear displacement for sawcut surface #4 at various normal pressure.

From testing results it is found that compressive failure and shear failure are two different modes of the failure mechanism of intact rock. Failure under compression is a matter of fracturing up to the point where a failure surface is initiated. After the formation of this surface, the failure process obeys the law of shear. Unfortunately, this shear process in compressive tests occurs extremely rapidly and cannot be easily observed. This is because the shear stress on the newly-formed fracture surface is much higher than the corresponding shear strength. Detailed analysis shows that this shear stress is up to three times higher than the strength for the specimens tested under compression. It is this extra shear stress that makes the failure violent. If this extra force can be extracted at the formation of the fracture surface, the failure can be reduced to non-violence. On the contrary, if a large shear load is suddenly applied to the shear specimen, violent failure can also be expected during shear process. This has been proven in the experiments by releasing the normal pressure instantly when slip began, resulting in bursting. Figure 6 shows the acoustic signals from one of these tests. The acoustic emission before slip has been completely shadowed by the peaking of signals at the instantaneous failure. Because the load is reduced to minima instantly, after shock is scarcely observed.

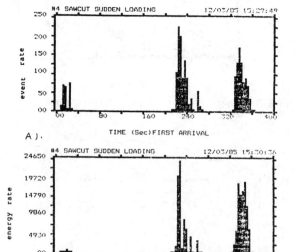

A).

B).

Fig. 6   AE from sawcut surface #4 under sudden loading, normal pressure 1, 2.5 and 4.5 ksi respectively.

3.   NUMERICAL SIMULATION OF ACOUSTIC ACTIVITY

3.1   Numerical model

From microseismic monitoring of rockbursts in situ and from acoustic emission tests of rock specimens in the laboratory, a large quantity of data of acoustic signals before rock failure is available. However, these data are only recordings of the acoustic signals. To date, the causes of these signals and how they actually occur remains unknown.

In this research, a numerical model is developed to simulate the acoustic activity prior to rock failure. This model is not based on any physical law of acoustic emissions or on any empirical formula from previous recordings. It is entirely based on the proposed shear failure mechanisms. Whether the failure is in the fracturing stage or at slip stage, any movement of rock particles at a local area will induce vibration among the surrounding rock particles. It is this vibration which causes acoustic signals. It is expected that, if this model

can produce results similar to the acoustic signals recorded during tests, this model can serve the following purposes:

1. To justify the shear process as a mechanism of rock failure and rockbursting.

2. To verify the acoustic emission as a useful precursive signal.

3. To provide a tool to study the acoustic activity prior to rock failure.

As in the finite element method of stress-strain analysis, the rock mass is discretized into individual elements. Because the shear process takes place on the contacting surfaces, the movement only occurs on the failure plane. Only two variables are needed to describe an exact location in a plane. However, this model is not involved in the exact descriptions of locations of the elements, only the behaviour of the elements during the movement is of interest. Therefore, only one degree of freedom is permitted.

The   model consists of a series of particles, N, connected together by springs, Figure 7. The mass of the material is concentrated on those particles and the spring represents the elasticity of the rock mass. Let the mass of particle i be m_i and the stiffness of spring i be λ_i. If we further assume that at the beginning all particles are at rest, by the force equilibrium of particle i as shown in Figure 7, the equation of motion of particle will be

$$m_i \ddot{x}_i = F_i - F_{i-1} - f_i^* \quad \ldots \ldots \ldots (2)$$
$$i = 1, 2 \ldots N$$

where F_i and F_{i-1} are the forces in the springs in front and behind particle i, and f_i^* is the total resistance including frictional resistance and seismic force.

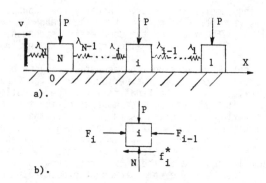

a).

b).

Fig. 7   Diagram of acoustic activity model.

After substituting all these forces, the above equation becomes

$$m_i \ddot{x}_i = \lambda_i (a + x_{i+1} - x_i) - \lambda_{i-1} (a + x_i - x_{i-1})$$
$$- f_i - E_o x_i \quad \ldots \ldots \ldots \ldots (3)$$
$$i = 1, 2 \ldots n$$

where a is the space between two particles, f_i the frictional force, E_o the seismic coefficient, x_i, $\dot{x}_i$ and $\ddot{x}_i$ the location, slip speed and acceleration of particle i, respectively.

When the initial conditions are considered, equation (3) can be solved for the unknowns x_i, $\dot{x}_i$ and $\ddot{x}_i$ for each particle at any time. Because f_i contains an unknown $\dot{x}_i$ in denominator, and explicit solution cannot be found. Therefore it is desired to search for a numerical solution, which is obtained by the Runge-Kuta method. The tedious work of calculation is left to the computer.

A computer program was written, which in addition to the above unknowns, can also calculate the kinetic energy, the work done against friction and the seismic energy radiated, at any moment. The event number is counted by checking the change of status of every particle continuously. Obviously, at the beginning only parts of the system will move under the load. As loading continues, the number of particles in movement will increase. If the onset of movement of all particles is considered as the final failure of the system, the energy changes and acoustic activity prior to the failure can be simulated.

3.2 Modelling results

This model generates excellent results. Some results from a typical analysis are given in Figure 8. As can be seen, the event rate increases sharply as the failure is approached and then drops to the previous low level immediately preceding the failure. Meanwhile, the seismic energy remains low when the event rate goes up, and increases dramatically prior to the failure. In the results from all runs of the program, the energy rate and energy ratio show a similar behaviour, although the energy ratio shows the anomaly more clearly.

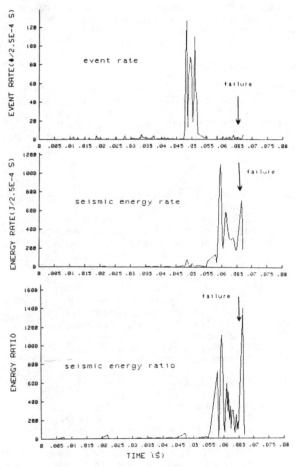

Fig. 8   Computer results from the numerical acoustic model.

All the results from this model are in very good agreement with the experimental results, even though the model itself has no direct relation to the acoustic emission. The increase of the event rate corresponds to fracture propagation. The drop of event rate and increase of acoustic energy indicates the formation of macrofractures. This shows that the postulated shear failure mechanism to interpret rock failure and the acoustic activity prior to the failure is justified. It also verifies that the acoustic emissions are indeed a precursive signal for rock

failure and can be used for predictive purposes. The problem is how to use this signal correctly and efficiently.

Consequently, this model provides us with a method to study the rock failure and the related acoustic activity. It can be used to simulate the acoustic emission under various conditions and to study the influence on the acoustic emission from the change of geological conditions and loading conditions. This part of the research is still in progress.

4.   COMPARISON WITH FIELD RESULTS

In order to check the acceptability of the above results from experiments and the numerical modelling, some field results of rockbursting monitoring have been studied. By comparison, it is found that the above results are in agreement with measurements made in situ in a deep level South African mine [7]. Two typical cases are presented:

4.1   Case 1:   Rockburst on May 15, 1983

A large rockburst of magnitude 3.4 occurred on 101W1 Panel, No. 3 shaft on May 15, 1983, at 03.37 hours. A concentration of microseismic events prior to the burst is apparent. In Figure 9 the number of microseismic events per hour originating from the panel for the period 8th to 15th May, 1983, is plotted. A steady increase can be seen, from approximately 60 events 6 days before the burst, to almost 300 events only 24 hours beforehand. A sharp drop in the rate of microseismic activity was measured immediately before the burst. For this particular case, the changes in the ratio between numbers of larger and smaller events provided the researcher with additional information to make a reliable prediction.

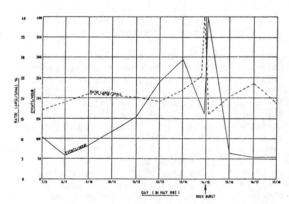

Fig. 9.   Event rate and relative energy one week and three days after the May 15 event (after Brink)

In this example, the agreement between the field data and the results from experiments and the model is apparent. In all cases, the event rate increases sharply at first and drops immediately preceding the failure. The abrupt increase of the ratio between the number of large and small events is equivalent to the increase of event energy, because this change of the ratio is due to both the decrease of the small event number and the increase of the large event number, with more energy being released.

4.2   Case 2:   Rockbursts on October 4 and 10, 1984

On October 4, a 2.6 magnitude rockburst occurred during shift time (10:31 hours) on the 110 level. Figure 10 shows the event rate, average corner frequency and average event energy as observed from that area for the time window 2200 to 0400 hours every night. On September 27th, influence from an external source made the measurement unreliable. On the basis of event rate alone, the rockburst would

not have been anticipate on October 4th, as the event
rate parameter is very sensitive to the mining activity and no blasting took place in that area the previous afternoon. However, the corner frequency
showed a steady drop for the preceding 11 days and a
further drop to below 600 Hz is indicated a few hours
before the burst. This behaviour of the corner frequency gave a clear precursive indication of a pending rockburst. The average event energy also confirmed what was expected. Five days later, regular
blasting started and was followed by a small burst
(magnitude 1.4) at 4:39 hours on October 10th.
Again, a relatively low corner frequency and a relatively high event energy preceded the burst. The
blasting the previous afternoon made the event rate
unusually high.

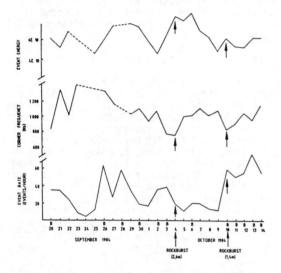

Fig. 10. Corner frequency, event rate and energy
over 25 days, covering 2 bursts(after Brink)

In this example, similar results as given previously were recorded. The energy goes up sharply
in both bursts. The event rate shows an increase
and drop prior to the burst, except the second
burst, which was influenced by blasting.

The corner frequency may be another important
parameter to be used. It is schematically defined
as f_O in Figure 11 [8], which shows that when $f < f_O$,
the amplitude spectrum is level, and when $f > f_O$, the
spectrum decays. In other words, higher frequency
corresponds to lower magnitude or to smaller event.
The drop of f_O may indicate a greater number of
events in the low frequency band and more energy released. The increase of magnitude of individual
events will be accompanied by a decrease of event
rate because of the coalescence of microfractures.
This hypothesis is supported by the empirical relation between the number of events and their magnitudes derived from years of observations of seismic
events [7].

$$\log N = a - b M \dots\dots\dots\dots\dots\dots (4)$$

where a and b are constants, M is the magnitude and
N the number of events of magnitude $\geq$ M. Therefore,
the downshift of the corner frequency also indirectly indicates the drop of the event rate and the
sharp increase of the energy release.

5. CONCLUSIONS

In this research, the source mechanism of rockbursts
is studied and the acoustic emission prior to the
bursting is analyzed. The shear failure is postulated as the basic mechanism of rock failure and
rockbursting. Important results were obtained from
experiments on acoustic emission from rock specimens.

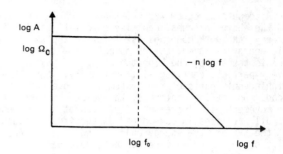

Fig. 11. Schematic far-field seismic spectrum (A=
amplitude spectral density, f = frequency),
clarifying the concepts of low-frequency
amplitude level (Ω_0), corner frequency
(f_O), and high-frequency amplitude decay
($\sim f^{-n}$ with n = 2 or 3). (after Bath).

A numerical model based on the shear failure mechanisms has produced results very similar to those from
experiments. All these results are in agreement with
measurements made in situ in a deep mine. This
suggests that the postulated mechanism is true and
the experimental results obtained clearly indicate
that the monitoring of acoustic emissions can be
used to reliably predict rockbursts.

6. ACKNOWLEDGEMENTS

Special thanks are given to Professor J. Nadeau in
the Department of Metallurgical Engineering at the
University of British Columbia for the loan of
acoustic monitoring equipment. Help from Mrs.
Melba Weber and Mark Stoakes during the preparation
of this paper is acknowledged.

REFERENCES

Jaeger, J.C. & Cook, N.G.W. 1969. "Fundamentals of
Rock Mechanics". Textbook.
Hoskins, E.R., Jaeger, J.C. & Rosengren, K.J. 1968.
"A Medium-Scale direct friction experiment". Int.
J. Rock Mech. Min. Sci. Vol. 5, p. 143-154.
Stesky, K.M. 1978. "Rock friction effect of con-
fining pressure, temperature and pore pressure".
Pageoph, Vol. 116, p. 690-703.
Brace, W.F. & Byerlee, J.D. 1966. "Stick-slip as
a mechanism of earthquakes". Science 153, p. 990.
Dieterich, J.H. 1978. "Time dependent friction
and the mechanism of stick-slip". Pageoph, Vol.
116, p. 790-805.
Zou, Daihua & Miller, Hamish D.S. "Acoustic
emissions from rock under unioxial compressive
test and direct shear test". Proc. 29th Acoustic
Emissions Working Group Meeting, Royal Military
College, Kingston, Ontario, June 23-26, 1986.
Brink, A.V.Z. & Mountfort, P.I. 1985. "Rockburst
prediction research at Western Deep Levels Ltd. -
A review report of the period 1981-1984". Inter-
nal report No. R, p. 21.
Bath, M. 1984. "Rockburst Seismology". Rockburst
and seismicity in mines, S. Afr. Inst. Min. & Met.
Symp. Series No. 6, p. 7-15.